5th International Conference on Topical Problems of Continuum Mechanics with a Special Session in Honor of Alexander Manzhirov's 60th Birthday

Journal of Physics: Conference Series Volume 991

Tsakhkadzor, Armenia
2 – 7 October 2017

ISBN: 978-1-5108-6163-3
ISSN: 1742-6588

Printed from e-media with permission by:

Curran Associates, Inc.
57 Morehouse Lane
Red Hook, NY 12571

Some format issues inherent in the e-media version may also appear in this print version.

Printed by Curran Associates, Inc. (2018)

For permission requests, please contact the Institute of Physics at the address below.

Institute of Physics
Dirac House, Temple Back
Bristol BS1 6BE UK

Phone: 44 1 17 929 7481
Fax: 44 1 17 920 0979

techtracking@iop.org

Additional copies of this publication are available from:

Curran Associates, Inc.
57 Morehouse Lane
Red Hook, NY 12571 USA
Phone: 845-758-0400
Fax: 845-758-2633
Email: curran@proceedings.com
Web: www.proceedings.com

Table of contents

Volume 991

5th International Conference on Topical Problems of Continuum Mechanics with a Special Session in Honor of Alexander Manzhirov's 60th Birthday

2–7 October 2017, Tsakhkadzor, Armenia

Accepted papers received: 15 March 2018
Published online: 13 April 2018

Preface

Papers

IOP Conf. Series: Journal of Physics: Conf. Series **991** (2018) 011001 doi:10.1088/1742-6596/991/1/011001

V International Conference on Topical Problems of Continuum Mechanics with a Special Session in Honor of Alexander Manzhirov's 60th Birthday October 2–7, 2017, Tsaghkadzor, Armenia

The V International Conference was held 2–7 October 2017 hosted by the Institute of Mechanics of the National Academy of Sciences of the Republic of Armenia. During the conference a Special Session was organized in honor of the Foreign Member of the National Academy of Sciences of the Republic of Armenia Professor Alexander Manzhirov on the occasion of his 60th birthday and to celebrate his contributions to mechanics and applied mathematics.

The conference aims to bring together leading academic scientists, researchers, and engineers to exchange and share their experience and results on all aspects of continuum mechanics. It also aims to provide the premier interdisciplinary forum for researchers, practitioners, and educators to present and discuss the most recent innovations, trends, and concerns, practical challenges encountered and solutions adopted in the field of continuum mechanics.

The conference program covers such topics as:

- General Problems
- Contact Mechanics
- Mechanics of Growing Solids
- Mechanics of Technological Processes
- Mechanics of Additive Manufacturing
- Thin-walled Structures
- Elasticity and Viscoelasticity
- Plasticity
- Tribology
- Fracture Mechanics
- Coupled Fields in Solids
- Composites
- Soil Mechanics
- Fluid Mechanics
- Optimal Design and Control Theory
- Mechanics of Nano-systems
- Structural Mechanics
- Biomechanics
- Hydraulics and Hydraulic Facilities
- Experimental Mechanics

During the conference a joint discussion of fundamental, computational, and applied problems of the new scientific area by leading mechanicians, mathematicians, physicists and technologists, exchange of the latest ideas and achievements, as well as working out the main directions of its further development were carried out. Representatives from many countries such as Russia (48), Armenia (37), Ukraine (3), USA (2), Germany (2), France (1), Great Britain (1), Hungary (1), India (1), South Africa (1), Georgia (1), Latvia (1) were participated in the conference. The geography of participants from Russia is quite wide: Moscow, Sankt-Peterburg, Kazan, Nizhny Novgorod, Novosibirsk, Perm, Rostov-on-Don, Saratov, Tomsk, Voronezh, Yakutsk, Yekaterinburg.

IOP Conf. Series: Journal of Physics: Conf. Series **991** (2018) 011001 doi:10.1088/1742-6596/991/1/011001

Figure 1. Conference session

Figure 2. Sightseeing tour to lake Sevan

The keynote lectures were delivered by the professors L.A. Aghalovyan (Armenia), H. Altenbach (Germany), A.S. Avetisyan (Armenia), S.Y. Dobrokhotov (Russia), V.N. Hakobyan (Armenia), I.G. Goryacheva (Russia), S. Joubert (RSA), I.V. Kirillova (Russia), I. Kudish (USA), A.V. Manzhirov (Russia), S.M. Mkhitaryan (Armenia), V.E. Nazaikinsky (Russia), M. Raous (France), A.V. Sahakyan (Armenia), V. Silbershmidt (UK), A. Spector (USA), M.A. Sumbatyan (Russia), R. Velmurugan (India).

The 84 oral and 31 poster addresses have been presented at the conference. The materials of the conference consisting of the 116 brief papers have been published at the beginning of the conference.

V. N. Hakobyan
A. V. Manzhirov
A. V. Sahakyan

IOP Conf. Series: Journal of Physics: Conf. Series **991** (2018) 011002 doi:10.1088/1742-6596/991/1/011002

Peer review statement

All papers published in this volume of *Journal of Physics: Conference Series* have been peer reviewed through processes administered by the proceedings Editors. Reviews were conducted by expert referees to the professional and scientific standards expected of a proceedings journal published by IOP Publishing.

IOP Conf. Series: Journal of Physics: Conf. Series **991** (2018) 012001 doi:10.1088/1742-6596/991/1/012001

On 60th Anniversary of Professor Alexander Vladimirovich Manzhirov

Yuri N. Radayev

[1]Ishlinsky Institute for Problems in Mechanics of the Russian Academy of Sciences, Moscow, Russia

[2]Department of Energy Conversion Science, Graduate School of Energy Science, Kyoto University, Kyoto, Japan

E-mail: y.radayev@gmail.com, radayev@ipmnet.ru, radaev.iurii.8e@kyoto-u.ac.jp

Abstract. On May, 24, 2017 the 60th jubilee of Prof. A.V. Manzhirov was celebrated at Ishlinsky Institute for Problems in Mechanics of the Russian Academy of Sciences. Prof. A.V. Manzhirov is known as a prominent scientist in the field of mechanics and applied mathematics. The principal directions of his academic activity are Mechanics of Growing Solids, Theory of Creep and Viscoelasticity, Contact Mechanics, Tribology, Integral Equations and their numerous applications.

1. Preliminary remarks

A.V. Manzhirov is a founder of a new scientific area (Mechanics of Growing Solids), which has been originated fulfilling to the needs of the applied mechanics in the study of complex technological and natural processes. The Mechanics of Growing Solids, in particular, can effectively describe a wide range of phenomena, such as concreting, solidification and polymerization, electroforming and pyrolytic deposition, laser spraying and fusing, melting curing and crystal growth. It also provides means for an adequate description of the glaciers formation, sedimentation and volcanic rock massif growth, gravitating objects accreting, volumetric and surface growth of biological tissues. From the beginning of 80th of the previous century until now he extensively developed the foundations of the Mathematical Theory of Growing Solids, proposed a classification of growing processes, derived models governing equations and resolved a number of boundary value problems. Early experimental studies have been carried out in order to identify the physical properties of growing solids and to predict the stress-strain state evolution. New mechanical effects attributed exceptionally to the growing solids have been discovered and studied under his supervision.

A.V. Manzhirov is also a leading Russian scientist in the area of Contact Mechanics. He developed the theory of mixed integral equations with constant and variable integration limits aimed at study of complex evolutionary processes of contact interaction and wearing. His projection method for solving mixed integral equations allows us to study the contact problems by employing experimental surface profilograms described by rapidly changing functions. He is also an outstanding expert in the field of Nonlinear Creep and Viscoelasticity of Inhomogeneous Aging Materials.

A.V. Manzhirov is an author of several fundamental high quality reference guides on Higher Mathematics and Integral Equations published in Russia and abroad. "Handbook on Integral Equations" is a unique sample of reference books. "Handbook of Mathematics for Engineers and Scientists" published by CRC Press in 2006 in English is the comprehensive one in the math world, embracing the huge data from different branches of modern mathematics and until now is unsurpassed with respect to collected methods for solving various equations arising in numerous applications of mathematics.

2. Education

A.V. Manzhirov was born on May 24, 1957 in the city of Rostov-on-Don.

In 1974 A.V. Manzhirov graduated with excellence *Specialized Secondary School No. 14* in Rostov-on-Don. In 1974 he entered *Faculty of Mechanics and Mathematics of Rostov State University*, where he was specialized in mechanics and applied mathematics at *Department of Theory of Elasticity*. *Mechanics and Mathematics Department of Rostov State University* in those years provided the highest level of education in various branches of mathematics and mechanics.

In 1979 A.V. Manzhirov graduated with excellence the Rostov State University and was awarded by Master of Mechanics Degree (M.Sc.). Title of his master thesis is "Dynamical Problem for a Stripe: Limiting Amplitude Method".

On the invitation of N.Kh. Arutyunyan and V.M. Alexandrov (August 22, 1979) A.V. Manzhirov came to *Institute for Problems in Mechanics of the Academy of Sciences of the USSR (IPMech AS USSR)*.

N.Kh. Arutyunyan is an outstanding scientist and statesman of the USSR and one of the founders of the scientific school of mechanics in Armenia. He and his students determined a number of new scientific trends in the Mechanics of Solids, in particular, in Mechanics of Growing Solids. By now A.V. Manzhirov is the head of the scientific school at *IPMech RAS* founded in early 1980s by N.Kh. Arutyunyan. At the same time Prof. V.M. Aleksandrov involved A.V. Manzhirov to the studies of contact and mixed problems of the Mechanics of Solids. Until now these studies are a significant part of his permanent scientific interests.

In November 10, 1983 A.V. Manzhirov successfully defended his Ph.D. thesis titled "Stress-Strain Analysis of Inhomogeneous Viscoelastic Solids and their Interaction with Stress Concentrators and Rigid Indentors" and was awarded by Ph.D. Degree in Physics and Mathematics from *Moscow Institute of Electronics and Mathematics.*

In October 7, 1993 he defended his doctor dissertation titled "Contact Problems of the Theory of Viscoelasticity of Growing Solids" at Dissertation Council of the *IPMech RAS.*

3. Work experience

Since 1995 he has been working as a leading researcher, since 2004 as the head of *Laboratory for Modeling in the Mechanics of Solids of IPMech RAS.* In September 2015 he was appointed to the position of deputy director for scientific work.

In 1988 the results related to contact problems were summarized in the joint monograph by N.Kh. Arutyunyan and A.V. Manzhirov "Contact Problems of Creep Theory", Yerevan, Publishing House of NAS RA, 1999. For the first time the foundations of the creep theory of inhomogeneous aging solids (including various types of constitutive equations and relaxation kernels, analysis of creeping structures, study of the correspondence principles) have been elucidated. Theory of Nonlinear Steady Creep, as well as the Mechanics of Growing Solids, and within the framework of the above theories the problems of contact interaction have been formulated and investigated. A significant place in this book is given to mathematical methods for constructing solutions of integral equations and systems of integral equations usable in the analysis of contact problems in the creep theory and algorithms for constructing exact and approximate solutions of nonlinear problems. One of the principal scientific achievements there has been a method of solving integral equations, which in application to contact problems allows us to obtain a sequence of independent Volterra's equations instead of infinite systems of integral Volterra's equations. The monograph includes solutions of a number of applied problems.

In 1991 the monograph by N.Kh. Arutyunyan, A.V. Manzhirov, and V.E. Naumov "Contact Problems of the Mechanics of Growing Solids", Moscow, Nauka was published and recognized as a an important step in the development of continuum mechanics. This book is entirely devoted to the new problems of mechanics of solids: contact problems of continuously and discretely built up solids with complex rheological properties (including aging and age-related inhomogeneity). In this study the theoretical foundations of the mechanics of the contact interaction of solids have been broadly discussed.

Since the mid of 1990s A.V. Manzhirov devoted himself to the development of the Theory of Integral Equations. Integral equations can be found in many areas of the continuum mechanics and physics (in Elasticity, Plasticity, Theory of Heat and Mass Transfer, Hydromechanics, Electrodynamics and the Theory of Wave Propagation). Since 1998 A.V. Manzhirov published a series of joint books devoted to exact solutions of integral equations and methods for their solution.

In 2006 A.V. Manzhirov with the co-author published a remarkable reference guide: A.D. Polyanin and A.V. Manzhirov "Handbook of Mathematics for Engineers and Scientists", Boca Raton, London, Chapman & Hall / CRC Press, 2006. In 2007 they released the most complete, extended and revised second edition of the handbook on integral equations: A.D. Polyanin and A.V. Manzhirov "Handbook of Mathematics for Engineers And Scientists", Chapman & Hall / CRC Press, Boca Raton, London, 2007.

A.V. Manzhirov actively participates in international scientific cooperation with Armenian, Indian, Polish, and South African scientists. During recent years A.V. Manzhirov is a coordinator of the Integrated Long Term Program of Scientific and Technical Cooperation between Russia and India in the field of mechanics. Together with the coordinator from India, Prof. N.K. Gupta he organized a permanent Russian-Indian seminar, which had been personally attended by the leading scientists from the two countries. The result of this activity was the publication of three collective Indo-Russian monographs.

Important studies devoted to thermomechanics of laser processing of materials have been carried out in cooperation with *Tshwane University of Technology* (Pretoria, South Africa) following a plan supported by a joint grant (*RFBR* and the *National Research Foundation of South Africa*). These studies are focused on the theoretical and experimental problems of laser deposition and fusing.

A.V. Manzhirov put into practice the fruitful cooperation with *Instytut Podstawowych Problemów Techniki Polskiej Akademii Nauk* under the Agreement of Cooperation between *RAS* and *Polish Academy of Sciences.*

Due to Prof. A.V. Manzhirov the studies in the Mechanics of Growing Solids has been internationally supported: in 2015 the IUTAM Symposium on Growing Solids took place in Moscow at *Institute for Problems in Mechanics of RAS.*

Since 1994 A.V. Manzhirov works as a professor of higher mathematics at *Moscow State University of Instrument Engineering and Computer Science*; since 1997 he is a professor of *Department of Applied Mathematics of Bauman Moscow State Technical University*; since 2014 he is a professor of *Department of Mathematics of the National Research Nuclear University.* He delivered original lecture courses on analytical geometry, higher algebra, calculus, differential equations, probability theory and mathematical statistics, queuing theory, mathematical physics. Special courses on the theory of creep of solids, mechanics of growing solids, non-classical problems of solid mechanics, mechanics of materials of constructions, integral equations have been also prepared.

4. Honours and awards

Professor Manzhirov is a winner of the First Competition of the Science Support Foundation in 2001, Moscow. He was awarded by the State scientific grant for outstanding Russian scientists (1997-2003).

Since 2015 A.V. Manzhirov is a foreign member of the *National Academy of Sciences of the Republic of Armenia.* 19 May 2017 he was elected as a full member of the *International Academy of Engineering.*

A.V. Manzhirov is a member of the editorial boards of scientific journals *Mechanics of Solids, Computational Continuum Mechanics, Izv. of Saratov State University. New Series. Mathematics. Mechanics. Computer Science, Proceedings of National Academy of Sciences of Armenia. Mechanics, Vestnik of Yakovlev Cheboksary State Pedagogical University. Series of Mechanics of Limit State.* He is a member of *Russian National Committee on Theoretical and Applied Mechanics, American Society of Mechanical Engineers (ASME), American Mathematical Society (AMS), International Association of Applied Mathematics and Mechanics (GAMM), International Association of Engineers (IAENG), European Society for Mechanics (EUROMECH).*

Concluding remarks

Prof. A.V. Manzhirov is a specialist of the highest qualification deeply involved in Russian educational and research systems. He is very kind to colleagues, associates and students. He always defends the principle scientific viewpoints in all professional and educational controversies. All these allows us to state that Prof. A.V. Manzhirov is a respectful professional in the Science of Mechanics.

Acknowledgments

The author is grateful to Ph.D. E.V. Murashkin and M.Sc. N.E. Stadnik for their invaluable assistance in preparing the present paper.

IOP Conf. Series: Journal of Physics: Conf. Series **991** (2018) 012001 doi:10.1088/1742-6596/991/1/012001

Books by Prof. A.V. Manzhirov

References

[1] Arutyunyan N Kh, Manzhirov A V and Naumov V E 1991 *Contact Problems in Mechanics of Growing Solids* (Moscow: Nauka Publishers) [in Russian]

[2] Polyanin A D and Manzhirov A V 1998 *Handbook of Integral Equations: Exact Solutions* (Moscow: Factorial Press) [in Russian]

[3] Polyanin A D and Manzhirov A V 1998 *Handbook of Integral Equations* (Boca Raton-London: Chapman & Hall/CRC Press)

[4] Manzhirov A V and Polyanin A D 1999 *Methods for the Solution of Integral Equations: A Handbook* (Moscow: Factorial Press) [in Russian]

[5] N. Kh. Arutyunyan and Manzhirov A V 1999 *Contact Problems in Creep Theory of Creep* (Erevan: National Academy of Sciences of Armenia) [in Russian]

[6] Polyanin A D and Manzhirov A V 1999 *Handbuch der Integralgleichungen: Exakte Lösungen* (Berlin: Spektrum Akademischer Verlag, Heidelberg)

[7] Manzhirov A V and Polyanin A D 2000 *Handbook of Integral Equations: Solution Methods* (Moscow: Factorial Press) [in Russian]

[8] Polyanin A D and Manzhirov A V 2003 *Handbook of Integral Equations* (Moscow: Nauka Publishers) [in Russian]

[9] Polyanin A D and Manzhirov A V 2007 *Handbook of Integral Equations. Second Edition* (Boca Raton-London: Chapman & Hall/CRC Press)

[10] Gupta N K and Manzhirov A V Eds. 2008 *Topical Problems in Solid Mechanics* (New Delhi: Elite Publishing)

[11] Polyanin A D and Manzhirov A V 2008 *Handbook of Mathematics for Engineers and Scientists* (Boca Raton–London: Chapman & Hall/CRC Press)

[12] Chernoutsan A I, Egorov A V, Manzhirov A V, etc. 2010 *A Concise Handbook of Mathematics, Physics, and Engineering Sciences. Ed. by A. D. Polyanin and A.I. Chernoutsan* (Boca Raton-London: Chapman & Hall/CRC Press)

[13] Manzhirov A V, Gupta N K, and Indeitsev D A Eds. 2013 *Topical Problems in Solid and Fluid Mechanics* (New Delhi: Elite Publishing)

[14] Gupta N K, Manzhirov A V, and Velmurugan R Eds. 2013 *Topical Problems in Theoretical and Applied Mechanics* (New Delhi: Elite Publishing)

[15] Polyanin A D, Manzhirov A V 2017 *Integral Equations in 2 volumes. Handbook for Higher Education. Part 1. Second Edition, revised and extended* (Moscow: Urait Publ. House) [in Russian]

[16] Polyanin A D, Manzhirov A V 2017 *Integral Equations in 2 volumes. Handbook for Higher Education. Part 2. Second Edition, revised and extended* (Moscow: Urait Publ. House) [in Russian]

IOP Conf. Series: Journal of Physics: Conf. Series **991** (2018) 012002 doi:10.1088/1742-6596/991/1/012002

Forced vibrations of a two-layered shell in the case of viscous resistance

L A Aghalovyan[1*] **and L G Ghulghazaryan**[1,2**]

[1]Institute of Mechanics of the National Academy of Sciences of Armenia, Yerevan, Armenia
[2]Kh. Abovyan Armenian State Pedagogical University, Yerevan, Armenia

E-mail: *lagal@sci.am, **lusina@mail.ru

Abstract. Forced vibrations of a two-layered orthotropic shell are studied in the case of viscous resistance in the lower layer of the shell. Two versions of spatial boundary conditions on the upper surface of the shell are posed, and the displacement vector is given on the lower surface. An asymptotic method is used to solve the corresponding dynamic equations and relations of the three-dimensional problem of elasticity. The amplitudes of the forced vibrations are determined, and the resonance conditions are established.

1. Basic equations and statement of the boundary-value problems

Forced vibrations of the two-layered orthotropic shell $\Omega = \{\alpha, \beta, \gamma; \alpha, \beta \in \Omega_0, -h_2 \leq \gamma \leq h_1\}$ are considered in the case of viscous resistance in the lower layer. Here Ω_0 is the surface of contact between the layers, α and β are the curvature lines of the surface of contact, and γ is the rectilinear axis directed perpendicularly to the surface of contact between the layers. In the chosen three-dimensional orthogonal coordinate system, it is required to find nonzero solutions of dynamic equations of elasticity which satisfy the boundary conditions of the second or mixed boundary-value problems of elasticity on the surfaces of the shell. To simplify the computations, the asymmetric stress tensor components τ_{ij} [1, 2] are denoted by $\tilde{\gamma}_i = 1+\gamma/R_i$ $(i = 1, 2)$, where R_1 and R_2 are the basic radii of curvature of the surface of contact between the layers.

We have:
the equations of motion

$$\frac{1}{AB}\frac{\partial}{\partial\alpha}(B\tau_{\alpha\alpha}^{(\mathrm{I})}) - k_\beta\tau_{\beta\beta}^{(\mathrm{I})} + \frac{1}{AB}\frac{\partial}{\partial\beta}(A\tau_{\beta\alpha}^{(\mathrm{I})}) + k_\alpha\tau_{\alpha\beta}^{(\mathrm{I})} + \tilde{\gamma}_1\frac{\partial\tau_{\alpha\gamma}^{(\mathrm{I})}}{\partial\gamma} + \frac{2\tau_{\alpha\gamma}^{(\mathrm{I})}}{R_1} = \rho^{\mathrm{I}}\tilde{\gamma}_1\tilde{\gamma}_2\frac{\partial^2 U^{(\mathrm{I})}}{\partial t^2},$$

$$\frac{1}{AB}\frac{\partial}{\partial\alpha}(B\tau_{\alpha\alpha}^{(\mathrm{II})}) - k_\beta\tau_{\beta\beta}^{(\mathrm{II})} + \frac{1}{AB}\frac{\partial}{\partial\beta}(A\tau_{\beta\alpha}^{(\mathrm{II})}) + k_\alpha\tau_{\alpha\beta}^{(\mathrm{II})} + \tilde{\gamma}_1\frac{\partial\tau_{\alpha\gamma}^{(\mathrm{II})}}{\partial\gamma} + \frac{2\tau_{\alpha\gamma}^{(\mathrm{II})}}{R_1}$$
$$- k_1\tilde{\gamma}_1\tilde{\gamma}_2\frac{\partial U^{(\mathrm{II})}}{\partial t} = \rho^{\mathrm{II}}\tilde{\gamma}_1\tilde{\gamma}_2\frac{\partial^2 U^{(\mathrm{II})}}{\partial t^2} \quad (A, B; \alpha, \beta; R_1, R_2; U, V),$$

$$\frac{\partial \tau_{\gamma\gamma}^{(\mathrm{I})}}{\partial \gamma} - \left(\frac{\tau_{\alpha\alpha}^{(\mathrm{I})}}{R_1} + \frac{\tau_{\beta\beta}^{(\mathrm{I})}}{R_2}\right) + \frac{1}{A}\frac{\partial \tau_{\alpha\gamma}^{(\mathrm{I})}}{\partial \alpha} + \frac{1}{B}\frac{\partial \tau_{\beta\gamma}^{(\mathrm{I})}}{\partial \beta} + k_\beta \tau_{\alpha\gamma}^{(\mathrm{I})} + k_\alpha \tau_{\beta\gamma}^{(\mathrm{I})} = \rho^{\mathrm{I}} \tilde{\gamma}_1 \tilde{\gamma}_2 \frac{\partial^2 W^{(\mathrm{I})}}{\partial t^2},$$

$$\frac{\partial \tau_{\gamma\gamma}^{(\mathrm{II})}}{\partial \gamma} - \left(\frac{\tau_{\alpha\alpha}^{(\mathrm{II})}}{R_1} + \frac{\tau_{\beta\beta}^{(\mathrm{II})}}{R_2}\right) + \frac{1}{A}\frac{\partial \tau_{\alpha\gamma}^{(\mathrm{II})}}{\partial \alpha} + \frac{1}{B}\frac{\partial \tau_{\beta\gamma}^{(\mathrm{II})}}{\partial \beta}$$
$$+ k_\beta \tau_{\alpha\gamma}^{(\mathrm{II})} + k_\alpha \tau_{\beta\gamma}^{(\mathrm{II})} - k_1 \tilde{\gamma}_1 \tilde{\gamma}_2 \frac{\partial W^{(\mathrm{II})}}{\partial t} = \rho^{\mathrm{II}} \tilde{\gamma}_1 \tilde{\gamma}_2 \frac{\partial^2 W^{(\mathrm{II})}}{\partial t^2},$$

where $\tilde{\gamma}_1 \tau_{\alpha\beta}^{(j)} = \tilde{\gamma}_2 \tau_{\beta\alpha}^{(j)}$, $j = \mathrm{I}, \mathrm{II}$ (symmetry conditions);
the equations of state (relations of elasticity)

$$\begin{aligned}
&\tilde{\gamma}_2\left(\frac{1}{A}\frac{\partial U^{(j)}}{\partial \alpha} + k_\alpha V^{(j)} + \frac{W^{(j)}}{R_1}\right) = \tilde{\gamma}_1 a_{11}^{(j)} \tau_{\alpha\alpha}^{(j)} + \tilde{\gamma}_2 a_{12}^{(j)} \tau_{\beta\beta}^{(j)} + a_{13}^{(j)} \tau_{\gamma\gamma}^{(j)}, \quad j = \mathrm{I}, \mathrm{II}, \\
&\qquad (A, B;\ \alpha \leftrightarrow \beta;\ R_1 \leftrightarrow R_2;\ U \leftrightarrow V;\ a_{11}^{(j)}, a_{22}^{(j)};\ a_{13}^{(j)}, a_{23}^{(j)}), \\
&\tilde{\gamma}_1 \tilde{\gamma}_2 \frac{\partial W^{(j)}}{\partial \gamma} = \tilde{\gamma}_1 a_{13}^{(j)} \tau_{\alpha\alpha}^{(j)} + \tilde{\gamma}_2 a_{23}^{(j)} \tau_{\beta\beta}^{(j)} + a_{33}^{(j)} \tau_{\gamma\gamma}^{(j)}, \\
&\tilde{\gamma}_1\left(\frac{1}{B}\frac{\partial U^{(j)}}{\partial \beta} - k_\beta V^{(j)}\right) + \tilde{\gamma}_2\left(\frac{1}{A}\frac{\partial V^{(j)}}{\partial \alpha} - k_\alpha U^{(j)}\right) = \tilde{\gamma}_1 a_{66}^{(j)} \tau_{\alpha\beta}^{(j)}, \\
&\tilde{\gamma}_1 \tilde{\gamma}_2 \frac{\partial U^{(j)}}{\partial \gamma} - \tilde{\gamma}_2 \frac{U^{(j)}}{R_1} + \frac{1}{A} \tilde{\gamma}_2 \frac{\partial W^{(j)}}{\partial \alpha} = \tilde{\gamma}_1 a_{55}^{(j)} \tau_{\alpha\gamma}^{(j)}, \\
&\qquad (A, B;\ \alpha, \beta;\ R_1 \leftrightarrow R_2;\ U, V;\ a_{55}^{(j)}, a_{44}^{(j)}),
\end{aligned} \tag{1}$$

where k_α, k_β are geodesic curvatures, A, B are coefficients of the first quadratic form, $\rho^{(j)}$ are the layer densities, $a_{ik}^{(j)}$ are elasticity constants, and j is the layer number.

On the face surface $\gamma = h_1$, the boundary conditions are

$$\tau_{\alpha\gamma}^{\mathrm{I}}(h_1) = 0, \quad \tau_{\beta\gamma}^{\mathrm{I}}(h_1) = 0, \quad \tau_{\gamma\gamma}^{\mathrm{I}}(h_1) = 0, \tag{2}$$

or

$$U^{\mathrm{I}}(h_1) = 0, \quad V^{\mathrm{I}}(h_1) = 0, \quad W^{\mathrm{I}}(h_1) = 0, \tag{3}$$

and on the surface $\gamma = -h_2$, the conditions are

$$U^{\mathrm{II}}(-h_2) = u^-(\alpha, \beta)\sin(\Omega t), \quad V^{\mathrm{II}}(-h_2) = v^-(\alpha, \beta)\sin(\Omega t), \quad W^{\mathrm{II}}(-h_2) = w^-(\alpha, \beta)\sin(\Omega t). \tag{4}$$

On the surface of the contact between the layers, the conditions of complete contact must be fulfilled:

$$\tau_{\alpha\gamma}^{\mathrm{I}}(\gamma = 0) = \tau_{\alpha\gamma}^{\mathrm{II}}(\gamma = 0), \quad \tau_{\beta\gamma}^{\mathrm{I}}(\gamma = 0) = \tau_{\beta\gamma}^{\mathrm{II}}(\gamma = 0), \quad \tau_{\gamma\gamma}^{\mathrm{I}}(\gamma = 0) = \tau_{\gamma\gamma}^{\mathrm{II}}(\gamma = 0), \tag{5}$$

$$U^{\mathrm{I}}(\gamma = 0) = U^{\mathrm{II}}(\gamma = 0), \quad V^{\mathrm{I}}(\gamma = 0) = V^{\mathrm{II}}(\gamma = 0), \quad W^{\mathrm{I}}(\gamma = 0) = W^{\mathrm{II}}(\gamma = 0). \tag{6}$$

The conditions on the lateral surface are not specified, and in the problems of this class, they imply origination of a boundary layer [1–3].

2. Solution of the outer problem

To solve above-formulated boundary-value problems (1)–(6), we pass to dimensionless coordinates and displacements in equations (1) by the formulas

$$\alpha = R\xi, \quad \beta = R\eta, \quad \gamma = \varepsilon R\zeta = h\zeta, \quad U = Ru, \quad V = Rv, \quad W = Rw,$$
$$h = \max\{h_1, h_2\}, \quad h \ll R,$$

where R is the characteristic dimension of the shell (the smallest value of the radii of curvature and the linear dimensions of the surface of contact between the layers) and $\varepsilon = h/R$ is a small parameter. The solution of the transformed equations [4, 5] is sought in the form

$$Q^{(j)}_{\alpha\beta}(x,y,z,t) = Q^{(j)}_1(\xi,\eta,\zeta)\sin(\Omega t) + Q^{(j)}_2(\xi,\eta,\zeta)\cos(\Omega t) \quad (\alpha,\beta,\gamma), \quad j = \mathrm{I},\mathrm{II}, \tag{7}$$

where $Q^{(j)}_{\alpha\beta}$ are any values of stresses and displacements and Ω is the frequency of the external forcing effect. As a result, a system of equations for $Q^{(j)}_1$, $Q^{(j)}_2$ singularly perturbed by the small parameter ε is obtained, and its solution is the sum of solutions of the outer problem and boundary layers: $I = Q^{\text{out}} + R_{\text{b}}$ [1, 3, 6].

The solutions of the outer problem are sought as the asymptotic expansion [7–12]

$$\begin{gathered}\tau^{\text{out}(j)}_{mk,i}(\xi,\eta,\zeta) = \varepsilon^{-1+s}\tau^{(j,s)}_{mk,i}(\xi,\eta,\zeta), \quad m,k = 1,2,3, \quad s = \overline{0,N}, \quad j = \mathrm{I},\mathrm{II}, \quad i = 1,2,\\ \left(u^{\text{out}(j)}_i(\xi,\eta,\zeta), v^{\text{out}(j)}_i(\xi,\eta,\zeta), w^{\text{out}(j)}_i(\xi,\eta,\zeta)\right) = \varepsilon^s\left(u^{(j,s)}_i(\xi,\eta,\zeta), v^{(j,s)}_i(\xi,\eta,\zeta), w^{(j,s)}_i(\xi,\eta,\zeta)\right).\end{gathered} \tag{8}$$

Here and further, $s = \overline{0,N}$ means that by the mute (repeated) summation index s ranges between 0 and N.

The solution of the problem must satisfy conditions (2)–(6). If the above-mentioned structure of the general solution is known, then to determine $\tau^{(j,s)}_{mk,i}$, $u^{(j,s)}_i$, and (u,v,w) in the outer problem from conditions (2)–(4), we must satisfy the following boundary conditions at $\zeta = \zeta_1$ ($\zeta_1 = h_1/h$):

$$\tau^{(\mathrm{I},s)}_{13,i}(\zeta_1) = -\tau^{(\mathrm{I},s)}_{13,ib}(\zeta = \zeta_1), \quad (13,23,33), \quad i = 1,2, \tag{9}$$

or

$$u^{(\mathrm{I},s)}_i(\zeta_1) = \bar{u}^{(\mathrm{I},s)}_{ib}(\zeta = \zeta_1), \quad (u,v,w), \quad i = 1,2, \tag{10}$$

and the conditions at $\zeta = -\zeta_2$ ($\zeta_2 = h_2/h$)

$$u^{(\mathrm{II},s)}_i = u^{-(\mathrm{II},s)}_i(\xi,\eta), \quad (u,v,w), \tag{11}$$

where

$$\begin{gathered}u^{-(\mathrm{II},0)}_1 = \frac{u^-}{R}, \quad u^{-(\mathrm{II},0)}_2 = 0, \quad \bar{\tau}^{(\mathrm{I},0)}_{m3,ib} = 0, \quad \bar{u}^{(\mathrm{I},0)}_{ib} = 0, \quad u^{-(\mathrm{II},s)}_i = -\bar{u}^{(\mathrm{II},s)}_{ib}(\zeta = -\zeta_2),\\ s \neq 0, \quad (u,v,w), \quad m = 1,2,3, \quad i = 1,2.\end{gathered} \tag{12}$$

The values $\tau^{(\mathrm{I},s)}_{13,ib}$, $\bar{u}^{(\mathrm{I},s)}_{ib}$, $\bar{u}^{(\mathrm{II},s)}_{ib}$, and (u,v,w) are determined after constructing the boundary layer solution.

Standardly substituting expressions (8) into the obtained singularly perturbed system of equations, we obtain

$$
\begin{aligned}
&\frac{1}{AB}\frac{\partial}{\partial\xi}(B\tau_{11,i}^{(\mathrm{I},s-1)}) - k_\beta R\tau_{22,i}^{(\mathrm{I},s-1)} + \frac{1}{AB}\frac{\partial}{\partial\eta}(A\tau_{21,i}^{(\mathrm{I},s-1)}) + k_\alpha R\tau_{12,i}^{(\mathrm{I},s-1)} + \frac{\partial\tau_{13,i}^{(\mathrm{I},s)}}{\partial\zeta} + r_1\zeta\frac{\partial\tau_{13,i}^{(\mathrm{I},s-1)}}{\partial\zeta}\\
&\quad + 2r_1\tau_{13,i}^{(\mathrm{I},s-1)} + \rho^{\mathrm{I}}\Omega_*^2u_i^{(\mathrm{I},s)} + (r_1+r_2)\zeta\rho^{\mathrm{I}}\Omega_*^2u_i^{(\mathrm{I},s-1)} + r_1r_2\zeta^2\rho^{\mathrm{I}}\Omega_*^2u_i^{(\mathrm{I},s-2)} = 0,\\
&\quad (A\leftrightarrow B;\ \alpha\leftrightarrow\beta;\ r_1,r_2;\ \xi\leftrightarrow\eta;\ u,v;\ \tau_{12}\leftrightarrow\tau_{21};\ \tau_{11}\leftrightarrow\tau_{22};\ \tau_{13},\tau_{23}),\\
&\frac{\partial\tau_{33,i}^{(\mathrm{I},s)}}{\partial\zeta} - r_1\tau_{11,i}^{(\mathrm{I},s-1)} - r_2\tau_{22,i}^{(\mathrm{I},s-1)} + \frac{1}{A}\frac{\partial\tau_{13,i}^{(\mathrm{I},s-1)}}{\partial\xi} + \frac{1}{B}\frac{\partial\tau_{23,i}^{(\mathrm{I},s-1)}}{\partial\eta} + k_\beta R\tau_{13,i}^{(\mathrm{I},s-1)}\\
&\quad + k_\alpha R\tau_{23,i}^{(\mathrm{I},s-1)} + \rho^{\mathrm{I}}\Omega_*^2w_i^{(\mathrm{I},s)} + (r_1+r_2)\zeta\rho^{\mathrm{I}}\Omega_*^2w_i^{(\mathrm{I},s-1)} + r_1r_2\zeta^2\rho^{\mathrm{I}}\Omega_*^2w_i^{(\mathrm{I},s-2)} = 0,\\
&\frac{1}{AB}\frac{\partial}{\partial\xi}(B\tau_{11,1}^{(\mathrm{II},s-1)}) - k_\beta R\tau_{22,1}^{(\mathrm{II},s-1)} + \frac{1}{AB}\frac{\partial}{\partial\eta}(A\tau_{21,1}^{(\mathrm{II},s-1)}) + k_\alpha R\tau_{12,1}^{(\mathrm{II},s-1)} + \frac{\partial\tau_{13,1}^{(\mathrm{II},s)}}{\partial\zeta}\\
&\quad + r_1\zeta\frac{\partial\tau_{13,1}^{(\mathrm{II},s-1)}}{\partial\zeta} + 2r_1\tau_{13,1}^{(\mathrm{II},s-1)} + 2K\Omega_*u_2^{(\mathrm{II},s)} + \rho^{\mathrm{II}}\Omega_*^2u_1^{(\mathrm{II},s)} = -(r_1+r_2)\zeta 2K\Omega_*u_2^{(\mathrm{II},s-1)}\\
&\quad - r_1r_2\zeta^2 2K\Omega_*u_2^{(\mathrm{II},s-2)} - (r_1+r_2)\zeta\rho^{\mathrm{II}}\Omega_*^2u_1^{(\mathrm{II},s-1)} - r_1r_2\zeta^2\rho^{\mathrm{II}}\Omega_*^2u_1^{(\mathrm{II},s-2)},\\
&\quad (A\leftrightarrow B;\ \alpha\leftrightarrow\beta;\ r_1,r_2;\ \xi\leftrightarrow\eta;\ u,v;\ \tau_{12}\leftrightarrow\tau_{21};\ \tau_{11}\leftrightarrow\tau_{22};\ \tau_{13},\tau_{23}),\\
&\frac{\partial\tau_{33,1}^{(\mathrm{II},s)}}{\partial\zeta} - r_1\tau_{11,1}^{(\mathrm{II},s-1)} - r_2\tau_{22,1}^{(\mathrm{II},s-1)} + \frac{1}{A}\frac{\partial\tau_{13,1}^{(\mathrm{II},s-1)}}{\partial\xi} + \frac{1}{B}\frac{\partial\tau_{23,1}^{(\mathrm{II},s-1)}}{\partial\eta} + k_\beta R\tau_{13,1}^{(\mathrm{II},s-1)}\\
&\quad + k_\alpha R\tau_{23,1}^{(\mathrm{II},s-1)} + 2K\Omega_*w_2^{(\mathrm{II},s)} + \rho^{\mathrm{II}}\Omega_*^2w_1^{(\mathrm{II},s)} = -(r_1+r_2)\zeta 2K\Omega_*w_2^{(\mathrm{II},s-1)}\\
&\quad - r_1r_2\zeta^2 2K\Omega_*w_2^{(\mathrm{II},s-2)} - (r_1+r_2)\zeta\rho^{\mathrm{II}}\Omega_*^2w_1^{(\mathrm{II},s-1)} - r_1r_2\zeta^2\rho^{\mathrm{II}}\Omega_*^2w_1^{(\mathrm{II},s-2)},\\
&\frac{1}{AB}\frac{\partial}{\partial\xi}(B\tau_{11,2}^{(\mathrm{II},s-1)}) - k_\beta R\tau_{22,2}^{(\mathrm{II},s-1)} + \frac{1}{AB}\frac{\partial}{\partial\eta}(A\tau_{21,2}^{(\mathrm{II},s-1)}) + k_\alpha R\tau_{12,2}^{(\mathrm{II},s-1)} + \frac{\partial\tau_{13,2}^{(\mathrm{II},s)}}{\partial\zeta}\\
&\quad + r_1\zeta\frac{\partial\tau_{13,2}^{(\mathrm{II},s-1)}}{\partial\zeta} + 2r_1\tau_{13,2}^{(\mathrm{II},s-1)} - 2K\Omega_*u_1^{(\mathrm{II},s)} + \rho^{\mathrm{II}}\Omega_*^2u_2^{(\mathrm{II},s)} = (r_1+r_2)\zeta 2K\Omega_*u_1^{(\mathrm{II},s-1)}\\
&\quad + r_1r_2\zeta^2 2K\Omega_*u_1^{(\mathrm{II},s-2)} - (r_1+r_2)\zeta\rho^{\mathrm{II}}\Omega_*^2u_2^{(\mathrm{II},s-1)} - r_1r_2\zeta^2\rho^{\mathrm{II}}\Omega_*^2u_2^{(\mathrm{II},s-2)},\\
&\quad (A\leftrightarrow B;\ \alpha\leftrightarrow\beta;\ r_1,r_2;\ \xi\leftrightarrow\eta;\ u,v;\ \tau_{12}\leftrightarrow\tau_{21};\ \tau_{11}\leftrightarrow\tau_{22};\ \tau_{13},\tau_{23}),\\
&\frac{\partial\tau_{33,2}^{(\mathrm{II},s)}}{\partial\zeta} - r_1\tau_{11,2}^{(\mathrm{II},s-1)} - r_2\tau_{22,2}^{(\mathrm{II},s-1)} + \frac{1}{A}\frac{\partial\tau_{13,2}^{(\mathrm{II},s-1)}}{\partial\xi} + \frac{1}{B}\frac{\partial\tau_{23,2}^{(\mathrm{II},s-1)}}{\partial\eta} + k_\beta R\tau_{13,2}^{(\mathrm{II},s-1)} \qquad (13)\\
&\quad + k_\alpha R\tau_{23,2}^{(\mathrm{II},s-1)} - 2K\Omega_*w_1^{(\mathrm{II},s)} + \rho^{\mathrm{II}}\Omega_*^2w_2^{(\mathrm{II},s)} = (r_1+r_2)\zeta 2K\Omega_*w_1^{(\mathrm{II},s-1)}\\
&\quad + r_1r_2\zeta^2 2K\Omega_*w_1^{(\mathrm{II},s-2)} - (r_1+r_2)\zeta\rho^{\mathrm{II}}\Omega_*^2w_2^{(\mathrm{II},s-1)} - r_1r_2\zeta^2\rho^{\mathrm{II}}\Omega_*^2w_2^{(\mathrm{II},s-2)},\\
&\frac{1}{A}\frac{\partial u_i^{(j,s-1)}}{\partial\xi} + k_\alpha Rv_i^{(j,s-1)} + r_1w_i^{(j,s-1)} + r_2\zeta\left(\frac{1}{A}\frac{\partial u_i^{(j,s-2)}}{\partial\xi} + k_\alpha Rv_i^{(j,s-2)} + r_1w_i^{(j,s-2)}\right)\\
&\quad = r_1\zeta a_{11}^j\tau_{11,i}^{(j,s-1)} + r_2\zeta a_{12}^j\tau_{22,i}^{(j,s-1)} + \sum_{1i}^{(j,s)},\\
&\quad \left(A,B;\ \alpha,\beta;\ r_1\leftrightarrow r_2;\ \xi,\eta;\ u\leftrightarrow v;\ \tau_{11}\leftrightarrow\tau_{22};\ a_{11},a_{22};\ \sum_{1i}^{(j,s)},\sum_{2i}^{(j,s)}\right),
\end{aligned}
$$

$$
\begin{aligned}
&\frac{\partial w_i^{(j,s)}}{\partial \zeta} + \zeta(r_1+r_2)\frac{\partial w_i^{(j,s-1)}}{\partial \zeta} + \zeta^2 r_1 r_2 \frac{\partial w_i^{(j,s-2)}}{\partial \zeta} = \sum_{3i}^{(j,s)} + r_1\zeta a_{13}^j \tau_{11,i}^{(j,s-1)} + r_2\zeta a_{23}^j \tau_{22,i}^{(j,s-1)},\\
&\frac{1}{B}\frac{\partial u_i^{(j,s-1)}}{\partial \eta} - k_\beta R v_i^{(j,s-1)} + r_1\zeta\left(\frac{1}{B}\frac{\partial u_i^{(j,s-2)}}{\partial \eta} - k_\beta R v_i^{(j,s-2)}\right) + \frac{1}{A}\frac{\partial v_i^{(j,s-1)}}{\partial \xi} - k_\alpha R u_i^{(j,s-1)}\\
&\qquad + r_2\zeta\left(\frac{1}{A}\frac{\partial v_i^{(j,s-2)}}{\partial \xi} - k_\alpha R u_i^{(j,s-2)}\right) = a_{66}^j \tau_{12,i}^{(j,s)} + r_1\zeta a_{66}^j \tau_{12,i}^{(j,s-1)},\\
&\frac{\partial u_i^{(j,s)}}{\partial \zeta} + \zeta(r_1+r_2)\frac{\partial u_i^{(j,s-1)}}{\partial \zeta} + \zeta^2 r_1 r_2 \frac{\partial u_i^{(j,s-2)}}{\partial \zeta} - r_1 u_i^{(j,s-1)} - \zeta r_1 r_2 u_i^{(j,s-2)}\\
&\qquad + \frac{1}{A}\frac{\partial w_i^{(j,s-1)}}{\partial \xi} + \frac{r_2\zeta}{A}\frac{\partial w_i^{(j,s-2)}}{\partial \xi} = a_{55}^j \tau_{13,i}^{(j,s)} + r_1\zeta a_{55}^j \tau_{13,i}^{(j,s-1)},\\
&\qquad (A,B;\ r_1 \leftrightarrow r_2;\ \xi,\eta;\ u,v;\ \tau_{13},\tau_{23};\ a_{55},a_{44}),\\
&\tau_{12,i}^{(j,s)} + r_1\zeta\tau_{12,i}^{(j,s-1)} = \tau_{21,i}^{(j,s)} + r_2\zeta\tau_{21,i}^{(j,s-1)},\\
&\sum_{ki}^{(j,s)} = a_{k1}\tau_{11,i}^{(j,s)} + a_{k2}\tau_{22,i}^{(j,s)} + a_{k3}\tau_{33,i}^{(j,s)}, \quad i=1,2, \quad k=1,2,3, \quad j=\mathrm{I},\mathrm{II}.
\end{aligned}
$$

Taking into account that $Q^{(j,m)} \equiv 0$ for $m < 0$, from system (13) we derive the following equations for determining the decomposition coefficients (8):

$$
\begin{aligned}
&\frac{\partial \tau_{13,i}^{(\mathrm{I},s)}}{\partial \zeta} + \rho^{\mathrm{I}}\Omega_*^2 u_i^{(\mathrm{I},s)} = P_{6\tau i}^{(\mathrm{I},s-1)}, \quad (13,23,33;\ u,v,w;\ 6\tau,5\tau,4\tau),\\
&\frac{\partial u_i^{(\mathrm{I},s)}}{\partial \zeta} - a_{55}^{\mathrm{I}}\tau_{13,i}^{(\mathrm{I},s)} = P_{ui}^{(\mathrm{I},s-1)},\\
&\frac{\partial v_i^{(\mathrm{I},s)}}{\partial \zeta} - a_{44}^{\mathrm{I}}\tau_{23,i}^{(\mathrm{I},s)} = P_{vi}^{(\mathrm{I},s-1)},\\
&\frac{\partial w_i^{(\mathrm{I},s)}}{\partial \zeta} - \sum_{3i}^{(\mathrm{I},s)} = P_{wi}^{(\mathrm{I},s-1)},\\
&\frac{\partial \tau_{13,1}^{(\mathrm{II},s)}}{\partial \zeta} + 2K\Omega_* u_2^{(\mathrm{II},s)} + \rho^{\mathrm{II}}\Omega_*^2 u_1^{(\mathrm{II},s)} = P_{6\tau 1}^{(\mathrm{II},s-1)}, \quad (13,23,33;\ u,v,w;\ 6\tau,5\tau,4\tau),\\
&\frac{\partial \tau_{13,2}^{(\mathrm{II},s)}}{\partial \zeta} - 2K\Omega_* u_1^{(\mathrm{II},s)} + \rho^{\mathrm{II}}\Omega_*^2 u_2^{(\mathrm{II},s)} = P_{6\tau 2}^{(\mathrm{II},s-1)}, \quad (13,23,33;\ u,v,w;\ 6\tau,5\tau,4\tau),\\
&\tau_{12,i}^{(j,s)} = P_{1\tau i}^{(j,s-1)}, \quad \tau_{21,i}^{(j,s)} = P_{1\tau i}^{(j,s-1)} - r_2\zeta\tau_{21,i}^{(j,s-1)} + r_1\zeta\tau_{12,i}^{(j,s-1)},\\
&\sum_{1i}^{(j,s)} = P_{2\tau i}^{(j,s-1)}, \quad \sum_{2i}^{(j,s)} = P_{3\tau i}^{(j,s-1)}, \quad j=\mathrm{I},\mathrm{II}, \quad i=1,2,\\
&\frac{\partial u_i^{(j,s)}}{\partial \zeta} - a_{55}^j\tau_{13,i}^{(j,s)} = P_{ui}^{(j,s-1)}, \quad \frac{\partial v_i^{(j,s)}}{\partial \zeta} - a_{44}^j\tau_{23,i}^{(j,s)} = P_{vi}^{(j,s-1)}, \quad \frac{\partial w_i^{(j,s)}}{\partial \zeta} - \sum_{3i}^{(j,s)} = P_{wi}^{(j,s-1)},
\end{aligned} \tag{14}
$$

where

$$
\begin{aligned}
P_{4\tau i}^{(\mathrm{I},s-1)} &= r_1\tau_{11,i}^{(\mathrm{I},s-1)} + r_2\tau_{22,i}^{(\mathrm{I},s-1)} - \frac{1}{A}\frac{\partial\tau_{13,i}^{(\mathrm{I},s-1)}}{\partial\xi} - \frac{1}{B}\frac{\partial\tau_{23,i}^{(\mathrm{I},s-1)}}{\partial\eta} \\
&- k_\beta R\tau_{13,i}^{(\mathrm{I},s-1)} - k_\alpha R\tau_{23,i}^{(\mathrm{I},s-1)} - (r_1+r_2)\zeta\rho^{\mathrm{I}}\Omega_*^2 w_i^{(\mathrm{I},s-1)} - r_1r_2\zeta^2\rho^{\mathrm{I}}\Omega_*^2 w_i^{(\mathrm{I},s-2)}, \\
P_{5\tau i}^{(\mathrm{I},s-1)} &= -\frac{1}{AB}\frac{\partial}{\partial\eta}(A\tau_{22,i}^{(\mathrm{I},s-1)}) + k_\alpha R\tau_{11,i}^{(\mathrm{I},s-1)} - \frac{1}{AB}\frac{\partial}{\partial\xi}(B\tau_{12,i}^{(\mathrm{I},s-1)}) - k_\beta R\tau_{21,i}^{(\mathrm{I},s-1)} \\
&- r_2\zeta\frac{\partial\tau_{23,i}^{(\mathrm{I},s-1)}}{\partial\zeta} - 2r_2\tau_{23,i}^{(\mathrm{I},s-1)} - (r_1+r_2)\zeta\rho^{\mathrm{I}}\Omega_*^2 v_i^{(\mathrm{I},s-2)} - r_1r_2\zeta^2\rho^{\mathrm{I}}\Omega_*^2 v_i^{(\mathrm{I},s-2)}, \\
&(5\tau,6\tau;\ A,B;\ u,v;\ \alpha,\beta;\ r_2,r_1;\ \xi,\eta;\ \tau_{11},\tau_{22};\ \tau_{12},\tau_{21};\ \tau_{23},\tau_{13}), \\
P_{4\tau 1}^{(\mathrm{II},s-1)} &= r_1\tau_{11,1}^{(\mathrm{II},s-1)} + r_2\tau_{22,1}^{(\mathrm{II},s-1)} - \frac{1}{A}\frac{\partial\tau_{13,1}^{(\mathrm{II},s-1)}}{\partial\xi} - \frac{1}{B}\frac{\partial\tau_{23,1}^{(\mathrm{II},s-1)}}{\partial\eta} \\
&- k_\beta R\tau_{13,1}^{(\mathrm{II},s-1)} - k_\alpha R\tau_{23,1}^{(\mathrm{II},s-1)} - (r_1+r_2)\zeta 2K\Omega_* w_2^{(\mathrm{II},s-1)} \\
&- r_1r_2\zeta^2 2K\Omega_* w_2^{(\mathrm{II},s-2)} - (r_1+r_2)\zeta\rho^{\mathrm{II}}\Omega_*^2 w_1^{(\mathrm{II},s-1)} - r_1r_2\zeta^2\rho^{\mathrm{II}}\Omega_*^2 w_1^{(\mathrm{II},s-2)}, \\
P_{5\tau 1}^{(\mathrm{II},s-1)} &= -\frac{1}{AB}\frac{\partial}{\partial\eta}(A\tau_{22,1}^{(\mathrm{II},s-1)}) + k_\alpha R\tau_{11,1}^{(\mathrm{II},s-1)} - \frac{1}{AB}\frac{\partial}{\partial\xi}(B\tau_{12,1}^{(\mathrm{II},s-1)}) \\
&- k_\beta R\tau_{21,1}^{(\mathrm{II},s-1)} - r_2\zeta\frac{\partial\tau_{23,1}^{(\mathrm{II},s-1)}}{\partial\zeta} - 2r_2\tau_{23,1}^{(\mathrm{II},s-1)} - (r_1+r_2)\zeta 2K\Omega_* v_2^{(\mathrm{II},s-1)} \\
&- r_1r_2\zeta^2 2K\Omega_* v_2^{(\mathrm{II},s-2)} - (r_1+r_2)\zeta\rho^{\mathrm{II}}\Omega_*^2 v_1^{(\mathrm{II},s-1)} - r_1r_2\zeta^2\rho^{\mathrm{II}}\Omega_*^2 v_1^{(\mathrm{II},s-2)}, \\
&(5\tau,6\tau;\ A,B;\ v,u;\ \alpha,\beta;\ r_2,r_1;\ \xi,\eta;\ \tau_{11},\tau_{22};\ \tau_{12},\tau_{21};\ \tau_{23},\tau_{13}), \\
P_{4\tau 2}^{(\mathrm{II},s-1)} &= r_1\tau_{11,2}^{(\mathrm{II},s-1)} + r_2\tau_{22,2}^{(\mathrm{II},s-1)} - \frac{1}{A}\frac{\partial\tau_{13,2}^{(\mathrm{II},s-1)}}{\partial\xi} - \frac{1}{B}\frac{\partial\tau_{23,2}^{(\mathrm{II},s-1)}}{\partial\eta} \\
&- k_\beta R\tau_{13,2}^{(\mathrm{II},s-1)} - k_\alpha R\tau_{23,2}^{(\mathrm{II},s-1)} + (r_1+r_2)\zeta 2K\Omega_* w_1^{(\mathrm{II},s-1)} \\
&+ r_1r_2\zeta^2 2K\Omega_* w_1^{(\mathrm{II},s-2)} - (r_1+r_2)\zeta\rho^{\mathrm{II}}\Omega_*^2 w_2^{(\mathrm{II},s-1)} - r_1r_2\zeta^2\rho^{\mathrm{II}}\Omega_*^2 w_2^{(\mathrm{II},s-2)}, \\
P_{5\tau 2}^{(\mathrm{II},s-1)} &= -\frac{1}{AB}\frac{\partial}{\partial\eta}(A\tau_{22,2}^{(\mathrm{II},s-1)}) + k_\alpha R\tau_{11,2}^{(\mathrm{II},s-1)} - \frac{1}{AB}\frac{\partial}{\partial\xi}(B\tau_{12,2}^{(\mathrm{II},s-1)}) \\
&- k_\beta R\tau_{21,2}^{(\mathrm{II},s-1)} - r_2\zeta\frac{\partial\tau_{23,2}^{(\mathrm{II},s-1)}}{\partial\zeta} - 2r_2\tau_{23,2}^{(\mathrm{II},s-1)} + (r_1+r_2)\zeta 2K\Omega_* v_1^{(\mathrm{II},s-1)} \\
&+ r_1r_2\zeta^2 2K\Omega_* v_1^{(\mathrm{II},s-2)} - (r_1+r_2)\zeta\rho^{\mathrm{II}}\Omega_*^2 v_2^{(\mathrm{II},s-1)} - r_1r_2\zeta^2\rho^{\mathrm{II}}\Omega_*^2 v_2^{(\mathrm{II},s-2)}, \\
&(5\tau,6\tau;\ A,B;\ v,u;\ \alpha,\beta;\ r_2,r_1;\ \xi,\eta;\ \tau_{11},\tau_{22};\ \tau_{12},\tau_{21};\ \tau_{23},\tau_{13}), \\
P_{1\tau i}^{(j,s-1)} &= \frac{1}{a_{66}^j}\left[\frac{1}{B}\frac{\partial u_i^{(j,s-1)}}{\partial\eta} - k_\beta R v_i^{(j,s-1)} + r_1\zeta\left(\frac{1}{B}\frac{\partial u_i^{(j,s-2)}}{\partial\eta} - k_\beta R v_i^{(j,s-2)}\right)\right. \\
&\left. + \frac{1}{A}\frac{\partial v_i^{(j,s-1)}}{\partial\xi} - k_\alpha R u_i^{(j,s-1)} + r_2\zeta\left(\frac{1}{A}\frac{\partial v_i^{(j,s-2)}}{\partial\xi} - k_\alpha R u_i^{(j,s-2)}\right) - r_1\zeta a_{66}^j\tau_{12,i}^{(j,s-1)}\right] \\
P_{2\tau i}^{(j,s-1)} &= \frac{1}{A}\frac{\partial u_j^{(j,s-1)}}{\partial\xi} + k_\alpha R v_i^{(j,s-1)} + r_1 w_i^{(j,s-1)} \\
&+ r_2\zeta\left(\frac{1}{A}\frac{\partial u_i^{(j,s-2)}}{\partial\xi} + k_\alpha R v_i^{(j,s-2)} + r_1 w_i^{(j,s-2)}\right) - r_1\zeta a_{11}^j\tau_{11,i}^{(j,s-1)} - r_2\zeta a_{12}^j\tau_{22,i}^{(j,s-1)}, \\
&(2\tau,3\tau;\ A,B;\ \alpha,\beta;\ r_1\leftrightarrow r_2;\ \xi,\eta;\ u\leftrightarrow v;\ \tau_{11}\leftrightarrow\tau_{22};\ a_{11},a_{22}),
\end{aligned} \tag{15}
$$

$$
\begin{aligned}
P_{ui}^{(j,s-1)} &= -\zeta(r_1+r_2)\frac{\partial u_i^{(j,s-1)}}{\partial\zeta} - \zeta^2 r_1 r_2 \frac{\partial u_i^{(j,s-2)}}{\partial\zeta} + r_1 u_i^{(j,s-1)} + \zeta r_1 r_2 u_i^{(j,s-2)} \\
&- \frac{1}{A}\frac{\partial w_i^{(j,s-1)}}{\partial\xi} - \frac{r_2\zeta}{A}\frac{\partial w_i^{(j,s-2)}}{\partial\xi} + r_1\zeta a_{55}^j \tau_{13,i}^{(j,s-1)}, \\
&(u,v;\ A,B;\ r_1 \leftrightarrow r_2;\ \xi,\eta;\ \tau_{13},\tau_{23};\ a_{55},a_{44}),\quad j=\mathrm{I,II},\quad i=1,2; \\
P_{wi}^{(j,s-1)} &= -\zeta(r_1+r_2)\frac{\partial w_i^{(j,s-1)}}{\partial\zeta} - \zeta^2 r_1 r_2 \frac{\partial w_i^{(j,s-2)}}{\partial\zeta} + r_1\zeta a_{13}^j \tau_{11,i}^{(j,s-1)} + r_2\zeta a_{23}^j \tau_{22,i}^{(j,s-1)}.
\end{aligned}
$$

We use relations (14) to express the stress tensor components in terms of $u_i^{(j,s)}$, $v_i^{(j,s)}$, and $w_i^{(j,s)}$:

$$
\begin{aligned}
\tau_{13,i}^{(j,s)} &= \frac{1}{a_{55}^{(j)}}\left[\frac{\partial u_i^{(j,s)}}{\partial\zeta} - P_{ui}^{(j,s-1)}\right], \quad \tau_{23,i}^{(j,s)} = \frac{1}{a_{44}^{(j)}}\left[\frac{\partial v_i^{(j,s)}}{\partial\zeta} - P_{vi}^{(j,s-1)}\right], \\
\tau_{12,i}^{(j,s)} &= P_{1\tau i}^{(j,s-1)}, \quad \tau_{21,i}^{(j,s)} = P_{1\tau i}^{(j,s-1)} - r_2\zeta\tau_{21,i}^{(j,s-1)} + r_1\zeta\tau_{12,i}^{(j,s-1)}, \\
\tau_{11,i}^{(j,s)} &= \frac{1}{\Delta^{(j)}}\left[\Delta_2^{(j)}\frac{\partial w_i^{(j,s)}}{\partial\zeta} + \Delta_{23}^{(j)}P_{2\tau i}^{(j,s-1)} + \Delta_1^{(j)}P_{3\tau i}^{(j,s-1)} - \Delta_2^{(j)}P_{wi}^{(j,s-1)}\right], \\
&(11,22,33;\ \Delta_2,\Delta_3,\Delta_{12};\ \Delta_{23},\Delta_1,\Delta_2;\ \Delta_1,\Delta_{13},\Delta_3),\quad j=\mathrm{I,II},\quad i=1,2,
\end{aligned}
\tag{16}
$$

where

$$
\begin{aligned}
&\Delta_1^{(j)} = a_{13}^j a_{23}^j - a_{33}^j a_{12}^j, \quad \Delta_2^{(j)} = a_{12}^j a_{23}^j - a_{22}^j a_{13}^j, \quad \Delta_3^{(j)} = a_{13}^j a_{12}^j - a_{11}^j a_{23}^j, \\
&\Delta^{(j)} = a_{11}^j \Delta_{23}^{(j)} + a_{13}^j \Delta_2^{(j)} + a_{12}^j \Delta_1^{(j)}, \quad \Delta_{nk}^{(j)} = a_{nn}^j a_{kk}^j - (a_{nk}^j)^2 \quad n,k=1,2,3.
\end{aligned}
$$

For determining the displacement vector components in the first layer, we obtain the equations

$$
\begin{aligned}
&\frac{\partial^2 u_i^{(\mathrm{I},s)}}{\partial\zeta^2} + a_{55}^{\mathrm{I}}\rho^{\mathrm{I}}\Omega_*^2 u_i^{(\mathrm{I},s)} = a_{55}^{\mathrm{I}}P_{6\tau i}^{(\mathrm{I},s-1)} + \frac{\partial P_{ui}^{(\mathrm{I},s-1)}}{\partial\zeta}, \quad (u,v;\ a_{55}^{\mathrm{I}},a_{44}^{\mathrm{I}};\ 6\tau,5\tau), \\
&\frac{\partial^2 w_i^{(\mathrm{I},s)}}{\partial\zeta^2} + \frac{\Delta^{(\mathrm{I})}}{\Delta_{12}^{(\mathrm{I})}}\rho^{\mathrm{I}}\Omega_*^2 w_i^{(\mathrm{I},s)} = F_{wi}^{(\mathrm{I},s-1)}, \quad i=1,2, \\
&F_{wi}^{(\mathrm{I},s-1)} = \frac{1}{\Delta_{12}^{(\mathrm{I})}}\left[\Delta^{(\mathrm{I})}P_{4\tau i}^{(\mathrm{I},s-1)} - \Delta_2^{(\mathrm{I})}\frac{\partial P_{2\tau i}^{(\mathrm{I},s-1)}}{\partial\zeta} - \Delta_3^{(\mathrm{I})}\frac{\partial P_{3\tau i}^{(\mathrm{I},s-1)}}{\partial\zeta} + \Delta_{12}^{(\mathrm{I})}\frac{\partial P_{wi}^{(\mathrm{I},s-1)}}{\partial\zeta}\right],
\end{aligned}
\tag{17}
$$

whose solutions have the form

$$
\begin{aligned}
u_i^{(\mathrm{I},s)} &= C_{1i}^{(\mathrm{I},s)}\sin(\chi^{(\mathrm{I},u)}\zeta) + C_{2i}^{(\mathrm{I},s)}\cos(\chi^{(\mathrm{I},u)}\zeta) + \bar{u}_i^{(\mathrm{I},s)}(\xi,\eta,\zeta), \\
v_i^{(\mathrm{I},s)} &= C_{3i}^{(\mathrm{I},s)}\sin(\chi^{(\mathrm{I},v)}\zeta) + C_{4i}^{(\mathrm{I},s)}\cos(\chi^{(\mathrm{I},v)}\zeta) + \bar{v}_i^{(\mathrm{I},s)}(\xi,\eta,\zeta), \\
w_i^{(\mathrm{I},s)} &= C_{5i}^{(\mathrm{I},s)}\sin(\chi^{(\mathrm{I},w)}\zeta) + C_{6i}^{(\mathrm{I},s)}\cos(\chi^{(\mathrm{I},w)}\zeta) + \bar{w}_i^{(\mathrm{I},s)}(\xi,\eta,\zeta), \quad i=1,2,
\end{aligned}
\tag{18}
$$

where

$$
\chi^{(\mathrm{I},u)} = \sqrt{a_{55}^{\mathrm{I}}\rho^{\mathrm{I}}}\,\Omega_*, \quad \chi^{(\mathrm{I},v)} = \sqrt{a_{44}^{\mathrm{I}}\rho^{\mathrm{I}}}\,\Omega_*, \quad \chi^{(\mathrm{I},w)} = \sqrt{\frac{\Delta^{(\mathrm{I})}\rho^{\mathrm{I}}}{\Delta_{12}^{(\mathrm{I})}}}\,\Omega_*,
$$

and $\bar{u}_i^{(\mathrm{I},s)}$, $\bar{v}_i^{(\mathrm{I},s)}$, $\bar{w}_i^{(\mathrm{I},s)}$ are particular solutions of equations (17). For determining the displacement vector components in the second layer, we obtain the equations

$$\begin{aligned}
&\frac{\partial^2 u_1^{(\mathrm{II},s)}}{\partial\zeta^2} + a_{55}^{\mathrm{II}}(\rho^{\mathrm{II}}\Omega_*^2 u_1^{(\mathrm{II},s)} + 2K\Omega_* u_2^{(\mathrm{II},s)}) = a_{55}^{\mathrm{II}} P_{6\tau 1}^{(\mathrm{II},s-1)} + \frac{\partial P_{u1}^{(\mathrm{II},s-1)}}{\partial\zeta}, \quad (u,v;\ a_{55},a_{44};\ 6\tau,5\tau),\\
&\frac{\partial^2 u_2^{(\mathrm{II},s)}}{\partial\zeta^2} + a_{55}^{\mathrm{II}}(\rho^{\mathrm{II}}\Omega_*^2 u_2^{(\mathrm{II},s)} - 2K\Omega_* u_1^{(\mathrm{II},s)}) = a_{55}^{\mathrm{II}} P_{6\tau 2}^{(\mathrm{II},s-1)} + \frac{\partial P_{u2}^{(\mathrm{II},s-1)}}{\partial\zeta}, \quad (u,v;\ a_{55},a_{44};\ 6\tau,5\tau),\\
&\frac{\partial^2 w_1^{(\mathrm{II},s)}}{\partial\zeta^2} + \frac{\Delta^{(\mathrm{II})}}{\Delta_{12}^{(\mathrm{II})}}(\rho^{\mathrm{II}}\Omega_*^2 w_1^{(\mathrm{II},s)} + 2K\Omega_* w_2^{(\mathrm{II},s)}) = F_{w1}^{(\mathrm{II},s-1)},\\
&\frac{\partial^2 w_2^{(\mathrm{II},s)}}{\partial\zeta^2} + \frac{\Delta^{(\mathrm{II})}}{\Delta_{12}^{(\mathrm{II})}}(\rho^{\mathrm{II}}\Omega_*^2 w_2^{(\mathrm{II},s)} - 2K\Omega_* w_1^{(\mathrm{II},s)}) = F_{w2}^{(\mathrm{II},s-1)},\\
&F_{wi}^{(\mathrm{II},s-1)} = \frac{1}{\Delta_{12}^{(\mathrm{II})}}\left[\Delta^{(\mathrm{II})} P_{4\tau i}^{(\mathrm{II},s-1)} - \Delta_2^{(\mathrm{II})}\frac{\partial P_{2\tau i}^{(\mathrm{II},s-1)}}{\partial\zeta} - \Delta_3^{(\mathrm{II})}\frac{\partial P_{3\tau i}^{(\mathrm{II},s-1)}}{\partial\zeta} + \Delta_{12}^{(\mathrm{II})}\frac{\partial P_{wi}^{(\mathrm{II},s-1)}}{\partial\zeta}\right], \quad i=1,2.
\end{aligned} \tag{19}$$

From (19),

$$\begin{aligned}
&u_2^{(\mathrm{II},s)} = \frac{1}{2K\Omega_* a_{55}^{\mathrm{II}}}\left[a_{55}^{\mathrm{II}} P_{6\tau 1}^{(\mathrm{II},s-1)} + \frac{\partial P_{u1}^{(\mathrm{II},s-1)}}{\partial\zeta} - \frac{\partial^2 u_1^{(\mathrm{II},s)}}{\partial\zeta^2} - a_{55}^{\mathrm{II}}\rho^{\mathrm{II}}\Omega_*^2 u_1^{(\mathrm{II},s)}\right],\\
&(u,v;\ a_{55},a_{44};\ 6\tau 1,5\tau 1);\\
&w_2^{(\mathrm{II},s)} = \frac{\Delta_{12}^{(\mathrm{II})}}{2\Delta^{(\mathrm{II})}K\Omega_*}\left[F_{w1}^{(\mathrm{II},s-1)} - \frac{\partial^2 w_1^{(\mathrm{II},s)}}{\partial\zeta^2} - \frac{\Delta^{(\mathrm{II})}}{\Delta_{12}^{(\mathrm{II})}}\rho^{\mathrm{II}}\Omega_*^2 w_1^{(\mathrm{II},s)}\right],
\end{aligned} \tag{20}$$

we derive the equations

$$\begin{aligned}
&\frac{\partial^4 u_1^{(\mathrm{II},s)}}{\partial\zeta^4} + 2a_{55}^{\mathrm{II}}\rho^{\mathrm{II}}\Omega_*^2\frac{\partial^2 u_1^{(\mathrm{II},s)}}{\partial\zeta^2} + a_{55}^{\mathrm{II}2}\Omega_*^2(\rho^{\mathrm{II}2}\Omega_*^2 + 4K^2)u_1^{(\mathrm{II},s)} = Q_{u1}^{(\mathrm{II},s-1)},\\
&(u,v,w;\ a_{55},a_{44},\Delta/\Delta_{12}),\\
&Q_{u1}^{(\mathrm{II},s-1)} = \Omega_* a_{55}^{\mathrm{II}}\Bigg(-a_{55}^{\mathrm{II}}2KP_{6\tau 2}^{(\mathrm{II},s-1)} - 2K\frac{\partial P_{u2}^{(\mathrm{II},s-1)}}{\partial\zeta} + \frac{1}{\Omega_*}\frac{\partial^2 P_{6\tau 1}^{(\mathrm{II},s-1)}}{\partial\zeta^2}\\
&\qquad + \frac{1}{\Omega_* a_{55}^{\mathrm{II}}}\frac{\partial^3 P_{u1}^{(\mathrm{II},s-1)}}{\partial\zeta^3} + \rho^{\mathrm{II}}\Omega_* a_{55}^{\mathrm{II}} P_{6\tau 1}^{(\mathrm{II},s-1)} + \Omega_*\rho^{\mathrm{II}}\frac{\partial P_{u1}^{(\mathrm{II},s-1)}}{\partial\zeta}\Bigg),\\
&(u,v;\ a_{55},a_{44};\ 6\tau,5\tau),\\
&Q_{w1}^{(\mathrm{II},s-1)} = \frac{\Delta^{(\mathrm{II})}\Omega_*}{\Delta_{12}^{(\mathrm{II})}}\left(\frac{\Delta_{12}^{(\mathrm{II})}}{\Delta^{(\mathrm{II})}\Omega_*}\frac{\partial^2 F_{w1}^{(\mathrm{II},s-1)}}{\partial\zeta^2} + \Omega_*\rho^{\mathrm{II}} F_{w1}^{(\mathrm{II},s-1)} - 2KF_{w2}^{(\mathrm{II},s-1)}\right),
\end{aligned} \tag{21}$$

The solutions of equations (21) are

$$u_1^{(\mathrm{II},s)} = u_{10}^{(\mathrm{II},s)}(\xi,\eta,\zeta) + u_{1\mathrm{p}}^{(\mathrm{II},s)}(\xi,\eta,\zeta), \quad (u,v,w), \tag{22}$$

where the index "0" denotes the general solution of the corresponding homogeneous equation, and the index "p" denotes a particular solution of the inhomogeneous equation.

The solutions of the corresponding homogeneous equations are

$$\begin{aligned}
u_{10}^{(\mathrm{II},s)}(\xi,\eta,\zeta) &= C_1^{(u,\mathrm{II},s)}(\xi,\eta)\varphi_{1u}(\zeta) + C_2^{(u,\mathrm{II},s)}(\xi,\eta)\varphi_{2u}(\zeta)\\
&+ C_3^{(u,\mathrm{II},s)}(\xi,\eta)\varphi_{3u}(\zeta) + C_4^{(u,\mathrm{II},s)}(\xi,\eta)\varphi_{4u}(\zeta), \quad (u,v,w),
\end{aligned} \tag{23}$$

where

$$\varphi_{1u}(\zeta) = \cosh(\gamma_u\zeta)\cos(\delta_u\zeta), \quad \varphi_{2u}(\zeta) = \sinh(\gamma_u\zeta)\sin(\delta_u\zeta),$$
$$\varphi_{3u}(\zeta) = \cosh(\gamma_u\zeta)\sin(\delta_u\zeta), \quad \varphi_{4u}(\zeta) = \sinh(\gamma_u\zeta)\cos(\delta_u\zeta),$$
$$\gamma_u = \sqrt{\frac{a_{55}^{\text{II}}\Omega_*}{2}\left(\sqrt{\rho^{\text{II}2}\Omega_*^2 + 4K^2} - \Omega_*\right)}, \quad \delta_u = \sqrt{\frac{a_{55}^{\text{II}}\Omega_*}{2}\left(\sqrt{\rho^{\text{II}2}\Omega_*^2 + 4K^2} + \Omega_*\right)},$$
$$(u, v, w;\ a_{55}a_{44}, \Delta/\Delta_{12}).$$

For the displacements with index "II" we have

$$\begin{aligned} u_2^{(\text{II},s)} &= u_{20}^{(\text{II},s)}(\xi,\eta,\zeta) + u_{2\text{p}}^{(\text{II},s)}(\xi,\eta,\zeta), \quad (u,v,w), \\ u_{20}^{(\text{II},s)}(\xi,\eta,\zeta) &= -C_1^{(u,\text{II},s)}(\xi,\eta)\varphi_{2u} + C_2^{(u,\text{II},s)}(\xi,\eta)\varphi_{1u} \\ &\quad + C_3^{(u,\text{II},s)}(\xi,\eta)\varphi_{4u} - C_4^{(u,\text{II},s)}(\xi,\eta)\varphi_{3u}, \quad (u,v,w), \\ u_{2\text{p}}^{(\text{II},s)}(\xi,\eta,\zeta) &= \frac{1}{a_{55}^{\text{II}}2K\Omega_*}\left(a_{55}^{\text{II}}P_{6\tau1}^{(\text{II},s-1)} + \frac{\partial P_{u1}^{(\text{II},s-1)}}{\partial\zeta} - \frac{\partial^2 u_{1\text{p}}^{(\text{II},s)}}{\partial\zeta^2} - a_{55}^{\text{II}}\rho^{\text{II}}\Omega_*^2 u_{1\text{p}}^{(\text{II},s)}\right). \end{aligned} \tag{24}$$

Satisfying conditions (9), (11), (5), (6) and (10), (11), (5), (6), we correspondingly obtain three algebraic systems of equations for $C_{ki}^{(\text{I},s)}(\xi,\eta)$, $C_m^{(u,\text{II},s)}$ ($k = \overline{1,6}$, $m = \overline{1,4}$, $i = 1,2$; u,v,w). The obtained systems have solutions if their determinants are different from zero. So the following conditions corresponding to boundary conditions (9), (11), (5), (6) must be satisfied:

$$\begin{aligned} \Delta_{u\tau} = \frac{\delta_u^2+\gamma_u^2}{2a_{55}^{\text{II}}}\cos^2(\chi^{(\text{I},u)}\zeta_1)[\cosh(2\gamma_u\zeta_2) + \cos(2\delta_u\zeta_2)] + \frac{(\chi^{(\text{I},u)})^2}{2a_{55}^{\text{I}2}}\sin^2(\chi^{(\text{I},u)}\zeta_1)[\cosh(2\gamma_u\zeta_2) \\ - \cos(2\delta_u\zeta_2)] - \frac{\chi^{(\text{I},u)}}{2a_{55}^{\text{I}}a_{55}^{\text{II}}}\sin(2\chi^{(\text{I},u)}\zeta_1)[\delta_u\sin(2\delta_u\zeta_2) + \gamma_u\sinh(2\gamma_u\zeta_2)] \neq 0, \quad (u,v,w), \end{aligned} \tag{25}$$

together with the conditions corresponding to boundary conditions (10), (11), (5), (6)

$$\begin{aligned} \Delta_{uu} = &\frac{\delta_u^2+\gamma_u^2}{2a_{55}^{\text{II}}}\sin^2(\chi^{(\text{I},u)}\zeta_1)[\cosh(2\gamma_u\zeta_2) + \cos(2\delta_u\zeta_2)] \\ &+ \frac{(\chi^{(\text{I},u)})^2}{2a_{55}^{\text{I}2}}\cos^2(\chi^{(\text{I},u)}\zeta_1)[\cosh(2\gamma_u\zeta_2) - \cos(2\delta_u\zeta_2)] \\ &+ \frac{\chi^{(\text{I},u)}}{2a_{55}^{\text{I}}a_{55}^{\text{II}}}\sin(2\chi^{(\text{I},u)}\zeta_1)[\delta_u\sin(2\delta_u\zeta_2) + \gamma_u\sinh(2\gamma_u\zeta_2)] \neq 0, \quad (u,v,w). \end{aligned} \tag{26}$$

In the case of one-layered shell [11] without viscous resistance, resonance (infinite amplitude) always occurs, and in the presence of viscous resistance [4], at resonance in the shell, the amplitude of forced vibrations is finite. In the case considered here, the presence of viscous resistance only in the lower layer of the shell results in resonance with an unlimited amplitude of forced vibrations, as the equations $\Delta_{u\tau} = 0$, $\Delta_{uu} = 0$, (u,v,w) have real roots which are the basic frequencies of natural vibrations, and resonance occurs when the imparted frequency of the forced effect concurs with the frequency of natural vibrations.

Conclusion

Dynamical three-dimensional problems for a two-layered orthotropic shell are solved asymptotically in the case of viscous resistance in one of the layers for two versions of the boundary conditions on the face surfaces. The amplitudes of forced vibrations are determined. It is shown that, for such a configuration, resonance with an infinite amplitude always occurs in the shell, in contrast to the case of a single-layered shell with viscosity, when the amplitude is finite.

References

[1] Aghalovyan L A 2015 *Asymptotic Theory of Anisotropic Plates and Shells* (Singapore - London: World Scientific Publishing) p 375
[2] Goldenveiser A L 1976 *Theory of Elastic Thin Shells* (Moscow: Nauka) p 512 [in Russian]
[3] Aghalovyan L A 2011 On the classes of problems for deformable one-layer and multilayer thin bodies solvable by the asymptotic method *Mekh. Kompoz. Mater.* **47** (1) 85–102 [*Mech. Compos. Mater.* (Engl. Transl.) **47** (1) 59–72]
[4] Ghulghazaryan L G 2015 Forced vibrations of orthotropic shells when there is viscous resistance *J. Appl. Math. Mech.* **79** (3) 281–92
[5] Azatyan G L 2007 Forced vibrations of an orthotropic plate when there is viscous resistance *Izv. Nats. Akad. Nauk Armenii. Mekh.* **60** (2) 29–40
[6] Aghalovyan L A 2000 To the asymptotic method of solution of dynamically mixed problems of anisotropic strips and plates *Izv. Vuzov. Severo-Kavkaz. Region. Est. Nauk: Natural sciences* No 3 8–11
[7] Aghalovyan L A, Gevorgyan R S and Ghulghazaryan L G 2010 The asymptotic solutions of 3D dynamic problems of orthotropic cylindrical and toroid shells *Izv. Nats. Akad. Nauk Armenii. Mekh.* **63** (1) 6–22
[8] Aghalovyan L A 2002 Asymptotics of solutions of classical and non-classical boundary-value problems of statics and dynamics of thin bodies *Appl. Mech.* **38** (7) 3–24
[9] Aghalovyan L A and Ghulghazaryan L G 2006 Asymptotic solutions of non-classical boundary-value problems of natural vibrations of orthotropic shells *J. Appl. Math. Mech.* **70** (1) 102–15
[10] Ghulghazaryan L G 2008 The own vibrations of two-layered orthotropic shells under the complete contact between the layers In *Proc. 6th Int. Conf. "The Problems of Dynamics of Interaction of Deformable Media," September 21–26, Goris - Stepanakert, 2008* (Goris - Stepanakert: Inst. Mekh. NAN RA) 202–11 [in Russian]
[11] Agalovyan L A and Gulgazaryan L G 2009 Forced vibrations of orthotropic shells: nonclassical boundary-value problems *Int. Appl. Mech* **25** (8) 888–903
[12] Ghulghazaryan L G 2013 Own vibrations of two-layered orthotropic shells under an incomplete contact between the layers *Vestnik Fonda Fund. Issled.* **65** (3) 51–69

IOP Conf. Series: Journal of Physics: Conf. Series **991** (2018) 012003 doi:10.1088/1742-6596/991/1/012003

Fundamentals of continuum mechanics – classical approaches and new trends[1]

H Altenbach

Institut für Mechanik, Otto-von-Guericke-Universität Magdeburg, Magdeburg, Germany

E-mail: holm.altenbach@ovgu.de

Abstract. Continuum mechanics is a branch of mechanics that deals with the analysis of the mechanical behavior of materials modeled as a continuous manifold. Continuum mechanics models begin mostly by introducing of three-dimensional Euclidean space. The points within this region are defined as material points with prescribed properties. Each material point is characterized by a position vector which is continuous in time. Thus, the body changes in a way which is realistic, globally invertible at all times and orientation-preserving, so that the body cannot intersect itself and as transformations which produce mirror reflections are not possible in nature. For the mathematical formulation of the model it is also assumed to be twice continuously differentiable, so that differential equations describing the motion may be formulated. Finally, the kinematical relations, the balance equations, the constitutive and evolution equations and the boundary and/or initial conditions should be defined. If the physical fields are non-smooth jump conditions must be taken into account.

The basic equations of continuum mechanics are presented following a short introduction. Additionally, some examples of solid deformable continua will be discussed within the presentation. Finally, advanced models of continuum mechanics will be introduced.

1. Introduction

The basic ideas with respect to continuum mechanics are given in a German textbook [1] which is written in the style of famous Russian school of profs' A. I. Lurie (1901–1980), V. A. Pal'mov and P. A. Zhilin (1942–2005). These professors presenting from the 60th of the last century up to the beginning of 2000s various courses at the Leningrad Polytechnic Institute/Peter the Great State Polytechnical University containing various elements of Continuum Mechanics. Lurie was more focused on elasticity problems. Palmov's idea was to combine continuum mechanics with rheological modeling. Zhilin presented, together with the three-dimensional approach, special courses devoted to one- and two-dimensional continua. Unfortunately, a part of monographs, related to various courses, have been published only in Russian [2–5] in which the following topics were typically discussed. The remaining monograph were translated [6–8].

The common idea of their courses was

- the use of the direct tensor notation,
- the clear split of material-independent and material-dependent equations,
- the introduction of five balance equations only,

[1] The paper is dedicated to Alexander Manzhirov's 60th birthday.

IOP Conf. Series: Journal of Physics: Conf. Series **991** (2018) 012003 doi:10.1088/1742-6596/991/1/012003

- the formulation of boundary conditions, and
- the formulation of initial conditions.

It should be noted that the courses were strongly based on mathematics. But the starting point was always mechanics (or physics) and only after that mathematics. This is partly in contrast to Clifford Truesdell (1919–2000) and his followers.

These items are discussed briefly as follows. At first, the elements are presented for the case of classical continua, sometimes named Cauchy (after Augustin-Louis Cauchy, 1789–1857) or Boltzmann (after Ludwig Boltzmann, 1844–1906) continua, and then some advanced models are introduced. The latter step is not final since many new models were introduced in recent years.

Continuum mechanics is a branch of mechanics that deals with the analysis of the mechanical behavior of materials and structures modeled as a continuous mass rather than as discrete particles. Maybe the French mathematician Augustin-Louis Cauchy was the first to formulate such models in the 19th century. Up to now, one has to understand that the starting point of continuum mechanics is not only the axioms of Isaac Newton (1643–1727), but also the ideas of Leonhard Euler (1707–1783). With the help of Newton's axiom, we will get only the simplest theory based on force interactions. Euler mentioned that there is more: for the creation of the beam theory it was necessary to assume the independence of forces and moments. Nowadays, research on developing new theories, based on Newton's axioms, is being carried out in several directions.

Some elements of continuum mechanics should be introduced from the very beginning. Firstly, one has to assume a space: let us consider here the Euclidean three-dimensional space — in special cases two-dimensional or one-dimensional spaces will be used. The four-dimensional space, discussed, for example, by Minkowski (1864–1909) and mentioned in [9, 10], is not applied here. Next step is the introduction of the (pseudo) time $t \geq 0$. After that, one has to define the body, with/without mass, which has a volume and a surface with an outward-pointing normal. Last but not least, homogeneity and isotropy should be assumed among other things. A continuum theory must be valid for solids and fluids. In addition, the theory should be scale-independent. Namely, not only the theory is applicable the the macro-scale but also it should provide proper results in the case of meso, micro or nano.

The details of the direct tensor notation will be not given here. However, a very good presentation of the tensor calculus is given in the appendices of [6, 7] and in the recent published textbook [11]. All these references followed the ideas of Josiah Willard Gibbs (1839–1903) and Max Lagally (1881–1945), see for example [12, 13].

2. Material-independent equations

In this section, the material-independent equations will be introduced, i.e., the equations are independent from the individual material response. The kinematical relations are based only on geometrical considerations and contain the changes of the geometry. Forces and stresses are the basic reasons for the geometrical changes in the classical continuum mechanics. The balances are physical principles formulated here in the general sense for the continuum volume. We have to distinguish between balance equation and inequality. There are four balance equations, the fifth one is an inequality necessary for the process direction estimation. Boundary and initial conditions allowing to solve initial-boundary problems. Jump conditions are necessary if the physical fields are non-smooth.

2.1. Kinematics

The starting point for any kinematical relation is the introduction of configurations. Generally, a minimum of two configurations is employed: the reference configuration at the initial moment t_0 and the actual configuration at the moment t. It is obvious that instead of time we have a (pseudo)time. It is our decision to fix the moment t_0 (the real moment of the beginning is unknown). In addition, it should be underlined that in many textbooks the initial configuration is related to the Lagrangian (named after Joseph-Louis Lagrange, 1736–1813) description and the reference to the Eulerian one. However historically, this is not correct: the Langrangian description was introduced by Leonhard Euler (1707–1783) in 1762 and the Eulerian by Jean-Baptiste le Rond d'Alambert (1717–1783) in 1752 [14, 15]. It should be noted that many other configurations may also exist: stress-free configuration, damaged configuration, etc. For solids, the Lagrangian description is preferred whereas the Eulerian description is used for fluids. But this restrictive split is under question if we deal, for example, with plasticity problems.

Any configuration can be introduced with the help of a reference point and a position vector. A coordinate system at the beginning is not necessary, but may be helpful in solving problems. At the same time, one has to select either a fixed- or a moving reference point. The latter allows to increase the number of distinct configurations.

In mechanics, geometrical changes are related to the motion and to the strains. At this stage, it is necessary to make a distinction by introducing a property of the medium: rigidity or deformability. In the first case, all material points of the body have the same motion (translations and rotations), while the distance between each two arbitrary points remains the same at all times. In the second case, we disregard the rigid motions and focus on the geometrical changes in the infinitesimal surrounding of the material points. It can be shown that these changes are related to extension/shorting and angle changes. In the case of extension/shortening, we look at the length of infinitesimal small radius-vector whereas in the second case, the focus is on the angle between two infinitesimal small radius-vectors. Note that in some applications, both approaches should be adopted concurrently.

The radius-vector approach for description of the material point positions results in two strain measures in the literature. After introducing the radius-vector in both configurations that means $\boldsymbol{R}$ for the reference configuration and $\boldsymbol{r}$ for the actual one, the displacement vector can be expressed as:

$$\boldsymbol{u}(\boldsymbol{R}, t) = \boldsymbol{r}(\boldsymbol{R}, t) - \boldsymbol{R}$$

with $\boldsymbol{r}(t = t_0) = \boldsymbol{R}$. This approach is the common way to present the Cauchy's strains in the geometrically-linear theory. The second approach is related to the deformation gradient — better gradient of the radius-vector, see [7]:

$$\boldsymbol{F} = (\boldsymbol{\nabla}_{\boldsymbol{R}}\boldsymbol{r})^{\mathrm{T}}$$

This property is more appropriate for the case of large strains since after some manipulation a more clear representation of the strains and rigid body motions is obtained. Note that linearizing each one of the relations coincides with the other one. More details about different strain measures are presented, for example, in [2, 7]. Concerning the linearization procedure, a excellent method is given in [16, 17].

With the strain measures and strain tensors one has only one part of the kinematics. In addition, the velocity vector, the acceleration vector, the strain rate tensor, and the strain gradient rate should be introduced. The calculation of time derivatives can be readily done in the case of geometrically-linear theory while it is more complicated in geometrically non-linear problems. One problem arises from the objectivity of these derivatives for instance. Different approaches are presented in the literature, see [5] and [6] among others. Some actual statements are presented in [18, 19].

2.2. Forces and stresses

The following classification of the external loading can be given:

(i) *Natural* loading models

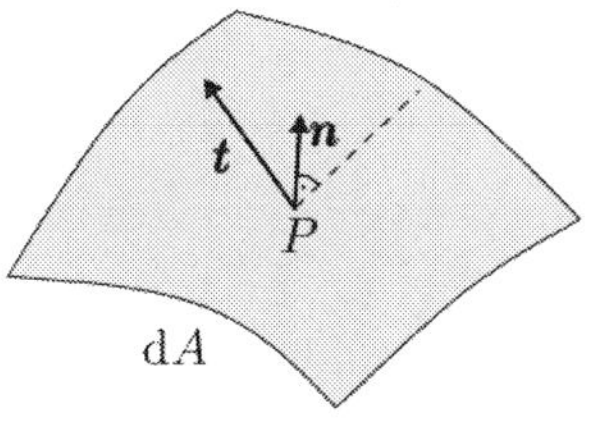

Figure 1. Force stress vector.

- body / mass / volume loading (forces, moments), and
- surface / contact loading (forces, moments).

(ii) *Artificial* loading models
- line loading (forces, moments), and
- single point loading (forces, moments).

Note that as in the Course of General Mechanics forces and moments are introduced. In addition, it should be mentioned that a moment is not only a moment of the force (torque) or a pair of forces — it can defined as an independent quantity. Dimensional analysis provides the following results for the single loads:

- dimension of force $[\mathbf{F}] = \mathrm{N}$ and
- dimension of moment $[\mathbf{M}] = \mathrm{Nm}$.

Hence, one can conclude that:

- body loads are expressed per unit volume,
- surface loads are expressed per unit area, and
- line loads are expressed per unit length.

Let us introduce body loadings:

- body force $\rho(\boldsymbol{r},t)\boldsymbol{k}(\boldsymbol{r},t) = \boldsymbol{k}^{\mathrm{V}}(\boldsymbol{r},t)$, and by analogues
- $\rho(\boldsymbol{r},t)\boldsymbol{l}(\boldsymbol{r},t) = \boldsymbol{l}^{\mathrm{V}}(\boldsymbol{r},t)$ the body moment.

Here $\rho = \rho(\boldsymbol{r},t)$ is the density. Examples of body forces are the weight force, the inertia force and potential force. The body moments in the classical theory will be ignored.

After introduction of forces and moments, we can define various stress vectors

- the force stress vector
$$\boldsymbol{t} = \lim_{\Delta A \to 0} \frac{\Delta \mathbf{F}}{\Delta A},$$
- and the couple stress vector
$$\boldsymbol{m} = \lim_{\Delta A \to 0} \frac{\Delta \mathbf{M}}{\Delta A}.$$

As usual, the second one is ignored in the textbooks of strength of materials. There are several arguments why the couple stresses can be ignored among which one is presented in [20].

Let us introduce an infinitesimal small surface element $\mathrm{d}A$ with the unit normal $\boldsymbol{n}$ (figure 1). The acting force stress vector does not have the same value over the surface but the difference of any two neighbouring magnitudes is negligible. From the definition of moment, it follows that the moment of the couple is infinitesimal since a small difference is multiplied by a small distance. Therefore, the couple stress vector can be ignored in this case.

A counterexample is the case of stress vectors on the crack tip: even over an infinitesimal small distance, the change in the force stress vector can be huge. Such cases initiate the arguments over

IOP Conf. Series: Journal of Physics: Conf. Series **991** (2018) 012003 doi:10.1088/1742-6596/991/1/012003

introducing couple stresses. Maybe this was the reason for extending the appendix in [20] and to publish another book on elasticity taking into account asymmetric stress tensors in addition to the symmetric ones [21].

Note Similar derivations can be established for both the actual and the reference configuration. Below we are performing the procedures for the actual configuration. That means writing $\boldsymbol{\nabla}$ implies $\boldsymbol{\nabla}_{\boldsymbol{R}}$, also $\boldsymbol{t}$ and $\boldsymbol{T}$ are the Cauchy stress vector and tensor in the actual configuration, respectively.

The components of the stress vector can be presented in Cartesian coordinates as

$$\boldsymbol{t} = t_{\boldsymbol{n}}\boldsymbol{n} + t_t\boldsymbol{e}_t = t_{\boldsymbol{n}}\boldsymbol{n} + t_{t_1}\boldsymbol{e}_{t_1} + t_{t_2}\boldsymbol{e}_{t_2},$$

where $\boldsymbol{e}_{t_1}$ and $\boldsymbol{e}_{t_2}$ are two arbitrary tangential directions on the surface $\mathrm{d}A$. Since we have a Cartesian coordinate system $\boldsymbol{n}\perp\boldsymbol{e}_{t_1}$, $\boldsymbol{n}\perp\boldsymbol{e}_{t_2}$, and $\boldsymbol{e}_{t_1}\perp\boldsymbol{e}_{t_2}$, so $\boldsymbol{n}$, $\boldsymbol{e}_{t_1}$, and $\boldsymbol{e}_{t_2}$ form an arbitrary orthonormal unit base.

Following Cauchy, a lemma can be written down

$$\boldsymbol{t}(\boldsymbol{r},\boldsymbol{n},t) = \boldsymbol{n}\cdot\boldsymbol{T}(\boldsymbol{r},t)$$

connecting the stress vector with the stress tensor. Cauchy got his lemma from equilibrium considerations [22]. Note that in many textbooks [23, for example] the lemma is presented in a different way

$$\boldsymbol{t}(\boldsymbol{r},\boldsymbol{n},t) = \boldsymbol{T}(\boldsymbol{r},t)\cdot\boldsymbol{n}.$$

This is equivalent to the linear mapping of two vector spaces with the help of a second rank tensor. Both formulations coincide if the stress tensor is symmetrical.

As in General Mechanics now the resultant force

$$\boldsymbol{F}^{\mathrm{R}} = \int_V \rho\boldsymbol{k}\,\mathrm{d}V + \int_A \boldsymbol{t}\,\mathrm{d}A$$

and the resultant moment

$$\boldsymbol{M}^{\mathrm{R}} = \int_V (\boldsymbol{r}\times\rho\boldsymbol{k} + \rho\boldsymbol{l})\,\mathrm{d}V + \int_A (\boldsymbol{r}\times\boldsymbol{t} + \boldsymbol{m})\,\mathrm{d}A$$

can be introduced. Taking into account only force actions the static equilibrium can be formulated for the continuum with $\boldsymbol{F}^{\mathrm{R}} = \boldsymbol{0}$ and $\boldsymbol{M}^{\mathrm{R}} = \boldsymbol{0}$

$$\int_V \rho\boldsymbol{k}\,\mathrm{d}V + \int_A \boldsymbol{t}\,\mathrm{d}A = \boldsymbol{0}, \qquad \int_V (\boldsymbol{r}\times\rho\boldsymbol{k})\,\mathrm{d}V + \int_A (\boldsymbol{r}\times\boldsymbol{t})\,\mathrm{d}A = \boldsymbol{0}.$$

With the divergence theorem (formulated independently by Carl Friedrich Gauß, 1777–1855, and Michail Wassiljewitsch Ostrogradski, 1801–1862) for the first equilibrium equation

$$\int_A \boldsymbol{t}\,\mathrm{d}A = \int_A \boldsymbol{n}\cdot\boldsymbol{T}\,\mathrm{d}A = \int_V \boldsymbol{\nabla}\cdot\boldsymbol{T}\,\mathrm{d}A$$

one gets

$$\int_V (\rho\boldsymbol{k} + \boldsymbol{\nabla}\cdot\boldsymbol{T})\,\mathrm{d}V = \boldsymbol{0}$$

and the local form can be obtained (if all fields are smooth)

$$\boldsymbol{\nabla}\cdot\boldsymbol{T} + \rho\boldsymbol{k} = \boldsymbol{0}$$

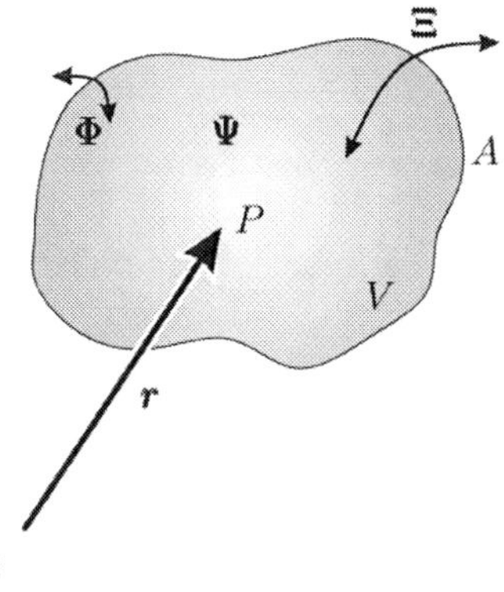

Figure 2. Formulation of the general balance equation.

Applying the divergence theorem to the second equilibrium equation the symmetry condition for the stress tensor ($\boldsymbol{T} = \boldsymbol{T}^{\mathrm{T}}$) can be established. Finally, with the d'Alambert's Principle one gets the dynamic equations (Euler's equation of motion)

$$\int_V \rho \boldsymbol{k} \,\mathrm{d}V + \int_A \boldsymbol{t} \,\mathrm{d}A - \int_V \rho \ddot{\boldsymbol{r}} \,\mathrm{d}V = \boldsymbol{0}.$$

Applying again the divergence theorem we obtain

$$\int_V (\rho \boldsymbol{k} + \boldsymbol{\nabla} \cdot \boldsymbol{T} - \rho \ddot{\boldsymbol{R}}) \,\mathrm{d}V = \boldsymbol{0}$$

and finally, if all fields are smooth, the local form is obtained

$$\boldsymbol{\nabla} \cdot \boldsymbol{T} + \rho \boldsymbol{k} = \rho \ddot{\boldsymbol{r}}.$$

2.3. Balances

2.3.1. General balance equation. Let us introduce some quantity named balance variable of the continuum $\Psi(\boldsymbol{r}, t)$ and defined in the actual configuration. The integration over all body points results in $Y(t)$

$$Y(t) = \int_V \Psi(\boldsymbol{r}, t) \,\mathrm{d}V.$$

With $\mathrm{d}V = \det\boldsymbol{F} \,\mathrm{d}V_0$ the recalculation of the properties for the reference configuration is possible.

The rate of $Y(t)$ is influenced by actions onto the volume and through the surface (flux)

$$\frac{\mathrm{D}}{\mathrm{D}t} Y(t) = \frac{\mathrm{D}}{\mathrm{D}t} \int_V \Psi(\boldsymbol{r}, t) \,\mathrm{d}V = \int_A \Phi(\boldsymbol{r}, t) \,\mathrm{d}A + \int_V \Xi(\boldsymbol{r}, t) \,\mathrm{d}V,$$

where $\mathrm{D}/\mathrm{D}t$ is the total time derivative, Φ and Ξ are named flux and supply, respectively. The visualization of the general balance equation is given in figure 2.

It should be underlined that the general balance equations can be introduced in a different form with three terms on the right-hand side [23]. Instead of one surface and one volume integral for the flux and the supply one surface and two volume integrals for the flux, the supply and the production are introduced. But there are only few applications for production. If Cauchy's lemma is valid for the general form after application of the divergence theorem one gets again the local form.

Note If the right-hand side of the general balance is equal to zero we have a conservation law. In other words, the first integral is a constant.

2.3.2. Balance of mass. In solid mechanics applications, the formulation of the balance equation for mass starts with

$$m = \int_V \rho(\boldsymbol{r}, t)\,\mathrm{d}V = \text{const},$$

where m is the mass. Then the integral form of the balance of mass is

$$\frac{\mathrm{D}m}{\mathrm{D}t} = \frac{\mathrm{D}}{\mathrm{D}t}\int_V \rho(\boldsymbol{r}, t)\,\mathrm{d}V = 0$$

and the local form is

$$\frac{\mathrm{D}}{\mathrm{D}t}(\mathrm{d}m) = \frac{\mathrm{D}}{\mathrm{D}t}(\rho\,\mathrm{d}V) = 0.$$

From this equation the continuity equation can be deduced

$$\frac{\mathrm{D}\rho}{\mathrm{D}t} + \rho\boldsymbol{\nabla}\cdot\boldsymbol{v} = 0$$

with the velocity $\boldsymbol{v} = \dot{\boldsymbol{r}}$. If $\rho = \text{const}$ with respect to the time the incompressibility condition is following as

$$\boldsymbol{\nabla}\cdot\boldsymbol{v} = 0.$$

2.3.3. Balance of momentum. By introducing the linear momentum of the continuum, the integral form can be written down as

$$\frac{\mathrm{D}}{\mathrm{D}t}\int_V \rho(\boldsymbol{r}, t)\boldsymbol{v}(\boldsymbol{r}, t)\,\mathrm{d}V = \int_A \boldsymbol{t}(\boldsymbol{r}, t)\mathrm{d}A + \int_V \rho(\boldsymbol{r}, t)\boldsymbol{k}(\boldsymbol{r}, t)\,\mathrm{d}V.$$

The local form is

$$\frac{\mathrm{D}}{\mathrm{D}t}\rho(\boldsymbol{r}, t)\boldsymbol{v}(\boldsymbol{r}, t) = \boldsymbol{\nabla}\cdot\boldsymbol{T}(\boldsymbol{r}, t) + \rho(\boldsymbol{r}, t)\boldsymbol{k}(\boldsymbol{r}, t).$$

2.3.4. Balance of moment of momentum. By introducing the angular momentum of the continuum, the integral form can be written down as

$$\frac{\mathrm{D}}{\mathrm{D}t}\int_V \boldsymbol{r}\times\rho(\boldsymbol{r}, t)\boldsymbol{v}(\boldsymbol{r}, t)\mathrm{d}V = \int_A \boldsymbol{r}\times\boldsymbol{t}(\boldsymbol{r}, t)\mathrm{d}A + \int_V \boldsymbol{r}\times\rho(\boldsymbol{r}, t)\boldsymbol{k}(\boldsymbol{r}, t)\,\mathrm{d}V.$$

If the balance of momentum is valid as given in the previous section, the local form results again in the symmetry condition for the stress tensor.

2.3.5. Balance of energy Let us assume at the beginning that we have only mechanical actions. In this case, the balance of energy, which is equivalent to the first law of thermodynamics, can be presented as

$$\frac{\mathrm{D}}{\mathrm{D}t}\int_V \left(\frac{1}{2}\boldsymbol{v}\cdot\boldsymbol{v} + u\right)\rho\,\mathrm{d}V = \int_A \boldsymbol{t}\cdot\boldsymbol{v}\,\mathrm{d}A + \int_V \boldsymbol{k}\cdot\boldsymbol{v}\rho\,\mathrm{d}V.$$

The kinetic and potential energy is influenced only by surface and volume forces. If the balance of moment The local form can be expressed as

$$\rho\dot{u} = \boldsymbol{T}\cdot\cdot(\boldsymbol{\nabla}\boldsymbol{v})^{\mathrm{T}} = \boldsymbol{T}\cdot\cdot\boldsymbol{D}.$$

The general first law of thermodynamics states that:

IOP Conf. Series: Journal of Physics: Conf. Series **991** (2018) 012003 doi:10.1088/1742-6596/991/1/012003

The changes in time of the total energy W within the volume is equal to the heat flux Q and the power of all external loadings P_{ext}

$$\frac{\mathrm{D}}{\mathrm{D}t}W = P_{\text{ext}} + Q,$$

where $W = U + K$ is the sum of the inner energy U and the kinetic energy K.

Let us introduce the four terms

$$K = \frac{1}{2}\int_V \boldsymbol{v}\cdot\boldsymbol{v}\rho\,\mathrm{d}V,$$
$$U = \int_m u\,\mathrm{d}m = \int_V \rho u\,\mathrm{d}V,$$
$$P_{\text{ext}} = \int_A \boldsymbol{t}\cdot\boldsymbol{v}\,\mathrm{d}A + \int_V \boldsymbol{k}\cdot\boldsymbol{v}\rho\,\mathrm{d}V,$$
$$Q = \int_V \rho r\,\mathrm{d}V - \int_A \boldsymbol{n}\cdot\boldsymbol{h}\,\mathrm{d}A.$$

Finally, we have

$$\frac{\mathrm{D}}{\mathrm{D}t}\int_V \left(u + \frac{1}{2}\boldsymbol{v}\cdot\boldsymbol{v}\right)\rho\,\mathrm{d}V = \int_A \boldsymbol{t}\cdot\boldsymbol{v}\,\mathrm{d}A + \int_V \boldsymbol{k}\cdot\boldsymbol{v}\rho\,\mathrm{d}V - \int_A \boldsymbol{n}\cdot\boldsymbol{h}\,\mathrm{d}A + \int_V \rho r\,\mathrm{d}V.$$

With the help of the Cauchy lemma and after some manipulations the local form can be deduced (it is assumed that the balance of momentum is valid):

$$\frac{\mathrm{D}u}{\mathrm{D}t} + \boldsymbol{T}\cdot\cdot\boldsymbol{D} - \boldsymbol{\nabla}\cdot\boldsymbol{h} + \rho r = 0.$$

2.3.6. Balance of entropy. The last balance cannot be formulated in the standard form since now we have two formulations: as an equation for non-dissipative processes or as an inequality for dissipative processes. The decision on the type of process can only be made after introducing statements concerning the material behavior.

Let us introduce the second law of thermodynamics in the integral form

$$\frac{\mathrm{D}}{\mathrm{D}t}\int_V \rho s\,\mathrm{d}V \geq \int_V \frac{r}{\Theta}\rho\,\mathrm{d}V - \int_A \frac{\boldsymbol{n}\cdot\boldsymbol{h}}{\Theta}\,\mathrm{d}A,$$

where s is the specific entropy, r is a energy source, Θ denotes the absolute temperature, and $\boldsymbol{h}$ is the heat flux vector. The changes in time of the entropy within the volume under consideration is not smaller then the rate of the outer entropy flux. Again after some manipulations

$$\int_A \frac{\boldsymbol{n}\cdot\boldsymbol{h}}{\Theta}\,\mathrm{d}A = \int_V \boldsymbol{\nabla}\cdot\frac{\boldsymbol{h}}{\Theta}\,\mathrm{d}V = \int_V \left(\frac{\boldsymbol{\nabla}\cdot\boldsymbol{h}}{\Theta} - \frac{\boldsymbol{h}\cdot\boldsymbol{\nabla}\Theta}{\Theta^2}\right)\mathrm{d}V$$

and taking into account

$$\frac{1}{\Theta}\boldsymbol{h}\cdot\boldsymbol{\nabla}\Theta = \boldsymbol{h}\cdot\boldsymbol{\nabla}\ln\Theta$$

we get the local form

$$\rho\theta\frac{\mathrm{D}s}{\mathrm{D}t} - \underline{\rho r - \boldsymbol{\nabla}\cdot\boldsymbol{h}} - \boldsymbol{h}\cdot\boldsymbol{\nabla}\ln\theta \geq 0.$$

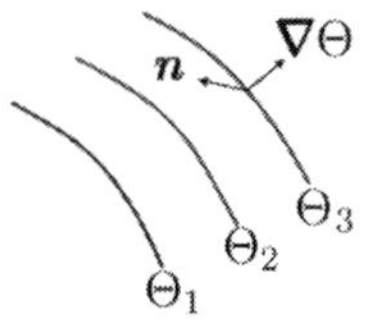

Figure 3. Temperature field $\Theta_1 < \Theta_2 < \Theta_3$.

The underlined terms are also in the first law of thermodynamics. Combining both equations we get

$$\rho\theta\frac{\mathrm{D}s}{\mathrm{D}t} - \frac{\mathrm{D}u}{\mathrm{D}t} + \boldsymbol{T}\cdot\cdot\boldsymbol{D} - \boldsymbol{h}\cdot\boldsymbol{\nabla}\ln\theta \geq 0.$$

With the Helmholtz free energy

$$u - \Theta s = f$$

we can define the dissipation function

$$\boldsymbol{T}\cdot\cdot\boldsymbol{D} - \rho(\dot{f} + s\dot{\Theta}) = \Phi \geq 0.$$

Finally, we obtain

$$\Phi - \boldsymbol{h}\cdot\boldsymbol{\nabla}\ln\theta \geq 0.$$

Since Φ is positive, the remaining part can be written

$$\boldsymbol{h}\cdot\boldsymbol{\nabla}\ln\theta \leq 0.$$

The limiting cases are defined by the following two processes:

- $\boldsymbol{h} = \boldsymbol{0}$ adiabatic process and
- $\boldsymbol{\nabla}\Theta = \boldsymbol{0}$ isothermal process.

Assuming a non-dissipative process $\Phi = 0$ with temperature field (figure 3).

The following solution is valid only for

$$\angle(\boldsymbol{h}, \boldsymbol{\nabla}\Theta) > \frac{\pi}{2}.$$

Note that the perpendicular case is excepted.

Going back to the first law

$$\rho\Theta\frac{\mathrm{D}s}{\mathrm{D}t} = \boldsymbol{T}\cdot\cdot\boldsymbol{D} - \rho\left(\frac{\mathrm{D}f}{\mathrm{D}t} + s\frac{\mathrm{D}\Theta}{\mathrm{D}t}\right) + \rho r - \boldsymbol{\nabla}\cdot\boldsymbol{h} = \Phi + \rho r - \boldsymbol{\nabla}\cdot\boldsymbol{h}$$

the following situations can be considered:

- non-dissipative process: $\Phi = 0$
$$\rho\Theta\frac{\mathrm{D}s}{\mathrm{D}t} = \rho r - \boldsymbol{\nabla}\cdot\boldsymbol{h},$$
which is a heat transfer equation,
- isothermal process: no heat transfer, mechanical and thermal processes are decoupled, and
- adiabatic process with $\boldsymbol{h} = \boldsymbol{0}$
$$\rho\Theta\frac{\mathrm{D}s}{\mathrm{D}t} = \rho r.$$

2.4. Boundary and initial conditions

The balance equations in the local form are a system of coupled differential equations in terms of the position-vector and time. That means for solving the system one should formulate boundary and initial equations.

There are different types of boundary conditions:

- A boundary condition which specifies the value of the function itself is a Dirichlet boundary condition (Peter Gustav Lejeune Dirichlet, 1805–1859).
- A boundary condition which specifies the value of the normal derivative of the function is a Neumann boundary condition (Carl Gottfried Neumann, 1832–1925).
- A boundary condition which specifies the sum of the value of the function itself multiplied by a constant and the value of the normal derivative of the function multiplied by another constant is a Robin boundary condition (Victor Gustave Robin, 1855–1897).
- If the boundary has the form of a curve or surface that gives a value to the normal derivative and the variable itself then it is a Cauchy boundary condition.
- Last but not least there are mixed boundary conditions that on one part of the boundary is given one condition, on another part — another condition.

If we have a pure mechanical problem, the simplest boundary conditions are related to the displacements or the stress vector on the boundary. In more complicated problems, the conditions should also be formulated for the temperature or the heat flux vector, etc.

With respect to the time, we have differential equations of second order in the case of a pure mechanical problem. That means one can formulate initial conditions for the function itself or the first time derivative. Examples are a prescribed displacement vector and/or velocity velocity at the beginning t_0. If both values are zero, we get the static solution only (no motion or constant velocity for all points of the continuum). In more complex problems, we have more initial conditions. The temperature problem, for example, is a first order differential equation with respect to time. That means the initial temperature can be prescribed.

2.5. Jump conditions

Up to now, it was assumed that the physical fields are continuously differentiable within the volume in both configurations. This assumption was also introduced for the Reynolds transport theorem (Osborne Reynolds, 1842–1912) necessary for the recalculation of the actual balances into balances in the reference configuration and the divergence theorem. If a surface within a material body exists with a discontinuous physical quantity, it is called a singular surface [23]. Examples are shock front in supersonic flows, surface between two different bodies (ski-snow, atmosphere-ocean) or the front between two phases. In all these cases, the balance can be formulated in the integral form but they should incorporate additional terms expressing the jump conditions. Details concerning the jump conditions are presented in [23–27] among others.

3. Material-dependent equations

Now our focus is on the specific (individual) response of the given material under an arbitrary load. We should include the information on material behavior since the number of governing equation is not equal to the number of unknowns in the set of governing equations. For example, consider a homogeneous three-dimensional solid which is only mechanically loaded. In this case, we have the following governing equations:

- balance of mass — one scalar equation,
- balance of momentum — one vectorial or three scalar equation,
- balance of moment of momentum — one vectorial or three scalar equation,

- balance of energy — one scalar equation, and
- balance of entropy — one scalar expression.

The unknown are:

- the density (one scalar),
- the displacement vector (one vector or three scalars),
- the stress tensor (one second rank tensor or 9 scalars), and
- the inner energy (one scalar).

It is obvious that we have only 8 governing equations, but 14 unknowns. To close the gap, we need more equations which can be established only with the help of information on the behavior of the materials. The material behavior itself is specific to each material, and thus a universal constitutive equation cannot be established.

3.1. Constitutive equations

The formulation of constitutive equations cannot be based on some physical principles (exception is that the requirements of the second of thermodynamics/balance of entropy should be fulfilled). The following modeling principles are established in the literature:

- inductive approach that is moving from the simplest to more complex models,
- deductive approach that is moving from a general frame to more specific cases, and
- rheological modeling which is a combination of both approaches.

The constitutive equations contain parameter, parameter functions, etc. which should be identified with the help of

- experimental observations (real or virtual),
- mathematical analysis, and
- theory of symmetry [28].

The last one should be used together with the Curie-Neumann's principle (Pierre Curie, 1859–1906, Fritz Ernst Neumann, 1798–1895). In addition, sometimes the models are restricted by some constraints (for example, the incompressibility condition). There are a lot of references in this matter among which is [17]. Herein, the deductive approach and the problems of identification are discussed. In addition, a classification of the material behavior can be done as follows:

- spontaneous material behavior with the subclasses of elastic and plastic behavior, and
- time-dependent material behavior with the subclasses of visco-elastic and visco-plastic behavior.

Formulation of suitable constitutive and evolution equations finally should be proved by

- checking the correctness of the formulation and the adequateness in terms of thermodynamic considerations and
- experimental validation or falsification, etc.

Let us introduce the following definition: *Constitutive equations connect all macroscopic phenomenological variables describing the behavior of the continuum* [29]. It is clear that this definition is very general and we need some restrictions with respect to the mathematical form of the constitutive equations. Such restrictions are introduced in [30] among others. One basic definition in this paper is: *Simple materials of the rank 1 are materials which are described by constitutive equations connecting local variables, e.g. the local strain tensor and the local*

IOP Conf. Series: Journal of Physics: Conf. Series **991** (2018) 012003 doi:10.1088/1742-6596/991/1/012003

heat flux vector with the local stress tensor and the local temperature gradient, respectively. All statements are related to the same material point and its differential neighborhood of rank 1. As a consequence only the first gradients are in the constitutive equations, higher order gradients are ignored. The next important definition is: *Process is the change of the constitutive parameters with respect of time* [2]. In addition, one can assume: *The behavior of the continuum in each material point is given as a set of constitutive variables which are operators with respect to time* [2].

If we want to distinguish solids and fluids, several definitions are presented in the literature:

- Under a given load, a solid will have non-zero components in the stress deviator, i.e., it shows resistance against any shape change.
- In contrast, under a given, a fluid will have only zero components in the stress deviator, i.e., it does not preserve its shape.

The starting point of the deductive approach in formulation of the constitutive equations is introducing of several axioms of the material theory [17]:

- causality,
- determinism,
- equipresence,
- material objectivity,
- local action,
- memory, and
- physical consistency.

A typical example of the set of constitutive equations is

$$\begin{aligned}
\boldsymbol{T}(\boldsymbol{R},\tau) &= \boldsymbol{P}\{\boldsymbol{R},\theta(\boldsymbol{R},\tau),\dot{\theta}(\boldsymbol{R},\tau),\boldsymbol{\nabla_r}\theta(\boldsymbol{R},\tau),\boldsymbol{\nabla_r}\dot{\theta}(\boldsymbol{R},\tau),\boldsymbol{\Gamma}(\boldsymbol{R},\tau)\},\\
\boldsymbol{h}(\boldsymbol{R},\tau) &= \boldsymbol{h}\{\boldsymbol{R},\theta(\boldsymbol{R},\tau),\dot{\theta}(\boldsymbol{R},\tau),\boldsymbol{\nabla_R}\theta(\boldsymbol{R},\tau),\boldsymbol{\nabla_r}\dot{\theta}(\boldsymbol{R},\tau),\boldsymbol{\Gamma}(\boldsymbol{R},\tau)\},\\
f(\boldsymbol{R},\tau) &= f\{\boldsymbol{R},\theta(\boldsymbol{R},\tau),\dot{\theta}(\boldsymbol{R},\tau),\boldsymbol{\nabla_R}\theta(\boldsymbol{R},\tau),\boldsymbol{\nabla_r}\dot{\theta}(\boldsymbol{R},\tau),\boldsymbol{\Gamma}(\boldsymbol{R},\tau)\},\\
s(\boldsymbol{R},\tau) &= s\{\boldsymbol{R},\theta(\boldsymbol{R},\tau),\dot{\theta}(\boldsymbol{R},\tau),\boldsymbol{\nabla_R}\theta(\boldsymbol{R},\tau),\boldsymbol{\nabla_r}\dot{\theta}(\boldsymbol{R},\tau),\boldsymbol{\Gamma}(\boldsymbol{R},\tau)\},
\end{aligned}$$

describing simple thermomechanical material behavior. The constitutive parameters are: the temperature, the temperature rate, the temperature gradient, the temperature rate gradient and a strain measure. Since $0 < \tau < t$, the constitutive equations are functionals. Other examples are given, for example, in [2, 14, 15].

Several simplifications can be made such as specification of the strain measure, neglecting aging effects and viscositywhich is equivalent to removing the explicit time dependency. Among such cases, the following simplified ones are worth mentioning:

- nonlinear elastic anisotropic material behavior

  $$\boldsymbol{T}(\boldsymbol{F}) = 2\rho\boldsymbol{F}\cdot u_{,\boldsymbol{C}}(\boldsymbol{C})\cdot\boldsymbol{F}^{\mathrm{T}}$$

  with $\boldsymbol{C}$ as the right Cauchy-Green strain tensor ($\boldsymbol{C} = \boldsymbol{F}^{\mathrm{T}}\cdot\boldsymbol{F}$).
- nonlinear elastic isotropic material behavior

  $$\boldsymbol{T}(\boldsymbol{F}) = 2\rho\boldsymbol{F}\cdot(\phi_0\boldsymbol{I} + \phi_1\boldsymbol{C} + \phi_2\boldsymbol{C}^2)\cdot\boldsymbol{F}^{\mathrm{T}}$$

  with the functions ϕ_i ($i = 0, 1, 2$) depending on the first, second and third invariants of the tensor $\boldsymbol{C}$.

- nonlinear elastic isotropic incompressible material behavior
$$\boldsymbol{T} = 2\rho \boldsymbol{F} \cdot \left[\left(\frac{\partial u}{\partial I_1} + I_1 \frac{\partial u}{\partial I_2} \right) \boldsymbol{I} - \frac{\partial u}{\partial I_2} \boldsymbol{C} \right] \cdot \boldsymbol{F}^{\mathrm{T}} - p\boldsymbol{I}$$
with the hydrostatic pressure p, and
- simple thermoelastic material behavior.

The inductive approach is the more engineering way of formulation constitutive equations. For example, we start we the Hooke's law

- tension with σ — normal stress, E — Young's modulus (named after Thomas Young, 1773–1829) and ε normal strain
$$\sigma = E\varepsilon$$
- or torsion with τ — shear stress, G — shear modulus and γ — shear strain
$$\tau = G\gamma$$

and generalize towards the following cases:

- three-dimensional isotropic case,
- three-dimensional anisotropic case,
- nonlinear behavior,
- ...

In each case, the thermodynamical consistency, which is guaranteed in the deductive approach, should be checked separately.

The last possibility to formulate constitutive equations is the rheological modeling [2, 8, 29]. This approach is founded on

- the introduction of some basic models, for example, related to
  - the elastic behavior,
  - the plastic behavior, and
  - the viscous behavior,
- and the assumption that the connection of basic models can be realized only
  - in parallel or
  - in series.

Then any complex behavior can be represented by these connection, for example,

- visco-elastic = elastic + viscous
- visco-plastic = plastic + viscous

The rheological modeling was introduced in [31] and discussed in [2, 8] for isotropic and anisotropic materials.

3.2. Evolution equations

In many cases it is enough to present the material behavior by constitutive equations discussed in the previous subsection. Sometimes the changes of the material behavior are related to evolution processes. Examples are

- the establishment of plastic zones,
- the development of damage processes,

IOP Conf. Series: Journal of Physics: Conf. Series **991** (2018) 012003 doi:10.1088/1742-6596/991/1/012003

- creep processes,
- the deformation induced anisotropy,
- the hardening and softening processes, or
- phase transformation processes.

The description of the evolution process is based on ordinary differential equations with respect to time derivatives of first order. Special initial conditions should be introduced. For example, if we have creep or plastic evolution the initial state is assumed to be elastic.

4. Advanced models

The last part of this paper presents some new directions in continuum mechanics. This overview is not comprehensive and only few comments are given. Nevertheless, the required references are given for further reading.

4.1. Lower dimensional continuum models

These models are interesting from the theoretical point of view. Commonly, structural models (beam, rod, plate, shell, etc.) are presented in the literature as sets of equations which are reduced from a general three-dimensional equation by means of hypotheses, mathematical simplification, etc. Examples of such hypotheses are:

- Euler-Bernoulli beam theory, for example, is based on the assumption that the beam cross-sections remain planar and perpendicular to the neutral axis before and after deformation, and
- Kirchhoff plate theory, for example, assumes that the line elements of the beam remain straight and perpendicular to the mid-plane before and after deformation.

Mathematical approaches are based on power series or asymptotic integration (in both case a small parameter is assumed).

A third way is more elegant and natural. At the beginning it is assumed that we have a deformable surface or line. Then, an exact continuum theory can be deduced. The disadvantage of this approach is the identification of the parameters in the constitutive equations.

Recently overviews concerning plate and shell theories and different approaches were published in [32, 33]. A micropolar plate model was presented in [34] and extended to micropolar shells in [35]. A direct theory of rods was suggested in [36, 37]. Further discussion concerning actual trends are given in [38, 39] among others.

4.2. Nanostructures

Nanostructures are a new class of structures with special properties. In many cases they are stiffer which is a result of high specific surface to volume ratio. In the case of classical structures the bulk behavior is dominant — in the case of nanostructures the stiffness properties are under the influence of the surface behavior.

A large number of publications were released in recent years which makes it almost impossible to provide a comprehensive overview. But within a more specific framework, the pioneering works of [40, 41] suggest considering the surface effects in the continuum mechanics of the bulk material — where it is relevant. Some publications in this context are [42–45]. References for further reading are given within these publications. An overview concerning surface effects is given also in [46].

4.3. Growing Solids

In June 2015, Alexander Manzhirov has organized an IUTAM[2] Symposium on growing solids in Moscow. From the first announcement, one can comprehend this class of solids by means of a nice description: *The vast majority of objects or solids which surround us arise from some growth processes. As an example, one can present such natural phenomena as growth of biological tissues, glaciers, blocks of sedimentary and volcanic rocks, as well as space objects. Similar processes determine the specific features of many technologies in industry, including well-known technologies of crystal growth, laser deposition, solidification of melts, electrolytic formation, pyrolytic deposition, polymerization and concreting. Recent research has shown that solids, which were formed due to the processes of growth, differ essentially in their properties from solids in the traditional view. Moreover, the classical approaches of solid mechanics to the modeling of growing solids behavior fail. They have to be replaced by new ideas and methods of modern mechanics, mathematics, physics, and engineering sciences. Thus, at present, a new area of solid mechanics, which deals with the construction of adequate models for solids growth processes is forming.*

There are world-wide different schools investigating growing solids. One of them was founded by N. Kh. Arutyunyan (1912–1993), who published a monograph in Russian on this topic [47]. Actual problems of the mechanics are the general theory of growth processes of solids, the numerical modeling of growth, moving boundaries and interfaces, surface effect, phase transitions, dislocations and disclinations in solids, crystal growth, growth of bone and soft tissues, laser deposition, solidification of melts, and electrolytic formation. Some of them are discussed, for example, in [48–50].

4.4. Reinforced materials

The final example of advanced materials is related to reinforced materials. By combining two or more different materials, one gets a new material with new properties. Considering the specific effective properties, it is obvious that, as a result a light-weight structural material is designed. The classical case is the are uni-directional fibre-reinforced layers. Finally, the layers can be combined and a laminate can be established. The classical ideas are published in various textbooks and monographs, for example in [51]. One of the actual problems is related to particle-reinforced matrix materials while taking into account the interphases [52–54].

5. Final Remarks

There are much more advanced models. At the moment we have a renaissance of the Cosserat theory (Eugìne Cosserat, 1866–1931, François Cosserat, 1852–1914) with applications to foams [34] or bones [55]. Further developments are related to micropolar continua [56]. Another direction is related to the application of higher gradients [57–60].

Summarizing up to now Continuum Mechanics is a actual branch of Mechanics with new directions. The limits are not clear since the continuum mechanics approach is applied to problems with smaller and smaller sizes. There is no need for new theories — only the classical continuum mechanics should be improved. There is only one important item: the requirements of the continuum definition should be fulfilled.

References

[1] Altenbach H 2015 *Kontinuumsmechanik - Eine elementare Einführung in die materialunabhängigen und materialabhangigen Gleichungen, 3rd edn.* (Berlin - Heidelberg: Springer)
[2] Palmov V A 2014 *Nonlinear Mechanics of Deformable Bodies* (St. Petersburg: Izdat. Politekh. Univ.) [in Russian]

[2] International Union of Theoretical and Appied Mechanics

[3] Zhilin P A 2006 *Applied Mechanics. Foundations of the Theory of Shells* (St. Petersburg: Izdat. Politekh. Univ.) [in Russian]
[4] Zhilin P A 2007 *Applied Mechanics. Theory of the Thin Elastic Rods* (St. Petersburg: Izdat. Politekh. Univ.) [in Russian]
[5] Zhilin P A 2012 *Rational Mechanics of Continua* (St. Petersburg: Izdat. Politekh. Univ.) [in Russian]
[6] Lurie A I 1990 *Nonlinear Theory of Elasticity* (Amsterdam: North-Holland)
[7] Lurie A I 2005 *Theory of Elasticity* (Berlin - Heidelberg: Springer)
[8] Palmov V A 1998 *Vibrations of Elasto-Plastic Bodies* (Berlin - Heidelberg: Springer)
[9] Sedov L I 1994 *Continuum Mechanics* vol 1 (Moscow: Nauka) [in Russian]
[10] Sedov L I 1994 Continuum Mechanics vol 2 (Moscow: Nauka) [in Russian]
[11] Lebedev L P, Cloud M J, Eremeyev V A *Tensor Analysis with Applications in Mechanics* 2nd ed (Singapore et al.: World Scientific)
[12] Wilson E B 1901 *Vector Analysis. A Text-Book for the Use of Students. Founded upon the Lectures of J. W. Gibbs* (New Haven, Connecticut, USA: Yale University Press)
[13] Legally M 1964 *Vorlesungen über Vektorrechnung* (Leipzig: Akademische Verlagsgesellschaft Geest und Portig)
[14] Truesdell C 1975 *A First Course in Rational Continuum Mechanics*, (Moscow: Mir) [in Russian]
[15] Truesdell C 1977 *A First Course in Rational Continuum Mechanics. Pure and Applied Mathematics* vol 1 (New York: Academic Press)
[16] Haupt P 1996 Konzepte der Materialtheorie *Technische Mechanik* **16** (1) 13–22
[17] Haupt P 2002 *Continuum Mechanics and Theory of Materials* 2nd ed (Berlin: Springer)
[18] Ivanova E A, Vilchevskaya E N, and Müller W H 2016 Time derivatives in material and spatial description — What are the differences and why do they concern us? In *Advanced Methods of Continuum Mechanics for Materials and Structures* ed Naumenko K and Aßmus M (Springer) pp 3–28
[19] Ivanova E, Vilchevskaya E, and Müller W H A study of objective time derivatives in material and spatial description In *Mechanics for Materials and Technologies* ed Altenbach H, Goldstein R V, and Murashkin E (Cham: Springer) 195–229
[20] Nowacki W 1975 *Theory of Elasticity* (Moscow: Mir) [in Russian]
[21] Nowacki W 1985 *Theory of Asymmetric Elasticity* (Oxford: Pergamon Press)
[22] Naumenko K and Altenbach H 2016 *Modeling High Temperature Materials Behavior for Structural Analysis. Part I: Continuum Mechanics Foundations and Constitutive Models. Advanced Structured Materials. Vol. 28* (Cham: Springer)
[23] Hutter K and Jöhnk K 2004 *Continuum Methods of Physikal Modeling - Continuum Mechanics, Dimensional Analysis, Turbulence* (Berlin: Springer)
[24] Altenbach H, Eremeyev, V A, Lebedev L P, and Rendón L A 2011 Acceleration waves and ellipticity in thermoelastic micropolar media ZAMM **80** (3) 217–27
[25] Casey J 2011 On the derivation of jump conditions in continuum mechanics *Int. J. Struct. Changes Solids* **3** (2) 61–84
[26] Müller I 1973 *Thermodynamik: die Grundlagen der Materialtheorie* (Bertelsmann Universitätsverlag)
[27] Šilhavý M 1997 *The Mechanics and Thermodynamics of Continuous Media* (Heidelberg: Springer)
[28] Nye J F 1985 *Physical Properties of Crystals* (Oxford: Oxford University Press)
[29] Krawietz A 1986 *Materialtheorie* (Berlin: Springer)
[30] Noll W 1972 A new mathematical theory of simple materials *Arch. Rat. Mech. Anal.*, **48** (1) 1–50
[31] Reiner M 1960 *Deformation, Strain and Flow: an Elementary Introduction to Rheology* (London: H. K. Lewis)
[32] Altenbach J, Altenbach H, and Eremeyev V A 2010 On generalized Cosserat-type theories of plates and shells: a short review and bibliography *Arch. Appl. Mech.* **80** (1) 73–92
[33] Altenbach H and Eremeyev V 2017 Thin-walled structural elements: classification, classical and advanced theories, new applications In *Shell-like Structures: Advanced Theories and Applications* ed Altenbach H and Eremeyev V (Cham: Springer) pp 1–62
[34] Altenbach H and Eremeyev V A 2009 On the linear theory of micropolar plates *ZAMM* **89** (4) 242–56
[35] Eremeyev V and Altenbach H Basics of mechanics of micropolar shells In *Shell-like Structures: Advanced Theories and Applications* ed Altenbach H and Eremeyev V (Cham: Springer) pp 63–111
[36] Birsan M and Altenbach H 2013 On the Cosserat model for thin rods made of thermoelastic materials with voids *Discr. Contin. Dyn. Syst. - Ser. S* **6** (6) 1473–85
[37] Altenbach H, Bîrsan M, and Eremeyev V A 2013 Cosserat-Type Rods In *Generalized Continua from the Theory to Engineering Applications* ed Altenbach H and Eremeyev V A (Wien: Springer Vienna) pp 178–248
[38] Altenbach H and Eremeyev V (Editors) 2017 *Shell-Like Structures. Advanced Theories and Applications*

(Wien: Springer)
[39] Altenbach H and Mikhasev G I (Editors) 2015 *Shell and Membrane Theories in Mechanics and Biology. From Macro- to Nanoscale Structures* (Cham: Springer)
[40] Gurtin M E and Murdoch A I 1975 A continuum theory of elastic material surfaces *Arch. Ration. Mech. Anal.* **57** (4) 291–323
[41] Gurtin Mn E and Murdoch A I 1978 Surface stress in solids *Int. J. Solids Struct.* **14** (6) 431–40
[42] Altenbach H, Eremeyev V A. and Morozov N F 2013 On the influence of residual surface stresses on the properties of structures at the nanoscale In *Surface Effects in Solid Mechanics* ed Altenbach H and Morozov N F (Heidelberg: Springer) pp 21–32
[43] Altenbach H, Eremeev V A, and Morozov N F 2010 On equations of the linear theory of shells with surface stresses taken into account *Mech. Solids* **45** (3) 331–42
[44] Altenbach H and Eremeyev V A 2014 Vibration analysis of non-linear 6-parameter prestressed shells *Meccanica* **49** (8) 1751–61
[45] Eremeyev V A 2016 On effective properties of materials at the nano-and microscales considering surface effects *Acta Mech.* **227** (1) 29–42
[46] Altenbach H and Morozov N F 2013 *Surface Effects in Solid Mechanics. Models, Simulations and Applications* (Berlin, Heidelberg: Springer)
[47] Arutyunyan N K, Drozdov A D, and Naumov V E 1987 *Mechanics of Growing Viscoelastoplastic Solids,* (Moscow: Nauka) [in Russian]
[48] Lychev S A 2011 Universal deformations of growing solids *Mech. Solids* **46** (6) 863–76
[49] Lychev S A and Manzhirov A V 2013 Reference configurations of growing bodies *Mech. Solids* **48** (5) 553–60
[50] Lychev S A and Manzhirov A V 2013 The mathematical theory of growing bodies. Finite deformations *J. Appl. Math. Mech.* **77** (4) 421–32
[51] Altenbach H, Altenbach J, and Kissing W 2018 *Mechanics of Composite Structural Elements* 2nd ed. (Heidelberg: Springer)
[52] Nazarenko L, Stolarski H, and Altenbach H 2017 A definition of equivalent inhomogeneity applicable to various interphase models and various shapes of inhomogeneity *Procedia IUTAM* **21** 63–70
[53] Nazarenko L, Stolarski H, and Altenbach H 2017 Thermo-elastic properties of random particulate nanomaterials for various models of interphase *Int. J. Mech. Sci.* **126** 130–41
[54] Nazarenko L, Stolarski H, and Altenbach H 2017 A model of cylindrical inhomogeneity with spring layer interphase and its application to analysis of short-fiber composites *Compos. Struct.* **160** 635–52
[55] Yang J F C and Lakes R S 1982 Experimental study of micropolar and couple stress elasticity in compact bone in bending *J. Biomech.* **15** (2) 91–8
[56] Eremeyev V A, Lebedev L P, and Altenbach H 2013 *Foundations of Micropolar Mechanics* (Berlin, Heidelberg: Springer)
[57] Abali B E, Müller W H, and dell'Isola F 2017 Theory and computation of higher gradient elasticity theories based on action principles *Archive of Applied Mechanics* doi: 10.1007/s00419-017-1266-5
[58] dell'Isola F, Andreaus U, and Placidi L 2015 At the origins and in the vanguard of peridynamics, non-local and higher-gradient continuum mechanics: An underestimated and still topical contribution of Gabrio Piola *Math. Mech. Solids* **20** (8) 887–928
[59] Eremeyev V A and Altenbach H 2014 Equilibrium of a second-gradient fluid and an elastic solid with surface stresses *Meccanica* **49** (11) 2635–43
[60] Eremeyev V A and Altenbach H 2015 On the direct approach in the theory of second gradient plates *Shell and Membrane Theories in Mechanics and Biology* ed Altenbach H and Mikhasev G I (Cham: Springer) pp 147–54

IOP Conf. Series: Journal of Physics: Conf. Series **991** (2018) 012004 doi:10.1088/1742-6596/991/1/012004

Some boundary-value problems for anisotropic quarter plane

K M Arkhypenko and O F Kryvyi*

National University "Odessa Maritime Academy," Odessa, Ukraine

E-mail: *krivoy-odessa@ukr.net

Abstract. To solve the mixed boundary-value problems of the anisotropic elasticity for the anisotropic quarter plane, a method based on the use of the space of generalized functions $\Im'(\mathrm{R}^2_+)$ with slow growth properties was developed. The two-dimensional integral Fourier transform was used to construct the system of fundamental solutions for the anisotropic quarter plane in this space and a system of eight boundary integral relations was obtained, which allows one to reduce the mixed boundary-value problems for the anisotropic quarter plane directly to systems of singular integral equations with fixed singularities. The exact solutions of these systems were found by using the integral Mellin transform. The asymptotic behavior of solutions was investigated at the vertex of the quarter plane.

1. Introduction

The plane problems of the theory of elasticity for isotropic and anisotropic wedges and the quarter plane have been considered by many authors. Here we should mention the papers [1, 2], where mixed, first and second basic problems for the isotropic wedge were solved using the Mellin integral transform. In [3], the mixed plane problem for the elastic quadrant was solved. A similar problem was also solved in [4] by reduction to systems of singular integral equations. In [5], the Mellin integral transform was used to solve the first fundamental problem for an anisotropic wedge. In [6], the first basic problem for a composite anisotropic wedge was considered and the stress behavior at the vertex of the wedge was investigated.

In this paper, the mixed boundary-value problems of anisotropic elasticity for the anisotropic quarter plane is solved bt an approach proposed in [7–13] for solving the problems for piecewise-homogeneous anisotropic unbounded media. The approach is implemented by constructing the system of fundamental solutions and singular integral relations connecting the boundary values of stresses and displacements on the faces of the anisotropic quarter plane. The latter permits reducing the mixed boundary-value problems for the anisotropic quarter plane directly to systems of singular integral equations (SIE).

2. Statement of the problem and construction of the fundamental system of solutions

The stress-strain state of the anisotropic quarter plane $(x, y) \in \mathrm{R}^2_+ = L_+ \times L_+$, $L_+ = (0, +\infty)$ is described by the vector $\mathrm{v} = \{v_p(x,y)\}_{p=\overline{1,5}} = \{\sigma_x, \sigma_y, \tau_{xy}, u, v\}$, whose boundary values of components on the faces of the quarter plane are denoted by

$$\chi_k(x) = v_{k+1}(x, +0), \quad k = \overline{1,4}, \quad x \in L_+, \tag{1}$$

$$\mu_1(y) = v_1(+0, y), \quad \mu_k(y) = v_{k+1}(+0, y), \quad k = \overline{2,4}, \quad y \in L_+. \tag{2}$$

The components of the vector v satisfy the equilibrium equations and the generalized Hooke's law, which in the absence of all forces can be written as

$$\mathrm{M}(\partial_1, \partial_2)v = 0, \quad (x, y) \in \mathrm{R}^2_+, \tag{3}$$

where

$$\mathrm{M}(\partial_1, \partial_2) \equiv \begin{pmatrix} \mathrm{D} & \mathrm{O}_{2\times 2} \\ \mathrm{A} & \mathrm{D}^{\mathrm{T}} \end{pmatrix}, \quad \mathrm{D} = \begin{pmatrix} \partial_1 & 0 & \partial_2 \\ 0 & \partial_2 & \partial_1 \end{pmatrix}, \quad \mathrm{A} = \{a_{kj}\}^3, \quad \partial_1 = \frac{\partial}{\partial x}, \quad \partial_2 = \frac{\partial}{\partial y},$$

a_{ij} are elastic constants of the anisotropic quarter plane [14], and $\mathrm{O}_{n\times m}$ is the zero $n \times m$ matrix.

In the statement of the problem for anisotropic elasticity, it is necessary to know two functions of representations (1) and (2) on each face of the quarter plane. The remaining functions are to be determined by solving the problem. Following [7–9], one must construct the system of fundamental solutions and establish integral relations between the functions (1) and (2) on the faces of the quarter plane.

To implement this approach, we pass to the space of generalized slowly growing functions $\Im'(\mathrm{R}^2_+)$. Using the boundary conditions (1), (2) and the connection formulas [8, 9] between the ordinary derivatives ∂_k $(k = 1, 2)$ and generalized derivatives $\tilde{\partial}_k$ $(k = 1, 2)$, instead of the matrix differential equation (3), we obtain the following matrix differential equation in the space of generalized functions $\Im'(\mathrm{R}^2_+)$:

$$(\mathrm{M}(\tilde{\partial}_1, \tilde{\partial}_2)\tilde{\mathrm{v}}, \zeta) = (\mathrm{f}, \zeta), \quad \tilde{\mathrm{v}} = \{\tilde{v}_j\}^5, \quad \tilde{v}_j \in \Im'(\mathrm{R}^2_+), \quad \zeta \in \Im(\mathrm{R}^2_+), \tag{4}$$

where

$$\mathrm{f} = \{f_j\}^5, \quad f_1 = \chi_2(x)\delta(L_1) + \mu_1(y)\delta(L_2), \quad f_2 = \chi_1(x)\delta(L_1) + \mu_2(y)\delta(L_2),$$
$$f_3 = \mu_3(y)\delta(L_2), \quad f_4 = \chi_4(x)\delta(L_1), \quad f_5 = \chi_3(x)\delta(L_1) + \mu_4(y)\delta(L_2),$$

$\delta(L_k) \in \Im'(L_k)$ is the generalized function concentrated on the semiaxis L_k, $L_1 = \{y = 0, x \in [0, +\infty)\}$, and $L_2 = \{x = 0, y \in [0, +\infty)\}$.

Following [7–13], we solve equation (4) using the *system of fundamental solutions* by which we mean the system of vectors $\mathrm{v}^0_j = \{v^0_{kj}\}^5$, $j = \overline{1,8}$, $v^0_{kj} \in \Im'(\mathrm{R}^2_+)$. In the space $\Im'(\mathrm{R}^2_+)$, these vectors satisfy the matrix differential equation

$$(\mathrm{M}(\tilde{\partial}_1, \tilde{\partial}_2)\mathrm{v}^0_j, \zeta) = (\mathrm{f}^0_j, \zeta), \quad \zeta \in \Im(\mathrm{R}^2_+) \quad (j = \overline{1,8}), \tag{5}$$

where

$$\mathrm{f}^0 = \{f^0_j\}^5, \quad f^0_1 = \delta_{2j}\delta(x - x_0, y) + \delta_{5j}\delta(x, y - y_0), \quad f^0_2 = \delta_{1j}\delta(x - x_0, y) + \delta_{6j}\delta(x, y - y_0),$$
$$f^0_3 = \delta_{7j}\delta(x, y - y_0), \quad f^0_4 = \delta_{4j}\delta(x - x_0, y), \quad f^0_5 = \delta_{3j}\delta(x - x_0, y) + \delta_{8j}\delta(x, y - y_0),$$

and δ_{kj} is the Kronecker symbol.

The support of generalized functions $\tilde{v}_j$, v^0_{kj} is the domain R^2_+, i.e., $\mathrm{supp}(\tilde{v}_j, v^0_{kj}) = \mathrm{R}^2_+$. The latter allows one to apply the two-dimensional Fourier transform to solve the matrix differential equation (5). As a result, we obtain explicit expressions for the vector components of the system of fundamental solutions

$$\begin{aligned} v_{pj}^0(x,y) &= \frac{1}{\pi}\mathrm{Im}\sum_{k,m=1}^{2}\frac{1}{q_k}\left[\frac{M_{pm}^0(z_k)\delta_{mj}}{x-x_0+z_ky}+\frac{M_{p,2+m}^0(z_k)\delta_{2+m,j}}{(x-x_0+z_ky)^2}\right. \\ &+\left.\frac{M_{p,4+m}^0(z_k)\delta_{4+m,j}}{x+z_k(y-y_0)}+\frac{M_{p,6+m}^0(z_k)\delta_{6+m,j}}{(x+z_k(y-y_0))^2}\right], \quad p=\overline{1,3}, \\ v_{pj}^0(x,y) &= \frac{1}{\pi}\mathrm{Im}\sum_{k,m=1}^{2}\frac{1}{q_k}\left\{M_{pm}^0(z_k)\delta_{mj}\left[\ln(x-x_0+z_ky)+\gamma-\frac{\pi i}{2}\right]+\frac{M_{p,2+m}^0(z_k)\delta_{2+m,j}}{x-x_0+z_ky}\right. \\ &+\left.M_{p,4+m}^0(z_k)\delta_{4+m,j}\left[\ln(x+z_k(y-y_0))+\gamma-\frac{\pi i}{2}\mathrm{sign}\,(y-y_0)\right]+\frac{M_{p,6+m}^0(z_k)\delta_{6+m,j}}{x+z_k(y-y_0)}\right\}, \quad p=4,5, \end{aligned} \tag{6}$$

$$\begin{aligned} &\mathrm{M}^0(z)=\{M_{pm}^0(z)\}_{p=\overline{1,5}}^{m=\overline{1,8}}, \\ &M_{11}^0(z)=M_{16}^0(z)=M_{57}^0(z)=zl_1(z), \quad M_{12}^0(z)=M_{15}^0(z)=M_{47}^0(z)=l_3(z), \\ &M_{13}^0(z)=M_{18}^0(z)=M_{37}^0(z)=z^3, \ M_{14}^0(z)=M_{27}^0(z)=M_{33}^0(z)=M_{38}^0(z)=-z^2, \ M_{17}^0(z)=-z^4, \\ &M_{21}^0(z)=M_{26}^0(z)=M_{54}^0(z)=-l_4(z), \quad M_{22}^0(z)=M_{25}^0(z)=M_{44}^0(z)=l_2(z), \quad M_{24}^0(z)=-1, \\ &M_{23}^0(z)=M_{28}^0(z)=M_{34}^0(z)=z, \quad M_{31}^0(z)=M_{36}^0(z)=M_{53}^0(z)=M_{58}^0(z)=-l_1(z), \\ &M_{32}^0(z)=M_{35}^0(z)=M_{43}^0(z)=M_{48}^0(z)=-zl_2(z), \quad M_{42}^0(z)=M_{45}^0(z)=-l_6(z), \\ &M_{41}^0(z)=M_{46}^0(z)=M_{52}^0(z)=M_{55}^0(z)=-l_5(z), \quad M_{51}^0(z)=M_{56}^0(z)=-l_7(z), \\ &q_k=a_{11}(z_k-z_{3-k})(z_k-\bar{z}_1)(z_k-\bar{z}_2), \quad k=1,2, \quad l_1(z)=a_{12}z^2-a_{26}z+a_{22}, \\ &l_2(z)=a_{11}z^2-a_{16}z+a_{12}, \quad l_3(z)=a_{16}z^3-(a_{12}+a_{66})z^2+2a_{26}z-a_{22}, \\ &l_4(z)=a_{11}z^3-2a_{16}z^2+(a_{12}+a_{66})z-a_{26}, \quad l_5(z)=-A_{23}z^2-(A_{12}+A_{33})z-A_{13}, \\ &l_6(z)=A_{22}z^2+2A_{23}z+A_{33}, \quad l_7(z)=A_{33}z^2+2A_{13}x+A_{11}, \end{aligned}$$

where z_k are roots of the characteristic equation [14] and A_{ij} is the cofactor of the matrix $\mathrm{A}=\{a_{kj}\}^3$.

The presence of explicit expressions (6) for the vector components $\mathrm{v}_j^0=\{v_{kj}^0\}^5$, $j=\overline{1,8}$, permits expressing the stresses and displacements in the space $\Im'(\mathrm{R}_+^2)$ in terms of the boundary values (1), (2) as the convolution

$$\tilde{v}_k(x,y)=\sum_{j=1}^{4}[\chi_j(x)*v_{kj}^0(x,y)+\mu_j(y)*v_{k,4+j}^0(x,y)], \quad k=\overline{1,5}. \tag{7}$$

Passung in the limit to the faces of the quarter plane and using the boundary values of displacements and stresses (1) and (2), we obtain the integral relations on the boundary:

$$\sum_{j=1}^{2}\left\{A_{pj}^1\chi_j(x)+\frac{B_{pj}^1}{\pi}\int_0^{+\infty}\frac{\chi_j(x_0)\,dx_0}{x_0-x}+A_{p,2+j}^1\chi_{2+j}'(x)+\frac{B_{p,2+j}^1}{\pi}\int_0^{+\infty}\frac{\chi_{2+j}'(x_0)\,dx_0}{x_0-x}\right.$$

$$+\operatorname{Im}\sum_{k=1}^{2}\left[C^1_{pj}(z_k)\int_0^{+\infty}\frac{\mu_j(y_0)\,dy_0}{x-z_ky_0}+C^1_{p,2+j}(z_k)\int_0^{+\infty}\frac{\mu'_{2+j}(y_0)\,dy_0}{x-z_ky_0}\right]\Bigg\}=0,\quad p=\overline{1,4},\quad x\in L_1,$$

$$\sum_{j=1}^{2}\Bigg\{A^2_{pj}\mu_j(y)+\frac{B^2_{pj}}{\pi}\int_0^{+\infty}\frac{\mu_j(y_0)\,dy_0}{y_0-y}+A^2_{p,2+j}\mu'_{2+j}(y)+\frac{B^2_{p,2+j}}{\pi}\int_0^{+\infty}\frac{\mu'_{2+j}(y_0)\,dy_0}{y_0-y}\qquad(8)$$

$$+\operatorname{Im}\sum_{k=1}^{2}\left[C^2_{pj}(z_k)\int_0^{+\infty}\frac{\chi_j(x_0)\,dx_0}{x_0-z_ky}+C^2_{p,2+j}(z_k)\int_0^{+\infty}\frac{\chi'_{2+j}(x_0)\,dx_0}{x_0-z_ky}\right]\Bigg\}=0,\quad p=\overline{1,4},\quad y\in L_2,$$

$A^1_{pj}=1+\operatorname{Re}\sum_{k=1}^{2}M^0_{p+1,j}(z_k),\ p,j=1,2,\ p,j=3,4,\quad A^1_{pj}=-\operatorname{Im}\sum_{k=1}^{2}M^0_{p+1,j}(z_k),\ p=1,2,\ j=3,4,$

$A^1_{pj}=\operatorname{Im}\sum_{k=1}^{2}M^0_{p+1,j}(z_k),\ p=3,4,\ j=1,2,\quad B^1_{pj}=\operatorname{Im}\sum_{k=1}^{2}M^0_{p+1,j}(z_k),\ p=\overline{1,4},\ j=1,2,\ p,j=3,4,$

$B^1_{pj}=-\operatorname{Im}\sum_{k=1}^{2}M^0_{p+1,j}(z_k),\ p=1,2,\ j=3,4,\quad C^1_{pj}(z)=-M^0_{p+1,4+j}(z),\ p=\overline{1,4},\ j=1,2,$

$C^1_{pj}(z)=z^{-1}M^0_{p+1,j}(z),\ p=1,2,\ j=3,4,\quad C^1_{pj}(z)=-z^{-1}M^0_{p+1,j}(z),\ p,j=3,4,$

$A^2_{pj}=1+\operatorname{Re}\sum_{k=1}^{2}z^{-1}M^0_{p,4+j}(z_k),\ p,j=1,2,\ p,j=3,4,\quad A^2_{pj}=\frac{3}{2}\operatorname{Im}\sum_{k=1}^{2}M^0_{p,4+j}(z_k),\ p=3,4,\ j=1,2,$

$A^2_{pj}=-\operatorname{Im}\sum_{k=1}^{2}z^{-2}M^0_{p,4+j}(z_k),\ p=1,2,\ j=3,4,\quad B^2_{pj}=\operatorname{Im}\sum_{k=1}^{2}z^{-1}M^0_{p,4+j}(z_k),\ p,j=1,2,\ p,j=3,4,$

$B^2_{pj}=-\operatorname{Im}\sum_{k=1}^{2}z^{-2}M^0_{p,4+j}(z_k),\ p=1,2,\ j=3,4,\quad B^2_{pj}=\operatorname{Im}\sum_{k=1}^{2}M^0_{p,4+j}(z_k),\ p=3,4,\ j=1,2,$

$C^2_{pj}(z)=M^0_{p,4+j}(z),\ p,j=1,2,\quad C^2_{pj}(z)=-M^0_{p,4+j}(z),\ p=1,2,\ j=3,4,$
$C^2_{pj}(z)=zM^0_{p,4+j}(z),\ p=3,4,\ j=\overline{1,4}.$

3. Solution of the mixed boundary-value problem and analysis of the obtaining results

Integral relations (8) allow one to reduce the boundary-value problems for the elastic anisotropic quarter plane directly to systems of SIE. In particular, let the conditions of complete cohesion be realized on one face and the stresses be known on the other:

$$\chi_k(x)=0,\quad x\in L_1,\quad k=3,4;\quad \mu_k(y)=P_k(y),\quad y\in L_2,\quad k=1,2.\qquad(9)$$

Having realized the first two conditions in (9) with the help of the third and fourth integral relations in (8) and the last two conditions in (9) and using the first and second integral relations, relatively to the unknown boundary values of the stresses $\chi_j(x)$, $x\in L_1$ $(j=1,2)$ and the unknown boundary values of the displacement derivatives $\mu'_j(y)$, $y\in L_2$ $(j=3,4)$, we obtain the following system of SIE with fixed singularities:

$$\begin{cases}\displaystyle\sum_{j=1}^{2}\left[A^1_{mj}\chi_j(x)+\frac{B^1_{mj}}{\pi}\int_0^{+\infty}\frac{\chi_j(x_0)\,dx_0}{x_0-x}+\operatorname{Im}\sum_{k=1}^{2}\frac{C^1_{m,2+j}}{\pi}\int_0^{+\infty}\frac{\mu'_{2+j}(y_0)\,dy_0}{x-z_ky_0}\right]=F_m(x),\quad m=1,2,\\ \displaystyle\sum_{j=3}^{4}\left[A^2_{mj}\mu'_j(y)+\frac{B^2_{mj}}{\pi}\int_0^{+\infty}\frac{\mu'_j(y_0)\,dy_0}{y_0-y}+\operatorname{Im}\sum_{k=1}^{2}\frac{C^2_{m,j-2}}{\pi}\int_0^{+\infty}\frac{\chi_j(x_0)\,dx_0}{x_0-z_ky}\right]=F_m(y),\quad m=3,4.\end{cases}\qquad(10)$$

The solution of system (10) was obtained by the Mellin transform

$$\tilde{\chi}_j(t) = \begin{cases} \chi_j(t), & j = 1, 2, \\ \mu_j'(t), & j = 3, 4, \end{cases} \tag{11}$$

$$\tilde{\chi}_j(t) = \frac{1}{2\pi i} \sum_{l=1}^{4} \int_{\sigma - i\infty}^{\sigma + i\infty} \frac{\sin(\pi s) M_{jl}(s) F_l(s)}{\det(\mathrm{M}(s))} t^{-s}\, ds, \quad 0 < \sigma < 1, \tag{12}$$

$$\mathrm{M}(s) = \{M_{jl}(s)\}^4, \quad M_{jl}(s) = \mathrm{Im} \sum_{k=1}^{2} C_{jl}^1(z_k)(-z_k)^{s-1},\ j = 1, 2,\ l = 3, 4;$$

$$M_{jl}(s) = \mathrm{Im}(\omega_{jl} e^{i\pi s}),\ j, l = 1, 2\ j, l = 3, 4, \quad M_{jl}(s) = \mathrm{Im} \sum_{k=1}^{2} C_{jl}^1(z_k)(-z_k)^{-s},\ j = 3, 4,\ l = 1, 2,$$

where $F_l(s)$ is the Mellin transform from the right-hand sides of system (10).

To calculate the integrals (12) and to determine the solution asymptotics at the vertex of the quarter plane, it is necessary to know the location of the roots of the transcendental equation

$$\det(M(s)) = 0, \tag{13}$$

in the half plane $\mathrm{Re}\, s < 1$. Let $\{s_k\}^\infty$ be the roots of equation (13) satisfying the conditions

$$1 > \mathrm{Re}\, s_1 > \mathrm{Re}\, s_2 > \mathrm{Re}\, s_3 > \ldots > \mathrm{Re}\, s_k > \ldots,$$

then

$$\tilde{\chi}_j(t) \simeq k_j t^{-\mathrm{Re}\, s_1}, \quad t \to 0. \tag{14}$$

The behavior of the roots of transcendental equation (13) was analyzed numerically. In particular, it was established that, for known anisotropic materials equation (13), has at least one root whose real part belongs to the interval $(0; 1)$. Therefore, the mixed boundary-value problem (9) has a unique solution in the class of functions with integrable singularity at zero. In addition, the asymptotic behavior of the roots at infinity is established. Namely, for each anisotropic material, one can find a number s_* such that all real roots of transcendental equation (13) belong to the interval $(-s_*; 1)$ and their number n_* is bounded. All other roots are complex and are arranged along two straight lines symmetric with respect to the real axis.

Table 1 shows the first ten roots of equation (5) for different angles between the coordinate axes of the plane and the main symmetry axes of anisotropic materials [14]: $m1$ – glass-fiber orthogonally reinforced plastic, $m2$ – glass-fiber plastic STET.

A numerical analysis has shown that, for the material $m1$ ($\varphi = 0$), transcendental equation (5) has four real roots: $n_* = 4$ and $s_* = 1.34$. All roots in $|s| > s_* = 1.34$ are arranged along the straight lines $z = \mathrm{Re}\, s \pm (0.034856 - 0.565527\mathrm{Re} s)i$ with step $h = 1.5137$. For the material $m2$ ($\varphi = 0$), the equation also has four real roots $n_* = 4$ and $s_* = 1.22$. All roots in $|s| > s_* = 1.22$ are arranged along the straight lines $z = \mathrm{Re}\, s \pm (0.076057 - 0.413467\mathrm{Re}\, s)i$ with step $h = 1.7067$. For the same materials but with $\varphi = \pi/4$, the localization domain of the real roots is wider, so for the material $m1$: $s_* = 37.5$, and for the material $m2$: $s_* = 26.5$. Figure 1 shows the root location for materials $m1$ and $m2$ $\varphi = 0$.

Conclusions

Thus, the received singular integral relations (8) allow the boundary mixed problems for the anisotropic quarter plane with any number of points of the boundary conditions' change to reduce directly to the SIE systems with fixed singularities. The kernels of these systems have a simple structure and make it possible to reveal not only the main solutions' asymptotics, but all up to any given order. This makes it possible to apply effective numerical methods to solving these SIE systems.

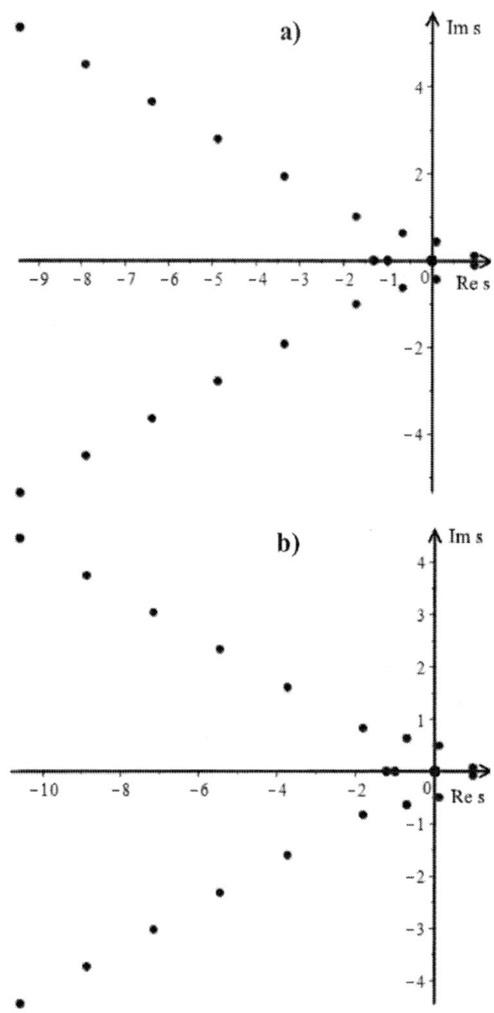

Figure 1. Location of roots of equation (13) for materials $m1$ (a) and $m2$ (b).

Table 1. Roots of transcendental equation (13).

	Material $m1$ ($\varphi = 0$)	Material $m1$ ($\varphi = \pi/4$)	Material $m2$ ($\varphi = 0$)	Material $m2$ ($\varphi = \pi/4$)
s_1	$0.96398 \pm 0.10714i$	0.68343	$0.96782 \pm 0.07313i$	$0.6501 \pm 0.43579i$
s_2	$0.09662 \pm 0.43387i$	0.47475	$0.10265 \pm 0.49705i$	$-0.40598 \pm 0.49585i$
s_3	-0.00336	0.07335	-0.00653	-0.89475
s_4	-0.03593	0.04851	-0.02948	-0.98229
s_5	$-0.66602 \pm 0.62959i$	$-0.46505 \pm 0.14226i$	$-0.7047 \pm 0.63859i$	$-1.41454 \pm 0.47715i$
s_6	-1.00312	$-1.48349 \pm 0.11995i$	-1.00558	$-2.47249 \pm 0.43114i$
s_7	-1.32116	-1.90011	-1.21599	-2.89316
s_8	$-1.7166 \pm 1.00599i$	-1.9648	$-1.80739 \pm 0.83147i$	-2.91223
s_9	$-3.35246 \pm 1.93375i$	-2.28077	$-3.72018 \pm 1.60498i$	$-3.46338 \pm 0.43883i$
s_{10}	$-4.87467 \pm 2.78832i$	-2.55499	$-5.44624 \pm 2.32672i$	$-4.39582 \pm 0.37442i$

IOP Conf. Series: Journal of Physics: Conf. Series **991** (2018) 012004 doi:10.1088/1742-6596/991/1/012004

References

[1] Uflyand Ya S 1959 The mixed problem of the theory of elasticity for a wedge *Izv. Akad. Nauk SSSR. OTN. Mekh. Mashinostr.* No. 2 156–8

[2] Belonosov S M 1960 A plane problem of the theory of elasticity for a wedge with stresses and displacements given at a boundary *Dokl. Akad. Nauk SSSR* **131** (5) 1042–5

[3] Kurshin L M 1959 Mixed plane boundary value problem of the theory of elasticity for a quadrant *J. Appl. Math. Mech.* **23** (5) 1403–8

[4] Hanson M T and Keer L M 1989 Stress analysis and contact problems for an elastic quarter-plane *Quarterly J. Mech. Appl. Math.* **42** (3) 364–83

[5] Aleksanyan R K and Gevorkyan S Kh 2000 To the first fundamental problem of the plane theory of elasticity for an anisotropic wedge *Izv. Nats. Akad. Nauk Armenii. Mekh.* **53** (2) 10–5

[6] Avanyan V T, Avanyan N V, and Babloyan A A 2000 The nature of the stress state in a small neighborhood of the wedge top made of two different anisotropic materials *Izv. Nats. Akad. Nauk Armenii. Mekh.* **53** (3) 5–11

[7] Krivoi A F and Radiolo M V 1984 Features of the stress field near inclusions in a composite anisotropic plane *Izv. Akad. Nauk SSSR. Mekh. Tverd. Tela* No. 3 84–92

[8] Krivoi A F 2001 Arbitrarily oriented defects in a composite anisotropic plane *Vest. Odessk. Nats. Univ. Fiz.-Mat. Nauki* **6** (3) 108–15

[9] Krivoi A F 2003 A fundamental solution for a four-component anisotropic plane *Vest. Odessk. Nats. Univ. Fiz.-Mat. Nauki* **8** (2) 140–9

[10] Krivoi A F and Popov G Ya 2008 Interface tunnel cracks in a composite anisotropic space *J. Appl. Math. Mech.* **72** (4) 499–507

[11] Krivoi A F and Popov G Ya 2008 Features of the stress field near tunnel inclusions in an inhomogeneous anisotropic space *Int. Appl. Mech.* **44** (6) 626–34

[12] Kryvyy O F 2014 Tunnel internal crack in a piecewise homogeneous anisotropic space *J. Math. Sci.* **198** (1) 62–74

[13] Kryvyi O F 2015 Mutual influence of an interface tunnel crack and an interface tunnel inclusion in a piecewise homogeneous anisotropic space *J. Math. Sci.* **208** (4) 409–16

[14] Lehnitsky S G 1977 *Theory of Elasticity of an Anisotropic Body* (Moscow: Nauka) p 415

IOP Conf. Series: Journal of Physics: Conf. Series **991** (2018) 012005 doi:10.1088/1742-6596/991/1/012005

Condition for transition to an unstable state (necking) of a specimen in tension

A R Arutyunyan and R A Arutyunyan*
Saint Petersburg State University, St. Petersburg, Russia
Mekhanobr-tekhnika Corp. (JSC), St. Petersburg, Russia
E-mail: *r.arutyunyan@spbu.ru

Abstract. On the stress-strain curve obtained in simple tension experiments, there is a region of instability due to the neck formation. In the theory of plasticity, conditions for the transition to an unstable state and appearance of the maximum point on the stress-strain diagram are determined. These condition were derived under the assumption of the material incompressibility. However, this assumption cannot be justified, since in the neck region, there are numerous damages (pores, micro-cracks), i.e., the material is compressible. With the current coefficient of lateral deformation, the authors formulate a condition for transition to an unstable state in a compressible plastic medium.

1. Introduction

The basic mechanical properties of materials are determined, in particular, in simple tension experiments at room temperature and constant strain rate.

The results of measurements of the force, the current length and diameter of the specimen were used to calculate the values of the true stress $\sigma = P/F = \sigma_0 F_0/F$, the relative $\varepsilon_0 = (l - l_0)/l_0$ and the logarithmic strains $\varepsilon = \ln l/l_0$ (P is force, $\sigma_0 = P/F_0$ is nominal stress, l_0 and F_0 are initial, l and F are the current length and the cross-section area of the specimen). A typical stress-strain curve received in a tension experiment is shown in figure 1.

If we neglect the change of the cross-section area of the specimen during the deformation, the stress-strain curves are constructed in the nominal stress–relative strain coordinates $\sigma_0 - \varepsilon$. In the case of large deformations (over twenty percent), the logarithmic deformation is used. At point M in figure 1, the nominal stress attains the maximum, a neck is formed on the specimen and the deformation becomes unstable. The drop-down part of the curve $\sigma_0 - \varepsilon$ is a result of a sharp decrease in the cross-section area of the specimen due to the necking. In the neck, the stresses and strains become non-uniform. The drop-down part of the nominal stress-strain curve is a consequence of the decrease in the load and not in the stress. According to the true stress-strain curve, the true stress does continue to rise until the fracture is reached. The metal continues to work-harden.

2. Condition for attaining the maximum in the stress-strain curve for incompressible and compressible plastic media

In the literature [1–5], the conditions for attaining the maximum at point M (figure 1) are determined under the incompressibility condition $l_0 F_0 = lF$, and then $P = \sigma F = \sigma F_0 e^{-\varepsilon}$.

IOP Conf. Series: Journal of Physics: Conf. Series **991** (2018) 012005 doi:10.1088/1742-6596/991/1/012005

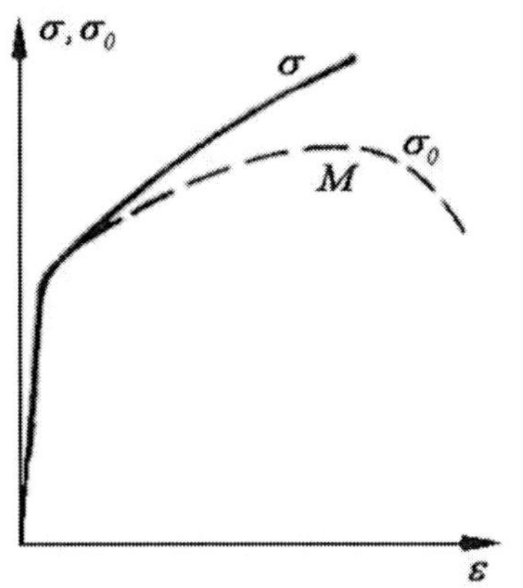

Figure 1. Stress–strain diagrams. Solid line corresponds to tension diagram plotted in true stresses and the dotted line is the diagram plotted in engineering stresses.

Differentiating the last expression with respect to ε, we have

$$\frac{dP}{d\varepsilon} = F_0 e^{-\varepsilon}\left(\frac{d\sigma}{d\varepsilon} - \sigma\right). \tag{1}$$

At the maximum point on the nominal stress-strain curve, the condition $dP/d\varepsilon = 0$ is satisfied and (1) implies the ratio

$$\frac{d\sigma}{d\varepsilon} = \sigma, \tag{2}$$

which determines a condition for transition to an unstable state and attainment of maximum at point M (figure 1).

Thus, relations (2) are derived under the incompressibility condition, which implies that the plastic deformation occurs by the viscous flow mechanism without formation of defect structures. At the same time, in real materials, for example, in metals, during the plastic deformation, particularly in the region of instability, there arise numerous defects (pores, cracks), and so the assumption of incompressibility cannot be considered reasonable in the general case.

We determine the compressibility condition using the current coefficient of the lateral deformation ν: $\nu = -\varepsilon_y/\varepsilon_x = -\varepsilon_z/\varepsilon_x$ ($\varepsilon_x = \ln(l/l_0)$ is the longitudinal deformation and $\varepsilon_y = \varepsilon_z = \ln(R/R_0)$ are the transverse deformations of a cylindrical specimen, R_0 is the initial radius, and R is the current radius of the specimen cross-section). In the elastic region, this coefficient is constant $\nu = \nu_0$ (ν_0 is the Poisson ratio).

With the transition to the plastic state, the current coefficient of the lateral deformation tends to $\nu = 0.5$, so we can assume that the material comes into the state of incompressibility. With further loading, the value ν decreases, which indicates a sharp increase in the irreversible bulk strain (loosening [6]) or a decrease in the density [7], which is characteristic of the pre-failure stage. To get to this point, the following two properties occur simultaneously: the process of destruction and the process of plastic deformation.

If the increase in tensile stresses is accompanied by a decrease in the current coefficient of the lateral deformation, then it is possible to assume that the destruction processes prevail over the processes of flow. Therefore, in formulating the condition of the specimen transition to an unstable state, it is necessary to consider the processes of loosening and, consequently, the compressibility of the material. These processes can be taken into account with the current

IOP Conf. Series: Journal of Physics: Conf. Series **991** (2018) 012005 doi:10.1088/1742-6596/991/1/012005

coefficient of the lateral deformation

$$\nu = -\frac{\ln(R/R_0)}{\ln(l/l_0)}. \tag{3}$$

From (3) it follows that $R_0/R = (l/l_0)^{2\nu}$. Taking into account the obvious relation $F_0/F = (R_0/R)^2$, we obtain [7]

$$\frac{F_0}{F} = \left(\frac{l}{l_0}\right)^{2\nu}. \tag{4}$$

Taking into account (4), we have

$$P = \sigma F_0 e^{-2\nu\varepsilon}. \tag{5}$$

Differentiating (5) with respect to ε, we obtain

$$\frac{dP}{d\varepsilon} = F_0 e^{-2\nu\varepsilon}\left(\frac{d\sigma}{d\varepsilon} - 2\sigma\varepsilon\frac{d\nu}{d\varepsilon} - 2\sigma\nu\right). \tag{6}$$

At the maximum point on the stress-strain curve $dP/d\varepsilon = 0$, it follows from (6) that

$$\frac{d\sigma}{d\varepsilon} = 2\sigma\nu + 2\sigma\varepsilon\frac{d\nu}{d\varepsilon}. \tag{7}$$

For $d\nu/d\varepsilon = 0$, relation (7) implies the maximum condition

$$\frac{d\sigma}{d\varepsilon} = 2\nu\sigma. \tag{8}$$

For an incompressible material, $\nu = 0.5$ and relation (8) coincides with formula (2). According to (8), the position of the maximum value on the stress-strain curve varies depending on the state of the material. Thus, the transition to instability with necking corresponds to the transition of a material from an incompressible state to a compressible state at the maximum point K on the ν–ε curve.

In the case of elastic-plastic media, $d\varepsilon = d\varepsilon^{\mathrm{e}} + d\varepsilon^{\mathrm{p}}$ (ε^{e} and ε^{p} are components of the elastic and plastic deformation, $\varepsilon^{\mathrm{e}} = \sigma/E$, and E is the Young modulus). The relation between stress and deformation can be determined by the equation

$$\frac{dl}{l} = \frac{d\sigma}{E} + \varphi(\sigma)\,d\sigma. \tag{9}$$

Integrating equation (9), we obtain

$$\varepsilon = \frac{\sigma}{E} + \int_0^{\sigma} \varphi(\sigma)\,d\sigma. \tag{10}$$

In the general case, substituting the relation $\sigma = \sigma_0 e^{2\nu\varepsilon}$ in (9), we obtain an equation written in terms of the lateral deformation ν. For different values of ν, we can plot nonmonotone σ_0–ε diagrams, and thus, can describe the experimental curves in engineering stresses-strain coordinates for metallic materials. Further, this approach is applied for the case of rigid-plastic Ludwig medium with nonlinear hardening

$$\sigma = \sigma_{\mathrm{T}} + b\varepsilon^m, \tag{11}$$

where σ_{T} is the yield stress and b, m are constants.

Let us write equation (11) in terms of σ_0 (see figure 2)

$$\sigma_0 = (\sigma_{\mathrm{T}} + b\varepsilon^m)e^{-2\nu\varepsilon}. \tag{12}$$

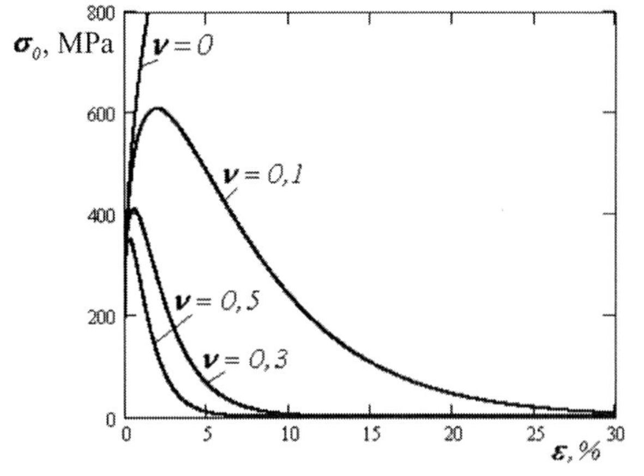

Figure 2. Theoretical curves σ_0–ε according formula (12).

Conclusions

In the mechanics of materials, the incompressibility condition (Hill, Nadai, Kachanov, etc.) is accepted in deriving conditions for a specimen transition to a state of instability and neck formation in tension with a constant speed. At the same time, under active loading, there are interrelated processes of plastic deformation and fracture. In the neck area, numerous defects (pores, cracks) are formed and the material passes into a state of compressibility. The compressibility effect is determined by the current coefficient of transverse deformation. The condition of transition to an unstable state is formulated with regard to the change in this coefficient. An analytical stress-strain relation for a rigid-plastic medium with nonlinear hardening is obtained and the corresponding theoretical curves are constructed for different values of the current coefficient of transverse deformation. In particular, it is shown that, for a compressible material, the maximum point is displaced in the region of instability. In the case of transition to an unstable state for an incompressible medium, there is a fixed maximum point on the stress-strain curve.

Acknowledgments

This work was supported by the Russian Scientific Foundation (Grant No. 17-79-30056).

References

[1] Hill R 1950 *The Mathematical Theory of Plasticity* (Oxford: Oxford Univ. Press)
[2] Kachanov L M 1974 *Fundamentals of Fracture Mechanics* (Moscow: Nauka) p 311 [in Russian]
[3] Nadai A 1931 *Plasticity and Fracture of Solids* (New York: McGraw Hill)
[4] Johnson W and Mellor P 1975 *Engineering Plasticity* (London: Van Nostrand Reinhold)
[5] Poirier J P 1985 *Creep of Crystals* (Cambridge: Cambridge Univ. Press)
[6] Novozhilov V V 1965 On plastic cavitation *J. Appl. Math. Mech.* **29** (4) 811–9
[7] Arutyunyan R A 2004 *Problem of Strain Aging and Long-Time Fracture in Mechanics of Materials* (St. Petersburg: Izdat. St. Petersburg Univ.) p 253 [in Russian]

IOP Conf. Series: Journal of Physics: Conf. Series **991** (2018) 012006 doi:10.1088/1742-6596/991/1/012006

Reflection of shear elastic waves from the interface of a ferromagnetic half–space

L H Atoyan* **and S H Terzyan****

Institute of Mechanics of the National Academy of Sciences of Armenia, Yerevan, Armenia

E-mail: *levous@mail.ru, **sat_and_21@yahoo.com

Abstract. In this paper, the problems of reflection and refraction of a pure elastic wave incident from a nonmagnetic medium on the surface of contact between two semi-infinite media of an infinite nonmagnetic/magnetic structure are considered. The resonance character of the interaction between elastic and magnetic waves is shown, and the dependence of the magnetoelastic wave amplitudes on the the incident elastic wave amplitude is also established.

1. Introduction

It is well known that there can be interconnected magneto-elastic (magneto-acoustic) waves in magnetically ordered media. The interrelated spin and elastic waves propagating through a ferromagnetic material are called magnetoelastic or elastic-spin waves. The theory of magnetoelastic coupling in ferromagnets was developed by Akhiezer A I et al. [1] and Kittel C [2]. The coupling effect is known as a magnetoacoustic effect. A number of monographs and almost infinitely many articles deal with this problem, and here are some of them [1–17].

In 1961, Damon R W and Eshbach J R [4] established the existence of a pure magnetostatic spin surface wave in ferromagnets, latter Maugin G A [5] and Hakmi A [6] proved the existence of shear magnetoelastic waves of Bleustein–Gulyaev-type in ferromagnets and obtained conditions under which the surface mode of these waves can be forced to propagate in a ferromagnetic medium.

Interest in these waves was caused by their great practical importance in engineering. Let us list some specific technical devices, where these waves find application: supersonic and hypersound generators, high-frequency magnetostrictive transducers, wave and frequency filters, delay lines, sensors and actuators for smart structures, radio transmitters, seismic devices, etc. Some of these devices operate due to conversion of the fluctuation energy of magnetic spin into the mechanical energy.

The coupled magnetoelastic problems are studied using equations of mechanical displacements, the Landau–Lifshits equations of motion of a magnetic moment, and Maxwell's equations of a magnetic field.

This paper studies the reflection and refraction of elastic shear waves incident from a nonmagnetic elastic medium on the surface of a ferromagnetic half-space bordering on a nonmagnetic medium.

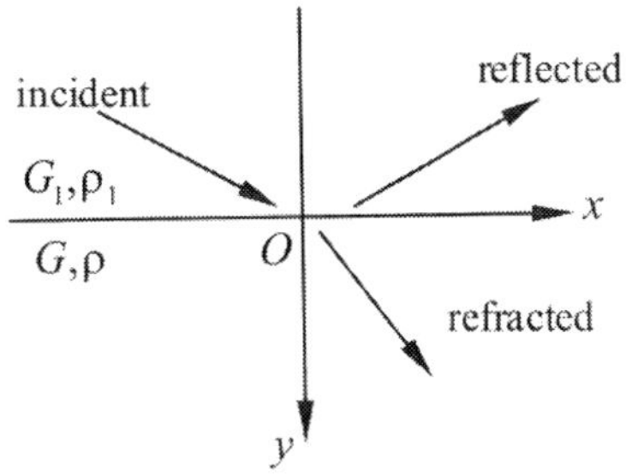

Figure 1. Scheme of incident, reflected and refracted waves.

2. Statement of the problem

We consider a structure consisting of two adjacent half-spaces one of which is not electromagnetically active and the other is ferromagnetic (figure 1). The elastic properties of both media are taken into account.

In figure 1, G_1, ρ_1 and G, ρ denote the shear moduli and the material densities of both media. It is assumed that an elastic shear wave is incident on the surface of the ferromagnetic half-space from the nonmagnetic medium ($y < 0$), where the displacement w_1 is directed along the axis Oz (the axis Oz is perpendicular to the plane of the figure).

The equation of mechanical motion in the nonmagnetic medium ($y < 0$) has the form

$$\frac{\partial^2 w_1}{\partial t^2} = c_{t1}^2 \Delta w_1, \quad c_{t1}^2 = \frac{G_1}{\rho_1}. \tag{1}$$

We assume that the magnetoelastic half-space ($y > 0$) is in a constant magnetic field $\vec{H}_0 = (0, 0, H_0)$. The saturation magnetization $\vec{M}_0$ is also directed along the axis Oz. The system comprising the equations of the mechanical displacement w in a ferromagnetic medium, the Landau–Lifshits equations of motion of a magnetic moment $\vec{\mu}(\mu, \nu, 0)$, and Maxwell's equations of a magnetic field becomes [1, 5, 7]

$$\begin{aligned}
&\frac{\partial^2 w}{\partial t^2} = c_t^2 \Delta w + M_0 f\left(\frac{\partial \mu}{\partial x} + \frac{\partial \nu}{\partial y}\right),\\
&\frac{\partial \mu}{\partial t} = \omega_M\left(\frac{1}{\rho}\frac{\partial \varphi}{\partial y} + \hat{b}\nu + \bar{b}\mu_0\frac{\partial w}{\partial y} - \lambda \Delta \nu\right),\\
&\frac{\partial \nu}{\partial t} = -\omega_M\left(\frac{1}{\rho}\frac{\partial \varphi}{\partial x} + \hat{b}\mu + \bar{b}\mu_0\frac{\partial w}{\partial x} - \lambda \Delta \mu\right),\\
&\Delta \varphi = \rho\left(\frac{\partial \mu}{\partial x} + \frac{\partial \nu}{\partial y}\right),
\end{aligned} \tag{2}$$

where $c_t^2 = G\rho$, $\hat{b} = b + H_0/M_0$, $\bar{b} = b + f$, H_0 is the intensity of the external magnetic field, M_0 is the bulk density of the saturation magnetization of the ferromagnetic medium, b and f are the coefficients of the piezomagnetic coupling, λ is an exchange constant, $\omega_M = \gamma M_0$, γ is the gyromagnetic ratio, $\mu_0 = \rho^{-1} M_0$ is the mass density of the saturation magnetization, w is the displacement, μ and ν are the respective components of the magnetization vector $\vec{\mu}$ along the coordinate axes Ox and Oy, and φ is the magnetostatic field potential.

In what follows, we do not take the excitation of the magnetostatic field and the exchange interaction into account; in this case, system (2) becomes

$$\begin{aligned}
\frac{\partial^2 w}{\partial t^2} &= c_t^2 \Delta w + M_0 f\left(\frac{\partial \mu}{\partial x} + \frac{\partial \nu}{\partial y}\right),\\
\frac{\partial \mu}{\partial t} &= \omega_M\left(\hat{b}\nu + \bar{b}\mu_0\frac{\partial w}{\partial y}\right),\\
\frac{\partial \nu}{\partial t} &= -\omega_M\left(\hat{b}\mu + \bar{b}\mu_0\frac{\partial w}{\partial x}\right).
\end{aligned} \tag{3}$$

The continuity conditions of the displacements and stresses on the boundary $y = 0$ are written as

$$w_1\big|_{y=0} = w\big|_{y=0}, \quad \sigma_{yz}^{(1)}\big|_{y=0} = \sigma_{yz}\big|_{y=0}, \tag{4}$$

where $\sigma_{yz}^{(1)} = G_1\partial w_1/\partial y$, $\sigma_{yz} = G\partial w/\partial y + \bar{b}M_0\nu$, $w_1 = w_{1n} + w_{10}$; w_{1n} and w_{10} are the incident and reflected waves, respectively [1, 7, 9].

The problem is to prove the existence of reflected and refracted waves, and to find the dependence of their parameters on the parameters of the incident wave.

3. The solution of the problem

It follows from equation (1) that the incident w_{1n} and reflected w_{10} waves are determined by the relations

$$\begin{aligned}
w_{1n} &= Ae^{i(\omega t - k_1 x - k_2 y)},\\
w_{10} &= Be^{i(\omega t - k_1 x + k_2 y)},
\end{aligned} \tag{5}$$

where k_1 and k_2 are components of the incident and reflected wave vectors and $k_2^2 = \omega^2/c_{t1}^2 - k_1^2$, A, and B are constants.

We seek the refracted elastic-spin wave in a magnetic medium $y > 0$ in the form

$$\begin{aligned}
w &= Ce^{i(\omega t - k_1 x - k_3 y)},\\
\mu &= Me^{i(\omega t - k_1 x - k_3 y)},\\
\nu &= Ne^{i(\omega t - k_1 x - k_3 y)},
\end{aligned} \tag{6}$$

where C, M, and N are constants and k_3 is a component of the wave vector of the refracted wave. In (6), we take into account the fact that the projections of the wave vectors of all three waves on the boundary must be the same, so that the boundary conditions are satisfied for all t and x [8].

We substitute solutions (6) in system (3) to obtain

$$\begin{aligned}
&[\omega^2 - c_t^2(k_1^2 + k_3^2)]C - ifM_0(k_1 M + k_3 N) = 0,\\
&i\omega M - \hat{b}\omega_M N + ik_3\bar{b}\mu_0\omega_M C = 0,\\
&i\omega N + \hat{b}\omega_M M - ik_1\bar{b}\mu_0\omega_M C = 0.
\end{aligned} \tag{7}$$

Equating the determinant of system (7) to zero, we arrive at the characteristic equation of system (7)

$$\begin{vmatrix}
-ifM_0k_1 & -ifM_0k_3 & \omega^2 - c_t^2(k_1^2 + k_1^3)\\
i\omega & -\hat{b}\omega_M & ik_3\bar{b}\mu_0\omega_M\\
b\omega_M & i\omega & -ik_1\bar{b}\mu_0\omega_M
\end{vmatrix} = 0.$$

After simple transformations, we obtain the the relationship between the components of the refracted wave vector

$$k_1^2 + k_3^2 = \frac{\omega^2(\omega^2 - \hat{b}^2\omega_M^2)}{f\bar{b}\hat{b}\mu_0\omega_M^2 M_0 + c_t^2(\omega^2 - \hat{b}^2\omega_M^2)}. \tag{8}$$

Further, we use system (7) to find the relationship between the amplitudes of the magnetization components and the amplitude of the elastic wave

$$M = \alpha C, \quad N = \beta C, \tag{9}$$

where

$$\alpha = \frac{\omega_M\bar{b}\mu_0(\omega k_3 + ik_1\hat{b}\omega_M)}{\hat{b}^2\omega_M^2 - \omega^2}, \quad \beta = \frac{\omega_M\bar{b}\mu_0(i\hat{b}k_3\omega_M - k_1\omega)}{\hat{b}^2\omega_M^2 - \omega^2}.$$

Formulas (9) show that the elastic oscillations in a ferromagnetic medium excite magnetic (spin) oscillations and vice versa, and the excitation of oscillations has a resonance character. In the literature, this phenomenon is called magnetoacoustic resonance. It follows from (9) that resonance occurs in our case at $\omega = \hat{b}\omega_M$.

We note that substituting (9) into the first equation of system (7), we again arrive at relation (8).

Now let us determine the dependence of the amplitudes of reflected and refracted waves on the amplitude of the incident wave. To this end, we use the boundary conditions, substitute solutions (5) and (6) in (3), and, after simplifying transformations, obtain

$$\begin{aligned} B &= \frac{k_2G_1 - k_3G + \bar{b}M_0\beta}{k_2G_1 - k_3G - \bar{b}M_0\beta}A, \\ C &= \frac{2k_2G_1}{k_3G + k_2G_1 - \bar{b}M_0\beta}A. \end{aligned} \tag{10}$$

For the amplitudes of the magnetization components, we have

$$\begin{aligned} M &= \frac{2\alpha k_2G_1}{k_3G + k_2G_1 - \bar{b}M_0\beta}A, \\ N &= \frac{2\beta k_2G_1}{k_3G + k_2G_1 - \bar{b}M_0\beta}A. \end{aligned} \tag{11}$$

If we equate the amplitude of the reflected wave to zero, we get the transparency condition for the boundary of our structure

$$k_2G_1 - k_3G + \bar{b}M_0\beta = 0.$$

In the particular case where the wave is incident perpendicularly to the surface ($k_1 = 0$), from (8) we derive

$$k_3^2 = \frac{\omega^2(\omega^2 - \hat{b}^2\omega_M^2)}{f\bar{b}\hat{b}\mu_0 M_0\omega_M^2 + c_t(\omega^2 - \hat{b}^2\omega_M^2)}.$$

Conclusions

Thus, a pure elastic shear wave incident from a nonmagnetic half-space on the interface between two media, one of which is magnetically ordered, is generally partially reflected as an elastic wave and partially refracted exciting an interrelated elastic-spin wave in the ferromagnetic medium. Relations (10) and (11) determine the relationship between the amplitudes of these waves.

Elastic oscillations generate magnetic (spin) oscillations in ferromagnetic media, and the oscillation excitation has a resonance character.

Acknowledgments
In conclusion, we would like to express our gratitude to Professor M. V. Belubekyan for the idea of the problem and useful discussions when solving it.

References

[1] Akhiezer A I, Baryakhtar V G, and Peletminskii S V 1967 *Spin waves* (Moscow: Nauka) p 368 [in Russian]
[2] Kittel C 1958 Interaction of spin waves and ultrasonic waves in ferromagnetic crystals *Phys. Rev. B* **110** 836–41
[3] Gurevich A G 1973 Magnetic resonance in ferrites and ferromagnets (Moscow: Nauka) p 591 [In Russian]
[4] Damon R W and Eshbach J R 1961 Magnetostatic modes of a ferromagnet slab *J. Phys. Chem. Solids* **19** 308–20
[5] Maugin G A 1981 Wave motion in magnetizable deformable solids *Int. J. Engng Sci.* **19** 321–88
[6] Maugin G A and Hakmi A 1985 Magnetoelastic surface waves in elastic ferromagnets *J. Acoust. Soc. Am.* **77** (3) 1010–26
[7] Hasanyan D J and Batra R C 2011 Antiplane shear waves in two contacting ferromagnetic half spaces *J. Elasticity* **103** 189–203
[8] Brekhovskikh L M and Goncharov V V 1982 *Introduction to the Mechanics of Continuum* (Moscow: Nauka) p 355 [in Russian]
[9] Danoyan Z N, Aghayan K L, and Atoyan L A 2015 Elastic-spin waves in a layered ferromagnetic-dielectric medium In *Topical Problems of Continuum Mechanics: Proc. of IV Int. Conf. 21–26 September 2015, Tsaghkadzor, Armenia* (Yerevan: Nats. Univ. Arkh. Stroit. Armenii) pp 185–9 [in Russian]
[10] Danoyan Z N, Ghazaryan K B, and Piliposyan G T 2009 Surface gap wave propagation in layered electro-magneto-elastic structures *Waves in Random and Complex Media* **19** (3) 521–34
[11] Danoyan Z N, Piliposyan G T, and Hasanyan D J 2009 Reflection of spin and spin-elastic waves at the interface of a ferromagnetic half-space *Waves in Random and Complex Media* **19** (4) 567–84
[12] Maugin G A 1988 *Continuum Mechanics of Electromagnetic Solids* (Amsterdam: North Holland) p 560
[13] Camley R E 1979 Magnetoelastic waves in a ferromagnetic film on a nonmagnetic substrate *J. Appl. Phys.* **50** 5272–84
[14] Mathews Hand Van de Vaart H 1969 Magnetoelastic Love waves *Appl. Phys. Lett* **15** (11) 373–5
[15] Baghdasaryan A and Belubekyan M 2009 On the problem of reflection of shear wave from a boundary of piezoelectric media of class 6 mm, In *Proc. of the 8th Int. Congr. on Thermal Stresses. TS2009. 1–4 July, Univ. of Illinois at Urbana-Champaing. Illinois, USA* vol 1 (Illinois) pp 183–6
[16] Bagdasaryan G E, Danoyan Z N, Atoyan L A, and Manukyan G A 2008 Reflection of spin waves from the boundary of a ferromagnetic half-space In *Proc. 6th Int. Conf. "The Problems of Dynamics of Interaction of Deformable Media," September 21-26, Goris-Stepanakert, 2008* (Goris - Stepanakert: Inst. Mekh. NAN RA) pp 115–25 [in Russian]
[17] Danoyan Z N and Atoyan L A 2010 Reflection of spin waves from the boundary of a ferromagnetic medium under generalized boundary conditions *Izv. Nats. Akad. Nauk Armenii. Mekh.* **63** (2) 34–9

IOP Conf. Series: Journal of Physics: Conf. Series **991** (2018) 012007 doi:10.1088/1742-6596/991/1/012007

Control of vibrations of a moving beam

N V Banichuk*, S Yu Ivanova, E V Makeev***, and A V Sinitsyn******

Ishlinsky Institute for Problems in Mechanics of the Russian Academy of Sciences, Moscow, Russia

E-mail: *banichuk@gmail.com, **syuivanova@yandex.ru, ***makeevev@yandex.ru, ****a@sinitsyn.info

Abstract. The translational motion of a thermoelastic beam under transverse vibrations caused by initial perturbations is considered. It is assumed that a beam moving at a constant translational speed is described by a model of a thermoelastic panel supported at the edges of the considered span. The problem of optimal suppression of vibrations is formulated when applying active transverse influences to the panel. To solve the optimization problem, modern methods developed in the theory of control of systems with distributed parameters described by partial differential equations are used.

The problem of suppressing vibrations of a deformable thermoelastic system moving in the longitudinal direction is of definite theoretical and practical interest. Previously arising problems in this direction were considered in [1–3]. The problems of elastic instability of vibrations, as well as the problem of finding the critical velocity parameters of longitudinal motion [4], material temperatures, structural parameters and other properties of the moving system, which determine its critical behavior including its static and dynamic forms of stability loss, were investigated.

A new formulation of the problem of optimal suppression of transverse vibrations of a continuous panel (elastic beam) moving in the longitudinal direction is described. Under the assumption of the presence of initial perturbations of the rectilinear movement of the beam and transverse velocities, an effective algorithm for suppressing perturbations is proposed. The above algorithm is based on obtaining and using necessary optimality conditions and the arising dynamic partial differential equations. These equations describe both the vibration processes of the moving beam (direct problems) and some processes for input conjugate variables (conjugate problems). It is shown that the solution of direct and conjugate problems is possible with the Galerkin method. An example is given that illustrates the main steps in the solution of the optimal damping problem.

Let us consider a continuous thermoelastic beam (panel) moving in the longitudinal direction x with constant velocity V_0 and simply supported at points $x = \pm l$. The transverse vibrations of this panel are described by the partial differential equation [4–7]

$$m\left(\frac{\partial^2 w}{\partial t^2} + 2V_0\frac{\partial^2 w}{\partial x \partial t} + V_0^2\frac{\partial^2 w}{\partial x^2}\right) - \left(T_0 - \frac{Eh}{1-\nu}\varepsilon_\theta\right)\frac{\partial^2 w}{\partial x^2} + D\frac{\partial^4 w}{\partial x^4} - g(x,t) = 0, \quad (1)$$
$$(x,t) \in \Omega\{-l \le x \le l,\ 0 \le t \le t_f\}$$

with boundary conditions

$$w\big|_{x=\pm l} = 0, \quad \left.\frac{\partial^2 w}{\partial x^2}\right|_{x=\pm l} = 0, \quad t \in [0, t_f] \quad (2)$$

where m is the mass per unit of panel area, $2l$ is the length of the span, h ($h \ll l$) is the panel thickness, t_s is the final time moment, $D = Eh^3/[12(1-\nu^2)]$ is the bending rigidity (cylindrical rigidity), E is Young's modulus, and ν is Poisson's ratio. In equation (1), the expressions $\partial^2 w/\partial t^2$, $2V_0\,\partial^2 w/(\partial x \partial t)$, $V_0^2\,\partial^2 w/\partial x^2$ are the local acceleration, Coriolis acceleration, and centripetal acceleration, consequently. The expressions $T_0\,\partial^2 w/\partial x^2$, $D\partial^4 w/\partial x^2$, $Eh\varepsilon/(1-v)\,\partial^2 w/\partial x^2$ are the resistance force of the given tension T_0 and the elastic resistance acting against the bending and the temperature action. The generalized thermal deformation ε_θ (we suppose that the temperature is homogeneous with respect to the panel thickness) is determined as [8, 9]

$$\varepsilon_\theta = \frac{1}{h}\int_{-h/2}^{h/2} \alpha_\theta \theta\, dz = \alpha_\theta \theta.$$

Here h is the panel thickness, $\theta = \theta_a - \theta_0$ is the temperature discrepancy, where θ_0 is the temperature when the thermal deformations are absent, and θ_a is the actual temperature (θ, θ_a, and θ_0 are measured).

The transverse vibrations in the moving panel are due to the initial perturbations $g_1(x)$,$g_2(x)$ of transverse displacements and its velocities, i.e., the initial conditions are determined as

$$w\big|_{t=0} = g_1(x), \quad \frac{\partial w}{\partial t}\bigg|_{t=0} = g_2(x), \quad x \in [-l,l], \quad g_1 \in H^1(-l,l), \quad g_2 \in L^2(-l,l). \tag{3}$$

For vibration suppression, the transverse dynamical actions $g = g(\alpha, t)$, $g \in L^2(\Omega)$ are applied and the following isoperimetric condition is added:

$$J_\mu = \int_{-l}^{l}\int_0^{t_f} g^2(x,t)\, dt\, dx = M_0. \tag{4}$$

Here $M_0 > 0$ is a given constant, L^2 is the space of square-integrable functions, H^1 is the space of Sobolev functions which together with their first derivatives are square integrable.

The minimizing quality criterion is written as

$$J_g = \int_{-l}^{l} \left[\alpha_1 w^2 + \alpha_2\left(\frac{\partial w}{\partial t}\right)^2\right]\bigg|_{t=t_f} dx, \tag{5}$$

where the integral J_g is considered at the final time moment t_f and α_1, α_2 are given positive constants.

The considered problem of optimization of vibrations suppression process consists in finding the control action $g(x,t)$ satisfying the isoperimetrical condition (4) and minimizing the quality functional (5) for initial boundary-value problem (1)–(3).

To obtain a necessary optimality condition, we use the equation in variations corresponding to equation (1) and the variated initial and boundary conditions (2), (3). We introduce the conjugate variable $v = v(x,t)$ and the notation $\tilde{D} = D/m$, $\tilde{g} = g/m$, $C^2 = T_0/m$, $e = Eh\theta\alpha_\theta/[(1-v)m]$; the tilde is omitted in what follows. We will derive the equation determining the conjugate variable with corresponding boundary conditions and conditions at the final time moment t_f of the considered process of vibration suppression. For this purpose, let us multiply the equation in variations by $v(x,t)$ satisfying the boundary conditions in the form (2) and integrate the obtain product in the domain Ω taking into account the considered relations in variations. We obtain an expression for the functional variation δJ_a accounting for the dynamics equation.

Then, we construct the augmented functional variation $\delta J = \delta J_g + \delta J_a + \mu\delta J_\mu$, where μ is the Lagrange factor corresponding to condition (4) and

$$\delta J_g = 2\int_{-l}^{l}\left[\alpha_1 w\delta w + \alpha_2\frac{\partial w}{\partial t}\delta\left(\frac{\partial w}{\partial t}\right)\right]\Big|_{t=t_f} dx, \qquad \delta J_\mu = 2\int_{-l}^{l}\int_0^{t_f} g\delta g\, dt\, dx,$$

$$\delta J_a = \int_{-l}^{l}\int_0^{t_f}\left\{\left[\frac{\partial^2 v}{\partial t^2} + 2V_0\frac{\partial^2 v}{\partial x\partial t} + (V_0^2 - C^2 + e)\frac{\partial^2 v}{\partial x^2} + D\frac{\partial^4 v}{\partial x^4}\right]\delta w - v\delta g\right\} dt\, dx$$
$$+ \int_{-l}^{l}\left[v\delta\left(\frac{\partial w}{\partial t}\right) - \left(\frac{\partial v}{\partial t} + 2V_0\frac{\partial v}{\partial x}\right)\delta w\right]\Big|_{t=t_f} dx.$$

Taking into account that $\delta J = 0$ and δg, δw are arbitrary variations for $x \in [-l, l]$, we obtain the necessary optimality condition

$$g_*(x,t) = \frac{1}{2\mu}v(x,t), \quad \mu^2 = \frac{1}{4M_0}\int_{-l}^{l}\int_0^{t_f} v^2\, dt\, dx, \quad (x,t) \in \Omega,$$

the homogeneous differential equation for the conjugate variable $V(x,t)$

$$\frac{\partial^2 v}{\partial t^2} + 2V_0\frac{\partial^2 v}{\partial x\partial t} + (V_0^2 - C^2 + e)\frac{\partial^2 v}{\partial x^2} + D\frac{\partial^4 v}{\partial x^4} = 0, \quad (x,t) \in \Omega \tag{6}$$

and the conditions at the final time moment $t = t_f (x \in [-l, l])$

$$v\Big|_{t=t_f} = -2\alpha_2\frac{\partial w}{\partial x}\Big|_{t=t_f}, \quad \frac{\partial v}{\partial t}\Big|_{t=t_f} = 2\left(\alpha_1 w + 2\alpha_2 V_0\frac{\partial^2 w}{\partial x^2}\right)\Big|_{t=t_f}. \tag{7}$$

We will use the dimensionless variable

$$\tilde{x} = \frac{x}{l}, \quad \tilde{t} = \frac{t}{\tau}, \quad \tilde{w}(\tilde{x},\tilde{t}) = \frac{w(l\tilde{x},\tau\tilde{t})}{h}, \quad \tilde{v}(\tilde{x},\tilde{t}) = \frac{v(l\tilde{x},\tau\tilde{t})}{h}, \quad \tilde{g}(\tilde{x},\tilde{t}) \equiv \frac{lg(l\tilde{x},\tau\tilde{t})}{mC^2} = \frac{g(l\tilde{x},\tau\tilde{t})}{T_0}$$

and the notation

$$\alpha = \frac{l}{\tau C}, \quad \beta = \frac{D}{ml^2C^2}, \quad v_0 = \frac{V_0}{C}, \quad \tilde{\Omega} = \{-1 \le \tilde{x} \le 1,\ 0 \le \tilde{t} \le \tilde{x}_f = t_f/\tau\},$$

where τ is a specific parameter with the dimension of time, and the tilde will be omitted in what follows. Then the vibration equation is

$$\alpha^2\frac{\partial^2 w}{\partial t^2} + 2\alpha v_0\frac{\partial^2 w}{\partial x\partial t} + (v_0^2 - 1 + e)\frac{\partial^2 w}{\partial x^2} + \beta\frac{\partial^4 w}{\partial x^4} = g(x,t). \tag{8}$$

The analogues equation for $v(x,t)$, we obtain with $g(t,x) = 0$ in (8).

For solving the nonhomogeneous partial differential equation (8) with conditions (2), (3) (the problem for determining the function w) and the corresponding homogeneous equation for v with conditions (7) and the boundary conditions in the form of (2), we apply the Galerkin method [10, 11] and represent the functions w, v in series form

$$w(x,t) = \sum_{n=1}^{n_0} f_n(t)\Psi_n(x), \quad v(x,t) = \sum_{n=1}^{n_0} q_n(t)\Psi_n(x), \quad \Psi_n(x) = \sin\left[\frac{n\pi}{2}(x+1)\right], \quad x \in [-1,1],$$

where $\Psi_n(x)$ are shape functions satisfying the pinned simply supported boundary conditions at $x = \pm 1$, n_0 is a given number of series members, $f_n(t)$ and $g_n(t)$ $(n = 1, \ldots, n_0)$ are unknown time functions.

Determining the function $f_n(t)$ is reduced to integrating a system of ordinary differential equations for $f_n(t)$ with the initial conditions

$$f_j(0) = \int_{-1}^{1} \Psi_j(x) g_1(x)\, dx, \quad \left.\frac{\partial f_j}{\partial t}\right|_{t=0} = \int_{-1}^{1} \Psi_j(x) g_2(x)\, dx, \quad j = 1, 2, \ldots$$

Determining the functions $q_n(t)$ is based on integrating the corresponding system of equations for $q_n(t)$ with the condition at the final time moment $t = t_f$

$$q_j(t_f) = -2\alpha_2 \sum_{n=1}^{n_0} \int_{-1}^{1} \Psi_j(x) \frac{d\Psi_n(x)}{dx}\, dx\, f_n(t_f), \quad \left.\frac{dq_f}{dt}\right|_{t=t_f} = 2\left[\alpha_1 - 2\alpha_2 v_0 \left(\frac{j\pi}{2}\right)^2\right] f_j(t_f).$$

As an example, we consider the case where

$$g_1(x) = \Psi_1(x), \quad g_2(x) = 0, \quad -1 \le x \le 1,$$

the thermal actions are ignored, and $g = g^0(x, 1) \equiv 1$, $(x, t) \in \Omega$ (non-optimal constant actions). In this case,

$$n_0 = 1, \quad G_1(t) = \frac{4}{\pi}, \quad f_1(0) = 1, \quad \left.\frac{df_1}{dt}\right|_{t=0} = 0$$

and the equation of transverse vibrations of the moving panel becomes

$$\begin{gathered} \frac{d^2 f_1}{dt^2} + c_1 f_1 + c_2 = 0, \quad 0 \le t \le t_f, \\ c_1 = \frac{\gamma_1}{\alpha^2}, \quad c_2 = \frac{G_1}{\alpha^2}, \quad \gamma_1 = \left(\frac{\pi}{2}\right)^2 \left[-(v_0^2 - 1) + \beta\left(\frac{\pi}{2}\right)^2\right]. \end{gathered} \tag{9}$$

Taking into account the initial conditions, we obtain the following solution

$$f_1(t) = \left(1 - \frac{c_2}{c_1}\right) \cos(\sqrt{c_1} t) + \frac{c_2}{c_1}, \quad 0 \le t \le t_f.$$

Then, we integrate the homogeneous equation

$$\frac{d^2 q_1}{dt^2} + c_1 q_1 = 0, \quad 0 \le t \le t_f, \tag{10}$$

from $t = t_f$ to $t = 0$ with the conditions

$$q_1(t_f) = -2\alpha_2 \int_{-1}^{1} \Psi_1 \frac{d\Psi_1}{dx}\, dx\, f_1(t_f) = 0, \quad \left.\frac{dq_1}{dt}\right|_{t=t_f} = 2\left[\alpha_1 - 2\alpha_2 v_0 \left(\frac{\pi}{2}\right)^2\right] f_1(t_f) \equiv Q,$$

and obtain the solution

$$q_1(t) = Q_1 \sin(\sqrt{c_1} t) + Q_2(\sqrt{c_1} t), \quad Q_1 = \frac{Q}{\sqrt{c_1}} \cos(\sqrt{c_1} t_f), \quad Q_2 = \frac{Q}{\sqrt{c_1}} \sin(\sqrt{c_1} t_f).$$

Thus, we have approximately found the extremal loading

$$d_*(x, t) = \frac{1}{2\mu} q_1(t) \Psi_1(x), \quad \mu^2 = \frac{1}{4\mu_0} \int_{-1}^{1} \int_{0}^{t_f} \Psi_1^2(x) q_1^2(t)\, dx\, dt.$$

For the obtained control loading, we solve equation (9), where

$$c_2 = \frac{G_1}{\alpha^2} = \frac{1}{\alpha^2}\int_{-1}^{1}\Psi_1(x)g_*(x,t)\,dx = \frac{q_1(t)}{2\mu\alpha^2}$$

and obtain the optimal solution

$$f_1^*(t) = -\frac{Q\cos(\sqrt{c_1}t_f)}{4\mu\alpha^2 C_1^{3/2}}\sin(\sqrt{c_1}t) + \left[-\frac{Q\cos(\sqrt{c_1}t_f)}{4\mu\alpha^2 C_1^{3/2}}\right]\cos(\sqrt{c_1}t)$$
$$+\frac{1}{4\mu\alpha^2 c_1}\left[(Q_1\sqrt{c_1}t - Q_2)\cos(\sqrt{c_1}t) - Q_2\sqrt{c_1}t\sin(\sqrt{c_1}t)\right].$$

In this case of optimal vibration suppression, the value of quality criterion (5) is

$$J_g^* = J_g\big|_{g=g_*} = \alpha_1\left(f_1^*\big|_{t=t_f}\right)^2 + \alpha_2\left(\frac{df_1^*}{dt}\bigg|_{t=t_f}\right)^2.$$

In the case of non-optimal action

$$g = g^0(x,t) = 1 \quad \left(c_2 = \frac{4}{\pi\alpha^2},\ \frac{c_2}{c_1} = \frac{4}{\pi\alpha^2\gamma_1}\right),$$

expression (5) has the form

$$J_g^* = J_g\big|_{g=g^0} = \alpha\left[\left(1-\frac{c_2}{c_1}\right)^2\cos^2(\sqrt{c_1}t_f) + 2\frac{c_2}{c_1}\left(1-\frac{c_2}{c_1}\right)\cos(\sqrt{c_1}t_f) + \left(\frac{c_2}{c_1}\right)^2\right] + \alpha_2 c_1\left(1-\frac{c_2}{c_1}\right)^2\sin^2(\sqrt{c_1}t_f).$$

In the case of absence of control action ($g = c_1 = c_2 = 0$), we have

$$J_g\big|_{g=g^0} = \alpha_1\cos^2(\sqrt{c_1}t_f) + \alpha_2 c_1\sin^2(\sqrt{c_1}t_f).$$

In figure 1, the dependence of functionals $J_{\text{opt}} = J_g^*(t_f)$, $J_1 = J_g^0(t_f)$ and $J_0 = J_g(t_f)\big|_{g=0}$ on the problem parameter t_f are presented for $\alpha_1 = \alpha_2 = 1$, $c_1 = 1$, $c_2 = 4/\pi$, $\alpha = \beta = 1$, $\gamma_1 = 1$, and $v_0 = 1.77$.

The solid line corresponds to the determined optimal control actions $g_{\text{opt}} = g_{\text{opt}}(x,t)$, the dotted line illustrates the case without control actions $g(T,t) = 0$, the dashed line characterizes the case with constant (non-optimal) actions $g(T,t) = 1$ applied to the moving beam.

Figure 1 shows that there is a time integral for which the applied optimal control actions are essentially effective.

In addition to a one-dimensional case of a moving beam, we could consider two-dimension plate models describing the longitudinal motions of wide bands in terms of tape systems. This case [4] can be based on the use of necessary conditions for the extremum and application of the Galerkin–Fourier methods too.

Conclusions

In this study a new formulation of the problem of optimal suppression of transverse vibrations of a continuous panel moving in the longitudinal direction is described.

To suppress the arising vibrations, control actions in the form $g = g(x,t)$ with non-separating independent variables x and t are used in the work. However, with this approach, the optimal control can be very complicated, and its practical implementation is difficult. In this case, it is possible to use the representation of the control action in the form $g(x,t) = f(t)\chi(x)$ with

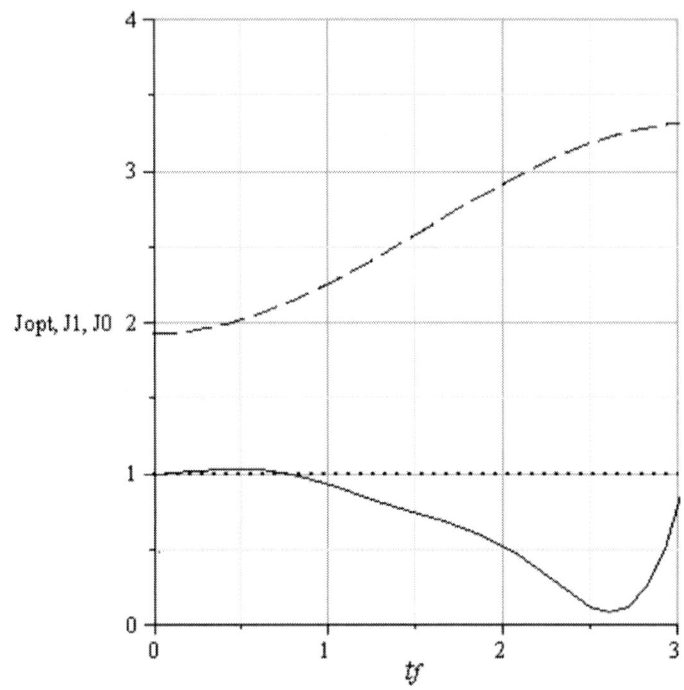

Figure 1. Dependence of the quality criterion on the problem parameter.

separated functions of position and time, describing both the specific geometric realization of location of the actions (actuators) set by $\chi(x)$ and the way of changing the time effects, denoted by $f(t)$. This approach allows us to consider and to compare the effectiveness of different methods to apply the effects to different parts of forces, and to compare such loadings of the beam applied at individual points of concentrated forces, and to compare such loadings in time as relay, harmonic, shock and other controls.

Acknowledgments
The work was supported by the Russian Science Foundation (project No. 17-19-01247).

References

[1] Banichuk N V, Ivanova S Yu, and Sharanyuk A V 1989 *Dynamics of Structures. Analysis and Optimization* (Moscow: Nauka)
[2] Banichuk N V and Bratus A C 1993 On optimal design of constructions equipped with actuators *Izv. Ross. Akad. Nauk. Tekh. Kibernet.* No. 1 24–33
[3] Banichuk N V, Bratus A C, and Posviansky V P 1994 Optimum positioning of actuator for structural control In *Fourth Int. Conf. on Adaptive Structures* ed E J Breitbach, B K Woda, and M Natori (Lancaster: Technomic Publ. Co.) pp 179-188
[4] Banichuk N, Jeronen J, Neittaanmaki P, et al. (Editors) 2014 *Mechanics of Moving Materials* (Switzerland: Springer) p 253
[5] Wickert J A 1992 Non-linear vibration of a traveling tensioned beam *Int. J. Non-Lin. Mech.* **27** (3) 503–17
[6] Marinowski K and Kapitaniak T 2014 Dynamics of axially moving continua *Int. J. Mech. Sci.* **81** 26–41
[7] Chen L-Q 2005 Analysis and control of transverse vibrations of axially moving strings *ASME Appl. Mech. Rev.* **58** 91–116
[8] Timoshenko S P and Woinowsky-Kriger S 1959 *Theory of Plates and Shells* 2nd ed. (New York - Tokyo: McGraw-Hill) p 595
[9] Kovalenko A D 1969 *Thermoelasticity* (Kiev: Vishcha Shkola) p 251 [in Russian]
[10] Washizy K 1982 *Variational Methods in Elasticity and Plasticity* (Oxford: Pergamon Press) p 540
[11] Michin S G 1964 *Variational Methods in Mathematical Physics* (New York: Pergamon Press) p 584

IOP Conf. Series: Journal of Physics: Conf. Series **991** (2018) 012008 doi:10.1088/1742-6596/991/1/012008

Equations of an elliptic boundary layer in shells of revolution under shock normal edge loading

G A Baregamyan*, I V Kirillova, and L Yu Kossovich**

Chernyshevsky Saratov State University, Saratov, Russia

E-mail: *bar.gaik@mail.ru, **nano-bio@sgu.ru

Abstract. The asymptotic equations describing an elliptic boundary layer in shells of revolution at shock normal edge loading are obtained. The considered type of a boundary layer is concentrated in the small vicinity of Rayleigh front. Asymptotic methods in combinations with a symbolical method of Lurye are used. The elliptical equations describing behavior of a boundary layer on thickness for this boundary layer and the hyperbolic equations describing boundary conditions on faces are received.

We consider the shell of revolution shown in figure 1. The following notation is introduced: α is the arc length along the meridian, θ is the angle in the circumferential direction, and z is the external normal to the middle surface. We denote the stresses and displacements by σ_{ij} and v_i $(i = 1, 2, 3)$, respectively. Let's consider a case of shock tangential impact on the edge $\alpha = 0$ at zero initial conditions

$$\sigma_{13} = IH(t), \quad v_1 = 0, \tag{1}$$

at free faces $z = \pm h$, where h is the shell half-thickness and t is the time.

Let us introduce the following dimensionless variables:

$$\xi = \frac{\alpha}{h}, \quad \zeta = \frac{z}{h}, \quad \tau = \frac{tc_2}{h}, \quad c_2 = \sqrt{\frac{E}{2(1+\nu)\rho}}, \tag{2}$$

where E, ν, and ρ are the Youngs modulus, Poisson coefficient, and the shell material density, respectively. The governing dynamic equations are given in the dimension coordinates in [1].

In the dimensionless coordinates boundary conditions take the following form

$$\frac{\partial v_3}{\partial \xi} = I^* H(\tau), \quad v_1 = 0, \tag{3}$$

where $I^* = I/\{E/[2(1+\nu)h]\}$. We will lower asterisks further.

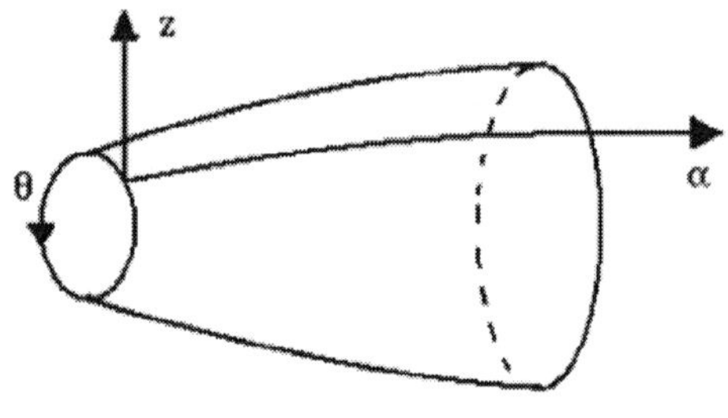

Figure 1. Shell of revolution.

Let us consider governing equations in the displacements in the dimensionless form

$$
\begin{aligned}
&æ^{-2}\frac{\partial^2 v_1}{\partial \xi^2}+\frac{\partial^2 v_1}{\partial \zeta^2}+\frac{1}{1-2\nu}\frac{\partial^2 v_3}{\partial \xi \partial \zeta}-\frac{\partial^2 v_1}{\partial \tau^2}+\eta æ^{-2}\frac{B'}{B}\frac{\partial v_1}{\partial \xi}=0,\\
&\frac{1}{1-2\nu}\frac{\partial^2 v_1}{\partial \xi \partial \zeta}+\frac{\partial^2 v_3}{\partial \xi^2}+æ^{-2}\frac{\partial^2 v_3}{\partial \zeta^2}-\frac{\partial^2 v_3}{\partial \tau^2}+\eta\left(\frac{1}{1-2\nu}\frac{B'}{B}\frac{\partial v_1}{\partial \zeta}+\frac{B'}{B}\frac{\partial v_3}{\partial \xi}\right)=0,\\
&\sigma_{11}=\frac{E}{2(1+\nu)æ^2 h}\left(\frac{\partial v_1}{\partial \xi}+\frac{\nu}{1-\nu}\frac{\partial v_3}{\partial \zeta}+\eta\frac{\nu}{1-\nu}\frac{B'}{B}v_1\right),\\
&\sigma_{33}=\frac{E}{2(1+\nu)æ^2 h}\left(\frac{\nu}{1-\nu}\frac{\partial v_1}{\partial \xi}+\frac{\partial v_3}{\partial \zeta}+\eta\frac{\nu}{1-\nu}\frac{B'}{B}v_1\right),\\
&\sigma_{13}=\frac{E}{2(1+\nu)h}\left(\frac{\partial v_1}{\partial \zeta}+\frac{\partial v_3}{\partial \xi}\right),
\end{aligned}
\tag{4}
$$

where B is the distance to the axis of revolution, $æ^2=(1-2\nu)/(2-2\nu)$, $\eta=h/R$ is the small parameter, and R is the typical value of the curvature radius.

In the work [1] the asymptotic model describing the nonstationary stress strain state (SSS) of the shell of revolution in the small vicinity of the Rayleigh wave front at shock face loading was describing. The purpose of this work is to generalize of the offered approach on a the case of shock edge loading.

Let's reduce the considered problem to the equivalent problem for the infinite shell. We use the method described in [2, 3] for the case of strips and cylindrical shell at action of shock face load. For this purpose at first we will construct the private solution which corresponds only the edge conditions (4). In the specified works this private solution was constructed from equations in partial derivatives with constant coefficients. In the case under consideration the equations for the shell of revolution have variable coefficients and the private solution could be constrcted by the front asymptotic method [4].

According to the most simple variant of a method of a front asymptotics which is widely used in [4] for definition of slowly changing components of the nonstationary SSS of shell of revolution, the required solution is presented in the decomposition form on deviation degrees from fronts of waves. In the case under consideration construction of the private solution satisfying to the equations (4) and the edge boundary conditions we will take private solution for semi-strip [2] as base and will present zero approximations of the required solution in the form (we will denote the private solution by a superscript (0))

$$
v_1^{(0)}=0,\quad v_3^{(0)}=-IF(\xi_0)(æ^{-1}\tau-\xi)H(æ^{-1}\tau-\xi),\quad \sigma_{11}^{(0)}=\sigma_{33}^{(0)}=0,\quad \sigma_{13}^{(0)}=IF(\xi_0)H(æ^{-1}\tau-\xi),\tag{5}
$$

where $F(\xi_0)$ is unknown function which will be found by means of decomposition in small parameter η. Substituting (5) in (4) and, neglecting values $O(\eta^2)$, we will receive the equation

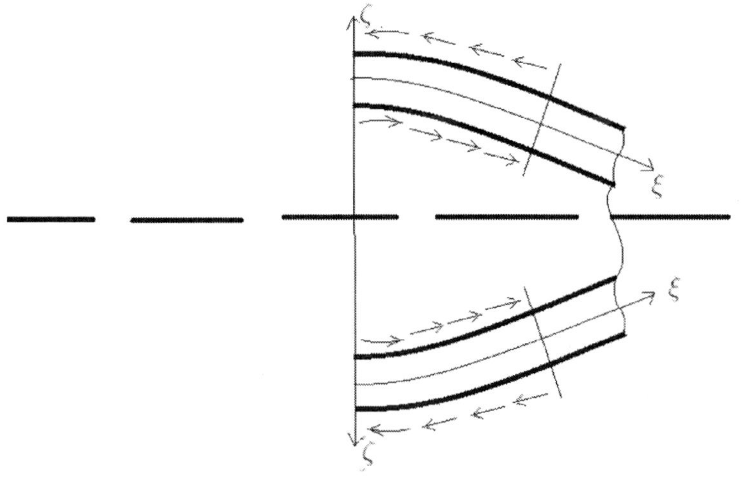

Figure 2. Loadings on front faces.

for $F(\xi_0)$:

$$2\frac{dF}{d\xi_0} + \frac{1}{B}\frac{dF}{d\xi_0}F = 0. \tag{6}$$

Solution of (6) gives expression for F

$$F = \frac{1}{\sqrt{B}}. \tag{7}$$

Then the private solution (5) takes the following final form

$$\begin{aligned} &v_1^{(0)} = 0, \quad v_3^{(0)} = -I\frac{1}{\sqrt{B}}(æ^{-1}\tau - \xi)H(æ^{-1}\tau - \xi), \\ &\sigma_{11}^{(0)} = \sigma_{33}^{(0)} = 0, \quad \sigma_{13}^{(0)} = I\left[\frac{1}{\sqrt{B}} - \eta\frac{B'}{B^{3/2}}(æ^{-1}\tau - \xi)\right]H(æ^{-1}\tau - \xi). \end{aligned} \tag{8}$$

Let's represent the solution of our problem, as well as in [2], in the form of the following sum

$$SSS = SSS^{(0)} + SSS^{(1)}, \tag{9}$$

where for a component with a superscript (1) the following boundary conditions are formulated

$$\begin{aligned} &\frac{\partial v_3^{(1)}}{\partial \xi} = 0, \quad v_1^{(1)} = 0, \quad \xi = 0, \\ &\sigma_{33}^{(1)} = 0, \quad \sigma_{13}^{(1)} = -I\left[\frac{1}{\sqrt{B}} - \eta\frac{B'}{B^{3/2}}(æ^{-1}\tau - \xi)\right]H(æ^{-1}\tau - \xi), \quad \zeta = \pm 1. \end{aligned} \tag{10}$$

The analysis shows that for solution (8) the deformed edge section is rectilinear and perpendicular to axis ξ (figure 2). Consequently, the component with a superscript (1) has the same properties. Therefore, the SSS of the infinite shell of revolution with the symmetric relatively ζ loading on the faces determined by boundary conditions (10) (figure 3) is equivalent to the SSS for the semi-infinite shell.

Let's consider the equivalent problem for the infinite shell of revolution. Let's use operators $\partial_\xi = \partial/\partial\xi$, $\partial_\tau = \partial/\partial\tau$.

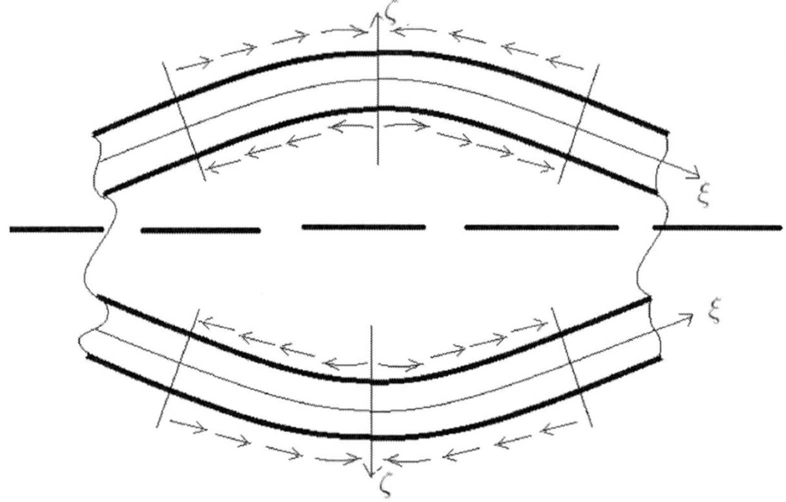

Figure 3. Problem for the infinite envelope.

Write down equations of motion in operatore form

$$\begin{gathered}\frac{\partial^2 v_1}{\partial\zeta^2} + \left(\ae^{-2}\partial_\xi^2 - \partial_\tau^2 + \eta\ae^{-2}\frac{B'}{B}\partial_\xi\right)v_1 + (\ae^{-2}-1)\partial_\xi\frac{dv_3}{d\zeta} = 0,\\ (\ae^{-2}-1)\left(\partial_\xi + \eta\frac{B'}{B}\right)\frac{dv_1}{d\zeta} + \ae^{-2}\frac{d^2v_3}{d\zeta^2} + \left(\partial_\xi^2 - \partial_\tau^2 + \eta\frac{B'}{B}\partial_\xi\right)v_3 = 0.\end{gathered} \tag{11}$$

The solution of the operator equations has the following form

$$\begin{gathered}v_1 = \partial_\xi e^{i(1+\zeta)\alpha_1}D_1 + \partial_\xi e^{i(1-\zeta)\alpha_1}D_2 + \beta_1 e^{i(1+\zeta)\beta_1}D_3 + \beta_1 e^{i(1-\zeta)\beta_1}D_4,\\ v_3 = i\alpha_1 e^{i(1+\zeta)\alpha_1}D_1 - i\alpha_1 e^{i(1-\zeta)\alpha_1}D_2 + i\left(\partial_\xi + \eta\frac{B'}{B}\right)e^{i(1+\zeta)\beta_1}D_3 - i\left(\partial_\xi + \eta\frac{B'}{B}\right)e^{i(1-\zeta)\beta_1}D_4,\end{gathered} \tag{12}$$

where D_i the arbitrary parameters, $\alpha_1^2 = \partial_\xi^2 - \ae^2\partial_\tau^2 + \eta(B'/B)\partial_\xi$, $\beta_1^2 = \partial_\xi^2 - \partial_\tau^2 + \eta(B'/B)$.

Expressions for stresses are written as:

$$\begin{aligned}\sigma_{33} = &-\frac{E}{(1+\nu)h}\Big[\gamma_1^2 e^{i(1+\zeta)\alpha_1}D_1 + \gamma_1^2 e^{i(1-\zeta)\alpha_1}D_2\\ &+ \left(\partial_\xi + \eta\frac{B'}{B}\right)\beta_1 e^{i(1+\zeta)\beta_1}D_3 + \left(\partial_\xi + \eta\frac{B'}{B}\right)\beta_1 e^{i(1-\zeta)\beta_1}D_4\Big],\\ \sigma_{13} = &\frac{E}{(1+\nu)h}\Big[i\partial_\xi\alpha_1 e^{1(1+\zeta)\alpha_1}D_1 - i\partial_\xi\alpha_1 e^{1(1-\zeta)\alpha_1}D_2 + i\gamma_1^2 e^{i(1+\zeta)\beta_1}D_3 - i\gamma_1^2 e^{i(1-\zeta)\beta_1}D_4\Big].\end{aligned} \tag{13}$$

Substituting solution (13) in boundary conditions (10) at $\zeta = \pm 1$, we will receive system of the algebraic equations for D_i:

$$\begin{gathered}\gamma_1^2 D_1 + \gamma_1^2 e^{i2\alpha_1}D_2 + \left(\partial_\xi + \eta\frac{B'}{B}\right)\beta_1 D_3 + \left(\partial_\xi + \eta\frac{B'}{B}\right)\beta_1 e^{i2\beta_1}D_4 = 0,\\ \gamma_1^2 e^{i2\alpha_1}D_1 + \gamma_1^2 D_2 + \left(\partial_\xi + \eta\frac{B'}{B}\right)\beta_1 e^{i2\beta_1}D_3 + \left(\partial_\xi + \eta\frac{B'}{B}\right)\beta_1 D_4 = 0,\\ \partial_\xi\alpha_1 D_1 - \partial_\xi\alpha_1 e^{i2\alpha_1}D_2 + i\gamma_1^2 D_3 - i\gamma_1^2 e^{i2\beta_1}D_4 = -S,\\ \partial_\xi\alpha_1 e^{i2\alpha_1}D_1 - \partial_\xi\alpha_1 D_2 + i\gamma_1^2 e^{i2\beta_1}D_3 - i\gamma_1^2 D_4 = -S,\end{gathered} \tag{14}$$

IOP Conf. Series: Journal of Physics: Conf. Series **991** (2018) 012008 doi:10.1088/1742-6596/991/1/012008

where

$$S=\frac{EI}{2(1+\nu)h}\left[\frac{1}{\sqrt{B}}-\eta\frac{B'}{B^{3/2}}(æ^{-1}\tau-\xi)\right]H(æ^{-1}\tau-\xi). \tag{15}$$

As well as in [1], we will receive the main part of approximation of the equations for boundary conditions, neglecting in equations (14) terms with factors $e^{i2\alpha_1}$, $e^{i2\beta_1}$

$$\begin{aligned}
&\gamma_1^2D_1+\left(\partial_\xi+\eta\frac{B'}{B}\right)\beta_1D_3=0,\\
&\gamma_1^2D_2+\left(\partial_\xi+\eta\frac{B'}{B}\right)\beta_1D_4=0,\\
&\partial_\xi\alpha_1D_1+\gamma_1^2D_3=-S,\\
&\partial_\xi\alpha_1D_2+\gamma_1^2D_4=-S.
\end{aligned}\tag{16}$$

Let's receive from (16) expression for D_1 and D_2 through D_3 and D_4:

$$\begin{aligned}
D_1&=-\frac{1}{\gamma_1^2}\left(\partial_\xi+\eta\frac{B'}{B}\right)\beta_1D_3,\\
D_2&=-\frac{1}{\gamma_1^2}\left(\partial_\xi+\eta\frac{B'}{B}\right)\beta_1D_4.
\end{aligned}\tag{17}$$

Governing equations for D_3 and D_4 have the form

$$\begin{aligned}
&\left[\gamma_1^4-\partial_\xi\left(\partial_\xi+\eta\frac{B'}{B}\right)\alpha_1\beta_1\right]D_3=-\gamma_1^2S,\\
&\left[\gamma_1^4-\partial_\xi\left(\partial_\xi+\eta\frac{B'}{B}\right)\alpha_1\beta_1\right]D_4=-\gamma_1^2S.
\end{aligned}\tag{18}$$

Let's consider great values of time (at $\tau\gg1$). We introduce variables y and τ_0:

$$y=\frac{\xi-æ_R\tau}{\varepsilon},\quad \tau_0=\varepsilon\tau,\quad \varepsilon=\frac{1}{T}. \tag{19}$$

We also introduce new operators ∂_y, ∂_{τ_0}:

$$\partial_y=\frac{\partial}{\partial y},\quad \partial_{\tau_0}=\frac{\partial}{\partial\tau_0}. \tag{20}$$

Then equations (18) are written as

$$\begin{aligned}
&\left(\varepsilon2æ_R\partial_y\partial_{\tau_0}+\eta\frac{B'}{B}\partial_y\right)D_3=\frac{æ_RB_\omega}{g}S,\\
&\left(\varepsilon2æ_R\partial_y\partial_{\tau_0}+\eta\frac{B'}{B}\partial_y\right)D_4=\frac{æ_RB_\omega}{g}S.
\end{aligned}\tag{21}$$

We transform the first and second equation in (16). We receive:

$$\begin{aligned}
&\left[g\partial_y^2+\varepsilon æ_R\partial_y\partial_{\tau_0}+\eta\frac{B'}{B}\partial_y\right]D_1+\left[b\partial_y^2+\varepsilon\frac{æ_R}{b}\partial_y\partial_{\tau_0}+\eta\left(b+\frac{1}{2b}\right)\frac{B'}{B}\right]D_3=0,\\
&\left[g\partial_y^2+\varepsilon æ_R\partial_y\partial_{\tau_0}+\eta\frac{B'}{B}\partial_y\right]D_2+\left[b\partial_y^2+\varepsilon\frac{æ_R}{b}\partial_y\partial_{\tau_0}+\eta\left(b+\frac{1}{2b}\right)\frac{B'}{B}\right]D_4=0.
\end{aligned}\tag{22}$$

Asymptotic representation of expressions for displacement and stresses will differ from representations from [1]: terms of higher order of smallness should be taking into account in representation of stress σ_{13}:

$$
\begin{aligned}
v_1 &= \partial_y e^{ia(1+\zeta)\partial_y} D_1 + \partial_y e^{ia(1-\zeta)\partial_y} D_2 + b\partial_y e^{ib(1+\zeta)\partial_y} D_3 + b\partial_y e^{ib(1-\zeta)\partial_y} D_4,\\
v_3 &= ia\partial_y e^{ia(1+\zeta)\partial_y} D_1 - ia\partial_y e^{ia(1-\zeta)\partial_y} D_2 + i\partial_y e^{ib(1+\zeta)\partial_y} D_3 - i\partial_y e^{ib(1-\zeta)\partial_y} D_4,\\
\sigma_{33} &= -\frac{E}{(1+\nu)h}\left[g\partial_y^2 e^{ia(1+\zeta)\partial_y} D_1 + g\partial_y^2 e^{ia(1-\zeta)\partial_y} D_2 + b\partial_y^2 e^{ib(1+\zeta)\partial_y} D_3 + b\partial_y^2 e^{ib(1-\zeta)\partial_y} D_4\right],\\
\sigma_{13} &= \frac{E}{(1+\nu)h}\left[i\left(a\partial_y^2 + \varepsilon\frac{æ^2}{a}æ_R\partial_y\partial_{\tau_0} + \eta\frac{1}{2a}\frac{B'}{B}\partial_y\right)e^{ia(1+\zeta)\partial_y} D_1\right.\\
&\quad - i\left(a\partial_y^2 + \varepsilon\frac{æ^2}{a}æ_R\partial_y\partial_{\tau_0} + \eta\frac{1}{2a}\frac{B'}{B}\partial_y\right)e^{ia(1-\zeta)\partial_y} D_2\\
&\quad \left. + i\left(g\partial_y^2 + \varepsilon æ_R\partial_y\partial_{\tau_0} + \eta\frac{B'}{B}\partial_y\right)e^{ib(1+\zeta)\partial_y} D_3 - i\left(g\partial_y^2 + \varepsilon æ_R\partial_y\partial_{\tau_0} + \eta\frac{B'}{B}\partial_y\right)e^{ib(1-\zeta)\partial_y} D_4\right].
\end{aligned}
\tag{23}
$$

We return in (21), (22) to initial operators

$$
\begin{aligned}
&\left(æ_R^2\partial_\xi^2 - \partial_\tau^2 + \eta\frac{B'}{B}\partial_\xi\right] D_3 = \frac{æ_R B_\omega}{g}S,\\
&\left[æ_R^2\partial_\xi^2 - \partial_\tau^2 + \eta\frac{B'}{B}\partial_\xi\right] D_4 = \frac{æ_R B_\omega}{g}S,
\end{aligned}
\tag{24}
$$

$$
\begin{aligned}
&\left[g\partial_\xi^2 + \frac{1}{2}(æ_R^2\partial_\xi^2 - \partial_\tau^2) + \eta\frac{B'}{B}\partial_\xi\right] D_1 + \left[b\partial_\xi^2 + \frac{1}{2b}(æ_R^2\partial_\xi^2 - \partial_\tau^2) + \eta\left(b + \frac{1}{2b}\right)\frac{B'}{B}\partial_\xi\right] D_3 = 0,\\
&\left[g\partial_\xi^2 + \frac{1}{2}(æ_R^2\partial_\xi^2 - \partial_\tau^2) + \eta\frac{B'}{B}\partial_\xi\right] D_2 + \left[b\partial_\xi^2 + \frac{1}{2b}(æ_R^2\partial_\xi^2 - \partial_\tau^2) + \eta\left(b + \frac{1}{2b}\right)\frac{B'}{B}\partial_\xi\right] D_4 = 0.
\end{aligned}
\tag{25}
$$

Also expressions (23) will be transformed similarly

$$
\begin{aligned}
v_1 &= \partial_\xi e^{ia(1+\zeta)\partial_\xi} D_1 + \partial_\xi e^{ia(1-\zeta)\partial_\xi} D_2 + b\partial_\xi e^{ib(1+\zeta)\partial_\xi} D_3 + b\partial_\xi e^{ib(1-\zeta)\partial_\xi} D_4,\\
v_3 &= ia\partial_\xi e^{ia(1+\zeta)\partial_\xi} D_1 - ia\partial_\xi e^{ia(1-\zeta)\partial_\xi} D_2 + i\partial_\xi e^{ib(1+\zeta)\partial_\xi} D_3 - i\partial_\xi e^{ib(1-\zeta)\partial_\xi} D_4,\\
\sigma_{33} &= -\frac{E}{(1+\nu)h}\left[g\partial_\xi^2 e^{ia(1+\zeta)\partial_\xi} D_1 + g\partial_\xi^2 e^{ia(1-\zeta)\partial_\xi} D_2 + b\partial_\xi^2 e^{ib(1+\zeta)\partial_\xi} D_3 + b\partial_\xi^2 e^{ib(1-\zeta)\partial_\xi} D_4\right],\\
\sigma_{13} &= \frac{E}{(1+\nu)h}\left[i\left(a\partial_\xi^2 + \varepsilon\frac{æ^2æ_R}{a}\partial_\xi\partial_{\tau_0} + \eta\frac{1}{2a}\frac{B'}{B}\partial_\xi\right)e^{ia(1+\zeta)\partial_\xi} D_1\right.\\
&\quad - i\left(a\partial_\xi^2 + \varepsilon\frac{æ^2æ_R}{a}\partial_\xi\partial_{\tau_0} + \eta\frac{1}{2a}\frac{B'}{B}\partial_\xi\right)e^{ia(1-\zeta)\partial_\xi} D_2\\
&\quad + i\left(g\partial_\xi^2 + \varepsilon æ_R\partial_\xi\partial_{\tau_0} + \eta\frac{B'}{B}\partial_\xi\right)e^{ib(1+\zeta)\partial_\xi} D_3 - i\left(g\partial_\xi^2 + \varepsilon æ_R\partial_\xi\partial_{\tau_0} + \eta\frac{B'}{B}\partial_\xi\right)e^{ib(1-\zeta)\partial_\xi} D_4.
\end{aligned}
\tag{26}
$$

Let's receive an approximate form of the equations for the small vicinity of the Rayleigh front. For this purpose we will divide Lame function potentials into two parts, according to directions of wave propagation:

$$
\varphi_i = e^{ia(1\pm\zeta)\partial_\xi} D_i, \quad \psi_i = e^{ib(1\pm\zeta)\partial_\xi} D_{i+2}, \quad i = 1, 2. \tag{27}
$$

These potential functions will be described by the equations

$$
\begin{aligned}
a^2\frac{\partial^2\varphi_i}{\partial\xi^2} + \frac{\partial^2\varphi_i}{\partial\zeta^2} &= 0,\\
b^2\frac{\partial^2\psi_i}{\partial\xi^2} + \frac{\partial^2\psi_i}{\partial\zeta^2} &= 0, \quad i = 1, 2.
\end{aligned}
\tag{28}
$$

and boundary conditions will be defined by the following equations, determined by the operator equations (24) and (25):

$$\begin{aligned}
&æ_R^2\frac{\partial^2\psi_1}{\partial\xi^2}-\frac{\partial^2\psi_1}{\partial\tau^2}+\eta\frac{B'}{B}\frac{\partial\psi_1}{\partial\xi}=\frac{æ_R B_\omega}{g}S, \quad \zeta=-1,\\
&æ_R^2\frac{\partial^2\psi_2}{\partial\xi^2}-\frac{\partial^2\psi_2}{\partial\tau^2}+\eta\frac{B'}{B}\frac{\partial\psi_2}{\partial\xi}=\frac{æ_R B_\omega}{g}S, \quad \zeta=1,\\
&\left(g+\frac{æ_R^2}{2}\right)\frac{\partial^2\varphi_1}{\partial\xi^2}-\frac{1}{2}\frac{\partial^2\varphi_1}{\partial\tau^2}+\eta\frac{B'}{B}\frac{\partial\varphi_1}{\partial\xi}+\left(b+\frac{æ_R^2}{2b}\right)\frac{\partial^2\psi_1}{\partial\xi^2}-\frac{1}{2b}\frac{\partial^2\psi_1}{\partial\tau^2}+\eta\left(b+\frac{1}{2b}\right)\frac{B'}{B}\frac{\partial\psi_1}{\partial\xi}=0, \quad \zeta=-1,\\
&\left(g+\frac{æ_R^2}{2}\right)\frac{\partial^2\varphi_2}{\partial\xi^2}-\frac{1}{2}\frac{\partial^2\varphi_2}{\partial\tau^2}+\eta\frac{B'}{B}\frac{\partial\varphi_2}{\partial\xi}+\left(b+\frac{æ_R^2}{2b}\right)\frac{\partial^2\psi_2}{\partial\xi^2}-\frac{1}{2b}\frac{\partial^2\psi_2}{\partial\tau^2}+\eta\left(b+\frac{1}{2b}\right)\frac{B'}{B}\frac{\partial\psi_2}{\partial\xi}=0, \quad \zeta=1.
\end{aligned} \tag{29}$$

Expressions for displacements and stresses through potential functions are written as:

$$\begin{aligned}
v_1&=\frac{\partial\varphi_1}{\partial\xi}+\frac{\partial\varphi_2}{\partial\xi}+b\frac{\partial\varphi_1}{\partial\xi}+b\frac{\partial\varphi_2}{\partial\xi},\\
v_3&=\frac{\partial\varphi_1}{\partial\zeta}+\frac{\partial\varphi_2}{\partial\zeta}+\frac{1}{b}\frac{\partial\psi_1}{\partial\zeta}+\frac{1}{b}\frac{\partial\psi_2}{\partial\zeta},\\
\sigma_{33}&=-\frac{E}{(1+\nu)h}\left(g\frac{\partial^2\varphi_1}{\partial\xi^2}+g\frac{\partial^2\varphi_2}{\partial\xi^2}+b\frac{\partial^2\psi_1}{\partial\xi^2}+b\frac{\partial^2\psi_2}{\partial\xi^2}\right),\\
\sigma_{13}&=\frac{E}{(1+\nu)h}\left[\frac{\partial^2\varphi_1}{\partial\xi\partial\zeta}+\varepsilon\frac{æ^2æ_R}{a}\frac{\partial^2\varphi_1}{\partial\zeta\partial\tau_0}+\eta\frac{1}{2a}\frac{B'}{B}\frac{\partial\varphi_1}{\partial\zeta}+\frac{\partial^2\varphi_2}{\partial\xi\partial\zeta}+\varepsilon\frac{æ^2æ_R}{a}\frac{\partial^2\varphi_2}{\partial\zeta\partial\tau_0}+\eta\frac{1}{2a}\frac{B'}{B}\frac{\partial\varphi_2}{\partial\zeta}\right.\\
&\left.+\frac{g}{b}\frac{\partial^2\psi_1}{\partial\xi\partial\zeta}+\varepsilon\frac{æ_R}{b}\frac{\partial^2\psi_1}{\partial\zeta\partial\tau_0}+\eta\frac{1}{b}\frac{\partial\psi_1}{\partial\zeta}+\frac{g}{b}\frac{\partial^2\psi_2}{\partial\xi\partial\zeta}+\varepsilon\frac{æ_R}{b}\frac{\partial^2\psi_2}{\partial\zeta\partial\tau_0}+\eta\frac{1}{b}\frac{\partial\psi_2}{\partial\zeta}\right].
\end{aligned} \tag{30}$$

Let's write down the governing equations in an initial dimensional form. The equations for potential functions keep a form

$$\begin{aligned}
&a^2\frac{\partial^2\varphi_i}{\partial\alpha^2}+\frac{\partial^2\varphi_i}{\partial z^2}=0,\\
&b^2\frac{\partial^2\psi_i}{\partial\alpha^2}+\frac{\partial^2\psi_i}{\partial z^2}=0, \quad i=1,2.
\end{aligned} \tag{31}$$

and boundary conditions and expressions for displacement and stresses are written as:

$$\begin{aligned}
&\frac{\partial^2\psi_1}{\partial\alpha^2}-\frac{1}{c_R^2}\frac{\partial^2\psi_1}{\partial t^2}+\frac{1}{æ_R^2}\frac{B'}{B}\frac{\partial\psi_1}{\partial\alpha}=\frac{B_\omega}{gæ_R}S, \quad z=-h,\\
&\frac{\partial^2\psi_2}{\partial\alpha^2}-\frac{1}{c_R^2}\frac{\partial^2\psi_2}{\partial t^2}+\frac{1}{æ_R^2}\frac{B'}{B}\frac{\partial\psi_2}{\partial\alpha}=\frac{B_\omega}{gæ_R}S, \quad z=h,\\
&\left(g+\frac{æ_R^2}{2}\right)\frac{\partial^2\varphi_1}{\partial\alpha^2}-\frac{æ_R^2}{2c_R^2}\frac{\partial^2\varphi_1}{\partial t^2}+\frac{B'}{B}\frac{\partial\varphi_1}{\partial\alpha}+\left(b+\frac{æ_R^2}{2b}\right)\frac{\partial^2\psi_1}{\partial\alpha^2}-\frac{æ_R^2}{2bc_R^2}\frac{\partial^2\psi_1}{\partial t^2}+\left(b+\frac{1}{2b}\right)\frac{B'}{B}\frac{\partial\psi_1}{\partial\alpha}=0, \quad \zeta=-h,\\
&\left(g+\frac{æ_R^2}{2}\right)\frac{\partial^2\varphi_2}{\partial\alpha^2}-\frac{æ_R^2}{2c_R^2}\frac{\partial^2\varphi_2}{\partial t^2}+\frac{B'}{B}\frac{\partial\varphi_2}{\partial\alpha}+\left(b+\frac{æ_R^2}{2b}\right)\frac{\partial^2\psi_2}{\partial\alpha^2}-\frac{æ_R^2}{2bc_R^2}\frac{\partial^2\psi_2}{\partial t^2}+\left(b+\frac{1}{2b}\right)\frac{B'}{B}\frac{\partial\psi_2}{\partial\alpha}=0, \quad \zeta=h,
\end{aligned} \tag{32}$$

$$
\begin{aligned}
v_1 &= h\left(\frac{\partial\varphi_1}{\partial\alpha} + \frac{\partial\varphi_2}{\partial\alpha} + b\frac{\partial\psi_1}{\partial\alpha} + b\frac{\partial\psi_2}{\partial\alpha}\right),\\
v_3 &= \frac{\partial\varphi_1}{\partial z} + \frac{\partial\varphi_2}{\partial z} + \frac{1}{b}\frac{\partial\psi_1}{\partial z} + \frac{1}{b}\frac{\partial\psi_2}{\partial z},\\
\sigma_{33} &= -\frac{Eh}{1+\nu}\left(g\frac{\partial^2\varphi_1}{\partial\alpha^2} + g\frac{\partial^2\varphi_2}{\partial\alpha^2} + b\frac{\partial^2\psi_1}{\partial\alpha^2} + b\frac{\partial^2\psi_2}{\partial\alpha^2}\right),\\
\sigma_{13} &= \frac{Eh}{1+\nu}\left[\frac{\partial^2\varphi_1}{\partial\alpha\partial z} + \frac{æ^2æ_R}{ac_2}\frac{\partial^2\varphi_1}{\partial z\partial t} + \frac{1}{2a}\frac{B'}{B}\frac{\partial\varphi_1}{\partial z} + \frac{\partial^2\varphi_2}{\partial\alpha\partial z} + \frac{æ^2æ_R}{ac_2}\frac{\partial^2\varphi_2}{\partial z\partial t} + \frac{1}{2a}\frac{B'}{B}\frac{\partial\varphi_2}{\partial z}\right.\\
&\left. + \frac{g}{b}\frac{\partial^2\psi_1}{\partial\alpha\partial z} + \frac{æ_R}{bc_2}\frac{\partial^2\psi_1}{\partial z\partial t} + \frac{1}{b}\frac{B'}{B}\frac{\partial\psi_1}{\partial z} + \frac{g}{b}\frac{\partial^2\psi_2}{\partial\alpha\partial z} + \frac{æ_R}{bc_2}\frac{\partial^2\psi_2}{\partial z\partial t} + \frac{1}{b}\frac{B'}{B}\frac{\partial\psi_2}{\partial z}\right].
\end{aligned}
\tag{33}
$$

References

[1] Kirillova I V and Kossovich L Yu 2016 Elliptic boundary layer in shells of revolution under normal shock surface loading *Vestn. St.-Pet. Univ. Ser. 1: Mat. Mekh. Astron.* **49** 139–46
[2] Kovalev V A, Kossovich L Yu, and Taranov O V 2005 The far field of a Rayleigh wave in an elastic half-strip under end loads *Mech. Solids* **40** (5) 66–72
[3] Kovalev V A and Taranov O V 2007 The distant field of a wave of Rayleigh for an elastic cylindrical cover at action of face loading *Vestnik SamGU. Estestvennonauch. Ser.* No. 4(54) 197–208
[4] Kossovich L Yu 1986 *Nonstationary tasks of the theory of resilient thin envelopes* (Saratov: Izdat. SGU) p 176

IOP Conf. Series: Journal of Physics: Conf. Series **991** (2018) 012009 doi:10.1088/1742-6596/991/1/012009

Experimental validation of thermo-chemical algorithm for a simulation of pultrusion processes

E Barkanov[1*], P Akishin[1], N L Miazza[2], S Galvez[2], and N Pantelelis[3]
[1]Institute of Materials and Structures, Riga Technical University, Riga, Latvia
[2]AIMPLAS, Paterna, Spain
[3]Synthesites Innovative Technologies, Piraeus, Greece
E-mail: *barkanov@latnet.lv

Abstract. To provide better understanding of the pultrusion processes without or with temperature control and to support the pultrusion tooling design, an algorithm based on the mixed time integration scheme and nodal control volumes method has been developed. At present study its experimental validation is carried out by the developed cure sensors measuring the electrical resistivity and temperature on the profile surface. By this verification process the set of initial data used for a simulation of the pultrusion process with rod profile has been successfully corrected and finally defined.

1. Introduction

Pultrusion is a continuous and cost-effective process for a production of composite structural components with a constant cross-sectional profile (figure 1). During pultrusion, the fibre reinforcements are saturated with the resin in a resin bath and then continuously pulled through a heated die by a puller. Inside the die, the resin gradually cures and solidifies to form a composite part with the same cross-section profile as in the die. At the final stage a travelling cut-off saw cuts the composite profile into desired lengths.

To provide better understanding of the pultrusion processes, to support the pultrusion tooling design and process control, a lot of numerical simulations have been done in the last thirty years [1–3]. Most of them are focusing on the analysis of the heat transfer and cure, on the pressure rise in the tapered zone of the die and on problems related to impregnation of reinforcing fibres to obtain a final product characterised by the desired mechanical properties. The effects of the processing parameters, like die temperature, pulling speed, post-cure temperature and time, filler and fibre type and content,

on the mechanical properties of composites produced with the pultrusion have been also intensively investigated [4–6]. It is necessary to note that numerical simulations of the pultrusion are carried out using non-commercial finite element codes and general-purpose finite element packages, and most of them are firmly connected with the final product and technology applied.

Also pultrusion processes have been investigated experimentally for various unidirectional composite profiles [7–9]. In these studies, experimental temperatures have been measured by using thermocouples inside the composite or die, but the degree of cure has been obtained from the differential scanning calorimetric (DSC) analysis of samples prepared from the pultruded

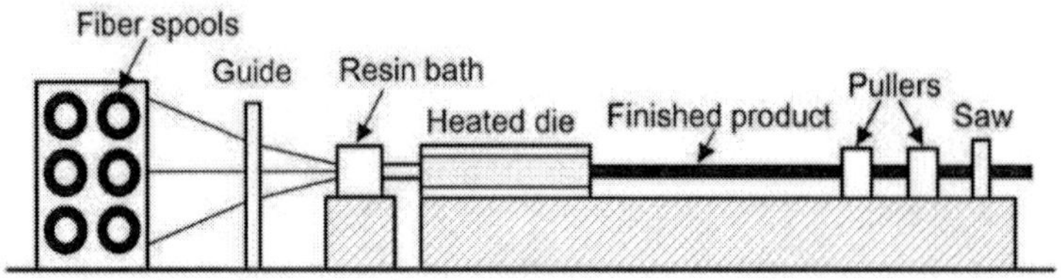

Figure 1. Pultrusion process.

profiles. The experimental results have been used in most cases for a validation of the developed numerical algorithms.

At present study the algorithm based on the mixed time integration scheme and nodal control volumes method developed in [10] for a solution of the coupled temperature state and degree of cure by using an iteration technique is validated experimentally. For this purpose new cure sensors measuring the electrical resistivity and temperature on the profile surface have been produced and applied for a real-time monitoring of the degree of cure in pultrusion process with a complex temperature control. Since the set of initial data is not clearly defined, the parameters of technological process are precised comparing the temperatures on thermocouples and sensors with the numerically obtained.

2. Formulation of thermo-chemical problem in pultrusion

To investigate numerically the pultrusion process, the following thermo-chemical problem consisting of the energy equation for the tool, the energy equation for the composite moving in the pull direction and the species equation for the resin should be solved

$$\begin{cases} \rho c_p \dfrac{\partial T}{\partial t} - \dfrac{\partial}{\partial x}\left(k_x \dfrac{\partial T}{\partial x}\right) - \dfrac{\partial}{\partial y}\left(k_y \dfrac{\partial T}{\partial y}\right) - \dfrac{\partial}{\partial z}\left(k_z \dfrac{\partial T}{\partial z}\right) - q_b = 0, \\ \bar{\rho}\bar{c}_p \left(\dfrac{\partial T}{\partial t} + u\dfrac{\partial T}{\partial x}\right) - \dfrac{\partial}{\partial x}\left(\bar{k}_x \dfrac{\partial T}{\partial x}\right) - \dfrac{\partial}{\partial y}\left(\bar{k}_y \dfrac{\partial T}{\partial y}\right) - \dfrac{\partial}{\partial z}\left(\bar{k}_z \dfrac{\partial T}{\partial z}\right) - q = 0, \\ \left(\dfrac{\partial \alpha}{\partial t} + u\dfrac{\partial \alpha}{\partial x}\right) = R_r, \end{cases} \tag{1}$$

where T is the temperature, ρ and c_p are the density and specific heat of the tooling materials, k_x, k_y, k_z are the thermal conductivities of the tooling materials in x, y, z directions, q_b is the rate of energy exchange at the boundary, u is the pull speed, $\bar{\rho}$ and $\bar{c}_p$ are the lumped density and specific heat for the composite material, $\bar{k}_x$, $\bar{k}_y$, $\bar{k}_z$ are the lumped thermal conductivities of the composite material in x, y, z directions, q is the generative term related to the internal heat generation due to the exothermic resin reaction, $\alpha = H(t)/H_{tr}$ is the degree of cure and $H(t)$ is the amount of heat evolved during the curing up to time t.

Heat transfer in the composite occurs as a result of conduction and the generation of heat resulting from the exothermic chemical reaction initiated by the die temperature. The generative term related to the internal heat generation due to the exothermic resin reaction could be written as:

$$q = V_r \rho_r H_{tr} R_r \tag{2}$$

where V_r is the resin volume fraction, ρ_r is the resin density, H_{tr} is the total heat of reaction and R_r is the rate of resin reaction determined as:

$$R_r(\alpha, T) = \frac{\partial \alpha}{\partial t} = \frac{1}{H_{tr}} \frac{dH(t)}{dt} = K(T) \cdot f(\alpha), \tag{3}$$

where $f(\alpha)$ depends on the resin properties and varies with the applied resin reaction model, and $K(T)$ is defined by the Arrhenius relationship:

$$K(T) = K_0 \exp\left(-\frac{E}{RT} \right) \tag{4}$$

where $R = 8.314\,\mathrm{J/mol{\cdot}K}$ is the universal gas constant. It is necessary to note that coefficients of the Arrhenius relationship: activation energy E and frequency factor K_0 are physical values and are determined by the Kissinger method or ASTM E 698 standard methodology from DSC tests [11]. Coefficients of the selected function $f(\alpha)$ could be obtained in a simple way by a fitting of the experimental heat flow curves applying the least squares method [11].

The reinforcement is saturated with the resin before entering the heated die in pultrusion process. Therefore, it is reasonable to assume that the resin does not flow. In most cases the continuous model with lumped material properties is used for a simulation of the pultrusion processes. These properties are evaluated by the rule of mixture:

$$\begin{aligned} \bar{\rho} &= (1 - V_r)\rho_f + V_r\rho_r, \\ \bar{c}_p &= \frac{(1 - V_r)\rho_f c_{pf} + V_r \rho_r c_{pr}}{\bar{\rho}}, \\ \bar{k} &= \frac{k_f k_r \bar{\rho}}{(1 - V_r)\rho_f k_r + V_r \rho_r k_f}, \end{aligned} \tag{5}$$

where indexes f and r relates to the fibres and resin, respectively.

The system of equations (1) could be solved by using the prescribed initial and boundary conditions. It is assumed that at time $t = 0$, for a curing composite the temperature is $T = T^0$ and the degree of cure is $\alpha = \alpha^0$, where index "0" denotes the initial values. The temperature of the composite at the die entrance and everywhere in the composite at the first step of numerical algorithm is prescribed as the pre-heat resin temperature. The value of the degree of cure at the die entrance and everywhere in the composite at the first step of numerical algorithm is taken as zero or very small value depending on the resin curing kinetic model. In some pultrusion processes this value could be much higher in relation to the ambient room temperature which is used also as the boundary condition for a die to describe the convection effect demonstrating the thermal energy transfer between a pultrusion die and an air surrounding it as a result of a temperature difference. It is important to note that the power flux at the entrance and exit of the composite in the die is specified as zero. The insulated boundary conditions could be applied also to reduce unnecessary heat losses and conserve energy in pultrusion processes.

3. Simulation of thermo-chemical problem

To solve the system of coupled energy and species equations (1), the numerical algorithm using ANSYS Mechanical environment and based on the mixed time integration scheme and nodal control volumes method has been developed in [10] and modified for a simulation of pultrusion processes with a temperature control (figure 2). It is seen in figure 2 that ANSYS software is applied only for a solution of the transient thermal problem.

4. An experimental set-up

An experimental set-up and simplified model of the pultrusion die developed for a validation of the numerical algorithm is presented in figure 3. The pultrusion process with a temperature control is investigated experimentally for a cylindrical rod with the diameter of 16 mm. Heating of the die is realised by 12 electrical heaters subdivided into 3 groups and controlled by the proportional-integral derivative (PID) controller and thermocouples located inside each heaters

IOP Conf. Series: Journal of Physics: Conf. Series **991** (2018) 012009 doi:10.1088/1742-6596/991/1/012009

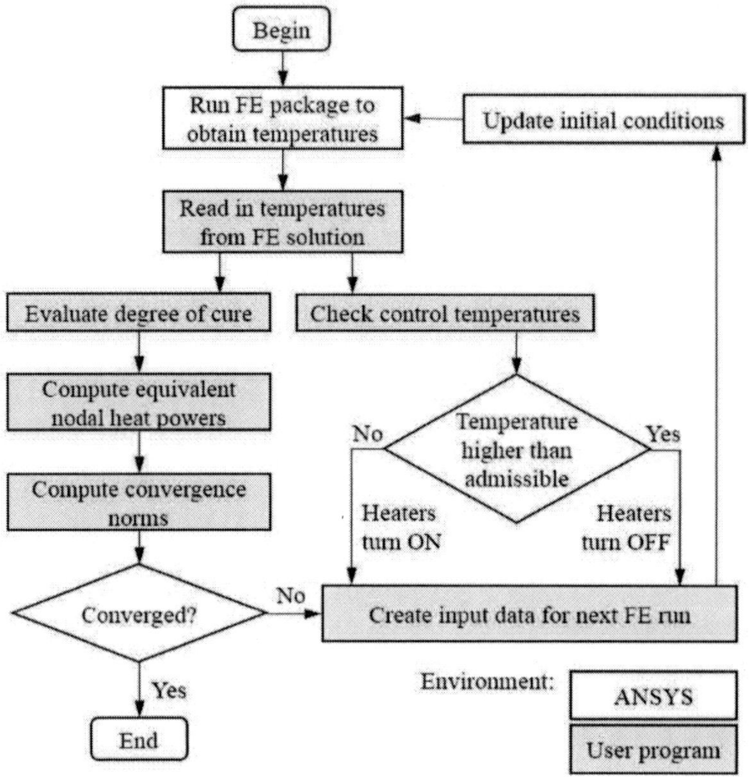

Figure 2. Thermo-chemical algorithm for a simulation of pultrusion.

Table 1. Control temperatures.

Thermocouple	Symbol (unit)	Initial	Corrected 1	Corrected 2	
				First turn OFF	Main control
1	T_1 (°C)	100	85	65	85
2	T_2 (°C)	120	105	85	105
3	T_3 (°C)	140	125	105	125

Table 2. Parameters of pultrusion process.

Parameter	Symbol (unit)	Value
Room temperature	T_{room} (°C)	17
Resin temperature in the bath	T_{resin} (°C)	17
Pull speed	V_{pull} (m/min)	0.18

group. The electrical power of each heater is 315 W. Initially it has been taken that the controller turns off the heaters groups when the temperature on the corresponding thermocouples reaches 100, 120, or 140°C (table 1). Other parameters of the pultrusion process are given in table 2. For a real-time monitoring of the degree of cure and temperature in pultrusion processes, new cure sensors measuring the electrical resistivity and temperature on the profile surface have been developed and located on the die top. For a better contact with the rod profile, they have been produced with a curved contact surface (figure 4).

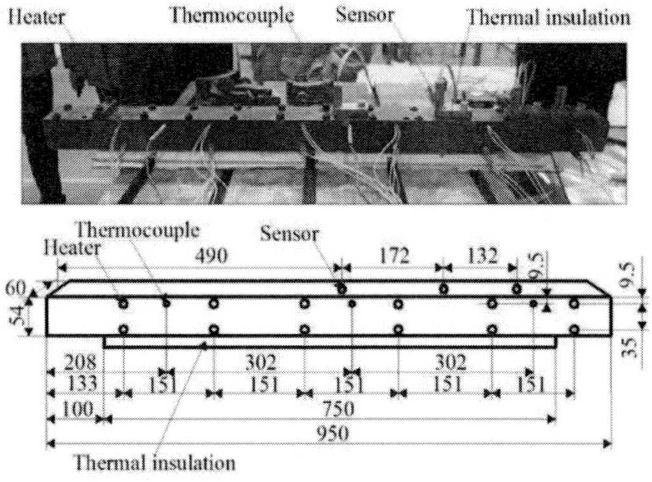

Figure 3. Pultrusion set-up.

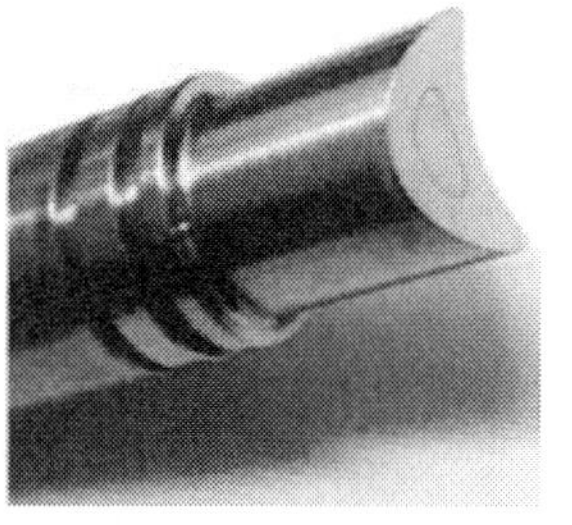

Figure 4. Cure sensor for the rod profile.

Table 3. Thermal properties of die material.

Property	Symbol (unit)	Steel 40Cr Initial	Steel 40Cr Corrected
Density	ρ (kg/m^3)	7850	7850
Specific heat	C (J/(kg·K))	460	460
Thermal conductivity	k_x (W/(m·K))	46	33

Table 4. Thermal properties of pultruded material.

Property	Symbol (unit)	Unifilo 4800 tex	POLRES 305BV	Lumped
Density	ρ (kg/m^3)	2500	1100	1870
Specific heat	C (J/(kg·K))	1235	1360	1268
Thermal conductivity	k_x (W/(m·K))	11	0.209	0.750
Thermal conductivity	k_y (W/(m·K))	1	0.209	0.500
Max allowable temperature	T (°C)	1200	170	—

The pultrusion die has been made of steel 40Cr with the thermal properties presented in table 3. Materials used for a production of the cylindrical rod are glass fibres Unifilo 4800 tex and polyester resin POLRES 305BV. The fibre volume content in the pultruded material is 55%. The thermal properties of constituent materials of the pultruded profile are given in table 4. The rate of resin reaction is described by using the Kamal-Sourour curing kinetic model:

$$\frac{\partial \alpha}{\partial t} = \left[K_1 \exp\left(-\frac{E_1}{RT}\right) + K_2 \exp\left(-\frac{E_2}{RT}\right)\alpha^m\right](1-\alpha)^n \tag{6}$$

with the parameters presented in table 5 and determined by using dynamic DSC tests.

5. Experimental validation of thermo-chemical algorithm

The finite element model for a simulation of the thermo-chemical problem in pultrusion of cylindrical rod has been created in ANSYS Mechanical by using 3-D thermal solid finite elements SOLID 70. The finite element has eight nodes with a single degree of freedom, temperature,

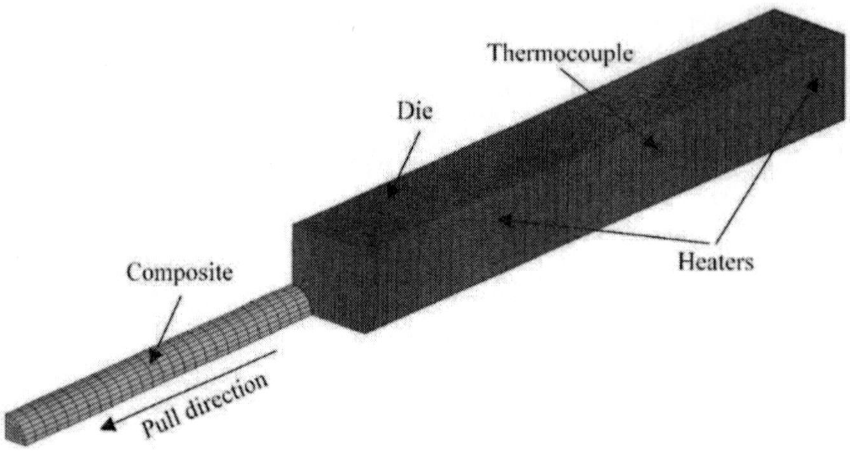

Figure 5. Fragment of the finite element model.

Table 5. Parameters of curing kinetic model.

Parameter	Symbol (unit)	POLRES 305BV
Heat reaction	H_{tr}, (J/kg)	323074
Frequency factor	K_1 (s^{-1})	14289310986
Frequency factor	K_2 (s^{-1})	285.870
Activation energy	E_1 (J/mol)	85573
Activation energy	E_2 (J/mol)	33141
Order of the reaction	n	2.342
Order of the reaction	m	0.519

at each node and the orthotropic material properties. Thermal insulation under the die is not taking into account in the finite element model. In this case symmetry of the simulated domain is used and only a quarter of the die is modelled. To analyse heat transfer and curing processes in the post-die region, the modelling of the profile is continued at the distance of 250 mm from the die exit. Fragment of the finite element model is presented in figure 5. The finite element mesh is regular in pull direction and has 240 elements along the profile. The total number of finite elements is 73040 and it consists of 7680 elements used for the composite modelling and 65360 elements used for the die including heaters and thermocouples. The time step of the solution is 1.67 s. During this time, composite travels in the pull direction on the distance equals to the dimension of one finite element (5 mm). Initial conditions are applied at time $t = 0$ when all nodal points of the composite have the room temperature and degree of cure equals $\alpha = 10^{-10}$.

Results of simulation, temperature and degree of cure obtained on sensors 1–3, together with the results of experimental measurements are presented in figure 6. This figure shows that close agreement between experimental and simulation results has not been obtained. If to suppose that experimental results are correct, a disagreement appeared should be identified. Smooth temperature curve at the end of measurement time on sensor 1 (figure 6 *a*) talks that this sensor is not screwed to the final position and for this reason the cavity appeared has been filled with uncured material. This is why sensor 1 not properly reacts on the heaters work. This phenomenon should be taken into account later analyzing the experimental and simulation results.

IOP Conf. Series: Journal of Physics: Conf. Series **991** (2018) 012009 doi:10.1088/1742-6596/991/1/012009

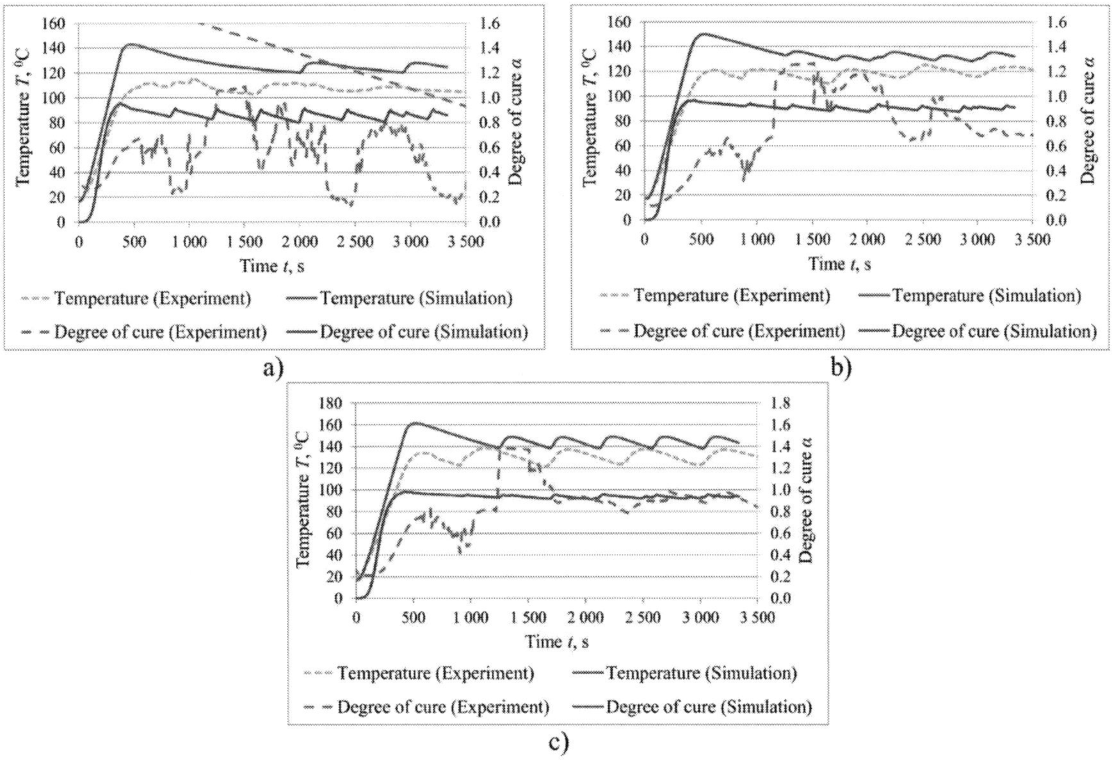

Figure 6. Temperature and degree of cure on sensors 1 (a), 2 (b), and 3 (c).

5.1. Correction of control temperatures

To identify disagreement between experimental and simulation results, the initial control temperatures (table 1) have been checked at first since the controller regulating these temperatures has been used by technologists as "the black box." It is seen from figure 7 that sensor 1 is located very closely to thermocouple 2 with the distance of 20 mm between thermocouple and sensor axes. It is reasonable to assume that temperatures on sensor 1 and thermocouple 2 are very close or even the same. The results of the finite element simulation confirm this assumption (figure 8). Moreover, it is seen that the experimental temperature on sensor 1 is lower than the controlled temperature on thermocouple 2 (figure 8) and simulated temperatures on all sensors are larger anytime than the temperatures obtained experimentally (figure 6). For this reason it is assumed that the temperatures really controlled by thermocouples during an experiment are lower than taken initially. After that additional calculations have been performed with the corrected control temperatures. For thermocouples 2 and 3 they have been taken from the experimental results for sensors 1 and 3 respectively. The control temperature for thermocouple 1 has been taken proportionally to the temperatures on thermocouples 2 and 3. The corrected values of the control temperatures are given in table 1.

Experimental measurements together with the results of simulation, temperature and degree of cure obtained on sensors 1–3 with the corrected control temperatures, are presented in figure 9. It is seen that simulation temperatures have moved down and now more close agreement between experimental and simulation results has been obtained. The largest disagreement is visible for the temperatures after the first turn off of the heaters at time about 400 s from the process start. At this time the simulation temperature continues to rise according to the thermal inertia effect.

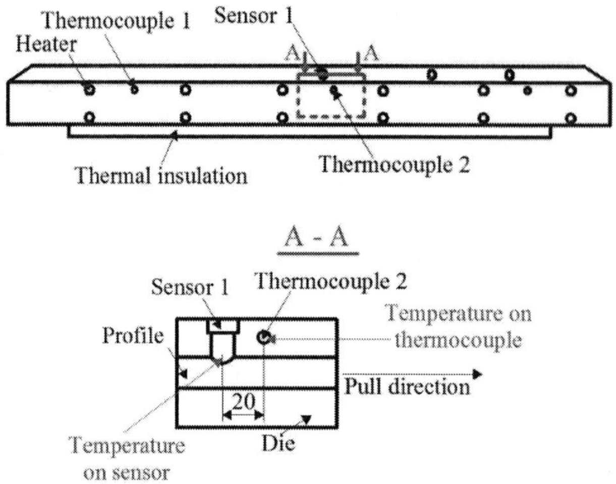

Figure 7. Location of sensor 1 and thermocouple 2 in the die.

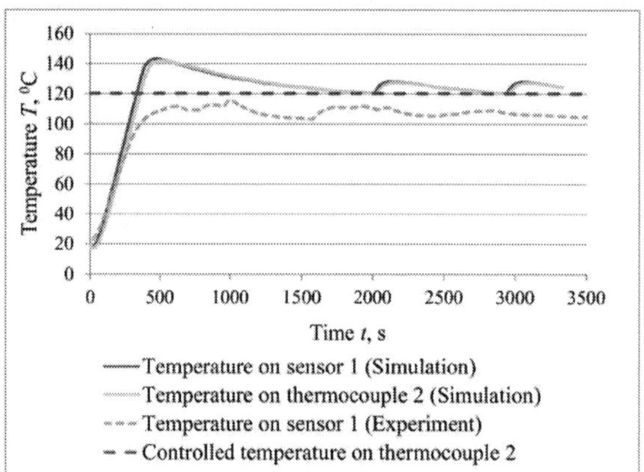

Figure 8. Simulated and experimental temperatures on sensor 1 and thermocouple 2.

5.2. Correction of temperature control algorithm

To bring the temperature control algorithm used in simulations closer to the experimental conditions, reduction of the control temperatures for the first turn off of the heaters has been executed. The control temperatures have been reduced by 20°C (table 1). Experimental measurements together with the results of simulation, temperature and degree of cure obtained on sensors 1–3 with the corrected first and main control temperatures, are presented in figure 10. It is seen that the first simulation peak has been reduced considerably and more close agreement between experimental and simulation results has been obtained. However, some disagreement is visible now for the duration of temperature cycles. Obviously, that this effect is dependent on the thermal conductivity of die material.

5.3. Correction of die thermal properties

To avoid last disagreement, the thermal properties of die material have been verified and it has been found that they differ considerably in different material data sheets. The corrected properties are given in table 3. Experimental measurements together with the results of simulation, temperature and degree of cure obtained on sensors 1–3 with the corrected control temperatures and die material properties, are presented in figure 11. It is seen that the duration

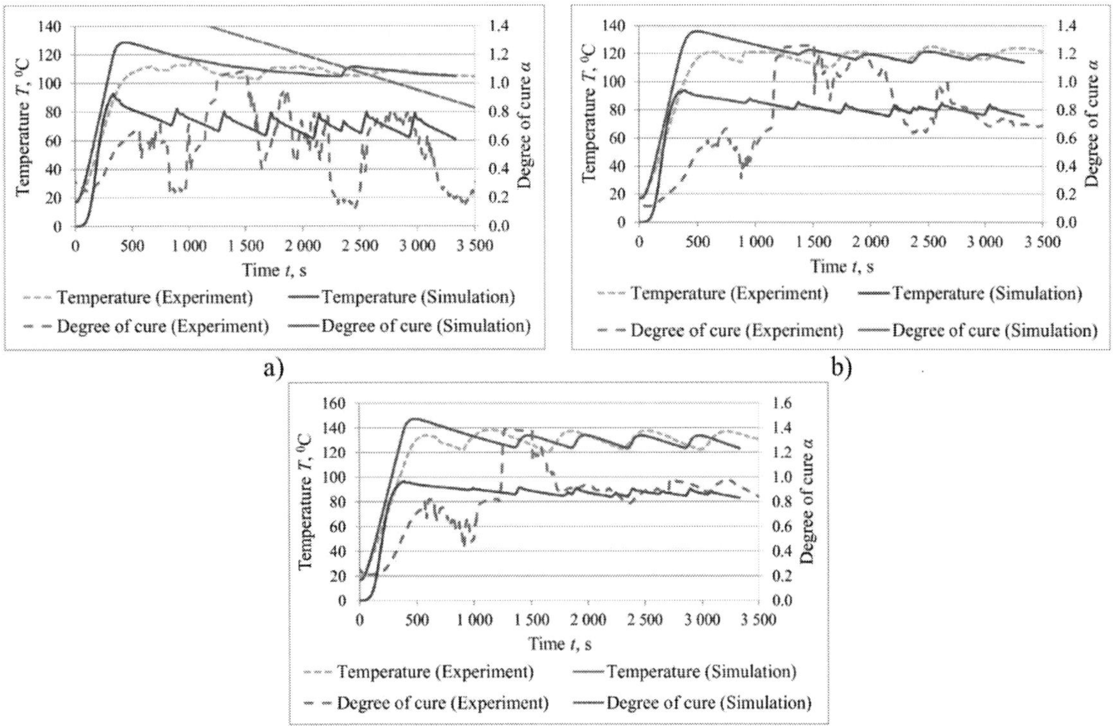

Figure 9. Temperature and degree of cure with corrected control temperatures on sensors 1 (a), 2 (b), and 3 (c).

of temperature cycles has been increased considerably and very good agreement between experimental and simulation temperatures has been obtained now. Final distribution of the temperature and degree of cure along the composite profile at centreline and on the surface is given in figure 12.

Figure 11 demonstrates that the degrees of cure predicted by the developed algorithm in general could be examined as a proper averaging of the values measured in real time by the developed sensors. Some noise in the online measurements appears due to not ideal contact between moved profile and electrodes. Comparing both results it is seen that difference between them is larger for sensor 1 (Figure 11 a), where the resin is at the gelation stage, but it is significantly reduced on sensor 2 (Figure 11 b). The difference between experimentally measured and simulation results is practically eliminated after the transient period on sensor 3 (figure 11 c) where the resin is almost fully cured.

Conclusions

The thermo-chemical algorithm based on the mixed time integration scheme and nodal control volumes method and developed for a simulation of pultrusion processes has been successfully validated by the pultrusion trial of the rod profile. For a real-time monitoring of the temperature and degree of cure, new cure sensors measuring the electrical resistivity and temperature on the profile surface have been developed. Comparing experimental and simulation temperatures in the validation process, very good agreement between them has been found correcting an inaccurate initial data set.

Numerical and experimental characterisation of the pultrusion process has shown that the

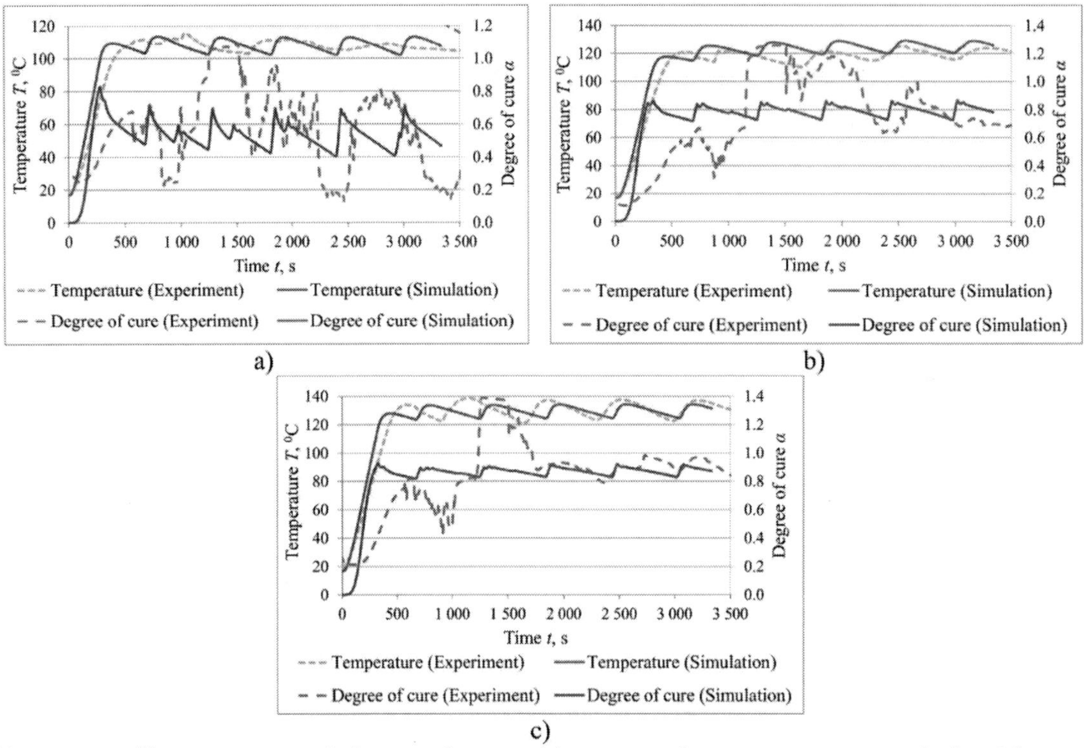

Figure 10. Temperature and degree of cure with corrected temperature control algorithm on sensors 1 (*a*), 2 (*b*), and 3 (*c*).

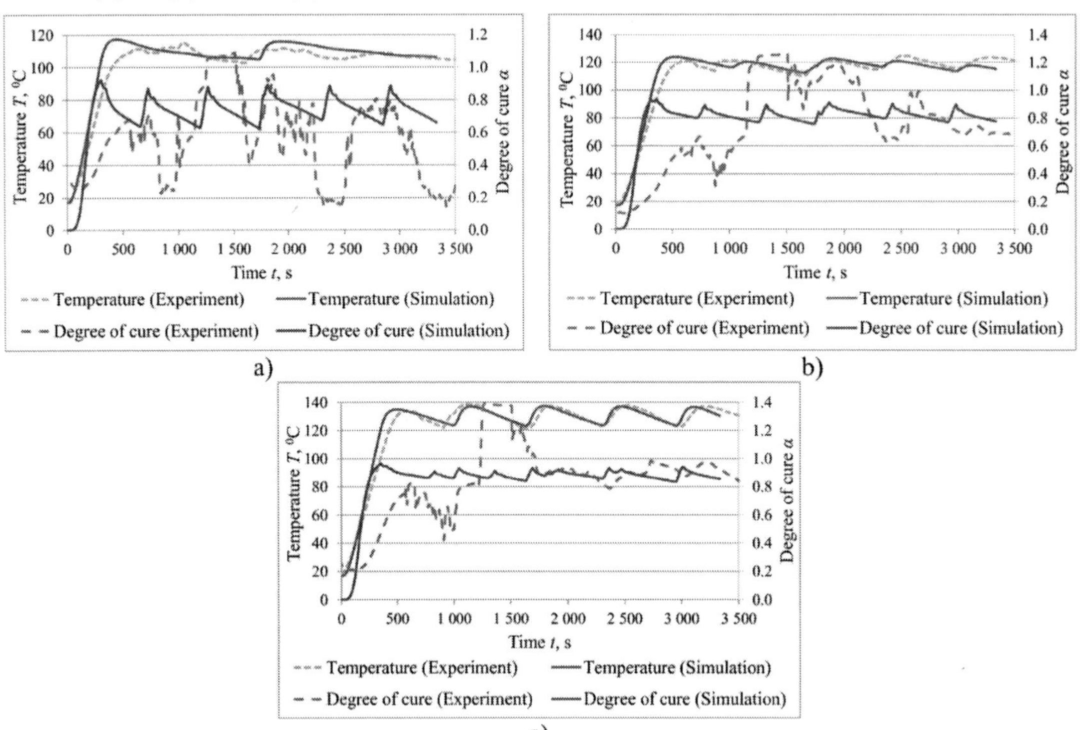

Figure 11. Temperature and degree of cure with corrected thermal properties of die material on sensors 1 (*a*), 2 (*b*), and 3 (*c*).

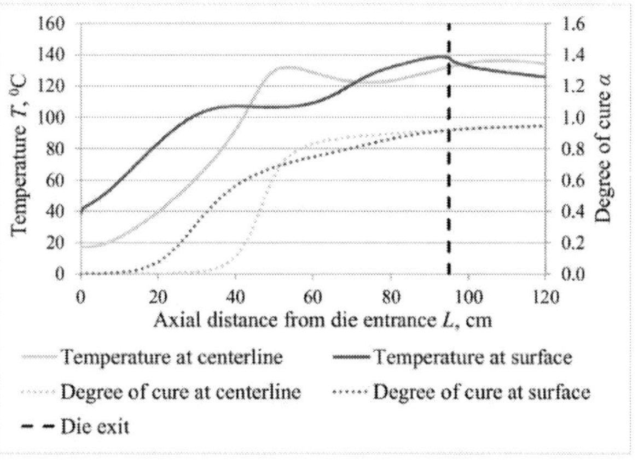

Figure 12. Distribution of temperature and degree of cure along the composite profile.

developed cure sensors require some improvements in the degree of cure measurements in the areas where the resin is at the gelation stage. The noise in the online measurements of the degree of cure appearing due to the not ideal contact between moved profile and electrodes should be removed from experimental results. However, it is necessary to note that in the areas where the resin is almost fully cured, negligible difference between experimentally measured and simulated degrees of cure has been obtained after the transient period.

Acknowledgments

The authors gratefully acknowledge the support of European Commission under FRAMEWORK 7 program, contract no. NMP2-SL-2013-609149, project "Development of an Innovative Manufacturing Process for the In-Line Coating of Pultruded Composites (COALINE)."

References

[1] Han C D and Lee D S 1986 Development of a mathematical model for the pultrusion process *Polym. Engng Sci.* **26** 393–404
[2] Liu X-L 2001 Numerical modelling on pultrusion of composite I beam *Composites: Part A* **32** 663–81
[3] Carlone P, Palazzo G S, and Pasquino R 2006 Pultrusion manufacturing process development by computational modelling and methods *Math. Comput. Model.* **44** 701–9
[4] Chen C-H and Wang W-S 1998 The development of glass fiber reinforced polystyrene for pultrusion. II: Effect of processing parameters for optimising the process *Polym. Compos.* **19** (4) 423–30
[5] Coelho R M L and Calado V M A 2002 An optimisation procedure for the pultrusion process based on a finite element formulation *Polym. Compos.* **23** (3) 329–41
[6] Carlone P, Palazzo G S, and Pasquino R 2007 Pultrusion manufacturing process development: Cure optimisation by hybrid computational methods *Comput. Math. Appl.* **53** 1464–71
[7] Roux J A, Vaughan J G, Shanku R, et al. 1998 Comparison of measurements and modelling for pultrusion of a fiberglass/epoxy I-beam *J. Reinforc. Plast. Compos.* **17** (17) 1557-79
[8] Liu X L and Hillier W 1999 Heat transfer and cure analysis for the pultrusion of a fiberglass-vinyl ester I beam *Compos. Struct.* **47** 581-8
[9] Liang G, Garg A, and Chandrashekhara K 2005 Cure characterization of pultruded soy-based composites *J. Reinforc. Plast. Compos.* **24** (14) 1509–20
[10] Barkanov E, Akishin P, Miazza N L, and Galvez S 2017 ANSYS-based algorithms for a simulation of pultrusion processes *Mech. Adv. Mater. Struct.* **24** (5) 377–84
[11] Akishin P, Barkanov E, Miazza N, and Galvez S 2017 Curing kinetic models of resins for microwave assisted pultrusion *Key Engng Mater.* **721** 92–6

Control problem for a system of linear loaded differential equations

V R Barseghyan and T V Barseghyan

Institute of Mechanics of the National Academy of Sciences of Armenia, Yerevan, Armenia
Yerevan State University, Yerevan, Armenia

E-mail: barseghyan@sci.am

Abstract. The problem of control and optimal control for a system of linear loaded differential equations is considered. Necessary and sufficient conditions for complete controllability and conditions for the existence of a program control and the corresponding motion are formulated. The explicit form of control action for the control problem is constructed and a method for solving the problem of optimal control is proposed.

Introduction

The study of different control processes in various fields of science and technology allows one to conclude that the future course of many control processes turns out to depend not only on the present but also on the prehistory of the process. A mathematical description of these dynamic processes can be carried out by using ordinary differential equations with memory of various types, also called equations with aftereffect or loaded differential equations. In the literature [1–5], the loaded differential equations are usually the equations whose coefficients or the right-hand side contain some functionals (functions) of the solution, in particular, the values of the solution at which the phase state of the process at a point and at a time moment can influence the dynamics of the process as a whole. In practice, such problems arise, for example, when, if it is necessary to make observations of a dynamic process, the phase states are measured at some time moments, and the information is continuously transmitted through the feedback.

The presence of a loaded summand in the dynamics of the system does not always allows one directly to apply the known methods developed in investigations of ordinary (unloaded) dynamic systems. This emphasizes both the theoretical and practical relevance of studying various control problems for loaded differential equations. We note that the presence of a loaded summand in differential equations leads to new problems that have not yet been investigated in the theory of control.

Over the last few years, the loaded differential equations related to various applied problems of mechanics, biology, ecology and chemistry modeled with loaded equations have been studied intensively. The interest of investigators in the control problems of loaded dynamical systems is also stimulated by the capabilities of modern computing and measuring technology, which permits using the most adequate mathematical models of the processes under study.

The loaded ordinary differential equations and boundary-value problems for such equations are considered in [1–5], and conditions for their solvability by various methods are established.

IOP Conf. Series: Journal of Physics: Conf. Series **991** (2018) 012010 doi:10.1088/1742-6596/991/1/012010

A significant contribution to the development of the theory of loaded equations was made by the article [1] (and in other works by the same author), where the definitions of loaded differential, loaded integro-differential, loaded functional equations, and their numerous applications are given. In the monograph [2], the loaded differential equations are interpreted as perturbations of differential equations. In [5], on the basis of the parametrization method, a linear multi-point boundary-value problem for a system of loaded differential equations is investigated and an algorithm for finding the solution is proposed. Many studies have been devoted to investigation of the existence of solutions of loaded linear differential equations, but relatively little attention has been paid to the problems of their control.

In this work, we consider the problem of control and optimal control of a system of linear loaded differential equations. Necessary and sufficient conditions for complete controllability are formulated. A constructive method for solving the control problem is proposed and conditions for the existence of program control and motion are formulated. The analytic forms of motion and control action for the control problem are constructed, as well as a method for solving the optimal control problem is proposed.

1. Statement of the problem

Let us consider a control process whose dynamics is described by loaded linear differential equations

$$\dot{x} = A_0(t)x + A_1(t)x(t_1) + A_2(t)x(t_2) + A_3(t)x(t_3) + B(t)u, \tag{1}$$

where $x(t) \in R^n$ is the phase vector of the system, $A_k(t)$ and $B(t)$ $(k = 0, 1, 2, 3)$ are $n \times n$ and $n \times r$ matrices of the system parameters (continuous on $[t_0, T]$) and $u(t)$ is the control action of dimension $r \times 1$.

In formula (1), the summands $A_k(t)x(t_k)$ $(k = 1, 2, 3)$ influence the system as functions starting from the moment $t \geq t_k$, since the value of the phase state $x(t_k)$, as a result of measurement, is determined at the time moment $t = t_k$ and, from that moment (when $t \geq t_k$), continuously affects the system in the form of summands $A_k(t)x(t_k)$.

Let the initial

$$x(t_0) = x_0 \tag{2}$$

and final

$$x(T) = x_T \tag{3}$$

states of system (1) be given.

It is assumed that the time moments of loading points and $0 \leq t_0 < t_1 < t_2 < t_3 < t_4 = T$ are given. The function $x(t)$ is continuous on the intervals $[t_{k-1}, t_k)$ $(k = 1, 2, 3, 4)$ and has finite left-hand limits $\lim\limits_{t \to t_k - 0} x(t) = x(t_k)$ at the loading points t_k.

We consider the following problems.

Problem 1. Determine the conditions under which there exists a program control action $u = u(t)$ and a program motion transferring the motion of system (1) from the initial state (2) to the final state (3) over the time interval $[t_0, T]$, as well as construct the control action and motion.

Let a quality criterion æ$[u]$, which can have the meaning of the norm of some normalized space, be given for the selection of the optimal solutions over the time interval $[t_0, T]$.

The problem of optimal control for system (1) with conditions (2), (3) and the quality criterion æ$[u]$ can be formulated as follows.

Problem 2. Determine the optimal control action $u^0(t)$, $t \in [t_0, T]$, which transfers the motion of system (1) from the initial state (2) to the final state (3) and takes the least possible value of the quality criterion æ$[u0]$.

IOP Conf. Series: Journal of Physics: Conf. Series **991** (2018) 012010 doi:10.1088/1742-6596/991/1/012010

2. Construction of the motion of the loaded system

To construct the motion of system (1), we divide the interval $[t_0, T]$ into parts by loading points: $[t_0, T) = \bigcup_{k=1}^{4} [t_{k-1}, t_k)$. Taking into account the sequence of loading points and the character of the influence of the corresponding loaded summands, we write equation (1) separately on the intervals $[t_{k-1}, t_k)$ $(k = 1, 2, 3, 4)$ in the form

$$\dot{x} = \begin{cases} A_0(t)x + B(t)u & \text{for } t \in [t_0, t_1), \\ A_0(t)x + A_1(t)x(t_1) + B(t)u & \text{for } t \in [t_1, t_2), \\ A_0(t)x + A_1(t)x(t_1) + A_2(t)x(t_2) + B(t)u & \text{for } t \in [t_2, t_3), \\ A_0(t)x + A_1(t)x(t_1) + A_2(t)x(t_2) + A_3(t)x(t_3) + B(t)u & \text{for } t \in [t_3, T], \end{cases} \tag{4}$$

which is the stage by stage changing of differential equations.

We assume that the stage by stage changing of the system of loaded differential equations (1) (or (4)) over the time interval $[t_0, T]$ is completely controllable [6–8].

To solve the above-formulated problems, we construct the motion of stage by stage changing of system (1) (or (4)) over the time interval $[t_0, T]$. For this, we write the solution of system (4) for the time interval $[t_0, t_1)$ in the form

$$x(t) = X[t, t_0]x(t_0) + \int_{t_0}^{t} H[t, \tau]u(\tau)\, d\tau, \tag{5}$$

where $H[t, \tau] = X[t, \tau]B(\tau)$ and $X[t, \tau]$ denotes the normalized fundamental matrix of the solution of the homogeneous part of $\dot{x} = A_0(t)x$.

For the time interval $[t_1, t_2)$, the solution of equation (4) is represented as

$$x(t) = X[t, t_1]x(t_1) + \int_{t_1}^{t} X[t, \tau]\big(A_1(\tau)x(t_1) + B(\tau)u(\tau)\big)\, d\tau. \tag{6}$$

Taking into account that $\lim_{t \to t_k - 0} x(t) = x(t_k)$, from formula (5), subtracting the value of $x(t_1)$ and substituting in (6), we obtain the motion of system (4) at the time moment $t \in [t_1, t_2)$ in the form

$$x(t) = \tilde{X}[t, t_1]X[t_1, t_0]x(t_0) + \tilde{X}[t, t_1] \int_{t_0}^{t_1} H[t_1, \tau]u(\tau)\, d\tau + \int_{t_1}^{t} H[t, \tau]u(\tau)\, d\tau. \tag{7}$$

Continuing this procedure, we obtain the formula for representing the motion of system (4) at the time moment $t \in [t_2, t_3)$ in the form

$$\begin{aligned} x(t) = {} & Y[t, t_2]X[t_1, t_0]x(t_0) + Y[t, t_2] \int_{t_0}^{t_1} H[t_1, \tau]u(\tau)\, d\tau \\ & + \tilde{X}[t, t_2] \int_{t_1}^{t_2} H[t_2, \tau]u(\tau)\, d\tau + \int_{t_2}^{t} H[t, \tau]u(\tau)\, d\tau, \end{aligned} \tag{8}$$

where

$$Y[t, t_2] = \tilde{X}[t, t_2]\tilde{X}[t_2, t_1] + \int_{t_2}^{t} X[t, \tau]A_1(\tau)\, d\tau,$$

and at the time moment $t \in [t_3, t_4]$,

$$\begin{aligned} x(t) &= Z[t,t_3]X[t_1,t_0]x(t_0) + Z[t,t_3]\int_{t_0}^{t_1} H[t_1,\tau]u(\tau)\,d\tau \\ &+ \left(\int_{t_3}^{t} X[t,\tau]A_2(\tau)\,d\tau + \tilde{X}[t,t_3]\tilde{X}[t_3,t_2]\right)\int_{t_1}^{t_2} H[t_2,\tau]u(\tau)\,d\tau \\ &+ \tilde{X}[t,t_3]\int_{t_2}^{t_3} H[t_3,\tau]u(\tau)\,d\tau + \int_{t_3}^{t} H[t,\tau]u(\tau)\,d\tau, \end{aligned} \tag{9}$$

where

$$\begin{aligned} \tilde{X}[t,t_j] &= X[t,t_j] + \int_{t_j}^{t} X[t,\tau]A_j(\tau)\,d\tau \quad (j=1,2,3), \\ Z[t,t_3] &= \int_{t_3}^{t} X[t,\tau]A_1(\tau)\,d\tau + \int_{t_3}^{t} X[t,\tau]A_2(\tau)\,d\tau\tilde{X}[t_2,t_1] + \tilde{X}[t,t_3]Y[t_3,t_2]. \end{aligned} \tag{10}$$

Thus, having the initial state $x(t_0)$ of system (1) and setting the control action $u(t)$, we use formulas (5) and (7)–(9) to determine the phase state $x(t)$ of system (1) (the solution of the loaded equation (1)) for the respective time intervals $[t_{k-1}, t_k)$ $(k = 2, 3, 4)$.

3. Solution of the problems

From formulas (9) for $t = t_4$, we have

$$\begin{aligned} x(t_4) &= Z[t_4,t_3]X[t_1,t_0]x(t_0) + Z[t_4,t_3]\int_{t_0}^{t_1} H[t_1,\tau]u(\tau)\,d\tau \\ &+ \left(\int_{t_3}^{t_4} X[t_4,\tau]A_2(\tau)\,d\tau + \tilde{X}[t_4,t_3]\tilde{X}[t_3,t_2]\right)\int_{t_1}^{t_2} H[t_2,\tau]u(\tau)\,d\tau \\ &+ \tilde{X}[t_4,t_3]\int_{t_2}^{t_3} H[t_3,\tau]u(\tau)\,d\tau + \int_{t_3}^{t_4} H[t_4,\tau]u(\tau)\,d\tau. \end{aligned} \tag{11}$$

Now, instead of functions $H[t_k, t]$, we introduce the functions $\bar{H}[t_k, t]$ as

$$\begin{aligned} &\bar{H}[t_1,t] = \begin{cases} H[t_1,t] & \text{for } t_0 \le t < t_1, \\ 0 & \text{for } t_1 \le t \le T, \end{cases} \qquad \bar{H}[t_2,t] = \begin{cases} 0 & \text{for } t_0 \le t < t_1, \\ H[t_2,t] & \text{for } t_1 \le t < t_2, \\ 0 & \text{for } t_2 \le t \le T, \end{cases} \\ &\bar{H}[t_3,t] = \begin{cases} 0 & \text{for } t_0 \le t < t_2, \\ H[t_3,t] & \text{for } t_2 \le t < t_3, \\ 0 & \text{for } t_3 \le t \le T, \end{cases} \qquad \bar{H}[t_4,t] = \begin{cases} 0 & \text{for } t_0 \le t < t_3, \\ H[t_4,t] & \text{for } t_3 \le t \le T. \end{cases} \end{aligned} \tag{12}$$

We use the functions introduced in (12) to write the integral relations (11) as

$$x(t_4) = Z[t_4,t_3]X[t_1,t_0]x(t_0) + \int_{t_0}^{T}\left(\sum_{i=1}^{4} H_i[t]\right)u(t)\,dt, \tag{13}$$

where

$$\begin{aligned} &H_1[t] = Z[t_4,t_3]\bar{H}[t_1,t], \quad H_2[t] = \left(\int_{t_3}^{t_4} X[t_4,\tau]A_2(\tau)\,d\tau + \tilde{X}[t_4,t_3]\tilde{X}[t_3,t_2]\right)\bar{H}[t_2,t], \\ &H_3[t] = \tilde{X}[t_4,t_3]\bar{H}[t_3,t], \quad H_4[t] = \bar{H}[t_4,t]. \end{aligned}$$

The integral relations (13) become

$$\int_{t_0}^{T} \left(\sum_{i=1}^{4} H_i[t] \right) u(t)\, dt = x(t_4) - Z[t_4, t_3] X[t_1, t_0] x(t_0) = \eta(t_0, \dots, T). \tag{14}$$

Thus, we see that system (1) over the time interval $[t_0, T]$ is completely controllable if and only if, for any vector $\eta(t_0, \dots, T) = x(T) - Z[T, t_3]X[t_1, t_0]x(t_0)$ in the space R^n, a control $u = u(t, \eta)$ satisfying condition (14) can be indicated.

The question of solvability is fundamental for any control problem and reduces to analyzing the controllability of the system.

We can formulate the assertion for the complete controllability of system (1) with conditions (2) and (3) as follows.

For the loaded system (1) (or the stage by stage changing of system (4)) to be completely controllable over the interval $[t_0, T]$, it is necessary and sufficient that the column vector of the matrix $\left(\sum_{i=1}^{4} H_i[t]\right)^T$ be linearly independent over this interval. Hereinafter "T" denotes the operation of transposition.

Following [6, 7, 9], we seek the function $u(t)$, $t \in [t_0, T]$, satisfying the integral relation (14) in the form

$$u(t) = \left(\sum_{i=1}^{4} H_i[t] \right)^T C + v(t), \tag{15}$$

where C is a constant vector to be determined and $v(t)$ is a vector function satisfying the orthogonality condition

$$\int_{t_0}^{T} \left(\sum_{i=1}^{4} H_i[t] \right) v(t)\, dt = 0. \tag{16}$$

Substituting (15) into (14), we obtain

$$Q(t_0, \dots, T)C = \eta(t_0, \dots, T), \tag{17}$$

where

$$Q(t_0, \dots, T) = \int_{t_0}^{T} \left(\sum_{i=1}^{4} H_i[t] \right) \left(\sum_{i=1}^{4} H_i[t] \right)^T dt. \tag{18}$$

Equation (17) present a system of n algebraic equations in the unknowns C_j $(j = 1, \dots, n)$. Equation (17) has a solution if $\det Q \neq 0$ or the rank of the matrix Q coincides with the rank of the extended matrix $\{Q, \eta\}$.

If $\det Q \neq 0$, then the solution of equation (17) is $C = Q^{-1}\eta$, and consequently, it follows from (15) that

$$u(t) = \left(\sum_{i=1}^{4} H_i[t] \right)^T Q^{-1} \left(x(T) - Z[T, t_3] X[t_1, t_0] x(t_0) \right) + v(t). \tag{19}$$

Therefore, the solution of Problem 1 can be formulated as a theorem similar to that proved in [6, 9].

Theorem. *For the existence of the program control* (15) *and its corresponding solution of system* (1) *satisfying conditions* (14), (16), *it is necessary and sufficient that the matrix* (18) *be nonsingular or that the ranks of the matrix* Q *and* $\{Q, \eta\}$ *coincide.*

With (12) taken into account, under the assumption that $\det Q \neq 0$ and $v(t) = 0$, it follows from (19) that the control action $u(t)$ is can be presented in the form

$$u(t) = \begin{cases} (Z[t_4, t_3]H[t_1, t])^T Q^{-1}(x(T) - Z[T, t_3]X[t_1, t_0]x(t_0)) & \text{for } t \in [t_0, t_1), \\ \left[\left(\int_{t_3}^{t_4} X[t_4, \tau]A_2(\tau)\, d\tau + \tilde{X}[t_4, t_3]\tilde{X}[t_3, t_2]\right)H[t_2, t]\right]^T & \\ \qquad \times Q^{-1}(x(T) - Z[T, t_3]X[t_1, t_0]x(t_0)) & \text{for } t \in [t_1, t_2), \\ (\tilde{X}[t_4, t_3]H[t_3, t])^T Q^{-1}(x(T) - Z[T, t_3]X[t_1, t_0]x(t_0)) & \text{for } t \in [t_2, t_3), \\ (H[t_4, t])^T Q^{-1}(x(T) - Z[T, t_3]X[t_1, t_0]x(t_0)) & \text{for } t \in [t_3, T]. \end{cases} \tag{20}$$

By substituting the control action from (20) into (5), (7)–(9), respectively, we obtain the program motion of system (1) over the time interval $[t_0, T]$ and satisfy conditions (2), (3).

To solve Problem 2, we note that the left-hand side of condition (14) is a linear operation generated by the function $u(t)$ over the time interval $[t_0, T]$. Consequently, if the functional æ$[u]$ is the norm of some linear normalized space, then the solution of Problem 2 should be constructed by the algorithm for solving the corresponding problem of moments [6, 8]. Then the solution of Problem 2 is given by the optimal control action $u^0(t)$, $t \in [t_0, T]$, minimizing the functional æ$[u]$.

Conclusions

A constructive approach to studying the problem of control for a system of linear loaded differential equations is proposed. A formula for determining the phase state of such dynamic systems at any time moment in a given initial state is obtained. Necessary and sufficient conditions for complete controllability and conditions for the existence of the program control and motion are formulated. The solution of the control problem for linear loaded differential equations is constructed, and a method for solving the problem of optimal control is proposed.

References

[1] Nakhushev A M 2012 *Loaded Equations and Their Applications* (Moscow: Nauka) p 232 [in Russian]
[2] Dzhenaliev M T and Ramazanov M I 2010 *Loaded Equations as Perturbations of Differential Equations* (Almaty: Fylym) p 336 [in Russian]
[3] Dzhenaliev M T 1989 Optimal control of linear loaded parabolic equations *Diff. Uravn.* **25** (4) 641–51
[4] Kozhanov A I 2004 Nonlinear Loaded Equations and Inverse Problems *Comput. Math. Math. Phys.* **44** (4) 657–75
[5] Bakirova E A and Kadirbayeva Zh M 2016 On a solvability of linear multipoint boundary-value problem for the loaded differential equations *Izv. Nats. Akad. Nauk Resp. Kazakhstan. Ser. Fiz.-Mat.* **5** (309) 168–75
[6] Barseghyan V R 2016 *Control of Compound Dynamic Systems and of Systems with Multipoint Intermediate Conditions* (Moscow: Nauka) p 230 [in Russian]
[7] Barseghyan V R 2012 Control of stage by stage changing linear dynamic systems *Yugoslav J. Operat. Res.* **22** (1) 31–9
[8] Krasovskii N N 1968 *The Theory of Motion Control* (Moscow: Nauka) p 476 [in Russian]
[9] Zubov V I 1975 *Lectures on the Theory of Control* (Moscow: Nauka) p 496 [in Russian]

Reflection of electromagnetic wave from the boundary of the piezoelectric half-space with cubic symmetry

A Kh Berberyan* **and V G Garakov****

Institute of Mechanics of the National Academy of Sciences of Armenia, Yerevan, Armenia

E-mail: *slavia555@yahoo.com, **garakov@yandex.ru

Abstract. A large number of works have been devoted to investigation of the influence of the piezoelectric properties of a material on the propagation of elastic waves [1–3]. Herewith, the quasi-static piezoelasticity model was mainly used. In the problem of an electromagnetic wave reflection from an elastic medium with piezoelectric properties, it is necessary to consider hyperbolic equations [4].

1. Governing equations and problem statement

In a rectangular Cartesian coordinate system (x, y, z), a half-space of a material with piezoelastic properties of cubic symmetry occupies the region $(-\infty < x < \infty,\ 0 \le y < \infty,\ -\infty < z < \infty)$.

In [5], a quasi-hyperbolic model was proposed to investigate the wave propagation in piezoelectric media with cubic symmetry. This model is a simplification of an exact system of equations preserving the property of being hyperbolic.

An analogous quasihyperbolic model for piezoelectrics of a hexagonal system was proposed earlier in [6].

According to [2], the equation of propagation of shear electroelastic waves, has the form

$$\begin{aligned} c_{44}\Delta w + 2e_{14}\frac{\partial^2\varphi}{\partial x\partial y} &= \rho\frac{\partial^2 w}{\partial t^2}, \\ \Delta\varphi - \frac{2e_{14}}{\varepsilon}\frac{\partial^2 w}{\partial x\partial y} &= \varepsilon\mu\frac{\partial^2\varphi}{\partial t^2}, \end{aligned} \tag{1}$$

where w is the elastic displacement, c_{44} is the shift module, ρ is the density of the material, e_{14} is the piezoelectric coefficient in the case of cubic symmetry, φ is the electric potential, ε and μ are the dielectric and magnetic permeability of the material.

The material equations have the form

$$\begin{aligned} \sigma_{13} &= c_{44}\frac{\partial w}{\partial x} + e_{14}\frac{\partial\varphi}{\partial y}, \quad \sigma_{23} = c_{44}\frac{\partial w}{\partial y} + e_{14}\frac{\partial\varphi}{\partial x}, \\ D_1 &= -\varepsilon\frac{\partial\varphi}{\partial x} + e_{14}\frac{\partial w}{\partial y}, \quad D_2 = -\varepsilon\frac{\partial\varphi}{\partial y} + e_{14}\frac{\partial w}{\partial x}, \end{aligned} \tag{2}$$

where σ_{13} and σ_{23} are components of the shifting tension, D_1 and D_2 are vectors of the electrical field induction. During the wave propagation, there also arises a plane magnetic field perpendicular to the propagation direction, which is determined by the equation

$$\frac{\partial H_3^{(1)}}{\partial x} = -\frac{1}{c_1}\frac{\partial D_2}{\partial t}, \quad c_1 = \frac{1}{\sqrt{\varepsilon\mu}}. \tag{3}$$

According to the above equations (1)–(3), an electromagnetic wave incident on the boundary of the piezoelectric half-space $y = 0$ from the region $y < 0$, which is identified with a vacuum, must satisfy the equations

$$\frac{\partial H_3}{\partial y} = \frac{1}{c}\frac{\partial E_1}{\partial t}, \quad \frac{\partial H_3}{\partial x} = -\frac{1}{c}\frac{\partial E_2}{\partial t}, \quad \frac{\partial E_2}{\partial x} - \frac{\partial E_1}{\partial y} = -\frac{1}{c}\frac{\partial H_3}{\partial t}. \tag{4}$$

Thus, the incident (and reflected) electromagnetic wave is a solution of equations (4), and the refracted wave is a solution of equations (1).

At the boundary of the section $y = 0$, it is necessary to satisfy the boundary conditions [7]

$$E_1 = E_1^{(1)}, \quad H_3 = H_3^{(1)}, \quad \sigma_{23} = 0. \tag{5}$$

Here, the values of the electromagnetic field without any upper index refer to the region $y < 0$, and those with index (1), to the region $y > 0$.

2. General solution constructing

It is known from (4) that the expressions for the incident and reflected waves of the magnetic field have the form

$$\begin{aligned} H_{3n} &= A\exp[i(\omega t - k_1 x - k_2 y)], \\ H_{30} &= B\exp[i(\omega t - k_1 x + k_2 y)], \end{aligned} \tag{6}$$

where

$$k_2 = \sqrt{\frac{\omega^2}{c^2} - k_1^2}. \tag{7}$$

The corresponding quantities for the electric field, in particular, for E_1, become

$$\begin{aligned} E_{in} &= -\frac{k_2 c}{\omega}A\exp[i(\omega t - k_1 x - k_2 y)], \\ E_{in} &= \frac{k_2 c}{\omega}B\exp[i(\omega t - k_1 x + k_2 y)]. \end{aligned} \tag{8}$$

The solution of the system of equations (1) is represented in the form

$$\begin{aligned} w &= C\exp[i(\omega t - k_1 x - k_3 y)], \\ \varphi &= F\exp[i(\omega t - k_1 x - k_3 y)]. \end{aligned} \tag{9}$$

Substituting (9) into (1), we obtain a system of homogeneous algebraic equations for the constants C and F:

$$\begin{aligned} &(s_1 - k_3^2)C - \frac{2e_{14}}{c_{44}}k_1 k_3 F = 0, \\ &\frac{2e_{14}}{\varepsilon}k_1 k_3 C + (s_2 - k_3^2)F = 0, \end{aligned} \tag{10}$$

with the notation

$$s_1 = \frac{\omega^2}{c_t^2} - k_1^2, \quad s_2 = \frac{\omega^2}{c_1^2} - k_1^2, \quad c_t^2 = \frac{c_{44}}{\rho}, \quad c_1^2 = \frac{1}{\varepsilon\mu}. \tag{11}$$

The condition that the determinant of system (10) is zero leads to the equation

$$k_3^4 - (s_1 + s_2 - \chi k_1^2)k_3^2 + s_1 s_2 = 0, \quad \chi = \frac{4\varepsilon_{14}^2}{c_{44}\varepsilon}. \tag{12}$$

Determining the positive roots k_{31} and k_{32} from (12), the general solution of (1) with regard to (9) have the form

$$\begin{aligned} w &= C_1 \exp[i(\omega t - k_1 x - k_{31} y)] + C_2 \exp[i(\omega t - k_1 x - k_{32} y)], \\ \varphi &= F_1 \exp[i(\omega t - k_1 x - k_{31} y)] + F_2 \exp[i(\omega t - k_1 x - k_{32} y)], \end{aligned} \tag{13}$$

where

$$k_{3i} = \frac{1}{\sqrt{2}}\left[s_1 + s_2 - \chi k_1^2 \pm \sqrt{(s_1 + s_2 - \chi k_1^2)^2 - 4 s_1 s_2}\right]^{1/2}, \quad i = 1, 2.$$

According to the equations in the system (10), the arbitrary constants C_1, C_2 and F_1, F_2 are connected by the relations

$$C_i = R_i F_i, \quad R_i = \frac{e_{14}}{c_{44}} k_1 k_{3i} (s_1 - k_{3i}^2)^{-1}, \quad i = 1, 2. \tag{14}$$

The components of the electric field are determined from the condition $\bar{E}^{(1)} = -\operatorname{grad}\varphi$. It follows from the equation that

$$\begin{aligned} H_3^{(1)} = \frac{i\omega}{KC_1}\{&(\varepsilon k_{31} - e_{14} k_1 R_1) F_1 \exp[i(\omega t - k_1 x - k_{31} y)] \\ &+ (\varepsilon k_{32} - e_{14} k_1 R_2) F_2 \exp[i(\omega t - k_1 x - k_{32} y)]\}. \end{aligned} \tag{15}$$

3. Boundary conditions and particular solution

The requirement that the solutions (6), (8), (13), (15) satisfy the boundary conditions (5) with regard to (14) and

$$H_3 = H_{3n} + H_{30}, \quad E_1 = E_{1n} + E_{10}, \tag{16}$$

leads to the following system of equations for the unknown constants B, F_1, and F_2:

$$\begin{aligned} A - B &= i\frac{k_1\omega}{k_2 c}(F_1 + F_2), \\ A + B &= i\frac{\omega}{k_1 c_1}[(\varepsilon k_{31} - e_{14} k_1 R_1) F_1 + (\varepsilon k_{32} - e_{14} k_1 R_2) F_2], \\ (k_{31} c_{44} R_1 &+ e_{14} k_1) F_1 + (k_{32} c_{44} R_2 + e_{14} k_1) F_2 = 0. \end{aligned} \tag{17}$$

The expressions for B, F_1, and F_2 determined from (17) and characterizing the amplitudes of the reflected electromagnetic and refracted electroelastic waves turn out to be quite cumbersome. Here these expressions are given in a certain approximation.

First, it is taken into account that the electromechanical coupling coefficient x is sufficiently small and hence, instead of (9), the following approximate expressions can be obtained:

$$\tilde{k}_{3i} \approx \left[s_i\left(1 - \frac{\chi k_1^2}{s_1 - s_2}\right)\right]^{1/2}, \quad (i = 1, 2). \tag{18}$$

Also using the inequality $c_t \ll c_1$, for constants B and F_2, we have

$$B = \frac{p_1 - p_2}{p_1 + p_2} A, \quad F_2 = -i \frac{2}{\omega(p_1 + p_2)} A, \tag{19}$$

where

$$p_1 = \frac{\varepsilon}{2k_1 c_1} [k_{32}(1 - 0.5\chi\eta) - \chi k_{31} \gamma (1 + \eta)], \quad \eta = \frac{k_1^2 c_t^2}{\omega^2},$$
$$p_2 = \frac{k_1}{k_2 c}(1 - \gamma), \quad \gamma = \frac{2 - \eta}{2\eta(1 + 0.5\chi\eta)(1 + \chi k_1^2 s_2)}. \tag{20}$$

According to (14) and the last equation in (17), the remaining unknown constants are determined by the formulas

$$C_i = R_i F_i, \quad F_1 = -\chi\gamma F_2. \tag{21}$$

4. Results discussion

It is shown in section 3 that a flexible wave arises when an electro-magnetic wave is incident on a tension-free surface. However, a flexible wave can also arise during the incidence of an electro-magnetic wave on a consolidated boundary [8].

Assume that, instead of condition (5), we have the following condition on the half-space boundary:

$$E_1 = E_1^1, \quad H_3 = H_3^1, \quad \omega = 0, \quad \text{when } y = 0. \tag{22}$$

Substituting (6), (8), (13), (15) into the boundary condition (22), we obtain an chain of equations of the form (170 for the required constants, where the third equation must be replaced by the following one:

$$R_1 F_1 + R_2 F_2 = 0. \tag{23}$$

Using (23) results in the first two equations of (17) of the form:

$$A - B = \frac{ik_1\omega}{k_2 c}\left(1 - \frac{R_2}{R_1}\right) F_2,$$
$$A + B = i \frac{\omega\varepsilon}{k_1 c_1}\left(k_{32} - k_{31}\frac{R_2}{R_1}\right) F_2. \tag{24}$$

From here, the desired reflection coefficient B and one of the refraction coefficients F_2 can be determined as

$$B = \frac{q_1 - q_2}{q_1 + q_2} A, \quad F_2 = -i \frac{2}{\omega(q_1 + q_2)} A, \tag{25}$$

where

$$q_1 = \frac{\varepsilon}{k_1 c_1}(k_{32} - k_{31})\frac{R_2}{R_1}, \quad q_2 = \frac{k_1}{k_2 c}\left(1 - \frac{R_2}{R_1}\right). \tag{26}$$

The reflection coefficient F_1 is determined from (23), while c_1 and c are determined from (14).

Conclusion

The electromagnetic wave incident on a half-space of a piezoelectric material with cubic symmetry excites an elastic wave. The amplitude of the elastic wave depends on the angle of incidence of the electromagnetic wave and the electromechanical coupling factor. No elastic wave is excited in normal incidence of an electromagnetic wave ($k_1 = 0$).

References

[1] Auld B A 1978 *Modern Problems in Elastic Wave Propagation* (New York: Wiley) p 459
[2] Balakirev M K and Gilinski I A 1982 *Waves in Piezocrystals* (Novosibirsk: Nauka) p 239
[3] Bardzokas D I, Kudryavtsev B A, and Senik N A 2003 *Propagation of Waves in Electromagnetoelastic Media* (Moscow: Editorial URSS) p 334
[4] Baghdasaryan A and Belubekyan M 2009 On the problem of reflection of shear wave from a boundary of piezoelectric media of class 6 mm, In *Proc. of the 8th Int. Congr. on Thermal Stresses. TS2009. 1–4 July, Univ. of Illinois at Urbana-Champaing. Illinois, USA* vol. 1 (Illinois) 183–6
[5] Belubekyan M V 2010 Separation of waves of an electromagnetic field for the antiplane problem of a piezoelectric of class 23 In *Topical Problems of Continuum Mechanics: Proc. of II Int. Conf. 4–8 October 2010, Dilijan, Armenia* vol 1 (Yerevan: EGUAS) pp 138–42 [in Russian]
[6] Schaofan L 1996 The electromagnetic–acoustic surface wave in piezoelectric medium: The Bleustein–Gulyaev mode. *J. Appl. Phys.* **80** (9) 5264–9
[7] Brekhovskikh L M 1973 *Waves in Layered Media* (Moscow: Nauka) p 343
[8] Belubekyan M V 2008 Screened shifted surface wave in piezo-active half-space of hexagonal symmetry In *Proc. 6th Int. Conf. "The Problems of Dynamics of Interaction of Deformable Media," September 21–26, Goris-Stepanakert, 2008* (Goris-Stepanakert: Institute of Mechanics NAS) 125–30

Physical nature of strain rate sensitivity of metals and alloys at high strain rates

E N Borodin[1,2,3*], A A Gruzdkov[4], A E Mayer[1], and N S Selyutina[3,5]
[1]Chelyabinsk State University, Chelyabinsk, Russia
[2]Ural Federal University, Ekaterinburg, Russia
[3]Institute of Problems of Mechanical Engineering Russian Academy of Sciences, St. Petersburg, Russia
[4]Saint Petersburg State Technological Institute (Technical University), St. Petersburg, Russia
[5]Saint Petersburg State University, St. Petersburg, Russia
E-mail: *elbor7@gmail.com

Abstract. The role of instabilities of plastic flow at plastic deformation of various materials is one of the important cross-disciplinary problems which is equally important in physics, mechanics and material science. The strain rate sensitivities under slow and high strain rate conditions of loading have different physical nature. In the case of low strain rate, the sensitivity arising from the inertness of the defect structures evolution can be expressed by a single parameter characterizing the plasticity mechanism. In our approach, this is the value of the characteristic relaxation time. In the dynamic case, there are additional effects of "high-speed sensitivity" associated with the micro-localization of the plastic flow near the stress concentrators. In the frames of mechanical description, this requires to introduce additional strain rate sensitivity parameters, which is realized in numerous modifications of Johnson–Cook and Zerilli–Armstrong models. The consideration of both these factors is fundamental for an adequate description of the problems of dynamic deformation of highly inhomogeneous metallic materials such as steels and alloys. The measurement of the dispersion of particle velocities on the free surface of a shock-loaded material can be regarded as an experimental expression of the effect of micro-localization. This is also confirmed by our results of numerical simulation of the propagation of shock waves in a two-dimensional formulation and analytical estimations.

1. Introduction

Some researches [1, 2], on the basis of a detailed consideration of microstructural processes, suppose that the instabilities such as the shear bands after the shock wave front make almost a dominant contribution to the plastic dissipation of wave energy. This means that the plastic flow instabilities are of sufficient mechanical importance like the strain rate sensitivity effect due to a decrease in the shock wave amplitude and, hence, an increase in the dynamical strength of material. Many other researchers tend to think about the micro-localization phenomenon mostly as about a feature of the plastic flow that is interesting only for material science but not for mechanics. Besides that, the nature of strain rate sensitivity parameter at high strain rates i s not clear and is a question to be discussed. Numerical simulations [3–5] with analytical consideration on the basis of the integral form for plasticity criterion [5–7] (which is appropriate for the high strain rate conditions) can shed new light on this problem. This type of criteria includes two original parameters: one of them is the characteristic relaxation time τ which has a clear physical meaning [5], but the second one is an empirical parameter α whose nature is unclear, and one can empirically find that $\alpha > 1$ only for some classes of alloys and steels. The parameter α gives an "addition" to the strain rate sensitivity for such materials, while the parameter τ always makes the main contribution to the dynamic effects.

One can compare this point of view with the commonly used definition of strain rate sensitivity at both low and high strain rates. The well-known relations for the yield stress are

$$\sigma_{\text{eff}} = C\dot{\varepsilon}^{\text{m}}, \tag{1}$$

where $\sigma_{\text{eff}} = \sigma - \sigma_y^0$ is an effective stress (without reference value of the yield stress) and the parameter $m = (\partial \ln \sigma / \partial \ln \dot{\varepsilon})_{T,\varepsilon}$ is the strain rate sensitivity.

Numerical simulation based on the dislocation plasticity model discovers the new possibilities for understanding the mechanisms of the material response to the external loading connected with the defect structure evolution. These models also allow one to perform a detailed investigation of the influence of the plastic flow instabilities on the mechanical behavior of the material. The aim of our current investigation is to give a clear interpretation of the mechanistic parameter of strain rate sensitivity at high strain rates by comparison of the simulation results with the experimental data and the analytical consideration.

2. Strain rate sensitivity at low strain rates

The dependence of strain rate sensitivity parameter on the microstructure of the material can be expressed in terms of the activation volume V^* as

$$m = \frac{\sqrt{3}k_{\text{b}}T}{\sigma V^*}, \tag{2}$$

where k_{b} is the Boltzmann constant and T is the temperature. For the activation volume one can write the well-known relation $V^* = b\xi l^*$ [8], which is connected with the overcoming of dislocation loops of obstacles located one from another at a distance l^*. Here ξ is the distance swept out by the gliding dislocation during one activation event, which is approximately equal to b. If the obstacles mainly have a dislocation nature (immobile dislocation clusters), then $l^* \sim \rho_{\text{D}}-1/2$ and $V^* = b\xi\rho_{\text{D}}^{-1/2}$. An experimental estimate of the activation volume for coarse-grained aluminum equal to $1800b^3$ is given in [8]. This value decreases by almost an order of magnitude for ultrafine-grained aluminum. In the quasi-static case, for the typical value of the parameter m [8] in pure metals, the activation volume is 0.005–0.01. For higher strain rates, but lesser than $10^2\,\text{s}^{-1}\text{m}$, it is equal to 0.035–0.055.

The effect of the strain rate sensitivity parameter is shown mechanically according to equation (1) as an increase in the yield strength of the material, which is the limiting value of the stress, after which the macroscopic plastic deformation of the material begins and is expressed as the classical Tresca–Guest yield criterion for quasi-stationary processes:

$$\Sigma(t) \leq \sigma_{\text{y}}^0, \tag{3}$$

where $\Sigma(t)/2$ is the current maximal shear stress, $\Sigma(t)$ is the longitudinal stress in the case of simple compression, and σ_{y}^0 is the static yield strength. Equation (3) together with equation (1) allows one to take into account the effect of the strain rate on the plastic flow of the material. In this case, the value of the strain rate sensitivity parameter is unambiguously interpreted from the thermo-activation point of view as the ratio of the internal energy to the energy of the field of external forces (2). This is an essentially "static view" of the deformation process, which does not imply an explicit consideration of temporal effects.

There is also another way to take into account the effect of strain rate on the yield strength of the material widely used in polymer mechanics. For this purpose, one can use the concept of fading memory [9]: the loading, which was applied at the previous time moments $s \ll \tau$, influences the current state of the material considerably less than the recent loading. Consequently, the current stress should be replaced by its "relaxed" value, and the earlier the stress acted in the material, the lesser its weight in the equation. In the general case, this concept leads to the following form of the integral inequality:

$$I(t) = \int_0^t \sigma(s) K(t-s)\, ds \le \sigma_{\rm y}^0, \tag{4}$$

where the kernel of the integral operator $K(t)$ is the memory fading function. Criterion (4) for large times transforms into (3). If one takes $K(t)$ in the form of step-wise function

$$K(t) = \begin{cases} \tau^{-1}, & 0 \le t \le \tau \\ 0, & t > \tau, \end{cases}$$

then the other well-known Maxwell model for very viscous liquid is obtained. In the latter case, one gets a plasticity criterion with the only parameter of the characteristic time of plastic relaxation [6, 10]:

$$\frac{1}{\tau}\int_{t-\tau}^{t} \Sigma(s)\, ds \le \sigma_{\rm y}^0. \tag{5}$$

Integration of (5) at constant strain rate $\dot\varepsilon = \text{const}$ for $\sigma(t) = 2G\dot\varepsilon t$ gives [5] the yield strength as a function of the strain rate:

$$\Sigma_{\rm d}(\dot\varepsilon) = \begin{cases} \sqrt{4G\dot\varepsilon\tau\sigma_{\rm y}^0}, & \dot\varepsilon \ge \dfrac{\sigma_{\rm y}^0}{G\tau}, \\ \sigma_{\rm y}^0 + G\dot\varepsilon\tau, & \dot\varepsilon < \dfrac{\sigma_{\rm y}^0}{G\tau}, \end{cases} \tag{6}$$

where G is the shear modulus. As one can see, this strain rate dependence on the yield strength consists of a linear part (Maxwell model) that is enough for all low strain rate loading conditions and a nonlinear part that is actual for high strain rate dynamic loading with the transition point between two modes at the strain rate $\sigma_{\rm y}^0/(G\tau)$. It is shown [5] that the characteristic time depends on both the material structure and the plasticity mechanism and reflects the inertness of the plastic relaxation. We have the following two parameters: the static yield as the strength of the material and the characteristic time as its strain rate sensitivity characteristic. These two parameters are enough for a unique determination of the dynamic yield stress in a wide range of strain rates.

One can interpret τ in terms of the strain rate sensitivity at low strain rates. In [10, 11], it was shown that, at high strain rates, the linear approximation of the dislocation velocity is valid. On the basis of the Kelvin–Voigt model [11], under steady-state conditions as $t \to \infty$, if the static yield stress is neglected in comparison with the acting stresses ($\Sigma_\tau = \sigma_{\rm y} + 2G\dot\varepsilon\tau$) $\Sigma_\tau \gg \sigma_{\rm y}$, then the yield strength Σ_τ is determined from the following condition at different strain rates $\dot\varepsilon$:

$$\Sigma_\tau = 2G\dot\varepsilon\tau. \tag{7}$$

The macroscopic plastic shear deformation in the motion of dislocations was introduced by Johnston and Gilman [12] and Johnston [13]:

$$\varepsilon = b\rho_{\rm m}\bar{x}. \tag{8}$$

Here we use the notation: b is the Bürgers vector, $\rho_{\rm m}$ is the density of mobile dislocations, and $\bar{x}$ is the mean dislocation path. Then the strain rate is

$$\dot\varepsilon = b\rho_{\rm m}\bar{v}, \tag{9}$$

where $\bar{v}$ is the average dislocation velocity. Replacing the deformation rate obtained in condition (7), we obtain the following expression for the relaxation time:

$$\tau = \frac{\Sigma_\tau}{2G}\frac{1}{b\rho_{\rm m}\bar{v}}. \qquad (10)$$

In the classical experiments by Johnston [13] on lithium fluoride crystals, an empirical dependence of the strain rate on the applied shear stresses σ_τ was observed:

$$\bar{v}(\sigma_\tau) = \left(\frac{\sigma_\tau}{D_{\rm m}}\right)^n, \qquad (11)$$

where $D_{\rm m}$ and n are the constant values for a fixed material structure. Note that the estimates of $D_{\rm m}$ for various materials are calculated for a certain dislocation velocity in the range ~ 0.1–$1000\,{\rm m/s}$ [12]. In particular, Johnston [13] estimated for lithium fluoride $D_{\rm m} = 5.3\,{\rm MPa}$ at $V_{\rm D} = 0.1\,{\rm m/s}$. Substituting (11) with the material flow condition $\sigma_\tau = \Sigma_\tau$ into equation (9) and assuming that $\bar{v} = V_{\rm D}(\Sigma_\tau)$, we get the characteristic relaxation time through the mobility parameters of dislocations ($D_{\rm m}$, m, $\rho_{\rm m}$):

$$\tau = \frac{1}{2b\rho_{\rm m}}\frac{\Sigma_\tau}{G}\left(\frac{D_{\rm m}}{\Sigma_\tau}\right)^n. \qquad (12)$$

In this sense, the introduction of τ completely reflects the effect of the material sensitivity on the stress level even at low strain rates. As one can see from the comparison of (1), (9), and (11), n must be inversely proportional to m.

3. Strain rate sensitivity at high strain rates

At high strain rates that are typical of stress waves propagation in metals, an over barrier motion regime occurs for the dislocation slip. Under these conditions, the thermal activation processes no longer have a noticeable effect on their slip, and the dislocation velocity is determined by the phonon drag $B_{\rm D}$ so that the expression for their velocity has a form linear in the velocity [14]. We have obtained the following expression for the relaxation time $\tau_{\rm D}$ connected with the dislocation motion [5]:

$$\tau_{\rm D} = \frac{8\chi B_{\rm D}}{3\rho_{\rm D}Gb^2}, \qquad (13)$$

Hence, the value of the characteristic relaxation time no longer explicitly reflects the high-strain-rate sensitivity given by relation (1) but still reflects the properties of the inertness of the dislocation structure of the material during the relaxation of the arising stresses, which can lead to their significant growth.

J. Campbell in 1953 proposed a dynamic yield criterion for mild steel [15]. He assumed the thermo-activation nature of the plasticity processes, so that the time $t' = Ce^{U/(k_{\rm b}T)}$ is required to release dislocations, where C is a constant with the dimension of time. For the activation energy, the approximation $U = -U_0 \ln[\sigma(t)/\sigma_0]$ proposed by Yokobori [16] was used, where $\sigma(t)$ is the value of the flow stress, σ_0 is the yield strength at temperature 0 K which can be replaced by the static yield strength, and U_0 is a constant. Combining the two relations, assuming that, for the onset of plastic flow, it is necessary to attain a certain critical density of the mobile dislocations, and taking into account that the flow stress can change in time, we obtain an integral criterion for the onset of plastic yield in the form:

$$\int_0^{t_*}\left[\frac{\sigma(t)}{\sigma_0}\right]^\alpha dt = C, \qquad (14)$$

where $\alpha = U_0/(k_b T)$ is a constant at a given temperature, t_* is the moment of transition to the plastic state (it is assumed that the time count starts from the moment when the load is applied). The thermo-activation nature of the plasticity process, assumed in deriving criterion (14), turned out to be erroneous in dynamics, as well as the inverse dependence of the parameter α on the temperature. Nevertheless, it is significant that the concept of parameter α characterizing the sensitivity of the material at a stress level was empirically introduced in the model [14]. Comparing (14) with (5), it is clear that they coincide with an accuracy up to α if one identifies C with τ. On the yield stress-strain rate curve, the characteristic relaxation time determines the beginning of the dynamic branch of the curve, i.e., the critical strain rate after which a sharp increase in the yield strength begins [5]. A decrease in the characteristic time leads to an increase in the value of this critical velocity $\dot{\varepsilon}_{tr} = \sigma_y^0/(G\tau)$. The parameter α determines the slope of the high-strain-rate branch of the yield stress-strain rate curve. Many experimental data cannot be described by criterion (5) with $\alpha = 1$. At the same time, for pure metals it is close to unity [5]. Processing of a large number of experimental data from many sources, performed by Yu.V. Suvorova and A.K. Pertsov [17, 18], as well as by other researcher, convincingly demonstrates the necessity of introducing the alpha parameter for most steels and alloys for which its values can reach several dozens, and it turns out to be mechanically significant. From the more recent developments, numerous extensions [19, 20] to the classical Johnson–Cook model [21] were obtained. A part of them, where the additional terms are introduced proportionally to the dimensionless deformation rate up to $k \sim 0.5$ [20], reflect the features described above. Similarly to Campbell [15], the parameter of the material sensitivity to the intensity level of the local stresses was introduced purely phenomenologically in the criterion (2) in [5–7, 22–25]. It can be shown that the relationship between the parameters m and α is expressed by the relation

$$m = \frac{1}{\alpha + 1}. \tag{15}$$

Thus, $\alpha + 1$ is equal to the reciprocal of the parameter of the strain rate sensitivity of stresses. From the number of calculations by criteria like (5), it is known that $\alpha > 1$ is only necessary for alloys and steels and only at high strain rates, when any τ in (5) give a markedly lowered strain rate sensitivity of stresses. The physical nature and mechanisms reflected by the alpha parameter remain a debatable issue.

4. Experimental investigations of the velocity dispersion

Determination of metal yield strength in dynamic experiments with high strain rates is usually performed by analyzing the rear surface velocity profiles. Besides, it is possible simultaneously to measure the second parameter, i.e., the dispersion of the rear surface velocity [1, 2]. If the signal of the laser beam is reflected from a perturbed surface in the presence of shear flow instabilities, then its amplitude decreases as compared to its counterpart with an ideal unperturbed surface. Their ratio gives an experimental value of the velocity dispersion [1]. The velocity dispersion value gives additional information about the micro-scale processes, which take place in the material under the shock wave loading conditions. Figure 1*a* shows experimental curves obtained by Yu.I. Meshcheryakov with coauthors [2] for aluminum alloy D16 (the impactor velocity is $V_{imp} = 160\,\mathrm{m/s}$, the target thickness is $h = 5\,\mathrm{mm}$). One can see three maximuma of the velocity dispersion V_D consequently corresponding to the elastic precursor, the compression wave, and the rarefaction wave. The values of dispersion maximums lay in the range from 20 to $40\,\mathrm{m/s}$. The maximal relative value of dispersion (V_{imp}/V_D) is about 0.23. It is obvious from other experiments that the relative value of dispersion is proportional to the solute concentration, and it can be achieved up to several tens of percents from V_{imp} value for D16 alloy.

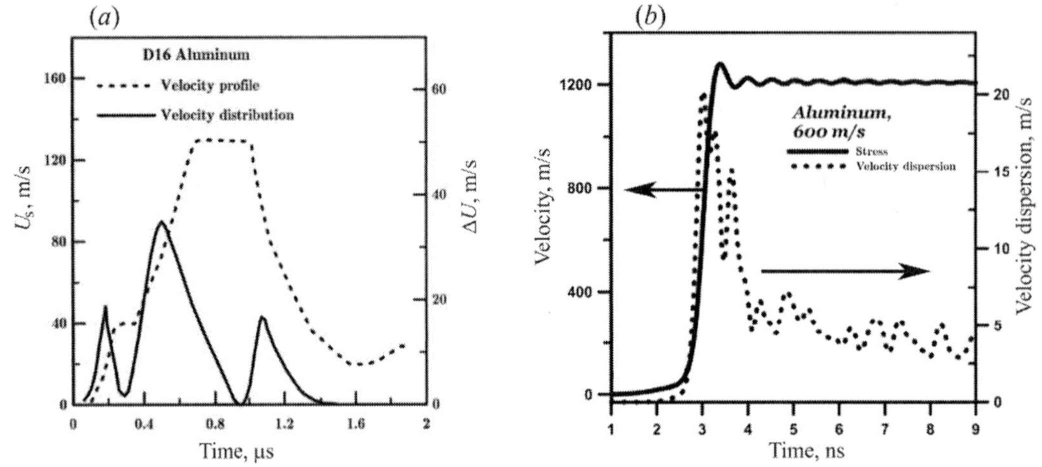

Figure 1. Average velocity of the target free surface and the velocity dispersion on the wave front: experimental [2] (*a*) and our numerical simulation data (*b*).

5. Numerical simulation and analytical consideration of the problem

Under real experimental conditions, the shock wave propagates through a material with initial random perturbations of the microstructure parameters (dislocation density, impurity concentration). to investigate the influence of these random perturbations, we performed 2D numerical simulation of the plastic flow under shock wave loading on the basis of the model for dynamics and kinetics of dislocations [3–5]. The initial dislocation density distribution was randomly perturbed. Formation of shear bands inclined at 45 degrees to the shock wave front was observed. The dispersion of the back surface velocity takes place similarly to the one observed by Yu.I. Meshcheryakov et al [1, 2], it is but weaker. To estimate the plastic flow heterogeneity, the normalized deviation $W = \Delta w/\langle w\rangle$ of the plastic deformation intensity $w = \sqrt{\frac{4}{3}(w_{xx}^2 + w_{yy}^2 + w_{xy}^2 + w_{xx}w_{yy})}$ was used, where $\Delta w = w_{\max} - w_{\min}$ and $\langle w\rangle$ is the average value of w. In simulations, the observed heterogeneities of the plastic flow correspond to the normalized deviation W lesser than 0.05, and according to our consideration, it corresponds to the dispersion of the scalar dislocation density $k = \rho_D^{\max}/\rho_D^{\min} \sim 10$. The velocity dispersion on the shock wave front in this case is about 2 m/s, which is an order of magnitude lower than the experimental findings for various aluminum alloys reaching the values about 20–40 m/s [2]. In this way, the instabilities caused by the heterogeneities of the only initial dislocation density cannot themselves provide a mechanically significant level of the plastic flow instability. One can approximate the numerical simulation data in the form of the dependence of the velocity deviation on the localization level by a logarithmic function $W = \Delta w/\langle w\rangle = 0.07\log k$.

It is important that this relation cannot changing the minimum initial density of dislocations in the range of orders of magnitude because of the intensive nucleation of dislocations during the deformation (see figure 2).

The introduction of impurities in the material can create plastic deformation with greater deviation as compared to the initial dislocation density perturbation. In the case of impurities, the deviation reaches about 0.09 and hence $k > 40$ which gives the velocity dispersion after the shock wave front of about 10–20 m/s. It better corresponds to the experimental data for aluminum alloys. In figure 3, one can see the calculated picture of micro-localization near the randomly distributed hard copper inclusions in a soft aluminum media.

It is interesting to compare these values of plastic deformation perturbations obtained in the

IOP Conf. Series: Journal of Physics: Conf. Series **991** (2018) 012012 doi:10.1088/1742-6596/991/1/012012

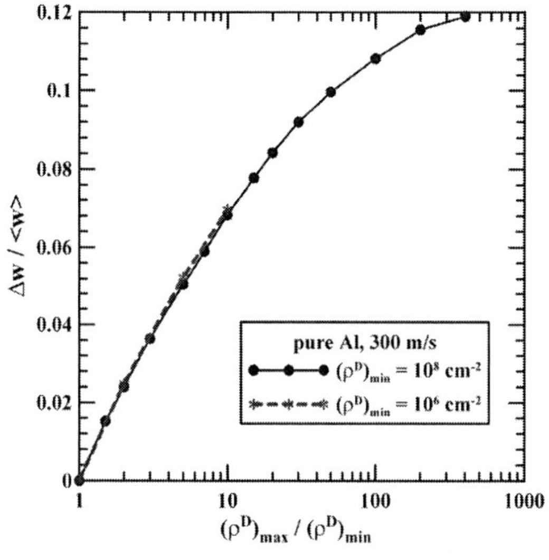

Figure 2. Dependence of the dispersion of the intensity of plastic deformation (associated with the dispersion of the free surface velocity of the target) on the variance of the initial dislocation density for two different minimum values of the dislocation densities.

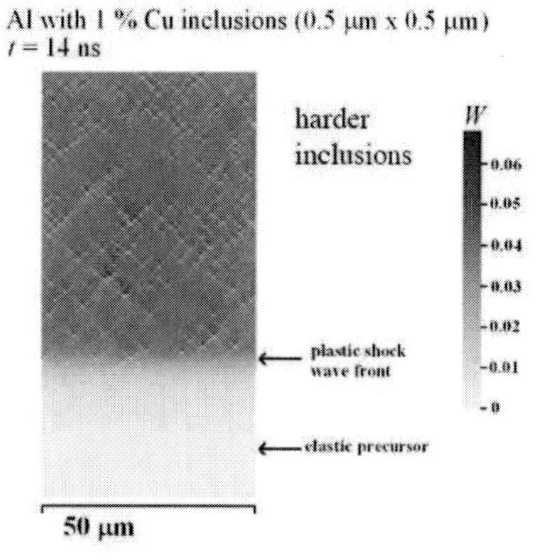

Figure 3. Two-dimensional numerical simulation of micro-localization behind the stress wave front under the impact loading of an aluminum alloy with randomly distributed stress concentrators.

numerical simulation and the experimental measurements with the empirically fitted values of the strain rate sensitivity parameter α from the integral criterion of plasticity (1). One can show that it is possible if we replace the parameter α in the integral criterion by the dispersion of the characteristic relaxation time parameter. It obviously follows from equation (13) that if there are any heterogeneities in the plastic flow in the material $\Delta\tau_\mathrm{D} = -[8\chi B_\mathrm{D}/(3\rho_\mathrm{D}^{(1)}\rho_\mathrm{D}^{(2)}Gb^2)]\Delta\rho_\mathrm{D}$ such as in the case of spatial micro-localization of the plastic flow, then we have different flow channels with different parameters of characteristic relaxation times that vary in a range $[\tau_1, \tau_2]$.

Let us consider how important is the effect of dispersion or "smoothing" of the real spectrum of the characteristic relaxation times on the mechanical characteristics of the material. For this purpose, we have to consider a continuous spectrum of relaxation times. Suppose that one can select a finite range of frequencies $p = 1/\tau$ which make the main contribution to the relaxation of stresses. A more general case of the kernel of integral operator (4) than the stepwise function (5) and exponential functions is its representation in the form of the Laplace integral [25]

$$K(t) = \int_0^{+\infty} \varphi(p)e^{-pt}\,dp, \tag{16}$$

where $\varphi(p)$ is the spectral density which must satisfy the normalization condition $\int_0^{+\infty}[\varphi(p)/p]\,dp = 1$. For the simplest case of bands of uniformly distributed characteristic relaxation times in the interval $\Delta\tau = \tau_2 - \tau_1$ $(\tau_1 < \tau_2)$, one can choose the spectral density $\varphi(p)$ in the form

$$\varphi(p) = \frac{H(p-\tau_1^{-1}) - H(p-\tau_2^{-1})}{\ln(\tau_2/\tau_1)}, \tag{17}$$

where $H(x)$ is the Heaviside function. Substituting equation (17) into equation (16) and integrating at a constant strain rate, we obtain the dynamic yield strength as a function of the time of onset of the plastic flow $\Sigma_{\mathrm{d}}(t_*)$ [25]:

$$\Sigma_{\mathrm{d}}(t_*) = \sigma_{\mathrm{y}}^0 \begin{cases} \dfrac{2\tau}{t_*}, & t_* < \tau_1, \\ \dfrac{\ln(\tau_2/\tau_1)}{\tau_1/(2t_*) - t_*/(2\tau_2) + \ln(t_*/\tau_1)}, & \tau_1 \le t_* < \tau_2, \\ 1 - \dfrac{\tau_1\tau_2}{2t_*\tau}, & t_* \ge \tau_2. \end{cases} \tag{18}$$

Here $\tau = \tau_1\tau_2(\ln\tau_2 - \ln\tau_1)/(\tau_2 - \tau_1)$. The relationship with the dispersion of the dislocation density is expressed in the form $\Delta\tau_{\mathrm{D}} = \tau(1 - k^2)/(2k)$, where we assume in (13) that $\rho_{\mathrm{D}} = \bar{\rho}_{\mathrm{D}} = \frac{1}{2}(\rho_{\mathrm{D}}^{(1)} - \rho_{\mathrm{D}}^{(2)})$. In the case where the band is narrow enough and $k = 1$, this model turns into the model with one relaxation time (6). Thus, we obtain a new formulation of the integral criterion, where a range of characteristic relaxation times is used instead of the phenomenological parameter of the strain rate sensitivity to the stress level. According to (13), this range has a clear physical meaning, which allows one directly to compare the spread of characteristic times with the spread of the dislocation densities. Our calculations show that the relation $\alpha \sim 5\log k$ is possible. Namely, the dispersion of the characteristic relaxation time and the dispersion of the dislocation density $k < 10$ corresponds to the values $\alpha < 5$. This is the case of numerical simulations with the dislocation plasticity model of pure material without impurities and the relative dispersion of the plastic deformation intensity of approximately 0.07. The case with $k > 100$ corresponds to $\alpha > 10$, which is typical of materials with strong stress concentrators such as steels [6].

Conclusions

The strain rate sensitivities under quasi-static and dynamic conditions of loading have different physical nature. In the low strain rate case, the sensitivity arises from inertness of the defect structures evolution and can be expressed by a single parameter characterizing the plasticity mechanism. In our approach, this parameter is the value of the characteristic relaxation time. In the dynamic case, the effects of inertness of defective microstructure evolution described by this parameter become dominant and determine the value of the dynamic yield strength of the material, but, in addition, new strain rate sensitivity effects arise that are associated with the micro-localization of the plastic flow near heterogeneities or stress concentrators. In the mechanical description, this requires the introduction of additional strain rate sensitivity parameters, which is realized in numerous refinements of the Johnson–Cook and Zerilli–Armstrong models [19, 20].

The consideration of both of these factors is required for an adequate description of the problems of dynamic deformation of highly inhomogeneous metallic materials such as steels and alloys. Many authors identify the values of the strain rate sensitivity in quasi-static with dynamics regimes having a different order of magnitude and different physical nature, which creates confusion in the interpretation of the research results. The measured dispersion of particle velocities on the free surface of a shock-loaded material can be considered as an experimental expression of the effect of micro-localization [1, 2]. This is also confirmed by the results of numerical simulation of the propagation of shock waves in a 2D-formulation and analytical estimates. A modification of the integral yield criterion [6, 22, 23] with regard to the real dispersion of the material properties in the shear bands in a certain range of parameters allows one to take into account all these effects in a natural way. It is possible to relate the width of the range of characteristic relaxation times corresponding to different regions of the heterogeneously deformed material to the value of the empirical parameter of its rate sensitivity at high strain rates.

Acknowledgments

The study was supported by the grants from the President of the Russian Federation (Projects Nos. MK-4649.2016.1 and MD-7481.2016.1) and by the grant from the Ministry of Education and Science of the Russian Federation (competitive part of State Task of NIR CSU No. 3.2510.2017/PP).

References

[1] Meshcheryakov Yu I, Divakov A K, Zhigacheva N I, et al. 2008 Dynamic structures in shock-loaded copper *Phys. Rev. B* **78** 64301–16
[2] Meshcheryakov Yu I, Vasilkov V B, and Atroshenko S A 1993 About rotational and translational modes of plasticity and fracture initiated by dynamic loading of materials *J. Mech. Behavior Mater.* **4** 255–66
[3] Mayer A E, Khishchenko K V, Levashov P R, and Mayer P N 2013 Modeling of plasticity and fracture of metals at shock loading *J. Appl. Phys.* **113** 193508
[4] Mayer A E, Borodin E N, and Mayer P N 2013 Localization of plastic flow at high-rate simple shear *Int. J. Plast.* **51** 188–99
[5] Selyutina N, Borodin E N, Petrov Y, and Mayer A E 2016 The definition of characteristic times of plastic relaxation by dislocation slip and grain boundary sliding in copper and nickel *Int. J. Plast.* **82** 97–111
[6] Gruzdkov A A, Petrov Yu V, and Smirnov V I 2002 An invariant form of the dynamic criterion for yield of metals *Phys. Solid State* **44** (11) 2080–2
[7] Gruzdkov A A and Petrov Yu V 1999 On temperature-time correspondence in high-rate deformation of metals *Dokl. Phys.* **44** (2) 114–6
[8] Suo T, Chen Y, Li Y, et al. 2013 Strain rate sensitivity and deformation kinetics of ECAPed aluminium over a wide range of strain rates *Mater. Sci. Engng A* **560** 545–51
[9] Rabotnov Yu N 1980 *Elements of Hereditary Solid Mechanics* (Moscow: MIR) p 392
[10] Petrov Yu V and Borodin E N 2015 Relaxation mechanism of plastic deformation and its justification using the example of the sharp yield point phenomenon in whiskers *Phys. Solid State* **57** (2) 353–9
[11] Borodin E N, Mayer A E, Petrov Yu V, and Gruzdkov A A 2014 Maximum yield strength under quasi-static and high-rate plastic deformation of metals *Phys. Solid State* **56** (12) 2470–9
[12] Hull D and Bacon D J 2011 *Introduction to Dislocations* (Amsterdam: Elsevier)
[13] Johnston W G 1962 Yield points and delay times in single crystals *J. Appl. Phys.* **33** 2716
[14] Krasnikov V S, Mayer A E, and Yalovets A P 2011 Dislocation based high-rate plasticity model and its application to plate-impact and ultra short electron irradiation simulations *Int. J. Plast.* **27** (8) 1294–308
[15] Campbell J D 1953 The dynamic yielding of mild steel *Acta Metall.* **1** 706–10
[16] Yokobori T 1986 *An Interdisciplinary Approach to Fracture and Strength of Solids* (New York: Gordon & Breach)
[17] Mnev E N and Pertsov A K 1970 *Hydroelasticity of Shells* (Leningrad: Sudostroenie) [in Russian]
[18] Suvorova Yu V 1968 Delayed yield in steels (review) *J. Appl. Math. Tech. Phys.* **9** (3) 270–4
[19] Couque H, Boulanger R, and Bornet F 2006 A modified Johnson-Cook model for strain rates ranging from 10^{-3} to $10^5\,s^{-1}$ *J. Phys. IV (Proc.)* **134** 87–94
[20] Couque H 2014 The use of the direct impact Hopkinson pressure bar technique to describe thermally activated and viscous regimes of metallic materials *Phil. Trans. Roy. Soc. A* **372** 20130218
[21] Johnson G R and Cook W H 1983 A constitutive model and data for metals subjected to large strains, high strain rates, and high temperatures In *Proc. of 7th Int. Symp. on Ballistics, The Hague, 19–21 April 1983* (Hague: The Netherlands) pp 541–7
[22] Petrov Yu V and Sitnikova Y V 2005 Temperature dependence of spall strength and the effect of anomalous melting temperatures in shock-wave loading *Tech. Phys.* **50** (8) 1034–7
[23] Gruzdkov A A and Petrov Yu V 2012 Incubation-time based approach for dynamic yielding of metals *J. Ningbo University* **25** (1) 78–82
[24] Gruzdkov A A, Sitnikova E V, Morozov N F, and Petrov Y V 2009 Thermal effect in dynamic yielding and fracture of metals and alloys *Math. Mech. Solids* **14** (1-2) 72–87

Investigation of the mechanical properties of organoplastic under shock wave loading conditions

A M Bragov*, L A Igumnov, A Yu Konstantinov, and A K Lomunov
Research Institute for Mechanics, Lobachevsky State University of Nizhny Novgorod, Nizhny Novgorod, Russia

E-mail: *bragov@mech.unn.ru

Abstract. The paper presents results of dynamic tests of a typical representative of new composite and damping materials: organoplastics. Compression testing was performed using the traditional Kolsky method and its original modification. The strength and deformation properties of organoplastics under conditions of uniaxial stress and uniaxial deformation were studied. When the organoplastic is compressed transversely to the Kevlar fabric layers under conditions of a uniaxial stress state, the material begins to break down (to lose the layer cohesion) at a stress of about 200 MPa, while under the conditions of uniaxial strain, it retains its apparent integrity at stresses up to 500 MPa. The small value of the lateral thrust factor indicates a large internal strength of the material in tension in the radial direction.

1. Introduction

It is known that objects of aerospace technology can be subjected to intensive dynamic impacts during operation due to various factors of explosive, shock or other nature. In modern designs, various composite materials are widely used, both as supporting structural elements and as damping materials.

To prevent the destruction of a construction of new technology under the impact of shock wave loads, a special layer of organoplastics (aramid–epoxy composite) is included in the structure of their shell, which reduces (dampens) the acting loads. Since the physical-mechanical properties of composite materials under conditions of intense dynamic effects depend on the strain rate, the data on the properties of the composite materials obtained at high strain rates are required to ensure and confirm the stability of such structures.

The investigated material (organoplastic) is a light polymer composite material consisting of a fibrous high-strength filler of organic nature (aramid fibers) distributed in a matrix of polymer (epoxy) resins. It is used in aircraft structures in which special strength and lightness is required, as well as in manufacturing of individual armour protection. In aeronautical engineering, the organoplastics ensure the fan case impenetrability when the blade breaks as a result of multi-cycle fatigue and in the case of birds or some foreign objects getting into the engine.

2. Experimental methods

To investigate the mechanical properties of organoplastics under compression, an apparatus [1] was developed, where the Kolsky method was used [2]. A gas gun of caliber of 20 mm was used as a loading device. The pressure bars of a diameter of 20 mm were made both of high-strength martensitic-aging steel and high-strength aluminum alloy D16T.

To study the properties under conditions of uniaxial deformation, a test sample was placed in a rigid jacket of original configuration [3] equipped with strain gauges to measure its deformation in the circular direction. Under conditions of the jacket elastic operation, this permits obtaining the radial stress component $\sigma_r(t)$ in the specimen [4, 5], which together with the longitudinal stress component $\sigma_x(t)$, determined by traditional formulas of the Kolsky technique, allowed one to obtain the pressure dependence in the specimen $P(t) = -\frac{1}{3}[\sigma_x(t) + 2\sigma_r(t)]$, the shear stress $\tau(t) = \frac{1}{2}[\sigma_x(t) - \sigma_r(t)]$, and the coefficient of lateral pressure (thrust) $\xi(t) = \sigma_r(t)/\sigma_x(t)$. All these quantities are parametric processes, so the relationship between them should be considered throughout the test.

The experimental setup allows one to vary the strain rate in a wide range. To cover the entire area, the strikers are fired at different initial velocities, and longer strikers are used to provide sufficient deformation at lower speeds. The strikers used in the experiments were made of steel and D16T alloy of length varying from 200 mm to 400 mm, which permits exciting the loading impulses of duration ($t = 2L_{\text{str}}/C_{\text{str}}$) from 80 to 160 microseconds.

3. Tested specimens

The specimens for testing shaped as cylinders of diameter of 20 mm and height of about 9 mm were cut from a pressed plate consisting of a great many layers of fabric of aramid fibers with equal strength of the weft and the base. Such dimensions (the ratio $L/D \approx 0.5$) correspond to the minimum error in the measurement of stresses introduced by inertia forces [6].

It should be noted that the end surfaces of the organoplastic specimens are not flat and smooth due to manufacturing technology. To reduce the influence of these errors and to improve the acoustic contact between the ends of the measuring bars and the ends of the specimen, a viscous lubricant layer was applied onto these surfaces before the test.

The specimens were loaded perpendicular to the fabric layers.

4. Results of the tests

Some of the tests were carried out using steel measuring bars (and steel confining jacket), which permits obtaining a large stress level in the specimen and, correspondingly, high strain rates. To obtain the properties at low strain rates, when the amplitude of the registered signal from the transmitting bar is small, we used bars (and a jacket) made from D16T alloy.

As a result of processing the experimental data according to the formulas of the Kolsky method, the dynamic stress-strain curves are obtained for compression under uniaxial stress conditions, as well as for uniaxial deformation (volumetric stress state). Analysis of these curves allowed us to determine the moduli of loading and unloading branches, strength characteristics, lateral expansion ratio, and parameters of shearing strength.

4.1. Investigation of the behavior in a uniaxial stress state

Figure 1 shows the characteristic diagrams of organoplastic deformation under the conditions of a uniaxial stress state obtained at different strain rates in order to determine the magnitude of the load that leads to destruction of the material. Here, the solid curves show the true dynamic σ_s–ε_s curves, and the dashed lines in the lower part of the graph show the corresponding strain rate change curves $\dot{\varepsilon}_s$–ε_s. By varying the striker speed, i.e., the amplitude of the loading wave, we could choose loading regimes in which the sample either retained its apparent integrity and strength (curves *1–3*) or was fractured (curves *4–5*). One can see that, in tests without a jacket (under the conditions of a uniaxial stress state) at a stress above 200 MPa, the fracture processes begin in the material, i.e., the stress decreases with a significant increase in the strain.

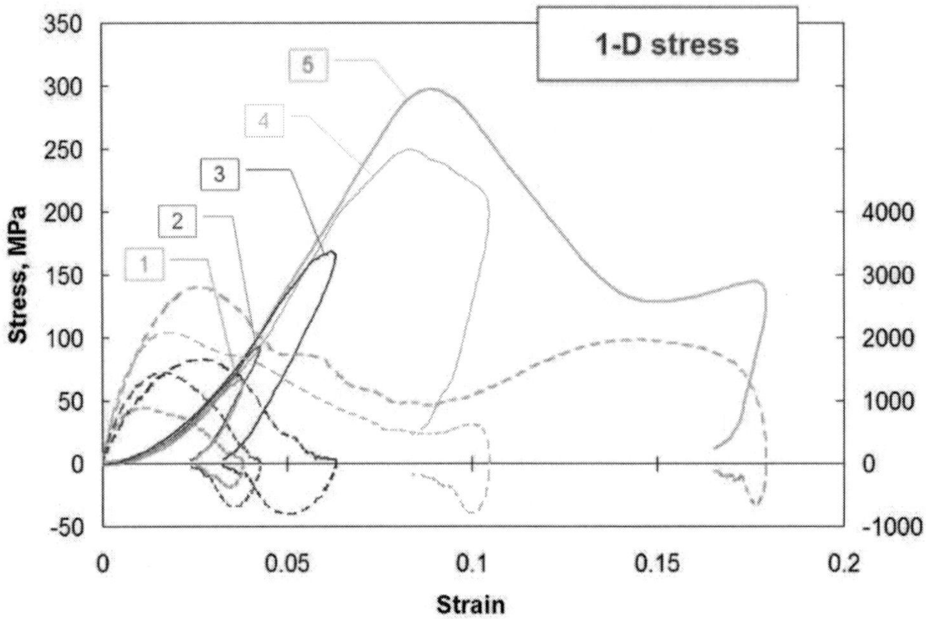

Figure 1. Effect of strain rate on the behavior of organoplastic.

The obtained stress-strain curves of organoplastic are nonlinear in the loading part. At the beginning of the diagrams, there is a significant (from 0.3 to 1.5%) section with practically zero stress. The main reason for this may be the non-flatness of the end faces of the specimen and, as a consequence, the presence of a small gap between the ends of the loading measuring bar and the specimen. On the other hand, the sample consists of Kevlar tissue layers impregnated with binder alternating with epoxy resin layers. At the initial moment of the specimen loading at a sufficiently low stress level, a nonlinear elastic deformation of the polymer binder occurs both inside the Kevlar layers and between these layers, which causes a delay in the appearance of the transmitted pulse. In addition, the velocity of the wave propagation in the composite sample is much lower than in the measuring bars, which is also the cause of the delay in the initial section of the diagram.

Next, the layers of the epoxy binder and Kevlar fabric layers are compressed. The material begins to resist the deformation more strongly, which leads to an increase in the stress according to a nearly linear law, up to values determined by the incident pulse amplitude. If this amplitude is insufficient to destroy the specimen, then after the loading pulse termination, the specimen is also unloaded according to a nearly linear law up to a stress level of $\sim 30\,\text{MPa}$ and then with a gradually decreasing deformation modulus.

It is clearly seen that when the specimen breaks down in testing under conditions of free extension, this leads to a drop in the stress level in the sample after a certain maximum and to a synchronous increase in the strain rate. With preservation of the the specimen integrity, the strain rate curve monotonically decreases after attaining its maximum on the loading branch. On the unloading branch, the strain rate has negative values.

The growth in the stress of the destroyed specimen (curve 5 in figure 1) after deformation of 0.15 can be explained by the exhaustion of the longitudinal deformation of relatively weak layers of the binder and the mutual compression of contiguous layers of the Kevlar fabric, which has a great strength.

The study of the material endurance under repeated loading at loading energies insufficient

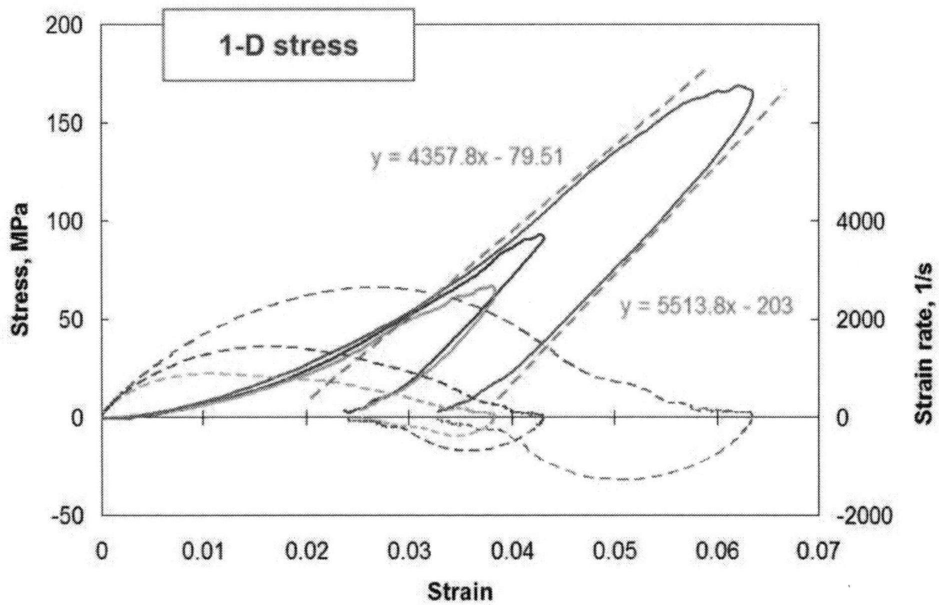

Figure 2. Study of the material endurance under repeated loading.

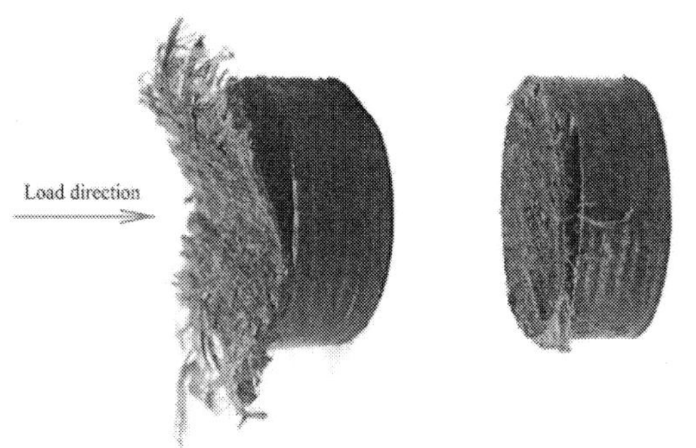

Figure 3. Appearance of organoplastic samples after loading under uniaxial stress conditions with different loading energies.

Figure 4. A specimen with strain gauges on the lateral surface for measuring the Poisson's ratio.

for visible fracture is illustrated in figure 2. The same specimen was loaded several times with different gradually increasing loading energies (strain rates). It can be seen that the loading branch of the chart, which has a nonlinear character, is repeated from experiment to experiment, and the increasing loading energy causes only an increase in the level of the stress achieved in the specimen. The average value of the modulus of the loading branch in the range of considered strain rates is ~ 4.3 GPa, the unloading modulus is 5.5 GPa.

The visual analysis of samples after testing indicates that the sample, after compression with insignificant loading energy, retains its apparent integrity and initial length, and only on the lateral surface, there is some extrusion of Kevlar fibers, apparently due to their straightening in the fabric when compressed. At loads greater than 200 MPa, the irreversible destruction occurs, i.e., the material is delaminated, and a greater delamination is observed at the end where the load is applied (figure 3).

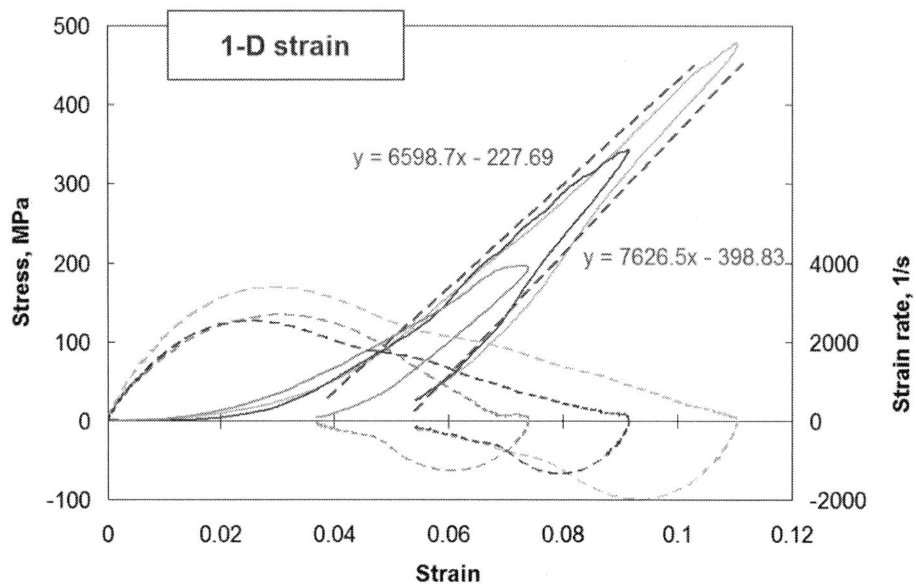

Figure 5. Diagrams of compression under uniaxial strain.

An attempt was made to measure the Poisson's ratio by a "direct" method, by attaching the strain gages directly to the lateral surface of the specimen (figure 4) and by its loading between the ends of the measuring rods under uniaxial stress conditions.

Certain difficulties arise when recording the signals from the strain gauges glued to the lateral surface of the specimen, since the sensors detect not only the integral peripheral (circular) deformation, but also the possible displacements and extrusion of individual layers of the fabric.

Based on the results of several experiments, the average Poisson's ratio was 0.05.

4.2. Investigation of the behavior under uniaxial deformation

The results of compression tests under conditions of uniaxial deformation (in a rigid jacket equipped with strain gages on the lateral surface) are presented below. Figure 5 shows curves in the "axial stress–axial strain" axes. Also, the corresponding strain rate curves are shown by dotted lines.

The average value of the module of the loading branch is $\sim$ 6.6 GPa, unloading — 7.6 GPa. Under conditions of radial expansion confining, the material retains its apparent integrity at stresses up to 500 MPa. This is also shown by the unloading branches. The appearance of the specimens is almost the same as that of the original, only on the lateral face of the specimen, there is a slight extrusion of the fibers from the Kevlar fabric layers.

An analysis of the loading branches of the charts showed that, with an exception of the initial section (delay of the curve appearance due to the gradual removal of the gap), the organoplastic has compression modulus almost independent of the strain rate. The material unloading is also nonlinear and has a slightly steeper slope. It is impossible to trace the course of the diagram until the specimen is completely unloaded, since the duration of the pulses reflected and transmitted through the specimen greatly exceeds the duration of the incident pulse due to the polymer binder viscosity and cannot be completely registered because of the insufficient length of the measuring bars.

Figure 6 shows the initial pulses in the measuring bars and on the circular surface of the jacket. It is clearly seen that the basic assumption of the Kolsky method that of forces are equal

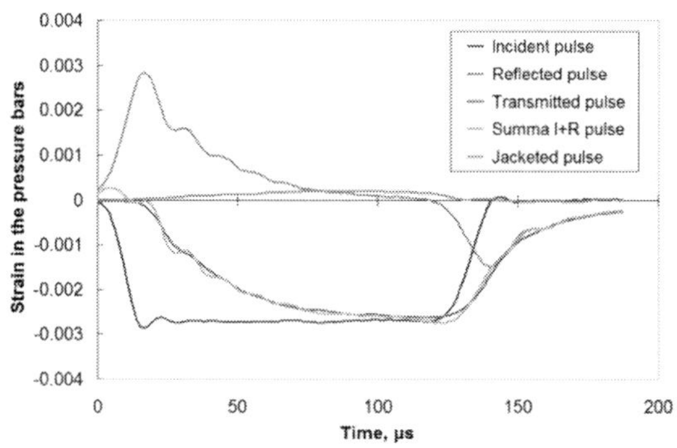

Figure 6. Strain pulses in the pressure bars and in the confining jacket.

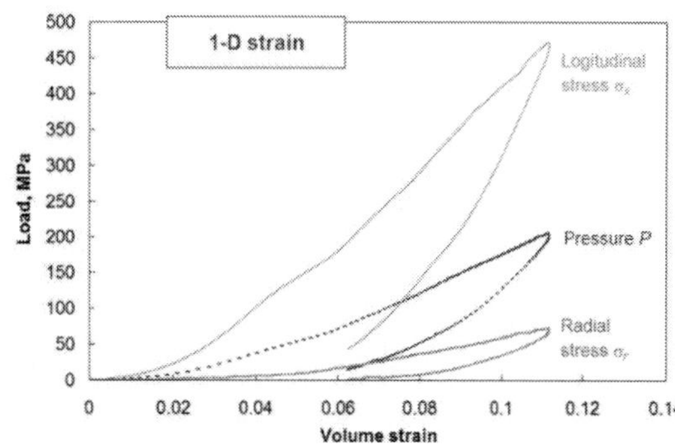

Figure 7. Load components in the specimen under uniaxial deformation.

at the end faces of the specimen, $\varepsilon^{\mathrm{I}}(t) + \varepsilon^{\mathrm{R}}(t) = \varepsilon^{\mathrm{T}}(t)$, is satisfied throughout the test, and the reflected and transmitted pulses have duration significantly longer than the loading wave. The transmitted pulse has a significant delay in appearance relative to the beginning of the reflected pulse.

The longitudinal compression of the specimen causes its transverse expansion according to the Poisson's ratio. However, since the radial deformation of the specimen is limited by the jacket, a near-uniaxial deformation state and a volumetric stress state arise in the specimen. For such a stress-strain state, the Poisson's ratio cannot be estimated, but only the coefficient of lateral thrust. The value of the radial stress component is determined by registering the signals from the confining jacket, which together with the signals from the strain gauges on the measuring bars permit obtaining the tangential stress and pressure in the specimen, and also to estimate the lateral thrust ratio.

For one of the experiments, figure 7 shows the curves of variation of the axial and radial stress components, as well as the pressure in the specimen as a function of the volume strain. One can see that the radial stress component is much smaller than the axial one. It should be noted that the results of measuring on the circular surface of the confining jacket determine the stress developed on the inner surface of the jacket. This stress is taken as the radial component of the stress in the specimen.

Another possible reason for the small value of the radial stress can be the inhomogeneity of the lateral surface of the specimen and the presence of a certain gap between the outer surface of the specimen and the inner surface of the jacket. Due to a gap between the specimen and the jacket, its participation in the formation of volumetric stress-strain state is inadequate.

The large internal tensile strength of the material (Kevlar fibers in the fabric) in the radial direction as compared to the compression strength (layers of the polymer binder) in the axial direction is also a cause of large difference between σ_r and σ_x. This respectively leads to a small value of the lateral thrust coefficient, whose curve in the loading and unloading sections is shown in figure 8. It should be noted that since the lateral thrust ratio is defined as the quotient of the radial stress component to the axial one and these quantities are sufficiently small at the initial and final loading stages, the quotient of two small quantities with possible errors and distortions can be sufficiently large.

In figure 8*a*, the dotted line additionally shows the curve of the time variations in the tangential stress. Depending on the pressure developing in the specimen (figure 8*b*), this curve practically coincides at the loading and unloading stages.

Figure 9 shows the curves of tangential stress changes as a function of the pressure developing in the specimen obtained for different loading energies. One can see that they almost coincide at

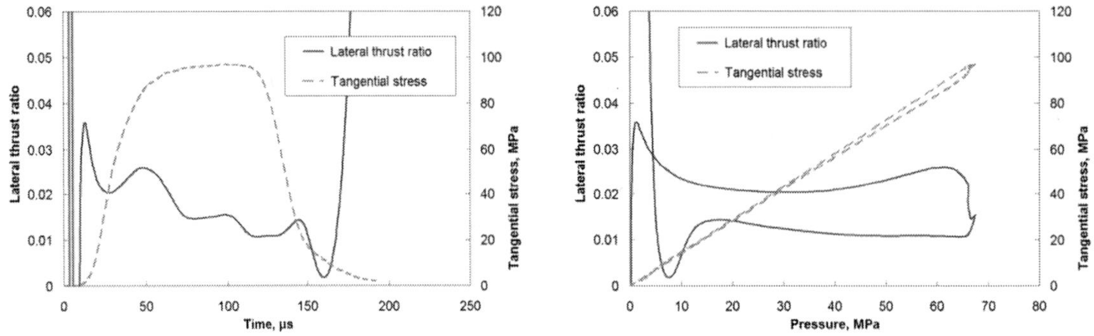

(a) (b)

Figure 8. Variation in the lateral thrust ratio and tangential stress depending on time (*a*) and on the pressure in the sample (*b*).

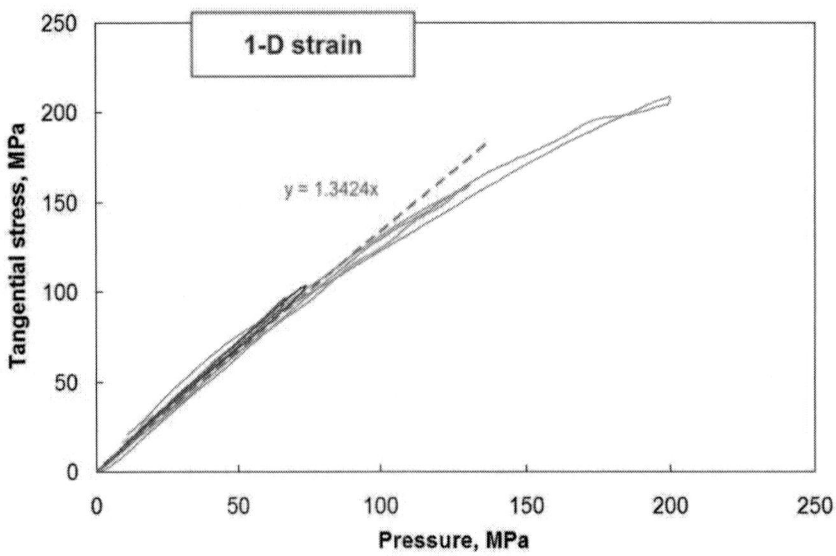

Figure 9. Changes in the tangential stress as a function of the pressure in the specimen.

loading and unloading of the specimen. For small load values, this curve can be approximated by a linear relation, whereas the linear dependence is violated for large load amplitudes. A reason for this may be a nonlinearly elastic deformation of the polymer binder.

Conclusions

The study has shown that when the organoplastic is compressed transversely to the Kevlar fabric layers under uniaxial stresses, the material begins to break down (to lose the layer cohesion) at a stress of about 200 MPa, while under the conditions of confining radial expansion, it retains its apparent integrity at stresses up to 500 MPa. The small value of the lateral thrust factor indicates a large internal strength of the material in tension in the radial direction. since the Kevlar fibers in the fabric have a high tensile strength.

Acknowledgments
Dynamic investigation was financially supported by the Russian Science Foundation (grant No. 15-19-10032).

References
[1] Bragov A M and Lomunov A K 1995 Methodological aspects of studying dynamic material properties using the Kolsky method *Int. J. Impact Engng* **16** (2) 321–30
[2] Kolsky H 1949 An investigation of the mechanical properties of materials at very high rates of loading *Proc. Phys. Soc. B* **62** (11) 676–700
[3] Bragov A M, Kotov V L, Lomunov A K, and Sergeichev I V 2004 Measurement of the dynamic characteristics of soft soils using the Kolsky method *J. Appl. Mech. Tech. Phys.* **45** (4) 580–5
[4] Bragov A, Grushevsky G, and Lomunov A 1994 Use of the Kolsky method for studying shear resistance of soils *DYMAT J.* **1** (3) 253–9
[5] Bragov A M, Lomunov A K, Sergeichev I V, et al. 2008 Determination of physicomechanical properties of soft soils from medium to high strain rates *Int. J Impact Engng* **35** (9) 967–76
[6] Davies E D H and Hunter S C 1963 The dynamic compression testing of solids by the method of the split Hopkinson pressure bar *J. Mech. Phys. Solids* **11** (3) 155–79

On multiple crack identification by ultrasonic scanning

M Brigante[1*] and M A Sumbatyan[2]**
[1]University of Naples – Federico II, Naples, Italy
[2]Institute of Mathematics, Mechanics, and Computer Science of the Southern Federal University, Rostov-on-Don, Russia
E-mail: *brigante@unina.it, **masumbatyan@sfedu.ru

Abstract. The present work develops an approach which reduces operator equations arising in the engineering problems to the problem of minimizing the discrepancy functional. For this minimization, an algorithm of random global search is proposed, which is allied to some genetic algorithms. The efficiency of the method is demonstrated by the solving problem of simultaneous identification of several linear cracks forming an array in an elastic medium by using the circular Ultrasonic scanning.

1. Introduction

Various optimization methods are widespread in the engineering science due to their efficiency when applied to both theoretical and practical problems, see, e.g., [1–3]. Classical regular methods are typically applied to optimize regular functionals, see [4]. If it is required to find the global minimum or maximum of a certain functional of a complex structure, then classical standard algorithms can hardly be used. However, some advanced algorithms of global random search demonstrate high efficiency just in applications to the problems with complicated target functionals, see, e.g., [5].

In the present paper, we demonstrate that many problems in engineering science can be reduced to certain classes of operator equations which can then be reduced to optimization problems. We construct an algorithm which permits efficient numerical treatment of the arising optimization problems and demonstrate the application of the proposed approach to the important problem of defect identification by using the Ultrasonic (US) scanning with an example, where the defects are represented by a finite array of linear cracks.

Many published works were currently dealing with inverse identification problems. This interest is caused by the importance of such investigations in many practical applications, in particular, as discussed above, in the US Nondestructive Testing. Chronologically, the first recognition methods were based on an approximation of weak wave interaction, namely on the theories of Born and Rytov [6]. The well-developed methods of acoustical tomography based on the Radon transformation are also related to the class of theories using the weak scattering hypothesis [7]. More strict diffraction theories are based on direct and inverse mathematical diffraction problems [8, 9]. The present authors have published a series of works on the image identification of single defects [10, 11]. Here we demonstrate that the optimization methods can also efficiently be applied to identify several defects simultaneously, on the basis of US methods.

IOP Conf. Series: Journal of Physics: Conf. Series **991** (2018) 012014 doi:10.1088/1742-6596/991/1/012014

2. Operator equations and optimization problems

Many problems in engineering science can mathematically be reduced to operator equations of the form

$$A\varphi = f, \quad \varphi \in D_\varphi \subset R^N, \quad f \in D_f \subset R^M, \tag{1}$$

where f is known but φ is unknown. In principle, Eq. (1) can be studied in a pair of Hilbert spaces, however for the applications discussed below it is sufficient to consider operator A acting from a finite-dimensional Euclidean space R^N of dimension N to another Euclidean space R^M of dimension M. To be more specific, we assume that D_φ and D_f are some simply connected convex bounded domains in R^N and R^M, respectively. Very often, the operator A is strongly nonlinear, and in such cases it is impossible to draw a conclusion about the existence and uniqueness of the solution of Eq. (1). However, in practice, a solution to Eq. (1) must be constructed even under the condition of such an uncertainty.

Let us assume that there is *a priori* information that there exists a certain solution φ_0 to Eq. (1). Then the key point is to construct an algorithm for determining this solution or finding at least a good approximation to φ_0. In the case where the operator A is continuous, even being nonlinear, an approximate solution φ_* close to φ_0 corresponds to the respective right-hand side f_* close to f. Let us introduce the following discrepancy functional:

$$\Omega(\varphi) = \|A\varphi - f\|^2. \tag{2}$$

Obviously, it is nonnegative: $\Omega(\varphi) \geq 0$, and $\Omega(\varphi_0) = 0$ since $A\varphi_0 = f$. Therefore, the minimization of the functional $\Omega(\varphi)$ may give a good approximation to the solution φ_0. In the case where the operator A has a continuous inverse operator in the considered domain, the lower is the attained minimal value $\Omega_{min} \geq 0$ of the functional, the more precise is the constructed approximation $\varphi_* \approx \varphi_0$. It should be noted that, in nonlinear problems, the existence of the inverse operator cannot, as a rule, be proved as a theorem, but sometimes this fact is evident from the physical point of view. The important question regarding the uniqueness of the solution is beyond the present study.

3. Key properties of the applied algorithm

The functional (2) can be minimized by any classical optimization method [4]. However, the main restriction of regular iterative schemes is that they give only a local minimum of the respective functional. Under such conditions, it is not evident that a local minimum is simultaneously the global minimum of the functional. In fact, it is well known that, for nonlinear equations, such values of local minima may be too far from the desired value $\Omega = 0$.

The described property is demonstrated with an example of a one-dimensional functional which, in the one-dimensional case, is simply a normal function of the variable φ:

$$\Omega(\varphi) = |\sin(10\varphi) + 0.8\cos(20\sqrt{2}\varphi) + 0.5\sin(15\sqrt{3}\varphi)|, \quad \varphi \in (0, 2\pi) \tag{3}$$

This function is plotted in figure 1. This is a strongly oscillating continuous nonnegative function, which has several minimal zero values. These points arise every time when the function inside the modulus symbol in (3) changes its sign as the argument φ monotonically increases.

If the analytical structure of function (3) is unknown, then one needs to apply an appropriate numerical algorithm for minimization. Obviously, if in the frames of any regular iteration process like the steepest descent method or Newton method, one starts from a certain initial point, then one arrives at a local minimum of the function at hand. Therefore, the global minimization cannot be resolved by such regular algorithms.

For this reason, in our numerical experiments, we use a version of the global random search method [5] contiguous to the one described in detail in [12]. This algorithm was developed to

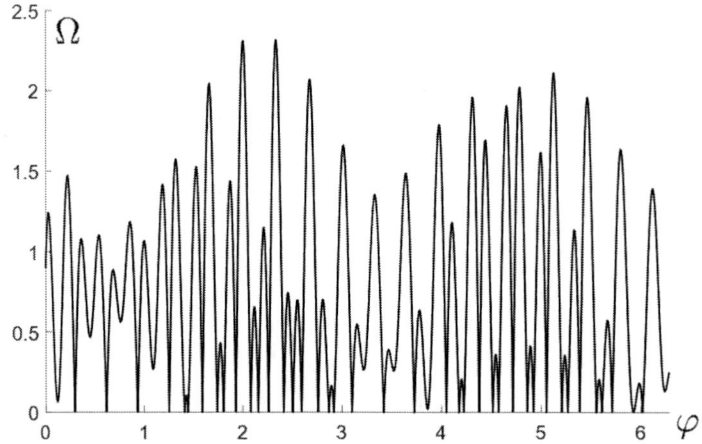

Figure 1. An example of the minimized function.

seek maxima, but it can also be used to seek minima. It is constructed so that it moves both up and downhill and, as the optimization process proceeds, focuses on the most promising area. As the first step, it randomly chooses a trial point within the step from the starting point selected by the user. The function is evaluated at this trial point, and its value is compared with its value at the initial point. In the minimization problem, all downhill moves are accepted and the algorithm continues from that trial point. The relationship between the initial value of Ω and the resulting step length is function dependent. This algorithm shows perfect convergence for many problems, also for our inverse identification problem, but unfortunately it sometimes arrives at a local extremum instead of the global one in the cases where there are a lot of global minima of the objective function.

The algorithm applied is a slight modification of these ideas. It possesses the following two specific features: (1) random sampling of values in a neighborhood of the points at which the values of the functional are smaller happens more frequently than in a neighborhood of worse points, and (2) the domains in which random values of variables are chosen are gradually contracted to small neighborhoods of the points with smaller values of the functional. This technique demonstrates remarkable convergence for many known multidimensional functions of irregular behavior, like Rosenbrock's function, Rastrigin function, Himmelblau's function, and many other test functions. This turns out very efficient also in the engineering problem discussed below and dealing with crack identification by the Ultrasonic (US) scanning echo-method.

4. Formulation of the crack identification problem

Let a finite array of linear cracks be located in a linear isotropic homogeneous elastic medium inside a certain bounded domain; l_n $(n = 1, \dots, N_c)$ denotes the surface of the nth crack, and $L = \bigcup_{n=1}^{N_c} l_n$ denotes the full set of cracks. Let us choose the origin of the Cartesian coordinate system somewhere in this domain, close to the cluster of cracks. In order to identify the geometry of each crack in this array, we apply a circular scanning by an US sensor which operates in the echo-regime, see figure 2. The current position of the sensor is given by the Cartesian coordinates $(R\cos\alpha, R\sin\alpha)$, and it is assumed that the amplitude of the back-scattered impulse is known for the full angular scanning: $\alpha \in (0, 2\pi)$.

Let us cite the governing equations in the problem under study. In the case of the so-

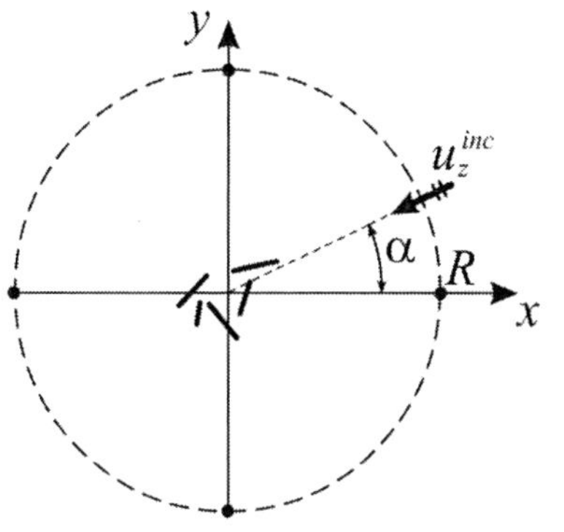

Figure 2. Circular scanning of the array of linear cracks by the US transducer in the echo-regime.

called "anti-plane" (or "shear-stress," SH) problem, this stress is directed perpendicularly to the considered plane (x, y), so that the deformation mode is identical for all cross-sections $z = \text{const}$. Then in the fixed rectangular Cartesian coordinate system $Oxyz$, the displacement vector components are $\bar{u} = \{0, 0, u_z(x, y)\}$, and this determines the two nontrivial components of the stress tensor

$$\tau_{xz}(x, y) = \mu \frac{\partial u_z(x, y)}{\partial x}, \quad \tau_{yz}(x, y) = \mu \frac{\partial u_z(x, y)}{\partial y}, \tag{4}$$

where μ is the shear elastic modulus. Under such conditions the equations of motion reduce to the single scalar Helmholtz equation for the function $u_z(x, y)$

$$\frac{\partial^2 u_z(x, y)}{\partial x^2} + \frac{\partial^2 u_z(x, y)}{\partial y^2} + k^2 u_z(x, y) = 0, \quad k = \frac{\omega}{c_s}, \tag{5}$$

where k is the wave number related to the transverse wave speed, the time-dependent factor $e^{-i\omega t}$ is omitted in all formulas, and c_s is the transverse wave speed.

The internal stress vector $\bar{T} = \{T_x, T_y, T_z\}$ over an arbitrary elemental area with the normal $\bar{n} = \{n_x, n_y, 0\}$ in the considered case of anti-plane deformation has the only nontrivial component

$$T_x = T_y = 0, \quad T_z = \tau_{xz} n_x + \tau_{yz} n_y = \mu \left(\frac{\partial u_z}{\partial x} n_x + \frac{\partial u_z}{\partial y} n_y \right) = \mu \frac{\partial u_z}{\partial n}. \tag{6}$$

Then the boundary conditions over the set of cracks L located in the medium should be satisfied over the faces of all defects which are free of load. Let us represent the full wave field as a sum of the incident and the scattered ones

$$\begin{aligned} &u_z = u_z^{inc} + u_z^{sc}(x, y), \quad u_z^{inc}(x, y) = e^{-ik(x\cos\alpha + y\sin\alpha)}, \quad w(x, y) = u_z^{sc}(x, y), \\ &T_z = T_z^{\text{inc}} + T_z^{\text{sc}}, \quad T_z^{\text{inc}} = -ik\mu(n_x \cos\alpha + n_y \sin\alpha)\, e^{-ik(x\cos\alpha + y\sin\alpha)}, \end{aligned} \tag{7}$$

where α is the angle of incidence.

It is obvious that the boundary condition for the scattered wave field implies that the crack faces are loaded by a certain tangential stress:

$$T_z\big|_L = 0 \quad \sim \quad T_z^{\text{sc}}\big|_L = -T_z^{\text{inc}}\big|_L \quad \Longrightarrow \quad \frac{\partial u_z}{\partial n}\bigg|_L = 0 \quad \sim \quad \frac{\partial w}{\partial n}\bigg|_L = -\frac{T_z^{\text{inc}}}{\mu}\bigg|_L. \tag{8}$$

In discrete formulation, in order to apply a numerical algorithm, all cracks are subdivided into I small elementary cracks. Then, by applying some known results of the classical potential theory, the diffraction problem can be reduced to a system of linear algebraic equations (SLAE) of the form

$$\sum_{j=1}^{I} K_{lj} g_j = ik(n_{lx}\cos\alpha + n_{ly}\sin\alpha)\exp[-i\,k(x_l\cos\alpha + y_l\sin\alpha)], \quad l = 1,2,\ldots,I, \tag{9}$$

where

$$K_{lj} = \tau''^{sc}_{xz}\bar{n}_l\cdot\bar{t}_j + \tau''^{sc}_{yz}\bar{n}_l\cdot\bar{n}_j, \tag{10}$$

$$\tau''^{sc}_{xz} = -\frac{iy''_l}{4}\left[K\left(\frac{\varepsilon_j}{2} - x''_l, y''_l\right) - K\left(-\frac{\varepsilon_j}{2} - x''_l, y''_l\right)\right],$$

$$\tau''^{sc}_{yz} = \frac{i}{4}\left\langle\frac{\varepsilon_j}{2}Q_y\left(\frac{\varepsilon_j}{2} - x''_l, y''_l\right) + Q_y\left(\frac{\varepsilon_j}{2} + x''_l, y''_l\right)\right. \tag{11}$$

$$\left. + \frac{2i}{\pi}\left[\frac{x''_l - \varepsilon_j/2}{(x''_l - \varepsilon_j/2)^2 + (y''_l)^2} + \frac{x''_l + \varepsilon_j/2}{(x''_l + \varepsilon_j/2)^2 + (y''_l)^2}\right]\right\rangle,$$

$$x''_l = (\bar{r}_j - \bar{r}_l)\cdot\bar{t}_j, \quad y''_l = (\bar{r}_j - \bar{r}_l)\cdot\bar{n}_j. \tag{12}$$

Here $\bar{r}_j$ is the radius vector of the jth elementary crack in the chosen Cartesian coordinate system, $\bar{n}_j$ and $\bar{\tau}_j$ are the unit normal and unit tangential vectors to this elementary crack, ε_j is its length, k is the wave number which is assumed to be fixed, and α is the angle of incidence with respect to the axis x. Besides,

$$K(x,y) = \frac{kH_1^{(1)}(k\rho)}{\rho}, \quad Q(x,y) = \frac{kH_1^{(1)}(k\rho)}{\rho} + \frac{2i}{\pi\rho^2}, \quad Q_y = \frac{\partial(yQ)}{\partial y}, \quad \rho^2 = x^2 + y^2, \tag{13}$$

where $H_1^{(1)}$ is the Hankel function. Once SLAE (9) is solved, i.e., all quantities g_j are found, the amplitude of the back-scattered far-field wave can be calculated as

$$A(\alpha) = \left|\sum_{j=1}^{I} g_j \tan[\arccos(t_{jx}\cos\alpha + t_{jy}\sin\alpha)]\sin\frac{k(t_{jx}\cos\alpha + t_{jy}\sin\alpha)\varepsilon_j}{2}\right|. \tag{14}$$

Let the system under identification be an array of linear cracks $l_n, n = 1,\ldots,N_c$. Each crack is defined by its central point with the Cartesian coordinates (a_n, b_n), by its length ζ_n, and by the angle of slope θ_n ($|\theta_n| \le \pi/2$) with respect to the axis x. If the nth crack contains J_n elementary cracks of length $\varepsilon_n = \zeta_n/J_n$, then the full set of elementary cracks is a union of ones when running over the given system of linear cracks:

$$\begin{cases} x_i = a_n + \left[\varepsilon_n(q-0.5) - \dfrac{\zeta_n}{2}\right]\cos\theta_n, \\ y_i = b_n + \left[\varepsilon_n(q-0.5) - \dfrac{\zeta_n}{2}\right]\sin\theta_n, \end{cases} \quad (q = 1,\ldots,J_n; n = 1,\ldots,N_c). \tag{15}$$

Coming to the formulated identification problem, let us estimate the total number of unknown parameters to be reconstructed. If the number of linear cracks is N_c, then one has four unknown parameters for each of them: a_n, b_n, ζ_n, θ_n ($n = 1,\ldots,N_c$). Therefore, the total number of the unknown parameters to be reconstructed is $4N_c$.

IOP Conf. Series: Journal of Physics: Conf. Series **991** (2018) 012014 doi:10.1088/1742-6596/991/1/012014

In order to find all these unknowns, we construct an objective functional, and reduce the reconstruction to an optimization problem for this functional, as discussed in section 2. For this, we represent the SLAE (9) in operator form

$$\mathbf{K}g = f, \quad \mathbf{K} = \{K_{lj}\}, \quad g = \{g_l\}, \quad f = \{f_l\}, \quad (l, j = 1, \ldots, I), \tag{16}$$

whose solution can be expressed in terms of the inverse matrix as

$$g = \mathbf{K}^{-1} f \quad \Longrightarrow \quad g_l = (\mathbf{K}^{-1} f)_l. \tag{17}$$

Obviously, the right-hand side of Eq. (17) depends on all $4N_c$ unknown parameters, as well as on the angle of incidence α: $\mathbf{K}^{-1}f = (\mathbf{K}^{-1}f)(a_n, b_n, \zeta_n, \theta_n, \alpha)$.

Now, let us pass to the question: What is the measured information which can be used as the input data for the inverse identification problem? We assume that the scanning US sensor can measure the amplitude of the echo-impulse at a fixed distance R, in a far zone, for M positions of the sensor corresponding to the values of the irradiation angle $\alpha = \alpha_m$ $(m = 1, \ldots, M)$. To be more specific, we assume that the values α_m are uniformly distributed over the full circular interval $(0, 2\pi)$. The registered values of the back-scattered amplitude form an array of the input data F_m, $(m = 1, \ldots, M)$. In our numerical experiments, the quantities F_m may be taken as the amplitude $A(\alpha)$ calculated from the solution of the respective direct problem: $F_m = A(\alpha_m)$, see Eq. (14). Then, by substituting (17) into (14), one comes to the system of nonlinear equations for the parameters a_n, b_n, ζ_n, θ_n written in discrete form

$$\left| \sum_{j=1}^{I} (\mathbf{K}^{-1}f)(a_n, b_n, \zeta_n, \theta_n, \alpha_m) \tan[\arccos(t_{jx}\cos\alpha_m + t_{jy}\sin\alpha_m)] \right.$$
$$\left. \times \sin\left[\frac{k(t_{jx}\cos\alpha_m + t_{jy}\sin\alpha_m)\varepsilon_j}{2}\right] \right| = F_m^* \quad (m = 1, \ldots, M). \tag{18}$$

5. Numerical algorithm

The system of equations (18) can be resolved by minimizing the discrepancy functional, as discussed in section 2:

$$\begin{aligned}
&\min[\Omega(a_n, b_n, \zeta_n, \theta_n)], \\
&\Omega(a_n, b_n, \zeta_n, \theta_n) = \left\| \left| \sum_{j=1}^{I} (\mathbf{K}^{-1}f)(a_n, b_n, \zeta_n, \theta_n, \alpha_m) \tan[\arccos(t_{jx}\cos\alpha_m + t_{jy}\sin\alpha_m)] \right.\right. \\
&\quad \left.\left. \times \sin\left[\frac{k(t_{jx}\cos\alpha_m + t_{jy}\sin\alpha_m)\varepsilon_j}{2}\right] \right| - F_m^* \right\|^2 \\
&= \sum_{m=1}^{M} \left\{ \left| \sum_{j=1}^{I} (\mathbf{K}^{-1}f)(a_n, b_n, \zeta_n, \theta_n, \alpha_m) \tan[\arccos(t_{jx}\cos\alpha_m + t_{jy}\sin\alpha_m)] \right.\right. \\
&\quad \left.\left. \times \sin\left[\frac{k(t_{jx}\cos\alpha_m + t_{jy}\sin\alpha_m)\varepsilon_j}{2}\right] \right| - F_m^* \right\}^2 .
\end{aligned} \tag{19}$$

The algorithm in such a form was realized in [13]. This shows a brilliant precision in reconstruction of the inclination angles θ and crack length ζ for each crack. The position of the cracks forming the array is identified incorrectly. This can easily be explained by the fact that the far-field scattered wave field, see expression (14), is free from the coordinates x_j, y_j and depends only on the angle of incidence. It is obvious that one cannot identify the real

Table 1. Identification of the single crack (1), see figure 3.

Configuration	a, mm	b, mm	ζ, mm	θ
Exact	6.000	−3.000	12.000	$-\pi/9 = -0.349$
One crack	**6.000**	**−3.003**	**11.984**	**−0.349**
Two cracks	5.998	−2.996	11.993	−0.349
	−18.596	15.354	17.611	0.521
Three cracks	6.002	−2.998	11.991	−0.348
	4.026	0.017	11.428	0.177
	10.337	−5.570	14.357	0.745
Four cracks	6.001	−3.011	11.999	−0.348
	−9.814	0.318	13.572	−0.149
	−10.578	8.882	10.376	0.918
	−7.056	−3.159	7.897	0.070

position of a reflector by using only the information from the far-field zone. In order to improve the identification of the position of cracks, i.e., the parameters a and b, let us involve the information about the "time-of-flight" of the US impulse. The latter is always reflected in the echo-regime from the point which is the nearest to the sensor; hence this information gives the value of the most distant point on the cracks in each direction of incidence, i.e., the quantity d_m,

$$d_m = \max_j\{x_j \cos\alpha_m + y_j \sin\alpha_m\}. \tag{20}$$

Therefore, it is reasonable to add an analogous discrepancy functional of the operator relation (20) to the functional (19):

$$\Omega_d = \sum_{m=1}^{M} \left(\max_j\{x_j \cos\alpha_m + y_j \sin\alpha_m\} - d_m \right)^2. \tag{21}$$

The final functional is thus constructed as a sum of expressions presented in equations (19) and (21) with appropriate weights. Numerous tests show that, to provide good precision for all geometric parameters under identification, the weight in front of functional (21) should be at least by an order of magnitude greater than the one for functional (19),

For multiple crack array, the algorithm operates so that it selects the most likely geometry, sequentially for the number of cracks $N_c = 1, 2, 3, 4$. Then, in the set of these four geometries, the algorithm chooses the best value of the functional Ω to come to the true solution including the recognition of the true number of cracks by itself.

Let us test the proposed method with some examples of the simultaneous identification of several cracks. If one applies US probes with the cyclic frequency $f = 1\,\mathrm{mHz}$ and the wave speed in the medium is 6 km/s, then the wave length is $\lambda = 6 \times 10^3/1 \times 10^6 = 6 \times 10^{-3}\,\mathrm{m} = 6\,\mathrm{mm}$.

For all examples listed below the input data for the reconstruction are taken from the solution of the respective direct problem. In our simulation, we always use $M = 72$ points of measurements uniformly distributed over the total circular interval of the polar angle $\alpha_m \in (0, 2\pi)$, $m = 1, \ldots, M$, of incidence, with the step $360°/72 = 5°$. For all examples demonstrated below, the maximum possible number of cracks in the cluster is taken $N_c = 4$. The real number of cracks is an additional parameter to be identified. Some examples of the

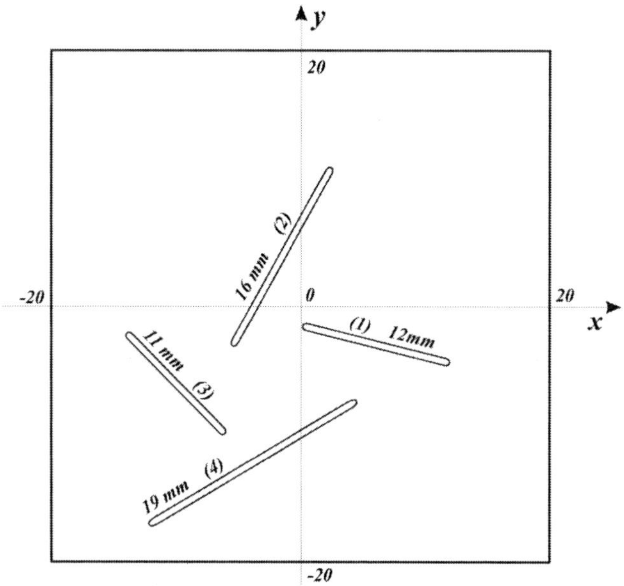

Figure 3. Array of four cracks to be identified in the $(-20, 20) \times (-20, 20)$ mm quadratic domain.

Table 2. Identification of two cracks (1) and (2), see figure 3.

Configuration	a, mm	b, mm	ζ, mm	θ
Exact	6.000	−3.000	12.000	$-\pi/9 = -0.349$
	−1.500	4.000	16.000	$\pi/3 = 1.047$
One crack	3.140	0.974	18.278	1.050
Two cracks	**6.101**	**−2.958**	**11.906**	**−0.348**
	−1.483	**4.000**	**15.990**	**1.047**
Three cracks	6.013	−3.013	11.930	−0.349
	−1.510	3.994	16.054	1.048
	−16.346	−13.390	16.681	−1.506
Four cracks	5.995	−2.986	12.030	−0.350
	−1.492	4.011	16.021	1.047
	−2.380	4.088	18.244	−0.976
	9.366	5.367	15.416	−0.006

reconstruction are presented in Tables 1–4. All sizes are given in mm. The configuration of the system of cracks is shown in figure 3.

Conclusions

1. The number of cracks, parameter N_c, for any configuration, as a rule, is identified correctly. In practice, this property takes place for both good and rough precision of the identification.
2. Following the conclusions of [13], one can see that, when ignoring the time-of-flight factor, i.e. with identification by minimizing functional (19) only and ignoring functional (21), the

Table 3. Identification of three cracks (1), (2), and (3), see figure 3.

Configuration	a, mm	b, mm	ζ, mm	θ
Exact	6.000	−3.000	12.000	$-\pi/9 = -0.349$
	−1.500	4.000	16.000	$\pi/3 = 1.047$
	−10.000	−6.000	11.000	$-\pi/4 = -0.785$
One crack	−0.711	−0.728	20.000	−0.157
Two cracks	7.808	−3.215	9.093	−0.540
	−7.253	2.445	19.996	1.015
Three cracks	**6.424**	**−3.022**	**11.126**	**−0.353**
	−1.393	**4.139**	**15.567**	**1.051**
	−9.710	**−5.988**	**11.729**	**−0.789**
Four cracks	6.088	−3.083	11.547	−0.333
	−0.985	4.700	14.254	1.045
	−10.079	−6.034	10.589	−0.775
	−4.629	3.940	13.068	0.594

Table 4. Identification of four cracks (1), (2), (3), and (4), see figure 3.

Configuration	a, mm	b, mm	ζ, mm	θ
Exact	6.000	−3.000	12.000	$-\pi/9 = -0.349$
	−1.500	4.000	16.000	$\pi/3 = 1.047$
	−10.000	−6.000	11.000	$-\pi/4 = -0.785$
	−4.000	−12.200	19.000	$\pi/6 = 0.524$
One crack	−2.290	−3.822	20.000	0.593
Two cracks	4.901	4.072	19.797	−0.690
	−5.382	−11.338	19.566	0.500
Three cracks	6.034	−0.260	14.171	−0.710
	−4.293	5.820	17.991	0.544
	−8.120	−10.664	15.425	1.007
Four cracks	**5.861**	**−3.063**	**12.193**	**−0.348**
	−1.652	**3.978**	**16.149**	**1.043**
	−10.162	**−5.988**	**10.616**	**−0.779**
	−4.183	**−12.294**	**18.555**	**0.525**

length of the crack (parameter ζ) and the inclination angle (parameter θ) are identified quite precisely. The Cartesian coordinates of the cracks cannot be correctly identified with this particular technique, which takes into account only the back-scattered amplitude of the US impulse. This is quite natural from the physical point of view when the impulse amplitude is measured in the far-field zone, where the value of the registered signal is insensitive to the real position of a defect.

3. It can also be seen from the presented Tables 1–4 that, by involving the time-of-flight

information, one can attain more precise identification of cracks' location too. This is also physically natural, since the time-of-flight information is directly connected with the distance between a defect and the sensor.

4. It is very interesting to notice that even with an incorrect number of cracks, the identification, as a rule, contains the information about some true cracks. This can easily be seen on example of table 2, where the first crack among others, for "Three cracks" identification, has all four reconstructed parameters very close to the one for the first of the two real cracks.
5. It should be finally noted that in some extreme situations the identification cannot in principle be performed precisely. We mean the cases when a certain crack is hidden in an acoustic "shadow" of other cracks, like a crack located between a pair of two long parallel cracks.

Acknowledgments

The second author thanks the Russian Science Foundation for the support by Project No. 15-19-10008.

References

[1] Borne P, Popescu D, Filip F G and Stefanoiu D 2012 *Optimization in Engineering Sciences: Exact Methods* (London: Wiley-ISTE)
[2] Stefanoiu D, Borne O, Popescu D, et al. 2014 *Optimization in Engineering Sciences: Metaheuristic, Stochastic Methods and Decision Support* (London: Wiley-ISTE)
[3] Parkinson A R, Balling R J and Hedengren J D 2013 *Optimization Methods for Engineering Design. Applications and Theory* (Brigham: Brigham Young University)
[4] Gill P E, Murray W and Wright M H 1981 *Practical Optimization* (Academic Press) p 402
[5] Zhigljavsky A A 1991 *Theory of Global Random Search* (Springer)
[6] Ramm A G 1986 *Scattering by Obstacles* (Dordrecht: Reidel Publish.)
[7] Natterer F 1986 *The Mathematics of Computerized Tomography* (Stuttgart: John Wiley)
[8] Colton D and Kress R 1992 *Inverse Acoustic and Electromagnetic Scattering Theory* (Springer-Verlag)
[9] Sumbatyan M A Scalia A 2005 *Equations of Mathematical Diffraction Theory* (CRC Press)
[10] Brigante M and Sumbatyan M A 2010 Application of the Inverse Diffraction Technique to the Problem of Reconstruction of Complexly Shaped Flaws by Ultrasonic Method *Russ. J. NDT* **46** 98–111
[11] Brigante M and Sumbatyan M A 2010 An efficient numerical algorithm for crack reconstruction in elastic media by the circular US scanning *Inv. Probl. Sci. Engng* **18** (3) 361–79
[12] Corana M, Marchesi M, Martini C, and Ridella S 1987 Minimizing Multimodal Functions of Continuous Variables with the "Simulated Annealing" Algorithm *ACM Trans. Math. Software* **13** (3) 262–80
[13] Sumbatyan M A and Brigante M 2017 Identification of Arrays of Cracks in the Elastic Medium by the Ultrasonic Scanning In *Wave Dynamics and Composite Mechanics for Microstructured Materials and Metamaterials. Advanced Structured Materials* **59** 71–89

IOP Conf. Series: Journal of Physics: Conf. Series **991** (2018) 012015 doi:10.1088/1742-6596/991/1/012015

Explosive compaction of aluminum oxide modified by multiwall carbon nanotubes

A E Buzyurkin[1*], E I Kraus[1], and Ya L Lukyanov[2]

[1]Khristianovich Institute of Theoretical and Applied Mechanics of the Siberian Branch of the Russian Academy of Sciences, Novosibirsk, Russia

[2]Lavrentyev Institute of Hydrodynamics of the Siberian Branch of the Russian Academy of Sciences, Novosibirsk, Russia

E-mail: *buzjura@itam.nsc.ru

Abstract. This paper presents experiments and numerical research on explosive compaction of aluminum oxide powder modified by multiwall carbon nanotubes (MWCNT) and modeling of the stress state behind the shock front at shock loading. The aim of this study was to obtain a durable low-porosity compact sample. The explosive compaction technology is used in this problem because the aluminum oxide is an extremely hard and refractory material. Therefore, its compaction by traditional methods requires special equipment and considerable expenses.

1. Introduction

Ceramics are usually chemically inert, hard, refractory and have a relatively low density in comparison with metals. Therefore, they form an interesting group of materials for high-tech applications. Hardness and refractivity exclude the possibility of their processing by casting, forging and machining, as is usually the case for metals and many polymers.

Instead, the technical ceramics are usually synthesized in the form of powders and further processed by precompiling them in green form and then sintering at high temperature in the furnace for their final compaction. A high-temperature process is usually required, since sintering is a process controlled by diffusion. The growth of grain and the shift of the chemical composition to a more thermodynamically stable one are inevitable during this process.

Thus, materials consisting of metastable phases and very small size particle (nanocrystalline, high-alloy and amorphous powders) cannot be processed using sintering without some loss of their original, often unique properties. Also, materials that have a low self-diffusion coefficient or decompose at high temperatures cannot be processed in this way. For these materials, as well as for unintentionally reacting material mixtures, alternative methods of compaction should be used. One of such methods is shock or explosive compaction of powders, in which the compaction of the powder occurs in a shock wave.

In a number of works, composite ceramic materials based on nano-sized oxides Al_2O_3, TiO_2, ZrO_2, MgO modified with carbon nanotubes were studied [1–6]. They showed that the insertion of small amounts of additions of carbon nanotubes (1–10 mass%) into dielectric ceramics drastically changes its electromagnetic properties leading to an increase in conductivity by several orders of magnitude. This opens the possibility of using such composite materials along with chemical, thermal and mechanical ones. It is necessary to provide certain electrophysical

IOP Conf. Series: Journal of Physics: Conf. Series **991** (2018) 012015 doi:10.1088/1742-6596/991/1/012015

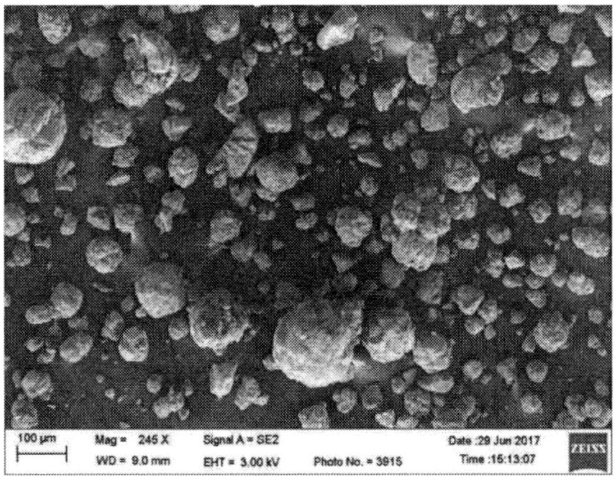

Figure 1. Original Al_2O_3 powder.

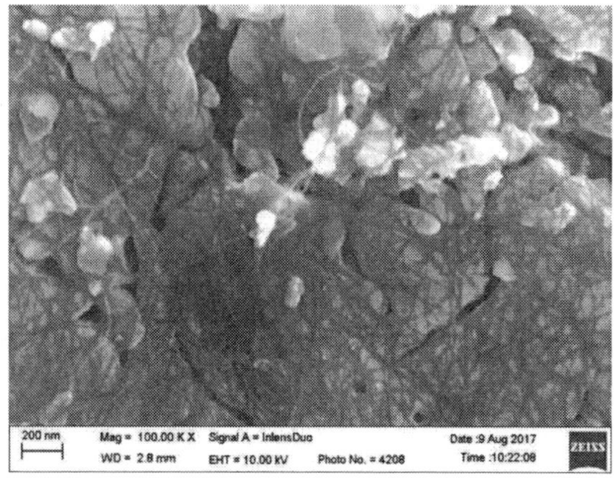

Figure 2. Initial mixture of MWCNT–Al_2O_3.

characteristics. At the same time, different values of conductivity are required for different applications. In this paper, an attempt was made to obtain a composite material based on Al_2O_3 ceramics modified by multiwall carbon nanotubes by explosive compaction.

2. Explosive compaction of Al_2O_3 modified by multiwall carbon nanotubes

Experiments have been carried out on explosive compaction of ceramic powder Al_2O_3, as well as its mixtures with carbon nanotubes. The powder, the mixture and the structure of the explosive compacts were examined with the MERLIN Compact scanning microscope. The original Al_2O_3 powder had granules of size of $10 \div 100\,\mu$m (figure 1).

The bulk density was 1.56 g/cm^3 or +40% of the theoretical maximum density (TMD). Multi-walled carbon nanotubes were supplied as an aqueous suspension containing 0.2% nanotubes by weight. For the experiment, 60 g of the powder were mixed with 40 ml of the suspension, then the mixture was dried and processed in a ball mill for 1 hour. Thus, the content of MWCNT in the mixture was 0.13% by weight. Figure 2 shows the mixture after drying.

Further, the pure Al_2O_3 powder and the mixture were compacted by explosion in ampoules of conservation according to the standard cylindrical scheme (figure 3) [7–9]. The explosive loading process using the standard cylindrical configuration is performed by plastic deformation of the container along with porous material therein at detonation of contact explosive charge in the mode of running load (figure 4). The shock wave propagates in the loaded sample with a great speed and determines the boundary between the compressed and the undisturbed material. At that, the position of the shock wave varies with time. Due to the radial component and the cylindrical geometry, the shock wave decays as it spreads in the powder. The area on which the shock wave acts reduces and hence the energy density increases.

As the explosive we used the mixture of ammonite with soda 2 : 1 by weight. The detonation velocity was measured by an electrocontact method and amounted to 2830 m/s. After compacting, the samples were heat-treated at 900°C for 3 hours. This sintering temperature was insufficient, so the compacts had a low strength and a porous structure. The compact density was 3.5 ± 0.1 g/cm^3 or 86% of the TMD. Figure 5 shows the microstructure of a compact of pure Al_2O_3 after explosive loading.

The microstructure of mixture of the Al_2O_3 with carbon nanotubes after explosive loading is shown in figure 6.

Also from compacts, samples of size of $5 \times 5 \times 5$ mm were cut out and their electrical resistance was measured by the LCR-meter Instec LCR-816. For a compact of pure Al_2O_3, it

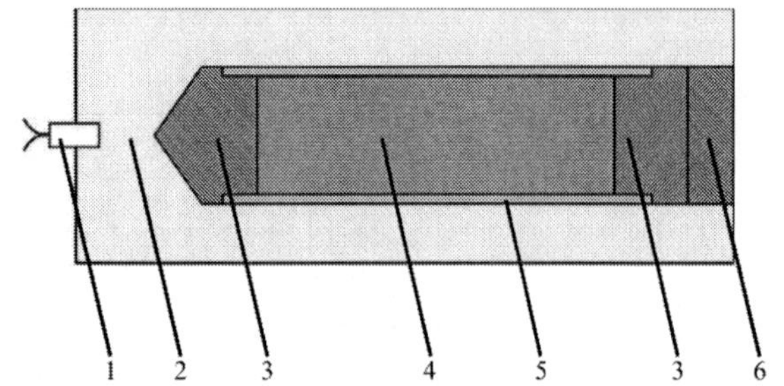

Figure 3. Standard cylindrical configuration: 1 — detonator, 2 — explosive, 3 — stoppers, 4 — container, 5 — powder, 6 — spall element.

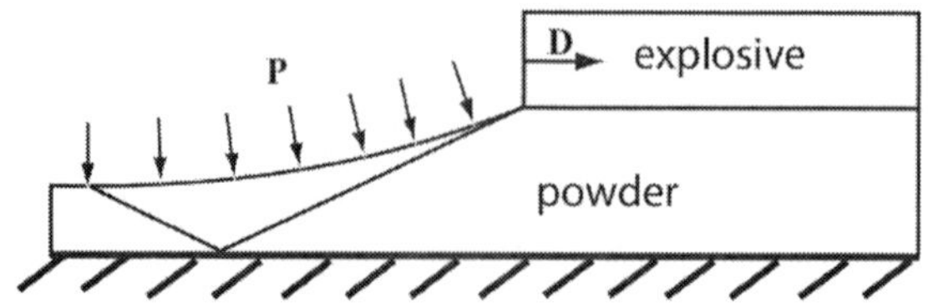

Figure 4. Statement of the problem according to experimental scheme.

Figure 5. The microstructure of a compact of pure Al_2O_3.

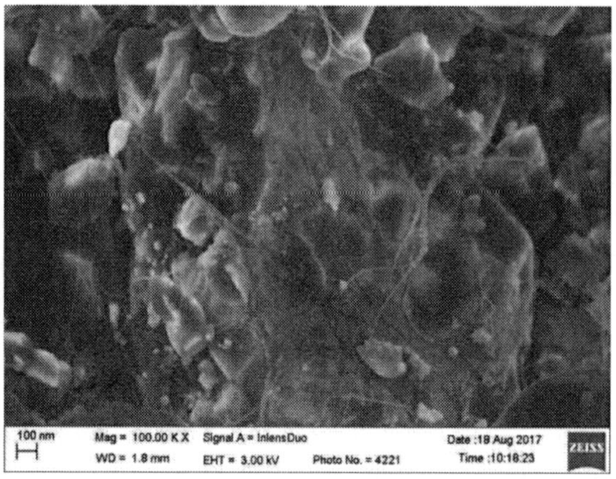

Figure 6. Microstructure of MWCNT–Al_2O_3 composite.

was 62 Megaohm, and for MWCNT–Al_2O_3 composite, it was 3.4 kiloohm, that is, four orders of magnitude smaller.

3. Calculation results and discussion

For the numerical simulation of propagation of shock waves, a complete system of equations of deformation of the porous elastic-plastic material was solved [10]. The methods for numerical modeling of explosive loading of porous materials are described in detail in [11,12]. In this paper, we used the few-parameter equation of state adequately describing the physics of collision at high pressures and temperatures [13–15], which made it possible to calculate the shock-wave processes with a minimal number of physical parameters as the initial data. The geometrical dimensions and the values of physical parameters correspond to the experimental data mentioned above.

With the results obtained in [13,14], the Hugoniot for aluminum oxide have been constructed (figure 7). The presented comparison of the results obtained in this study with known experimental data [16–23] shows that the error in calculating of the parameters of the Hugoniot does not exceed 5–10%.

Joint theoretical and experimental studies allowed us to implement an approach based on mathematical and physical simulation of shock-wave loading of powdered materials. The Hugoniot curve of the aluminum oxide was constructed. Thus, the technology of explosive compaction was used to obtain compact samples of aluminum oxide. The above research allows us to draw the following conclusions:

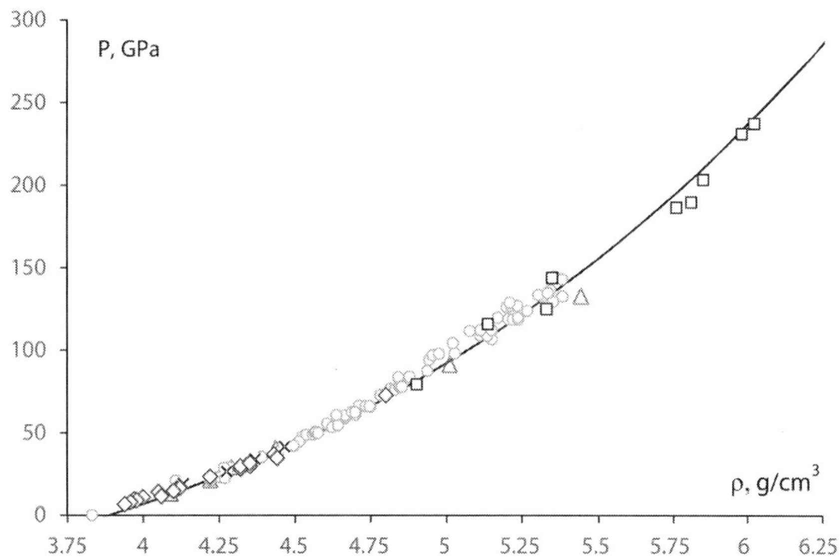

Figure 7. The Hugoniot curve for aluminum oxide. Experimental data: △ – [22], □ – [24], ○ – [25], × – [26], ◊ – [27].

- When the aqueous suspension is a mixture of multiwall carbon nanotubes and Al_2O_3 powder, the distribution of nanotubes is fairly uniform.
- It is possible to obtain a composite Al_2O_3 modified by multiwall carbon nantubes by the explosive method followed by sintering.
- Even in the case of a low content of multiwall carbon nantubes in compacts (0.13%), the electrical conductivity increases by four orders of magnitude.

Acknowledgments

This work is supported by the Russian Foundation for Basic Research (grant No. 16-08-00244). The Hugoniot for aluminum oxide have been constructed with support by the Russian Science Foundation (grant No. 16-19-10300).

References

[1] Sarkar S and Das P K 2014 Processing and properties of carbon nanotube/alumina nanocomposites: A review *Rev. Adv. Mater. Sci.* **37** (1–2) 53–82
[2] Kim D I, Vu D-T, Zhang T, et al. 2014 Transition in micro/nano-scale mechanical properties of ZrO2/multi-wall carbon nanotube composites *J. Ceramic Soc. Japan* **122** (1432) 1028–31
[3] Berki P, Reti B, Terzi K, et al. 2014 The effect of titania precursor on the morphology of prepared TiO2/MWCNT nanocomposite materials *Physica Status Solidi B* **251** (12) 2384–8
[4] Taleshi F and Hosseini A A 2012 Synthesis of uniform MgO/CNT nanorods by precipitation method *J. Nanostruct. Chem.* **3** (1) 4
[5] Yamamoto G, Shirasu K, Nozaka Y, et al. 2014 Microstructure-property relationships in pressureless-sintered carbon nanotube/alumina composites *Mater. Sci. Engng A* **617** 179–86
[6] Shutilov R A, Myz A L, Kuznetsov V L, and Karagedov G R 2016 Current conductive Al2O3 ceramic composites modified by multiwall carbon nanotubes *Perspektivn. Mater.* **8** 64–73
[7] Buzyurkin A E, Kraus E I, and Lukyanov Ya L 2015 Explosive compaction of WC+Co mixture by axisymmetric scheme *J. Phys. Conf. Ser.* **653** 012036
[8] Buzyurkin A E, Kraus E I, and Lukyanov Ya L 2016 Dynamic compaction of boron carbide by a shock wave *AIP Conf. Proc.* **1770** 030091
[9] Buzyurkin A E, Kraus E I, and Lukyanov Ya L 2013 Study of the conditions of fracture at explosive compaction of powders *Frattura ed Integrita Strutturale* **24** 102–11

[10] Fomin V M and Kiselev S P 1997 Elastic-plastic waves in porous materials In *High-Pressure Shock Compression of Solids. Response of Highly Porous Solids to Shock Loading* ed L Davison, Y Horie, and M Shahinpoor (New York - Berlin - Heidelberg: Springer-Verlag) pp 205–232

[11] Buzjurkin A E and Kiselev S P 2000 On appearance of "cold" layer in explosive consolidation of powders *Shock Waves* **10** (3) 159–65

[12] Buzyurkin A E and Kiselev S P 2000 Appearance of a "cold" layer upon explosive compacting of powders *J. Appl. Mech. Tech. Phys.* **41** 176–81

[13] Fomin V M, Kraus E I, and Shabalin I I 2004 An equation of state for condensed matter behind intense shockwaves *Mater. Phys. Mech.* **7** (1) 23–8

[14] Kraus E I and Shabalin I I 2013 Impact loading of a space nuclear powerplant *Frattura ed Integrita Strutturale* **24** 138–50

[15] Fomin V M, Kraus E I, and Shabalin I I 2010 A few-parameter equation of state of the condensed matter and its application to the impact problems In *EPJ Web Conf.* vol 10 p 00027

[16] Wilkins M L 1968 Third progress report of light armor program In *Tech. Rep. UCRL-50460 Lawrence Livermore National Laboratory* (Livermore, CA)

[17] Taniguchi T, Kondo K, and Sawaoka A 1986 Shock waves profiles in powder compacts of α-Al_2O_3 In *Shock Waves in Condensed Matter* ed Y M Gupta (Boston, MA: Springer US) pp 773–8

[18] Grady D E 1994 Shock-wave strength properties of boron carbide and silicon carbide *J. Phys. IV* **4** (C8) 385–91

[19] McQueen R G, Marsh S P, Taylor J W, et al. 1970 The equation of state of solids from shock wave studies In *High-Velocity Wave Phenomena* ed R Kinslow (New York: Academic Press) pp 293–417

[20] Zhang Y, Mashimo T, Uemura Y, et al. 2006 Shock compression behaviors of boron carbide (B_4C) *J. Appl. Phys.* **100** (11) 113536

[21] Vogler T J, Reinhart W D, and Chhabildas L C 2004 Dynamic behavior of boron carbide *J. Appl. Phys.* **95** (8) 4173

[22] Pavlovskii M N 1971 Shock compressibility of six very hard substances *Sov. Phys. Solid State* **12** (7) 1736–7

[23] Gust W H and Royce E B 1971 Dynamic yield strengths of B_4C, BeO, and Al_2O_3 Ceramics *J. Appl. Phys.* **42** (1) 276

[24] Erskine D 1994 High pressure Hugoniot of sapphire In *AIP Conf. Proc.* vol 309 (AIP) pp 141–143

[25] Marsh S P 1980 *LASL shock Hugoniot data* (Univ of California Press)

[26] Grady D E 1995 Dynamic properties of ceramic materials, Sendia National Laboratories Report, SAND 94-3266 *Sendia National Laboratories, Albuquerque, NM*

[27] Mashimo T, Hanaoka Y, and Nagayama K 1988 Elastoplastic properties under shock compression of Al_2O_3 single crystal and polycrystal *J. Appl. Phys.*, **63** (2) 327

IOP Conf. Series: Journal of Physics: Conf. Series **991** (2018) 012016 doi:10.1088/1742-6596/991/1/012016

Determination of residual stresses in objects at their additive manufacturing by layer-by-layer photopolymerization method

P S Bychkov, A V Chentsov, V M Kozintsev, and A L Popov*

Ishlinsky Institute for Problems in Mechanics of the Russian Academy of Sciences, Moscow, Russia

E-mail: *popov@ipmnet.ru

Abstract. A calculation-experimental technique is developed for identification of the shrinkage stresses generated in objects after their additive manufacturing by layer-by-layer photopolymerization. The technique is based on the analysis of shrinkage deformations at bending occurring in a series of samples in the form of plates-stripes with identical sizes, but with different time of polymerization which is predetermined during their production on the 3D printer.

1. Introduction

Today we witness the rapid development of additive manufacturing technologies as they enable quick and relatively cheap reproduction of spatial objects of as much as irregular shape (and theoretically from any material) with high precision [1–3]. Additive manufacturing technologies include stereolithography, electrolytic precipitation, winding, selective laser agglomeration, electron beam melting, sputtering, 3-dimensional printing, lamination and some other. Thousands of works on this topic has been published by technologists, physicists and chemists. As a rule, the technological aspects of production and properties of final objects are discussed in these works. However, it is difficult to find works describing the reasons why the products gain these or those properties, other than as-designed and how these properties can be corrected, having changed parameters of the manufacturing process. From the point of view of mechanics of additive manufacturing, main problems still remain — shrinkage stresses, deviation of objects final form from the expected design, and also – strength, stability, and durability of final products.

Special problem of additive manufacturing processes is the experimental verification of mathematical models of the produced parts and constructions. Numerous theoretical studies, in particular, show that a necessary condition for formulation of the self-contained model of layer-by-layer (discrete) or continuous process of additive manufacturing is knowing the deformations in the developed elementary layer of finite or infinitesimal thickness (see, for example, [4–8]). Now the most clear and reliable way to determine these data is performing experimental analysis.

In this paper we study shrinkage of objects made of polymeric materials made layer-by-layer using 3D printer. One of the main problems of such process is the essential deviation of final form of the objects from expected design. To overcome this, the companies-producers of 3D-printers use special substrates which prevent of shrinkage of the material when printing the object. However, at the same time, shrinkage stresses are formed in the object, which cause deformations after separating the object from the substrate, distorting its designed dimensions. A simple mechanical-mathematical model for calculation the shrinkage stresses formed during layer-by-layer photopolymerization of a liquid substance at its irradiation by an ultra-violet light source through a projector matrix is given below.

2. Formation of shrinkage stresses at 3D printing

We use Master EV series 3D printer which is a device of a widespread DLP type [2]. This type of printers has a high resolution — 1920 by 1080 points. The consumable material for DLP technology printing is liquid photopolymer. The process goes on as follows. The DLP printer projects the image of a digital model onto a thin layer of photopolymer and exposes it to light for the time predetermined by the operator. During exposure time the polymerization takes place, curing the lit sites of photopolymer. Then the substrate with the hardened layer of photopolymer rises by a height of a layer, a new layer of a liquid substance is put and the image of the following layer of an object is projected onto the object. This process is repeated until the last layer of the object is deposited and hardened.

Owing to the fact that during polymerization of a liquid substance the polymeric chains are formed that reduces the distance between monomer molecules, the hardening material tends to shrinkage which is prevented by adhesion forces between the material and substrate. As a result, the shrinkage stresses are formed in the object. Upon separation of the printed object from the substrate the shrinkage stresses can create the deformations distorting the design sizes of an object.

3. Modeling shrinkage stresses

A theoretical description of shrinkage deformations relation to the stresses causing them at 3D printing can be obtained using a model of a non-central bending of an elastic plate-stripe [9–11]. Such representation is possible due to the fact that each subsequent layer of the object obtains unequal radiation dose, as compared to the previous layer, that is caused by light-transparency of liquid substance and the solid material, formed from it. As a result, the first layer adjoining the substrate gets harder than the second one, second layer gets harder than the third, etc. We observe that the elastic modulus of the cured material increases during its exposure to light. At the same time elastic modulus increase leads to intense shrinkage of material [13, 14].

Thus, considering process of 3D printing by layer-by-layer hardening of the object in the form of a plate-stripe, rigidly bonded with substrate, the non-uniform stress field will be formed through the thickness of the plate which will force it to be bent after plate's removal from the substrate. Let's assume that shrinkage of such plate is described by a model of pure bending of the stripe in the direction of a normal to the substrate under the influence of the boundary moments which occur because of uneven distribution of shrinkage stresses along stripe's thickness. Figure 1 qualitatively illustrates this idea of shrinkage stresses action.

In Figure 1 *a* there is a scheme of the process of multilayered plate manufacturing on a 3D printer: the plate is limited to ABCD rectangle, slanted dashed lines above the plate denote the substrate; below the growing layered plate there is a tank with liquid photopolymer which surface is adjoining to the last hardened layer of the plate; the direction of light radiation is shown by a dark sector and arrows. In figure 1 *b* there is a plate after separation from the substrate being bent by shrinkage stresses in the direction of a normal to the substrate with the triangular distribution of shrinkage stresses σ_{sm} along the thickness of a plate on its edges, as assumed according to manufacturing technique, causing its bending.

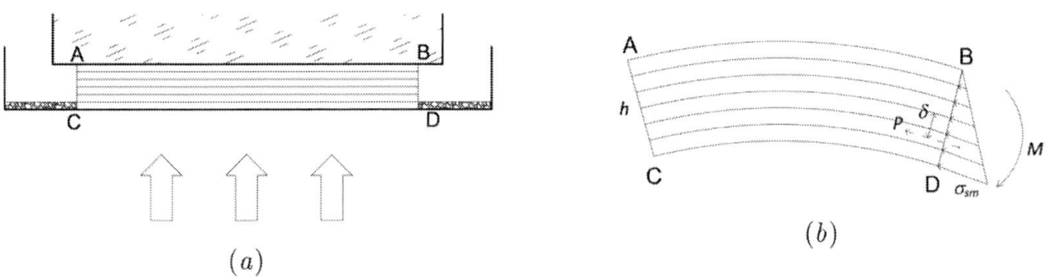

Figure 1.

To determine the stresses $\sigma_s(y)$ we will consider shrinkage deformation of the plate-stripe removed from the substrate by a model of pure bending of a beam. The equation relating curvature $1/\rho$ of a neutral layer of a beam of rectangular transverse section with the edge bending moments M has the form of

$$\frac{1}{\rho} = \frac{M}{EJ}, \quad J = \frac{bh^3}{12}, \tag{1}$$

where E is the elastic modulus, b and h are the width and thickness of transverse section of a beam, respectively [14].

Value of the bending moment M can be related to the maximum value of the given shrinkage stresses $\sigma_{sm} = \max \sigma_s(y) = \sigma_s(h/2)$ through the resulting axial force P and arm δ, defining the displacement of a point of application of the resulting force (the center of gravity of a rectangular triangle with legs h and σ_{sm}) from the location of a beams neutral axis (figure 1 c):

$$M = P\delta, \quad P = \frac{1}{2}\sigma_{sm}bh, \quad \delta = \frac{h}{6}. \tag{2}$$

Comparing 1 and 2, we will obtain simple expression for the maximum value of the given shrinkage stress containing only the elastic modulus, thickness and curvature radius of a beam's central line

$$\sigma_{sm} = E\frac{h}{\rho}. \tag{3}$$

4. Experimental modeling of shrinkage deformations

For experimental modeling of deformations caused by shrinkage stresses at layer-by-layer 3D printing several series of rectangular strips with identical in-plane sizes were made: 50 mm long, 10 mm wide, differing in number of layers — 2 or 7 and time of a light exposure of separate layers — 5, 10 and 15 seconds. To evaluate the mechanical characteristics of the produced materials the standard H-shaped samples with working part 2 mm wide and 20 mm long with the number of layers and times of light exposure being same as the one for the rectangular stripes, were printed at the same time. In-plane photos of one of H-shaped test sample with reference scale on the ruler is given in figure 2.

For the chosen layer thickness of about 0.1 mm, single-layer samples could not be separated from the substrate without getting damaged. Therefore two-layer samples were the samples of minimum thickness. Samples with seven layers were also chosen for study considering the fact that their cylindrical bending due to shrinkage deformation, remaining visually noticeable, occurred along the axis parallel to short side of a strip.

After separation from substrate the samples immediately deformed due to shrinkage stresses action, and further polymerization at sunlight or under special UV-lamp only enhanced this

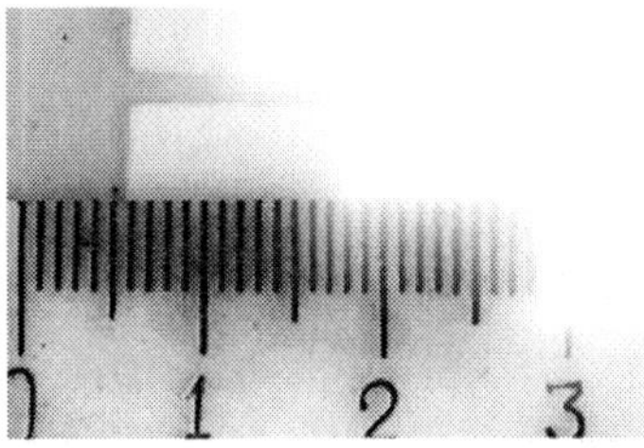

Figure 2.

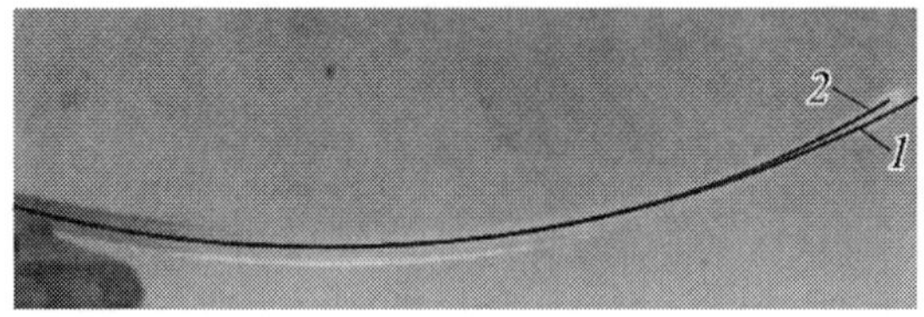

Figure 3.

deformation. This observation corresponds to observations of other authors at a bit different kind of photopolymerization — volumetric polymerization [14].

We found that most evident cases for determining curvature of the strips bent by shrinkage deformations are the samples with minimum time of light exposure. One of such bent strips is given in figure 3 where 7-layers sample gets bent after 5-second light exposure of each layer.

Lines *1* and *2* in figure 3 approximate the bending form of the strip by a circle and a curve of the fourth order respectively. Apparently, approximation by a circle which follows deflection at a pure bending of a beam describes the real bending of the hardened photopolymer strip removed from a substrate due to shrinkage stresses quite adequately. As a result, by the known arch length of a curved strip we can find its curvature radius. In particular, for arch length of 50 mm the radius of approximating circle turns out to be 55 mm. Average thickness of a 7-layer strip is 0.7 mm.

Note that the material obtained as a result of polymerization process has properties of elasticity and creep. Moreover, it ages over time, i.e. its properties depend on intensity and duration of light radiation (see, for example, [4, 8]). As long as we consider a simple model which only accounts for elastic behavior of material, for measurement of bending and evaluation of the elastic modulus of plates-stripes we used the samples made in the same technological process, and the age of the compared samples (time which passed since the end of manufacturing) was chosen almost identical.

Evaluation of elastic modulus for the samples given in figure 2 was performed on the MTS Synergie 400 universal test system. Figure 4 contains tension diagrams of three 7-layer samples, first of which had 5-second light exposure of each of layer and was tested immediately after manufacturing (curve 1), the second one was also tested just after manufacturing, but had a light exposure time per layer — 10 seconds (curve 2), the third one, having a 10-second light exposure time per layer, was additionally lit by the UV-lamp for 4 minutes (curve 3) to promote polymerization. The tensile tests showed that curves strongly differ from each other, and values of the maximum elastic modulus for the considered samples essentially differ. So, for the first sample elastic modulus $E = 5\,\mathrm{MPa}$ (curve 1), for the second $E = 50\,\mathrm{MPa}$ (curve 2), for the third $E = 500\,\mathrm{MPa}$ (curve 3).

Using the test results we can calculate estimate of the maximum value of shrinkage stresses which caused distortion of a 7-layer plate-strip with 5 second light exposure time per layer given in Figure 3. Substitution to equation (3) of values $E = 5\,\mathrm{MPa}$, $h = 0.7\,\mathrm{mm}$, and $\rho = 55\,\mathrm{mm}$ yields maximum value of shrinkage stresses $\sigma_{sm} = 63.6\,\mathrm{kPa}$.

Conclusion

The suggested experimental-theoretical approach enables estimation of the residual stresses that accumulate at additive manufacturing of a plate-strip by layer-by-layer photopolymerization method. It gives the chance to correct the technological process directly for improvement of quality of particular finial object, and to use the self-contained theoretical model of additive

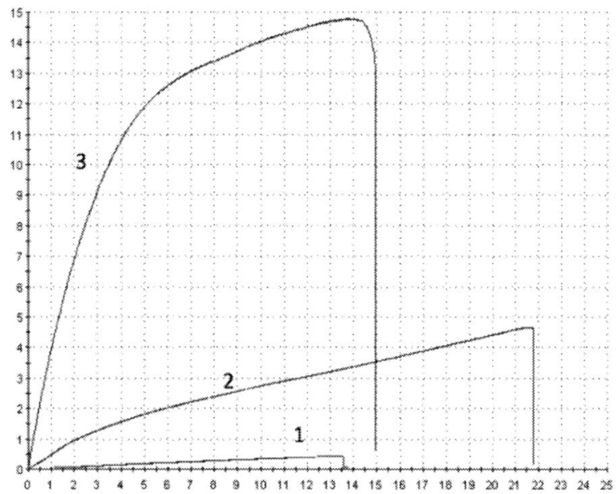

Figure 4.

manufacturing process (surface growth) for design of objects of very irregular shape, with increased accuracy.

Acknowledgments

The work is performed with financial support of the Russian Science Foundation (RSF project No. 17-19-01257).

References

[1] Gebhardt A and Hoetter J S 2016 *Additive Manufacturing: 3D Printing for Prototyping and Manufacturing* (Munich: Carl Hanser Verlag) p 611
[2] Zlenko M A, Nagaytsev M V, and Dovbysh V M 2015 *Additive Technologies in Mechanical Engineering* (Moscow: FGUP NAMI) p 220 [in Russian]
[3] Arutyunyan N H, Manzhirov A V, and Naumov V E 1991 *Contact Problems of Mechanics of the Growing Bodies* (Moscow: Nauka) p 172 [in Russian]
[4] Manzhirov A V 1995 The general non-inertial initial-boundary value problem for a viscoelastic ageing solid with piecewise-continuous accretion *J. Appl. Math. Mech.* **59** 805–16
[5] Lychev S A, Manzhirov A V, and Joubert S V 2010 Closed solutions of boundary-value problems of coupled thermoelasticity *Mech. Solids* **45** (4), 610–23
[6] Lychev S A, Lycheva T N, and Manzhirov A V 2011 Unsteady vibration of a growing circular plate *Mech. Solids* **46** (2), 325–333
[7] Manzhirov A V 2013 Mechanics of growing solids and phase transitions *Key Engng Mater.* **535–536** 89–93
[8] Manzhirov A V 2015 Mechanical design of viscoelastic parts fabricated using additive manufacturing technologies In *Lect. Notes Engng Comp. Sci.: Proc. of The World Congr. on Engineering* (London: IAENG) pp 710–4
[9] Manzhirov A V 2017 Advances in the theory of surface growth with applications to additive manufacturing technologies *Proc. Engng* **173** 11–6
[10] Birger I A 1963 *Residual Stressess* (Moscow: Mashgiz) p 232 [in Russian]
[11] Chernyshev G N, Popov A L, Kozintsev V M, and Ponomarev I I 1996 *Residual Stresses in Deformable Solid Bodies* (Moscow: Fizmatlit) p 240 [in Russian]
[12] Goldstein R V, Popov A L, Kozintsev V M, and Chelubeev D A 2015 Some new applications of ESPI at the mechanical tests *Meccanica* **50** 389–99
[13] Chen Z, Huang G S, Trase I, et al. 2016 Mechanical self-assembly of a strain-engineered flexible layer: Wrinkling, rolling and twisting *Phys. Rev. Appl.* **5** 017001–33
[14] Zhao Z, Wu J, Mu X, et al. 2017 Origami by frontal photo polymerization *Sci. Adv.* **3** e1602326
[15] Feodosyev V I 1999 *Resistance of Materials: Textbook* (Moscow: Izdat. MGTU im. N.E. Baumana) p 592 [in Russian]

Experimental study of auxetic behavior of cellular structure

A V Chentsov* **and D S Lisovenko**

Ishlinsky Institute for Problems in Mechanics of the Russian Academy of Sciences, Moscow, Russia

E-mail: *chentsov@ipmnet.ru

Abstract. The uniaxial tension of two-dimensional auxetic cellular constructions is studied experimentally. Samples were made of nonauxetic polyethylene terephthalate (PET-A amorphous) and subjected to monotonous uniaxial tension until the last moment when they still remained plane. As a result of the experimental data analysis, comparison of the mechanical properties is given for a faultless sample and constructions in which one horizontal or vertical element in the central area of the sample was removed. It is shown that the lack of one horizontal element of the construction has little influence on the auxetic properties of these constructions unlike the structures with one vertical element being absent.

1. Introduction

Poisson's ratio for isotropic materials varies in the range from -1 to 0.5 [1]. This ratio characterizes the relation of transverse deformations to longitudinal deformations. The majority of the known materials have positive values of Poisson's ratio. In [2], the polymeric foam with negative Poisson's ratio was obtained for the first time. The very name of auxetics was attributed to materials with negative Poisson's ratio [3]. Recently, we observe a certain increase in the number of studies on the physical and mechanical properties of auxetics [4–6]. The main directions of these studies are the studies of auxetic crystal materials [7–23], creation of auxetic structures [24–29], and production of composites with effective negative Poisson's ratio [30, 31]. Note that there are over four hundred crystal auxetics known at the moment. The greatest number (more than three hundred) were found among cubic crystals. For example, monocrystals of lithium, potassium, calcium, nickel, iron, gold, silver, etc. possess auxetic properties (see [16 17]). It was for the first time in [24, 25], where the concave hexagon was suggested as a cell of the auxetic two-dimensional plane for creation of auxetic structures. In [24], the auxetic two-dimensional framework consisting of concave hexagons was considered, and calculations of the average elastic characteristics were given. In [25], a two-dimensional construction comprised of concave hexagons was also proposed, and the geometrical parameters at which Poisson's ratio is observed to be equal to -1, were calculated. In [27, 28], the concave hexagon in which a part of straight elements is replaced with curvilinear ones was suggested. An experimental analysis of such two-dimensional construction showed that Poisson's ratio can reach -0.6. The two-dimensional and three-dimensional auxetic polymeric planes were experimentally studied in [29], where the dependence of their Poisson's ratio on the real deformation was calculated.

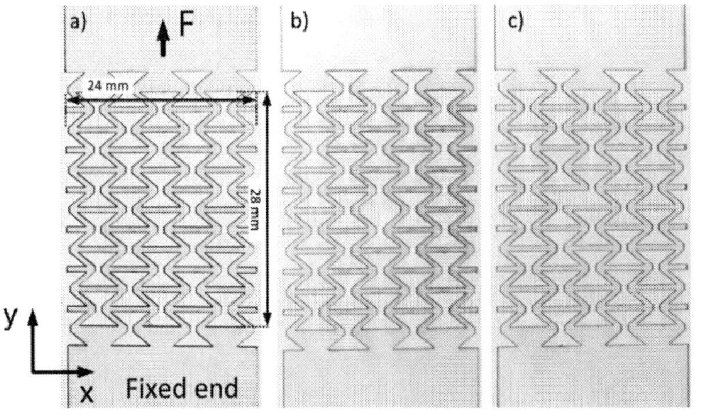

Figure 1. Three plane cellular samples with central area 28 × 24 × 0.7 mm: faultless sample zz (a); sample with one horizontal element being absent zzH (b); sample with one vertical element being absent zzV (c).

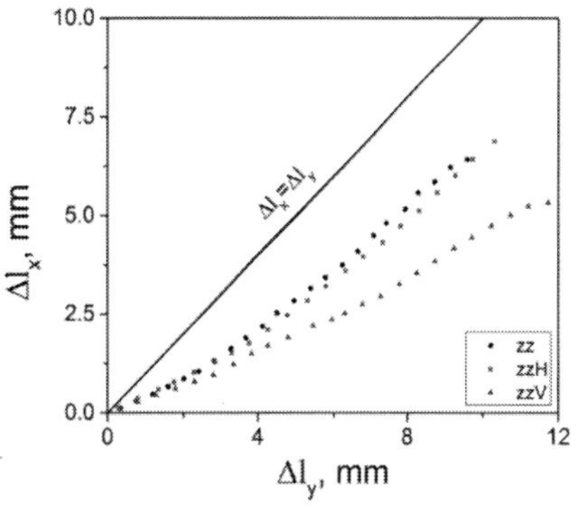

Figure 2. Variation in the transverse displacement Δl_x versus the longitudinal displacement Δl_y.

The present paper considers an experimental study of uniaxial tension of two-dimensional auxetic cellular constructions consisting of concave hexagons which contain defects in the central part of the samples.

2. Methods

The mechanical tests were performed on samples which were obtained from nonauxetic polyethylene terephthalate (PET-A amorphous) 0.7 mm thick by the femtosecond laser cutting method. The laser machine carried out displacements of the billet with micron accuracy. The duration of radiation impulses was 500 fs. The energy of an impulse was 120μJ, and the radiation power in an impulse reached 240 MW. The average power of radiation was 3 W. As a result of laser cutting, three plane cellular samples with dimensions 112.5 × 24 × 0.7 mm with central area 28 × 24 × 0.7 mm were obtained (figure 1): a faultless zz sample (figure 1a), zzH sample with one horizontal element being absent (figure 1b), and zzV sample with one vertical element being absent (figure 1c). The transverse size of elements of the cellular construction was equal to the thickness of the initial plate.

All three samples were subjected to monotonous uniaxial tension on a universal uniaxial setup (MTS Synergie 400) at 1 mm/min loading rate, with simultaneous registration of displacements and forces. The samples were stretched until the last moment when they still remained plane. The changes of the sample geometry were registered in the 12-Megapixel photo and video images taken during the experiment by Dahua IPC-HF81200E camera. The force and displacement of the cantilever-moving beam of the upper grip were also registered by the built-in sensors of the setup. The displacements and deformations in the plane of the sample were calculated later after decoding the video into separate frames by digital image correlation (DIC) method using Ncorr software package [32]. A random two-color pattern was applied to the initially transparent samples to enable tracking of specific areas by DIC post-processing.

Based on the force-displacement data, the experimental dependence of equivalent longitudinal stresses acting in the transverse section of the sample on longitudinal deformation can be calculated. The initial stage of almost linear deformation is related to the elastic reaction of the structure elements. At the next stage, the plastic deformation develops in curved elements and straight parts connecting them, the in-plane rotations is performed. At the last stage, the rigidity of the structure is restored.

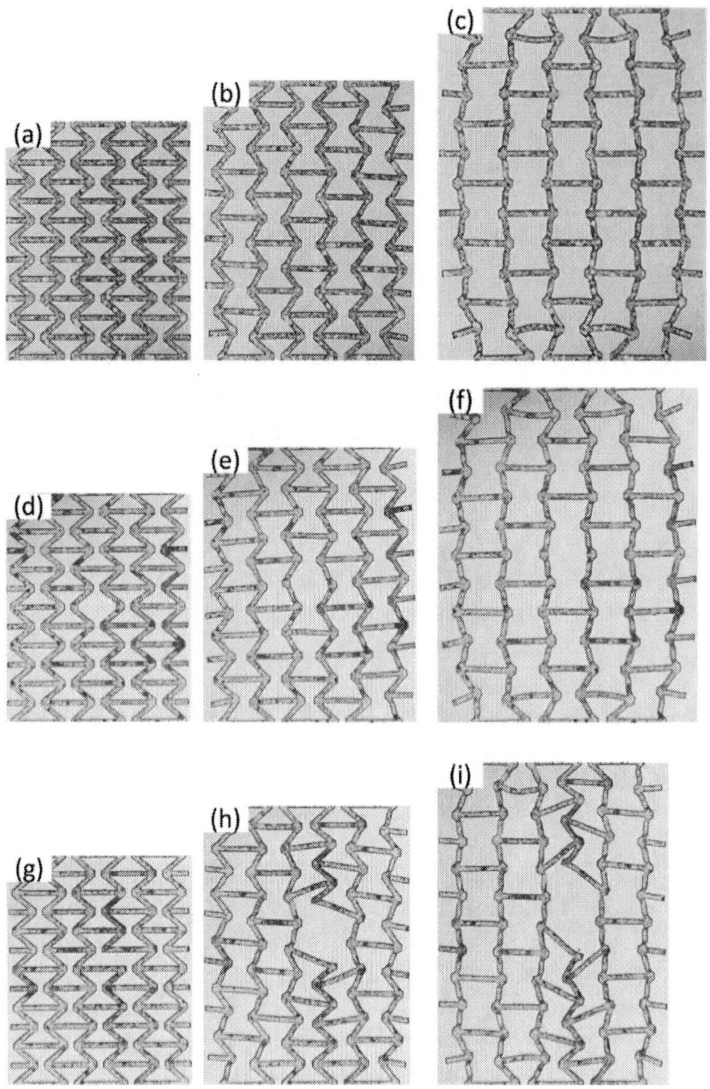

Figure 3. Photographs of sample zz at different stages of deformation $\varepsilon_x = 0$, $\varepsilon_y = 0$ (a), $\varepsilon_x = 0.110$, $\varepsilon_y = 0.172$ (b), $\varepsilon_x = 0.319$, $\varepsilon_y = 0.400$ (c); sample zzH at different stages of deformation $\varepsilon_x = 0$, $\varepsilon_y = 0$ (d), $\varepsilon_x = 0.118$, $\varepsilon_y = 0.188$ (e), $\varepsilon_x = 0.287$, $\varepsilon_y = 0.365$ (f); sample ssV at different stages of deformation $\varepsilon_x = 0$, $\varepsilon_y = 0$ (g), $\varepsilon_x = 0.098$, $\varepsilon_y = 0.210$ (h), $\varepsilon_x = 0.218$, $\varepsilon_y = 0.395$ (i); the axis y corresponds to the tension direction, and the axis x corresponds to the transverse direction.

3. Result and discussion

Variations in the transverse displacement Δl_x versus the longitudinal displacement Δl_y for three cellular samples are shown in figure 2. The sample with one vertical element being absent (zzV sample) is extended much stronger as compared to other samples. The maximum change of longitudinal displacement was 9.6 mm for sample zz, 10.3 mm for sample zzH, and 11.7 mm for sample zzV. The lack of one horizontal element in sample zzH led to a longer time till the loss of stability as compared to sample zz. Thus, in the monotonous tension, sample zzH was extended slightly stronger as compared to the faultless construction. The lack of one vertical element in sample zzV leads to its poor extension in the transverse direction during the experiment

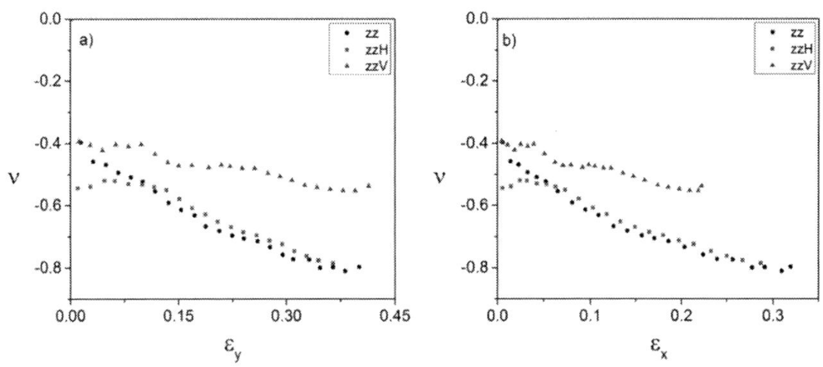

Figure 4. Poisson's ratio as a function of (*a*) longitudinal deformations ε_y and (*b*) transverse deformations ε_x for three samples.

as compared to the two other samples. The maximum changes of transverse displacements reached 6.4 mm for sample zz, 6.9 mm for sample zzH, and 5.3 mm for sample zzV. The pictures of three studied samples at different stages of deformation are given in figure 3.

Poisson's ratio in elasticity theory in the case of small deformations is determined by the formula

$$\nu = -\frac{\varepsilon_x}{\varepsilon_y},$$

where ε_x is the transverse deformation and ε_y is the longitudinal deformation. The variability of Poisson's ratio for three cellular samples was studied by analogy with this formula. In this case, $\varepsilon_y = \Delta l_y/L_0$ is the longitudinal deformation, $L_0 = 28\,\mathrm{mm}$ is the length of the central area, $\varepsilon_x = \Delta l_x/l_0$ is the transverse deformation, and $l_0 = 24\,\mathrm{mm}$ is the width of the central area.

The dependence of Poisson's ratio on longitudinal and transverse deformations for three cellular samples is given in figure 4. Figure 4*a* shows that samples zz and zzV experience large longitudinal deformations as compared to sample zzH. In the case of transverse deformations (see figure 4*b*), the lack of one vertical element leads to the fact that sample zzV has significantly smaller transverse deformation as compared to the two other samples. The maximum longitudinal and maximum transverse deformations reached 40% and 32% for sample zz, 36% and 29% for sample zzH, and 41% and 22% for sample zzV, respectively).

Conclusions

Poisson's ratio for sample zz varies in a wider range (from −0.40 to −0.81) as compared to the two other samples. In the case of sample zzH, the range of Poisson's ratio is reduced to values from −0.54 to −0.79. The lack of one horizontal element in the cellular plane (sample zzH) has weak influence on the mechanical properties. Distinctions in Poisson's ratio take place at longitudinal deformations up to 13%. The narrowest range of Poisson's ratio variations is for sample zzV (from −0.40 to −0.54). This is related to the fact that this sample extends stronger with less widening as comparison to sample zz and sample zzH (figure 2).

Acknowledgments

The study was supported by the Russian Foundation for Basic Research under grants Nos. 16-01-00325 and 17-01-00526.

References

[1] Landau L D and Lifshitz E M 1986 *Theory of Elasticity. Vol. 7 Course of Theoretical Physics* (Oxford: Butterworth-Heinemann)

[2] Lakes R 1987 Foam structures with a negative Poisson's ratio *Science* **235** (4792) 1038–40
[3] Evans K E, Nkansah M A, Hutchinson I J, and Rogers S C 1991 Molecular network design *Nature* **353** 124–5
[4] Lim T-C 2015 *Auxetic Materials and Structures* (Singapore: Springer)
[5] Kolken H M A and Zadpoor A A 2017 Auxetic mechanical metamaterials *RSC Adv.* **7** 5111–29
[6] Carneiro V H, Meireles J, and Puga H 2013 Auxetic materials – A review *Mater. Sci. - Pol.* **31** (4) 561–71
[7] Ballato A 1996 Poisson's ratio for tetragonal, hexagonal, and cubic crystals *IEEE Trans. Ultrasonics, Ferroelectr. Freq. Control* **43** (1) 56–62
[8] Baughman R H, Shacklette J M, Zakhidov A A, and Stafstrom S 1998 Negative Poisson's ratios as a common feature of cubic metals *Nature* **392** 362–5
[9] Lubarda V A and Meyers M A 1999 On the negative Poisson ratio in monocrystalline zinc *Scripta Mater.* **40** 975–7
[10] Tokmakova S P 2005 Stereographic projections of Poisson's ratio in auxetic crystals *Phys. Status Solidi B* **242** (3) 721–9
[11] Brańka A C, Heyes D M, and Wojciechowski K W 2009 Auxeticity of cubic materials *Phys. Status Solidi B* **246** (9) 2063–71
[12] Goldstein R V, Gorodtsov V A, and Lisovenko D S 2010 Auxetic mechanics of crystalline materials *Mech. Solids* **45** (4) 529–45
[13] Ballato A 2010 Poisson's ratios of auxetic and other technological materials *IEEE Trans. Ultrasonics, Ferroelectr. Freq. Control* **57** (1) 7–15
[14] Lethbridge Z A D, Walton R I, Marmier A S H, et al. 2010 Elastic anisotropy and extreme Poisson's ratios in single crystals *Acta Mater.* **58** (19) 6444–51
[15] Belomestnykh V N and Soboleva E G 2012 Unconventional approach to determining anisotropic Poisson's ratios in cubic crystals *Letters on Materials* **2** (1) 13–6
[16] Goldstein R V, Gorodtsov V A, and Lisovenko D S 2013 Classification of cubic auxetics *Phys. Status Solidi B* **250** (10) 2038–43
[17] Goldstein R V, Gorodtsov V A, Lisovenko D S, and Volkov M A 2014 Negative Poisson's ratio for cubic crystals and nano/microtubes *Phys. Mesomech.* **17** (2) 97–115
[18] Goldstein R V, Gorodtsov V A, and Lisovenko D S 2015 Young's modulus and Poisson's ratio for seven-constant tetragonal crystals and nano/microtubes *Phys. Mesomech.* **18** (3) 213–22
[19] Volkov M A 2014 Extreme values of Poisson's ratio for triclinic and monoclinic crystals *Letters on Materials* **4** (3) 167–70
[20] Krasavin V V and Krasavin A V 2014 Auxetic properties of cubic metal single crystals *Phys. Status Solidi B* **251** (11) 2314–20
[21] Goldstein R V, Gorodtsov V A, Lisovenko D S, and Volkov M A 2015 Auxetics among 6-constant tetragonal crystals *Letters on Materials* **5** (4) 409–13
[22] Goldstein R V, Gorodtsov V A, Lisovenko D S, and Volkov M A 2016 Mechanical characteristics for seven-constant rhombohedral crystals and their nano/microtubes *Letters on Materials* **6** (2) 93–7
[23] Epishin A I and Lisovenko D S 2016 Extreme values of the Poisson's ratio of cubic crystals *Tech. Phys.* **61** (10) 1516–24
[24] Kolpakov A G 1985 Determination of the average characteristics of elastic frameworks *J. Appl. Math. Mech.* **49** (6) 739–45
[25] Almgren R F 1985 An isotropic three-dimensional structure with Poisson's ratio = −1 *J. Elasticity* **15** (4) 427–30
[26] Li D, Dong L, and Lakes R S 2016 A unit cell structure with tunable Poisson's ratio from positive to negative *Mater. Lett.* **164** 456–9
[27] Goldstein R V, Lisovenko D S, Chentsov A V, and Lavrentyev S Yu 2017 Experimental study of auxetic behavior of re-entrant honeycomb with curvilinear elements *Letters on Materials* **7** (2) 81–4
[28] Goldstein R V, Lisovenko D S, Chentsov A V, and Lavrentyev S Yu 2017 Experimental study of defects influence on auxetic behavior of cellular structure with curvilinear elements *Letters on Materials* **7** (4) 355–8
[29] Fozdar D Y, Soman P, Lee J W, et al. 2011 Three-dimensional polymer constructs exhibiting a tunable negative poisson's ratio *Adv. Funct. Mater.* **21** (14) 2712–20
[30] Lim T-C 2012 Mixed auxeticity of auxetic sandwich structures *Phys. Status Solidi B* **249** (7) 1366–72
[31] Goldstein R V, Gorodtsov V A, and Lisovenko D S 2017 Longitudinal elastic tension of two-layered plates from isotropic auxetics-nonauxetics and cubic crystals *Eur. J. Mech. A Solids* **63** 122–7
[32] Blaber J, Adair B, and Antoniou A 2015 Ncorr: open-source 2d digital image correlation Matlab software *Exp. Mech.* **55** (6) 1105–22

IOP Conf. Series: Journal of Physics: Conf. Series **991** (2018) 012018 doi:10.1088/1742-6596/991/1/012018

Lame problem for a multilayer viscoelastic hollow ball with regard to inhomogeneous aging

Z A Davtyan, S Y Mirzoyan, and A V Gasparyan*
Institute of Mechanics of the National Academy of Sciences of Armenia, Yerevan, Armenia
E-mail: *anush@mechins.sci.am

Abstract. Determination of characteristics of the stress strain state of compound viscoelastic bodies is of both theoretical and practical interest. In the present paper, the Lamé problem is investigated for an uneven-aged multilayer viscoelastic hollow ball in the framework of N. Kh. Arutyunyan's theory of creep of nonhomogeneously aging bodies [1, 2]. Solving this problem reduces to solving an inhomogeneous finite-difference equation of second order that contains operators with coordinates of time and space. The obtained formulas allow one to determine the required contact stresses and other mechanical characteristics of the problem related to uneven age of contacting balls.

1. Problem formulation and derivation of basic equations

In a spherical coordinate system (r, φ, θ), consider a composite shaped as a package of an arbitrary n number of viscoelastic hollow balls

$$(r_{k-1} \le r \le r_k,\ 0 < \varphi \le 2\pi,\ 0 < \theta \le 2\pi) \quad (k = \overline{1, n})$$

with instant deformation moduli $E_k(t)$ soldered together through their surfaces. We assume that the contacting balls possess the property of creep and are of different ages. Assume further that, at time $t = \tau_0$, uniformly distributed forces of intensities $P_0(t)$ and $P_n(t)$ are respectively applied to the composite surfaces $r = r_0$ and $r = r_n$, i.e.,

$$\sigma_r(r,t)\big|_{r=r_0} = -P_0(t), \quad \sigma_r(r,t)\big|_{r=r_n} = -P_n(t), \tag{1}$$

where $\sigma_r(r, t)$ is the radial stress component, and the tangential stresses are equal to zero due to symmetry.

To derive the governing equations of the problem, we consider the kth layer of the package. Then to determine the elastic instant stresses and displacements of the layers, according to Lamé formulas, we have the expressions [3]

$$\begin{aligned}
&\sigma_r^{(k)}(r,t) = \frac{\lambda_k(t)}{\nu_k} A_k(t) + \frac{\lambda_k(t)(3\nu_k - 2)}{\nu_k}\frac{B_k(t)}{r^3},\\
&\sigma_\varphi^{(k)}(r,t) = \sigma_\theta^{(k)}(r,t) = \frac{\lambda_k(t)}{\nu_k}\left[A_k(t) + (1 - 3\nu_k)\frac{B_k(t)}{r^3}\right],\\
&U_r^{(k)}(r,t) = A_k(t) + \frac{B_k(t)}{r^2}, \quad \lambda_k(t) = \frac{E_k(t)\nu_k}{(1+\nu_k)(1-2\nu_k)} \quad (k = \overline{1, n}),
\end{aligned} \tag{2}$$

$$\sigma_r^{(k)}(r,t)\big|_{r=r_{k-1}} = -P_{k-1}(t), \quad \sigma_r^{(k)}(r,t)\big|_{r=r_k} = -P_k(t). \tag{3}$$

Here $A_k(t)$ and $B_k(t)$ are unknown coefficients to be determined, $U_r^{(k)}(r,t)$ is the elastic instant displacement of the layer, $P_{k-1}(t)$ and $P_k(t)$ are unknown radial contact stresses on the surfaces $r = r_{k-1}$ and $r = r_k$ of the layer, respectively, ν_k is Poisson ratio of the material of the layer, and $E_k(t)$ is the elastic instant modulus of deformation of the layer.

Satisfying the boundary-value conditions in equation (3), for the radial displacement from equation (2) we obtain

$$\begin{aligned} U_r^{(k)}(r,t) &= \frac{r_k^3 r_{k-1}^3 [P_k(t) - P_{k-1}(t)]}{d_k(t) l_k} \frac{1}{r^2} - \frac{r_k^3 P_k(t) - r_{k-1}^3 P_{k-1}(t)}{c_k(t) l_k} r, \\ U_r^{(k+1)}(r,t) &= \frac{r_{k+1}^3 r_k^3 [P_{k+1}(t) - P_k(t)]}{d_{k+1}(t) l_{k+1}} \frac{1}{r^2} - \frac{r_{k+1}^3 P_{k+1}(t) - r_k^3 P_k(t)}{c_{k+1}(t) l_{k+1}} r, \end{aligned} \tag{4}$$

where

$$c_k(t) = \frac{\lambda_k(t)}{\nu_k}, \quad d_k(t) = \frac{\lambda_k(t)(3\nu_k - 2)}{\nu_k}, \quad l_k = r_k^3 - r_{k-1}^3.$$

Now the condition of displacement continuity on the surface of contact $r = r_k$ of the two successive balls, with regard to the creep, can be written as

$$(I - L_k) U_r^{(k)}(r,t) = (I - L_{k+1}) U_r^{(k+1)}(r,t) \quad (k = \overline{1, n-1}). \tag{5}$$

Here

$$L_i[Y(t)] = \int_0^t E_i(u) K_i(t + \rho_i, u + \rho_i) Y(u) du, \quad \rho_i = \tau_i - \tau_0,$$

$$K_i(t,u) = \frac{\partial}{\partial u}\left[\frac{1}{E_i(u)} + C_i(t,u)\right] \quad (i = \overline{1,n}),$$

where τ_i are the ages and $C_i(t,u)$ are the amounts of creep ($i = \overline{1,n}$) of the layers.

We use equation (4) to rewrite contact condition (5) as

$$\begin{aligned} &\frac{1}{l_k}(I - L_k)\left[\frac{r_{k-1}^3 P_k(t) - r_{k-1}^3 P_{k-1}(t)}{d_k(t)} - \frac{r_k^3 P_k(t) - r_{k-1}^3 P_{k-1}(t)}{c_k(t)}\right] \\ &\quad = \frac{1}{l_{k+1}}(I - L_{k+1})\left[\frac{r_{k+1}^3 P_{k+1}(t) - r_{k+1}^3 P_k(t)}{d_{k+1}(t)} - \frac{r_{k+1}^3 P_{k+1}(t) - r_k^3 P_k(t)}{c_{k+1}(t)}\right] \quad (k = \overline{1, n-1}). \end{aligned}$$

The latter can be reduced to the following second-order finite-difference equations [4]:

$$(I - L_k)[a_k(t) q_{k-1}(t) + b_k(t) q_k(t)] = f_k(t), \tag{6}$$

$$(I - L_{k+1})[\delta_{k+1}(t) q_k(t) + a_{k+1}(t) q_{k+1}(t)] = g_{k+1}(t). \tag{7}$$

Under the conditions

$$f_k(t) + g_{k+1}(t) = 0 \quad (k = \overline{1, n-1}), \tag{8}$$

equations (6) and (7) can be rewritten as

$$a_k(t) q_{k-1}(t) + b_k(t) q_k(t) = (I + R_k) f_k(t), \tag{9}$$

$$\delta_{k+1}(t) q_k(t) + a_{k+1}(t) q_{k+1}(t) = (I + R_{k+1}) g_{k+1}(t). \tag{10}$$

Here $R_s(t,u)$ is the resolvent of the kernel $K_s(t,u)$ ($s = \overline{1,n}$), and the following notation is introduced:

$$a_k(t) = \frac{1}{c_k(t) l_k} - \frac{1}{d_k(t) l_k}, \quad q_k(t) = r_k^3 P_k(t), \quad \varphi_k = \frac{r_{k-1}^3}{r_k^3},$$

$$b_k(t) = \frac{\varphi_k}{d_k(t) l_k} - \frac{1}{c_k(t) l_k}, \quad \delta_k(t) = \frac{1}{\varphi_k d_k(t) l_k} - \frac{1}{c_k(t) l_k}.$$

IOP Conf. Series: Journal of Physics: Conf. Series **991** (2018) 012018 doi:10.1088/1742-6596/991/1/012018

2. Solution of the finite-difference equations

The solution of the first-order finite-difference equations (9) and (10) is constructed according to the well-known method [4, 5]. It can be represented in the form

$$q_k(t) = [1 - F(k,t)]q_{k-1}(t) + \frac{(I + R_k)f_k(t)}{b_k(t)}. \tag{11}$$

Here

$$F(k,t) = \frac{[a_k(t) + b_k(t)]}{b_k(t)} \quad (k = \overline{1, n-1}).$$

By analogy with [4], we find the general solution of equation (11)

$$q_k(t) = \prod_{j=1}^{k}[1 - F(j,t)]\left\{ \sum_{i=1}^{k}(I + R_i)f_i(t)\left[b_i(t)\prod_{r=1}^{i}(1 - F(j,t))\right]^{-1} + P_0(t)\right\} \quad (k = \overline{1, n-1}). \tag{12}$$

Quite similarly, the solution of equation (10) can be represented as

$$q_k(t) = \prod_{j=k}^{n-1}[1 - \tilde{F}(j,t)]^{-1}\left\{ \sum_{i=k}^{n-1}(I + R_{i+1})g_{i+1}(t)\left[a_{i+1}(t)\prod_{r=i+1}^{n-1}(1 - \tilde{F}(r,t))^{-1}\right]^{-1} + P_n(t)\right\}. \tag{13}$$

Here

$$\tilde{F}(k,t) = \frac{[a_{k+1}(t) + \delta_{k+1}(t)]}{a_{k+1}(t)} \quad (k = \overline{1, n-1}).$$

Now equating expressions (12) and (13) and taking (8) into account, we obtain the following system of Volterra integral equations of the second kind with respect to $f_i(t)$ $(i = \overline{1, n-1})$:

$$\sum_{i=1}^{k}\frac{1}{b_i(t)}\prod_{j=1}^{k}\frac{a_j(t)}{b_j(t)}\prod_{r=1}^{i}\frac{b_r(t)}{a_r(t)}(I + R_i)f_i(t) + \sum_{i=k}^{n-1}\frac{1}{a_{i+1}(t)}\prod_{j=k}^{n-1}\frac{a_{j+1}(t)}{\delta_{j+1}(t)}\prod_{r=i+1}^{n-1}\frac{\delta_{r+1}(t)}{a_{r+1}(t)}(I + R_{i+1})f_i(t)$$
$$= P_0(t)\prod_{j=1}^{k}\frac{a_j(t)}{b_j(t)} - P_n(t)\prod_{j=k}^{n-1}\frac{a_{j+1}(t)}{b_{j+1}(t)} \quad (k = \overline{1, n-1}) \tag{14}$$

or

$$Q_k(t)(I + R_k)f_k(t) + \sum_{i=1}^{k-1}Q_i(t)\frac{M_i(t)}{M_k(t)}(I + R_i)f_i(t)$$
$$+ \sum_{i=k}^{n-1}H_i(t)\frac{N_{i+1}(t)}{N_k(t)}(I + R_{i+1})f_i(t) = \frac{P_0(t)}{M_k(t)} - \frac{P_n(t)}{N_k(t)} \quad (k = \overline{1, n-1}). \tag{15}$$

Here the following notation is used:

$$Q_i(t) = \frac{1}{b_i(t)}, \quad H_i(t) = \frac{1}{a_{i+1}(t)},$$
$$M_i(t) = \prod_{r=1}^{i}\frac{b_r(t)}{a_r(t)} \quad (i = \overline{1, k}), \quad N_i(t) = \prod_{r=i}^{n-1}\frac{\delta_{r+1}(t)}{a_{r+1}(t)} \quad (i = \overline{k, n-1}).$$

Once the governing system of integral equations (15) is solved with respect to $f_i(t)$, the radial contact stresses $q_k(t)$ $(k = \overline{1, n-1})$ are determined from equations (9) and (10).

3. Particular case

Let us consider a particular case of the problem where the composite consists of two heterogeneous balls, i.e., $n = 2$. Note that a similar problem is considered in [6]. In our case of two layers, we determine $f_1(t)$ from equation (11). Next, we find the radial contact stress $q_1(t)$ from equations (12) and/or (13) through $P_0(t)$ and $P_2(t)$. The expression for $q_1(t)$ can be obtained as well from equations (6) and (7) with $n = 2$. In this case, to determine the unknown pressure $q_1(t)$, according to equation (5), we obtain the following Volterra integral equations of the second kind

$$\alpha(t)q_1(t) - \int_{\tau_0}^{t} [\alpha_1(\tau)E_1(\tau)K_1(t,\tau) + \alpha_2(\tau)E_2(\tau)K_2(t,\tau)]q_1(\tau)\,d\tau = F(t). \tag{16}$$

Here the following notation is used:

$$F(t) = \beta_1(t)P_0(t) + \beta_1(t)P_2(t) - \int_{\tau_0}^{t} [\beta_1(\tau)E_1(\tau)K_1(t,\tau)P_0(t) - \beta_2(\tau)E_2(\tau)K_2(t,\tau)P_2(\tau)]\,d\tau,$$

$$\alpha_1(t) = \frac{[r_0^3 - (3\nu_1 - 2)r_1^3](1+\nu_1)(1-2\nu_1)}{r_1^2(r_1^3 - r_0^3)(3\nu_1 - 2)E_1(t)}, \quad \alpha_2(t) = \frac{[r_2^3 - (3\nu_2 - 2)r_1^3](1+\nu_2)(1-2\nu_2)}{r_1^3(r_2^3 - r_1^3)(3\nu_2 - 2)E_2(t)},$$

$$\beta_1(t) = \frac{3r_1r_0^3(1-\nu_1^2)(1-2\nu_1)}{(r_1^3 - r_0^3)(3\nu_1 - 2)E_1(t)}, \quad \beta_2(t) = \frac{3r_1r_2^3(1-\nu_2^2)(1-2\nu_2)}{(r_2^3 - r_1^3)(3\nu_2 - 2)E_2(t)},$$

$$\alpha(t) = \alpha_1(t) + \alpha_2(t).$$

To solve integral equation (16), we assume that the kernels of the creep of materials have the form [1]

$$K_i(t,\tau) = \frac{\partial}{\partial\tau}\left[\frac{1}{E_i(\tau)} + C_i(t,\tau)\right], \quad C_i(t,\tau) = \varphi_i(\tau)(1 - e^{-\gamma(t-\tau)}) \quad (i = 1,2), \tag{17}$$

where $\varphi_1(\tau)$ and $\varphi_2(\tau)$ are the functions of aging for the materials of the outer and inner layers, respectively. The meaning of the other parameters well-known in the theory of creep is given in [1]. Later, we also assume that $E_1(t) = E_1 = \text{const}$ and $E_2(t) = E_2 = \text{const}$.

Since the kernels determined by formula (17) are singular, the Volterra integral equations of the second kind (16) can be reduced to a differential equation with variable coefficients.

For simplicity, let us consider a particular case of external load. Assume that, at time $t = \tau_0$, a concentrated radial force of intensity T_0 is applied to the outer sphere of the compound hollow ball. Afterwards, the applied force remains constant in time. Then $P_0(t) = P_2(t) = T_0H(t-\tau_0)$. Here $H(t)$ is the unit Heaviside function. Also, assume that $\rho_1 = 0$. Under the applied load, the solution of integral equation (16) becomes

$$q_1(t) = T_0\left\{(\beta_1 + \beta_2)\lambda_0 + \lambda_0^2(\beta_1\alpha_2 - \beta_2\alpha_1)[E_1\varphi_1(\tau_0) - E_2\varphi_2(\tau_2)]\int_{\tau_0}^{t} e^{-\eta(z)}\,dz\right\}, \tag{18}$$

where

$$\eta(t) = \gamma\int_{\tau_0}^{t} [1 + \lambda_0\alpha_1E_1\varphi_1(\tau) + \lambda_0\alpha_2E_1\varphi_2(\tau + \rho_2)]\,d\tau, \quad \lambda_0 = \frac{1}{\alpha_1 + \alpha_2}$$

It is easy to see that when the creep deformation of two contacting bodies is caused by constant stresses applied to the layers from inside and outside simultaneously (at the age of τ_0 and τ_2, respectively) and proportionally to their elastic deformation, i.e., $E_1\varphi_1(\tau_0) = E_2\varphi_2(\tau_2)$, then the solution of the above-stated problem coincides with the solution of the corresponding elastic problem.

IOP Conf. Series: Journal of Physics: Conf. Series **991** (2018) 012018 doi:10.1088/1742-6596/991/1/012018

If we assume that $P_2(t) = 0$, then from the general solution of equation (18) we obtain

$$q_1(t) = T_0\beta_1\lambda_0\left\{1 + \lambda_0\alpha_2[E_1\varphi_1(\tau_0) - E_2\varphi_2(\tau_2)]\int_{\tau_0}^{t} e^{-\eta(z)}\, dz\right\}. \quad (19)$$

Meanwhile, equation (19) implies that, as a function of time, the radial pressure $q_1(t)$ increases for $E_1\varphi_1(\tau_0) > E_2\varphi_2(\tau_2)$ and, vice versa, it decreases for $E_1\varphi_1(\tau_0) < E_2\varphi_2(\tau_2)$.

Conclusions

In the paper an inhomogeneously aging multilayer viscoelastic hollow ball was investigated when stressed uniaxially. In particular, the problem was considered for two-layered composite with the creep property. Simple formulas were obtained for unknown contact stresses. It was shown that if the creep deformations and the elastic deformations of the layers are proportional, then the solution of the formulated problem coincides with that of the corresponding elastic problem.

References

[1] Arutyunyan N Kh and Kolmanovski W B 1983 *Creep Theory of Inhomogeneous Solids* (Moscow: Nauka) p 336 [in Russian]
[2] Arutyunyan N Kh and Manzhorov AV 1999 *Contact Problems of Theory of Creep* (Yerevan: Izdat. Inst. Mekaniki NAN RA) p 318 [in Russian]
[3] Rekach W G 1984 *Guide into the Solution of Problems of the Theory of Creep* (Moscow: Vysshaya Shkola) p 287 [in Russian]
[4] Gelfond A O 1967 *Calculus of Finite Differences* (Moscow: Nauka) p 375 [in Russian]
[5] Gasparyan A V and Mkhitaryan S M 2010 On Stressed State of a Composite in Form of a Package Composed of an Arbitrary Number of Elastic Layers under Antiplane Deformation, *Izv. Nats. Akad. Nauk Armenii. Mekh.* **63** (2) 10–20
[6] Mirzoyan S E 1983 Lamé problem for a two-layer construction In *Studies on Mechanics of Deformable Solids* (Yerevan: Izdat. NAN RA) pp 112–9 [in Russian]

Investigation of strength characteristics of aluminum alloy under dynamic tension

A D Evstifeev
Saint Petersburg State University, St. Petersburg, Russia
E-mail: ad.evstifeev@gmail.com

Abstract. The study presents the results of experimental-theoretical analysis for aluminum alloy subjected to static and dynamic tension on samples of different types. The material was tested under initial coarse-grained (CG) and in ultrafine-grained (UFG) condition. The time dependence of the tensile strength is calculated using an incubation time fracture criterion based on a set of fixed constants of the material.

1. Introduction

In the second half of the 20th century, interest in studying the behavior of solids under conditions of dynamic effects increased significantly. The area of fracture mechanics was formed. The appearance of a relatively new field of science and a large number of theoretical and experimental works did not solve all problems associated with effects of dynamic destruction, but only formed new tasks. Additional features were also revealed when new materials appeared. The main source of problems was the lack of a single fracture criterion capable of solving the problems posed in both the static and dynamic domains.

The research of behavior of materials under high rate pulse loads is of great scientific and practical interest especially considering fast development of such areas as aerospace industry and recent appearance of new materials and new methods of material treatment and processing. The identification of regularities and prediction of the behavior of materials under dynamic and static pulsed loads is the main goal of the project. Special attention in industry now paid to nanostructured materials obtained by Severe Plastic Deformation (SPD) [1], because the reliability and efficiency of modern equipment operating under conditions of high rate dynamic loads (for example, elements of gas turbine engines, parts of aircraft and spacecraft) depend on the completeness and quality of the experimental and theoretical studies carried out in the entire range of variations in the parameters of the external impact. The investigation of such materials is complicated by the specific features of small-scale samples currently obtained by SPD methods (for example, disks obtained using the method of High-Pressure Torsion (HPT) are usually up to 20 mm in diameter and up to 1–2 mm in thickness). A lot of studies of the properties of UFG materials including static strength [2, 3], fatigue [4], wear and coating adhesion [5] have been recently carried out by various research groups. Conducting experiments on high rate loading requires to develop and probe the test procedure with regard to the dimensional features of samples of nanostructured materials. To achieve the goal, a methodology was developed and experiments on dynamic tension were carried out using droptower impact systems with an accelerator.

IOP Conf. Series: Journal of Physics: Conf. Series **991** (2018) 012019 doi:10.1088/1742-6596/991/1/012019

This study presents the results of experimental-theoretical analysis of materials subjected to static and dynamic tension. The experiments were carried out on model material aluminum alloy 1230 (99.3 Al). The strength properties of these materials under high strain rates were obtained and used to analyze the tensile strengths by the incubation time fracture criterion [6, 7].

2. Materials

Experiments were carried out on model material aluminum alloy 1230 (99.3 Al). The alloy has the following composition: 99.3% of Al with additions of Fe, Si, Mn, Ti, Cu, Mg, and Zn. This material and other aluminum alloy have a wide range of practical applications in aerospace industry. The nano-materials were made by HPT processing (disks of diameter 20 mm, 6 GPa pressure, and 10 revolutions at room temperature).

The Vickers microhardness of samples was assessed for both types of the materials. The measurements were performed using Shimadzu HMV-G machine and ISO 6507-1:2005 standard. The applied load was 50 g with dwell time of 15 s. The microhardness measurements showed on average 28.6 HV for CG alloy and 53.7 HV for UFG alloy.

3. Experimental Techniques

The area of dynamic loads is of a great interest, as it is well known that the materials behavior under dynamic loading conditions significantly differs from operation under static conditions. The actual task in this area is an experimental and theoretical study of the materials properties in both the static and the dynamic regime, where the main loads are high rate pulse loads.

The earliest works in this area belong to Hopkinson [8] and Manson [9]. They used a dropping tip to produce a tension pulse. Ginns [10], using a spring mechanism to apply a dynamic load and a strain gage to measure the voltage, was one of the first who tried directly to record the dynamic stress-strain curve. Later, a lot of equipments and methods were developed.

At present, fundamental and experimental studies in the field of high-speed impulse actions are conducted at many Russian and world scientific centers. Great results have been achieved at University of Applied Sciences of Southern Switzerland, University of Texas, Stanford Research Institute, California Institute of Technology, Johns Hopkins University, and others. A lot of experience has been accumulated in experimental work. One of the main tools for implementing high-speed impulse impact is the Kolsky method using Hopkinson's split rods [11]. There are a lot of various modifications of this method, which permits realizing compression, tension, torsion, etc. The method is based on the one-dimensional theory of propagation of elastic waves along elastic rods. According to formulas of the Kolsky method, one can determine the parametric dependence of development of stress, deformation, and strain rate in the sample during the test. However, this requires a special experimental and diagnostic knowledge, as well as significant staff experience in fixing and interpreting the received signals. These difficulties can partially be overcome using standard certified installations.

In the present study, the dynamic characteristics of the material were performed on the Instron CEAST 9350 droptower at a high strain rate. A high-energy configuration increases the velocity from 4.65–24.0 m/s. This allows one to carry out the experiment for different type of samples with automated result calculation. The samples of the first type (figure 1) were flat samples of ISO 8256 with the work part of length 10 mm and width 3 mm (Sample I). The samples of the other type (figure 2) were adapted to specific features of small-scale samples currently obtained by SPD methods. The samples has the work part of length 5 mm and width 2 mm (Sample II). The samples were cut by the electrical discharge machine ARTA 123 PRO with high precision. A special holder for small samples was also made for dynamic tension experiment.

Figure 3 shows the pulses obtained in the sensor during tensile tests under different strain rates. The signal was stable and may be used for obtaining threshold characteristic of the materials.

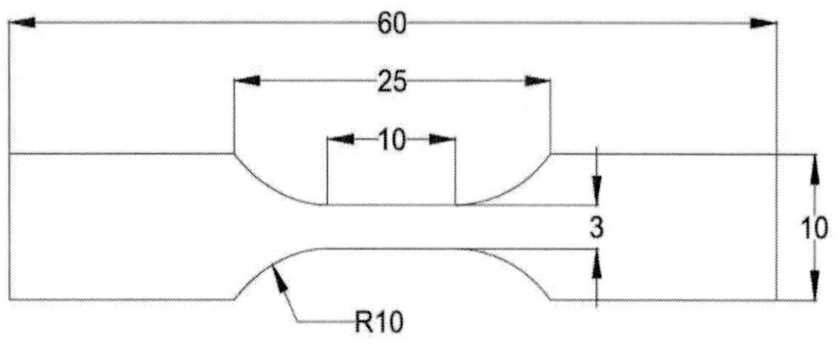

Figure 1. Type of samples I.

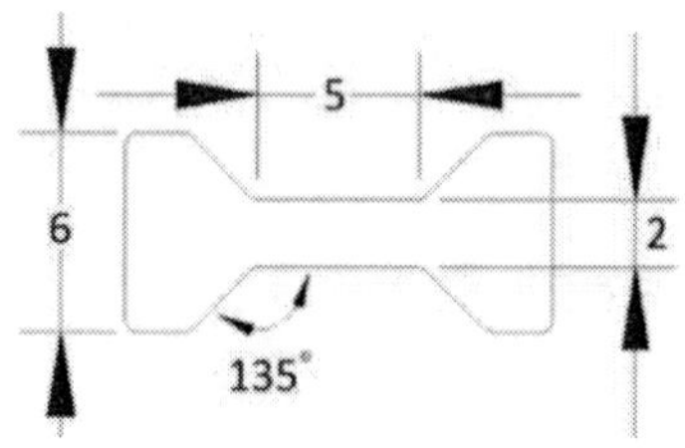

Figure 2. Type of samples II.

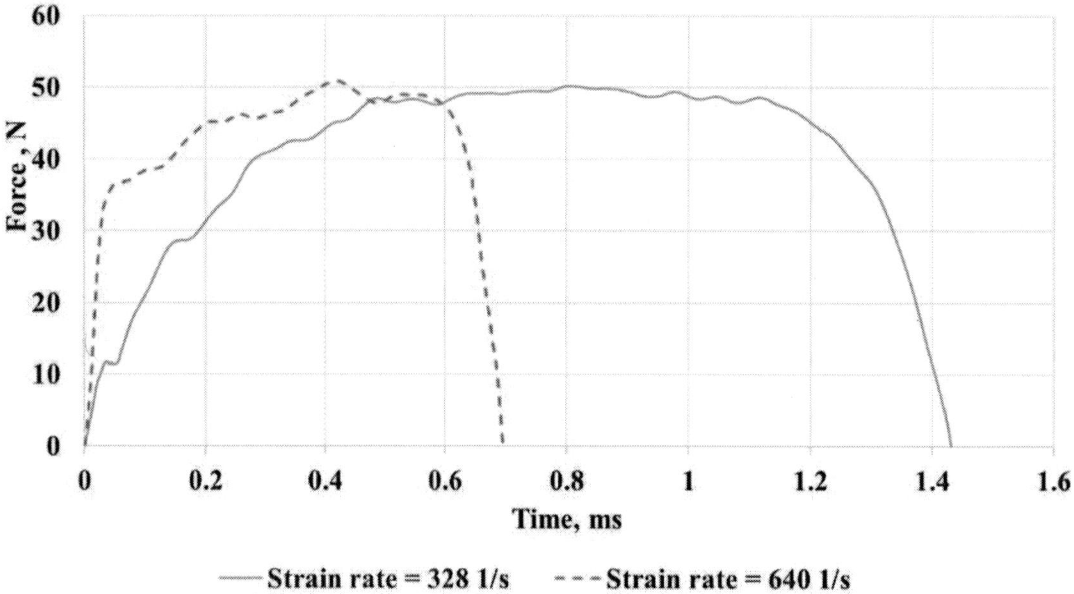

Figure 3. Force obtained in the machine during tensile test for different strain rates.

Experiments in static regime of deformation by tension were carried out on the Shimadzu tensile machine AG-50kNX.

4. Experimental results and theoretical investigation

In the case of slowly applied load, there occurs a destruction under the threshold value of the load amplitude. In dynamics, the strength characteristics of materials greatly differ from those obtained in the case of static loading. It can be explained by the fact that the time of loading is of the same order as the typical time of certain processes on the micro-level. This time can be described by the material constant as the incubation time parameter. For this task, for all ranges of loading an efficient criterion can be formulated in the following form [6, 7]:

$$\frac{1}{\tau}\int_{t-\tau}^{t} \sigma(s)\,ds \le \sigma_{st}, \qquad (1)$$

where τ is the incubation time, σ_{st} is the tensile strength for quasi-static loading, and $\sigma(t)$ is the applied stress. The incubation time characterizes the material sensitivity to the rate of loading. The fracture moment t^* corresponds to the earliest realization of equality in criterion (1).

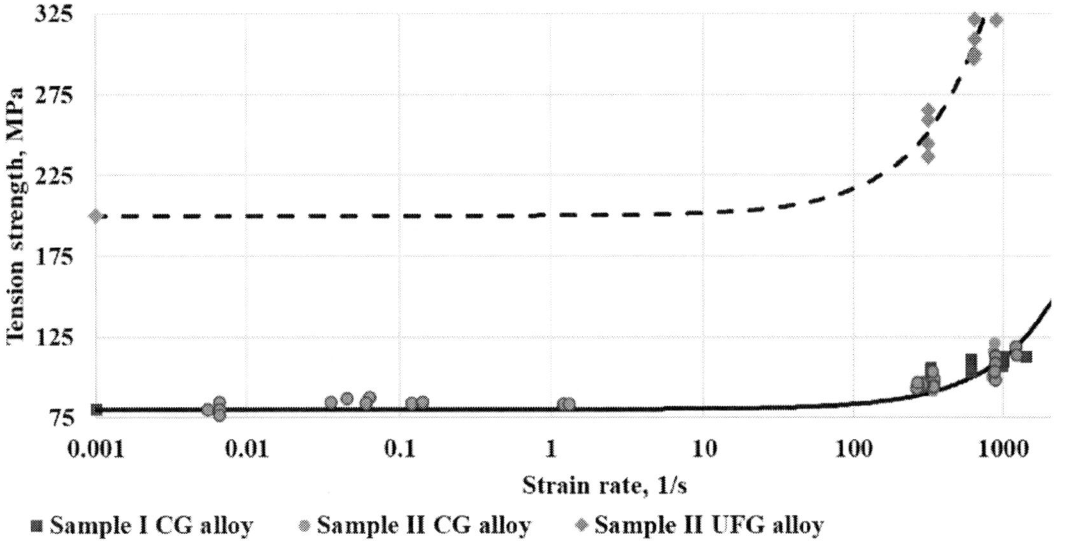

Figure 4. Tension strength versus strain rate for CG and UFG aluminum alloy for different type of samples. The theoretical line was modelled using the concept of incubation time with material parameters: for CG alloy $E = 72\,\text{GPa}$, $\tau = 0.8\,\mu\text{s}$, $\sigma_{st} = 80\,\text{MPa}$; for UFG alloy $E = 72\,\text{GPa}$, $\tau = 4.4\,\mu\text{s}$, $\sigma_{st} = 200\,\text{MPa}$.

The experimental results for CG aluminum alloy for Samples I and II are shown in figure 4. It shows a strain rate effect at the dynamic tensile strength. The points indicate the experimental data, whereas the curves show the theoretical line constructed by formula (1) using the material parameters $E = 72\,\text{GPa}$, $\tau = 0.8\,\mu\text{s}$, $\sigma_{st} = 80\,\text{MPa}$. The obtained experimental data can serve as a basis for further studies with nano-materials because they exhibited a good compliance between the theoretical line and the experimental data for all sizes of specimens.

The experimental results for UFG materials are also shown in figure 4. The experiment was carried out with a small sample. The theoretical line is constructed by formula (1) using the material parameters $E = 72\,\text{GPa}$, $\tau = 4.4\,\mu\text{s}$, $\sigma_{st} = 200\,\text{MPa}$. The obtained data showed an increase in the strength of the UFG material in the static and dynamic regimes of investigation compared with the properties of the CG material.

Conclusion

The test procedure for tensile testing of small samples under conditions of dynamic tension was developed using the Instron CEAST 9350 droptower. The approbation of the methodology was carried out on the model material aluminum alloy, including the comparison of the results obtained on small nontypical samples with the results obtained on standard samples. The high rate dependence of the tensile strength of samples for the model material was obtained. The experimental data after static and dynamic tension of the samples were analyzed by structural time fracture criteria. The experimental data obtained in the experiment on two types of samples agree well with the theoretical line. This allows one to carry out the further study on small-scale samples of nano-materials.

The samples were made from the aluminum alloy by the method of severe plastic deformation using high-pressure torsion procedure and then were tested like a coarse-grained analog. The UFG materials also depend on the strain rate. The severe plastic deformation significantly changed the static and dynamic strength of the investigated material. The dynamic strength τ of UFG alloy is more than 5 times greater than the same property of CG alloy.

The experimental methods can further be used in experimental research in the field of dynamic loads. The analysis of the results can be useful in the future study of a whole range of structural materials required in the transport and aerospace industry, as well as for introducing structural temporal fracture criteria in numerical schemes.

The studies of the strain rate features using the incubation time approach provide an effective instrument for examining the fracture process that is crucial for predicting the critical parameters of external loads under various loading conditions. The theoretical analysis showed that the application of the incubation time concept allows one to determine the parameters of materials necessary to predict the strength properties under static and dynamic load conditions and permits selecting suitable materials for specific applications.

Acknowledgments
The work was supported by the Russian Science Foundation (grant No. 17-79-10145).

References

[1] Valiev R 2004 Nanostructuring of metals by severe plastic deformation for advanced properties *Nature Mater.* **3** (8) 511–6
[2] Wei Q, Zhang H T, Schuster B E, et al. 2006 Microstructure and mechanical properties of super-strong nanocrystalline tungsten processed by high-pressure torsion *Acta Mater.* **54** (15) 4079–89
[3] Mishra A, Martin M, Thadhani N N, et al. 2008 High-strain-rate response of ultra-fine-grained copper *Acta Mater.* **56** (12) 2770–83
[4] Witney A B, Sanders P G, Weertman J R, and Eastman J A 1995 Fatigue of nanocrystalline copper *Scr. Metall. Mater.* **33** (12) 2025–30
[5] Wang C T, Gao N, Gee M G, et al. 2013 Tribology testing of ultrafine-grained Ti processed by high-pressure torsion with subsequent coating *J. Mater. Sci.* **48** (13) 4742–8
[6] Morozov N F and Petrov Y V 2000 *Dynamics of Fracture* (Berlin: Springer) p 99
[7] Petrov Y V and Morozov N F 1994 On the modeling of fracture of brittle solids *ASME J. Appl. Mech.* **61** (3) 710–2
[8] Hopkinson B 1905, The effects of momentary stresses in metals *Proc. Roy. Soc. A* **74** 498–506
[9] Mason W 1934 The Yield of Steel Wire under Stresses of Very Small Duration *Proc. Inst. Mech. Eng.* **128** (1) 409–38
[10] Ginns D W 1937 The mechanical properties of some metals and alloys broken at ultra-high speeds *J. Inst. Metals* **61** 61–78
[11] Bragov A M and Lomunov A K 1995 Methodological aspects of studying dynamic material properties using the Kolsky method *Int. J. Impact Engng* **16** (2) 321–30

Determination of the mechanical characteristics of nanomaterials under tension and compression

A A Filippov[1*] **and V M Fomin**[1,2,3]
[1]Khristianovich Institute of Theoretical and Applied Mechanics, Siberian Branch of the Russian Academy of Sciences, Novosibirsk, Russia
[2]Novosibirsk State Technical University, Novosibirsk, Russia
[3]Novosibirsk State University, Novosibirsk, Russia

E-mail: *filippov@itam.nsc.ru

Abstract. In this paper, new method for determining the mechanical characteristics of nanoparticles in a heterogeneous mixture is proposed. The heterogeneous mixture consists of a thermosetting epoxy resin and silicon dioxide powder of different dispersity. The mechanical characteristics of such a material at a constant concentration for nanopowder are experimentally determined. Using existing formulas for obtaining effective characteristics, the Lame coefficients for nanoparticles of various sizes are calculated. The dependence of the elastic characteristics on the particle size is obtained.

1. Introduction

The modern material science assumes the widespread use of composite materials to reduce the weight of structures and to increase the elastic-strength characteristics. The use of filled polymer materials has proved itself in aviation and rocket engineering, automotive industry and others. One of the simplest of these composites is the polymer filled with dispersed particles to change the strength, electrically conductive, heat-conducting properties. Advantages of using dispersed particles in polymers can be called the low cost, high wear resistance of the resulting composites, the possibility of being used together with other reinforcement technologies. Prediction of the properties of heterogeneous materials is a problem whose solution is not possible when taking into account the properties of each particle in the volume. However, it is quite possible to determine the macroscopic characteristics of a heterogeneous material through its physical and mechanical components using models of homogenization. Such approaches have been developed very successfully since the beginning of the 20th century and up to the present, covering an increasing number of features of heterogeneous materials deformation behavior. Using the nanoobjects like nanoparticles or nanotubes in composite materials as filler permits understanding their mechanical properties. The results of prediction by analytical models do not agree with experiments, because they do not take the average size of the filler particles into account [1, 2]. The purpose of this paper is to determine the mechanical properties of nanoparticles for describing the deformation behavior of heterogeneous materials. Experiments to determine the mechanical characteristics were performed on an epoxy resin with the addition of ultra-disperse silicon dioxide powders, namely, the epoxy resin DER-330 (Dow Chemical, USA) actively used in many industries and the Tarkosil silica fume powder (ITAM SB RAS, Novosibirsk, Russia) with different specific surface area (table 1).

Table 1. Roots of transcendental equation (13).

Filler type	Specific surface area (m^2/g)	Bulk density (kg/m^3)	Agglomerates diameter (μm)
Tarkosil T-05	50	80–100	20
Tarkosil T-15	96	70–80	10–12
Tarkosil T-20	134	60–70	5–7

Figure 1. Agglomerates of Tarkosil T-20 nanopowder.

The powder was manufactured by using a special equipment. This is an industrial electron accelerator with power capacity of 100 kW, a reactor for evaporating matter, a coagulator, a series of cyclones and filters for purification and isolation of nanoparticles [3]. Due to the high electron energy, the evaporated substance is heated at a rate of about 1000° C/sec in the reactor. Then the matter is cooled and moved by air (or other accelerated gas). Passing through the coagulator, a system of filters and cyclones, the vapor condenses into nanoparticles and trapped as a powder. The resulting white powder has a very low bulk density, X-ray amorphous and has a high specific surface area controlled by parameters of equipment. Because of the high surface energy, the particles tend to stick together and form agglomerates of a few microns in size (figure 1).

2. Experimental moduli of epoxy resin with a nanofiller

The preparation and production of heterogeneous materials on the basis on epoxy resin and silica fume powder including the dispersion of particles in the matrix and the preparation of test specimens are described in detail in [4]. The curing agent (IMTHPA) is then added to the epoxy resin mixed with nanopowder in ratio 100 : 2 : 80, respectively. The curing of the mixture was carried out by furnace in step wise mode. Mechanical test parameters were chosen in accordance with ASTM D 790: the loading speed 400 N/min, and the preload 50 ± 10 N. The modulus were determined by the secant method on stress-strain curve in the 1–2% deformation section. The modulus of elasticity of pure epoxy resin was 1450 ± 50 MPa. The test results are graphically shown in figure 2. The obtained data indicate a significant change in the properties

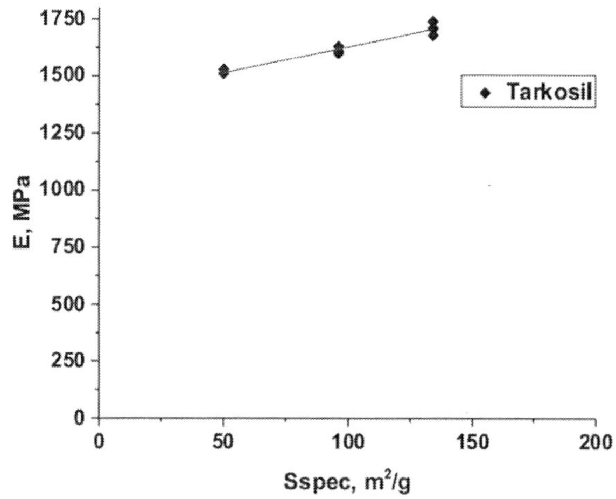

Figure 2. Dependence of the tensile modulus on the average particle size of the filler.

at low filler concentrations. It is especially worth noting the difference in the elastic properties of heterogeneous materials with a change in the specific surface of the filler. For Tarkosil, the growth of the elastic modulus from 1500 to 1750 MPa is observed with an increase in the surface area. For a larger specific surface more significant, the Tarkosil powders showed less effect on the elastic characteristics of the material.

Despite the same amount of the filler, the elastic properties are higher when the particle size is smaller. Such an experiment may show a change in the elastic properties from a macro object to a nano object. The reasons of this phenomenon are not clearly understandable. This could be the properties of the filler surface and the chemical reactions with epoxy resin or curing agent. The hypothesis advanced by the authors is that when the size decreases, the number of internal defects in filler decreases and the modulus of the nanoparticle becomes closer to the modulus of the crystalline matter.

3. Models of homogenization used to predict the moduli of nanofiller

The elastic properties of heterogeneous materials can be described by different existing models of homogenization using different hypotheses for averaging properties [5, 6]:

$$\text{Voigt:}\quad K = c_1K_1 + c_2K_2; \tag{1}$$

$$\text{Reuss:}\quad \frac{1}{K} = \frac{c_1}{K_1} + \frac{c_2}{K_2}; \tag{2}$$

$$\text{Hashin:}\quad K = K_2 + \frac{c_1}{1/(K_1-K_2) + 3c_2/(3K_2+4\mu_2)}; \tag{3}$$

$$\text{Shtrikman:}\quad K = K_1 + \frac{c_2}{1/(K_2-K_1) + 3c_1/(3K_1+4\mu_1)}; \tag{4}$$

$$\text{Christensen low volume fraction of filler:}\quad K = K_2 + \frac{(K_1-K_2)c_1}{1+(K_1-K_2)/(K_2+\frac{4}{3}\mu_2)}; \tag{5}$$

$$\text{Christensen polydispersity model:}\quad K = K_2 + \frac{(K_1-K_2)c_1}{1+[c_2(K_1-K_2)/(K_2+\frac{4}{3}\mu_2)]}; \tag{6}$$

$$\text{Cuento:}\quad \frac{1}{K} = \frac{1-\sqrt{c_2}}{K_1} + \left(\frac{1-\sqrt{c_2}}{\sqrt{c_2}}K_1 + K_2\right)^{-1}; \tag{7}$$

$$\text{Bache–Nepper–Chestenson:}\quad K = K_1^{c_1}\cdot K_2^{c_2}; \tag{8}$$

$$\text{Krivoglazov:} \quad K = K_2 \cdot \left[1 + \frac{3(1-\nu_2)(K_1-K_2)c_1}{2K_2(1-2\nu_2)+K_1(1+\nu_2)}\right]; \tag{9}$$

$$\text{Berryman:} \quad K = K_2 - c_1(K_2-K_1)\left[1 - \frac{c_2(3K_2-3K_1)}{3\mu_1+3K_2+\mu_2}\right]; \tag{10}$$

$$\text{Guz:} \quad K = c_1K_1 + c_2K_2 - \frac{3c_1c_2(K_1-K_2)^2}{3c_1K_2+3c_2K_1+4\mu_1\mu_2/(c_1\mu_1+c_2\mu_2)}; \tag{11}$$

Kulak (in assumption that the structural parameter $k_{ai} = k_1$):

$$K = K_2 + \frac{(K_{ai}-K_2)c_1}{1+(K_{ai}-K_2)c_2/(K_2-\frac{4}{3}\mu_2)}, \quad K_{ai} = K_1 + \frac{(K_2-K_1)c_1}{1+(K_2-K_1)c_2/(K_1-\frac{4}{3}\mu_1)}. \tag{12}$$

The combination of Voigt and Reuss models (Takayangi models):

$$\text{Hirsh–Dugill:} \quad K = \frac{1}{2}\left(\frac{1}{c_1K_1+c_2K_2} + \frac{c_1}{K_1} + \frac{c_2}{K_2}\right); \tag{13}$$

$$\text{Popovich:} \quad K = \frac{1}{2}\left(c_1K_1+c_2K_2 + \frac{1}{c_1/K_1+c_2/K_2}\right) \tag{14}$$

It should be noted, that the sum of volume concentrations is 1

$$c_1 + c_2 = 1.$$

Relation between K, μ, and ν is well-known

$$K = \frac{2\mu(1+\nu)}{3(1-2\nu)}$$

where where K is bulk modulus, μ is shear modulus, ν is Puasson ratio, c is volume concentration. Elastic constants without an index correspond to a heterogeneous material, index 1 is to a filler, index 2 is to a matrix.

For the above models, the elastic modulus was plotted against the filler concentration. The mechanical properties for the matrix and the filler are

$$K_1 = 1\,\text{GPa}, \quad K_2 = 100\,\text{GPa}, \quad \nu_1 = 0.3, \quad \nu_2 = 0.3.$$

The basic principles of constructing such dependencies are

- a monotonous increase in the characteristic from smaller to larger values;
- at a concentration of 0% of one of the components, the composite characteristic is equal to the characteristic of the second component, i.e., $K = K_2$ for $c = 0$ and $K = K_1$ for $c = 1$;
- the curves are located inside the Voigt–Reuss fork;
- the compliance with experimental data is attained.

These principles could be ignored like the Berryman model or Christensen low volume fraction, but the main criteria for the prediction of effective modulus of the mixture for different combinations of materials are always required. The application of these models is used to predict the properties of mixtures at a certain concentration, but in our case, when the properties of the filler are not precisely determined, their use is difficult. The approach proposed by the authors is to convert these models to the form $K_1 = f(K, K_2, c_1, \nu_1, \nu_2)$ and to calculate the modulus of elasticity for the filler nanoparticles. This dependence for the moduli gives only estimates of the contribution of the mechanical properties of the filler in a particular model and their use should be carefully analyzed. In addition, the prediction of the elastic properties of a new material

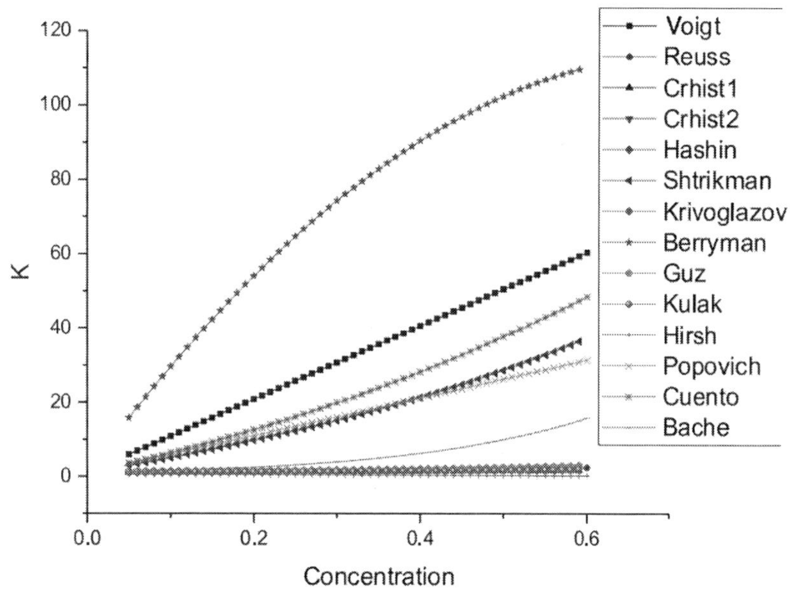

Figure 3. Dependence of the effective modulus of elasticity on the filler concentration for various models of homogenization.

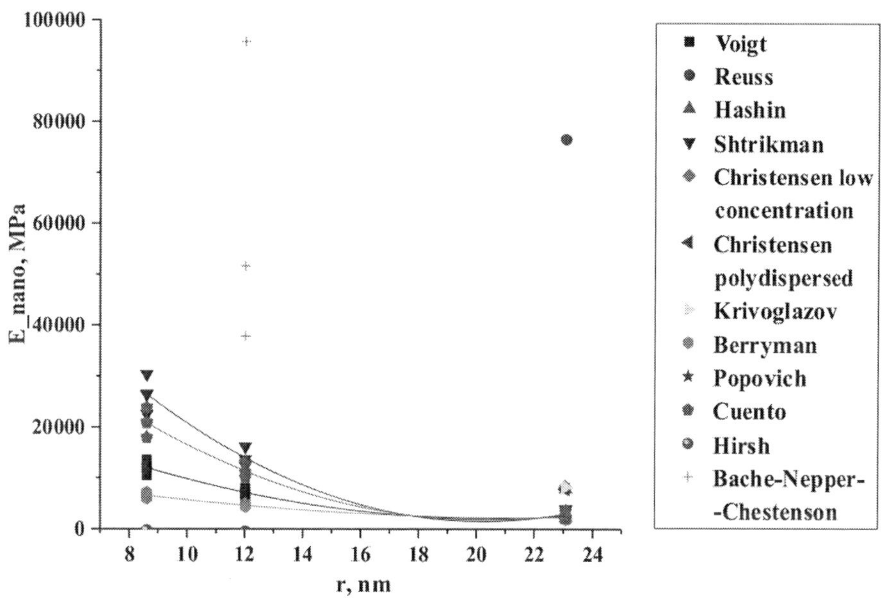

Figure 4. Estimated modulus of elasticity of the filler in various models of homogenization.

requires the determination of at least two independent elastic constants. In the present paper, it is assumed that Poisson's ratios of nanoparticles are the same as for the macro material, i.e., only one constant changes. To calculate the modulus of nanoparticles, formulas (1)–(14) used the Poisson coefficients (for nanoparticles $\nu_1 = 0.3$, for resins $\nu_2 = 0.3$). The specific surface area of the particles was recalculated for the radius by the formula $r = 3/(S_{\text{spec}} * \rho)$, where ρ is the density of quartz (2400 kg/m^3).

The calculation results for the presented models are shown in the figure. The range of the moduli is artificially specified and has limitations: a low level corresponds to the positive modulus of elasticity by definition, and the high level corresponds to the elasticity modulus for the quartz crystal (100 GPa). It should be noted that, for some homogenization models due to the nonlinear dependence, two values can be obtained at once. For models predicting a small increase with increasing concentration, such as the Reuss, Guz, Hirsch, Kulak and others models, the filler modulus values could be very large or negative. For example, for the Reuss model, from $1/K = c_1/K_1 + c_2/K_2$ we can express $K_1 = f(K, K_2, c_1, \nu_1, \nu_2)$:

$$K_1 = \frac{c_1 K K_2}{K_2 - (1 - c_1)K}.$$

The denominator must be greater than zero:

$$K_{\text{exp}} < \frac{K_2}{1 - c_1}.$$

This restriction in the Reuss model leads to high estimates of the filler modulus in the case where the inequality is satisfied and to negative values in the case where the experimental values of the modulus of the mixture do not satisfy the inequality. But these models are quite suitable for predicting the properties of fillers if the Young modulus of the filler is smaller than the matrix modulus.

A number of dependencies can predict the properties of the nanofiller in the composition of heterogeneous materials with an increase in their mechanical properties. The models of Voigt, Shtrikman, Popovich, and Cuento give satisfactory results for the entire set of presented experimental data. The Voigt model is applicable for determining the mechanical properties of the components in the experimental region of the experimental data. The applicability of the Voigt model is limited by the condition $E_{\text{exp}} > E_2 - c_1 E_2$.

According to the listed models, the elastic modulus increases with decreasing particle size. The averaging models do not take into account the size of the filler particles, but only the volume fraction in the mixture. However, even at this level, it is possible to compare the values of the moduli with the experimental data on nanoindentation or model the properties of particles by the molecular dynamics method [7].

Conclusions

The Young modulus for heterogeneous materials based on epoxy resin with addition of nano-sized silicon dioxide powders with different mean particle sizes at constant concentration was obtained. It is shown that as the average size decreases, the modulus of elasticity of the heterogeneous material increases significantly. The method proposed by the authors allows one, using the existing models of homogenization of the elastic properties of a heterogeneous material, to evaluate the characteristics of the filler which cannot be measured by direct experimental methods. Applying the above homogenization models to the obtained set of experimental data, it is shown that the modulus of elasticity of the filler increases with decreasing average particle size.

Acknowledgments

This work was supported by the Russian Scientific Foundation "Physical basis for creation of heterogeneous materials using additive technologies" under grant No. 16-19-10300.

IOP Conf. Series: Journal of Physics: Conf. Series **991** (2018) 012020 doi:10.1088/1742-6596/991/1/012020

References

[1] Ni Q-Q, Fu Y, and Iwamoto M 2004 Evaluation of elastic modulus of nano particles in PMMA/silica nanocomposites *J. Soc. Mater. Sci. Jap.* **53** (9) 956–61

[2] Duan H L, Wang J, Huang Z P, and Karihaloo 2005 Size-dependent effective elastic constants of solids containing nano-inhomogeneities with interface stress *J. Mech. Phys. Solids* **53** (7) 1574–96

[3] Bardakhanov S P, Korchagin A I, Kuksanov N K, et al. 2006 Nanopowder production based on technology of solid raw substances evaporation by electron beam accelerator *Mater. Sci. Engng B* **132** (1-2) 204–8

[4] Brusentseva T A, Filippov A A, Fomin V M, et al. 2014 Influence of the nanosized filler nature on the mechanical properties of epoxy-anhydride polymer composites *Nanotechnol. Russ.* **9** (11-12) 638–44

[5] Christensen R M 1979 *Mechanics of Composite materials* (New York: John Wiley & Sons, Inc.)

[6] Torquato S 2002 *Random Heterogeneous Materials. Microstructure and Macroscopic properties. Interdisciplinary applied mathematics, vol 16* (New York: Springer)

[7] Vakhrushev A V, Fedotov A Y, and Golubchikov V B 2016 Theoretical bases of modeling of nanostructures formed from the gas phase *Int. J. Math. Comput. Simulat.* **10** 192–201

IOP Conf. Series: Journal of Physics: Conf. Series **991** (2018) 012021 doi:10.1088/1742-6596/991/1/012021

Use of various versions of Schwarz method for solving the problem of contact interaction of elastic bodies

M P Galanin*, V V Lukin, and A S Rodin
Keldysh Institute of Applied Mathematics of the Russian Academy of Sciences, Moscow, Russia
Bauman Moscow State Technical University, Moscow, Russia
E-mail: *galan@keldysh.ru

Abstract. A definition of a sufficiently common problem of mechanical contact interaction in a system of elastic bodies is given. Various versions of realization of the Schwarz method for solving the contact problem numerically are described and the results of solution of a number of problems are presented. Special attention is paid to calculations where the grids in the bodies significantly differ in steps.

1. Introduction

Taking the contact interaction between various structural elements of constructions into account is an important component of assessment of the strain stress distribution (SSD) in bodies. Analytical solutions of contact problems have been received for a very restricted number of contact interactions and forms of the contacting surfaces. In almost all important situations, it is necessary to apply numerical methods among which, for solving the problem of the deformable solid mechanics (DSM), the leading position is held by the finite element method (FEM) [1–4].

2. Definition of the problem of contact interaction in a system of thermo-elastic bodies

In the three-dimensional space R^3 with a Cartesian coordinate system $Ox_1x_2x_3$, we consider a group of bodies occupying the area $G = \bigcup\limits_{\alpha} G_\alpha$ with a piecewise smooth border $\partial G = \bigcup\limits_{\alpha} \partial G_\alpha$. We assume that the relation between temperatures and deformation can be neglected. Therefore, the problem of heat conductivity can be solved separately, and the received temperature profile can be applied at the solution of a quasi-static problem of equilibrium of bodies.

We use the classical problem definition of heat conductivity taking into account the boundary conditions of second and third kinds [3, 4]. To solve the contact problem, we assume that the temperature at the corresponding points in the neighborhood of contacting bodies is the environment temperature on the site boundary where the heat exchange condition is set.

The mathematical statement of a quasi-static untied problem of DSM in the considered thermo-elastic statement includes the following ratios (for each body G_α, $i, j = \overline{1,3}$):

the equilibrium equations

$$\sigma_{ji,j}(\mathbf{u}, T) + Q_i(\mathbf{x}, t) = 0, \quad \mathbf{x} \in G_\alpha; \tag{1}$$

the boundary conditions (kinematic and stress)

$$\mathbf{u}(\mathbf{x},t)\big|_{S_1^\alpha} = \mathbf{u}^0(\mathbf{x},t), \quad \mathbf{x} \in S_1^\alpha \subset \partial G_\alpha, \tag{2}$$

$$\sigma_{ji}(\mathbf{u},T)n_j\big|_{S_2^\alpha} = p_i(\mathbf{x},t), \quad \mathbf{x} \in S_2^\alpha \subset \partial G_\alpha; \tag{3}$$

and the governing equations (in this case, Hooke's law) for stress tensor components

$$\sigma_{ij} = C_{ijkl}(\varepsilon_{kl} - \varepsilon_{kl}^{\text{th}}), \tag{4}$$

where $\mathbf{u}(\mathbf{x},t)$ is the displacement vector of the point determined by the radius vector $\mathbf{x} = x_i\mathbf{e}_i$, $\mathbf{u}^0(\mathbf{x},t)$ is the displacement vector of a point located on a surface S_1^α, $\mathbf{Q}(\mathbf{x},t) = Q_i(\mathbf{x},t)\mathbf{e}_i$ is the vector of mass forces, $\mathbf{p}(\mathbf{x},t) = p_i(\mathbf{x},t)\mathbf{e}_i$ is the vector of the external loading acting on a surface S_2^α, ε_{kl} are components of a tensor of the total strain (set by Cauchy relations), $\varepsilon_{kl}^{\text{th}}$ are components of a tensor of temperature strain, C_{ijkl} are components of a tensor of elastic coefficients.

When solving the contact problem on the contact surfaces of bodies, it is necessary to satisfy the contact conditions for interaction between displacements and stresses. For simplicity, we consider only the case of two bodies with one couple of contact surfaces. Let us consider two thermo-elastic contacting bodies A and B occupying the areas G_A and G_B with piecewise smooth boundaries ∂G_A and ∂G_B in the space R^3.

On the surface of contact $S_k = S_k^A = S_k^B$, the following conditions must be satisfied: on the displacements (kinematic conditions)

$$u_n^A(\mathbf{x},t) - u_n^B(\mathbf{x},t) = \delta_n(\mathbf{x}), \quad \mathbf{x} \in S_k, \tag{5}$$

and on the stresses (stress condition)

$$\sigma_n^A(\mathbf{x},t) = \sigma_n^B(\mathbf{x},t) \le 0, \quad \mathbf{x} \in S_k, \tag{6}$$

where u_n^A and u_n^B are projections of the displacement vectors of boundary points on the direction of the external normal $\mathbf{n}_A$ to the boundary of a body L, δ_n is the initial distance (gap) along the normal between boundary points of bodies A and B, σ_n^A and σ_n^B are projections of the stress vectors $\boldsymbol{\sigma}^A$ and $\boldsymbol{\sigma}^B$ on the external normals $\mathbf{n}_A$ and $\mathbf{n}_B$ respectively, $S_1^\alpha \bigcup S_2^\alpha \bigcup S_k^\alpha = \partial G_\alpha$, $\alpha \in \{A, B\}$.

The tangential contact stresses $\sigma_\tau^\alpha = \boldsymbol{\sigma}^\alpha \cdot \boldsymbol{\tau}_\alpha$ ($\boldsymbol{\tau}_\alpha$ is tangent to the contact boundary of the corresponding body) are calculated by the formula (Coulomb's law)

$$|\sigma_\tau^\alpha| = \mu|\sigma_n^\alpha|, \tag{7}$$

where μ is the friction coefficient (sliding friction).

If we pose adhesion conditions on the contact bundary, then not only normal components of the vectors of displacements and stresses, but all components of these vectors, appear in formulas (5), (6).

The set of relations (1)–(7) makes the mathematical statement of a contact problem of DSM. It is supposed that all functions entering this formulation have sufficient smoothness.

The multi-contact character of the considered problem is determined, first of all, by the geometry of the interacting bodies. Two bodies can interact but have several incoherent surfaces of contact. On the other hand, several bodies also can participate in interaction. These circumstances need to be considered when developing an algorithm for solving problem (1)–(7) numerically.

3. Application of the alternating Schwarz method for solving the contact problem numerically

The stated problem is solved numerically by FEM [1–4]. In particular, the applied version of FEM for this problem is described in [3, 5]. As a result of FEM application, the problem of MDSB (1)–(7) reduces to solving the linear matrix equation [2, 4]

$$[K]\{U\} = \{R\}. \tag{8}$$

Here the following notation is used: $[K]$ is global stiffness matrix, $\{U\}$ is global displacements vector, $\{R\}$ is global vector of loading.

Various iterative methods are used to solve the contact problems, for example, the penalty method, the Lagrange multiplier method, the combined method of penalties and Lagrange, the pseudo-medium method, an alternating Schwarz method and others [2, 4, 5]. In this work, the application of Schwarz method is considered (a version of decomposition method – see [6, p. 412] and [7] for a general case).

The essence of the method is as follows: at the first step, on the contact surfaces of the bodies, the initial approximation for components of the vector of displacements is set (the approximation is chosen from the range of expected values for the region of contact interaction). After solving this task, kinematic condition (5) on the contact surface is satisfied, but the calculated contact pressures on the opposite contact surfaces of the interacting bodies are not equal (condition (6) is violated). At the next step, by means of in a specially executed correction, we can obtain equal contact stresses but the obtained displacements do not satisfy condition (5). Further, at the next iteration step, the corrected kinematic conditions are again applied (combining the contacting surfaces). The stress and kinematic iterations are alternated to attain the convergence when both kinematic (5) and stress conditions (6) on the contact surface are satisfied with a prescribed accuracy. This method is described in more detail in [5, 8, 9]. Thus, the alternating Schwarz method is an iterative method and its essence within the finite element technology is as follows. At even iterations, the components of vectors of displacements of contact clusters $\{U_k\}_{(A)}$ and $\{U_k\}_{(B)}$ of bodies A and B is corrected. For body A, the correcting expressions have a form

$$\{U_k\}^{2n}_{(A),m} = \{U_k\}^{2n-1}_{(A),m} + \alpha^{2n-1}_{(A),m}(\{U_k\}^{2n-1}_{(B),s} - \{U_k\}^{2n-1}_{(A),m}), \quad n = 1, 2, \ldots, \tag{9}$$

for some initial displacement. Here $\alpha^{2n-1}_{(A),m}$ is an iteration parameter, m $(1 \leq m \leq M_A)$ is the number of the current node lying on the contact surface S^A_k of body A, M_A is the number of contact nodes on a surface S^A_k, $\{U\}^{2n-1}_{(B),s}$ is the displacement vector of the corresponding point s lying on the contact surface S^B_k of body B.

Similar relations are used to correct the components of displacement vectors of contact nodes of body B. Further, we solve two similar tasks of the elastic theory where the vectors $\{U_k\}_{(A)}$ and $\{U_k\}_{(B)}$ also play the role of additional kinematic boundary conditions.

For the Schwarz method used for odd iterations, the vector of global loading is the sum of two parts: the vector $\{R_{nk}\}$, which describes the impact of all forces on a body except for the contact ones, and the vector of nodal contact forces $\{R_k\}$ in which only the components corresponding to the clusters on the contact surface are nonzero.

The calculation of components of nodal forces $\{R_k\}_{(A)}$ and $\{R_k\}_{(B)}$, arising at contact nodes of bodies A and B, is also based on the assumption that, for any kinematic restriction imposed on the surface site, there is a stress loading such that it makes an equivalent impact on the body. Therefore, after carrying out the kinematic iteration with number $2n$, the values $\{R_k\}^{2n}_{(A),m}$ and $\{R_k\}^{2n}_{(B),m}$ can also be calculated by the formula (j is the global number of the variable corresponding to the node on the contact interface with number m):

$$\{R_k\}^{2n}_j = [K]_j\{U\}^{2n} - \{R_{nk}\}^{2n}_j. \tag{10}$$

In the simplest case of the stress conditions, corrections of nodal vectors of forces $\{R_k\}_{(A)}$ and $\{R_k\}_{(B)}$ are calculated. For body A, the correction has the form

$$\{R_k\}^{2n+1}_{(A),m} = \{R_k\}^{2n}_{(A),m} - \alpha^{2n}_{(A),m}(\{R_k\}^{2n}_{(B),s} + \{R_k\}^{2n}_{(A),m}), \quad n = 0, 1, 2, \dots . \tag{11}$$

Here $\alpha^{2n}_{(A),m}$ is the iteration parameter, m $(1 \le m \le M_A)$ is the number of the current node on the contact surface S^A_k of body A, $\{R_k\}^{2n}_{(B),s}$ is the vector of contact nodal force at the point s corresponding to the node m and lying on the contact surface S^B_k of body B. A similar relation is used to correct the components of vectors of the nodal forces arising at contact nodes of body B.

The vectors $\{R_k\}^{2n+1}_{(A)}$ and $\{R_k\}^{2n+1}_{(B)}$ are also used to form the global vectors of nodal loading $\{R\}_{(A)}$ and $\{R\}_{(B)}$ for bodies A and B. Then matrix equation (8) is solved for each of the two considered bodies.

Convergence of the described iteration scheme is considered in [8, 9].

The use of relations (11) for grids whose nodes on the contact surfaces of the two bodies coincide or are close to each other results in rather good results. But in the case of distinct nodes, especially when the grid steps are significantly different for the two bodies, the immediate application of (11) significantly worsens the convergence of the repetition process (when the second or higher order FE are used, the situation becomes even worse).

Because, in reality, not the concentrated forces carrying to grid clusters but the distributed contact forces operating on all contact boundaries act on the contacting sites of the surfaces of two bodies (by analogy with the vector of surface forces), the following expression can be written for the jth component of the vector $\{R_k\}$:

$$\{R_k\}_j = \int_{S_k} N^S_j p_k(\mathbf{x})\, dS, \tag{12}$$

where N^S_j is the basic function defined on a surface and corresponding to the node with number j and $p_k(x)$ is the contact pressure.

It is obvious that, when using distinct grids for stress iterations, it is necessary to correct not the values of nodal forces $\{R_k\}_{(A)}$ and $\{R_k\}_{(B)}$, which experience not only the action of pressure, and integrals of basic functions, but the values of contact pressure $p^{(A)}_k$ and $p^{(B)}_k$ in the corresponding clusters of the grid.

Further, we consider various realizations of the Schwarz method, which differ only in relations for stress iterations (for kinematic ones, the displacements are calculated by formula (9)).

The first realization of the Schwarz method is based on the following steps:

1. We replace the unknown contact pressure in (12) with an approximation based on use of basic functions N^S which are falling into clusters of the surface grid (for the first-order FE, they are piecewise linear functions)

$$p_k(x) \approx \sum_{m=1}^{M} p_{k,m} N^S_m, \tag{13}$$

where M is number of the nodes which are on the considered surface of contact.

2. After each kinematic iteration with number $2n$ $(n = 0, 1, 2, \dots)$, we determine the values of components of the vector of nodal forces $\{R_k\}^{2n}_m$, $m = 1, \dots, M$.

3. Substituting (13) in (12), we obtain a system of M equations for M unknown pressures

$$\int_{S_k} [N^S]^T [N^S]\, dS \{p_k\}^{2n} = \{R_k\}^{2n}. \tag{14}$$

Solving (14), we obtain the values of contact pressures after kinematic iteration. These pressures do not satisfy condition (6).

4. We carry out corrections of the received pressures by a formula similar to (11):

$$\{p_k\}^{2n+1}_{(A),m} = \{p_k\}^{2n}_{(A),m} - \alpha^{2n}_{(A),m}(\{p_k\}^{2n}_{(B),s} + \{p_k\}^{2n}_{(A),m}). \tag{15}$$

5. Knowing the corrected pressures, we calculate new values of nodal forces:

$$\{R_k\}^{2n+1} = \int_{S_k} [N^S]^T [N^S]\, dS \{p_k\}^{2n+1}. \tag{16}$$

6. Considering the obtained vector of nodal forces in the vector of global loading, we solve matrix equation (8) for each body.

Further, the method defined by points 1–6 will be called the "first method."

In [10], a generalization of the given version of the Schwarz method to the case of multi-contact interaction is considered. The influence of the choice of an initial approximation on the speed of convergence of the iteration process is investigated. For a particular class of problems, a modification of the Schwarz method is proposed which allows one to receive an initial approximation by solving a number of auxiliary problems by direct consideration of various loadings acting on the bodies.

4. Application of the Schwarz method to the problem of contact of a large number of bodies

The numerical algorithms described above are realized as computational procedures which, as a component of the central computing core, enter the prototype of the integrated program platform TEMETOS for carrying out computing experiments in complex problems of mathematical model operation [12].

The authors executed a number of calculations in which some thermo-mechanical effects occurring in a fuel element (TVEL) [10, 11] are simulated.

For example, for calculating the SSD parameters in TVEL at the exit to a rated duty, the following problem is considered: there is a thick-walled pipe (TVEL cladding) containing a column of identical cylinders put over each other; the cylinders have an internal opening and flats on both end faces (fuel pellets). In the cylinders, the uniform thermal emission obeying the following law is assumed: there is a linear rise in power of thermal emission in time till the limit value of $q_{l\,\max}$, and then the power remains constant. It is necessary to maintain the temperature of the external surface of the pipe at a constant value T_1. Between the external surfaces of the cylinders and the internal surface of the pipe, there is a heat exchange. The lower end face of the pipe and the lower end face of the lower cylinder are fixed. On the external surface of the pipe, the constant pressure p_1 is set. On the upper face of the top cylinder, the constant pressure p_2 is set.

The definition of the problem allows one to use a modified version of the Schwarz method. In the course of solving the problem, the dynamic temperature problem is solved first, and then the received temperature fields are used to solve a quasi-stationary equilibrium equation.

In [10], a series of calculations in the axially symmetric statement, where the number of cylinders varies from several tens to several hundreds (as contact conditions, the sliding

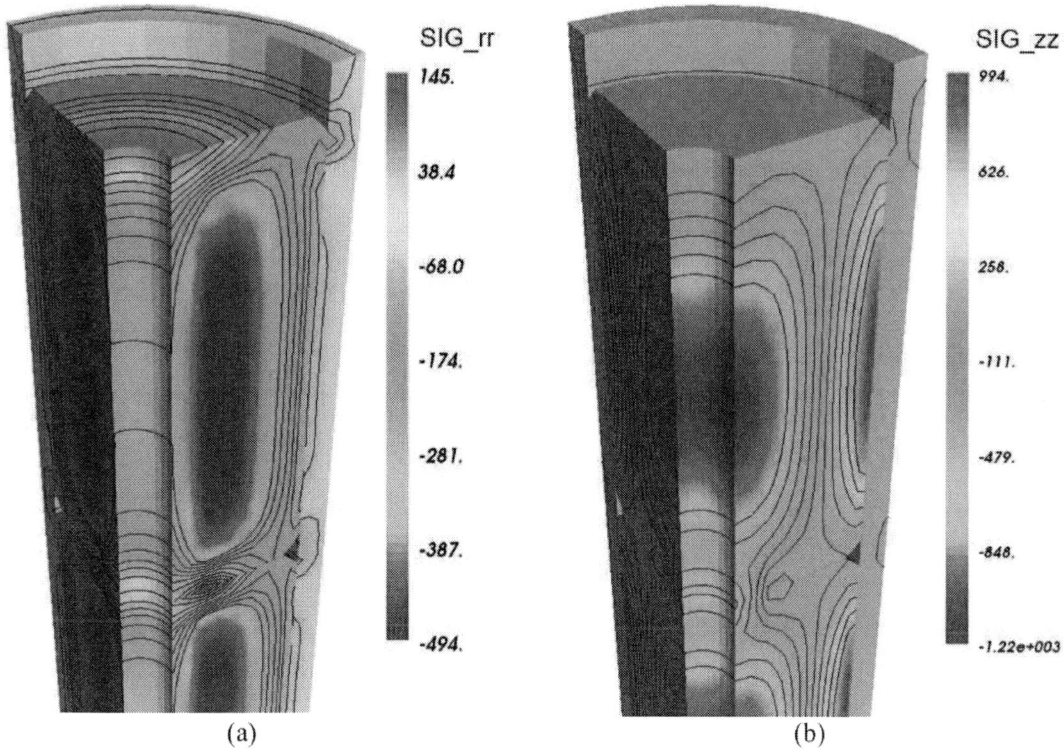

Figure 1. Radial (a) and axial (b) stresses.

conditions without sliding friction are chosen) is described. At an initial instant, there was a gap between the cylinders and pipe. Then the increase in the body temperature strains was a result of the contact between the cylinders and the pipe. The iterations stopped when the maximal relative change of displacements (compared with the two next iterations) did not exceed 1%. To achieve a similar accuracy, it is required, on the average, 10–15 iterations at one step in time (for any number of cylinders).

Let us note that, in these problems, the total system of linear equations for each body was solved self-contained which allowed one significantly to reduce the calculation time.

A similar problem is also solved in the three-dimensional statement (the number of cylinders varied from two to ten). In [11], the results of three-dimensional calculations for 4 pellets are given. The case of an axially symmetric loading where the calculation area corresponds to the sector 90° is therefore considered. As the contact conditions, the sliding condition with sliding friction is chosen.

The obtained results are compared with the results of two-dimensional (axisymmetric) calculations by means of the ANSYS code by using the Lagrange multiplier method. The comparison showed a rather good qualitative and quantitative compliance between the considered sizes.

Figure 1 shows the distribution of radial and axial stresses for the three-dimensional calculation area consisting of two cylinders and a pipe at one of the time instants.

At the moment, the used version of the Schwarz method has an essential drawback: it cannot be applied directly if there are cells of the surface grid which are only in partial contact (this will be called "partial contact of a cell"). In this case, for some clusters located on the boundaries of the contact surface, it is more difficult to interpret the concept of "corresponding point," which is of key significance for formulas (9), (11). In the course of heating, considerable axial displacements are observed on the cylinders, and similar situations with partial contact of cells arise constantly. To resolve this problem, a local reorganization of the grid in the pipe near the boundaries of the surface of contact between the pipe and each of cylinders was carried out at each iteration step. Because of such reorganizations of the grid, it was possible to avoid problems with partial contact of cells, but there were situations where the grid steps in special subareas from the cylinders and from the pipe differed significantly (sometimes by more than twice).

At the same time, in all carried-out calculations (three-dimensional and two-dimensional), there were stress concentrators near the contact of the corners of the cylinder facet and the pipe. The numerical values of contact pressure in these areas are very sensitive to the used calculation grid, and therefore such differences in the grid steps led to oscillations on the curves of contact pressure and decelerated the convergence of the iteration process.

These specified problems led to the comprehension that it is necessary to develop a more perfect realization of the Schwarz method. The main ideas which are the cornerstone of new version are explained below.

5. New version of the Schwarz method

The disadvantages of the previous version of the Schwarz method can be explained by the fact that when the grid is too rough for describing the contact pressure, the total force applied to the contact surface of the first body may differ from the total force applied to the contact surface of the second body, which leads to deceleration of the iteration process convergence.

To avoid this situation, it is possible to supplement the applied algorithm with the requirement that the forces acting on the contacting surfaces coincide. This restriction can be considered as follows: we divide the contact surface of each body into non-itersecting sites and, for the power iteration, demand that the equality hold not for the values of pressure at grid nodes but for the values of contact forces obtained after integration of the pressure on each of such sites for one body and for the corresponding site of the other body.

Let us note that a similar approach potentially allows one to solve a problem with partial contact of cells if, instead of the term "corresponding point," one uses the more universal concept "corresponding site."

Let us describe one of possible realizations of the given algorithm:

1. We replace the unknown contact pressure in (12) with an approximation based on use of basic functions χ^S which fall into the surface grid but optionally coincide with the basic functions N^S:
$$p_k(x) \approx \sum_{m=1}^{L} p_{k,m}\chi_m^S, \tag{17}$$
where L is the number of introduced basic functions. We see that the contact surface of the considered body consists of L non-intersecting sites $S_k = \bigcup_{j=1}^{L} S_{k,j}$.
2. After each kinematic iteration with number $2n$ ($n = 0, 1, 2, \ldots$), we find the values of components of the vector of nodal forces $\{R_k\}_m^{2n}$, $m = 1, \ldots, M$.
3. We obtain the system of M equations for L unknown pressurs
$$\int_{S_k} [N^S]^T [\chi^S]\, dS \{p_k\}^{2n} = \{R_k\}^{2n}. \tag{18}$$
Solving (18), we find the values of the contact pressure after kinematic iteration (if L does not coincide with M, then it is necessary to use the methods focused on solving systems with a rectangular matrix). The obtained pressure does not satisfy the stress condition (6).

IOP Conf. Series: Journal of Physics: Conf. Series **991** (2018) 012021 doi:10.1088/1742-6596/991/1/012021

4. We calculate the contact forces corresponding to each site $S_{k,j}$:

$$P_{k,j}^{2n} = \sum_i \int_{S_{k,j}} \chi_i^S \, dS p_{k,i}^{2n}. \tag{19}$$

5. We adjust the received contact forces by the formula similar to (11):

$$\{P_k\}_{(A),j}^{2n+1} = \{P_k\}_{(A),j}^{2n} - \alpha_{(A),j}^{2n}(\{P_k\}_{(B),s}^{2n} + \{P_k\}_{(A),j}^{2n}). \tag{20}$$

   Here $\{P_k\}_{(B),s}^{2n}$ is contact force calculated on the site of the surface of body B which corresponds to the considered part $S_{k,j}$ of body A.

6. After calculation of the contact forces, we determine the contact pressure. For this, it is necessary to solve a set of equations:

$$\sum_i \int_{S_{k,j}} \chi_i^S \, dS p_{k,i}^{2n+1} = P_{k,j}^{2n+1}, \quad j = 1, \ldots, L. \tag{21}$$

7. When the corrected pressures are known, we calculate new values of the nodal forces:

$$\{R_k\}^{2n+1} = \int_{S_k} [N^S]^T [\chi^S] \, dS \{p_k\}^{2n+1}. \tag{22}$$

8. Considering the obtained vector of nodal forces as the vector of global loading, we solve for each body the matrix equation (8) for each body.

It is possible to specify functions, piecewise constant on the surface elements, as one of possible choices of basic functions χ^S. Then L is the number of surface elements which are in contact (L differs from M, which is the number of nodes in contact).

For further application, we consider the case $L = M$. As non-intersecting sites of the contact surface $S_{k,j}$, we take the Dirichlet cells corresponding to nodes of the surface grid. As basic functions χ^S, we choose the functions N^S (for FE of first order, they are piecewise linear functions).

Further, the method given by points 1–8 is called the "second method."

6. Comparison of various versions of the Schwarz method for a test problem

Let us analyze the results of calculations with application of two realizations of the Schwarz method described above for the following two-dimensional test problem used as an exmple: the second body of width 6 and height 2 rests on the first body of height 4 and width 9. On the left side, the bodies are fixed across, and the lower body is fixed down. The upper body is affected by the distributed load given by the formula $p(x) = p_0[1 - \cos(2\pi x/l)]$, where $p_0 = 10$, $l = 3$.

As the conditions on the contact surface, we choose the adhesion condition, and therefore the solution of the considered contact problem must coincide with the solution of the problem where the considered design is a uniform body. Therefore, to estimate the accuracy of the obtained results, it is advisable to carry out comparison with the calculations corresponding to the case of one body under the same loading.

In all calculations, the FE of first order were applied. All values are given in dimensionless form.

The chosen initial approximation can affect the speed of convergence of the Schwarz method. But for the problems considered in this work, this influence is rather restrictive, and therefore, in all calculations, the following values of displacements on all surfaces of contact were chosen as the initial approximation: $u_x = 0$, $u_y = -10^{-4}$.

Table 1. Maximal relative errors for calculation with various grids.

Method number	$u_x(2)$	$u_y(2)$	$u_x(2)$	$u_y(1)$	$\sigma_{xx}(1)$	$\sigma_{yy}(1)$
			$h_1 = 0.25,\ h_2 = 0.25,\ h = 0.25$			
1	2.78×10^{-5}	2.39×10^{-5}	1.67×10^{-5}	1.79×10^{-5}	1.38×10^{-5}	2.08×10^{-5}
2	2.63×10^{-5}	2.38×10^{-5}	1.69×10^{-5}	1.79×10^{-5}	1.44×10^{-5}	2.08×10^{-5}
			$h_1 = 0.125,\ h_2 = 0.25,\ h = 0.125$			
1	4.9×10^{-2}	6.9×10^{-2}	1.5×10^{-2}	3.0×10^{-2}	1.8×10^{-1}	3.3×10^{-2}
2	1.2×10^{-2}	1.0×10^{-2}	2.0×10^{-2}	1.6×10^{-2}	1.9×10^{-1}	4.8×10^{-2}
			$h_1 = 0.125,\ h_2 = 0.375,\ h = 0.125$			
1	2.4×10^{-1}	5.1×10^{-2}	4.7×10^{-2}	7.3×10^{-2}	1.6×10^{-1}	1.5×10^{-1}
2	1.8×10^{-2}	1.3×10^{-2}	2.2×10^{-2}	2.8×10^{-2}	2.4×10^{-1}	6.2×10^{-2}

Let us consider the numerical results obtained for the solution by using two methods on various grids. The maximal relative errors of the numerical solution of the contact problem (compared with the case of one body) after 50 iterations (stress iteration) are presented in table 1. Calculations with the following steps of the grid are considered: $h_1 = 0.25$ and $h_2 = 0.25$, $h_1 = 0.125$ and $h_2 = 0.25$ (the grid steps differ by a factor of two), and $h_1 = 0.125$ and $h_2 = 0.375$ (the grids steps are differ by a factor of three).

To estimate the accuracy of the numerical solution of the contact problem, the following values were considered: the stresses (σ_{xx}, σ_{yy}) obtained near the contact boundary (at the center of the cells located along the contact surface of the first body), and the displacements (u_x, u_y) at the nodes located on the contact boundary (for both bodies). The number of a body is given in brackets. These values were compared with the corresponding values obtained in calculations for one body with h step (for the similar section).

For calculations where a uniform grid with an identical step was used in both bodies, the results obtained by different methods are almost identical. They coincide with the values obtained in calculations for one body with a relative accuracy of about 10^{-5}. When the grid steps differ rather strongly, then the results of calculations become different. In calculations on similar grids, all considered sizes are calculated with a significantly smaller accuracy. Moreover, one can observe that a larger discrepancy of displacements on the contact surfaces of two bodies is essential (after the power iteration): the previously obtained relative error of displacements (when comparing the values on the surfaces of the first and second body) was approximately 10^{-5}, but the displacements obtained now differ by percents. At the same time, the further decrease in the accuracy is not observed: at the power and kinematic iterations, we have two sufficiently steady solutions, which do not coincide with each other. At the power iterations, the nodal values of the contact pressure coincide, but there is an overlap of the displacements. At the kinematic iterations, the displacements coincide, but there is a gap of the contact pressure.

For descriptional reasons, we present the curves of displacements and stresses after 50 iterations for the maximal difference of steps of grids ($h_1 = 0.125$, $h_2 = 0.375$). Figure 2 shows the curves of displacement of u_x and u_y on the contact surfaces of both bodies calculated by various methods (curves with markers) and calculated for one body (the grid step is $h = 0.125$, curves without markers). The curves show that, for the first method, on the given grids at the considered power iteration, there is a very big divergence of displacements (for u_x, the discrepancy attains 25%, and for u_y, 7%). For the second method, the gap of displacements at the power iteration, is significantly less.

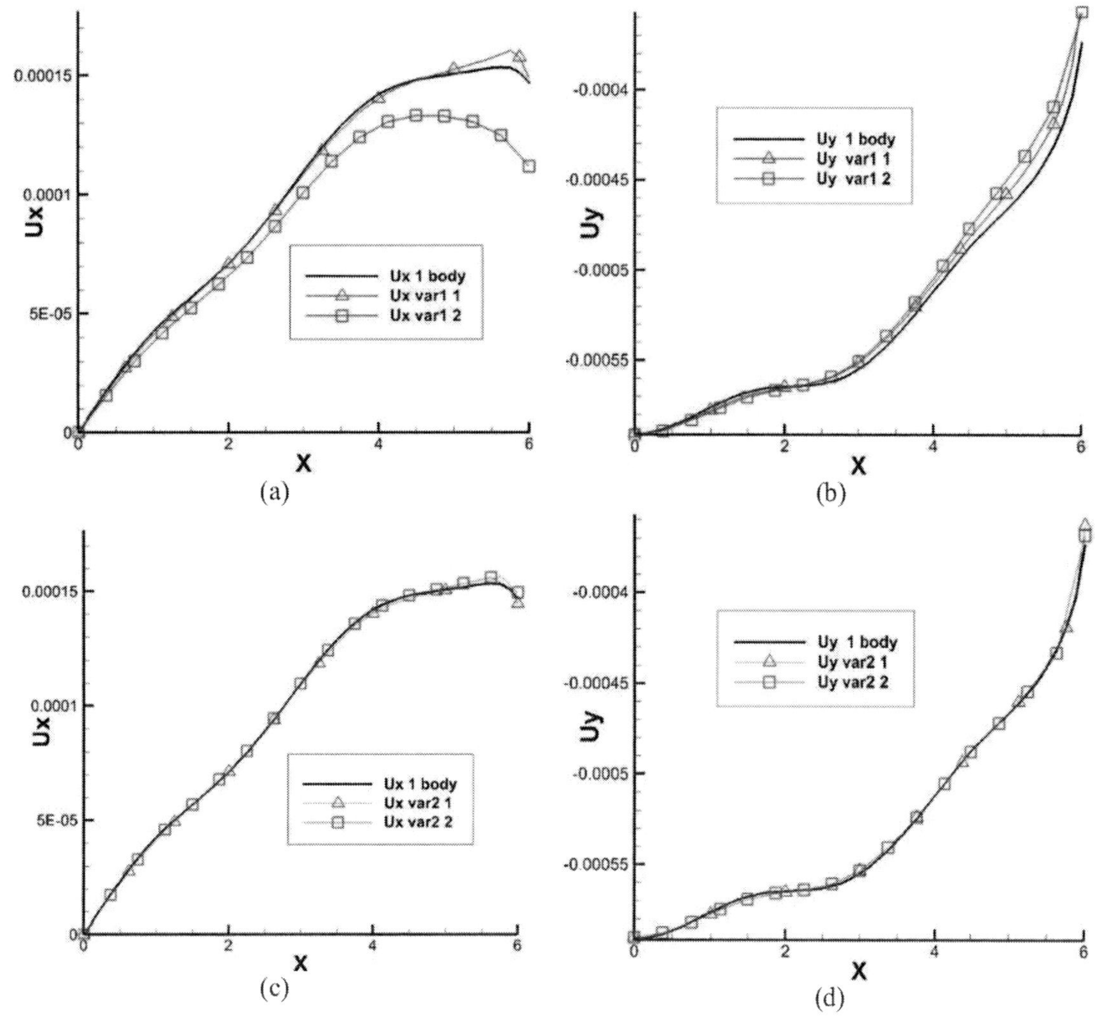

Figure 2. Components of the displacement vector: (a), (b) — "first method"; (c), (d) — "second method".

Figure 3 shows the curves of stresses (σ_{xx}, σ_{yy}) near the contact boundary of the first body for calculations by various contact methods (curves with markers) and for calculations for one body (the grid step is $h = 0.125$, curves without markers). For the considered problem, on the contact surface, there is a singular point, the edge of the upper body. As the grid becomes finer, the contact pressure grows at this point without bounds. Therefore, the stresses calculated in a neighborhood of this point are most sensitive to the grid steps. Therefore, in cases where grids with various steps are constructed, the greatest divergence of the solutions is observed.

Proceeding from the above analysis of the obtained data, it is possible to conclude that, for the considered test problem where significantly different grids were used, the second method allows one to obtain more precise results. Its application for the multi-contact problem considered in section 3 is planned in the near future.

IOP Conf. Series: Journal of Physics: Conf. Series **991** (2018) 012021 doi:10.1088/1742-6596/991/1/012021

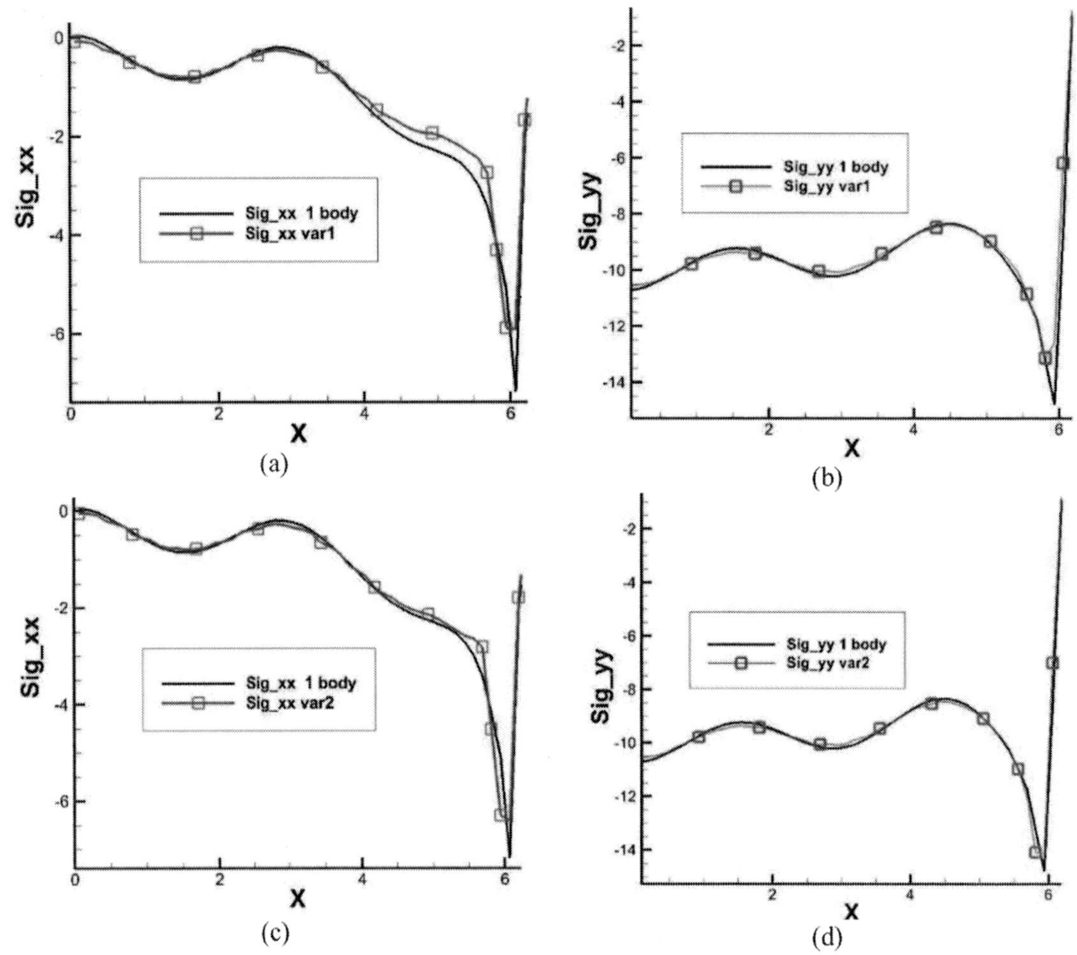

Figure 3. Components of the stress tensor: (a), (b) — "first method"; (c), (d) — "second method".

Conclusions

An algorithm for numerical solution of the contact problem for a system of interacting elastic bodies by using the alternating Schwarz method is explained. The results of application of this method for solving a multidimensional problem of contact interaction between a large number of bodies are given. A new version of the Schwarz method realization, which is focused on the case where the grid steps are significantly different in different bodies is proposed. An example of a two-dimensional test problem of contact is used to compare two methods for a series of calculations with more and more differing grids. It is shown that the new method allows one to obtain better results compared with the first.

Acknowledgments

The work was partially supported by the Russian Foundation for Basic Research (project No. 15-01-03073).

The authors thank professor I. V. Stankevich for great help in formulating the problem and choosing methods for solving it.

References

[1] Bourago N G and Kukudzhanov V N 2005 A review of contact algorithms *Mech. Solids* **40** (1) 35–71
[2] Zienkiewicz O 1975 *The Finite Element Method in Engineering Science* (New York: McGraw-Hill)
[3] Zarubin V S and Stankevich I V 2005 *Computations of Thermal Stress Structures* (Moscow: Mashinostroenie) p 352 [in Russian]
[4] Bathe K-J 1996 *Finite Element Procedures* (New York: Prentice Hall)
[5] Stankevich I V, Yakovlev M E, and Si Tu Khtet 2011 Development of contact interaction algorithm on the basis of Schwarz alternating method *Vestnik MGTU im. N.E. Baumana. Estestv. Nauki* No. S 134–41
[6] Marchuk G I 1989 *Methods of Numerical Mathematics* (Moscow: Nauka) p 608 [in Russian]
[7] Toselli A and Widlund O 2005 *Domain Decomposition Methods – Algorithms and Theory* (Berlin - Heidelberg - New York: Springer) p 450
[8] Tsvik L B 1978 Principle of alternating continuity in piecewise solutions of problems in field theory *Sov. Phys. Dokl.* **23** 824–6
[9] Tsvik L B 1980 Principle of alternation in problems of the conjugation and contact of solid deformable bodies *Int. Appl. Mech.* **16** (1) 9–13
[10] Galanin M P, Lukin V V, Rodin A S, and Stankevich I V 2015 Application of the Schwarz alternating method for simulating the contact interaction of a system of bodies *Comput. Math. Math. Phys.* **55** (8) 1393–406
[11] Galanin M P, Krupkin A V, Kuznetsov V I, et al. 2016 Modeling of contact interaction of a thermoelastic body system using Schwartz method for a multidimensional case *Izv. VUZov. Mashinostroenie* No. 12 9–20
[12] Galanin M P, Gorbunov-Possadov M M, Ermakov A V, et al. 2014 Prototype of an integrated software platform for tracking computer simulations to solve complex problems of mathematical modeling *Trudy ISP RAN* **26** (3) 51–68

IOP Conf. Series: Journal of Physics: Conf. Series **991** (2018) 012022 doi:10.1088/1742-6596/991/1/012022

Magneto-electric transition in nickel-gallium arsenide-nickel multiferroic structure

T A Galichyan[1*], D A Filippov[2], V M Laletin[3], T O Firsova[2], and N N Poddubnaya[3]

[1]Institute of Mechanics of the National Academy of Sciences of Armenia, Yerevan, Armenia
[2]Novgorod State University, Velikiy Novgorod, Russia
[3]Institute of Technical Acoustics, National Academy of Sciences of Belarus, Vitebsk, Belarus

E-mail: *tigrangalichyan@yahoo.com

Abstract. Experimental studies of the magnetoelectric effect are presented in structures manufactured by electrolytic deposition of nickel on a substrate of gallium arsenide. It is shown that the use of gold-germanium-nickel sublayer, when sprayed on a substrate, significantly improves the adhesion between electrolytically deposited nickel and substrate. Linear and nonlinear magnetoelectric effects on the alternating magnetic field are observed in these structures. Both effects have resonant character and the resonance frequency of the nonlinear effect is twice less than that of the linear effect. In weak fields, the value of the nonlinear magnetoelectric effect is in quadratic dependence on the alternating magnetic field and unlike the linear magnetoelectric effect, it does not depend on the bias field.

1. Introduction

Magnetoelectric (ME) multiferroic structures, simultaneously exhibiting ferromagnetism and ferroelectricity, have recently motivated an increasing number of research activities for their scientific interest and have drawn significant interest due to their potential applications in many multifunctional devices, such as transducers, passive magnetic field sensors, electric-write magnetic-read memory devices, microwave filters, energy harvest devices, actuators, etc. [1–3]. Natural multiferroic single-phase compounds are rare, and their ME responses are either relatively weak or occur at temperatures too low for practical applications. In contrast, the multiferroic composites, which incorporate both ferroelectric and ferri-ferromagnetic phases, typically yield giant ME coupling response above room temperature, which makes them ready for technical applications.

The historical perspectives of the multiferroic ME composite materials appeared in 1972. In such composites, the ME effect is generated as a product property of a magnetostrictive and a piezoelectric material. The multiferroic materials with coexistence of at least two ferroic orders (ferroelectric, ferromagnetic, or ferroelastic) have drawn growing attention due to their potential for applications as multifunctional devices. In multiferroic materials, the coupling interaction between different order parameters could produce new effects, such as the ME effect [4–7].

The layered magnetoelectric (ME) composites are structures consisting of interlaced, mechanically interacting magnetostrictive and piezoelectric plates. They have certain advantages over bulk composites: they are better polarized, and metals having a high coefficient of magnetostriction can be used as magnetostrictive phase [8]. However, the greatest disadvantages of layered structures are the poor mechanical durability and the samples fibering on the phase boundaries. Most of the laminated structures are prepared using an interlayer polymer phase (glue), which impairs the parameters, leads to an undesirably high temperature dependence, and reduces the quality factor of the structure. The magnetoelectric effect in a structure, where the magnetostrictive phase was applied to the piezoelectric substrate by the spraying method was studied [9]. This provides a good mechanical contact between the phases, but does not allow obtaining high values of the effect. The maximum effect is achieved under the condition ${}^{\mathrm{p}}t\sqrt{{}^{\mathrm{p}}Y} = {}^{\mathrm{m}}t\sqrt{{}^{\mathrm{m}}Y}$, where ${}^{\mathrm{m}}Y$ and ${}^{\mathrm{p}}Y$ are the Young's moduli of magnet and piezoelectric, ${}^{\mathrm{m}}t$ and ${}^{\mathrm{p}}t$ are the thicknesses, respectively [10]. Young's moduli of piezoelectric and magnet materials generally differ less than two times, so the maximum effect is obtained at about the same thickness of the magnet and piezoelectric. Using the electrodeposition method allows getting magnetostrictive layers which have thickness comparable with the thickness of piezoelectric substrate. However, a problem of adhesion arises. The previously obtained samples with thickness of magnetostrictive phase of $100\,\mu$m showed significant reduction in the adhesion properties. Consequently, the value of the effect in such structures is considerably lower than that of the adhesive structures [11]. An enhanced converse ME effect has been experimentally observed in cylindrical PZTp-Terfenolp-D piezoelectric–magnetostrictive bilayered composites [12]. In this paper, sublayers are used to improve the adhesion between the phases, which are deposited on the substrate by electron-beam evaporation.

2. Material and methods

According to the results of numerous performed experiments, the best results for GaAs substrate are achieved by using sublayers of Au–Ge–Ni composition. The thickness of each sublayer is of order $0.15\,\mu$m. In manufacturing the layered structures, a material with highest piezoelectric value is usually selected as a piezoelectric, for example, lead zirconate titanate (PZT). However, the results of calculations in [10, 13] show that the value of ME effect is directly proportional to the piezoelectric modulus and inversely proportional to the dielectric constant of the piezoelectric. Although the piezoelectric value of GaAs is 37 times less than that of PZT (GaAs $d = -2.69$ pC/m, PZT $d = 100$ pC/m), its dielectric constant is 135 times less than that of PZT (GaAs $\varepsilon = 12.9$, PZT $\varepsilon = 1750$). As a result, under the other equal conditions, we may expect the value of the ME effect in structures based on GaAs to be 3.5 times greater than in structures based on PZT.

Another challenge in creating structures is based on the effects that occur at the interface due to the disproportion between the phases. The lattice constant of GaAs is 0.56533 nm and 0.3524 nm for Ni. As a result, large mechanical stresses arise during the growth of thick layers in the structure. The structure is destroyed even under small mechanical influences due to the warping of the sample if the Ni layer is deposited on one side of the substrate. Therefore, the trilayer system Ni–GaAs–Ni is used to manufacture the structures. This increases the durability of the structure, on one hand, and in turn prevents the samples to change their initial shape, on the other hand. Gallium arsenide belongs to the hextetrahedral class (the international symbol is $\bar{4}3$m), and the nonzero components of piezoelectric moduli are $d_{14} = d_{25} = d_{36}$. Therefore, the ME effect in GaAs occurs as a result of transverse oscillations propagation in the structure, unlike the PZT, where the ME effect is a result of longitudinal oscillations. Shear deformations occur at the XY plane in the magnetostrictive component when the sample is placed in an alternating magnetic field having Y projection. These deformations are transmitted to the piezoelectric phase by the stress tensor T_{yz}, which leads to a change in polarization and the appearance of voltage between the electrodes.

Magnetostriction is an effect, generally nonlinearly dependent on the magnetization. Its value depends on the size of the sample and the thickness of the film. The dependence of

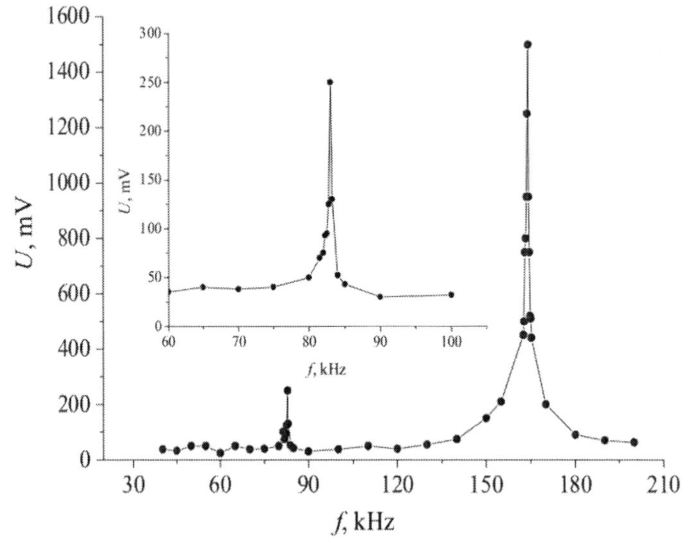

Figure 1. Frequency dependence of the ME effect for Ni-GaAs-Ni structure. The alternating magnetic field is $H = 0.1\,\mathrm{Oe}$ and the bias magnetic field is $H_{\mathrm{bias}} = 10\,\mathrm{Oe}$.

magnetostriction on the magnetic field is shown in [14] for samples of length of about 14 mm and with deposited Ni films. As was shown in the figure [14], in weak fields (up to 50 Oe), the dependence of the magnetostriction $\lambda(H)$ can be represented as

$$\lambda(H) = qH + gH^2. \tag{1}$$

Here the following designations are used: $q = \partial\lambda/\partial H|_{H=H_{\mathrm{bias}}}$ is the piezomagnetic coefficient, $g = \frac{1}{2}\partial^2\lambda/\partial H^2|_{H=H_{\mathrm{bias}}}$ is the magnetostrictive coefficient, and H_{bias} is the bias magnetic field. It is clear that in the absence of the bias field, i.e., for $H_{\mathrm{bias}} = 0$, the piezomagnetic coefficient is $q = 0$, whereas the magnetostriction coefficient is $g \neq 0$. This leads to the linear ME effect which is observed only for a bias magnetic field, except for the samples with hysteresis [15]. The nonlinear ME effect is observed for both $H_{\mathrm{bias}} = 0$ and in the case where the magnetizing field is nonzero. Since the oscillation frequency is proportional to the squared harmonic signal, the oscillations of the medium are excited with twice the frequency of the alternating magnetic field. As a result, the resonant excitation occurs at half the magnetic field frequency of the main resonance.

3. Experimental section

The experimentally obtained frequency dependence of the ME effect is shown in figure 1 for Ni–GaAs–Ni structure. The structures were prepared by the method of electrodeposition of nickel on a substrate of gallium arsenide with sublayers Au–Ge–Ni. The samples were cut from plates with surface orientation (100) as parallelepipeds of thickness ${}^{\mathrm{p}}t = 0.4\,\mathrm{mm}$, length $L = 15\,\mathrm{mm}$, and width $W = 5\,\mathrm{mm}$. The thickness of each nickel layer was ${}^{\mathrm{m}}t = 20\,\mu\mathrm{m}$. The long side of the sample had direction $< 110 >$. The samples were placed in a solenoid so that the long boundaries of the samples and the solenoid axis were at 45 degrees with respect to each other. The value of the alternating magnetic field was $H = 0.1\,\mathrm{Oe}$.

Figure 1 shows that, along with the main resonance at a frequency $f_1 = 165\,\mathrm{kHz}$, an additional resonance is observed at a frequency $f_2 = 82.5\,\mathrm{kHz}$, which is half the frequency of the main resonance. The frequency dependence of the ME effect in the second resonance region is additionally shown in the insertion. The value of this resonance did not changed as the magnetizing field changed. The peak value of the voltage for linear ME effect is $U_{\mathrm{res}} = 1.5\,\mathrm{V}$. This gives a value of the ME coefficient equal to $\alpha_{\mathrm{E}} = U/{}^{\mathrm{P}}t \cdot H = 375\,\mathrm{V/cm{\cdot}Oe}$ or $\alpha_{\mathrm{E}} = 470\,\mathrm{V/A}$ when the piezoelectric thickness is ${}^{\mathrm{P}}t = 0.4\,\mathrm{mm}$ and the magnet field value is $H = 0.1\,\mathrm{Oe}$.

The obtained value for the ME coefficient is 5 times greater than the same value for structures manufactured by the method of spraying Ni on GaAS substrate (the ME coefficient for such structures is $\alpha_{\mathrm{E}} = 85\,\mathrm{V/A}$). It is also greater than the value of the ME coefficient ($\alpha_{\mathrm{E}} = 113\,\mathrm{V/A}$) in Ni–PZT–Ni adhesive structures (Ni and PZT layers are glued).

Conclusions

Structures with good adhesion between the layers and with thickness of magnetostrictive layer 20–30 μm can be obtained using Au–Ge–Ni sublayers deposited on GaAs substrate and followed by Ni electrolytic deposition. The linear ME effect in an alternating magnetic field and a nonlinear ME effect are observed in the structure. In the obtained structures, the resonance value of the ME coefficient is almost 5 times greater than this value for the structures obtained by Ni deposition on GaAs substrate and 4 times greater than this value for the adhesive structure Ni–PZT–Ni.

References

[1] Nan C W, Bichurin M I, Dong S, et al. 2008 Multiferroic magnetoelectric composites: Historical perspective, status, and future directions *J. Appl. Phys.* **103** (3) 031101

[2] Wang Y, Gray D, Berry D, et al. 2011 An extremely low equivalent magnetic noise magnetoelectric sensor *Adv. Mater.* **23** (35) 4111–4

[3] Li P, Wen Y, Liu P, et al. 2010 A magnetoelectric energy harvester and management circuit for wireless sensor network *Sensors Actuat. A. Phys.* **157** (1) 100–6

[4] Schmid H (Editor) 1994 *Magnetoelectric Interaction Phenomena in Crystals, Proceedings of the MEIPIC-2 Ascona, Switzerland, 13–18 September 1993* [*Ferroelectrics* **161–162**]

[5] Bichurin M (Editor) 1997 *Proceedings of the MEIPIC-3, Novgorod, Russia, 16–20 September 1996* [*Ferroelectrics* **204**]

[6] Bichurin M *Proceedings of the MEIPIC-4, Novgorod, Russia, 16–19 October 2001* [*Ferroelectrics* **279–280**]

[7] Fiebig M, Eremenko V V, and Chupis I E 2004 Magnetoelectric interaction phenomena in crystals In *Proceedings of the MEIPIC-5, Sudak, Ukraine, 21–24 September 2003* (Dordrecht: Kluwer)

[8] Filippov D A, Laletin V M, and Srinivasan G 2012 Low-frequency and resonance magnetoelectric effects in nickel ferrite-PZT bulk composites *Tech. Phys.* **57** (1) 44–7

[9] Laletin V M, Stognii A I, Novitskii N N, and Poddubnaya N N 2014 The magnetoelectric effect in structures based on metallized gallium arsenide substrates *Tech. Phys. Lett.* **40** (11) 969–71

[10] Filippov D A, Laletin V M, and Galichyan T A 2013 Magnetoelectric effect in a magnetostrictive-piezoelectric bilayer structure *Phys. Solid State* **55** (9) 1840–5

[11] Poddubnaya N N, Laletin V M, Stognij A I, and Novitskii N N 2010 Dependence of magnetoelectric effect in layered lead zirconate-titanate / nickel heterostructures on the interface type *Funct. Mater.* **17** (3) 329–34

[12] Wu G, Zhang R, and Zhang N 2016 Enhanced converse magnetoelectric effect in cylindrical piezoelectric-magnetostrictive composites *Eur. Phys. J. Appl. Phys.* **76** (1) 10602

[13] Filippov D A, Galichyan T A, and Laletin V M 2014 Magnetoelectric effect in bilayer magnetostrictive-piezoelectric structure. Theory and experiment *Appl. Phys. A* **115** (3) 1087–91

[14] Burdin D A, Chashin D V, Ekonomov N A, et al. 2014 Nonlinear magneto-electric effects in ferromagnetic-piezoelectric composites *J. Magn. Magn. Mater.* **358–359** 98–104

[15] Burdin D A, Fetisov L Y, Fetisov Y K, et al. 2014 Resonance magnetoelectric effect without a bias field in a piezoelectric langatate-hysteretic ferromagnet monolithic structure *Tech. Phys.* **59** (9) 1350–5

IOP Conf. Series: Journal of Physics: Conf. Series **991** (2018) 012023 doi:10.1088/1742-6596/991/1/012023

Magnetoelectric effect in a sandwich structure of gallium arsenide–nickel–tin–nickel

T A Galichyan[1*], D A Filippov[2], A A Tihonov[2], V M Laletin[3], T O Firsova[2], and I N Manicheva[2]

[1]Institute of Mechanics of the National Academy of Sciences of Armenia, Yerevan, Armenia
[2]Novgorod State University, Velikiy Novgorod, Russia
[3]Institute of Technical Acoustics, National Academy of Sciences of Belarus, Vitebsk, Belarus

E-mail: *tigrangalichyan@yahoo.com

Abstract. The results of investigation of the magnetoelectric effect in a nickel-tin-nickel sandwich structure obtained by galvanic deposition of gallium arsenide on a substrate are presented. The technology of constructing such structures is described and the experimental results of the frequency dependence of the effect are presented. It is shown that the use of tin as an intermediate layer reduces the mechanical stresses resulting from the incommensurability of the phases, which permits obtaining qualitative structures with the nickel thickness of about 70 μm. The resulting structures exhibit good adhesion between the layers and have a high quality factor.

1. Introduction

The magnetoelectric (ME) effect, discovered more than half a century ago, recently attracts more attention of the researchers which is shown by an increasing number of publications on this subject [1]. This is due to the fact that rather many materials with the effect magnitude sufficient for practical applications have recently been discovered. This permits creating various devices of solid-state electronics such as magnetic field sensors or memory cells on the basis of the ME effect, which, by their parameters, are not only inferior but, in some cases, superior to the traditional devices. The composite materials consisting of magnetostrictive and piezoelectric phases are widely used to create such devices. These materials can be divided into two classes, bulk and layered. The bulk composites are manufactured by the ceramic technology and represent mechanically coupled mixtures of powders of magnetostrictive and piezoelectric phases [2, 3]. The layered composites consist of mechanically coupled magnetostrictive and piezoelectric layers [4–6], where the layers can be arranged both in parallel and in series. The structures can be shaped as plates, disks or rings with radial polarization and as cylinders. Both bulk and layered composites have their advantages and disadvantages. The bulk composite materials obtained by sintering the mixtures of ferrite and piezoelectric powders are easily manufactured and have good mechanical properties, but they have smaller ME parameters compared to the layered composites. The advantages of layered structures is a high degree of the piezoelectric phase polarization and small leakage currents, since the magnetostrictive phase with higher conductivity is isolated by the piezoelectric phase with high resistivity. In this case, the metals with a large magnetostriction coefficient can be used as a magnetostrictive phase. However, great disadvantages of layered structures are the poor mechanical strength, stratification of the samples along the phase boundaries, and a low quality factor. Most of the layered structures are obtained using an intermediate polymer layer (glue) which degrades the parameters and leads to undesirably high temperature dependence and a decrease in the quality factor of the structure. In [7], the ME effect was studied in the structures where the magnetostrictive phase was applied to a piezoelectric substrate by the deposition method. This provides a good mechanical contact between the phases, but does not allow obtaining a large value of the effect. As was shown in [8], the maximum value of the effect is obtained under the condition ${}^{\mathrm{p}}t\sqrt{{}^{\mathrm{p}}Y} = {}^{\mathrm{m}}t\sqrt{{}^{\mathrm{m}}Y}$, where ${}^{\mathrm{p}}Y$, ${}^{\mathrm{m}}Y$ are the Young's moduli of the piezoelectric and the magnet, respectively, and ${}^{\mathrm{p}}t$, ${}^{\mathrm{m}}t$ are the thicknesses. Young's moduli of a piezoelectric and a magnetic material, as a rule, differ at most by two times, so the maximum value of the effect is obtained for approximately equal thicknesses of the magnet and piezoelectric. The use of the method of electrolytic deposition makes permits obtaining magnetostrictive layers whose thickness is commensurable with the thickness of the piezoelectric substrate. However, this raises the problem of adhesion. The better adhesion between the magnetostrictive and piezoelectric phases can be obtained by using pre-sprayed sublayers. As was shown in [9], the use of Au–Ge–Ni sublayers deposited on a GaAs substrate with subsequent electrolytic deposition of Ni allows obtaining structures with good adhesion between the layers. However, mechanical stresses arise due to the disparity of the lattice parameters, which, at large layer thicknesses, lead to distortion and destruction of the structure. In this paper, we propose a method for eliminating these stresses by creating a sandwich structure in which the electrolytically deposited layer of nickel alternates with an electrolytically deposited layer of tin.

2. Technique for manufacturing the structure

The initial samples were cut from plates with surface orientation (100) as parallelepipeds of dimensions $14 \times 4 \times 0.4$ mm so that their long side coincided with the $< 110 >$ direction of the crystal. In the production of layered structures, as a rule, the piezoelectric material is a material with the highest piezoelectric modulus value d. However, the calculations show [10] that the ME effect is directly proportional to the piezoelectric modulus and is inversely proportional to the dielectric constant of the piezoelectric material. Although the piezoelectric modulus of GaAs is 37 times smaller than that of a PZT ($d = -2.69$ pC/m for GaAs and $d = 100$ pC/m for PZT), its dielectric constant is 135 times smaller than that of a PZT ($\varepsilon = 12.9$ for GaAs and $\varepsilon = 1750$ for PZT). This implies that, under the other equal conditions, it is expected that the magnitude of the ME effect in GaAs based structures is almost the same as in the structures based on the PZT. However, unlike the PZT, gallium arsenide is a single crystal with more stable properties. It should not be preliminarily polarized and, in addition, a well-tuned semiconductor technology can be used in the production of structures.

To improve the adhesion to the samples, Au–Ge–Ni sublayers were previously deposited. Before plating, all samples were first contacted with a 0.2 mm diameter nickel wire or with a clamp. The samples were then degreased with Vienn lime or lambomide solution 203.

Vienna lime is a mixture of calcium oxide and magnesium oxide. Degreasing with Viennese lime is done in small-scale production by hand using a brush. Thin slurry of Viennese lime is rubbed by brush onto the surface of the product, and then the product is washed with water. These operations were performed three times until the surface of the product was completely wetted with water. More efficient, less laborious, and allowing one to obtain high quality degreasing is the operation with the help of labomide 203. For this purpose, a solution of labomide 203 concentration from 30 to 40 g/l was used, and the solution was heated to 70–80°C. The degreasing time was 10–20 minutes. After degreasing, the samples were washed in hot water at a temperature of 60–80°C for 1 minute.

Electrolytic deposition of nickel generates high internal stresses and it is impossible to obtain

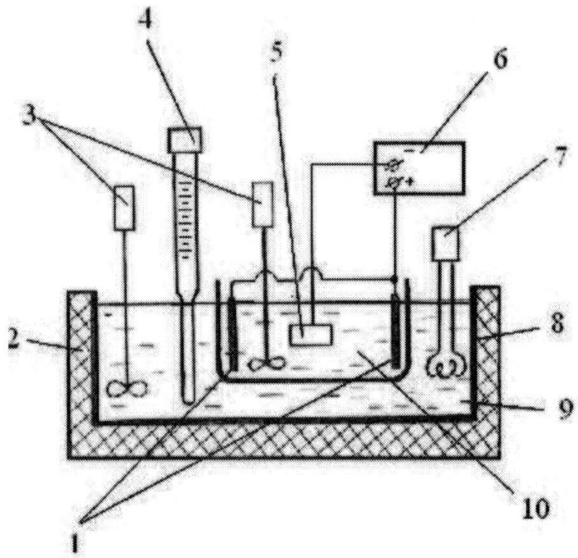

Figure 1. Coating installation 1 — anodes, 2 — heat insulation, 3 — device for mixing, 4 — contact thermometer, 5 — cathode, 6 — DC source, 7 — tubular electric heater, 8 — thermostat, 9 — water, 10 — electrolyte.

Table 1. Composition of electrolytes used to create the structures.

Components of electrolyte	Electrolyte No. 1, g/l	Electrolyte No. 2, g/l
Nickel sulfate heptahydrate	250	—
Nickel chloride hexahydrate	50	—
Tin sulfate	—	60
Boric acid	25	—
Sulfuric acid	—	105
Preparation OS-20	—	4.5

thick (more than 50 μm) layers as a result. In order to increase the plasticity of nickel-based coatings, it was decided to introduce thin interlayers of tin into the coatings. The electrolytes shown in table 1 were used to apply galvanic coatings.

A current pulse was applied for all electrolytes, i.e., at the beginning of electrolysis, the cathode current density (briefly up to 1 min) was increased by a factor of 2 compared to the main working cathode current density.

For the application of coatings, the installation shown in figure 1 was used.

In manufacturing a multilayer structure, the electrolytic deposition was alternately used in the sulfuric acid electrolyte of nickel No. 1 with the cathode current density of 1 A/dm^2 and the electrolyte temperature of 55–65°C, and then in the tin electrolyte No. 2 with the cathode current density of 2 A/dm^2 at room temperature. As a result, we obtained a multilayer structure consisting of the following layers: tin (8 μm), nickel (12 μm), tin (9.6 μm), nickel (12 μm), tin (38.4 μm), nickel (6 μm), tin (8 μm), nickel (12 μm), tin (7.2 μm), nickel (12 μm), tin (9.6 μm, nickel (12.6 μm, tin (9.6 μm). Thus, a sandwich structure was obtained consisting of six layers of nickel, of total thickness of 66.6 μm, and seven layers of tin, of total thickness of 90.4 μm. The total thickness of the multilayer structure was 157 μm. The coating on the gallium arsenide turned out to be smooth, frosted and without visible defects.

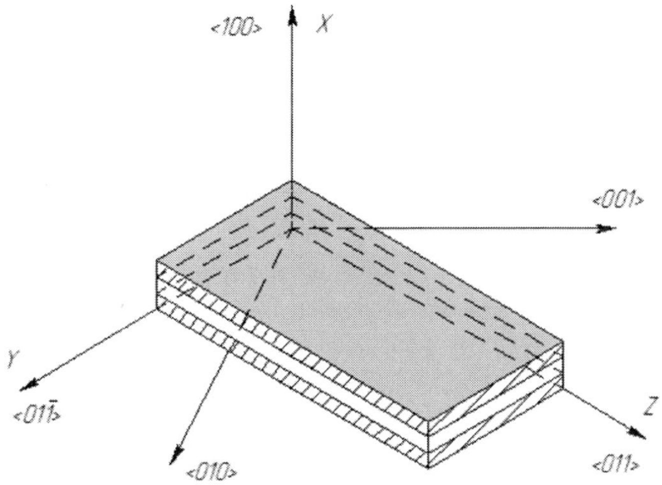

Figure 2. Sample placed in a constant (magnetizing) and an alternating magnetic field.

3. Magnetoelectric effect

The magnetoelectric (ME) effect in the structure was studied by measuring the voltage on a sample placed in a constant (magnetizing) and an alternating magnetic field (figure 2).

First, the field dependence of the low-frequency ME signal was investigated. At a constant value of the strength of the alternating magnetic field of 1 Oe, the dependence of the ME coefficient on the strength of the bias field was measured. Then, with the magnetization field strength corresponding to the maximum of the effect, the frequency dependence of the magnetoelectric coefficient in the electromechanical resonance region was investigated. The measurements were carried out when the permanent and alternating magnetic fields were directed along the long side of the sample. The singularity of the ME effect, when gallium arsenide is used as a piezoelectric, is that the nonzero components of the piezoelectric tensor are $d_{14} = d_{25} = d_{36}$, and therefore, the electric voltage induced on the sample plates arises due to shear deformations rather than to tension-compression deformations as in PZT. In our case, an alternating magnetic field directed along the long side of the sample (the Z axis) induces tension-compression strains in the magnetic component whose tensor, in the coordinate system associated with the sample, will be denoted by S_{zz}. These deformations are transferred to a piezoelectric which, as a result of piezoelectric effect, induced by the electric field. The equation of motion for the z projection of the displacement vector of the medium can be written as

$$\bar{\rho}\frac{\partial^2 u}{\partial t^2} = \frac{\partial \bar{T}_{zz}}{\partial z}, \tag{1}$$

where $\bar{\rho}$ is the average value of the density of the sample, and $\bar{T}_{zz}$ is the average value of the stress tensor of the sample.

The equations for the component of the strain tensors ${}^{\mathrm{p}}S_{zz}$ and the x projection of the electric induction ${}^{\mathrm{p}}D_x$ of the polarized piezoelectric phase take the form

$${}^{\mathrm{p}}S_{zz} = \frac{1}{{}^{\mathrm{p}}Y}{}^{\mathrm{p}}T_{zz} + {}^{\mathrm{p}}d_{x,zz}{}^{\mathrm{p}}E_x, \tag{2}$$

$${}^{\mathrm{p}}D_x = {}^{\mathrm{p}}\varepsilon_{xx}{}^{\mathrm{p}}E_x + {}^{\mathrm{p}}d_{x,zz}{}^{\mathrm{p}}T_{zz}, \tag{3}$$

where ${}^{\mathrm{p}}T_{zz}$ is the component of the stress tensor in the piezoelectric phase, ${}^{\mathrm{p}}Y$ is the Young's modulus along the direction $< 011 >$ (Z axis), ${}^{\mathrm{p}}d_{x,zz}$ is the piezoelectric tensor in the coordinate system XYZ (see figure 2), ${}^{\mathrm{p}}\varepsilon_{xx}$ is the dielectric tensor, and ${}^{\mathrm{p}}E_x$ is the x component of the electric field.

The components of the piezoelectric tensor ${}^{\mathrm{p}}d_{x,zz}$ in the coordinate system XYZ connect with the components of the piezoelectric tensor ${}^{\mathrm{p}}d_{\alpha,\beta}$ in a coordinate system associated with crystallographic axes by the following relation

$$d_{x,zz} = d_{14}\beta_{z2}\beta_{z3}, \tag{4}$$

where β_{z2} and β_{z4} are the cosiness of the angle between the axes Z and the axes 2 and 3 (the $< 010 >$ and $< 001 >$ directions).

Equations for the magnetostrictive phase can be written in the form

$${}^{\mathrm{m}}S_{zz} = \frac{1}{{}^{\mathrm{m}}Y}{}^{\mathrm{m}}T_{zz} + {}^{\mathrm{m}}q_{z,zz}{}^{\mathrm{m}}H_z, \tag{5}$$

where ${}^{\mathrm{m}}T_{zz}$ is the component of the stress tensor in the magnetostrictive phase, ${}^{\mathrm{m}}Y$ is the Young modulus, ${}^{\mathrm{m}}q_{z,zz}$ is the longitudinal piezomagnetic coefficient, and ${}^{\mathrm{m}}H_z$ is the magnetic field.

Let us represent the solution of the equation for the displacement vector of the medium in following form:

$$u(z) = A\cos(kz) + B\sin(kz), \tag{6}$$

where A and B are the integration constants.

Using the Eq. (6), we obtain the equation for the dependence of the frequency on the wave vector at the propagation of elastic waves in the magnetostrictive-piezoelectric multilayer structure in following form

$$\omega = \sqrt{\frac{\bar{Y}}{\bar{\rho}}}k = \sqrt{\frac{{}^{\mathrm{m}}Y{}^{\mathrm{m}}t + {}^{\mathrm{l}}Y{}^{\mathrm{l}}t + {}^{\mathrm{p}}Y{}^{\mathrm{p}}t}{{}^{\mathrm{m}}\rho{}^{\mathrm{m}}t + {}^{\mathrm{l}}\rho{}^{\mathrm{l}}t + {}^{\mathrm{p}}\rho{}^{\mathrm{p}}t}}k. \tag{7}$$

4. Magnetoelectric voltage coefficient

The magnetoelectric voltage coefficient is defines as the ratio of the electric field E induced by the magnetic field to the magnetic field H; i.e.,

$$\alpha_{\mathrm{E}} = \frac{E}{H}. \tag{8}$$

However, in contrast to bulk composites, the magnetoelectric voltage coefficient of layered magnetostrictive-piezoelectric structures can be defined in two ways. The first one is the ratio of the electric field induced in the piezoelectric to the magnetic field in the magnet producing this electric field; i.e.,

$${}^{\mathrm{p}}\alpha_{\mathrm{E}} = \frac{{}^{\mathrm{p}}E}{{}^{\mathrm{m}}H}. \tag{9}$$

The magnetoelectric coefficient defined in this ways characterizes well the efficiency of the magnetoelectric field transformation, i.e., the transformation of the magnetic field in the magnet into the electric field in the piezoelectric. However, under this definition, the magnetoelectric coefficient characterizes rather poorly the efficiency of the magnetoelectric transformation of the entire structure. In the experimental determination of the magnetoelectric coefficient, the voltage that appears across the capacitor plates is measured. The magnetoelectric coefficient defined by Eq. (9) is the higher the thinner is the piezoelectric layer, although in this case the voltage induced across the plates decreases. The second way of defining the magnetoelectric voltage coefficient is the ratio of the average electric field in the structure to the magnetic field; i.e.,

$$\langle \alpha_{\rm E} \rangle = \frac{\langle E \rangle}{H}, \tag{10}$$

where $\langle E \rangle = U/({}^{\rm m}t + {}^{\rm l}t + {}^{\rm p}t)$ is the average value of the electric field in the structure and U is the voltage induced between the electrodes. The magnetoelectric voltage coefficient defined in this way quantifies the efficiency of the magnetoelectric transformation of the entire structure.

To derive the theoretical expression for the magnetoelectric voltage coefficient we use the method elaborated in our previous works [2–4]. First, the integration constants A and B entering Eq. (6) for the displacement vector of the medium are determined from the boundary conditions at the sidewalls of the sample. Next, the found values are substituted into Eq. (2) for the strain tensor of the piezoelectric, from which the component ${}^{\rm p}T_{zz}$ of the stress tensor is expressed. Finally, found expression is substituted into Eq. (3) for the electric induction and the electric field in the piezoelectric and the magnetoelectric voltage coefficient are determined from the condition of an open circuit.

The conditions of mechanical equilibrium at the free sidewalls of the sample yield the following boundary conditions

$$\int_0^{{}^{\rm p}t} {}^{\rm p}T_{zz}(\pm\tfrac{1}{2}L, x)\,dx + \int_{{}^{\rm p}t}^{{}^{\rm l}t} {}^{\rm l}T_{zz}(\pm\tfrac{1}{2}L, x)\,dx + \int_{{}^{\rm l}t}^{{}^{\rm m}t} {}^{\rm m}T_{zz}(\pm\tfrac{1}{2}L, x)\,dx = 0. \tag{11}$$

Using these boundary conditions, we find the expressions for the integration constants

$$A = 0, \quad B = \frac{{}^{\rm m}Y{}^{\rm m}t{}^{\rm m}q_{z,zz}\langle {}^{\rm m}H_z \rangle + {}^{\rm p}Y{}^{\rm p}t{}^{\rm p}d_{x,zz}\langle {}^{\rm p}E_x \rangle}{k\cos\kappa} \frac{1}{{}^{\rm m}t{}^{\rm m}Y + {}^{\rm l}t{}^{\rm l}Y + {}^{\rm p}t{}^{\rm p}Y}, \tag{12}$$

where we have introduced the dimensionless parameter $\kappa = kL/2$.

Having expressed the component of the stress tensor in terms of the components of the strain tensor from Eq. (2) and substituting the resulting expression into the equation for the normal component of the electric induction vector, we come to the equation of the form

$${}^{\rm p}D_z = {}^{\rm p}\varepsilon_{xx}{}^{\rm p}E_x + {}^{\rm p}Y{}^{\rm p}d_{x,zz}\frac{\partial {}^{\rm p}u_z}{\partial z} - {}^{\rm p}Y({}^{\rm p}d_{x,zz})^2\,{}^{\rm p}E_x. \tag{13}$$

The electric field $\langle {}^{\rm p}E_x \rangle$ induced in the piezoelectric is found from Eq. (10) with the use of the open-circuit condition $I = 0$. Taking into account this condition and the expression for B we find from Eq. (12) the following equation

$$\langle {}^{\rm p}E_x \rangle = \frac{{}^{\rm p}Y{}^{\rm p}d_{x,zz}{}^{\rm m}q_{x,zz}}{{}^{\rm p}\varepsilon_{xx}} - \frac{\langle {}^{\rm m}H_x \rangle}{\Delta}\frac{{}^{\rm m}Y{}^{\rm m}t}{{}^{\rm m}t{}^{\rm m}Y + {}^{\rm l}t{}^{\rm l}Y + {}^{\rm p}t{}^{\rm p}Y}\frac{\tan\kappa}{\kappa}, \tag{14}$$

$$\Delta = 1 - k_{\rm p}^2 \frac{\langle {}^{\rm m}H_x \rangle}{({}^{\rm m}t{}^{\rm m}Y + {}^{\rm l}t{}^{\rm l}Y + {}^{\rm p}t{}^{\rm p}Y)\tan\kappa/\kappa} \tag{15}$$

We will use the definition of the magnetoelectric coefficient given by Eq. (10) taking into account the fact that the voltage induced between the electrodes is $U = \langle {}^{\rm p}E_x \rangle {}^{\rm p}t$. Thus, we finally find the magnetoelectric voltage coefficient

$$\alpha_{\rm E} = \frac{{}^{\rm p}Y{}^{\rm p}d_{x,zz}{}^{\rm m}q_{z,zz}}{{}^{\rm p}\varepsilon_{xx}({}^{\rm m}t{}^{\rm m}Y + {}^{\rm l}t{}^{\rm l}Y + {}^{\rm p}t{}^{\rm p}Y)}\frac{{}^{\rm m}Y{}^{\rm m}t}{\Delta}\frac{\tan\kappa}{\kappa}. \tag{16}$$

According to Eq. (16), the frequency dependence of the magnetoelectric voltage coefficient is of the resonance type.

IOP Conf. Series: Journal of Physics: Conf. Series **991** (2018) 012023 doi:10.1088/1742-6596/991/1/012023

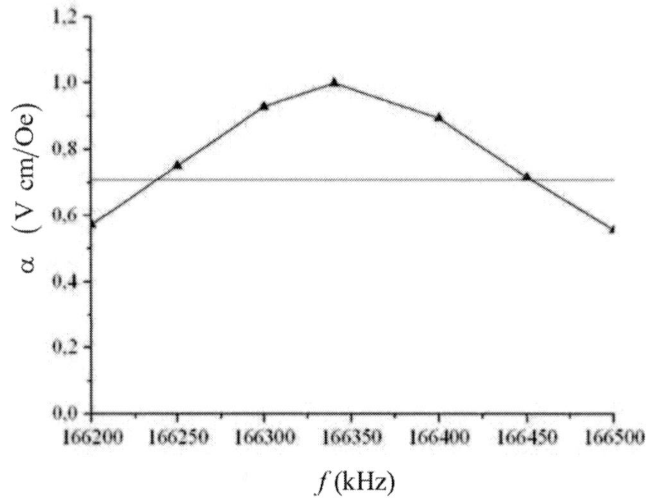

Figure 3. Frequency dependence of the sandwich structure of gallium arsenide-nickel-tin-nickel. The bias field is $H_{\text{bias}} = 360\,\text{Oe}$.

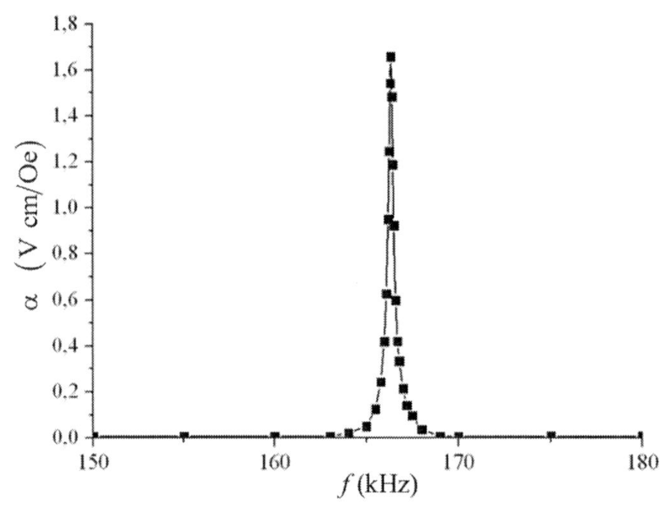

Figure 4. Frequency dependence of the effect in the resonance region.

The experimental measurements of the frequency dependence of the structure are illustrated in figure 3.

As can be seen from the figure, the frequency dependence has a sharp resonance character. The ME value of the voltage coefficient is much lower than that in the pure structure of nickel-gallium arsenide [9], which is explained by the presence of a passive tin buffer layer. However, this structure has a very high quality factor. The frequency dependence of the effect in the resonance region is shown in figure 4.

As follows from the figure, the quality factor of the system is $Q = 800$, which is much better than the quality factor of the samples obtained by the gluing method and is comparable with the quality factor of bulk composites [3].

Conclusions

The use of an intermediate tin layer in the electrolytic deposition of nickel on an arsenide-gallium substrate permits obtaining structures with the nickel layer of thickness up to 70 μm. These structures have good adhesion between layers, good mechanical strength, high quality factor and are very promising for creating devices based on the magnetoelectric effect.

References

[1] Pyatakov A P and Zvezdin A K 2012 Magnetoelectric and multiferroic media *Phys.-Usp.* **55** (6) 557–81
[2] Laletin V M and Srinivasan G 2002 Magnetoelectric effects in composites of nickel ferrite and barium lead zirconate titanate *Ferroelectrics* **280** (1) 177–85
[3] Filippov D A, Laletin V M, and Srinivasan G 2012 Low-frequency and resonance magnetoelectric effects in nickel ferrite–PZT bulk composites *Tech. Phys.* **57** 44–7
[4] Ryu J, Carazo A V, Uchino K, and Kim H E 2001 Magnetoelectric properties in piezoelectric and magnetostrictive laminate composites *Japan. J. Appl. Phys.* **40** (8R) 4948–51
[5] Srinivasan G, Rasmussen E T, Gallegos J, et al. 2001 Novel magnetoelectric bilayer and multilayer structures of magnetostrictive and piezoelectric oxides *Phys. Rev. B* **64** 214408
[6] Kalgin A V, Gridnev S A, and Gribe Z H 2014 Specific features of the inverse magnetoelectric effect in two-layered $Tb_{0.12}Dy_{0.2}Fe_{0.68}$–$PbZr_{0.53}Ti_{0.47}O_3$ composites *Phys. Solid State* **56** (7) 1327–33
[7] Laletin V M, Stognii A I, Novitskii N N, and Poddubnaya N N 2014 The magnetoelectric effect in structures based on metallized gallium arsenide substrates *Tech. Phys. Lett.* **40** (11) 969–71
[8] Filippov D A, Laletin V M, and Galichyan T A 2013 Magnetoelectric effect in a magnetostrictive-piezoelectric bilayer structure *Phys. Solid State* **55** (9) 1840–5
[9] Filippov D A, Firsova T O, Laletin V M, and Poddubnaya N N 2017 The magnetoelectric effect in nickel–GaAs–nickel structures *Tech. Phys. Lett.* **43** (3) 313–5
[10] Filippov D A, Galichyan T A and Laletin V M 2014 Magnetoelectric effect in bilayer magnetostrictive-piezoelectric structure. Theory and experiment *Appl. Phys. A* **115** (3) 1087–91

IOP Conf. Series: Journal of Physics: Conf. Series **991** (2018) 012024 doi:10.1088/1742-6596/991/1/012024

On the dynamic instability of a rectangular plate with one free edge

G Z Gevorgyan and A A Sahakyan*

Institute of Mechanics of the National Academy of Sciences of Armenia, Yerevan, Armenia

E-mail: *areg1992@gmail.com

Abstract. The problem of dynamic stability of a rectangular plate is considered. It is supposed that two opposite sides of the plate are freely supported, the third side is either hingedly supported or rigidly clamped, and the fourth side is free and carries a finite mass. The plate is compressed by a tracking load applied to the last side. The dependence of the critical load on the relative length of the plate, i.e., the distance between sides with different conditions, is obtained.

1. Introduction

A.Yu.Ishlinskii [1] investigated the stability problem of a thin rectangular plate with two opposite free and two freely supported edges when a compressive load is applied to the freely supported sides. In fact, this work for the first time showed the possibility of the existence of a localized instability near the free edges of the plate. Later, this question was studied in a number of articles [2–6]. In the case where a compressive load is applied to the freely supported sides, there is a complete analogy with the problem of plate vibrations [7]. The equations determining the critical loads and the vibration frequencies of localized instability and localized vibrations coincide. But when the load is applied to a free edge, there is no analogy with the problem of vibrations. This raises the natural question of determining the conditions for the onset of localized instability in the case where a compressive load is applied to free edges. In this aspect, we can, for example, refer to [8]. In [9], the problem of localized instability of a rectangular plate is considered, where a compressive load is applied to the free side, the two opposite sides of the plate are freely supported, and the fourth side is either freely supported or rigidly clamped. It is shown that if a uniformly distributed tracking load acts on the free side, then no localized instability can arise in any of the cases considered.

In this paper, we consider the problem stated in [9], where the compressive load is tracking and the possibility of the appearance of dynamic instability of an elastic rectangular plate is investigated in Bolotin's formulation [10].

2. Statement of the problem

Let a thin elastic plate of constant thickness $2h$ in the Cartesian coordinate system (x, y, z) occupy the area $0 \le x \le a$, $0 \le y \le b$, $-h \le z \le h$. According to [10], the equation of plate stability can be written as

$$\Delta^2 w(x,y,t) + \alpha^2 \frac{\partial^2 w(x,y,t)}{\partial x^2} = 0, \quad \alpha^2 = \frac{P}{D}. \tag{1}$$

Here $w(x, y, t)$ is the deflection whose dependence on the time t is contained only in the boundary conditions, P is the intensity of uniformly distributed compressive tracking load, and D is the bending rigidity of the plate.

Assuming that the plate edges $y = 0, b$ are freely supported, we have the following boundary conditions on this edges:

$$w = 0, \quad \frac{\partial^2 w}{\partial y^2} = 0 \quad \text{for } y = 0, b. \tag{2}$$

On the free edge ($x = 0$), where a uniform distributed mass m is assumed to be located, we have the conditions

$$\begin{gathered} \frac{\partial^2 w(x,y,t)}{\partial x^2} + \nu \frac{\partial^2 w(x,y,t)}{\partial y^2} = 0, \\ \frac{\partial}{\partial x}\left[\frac{\partial^2 w(x,y,t)}{\partial x^2} + (2-\nu)\frac{\partial^2 w(x,y,t)}{\partial y^2}\right] = -\frac{m}{D}\frac{\partial^2 w(x,y,t)}{\partial t^2}. \end{gathered} \tag{3}$$

The last conditions are written in the case where the external load is tracking. The conditions on the edge $x = a$ will be given below.

The solution of equation (1), satisfying conditions (2) and taking into account the presence of an inertial term in (3), can be written as

$$w(x,y,t) = e^{i\omega t}\sum_{n=1}^{\infty} f_n(x)\sin\lambda_n y, \quad \lambda_n = \frac{n\pi}{b}. \tag{4}$$

Substituting (4) in equation (1), we obtain a sequence of ordinary differential equations for the functions $f_n(x)$, whose solutions can be written in the form [9]

$$\begin{aligned} f_n(x) &= [A_n \sin(\lambda_n s_2 x) + B_n \cos(\lambda_n s_2 x)]\sinh(\lambda_n s_1 x) \\ &+ [D_n \sin(\lambda_n s_2 x) + C_n \cos(\lambda_n s_2 x)]\cosh(\lambda_n s_1 x), \end{aligned} \tag{5}$$

where A_n, B_n, C_n, D_n are arbitrary constants,

$$s_1 = \sqrt{1 - \frac{\alpha_n^2}{2}}, \quad s_2 = \frac{\alpha_n}{\sqrt{2}}, \quad \alpha_n^2 = \frac{P}{2D\lambda_n^2}.$$

Conditions (3) for the functions $f_n(x)$ become

$$\begin{gathered} f_n''(x) - \nu\lambda_n^2 f_n(x) = 0, \\ f_n'''(x) - (2-\nu)\lambda_n^2 f_n'(x) - \frac{m\omega^2}{D} f_n(x) = 0. \end{gathered}$$

Substituting representation (5) in these conditions, we obtain

$$\begin{gathered} \lambda_n^2[C_n(s_1^2 - s_2^2) + 2A_n s_1 s_2] - \nu\lambda_n^2 C_n = 0, \\ B_n s_1(s_1^2 - 3s_2^2) - D_n s_2(s_2^2 - 3s_1^2) - (2-\nu)(B_n s_1 + D_n s_2) - \eta C_n = 0, \quad \eta = \frac{m\omega^2}{D\lambda_n^3}. \end{gathered} \tag{6}$$

From (6) we express A_n and D_n through B_n and C_n to write the desired function in the form

$$\begin{aligned} f_n(x) &= B_n\left[\cos(\lambda_n s_2 x)\sinh(\lambda_n s_1 x) + \frac{s_1(1-\nu+4s_2^2)}{s_2(1+\nu-4s_2^2)}\sin(\lambda_n s_2 x)\cosh(\lambda_n s_1 x)\right] \\ &+ C_n\left\{-\frac{1-\nu-2s_2^2}{2s_1 s_2}\sin(\lambda_n s_2 x)\sinh(\lambda_n s_1 x)\right. \\ &+ \left.\left[\frac{\eta}{s_2(1+\nu-4s_2^2)}\sin(\lambda_n s_2 x) + \cos(\lambda_n s_2 x)\right]\cosh(\lambda_n s_1 x)\right\}. \end{aligned} \tag{7}$$

Thus, we obtain a general solution satisfying the boundary conditions on three edges of the plate and containing two arbitrary constants whose choice allows us to satisfy the conditions on the fourth side. According to the title of the paper, different boundary conditions will be considered on this side, and hence we consider each of these cases separately.

3. Freely supported edge

In this case, on the edge $x = a$, we have the conditions

$$w = 0, \quad \frac{\partial^2 w}{\partial x^2} = 0, \quad \text{for } x = a. \tag{8}$$

Substituting expression (7) in conditions (8), we obtain a homogeneous system of linear algebraic equations. For the existence of a nontrivial solution, i.e., nonzero values of the constants B_n and C_n, it is necessary that the determinant of the matrix of the obtained system be equal to zero. Satisfying this condition, we come to the equation

$$\begin{aligned} &- s_1[1 + \beta - (2 + \beta)\nu + \nu^2 - 4s_2^2]\sin(2\zeta_2) \\ &\qquad - s_2[3 + \beta - (2 - \beta)\nu - \nu^2 - 4s_2^2]\sinh(2\zeta_1) + \eta\{2s_1s_2[\cosh(2\zeta_1) - \cos(2\zeta_2)]\} = 0, \end{aligned}$$

where the following notation is adopted

$$\zeta_1 = \lambda_n s_1 a, \quad \zeta_2 = \lambda_n s_2 a.$$

From here we derive

$$\eta = \frac{s_1[(1 - \nu)^2 - 4s_2^2\nu]\sin(2\zeta_2) - s_2[(1 + \nu)^2 - 4 - 4s_2^2\nu]\sinh(2\zeta_1)}{2s_1s_2[\cosh(2\zeta_1) - \cos(2\zeta_2)]}. \tag{9}$$

If the representation $\eta = m\omega^2/(D\lambda_n^3)$ is taken into account, then the implementation of the condition $\eta > 0$ is necessary for the dynamic stability of the plate. It is obvious that if the external load is absent ($\alpha_n = 0$), then the plate is stable and, consequently, as the parameter α_n increases, the loss of stability can occur when either the numerator or the denominator in expression (9) vanishes for the first time. In this case, the condition that the numerator is zero coincides with the equation for determining the critical load for the static loss of stability, which was obtained in [9].

In order to determine the critical load leading to the loss of stability, we equate the numerator and the denominator of expression (9) to zero to obtain the system of equations

$$s_1[(1 - \nu)^2 - 4s_2^2\nu]\sin(2\zeta_2) + s_2[4 - (1 + \nu)^2 + 4s_2^2\nu]\sinh(2\zeta_1) = 0, \tag{10}$$

$$\cosh(2\zeta_1) - \cos(2\zeta_2) = 0. \tag{11}$$

It is easy to verify that none of the equations has roots satisfying the condition $\alpha_n^2 < 2$, and therefore the localized instability can not occur. For $\alpha_n^2 > 2$, equations (10)–(11) become

$$\begin{aligned} &\sqrt{\frac{\alpha_n^2}{2} - 1}[(1 - \nu)^2 - 2\alpha_n^2\nu]\sin(\sqrt{2}\pi n a^* \alpha_n) \\ &\qquad + \frac{\alpha_n}{\sqrt{2}}[4 - (1 + \nu)^2 + 2\alpha_n^2\nu]\sin\left(2\pi n a^*\sqrt{\frac{\alpha_n^2}{2} - 1}\right) = 0, \end{aligned} \tag{12}$$

$$\cos\left(2\pi n a^*\sqrt{\frac{\alpha_n^2}{2} - 1}\right) - \cos(\sqrt{2}\pi n a^* \alpha_n) = 0, \tag{13}$$

where $a^* = a/b$.

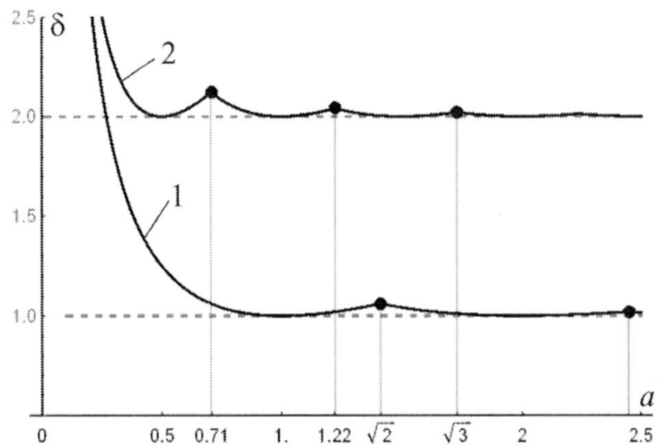

Figure 1. Dependence of the critical tracking load on the relative length of the plate: curve 1 corresponds to the first form of the loss of stability, and curve 2 corresponds to the second form of the loss of stability.

The solution of equation (13) is the sequence

$$\frac{\alpha_n}{\sqrt{2}} = \frac{k^2 + (a^* n)^2}{2a^* nk} \quad (k = 1, 2, 3 \dots), \tag{14}$$

which is not arranged in ascending order, and the ordinal number of its smallest term depends essentially on the parameter a^*.

The roots of equation (12) exceed the roots (14).

We introduce a dimensionless parameter $\delta \equiv \sqrt{Pb^2/(4\pi^2 D)} = (\alpha_n n)/\sqrt{2}$ characterizing the external load. Then, from (14), for each form ($n = 1, 2, 3 \dots$), we have

$$\delta = \min_k \frac{k^2 + (a^* n)^2}{2a^* k}.$$

It is obvious that we have the smallest critical load at $n = 1$, but, depending on a^*, for different values of the parameter k.

The graphs of the dependence of δ on the parameter a^* for the first two forms of the loss of stability are shown in figure 1.

It is easy to verify that the transition from curve k to curve $k+1$ occurs at the following values of the parameter a^*:

$$a^* = \frac{\sqrt{k(k+1)}}{n} \quad (k, n = 1, 2, 3 \dots),$$

and at these points, we have

$$\delta = \frac{2k+1}{2\sqrt{k(k+1)}} n.$$

It is interesting to note that, for $n = 1$, the value a^* is the geometric mean of two successive natural numbers, and δ is the ratio of their arithmetic to the geometric mean.

4. Rigidly clamped edge

In this case, on the edge $x = a$, we have the conditions

$$w = 0, \quad \frac{\partial w}{\partial x} = 0, \quad \text{for } x = a,$$

Satisfying the last conditions, we obtain the homogeneous system of linear algebraic equations

$$\begin{aligned}
&B_n\left[\sinh\zeta_1\cos\zeta_2 + \frac{s_1}{s_2}\frac{1-\nu+4s_2^2}{1+\nu-4s_2^2}\cosh\zeta_1\sin\zeta_2\right] \\
&\quad + C_n\left[\eta\frac{\cosh\zeta_1\sin\zeta_2}{s_2(1+\nu-4s_2^2)} + \cosh\zeta_1\cos\zeta_2 - \frac{1-\nu-2s_2^2}{2s_1s_2}\sinh\zeta_1\sin\zeta_2\right] = 0, \\
&B_n\left[\frac{2(1-\nu+2s_2^2)}{1+\nu-4s_2^2}\sin\zeta_2\sinh\zeta_1 + \frac{4s_1s_2}{1+\nu-4s_2^2}\cos\zeta_2\cosh\zeta_1\right] + C_n\left[(1+\nu)\frac{s_2}{s_1}\cos\zeta_2\sinh\zeta_1\right. \\
&\quad \left. -(1-\nu)\sin\zeta_2\cosh\zeta_1 + \frac{2\eta s_2}{1+\nu-4s_2^2}(\cos\zeta_2\cosh\zeta_1 + \frac{s_1}{s_2}\sin\zeta_2\sinh\zeta_1)\right] = 0.
\end{aligned}$$

From the condition that the determinant of the matrix of this system is zero we obtain the equation

$$\begin{aligned}
&2\eta s_1 s_2^3[s_1\sin(2\zeta_2) - s_2\sinh(2\zeta_1)] + (1-\nu)^2 + 4s_2^2(1-2s_2^2) \\
&\quad - s_1^2[(1-\nu)^2 - 4s_2^2\nu]\cos(2\zeta_2) - s_2^2[(1+\nu)^2 - 4 - 4\nu s_2^2]\cosh(2\zeta_1) = 0.
\end{aligned}$$

As in the previous case, to determine the critical load leading to the loss of stability, we obtain the system of equations

$$\begin{aligned}
&(1-\nu)^2 + 4s_2^2(1-2s_2^2) - s_1^2[(1-\nu)^2-4s_2^2\nu]\cos(2\zeta_2) - s_2^2[(1+\nu)^2-4-4\nu s_2^2]\cosh(2\zeta_1) = 0, \\
&s_2\sinh(2\zeta_1) - s_1\sin(2\zeta_2) = 0.
\end{aligned} \tag{15}$$

The first equation in (15) is a dispersion equation in static formulation and has a unique positive root $\alpha_n = \sqrt{2}$ which, as is easy to verify, leads only to the trivial, zero solution of the problem under study. The second equation in (15) does not have roots satisfying the condition $\alpha_n^2 < 2$; under the assumption that $\alpha_n^2 > 2$, it becomes an equation that, as an expression for the parameter δ, has the form

$$\delta\sin(2\pi a^*\sqrt{\delta^2 - n^2}) - \sqrt{\delta^2 - n^2}\sin(2\pi a^*\delta) = 0$$

The dependence of the parameter δ on the relative length a^* was found by solving this equation numerically. For a comparative analysis, in figure 2, this dependence for the first form $n = 1$ is shown together with a similar curve in the case of free support of the edge $x = a$.

As can be seen from the graphs in figure 2, the critical value of the compressive load in the case where the fourth side is rigidly clamped, as expected, exceeds the analogous value for the freely supported side, and this excess is most significant at small values of the relative length. Of special interest are the values of the relative length a^* equal to the geometric mean of two successive natural numbers, when the critical values of the compressive load coincide regardless of the type of the fourth side fixation.

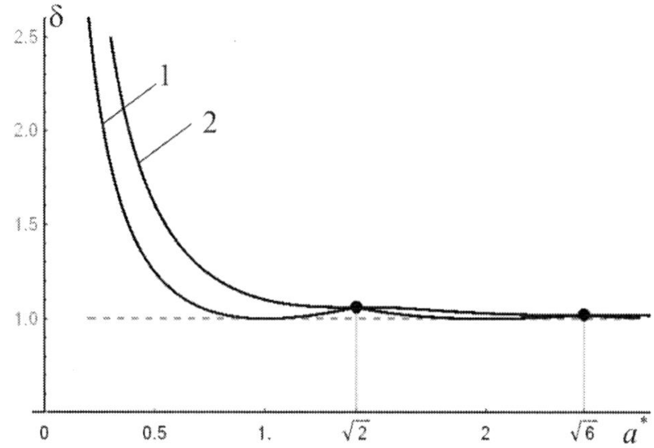

Figure 2. Dependence of the critical tracking load on the relative length of the plate: curve 1 corresponds to the free support, and curve 2 corresponds to the rigid clamping of the edge $x = a$.

Conclusions

The possibility of the occurrence of instability of a rectangular plate is investigated in dynamic formulation. It is assumed that one side of the plate is not supported, two opposite sides are freely supported, and various fixation conditions can be posed on the fourth side, – free support or rigid clamping. It is shown that when a uniformly distributed compressive tracking load acts on the free edge, the loss of stability occurs in both cases, while the critical value of the compressive load in the second case exceeds the corresponding value in the first case, and these values coincide only for certain values of the relative length. The minimum values of the relative length of the plate, at which the loss of stability may begin, are determined by the strength characteristics of the plate material.

References

[1] Ishlinskii A Yu 1954 On a limit transition in the theory of stability of elastic rectilinear plates *Dokl. Akad. Nauk SSSR* **95** (3) 477–9
[2] Belubekyan M V 1997 Problems of localized instability of a plate In *Proc. Int. Conf. on Optimal Control, Stability and Strength of Mechanical Systems* (Yerevan: Yerevan State University) 95–9 [In Russian]
[3] Tovstik P E 1995 *Stability of Thin Shells (Asymptotic Methods)* (Moscow: Nauka) p 308 [In Russian]
[4] Belubekyan M V and Chil-Hakobyan E O 2004 Problems of localized instability of a plate with a free edge *Izv. Nats. Akad. Nauk Armenii. Mekh.* **57** (2) 34–3
[5] Banichuk N V and Ishlinskii A Yu 1995 Some special features of problems of the stability and vibrations of rectangular plates *J. Appl. Math. Mech.* **59** (4) 593–7
[6] Banichuk N V and Barsuk A A 2008 Localization of eigenforms and limit transformations in problems of the stability of rectangular plates *J. Appl. Math. Mech.* **72** (2) 302–7
[7] Konenkov Yu K 1960 A Rayleigh-type flexural wave *Sov. Phys. Acoust.* **6** (1) 122–3
[8] Sharifian R and Belubekyan V M 2012 Stability of a rectangular plate axially compressed on its two opposite free edges *ZAMM* **92** (7) pp 558–64
[9] Belubekyan M V and Sahakyan A A 2016 Localized instability of a rectangular plate under the action of a compressive load on the free edge In *MECHANICS 2016: Proc. of Int. School-Conf. of Young Scientists, 3–7 October 2016, Tsaghkadzor, Armenia* (Yerevan: Nats. Univ. Arkh. Stroit. Armenii) 43–6 [in Russian]
[10] Bolotin V V 1963 *Nonconservative Problems of the Theory of Elastic Stability* (New York: Macmillan) p 324

Reflection, refraction, and transmission of SH waves at a micropolar layer separating two elastic media

K B Ghazaryan, A A Papyan, and S K Ohanyan

Institute of Mechanics of the National Academy of Sciences of Armenia, Yerevan, Armenia

Abstract. In this paper, we study the reflection, refraction, and transmission of SH-waves in media separated by a layer whose behavior is described by a simplified model of Cosserat (micropolar) elasticity. The effect of a micropolar medium on the reflection and refraction of SH-waves is obtained. The amplitude ratios and energy fluxes of reflected and transmitted waves are graphically represented.

1. Introduction

A simplification of the Cosserat elasticity model was first mentioned by Ugodchikov A. G. in [1], and it was used by Manookian V. F. to study the existence of surface shear waves in a micropolar medium [2]. The term simplified was first used by Belubekyan M. V. to study SH-waves in periodic stratified media [3]. Then a simplified model of the Cosserat continuum was considered by Avetisyan A. S. in [4]. We use a simplified model of Cosserat elasticity to describe the boundary layer separating two homogeneous elastic half-spaces described by the classical model of elasticity.

2. Formulation of the problem

The geometry of the problem under consideration is depicted in figure 1. Homogeneous elastic half-spaces with rigidity μ_s, mass density ρ_s (where $s = 1, 2$ denotes the lower and upper half-spaces, respectively) are separated by a micropolar layer with rigidity μ_0, mass density ρ_0, rotational inertia J, and thickness h. The structure is referred to a rectangular coordinate system with positive Y-axis directed vertically upwards and origin at the middle of the micropolar layer. The SH-waves from the lower half-space reflect and transmit through the layer into the upper half-space.

At the interface $Y = -h/2$, the incoming SH-wave propagates back partially reflected as a SH-wave in the same medium and partially transmitted as a SH-wave into the upper media. Let the incident wave and the transmitted wave make respective angles φ_1 and φ_2 with the normal to the boundary.

By assumptions of small deformations and neglect of the body forces, the equation of motion for SH-wave is given by

$$\frac{\partial \sigma_{xz}^{(s)}}{\partial x} + \frac{\partial \sigma_{yz}^{(s)}}{\partial x} = \rho \frac{\partial^2 u^{(s)}}{\partial t^2}, \quad s = 1, 2, \tag{1}$$

where s denotes lower (1) and upper (2) half-spaces and $u \equiv u(x, y, t)$ is the displacement component along Z-direction for anti-plane state of the problem.

IOP Conf. Series: Journal of Physics: Conf. Series **991** (2018) 012025 doi:10.1088/1742-6596/991/1/012025

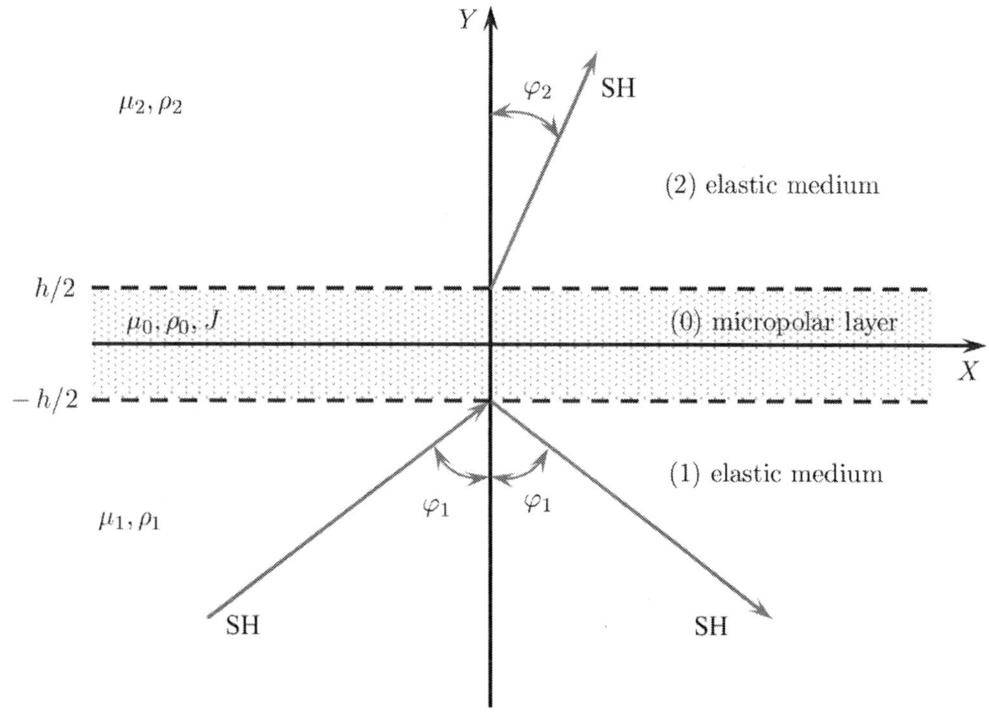

Figure 1. Reflection and transmission of SH–waves.

2.1. Relations for the upper and lower half-spaces

The non-zero stress-strain relations are $\sigma_{xz}^{(s)} = 2\mu\varepsilon_{xz}^{(s)}$ and $\sigma_{yz}^{(s)} = 2\mu\varepsilon_{yz}^{(s)}$, where $\varepsilon_{ij} = \frac{1}{2}(u_{i,j}+u_{j,i})$, $i, j = x, y$ are the strain tensor components. For half-spaces, the stress–displacement relations are

$$\begin{aligned} \sigma_{yz}^{(s)} &= \mu\frac{\partial u^{(s)}}{\partial y}, \\ \sigma_{xz}^{(s)} &= \mu\frac{\partial u^{(s)}}{\partial x}. \end{aligned} \tag{2}$$

2.2. Relations for the micropolar layer

According to the simplified Cosserat elasticity model, the constitutive equations (generalized Hooke's law) for bound layers are $\sigma_{ij} = 2\mu\varepsilon_{ij} + \delta_{ij}\lambda\varepsilon_{kk} + J\partial^2\omega_{ij}/\partial t^2$, where $\omega_{ij} = \frac{1}{2}(u_{j,i} - u_{i,j})$ are components of the rotation tensor, λ and μ are Lame's coefficients, and J is a dynamic characteristic of the medium (rotational inertia).

The nonzero stress components for the layers are

$$\begin{aligned} \sigma_{yz}^{(s)} &= \mu\frac{\partial u^{(s)}}{\partial y} + J\frac{\partial^3 u^{(s)}}{\partial t^2 \partial y}, \\ \sigma_{xz}^{(s)} &= \mu\frac{\partial u^{(s)}}{\partial x} + J\frac{\partial^3 u^{(s)}}{\partial t^2 \partial x}. \end{aligned} \tag{3}$$

IOP Conf. Series: Journal of Physics: Conf. Series **991** (2018) 012025 doi:10.1088/1742-6596/991/1/012025

3. Solutions of the equations of motion

After substituting stress-strain relations (2) and (3) in (1) we obtain

$$\mu\Delta u + J\frac{\partial^2}{\partial t^2}(\Delta u) = \rho\frac{\partial^2 u}{\partial t^2}, \tag{4}$$

$$\mu\Delta u^{(s)} = \rho\frac{\partial^2 u^{(s)}}{\partial t^2}, \tag{5}$$

where $\Delta = \frac{\partial^2}{\partial x^2} + \frac{\partial^2}{\partial y^2}$ is the two-dimensional Laplace delta operator.

Substituting the solution $u(x, y, t) = f(y)e^{i(\omega t - kx)}$ in (4), we obtain

$$f''(y) + p^2 f(y) = 0, \tag{6}$$

where $p^2 = (\rho_0\omega^2 - \beta k^2)/\beta$ and $\beta = \mu_0 - J\omega^2$. The solution of differential equation (6) is $f(y) = A_1e^{ipy} + A_2e^{-ipy}$. For (5), the characteristic equation is

$$g''_s(y) + q_s^2 g_s(y) = 0, \quad s = 1, 2, \tag{7}$$

where $q_s^2 = \rho_s\omega^2/\mu_s - k^2$ and $g_s(y)$ is the amplitude function. At the half-spaces, the displacement components of incident, reflected, and refracted waves are

$$\begin{aligned} u^{(1)} &= B_1e^{iq_1y} + B_2e^{-iq_1y}, \\ u^{(2)} &= Ce^{iq_2y}, \end{aligned} \tag{8}$$

and the displacement and stress components for the micropolar layer are

$$\begin{aligned} u(x, y, t) &= (A_1e^{ipy} + A_2e^{-ipy})e^{i(\omega t - kx)}, \\ \sigma_{yz} &= ip\beta(A_1e^{ipy} - A_2e^{-ipy})e^{i(\omega t - kx)}, \end{aligned} \tag{9}$$

where A_1 and A_2 are the amplitudes of the incident and reflected waves, respectively.

4. Matrix manipulations

In matrix form, expressions (9) become

$$\begin{pmatrix} u^{(0)} \\ \sigma_{yz}^{(0)} \end{pmatrix} = \begin{pmatrix} e^{ipy} & e^{-ipy} \\ ip\beta e^{ipy} & -ip\beta e^{-ipy} \end{pmatrix} \begin{pmatrix} A_1 \\ A_2 \end{pmatrix} e^{i(\omega t - kx)}.$$

We rewrite (omitting $e^{i(\omega t - kx)}$) this equation as

$$U^{(0)}(y) = M(y)A,$$

where

$$U^{(0)}(y) = \begin{pmatrix} u^{(0)} \\ \sigma_{yz}^{(0)} \end{pmatrix}, \quad M(y) = \begin{pmatrix} e^{ipy} & e^{-ipy} \\ ip\beta e^{ipy} & -ip\beta e^{-ipy} \end{pmatrix}, \quad A = \begin{pmatrix} A_1 \\ A_2 \end{pmatrix}.$$

For two arbitrary points y_0 and y_1 within the layer, we have

$$U^{(0)}(y_0) = M(y_0)A, \quad U^{(0)}(y_1) = M(y_1)A,$$

and subsequently, the stress strain state for these points can be represented by the relation

$$U^{(0)}(y_0) = N(y_0, y_1)U^{(0)}(y_1),$$

where

$$N(y_0, y_1) = M(y_0)M^{-1}(y_1) = \begin{pmatrix} \cos[p(y_0 - y_1)] & \dfrac{\sin[p(y_0 - y_1)]}{p\beta} \\ -p\beta\sin[p(y_0 - y_1)] & \cos[p(y_0 - y_1)] \end{pmatrix}$$

is the transfer matrix.

5. Boundary conditions

According to the boundary conditions $U^{(1)}(-h/2) = N(-h/2; h/2)U^{(2)}(h/2)$ at $(-h/2, h/2)$ of ideally contact, we obtain the following boundary relations for the displacement and stress components of the lower and upper half-spaces:

$$\begin{aligned} u^{(1)}(-h/2) &= \cos(ph)u^{(2)}(h/2) - \frac{\sin(ph)}{\beta p}\tau^{(2)}(h/2), \\ \tau^{(1)}(-h/2) &= \beta p \sin(ph) u^{(2)}(h/2) - \cos(ph)\tau^{(2)}(h/2). \end{aligned} \tag{10}$$

After substituting displacements (8) in (10), we obtain

$$\begin{aligned} B_1 + B_2 e^{ihq_1} &= C\frac{e^{ih(q_1+q_2)/2}}{p\beta}[p\beta\cos(hp) - i\mu_2 q_2 \sin(hp)], \\ (B_1 + B_2)\sin(hq_1/2) + i(B_1 - B_2)\cos(hq_1/2) &= C\frac{e^{ihq_2/2}}{\mu_1 q_1}[i\mu_2 q_2 \cos(hp) + p\beta \sin(hp)]. \end{aligned} \tag{11}$$

6. Reflection and refraction coefficients

The reflection and refraction amplitude ratios from (11) are respectively $R = B_2/B_1$ and $T = C/B_1$:

$$\begin{aligned} R &= \frac{e^{-ihq_1}[(\mu_1\mu_2 q_1 q_2 - p^2\beta^2)\sin(ph) - ip\beta(\mu_2 q_2 - \mu_1 q_1)\cos(ph)]}{(\mu_1\mu_2 q_1 q_2 + p^2\beta^2)\sin(ph) + ip\beta(\mu_2 q_2 + \mu_1 q_1)\cos(ph)}, \\ T &= \frac{2ie^{-ih(q_1+q_2)/2}\mu_1 q_1 p\beta}{(\mu_1\mu_2 q_1 q_2 + p^2\beta^2)\sin(ph) + ip\beta(\mu_2 q_2 + \mu_1 q_1)\cos(ph)}, \end{aligned} \tag{12}$$

where $q_s = k_s\sqrt{c^2/c_s^2 - 1} = k_s \tan(\varphi_s)$. The dependence of the angles of reflection and refraction is given by Snell's law:

$$\begin{aligned} c_1 \sin\varphi_1 &= c_2 \sin\varphi_2, \\ c\cos\varphi_s &= c_s. \end{aligned}$$

7. Frequency ranges

From the existence condition $p^2 > 0$ for harmonic waves, we obtain $\mu_0 k^2/(\rho_0 + Jk^2) < \omega^2 < \mu_0/J$; in this case, the solution of (6) is real and the condition $\mu_0 k^2/(\rho_0 + Jk^2) < \mu_0/J$ is always true. For the elastic half-spaces (7), from the condition $q_s^2 > 0$, we obtain $\omega^2 > \mu_s k^2/\rho_s$. So we have the frequency range $\sqrt{\mu_0 k_0^2/(\rho_0 + Jk_0^2)} < \omega < \sqrt{\mu_0/J}$, where $p^2 > 0$, and the solution is real. In the case where $\omega < \sqrt{\mu_0 k_0^2/(\rho_0 + Jk_0^2)}$ or $\omega > \sqrt{\mu_0/J}$, $p^2 < 0$, the solution is complex and can be represented as ip_0 for $p_0 > 0$.

The SH-wave propagates throughout the half-spaces and $q^2 > 0$, when

$$\text{a)}\quad \alpha_s < \alpha_l, \quad \begin{cases} p^2 > 0 & \text{if } \omega \in (\alpha_l, \alpha_u), \\ p^2 < 0 & \text{if } \omega \in (\alpha_s, \alpha_l) \cup (\alpha_u, +\infty), \end{cases}$$

$$\text{b)}\quad \alpha_l < \alpha_s < \alpha_u, \quad \begin{cases} p^2 > 0 & \text{if } \omega \in (\alpha_s, \alpha_u), \\ p^2 < 0 & \text{if } \omega \in (\alpha_u, +\infty), \end{cases}$$

$$\text{c)}\quad \alpha_u < \alpha_s, \quad p^2 < 0 \text{ if } \omega \in (\alpha_s, +\infty),$$

where $\alpha_s = \sqrt{\mu_s k_s^2/\rho_s}$ is the frequency limit for the half-spaces, $\alpha_l = \sqrt{\mu_0 k_0^2/(\rho_0 + Jk_0^2)}$ and $\alpha_u = \sqrt{\mu_0/J}$ are the respective lower and upper limits for the frequency range of the micropolar layer. These cases are graphically presented in figure 2.

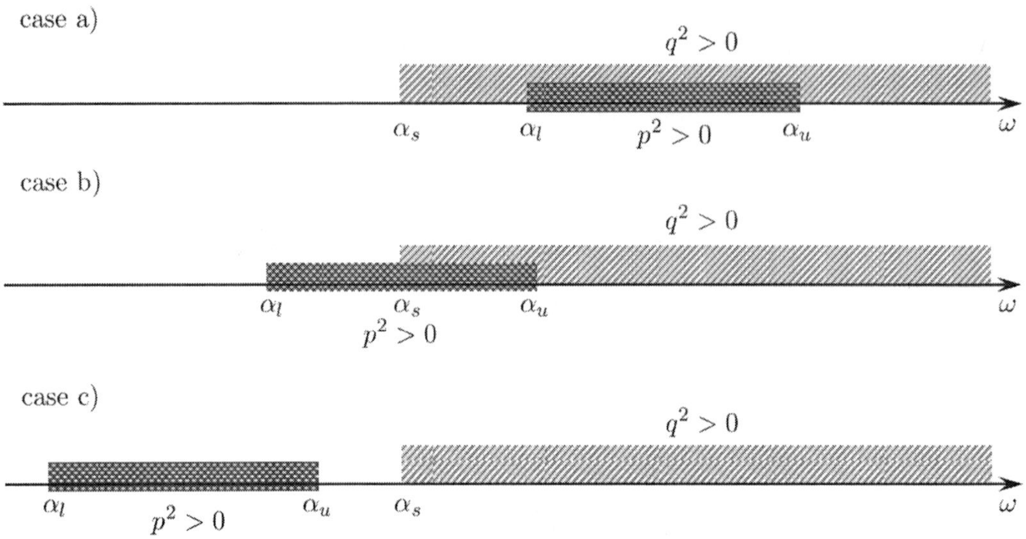

Figure 2. Graphical representation of frequency ranges.

8. Particular case: half-spaces with identical material properties

There are two possible cases for reflection and refraction amplitude ratios (12): $p^2 > 0$ and $p^2 < 0$.

In the case $p^2 > 0$, $\omega \in (\alpha_l, \alpha_u)$ for $\alpha_s < \alpha_l$, and $\omega \in (\alpha_s, \alpha_u)$ for $\alpha_l < \alpha_s < \alpha_u$:

$$\begin{aligned} R &= \frac{e^{-ihq}(q^2\mu^2 - p^2\beta^2)\sin(hp)}{2ipq\beta\mu\cos(hp) + (p^2\beta^2 + q^2\mu^2)\sin(hp)}, \\ T &= \frac{2ie^{-ihq}pq\beta\mu}{2ipq\beta\mu\cos(hp) + (p^2\beta^2 + q^2\mu^2)\sin(hp)}. \end{aligned} \tag{13}$$

In the case $p^2 < 0$, $p = ip_0$, $p_0 > 0$, $\omega \in (\alpha_s, \alpha_l) \cup (\alpha_u, +\infty)$ for $\alpha_s < \alpha_l$ $\omega \in (\alpha_u, +\infty)$ for $\alpha_l < \alpha_s < \alpha_u$, and $\omega \in (\alpha_s, +\infty)$ for $\alpha_u < \alpha_s$:

$$\begin{aligned} R &= \frac{ie^{-ihq}\sinh(hp_0)(q^2\mu^2 + \beta^2 p_0^2)}{-2q\beta\mu p_0\cosh(hp_0) + i\sinh(hp_0)(q^2\mu^2 - \beta^2 p_0^2)}, \\ T &= \frac{-2e^{-ihq}q\beta\mu p_0}{-2q\beta\mu p_0\cosh(hp_0) + i\sinh(hp_0)(q^2\mu^2 - \beta^2 p_0^2)}. \end{aligned} \tag{14}$$

The energy flux conservation is then expressed via the reflection and transmission amplitudes by the algebraic identity $|R|^2 + q_2\mu_2/(q_1\mu_1)|T|^2 = 1$, in the case where the half-spaces are identical

$$|R|^2 + |T|^2 = 1, \tag{15}$$

for $p^2 > 0$,

$$\begin{aligned} |R|^2 &= 1 - \frac{4p^2q^2\beta^2\mu^2}{4p^2q^2\beta^2\mu^2\cos^2(hp) + (p^2\beta^2 + q^2\mu^2)^2\sin^2(hp)}, \\ |T|^2 &= \frac{4p^2q^2\beta^2\mu^2}{4p^2q^2\beta^2\mu^2\cos^2(hp) + (p^2\beta^2 + q^2\mu^2)^2\sin^2(hp)}, \end{aligned} \tag{16}$$

and for $p^2 < 0$,

$$|R|^2 = 1 - \frac{4q^2\beta^2\mu^2 p_0^2}{q^4\mu^4\sinh^2(hp_0) + q^2\beta^2\mu^2[3+\cosh^2(2hp_0)]p_0^2 + \beta^4\sinh^2(hp_0)p_0^4},$$
$$|T|^2 = \frac{4q^2\beta^2\mu^2 p_0^2}{q^4\mu^4\sinh^2(hp_0) + q^2\beta^2\mu^2[3+\cosh^2(2hp_0)]p_0^2 + \beta^4\sinh^2(hp_0)p_0^4}, \tag{17}$$

where $q = k\tan\varphi$; we introduce the following notation to obtain the dimensionless parameters: $p = k\sqrt{\eta/(1-\gamma\eta)-1}$, $\beta = \rho_0 c^2(1/\eta - \gamma)$.

So for $p^2 > 0$,

$$|R|^2 = 1 - \frac{-4\zeta^2 a\tan^2\varphi}{-4\zeta^2 a\cos^2\theta\tan^2\varphi + \sin^2\theta(a - \zeta^2\tan^2\varphi)^2},$$
$$|T|^2 = \frac{-4\zeta^2 a\tan^2\varphi}{-4\zeta^2 a\cos^2\theta\tan^2\varphi + \sin^2\theta(a - \zeta^2\tan^2\varphi)^2},$$

and for $p^2 < 0$,

$$|R|^2 = 1 - \frac{-4a\eta^4\zeta^2\tan^2\varphi}{a\eta^4\sinh^2\theta - \zeta^2 a\eta^2(3+\cosh\theta)\tan^2\varphi + \zeta^4\sinh^2\theta\tan^4\varphi},$$
$$|T|^2 = \frac{-4a\eta^4\zeta^2\tan^2\varphi}{a\eta^4\sinh^2\theta - \zeta^2 a\eta^2(3+\cosh\theta)\tan^2\varphi + \zeta^4\sinh^2\theta\tan^4\varphi},$$

where $a = \eta^2(\gamma\eta - 1)(\gamma\eta + \eta - 1)$, $\theta = \alpha\sqrt{\eta/(1-\gamma\eta)-1}$, $\alpha = hk$, $\eta = c/c_0$, $\gamma = Jk^2/\rho_0$, and $\zeta = \mu/\mu_0$; here $\eta = c/c_0$ is dimensionless phase velocity.

9. Numerical calculations

The comparison of the micropolar elastic layer case, which is described by the *simplified Cosserat model* of elasticity, with classical case is presented in figure 3. In this case, the shear moduli ratio ζ and the dimensionless phase velocity η are equal to 0.5, which means that the half-spaces are softer than the layer. Figure 3 shows that taking into account the effect of micro-rotations according to *simplified Cosserat model* of elasticity for a thin layer (there and below, *thin* means that the thickness of the layer is less than the wavelength) has a small impact on the dependence of the energy fluxes of reflected and transmitted waves on the angle of incidence. For the layer thickness equal to the wavelength of the incident wave, the influence of micro-rotations is sufficient, in the *classical* case, the wave is completely reflected for any angle of incidence, but in the *micropolar layer* case, the wave incident perpendicularly to the layer boundary is completely transmitted, and then the energy flux of the transmitted wave decreases until, at 30 degrees, the waves became almost completely reflected.

The dependence of energy fluxes and amplitude ratios on the angle of incidence are presented in figures 4–9 for $p^2 > 0$, and in figures 10–15 for $p^2 < 0$.

Figures 4–6 graphically present the effect of micropolar layer dependent on its thickness in the case where the shear moduli ratio ζ and dimensionless phase velocity η are equal to 1. A thin layer does not influence the shear wave reflection and transmission, and waves are completely transmitted. But as the layer thickness increases to the wavelength dimensions (see figure 7), a significant impact is obtained, so that the changes in the shear moduli ratio (see figures 8 and 9) result in the complete reflection of waves.

The influence of increasing thickness of the micropolar layer described above is more visible in the complex case $p^2 < 0$ (see figures 10–12). Figures 13–15 show that the changes in rigidity ratio ζ make a smaller impact in the case $p^2 < 0$.

Graphs for the SH-waves represented by trigonometric functions are shown in figures 13–15.

IOP Conf. Series: Journal of Physics: Conf. Series **991** (2018) 012025 doi:10.1088/1742-6596/991/1/012025

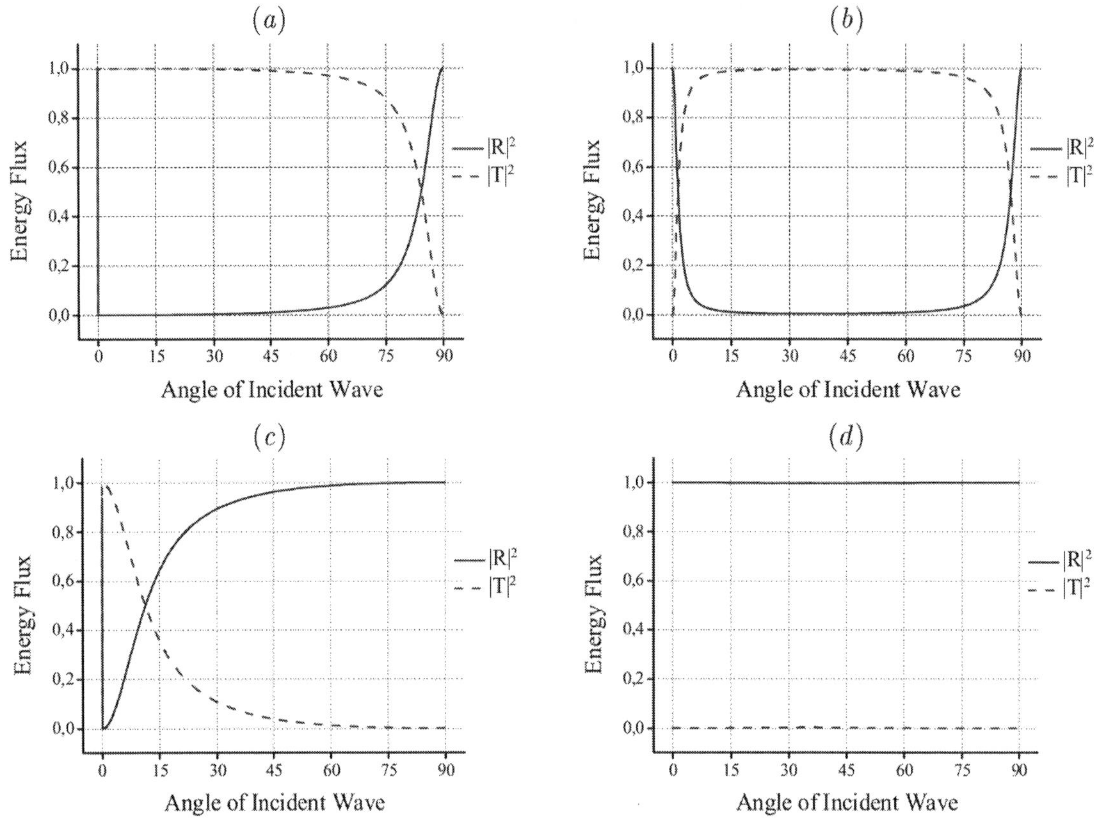

Figure 3. Simplified Cosserat elastic layer case ((a) for $\alpha = 0.1$, $\eta = 0.5$, $\gamma = 1$, $\zeta = 0.5$ and (c) for $\alpha = 5$, $\eta = 0.5$, $\gamma = 1$, $\zeta = 0.5$) and the classical elastic layer case ((b) for $\alpha = 0.1$, $\eta = 0.5$, $\gamma = 0$, $\zeta = 0.5$ and (d) for $\alpha = 5$, $\eta = 0.5$, $\gamma = 0$, $\zeta = 0.5$).

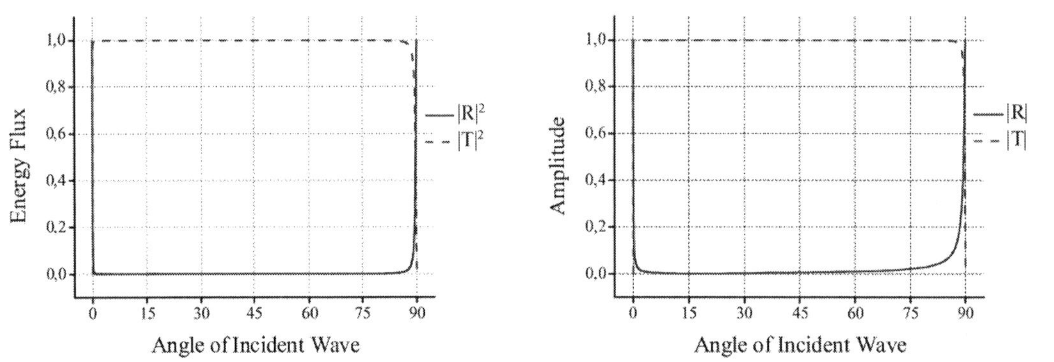

Figure 4. Values of dimensionless parameters: $\alpha = 0.01$, $\eta = 1$, $\gamma = 0.1$, $\zeta = 1$. The layer is thin enough to make no impact on the SH-wave transmission.

Conclusions

This paper presents an analytical solution to study of reflection and transmission of shear SH elastic waves in a multilayered structure composed of elastic micropolar layer sandwiched between two elastic materials on either side. The micropolar layer material obeys to the equation of the simplified model of Cosserat (micropolar) elasticity theory. The transfer matrix method

IOP Conf. Series: Journal of Physics: Conf. Series **991** (2018) 012025 doi:10.1088/1742-6596/991/1/012025

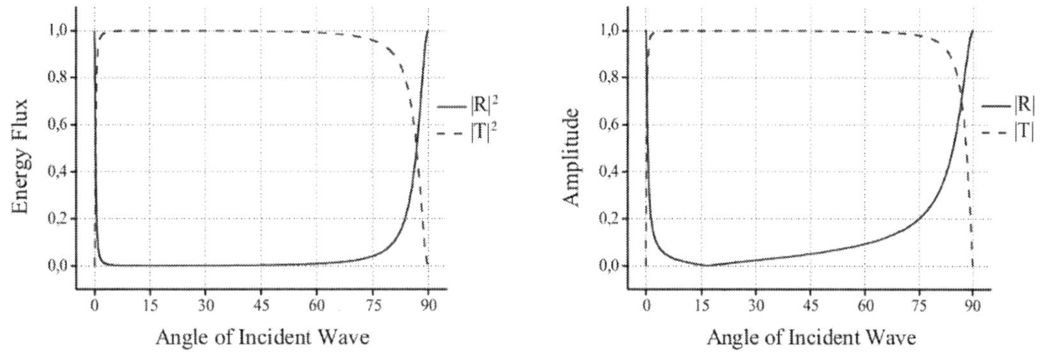

Figure 5. Values of dimensionless parameters: $\alpha = 0.1$, $\eta = 1$, $\gamma = 0.1$, $\zeta = 1$. Here we can observe the appearance of a cross point at 85 degrees when only half the energy is transmitted, which is a result of increasing thickness of the layer.

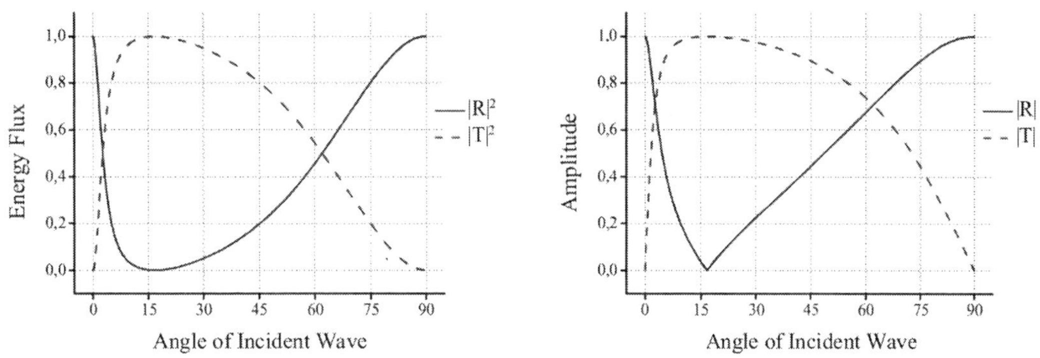

Figure 6. Values of dimensionless parameters: $\alpha = 1$, $\eta = 1$, $\gamma = 0.1$, $\zeta = 1$. In this figure, we can see a determined point at 60 degrees when half the energy of SH-waves is reflected and the other half is transmitted. The cusp at 15 degrees indicates the angle of incidence at which the SH-waves are completely transmitted.

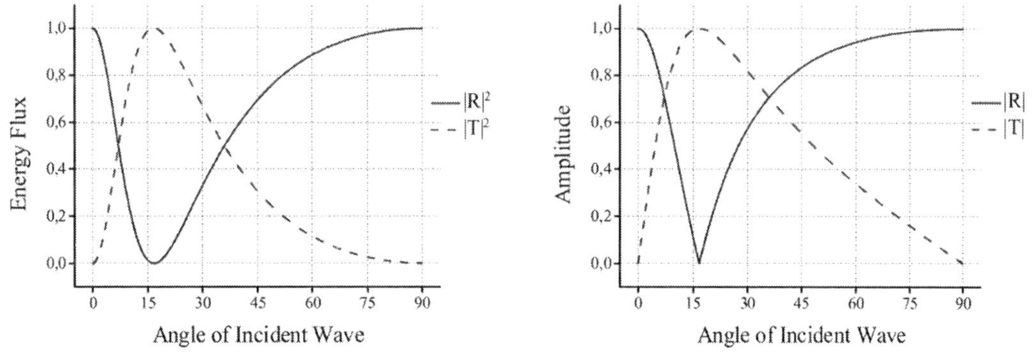

Figure 7. Values of dimensionless parameters: $\alpha = 5$, $\eta = 1$, $\gamma = 0.1$, $\zeta = 1$. The change in the thickness (equal to the wave number) results in two more visible points (7 and 37 degrees) at which the graphs for the transmitted and reflected waves intersect, so only half the energy is transmitted.

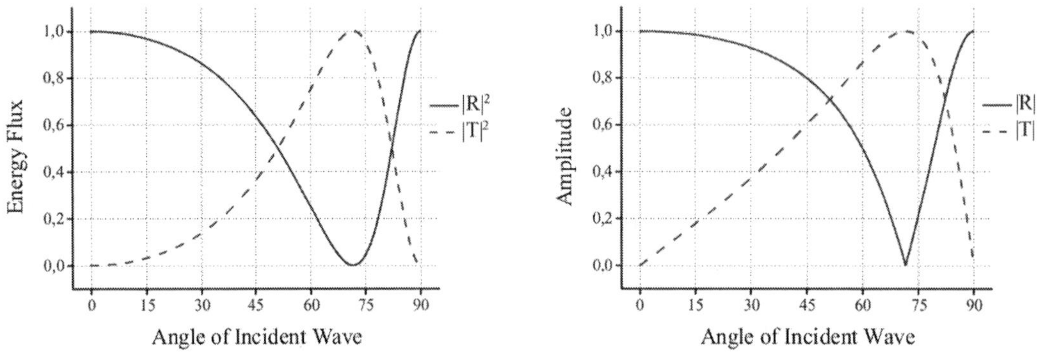

Figure 8. Values of dimensionless parameters: $\alpha = 5$, $\eta = 1$, $\gamma = 0.1$, $\zeta = 0.1$. The changes in the rigidity ratio leads to changes in the position (degree) of cross points (and the cusp point). Subsequent decrease of rigidity ratio results to complete reflection.

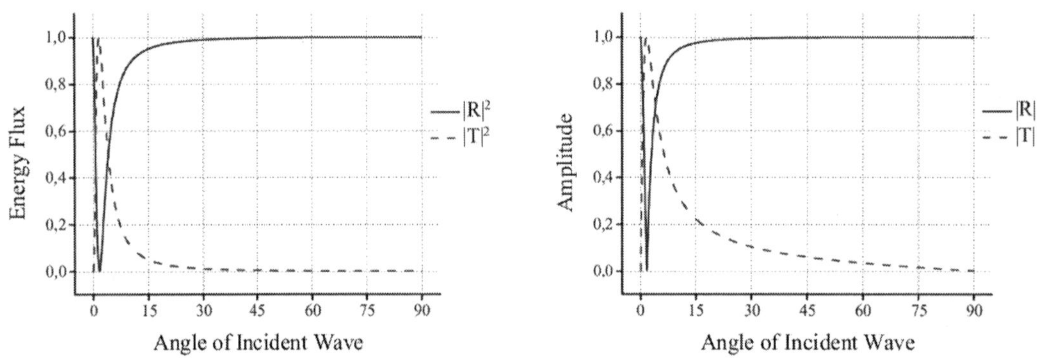

Figure 9. Values of dimensionless parameters: $\alpha = 5$, $\eta = 1$, $\gamma = 0.1$, $\zeta = 10$. The result of increase in the rigidity ratio is similar to that of decrease, so it leads to complete reflection.

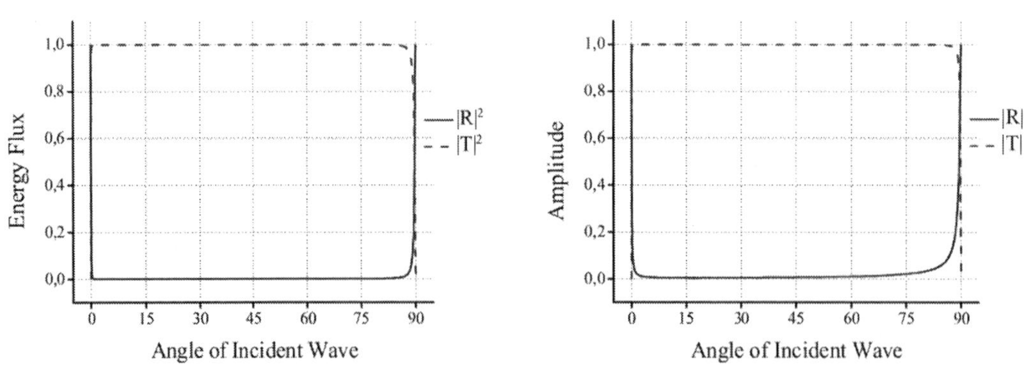

Figure 10. Values of dimensionless parameters: $\alpha = 0.01$, $\eta = 1$, $\gamma = 0.1$, $\zeta = 1$. As in figure 4, the micropolar layer is thin enough to make no changes in the SH-waves transmission.

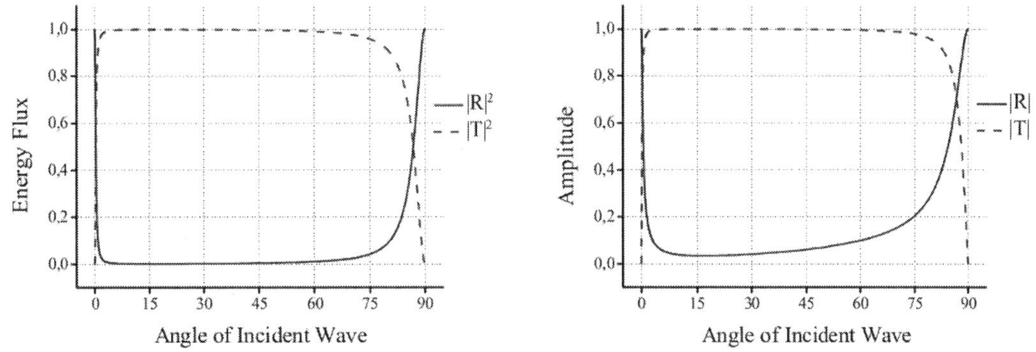

Figure 11. Values of dimensionless parameters: $\alpha = 0.1$, $\eta = 1$, $\gamma = 0.1$, $\zeta = 1$. Here is no difference from figure 5, the cross point is at 85 degrees.

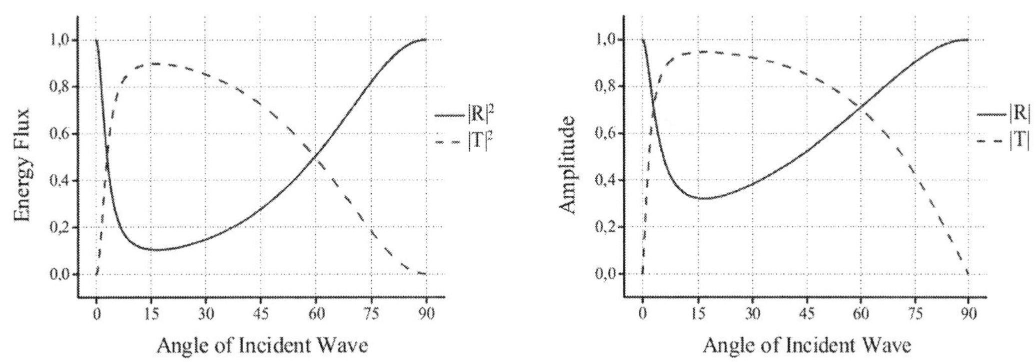

Figure 12. Values of dimensionless parameters: $\alpha = 1$, $\eta = 1$, $\gamma = 0.1$, $\zeta = 1$. The main difference from figure 6 is the absence of a cusp point (degree) at which the SH-waves are completely transmitted.

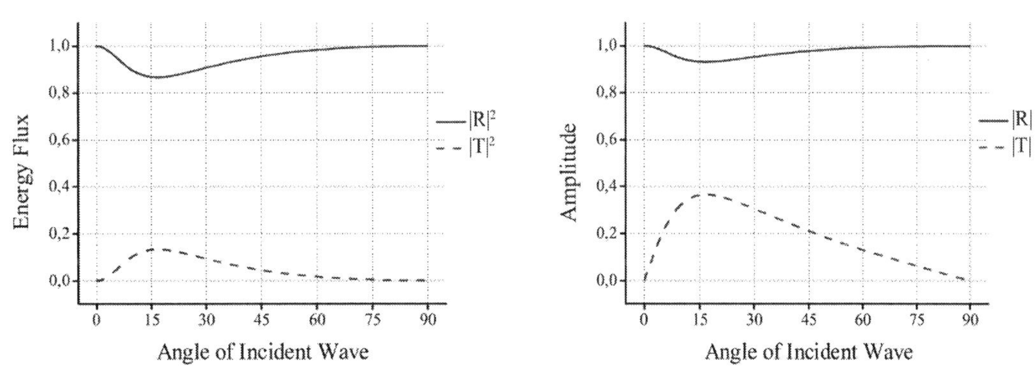

Figure 13. Values of dimensionless parameters: $\alpha = 5$, $\eta = 1$, $\gamma = 0.1$, $\zeta = 1$. This figure illustrates the difference from figure 7, there are no cross points for graphs and there is no point (degree) at which the SH-waves can be completely transmitted.

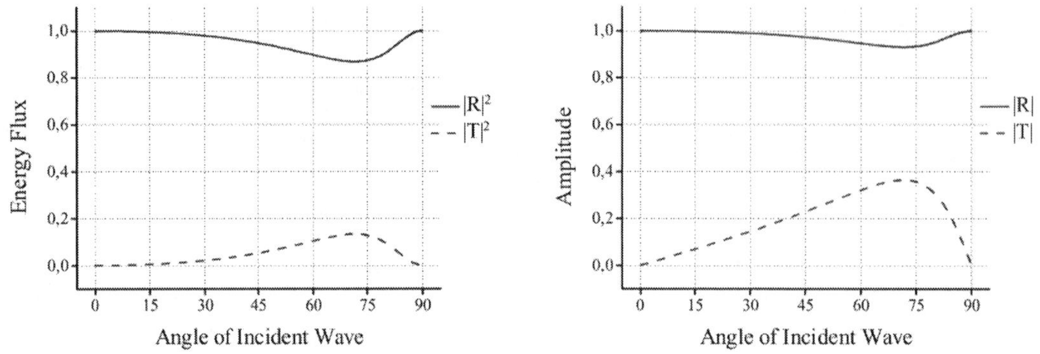

Figure 14. Values of dimensionless parameters: $\alpha = 5$, $\eta = 1$, $\gamma = 0.1$, $\zeta = 0.1$. As in figure 8, the changes in the rigidity ratio lead to changes in the position (degree) of cusp point of the graphs. A subsequent decrease in the rigidity ratio results in the complete reflection.

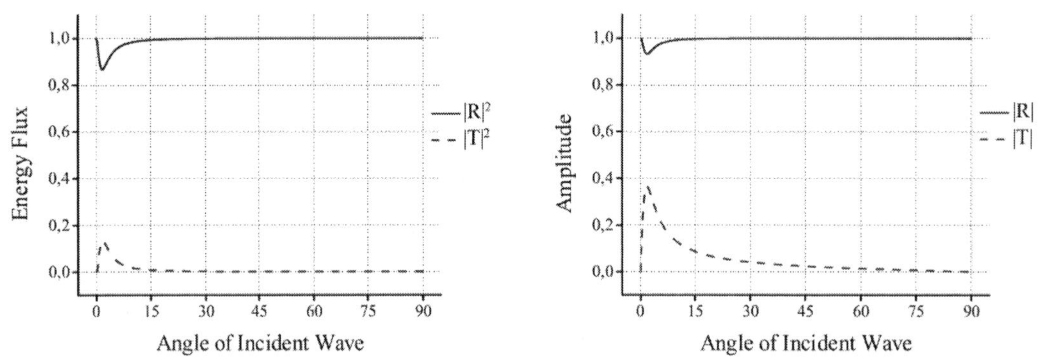

Figure 15. Values of dimensionless parameters: $\alpha = 5$, $\eta = 1$, $\gamma = 0.1$, $\zeta = 10$. As in figure 9, the result of increase in the rigidity ratio is similar to that of decrease, so it leads to complete reflection.

is used to find the amplitudes and energy fluxes of reflected and transmitted waves in different frequency bands of incident wave. On the basis of analytical solutions it is concluded that micropolar properties of the layer produce a strong effect on the energy fluxes of reflected and transmitted waves when the layer thickness is compared with wavelength of the incident wave.

References

[1] Ugodchikov A G 1995 Torque dynamics of linear elastic bodies *Dokl. Akad. Nauk* **340** (1) 50–8

[2] Manukyan V F 1997 On the existence of surface shear waves in a micropolar medium *Izv. Nats. Akad. Nauk Armenii. Mekh.* **50** (2) 75–9

[3] Ambartsumian S A, Belubekyan M V, and Ghazaryan K B 2014 Shear elastic waves in a periodic medium with the Cosserat simplified model properties *Izv. Nats. Akad. Nauk Armenii. Mekh.* **67** (4) 3–9

[4] Ambartsumyan S A, Avetisyan A S, and Belubekyan M V 2017 Propagation of elastic waves in a plane waveguide layer on the basis of a simplified model of the Cosserat continuum *Izv. Nats. Akad. Nauk Armenii. Mekh.* **70** (2) 15–27

On a model of the processes of maintaining a technological area by a manipulator

A A Ghukasyan[1,2*] **and A Ya Ordyan**[2**]

[1]Institute of Mechanics of the National Academy of Sciences of Armenia, Yerevan, Armenia
[2]Goris State University, Goris, Armenia

E-mail: *ghukasyan10@yandex.ru, **oarpine@rambler.ru

Abstract. The research refers to the results of mathematical modeling of the process of maintaining a technological area which consists of unstable or fixed objects (targets) and a controlled multi-link manipulator [1–9]. It is assumed that, in the maintenance process, the dynamic characteristics and the phase vector of the manipulator state can change at certain finite times depending on the mass of the cargo or instrument [10, 11]. Some controllability problems are investigated in the case where the manipulator motion on each maintenance interval is described by linear differential equations with constant coefficients and the motions of the objects are given.

1. Description of the maintenance model

The equations describing the manipulator motion on each maintenance interval are generally represented as linear differential equations with constant coefficients of various size

$$\dot{\mathbf{x}}^i = \mathbf{A}_i\mathbf{x}^i + \mathbf{b}^i u, \quad \text{where} \quad \mathbf{A}_i = \mathbf{A}(\boldsymbol{\omega}^i) \quad (i = 1, 2, \ldots, k), \tag{1}$$

where k is the number of maintenance stages to which the time intervals $[t_{i-1}, t_i]$ $(i = 1, 2, \ldots, k)$ correspond, $\mathbf{x}_i$ is the n_i-dimensional phase vector of the manipulator state on the time interval $t \in [t_{i-1}, t_i]$, $\mathbf{A}_i$ is the $n_i \times n_i$ dimensional matrix with constant elements on the interval $[t_{i-1}, t_i]$ depending on the parameter $\boldsymbol{\omega}^i$, $\mathbf{b}^i$ is an n_i-dimensional constant vector, $u(t)$ is the scalar control function, $t \in [t_0, T]$ (t_0 is the initial time and $t_k = T$ is the terminal time). The intermediate times t_i $(i = 1, 2, \ldots, k,\ t_0 < t_1 < t_2 < \ldots < t_{k-1} < t_k = T)$ can be fixed or determined in additional conditions.

Suppose that, in the state space, arbitrary initial (for $t = t_0$, $i = 1$) and finite (for $t = T$, $i = k$) positions of the system (1) are given in the form of

$$\mathbf{x}^1(t_0) = \mathbf{x}^1_{t_0}, \quad \mathbf{x}^k(t_k) = \mathbf{x}^k(T) = \mathbf{x}^k_T, \tag{2}$$

where the phase vector $\mathbf{x}^1_{t_0}$ has dimension n_1 and $\mathbf{x}^k_T$ has dimension n_k.

In the general case, it is assumed that $n_1 \neq n_k$ and, at intermediate times t_i $(i = 1, 2, \ldots, k-1)$, when there is a transition from one maintenance stage to another, the vectors of composite systems (1) must correspond to the conditions

$$\mathbf{x}^i(t_i) = \mathbf{z}^i(t_i) = \mathbf{x}^{i+1}(t_i), \quad (i = 1, 2, \ldots, k-1), \tag{3}$$

where $\mathbf{z}^i(t_i)$ is the phase state of the objects at times t_i $(i = 1, 2, \ldots, k-1)$.

Conditions (3) ensure the continuity of the process and mean [12] that the end of the manipulator movement at time t_i $(i = 1, 2, \ldots, k)$ is the beginning of motion on the next time interval $t \in [t_i, t_{i+1}]$.

Without loss of generality, we assume that $n_{i+1} > n_i$ $(i = 1, 2, \ldots, k-1)$. To match the dimension of the phase vectors with different stages of motion, we formally assume that the motion at the stage i occurs in the n_{i+1}-dimensional space $R^{n_{i+1}}$, where the last $(n_{i+1} - n_i)$ components of the phase vector $\mathbf{x}^i(t)$ are identically zero, that is,

$$\begin{gathered} \mathbf{x}^i(t) = (x_1^i, x_2^i, \ldots, x_{n_i}^i, 0_{n_i+1}, 0_{n_i+2}, \ldots, 0_{n_{i+1}})^T, \\ \mathbf{x}^{i+1}(t) = (x_1^{i+1}, x_2^{i+1}, \ldots, x_{n_i}^{i+1}, x_{n_i+1}^{i+1}, x_{n_i+2}^{i+1}, \ldots, x_{n_{i+1}}^{i+1})^T, \\ (x_1^{i+1}(t_i) = x_1^i(t_i),\ x_2^{i+1}(t_i) = x_2^i(t_i),\ \ldots,\ x_{n_i}^{i+1}(t_i) = x_{n_i}^i(t_i), \\ x_{n_i+1}^{i+1}(t_i) \equiv 0,\ x_{n_i+2}^{i+1}(t_i) \equiv 0,\ \ldots,\ x_{n_{i+1}}^{i+1}(t_i) \equiv 0). \end{gathered} \tag{4}$$

Similarly, it is possible formally to increase the dimension of the state space at the $i+1$st stage of motion by an amount $(n_i - n_{i+1})$, when $n_{i+1} < n_i$ $(i = 1, 2, \ldots, k-1)$.

The complete set of systems of differential equations describing the stepwise motion of the manipulator on the time intervals $[t_{i-1}, t_i]$ $(i = 1, 2, \ldots, k)$, with allowance for (2)–(4), can be represented as

$$\left[\begin{array}{ll} \dot{\mathbf{x}}^1 = \mathbf{A}_1\mathbf{x}^1 + \mathbf{b}^1 u & \text{for } t_0 \le t \le t_1, \\ \dot{\mathbf{x}}^2 = \mathbf{A}_2\mathbf{x}^2 + \mathbf{b}^2 u & \text{for } t_1 \le t \le t_2, \\ \ldots\ldots\ldots\ldots & \\ \dot{\mathbf{x}}^{k-1} = \mathbf{A}_{k-1}\mathbf{x}^{k-1} + \mathbf{b}^{k-1} u & \text{for } t_{(k-2)} \le t \le t_{(k-1)}, \\ \dot{\mathbf{x}}^k = \mathbf{A}_k\mathbf{x}^k + \mathbf{b}^k u & \text{for } t_{(k-1)} \le t \le t_k = T \end{array}\right. \tag{5}$$
$$(t_0 < t_1 < t_2 < \ldots < t_{k-1} < t_k = T).$$

Since the manipulator motion is described by the system of linear differential equations (1) on each time interval $[t_{i-1}, t_i]$ $(i = 1, 2, \ldots, k)$, it follows that, with a proper choice of the control function $u = u(t)$, $t \in [t_{i-1}, t_i]$ $(i = 1, 2, \ldots, k)$, in the domain of admissible controls, we have a unique trajectory of the motion of the characteristic points of the mechanical system satisfying conditions (2) and (3). The obtained trajectory $\mathbf{x}(t) = \{\mathbf{x}^i(t)\}$ $(i = 1, 2, \ldots, k)$ of motion, combining the trajectories on the intervals $[t_{i-1}, t_i]$ $(i = 1, 2, \ldots, k)$, is continuous and piecewise differentiable, i.e., the solution $x(t)$ is always continuously differentiable [13], except for the times t_i $(i = 1, 2, \ldots, k)$.

2. Controllability of the maintenance process

Important concepts for the further investigation of the process of maintaining the objects (technological area) by a manipulator are the concepts of controllability, not only at separate stages of maintenance, but also at those of the entire system over the entire time interval.

As is known [14–16], each system in the set (5)

$$\dot{\mathbf{x}}^i = \mathbf{A}_i\mathbf{x}^i + \mathbf{b}^i u \quad (i = 1, 2, \ldots, k) \tag{6}$$

is completely controllable on the time interval $t \in [t_{i-1}, t_i]$ $(i = 1, 2, \ldots, k)$ if the rank of the controllability matrix is equal to n_i, that is,

$$\operatorname{rang}\begin{pmatrix} \mathbf{b}^i & \mathbf{A}_i\mathbf{b}^i & \mathbf{A}_i^2\mathbf{b}^i & \cdots & \mathbf{A}_i^{n-1}\mathbf{b}^i \end{pmatrix} = n_i \quad (i = 1, 2, \ldots, k), \tag{7}$$

and not controllable on the time interval $t \in [t_{i-1}, t_i]$ if

$$\operatorname{rang}\begin{pmatrix} \mathbf{b}^i & \mathbf{A}_i\mathbf{b}^i & \mathbf{A}_i^2\mathbf{b}^i & \cdots & \mathbf{A}_i^{n-1}\mathbf{b}^i \end{pmatrix} < n_i \quad (i = 1, 2, \ldots, k). \tag{8}$$

Following [14–16], we define completely controllable autonomous dynamical systems (6). An autonomous system (6) on a time interval $[t_{i-1}, t_i]$ is considered to be completely controllable (possesses the controllability property) if, for any pair of points $\mathbf{x}^i(t_{i-1})$ and $\mathbf{x}^i(t_i)$ $(i = 1, 2, \ldots, k)$, there is a bounded measurable control $u = u(t)$, $t \in [t_{i-1}, t_i]$, that takes system (6) from point $\mathbf{x}^i(t_{i-1})$ to point $\mathbf{x}^i(t_i)$ $(i = 1, 2, \ldots, k)$. The definition of the complete controllability of systems (6) and conditions (2), (3) implies a generalized definition of completely controllable set (5) on the time interval $t \in [t_0, T]$.

Definition. The set (5) on the time interval $[t_0, T]$ is completely controllable if, for any initial $\mathbf{x}^1(t_0)$ and finite states $\mathbf{x}^k(T)$, there is an admissible control $u = u(t)$, $t \in [t_0, T]$, that takes (5) from the initial state to the final state so that, at intermediate time t_i $(t_0 < t_1 < t_2 < \ldots < t_{k-1} < t_k = T)$, the condition $\mathbf{x}^i(t_i) = \mathbf{x}^{i+1}(t_i)$ $(i = 1, 2, \ldots, k-1)$ is satisfied (the control $u = u(t)$ is constructed as a combination of controls by the manipulator at each maintenance stage). Since the aggregate (5) describes the whole manipulator maintenance process of the technological area and each system (6) is the equation of the controlled motion on the time interval $t \in [t_i, t_{i+1}]$ $(i = 1, 2, \ldots, k-1)$, the controllability of the technological process as a whole depends on the controllability on each interval.

By the method of the mathematical induction, it is easy to show that the maintenance process is fully controllable if it is completely controllable at each stage and not controllable if at least one of the systems in (6) is not completely controllable on its interval of definition.

We assume that, on the time interval $[t_0, T]$, there is only one intermediate moment t_1, where not only the dynamic characteristics of the mechanical system of the manipulator but also the dimension of the phase vector can change.

The manipulator motion on each interval $[t_0, t_1]$, $[t_1, T]$ $(k = 2)$ is described by the equations

$$\left[\begin{array}{ll} \dot{\mathbf{x}}^1 = \mathbf{A}_1\mathbf{x}^1 + \mathbf{b}^1 u & \text{at } t_0 \le t \le t_1, \\ \dot{\mathbf{x}}^2 = \mathbf{A}_2\mathbf{x}^2 + \mathbf{b}^2 u & \text{at } t_1 \le t \le t_2 = T \end{array}\right. \tag{9}$$

with the conditions

$$\mathbf{x}^1(t_0) = \mathbf{x}^1_{t_0}, \quad \mathbf{x}^1(t_1) = \mathbf{z}^1(t_1) = \mathbf{x}^2(t_1), \quad \mathbf{x}^2(t_2) = \mathbf{x}^2_T. \tag{10}$$

According to the controllability criterion (7), the autonomous systems (9) are completely controllable on the time intervals $[t_0, t_1]$, $[t_1, t_2]$ if the corresponding controllability matrices have maximum rank, that is,

$$\operatorname{rang}\left(\begin{array}{ccccc} \mathbf{b}^i & \mathbf{A}_i\mathbf{b}^i & \mathbf{A}_i^2\mathbf{b}^i & \cdots & \mathbf{A}_i^{n-1}\mathbf{b}^i \end{array}\right) = n_i \quad \text{at } t \in [t_{i-1}, t_i] \quad (i = 1, 2). \tag{11}$$

Here, in general, the matrices $\mathbf{A}_1$, $\mathbf{A}_2$ and the vectors $\mathbf{b}^1$, $\mathbf{b}^2$ have the structures

$$\mathbf{A}_i = \begin{bmatrix} a^i_{11} & a^i_{12} & \cdots & a^i_{1n_i} \\ a^i_{21} & a^i_{22} & \cdots & a^i_{2n_i} \\ \vdots & \vdots & \ddots & \vdots \\ a^i_{n_i1} & a^i_{n_i2} & \cdots & a^i_{n_in_i} \end{bmatrix}, \quad \begin{array}{l} \mathbf{b}^i = (b^i_1,\ b^i_2,\ \ldots,\ b^i_{n_i})^T, \\ n_1 \ne n_2, \end{array} \quad (i = 1, 2). \tag{12}$$

For the further investigation of the problems of the completely controllable maintenance process described by the combining systems (9) for $n_2 > n_1$ in the whole period of time $[t_0, T]$, we consider the motion on a state area $R^{n_1+n_2}$ of dimension $(n_1 + n_2)$, where the motion at the first stage, that is, for $[t_0, t_1]$, occurs in the subspace R^{n_1}, and at the second stage, in the subspace R^{n_2}. The subspaces R^{n_1} and R^{n_2} ($R^{n_1} \bigcup R^{n_2} = R^{n_1+n_2}$) have no interior points, while at $t = t_1$, they have common boundary points.

To describe the motion in $R^{n_1+n_2}$ during the time $t \in [t_0, t_2]$, we derive the (n_1+n_2)-dimensional state vector y as

$$\mathbf{y} = (x_1^1,\ x_2^1,\ \ldots,\ x_{n_1}^1,\ x_1^2,\ x_2^2,\ \ldots,\ x_{n_2}^2)^T, \tag{13}$$

where

$$\mathbf{y}^1 = (x_1^1,\ x_2^1,\ \ldots,\ x_{n_1}^1,\ 0,\ 0,\ \ldots,\ 0)^T \quad \text{for } t_0 \le t \le t_{1-} \tag{14}$$

and

$$\mathbf{y}^2 = (0,\ 0,\ \ldots,\ 0,\ x_1^2,\ x_2^2,\ \ldots,\ x_{n_2}^2)^T \quad \text{for } t_{1+} \le t \le t_2. \tag{15}$$

At time t_1, it follows from (10) that $|\mathbf{y}^1(t_1)| = |\mathbf{y}^2(t_1)|$.

Conditions (14) and (15) can also be represented as

$$\lim_{t\to t_1-} \mathbf{y} = \lim_{t\to t_1-} (\mathbf{x}^1, \mathbf{x}^2)^T = (\mathbf{x}^1, 0)^T = \mathbf{y}^1(t_1), \quad \lim_{t\to t_1+} \mathbf{y} = \lim_{t\to t_1+} (\mathbf{x}^1, \mathbf{x}^2)^T = (0, \mathbf{x}^2)^T = \mathbf{y}^2(t_1).$$

We also introduce an $(n_1+n_2)\times(n_1+n_2)$ block matrix $\mathbf{C}$ and an (n_1+n_2)-dimensional vector $\mathbf{d}$:

$$\mathbf{C} = \begin{bmatrix} \mathbf{A}_1 & \mathbf{0}_1 \\ \mathbf{0}_2 & \mathbf{A}_2 \end{bmatrix} \tag{16}$$

where $\mathbf{A}_i$ $(i = 1, 2)$ are determined from (12), and the vector $\mathbf{d}$ has the form

$$\mathbf{d} = (b_1^1,\ b_2^1,\ \ldots,\ b_{n_1}^1,\ b_1^2,\ b_2^2,\ \ldots,\ b_{n_2}^2)^T. \tag{17}$$

Similarly, with (13)–(15) taken into account, the matrix $\mathbf{C}$ (16) and the vector $\mathbf{d}$ (17) satisfy the conditions

$$\mathbf{C}_1 = \begin{bmatrix} \mathbf{A}_1 & \mathbf{0}_1 \\ \mathbf{0}_2 & \mathbf{0} \end{bmatrix}, \quad \mathbf{d}^1 = (b_1^1,\ b_2^1,\ \ldots,\ b_{n_1}^1,\ 0,\ 0,\ \ldots,\ 0)^T \quad \text{for } t_0 \le t \le t_{1-}, \tag{18}$$

$$\mathbf{C}_2 = \begin{bmatrix} \mathbf{0} & \mathbf{0}_1 \\ \mathbf{0}_2 & \mathbf{A}_2 \end{bmatrix}, \quad \mathbf{d}^2 = (0,\ 0,\ \ldots,\ 0,\ b_1^2,\ b_2^2,\ \ldots,\ b_{n_2}^2)^T \quad \text{for } t_{1+} \le t \le t_2. \tag{19}$$

The matrices $\mathbf{0}_1$ and $\mathbf{0}_2$ with zero elements in (18), (9) have dimensions $n_1 \times n_2$ and $n_2 \times n_1$, respectively, and the matrix $\mathbf{0}$ with zero elements in (18) has dimension $n_2 \times n_2$, and $\mathbf{0}$ in (19) has dimension $n_1 \times n_1$.

In the general case, conditions (10), on account of (13)–(19), allow one to represent (9) as

$$\dot{\mathbf{y}}^1 = \mathbf{C}_1\mathbf{y}^1 + \mathbf{d}^1 u \quad \text{at } t \in [t_0, t_1], \tag{20}$$

$$\dot{\mathbf{y}}^2 = \mathbf{C}_2\mathbf{y}^2 + \mathbf{d}^2 u \quad \text{at } t \in [t_1, t_2]. \tag{21}$$

Equation (20) can be considered on the whole time interval $t \in [t_0, t_2]$ with zero elements at $t \in [t_1, t_2]$, and (21) can be considered at $t \in [t_0, t_2]$ with zero elements at $t \in [t_0, t_1]$.

By combining the systems (20) and (21), we can formally describe the maintenance process by the equation

$$\dot{\mathbf{y}} = \mathbf{C}\mathbf{y} + \mathbf{d}u, \quad t \in [t_0, t_2]. \tag{22}$$

Since, at time $t \in [t_0, t_2]$, system (22) either coincides with system (20) or (21), it follows that, in the variables and parameters (13)–(21), the controllability matrixes (7) for systems (20) and (21) have the following respective structures,

$$\mathbf{M}_1 = (\mathbf{d}^1, \mathbf{C}_1\mathbf{d}^1, \mathbf{C}_1^2\mathbf{d}^1, \ldots, \mathbf{C}_1^{n_1-1}\mathbf{d}^1) = \begin{bmatrix} \mathbf{b}^1 & \mathbf{A}_1\mathbf{b}^1 & \mathbf{A}_1^2\mathbf{b}^1 & \cdots & \mathbf{A}_1^{n_1-1}\mathbf{b}^1 \\ 0 & 0 & 0 & \cdots & 0 \\ \vdots & \vdots & \vdots & \ddots & \vdots \\ 0 & 0 & 0 & \cdots & 0 \end{bmatrix} \quad \text{for } t \in [t_0, t_1] \tag{23}$$

and

$$\mathbf{M}_2 = (\mathbf{d}^2, \mathbf{C}_2\mathbf{d}^2, \mathbf{C}_2^2\mathbf{d}^2, \ldots, \mathbf{C}_2^{n_2-1}\mathbf{d}^2) = \begin{bmatrix} 0 & 0 & 0 & \cdots & 0 \\ \vdots & \vdots & \vdots & \ddots & \vdots \\ 0 & 0 & 0 & \cdots & 0 \\ \mathbf{b}^2 & \mathbf{A}_2\mathbf{b}^2 & \mathbf{A}_2^2\mathbf{b}^2 & \cdots & \mathbf{A}_2^{n_2-1}\mathbf{b}^2 \end{bmatrix} \text{ for } t \in [t_1, t_2]. \quad (24)$$

Here, the matrices $\mathbf{M}_1$ and $\mathbf{M}_2$ have dimensions $(n_1+n_2)\times n_1$ and $(n_1+n_2)\times n_2$, respectively.

Taking (11) into account, we obtain $\operatorname{rang}\mathbf{M}_1 = n_1$ and $\operatorname{rang}\mathbf{M}_2 = n_2$.

Combining the matrices (23) and (24), we obtain the complete controllability matrix for the union (22) on the entire time interval $[t_0, T]$

$$\begin{aligned} \mathbf{M}^1 &= \{\mathbf{M}_1, \mathbf{M}_2\} = \{(\mathbf{d}^1, \mathbf{C}_1\mathbf{d}^1, \mathbf{C}_1^2\mathbf{d}^1, \ldots, \mathbf{C}_1^{n_1-1}\mathbf{d}^1), (\mathbf{d}^2, \mathbf{C}_2\mathbf{d}^2, \mathbf{C}_2^2\mathbf{d}^2, \ldots, \mathbf{C}_2^{n_2-1}\mathbf{d}^2)\} \\ &= \begin{bmatrix} \mathbf{M}_{11} & \mathbf{0}_{12} \\ \mathbf{0}_{21} & \mathbf{M}_{22} \end{bmatrix}. \end{aligned} \quad (25)$$

The block matrix $\mathbf{M}^1$ has dimension $(n_1+n_2)\times(n_1+n_2)$, $\mathbf{M}_{11} = (\mathbf{b}^1, \mathbf{A}_1\mathbf{b}^1, \mathbf{A}_1^2\mathbf{b}^1, \ldots, \mathbf{A}_1^{n_1-1}\mathbf{b}^1)$ coincides with the controllability matrix (11) for $i = 1$, $\mathbf{M}_{22} = (\mathbf{b}^2, \mathbf{A}_2\mathbf{b}^2, \mathbf{A}_2^2\mathbf{b}^2, \ldots, \mathbf{A}_2^{n_2-1}\mathbf{b}^2)$ coincides with the controllability matrix (11) for $i = 2$. The matrices $\mathbf{0}_{12}$ and $\mathbf{0}_{21}$ with zero elements have dimensions $n_1 \times n_2$ and $n_2 \times n_1$, respectively.

The determinant of the bakery-diagonal matrix $\mathbf{M}^1$ is calculated as [17]

$$\det \mathbf{M}^1 = \det \begin{bmatrix} \mathbf{M}_{11} & \mathbf{0}_{12} \\ \mathbf{0}_{21} & \mathbf{M}_{22} \end{bmatrix} = \det(\mathbf{M}_{11}\mathbf{M}_{22}) = \det \mathbf{M}_{11} \det \mathbf{M}_{22} \neq 0. \quad (26)$$

Consequently, $\operatorname{rang}\mathbf{M}^1 = n_1 + n_2$.

It follows from (26) that the motion of the manipulator mechanical system described by the union (22) and including systems (20) and (21) is completely controllable on the time interval $t \in [t_0, T]$ only if subsystems in (9) are controllable on the time intervals $[t_0, t_1]$ and $[t_1, t_2]$, respectively, and are completely uncontrollable if one of the subsystems in (9) is not completely controllable on its interval of definition, i.e.,

$$\det \mathbf{M}^1 = \det \mathbf{M}_{11} \det \mathbf{M}_{22} = 0 \quad \text{if } \operatorname{rang}\mathbf{M}_{11} < n_1 \text{ or } \operatorname{rang}\mathbf{M}_{22} < n_2.$$

Suppose that the assertion about controllability is true for $(k-1)$ stages of motion. That is, the controllability matrix for the combined systems (5) with $(k-1)$ equation has determinant of the form

$$\begin{aligned} \det \mathbf{M}^{k-1} &= \det \begin{bmatrix} \mathbf{M}_{11} & \mathbf{0}_{12} & \mathbf{0}_{13} & \cdots & \mathbf{0}_{1\,k-1} \\ \mathbf{0}_{21} & \mathbf{M}_{22} & \mathbf{0}_{23} & \cdots & \mathbf{0}_{2\,k-1} \\ \vdots & \vdots & \vdots & \ddots & \vdots \\ \mathbf{0}_{k-1\,1} & \mathbf{0}_{k-1\,2} & \mathbf{0}_{k-1\,3} & \cdots & \mathbf{M}_{k-1\,k-1} \end{bmatrix} \\ &= \det(\mathbf{M}_{11}\mathbf{M}_{22} \cdots \times \mathbf{M}_{k-1k-1}) = \det \mathbf{M}_{11} \det \mathbf{M}_{22} \det \mathbf{M}_{33} \cdots \det \mathbf{M}_{k-1k-1} \neq 0. \end{aligned} \quad (27)$$

Here the matrix $\mathbf{M}_{jj}$ $(j = 1, 2, \ldots, k-1)$ is the controllability matrix of the system with number j $(j = 1, 2, \ldots, k-1)$ in (5) on the time interval $[t_{j-1}, t_j]$ and has dimension $(n_j \times n_j)$, and the controllability matrix $\mathbf{M}^{k-1}$ of the combined system has dimension $\left(\sum_{j=1}^{k-1} n_j\right) \times \left(\sum_{j=1}^{k-1} n_j\right)$ on the time interval $[t_0, t_{k-1}]$.

Since each $\det \mathbf{M}_{jj} \neq 0$ $(j = 1, 2, \ldots, k-1)$, it follows that $\operatorname{rang} \mathbf{M}^{k-1} = \sum_{j=1}^{k-1} n_j$. The matrices $\mathbf{0}_{12}$, $\mathbf{0}_{13}$, …, $\mathbf{0}_{1k-1}$ have dimensions $n_1 \times n_2$, $n_1 \times n_3$, …, $n_1 \times n_{k-1}$, respectively. Similarly, $\mathbf{0}_{k-11}$, $\mathbf{0}_{k-12}$, $\mathbf{0}_{k-13}$, …, $\mathbf{0}_{k-1k-2}$ have dimensions $n_{k-1} \times n_1$, $n_{k-1} \times n_2$, …, $n_{k-1} \times n_{k-2}$, respectively.

To combine the system of equations (5) describing $(k-1)$ stages of motion, let us consider the state spaces of dimension $\sum_{j=1}^{k-1} n_j$ $(R^{\sum_{j=1}^{k-1} n_j})$. The state vector $\mathbf{y}$ is presented as

$$y = (x_1^1,\ x_2^1,\ \ldots,\ x_{n_1}^1,\ x_1^2,\ x_2^2,\ \ldots,\ x_{n_2}^2,\ \ldots,\ x_1^{k-1},\ x_2^{k-1},\ \ldots,\ x_{n_{k-1}}^{k-1})^T. \tag{28}$$

The vector y has dimension $\sum_{j=1}^{k-1} n_j$. On each interval of motion, we have

$$\begin{aligned}
&\mathbf{y}^1 = (x_1^1,\ x_2^1,\ \ldots,\ x_{n_1}^1,\ 0,\ 0,\ \ldots,\ 0,\ \ldots,\ 0,\ 0,\ \ldots,\ 0)^T \quad \text{for } t_0 \le t \le t_{1-},\\
&\mathbf{y}^2 = (0,\ 0,\ \ldots,\ 0,\ x_1^2,\ x_2^2,\ \ldots,\ x_{n_2}^2,\ \ldots,\ 0,\ 0,\ \ldots,\ 0)^T \quad \text{for } t_{1+} \le t \le t_{2-},\\
&\cdots\cdots\cdots\cdots\\
&\mathbf{y}^{k-1} = (0,\ 0,\ \ldots,\ 0,\ 0,\ 0,\ \ldots,\ 0,\ \ldots,\ x_1^{k-1},\ x_2^{k-1},\ \ldots,\ x_{n_{k-1}}^{k-1})^T \quad \text{for } t_{(k-2)+} \le t \le t_{k-1}.
\end{aligned} \tag{29}$$

Similarly to (13)–(15), at intermediate time t_j $(j = 1, 2, \ldots, k-2)$, the state vector $\mathbf{y}$ satisfies the conditions

$$\begin{aligned}
\lim_{t \to t_j -} \mathbf{y} &= \lim_{t \to t_j -} (\mathbf{x}^1,\ \mathbf{x}^2,\ \ldots,\ \mathbf{x}^j,\ \mathbf{x}^{j+1},\ \ldots,\ \mathbf{x}^{k-1})^T\\
&= (\mathbf{0},\ \mathbf{0},\ \ldots,\ \mathbf{0},\ \mathbf{x}^j,\ \mathbf{0},\ \ldots,\ \mathbf{0})^T = \mathbf{y}^j(t_j), \quad (j = 1, 2, \ldots, k-2).
\end{aligned}$$

In the case under consideration, parameters (16)–(19) have the form

$$\begin{aligned}
\mathbf{C} &= \begin{bmatrix} \mathbf{A}_1 & \mathbf{0}_{12} & \mathbf{0}_{13} & \cdots & \mathbf{0}_{1k-1}\\ \mathbf{0}_{21} & \mathbf{A}_2 & \mathbf{0}_{23} & \cdots & \mathbf{0}_{2k-1}\\ \vdots & \vdots & \vdots & \ddots & \vdots\\ \mathbf{0}_{k-11} & \mathbf{0}_{k-12} & \mathbf{0}_{k-13} & \cdots & \mathbf{A}_{k-1} \end{bmatrix},\\
\mathbf{d} &= (b_1^1,\ b_2^1,\ \ldots,\ b_{n_1}^1,\ b_1^2,\ b_2^2,\ \ldots,\ b_{n_2}^2,\ \ldots,\ b_1^{k-1},\ b_2^{k-1},\ \ldots,\ b_{n_{k-1}}^{k-1})^T
\end{aligned} \tag{30}$$

on the time interval $t \in [t_0, t_{k-1}]$. The matrix $\mathbf{C}$ and the vector $\mathbf{d}$ satisfy the conditions

$$\begin{aligned}
\mathbf{C}_1 &= \begin{bmatrix} \mathbf{A}_1 & \mathbf{0}_{12} & \mathbf{0}_{13} & \cdots & \mathbf{0}_{1k-1}\\ \mathbf{0}_{21} & \mathbf{0}_{22} & \mathbf{0}_{23} & \cdots & \mathbf{0}_{2k-1}\\ \vdots & \vdots & \vdots & \ddots & \vdots\\ \mathbf{0}_{k-11} & \mathbf{0}_{k-12} & \mathbf{0}_{k-13} & \cdots & \mathbf{0}_{k-1k-1} \end{bmatrix},\\
\mathbf{d}^1 &= (b_1^1,\ b_2^1,\ \ldots,\ b_{n_1}^1,\ 0,\ 0,\ \ldots,\ 0,\ \ldots,\ 0,\ 0,\ \ldots,\ 0)^T \quad \text{for } t \in [t_0, t_1],
\end{aligned} \tag{31}$$

$$\begin{aligned}
\mathbf{C}_2 &= \begin{bmatrix} \mathbf{0}_{11} & \mathbf{0}_{12} & \mathbf{0}_{13} & \cdots & \mathbf{0}_{1k-1}\\ \mathbf{0}_{21} & \mathbf{A}_2 & \mathbf{0}_{23} & \cdots & \mathbf{0}_{2k-1}\\ \vdots & \vdots & \vdots & \ddots & \vdots\\ \mathbf{0}_{k-11} & \mathbf{0}_{k-12} & \mathbf{0}_{k-13} & \cdots & \mathbf{0}_{k-1k-1} \end{bmatrix},\\
\mathbf{d}^2 &= (0,\ 0,\ \ldots,\ 0,\ b_1^2,\ b_2^2,\ \ldots,\ b_{n_2}^2,\ \ldots,\ 0,\ 0,\ \ldots,\ 0)^T \quad \text{for } t \in [t_1, t_2]
\end{aligned} \tag{32}$$

On the interval $t \in [t_{k-2}, t_{k-1}]$, we have

$$\mathbf{C}_{k-1} = \begin{bmatrix} \mathbf{0}_{11} & \mathbf{0}_{12} & \mathbf{0}_{13} & \cdots & \mathbf{0}_{1k-1} \\ \mathbf{0}_{21} & \mathbf{0}_{22} & \mathbf{0}_{23} & \cdots & \mathbf{0}_{2k-1} \\ \vdots & \vdots & \vdots & \ddots & \vdots \\ \mathbf{0}_{k-11} & \mathbf{0}_{k-12} & \mathbf{0}_{k-13} & \cdots & \mathbf{A}_{k-1} \end{bmatrix}, \tag{33}$$
$$\mathbf{d}^{k-1} = (0,\ 0,\ \ldots,\ 0,\ 0,\ 0,\ \ldots,\ 0,\ \ldots,\ b_1^{k-1},\ b_2^{k-1},\ \ldots,\ b_{n_{k-1}}^{k-1})^T.$$

The matrices $\mathbf{M}_1$, $\mathbf{M}_2$, …, $\mathbf{M}_{k-1}$ have structures similar to (23), (24) of dimensions $(n_1 + n_2 + \cdots + n_{k-1}) \times n_j$, $(j = 1, 2, \ldots, k-1)$. In particular,

$$\mathbf{M}_1 = (\mathbf{d}^1, \mathbf{C}_1\mathbf{d}^1, \mathbf{C}_1^2\mathbf{d}^1, \ldots, \mathbf{C}_1^{n_1-1}\mathbf{d}^1)$$
$$= \begin{bmatrix} \mathbf{b}^1 & \mathbf{A}_1\mathbf{b}^1 & \mathbf{A}_1^2\mathbf{b}^1 & \cdots & \mathbf{A}_1^{n_1-1}\mathbf{b}^1 \\ 0 & 0 & 0 & \cdots & 0 \\ 0 & 0 & 0 & \cdots & 0 \\ \vdots & \vdots & \vdots & \ddots & \vdots \\ 0 & 0 & 0 & \cdots & 0 \end{bmatrix} \quad \text{for } t \in [t_0, t_1], \tag{34}$$

$$\mathbf{M}_2 = (\mathbf{d}^2, \mathbf{C}_2\mathbf{d}^2, \mathbf{C}_2^2\mathbf{d}^2, \ldots, \mathbf{C}_2^{n_2-1}\mathbf{d}^2)$$
$$= \begin{bmatrix} 0 & 0 & 0 & \cdots & 0 \\ \vdots & \vdots & \vdots & \ddots & \vdots \\ 0 & 0 & 0 & \cdots & 0 \\ \mathbf{b}^2 & \mathbf{A}_2\mathbf{b}^2 & \mathbf{A}_2^2\mathbf{b}^2 & \cdots & \mathbf{A}_2^{n_2-1}\mathbf{b}^2 \\ 0 & 0 & 0 & \cdots & 0 \\ \vdots & \vdots & \vdots & \ddots & \vdots \\ 0 & 0 & 0 & \cdots & 0 \end{bmatrix} \quad \text{for } t \in [t_1, t_2], \tag{35}$$

$$\mathbf{M}_{k-1} = (\mathbf{d}^{k-1}, \mathbf{C}_{k-1}\mathbf{d}^{k-1}, \mathbf{C}_{k-1}^2\mathbf{d}^{k-1}, \cdots, \mathbf{C}_{k-1}^{n_{k-1}-1}\mathbf{d}^{k-1})$$
$$= \begin{bmatrix} 0 & 0 & 0 & \cdots & 0 \\ 0 & 0 & 0 & \cdots & 0 \\ \vdots & \vdots & \vdots & \cdots & \vdots \\ \mathbf{b}^{k-1} & \mathbf{A}_{k-1}\mathbf{b}^{k-1} & \mathbf{A}_{k-1}^2\mathbf{b}^{k-1} & \cdots & \mathbf{A}_{k-1}^{n_{k-1}-1}\mathbf{b}^{k-1} \end{bmatrix} \quad \text{for } t \in [t_{k-2}, t_{k-1}]. \tag{36}$$

At the last stage of the manipulator movement, the dynamic characteristics of the mechanical system change at time $t = t_{k-1}$ and the manipulator moves from point $\mathbf{x}^k(t_{k-1})$ to the final point $\mathbf{x}^k(T)$ (remember that $\mathbf{x}^{k-1}(t_{k-1}) = \mathbf{x}^k(t_{k-1})$ is condition (3)). The system of the equations of motion on the time interval $t \in [t_{k-1}, t_k]$ coincides with the last system in (5), that is,

$$\dot{\mathbf{x}}^k = \mathbf{A}_k\mathbf{x}^k + \mathbf{b}^k u \tag{37}$$

for $\mathbf{x}^{k-1}(t_{k-1}) = \mathbf{x}^k(t_{k-1})$. On the interval $t \in [t_{k-1}, t_k]$, system (37) is completely controllable if

$$\operatorname{rang}\mathbf{M}_{kk} = \operatorname{rang}(\mathbf{b}^k,\ \mathbf{A}_k\mathbf{b}^k,\ \mathbf{A}_k^2\mathbf{b}^k,\ \ldots,\ \mathbf{A}_k^{n_k-1}\mathbf{b}^k) = n_k. \tag{38}$$

On the entire interval $[t_0, T]$, the determinant of the controllability matrix $\mathbf{M}^k$ of the combined systems (5), on the account of (34), (38), has the structure

$$\begin{aligned}\mathbf{M}^k &= \{\mathbf{M}_1, \mathbf{M}_2, \ldots, \mathbf{M}_{k-1}, \mathbf{M}_k\}\\ &= \{(\mathbf{d}^1,\ \mathbf{C}_1\mathbf{d}^1,\ \mathbf{C}_1^2\mathbf{d}^1,\ \ldots,\ \mathbf{C}_1^{n_1-1}), (\mathbf{d}^2,\ \mathbf{C}_2\mathbf{d}^2,\ \mathbf{C}_2^2\mathbf{d}^2,\ \ldots,\ \mathbf{C}_2^{n_2-1}\mathbf{d}^2),\ \ldots,\\ &(\mathbf{d}^{k-1},\ \mathbf{C}_{k-1}\mathbf{d}^{k-1},\ \mathbf{C}_{k-1}^2\mathbf{d}^{k-1},\ \ldots,\ \mathbf{C}_{k-1}^{n_{k-1}-1}\mathbf{d}^{k-1}),\ (\mathbf{d}^k,\ \mathbf{C}_k\mathbf{d}^k,\ \mathbf{C}_k^2\mathbf{d}^k,\ \ldots,\ \mathbf{C}_k^{n_k-1}\mathbf{d}^k)\}\\ &= \begin{bmatrix} \mathbf{M}_{11} & \mathbf{0}_{12} & \mathbf{0}_{13} & \cdots & \mathbf{0}_{1k-1} & \mathbf{0}_{1k}\\ \mathbf{0}_{21} & \mathbf{M}_{22} & \mathbf{0}_{23} & \cdots & \mathbf{0}_{2k-1} & \mathbf{0}_{2k}\\ \vdots & \vdots & \vdots & \ddots & \vdots & \vdots\\ \mathbf{0}_{k-11} & \mathbf{0}_{k-12} & \mathbf{0}_{k-13} & \cdots & \mathbf{M}_{k-1k-1} & \mathbf{0}_{k-1k}\\ \mathbf{0}_{k1} & \mathbf{0}_{k2} & \mathbf{0}_{k3} & \cdots & \mathbf{0}_{kk-1} & \mathbf{M}_{kk}\end{bmatrix}.\end{aligned} \tag{39}$$

The matrix $\mathbf{M}^k$ has dimension $\left(\sum_{j=1}^{k} n_j\right) \times \left(\sum_{j=1}^{k} n_j\right)$. The determinant of the controllability matrix (39), on the account of (26), (27), is defined as $\det \mathbf{M}^k = \det \mathbf{M}_{kk} \det \mathbf{M}^{k-1}$, where $\det \mathbf{M}_{kk} \neq 0$ and $\det \mathbf{M}^{k-1} \neq 0$, i.e.,

$$\operatorname{rang} \mathbf{M}^k = n_1 + n_2 + \cdots + n_k.$$

Hence, from (26), (27), and (39) we conclude that inclusion (5) in the intermediate states (3) over the entire time interval $[t_0, T]$ is completely controllable if all systems in (6) are controllable at their intervals of definition and not completely controllable if at least one system in (6) is not completely controllable on its interval of definition. This means that the manipulator maintenance process on the whole time interval is completely controllable if each stage of the maintenance is completely manageable and unmanageable if the maintenance process is uncontrollable at least on one interval.

Conclusions

A model of a technological maintenance process consisting of a controlled ladder manipulator and mobile or fixed objects (targets) is described. The manipulator must automatically support continuous operation of the objects. It is assumed that in the the maintenance process, the dynamic characteristics and the phase vectors of the manipulator state can change at some finite instants of time. In a particular case, when the manipulator movement at each maintenance interval is described by linear differential equations with constant coefficients and the motions of the objects are specified, the problems of controllability of the entire process are investigated depending on the controllability at each stage of the motion. It was shown that the maintenance process on the entire time interval is completely manageable if it is controllable on each maintenance interval and is not manageable if it is not controllable at least on one interval. The controllability matrix of the entire system that unites finitely many composite systems is presented.

References

[1] Chernousko F L, Bolotnik N N, and Gradetsky V G 1989 *Manipulation Robots: Dynamics, Control, and Optimization* (Moscow: Nauka) p 368 [in Russian]

[2] Akulenko L D and Bolotnik N N 1986 Synthesis of optimal control of transport motions of manipulation robots *Mech. Solids* **21** (4) 18–26

[3] Gradetsky V G, Ghukasyan A A, Grudev A I and Chernousko F L 1985 On the effect of the elastic compliance of the robot design on its dynamics *Izv. AN SSSR. Mekh. Tverd. Tela* No. 3 63–71

[4] Ghukasyan A A and Machkalyan R N 2007 Kinematics of the motion of a manipulator with elastic connecting nodes in a curvilinear coordinate system *Izv. Nats. Akad. Nauk Armenii. Mekh.* **60** (3) 114–20

[5] Ghukasyan A A 2014 On the kinematics of a multi-link manipulator with elastic connecting nodes and elastic links *Izv. Nats. Akad. Nauk Armenii. Mekh.* **67** (3) 68–83
[6] Ghukasyan A A 2014 Kinematic control of the movement of the grip of a two-link elastic manipulator *Dokl. Nats. Akad. Nauk Armenii* **114** (4) 316–24
[7] Ghukasyan A A 2014 On the spatial position and deformation of the elastic links of the manipulator *Izv. Nats. Akad. Nauk Armenii. Mekh.* **67** (4) 53–64
[8] Ghukasyan A A 2015 Two approaches to the study of the kinematics of elastic manipulators *Izv. Nats. Akad. Nauk Armenii. Mekh.* **68** (2) 53–67
[9] Ghukasyan A A 1986 A problem of optimal modelling of technological processes served by a manipulator robot *Izv. Nats. Akad. Nauk Arm. SSR. Mekh.* **39** (6) 39–49
[10] Ghukasyan A A and Matevosyan A G 2002 On the controllable motion of a material point with a variable mass *Izv. Nats. Akad. Nauk Armenii. Mekh.* **55** (1) 75–81
[11] Ghukasyan A A and Matevosyan A G 2003 The optimal control problem of the motion of a point of the variable mass In *Mathematical Analysis and Its Applications. Collection of Scientific Papers: ASPU* issue 3 (Yerevan) pp 29–40 [in Russian]
[12] Velichenko V V 1967 Optimality conditions for control problems with intermediate conditions *Dokl. Akad. Nauk SSSR* **174** (5), 1011–3
[13] Pontryagin L S, Boltyansky V G, Gamkrelidze R V, and Mishchenko E F 1983 *Mathematical Theory of Optimal Processes* (CRC Press) p 360
[14] Krasovskii N N 1968 *Theory of Motion Control* (Moscow: Nauka) p 475 [in Russian]
[15] Roitenberg Ya N 1978 *Automatic Control* (Moscow: Nauka) p 552 [in Russian]
[16] Lee E B and Markus L 1972 *Foundations of Optimal Control Theory* (Malabar, FL: Krieger Pub Co) p 576
[17] Gantmakher F R 2004 *The Theory of Matrices* (Moscow: Nauka) p 560 [in Russian]

IOP Conf. Series: Journal of Physics: Conf. Series **991** (2018) 012027 doi:10.1088/1742-6596/991/1/012027

Symmetric vibrations of a liquid in a vessel with a separator and an elastic bottom

D A Goncharov* **and A A Pozhalostin****

Bauman Moscow State Technical University, Moscow, Russia

E-mail: *goncharov@bmstu.ru, **a.pozhalostin@mail.ru

Abstract. The paper considers the problem of small axisymmetric vibrations of an ideal fluid filling a vessel with rigid walls and an elastic bottom. The liquid is divided into two layers by an elastic septum. The elastic baffle and the vessel elastic bottom are modeled by elastic membranes. The Neumann boundary-value problem is posed for the fluid. The equations of motion of the membranes are integrated with boundary conditions.

1. Introduction

The problem originates in practice. In rocket technology, various intra-tank structures are widely used to launch a propulsion systems under the space flight conditions [1]. These designs are significantly complicated for the frequency analysis. At the same time, the frequency analysis is very important, since the elastic tank, the propulsion path of the rocket engine, and the engine itself form a closed system with feedback in which self-oscillations are possible [2].

Since these phenomena and their influence on the spacecraft and launch vehicle operation are of great importance, the symmetric liquid oscillations in vessels of various shapes and configurations were studied in numerous works [3–8]. Some generalizations are given in the monograph [9].

2. Formulation of the problem

We consider small axisymmetric oscillations of an ideal fluid filling a vessel of cylindrical shape of radius R with rigid walls. The vessel is divided by an elastic membrane. The bottom of the vessel is also an elastic membrane. We introduce cylindrical coordinates $O_1x_1\eta r$ and $O_2x_2\eta r$ with origins at the respective centers of the partition and the bottom. The liquid fills the areas ϑ_1 and ϑ_2. The motion of the fluid is assumed to be potential with velocity potential $\Phi_i = \Phi_i(x_i, r, t)$.

In the areas ϑ_1 and ϑ_2, the Laplace equation has the form

$$\nabla^2\Phi_i = 0, \quad i = 1, 2, \tag{1}$$

where ∇ is the Hamilton operator and Φ_i is the velocity potential in the area ϑ_1. On the free surface, we pose the linearized boundary condition in simplified form

$$\Phi_1\big|_{x_1=h_1} = 0. \tag{2}$$

If the inertia of membranes is neglected, then their equations of motion take the following forms for the separating and bottom membranes:

$$\nabla^2 \dot{w}_1 = \frac{\varrho}{\tau}\left(\frac{\partial^2 \Phi_1}{\partial t^2}\bigg|_{x_1=0} - \frac{\partial^2 \Phi_2}{\partial t^2}\bigg|_{x_2=h_2}\right), \quad w_1\big|_{r=R} = 0, \quad w_1\big|_{r=0} < \infty, \tag{3}$$

$$\nabla^2 \dot{w}_2 = \frac{\varrho}{\tau}\frac{\partial^2 \Phi_2}{\partial t^2}\bigg|_{x_2=0}, \quad w_2\big|_{r=R} = 0, \quad w_2\big|_{r=0} < \infty, \tag{4}$$

where w_1 and w_2 are deflections of the separating and bottom membranes, $\dot{w}_1$ and $\dot{w}_2$ are velocities of the membrane deflections, and τ is the tension of the membranes.

The kinematic conditions are also satisfied. In particular, the condition that the liquid velocities are the same in the areas ϑ_1 and ϑ_2:

$$\frac{\partial \Phi_1}{\partial x_1}\bigg|_{x_1=0} = \frac{\partial \Phi_2}{\partial x_2}\bigg|_{x_2=h_2}, \tag{5}$$

the condition that the fluid speed is equal to the rate of deflection of the separating membrane:

$$\frac{\partial \Phi_2}{\partial x_2}\bigg|_{x_2=h_2} = \dot{w}_1, \quad \frac{\partial \Phi_1}{\partial x_1}\bigg|_{x_1=0} = \dot{w}_1, \tag{6}$$

and the condition that the fluid speed is equal to the rate of deflection of the bottom membrane:

$$\frac{\partial \Phi_2}{\partial x_2}\bigg|_{x_2=0} = \dot{w}_2. \tag{7}$$

In addition, the non-flow conditions are satisfied:

$$\frac{\partial \Phi_1}{\partial r}\bigg|_{r=R} = 0, \quad \frac{\partial \Phi_2}{\partial r}\bigg|_{r=R} = 0, \tag{8}$$

3. Solution of the boundary-value problem

We consider the potential for the liquid velocities in the form

$$\Phi_i = \Phi_{i0} e^{i\omega t}, \tag{9}$$

where i is the imaginary unit, ω is the natural frequency, and t denotes the time. Satisfying the Laplace equation, the non-flow conditions, and the condition on the free surface, we obtain expressions for the potentials of fluid velocities in the form

$$\Phi_{10} = B_{10}\left(\frac{x_1}{h_1} - 1\right) + \sum_{i=1}^{\infty} C_{1i} J_0\left(\xi_i \frac{r}{R}\right)\left[\cosh\left(\xi_i \frac{x_1}{R}\right) - \coth\left(\xi_i \frac{h_1}{R}\right)\sinh\left(\xi_i \frac{x_1}{R}\right)\right], \tag{10}$$

$$\Phi_{20} = A_{20} + B_{20}\frac{x_2}{h_2} + \sum_{i=1}^{\infty} J_0\left(\xi_i \frac{r}{R}\right)\left[C_{2i}\cosh\left(\xi_i \frac{x_2}{R}\right) + C_{3i}\sinh\left(\xi_i \frac{x_2}{R}\right)\right]. \tag{11}$$

Here B_0, A_{20}, B_{20}, C_{1i}, C_{2i}, C_{3i} are constants of integration, $J_0(\xi_i\, dr/dR)$ is the Bessel function of the first kind of order zero, and ξ_i is the root of equation $J_0'(x) = 0$ obtained from the non-flow conditions. In this case, the equations of membrane motions take the following forms for the separating and bottom membranes:

$$\nabla^2 \dot{w}_1 = \frac{\varrho\omega^2}{\tau}\Bigg[-\Bigg\{A_{20} + B_{20} + \sum_{i=1}^{\infty} J_0\Big(\xi_i\frac{r}{R}\Big)\Big[C_{2i}\cosh\Big(\xi_i\frac{x_2}{R}\Big) + C_{3i}\sinh\Big(\xi_i\frac{x_2}{R}\Big)\Big]\Bigg\}$$
$$- B_0 + \sum_{i=1}^{\infty} C_{1i}J_0\Big(\xi_i\frac{r}{R}\Big), \tag{12}$$

$$\nabla^2 \dot{w}_2 = \frac{\varrho\omega^2}{\tau}\Big[A_{20} + \sum_{i=1}^{\infty} C_{2i}J_0\Big(\xi_i\frac{r}{R}\Big)\Big]. \tag{13}$$

Integrating the above equations, we obtain the following expressions for the rates of deflection of the separate and bottom membrane, respectively

$$\dot{w}_1 = \frac{\varrho\omega^2}{\tau}\Bigg\{C_{10} - (A_{20}+B_{20})\frac{r^2}{4} + \sum_{i=1}^{\infty}\frac{R^2}{\xi_i^2}\Big[C_{2i}\cosh\Big(\xi_i\frac{x_2}{R}\Big) + C_{3i}\sinh\Big(\xi_i\frac{x_2}{R}\Big)\Big]$$
$$- B_0\frac{r^2}{4} + \sum_{i=1}^{\infty}\frac{R^2}{\xi_i^2}C_{1i}J_0\Big(\xi_i\frac{r}{R}\Big)\Bigg\}, \tag{14}$$

$$\dot{w}_2 = C_0 + \frac{\varrho\omega^2}{\tau}A_{20}\frac{r^2}{4} + \frac{\varrho\omega^2}{\tau}\sum_{i=1}^{\infty}\frac{R^2}{\xi_i^2}C_{2i}J_0\Big(\xi_i\frac{r}{R}\Big). \tag{15}$$

To complete the system of arbitrary constants of integration in expressions (14) and (15), we use the following procedure [10]. We represent $r^2/4$ as a series in Bessel eigenfunctions

$$\frac{r^2}{4} = \sum_{i=1}^{\infty}\beta_{1i}J_0\Big(\xi_i\frac{r}{R}\Big) + \beta_0. \tag{16}$$

The coefficients are determined from the following relations

$$\int_0^R \frac{r^3}{4}\,dr = \int_0^R \sum_{i=1}^{\infty}\beta_{1i}J_0\Big(\xi_i\frac{r}{R}\Big) r\,dr + \int_0^R \beta_0 r\,dr, \tag{17}$$

$$\int_0^R \frac{r^3}{4}J_0\Big(\xi_i\frac{r}{R}\Big) r\,dr = \int_0^R \sum_{i=1}^{\infty}\beta_{1i}J_0^2\Big(\xi_i\frac{r}{R}\Big) r^2\,dr + \int_0^R \beta_0 J_0^2\Big(\xi_i\frac{r}{R}\Big) r\,dr. \tag{18}$$

In this case, the expressions for the deflections rates of the separating and bottom membranes become

$$\dot{w}_1 = \frac{\varrho\omega^2}{\tau}\Bigg\{C_{10} - (A_{20}+B_{20})\Big[\sum_{i=1}^{\infty}\beta_{1i}J_0\Big(\xi_i\frac{r}{R}\Big) + \beta_0\Big] + \sum_{i=1}^{\infty}\frac{R^2}{\xi_i^2}\Big[C_{2i}\cosh\Big(\xi_i\frac{h_2}{R}\Big)$$
$$+ C_{3i}\sinh\Big(\xi_i\frac{h_2}{R}\Big)\Big]J_0\Big(\xi_i\frac{r}{R}\Big) - B_0\Big(\sum_{i=1}^{\infty}\beta_{1i}J_0\Big(\xi_i\frac{r}{R}\Big) + \beta_0\Big) + \sum_{i=1}^{\infty}\frac{R^2}{\xi_i^2}C_{1i}J_0\Big(\xi_i\frac{r}{R}\Big)\Bigg\}, \tag{19}$$

$$\dot{w}_2 = C_0 + A_{20}\frac{\varrho\omega^2}{\tau}\Big[\sum_{i=1}^{\infty}\beta_{1i}J_0\Big(\xi_i\frac{r}{R}\Big) + \beta_0\Big] + \frac{\varrho\omega^2}{\tau}\sum_{i=1}^{\infty}\frac{R^2}{xi_i^2}C_{2i}J_0\Big(\xi_i\frac{r}{R}\Big). \tag{20}$$

With the last expression taken account, the kinematic conditions can be written as

$$\begin{aligned}&\frac{\varrho\omega^2}{\tau}\Bigg\{C_{10}-(A_{20}+B_{20})\Bigg[\sum_{i=1}^{\infty}\beta_{1i}J_0\Big(\xi_i\frac{r}{R}\Big)+\beta_0\Bigg]\\&\quad+\sum_{i=1}^{\infty}\frac{R^2}{xi_i^2}\Bigg[C_{2i}\cosh\Big(\xi_i\frac{h_2}{R}\Big)+C_{3i}\sinh\Big(\xi_i\frac{h_2}{R}\Big)\Bigg]J_0\Big(\xi_i\frac{r}{R}\Big)\\&\quad-B_0\Bigg(\sum_{i=1}^{\infty}\beta_{1i}J_0\Big(\xi_i\frac{r}{R}\Big)+\beta_0\Bigg)+\sum_{i=1}^{\infty}\frac{R^2}{\xi_i^2}C_{1i}J_0\Big(\xi_i\frac{r}{R}\Big)\Bigg\}\\&\quad=\frac{B_{20}}{h_2}+\sum_{i=1}^{\infty}J_0\Big(\xi_i\frac{r}{R}\Big)\Bigg[C_{2i}\sinh\Big(\xi_i\frac{h_2}{R}\Big)+C_{3i}\cosh\Big(\xi_i\frac{h_2}{R}\Big)\Bigg]\frac{\xi_i}{R},\end{aligned}\tag{21}$$

$$\begin{aligned}&\frac{B_{20}}{h_2}+\sum_{i=1}^{\infty}J_0\Big(\xi_i\frac{r}{R}\Big)\Bigg[C_{3i}\frac{\xi_i}{R}\Bigg]\\&\quad=C_0+A_{20}\frac{\varrho\omega^2}{\tau}\Bigg[\sum_{i=1}^{\infty}\beta_{1i}J_0\Big(\xi_i\frac{r}{R}\Big)+\beta_0\Bigg]+\frac{\varrho\omega^2}{\tau}\frac{R^2}{\xi_i^2}\sum_{i=1}^{\infty}C_{2i}J_0\Big(\xi_i\frac{r}{R}\Big).\end{aligned}\tag{22}$$

In addition, in view of (5), we have

$$\begin{aligned}&\frac{B_0}{h_1}+\sum_{i=1}^{\infty}C_{1i}J_0\Big(\xi_i\frac{r}{R}\Big)\frac{\xi_i}{R}\coth\Big(\xi_i\frac{h_1}{R}\Big)\\&\quad=\frac{B_{20}}{h_2}+\sum_{i=1}^{\infty}J_0\Big(\xi_i\frac{r}{R}\Big)\Bigg[C_{2i}\frac{\xi_i}{R}\sinh\Big(\xi_i\frac{h_2}{R}\Big)+C_{3i}\frac{\xi_i}{R}\cosh\Big(\xi_i\frac{h_2}{R}\Big)\Bigg].\end{aligned}\tag{23}$$

Thus, we obtained a system of equations that allows us to close the system of arbitrary constants. In particular, we have

$$\begin{aligned}&B_{20}=B_0\frac{h_1}{h_2},\quad C_0=\frac{B_0}{h_1}-A_{20}\beta_0,\quad C_{10}=\frac{B_0}{h_1}+A_{20}\beta_0+B_0\beta_0,\\&C_{1i}=\frac{1}{\coth(\xi_ih_1/R)}\Bigg[C_{2i}\sinh\Big(\xi_i\frac{h_2}{R}\Big)+C_{3i}\cosh\Big(\xi_i\frac{h_2}{R}\Big)\Bigg],\\&C_{2i}=a_{10}^{(i)}B_0+a_{20}^{(i)}A_{20},\quad C_{3i}=b_{10}^{(i)}B_0+b_{20}^{(i)}A_{20},\end{aligned}$$

where, in turn,

$$a_{10}^{(i)}=-\frac{\varepsilon_{3i}}{\varepsilon_{1i}},\quad a_{20}^{(i)}=-\frac{\varepsilon_{2i}}{\varepsilon_{1i}},\quad b_{20}^{(i)}=\frac{\varrho\omega^2}{\tau}\beta_{1i}\frac{R}{\xi_i}+\frac{\varrho\omega^2}{\tau}\frac{R^3}{\xi_i^3}a_{20}^{(i)},\quad b_{10}^{(i)}=\frac{\varrho\omega^2}{\tau}\frac{R^2}{\xi_i^2}a_{10}^{(i)}.$$

Here

$$\varepsilon_{1i} = \frac{\varrho\omega^2}{\tau}\frac{R^2}{\xi_i^2}\coth\left(\xi_i\frac{h_2}{R}\right) + \frac{\varrho^2\omega^4}{\tau^2}\frac{R^5}{\xi_i^5}\sinh\left(\xi_i\frac{h_2}{R}\right) + \frac{\varrho\omega^2}{\tau}\frac{R^4}{\xi_i^4}\sinh\left(\xi_i\frac{h_2}{R}\right)\tanh\left(\xi_i\frac{h_1}{R}\right)$$
$$+ \left(\frac{\varrho^2\omega^4}{\tau^2}\right)\frac{R^6}{\xi_i^6}\cosh\left(\xi_i\frac{h_2}{R}\right)\tanh\left(\xi_i\frac{h_1}{R}\right) - \cosh\left(\xi_i\frac{h_1}{R}\right) - \frac{\varrho\omega^2}{\tau}\frac{R^3}{\xi_i^3}\sinh\left(\xi_i\frac{h_1}{R}\right),$$
$$\varepsilon_{2i} = -\frac{\varrho\omega^2}{\tau}\beta_{1i} + \left(\frac{\varrho\omega^2}{\tau}\right)^2\frac{R^3}{\xi_i^3}\beta_{1i}\sinh\left(\xi_i\frac{h_2}{R}\right)$$
$$+ \left(\frac{\varrho\omega^2}{\tau}\right)^2\frac{R^5}{\xi_i^5}\beta_{1i}\cosh\left(\xi_i\frac{h_2}{R}\right)\tanh\left(\xi_i\frac{h_1}{R}\right) - \frac{\varrho\omega^2}{\tau}\frac{R}{\xi_i}\beta_{1i}\sinh\left(\xi_i\frac{h_1}{R}\right),$$
$$\varepsilon_{3i} = -\frac{\varrho\omega^2}{\tau}\beta_{1i} - \frac{\varrho\omega^2}{\tau}\frac{R^2}{\xi_i^2}\beta_{1i}.$$

A closed system of arbitrary constants makes it possible to construct the system

$$\begin{cases} a_{11}B_0 + a_{12}A_{20} = 0 & \text{from } \dot{w}_1\big|_{r=R} = 0, \\ a_{21}B_0 + a_{22}A_{20} = 0 & \text{from } \dot{w}_2\big|_{r=R} = 0. \end{cases} \tag{24}$$

From the this system we can obtain obtain the frequency equation in the form

$$a_{11}a_{22} - a_{12}a_{21} = 0. \tag{25}$$

Conclusions

Thus, we showed that, adopting a number of simplifying assumptions, it is possible to obtain an analytical solution of the problem of fluid oscillations in a vessel with an elastic bottom and a separating membrane. In further studies, it will be possible to modify the condition on the free surface, since the above solution method works for various conditions on the free surface. In addition, it is possible to consider membranes without neglecting their inertia, but then the expressions become much more complicated. The frequency equation is very important for studying various technical systems, since it allows modeling of real systems by using mechanical analogs and laboratory experiments.

References

[1] Bagrov V V, Kurpatenkov A V, and Polyaev V M 1997 *Capillary Systems for Fluid Extraction from Tanks of Spacecrafts* (Moscow: UNPTs Energomash) [in Russian]
[2] Mikishev G N and Rabinovich B I 1971 *Dynamics of Thin-Walled Structures with Compartments Containing Liquid* (Moscow: Mashinostroyenie) [in Russian]
[3] Okhotsimskii D E 1956 Theory of motion of a body with cavities partially filled with liquid *Prikl. Mat. Mekh.* **20** (1) 3–20
[4] Chernousko F L 1966 On free oscillations of a viscous fluid in a vessel *J. Appl. Math. Mech.* **30** (5) 990–1003
[5] Bauer H F 1981 Hydroelastische schwingungen einer oberflächenspannungsstruktur in einem sateliiten-behälter *Z. Flugwiss. Weltraumforsch.* **5** (5) 303–13
[6] Kononov Yu N and Tatarenko E A 2006 Free vibrations of elastic membranes that separate a multilayer liquid in a cylindrical vessel with an elastic bottom *Dinamich. sist.* No 21 7–13
[7] Kravtsov A V and Sekerzh-Zen'kovich S Ya 2004 On free vibrations of a low-viscosity liquid in a vessel partially filled with a porous medium *Zh. Vych. Mat. Mat. Fiz.* **44** (4) 746–51
[8] Bukreev V I, Sturova I V, and Chebotnikov A V 2014 Seiche oscillations in a reservoir filled with a double-layer fluid *Fluid Dynamics* **49** (3) 395–402
[9] Ibrahim R A 2005 *Liquid Sloshing Dynamics: Theory and Applications* (Cambridge: Cambridge Univ. Press)
[10] Pozhalostin A A and Goncharov D A 2015 Free axisymmetric oscillations of a two-layer liquid with an elastic separator between layers *Russ. Aeronaut.* **58** (1) 37–41

IOP Conf. Series: Journal of Physics: Conf. Series **991** (2018) 012028 doi:10.1088/1742-6596/991/1/012028

Modeling of normal contact of elastic bodies with surface relief taken into account

I G Goryacheva and I Yu Tsukanov*

Ishlinsky Institute for Problems in Mechanics of the Russian Academy of Sciences, Moscow, Russia

E-mail: *ivan.yu.tsukanov@gmail.com

Abstract. An approach to account the surface relief in normal contact problems for rough bodies on the basis of an additional displacement function for asperities is considered. The method and analytic expressions for calculating the additional displacement function for one-scale and two-scale wavy relief are presented. The influence of the microrelief geometric parameters, including the number of scales and asperities density, on additional displacements of the rough layer is analyzed.

1. Introduction

Contact problems accounting surface microrelief are used in many engineering applications. Since the contact stresses and displacements can reach significant values, they should be accounted in design of curvilinear bearings and liners in machines and instruments. This problem is also actual for thermal and electrical contacts between curvilinear surfaces, where the thermal and electric conductivities depend on the real contact area and the value of the gap between the surfaces. Another field of application deals with probe methods of materials testing, where it is necessary to indent unprepared surfaces.

Many experimental investigations of the contact stiffness of the machine and instrument parts showed that there is a certain difference between the experimental and computational results obtained for smooth elastic bodies. To remove this discrepancy, it is required to formulate and solve contact problems for rough elastic bodies. The first formulation of such problem was done in [1]. The further research was related to both experimental determination of the influence of the roughness parameters on the contact characteristics [2, 3] and formulation of new contact problems for bodies of various macro- and microgeometry interacting under specified boundary conditions [4–8].

The influence of rough layer on the contact interaction of elastic bodies can be quantitatively estimated on the basis of discrete contact mechanics. Here the two basic approaches – statistical and deterministic – can be applied. The statistical approach assumes interaction of two single asperities of regular shape with a number of contacting asperities depending on the probability of their heights. The displacements of asperities are summarized with the displacements of the whole curvilinear elastic body. Then, as a rule, the contact problem is solved neglecting the elastic interaction between asperities. One of the first study in this direction was published in [9]. A comparative review of statistical approach is given in [10]. In the present paper, we consider the deterministic approach for evaluation of additional displacement due to the surface roughness in contact problems, which allows us to take the elastic interaction between asperities into account. On the basis of this approach, the additional displacement as a function of nominal pressure is analyzed for various types of the surface microgeometry.

2. Modeling of the surface roughness effect by an additional displacement function

The plane contact problem for rough elastic bodies was first posed by Schtaierman [1]. In his work, it was assumed that under the action of a rough body of specified macrogeometry $f(x)$, the elastic displacements at each point of the nominal contact area are defined by a sum of basic displacements u_{z1} of the half-plane under the action of nominal pressure $p(x)$ and additional displacements u_{z2} related to local deformations of roughness. In [1], it is also assumed that this additional displacement is proportional to the acting nominal pressure, i.e., $u_{z2} = kp(x)$, where k is a coefficient depending on the surface microgeometry. The nominal contact pressure $p(x)$ during the penetration of a body of a given macroshape $f(x)$ into the elastic half-plane is determined by the integral equation [1]

$$-\frac{\partial f(x)}{\partial x} = k\frac{\partial p(x)}{\partial x} + \frac{2(1-\nu^2)}{\pi E}\int_{-a}^{a}\frac{p(\xi)}{x-\xi}\,d\xi. \quad (1)$$

Schtaierman proposed to define the coefficient k in equation (1) by tests. Attempts to determine the additional displacement experimentally were made in [2]. The experiments were carried out on the developed test bench for determining the contact separation of circular specimens under the first and successive loadings. The bench design allows avoiding the influence of the specimen bulk deformation and determining only the displacements of the surface asperities. However, no general technique of experimental determination of the additional displacement function has yet been developed. The research results [2] show that, at small displacements, this dependence is better described by the power function $u_{z2} = k(p(x,y))^{\alpha}$ than by a linear one.

Goryacheva [11–15] proposed derivation of an equation similar to (1) on the basis of modeling the discrete contact of bodies with specified macro- and microgeometry. The contact problem for a rigid punch, having a specified macrogeometry $f(x,y)$ and a rough surface, on an elastic half-space was formulated and solved in the linear statement under the assumption that elastic deformations were infinitesimally small. This assumption is satisfied if the heights of asperities are much smaller than their dimensions in the longitudinal and transverse directions. The real contact area consists of a finite number n of discrete subareas ω_i ($i = 1, 2\ldots, n$). The displacements of the elastic half-space boundary are written as [11, 12]

$$u_z(x,y) = \frac{1-\nu^2}{\pi E}\sum_{i=1}^{n}\iint_{\omega_i}\frac{p_i(x,y)\,d\xi\,d\eta}{\sqrt{(x-\xi)^2+(y-\eta)^2}}. \quad (2)$$

Here $p_i(x,y)$ are real contact pressures which satisfy the equilibrium equation

$$P = \sum_{i=1}^{n}\iint_{\omega_i} p_i(x,y)\,dx\,dy. \quad (3)$$

Here P is the total load applied to the rigid body which penetrates into the elastic half-space.

The idea of this approach consists in the following. Let us assume that we have many contact spots and, in an arbitrarily small vicinity Ω_0 of each point (x,y) of the nominal contact area, the microgeometry of the surface can be described by the function $F(x,y)$. There are two basic length scales in the problem under consideration (figure 1): the macrolevel which corresponds to the nominal contact area Ω and to the nominal (average) pressure $p(x,y)$ and the microlevel which corresponds to a subarea Ω_0 containing N real contact spots ω_i ($N \ll n$).

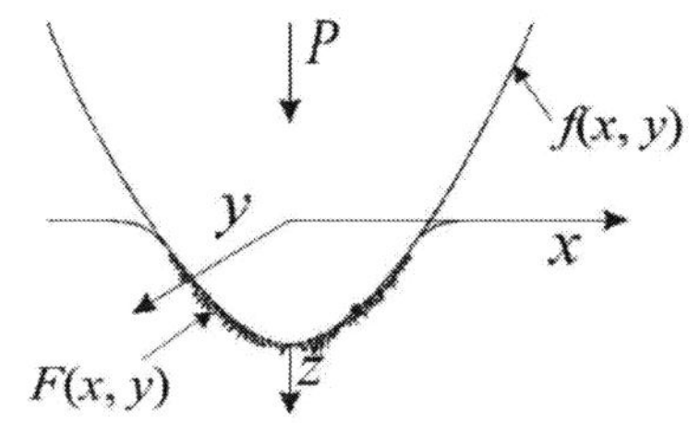

Figure 1. Scheme of contact of a body with a given macroshape $f(x, y)$ and microgeometry $F(x, y)$.

As $\Omega_0 \ll \Omega$, the curvature of the surface macrogeometry $f(x, y)$ can be neglected if the discrete contact problem is solved within Ω_0. It can be assumed that the uniformly distributed pressure $p(x, y)$ is applied in the subarea Ω_0. Then, as was proved in [13] using the localization principle [11], the additional displacements due to the rough layer can be calculated by the formula

$$C[p(x,y)] = -F(x_0, y_0) + \frac{1-\nu^2}{\pi E}\left[\sum_{i=1}^{N}\iint_{\omega_i}\frac{p_i(\xi,\eta)\,d\xi\,d\eta}{\sqrt{(x-\xi)^2+(y-\eta)^2}} - \iint_{\Omega_0}\frac{p(x,y)\,d\xi\,d\eta}{\sqrt{(x-\xi)^2+(y-\eta)^2}}\right]. \quad (4)$$

Here $F(x_0, y_0)$ is the height of the highest asperity in the subarea Ω_0.

As follows from equation (4), the additional displacement function depends on the distribution of the contact pressure at real contact spots, which is calculated with regard to the shape and location of asperities in the subarea Ω_0 and to the coupling of contact spots.

Then the equation for determining the nominal contact pressure distribution $p(x, y)$ and the nominal area of contact Ω with regard to the effect of microgeometry parameters has the form [13]

$$C[p(x,y)] + \frac{1-\nu^2}{\pi E}\iint_{\Omega}\frac{p(x,y)\,d\xi\,d\eta}{\sqrt{(x-\xi)^2+(y-\eta)^2}} = \delta - f(x,y). \quad (5)$$

Here δ is the punch penetration into the elastic half-space. To determine this value and the nominal contact area Ω, the equilibrium equation should be added to equation (5):

$$\iint_{\Omega} p(x,y)\,dx\,dy = P. \quad (6)$$

In addition, for a smooth function $f(x, y)$, the condition of a continuity of nominal pressures on the boundary $\partial\Omega$ of the nominal contact area Ω must be taken into account:

$$p(x,y)\big|_{(x,y)\in\partial\Omega} = 0. \quad (7)$$

It should be noted that equation (5) is a generalization of equation (1), proposed by Schtaierman, to the case of an arbitrary configuration of the body (three-dimensional problem) and to the case of a nonlinear additional displacement function.

Methods for solving integral equations like (1) and (5) are developed in [6–8]. Particular cases of linear and power function $C[p(x, y)]$ are considered in these papers.

For certain types of surfaces with regular microrelief, the function of additional displacement can be determined analytically or by approximating the numerical results of solving the corresponding contact problems on microscale level.

Goryacheva [13] developed an approach for determining the additional displacement function by modeling the contact interaction of a rough nominally flat body, having a regular periodic surface relief, with an elastic half-space. The contact problems of interaction of a system of periodically located asperities, having regular shape (spherical, cylindrical, etc.) and various heights, with an elastic half-space were solved in [12, 14]. Such model well describes the microrelief if the longitudinal and transversal roughnesses are commensurable (e.g., at grinding, polishing).

3. Additional displacement functions for surfaces with regular waviness

In another limiting case where the surface has clearly defined direction and periodicity of roughness (e.g., at finishing milling or turning), the surface is modeled by regular waviness of various shape neglecting the transverse roughness. Meanwhile, in finishing processing, the period of waviness is much higher than its amplitude, and so the linear elasticity can be applied.

For determining the additional displacement function for this type of model roughness, the periodic contact problem of plane elasticity should be solved. In that problem, a rigid body with regular relief, described by smooth function $F(x)$, penetrates into an elastic half-plane. The basic integral equation of the considered problem for the normal contact without friction is [1, 16]:

$$-\frac{E}{2(1-\nu^2)}\frac{\partial F(x)}{\partial x} = \frac{1}{2\pi}\int_{-a}^{a} p(\xi)\cot\frac{x-\xi}{2}\,d\xi, \tag{8}$$

where a is the contact zone half-length and $p(x)$ is the contact pressure.

The wavy surface penetrates into the elastic half-plane under the nominal pressure p_∞. Its value is determined by the ratio of linear load P on a single period to the value λ of the period, $p_\infty = P/\lambda$. It follows from the equilibrium equation that

$$P = \int_{-a}^{a} p(x)\,dx. \tag{9}$$

According to the theory described in Section 1, the additional displacement function for the wavy surface is determined by the expression

$$C[p_\infty] = -F(x_0) + \frac{2(1-\nu^2)}{\pi E}\left[\int_{-\lambda/2}^{\lambda/2} p(\xi)\ln|x-\xi|\,d\xi - p_\infty\int_{-\lambda/2}^{\lambda/2}\ln|x-\xi|\,d\xi\right] + C_1 - C_2. \tag{10}$$

The constants C_1 and C_2 depend on the selected displacement datum point. If the surface relief has a periodic distribution of asperities heights and the highest point is located at the origin of coordinates, then the additional displacements should be determined at the point $x_0 = 0$.

Thus, the additional displacement function can be written in the form

$$C[p_\infty] = -[\bar{u}_{z\lambda}(0) - \bar{u}_{z\lambda\infty}(0)], \tag{11}$$

where $\bar{u}_z(0)$ is the displacement of the asperities peak under the pressure distribution $p(x)$ acting on the period λ. The value of $\bar{u}_z$ for an arbitrary coordinate x is defined by the expression

$$\bar{u}_{z\lambda}(x) = \frac{2(1-\nu^2)}{\pi E}\int_{-\lambda/2}^{\lambda/2} p(\xi)\ln|x-\xi|\,d\xi + C_1, \tag{12}$$

and $\bar{u}_{z\infty}(0)$ is the displacement of the asperities peak from the nominal pressure p_∞. For an arbitrary coordinate x, it is given by the expression

$$\bar{u}_{z\lambda\infty}(x) = \frac{2(1-\nu^2)}{\pi E}p_\infty\int_{-\lambda/2}^{\lambda/2}\ln|x-\xi|\,d\xi + C_2. \tag{13}$$

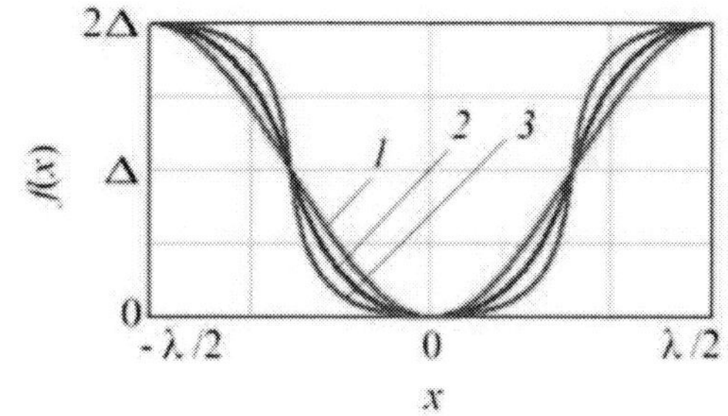

Figure 2. Waviness profiles on a single period for different values of m: $m = 0$ (curve *1*), $m = 1$ (curve *2*), and $m = 4$ (curve *3*).

To calculate equation (11) numerically, we use the following additional condition, which is implemented in numerical integration of equations (12) and (13):

$$\lim_{x\to\infty} [\bar{u}_z(x) - \bar{u}_{z\infty}(x)] = 0. \tag{14}$$

It follows from equation (10) that the additional displacement function is significantly influenced by the contact pressure distribution $p(x)$ on each period. The pressure distribution is, in turn, determined by the shape $F(x)$ of the wavy surface, as it follows from equation (8).

The surface waviness can generally be represented as a Fourier series. It can have different number of scales (harmonics of specified frequencies). In the simplest case, the waviness is defined by the sinusoid, i.e., $F(x) = \Delta[1 - \cos(2\pi x/\lambda)]$, where Δ is an amplitude and λ is a period. The contact problem for such a surface penetrating into an elastic half-plane was first solved by Westergaard [17].

To describe the waviness, other than the sinusoidal one, the close-form solution of the plane contact problem was found in [18] for the waviness with arbitrary shape of waveform defined by the function

$$f(x, m) = \Delta\left[1 - \frac{(m+1)\cos(2\pi x/\lambda)}{|m\cos(2\pi x/\lambda)| + 1}\right], \tag{15}$$

where m is a shape parameter. At $m = 0$, equation (15) becomes a simple sinusoid $f(x) = \Delta[1 - \cos(2\pi x/\lambda)]$; the radii of curvature of the profile peak and valley increase with increasing m (figure 2).

For determining the normal pressure in contact between the surface described by equation (15) and an elastic half-space, the following close-form solution was found [18]:

$$p(x, m) = -J(x, m)\frac{\sqrt{2}\pi\Delta(1-\nu^2)}{\lambda E}\left|\cos\frac{\pi x}{\lambda}\right|\sqrt{\cos\frac{2\pi x}{\lambda} - \cos\frac{2\pi a}{\lambda}}, \tag{16}$$

where $J(x, m)$ is defined by

$$J(x, m) = (m+1)^2\left(\left|m\cos\frac{2\pi x}{\lambda}\right| + 1\right)^{-2}\left(m\cos\frac{2\pi a}{\lambda} + 1\right)^{-1}. \tag{17}$$

Substituting (16) in (11)–(14) and calculating the integrals numerically, we determine the additional displacement function for the surface waviness described by the periodic function (15).

The graphs of the dimensionless additional displacement function for waviness of various shape m at $\Delta = 1$ mm and $\lambda = 10$ mm as a function of the dimensionless nominal pressure p_∞/p^*, where $p^* = -\pi\Delta_1 E/[(1-\nu^2)\lambda_1]$, are presented in figure 3.

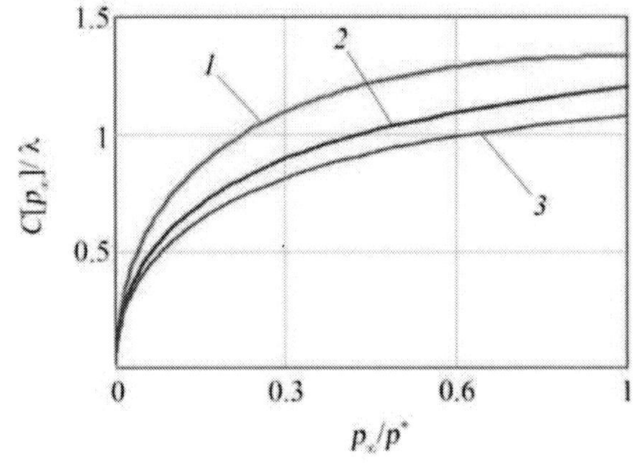

Figure 3. Additional displacement function, related to λ, for wavy profile (15) for $m = 0$ (curve *1*), $m = 0.5$ (curve *2*), and $m = 0.8$ (curve *3*).

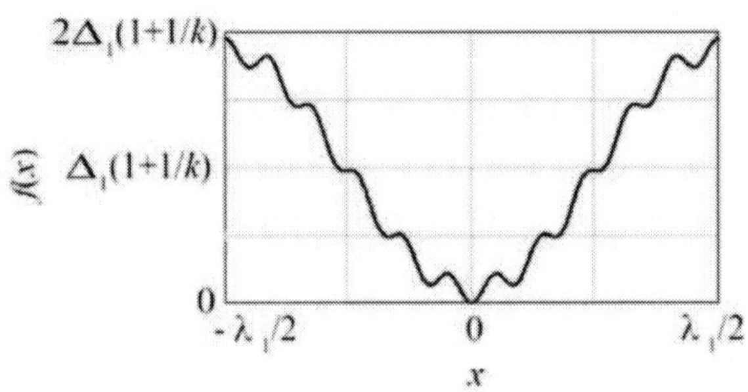

Figure 4. Example of profile with two-scale waviness.

The results illustrate that the additional displacement of asperities decreases as the parameter m increases. It should also be noted that the additional displacement increases linearly with the amplitude Δ and decreases nonlinearly as the period λ increases. The influence of the waviness period on the additional displacements is stronger than that of the amplitude. Meanwhile, the dependence $C[p_\infty]$ can be approximated by a power function for small values of p_∞.

The considered model is a one-scale model, however the actual waviness (roughness) profile is multiscale. As a first approximation, it can be represented by a sum of sinusoids with various amplitudes and periods. In [19], a contact problem for a rigid body with two-scale surface wavy profile and an elastic half-plane is considered. The two-scale profile was modeled by the function

$$f(x) = \Delta_1\left[1 - \cos\frac{2\pi x}{\lambda_1} + \frac{1}{k}\left(1 - \cos\frac{2\pi n x}{\lambda_1}\right)\right], \tag{18}$$

Here $k = \Delta_1/\Delta_2$ is the ratio of the long-wave to short-wave harmonic amplitude of waviness; $n = \lambda_1/\lambda_2$ is the ratio of their periods. The graph of waviness profile at $k = 10$ and $n = 11$ is presented in figure 4.

For $n/k < 0.5$, for determining of normal pressure in periodic contact of the surface described by equation (18) with an elastic half-plane, the integral equation was reduced and solved in [19] using a variable transform. The analytical expression for the contact pressure obtained in [19]

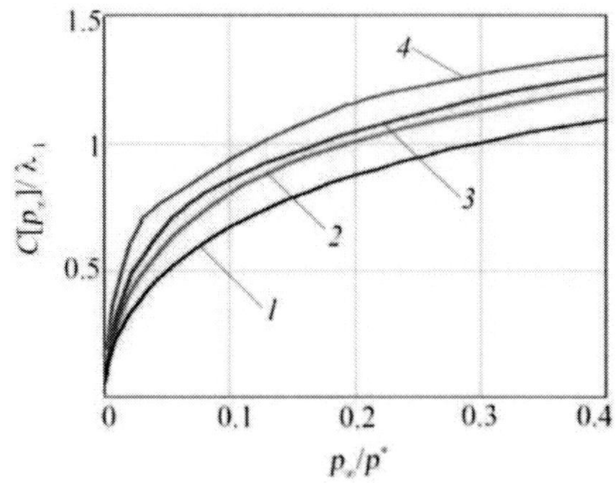

Figure 5. Additional displacement function related to λ_1 for wavy profile (18): single-scale profile (curve *1*), $n = 5$ (curve *2*), $n = 7$ (curve *3*), and $n = 11$ (curve *4*).

has the form

$$p(x) = p_1(x) + p_2(x), \tag{19}$$

$$p_1(x) = -\frac{\sqrt{2}\pi E\Delta_1}{(1-\nu^2)\lambda_1}\left|\cos\frac{\pi x}{\lambda_1}\right|\sqrt{\cos\frac{2\pi x}{\lambda_1} - \cos\frac{2\pi a}{\lambda_1}}, \tag{20}$$

$$p_2(x) = -\frac{2E}{(1-\nu^2)}\frac{\Delta_1 n}{\lambda_1 k}\sqrt{1-\left[\frac{\tan(\pi x/\lambda_1)}{\tan(\pi a/\lambda_1)}\right]^2}\sum_{j=1}^{\infty}U_{j-1}\left[\frac{\tan(\pi x/\lambda_1)}{\tan(\pi a/\lambda_1)}\right]\int_{-1}^{1}\frac{\varphi(s,a)T_j(s)}{\sqrt{1-s^2}}ds, \tag{21}$$

$$\varphi(s,a) = \frac{\tan(\pi a/\lambda_1)s}{1+[\tan(\pi a/\lambda_1)s]^2}U_{n-1}\left\{\frac{1-[\tan(\pi a/\lambda_1)s]^2}{1+[\tan(\pi a/\lambda_1)s]^2}\right\}, \tag{22}$$

where U_n are Chebyshev polynomials of the second kind of degree n.

The additional displacement function in dimensionless form for the two-scale waviness at $\Delta_1 = 1\,\mathrm{mm}$, $\lambda_1 = 10\,\mathrm{mm}$, $k = 40$, calculated using equations (11)–(14) and (19–22) is shown in figure 5 for various values of the ratio n of long to short harmonic period.

It follows from the results (figure 5), that the main features of the additional displacement function for two-scales and single scale profiles are similar. The influence of amplitude and period of long-wave asperities also has the same character, as for single-scale waviness. However, the existence of the short wave harmonics leads to discontinuity of the derivative of the additional displacement function $C[p_\infty]$. This effect increases with the increasing ratio n of long to short harmonic period. The ratio of amplitudes k linearly influences the additional displacements, as it follows from equations (20) and (21).

The additional displacement function calculated for various waviness parameters can be used in equations of the macroscale contact problem after substituting the function $C[p(x,y)] = C[p_\infty]$ in (5).

Conclusions

The approach for calculating the contact characteristics (pressure distribution, nominal contact area, and penetration) in normal contact of elastic bodies with regard to their surface roughness is discussed. The approach is based on introducing an additional displacement function which describes the additional compliance due to the surface roughness.

The method for calculating the additional displacement function is presented and used to analyze the influence of the surface waviness parameters on the dependence of the additional displacement due to the surface microgeometry on the nominal pressure in periodic contact problems for elastic half-planes. The analytical determination of the additional displacement function allows us to take into account the effect of the asperities of the contacting surfaces on the nominal contact characteristics, providing a unified approach to contact problems for rough bodies with microgeometry of various types and avoiding labor-consuming experiments. The form of the additional displacement function obtained in the present work for various types of the wavy surfaces, is comparable with the results for a three-dimensional system of spherical asperities [11, 12]. An analysis of the additional displacement functions for some particular types of surface microgeometry shows that as the nominal pressure p_∞ and hence the real contact area increase, the rate of change of additional displacement function decreases. The growth of the density of contact spots and the number of scales of asperities leads to an increase in the additional displacements. It should also be noted that the power function approximation of the additional displacement function describes its behavior with an adequate accuracy only at small nominal pressures and does not reflect the transition from partial to full contact.

Acknowledgments

The research was supported by Russian Science Foundation (project No. 14-29-00198).

References

[1] Schtaierman I Y 1949 *Contact Problem of Theory of Elasticity* (Moscow: Gostekhizdat) [in Russian]
[2] Kragelsky I V 1965 *Friction and Wear* (London: Butterworths)
[3] Kragelsky I V, Dobychin M N, and Kombalov V S 1982 *Friction and Wear: Calculation Methods* (Pergamon)
[4] Rabinovich A S 1979 On solving the contact problems for rough bodies *Izv. Akad. Nauk SSSR. Mekh. Tverd. Tela* No. 1 52–7
[5] Teplyi M I 1981 The problem of internal compression of cylindrical bodies with a surface layer of increased pliability *Sov. Mater. Sci.* **17** (2) 185–8
[6] Galanov B A 1985 The method of boundary equations of the hammerstein-type for contact problems of the theory of elasticity when the regions of contact are not known *J. Appl. Math. Mech.* **49** (5) 634–40
[7] Galanov B A 1984 Spatial contact problems for rough elastic bodies under elastoplastic deformations of the unevenness *J. Appl. Math. Mech.* **48** (6) 750–7
[8] Goryacheva I G 1979 Plane and axisymmetric contact problems for rough elastic bodies *J. Appl. Math. Mech.* **43** (1) 99–105
[9] Greenwood J A and Tripp J H 1967 The elastic contact of rough spheres *Trans. ASME. Ser. E. J. Appl. Mech.* **34** 153–9
[10] Sviridenok A I, Chizhik S A, and Petrokovets M I 1990 *Mechanics of discrete frictional contact* (Minsk: Science and Engineering) [in Russian]
[11] Goryacheva I 1997 *Contact Mechanics in Tribology* (Kluwer)
[12] Goryacheva I G 2001 *Mechanics of frictional interaction* (Moscow: Nauka) [in Russian]
[13] Goryacheva I G 2006 Mechanics of discrete contact *Trib. Int.* **39** (5) 381–6
[14] Goryacheva I G 1998 The periodic contact problem for an elastic half-space *J. Appl. Math. Mech.* **62** (6) 959–66
[15] Goryacheva I G 1999 Determination of contact characteristics with the account of the parameters of surface macro- and microgeometries *Frict. Wear* **20** (3) 239–48
[16] Block J M and Keer L M 2008 Periodic contact problems in plane elasticity *J. Mech. Mater. Struct.* **3** (7) 1207–37
[17] Westergaard H M 1939 Bearing pressures and cracks *Trans. ASME. J. Appl. Mech* **6** 49–53
[18] Tsukanov I Yu 2017 Effects of shape and scale in mechanics of elastic interaction of regular wavy surfaces *Proc. Inst. Mech. Engineers. Part J. J. Engng Trib.* **231** (3) 332–40
[19] Tsukanov I Yu 2018 Periodic contact problem for surface with two-scale waviness *Prikl. Mat. Mekh.* (in print)

IOP Conf. Series: Journal of Physics: Conf. Series **991** (2018) 012029 doi:10.1088/1742-6596/991/1/012029

An analytical method for the inverse Cauchy problem of Lame equation in a rectangle

Yu Grigor'ev
Academy of Sciences of Republic Sakha (Yakutia), Yakutsk, Russia
North-Eastern Federal University, Yakutsk, Russia
E-mail: grigyum@yandex.ru

Abstract. In this paper, we present an analytical computational method for the inverse Cauchy problem of Lame equation in the elasticity theory. A rectangular domain is frequently used in engineering structures and we only consider the analytical solution in a two-dimensional rectangle, wherein a missing boundary condition is recovered from the full measurement of stresses and displacements on an accessible boundary. The essence of the method consists in solving three independent Cauchy problems for the Laplace and Poisson equations. For each of them, the Fourier series is used to formulate a first-kind Fredholm integral equation for the unknown function of data. Then, we use a Lavrentiev regularization method, and the termwise separable property of kernel function allows us to obtain a closed-form regularized solution. As a result, for the displacement components, we obtain solutions in the form of a sum of series with three regularization parameters. The uniform convergence and error estimation of the regularized solutions are proved.

1. Introduction

Let us have an element of construction with boundary $S = S_1 \cup S_2$ consisting of two parts S_1 and S_2. The part S_1 is accessible for all measurements and the part S_2 is not accessible. In engineering practice it is often necessary to reconstruct the stress-strain state of such structural element by means of measured data on S_1. In this case within an elastic model, we have the Cauchy problem for the Lame equation of the theory of elasticity. The Cauchy problem for elliptic equations is a so-called ill-posed problem. The main difficulty for solving such problems is numerical instability. For the Cauchy problem in linear elasticity, there exist numerous ways for numerical solving. Many of these methods can be classified as Tikhonov-type regularization methods, the finite element method, the boundary element method, etc. (sf. [1, 2]). With regard to analytical solutions, only the usage of Carleman's method is known in the elasticity theory. In this paper, we present an analytical method for the Cauchy problem solution of the Lame equation in a rectangle. This method generalizes the Liu method [3] for an analytical solution of the Cauchy problem of the Laplace equation in a rectangle.

2. Formulation of the problem

In numerous publications under the name of Cauchy problem for the Laplace equation, there exist two types of problems. In them, it is necessary to find a harmonic function in $\Omega \in \mathbb{R}^3$. At first one has a cylindrical domain $\Omega = S \times (0, T)$ with lower base $S \in \mathbb{R}^2$ and lateral surface S_1. On S, the Cauchy conditions are given: $u = f$, $\partial u/\partial n = g$, and some boundary conditions are posed on the lateral surface S_1, for example, $u = 0$, and on the upper base, there are no given boundary conditions. Such problems can be named as initial boundary-value problems for the Laplace equation (see Ch. 9 in [1]). For another type, we have a bounded domain $\Omega \in \mathbb{R}^3$ with a smooth boundary $S = S_1 \cup S_2$ and the Cauchy conditions are given on the part S_1 of the boundary S. We deal with the first type of the Cauchy problem, and it is analogous in the elasticity theory. It is necessary to note that the first type of Cauchy problem is not considered for analytical solutions in the elasticity theory. All analytical results in the elasticity theory concern the second-type problem and use Carleman's method.

Let x, y, z be the Cartesian coordinates. For convenience, we also simultaneously use the index notation: x_i $(i = 1, 2, 3)$. An equilibrium equation in the theory of elasticity is the Lame equation

$$L\mathbf{u} \equiv \mu\Delta\mathbf{u} + (\lambda + \mu)\nabla(\nabla \cdot \mathbf{u}) = 0. \tag{1}$$

Hooke's law expresses connections between the components σ_{ij} of stress tensor and the components ε_{ij} of deformation tensor and for the linear elasticity has the form

$$\sigma_{ij} = (\lambda + 2\mu)(\nabla \cdot \mathbf{u})\delta_{ij} + 2\mu\varepsilon_{ij}, \quad \varepsilon_{ij} = \frac{1}{2}(u_{i,j} + u_{j,i}) \quad (i, j = 1, 2, 3). \tag{2}$$

Here and below, the comma and an index (or a variable) denote the corresponding partial derivative. For simplicity we consider the plane deformation when the components of displacement $\mathbf{u}$ have the form $u_x = u(x, y)$, $u_y = v(x, y)$, $u_z = 0$. In such a case, an equilibrium equation in stresses components has the form

$$\begin{aligned} \sigma_{xx,x} + \sigma_{xy,y} &= 0, \\ \sigma_{xy,x} + \sigma_{yy,y} &= 0. \end{aligned} \tag{3}$$

Let $\Omega \in \mathbb{R}^2$ be the rectangle $\mathbf{r} \in (0, a) \times (0, b)$. The problem to be solved is

$$\begin{aligned} &L\mathbf{u}(\mathbf{r}) = 0, \quad \mathbf{r} \in \Omega; \\ &\sigma_{yy} = u_x = u_y = 0, \quad \sigma_{xy} = \sigma(x), \quad u_y = 0 \quad \text{on } y = b; \\ &\sigma_{xx} = u_y = 0 \quad \text{on } x = 0 \text{ and } x = a. \end{aligned} \tag{4}$$

Thus, we have the Cauchy problem for the Lame equation with the Cauchy data on the upper base, with zero boundary conditions on the lateral sides of the rectangle, and with no boundary conditions on the lower base of the rectangle. On the upper base, only one boundary function $\sigma(x)$ of the four conditions is taken not to be equal to zero only for simplicity.

3. Solution method

3.1. Cauchy problem for the auxiliary harmonic function

Let us introduce an auxiliary function $f_0 = (\lambda + 2\mu)\nabla \cdot \mathbf{u}$. It is well known that this function is harmonic. From Hook's law (2), we have

$$\begin{aligned} \sigma_{xx} &= \lambda\nabla \cdot \mathbf{u} + 2\mu\varepsilon_{xx} = \lambda\nabla \cdot \mathbf{u} + 2\mu u_{x,x} = \lambda\nabla \cdot \mathbf{u} + 2\mu(\nabla \cdot \mathbf{u} - u_{y,y}), \\ \sigma_{yy} &= \lambda\nabla \cdot \mathbf{u} + 2\mu\varepsilon_{yy} = \lambda\nabla \cdot \mathbf{u} + 2\mu u_{y,y} = \lambda\nabla \cdot \mathbf{u} + 2\mu(\nabla \cdot \mathbf{u} - u_{x,x}), \end{aligned} \tag{5}$$

and for our auxiliary function, we have the expressions

$$\begin{aligned} f_0 &= \sigma_{xx} + 2\mu u_{y,y}, \\ f_0 &= \sigma_{yy} + 2\mu u_{x,x}. \end{aligned} \tag{6}$$

Then, after differentiating these expressions and using equilibrium equations (3) and (1), we obtain

$$\begin{aligned} f_{0,x} &= \sigma_{xx,x} + 2\mu u_{y,yx} = -\sigma_{xy,y} + 2\mu(\nabla\cdot\mathbf{u} - u_{x,x})_{,x} = -\sigma_{xy,y} + 2\mu\nabla\cdot\mathbf{u}_{,x} - 2\mu u_{x,xx} \\ &= -\sigma_{xy,y} + 2\mu\nabla\cdot\mathbf{u}_{,x} - 2[-\mu u_{x,yy} - (\lambda+\mu)\nabla\cdot\mathbf{u}_{,x}] = -\sigma_{xy,y} + 2\mu u_{x,yy} + 2f_{0,x}, \end{aligned}$$

which implies

$$f_{0,x} = \sigma_{xy,y} - 2\mu u_{x,yy}. \tag{7}$$

Analogically, we can obtain a similar formula,

$$f_{0,y} = \sigma_{xy,x} - 2\mu u_{y,xx}. \tag{8}$$

Let the boundary conditions of the Cauchy problem (4) provide the solution smoothness, $\mathbf{u} \in C^2(\overline{\Omega})$. In this case, all formulas (6)–(8) hold till the rectangle boundary. Then from formulas (6)–(8) and boundary conditions in (4), we obtain boundary conditions for the function f_0 and thus have the following Cauchy problem for the harmonic function f_0:

$$\begin{aligned} &\Delta f_0 = 0, \quad \mathbf{r}\in\Omega, \\ &f_0 = 0, \quad f_{0,y} = -h(x) \quad \text{on } y = b, \\ &f_0 = 0 \quad \text{on } x = 0 \text{ and } x = a, \end{aligned} \tag{9}$$

where the function $h(x)$ is expressed in terms of the initial boundary functions from (4)

$$h(x) = \sigma'(x). \tag{10}$$

The regularized solution f_0^α of this problem is obtained by using the Liu method [3] in the form

$$\begin{aligned} f_0^\alpha(x,y) &= \sum_{k=1}^{\infty} a_k^\alpha \frac{\sinh[k\pi(b-y)/a]}{\sinh(k\pi b/a)} \sin\frac{k\pi x}{a}, \\ a_k^\alpha &= \frac{2\sinh(k\pi b/a)}{k\pi + \alpha a\sinh(k\pi b/a)} \int_0^a h(\xi)\sin\frac{k\pi\xi}{a}d\xi, \end{aligned} \tag{11}$$

where α is a parameter of regularization. In this method, the integral equation of the first kind is solved by Lavrent'ev's regularization method using the Fourier method. The kernel in this integral equation has a termwise separability property. This property is the main reason for a solution to be obtained in closed form.

For solution (11), the following results presented in [3] are valid including the error estimation.

Theorem 1. If the Neumann datum $h(x)$ is bounded in the interval $x \in [0,a]$, then for any $\alpha > 0$ and $y_0 > 0$ for all $x \in [0,a]$ and $y \in [y_0, b]$, the regularized solution (11) converges uniformly, and this series can be termwise differentiated any number of times.

Theorem 2. Assume that the Neumann datum $h(x) \in L^2(0,a)$. Then a sufficient and necessary condition for the Cauchy problem (9) to have a solution is

$$\sum_{k=1}^{\infty} \frac{\sinh^2(k\pi b/a)}{k^2\pi^2}\left(\int_0^a h(\xi)\sin\frac{k\pi\xi}{a}d\xi\right)^2 < \infty. \tag{12}$$

Theorem 3. If the Neumann datum $h(x)$ satisfies condition (12) and there exists an $\varepsilon \in (0,1)$ such that

$$4\sum_{k=1}^{\infty}\left[\frac{a^2\sinh^2(k\pi b/a)}{k^2\pi^2}\right]^{1+\varepsilon}\left(\int_0^a h(\xi)\sin\frac{k\pi\xi}{a}d\xi\right)^2 := M^2(\varepsilon) < \infty, \tag{13}$$

then for any $\alpha > 0$, the regularized solution $f^\alpha(x,y)$ satisfies the following error estimation:

$$\|f^\alpha(x,y) - f(x,y)\|_{L^2(0,a)} \le \alpha^\varepsilon M(\varepsilon). \tag{14}$$

3.2. Cauchy problems for the displacement components

After determining $f_0^\alpha(x,y)$, we have Poisson equations for the components of a displacement from the equilibrium equations (1). It can be shown that, for all components $u_x(x,y)$ and $u_y(x,y)$, we can find Cauchy boundary conditions. And for the $u_x(x,y)$, we have the following Cauchy problem for the Poisson equations:

$$\begin{aligned}
&\Delta u_x = -\frac{\lambda+\mu}{\mu(\lambda+2\mu)} f^\alpha_{0,x}, \quad \mathbf{r}\in\Omega,\\
&u_x = 0, \quad u_{x,y} = \frac{1}{\mu}\sigma(x) \quad \text{on } y=b,\\
&u_{x,x} = 0 \quad \text{on } x=0 \text{ and } x=a.
\end{aligned} \tag{15}$$

For the component $u_y(x,y)$, we have another Cauchy problem

$$\begin{aligned}
&\Delta u_y = -\frac{\lambda+\mu}{\mu(\lambda+2\mu)} f^\alpha_{0,y}, \quad \mathbf{r}\in\Omega,\\
&u_y = u_{y,y} = 0 \quad \text{on } y=b,\\
&u_y = 0 \quad \text{on } x=0 \text{ and } x=a.
\end{aligned} \tag{16}$$

These problems are solved by the method similar to the Liu method. Thus, for the solution u_x and u_y of the Cauchy problem (4), we have regularized expressions similar to (11) but with additional regularization parameters β and γ. Now we give the final formulas for these solutions:

$$\begin{aligned}
&u_x^{\alpha,\beta}(x,y) = w_x^\alpha(x,y) + v_x^{\alpha,\beta}(x,y),\\
&u_y^{\alpha,\gamma}(x,y) = w_y^\alpha(x,y) + v_y^{\alpha,\gamma}(x,y);
\end{aligned} \tag{17}$$

$$w_x^\alpha(x,y) = \frac{\lambda+\mu}{\mu(\lambda+2\mu)} \sum_{k=1}^\infty w_{xk}(y)\cos\frac{k\pi x}{a},$$

$$w_{xk}(x,y) = \frac{a_k^\alpha}{\sinh(k\pi b/a)}\left[y\cosh\frac{k\pi(b-y)}{a} - b\frac{\sinh(k\pi y/a)}{\sinh(k\pi b/a)}\right];$$

$$v_x^\beta(x,y) = b_0(b-y) + \sum_{k=1}^\infty b_k^\beta \frac{\sinh[k\pi(b-y)/a]}{\sinh(k\pi b/a)}\cos\frac{k\pi x}{a},$$

$$b_k^\beta = \frac{k\pi\sinh(k\pi b/a)}{k\pi+\beta\sinh(k\pi b/a)}\frac{a}{\mu}\left[\frac{\lambda+\mu}{2(\lambda+2\mu)}\frac{a_k^\alpha}{\sinh(k\pi b/a)}\left(1-\frac{k\pi b}{a}\coth\frac{k\pi b}{a}\right) - \sigma_k\right];$$

$$w_y^\alpha(x,y) = -\frac{\lambda+\mu}{\mu(\lambda+2\mu)}\sum_{k=1}^\infty a_k^\alpha y\frac{\sinh[k\pi(b-y)/a]}{\sinh(k\pi b/a)}\sin\frac{k\pi x}{a};$$

$$v_y^\gamma(x,y) = \sum_{k=1}^\infty c_k^\gamma\frac{\sinh[k\pi(b-y)/a]}{\sinh(k\pi b/a)}\sin\frac{k\pi x}{a},$$

$$c_k^\gamma = \frac{\lambda+\mu}{2\mu(\lambda+2\mu)}\cdot\frac{k\pi b a_k^\alpha}{k\pi+\gamma\sinh(k\pi b/a)},$$

$$\sigma_k = \int_0^a \sigma(\xi)\cos\frac{k\pi\xi}{a}d\xi, \quad k\geq 1; \quad b_0 = -\frac{1}{a\mu}\int_0^a\sigma(\xi)d\xi.$$

If the boundary functions in (4) are bounded in closed intervals, then all of these regularized solutions converge uniformly and can be differentiated termwise any number of times. When the parameters $\alpha,\beta,\gamma\to 0$, then the regularized solutions $u_x^{\alpha,\beta}(x,y)$, $u_y^{\alpha,\gamma}(x,y)$ tend to the exact solutions. Error estimations for these solutions can be obtained analogically to Theorem 3. Effective numerical calculations can be realized by means of the formulas presented above.

Conclusions

The Liu method for solving the Cauchy problem for the Laplace equation is generalized to the Cauchy problem of the linear elasticity. This allows us to obtain a regularized solution of the problem in closed form by using the Fourier and Lavrent'ev methods. The obtained regularized solution converges uniformly and the error estimation of the regularized solution is proved. For simplicity, we only consider a two–dimensional rectangular domain. The method can be used for other boundary conditions, it can be generalized to the three-dimensional case. It is also interesting to note that the above-used auxiliary function $f_0 = (\lambda + 2\mu)\nabla \cdot \mathbf{u}$ is a scalar part of a regular quaternion function connected with the solution $\mathbf{u}$ of Lame equation (1) [4–6]. Thus it may be possible to use the above-presented method in the problem of regular extension to a domain of the quaternion function defined on a part of its boundary.

Acknowledgments

This work was supported by grant 15-41-05081 from the Russian Foundation for Basic Research.

References

[1] Kabanikhin S I 2012 *Inverse and Ill-Posed Problems: Theory and Applications* (Berlin - Boston: Walter de Gruyter GmbH & Co. KG)

[2] Durand B, Delvare F, and Bailly P 2011 Numerical solution of Cauchy problems in linear elasticity in axisymmetric situations *Int. J. Solid Struct.* **48** (21) 3041–53

[3] Liu C S 2011 An analytical method for the inverse Cauchy problem of Laplace equation in a rectangular plate *J. Mech.* **27** (4) 575–84

[4] Grigor'ev Yu M 2014 Three-dimensional Quaternionic Analogue of the Kolosov–Muskhelishvili Formulae Hypercomplex Analysis In *New Perspectives and Applications, Trends in Mathematics* ed S Bernstein, U Kaehler, I Sabadini, and F Sommen (Basel: Birkhauser) pp 145–166

[5] Grigor'ev Yu 2016 Three-dimensional Analogue of Kolosov-Muskhelishvili Formulae In *Modern Trends in Hypercomplex Analysis, Trends in Mathematics* ed S Bernstein, U Kaehler, I Sabadini, and F Sommen (Basel: Birkhauser) pp 203–215

[6] Grigor'ev Yu M 2017 Regular quaternionic polynomials and their properties *Compl. Variabl. Ellipt. Equat.* **62** (9) 1343–63

On the contact interaction of two identical stringers with an elastic semi-infinite continuous or vertically cracked plate

M S Grigoryan

Institute of Mechanics of the National Academy of Sciences of the Republic of Armenia, Yerevan, Armenia

E-mail: marinegrigoryan17@gmail.com

Abstract. This paper considers two connected contact problems on the interaction of stringers with an elastic semi-infinite plate. In the first problem, an elastic half-infinite continuous plate is reinforced on its boundary by two identical stringers exposed to a tensile external force. In the second problem, in the presence of the same stringers, the plate contains a collinear system of cracks on its vertical axis. The solution of both problems is reduced to the solution of singular integral equations (SIE) that are solved by a known numerical-analytical method.

1. Introduction

The problems of contact interaction of thin-walled elements such as stringers with massive deformable bodies generalize and develop the classical contact problems of the theory of elasticity. Due to their theoretical and practical significance, these problems are the subject of many studies [1–6]. A brief review of the main works and results is given in [7]. The problems of interaction between stringers and massive elastic bodies with cracks [8–12] are of significant interest.

The present paper considers two contact problems related to each other. In the first problem, an elastic half-infinite continuous plate which, in the right-hand rectangular coordinate system, occupies the lower half-plane is reinforced on its boundary by two identical stringers located symmetrically with respect to the origin of the coordinate system. At the far ends of the stringers, horizontal tensional concentrated forces, equal in value but opposite in direction, are applied. In addition, tensional distributed tangential forces of given intensities act on the upper faces of the stringers along their middle lines. The plate itself is exposed to the action of uniformly distributed horizontal forces tensile in the horizontal direction at infinity. This problem is mathematical formulated as the Prandtl integro-differential equation for stringers in the well-known Melan physical model [1–2]. The procedure of solving the latter by the method of Chebyshev polynomials [6] is reduced to solving a regular infinite system of linear algebraic equations (SLAE). Here the solution of the governing SIE of this problem is constructed by a method which is simpler from the standpoint of computational mathematics. Namely, the solution is represented in the form of a product of a weight function that takes into account the characteristic singularities of the contact stresses under the stringers and some regular function of the Hölder class. In a particular case where the stringers are not loaded or are under the influence of a self-balanced system of external forces, we also use the Multopp method [13].

Further, we use the Kolosov–Muskhelishvili complex potentials [14] to determine the normal breaking stresses on the vertical axis of the plate at the origin of the coordinate system. An analysis of the formula for these stresses shows that, as the near ends of the stringers tend to zero, i.e., when they infinitely approach each other, these stresses increase infinitely attaining the value of the brittle fracture limit for a given material. As a result, the vertical cracking of the elastic plate occurs.

Proceeding from this fact, the second problem closely related to the first problem is considered. We have the first problem configuration, but there is a crack of finite length on the vertical axis and the crack edges can be loaded with vertical forces. The following two cases of the crack location are discussed:

(a) when it is located inside the plate at some distance from its boundary;
(b) when the crack tip vertically reaches the boundary of the plate.

In both cases, the discussed problem is described by a system of two SIE.

In the case (a), the dislocation densities on the crack edges are unbounded at its tips, and they have ordinary root singularities. In the case (b), the dislocation densities are unbounded at the tip of the lower internal crack and, in general, are bounded at the upper tip. The solution of the governing system of the SIE is constructed by a known numerical-analytical method [15–17].

2. Solution of the first boundary-value problem for an elastic quarter-plane

Let the elastic semi-infinite plate with a rectangular coordinate system Oxy, which occupies the lower half-plane $y \leq 0$, have modulus of elasticity E, Poisson's ratio ν, and width d_1. Further, suppose that such a plate, on its boundary $y = 0$, on the segments $[-a, -b]$ and $[b, a]$ symmetrically located about the origin, is reinforced by two identical elastic stringers of modulus of elasticity E_s; d_2 is the width and h is the height. We assume that the stringers are also loaded symmetrically, and horizontal concentrated forces P equal in values and opposite in direction act at the centers of their end-sections $y = \pm a$, while the sections $y = \pm b$ are free from external forces and tensile tangential forces of intensity $\tau_+(y)$ $(y \in (-a, -b) \cup (b, a);\ \tau_+(-y) = -\tau_+(y))$ act along the middle lines of the upper edges of the stringers. At infinity, the plate is subject to the action of uniformly distributed tensile horizontal forces of intensity P_0.

Under these assumptions, within the physical model of the one-dimensional elastic continuum of stringers [1–2], it is required to determine the distribution of tangential contact stresses $\tau_-(y)$ referred to the unit width d_1 of the plate, and in addition $\tau_-(-y) = -\tau_-(y)$.

We introduce the governing SIE of the problem. Following [6], we consider the equilibrium of the part $[b, r]$ of the right stringer. We obtain

$$\sigma_s(r) = \frac{1}{A_s}\left[d\int_b^r \tau_-(s)\,ds - \int_b^r \tau_+(s)\,ds\right] \quad (r = y \in [b, a]). \tag{1}$$

Here $\sigma_s(r)$ $(r = y)$ is the axial stress in the section r of the stringer, $d = \min(d_1, d_2)$ is the effective width of the contact zone, and $A_s = d_2 h$ is the cross-section area of the stringer. According to Hooke's law, from (1), for the axial deformation $\varepsilon_s(r)$ of the stringer, we obtain

$$\varepsilon_s(r) = \frac{1}{A_s E_s}\left[d\int_b^r \tau_-(s)\,ds - \int_b^r \tau_+(s)\,ds\right] \quad (y = r \in (b, a)). \tag{2}$$

Similarly considering the same equilibrium of the part $[r, a]$ of the right stringer, we obtain

$$\varepsilon_s(r) = \frac{1}{A_s E_s}\left[-d\int_r^a \tau_-(s)\,ds + \int_r^a \tau_+(s)\,ds\right] \quad (y = r \in (b, a)). \tag{3}$$

Now, combining (2) and (3), we find

$$\varepsilon_s(r) = \frac{1}{2A_sE_s}\left[d\int_b^a \operatorname{sign}(r-s)\tau_-(s)\,ds - \int_b^a \operatorname{sign}(r-s)\tau_+(s)\,ds\right] \quad (y = r \in (b,a)). \qquad (4)$$

Note that setting $r = a$ in (1), we obtain the equilibrium condition for the right stringer:

$$d\int_b^a \tau_-(s)\,ds = Q; \quad Q = P + T_+ = P + \int_b^a \tau_+(s)\,ds. \qquad (5)$$

On the other hand, for the derivative of the horizontal displacements $u_y(y,0)$ of points of the elastic semi-infinite plate, we have [6]

$$u_y'(y,0) = \frac{4}{\pi E}\frac{d}{d_1}\int_b^a \frac{s\tau_-(s)\,ds}{s^2-y^2} + \frac{P_0}{E} \quad (-\infty < y < \infty). \qquad (6)$$

Next, we write the condition of contact between the right stringer and the plate

$$u_y'(y,0) = \varepsilon_s(y) \quad (-\infty < y < \infty). \qquad (7)$$

Substituting (4) and (6) into (7), we obtain the following governing SIE of the problem:

$$\frac{4}{\pi E}\frac{d}{d_1}\int_b^a \frac{s\tau_-(s)\,ds}{s^2-r^2} + \frac{P_0}{E} = \frac{1}{2A_sE_s}\left[d\int_b^a \operatorname{sign}(r-s)\tau_-(s)\,ds - \int_b^a \operatorname{sign}(r-s)\tau_+(s)\,ds\right] \qquad (8)$$
$$(b < y = r < a).$$

Now we introduce dimensionless quantities in (8) and in (5):

$$\xi = \frac{r}{a}, \quad \eta = \frac{s}{a}, \quad \tau(\xi) = \frac{d\tau_-(a\xi)}{d_1E}, \quad \tau_0(\xi) = \frac{\tau_+(a\xi)}{d_1E}, \quad p_0 = \frac{P_0}{E}, \quad \rho = \frac{b}{a},$$
$$\alpha_k = \frac{a_k}{a}, \quad \beta_k = \frac{b_k}{a} \quad (k = \overline{1,n}), \quad L_0 = \bigcup_{k=1}^{n}(\alpha_k,\beta_k), \quad \Lambda = \frac{ad_1E}{2A_sE_s}.$$

As a result, SIE (8) becomes

$$\frac{2}{\pi}\int_\rho^1 \frac{2\eta\tau(\eta)\,d\eta}{\eta^2-\xi^2} + p_0 = \Lambda\left[\int_\rho^1 \operatorname{sign}(\xi-\eta)\tau(\eta)\,d\eta - \int_\rho^1 \operatorname{sign}(\xi-\eta)\tau_0(\eta)\,d\eta\right] \quad (\rho < \xi < 1), \quad (9)$$

and condition (5) takes the form

$$\int_\rho^1 \tau(\eta)\,d\eta = Q_0, \quad Q_0 = \frac{Q}{ad_1E}. \qquad (10)$$

Further, in (9) and in (10), the interval $(\rho,1)$ is transformed into the interval $(-1,1)$ [6] for which we set

$$\xi = \sqrt{ct+d}, \quad \eta = \sqrt{cu+d} \quad \left(c = \frac{1-\rho^2}{2}, \quad d = \frac{1+\rho^2}{2}, \quad -1 < t,u < 1\right).$$

As a result, SIE (9) is transformed to the form

$$\frac{2}{\pi}\int_{-1}^1 \left[\frac{1}{u-t} + \frac{\pi\Lambda c}{4\sqrt{cu+d}}\operatorname{sign}(u-t)\right]\chi(u)\,du = f_0(t) \quad (-1 < t < 1),$$
$$\chi(u) = \tau(\sqrt{cu+d}), \quad f_0(t) = f_0(\sqrt{ct+d}), \quad f(\xi) = -p_0 + \Lambda\int_\rho^1 \operatorname{sign}(\eta-\xi)\tau_0(\eta)\,d\eta, \qquad (11)$$

and condition (10), to the form

$$\int_{-1}^{1} \frac{\chi(u)\,du}{\sqrt{cu+d}} = \frac{Q_0}{c}. \tag{12}$$

Now let us solve SIE (11)–(12). For this purpose, following the well-known procedure [15–17], we set

$$\chi(u) = \frac{\mathrm{X}(u)}{\sqrt{1-u^2}} \quad (-1 < u < 1),$$

where $\mathrm{X}(u)$ is the Hölder function on the interval $[-1, 1]$. As a result, the process of solving SIE (11)–(12) reduces to solving the SLAE

$$\sum_{n=1}^{M} L_{mn}\mathrm{X}_n = c_m \quad (m = \overline{1, M}), \tag{13}$$

$$L_{mn} = \begin{cases} \dfrac{2}{M}\left[\dfrac{1}{u_n - t_m} + \dfrac{\pi\Lambda c}{4\sqrt{cu_n + d}}\operatorname{sign}(u_n - t_m)\right], & n = \overline{1, M}, \quad m = \overline{1, M-1}, \\ \dfrac{\pi}{M}\dfrac{1}{\sqrt{cu_n + d}}, & n = \overline{1, M}, \quad m = M, \end{cases}$$

$$\mathrm{X}_n = \mathrm{X}(u_n) \quad (n = \overline{1, M}), \quad c_m = \begin{cases} f_0(t_m), & m = \overline{1, M-1}, \\ \dfrac{Q_0}{c}, & m = M, \end{cases}$$

$$u_n = \cos\frac{(2n-1)\pi}{2M} \quad (n = \overline{1, M}), \quad t_m = \cos\frac{\pi m}{M} \quad (m = \overline{1, M-1}).$$

Here u_n and t_m are Chebyshev nodes, i.e., the roots of the Chebyshev polynomials of the first kind $T_m(u)$ and of the second kind $U_{M-1}(t)$, respectively.

After solving SLAE (13), the reduced dimensionless tangential stresses is calculated by the formula

$$\chi(u_n) = \frac{\mathrm{X}(u)}{\sqrt{1-u^2}} \quad \left(n = \overline{1, M}, \quad u_n = \cos\frac{(2n-1)\pi}{2M}\right).$$

Writing formula (1) for the axial stresses in the form

$$\sigma_s(r) = \frac{1}{2A_s}\left[d\int_b^a \operatorname{sign}(r-s)\tau_-(s)\,ds - \int_b^a \operatorname{sign}(r-s)\tau_+(s)\,ds\right] \quad (r = y \in [b, a]),$$

turning to the interval $(-1, 1)$, and using the formulas given above, we can obtain the dimensionless axial stresses by solving SLAE (13).

3. Stress state of an elastic half-plane with a collinear system of vertical cracks

The problem considered in the previous section can be solved exactly if the stringers are absolutely rigid in tension and compression and absolutely flexible in bending. On the basis of this solution, using the normal Kolosov–Muskhelishvili complex potentials [14], one can determine the normal breaking stresses on the vertical axis of the elastic semi-infinite plate. An analysis of the formula for these stresses shows that they increase infinitely as the near ends of the stringers approach each other ($b \to +0$) and, consequently, attain the value of the brittle fracture limit for a given material. As a result, the elastic semi-infinite plate cracks, and a vertical crack occurs on the edge. Therefore, to understand the further development of a tensile elastic semi-infinite plate with two identical symmetrically disposed stringers, it is necessity to study the contact interaction between the stringers and the elastic semi-infinite plate with a vertical edge crack or, in general, with a collinear system of vertical cracks.

Proceeding from this fact, in the contact problem on two identical stringers on an elastic semi-infinite plate described in the previous section, we assume that the plate contains a collinear system of cracks $L = \bigcup_{k=1}^{n}[a_k, b_k]$ on its vertical axis Ox directed inwards. In addition, we assume that, on the edges of the system of cracks L, forces of intensity $p(x)$ normal to the edge are distributed which leads to normal separation, but there are no tangential forces.

Under these assumptions and again within the physical model of the one-dimensional elastic set of stringers [1–2], it is required to determine the distribution of tangential contact stresses $\tau_-(y)$ referred to unit width d_1 of the plate and the following characteristics of the cracks: the breaking stresses outside the cracks and along the line of their location, the SIF, and the crack openings.

We introduce the governing SIE of the problem. For this purpose, we first construct the solution of the following first boundary-value problem for the right elastic quarter-plane-wedge $\Omega_+ = \{0 \le r < \infty, 0 \le \vartheta \le \pi/2\}$ with the right angle in the polar coordinates r, ϑ:

$$\begin{gathered}\tau_{r\vartheta}\big|_{\vartheta=+0} = 0 \quad (0 < r < \infty), \quad \sigma_\vartheta\big|_{\vartheta=+0} = -q_0^+(r) = \begin{cases} -p(r), & r \in L, \\ -\sigma(r), & r \in [0,\infty)\backslash L, \end{cases} \\ \tau_{r\vartheta}\big|_{\vartheta=\pi/2-0} = \tau_-(r) \quad (0 < r < \infty), \quad \sigma_\vartheta\big|_{\vartheta=\pi/2-0} = 0 \quad (0 < r < \infty),\end{gathered} \tag{14}$$

where σ_ϑ and $\tau_{r\vartheta}$ are, respectively, the components of the normal and tangential stresses in the polar coordinates. In this coordinate system, we also write the equations without volume forces [18]:

$$\begin{gathered}(\lambda + 2\mu)\frac{\partial e}{\partial r} - \frac{\mu}{r}\frac{\partial \chi}{\partial \vartheta} = 0, \quad (\lambda + 2\mu)\frac{1}{r}\frac{\partial e}{\partial \vartheta} + \mu\frac{\partial}{\partial r}\left(\frac{\chi}{r}\right) = 0, \\ e = \frac{\partial u_r}{\partial r} + \frac{1}{r}\left(u_r + \frac{\partial u_\vartheta}{\partial \vartheta}\right), \quad \chi = r\frac{\partial u_\vartheta}{\partial r} + u_\vartheta - \frac{\partial u_r}{\partial \vartheta} = 2r\omega_z,\end{gathered} \tag{15}$$

and Hooke's law

$$\sigma_r = 2\mu\frac{\partial u_r}{\partial r} + \lambda e, \quad \sigma_\vartheta = \frac{2\mu}{r}\left(\frac{\partial u_\vartheta}{\partial \vartheta} + u_r\right) + \lambda e, \quad \tau_{r\vartheta} = \mu\left(\frac{\partial u_\vartheta}{\partial r} - \frac{u_\vartheta}{r} + \frac{1}{r}\frac{\partial u_r}{\partial \vartheta}\right). \tag{16}$$

Here u_r and u_ϑ are the respective components of displacements of the elastic body point in the radial and circular directions, e is the relative volume expansion, ω_z is the angle of rotation about the axis Oz perpendicular to the plane of the wedge Ω_+, λ and μ are the Lame constants of the elastic body expressed in terms of Young's modulus and Poisson's ratio:

$$\lambda = \frac{E\nu}{(1+\nu)(1+2\nu)}, \quad \mu = G = \frac{E}{2(1+\nu)}.$$

The solution of boundary-value problem (14)–(16) is constructed using the Mellin integral transformation with respect to the variable r. As a result, we have

$$\frac{du_r(r,\pi/2)}{dr} = \frac{\lambda + 2\mu}{\pi\mu(\lambda+\mu)}\int_0^\infty \frac{s\tau_-(s)\,ds}{s^2 - r^2} + \frac{2}{\pi}\int_0^\infty \frac{sr^2}{(s^2+r^2)^2}\varphi(s)\,ds \quad (0 < r < \infty), \tag{17}$$

$$q_0^+(r) = -\frac{4}{\pi}\int_0^\infty \frac{s^3\tau_-(s)\,ds}{(s^2+r^2)^2} - \frac{\mu(\lambda+\mu)}{\pi(\lambda+2\mu)}\int_0^\infty \left[\frac{1}{s-r} - \frac{1}{s+r} + \frac{6r}{(s+r)^2} - \frac{4r^2}{(s+r)^3}\right]\varphi(s)\,ds. \tag{18}$$

Here we introduced the dislocation density on the crack edges

$$\varphi(r) = 2\frac{du_\vartheta(r,+0)}{dr} \quad (r = x),$$

IOP Conf. Series: Journal of Physics: Conf. Series **991** (2018) 012030 doi:10.1088/1742-6596/991/1/012030

and $\varphi(r) = 0$ outside the cracks on $L' = [0,\infty)\backslash L$.

Now we can derive the system of the governing SIE of the posed problem. For this, we write the condition of contact between the right stringer and the plate

$$\frac{du_r(r,\pi/2)}{dr} = \varepsilon_s(r) \quad (b < r < a) \tag{19}$$

and the condition which follows from the second equality in (14)

$$q_0^+(r) = p(r) \quad (r \in L). \tag{20}$$

In addition, according to (6), the term P_0/E, which takes into account the longitudinal deformation of the plate due to tensile forces at infinity, must be added to the right-hand side of (17) and the function $p(r)$ must be replaced by a function $p(r) + P_0$ in (20). Then we use (17)–(18) and (19)–(20) to arrive at the following governing system of SIE of the posed problem:

$$\begin{aligned}
&\frac{4}{\pi E}\frac{d}{d_1}\int_b^a \frac{s\tau_-(s)\,ds}{s^2 - r^2} + \frac{2}{\pi}\int_L \frac{sr^2\varphi(s)\,ds}{(s^2+r^2)^2} + \frac{P_0}{E} \\
&\quad = \frac{1}{2A_sE_s}\left[d\int_b^a \operatorname{sign}(r-s)\tau_-(s)\,ds - \int_b^a \operatorname{sign}(r-s)\tau_+(s)\,ds\right] \quad (b < r < a). \\
&\frac{1}{\pi}\int_L\left[\frac{2r}{s^2-r^2} + \frac{6r}{(s+r)^2} - \frac{4r^2}{(s+r)^3}\right]\varphi(s)\,ds + \frac{16}{\pi E}\frac{d}{d_1}\int_b^a \frac{s^3\tau_-(s)\,ds}{(s^2+r^2)^2} = -\frac{4}{E}q(r) \\
&\quad (r \in L;\;\; q(r) = p(r) + P_0).
\end{aligned} \tag{21}$$

The solution of the SIE system of two equations (21) must satisfy the equilibrium condition of the right stringer (5) and the continuity conditions for the displacements at the tips of the cracks:

$$\int_{a_k}^{b_k} \varphi(s)\,ds = 0 \quad (k = \overline{1,n}). \tag{22}$$

We introduce the above-mentioned dimensionless variables in (21), (5) and (22), supplementing them with the following ones:

$$p_0(\xi) = \frac{p(a\xi)}{E}, \quad q_0(\xi) = \frac{q(a\xi)}{E} = p_0 + p_0(\xi), \quad \psi(\xi) = \varphi(a\xi).$$

As a result, system (21) becomes

$$\begin{aligned}
&\frac{2}{\pi}\int_\rho^1 \frac{2\eta\tau(\eta)\,d\eta}{\eta^2-\xi^2} + \frac{2}{\pi}\int_{L_0}\frac{\eta\xi^2\psi(\eta)\,d\eta}{(\xi^2+\eta^2)^2} + p_0 \\
&\quad = \Lambda\left[\int_\rho^1 \operatorname{sign}(\xi-\eta)\tau(\eta)\,d\eta - \int_\rho^1 \operatorname{sign}(\xi-\eta)\tau_0(\eta)\,d\eta\right] \quad (\rho < \xi < 1), \\
&\frac{1}{\pi}\int_{L_0}\left[\frac{2\xi}{\eta^2-\xi^2} + \frac{6\xi}{(\eta+\xi)^2} - \frac{4\xi^2}{(\eta+\xi)^3}\right]\psi(\eta)\,d\eta + \frac{16}{\pi}\int_0^1 \frac{\eta^3\tau(\eta)\,d\eta}{(\eta^2+\xi^2)^2} = -4q_0(\xi) \quad (\xi \in L_0),
\end{aligned} \tag{23}$$

condition (5) takes the form (10), and conditions (22) become

$$\int_{\alpha_k}^{\beta_k} \psi(\eta)\,d\eta = 0 \quad (k = \overline{1,n}). \tag{24}$$

Then, turning again to (23), we consider this equation on $L' = [0,\infty)\backslash L$ and take into account the second condition in (14). After passing to the above-mentioned dimensionless variables, we have

$$\sigma_0(\xi) = \frac{\sigma(a\xi)}{E} = -\frac{4}{\pi}\int_\rho^1 \frac{\eta^3\tau(\eta)\,d\eta}{(\eta^2+\xi^2)^2} - \frac{1}{4\pi}\int_{L_0}\left[\frac{2\xi}{\eta^2-\xi^2} + \frac{6\xi}{(\eta+\xi)^2} - \frac{4\xi^2}{(\eta+\xi)^3}\right]\psi(\eta)\,d\eta \quad (\xi \in L_0' = [0,\infty)\backslash L_0). \tag{25}$$

Further, for simplicity, we can confine to one crack and replace the interval $L_0 = (\alpha_1, \beta_1)$ by an interval (α, β). Then system (23) takes the form

$$\begin{gathered}\frac{2}{\pi}\int_\rho^1\left[\frac{2\eta}{\eta^2-\xi^2} + \frac{\pi\Lambda}{2}\operatorname{sign}(\eta-\xi)\right]\tau(\eta)\,d\eta + \frac{2}{\pi}\int_\alpha^\beta \frac{\eta\xi^2\psi(\eta)\,d\eta}{(\xi^2+\eta^2)^2} = f_0(\xi) \quad (\xi\in(\rho,1)),\\ \frac{1}{\pi}\int_\alpha^\beta\left[\frac{2\xi}{\eta^2-\xi^2} + \frac{6\xi}{(\eta+\xi)^2} - \frac{4\xi^2}{(\eta+\xi)^3}\right]\psi(\eta)\,d\eta + \frac{16}{\pi}\int_\rho^1\frac{\eta^3\tau(\eta)\,d\eta}{(\eta^2+\xi^2)^2} = -4q_0(\xi) \quad (\xi\in(\alpha,\beta)),\\ f_0(\xi) = -p_0 + \Lambda\int_\rho^1 \operatorname{sign}(\eta-\xi)\tau_0(\eta)\,d\eta.\end{gathered} \tag{26}$$

Condition (10) remains unchanged, and conditions (24) are reduced to the condition

$$\int_\alpha^\beta \psi(\eta)\,d\eta = 0. \tag{27}$$

Further in (26), as well as in (10) and (27), we transform all intervals into an interval $(-1,1)$ for which:

(1) in the first integral (26), we set

$$\xi = \sqrt{ct+d}, \quad \eta = \sqrt{cu+d} \quad \left(c = \frac{1-\rho^2}{2}, \quad d = \frac{1+\rho^2}{2}, \quad -1 < t, u < 1\right);$$

(2) in the second integral (26), we set

$$\xi = \sqrt{ct+d}, \quad \eta = \sqrt{\gamma u+\delta}, \quad \gamma = \frac{\beta^2-\alpha^2}{2}, \quad \delta = \frac{\beta^2+\alpha^2}{2} \quad (-1 < t, u < 1);$$

(3) in the third integral, we set

$$\xi = \sqrt{\gamma t+\delta}, \quad \eta = \sqrt{\gamma u+\delta} \quad (-1 < t, u < 1);$$

(4) in the fourth integral (25), we set

$$\eta = \sqrt{cu+d}, \quad \xi = \sqrt{\gamma t+\delta} \quad (-1 < t, u < 1);$$

(5) in condition (10), we set $\eta = \sqrt{cu+d}$;

(6) in condition (26), we set $\eta = \sqrt{\gamma u+\delta}$.

As a result, SIE system (26) is transformed to the form

$$\begin{gathered}\frac{2}{\pi}\int_{-1}^1\left[\frac{1}{u-t} + \frac{\pi\Lambda c}{4\sqrt{cu+d}}\operatorname{sign}(u-t)\right]\chi(u)\,du + \frac{\gamma}{\pi}\int_{-1}^1\frac{(ct+d)\omega(u)\,du}{(ct+\gamma u+d+\delta)^2} = h_0(t),\\ \frac{1}{\pi}\int_{-1}^1\left[\frac{1}{u-t} + \frac{3\gamma}{(\sqrt{\gamma u+\delta}+\sqrt{\gamma t+\delta})^2} - \frac{2\gamma\sqrt{\gamma t+\delta}}{(\sqrt{\gamma u+\delta}+\sqrt{\gamma t+\delta})^3}\right]\sqrt{\frac{\gamma t+\delta}{\gamma u+\delta}}\,\omega(u)\,du\\ + \frac{8c}{\pi}\int_{-1}^1\frac{(cu+d)\chi(u)\,du}{(cu+\gamma t+d+\gamma)^2} = -4g_0(t) \quad (-1<t<1),\\ \chi(u) = \tau(\sqrt{cu+d}), \quad \omega(u) = \psi(\sqrt{\gamma u+\delta}), \quad h_0(t) = f_0(\sqrt{ct+d}), \quad g_0(t) = q_0(\sqrt{\gamma t+\delta});\end{gathered} \tag{28}$$

condition (10) is transformed to the form (12), and condition (27) is transformed to the form

$$\int_{-1}^{1} \frac{\omega(u)\,du}{\sqrt{\gamma u+\delta}} = 0. \tag{29}$$

In (25), where $L_0 = (\alpha, \beta)$, we also move to the interval $(-1, 1)$. As a result,

$$\Sigma_0(t) = \sigma_0(\sqrt{\gamma t+\delta}) = -\frac{2c}{\pi}\int_{-1}^{1} \frac{(cu+d)\chi(u)\,du}{(cu+\gamma t+d+\gamma)^2} - \frac{1}{4\pi}\int_{-1}^{1}\left[\frac{1}{u-t} + \frac{3\gamma}{(\sqrt{\gamma u+\delta}+\sqrt{\gamma t+\delta})^2}\right.$$
$$\left. - \frac{2\sqrt{\gamma t+\delta}}{(\sqrt{\gamma u+\delta}+\sqrt{\gamma t+\delta})^3}\right]\sqrt{\frac{\gamma t+\delta}{\gamma u+\delta}}\,\omega(u)\,du \quad (t \in (-\delta/\gamma, -1) \cup (1, \infty)). \tag{30}$$

4. On the stress state of an elastic half–plane with a collinear system of absolutely rigid thin inclusions on the vertical axis

As was mentioned above, to solve this SIE system, we apply the numerical-analytic method for solving SIE on the basis of the Gauss quadrature formulas for Chebyshev nodes for calculating Cauchy-type integrals and ordinary integrals in [15–17]:

$$\chi(u) = \frac{\mathrm{X}(u)}{\sqrt{1-u^2}}, \quad \omega(u) = \frac{\Omega(u)}{\sqrt{1-u^2}} \quad (-1 < u < 1), \tag{31}$$

where $\mathrm{X}(u)$ and $\Omega(u)$ are the Hölder functions on the interval $[-1, 1]$, and using the procedure of this method, we reduce SIE system (28), (12), (29) to the finite system of linear algebraic equations

$$\sum_{n=1}^{2M} K_{mn} Z_n = a_m \quad (m = \overline{1, 2M}) \tag{32}$$

$$Z_n = \begin{cases} \mathrm{X}_n, & n = \overline{1, M}, \\ Y_{n-M} & n = \overline{M+1, 2M}, \end{cases} \qquad \mathrm{X}_n = \mathrm{X}(u_n), \quad Y_n = \Omega(u_n) \quad (n = \overline{1, M}),$$

$$u_n = \cos\frac{(2n-1)\pi}{2M} \quad (n = \overline{1, M}), \quad t_m = \cos\frac{\pi m}{M} \quad (m = \overline{1, M-1}),$$

$$a_m = \begin{cases} f_0(t_m), & m = \overline{1, M-1}, \\ \dfrac{Q_0}{c}, & m = M, \\ -4g_0(t_{m-M}), & m = \overline{M+1, 2M-1}, \\ 0, & m = 2M, \end{cases}$$

$$K_{mn} = \begin{cases} \dfrac{2}{M}\left[\dfrac{1}{u_n - t_m} + \dfrac{\pi\Lambda c}{4\sqrt{cu_n + d}}\operatorname{sign}(u_n - t_m)\right], & m = \overline{1, M-1}, \quad n = \overline{1, M}, \\ \dfrac{\pi}{M}\dfrac{1}{\sqrt{cu_n + d}} & m = M, \quad n = \overline{1, M}, \\ \dfrac{\gamma}{M}\dfrac{ct_m + d}{(ct_m + \gamma u_{n-M} + d + \delta)^2}, & m = \overline{1, M-1}, \quad n = \overline{M+1, 2M}, \\ \dfrac{8c}{M}\dfrac{cu_n + d}{(cu_n + \gamma t_{m-M} + d + \gamma)^2}, & m = \overline{M+1, 2M-1}, \quad n = \overline{1, M}, \\ \dfrac{1}{M}\left[\dfrac{1}{u_{n-M} - t_{m-M}} + \dfrac{3\gamma}{(\sqrt{\gamma u_{n-M}+\delta} + \sqrt{\gamma t_{m-M}+\delta})^2} \right. & \\ \left. \quad - \dfrac{2\gamma\sqrt{\gamma t_{m-M} + \delta}}{(\sqrt{\gamma u_{n-M}+\delta} + \sqrt{\gamma t_{m-M}+\delta})^3}\right]\sqrt{\dfrac{\gamma t_{m-M}+\delta}{\gamma u_{n-M}+\delta}}, & m = \overline{M+1, 2M-1}, \; n = \overline{M+1, 2M}, \\ \dfrac{\pi}{M}\dfrac{1}{\sqrt{\gamma u_m + \delta}}, & m = 2M, \quad n = \overline{M+1, 2M}. \end{cases}$$

Here M is any natural number, and u_n and t_m are Chebyshev nodes, i.e., the roots of the Chebyshev polynomials of the first and second kinds.

Let us calculate the characteristics of the problem under study. For the SIF at the upper end point $x = a_1$ of the crack, we have

$$K_I = \lim_{x \to a_1 - 0}[\sqrt{2\pi(a_1 - r)}\sigma_\vartheta] = -\sqrt{a}E \lim_{\xi \to \alpha - 0}[\sqrt{2\pi(\alpha - \xi)}\sigma_0(\xi)],$$

from which, for the dimensionless SIF, we obtain

$$K_I^{(0)} = \frac{K_I}{\sqrt{a}E} = \lim_{\xi \to \alpha - 0}[\sqrt{2\pi(\alpha - \xi)}\sigma_0(\xi)]. \tag{33}$$

We use (31), (32) to transform formula (33) and take into account the expression of the well-known integral from [19] (p. 175, formula (21)). After simple transformations, we have

$$K_I^{(0)} = \frac{\sqrt{\pi}}{4\sqrt{2}}\sqrt{\frac{\gamma}{\alpha}}\Omega(-1) \tag{34}$$

or, substituting the value $\Omega(-1)$ for the Lagrange interpolation polynomial and using the solution of the system of equations (32), we obtain

$$K_I^{(0)} = \frac{\sqrt{\pi}}{4\sqrt{2}}\sqrt{\frac{\gamma}{\alpha}}\frac{1}{M}\sum_{n=M+1}^{2M}(-1)^{M+n}Z_n \tan\frac{(2n-1)\pi}{4M}. \tag{35}$$

It follows from (34) or (35) that, as $\alpha \to +0$ ($a_1 \to +0$), the SIF $K_I^{(0)}$ infinitely increases and, as a result, a crack spreads out onto the plate boundary. To prevent the crack growth, it is possible to connect crack edges near the point $x = a_1$ by an inclusion, which is modelled by continuously distributed linear springs [20].

We note that, in the case of an edge crack ($a_1 = 0$), the solution of SIE system (26) or (28) must be constructed in the class of functions bounded at the point $x = a_1$ ($\xi = \alpha$) and unbounded at the point $x = b_1$ ($\xi = \beta$).

We also note that the dimensionless breaking stresses $\Sigma_0(t)$ from (30) are expressed in terms of the solution of the system of equations (32):

$$\Sigma_0(t) = -\frac{2c}{M}\sum_{n=1}^{M}\frac{(cu_n + d)Z_n}{(cu_n + \gamma t + d + \delta)^2} - \frac{\pi}{4M}\sum_{n=M+1}^{2M}\left[\frac{1}{u_{n-M} - t} + \frac{3\gamma}{(\sqrt{\gamma u_{n-M} + \delta} + \sqrt{\gamma t + \delta})^2}\right.$$
$$\left. - \frac{2\sqrt{\gamma t + \delta}}{(\sqrt{\gamma u_{n-M} + \delta} + \sqrt{\gamma t + \delta})^3}\right]\sqrt{\frac{\gamma t + \delta}{\gamma u_{n-M} + \delta}}Z_n \quad (t \in (-\delta/\gamma, -1) \cup (1, \infty)).$$

The crack opening is determined by the formula

$$\Phi(r) = \int_{a_1}^{b_1} \operatorname{sign}(r-s)\varphi(s)\,ds \quad (a_1 \le r \le b_1).$$

Transforming this integral to the interval $(-1,1)$, we obtain

$$\Phi_0(t) = \frac{\Phi(a\sqrt{ct+d})}{a} = \frac{\gamma}{4}\int_{-1}^{1} \operatorname{sign}(t-u)\frac{\omega(u)\,du}{\sqrt{\gamma u+\delta}}$$
$$= \frac{\pi\gamma}{4M}\sum_{n=M+1}^{2M} \frac{Z_n}{\sqrt{\gamma u_{n-M}+\delta}}\operatorname{sign}(t-u_{n-M}) \quad (-1 \le t \le 1).$$

The dimensionless tangential contact stresses under the right stringer, according to the first formula in (31) through the solution of the system (32), are determined by the formula

$$\chi(u_n) = \frac{Z_n}{\sqrt{1-u_n^2}} \quad \left(n=\overline{1,M}, \quad u_n = \cos\frac{(2n-1)\pi}{2M}\right).$$

Let us consider the axial directions in the right stringer. By formula (1) and by the formula related to it, which is obtained from (1) by replacing the interval (b,r) with the interval (r,a), we have

$$\sigma_s(r) = \frac{1}{2A_s}\left[d\int_b^a \operatorname{sign}(r-s)\tau_-(s)\,ds - \int_b^a \operatorname{sign}(r-s)\tau_+(s)\,ds\right] \quad (r=y\in[b,a]).$$

Hence, after passing to the interval $(-1,1)$ and to dimensionless variables, we obtain

$$\Sigma_s^{(0)}(t) = \frac{\Lambda_0 c}{2}\int_{-1}^{1} \operatorname{sign}(t-u)\frac{[\mathrm{X}(u)-\mathrm{X}_0(u)]\,du}{\sqrt{cu+d}}$$
$$= \frac{\pi\Lambda_0 c}{2M}\sum_{n=1}^{M} \operatorname{sign}(t-u_n)\frac{Z_n-\sqrt{1-u_n^2}\mathrm{X}_0(u)}{\sqrt{cu_n+d}} \quad (-1\le t\le 1),$$
$$\Sigma_s^{(0)}(t) = \sigma_s^{(0)}(\sqrt{ct+d}), \quad \sigma_s^{(0)}(\xi) = \frac{\sigma_s(a\xi)}{E}, \quad \Lambda_0 = \frac{ad_1}{2A_s},$$
$$\mathrm{X}(u) = \tau(\sqrt{cu+d}), \quad \mathrm{X}_0(u) = \tau_0(\sqrt{cu+d}).$$

In special cases when there are only stringers or only cracks, system of SIE (21), (26) or (28) splits into separate SIEs. We note that the problem of the stress state of an elastic semi-infinite plate containing only internal cracks on its vertical axis in the absence of stringer can be constructed by the Multopp method [16].

In the general case, different combinations of mutual locations of stringers and cracks are possible, which leads to various problems.

Conclusions

The above-considered problems directly related to the interaction of stress concentrators (such as stringers and cracks) with massive deformable bodies, which are often encountered in applications, are of great theoretical and practical interest. Different combinations of mutual locations of stringers and cracks lead to various problems which can be studied by using the results and approaches proposed in this paper.

References

[1] Melan E 1932 Ein Beitrag zur Theorie geschweiter Verbindungen *Ingr. Arch.* **3** (2) 123–9
[2] Bufler H 1964 Zur Krafteinleitung in Scheiben über geschweißte oder geklebte Verbindungen *Osterr. Ing.-Arch.* **18** (3–4) 284–92
[3] Bufler H 1961 Scheibe mit endlicher, elastischer Versteifung *VDI-Forschungsheft* **485** (B27) 5–44
[4] Koiter W T 1955 On the diffusion of load from a stiffener into a sheet *Quart. J. Mech. Appl. Math.* **8** (2) 164-78
[5] Muki R and Sternberg E 1967 Transfer of Load From an Edge-Stiffener to a Sheet – A Reconsideration of Melan's Problem *J. Appl. Mech.* **34** (3) 679–86
[6] Alexandrov V M and Mkhitaryan S M 1983 *Contact Problems for Bodies with Thin Coatings and Interlayers* (Moscow: Nauka) p 483 [in Russian]
[7] Galin L A (Editor) 1976 *Development of Contact Problems Theory in USSR* (Moscow: Nauka) p 493 [in Russian]
[8] Greif R and Sanders J L Jr 1965 The effect of a stringer on the stress in a cracked sheet *J. Appl. Mech.* **32** (1) 59–66
[9] Delale F and Erdogan F 1982 The Crack Problem for a Half-Plane Stiffened by Elastic Cover Plates *Int. J. Solids Struct.* **18** (3) 381-95
[10] Mkhitaryan S M and Aghayan K L 1978 On a periodic contact problem for an elastic strip weakened by cracks and reinforced by elastic stringers *Izv. Nats. Akad. Nauk Armenii. Mekh.* **31** (3) 3–17
[11] Mkhitaryan S M 2014 Contact interaction between an infinite or finite stringers and elastic semi-infinite plate with a finite vertical crack coming out to the stringer by one end *Adv. Mater. Res.* **1020** 253–7
[12] Hoakbyan V N, Mirzoyan S E, and Mkhitaryan S M 2015 The problem of the contact between a broken stringer and an elastic infinite strip containing a vertical edge crack *Int. Appl. Mech.* **51** (2) 176–86
[13] Multhopp H 1938 Die Berechnung der Auftriebsverteilung von Tragflugeln, *Luftfahrtforschung* **15** (4) 153–66
[14] Muskhelishvili N I 1966 *Some Basic Problems of the Mathematical Theory of Elasticity* (Moscow: Nauka) p 708 [in Russian]
[15] Erdogan F, Gupta G D, and Cook T S 1973 The numerical solutions of singular integral equations. In *Mechanics of Fracture, Vol. 1, Methods of Analysis and Solutions of Crack Problems* (Leyden: Noordhoff Intern. Publ.) pp 368–425
[16] Theocaris P S and Ioakimidis N I 1977 Numerical integration methods for the solution of singular integral equations *Quart. Appl. Math.* **XXXV** (1) 173–85
[17] Panasiuk V V, Savruk M P, and Datsishin A P 1976 *Stress Distribution near Cracks in Plates and Shells* (Kiev: Naukova dumka) p 443
[18] Novozhilov V V 1958 *Theory of Elasticity* (Leningrad: Sudpromgiz) p 372 [in Russian]
[19] Bateman H and Erdalyi A 1970 *Tables of Integral Transforms, Vol. 2* (Moscow: Nauka) p 328 [in Russian]
[20] Rose R L 1987 Crack reinforcement by distributed springs *J. Mech. Phys. Solids* **35** (4) 363–405

IOP Conf. Series: Journal of Physics: Conf. Series **991** (2018) 012031 doi:10.1088/1742-6596/991/1/012031

Stress state of a piecewise uniform layered space with doubly periodic internal cracks

V N Hakobyan* **and L L Dashtoyan**

Institute of Mechanics of the National Academy of Sciences of Armenia, Yerevan, Armenia

E-mail: *vhakobyan@sci.am

Abstract. The present paper deals with the stress state of a piecewise homogeneous plane formed by alternation junction of two distinct strips of equal height manufactured of different materials. There is a doubly periodic system of cracks on the plane. The governing system of singular integral equations of the first kind for the density of the crack dislocation is derived. The solution of the problem in the case where only one of the repeated strips contains one doubly-periodic crack is obtained by the method of mechanical quadratures.

Introduction

A large number of studies deal with periodic and doubly-periodic problems for massive homogeneous bodies with cracks. The monographs [1, 2] summarizing the basic results obtained in this field deserve our attention. In the recent years, the periodic and doubly-periodic problems for piecewise uniformly homogeneous layered bodies with interphase crack-type defects namely, foreign or absolutely rigid inclusions, are currently actual in the light of layered composites studied in [3–6]. Our review of the relevant literature showed that the research of the stress state of piecewise uniformly homogeneous layered bodies with internal doubly-periodic cracks is carried out for the first time in the present paper.

1. Statement of the problem and derivation of the governing equations

Assume that, in the Cartesian coordinates Oxy, a piecewise homogeneous elastic plane, which is formed by alternate junction of strips of thickness $2h$ manufactured of different materials with Young moduli E_1 and E_2 and Poisson ratios ν_1 and ν_2, contains the same periodic internal cracks with period $2l$ on the middle lines of the distinct strips $y = (2n+1)h$ $(n \in Z)$ along the lines $L_1 = \bigcup_{i=1}^{N}(a_i, b_i)$ and $L_2 = \bigcup_{i=1}^{M}(c_i, d_i)$. Let us further assume that the plane is deformed under the action of identical loads, which are symmetric with respect to the lines $x = 2nl$ $(n \in Z)$ and are applied on the crack edges so that the lines $x = (2n+1)l$ and $y = (2n+1)h$ $(n \in Z)$ are the lines of symmetry. Taking these assumptions into account, we can define our problem as a problem for the piecewise homogeneous rectangle (basic cell) occupying the region $\Omega\{|x| \le l; |y| \le H\}$ on the plane with symmetry conditions on the boundaries and on the lines $y = \pm h$ beyond the cracks; the normal stresses of intensity $P_0^{(j)}(x)$ are given on the crack edges and the shear stresses are absent.

The goal in this paper is to construct the solutions of the above-defined problem and to reveal how the laws of changes in the stress intensity factors at the crack end-points and the crack opening depend on the elastic and geometrical characteristics of the problem.

The problem is equivalent to the following boundary-value problem for the piecewise homogeneous rectangle $\Omega\{-l \le x \le l; -h \le y \le h\}$:

$$\begin{cases} \tau_{xy}^{(1)}(\pm l, y) = U_1(\pm l, y) = 0, & 0 < y < h, \\ \tau_{xy}^{(2)}(\pm l, y) = U_2(\pm l, y) = 0, & -h < y < 0, \\ \tau_{xy}^{(j)}(x, (-1)^{j+1}h) = V_j(x, (-1)^{j+1}h) = 0, & x \notin L_j, \\ \tau_{xy}^{(j)}(x, (-1)^{j+1}h) = 0, & x \in L_j, \quad j = 1, 2, \\ \sigma_y^{(j)}(x, (-1)^{j+1}h) = P_0^{(j)}(x), & x \in L_j, \end{cases} \tag{1a}$$

$$\begin{cases} \sigma_y^{(1)}(x, 0) = \sigma_y^{(2)}(x, 0), & \tau_{xy}^{(1)}(x, 0) = \tau_{xy}^{(2)}(x, 0), \\ U_1(x, 0) = U_2(x, 0), & V_1(x, 0) = V_2(x, 0), \end{cases} \quad |x| \le l. \tag{1b}$$

The functions $P_0^{(j)}(x)$ are even.

Here $U_j(x, y)$ and $V_j(x, y)$ $(j = 1, 2)$ are horizontal and vertical components of the displacement vector of the rectangle, which satisfy the Lame equations, while $\sigma_y^{(j)}(x, y)$ and $\tau_{xy}^{(j)}(x, y)$ $(j = 1, 2)$ are components of normal and shear stresses acting in the respective rectangle.

To solve the boundary-value problem, we should initially solve an auxiliary problem, which differs from problem (1) in the following relations considered instead of (1a):

$$V_j(x, (-1)^{j+1}h) = V_j(x) \quad (j = 1, 2; x \in L). \tag{2}$$

Here the functions $V_j(x)$ are the unknown displacements of the crack edges.

Just as in [3, 4], we use biharmonic functions of stresses (Airy functions) to construct solutions of the auxiliary problem. For the heterogeneous rectangles, these functions are represented as the series

$$\begin{aligned} F_j(x, y) = \sum_{k=1}^{\infty} \{A_k^{(j)} \cosh(\alpha_k y) + B_k^{(j)} \sinh(\alpha_k y) \\ + \alpha_k y [C_k^{(j)} \cosh(\alpha_k y) + D_k^{(j)} \sinh(\alpha_k y)]\} \cos(\alpha_k x) + a_j x^2 + b_j y^2 \quad (j = 1, 2), \end{aligned} \tag{3}$$

where $\alpha_k = \pi k / l$ $(-l \le x \le l; 0 \le y \le h)$ for $j = 1$ and $(-l \le x \le l, -h \le y \le 0)$ for $j = 2$. The coefficients $A_k^{(j)}$, $B_k^{(j)}$, $D_k^{(j)}$, $C_k^{(j)}$, a_j and b_j $(j = 1, 2)$ are unknown constants to be determined.

We take into account the relations

$$\sigma_x(x, y) = \frac{\partial^2 F(x, y)}{\partial y^2}, \quad \sigma_y(x, y) = \frac{\partial^2 F(x, y)}{\partial x^2}, \quad \tau_{xy}(x, y) = -\frac{\partial^2 F(x, y)}{\partial x \partial y},$$

$$\frac{E_j}{1 - \nu_j^2} U_j(x, y) = \int \frac{\partial^2 F_j(x, y)}{\partial y^2}\, dx - \frac{\nu_j}{(1 - \nu_j)} \frac{\partial F_j(x, y)}{\partial x} + a_0^{(j)} y + c_0^{(j)},$$

$$\frac{E_j}{1 - \nu_j^2} V_j(x, y) = \int \frac{\partial^2 F_j(x, y)}{\partial x^2}\, dy - \frac{\nu_j}{(1 - \nu_j)} \frac{\partial F_j(x, y)}{\partial y} - a_0^{(j)} x + b_0^{(j)}$$

connecting the stress and displacement components with the stress function, and satisfy conditions (1), besides the last relations in (1a), and relations (2) to obtain the unknown coefficients in terms of the Fourier coefficients $V_j(x)$ $(j = 1, 2)$.

The following is obtained:

$$
\begin{aligned}
&C_k^{(1)} = -\tanh\beta_k D_k^{(1)} + \frac{\nu_k^{(1)}}{\cosh\beta_k}, \quad C_k^{(2)} = \tanh\mu_k D_k^{(2)} + \frac{\nu_k^{(2)}}{\cosh\beta_k},\\
&D_k^{(1)} = \frac{\nu_k^{(1)}[\vartheta_2 + \vartheta_1(1-2\nu_1)] - 2\vartheta_1\nu_k^{(2)}(1-\nu_2) + (\vartheta_2 - \vartheta_1)\nu_k^{(3)}}{\Delta_k^{(1)}\sinh\beta_k},\\
&D_k^{(2)} = \frac{-2\vartheta_1\nu_k^{(1)}(1-\nu_1) + \nu_k^{(2)}[\vartheta_1 + \vartheta_2(1-2\nu_2)] + (\vartheta_2 - \vartheta_1)\nu_k^{(4)}}{\Delta_k^{(2)}\sinh\beta_k},\\
&A_k^{(1)} = A_k^{(2)} = -\left[\frac{\beta_k}{\sinh(2\beta_k)} + \frac{1}{2}\right](D_k^{(1)} + D_k^{(2)}) + \frac{\nu_k^{(1)} - \nu_k^{(2)} + \nu_k^{(3)} + \nu_k^{(4)}}{2\sinh\beta_k},\\
&B_k^{(1)} = -A_k^{(1)}\tanh\beta_k - \frac{\beta_k D_k^{(1)}}{\cosh^2\beta_k} + \frac{\nu_k^{(3)}}{\cosh\beta_k}, \quad B_k^{(2)} = A_k^{(2)}\tanh\beta_k + \frac{\beta_k D_k^{(2)}}{\cosh^2\beta_k} - \frac{\nu_k^{(4)}}{\cosh\beta_k},\\
&b_0^{(j)} = 2h(-1)^j\frac{(1-2\nu_j)}{(1-\nu_j)^2}a_j + \frac{E_j}{(1-\nu_j^2)}, \quad a_1 = a_2 = \frac{(V_0^{(1)} - V_0^{(2)})}{2h(m_1+m_2)}, \quad b_j = \frac{\nu_j a_j}{1-\nu_j},\\
&a_0^{(j)} = c_0^{(j)} = 0 \quad (j=1,2),
\end{aligned}
$$

where

$$
\begin{aligned}
&\Delta_k^{(1)} = \frac{2(\vartheta_2-\vartheta_1)\beta_k}{\sinh(2\beta_k)} + \Delta_1, \quad \Delta_k^{(2)} = \frac{2(\vartheta_2-\vartheta_1)\beta_k}{\sinh(2\beta_k)} - \Delta_2, \quad \Delta_1 = \vartheta_2 + \vartheta_1\varkappa_1, \quad \Delta_2 = \vartheta_1 + \vartheta_2\varkappa_2,\\
&\nu_k^{(j)} = \frac{V_k^{(j)}}{2\vartheta_j\alpha_k(1-\nu_j)} \quad (j=1,2), \quad \nu_k^{(j)} = (-1)^j(\beta_k\tanh\beta_k + 1)\nu_k^{(j-2)} \quad (j=3,4),\\
&m_j = \frac{\vartheta_j(1-2\nu_j)}{(1-\nu_j)}, \quad \beta_k = \frac{\pi k}{l}h, \quad \vartheta_j = \frac{1+\nu_j}{E_j}, \quad \varkappa_j = 3-4\nu_j, \quad V_0^{(j)} = \frac{1}{2l}\int_{L_j} V_j(x)\,dx,\\
&V_k^{(j)} = \frac{1}{l}\int_{L_j} V_j(x)\cos(\alpha_k x)\,dx \quad (j=1,2; k=1,2,\ldots).
\end{aligned}
\tag{4}
$$

Then, substituting the obtained expressions for the coefficients into (3) and taking (4) into account, we can express the stress function, and hence all values characterizing the stress state in the piecewise homogeneous rectangle in terms of the functions $V_j(x)$ $(j=1,2)$.

In particular, the following expressions are obtained for the components of stresses acting on the crack edges:

$$
\begin{aligned}
\sigma_y^{(1)}(x,h) = &-\frac{1}{2l(1-\nu_1)\vartheta_1}\left[\int_{L_1}\cot\frac{\pi(s-x)}{2l}V_1'(s)\,ds + \int_{L_1}K_{11}(s-x)V_1'(s)\,ds\right]\\
&+ \frac{1}{2l(1-\nu_2)\vartheta_2}\int_{L_2}K_{12}(s-x)V_2'(s)\,ds + 2a_1 \quad (-l<x<l),
\end{aligned}
\tag{5}
$$

$$
\begin{aligned}
\sigma_y^{(2)}(x,-h) = &-\frac{1}{2l(1-\nu_2)\vartheta_2}\left[\int_{L_2}\cot\frac{\pi(s-x)}{2l}V_2'(s)\,ds + \int_{L_2}K_{22}(s-x)V_2'(s)\,ds\right],\\
&+ \frac{1}{2l(1-\nu_1)\vartheta_1}\int_{L_1}K_{21}(s-x)V_1'(s)\,ds + 2a_1 \quad (-l<x<l),
\end{aligned}
\tag{6}
$$

$$
K_{ij}(x) = \frac{1}{l}\sum_{k=1}^{\infty}A_k^{(i,j)}\sin(\alpha_k x) \quad (i,j=1,2),
$$

$$A_k^{(1,1)} = \left\{ \frac{(\beta_k \tanh\beta_k + \beta_k \coth\beta_k + 1)[2\vartheta_1(1-\nu_1) - (\vartheta_2 - \vartheta_1)\tanh\beta_k]}{\Delta_k^{(1)}\sinh(2\beta_k)} \right.$$
$$\left. + \frac{2(1-\nu_1)\vartheta_1}{\Delta_k^{(1)}\sinh(2\beta_k)}\left[\frac{2\beta_k}{\sinh(2\beta_k)} + 1\right] - \frac{\beta_k}{2\cosh^2\beta_k} + \tanh\beta_k - 1 \right\},$$
$$A_k^{(1,2)} = \left\{ \frac{2\vartheta_2(1-\nu_2)(\beta_k \tanh\beta_k + \beta_k \coth\beta_k + 1)}{\Delta_k^{(1)}\sinh(2\beta_k)} \right.$$
$$\left. - \frac{2(1-\nu_2)\vartheta_2 + (\vartheta_2-\vartheta_1)\beta_k\tanh\beta_k}{\Delta_k^{(2)}\sinh(2\beta_k)}\left[\frac{2\beta_k}{\sinh(2\beta_k)} + 1\right] - \frac{\beta_k}{2\cosh^2\beta_k} \right\};$$
$$A_k^{(2,2)} = \left\{ - \frac{(\beta_k \tanh\beta_k + \beta_k \coth\beta_k + 1)[2\vartheta_2(1-\nu_2) + (\vartheta_2 - \vartheta_1)\tanh\beta_k]}{\Delta_k^{(2)}\sinh(2\beta_k)} \right.$$
$$\left. + \frac{2(1-\nu_2)\vartheta_2}{\Delta_k^{(1)}\sinh(2\beta_k)}\left[\frac{2\beta_k}{\sinh(2\beta_k)} + 1\right] - \frac{\beta_k}{2\cosh^2\beta_k} + \tanh\beta_k - 1 \right\};$$
$$A_k^{(2,1)} = \left\{ \frac{2\vartheta_1(1-\nu_1)(\beta_k \tanh\beta_k + \beta_k \coth\beta_k + 1)}{\Delta_k^{(2)}\sinh(2\beta_k)} \right.$$
$$\left. - \frac{2(1-\nu_1)\vartheta_1 - (\vartheta_2-\vartheta_1)\beta_k\tanh\beta_k}{\Delta_k^{(1)}\sinh(2\beta_k)}\left[\frac{2\beta_k}{\sinh(2\beta_k)} + 1\right] - \frac{\beta_k}{2\cosh^2\beta_k} \right\}.$$

Using the obtained expressions we satisfy the last relations in (1a). After some simplifications, the following system of singular integral equations of the first kind is obtained to determine the unknown derivatives of the functions $V_j(x)$ $(j = 1, 2)$:

$$\begin{cases} \dfrac{1}{\pi}\displaystyle\int_{L_1} \frac{V_1'(s)\,ds}{s-x} + \int_{L_1} R_{11}(s-x)V_1'(s)\,ds + \int_{L_2} R_{12}(s-x)V_2'(s)\,ds = f_1(x) \quad (x \in L_1), \\ \dfrac{1}{\pi}\displaystyle\int_{L_2} \frac{V_2'(s)\,ds}{s-x} + \int_{L_1} R_{21}(s-x)V_1'(s)\,ds + \int_{L_2} R_{22}(s-x)V_2'(s)\,ds = f_2(x) \quad (x \in L_2). \end{cases} \tag{7}$$

This system should be considered under the continuity conditions for the displacements at the crack end-points

$$\int_{a_j}^{b_j} V_1'(s)\,ds = 0, \quad \int_{c_i}^{d_i} V_2'(s)\,ds = 0 \quad (j = \overline{1,N}; i = \overline{1,M}). \tag{8}$$

Here

$$f_j(x) = (1-\nu_j)\vartheta_j P_0^{(j)}(x) \quad (j = 1, 2),$$
$$R_{jj}(t) = \frac{1}{2l}\cot\frac{\pi t}{2l} - \frac{1}{\pi t} + \frac{1}{2l}K_{jj}(t) + \frac{2(1-\nu_j)\vartheta_j t}{hl(m_1+m_2)} \quad (j = 1, 2),$$
$$R_{ij}(t) = (-1)^i \frac{m_{12}}{2l}K_{ij}(t) - (-1)^j \frac{2(1-\nu_i)\vartheta_i t}{hl(m_1+m_2)} \quad (i, j = 1, 2; i \neq j).$$

Note that, in the general case, system (7) with conditions (8) can be represented as a system of $N + M$ equations for the displacement dislocations on each of the intervals (a_j, b_j) $(j = \overline{1,M})$ and (c_j, d_j) $(j = \overline{1,N})$. This system can be solved either by using the apparatus of Chebyshev orthogonal polynomials reducing it to a quasi-complete regular system of infinite algebraic equations [7] or by the method of mechanical quadratures [8].

2. A piecewise homogeneous plane with a doubly-periodic system of internal cracks located in one of the repeated strips

As an example, we consider the case where there is a periodically repeated crack of length $2a$ occupying the interval $(a_1, b_1) = (-a, a)$ on the boundary $y = h$ of the basic cell on whose edges the symmetric normal stresses of intensity $P_0(x)$ are acting, i.e., $M = 1$, $N = 0$ and $P_0^{(1)}(x) = P_0(x)$.

In this special case, the following integral equation is obtained from system (7):

$$\frac{1}{\pi}\int_{-a}^{a}\frac{V_1'(s)\,ds}{s-x}+\int_{-a}^{a}R_{11}(s-x)V_1'(s)\,ds=f_1(x)\quad(|x|<a) \tag{9}$$

under the condition

$$\int_{-a}^{a}V_1'(s)\,ds=0. \tag{10}$$

The solution of the equation is constructed by the method of mechanical quadratures [8]. For this, after the change $s = a\xi$ and $x = a\eta$, the equation is considered on the interval $(-1, 1)$. We use the notation

$$V_1'(\eta a)=\varphi(\eta),\quad aR_{11}(a\eta)=R(\eta),\quad f_1(\eta a)=f_*(\eta)$$

to write the equation in the form

$$\frac{1}{\pi}\int_{-1}^{1}\frac{\varphi(\xi)}{\xi-\eta}\,d\xi+\int_{-1}^{1}R(\xi-\eta)\varphi(\xi)\,d\xi=f_*(\eta). \tag{11}$$

Condition (10) becomes

$$\int_{-1}^{1}\varphi(\xi)\,d\xi=0. \tag{12}$$

It is not difficult to state that the unknown function at the end-points of the interval of integration has a root singularity, which can be represented as

$$\varphi(\eta)=\frac{\varphi^*(\eta)}{\sqrt{(1-\eta)(1+\eta)}}. \tag{13}$$

Here $\varphi^*(\eta)$ is a continuous smooth function bounded up to the ends of the interval $[-1, 1]$.

Substituting the function $\varphi(\eta)$ from (13) into (11), (12) and using relations from [10], we standardly obtain a system of algebraic equations for the values $\varphi^*(\xi_i)$ $(i = \overline{1, n})$, where ξ_i are roots of the Chebyshev polynomial of the first kind $T_n(\xi_i) = 0$.

Using the Lagrange interpolation polynomial, we can easily reconstruct the function $\varphi^*(\eta)$ $(-1 < \eta < 1)$ and determine all necessary values characterizing the stress state in the two-component cell. In particular, the stress intensity factor at the crack end-points is determined by formula (5), which has the following form for $|\eta| > 1$ in the new notation:

$$\sigma_y^{(1)}(a\eta,h)=\frac{1}{\vartheta_1(1-\nu_1)}\left[\frac{1}{\pi}\int_{-1}^{1}\frac{\varphi(\xi)}{\xi-\eta}\,d\xi+\int_{-1}^{1}R(\eta-\xi)\varphi(\xi)\,d\xi\right]\quad(|\eta|>1). \tag{14}$$

Substituting $\varphi(\eta)$ from (13) and using the relation [9]

$$\frac{1}{\pi}\int_{-a}^{a}\frac{ds}{\sqrt{a^2-s^2}(s-x)}=-\frac{\operatorname{sign}(x)}{\sqrt{x^2-a^2}}\quad(x<a;x>b)$$

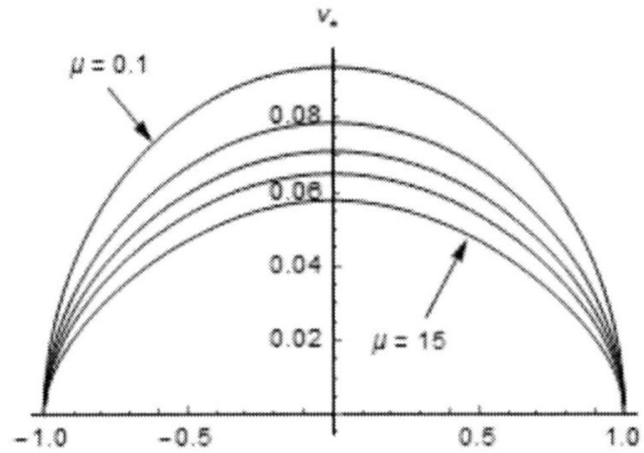

Figure 1. Crack opening for $\mu = 0.1, 0.5, 1, 2, 15$.

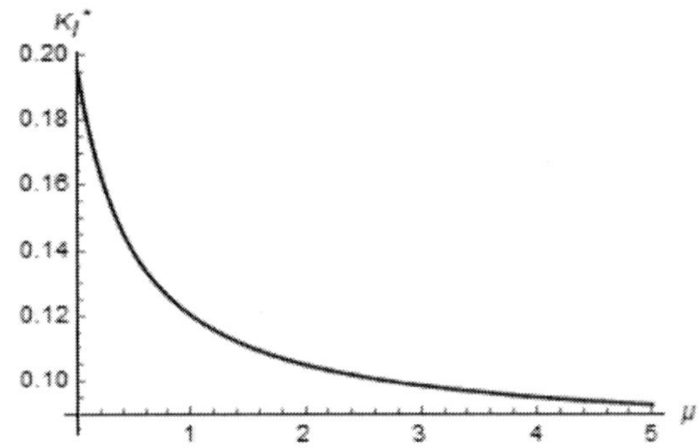

Figure 2. Dimensionless intensity stress factor $K_{\rm I}^*(a)$.

for the dimensionless stress factor at the crack end-points, we obtain

$$K_{\rm I}^*(\pm a) = \frac{K_{\rm I}(\pm a)}{E_1} = \sqrt{2\pi} \lim_{\eta\to\pm1\pm0} \sqrt{|\eta \mp 1|}\frac{\sigma_y^{(1)}(a\eta, 0)}{E_1} = \pm\frac{\sqrt{\pi}\varphi^*(\pm1)}{(1-\nu_1^2)}.$$

The following formula is obtained to determine the dimensionless crack opening:

$$V_*(\eta) = \frac{V_1(a\eta)}{a} = \int_{-1}^{\eta} \varphi(\xi)\, d\xi.$$

3. Numerical analysis and conclusions

The numerical analysis of above-stated problem was carried out for $P_0(x) = P_0 = \text{const}$. The laws of changes in the dimensionless stress intensity factor at the crack end-points and the crack opening depending on the Young moduli ratio $\mu = E_2/E_1$ of a separate strip were studied. The following values of Poisson's ratios and geometrical characteristics of the strips were taken: $\nu_1 = 0.28$, $\nu_2 = 0.3$, $\lambda_1 = a/l = 0.2$, $\lambda_2 = h/l = 0.5$, and $E_0 = P_0/E_1 = 0.1$.

Figures 1 and 2 show the graphs of the crack opening $V_*(\eta)$ and the dimensionless intensity stress factor $K_{\rm I}^*(a)$ versus the parameter μ.

It follows from these graphs that both the crack opening and the stress intensity factor at the crack end-points decrease tending to a certain limit as the parameter μ increases.

Conclusions

The governing system of equations of elasticity for uniform piecewise homogeneous plane consisting of two periodically repeated distinct strips with doubly periodic internal cracks is obtained. In the general case, the ways of solving the problem are indicated. Numerical-analytical solutions of the equations are obtained in the special case where only one of the strips contains a periodically repeated crack of length $2a$. It is shown that as the rigidities of the strips without cracks increase, while the other parameters remain unchanged, the crack opening and the stress intensity factors at the crack end-points decrease and tend to the values corresponding to the case where the strips with no cracks are absolutely rigid.

Acknowledgments

This work was supported by the RA MES State Committee of Science (SCS) and Russian Foundation for Basic Research (RFBR) in the frames of the joint research project SCS 18RF-061 and RFBR 18-51-05012 accordingly.

References

[1] Panasyuk V, Savruk M, and Datsyshyn O 1976 *Distribution of Stresses near Cracks in Plates and Shells* (Kiev: Naukova Dumka) p 443 [in Russian]
[2] Bardzokas D, Filshtinsky L, and Filshtinsky M 2010 *Actual Problems of Coupled Physical Fields in Deformable Solids* vol. 1 (Moscow - Izhevsk: NITs "RKhD") p 864 [in Russian]
[3] Dashtoyan L and Hakobyan L 2011 Stress state of compound plane with periodic system of the absolutely rigid thin inclusions In *The Problems of Dynamics of Interaction of Deformable Media. Proc. of the VII Int. Conf. September 19–23, 2011, Goris - Stepanakert* (Yerevan: In-t Mekh. NAN RA) pp 172–6 [in Russian]
[4] Hakobyan V and Hakobyan L 2015 The stress state of piecewise homogeneous space with periodic system of parallel interphase defects In
textitTopical Problems of Continuum Mechanics: Proc. of the IV Int. Conf. September 21–26, 2015, Tsaghkadzor, Armenia (Yerevan: Nats. Univ. Arkh. Stroit. Armenii) pp 28–32 [in Russian]
[5] Hakobyan V and Dashtoyan L 2015 *Doubly periodic problem for piecewise homogeneous plane with cracks* In *Topical Problems of Continuum Mechanics: Proc. of the IV Int. Conf. September 21–26, 2015, Tsaghkadzor, Armenia* (Yerevan: Nats. Univ. Arkh. Stroit. Armenii) pp 33–7 [in Russian]
[6] Hakobyan V and Dashtoyan L 2016 Doubly periodic problem for piecewise homogeneous plane with absolutely rigid inclusions In *Proc. of the 8th Int. Conf. Contemporary Problems of Architecture and Construction, Yerevan* (Yerevan) pp 125–8 [in Russian]
[7] Alexandrov V and Mkhitaryan S 1983 *Contact Problems for Bodies with Thin Coatings and Interlayers* (Moscow: Nauka) p 488 [in Russian]
[8] Sahakyan A 2000 Application of the method of discrete singularities for solving the singular integral equations with unmoved singularities *Izv. Nats. Akad. Nauk Armenii. Mekh.* **53** 12–9
[9] Prudnikov A P, Brychkov Yu A, and Marichev O I 1981 *Integrals and Series* (Moscow: Nauka) p 738 [in Russian]

Stress state of uniformly piecewise homogeneous space with a periodic system of parallel internal cracks

V N Hakobyan* **and L V Hakobyan**
Institute of Mechanics of the National Academy of Sciences of Armenia, Yerevan, Armenia
E-mail: *vhakobyan@sci.am

Abstract. This paper deals with the antiplane stress state of a uniformly piecewise homogeneous space formed by alternate junction of two distinct layers of the same thickness manufactured of different materials which are weakened by periodic parallel tunnel cracks in their middle planes. The governing system of singular integral equations and the ways of solving the system are are given in general form. The closed solution of the problem is obtained in the special case, where there are central cracks of length $2a$ in heterogenous layers.

Introduction

The periodic and doubly-periodic problems for two elastic homogeneous massive bodies with cracks form one of the developing directions of the theory of contact and mixed boundary-value problems of the mathematical theory of elasticity. The monographs [1, 2] outline the obtained results in this direction. The number of similar problems studied for piecewise homogeneous bodies is very small. This study considers several antiplane and plane periodic problems for a compound two-component space as well as periodic and doubly-periodic problems for uniformly piecewise homogeneous bodies with interphase cracks, which are important for the layered composites. We mention the research studies [3–7] which are more closely related to the problem.

1. Statement of the problem and derivation of discontinuous solutions

Let us consider the antiplane stress state of a piecewise homogeneous space formed by alternate junction of two distinct layers of thickness $2h$ with shear moduli G_1 and G_2. On the middle planes $y = (2n+1)h$ $(n \in Z)$ of the layers along the lines L_1 and L_2 $(-\infty < z < \infty)$ consisting of disjoint intervals (a_j, b_j) $(j = \overline{1, N})$ and (c_i, d_i) $(i = \overline{1, M})$, there is a periodic system of tunnel parallel cracks. It is assumed that the space is deformed by the action of identical but oppositely directed distributed loads $\tau_0^{(1)}(x)$ and $\tau_0^{(2)}(x)$ acting on the edges of cracks in heterogenous layers.

The problem is to construct the solution and to find the laws of changes in the intensity stress factor at the crack end-points and the crack opening depending on the elastic characteristics of heterogenous layers.

It is obvious that, in this case, the planes $y = (2n+1)h$ $(n \in Z)$ are antisymmetric and the stress state in the compound layers located between the symmetry planes $y = (2k-1)h$ and $y = (2k+1)h$ is the same. Therefore, we can consider only the two-component layer (basic layer) between the symmetry planes $y = \pm h$. For the basic layer, the problem can mathematically be formulated as the boundary-value problem

$$\begin{cases} W_j(x,(-1)^{j+1}h) = 0, & x \notin L_j,\ j = 1,2, \\ \tau_{yz}^{(1)}(x,0) = \tau_{yz}^{(2)}(x,0), & -\infty < x < \infty, \\ W_1(x,0) = W_2(x,0), & -\infty < x < \infty, \\ \tau_{yz}^{(j)}(x,(-1)^{j+1}h) = \tau_0^{(j)}(x), & (x \in L_j). \end{cases} \tag{1}$$

Here $W_j(x,y)$ $(j = 1,2)$ are components of displacements of points of the corresponding layers each of which, in the domain of its definition, satisfies the Laplace equation and is related to the stress components $\tau_{yz}^{(j)}(x,y)$ by the formula

$$\tau_{yz}^{(j)}(x,y) = G_j \frac{\partial W_j(x,y)}{\partial y}. \tag{2}$$

To solve the problem, let us introduce the unknown displacements of the crack edges in the heterogenous layers $W_j(x,(-1)^{j+1}h) = W_j(x)$ $(j = 1,2)$. First, we solve an auxiliary problem with the last condition in (1) replaced by this relation.

To this end, we represent the solutions of the Laplace equation as Fourier integrals

$$W_j(x,y) = \frac{1}{2\pi}\int_{-\infty}^{\infty} [A_j(\lambda)\sinh(\lambda y) + B_j(\lambda)\cosh(\lambda y)]e^{-i\lambda x}\, d\lambda \quad (j = 1,2). \tag{3}$$

Satisfying the conditions of the auxiliary boundary-value problem, we express the unknown coefficients $A_j(\lambda)$ and $B_j(\lambda)$ $(j = 1,2)$ in terms of the Fourier transforms of the functions $W_j(x)$ $(j = 1,2)$ and obtain

$$A_1 = \frac{A_2}{G} = \frac{\bar{w}_1(\lambda) - \bar{w}_2(\lambda)}{(1+G)\sinh(\lambda h)}, \quad B_1 = B_2 = \frac{G\bar{w}_1(\lambda) + \bar{w}_2(\lambda)}{(1+G)\cosh(\lambda h)},$$

where

$$\bar{w}_j(\lambda) = \frac{1}{2\pi}\int_{-\infty}^{\infty} W_j(x)e^{i\lambda x}\, dx.$$

After substitution of the obtained expressions into (3) and then into (2), the stress state field in the heterogenous layers can be determined by the functions $W_j(x)$ $(j = 1,2)$. In particular, the shear stresses on the boundaries of the two-component layer can be written in terms of $W_j(x)$ as

$$\tau_{yz}^{(1)}(x,h) = -\frac{G_1}{2h(G+1)}\left\{ \int_{L_1} \frac{G + \cosh[\mu(s-x)]}{\sinh[\mu(s-x)]} W_1'(s)\, ds + \int_{L_2} \frac{1 - \cosh[\mu(s-x)]}{\sinh[\mu(s-x)]} W_2'(s)\, ds \right\}, \tag{4}$$

$$\tau_{yz}^{(2)}(x,-h) = \frac{G_2 G}{2h(G+1)}\left\{ \int_{L_1} \frac{1 - \cosh[\mu(s-x)]}{\sinh[\mu(s-x)]} W_1'(s)\, ds + \int_{L_2} \frac{1 + G\cosh[\mu(s-x)]}{G\sinh[\mu(s-x)]} W_2'(s)\, ds \right\} \quad \left(G = \frac{G_1}{G_2},\ \mu = \frac{\pi}{2h},\ -\infty < x < \infty\right). \tag{5}$$

The obtained relations are used to satisfy the last two relations in (1). Finally, we obtain the following governing system of singular integral equations for the derivative of the displacement of the crack edges:

$$\begin{aligned}
&\frac{1}{2h}\int_{L_1}\frac{G+\cosh[\mu(s-x)]}{\sinh[\mu(s-x)]}W_1'(s)\,ds\\
&\qquad+\frac{1}{2h}\int_{L_2}\frac{1-\cosh[\mu(s-x)]}{\sinh[\mu(s-x)]}W_2'(s)\,ds=-\frac{G+1}{G_1}\tau_0^{(1)}(x)\quad(x\in L_1)\\
&\frac{G}{2h}\int_{L_1}\frac{1-\cosh[\mu(s-x)]}{\sinh[\mu(s-x)]}W_1'(s)\,ds\\
&\qquad+\frac{1}{2h}\int_{L_2}\frac{1+G\cosh[\mu(s-x)]}{\sinh[\mu(s-x)]}W_2'(s)\,ds=\frac{G+1}{G_2}\tau_0^{(2)}(x)\quad(x\in L_2).
\end{aligned}\tag{6}$$

This system should be considered with the continuity conditions for the displacements at the crack end-points.

$$\int_{a_j}^{b_j}W_1'(s)\,ds=0,\quad\int_{c_i}^{d_i}W_2'(s)\,ds=0\quad(j=\overline{1,N};i=\overline{1,M}).\tag{7}$$

Solving the problem is reduced to solving the governing system (6) under conditions (7). Note that, in the general case, for arbitrary L_1 and L_2, the governing system can be solved either by the method of Chebyshev orthogonal polynomials or by the method of mechanical quadratures, where the equations are written on each interval contained in L_1 and L_2. However, we shall continue our research by considering a particular problem which allows a closed solution.

2. Uniformly piecewise homogenous space with a periodic system of identical parallel internal cracks

Let us consider a special case of the problem, where the cracks in the heterogenous layers are identical, i.e., $L_1 = L_2 = L$. In this case, we use some simple calculations and introduce the functions

$$\begin{aligned}
&\varphi_1(x)=W_1(x)-W_2(x),\quad\varphi_2(x)=GW_1(x)+W_2(x),\\
&f_1(x)=-\frac{\tau_0^{(1)}(x)+G\tau_0^{(2)}(x)}{G_1},\quad f_2(x)=-\frac{\tau_0^{(1)}(x)-\tau_0^{(2)}(x)}{G_2}.
\end{aligned}\tag{8}$$

The governing system of equations (6) can be written as the following independent singular integral equations of the first kind:

$$\frac{1}{2h}\int_L\coth[\mu(s-x)]\varphi_1'(s)\,ds=f_1(x)\quad(x\in L),\tag{9}$$

$$\frac{1}{2h}\int_L\frac{\varphi_2'(s)}{\sinh[\mu(s-x)]}\,ds=f_2(x)\quad(x\in L).\tag{10}$$

Conditions (7) become

$$\int_{a_j}^{b_j}\varphi_1'(s)\,ds=0,\quad\int_{a_j}^{b_j}\varphi_2'(s)\,ds=0\quad(j=\overline{1,N}).\tag{11}$$

The closed solutions of equations (9) and (10) can be obtained for any finite value of N. We construct these solutions for $N = 1$ and $a_1 = -a$, $b_1 = a$, i.e., when the distinct layers are weakened by one finite tunnel crack of length $2a$ occupying the interval $(-a, a)$. In this case, introducing new variables by formulas $\xi = \exp(2\mu s)$, $\eta = \exp(2\mu x)$ and using the notation

$$\varphi_1^*(\eta) = \frac{1}{\eta}\varphi_1'\left(\frac{\ln\eta}{2\mu}\right), \quad f_1^*(\eta) = \frac{1}{\eta}f_1\left(\frac{\ln\eta}{2\mu}\right), \quad a_* = \exp(-2\mu a),$$
$$\varphi_2^*(\eta) = \frac{1}{\sqrt{\eta}}\varphi_2'\left(\frac{\ln\eta}{2\mu}\right), \quad f_2^*(\eta) = \frac{1}{\sqrt{\eta}}f_2\left(\frac{\ln\eta}{2\mu}\right), \quad b_* = \exp(2\mu a),$$

we obtain the following system of singular integral equations with the Cauchy kernel:

$$\frac{1}{\pi}\int_{a_*}^{b_*}\frac{\varphi_j^*(\xi)\,d\xi}{\xi-\eta} = f_j^*(\eta) \quad (j=1,2; \quad a_* < \eta < b_*). \tag{12}$$

Conditions (11) become

$$\int_{a_*}^{b_*}\varphi_j^*(s)\,ds = 0 \quad (j=1,2). \tag{13}$$

The solution of equations (12) with conditions (13) is given by the formulas [8]

$$\varphi_1^*(\eta) = -\frac{1}{\pi\omega(\eta)}\int_{a_*}^{b_*}\frac{\omega(\xi)f_1^*(\eta)\,d\xi}{\xi-\eta} \quad (a_* < \eta < b_*), \tag{14}$$
$$\varphi_2^*(\eta) = -\frac{1}{\omega(\eta)}\left[\frac{1}{\pi}\int_{a_*}^{b_*}\frac{\omega(\xi)f_1^*(\eta)\,d\xi}{\xi-\eta} - C\right] \quad (a_* < \eta < b_*). \tag{15}$$

Here

$$C = \frac{I_1}{I_2}, \quad I_1 = \frac{1}{\pi}\int_{a_*}^{b_*}\frac{d\eta}{\sqrt{\eta}\omega(\eta)}\int_{a_*}^{b_*}\frac{\omega(\xi)f_2^*(\xi)\,d\xi}{\xi-\eta}, \quad I_2 = \frac{2}{\sqrt{b_*}}F\left(\frac{\pi}{2}, k_1\right),$$
$$\omega(\eta) = \sqrt{(\eta-a_*)(b_*-\eta)}, \quad k_1 = \sqrt{1-\frac{a_*}{b_*}}.$$

The function $F(\alpha, k_1)$ is the elliptic integral of the first kind. Returning to the initial variables, we obtain

$$\varphi_1'(x) = -\frac{1}{2h\omega_*(x)}\int_{-a}^{a}\frac{\omega_*(s)f_1(s)\,ds}{\sinh[\mu(s-x)]}, \tag{16}$$
$$\varphi_2'(x) = -\frac{1}{2\omega_*(x)}\left\{\frac{2\mu}{\pi}\int_{-a}^{a}\frac{\omega_*(\xi)e^{\mu(s-x)}f_2(s)\,ds}{\sinh[\mu(s-x)]} - C\right\} \tag{17}$$
$$\omega_*(x) = \sqrt{\cosh(2\mu a)-\cosh(2\mu x)}, \quad |x| < a.$$

Then, using formulas (8), we obtain the following expressions for the derivative of displacements of the crack edges:

$$W_1'(x) = -\frac{1}{2(G+1)\omega_*(x)}\left\{\frac{1}{h}\int_{-a}^{a}\frac{\omega_*(\xi)[f_1(s)+e^{\mu(s-x)}f_2(s)]\,ds}{\sinh[\mu(s-x)]} - C\right\}, \tag{18}$$
$$W_2'(x) = \frac{1}{2(G+1)\omega_*(x)}\left\{\frac{1}{h}\int_{-a}^{a}\frac{\omega_*(\xi)[Gf_1(s)-e^{\mu(s-x)}f_2(s)]\,ds}{\sinh[\mu(s-x)]} + C\right\}. \tag{19}$$

Note that, in the special case, where $\tau_0^{(1)}(x) = \tau_0^{(2)}(x) = \tau_0(x)$, we have

$$f_1(x) = -\frac{1+G}{G_1}\tau_0(x), \quad f_2(x) = 0.$$

Then from formulas (16)–(19) we derive

$$C = 0, \quad \varphi_2(x) \equiv 0,$$
$$W_j'(x) = \frac{(-1)^j}{2hG_j\omega_*(x)} \int_{-a}^{a} \frac{\omega_*(\xi) f_1(s)\, ds}{\sinh[\mu(s-x)]}.$$

Now, let us determine the stress intensity factors at the crack end-points. First, we determine the stress intensity factors at the end-point of the first crack located on the line $y = h$. For this, we use formula (4) outside the interval $(-a, a)$. Substituting the values of the function $W_1'(x)$ from (18) into (4) and taking into account that the second term in (4) is a bounded function, we can write

$$\tau_{yz}^{(1)}(x, h) = \frac{G_1 q_1(x)}{4h(G+1)} \int_{-a}^{a} \frac{ds}{\omega_*(s)\sinh[\mu(s-x)]} + R_1(x) \quad (|x| > a).$$

Here

$$q_1(x) = \frac{1}{h}\int_{-a}^{a} \frac{\omega_*(s)[f_1(s) + e^{\mu(s-x)} f_2(s)]\, ds}{\sinh[\mu(s-x)]} - C,$$
$$R_1(x) = \frac{G_1}{4h(G+1)^2}\int_{-a}^{a} \frac{(G+1)[q_1(s) - q_1(x)] + \{\cosh[\mu(s-x)] - 1\} q_1(s)}{\sinh[\mu(s-x)]}\, ds$$
$$- \frac{G_1}{2h(G+1)}\int_{-a}^{a} \frac{1 - \cosh[\mu(s-x)]}{\sinh[\mu(s-x)]} W_2'(s)\, ds.$$

Then, taking into account the value of the integral

$$\int_{-a}^{a} \frac{ds}{\omega_*(s)\sinh[\mu(s-x)]} = \frac{\mathrm{sign}(x)}{\mu\sqrt{|\sinh[\mu(a+x)]\sinh[\mu(a-x)]|}},$$

which, after the change of variables $\xi = \exp(2\mu s)$, $\eta = \exp(2\mu x)$, can be reduced to calculating the integral [9]

$$\int_a^b \frac{d\xi}{\sqrt{(\xi - a)(b - \xi)}} = \begin{cases} \frac{1}{\sqrt{(a-\eta)(b-\eta)}}, & \eta < a, \\ -\frac{1}{\sqrt{(\eta-a)(\eta-b)}}, & \eta > b, \end{cases}$$

we obtain the following formulas for determining the reduced stress intensity factors:

$$K_{\mathrm{III}}^{(1)}(\pm a) = \sqrt{2\pi} \lim_{x \to \pm a \pm 0} \sqrt{|x \mp a|}\, \tau_{yz}^{(1)}(x, h)$$
$$= \pm \frac{G_1\sqrt{h}}{\pi(G+1)\sqrt{\sinh(2\mu a)}} \left\{ \frac{1}{h}\int_{-a}^{a} \frac{\omega_*(s)[f_1(s) + e^{\mu(s-x)} f_2(s)],\, ds}{\sinh[\mu(s \mp a)]} - C \right\}.$$

Similarly, for the stress intensity factors at the end-points of the second crack located on the line $y = -h$, we have

$$K_{\mathrm{III}}^{(2)}(\pm a) = \sqrt{2\pi} \lim_{x \to \pm a \pm 0} \sqrt{|x \mp a|}\, \tau_{yz}^{(2)}(x, h)$$
$$= \pm \frac{G_2\sqrt{h}}{\pi(G+1)\sqrt{\sinh(2\mu a)}} \left\{ \frac{1}{h}\int_{-a}^{a} \frac{\omega_*(s)[G f_1(s) - e^{\mu(s-x)} f_2(s)]\, ds}{\sinh[\mu(s \mp a)]} + C \right\}.$$

Conclusions

The governing equations for antiplane problem of the elasticity theory for uniformly piecewise homogeneous space with tunnel parallel internal cracks are constructed. It is shown that, in the case where the cracks in heterogenous layers are identical, the closed solutions of the equations can be obtained. These solutions are constructed in the case where the identical central cracks are located in heterogenous layers. Simple formulas are obtained for the determining the stress intensity factors at the crack end-points.

References

[1] Panasyuk V, Savruk M, and Datsyshyn O 1976 *Distribution of Stresses near Cracks in Plates and Shells* (Kiev: Naukova Dumka) p 443 [in Russian]

[2] Bardzokas D, Filshtinsky L, and Filshtinsky M 2010 *Actual Problems of Coupled Physical Fields in Deformable Solids* vol 1 (Moscow, Izhevsk: NITs "RKhD") p 864 [in Russian]

[3] Mkrtchyan M and Mkhitaryan S 1993 On a stress state of compound elastic infinite body with periodic system of collinear cracks under longitudinal shear *Dokl. Nats. Akad. Nauk Armenii* **94** 104–10

[4] Hakobyan V 2002 The mixed boundary value problem for compound plane, weakened by periodic system of cracks *Dokl. Nats. Akad. Nauk Armenii* **102** 29–34

[5] Dashtoyan L and Hakobyan L 2011 Stress state of compound plane with periodic system of the absolutely rigid thin inclusions In *The Problems of Dynamics of Interaction of Deformable Media. Proc. of the VII Int. Conf. September 19–23, 2011, Goris - Stepanakert* (Yerevan: In-t Mekh. NAN RA) 172–6 [in Russian]

[6] Hakobyan V and Hakobyan L 2015 The stress state of piecewise homogeneous space with periodic system of parallel interphase defects In *Topical Problems of Continuum Mechanics: Proc. of the IV Int. Conf. September 21–26, 2015, Tsaghkadzor, Armenia* (Yerevan: Nats. Univ. Arkh. Stroit. Armenii) 28–32 [in Russian]

[7] Hakobyan V and Dashtoyan L 2015 *Doubly periodic problem for piecewise homogeneous plane with cracks* In *Topical Problems of Continuum Mechanics: Proc. of the IV Int. Conf. September 21–26, 2015, Tsaghkadzor, Armenia* (Yerevan: Nats. Univ. Arkh. Stroit. Armenii) 33–7 [in Russian]

[8] Muskhelishvili N 1966 *Some Fundamental Problems of Mathematical Elasticity Theory* (Moscow: Nauka) p 708 [in Russian]

[9] Prudnikov A P, Brychkov Yu A, and Marichev O I 1981 *Integrals and Series* (Moscow: Nauka) p 738 [in Russian]

IOP Conf. Series: Journal of Physics: Conf. Series **991** (2018) 012033 doi:10.1088/1742-6596/991/1/012033

Estimation of mountain slope stability depending on ground consistency and slip-slide resistance changes on impact of dynamic forces

H S Hayroyan[1*], S H Hayroyan[1], and K A Karapetyan[2***]**
[1]Yerevan State University, Yerevan, Armenia
[2]Institute of Mechanics of the National Academy of Sciences of Armenia, Yerevan, Armenia
E-mail: *hayroyan21@gmail.com, **hairoyan@ysu.am, ***koryan@sci.am

Abstract. In this paper, three types of clayish soils with different consistency and humidity properties and slip-slide resistance indexes are considered on impact of different cyclic shear stresses. The side-surface deformation charts are constructed on the basis of experimental data obtained testing cylindrical soil samples. It is shown that the fluctuation amplitude depends on time and the consistency index depends on the humidity condition in the soil inner contact and the connectivity coefficients. Consequently, each experiment is interpreted. The main result of this research is that it is necessary to make corrections in the currently active schemes of slip-hazardous slopes stability estimation, which is a crucial problem requiring ASAP solution.

1. Introduction

In consequent calculation systems, the real slope stability estimation can significantly be affected by changes of the slope soils resistance under the action of static or dynamic forces. One of the most important issues in engineering geology and soil mechanics is the determination of slip-slide resistance indexes of clayish soils on impact of static and dynamic forces. Note that the above-mentioned problem has not been illustrated properly in the few currently existing scientific papers [1–3]. The current construction norms define that, when estimating the slope stability in consequent calculation systems [4], the dynamic and seismic impacts should be considered together with additional inertial forces, but no possible deviations of the soil slip-slide resistance are mentioned in these norms.

The main goal of this work is experimentally to justify that, in some cases, it would be useful to take into account the changes in the soil slip-slide resistance initiated on dynamic impact when the real stability of the slope is estimated. To realize the above-mentioned goal, we shall investigate the sliding resistance characteristics in a wide range of conditions of cyclic changes in the slipping relative stress ($\tau/\tau_{f,st}$) and soil consistency (I_L).

2. Testing method

Three samples of clayish soils with different consistency and humidity properties taken at the Dilijan town International school mountain slope in Armenia were used. Comparing the hydro-physical properties of soils (table 1), one can see that the clayish soil samples consequently have

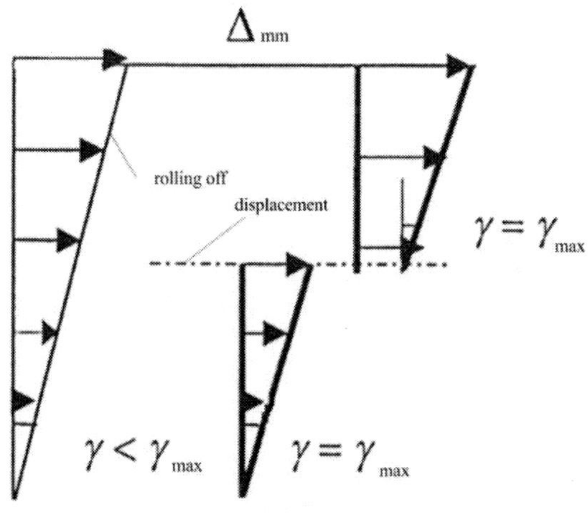

Figure 1. Dynamics of soil slip-deformation in the torsion test tool [5, 6].

Table 1. Experimentally determined hydro-physical indexes of soil samples.

Types	Specific gravity ρ g/cm^3	Soil humidity W	Mineral particles specific gravity ρ_s g/cm^3	Porosity coefficient e	Limits of plasticity			Consistency index $I_L = \frac{W-W_p}{W_L-W_p}$
					Liquid limit W_L	Plastic limit W_p	Plasticity index I_p	
1	1.98	0.155	2.67	0.579				0.045
2	1.95	0.186	2.67	0.644	0.281	0.149	0.132	0.280
3	1.89	0.245	2.67	0.759				0.727

1 — half-hard ($I_L = 0.045$), 2 — tight-plastic ($I_L = 0.280$), and 3 — soft-plastic ($I_L = 0.727$) consistency.

The soil tests were carried out for cylindrical samples with $d = 101$ mm and $h = 24$ mm. The samples were fixed in protecting rings of 2–4 mm height, which allowed tracing the process of slip-deformation.

Each sample participated in a specific dynamic process of slip-deformation during the testing [3, 5]:

- The soil hard, half-hard, tight-plastic ($I_L < 0.5$) consistency deformation under slip stresses up to τ_0 is manifested as aside-inclination (figure 1), but starting from $\tau_0 > \tau$, the critical deformation starts formation of a localized slip-zone, where the further process of slipping evolves (figure 1). Moreover, the soil does not deform above and below this zone.
- In the case of soft-plastic, flow-plastic ($I_L > 0.5$) consistency, at any value of the clayish soil deformation, the slip-stress takes place along the whole height in the form of aside-inclination according to Bingham–Shvedov viscous-plastic or Newton viscous body models [5, 6] (figure 1).

Experiments aimed at determining the clayish soil slip-stress resistance under static and dynamic forces were carried out using the upgraded M-5 torsion test tool (figure 2). In

Figure 2. Upgraded M-5 torsion test tool.

addition to the scrolling static moment, the tool also provides simultaneous action of stresses of frequency $f = 1$–$20\,\text{Hz}$ and shear alternating cyclic stresses of several values.

Five cyclic experiments with cylindrical test-samples of clayish soil were carried out according to the method described below.

First, the samples experienced the action of condensing normal stresses of magnitude $\sigma = 0.1$, 0.2, and $0.3\,\text{MPa}$, respectively, and then the initial shear τ_0 stresses are applied of the following value for each experiment:

1-series (soil type: Nos. 1 and 2, half-hard, tight-plastic): $\tau_0 = 0.5\tau_{f,st}$ ($\tau_{f,st}$ is the soil shear standard resistance);

2-series (soil type: Nos. 1 and 2, half-hard, tight-plastic): $\tau_0 = 0.77\tau_{f,st}$;

3-series (soil type: Nos. 1 and 2, half-hard, tight-plastic): $\tau_0 = 0.81\tau_{f,st}$;

4-series (soil type: No. 3, soft-plastic): $\tau_0 = 0.4\tau_{f,st}$;

5-series (soil type: No. 3, soft plastic): $\tau_0 = 0.650\tau_{f,st}$.

In the whole, 45 experiments were carried out in 5 series, and each of them consisted of 9 experiments.

After stabilization of the slipping process, the samples undergo the action of shear dynamic alternating stresses of several frequencies whose values were equal to $\Delta\tau = \pm(\tau_{f,st} - \tau_0)$, and then the obtained data were registered after compilation in each 10, 30, 60, 120, and 360 seconds.

3. Interpretation and discussion of obtained data

Now let us interpret the experimental data only in 6 cases where the cylindrical soil test-samples experience the action of $\sigma = 0.2\,\text{MPa} = \text{const}$ (half-hard and tight-plastic soils) and $\sigma = 0.1\,\text{MPa} = \text{const}$ (soft-plastic soils) static normal stresses.

Figure 3 presents 6 charts of the time (t) dependence of the aside-deformation fluctuation amplitude of cylindrical test-samples. The data were registered after stabilization of the initial deformation on impact of several frequency cyclic $\pm\Delta\tau$ relative stresses ($\tau_0/\tau_{f,st}$).

According to figures 3*a* and 3*c*, which are plotted for half-hard soil No. 1 (table 1) and plastic type No. 2 (table 1), when $\sigma = 02\,\text{MPa}$ and $\tau_0/\tau_{f,st} = 0.5$, during the initial time of cyclic changing stresses, it is observed that the slipping deformation amplitude gradually increases. This means that the soil's ability to resist the shear decreases. Moreover, as the amplitude growth indicates, its intensity and the time also significantly depend on the soil type, and dynamic shear $\pm\Delta\tau$ stresses, on the amplitude influencing the test sample. Parallel to the

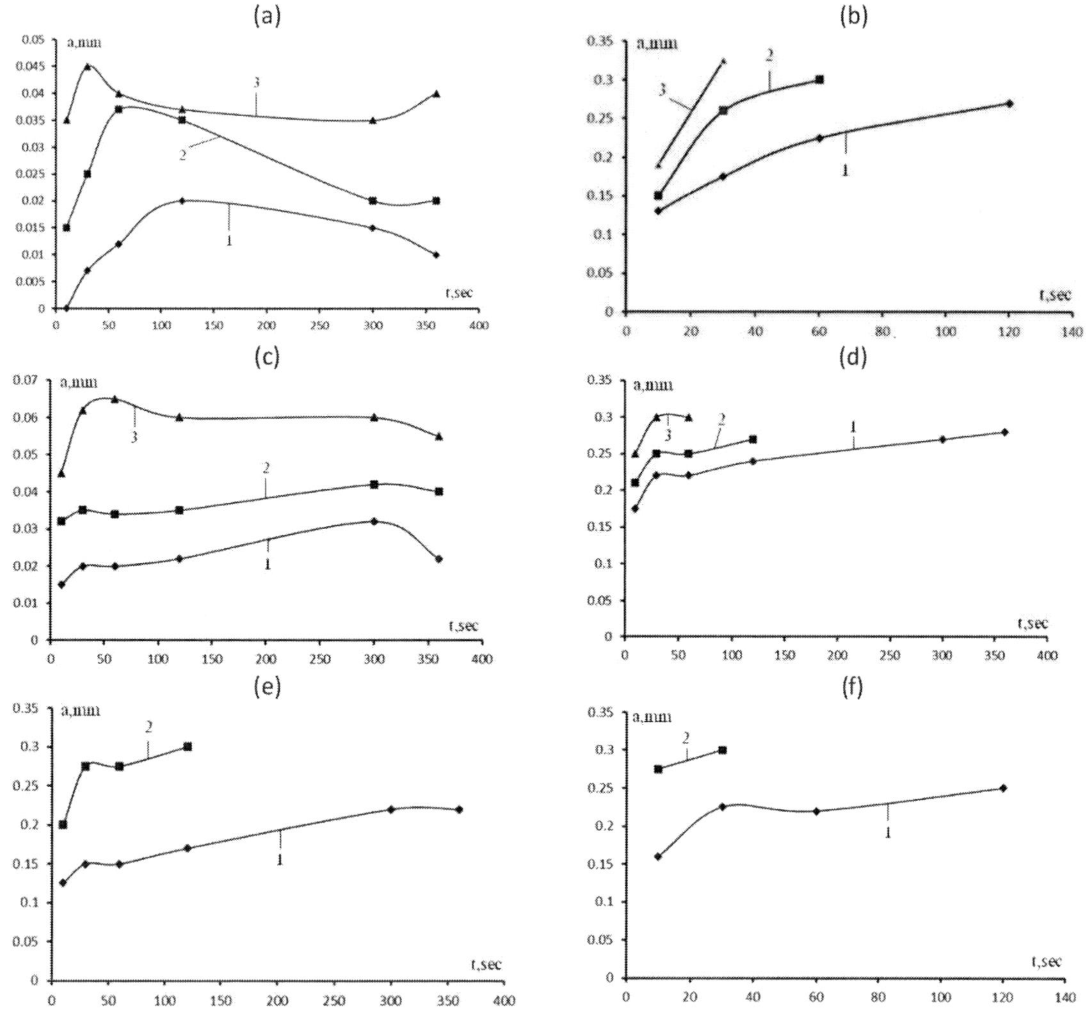

Figure 3. (a) and (b): half-hard soil's physical properties for $I_L = 0.0465$ ($w = 0.155$), $\sigma = 0.2\,\text{MPa}$ (*1* — $f = 8\,\text{Hz}$, *2* — $f = 10\,\text{Hz}$, *3* — $f = 12\,\text{Hz}$); (c) and (d): tight-plastic consistency soil with $I_L = 0.280$ ($w = 0.186$), $\sigma = 0.2\,\text{MPa}$ (*1* — $f = 8\,\text{Hz}$, *2* — $f = 10\,\text{Hz}$, *3* — $f = 12\,\text{Hz}$); (e) and (f): soft-plastic consistency soil with $I_L = 0.720$ ($w = 0.255$), $\sigma = 0.1\,\text{MPa}$ (*1* — $f = 6\,\text{Hz}$, *2* — $f = 8\,\text{Hz}$).

increase in the dynamic influence, a time decrease in slipping deformation amplitude is observed here, which is independent of the dynamic shear stress amplitude. Then, in the case of half-hard test-sample, when $f = 8\,\text{Hz}$, the deformation decreasing process persists, but when $f = 10$ and $12\,\text{Hz}$, the deformation ascending trend is observed (figure 3a). The initial growth of the deformation amplitude is observed after the above-mentioned deformation value decreases, as the time increases, and then it decreases independently of the value of the frequency of shear stresses in the case of a tight-plastic soil (figure 3c). The shear deformation is expressed via the aside-inclination without formation of the localized shear zone and, at the end of experiment, no decrease in shear strength is observed on the surfaces of the above-mentioned two test samples.

Let us consider the cases where the half-hard and tight-plastic soil samples were tested for $\sigma = 0.2\,\text{MPa}$ under normal stresses. First, the stresses $\tau_0/\tau_{f,st} = 0.81$, $\tau_0/\tau_{f,st} = 0.77$

Table 2. Soil slipping resistance indexes obtained by processing the data in figure 4.

Soil Type No.	Humidity W	Index of consistency I_L	Coefficient of internal friction: Under static conditions, $\tan\varphi_{st}$	Under dynamic conditions, $\tan\varphi_{dyn}$	Cohesion: Under static conditions, C_{st}, MPa	Under dynamic conditions, C_{dyn}, MPa
1	0.155	0.045	0.421	0.379	0.018	0.015
2	0.186	0.280	0.338	0.257	0.016	0.013
3	0.245	0.727	0.224	0.124	0.018	0.005

were respectively implemented on these test-samples, then after stabilization of shear deformation $\pm\Delta\tau$, the cyclic shear strains were transmitted (figures 3*b* and 3*d*). According to these charts, the shear deformation amplitudes for these samples have the same qualitative character regardless of the frequency of the used transmitter. An increase in the cyclic tension can result in a monotonous increase in the dynamic deformation amplitudes of slide, leading to a decrease in the test-sample resistance to slip. Quantitative comparisons show that both the intensities of variations in the time of dynamic deformation and the durability of resistance to slip substantially depend on the range of soil and the frequency f of the dynamic torque (figures 1*b* and 1*d*).

Now let us consider the experimental data on the soft-plastic consistency of clayish soils (soil type No. 3; table 1; figures 3*e* and 3*f*). According to these charts in the case of above-mentioned two cyclic conditions due to the increased duration of dynamic stress, the slide amplitude grows practically monotonically, which finally leads to the end of the soil resistance to slip without formation of a localized shear zone. In the course of time, the intensity of the slide amplitude growth and the time of decrease in the resistance depend on the characteristic values of $\tau_0/\tau_{f,st}$ and $\pm\Delta\tau$, and f. In this case, the test-values $\sigma = 0.1$ MPa, the ratios $\tau_0/\tau_{f,st} = 0.4$ and 0.6, and the frequencies of fluctuations of dynamic shear stresses $\pm\Delta\tau$ were, respectively, equal to $f = 6$ Hz and $f = 8$ Hz.

In the tests in the case of sliding resistance–normal tension dependence, all test-samples practically show linear functional independence of static or dynamic characteristics of the slip-stresses impact, as this is obvious from the charts in figure 4.

It is obvious that the values of the soil inner coefficient of internal friction and the connectivity values of slipping dynamic stress impact are significantly smaller than the same indexes under static conditions. This difference becomes more profound when the consistency index increases. So the differences for the inner contact indexes of half-hard, tight-plastic, and soft-plastic soils are, respectively, approximately about 21%, 24%, and 47%, and for connectivity, 17%, 19%, and 72%.

The data in table 2 and figure 4 allow one to obtain the following empirical expressions for the soil slipping resistance ($\tau_{f,st} = \sigma\tan\varphi + c$).

1. Half-hard consistency of soil: when ($I_L = 0.045$), then

$$\tau_{f,st} = 0.481\sigma + 0.018, \quad \tau_{f,dyn} = 0.379\sigma + 0,015.$$

2. Tight-plastic consistency of soil: when ($I_L = 0.280$), then

$$\tau_{f,st} = 0.338\sigma + 0.016, \quad \tau_{f,dyn} = 0.2657\sigma + 0.013.$$

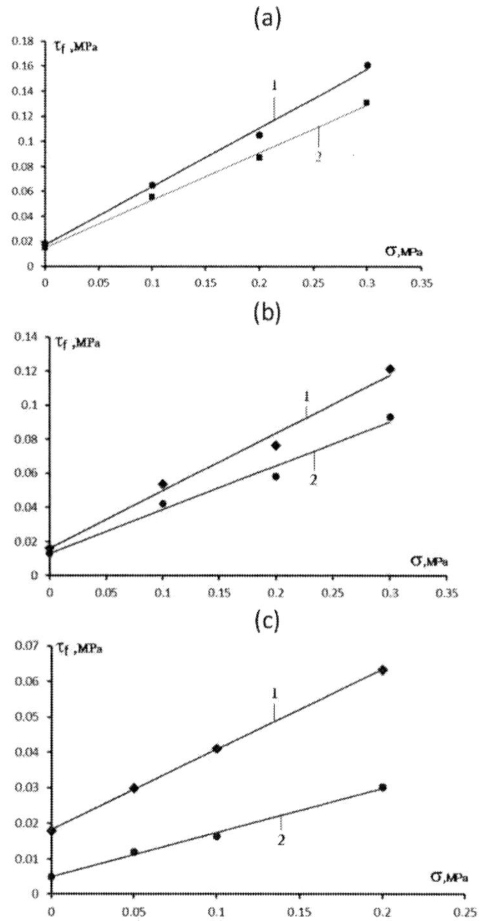

Figure 4. Relationship between the soil slip-resistance and condensing normal stresses. (a), (b), and (c) charts subject to half-hard, tight-plastic and soft-plastic consistent soils, respectively (*1* — static experiments data, *2* — dynamic experiments data).

3. Soft-plastic consistency of soil: when ($I_L = 0.727$), then

$$\tau_{f,st} = 0.224\sigma + 0.018, \quad \tau_{f,dyn} = 0.124\sigma + 0.005.$$

Under the conditions of static and dynamic slipping stress impact, the charts of functional dependence between the clayish soil inner coefficient of internal friction ($\tan\varphi$), cohesion (C), humidity (W), and consistency (I_L) are presented in figure 5. These charts are based on the data given in table 2.

The obtained data, which are illustrated in figure 5, can be described sufficiently well by the following empirical expressions.

1. Inner coefficient of internal friction–consistency index dependence:

$$\tan\varphi_{st} = 0.4865e^{-1.094IL}, \quad \tan\varphi_{dyn} = 0.4074e^{-1.637IL}.$$

2. Inner coefficient of internal friction–humidity dependence:

$$\tan\varphi_{st} = 0.0216W^{-1.665}, \tan\varphi_{dyn} = 0.004W^{-2.257}.$$

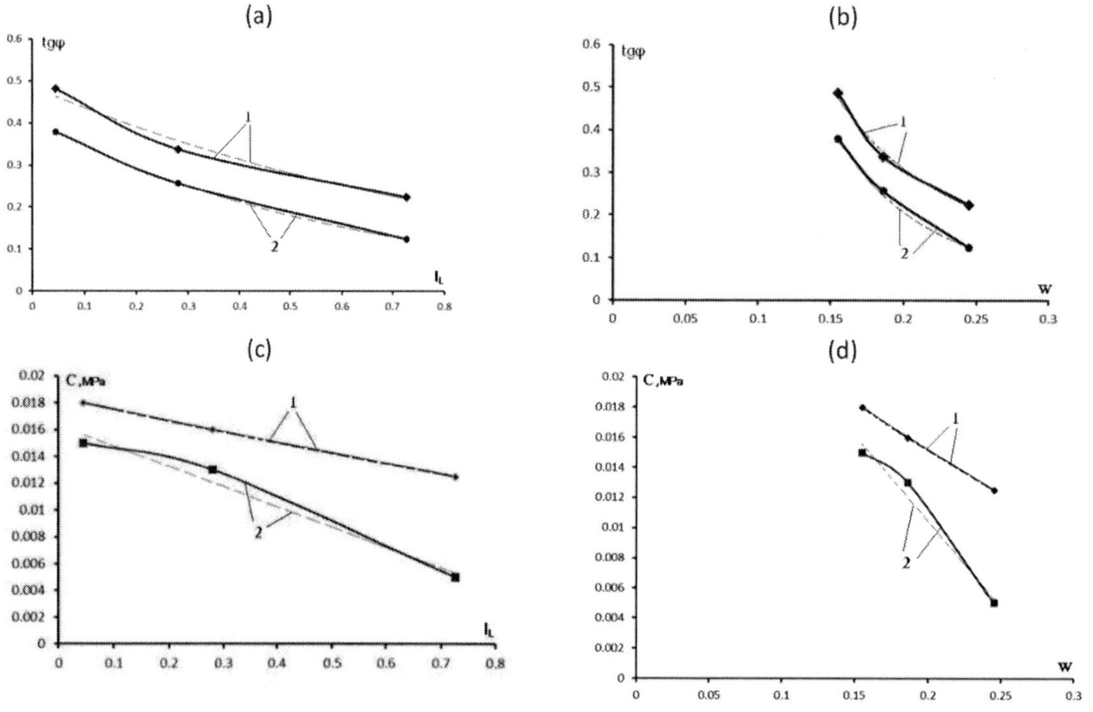

Figure 5. Dependence between: (*a*) soil inner coefficient of internal friction C and consistency I_L, (*b*) soil inner coefficient of internal friction $\tan\varphi$ and humidity W, (*c*) soil inner coefficient of internal friction $\tan\varphi$ and consistency I_L, and (*d*) cohesion C and humidity W under static (*1*) and dynamic (*2*) loadings.

3. Cohesion index–consistency index:

$$C_{st} = -0.0088 I_L + 0.0184, \quad C_{dyn} = -0.0151 I_L + 0.163.$$

4. Cohesion index–humidity index:

$$C_{st} = -0.0668 W + 0.0284, \quad C_{dyn} = -0.1143 W + 0.0333.$$

Conclusion

The variations in the slope fluctuations due to different causes can lead to a decrease in the ability to resist the slope whose value depends on the soil consistency index (IL), the value of slide-relative static stresses ($\tau_0/\tau_{f,st}$), and the frequency (f) of the shear dynamic stress fluctuation ($\pm\Delta\tau$). Moreover, for

- $\tau_0/\tau_{f,st} < 0.7$ and $f < 12$ Hz, half-hard ($0 < I_L < 0.25$) and tight-plastic ($0.25 < I_L < 0.50$) soils, no consumption of resistance occurs in consistent soils;
- $\tau_0/\tau_{f,st} > 0.7$ and $f < 12$ Hz, consumption of resistance occurs in the above-mentioned soils, and the time of attaining this point decreases $\tau_0/\tau_{f,st}$ while the frequency (f) increases;
- $\tau_0/\tau_{f,st} > 0.4$ and $f > 6$ Hz, consumption of resistance occurs in the soft-plastic consistency soils ($I_L > 0.50$) and, in this case, the higher the values of ($\tau_0/\tau_{f,st}$) and f, the smaller the soil resistance to sliding.

IOP Conf. Series: Journal of Physics: Conf. Series **991** (2018) 012033 doi:10.1088/1742-6596/991/1/012033

Thus, on the basis of the results obtained in direct experiments and the above-described results, it can be concluded that to adjust the accuracy of calculations of the landslide slopes stability is a crucial problem which urgently requires to be solved. This problem can be solved by coordinating the well-known data and new researches aimed at this target.

References

[1] Voznesensky E A 1997 *The Behavior of Soils under Dynamic Loads* (Moscow: Izdat. MGU) p 286
[2] Krasnikov N D 1970 *Dynamic Properties of Soil and Methods of Their Determination* (Moscow: Stroyizdat) p 240
[3] Trofimov V T, Korolev V A, Voznesensky E A, et al. 2005 *Soil Science* (Moscow: Izdat. MGU) p 1024
[4] *Building Regulations RA (SNRA) II-6.02.-2006. Seismic Resistant Construction.*
[5] Meschyan S R 2008 *Experimental Rheology of Clayey Soil* (Yerevan: Gitutyun NAN RA) p 495
[6] Ter-Martirosyan Z G and Proshin M V 2002 Short- and long-term slope stability *Soil Mech. Foundat. Engng* **39** (2) 45–49

IOP Conf. Series: Journal of Physics: Conf. Series **991** (2018) 012034 doi:10.1088/1742-6596/991/1/012034

Boundary-element modelling of dynamics in external poroviscoelastic problems

L A Igumnov[1], S Yu Litvinchuk[1*], A A Ipatov[1], and A N Petrov[2]
[1]Research Institute for Mechanics, Lobachevsky State University of Nizhni Novgorod, Nizhny Novgorod, Russia
[2]Research and Education Center "Materials," Don State Technical University, Rostov-on-Don, Russia

E-mail: *litvinchuk@mech.unn.ru

Abstract. A problem of a spherical cavity in porous media is considered. Porous media are assumed to be isotropic poroelastic or isotropic poroviscoelastic. The poroviscoelastic formulation is treated as a combination of Biot's theory of poroelasticity and elastic-viscoelastic correspondence principle. Such viscoelastic models as Kelvin–Voigt, Standard linear solid, and a model with weakly singular kernel are considered. Boundary field study is employed with the help of the boundary element method. The direct approach is applied. The numerical scheme is based on the collocation method, regularized boundary integral equation, and Radau stepped scheme.

1. Introduction

Various types of interactions in advanced dispersed media, such as porous or viscous media, are of great interest in many disciplines. The wave propagation in porous/viscous media is an important field of geophysics, geomechanics, geotechnical engineering, etc. The study of wave propagation processes in saturated porous media originates from Y. I. Frenkel [1] and M. Biot [2, 3]. The implementation of the solid viscoelastic effects in the theory of poroelasticity was first introduced by Biot [4].

The ever-growing interest of the research community in the wave propagation in porous media requires extension of Biot's theory to the case of porous media saturated with several fluid fillers or porous media with viscous effects.

There are two major approaches to dynamic processes modeled by the boundary element method (BEM): solving the system of boundary integral equations (BIE) directly in time or in the Laplace or Fourier transform domain followed by the respective transform inversion. The Laplace transform is the main method for dealing with the transient response of porous media. A classical three-dimensional formulation is employed for the study of the influence of viscoelastic parameters on dynamic responses in poroviscoelastic solids. The present paper is dedicated to the development of numerical modeling technique based on the BEM usage in the Laplace domain for solving 3D poroviscodynamic problems.

2. Mathematical model

The mathematical model can be written according to the Schanz and Li [5–7] approach. A poroelastic medium is represented by using the following mathematical model of heterogeneous material: an elastic matrix phase and two filler phases — a liquid and a gas filling the pore system. All three phases are assumed to be compressible. The temperature variations are neglected. Such a poroelastic material is classified as partially saturated, and its model is called a "three-phase" model. The porosity φ and the degree of the material saturation by fluid S_{w} and by gas S_{a} are defined as

$$\varphi = \frac{V_{\mathrm{void}}}{V}, \quad S_{\mathrm{w}} = \frac{V_{\mathrm{w}}}{V_{\mathrm{void}}}, \quad S_{\mathrm{a}} = \frac{V_{\mathrm{a}}}{V_{\mathrm{void}}}, \tag{1}$$

where V_{void} is the volume of interconnected pores in the specimen, V is the total volume of the material, V_{w} is the fluid fuller volume, and V_{a} is the gas filler volume. In the present paper, we consider the case where the pores are filled completely with the filler:

$$S_{\mathrm{a}} + S_{\mathrm{w}} = 1. \tag{2}$$

An important characteristic of the three-phase media state is the capillary pressure p^{c}. The capillary pressure is determined as a difference between the pore pressure of gas p^{a} and the pore pressure of liquid p^{w} and can be represented as a function of saturation degree

$$p^{\mathrm{c}} = p^{\mathrm{a}} - p^{\mathrm{w}} = p^{\mathrm{d}} S_{\mathrm{e}}^{-1/\theta}, \quad S_{\mathrm{e}} = \frac{S_{\mathrm{w}} - S_{\mathrm{rw}}}{S_{\mathrm{ra}} - S_{\mathrm{rw}}}, \tag{3}$$

where S_{e} is the effective water-saturation, S_{rw} is the residual water-saturation, S_{ra} is the residual gas saturation, p^{d} is the gas pressure required for driving the liquid out of the pores, and $\theta \in [0.2, 3]$ is the pore size distribution coefficient. The phase permeabilities κ_{w} and κ_{a} are defined as

$$\kappa_{\mathrm{w}} = \frac{K_{\mathrm{rw}} k}{\eta_{\mathrm{w}}}, \quad \kappa_{\mathrm{a}} = \frac{K_{\mathrm{ra}} k}{\eta_{\mathrm{a}}}, \tag{4}$$

where k is the poroelastic material permeability, K_{rw}, K_{ra} is the relative phase permeability, and η_{w}, η_{a} are the filler viscosities. The constitutive relations for the water saturation, as well as the water and air relative permeabilities, are assumed to be based on that proposed by Brooks and Corey [8] as

$$K_{\mathrm{rw}} = S_{\mathrm{e}}^{(2+3\theta)/\theta}, \quad K_{\mathrm{ra}} = (1 - S_{\mathrm{e}})^2 (1 - S_{\mathrm{e}}^{(2+\theta)/\theta}). \tag{5}$$

The governing equations for partially saturated poroelastic material, according to the effective stress principal, are following:

$$\sigma_{ij} = \left(K - \frac{2}{3}G\right)\delta_{ij} u_{k,k} + G(u_{i,j} + u_{j,i}) - \delta_{ij}\alpha(S_{\mathrm{w}} p^{\mathrm{w}} + S_{\mathrm{a}} p^{\mathrm{a}}), \tag{6}$$

where $\alpha = 1 - K/K_{\mathrm{s}}$ is the effective stress coefficient, K and G are elastic moduli of the material, and K_{s} is the bulk modulus of skeleton grains. The strain tensor ε_{ij} and the displacements u_i are connected by the relation

$$\varepsilon_{ij} = \frac{1}{2}(u_{i,j} + u_{j,i}). \tag{7}$$

The use of the Laplace transform allows us to rewrite the equations of porous media dynamics in terms of solid displacement $\hat{u}_i$ and pore fillers $\hat{p}^{\mathrm{w}}$ and $\hat{p}^{\mathrm{a}}$ in the Laplace domain with the transform parameter s

$$G\hat{u}_{i,jj} + \left(K + \frac{G}{3}\right)\hat{u}_{j,ij} - (\rho - \beta S_{\mathrm{w}}\rho_{\mathrm{w}} - \gamma S_{\mathrm{a}}\rho_{\mathrm{a}})s^2\hat{u}_i - (\alpha - \beta)S_{\mathrm{w}}\hat{p}^{\mathrm{w}}_{,i} - (\alpha - \gamma)S_{\mathrm{a}}\hat{p}^{\mathrm{a}}_{,i} = 0, \tag{8}$$

$$-(\alpha - \beta)S_{\mathrm{w}} s\hat{u}_{i,i} - \left(\zeta S_{\mathrm{ww}} S_{\mathrm{w}} + \frac{\varphi}{K_{\mathrm{w}}} S_{\mathrm{w}} - S_{\mathrm{u}}\right) s\hat{p}^{\mathrm{w}} + \frac{\beta S_{\mathrm{w}}}{\rho_{\mathrm{w}} s}\hat{p}^{\mathrm{w}}_{,ii} - (\zeta S_{\mathrm{aa}} S_{\mathrm{w}} + S_{\mathrm{u}}) s\hat{p}^{\mathrm{a}} = 0, \tag{9}$$

$$-(\alpha - \gamma)S_{\mathrm{a}} s\hat{u}_{i,i} - (\zeta S_{\mathrm{ww}} S_{\mathrm{a}} + S_{\mathrm{u}}) s\hat{p}^{\mathrm{w}} - \left(\zeta S_{\mathrm{aa}} S_{\mathrm{a}} + \frac{\varphi}{K_{\mathrm{a}}} S_{\mathrm{a}} - S_{\mathrm{u}}\right) s\hat{p}^{\mathrm{a}} + \frac{\gamma S_{\mathrm{a}}}{\rho_{\mathrm{a}} s}\hat{p}^{\mathrm{a}}_{,ii} = 0, \tag{10}$$

where

$$\zeta = \frac{\alpha-\varphi}{K_{\rm s}}, \quad S_{\rm ww} = S_{\rm w} - \theta(S_{\rm w}-S_{\rm rw}), \quad S_{\rm aa} = S_{\rm a} + \theta(S_{\rm w}-S_{\rm rw}), \quad b_{\rm w} = \frac{\varphi}{K_{\rm w}}, \tag{11}$$

$$b_{\rm a} = \frac{\varphi}{K_{\rm a}S_{\rm u}} = -\frac{\theta(S_{\rm ra}-S_{\rm rw})}{p^{\rm d}}\left(\frac{S_{\rm w}-S_{\rm rw}}{S_{\rm ra}-S_{\rm rw}}\right)^{(\theta+1)/\theta}, \quad \rho = (1-\varphi)\rho_{\rm s} + \varphi S_{\rm w}\rho_{\rm w} + \varphi S_{\rm a}\rho_{\rm a}, \tag{12}$$

$$\beta = \frac{\kappa_{\rm w}\varphi\rho_{\rm w}s}{\varphi S_{\rm w} + \kappa_{\rm w}\rho_{\rm w}s}, \quad \gamma = \frac{\kappa_{\rm a}\varphi\rho_{\rm a}s}{\varphi S_{\rm a} + \kappa_{\rm a}\rho_{\rm a}s}. \tag{13}$$

In the case of full saturation, equation (10) is eliminated, and equations (8) and (9) become

$$G\hat{u}_{i,jj} + \left(K + \frac{G}{3}\right)\hat{u}_{j,ij} - (\alpha-\lambda)\hat{p}^{\rm w}_{,i} - s^2(\rho - \lambda\rho_{\rm w})\hat{u}_i = 0, \tag{14}$$

$$\frac{\lambda}{s\rho_{\rm f}}\hat{p}^{\rm w}_{,ii} - \frac{\varphi^2 s}{R}\hat{p}^{\rm w} - (\alpha-\lambda)s\bar{u}^{\rm w}_{i,i} = 0, \tag{15}$$

$$\lambda = \frac{\kappa_{\rm w}\rho_{\rm w}\varphi^2 s^2}{\varphi^2 s + s^2\kappa_{\rm w}(\rho_{\rm g} + \varphi\rho_{\rm w})}. \tag{16}$$

Equations (14)–(15) describe a dynamic boundary-value problem of three dimensional isotropic poroelasticity in the Laplace domain with the boundary conditions

$$\mathbf{v}(\mathbf{x},s) = \tilde{\mathbf{v}} \quad (\mathbf{x}\in S^{\rm u}); \quad \mathbf{t}(\mathbf{x},s) = \tilde{\mathbf{t}} \quad (\mathbf{x}\in S^{\sigma}), \tag{17}$$

where $\mathbf{v} = (\hat{u}_1, \hat{u}_2, \hat{u}_3, \hat{p}^{\rm w}, \hat{p}^{\rm a})$ is the generalized displacement vector, $\mathbf{t} = (\hat{t}_1, \hat{t}_2, \hat{t}_3, -\hat{q}^{\rm w}, -\hat{q}^{\rm a})$ is the generalized traction vector, $\hat{t}_i = \hat{\sigma}_{ij}n_j$ are tractions, $\hat{q}^{\rm w} = -\beta\rho_{\rm w}^{-1}(\hat{p}^{\rm w}_{,i} + \rho_{\rm w}s^2\hat{u}_i)n_i/s$ is the fluid flow, $\hat{q}^{\rm a} = -\gamma\rho_{\rm a}^{-1}(\hat{p}^{\rm a}_{,i} + \rho_{\rm a}s^2\hat{u}_i)n_i/s$ is the gas flow, $\mathbf{n}$ is the outer normal vector, $S^{\rm u}$ denotes the Dirichlet boundary, and S^{σ} denotes the Neumann boundary.

The poroviscoelastic solution is obtained from the poroelastic solution by using the elastic-viscoelastic correspondence principle applied to the skeleton elastic constants K and G in the Laplace domain. The quantities $\hat{K}(s)$ and $\hat{G}(s)$ are calculated by the formulas

- $\hat{K}(s) = K\left(1 + \frac{s}{\gamma_1}\right)$, $\hat{G}(s) = G\left(1 + \frac{s}{\gamma_1}\right)$ — Kelvin–Voigt model;
- $\hat{K}(s) = K\left[(\beta_0 - 1)\frac{s}{s+\gamma_2} + 1\right]$, $\hat{G}(s) = G\left[(\beta_0 - 1)\frac{s}{s+\gamma_2} + 1\right]$ — standard linear solid model;
- $\hat{K}(s) = \frac{K}{1 + hs^{\alpha_0 - 1}}$, $\hat{G}(s) = \frac{G}{1 + hs^{\alpha_0 - 1}}$ — model with weakly singular kernel of Abel type;

where γ_1, γ_2, β_0, h, and α_0 are model parameters [9, 10].

3. Boundary-element approach

The integral representation and BIE with integral Laplace transform are used. The fundamental and singular solutions are considered in terms of singularity isolation. Numerical scheme is based on the Green–Betti–Somigliana formula.

Boundary-value problem (14) can be reduced to the BIE system as follows [10, 11]:

$$\frac{1-\alpha_\Omega}{2}v_i(\mathbf{x},s) + \int_\Gamma [T_{ij}(\mathbf{x},\mathbf{y},s)v_i(\mathbf{y},s) - T^0_{ik}(\mathbf{x},\mathbf{y},s)v_i(\mathbf{x},s) - U_{ij}(\mathbf{x},\mathbf{y},s)t_i(\mathbf{y},s)]\,d\Gamma = 0, \tag{18}$$

$\mathbf{x},\mathbf{y}\in\Gamma$,

where U_{ij}, T_{ij} are fundamental and singular solutions, T_{ij}^0 contains the isolated singularities, and $\mathbf{x} \in \Gamma$ is an arbitrary point. The coefficient α_Ω equals 1 in the case of finite domain and -1 in the case of infinite domain.

The boundary surface of our homogeneous solid is discretized by quadrangular and triangular elements, and the triangular elements are assumed to be singular quadrangular elements. We use the following reference elements: square $\xi = (\xi_1, \xi_2) \in [-1, 1]^2$ and triangle $0 \le \xi_1 + \xi_2 \le 1$, $\xi_1 \ge 0$, $\xi_2 \ge 0$, and each boundary element is mapped to a reference one by the formula

$$y_i(\xi) = \sum_{l=1}^{8} N^l(\xi) y_i^{\beta(k,l)}, \quad i = 1, 2, 3, \tag{19}$$

where l is the local node number in the element k, $\beta(k,l)$ is the global node number, and $N^l(\xi)$ are shape functions. Goldstein's displacement-stress mixed model is performed. To discretize the boundary surface, eight-node biquadratic quadrilateral elements are used, and generalized displacements and tractions are approximated by linear and constant shape functions, respectively.

Subsequent application of the collocation method leads to a system of linear equations. As the collocation nodes, we take the approximation nodes of boundary functions. The Gaussian quadratures are used to calculate integrals on regular elements. But if an element contains a singularity, then the algorithm of singularity avoiding or order reducing is applied. When the singularity is eliminated, we use an adaptive integration algorithm. An appropriate order of Gaussian quadrature is chosen from the primarily known necessary precision; if this is impossible, then the element is recursively subdivided into smaller elements.

Solving the system of linear equations leads to solving the initial boundary-value problem in the Laplace domain. The obtained solution should be inverted to the time domain. For this purpose, we use a stepped scheme based on Runge-Kutta nodes.

4. Stepped scheme based on Runge-Kutta nodes

For the Laplace transform

$$\hat{f}(s) = \int_0^\infty f(t) e^{-st}\, dt, \tag{20}$$

we consider a stepped scheme of numerical Laplace transform inversion based on Runge-Kutta nodes. Substituting the basic linear multistep method for the Runge-Kutta method and taking Butcher tableau into account, we have

$$\frac{\mathbf{c}|\mathbf{A}^T}{\mathbf{b}^T}, \quad \mathbf{A} \in R^{m\times m}, \quad \mathbf{b}, \mathbf{c} \in R^m. \tag{21}$$

The stepped method of numerical Laplace inversion is based on the integration theorem:

$$w(t) = \int_0^t f(\tau)\, d\tau. \tag{22}$$

In order to acquire a function in the time domain, we consider the following system of relations [12]:

$$w(0) = 0, \quad w(n\Delta t) = \mathbf{b}^T \mathbf{A}^{-1} \sum_{k=1}^{n} \omega_k(\Delta t), \quad n = 0, 1, \ldots, N, \tag{23}$$

$$\omega_k(\Delta t) = \frac{R^{-n}}{L} \sum_{l=0}^{L-1} \bar{f}(s) s e^{-inl\frac{2\pi}{L}}, \quad s = \frac{\chi(z)}{\Delta t}, \quad z = Re^{il\frac{2\pi}{L}}, \tag{24}$$

$$\chi(z) = \mathbf{A}^{-1} - z\mathbf{A}^{-1}\mathbf{I}\mathbf{b}^T\mathbf{A}^{-1}, \quad \mathbf{I} = (1, \ldots, 1)^T. \tag{25}$$

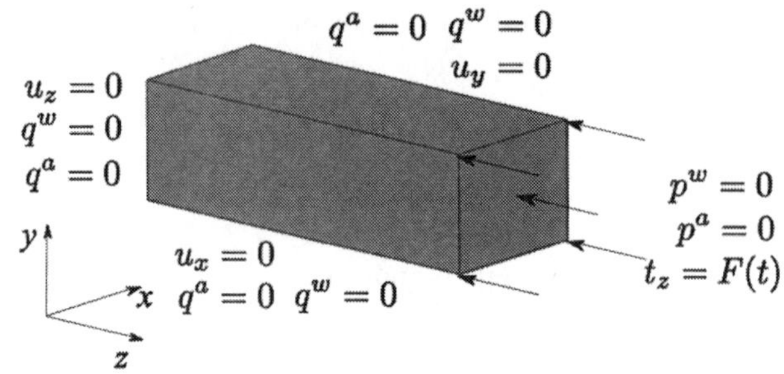

Figure 1. Geometry and boundary conditions of a partially saturated poroelastic column.

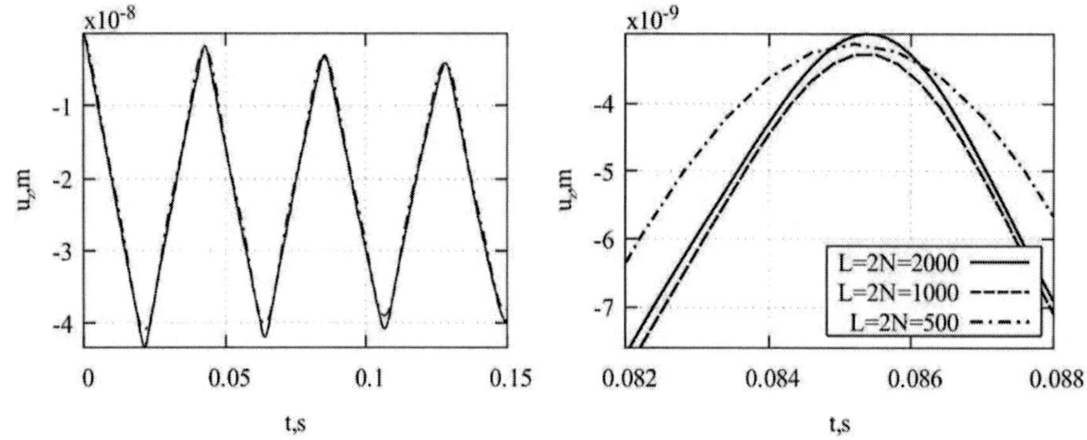

Figure 2. Displacement u_z for different L and N at the loaded end.

The Radau (Radau IIA) scheme is chosen for the further representation of the Runge-Kutta method.

5. Numerical results

5.1. Partially saturated poroelastic column

A problem of axial force $F(t) = 1\text{N/m}^2$ acting on a 3m-long 1m-wide prismatic poroelastic body with a rigidly fixed end is considered [13]. The boundary-value problem is presented in figure 1. The parameters of the partially saturated porous material are: $\varphi = 0.3$, $\rho_s = 2000\,\text{kg/m}^3$, $\rho_w = 1000\,\text{kg/m}^3$, $\rho_a = 1.22\,\text{kg/m}^3$, $K = 1\times10^8\,\text{N/m}^2$, $G = 2.14\times10^7\,\text{N/m}^2$, $K_s = 1.4\times10^{10}\,\text{N/m}^2$, $K_w = 4.3\times10^9\,\text{N/m}^2$, $K_a = 1\times10^5\,\text{N/m}^2$, $k = 4.6\times10^{-12}\,\text{m}^2$, $\eta_w = 1\times10^{-3}\,\text{N·s/m}^2$, $\eta_a = 1\times10^{-3}\,\text{N·s/m}^2$, $p^d = 2.25\times10^5\,\text{N/m}^2$, $\theta = 3$, $S_w = 0.9$, $S_{rw} = 0.3966$, and $S_{ra} = 1$.

A boundary-element mesh is employed for calculations. The BE-mesh consists of 504 quadrangular elements. The dynamic responses of displacements, fluid pore pressure, and air pore pressure in the case of different parameters of the stepped scheme are presented in figures 2 and 3.

5.2. Problem of a cavity in poroviscoelastic media

A problem of cavity in poroelastic/poroviscoelastic media is considered (figure 4). The cavity is subjected to the unit normal internal pressure. The cavity radius is $R = 5\,\text{m}$, the origin of the coordinate system is in the center of the cavity. The unknown functions are the displacements

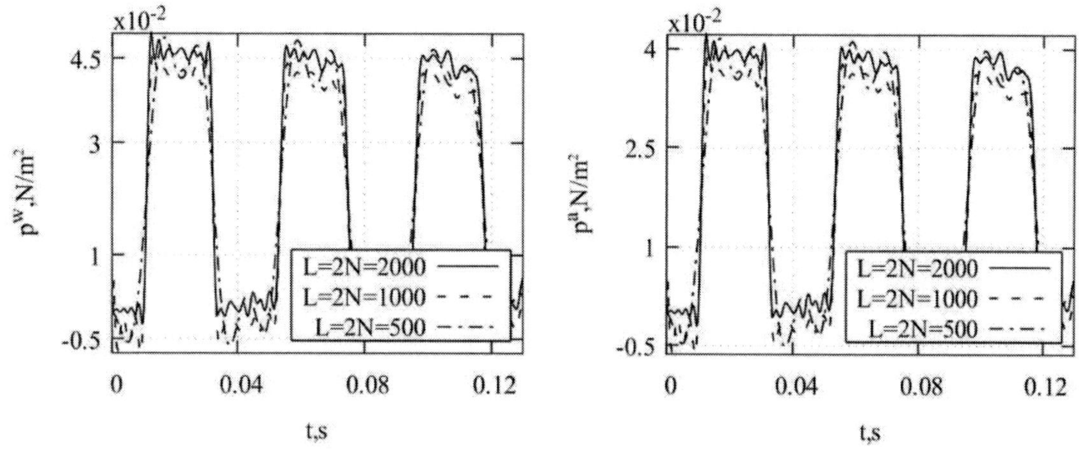

Figure 3. Pressures p^{w}, p^{a} for different L and N at the fixed end.

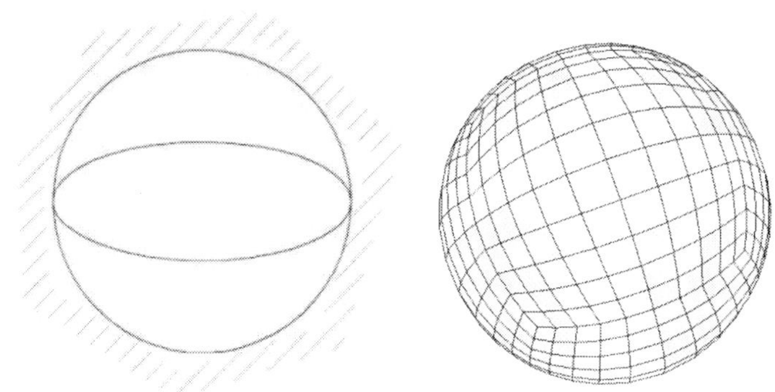

Figure 4. Geometry and boundary element mesh visualization.

and pore pressure on the cavity boundary. Load type:

$$P(t) = \begin{cases} 0 & \text{for } t \leq 0\,\mathrm{s}, \\ P_0 & \text{for } t > 0\,\mathrm{s}, \end{cases} \qquad P_0 = 1\,\mathrm{Pa}.$$

The poroelastic material parameters are: $K = 8 \times 10^9\,\mathrm{N/m^2}$, $G = 6 \times 10^9\,\mathrm{N/m^2}$, $\rho = 2458\,\mathrm{kg/m^3}$, $\phi = 0.66$, $K_s = 3.6 \times 10^{10}\,\mathrm{N/m^2}$, $\rho_f = 1000\,\mathrm{kg/m^3}$, $K_f = 3.3 \times 10^9\,\mathrm{N/m^2}$, $k = 1.9 \times 10^{-10}\,\mathrm{m^4/(N{\cdot}s)}$ (Berea sandstone).

The BE-mesh consists of 736 boundary elements (figure 4). The problem is solved with the symmetry taken into account, the real calculations are made on a quarter of the sphere.

Figures 5–7 shows the transient response of the displacements u and pore pressure p at the nodal point with the coordinates $(0, 0, 5)$ m. The influence of viscoelastic parameters on the dynamic responses is demonstrated in the figures below.

The effect of restructuring the wave field of inner displacements is demonstrated in the case of standard linear solid model, where the properties of poroviscoelastic material are changing from instantaneous to equilibrium (figure 6).

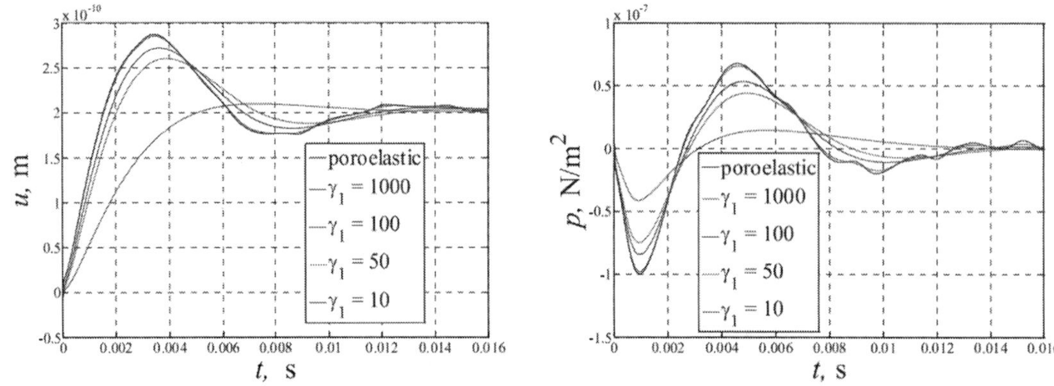

Figure 5. Displacement u and the pore pressure p for poroelastic (black line) and poroviscoelastic (colored lines) solution in the case of Kelvin–Voigt model.

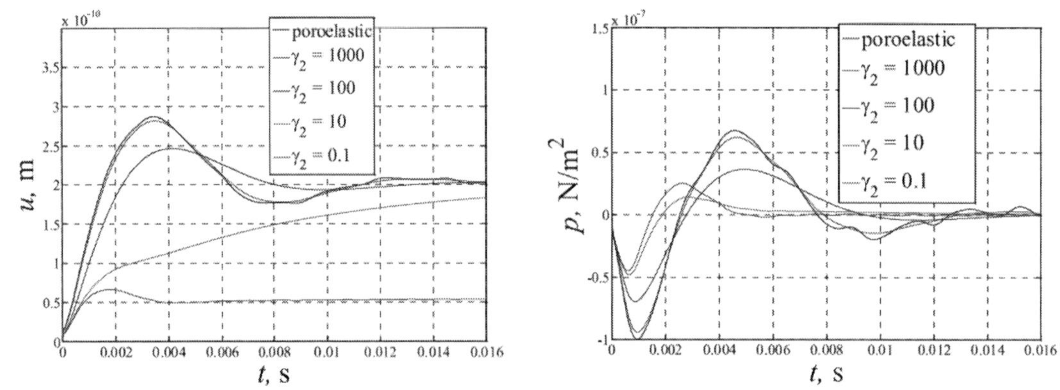

Figure 6. Displacement u and the pore pressure p for poroelastic (black line) and poroviscoelastic (colored lines) solution in the case of standard linear solid model ($\beta_0 = 4$).

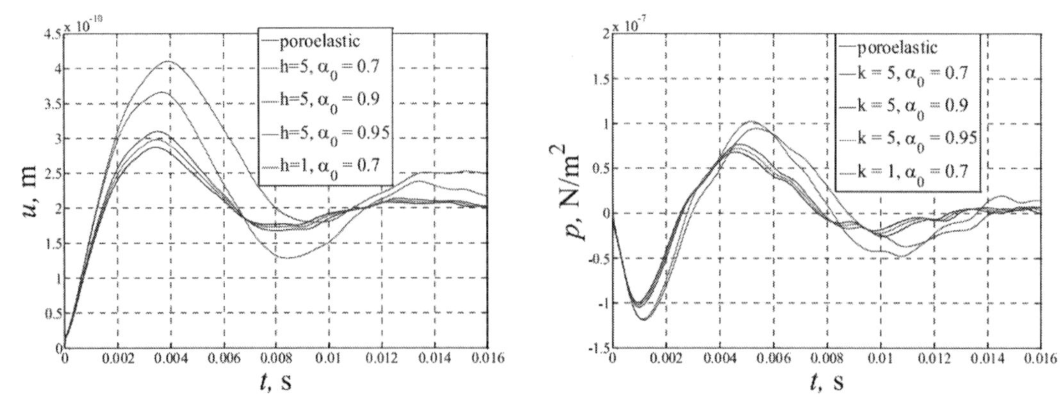

Figure 7. Displacement u and the pore pressure p for poroelastic (black line) and poroviscoelastic (colored lines) solution in the case of model with weakly singular kernel.

Conclusions

The problem of a spherical cavity considered in present paper is a classical external model boundary-value problem. The media are treated as poroelastic and poroviscoelastic, and the dynamic responses are compared for different poroviscoelastic properties of the considered viscoelastic models. The poroviscoelastic media modelling is based on Biot's theory of poroelastic material combined with the elastic-viscoelastic correspondence principle. The Kelvin–Voigt, standard linear solid, and weakly singular models are considered as viscoelastic constitutive models.

Acknowledgments

The work is financially supported by the Russian Foundation for Basic Research (RFBR) under grants Nos. 15-08-02817 and 15-48-02333.

References

[1] Frenkel J 1944 On the theory of seismic and seismoelectric phenomena in a moist soil *J. Phys. (Soviet)* **8** (4) 230–41
[2] Biot M A 1941 General theory of three-dimensional consolidation *J. Appl. Phys.* **12** (2) 155–64
[3] Biot M A 1956 Theory of deformation of a porous viscoelastic anisotropic solid *J. Appl. Phys.* **27** (5) 459–67
[4] Biot M A 1955 Theory of elasticity and consolidation for a porous anisotropic solid *J. Appl. Phys.* **26** (2) 182–5
[5] Li P and Schanz M 2011 Wave propagation in a one dimensional partially saturated poroelastic column *Geophys. J. Int.* **184** (3) 1341–53
[6] Li P 2012 Boundary element method for wave propagation in partially saturated poroelastic continua In *Computation in Engineering and Science* vol. 15 (Graz: Verlag der Technischen Universität Graz) p 131
[7] Li P and Schanz M 2013 Time domain boundary element formulation for partially saturated poroelasticity *Engng Anal. Bound. Elem.* **37** (11) 1483–98
[8] Brooks R H and Corey A T 1964 Hydraulic properties of porous media In *Hydrology papers* issue 3 (Fort Collins: Colorado State University) p 30
[9] Christensen R M 1971 *Theory of Viscoelasticity: an Introduction* (New York: Academic Press)
[10] Ugodchikov A G and Hutoryanskii N M 1986 *Boundary Element Method in Mechanics of Solid* (Kazan: Izdat. Kazan. Univ.) p 295 [in Russian]
[11] Igumnov L A, Belov A A, and Petrov A N 2015 Boundary-element modeling of the dynamics of elastic and viscoelastic bodies and media In *Advanced Materials — Studies and Applications* ed I A Parinov, S-H Chang, and S Theerakulpisut (New York: Nova Science Publishers) pp 301-317
[12] Banjai L, Messner M, and Schanz M 2012 Runge-Kutta convolution quadrature for the boundary element method *Comput. Methods Appl. Mech. Engng* **245–246** 90–101
[13] Igumnov L A and Petrov A N 2016 Dynamics of partially saturated poroelastic solids by boundary-element method *Vestnik PNIPU. Mekh.* **3** 47–61

IOP Conf. Series: Journal of Physics: Conf. Series **991** (2018) 012035 doi:10.1088/1742-6596/991/1/012035

Boundary element modelling of dynamic behavior of piecewise homogeneous anisotropic elastic solids

L A Igumnov, I P Markov, and S Yu Litvinchuk*

Research Institute for Mechanics, Lobachevsky State University of Nizhni Novgorod, Nizhny Novgorod, Russia

E-mail: *litvinchuk@mech.unn.ru

Abstract. A traditional direct boundary integral equations method is applied to solve three-dimensional dynamic problems of piecewise homogeneous linear elastic solids. The materials of homogeneous parts are considered to be generally anisotropic. The technique used to solve the boundary integral equations is based on the boundary element method applied together with the Radau IIA convolution quadrature method. A numerical example of suddenly loaded 3D prismatic rod consisting of two subdomains with different anisotropic elastic properties is presented to verify the accuracy of the proposed formulation.

1. Introduction

The analytical solutions of the dynamic problems of three-dimensional anisotropic and linear elasticity are severely restricted to a limited number of specific geometrical configurations and boundary conditions. More sophisticated problems, e.g., when nonhomogeneous regions are considered, can only be solved by suitable numerical tools.

Nowadays, the most popular domain-type numerical method is the finite element method (FEM). However, the boundary element method (BEM) is an interesting and highly-accurate alternative to FEM, due to its semi-analytical nature and the reduction of the problem dimension by one. In the recent decades, application of BEM for elastic wave propagation in isotropic elastic solids and media have received considerable attention. Two- and three-dimensional, time-domain, frequency- and Laplace-domain BEM formulations for transient isotropic elastodynamic problems are reported in abundance in the literature (see, e.g., [1–7]). Certain difficulties of development of the boundary element techniques for anisotropic and linear elastic solids are usually associated with the mathematical complexity of the corresponding dynamic fundamental solutions. They are not available in the explicit closed forms.

In this paper, a multi-domain three-dimensional boundary element formulation is applied for transient dynamic anisotropic and linear elastic multi-material, i.e., piecewise homogeneous, problems. The employed boundary element technique is formulated in the Laplace domain. The transient time-domain response is calculated via an inverse transformation by Radau IIA Runge-Kutta convolution quadrature method (CQM). Three-dimensional anisotropic elastodynamic fundamental solutions are expressed as surface integrals over half a unit sphere. Different interpolation functions are used to approximate the boundary variables. To demonstrate the reliability and accuracy of the proposed boundary element formulation, a numerical example is presented.

IOP Conf. Series: Journal of Physics: Conf. Series **991** (2018) 012035 doi:10.1088/1742-6596/991/1/012035

2. Basic equations and BEM formulation

We consider a homogeneous anisotropic and linear elastic solid $\Omega \in R^3$ with boundary $\Gamma = \partial\Omega$ which, in the absence of the body forces, satisfies the equations of motion

$$\sigma_{ij,j}(\mathbf{x},t) - \rho\ddot{u}_i(\mathbf{x},t) = 0, \quad \mathbf{x} \in \Omega, \quad i,j = \overline{1,3}, \tag{1}$$

the generalized Hooke's law

$$\sigma_{ij}(\mathbf{x},t) = C_{ijkl}\varepsilon_{kl}(\mathbf{x},t), \quad k,l = \overline{1,3}, \tag{2}$$

the elastic strain-displacement equations

$$\varepsilon_{ij}(\mathbf{x},t) = \frac{1}{2}[u_{i,j}(\mathbf{x},t) + u_{j,i}(\mathbf{x},t)], \tag{3}$$

the zero initial conditions

$$u_i(\mathbf{x},t) = \dot{u}_i(\mathbf{x},t) = 0, \quad t \leq 0, \tag{4}$$

and the boundary conditions

$$u_i(\mathbf{x},t) = u_i^*(\mathbf{x},t), \quad \mathbf{x} \in \Gamma_u, \tag{5}$$

$$t_i(\mathbf{x},t) = t_i^*(\mathbf{x},t), \quad \mathbf{x} \in \Gamma_t. \tag{6}$$

In equations (1)–(6), σ_{ij} denote the Cauchy stress tensor, u_i are displacement components, ρ is the mass density, C_{ijkl} is the elastic stiffness tensor, and ε_{ij} is the linear strain tensor; u_i^* and t_i^* denote the known displacement and traction vectors on the external boundaries Γ_u and Γ_t, respectively. For the outward unit normal vector n_i, the traction vector t_i is expressed as

$$t_i(\mathbf{x},t) = \sigma_{ik}(\mathbf{x},t)n_k(\mathbf{x}). \tag{7}$$

For subsequent derivations, we employ the Laplace transform in order to eliminate the time variable

$$\bar{f}(\mathbf{x},s) = L\{f(\mathbf{x},t)\} = \int_0^\infty f(\mathbf{x},t)\exp(-st)\,dt, \quad \mathrm{Re}\,(s) \geq 0, \tag{8}$$

with s denoting the Laplace transform parameter.

Applying Laplace transform (8) to equations (1)–(7) and combining the corresponding transforms of equations (1)–(3), we obtain a set of decoupled Laplace domain problems

$$C_{ijkl}\bar{u}_{k,lj}(\mathbf{x},s) - \rho s^2\bar{u}_i(\mathbf{x},s) = 0, \quad \mathbf{x} \in \Omega, \tag{9}$$

$$\bar{u}_i(\mathbf{x},s) = \bar{u}_i^*(\mathbf{x},s), \quad \mathbf{x} \in \Gamma_u, \tag{10}$$

$$\bar{t}_i(\mathbf{x},s) = \bar{t}_i^*(\mathbf{x},s), \quad \mathbf{x} \in \Gamma_t. \tag{11}$$

To solve the boundary-value problem defined in equations (9)–(10), the problem is reformulated as boundary integral equations (BIEs) for the displacements:

$$c_{ij}(\mathbf{x})\bar{u}_j(\mathbf{x},s) = \int_\Gamma \bar{g}_{ij}(\mathbf{r},s)\bar{t}_j(\mathbf{y},s)\,d\Gamma(\mathbf{y}) - \mathrm{p.v.}\int_\Gamma \bar{h}_{ij}(\mathbf{r},s)\bar{u}_j(\mathbf{y},s)\,d\Gamma(\mathbf{y}), \quad \mathbf{x} \in \Gamma, \tag{12}$$

where $\mathbf{r} = \mathbf{y} - \mathbf{x}$, with $\mathbf{y}$ and $\mathbf{x}$ being the field and source points, respectively; $c_{ij}(\mathbf{x})$ are the jump terms, p.v. stands for the Cauchy principal-value integral, $\bar{g}_{ij}$ and $\bar{h}_{jk}$ are the Laplace domain fundamental solutions for displacement and traction, respectively.

Following the usual boundary element procedure, the external boundary $\Gamma = \partial\Omega$ is discretized into quadrangular boundary elements. Quadratic shape functions are adopted to approximate the geometry of the elements. For the spatial discretization of the displacements and tractions over each boundary element, the linear and constant interpolation functions are used, respectively. The nodal collocation scheme is employed for the spatial discretization of boundary integral equations (12). For a piecewise homogeneous solid, which consists of a finite number distinct homogeneous subdomains, the discretization process of the BIEs (12) is performed for each homogeneous subdomain. Under the assumption of continuity of the displacements and equilibrium of tractions on the each interface of adjacent subdomains, the resulting complex-valued system of linear algebraic equations is obtained for a fixed value of Laplace transform parameter s

$$\bar{\mathbf{A}}(s)\bar{\mathbf{f}}(s) = \bar{\mathbf{b}}(s),$$

where $\bar{\mathbf{A}}(s)$ is the coefficient matrix, the vector $\bar{\mathbf{f}}(s)$ combines unknown boundary field variables (displacements and tractions), and $\bar{\mathbf{b}}(s)$ is a known vector.

3. Fundamental solutions

The elastodynamic fundamental solutions for the anisotropic and homogeneous solids can be split into the sum of a singular static and a regular dynamic part

$$\bar{g}_{ij}(\mathbf{r},s) = g^S_{ij}(\mathbf{r}) + \bar{g}^D_{ij}(\mathbf{r},s),$$
$$\bar{h}_{ij}(\mathbf{r},s) = h^S_{ij}(\mathbf{r}) + \bar{h}^D_{ij}(\mathbf{r},s).$$

For the three-dimensional displacement fundamental solution, $g^S_{ij}(\mathbf{r})$ and $\bar{g}^D_{ij}(\mathbf{r},s)$ have the following expressions [8, 9]

$$g^S_{ij}(\mathbf{r}) = \frac{1}{8\pi^2|\mathbf{r}|}\int_{|\mathbf{d}|=1} \Gamma^{-1}_{ij}(\mathbf{d})\, dL(\mathbf{d}),$$

$$\bar{g}^D_{ij}(\mathbf{r},s) = -\frac{1}{8\pi^2}\int_{\substack{|\mathbf{n}|=1\\ \mathbf{n}\cdot\mathbf{r}>0}} \sum_{m=1}^{3} \frac{k_m E_{im} E_{jm}}{\rho c_m^2} \exp(-k_m|\mathbf{n}\cdot\mathbf{r}|)\, dS(\mathbf{n}),$$

with

$$c_m = \sqrt{\frac{\lambda_m}{\rho}}, \quad k_m = \frac{s}{c_m}, \quad \Gamma_{ij}(\mathbf{d}) = C_{kijl} d_k d_l, \quad \Gamma_{ij}(\mathbf{n}) = C_{kijl} n_k n_l,$$
$$dL(\mathbf{d}(\varphi)) \in D^S = \{0 \le \varphi \le 2\pi\}, \quad dS(\mathbf{n}(b,\varphi)) \in D^D = \{0 \le b \le 1;\ 0 \le \varphi \le 2\pi\},$$
$$\mathbf{n}(b,\varphi) = \sqrt{1-b^2}\mathbf{d} + b\mathbf{e}, \quad \mathbf{e} = \frac{\mathbf{r}}{|\mathbf{r}|}, \quad \mathbf{e} = [e_1, e_2, e_3],$$
$$\mathbf{d}(\varphi) = \frac{[e_2\cos\varphi + e_1 e_3 \sin\varphi, -e_1\cos\varphi + e_2 e_3\sin\varphi, -(1-e_3^2)\sin\varphi]}{\sqrt{1-e_3^2}},$$

where λ_m and E_{jm} are eigenvalues and the corresponding eigenvectors of $\Gamma_{jk}(\mathbf{n})$.

The corresponding traction fundamental solution can be obtained as

$$\bar{h}_{jp}(\mathbf{r},s) = C_{ijkl}\bar{g}_{kp,l}(\mathbf{r},s) n_i(\mathbf{y}), \quad p = \overline{1,3},$$

where $n_i(\mathbf{y})$ is the outward unit normal vector to the boundary at the point $\mathbf{y}$.

IOP Conf. Series: Journal of Physics: Conf. Series **991** (2018) 012035 doi:10.1088/1742-6596/991/1/012035

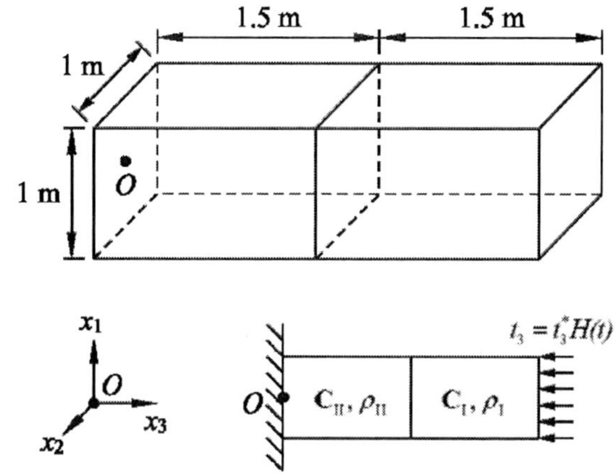

Figure 1. A piecewise homogeneous rod.

4. Numerical inversion of the Laplace transform

The convolution quadrature method was developed by Lubich [10, 11] for numerical approximation of the convolution integral, and it can also be used as a numerical inverse Laplace transform method. In this paper, we use CQM based on a Radau IIA Runge-Kutta method [12] given by Butcher's table $\frac{\mathbf{c}|\mathbf{A}}{\mathbf{b}^T}$ with $\mathbf{A} \in R^{m\times m}$, $\mathbf{b}$, $\mathbf{c} \in R^m$. For the transformed function $\bar{f}(s)$, the CQM approximation of $f(t)$ on the time interval $[0, N\Delta t]$ is

$$f(0) = 0, \quad f((n+1)\Delta t) = \mathbf{b}^T\mathbf{A}^{-1}\sum_{k=0}^{n}\omega_{n-k}(s\bar{f}(s)),$$

$$\omega_n(s\bar{f}(s)) = \frac{R^{-n}}{L}\sum_{p=0}^{L-1} s_p\bar{f}(s_p)e^{-in\phi_p}, \quad n = \overline{0, N-1},$$

$$s_p = \frac{\gamma(z_p)}{\Delta t}, \quad z_p = Re^{i\phi_p}, \quad \phi_p = 2\pi\frac{p}{L}, \quad \gamma(z_p) = \mathbf{A}^{-1} - z_p\mathbf{A}^{-1}\mathbf{I}\mathbf{b}^T\mathbf{A}^{-1}, \quad \mathbf{I} = (1,\dots,1)^T,$$

where Δt is the time step size, N is total number of time steps, and $0 < R < 1$ is a CQM parameter.

5. Numerical example

A three-dimensional piecewise homogeneous rod of total size $1 \times 1 \times 3$ m, as shown in figure 1, is considered. The rod is subjected to a Heaviside-type loading $t_3 = t_3^* H(t)$, $t_3^* = -10^5$ Pa at one end and fixed at another. The rest of the surface is free of tractions.

Four different uniform boundary-element meshes with 128 (mesh 1), 190 (mesh 2), 288 (mesh 3), and 378 (mesh 4) elements for each homogeneous part are used in computations. The obtained boundary element results for the displacements $u_3(t)$ at the points $A(0, 0, 3)$ m and $B(0, 0, 1.5)$ m are shown in figures 2 and 3 along with the corresponding FEM solutions. The traction results $t_3(t)$ at the points B and $C(0, 0, 0)$ m are displayed in figures 4 and 5. The mass densities of the respective homogeneous parts of the rod are $\rho_I = 2788\,\text{kg/m}^3$ and

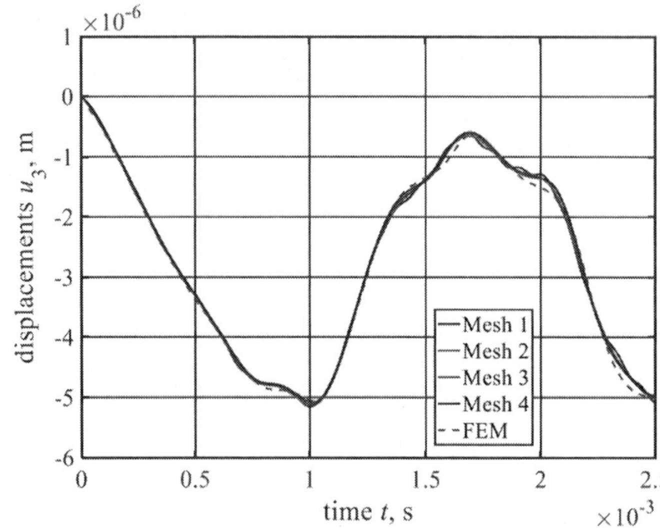

Figure 2. Displacement solution $u_3(t)$ at point A.

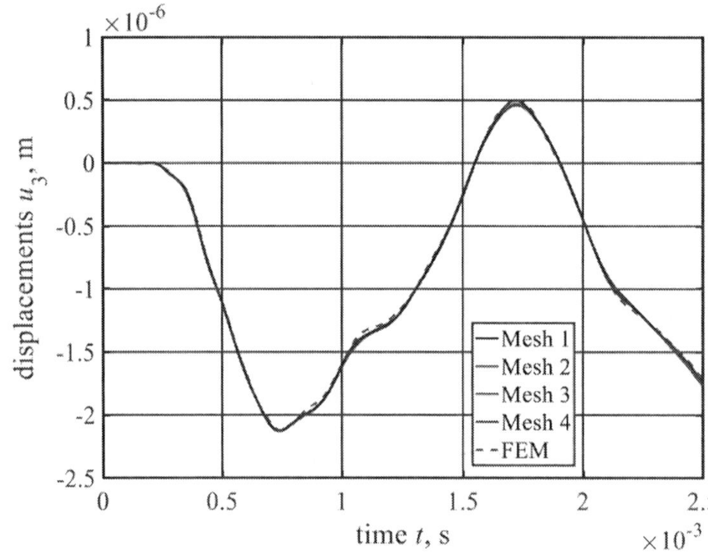

Figure 3. Displacement solution $u_3(t)$ at point B.

$\rho_{\mathrm{II}} = 3112\,\mathrm{kg/m^3}$, and the elastic stiffness tensors are given as

$$C_{\mathrm{I}} = \begin{bmatrix} 120 & 62 & 56 & 0 & 0 & 0 \\ 62 & 124 & 57 & 0 & 0 & 0 \\ 56 & 57 & 112 & 0 & 0 & 0 \\ 0 & 0 & 0 & 30 & 0 & 0 \\ 0 & 0 & 0 & 0 & 32 & 0 \\ 0 & 0 & 0 & 0 & 0 & 33 \end{bmatrix} \mathrm{GPa}, \quad C_{\mathrm{II}} = \begin{bmatrix} 188 & 84 & 78 & 0 & 0 & 0 \\ 84 & 195 & 80 & 0 & 0 & 0 \\ 78 & 80 & 191 & 0 & 0 & 0 \\ 0 & 0 & 0 & 57 & 0 & 0 \\ 0 & 0 & 0 & 0 & 55 & 0 \\ 0 & 0 & 0 & 0 & 0 & 55 \end{bmatrix} \mathrm{GPa}.$$

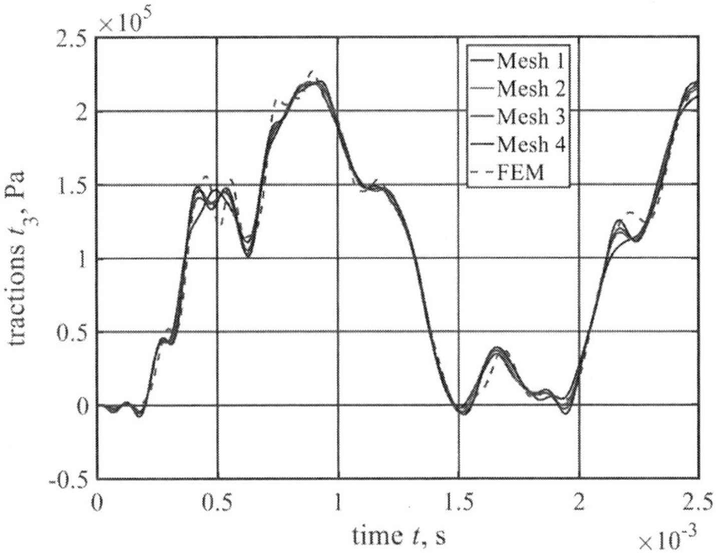

Figure 4. Traction solution $t_3(t)$ at point B.

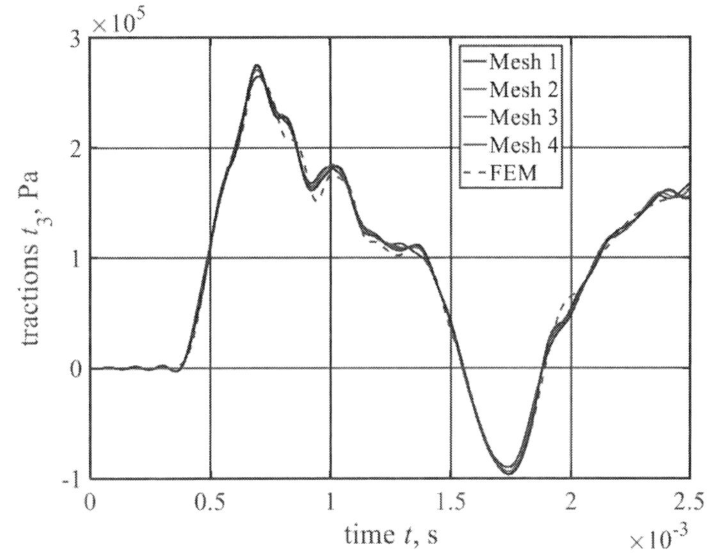

Figure 5. Traction solution $t_3(t)$ at point C.

Conclusions

In this study, we applied a Radau IIA Runge-Kutta CQM based boundary element method for transient dynamic analysis of generally anisotropic and linear elastic piecewise homogeneous three-dimensional solids. For the spatial discretization, a collocation method with mixed boundary elements is employed. The singular and regular parts of the anisotropic dynamic fundamental solutions are expressed as integrals over a unit circle and a half of a unit sphere, respectively. The comparison of the obtained boundary element results with the corresponding finite element solutions of the provided numerical example shows that the present boundary element formulation for multi-domain anisotropic elastodynamic problems is accurate and stable.

IOP Conf. Series: Journal of Physics: Conf. Series **991** (2018) 012035 doi:10.1088/1742-6596/991/1/012035

Acknowledgments

The work is financially supported by the Russian Foundation for Basic Research (RFBR) under grant No. 15-08-02814.

References

[1] Niwa Y, Fukui T, Kato S, and Fujiki K 1980 An application of the integral equation method to two-dimensional elastodynamics *Theor. Appl. Mech.* **28** 281–90

[2] Antes H 1985 A boundary element procedure for transient wave propagations in two-dimensional isotropic elastic media *Finite Elem. Anal. Des.* **1** (4) 313–22

[3] Israil A S M and Banerjee P K 1990 Advanced time-domain formulation of BEM for two-dimensional transient elastodynamics *Int. J. Numer. Meth. Engng* **29** (7) 1421–40

[4] Israil A S M and Banerjee P K 1990 Two-dimensional transient wave-propagation problems by time-domain BEM *Int J. Solids Struct.* **26** (8) 851–64

[5] Ahmad S and Banerjee P K 1988 Time-domain transient elastodynamic analysis of 3-D solids by BEM *Int. J. Numer. Meth. Engng* **26** (8) 1709–28

[6] Beskos D E 1997 Boundary element methods in dynamic analysis: Part II (1986–1996) *Appl. Mech. Rev.* **50** 149–97

[7] Dominguez J 1993 *Boundary Elements in Dynamics* (Southampton - Boston: Computational Mechanics Publication)

[8] Wang C Y and Achenbach J D 1994 Elastodynamic fundamental solutions for anisotropic solids *Geophys. J. Int.* **118** (2) 384–92

[9] Wang C Y and Achenbach J D 1995 Three-Dimensional Time-Harmonic Elastodynamic Green's Functions for Anisotropic Solids *Proc. Roy. Soc. London. A* **449** (1937) 441–58

[10] Lubich C 1988 Convolution quadrature and discretized operational calculus. I *Numer. Math.* **52** (2) 129–45

[11] Lubich C 1988 Convolution quadrature and discretized operational calculus. II *Numer. Math.* **52** (4) 413–25

[12] Banjai L, Messner M, and Schanz M 2012 Runge-Kutta convolution quadrature for the boundary element method *Comput. Method. Appl. Mech. Engng* **245-246** 90–101

IOP Conf. Series: Journal of Physics: Conf. Series **991** (2018) 012036 doi:10.1088/1742-6596/991/1/012036

Biomechanics of compensatory mechanisms in spinal-pelvic complex

D V Ivanov[1], **V V Hominets**[2], **I V Kirillova**[1*], **L Yu Kossovich**[1], **A L Kudyashev**[2], **and A V Teremshonok**[2]

[1]Chernyshevsky Saratov State University, Saratov, Russia
[2]S M Kirov Military Medical Academy, St. Petersburg, Russia

E-mail: *nano-bio@sgu.ru

Abstract. 3D geometric solid computer model of spinal-pelvic complex was constructed on the basis of computed tomography and full body X-ray in standing position data. The constructed model was used for biomechanical analysis of compensatory mechanisms arising in the spine with anteversion and retroversion of the pelvis. The results of numerical biomechanical 3D modeling are in good agreement with the clinical data.

1. Materials and methods

With the help of biomechanics methods, the compensatory changes in the sagittal profile of the spine were simulated, resulting from a change in the normal position of the pelvis in the sagittal plane due to its rotation around the bicocoxoforal axis. The change in the position of the pelvis was modeled by increasing and then decreasing the initial value of pelvic tilt (PT, pelvic tilt is an orientation of the pelvis in respect to the thighbones and the rest of the body) in the range from 5° to 25° in steps of 5°. The presented biomechanical model allowed reproducing the changes in sagittal vertebral-pelvic interrelations, which are characteristic for patients with deforming arthrosis of the hip joint of the third stage, accompanied by the formation of its flexion-leading contracture, and also for patients with degenerative-dystrophic spinal disease. When simulating the compensatory changes in sagittal profile of the spinal column, the rotated pelvis was fixed, and the seventh cervical vertebra was returned to its original position, at which the plumb line, drawn from the center of the C7 (C7 plumb line), passed through the cape of the sacrum. This mechanism of maintaining the sagittal balance of the trunk, allowing to minimize muscular efforts aimed at maintaining the vertical position of the body, was accompanied by a change in the physiological curves of the spine – lumbar lordosis and thoracic kyphosis. The stresses appearing in the vertebrae, arched joints and intervertebral discs were examined with the help of finite-element method and their localization was also determined.

1.1. *Initial data*

Computed tomography (CT) scans of a pelvis-spine complex of a healthy person (volunteer), performed from the level of the seventh cervical vertebra to the proximal parts of the femur, as well as the full body X-ray of the same subject executed in a standing position [1].

1.2. 3D solid models of pelvis-spine complex creation

Surface models obtained by semi-automatic processing of CT in special software products are only surfaces in the form of triangles that limit the volume of the objects under study, and are not suitable for performing computer biomechanical modeling. In this connection, for the implementation of biomechanical computer finite element modeling on the basis of the initial data (CT and X-ray diffraction), a solid 3D geometrical model of the pelvis-spine complex was constructed, containing the volumes of the objects under study.

CT data were used at first stage. CT data was processed using Mimics software, in which the models of vertebrae, pelvis and upper thirds of femurs were constructed. Since the CT scan was performed in the position of the patient lying on the back, the lumbar lordosis and thoracic kyphosis on the constructed model were smoothed out, and the location of the vertebrae relative to each other did not correspond to that in the standing position. At the second stage in SolidWorks software the constructed vertebral models were moved in the sagittal plane in exact accordance with their location on the roentgenogram performed to the subject in standing position. As a result, a pelvis-spine model was obtained containing vertebras, pelvic bones and upper thirds of the femurs, the spatial arrangement of which corresponded to the true sagittal spine-and-pelvic relationship of the subject in the standing position. Then solid 3D models of intervertebral discs, articular joints and ligaments of the spine have been completed with the help of the SolidWorks software product on the basis of known information about the anatomical structure of the lumbar-motor segments of the examined spine sections.

1.3. Finite-element modeling

Constructed 3D solid models of pelvis-spine complex were loaded into the ANSYS 18 finite element analysis system. It was used to model the rotation of pelvis around the bicoxo-femoral axis anteriorly (anteversion) and then backward (retroversion). In this case, the value of the PT parameter, equal 10°, was successively changed between +25° and −25 with a step of 5° (figures 2 and 3). Then the position of pelvis was fixed and the seventh cervical vertebra was returned to its normal position, keeping the original plumb line drawn from the center of C7 body through the cape of the sacrum.

Mechanical parameters of the vertebrae bone tissue, pelvis and femurs, as well as the parameters of intervertebral discs, arched joints and ligaments, were modeled on the basis of literature data. All pelvis-spine complex elements were considered to be isotropic, homogeneous and linearly elastic. Large deformations of pelvis-spine complex elements were assumed during calculations. Bonded contact was used between the contacting elements of pelvis-spine complex. Static formulation was used in calculations.

Tetrahedral finite-element computational mesh was created, the characteristic size of elements was calculated on the basis of the mesh convergence analysis and was assumed to be 2.5 mm. Displacements and equivalent von Mises stresses of the pelvis-spine complex were analyzed.

2. Results

2.1. 3D solid models of pelvis-spine complex

Figure 1 *a* shows 3D solid model of pelvis-spine complex, constructed on the basis of CT and full body X-ray data. Blue color shows pelvic bones, brown – proximal femur bones, green – vertebrae, red – intervertebral disks, lilac – sacrum, yellow – ligaments. Figure 1 *b* shows the lumbosacral and thoracic vertebrae, while their relationships are modeled on the basis of the full body X-ray in standing position. Red marks show position of the vertebrae on roentgenogram. Figure 1 *c* shows sagittal and frontal projections of the spine. In sagittal projection, spine model is superimposed on the X-ray. The plumb line from the center of the C7 vertebra body is shown in orange, the vertebrae of the lumbosacral segment are green, the thoracic vertebrae are yellow, the seventh cervical vertebra is red and the sacrum vertebra is lilac.

IOP Conf. Series: Journal of Physics: Conf. Series **991** (2018) 012036 doi:10.1088/1742-6596/991/1/012036

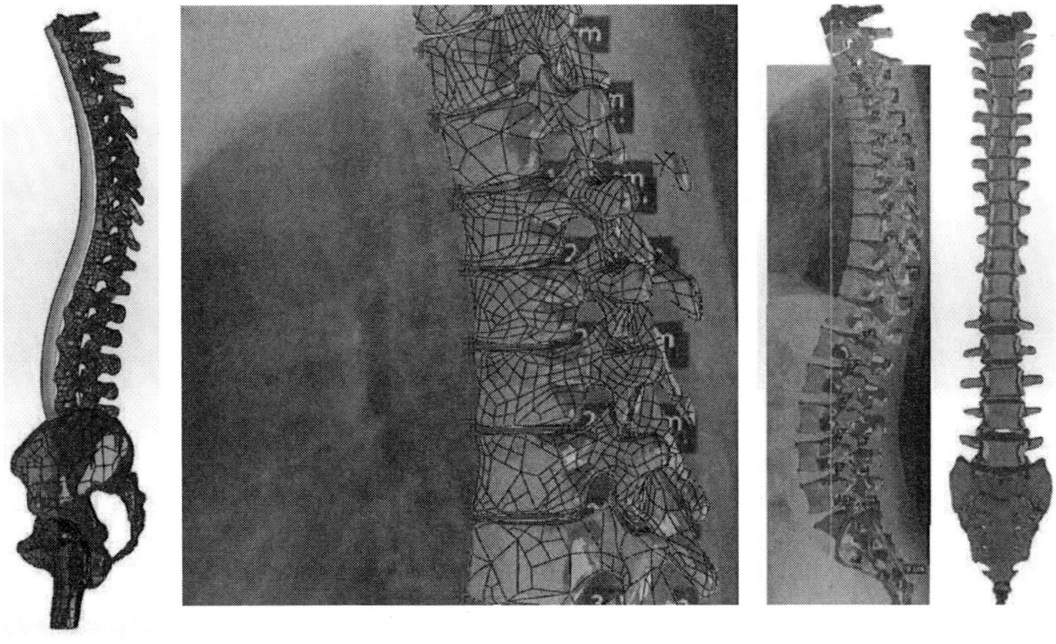

a *b* *c*

Figure 1. 3D solid model of: *a* – general view of 3D solid model of pelvis-spine complex; *b* – superposition of pelvis-spine complex 3D model segment on full body X-ray of spine; *c* – frontal and sagittal projection of the spine with superposition on the radiograph. A plumb line was constructed from the center of the body of the C7 vertebra.

2.2. Finite-element modeling

Finite-element modeling allowed calculating stress-strain state of the pelvis-spine complex in following ways of fixing and loading: 1) PT increased to 25° from the initial with a step of 5° and subsequent fixation of pelvis in this position, returning the seventh cervical vertebra to its normal position to restore the plumb line; 2) PT reduced to 25° from the original in 5° increments and subsequent fixation of pelvis in this position, returning the seventh cervical vertebra to its normal position to restore the plumb line.

Figure 2 shows 3D models of spine: normal and with the described changes. Vertical axis in each illustration is saved. Changes in physiological curves of spine are demonstrated. For each model of spine a plumb line is shown. Displacements are shown on scale from blue (zero displacement) to red (maximum displacement). Boundaries of undeformed spine are shown by thin black lines for each model.

Figure 2 demonstrates that with the increase in pelvic anteversion angle and subsequent return of seventh cervical vertebra to normal position, lumbar lordosis and thoracic kyphosis increase. In this case, plumb line practically does not change its position and passes slightly anterior to the cape of sacrum.

Figure 3 shows the results of modeling changes in the sagittal profile of the spine with increasing retroversion of the pelvis in steps of 5°.

Figure 3 illustrates the increase in retroversion of the pelvis and an increase in the distance from the plumb line drawn from the center of the body of C7 to the cape of the sacrum. There is also a significant increase in thoracic kyphosis and a flattening of the lumbar lordosis, leading to an inclination of the upper half of the trunk anteriorly and a sagittal imbalance.

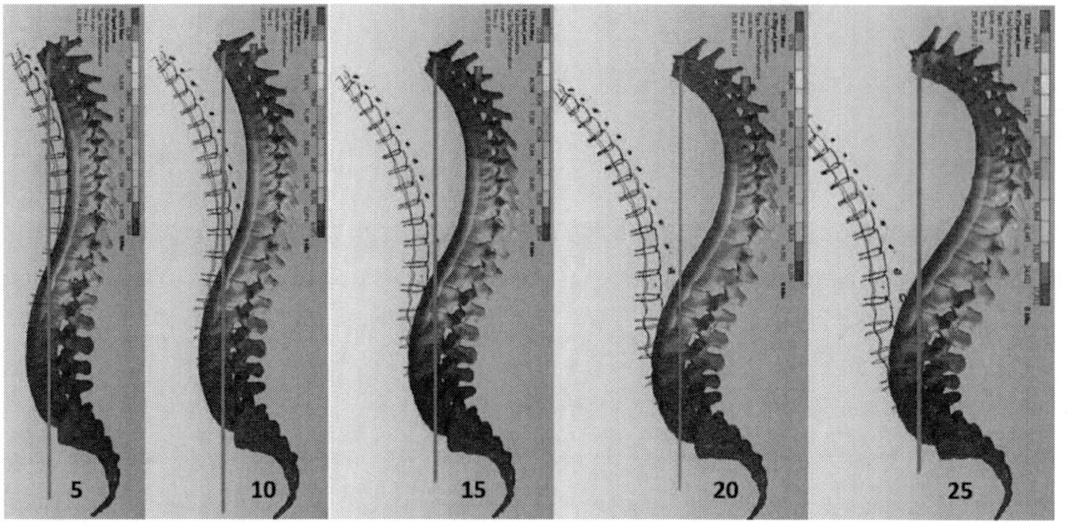

Figure 2. Changes in the sagittal profile of the spine with an increase in the initial pelvic anteversion in steps of 5°.

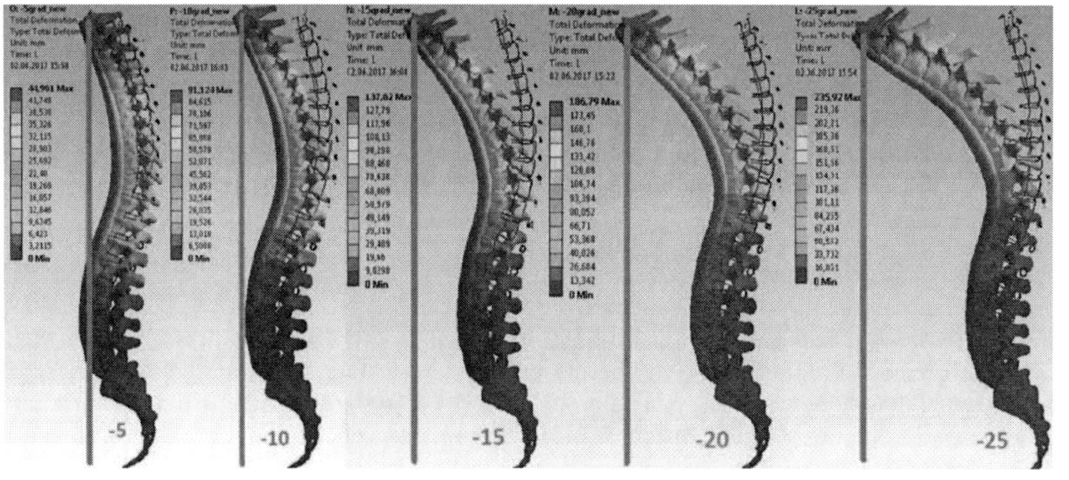

Figure 3. Changes in the sagittal profile of the spine with a decrease in the initial pelvic anteversion in steps of 5°.

Figure 4 shows distribution of effective stresses in various parts of spine and in various parts of vertebrae. Such stress field is typical for sagittal vertebral-pelvic relationships with an increase in pelvic anteversion.

Figure 4 shows that highest stress values for this variant of the sagittal profile occur in the region of the articular joints of the lumbar vertebrae, as well as in the anterior parts of the bodies of the upper thoracic vertebrae. With further increase in pelvic anteversion, stress values also increase. Analysis of deformations of intervertebral discs indicated the greatest changes in the lumbosacral spine.

Effective von Mises stresses for the vertebrae for pelvis retroversion are shown in figure 5.

With this variant of static deformation of pelvis-spine complex, the highest stresses occur in the posterior parts of the lumbar vertebrae, in the sacrum, and also in the anterior sections of

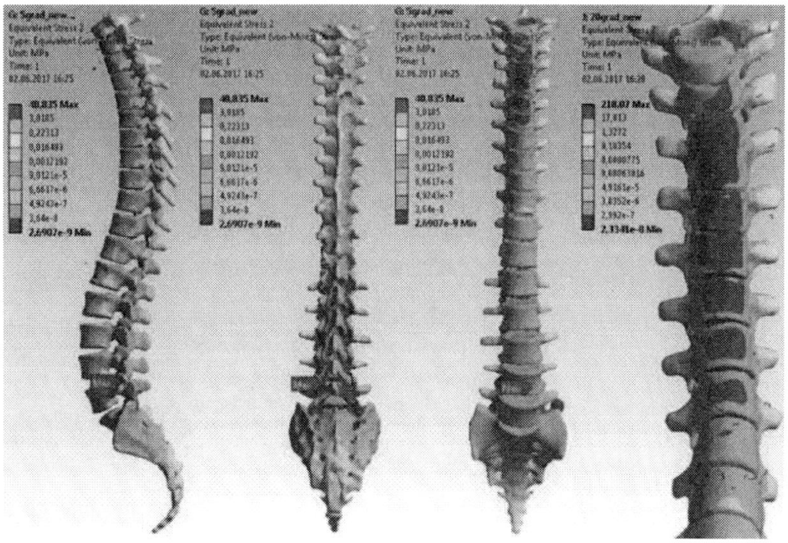

Figure 4. Effective stresses in the vertebrae with increasing pelvic anteversion (biomechanical model of a patient with coxarthrosis of III st., accompanied by a flexion-reducing contracture of hip joint.

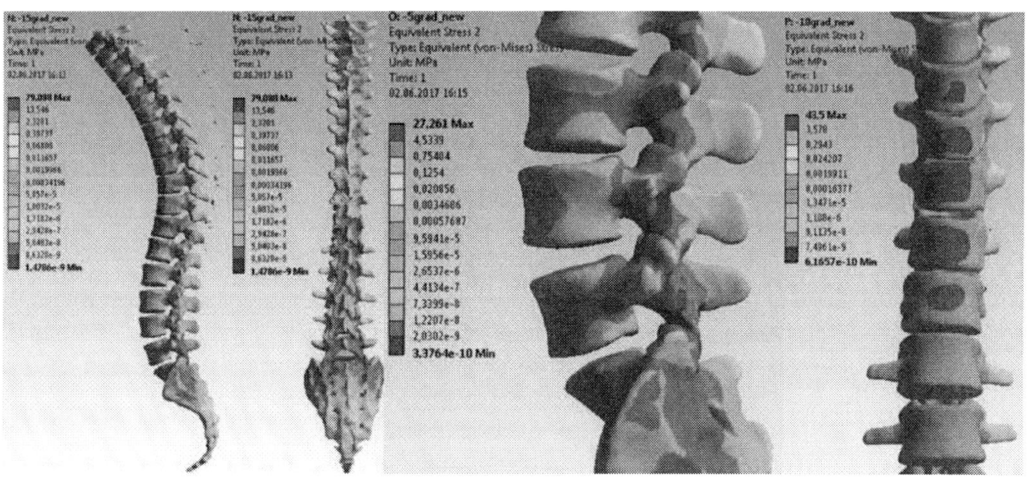

Figure 5. Effective Mises stresses for vertebrae at tilting the pelvis back.

the thoracic vertebrae. The largest deformations are characteristic of the spinous processes of the thoracic vertebrae.

3. Discussion

The data obtained as a result of biomechanical modeling of various variants of static deformations of pelvis-spine complex are fully confirmed by the results of clinical and radiological examination and assessment of the sagittal balance of patients suffering from both deforming arthrosis of the hip joint and degenerative-dystrophic disease of the spine.

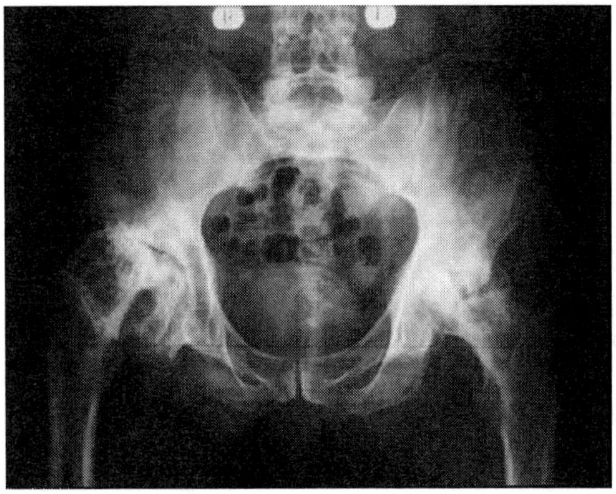

Figure 6. Review X-ray of the patient's pelvis D. Aseptic necrosis of the head of the femoral bones of the fourth degree, secondary bilateral coxarthrosis of the third degree.

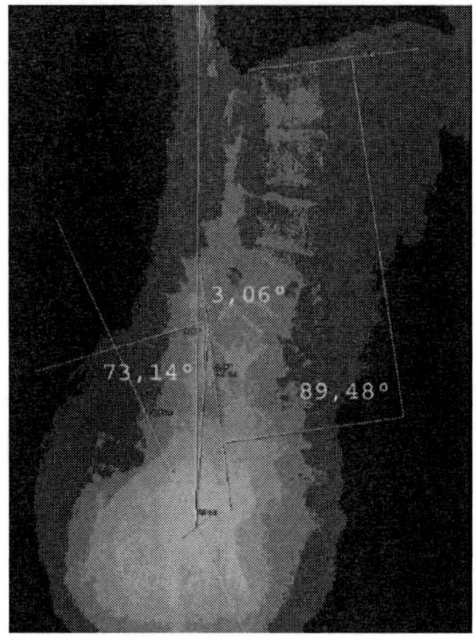

Figure 7. Sagittal X-ray of the patient's D. pelvis-spine complex. Excessive pelvic anteversion and compensatory lumbar lordosis enlargement.

3.1. Clinical example 1

Patient D. was 56 years old on examination and treatment in the clinic of military traumatology and orthopedics of the Military Medical Academy about aseptic necrosis of the femoral bones head of the IV degree, secondary bilateral coxarthrosis of the third degree, flexor-leading contractures of the hip joints (figure 6). In an objective clinical examination, lumbar lordosis and a positive symptom of Thomas were diagnosed.

Analysis of the sagittal profile of the patient D. testified to excessive pelvic anteversion, which arose as a result of the flexion element of the hip joint contracture, as well as the compensatory increase in lumbar lordosis (figure 7) with the formation of pain vertebrogenic and discogenic syndromes.

The presented variant of the sagittal profile is typical for patients with the phenomena of deforming arthrosis of the hip joint, which is accompanied by the formation of the flexion-leading contracture. It completely coincides with the results of 3D modeling of this variant of static deformation of pelvis-spine complex, presented in figure 2.

3.2. Clinical example 2

Patient Z. 53 years old was on treatment at the clinic of military traumatology and orthopedics of the Military Medical Academy in 2011 for degenerative-dystrophic disease of the lumbosacral spine, degenerative stable spondylolisthesis L4 of the vertebra of I st., degenerative subcompensated stenosis of the vertebral canal with neurogenic syndrome "Intermittent claudication." As a result of the study of the sagittal profile, a decrease in pelvic anteversion and lumbar lordosis has been diagnosed (figure 8).

This type of sagittal profile is typical for patients with degenerative-dystrophic disease of the spine, accompanied by the formation of antesponilolisthesis and stenosis of the spinal canal. It completely coincides with the results of 3D modeling of the considered variant of pelvis-spine complex static deformation, presented in figure 3.

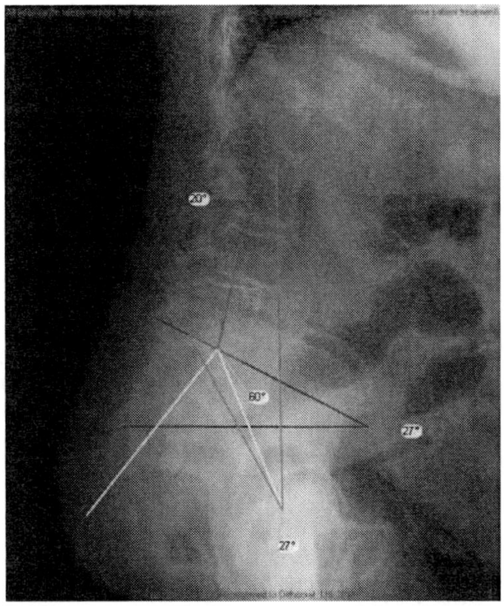

Figure 8. Sagittal X-ray of the patient's pelvis-spine complex. Reduced pelvic anteversion and lumbar lordosis, L4 antepspondilolisthesis on the I-st stage, degenerative-dystrophic disease of the spine.

Thus, the obtained results of biomechanical 3D modeling of the pelvis-spine complex, as well as the study of the behavior of the constructed model with a change in the spatial location of the pelvis, fully coincided with the data of additional examination of patients with diseases of the musculoskeletal system resulting in the formation of static deformations of the pelvis-spine complex, and also with the description of changes in the sagittal profile of the spine available in the scientific literature for various diseases [2–4].

Conclusions

Accurate 3D solid geometric model of the human spinal-pelvic complex was constructed. Full body x-ray in the standing position and a computer tomogram in the supine position were used as the initial data. The constructed 3D model was used in the biomechanical analysis of compensatory mechanisms arising in the spine with various pathologies.

References

[1] Shnayder L S, Pavlov V V, Krutko A V, and Golenkov O I 2016 Sagittal alignment of the spine in patients with hip dysplasia crowe IV stage based on the sagittal radiographs analysis *Sovrem. Probl. Nauki Obraz.* **6** 39

[2] Sato K, Itoi E, and Kasama F 1989 Abnormal posture associated with osteoporosis *J Muskuloskeletal Syst.* **2** 1451–62

[3] Nakamura Y, Funayama K, and Kita J 1996 Hipe spine syndrome: tracing of coxarthropathy and the relationship between the lumbar and sacral angles *Sendai Red. Cross. Med. J.* **5** 73–6

[4] Kudyashev A L, Shapovalov V M, Averkiev V A, et al. 2013 Radiographic evaluation of sagittal spinopelvic relations in patients with arthrosis deformans of the hip *Genii Ortopedii* **3** 27–32

Contact interaction of thin-walled elements with an elastic layer and an infinite circular cylinder under torsion

E G Kanetsyan*, M S Mkrtchyan, and S M Mkhitaryan

Institute of Mechanics of the National Academy of Sciences of Armenia, Yerevan, Armenia

E-mail: *ekanetsyan@gmail.com

Abstract. We consider a class of contact torsion problems on interaction of thin-walled elements shaped as an elastic thin washer – a flat circular plate of small height – with an elastic layer, in particular, with a half-space, and on interaction of thin cylindrical shells with a solid elastic cylinder, infinite in both directions. The governing equations of the physical models of elastic thin washers and thin circular cylindrical shells under torsion are derived from the exact equations of mathematical theory of elasticity using the Hankel and Fourier transforms. Within the framework of the accepted physical models, the solution of the contact problem between an elastic washer and an elastic layer is reduced to solving the Fredholm integral equation of the first kind with a kernel representable as a sum of the Weber–Sonin integral and some integral regular kernel, while solving the contact problem between a cylindrical shell and solid cylinder is reduced to a singular integral equation (SIE). An effective method for solving the governing integral equations of these problems are specified.

1. Introduction

The theory of torsion of elastic bodies and methods for solving torsion problems present one of the vast areas of mathematical theory of elasticity. The theoretical and applied significance of contact torsion problems lies in the fact that they, on the one hand, generalize and develop the classical contact problems of the theory of elasticity. On the other hand, they are directly related to important practical engineering issues on the transfer of loads from thin-walled elements to massive deformable bodies, often encountered in construction, machine building, especially in aircraft building, in mechanics of growing bodies, in measurement technology, in composites mechanics, and in other fields of applied mechanics. These problems are the subject of numerous studies [1–6]. The contact problem of torsion of an elastic half-space by an absolutely rigid circular washer is considered in [7]. An overview of the main results and papers on contact problems of torsion of elastic bodies is given in [8].

In the present paper, we consider a class of contact torsion problems on the interaction of thin-walled elements shaped as an elastic thin washer – a flat circular plate of small height – with an elastic layer, in particular, with a half-space and on the interaction of the thin cylindrical shells with a solid elastic cylinder, infinite in both directions.

The mechanical behavior of a thin circular elastic washer and a thin cylindrical shell under torsion is described by physical models that are completely analogous to the well-known Melan model of a rectilinear stringer [9, 10]. The deformation of an elastic layer and an elastic infinite solid cylinder is described by differential equations of axisymmetric torsion of the theory of elasticity.

In problems for the layer, here, in contrast to other studies, the corresponding equation of deformation of a thin circular washer is first derived from the exact differential equation of axisymmetric torsion.

Next, the contact problem of torsion is considered when the elastic layer, whose shear modulus varies exponentially with depth, is rigidly clamped along its lower face, whereas a thin elastic circular washer is fastened on its upper face. The torsional tangential forces act on the upper face of the washer, as well as on its lateral surface. It is required to determine the tangential contact stresses under the washer, as well as other deformation characteristics of the washer. From the condition of contact between the washer and the layer, the governing integral equation (GIE) of the problem is obtained. The kernel of this equation is the sum of its principal part in the form of the Weber–Sonin integral and the regular part in the form of a fairly rapidly convergent integral of the product of Bessel functions of the first kind. The GIE of the problem is solved by the method of spectral relationships containing the associated Legendre polynomials, or by the method of collocations using Gaussian quadrature formulas. Some particular cases are discussed.

In problems for an infinite cylinder, we first consider the boundary-value problem for an infinite hollow cylinder whose inner and outer surfaces are loaded with tangential torsional forces. Using the integral Fourier transform, we construct an exact solution of this problem containing modified Bessel functions and then, taking into account that the cylinder is thin, we derive an ordinary differential equation, as above, describing the deformation of a thin cylindrical shell. Further, within the framework of the obtained physical model, the problem of contact of a thin cylindrical shell of finite length with a solid infinite circular cylinder under torsion is considered. Eventually, solving this contact problem is reduced to solving the SIE with a kernel represented by the sum of the Cauchy kernel and a regular kernel. A well-known numerical-analytical method is used to solve the governing SIE [11].

2. Solution of two boundary-value problems on torsion of an exponentially inhomogeneous elastic layer under an axisymmetric deformation

Let an elastic layer $\Omega_0 = \{0 < r < \infty;\ 0 \le \vartheta < 2\pi;\ 0 \le z \le h\}$, referred to the cylindrical coordinate system r, ϑ, z, have a shear modulus G varying in depth by an exponential law

$$G = G_0 e^{\gamma z} \quad (0 \le z \le h,\ \gamma = \text{const}). \tag{1}$$

It is well known [6] that, under an axisymmetric deformation, the Lame equations in a cylindrical coordinate system fall into two groups. The first group consists of two equations and describes the axisymmetric deformation of an elastic body, and the second group consists of a single equation and describes the torsion of the elastic body. Under torsion we have $u_r = u_z \equiv 0$, $u_\vartheta = v(r, z)$ for elastic displacement components and

$$\begin{gathered} e_{r\vartheta} = \frac{\partial v}{\partial r} - \frac{v}{r}, \quad e_{\vartheta z} = \frac{\partial v}{\partial z}, \quad e_r = e_\vartheta = e_z = e_{rz} \equiv 0, \\ 2\omega_r = -\frac{\partial v}{\partial z}, \quad \omega_\vartheta = 0, \quad 2\omega_z = \frac{\partial v}{\partial r} + \frac{v}{r} \end{gathered} \tag{2}$$

for components of deformations and angles of rotation [6], respectively.

In addition, the differential equations of equilibrium in the absence of volume forces reduce to a single equation,

$$\frac{\partial \tau_{r\vartheta}}{\partial r} + \frac{\partial \tau_{\vartheta z}}{\partial z} + \frac{2}{r}\tau_{r\vartheta} = 0, \tag{3}$$

where $\tau_{r\vartheta}$ and $\tau_{\vartheta z}$ are components of tangential stresses. The Hooke's law has the form

$$\tau_{r\vartheta} = Ge_{r\vartheta}, \quad \tau_{\vartheta z} = Ge_{\vartheta z}. \tag{4}$$

We substitute the expressions for stresses (4) into equation (3) taking into account (1). As a result, we arrive at a differential equation for the function $v(r, z)$,

$$\frac{\partial^2 v}{\partial r^2} + \frac{1}{r}\frac{\partial v}{\partial r} + \frac{\partial^2 v}{\partial z^2} + \gamma\frac{\partial v}{\partial z} - \frac{v}{r^2} = 0 \quad (r, z \in \Omega_0). \tag{5}$$

Now for an elastic layer Ω_0, we consider the first boundary-value problem when the tangential stress components are given on its boundary planes $z = 0$ and $z = h$:

$$\tau_{\vartheta z}\big|_{z=+0} = \tau(r), \quad \tau_{\vartheta z}\big|_{z=h-0} = \tau_0(r) \quad (0 < r < \infty). \tag{6}$$

Using (5), (6), (4), and (1), this problem is formulated mathematically as the boundary-value problem

$$\begin{cases} \dfrac{\partial^2 v}{\partial r^2} + \dfrac{1}{r}\dfrac{\partial v}{\partial r} - \dfrac{v}{r^2} + \dfrac{\partial^2 v}{\partial z^2} + \gamma\dfrac{\partial v}{\partial z} = 0 \quad (0 < r < \infty,\ 0 < z < h), \\ \dfrac{\partial v}{\partial z}\bigg|_{z=+0} = \dfrac{\tau(r)}{G_0}, \quad \dfrac{\partial v}{\partial z}\bigg|_{z=h-0} = \dfrac{e^{-\gamma h}}{G_0}\tau_0(r) \quad (0 < r < \infty), \\ \tau_{r\vartheta}^2 + \tau_{\vartheta z}^2 \to 0 \ \text{ when } r^2 + z^2 \to \infty. \end{cases} \tag{7}$$

We construct the solution of boundary-value problem (7) by the method of the integral Hankel transform, for which we introduce the Hankel transformants

$$\{\bar{v} = \bar{v}(z, \lambda),\ \bar{\tau}_0 = \bar{\tau}_0(\lambda),\ \bar{\tau} = \bar{\tau}(\lambda)\} = \int_0^\infty \{v(r, z),\ \tau_0(r),\ \tau(r)\} r J_1(\lambda r)\, dr,$$

where $J_1(r)$ is the Bessel function of the first kind of order 1, and λ $(0 < \lambda < \infty)$ is a spectral parameter.

Further, we multiply both sides of the differential equation in (7) and boundary conditions by $rJ_1(\lambda r)$ and integrate them with respect to r over the interval $(0, \infty)$. Taking into account the well-known relation from [12] (p. 79, formula (2.32) for $\nu = 1$)

$$\int_0^\infty r\left(\frac{d^2 f}{dr^2} + \frac{1}{r}\frac{df}{dr} - \frac{f}{r^2}\right) J_1(\lambda r)\, dr = -\lambda^2 \bar{f},$$

we reduce the two-dimensional boundary-value problem (7) to the one-dimensional boundary-value problem

$$\begin{cases} \dfrac{d^2\bar{v}}{dz^2} + \gamma\dfrac{d\bar{v}}{dz} - \lambda^2\bar{v} = 0 \quad (0 < z < h), \\ \dfrac{d\bar{v}}{dz}\bigg|_{z=+0} = \dfrac{\bar{\tau}}{G_0}, \quad \dfrac{d\bar{v}}{dz}\bigg|_{z=h-0} = \dfrac{e^{-\gamma h}}{G_0}\bar{\tau}_0. \end{cases} \tag{8}$$

The differential equation in (8) has a general solution

$$\bar{v} = \bar{v}(z, \lambda) = Ae^{\alpha_1 z} + Be^{\alpha_2 z} \quad (0 \le z \le h),$$
$$\alpha_1 = -\frac{\lambda_* + \gamma}{2}, \quad \alpha_2 = \frac{\lambda_* - \gamma}{2}, \quad \lambda_* = \sqrt{\gamma^2 + 4\lambda^2}.$$

IOP Conf. Series: Journal of Physics: Conf. Series **991** (2018) 012037 doi:10.1088/1742-6596/991/1/012037

After determining the constants A and B from the boundary conditions of problem (8), we obtain

$$\bar{v}(z,\lambda) = \frac{e^{-\gamma z/2}}{2G_0\lambda^2\sinh(\lambda_* h/2)}\left\{e^{-\gamma h/2}\left(\lambda_*\cosh\frac{\lambda_* z}{2} - \gamma\sinh\frac{\lambda_* z}{2}\right)\bar{\tau}_0(\lambda)\right.$$
$$\left. - \left[\lambda_*\cosh\frac{\lambda_*(z-h)}{2} + \gamma\sinh\frac{\lambda_*(z-h)}{2}\right]\bar{\tau}(\lambda)\right\} \quad (0 \le z \le h,\ \lambda_* = \sqrt{\gamma^2+4\lambda^2}). \quad (9)$$

Now we will find the desired function $v(r,z)$ with the help of the inverse Hankel transform,

$$v(r,z) = \int_0^\infty \bar{v}(z,\lambda)\lambda J_1(\lambda r)\,d\lambda.$$

Next, we consider an important special case, where $\gamma = 0$, i.e., where the elastic layer Ω_0 is homogeneous. In this case, from (9) we have

$$\bar{v}(z,\lambda) = \frac{\cosh(\lambda z)\bar{\tau}_0 - \cosh[\lambda(z-h)]\bar{\tau}}{\lambda G_0\sinh(\lambda h)} \quad (0 \le z \le h). \quad (10)$$

Proceeding from (10), we obtain a differential equation for the deformation of an elastic thin layer or an elastic thin circular washer under torsion. For this, we write (10) in the form

$$\lambda G_0\sinh(\lambda h)\bar{v}_0(z,\lambda) = \cosh(\lambda z)\bar{\tau}_0 - \cosh[\lambda(z-h)]\bar{\tau} \quad (\bar{v}_0(z,\lambda) = \bar{v}(z,\lambda))$$

and expand the entire functions of $\lambda\sinh(\lambda h)$, $\cosh(\lambda z)$, and $\cosh[\lambda(z-h)]$ in power series restricting ourselves to terms of the order of λ^2. We get

$$G_0h\lambda^2\bar{v}_0 = G_0h\lambda^2\bar{v}_0(z,\lambda) = \bar{\tau}_0(\lambda) - \bar{\tau}(\lambda) + \frac{\lambda^2z^2}{2}\bar{\tau}_0(\lambda) - \frac{\lambda^2}{2}(z-h)^2\bar{\tau}(\lambda) \quad (0 \le z \le h). \quad (11)$$

Let us discuss two cases.

(1) $z = 0$. Then from (11) we obtain

$$G_0h\lambda^2\bar{v}_0(0,\lambda) = \bar{\tau}_0(\lambda) - \bar{\tau}(\lambda) - \frac{\lambda^2h^2}{2}\bar{\tau}(\lambda) \quad (0 < \lambda < \infty),$$

from which using the inverse Hankel transform, we arrive at the differential equation

$$G_0h\left(\frac{d^2v_0}{dr^2} + \frac{1}{r}\frac{dv_0}{dr} - \frac{v_0}{r^2}\right) = \tau(r) - \tau_0(r) - \frac{h^2}{2}\left(\frac{d^2\tau}{dr^2} + \frac{1}{r}\frac{d\tau}{dr} - \frac{\tau}{r^2}\right) \quad (z=0,\ 0<r<\infty). \quad (12)$$

Thus, the axisymmetric deformation of an elastic thin material layer $z = 0$ under torsion, within the framework of the accepted accuracy, is described by differential equation (12). Now in (12) we pass to the limits $G_0 \to \infty$ and $h \to 0$ in such a way that G_0h – the rigidity of the thin layer – remains constant. As a result, by analogy with the known results from [10], we come to the physical model of an elastic thin layer in torsion:

$$G_0h\left(\frac{d^2v_0}{dr^2} + \frac{1}{r}\frac{dv_0}{dr} - \frac{v_0}{r^2}\right) = \tau(r) - \tau_0(r) \quad (0<r<\infty). \quad (13)$$

If, in particular, we set in (12) $\tau_0(r) = \tau(r)$ $(0 < r < \infty)$, we obtain a homogeneous differential equation with respect to the function $v_0(r) + h\tau(r)/(2G_0)$. Since the displacements and stresses are bounded at $r = 0$ and $r = \infty$, this equation has only a trivial zero solution. As a result, we arrive at the Winkler model for a thin elastic layer under torsion,

$$v_0 = -\frac{h\tau}{2G_0} \quad (0<r<\infty) \quad (14)$$

with the bed coefficient $h/(2G_0)$.

(2) $z = h$. Then from (11) we get

$$G_0 h\lambda^2 \bar{v}_0(h,\lambda) = \bar{\tau}_0(\lambda) - \bar{\tau}(\lambda) + \frac{\lambda^2 h^2}{2}\bar{\tau}_0(\lambda) \quad (0<\lambda<\infty),$$

from which again with the help of the inverse Hankel transform, we obtain

$$G_0 h\left(\frac{d^2 v_0}{dr^2}+\frac{1}{r}\frac{dv_0}{dr}-\frac{v_0}{r^2}\right)=\tau(r)-\tau_0(r)+\frac{h^2}{2}\left(\frac{d^2\tau_0}{dr^2}+\frac{1}{r}\frac{d\tau_0}{dr}-\frac{\tau_0}{r^2}\right) \quad (z=h,\ 0<r<\infty). \quad (15)$$

Hence, as above, we arrive at the deformation model (13) of an elastic thin material layer $z = h$ in torsion and at the Winkler model

$$v_0 = \frac{h\tau_0}{2G_0} \quad (0<r<\infty) \quad (16)$$

with the bed coefficient $h/(2G_0)$.

Thus, physical models of deformation of a thin elastic layer under torsion, which are described by equations (12)–(14) and (15)–(16), were revealed from the exact equations of the theory of elasticity of axisymmetric torsion with a practically acceptable accuracy.

Now we consider the elastic equilibrium of an elastic thin circular washer of radius a and height h ($h \ll a$) assuming that the washer is loaded on its upper and lower faces, as above, by tangential forces of intensities $\tau_0(r)$ and $\tau(r)$ ($0 < r < a$), respectively, and on the lateral surface, it is also loaded by tangential forces with the resultant force $T = T(a)$,

$$T = T(a) = \int_0^{2\pi} d\vartheta \int_0^h \tau_{r\vartheta}(a,z)\,dz = 2\pi\int_0^h \tau_{r\vartheta}(a,z)\,dz.$$

Since h ($h \ll a$) is small, we can assume that these forces are uniformly distributed over the side surface of the washer, and we set $T = T(a) = T_0 h$.

Under these assumptions, we determine the required characteristics of the washer. First, proceeding from (13) for $0 < r < a$, we determine the circular displacements of the washer. For this, in (13) we set $r = e^t$ ($-\infty < t < \ln a$), and as a result of this, it becomes

$$\frac{d^2u_0}{dt^2} - u_0 = \frac{f_0(t)}{G_0 h}, \quad f_0(t) = e^{2t}[\tau(e^t) - \tau_0(e^t)], \quad u_0(t) = v_0(e^t).$$

The general solution of this differential equation consists of the sum of the general solution of the homogeneous equation and a particular solution of the inhomogeneous equation, which is expressed by the well-known Cauchy formula. As a result, we have

$$u_0(t) = C_1 e^t + D_1 e^{-t} + \frac{1}{G_0 h}\int_{-\infty}^{t} \sinh(t-\xi) f_0(\xi)\,d\xi \quad (-\infty < t < \ln a).$$

Since the function $v_0(r)$ is bounded at the center of the washer $r = 0$, we should set $D_1 = 0$. Then returning to the previous variable r, we have

$$v_0(r) = C_1 r + \frac{1}{2G_0 hr}\int_0^r (r^2 - s^2)[\tau(s) - \tau_0(s)]\,ds \quad (0 \le r \le a). \quad (17)$$

The solution of equation (13) for $0 \le r \le a$ can also be represented by the formula

$$v_0(r) = C_2 r + \frac{1}{2G_0 hr}\int_r^a (s^2 - r^2)[\tau(s) - \tau_0(s)]\,ds \quad (0 \le r \le a). \quad (18)$$

IOP Conf. Series: Journal of Physics: Conf. Series **991** (2018) 012037 doi:10.1088/1742-6596/991/1/012037

Finally, using (17) and (18), we can write

$$v_0(r) = Cr + \frac{1}{4G_0hr}\int_0^a |r^2 - s^2|[\tau(s) - \tau_0(s)]\,ds \quad \left(0 \le r \le a,\ C = \frac{C_1 + C_2}{2}\right). \tag{19}$$

To find the physical meaning of the constant C, we calculate the angle of rotation of the washer around the vertical axis Oz. According to (2) and (19) we have

$$\omega_z = \frac{1}{2}\left(\frac{dv_0}{dr} + \frac{v_0}{r}\right) = C + \frac{1}{4G_0h}\int_0^a \operatorname{sign}(r-s)[\tau(s) - \tau_0(s)]\,ds \quad (0 \le r \le a).$$

Whence it follows that $\omega_z = C = \omega$ for an absolutely rigid washer ($G_0 = \infty$). Therefore, the constant $C = \omega$ represents the angle of rotation of the absolutely rigid washer around the axis Oz.

So, finally

$$\begin{aligned} v_0(r) &= \omega r + \frac{1}{4G_0hr}\int_0^a |r^2 - s^2|[\tau(s) - \tau_0(s)]\,ds, \\ \varkappa(r) &= \omega + \frac{1}{4G_0h}\int_0^a \operatorname{sign}(r-s)[\tau(s) - \tau_0(s)]\,ds \quad (\varkappa(r) = \omega_z = \omega_z(r),\ 0 \le r \le a). \end{aligned} \tag{20}$$

Let us determine the radial force $T(r) = h\tau_{r\vartheta}(r)$ in the current washer lateral surface r. Proceeding from the Hooke's law (4), one can write

$$\tau_{r\vartheta} = G_0 e_{r\vartheta} = G_0\left(\frac{dv_0}{dr} - \frac{v_0}{r}\right),$$

whence

$$\frac{d\tau_{r\vartheta}}{dr} = G_0\left(\frac{d^2v_0}{dr^2} - \frac{1}{r}\frac{dv_0}{dr} + \frac{v_0}{r^2}\right).$$

Transforming this equation further with the help of equation (13) for $0 < r < a$, we obtain

$$\frac{dT}{dr} = 2G_0h\frac{d^2v_0}{dr^2} - [\tau(r) - \tau_0(r)] \quad (0 < r < a). \tag{21}$$

On the other hand, equation (13) itself can be represented as

$$G_0h\frac{d^2v_0}{dr^2} + \frac{T(r)}{r} = \tau(r) - \tau_0(r) \quad (0 < r < a). \tag{22}$$

Further, from (21) and (22), eliminating the second-order derivative with respect to v_0, we arrive at the differential equation

$$\frac{dT}{dr} + \frac{2}{r}T = \tau(r) - \tau_0(r) \quad (0 < r < a). \tag{23}$$

The solution of equation (23), bounded for $r = 0$, is given by formula

$$T(r) = \frac{1}{r^2}\int_0^r s^2[\tau(s) - \tau_0(s)]\,ds \quad (0 \le r \le a). \tag{24}$$

Let us also write the moment equilibrium condition of the washer

$$\int_0^{2\pi} d\vartheta \int_0^a \tau_0(r)r\cdot r\,dr - \int_0^{2\pi} d\vartheta \int_0^a \tau(r)r\cdot r\,dr + 2\pi a T(a)\cdot a = 0.$$

IOP Conf. Series: Journal of Physics: Conf. Series **991** (2018) 012037 doi:10.1088/1742-6596/991/1/012037

Hence we arrive at the condition

$$\int_0^a s^2\tau(s)\,ds = \frac{M_0}{2\pi} + a^2 T(a), \quad M_0 = 2\pi\int_0^a s^2\tau_0(s)\,ds, \quad T(a) = T_0 h. \tag{25}$$

Note that if we set $r = a$ in (24), then we obtain condition (25).

Now we turn to the second boundary-value problem. Let us again consider an elastic layer $\Omega_1 = \{0 < r < \infty,\ 0 \le \vartheta < 2\pi,\ -H \le z \le 0\}$ whose shear modulus G, as in the first problem, varies in depth by the exponential law

$$G = G_1 e^{\gamma z} \quad (-H \le z \le 0,\ G_1 = \text{const},\ \gamma = \text{const}).$$

Let then the tangential forces of intensity $\tau(r)$ act on the upper boundary plane $z = 0$ of the layer Ω_1, whereas the layer lower face $z = -H$ is rigidly clamped. This mixed boundary-value problem is formulated mathematically as

$$\begin{cases} \dfrac{\partial^2 v_1}{\partial r^2} + \dfrac{1}{r}\dfrac{\partial v_1}{\partial r} + \dfrac{\partial^2 v_1}{\partial z^2} + \gamma\dfrac{\partial v_1}{\partial z} - \dfrac{v_1}{r^2} = 0 \quad (0 < r < \infty,\ -H < z < 0), \\ \tau_{\vartheta z}\big|_{z=-0} = \tau(r), \quad v_1\big|_{z=-H+0} = 0 \quad (0 < r < \infty), \\ \tau_{r\vartheta}^2 + \tau_{\vartheta z}^2 \to 0 \quad \text{when } r^2 + z^2 \to \infty. \end{cases} \tag{26}$$

Here $v_1 = v_1(r, z)$ is the component of the circular displacements of points of the layer Ω_1.

Here, too, we again construct the solution of problem (26) by the method of the integral Hankel transform. As a result, for the Hankel transformants $\bar{v}_1(z, \lambda)$, we have

$$\bar{v}_1(z,\lambda) = \frac{2e^{-\gamma z/2}\sinh[\lambda_*(z+H)/2]}{\lambda_*\cosh[\lambda_* H/2] - \gamma\sinh(\lambda_* H/2)}\frac{\bar{\tau}(\lambda)}{G_1} \quad (-H \le z \le 0,\ 0 < \lambda < \infty),$$

where the notation is the same. Whence for $z = 0$, we have

$$\bar{v}_1(0,\lambda) = \frac{2\tanh(\lambda_* H/2)}{\lambda_* - \gamma\tanh(\lambda_* H/2)}\frac{\bar{\tau}(\lambda)}{G_1} \quad (0 < \lambda < \infty).$$

Now, using the inverse Hankel transform formula, we get

$$v_1(r,0) = \int_0^\infty \bar{v}_1(0,\lambda)J_1(\lambda r)\lambda d\lambda = \frac{2}{G_1}\int_0^\infty \frac{\tanh(\lambda_* H/2)\lambda\bar{\tau}(\lambda)}{\lambda_* - \gamma\tanh(\lambda_* H/2)}J_1(\lambda r)\,d\lambda.$$

After simple transformations, we have

$$\begin{gathered} v_1(r,0) = \frac{1}{G_1}\int_0^\infty K(r,\rho)\tau(\rho)\rho\,d\rho \quad (0 < r < \infty), \\ K(r,\rho) = \int_0^\infty J_1(\lambda r)J_1(\lambda\rho)h(\lambda)\,d\lambda, \quad h(\lambda) = \frac{2\tanh(\lambda_* H/2)\lambda}{\lambda_* - \gamma\tanh(\lambda_* H/2)}. \end{gathered} \tag{27}$$

Since

$$\lambda_* = \sqrt{\gamma^2 + 4\lambda^2} = 2\lambda\left(1 + \frac{\gamma^2}{4\lambda^2}\right)^{1/2} = 2\lambda + \frac{\gamma^2}{4\lambda} + O\left(\frac{1}{\lambda^3}\right) \quad (\lambda \to \infty),$$

we can see that

$$h(\lambda) \sim 1 \quad \text{when } \lambda \to \infty.$$

Hence the function $h(\lambda)$ can be transformed as

$$h(\lambda)=\left[\frac{2\lambda\tanh(\lambda_*H/2)}{\lambda_*-\gamma\tanh(\lambda_*H/2)}-1\right]+1=1+h_0(\lambda),\quad h_0(\lambda)=\frac{(2\lambda+\gamma)\tanh(\lambda_*H/2)-\lambda_*}{\lambda_*-\gamma\tanh(\lambda_*H/2)},\quad (28)$$

and $h_0(\lambda)=O(1/\lambda)$ as $\lambda\to\infty$, $h_0(0)=-1$.

Taking into account (28), the kernel $K(r,\rho)$ can be represented as

$$\begin{aligned}K(r,\rho)&=K_0(r,\rho)+R_0(r,\rho),\\ K_0(r,\rho)&=\int_0^\infty J_1(\lambda r)J_1(\lambda\rho)\,d\lambda,\\ R_0(r,\rho)&=\int_0^\infty J_1(\lambda r)J_1(\lambda\rho)h_0(\lambda)\,d\lambda.\end{aligned}\quad (29)$$

Note that as we pass to the limit as $H\to\infty$, then the elastic layer Ω_1 becomes the lower elastic half-space $z\le 0$, and from (27) we obtain

$$v_1(r,0)=\frac{1}{G_1}\int_0^\infty L(r,\rho)\tau(\rho)\rho\,d\rho\quad(0<r<\infty),\quad L(r,\rho)=2\int_0^\infty\frac{1}{\lambda_*-\gamma}J_1(\lambda r)J_1(\lambda\rho)\,d\lambda.\quad (30)$$

Using simple transformations and separating the principle part of the kernel, we find

$$\begin{aligned}&L(r,\rho)=K_0(r,\rho)+L_0(r,\rho),\quad L_0(r,\rho)=\int_0^\infty J_1(\lambda r)J_1(\lambda\rho)g_0(\lambda)\,d\lambda,\\ &g_0(\lambda)=\frac{2\gamma}{\sqrt{\gamma^2+4\lambda^2}+2\lambda-\gamma}\quad\left(g_0(\lambda)\sim\frac{\gamma}{\lambda}\text{ when }\lambda\to\infty\right).\end{aligned}\quad (31)$$

For a homogeneous half-space ($z\le 0$), we have $L(r,\rho)=K_0(r,\rho)$.

We express the kernel $K_0(r,\rho)$ from (29), which is a well-known particular Weber–Sonin integral, in terms of a hypergeometric function. To this end, we use the well-known formulas from [13] (p. 49, Eq. 8.11(9)) and from [14] (p. 113, Eq. (36)). As a result, for kernel $K_0(r,\rho)$ from (31) we get

$$K_0(r,\rho)=\frac{r\rho}{2(r+\rho)^3}F\left(\frac{3}{2},\frac{1}{2};3;\frac{4r\rho}{(r+\rho)^2}\right)\quad(0<r,\ \rho<\infty),\quad (32)$$

where $F(a,b;c;x)$ is a hypergeometric function; the series corresponding to the function absolutely converges for $r=\rho$.

3. Contact problem of torsion of an elastic exponentially inhomogeneous layer by means of an elastic circular washer adhered to it

Let the elastic layer Ω_1 considered above, whose shear modulus varies exponentially with depth and whose lower face $z=-H$ is rigidly clamped, be subject to torsion by means of an elastic thin circular washer of shear modulus G_0, radius a, and height h adhered to the layer along its upper face $z=0$. Torsional tangential forces of intensity $\tau_0(r)$ act on the upper face of the washer. In addition, the washer on its lateral surface along the mid-height circle is loaded with torsional tangential forces of constant intensity $T=T(a)$. It is required to determine the tangential contact stresses on the contact circle of radius a.

We reduce solving the posed problem to solving an integral equation (IE), which we supplement with the contact condition

$$v_0(r)=v_1(r,0)\quad(0\le r\le a).$$

Setting the expressions for $v_0(r)$ and $v_1(r,0)$ from (20) and (30) and passing to dimensionless quantities

$$\xi = \frac{r}{a}, \quad \eta = \frac{\rho}{a}, \quad \varphi(\xi) = \frac{\tau(a\xi)}{G_1}, \quad f_0(\xi) = \frac{\tau_0(a\xi)}{G_1},$$
$$\alpha = a\lambda, \quad H_0 = \frac{H}{a}, \quad \gamma_0 = a\gamma, \quad \Lambda = \frac{G_1 a}{4G_0 h}, \quad \lambda_0 = \sqrt{\gamma_0^2 + 4\alpha^2},$$

after simple transformations, we arrive at the governing Fredholm IE of the first kind

$$\int_0^1 \{[M_0(\xi,\eta) + N_0(\xi,\eta)]\xi\eta - \Lambda|\xi^2 - \eta^2|\}\varphi(\eta)\,d\eta = \omega\xi^2 - \Lambda h_0(\xi) \quad (0 < \xi < 1), \tag{33}$$
$$M_0(\xi,\eta) = aK_0(a\xi, a\eta) = \int_0^\infty J_1(\alpha\xi)J_1(\alpha\eta)\,d\alpha, \quad h_0(\xi) = \int_0^1 |\xi^2 - \eta^2| f_0(\eta)\,d\eta,$$
$$N_0(\xi,\eta) = aR_0(a\xi, a\eta) = \int_0^\infty J_1(\alpha\xi)J_1(\alpha\eta)n_0(\alpha)\,d\alpha, \quad n_0(\alpha) = \frac{(2\alpha + \gamma_0)\tanh(\lambda_0 H_0/2) - \lambda_0}{\lambda_0 - \gamma_0 \tanh(\lambda_0 H_0/2)},$$

from which the dimensionless tangential contact stresses $\varphi(\xi)$ are determined.

In terms of the above-introduced dimensionless quantities, we also write the angle of rotation of the washer about the vertical axis $\varkappa(r)$ from (20), radial force in the washer $T(r)$ from (24), and the moment condition of the washer (25). We get, respectively,

$$\varkappa_0(\xi) = \omega + \Lambda \int_0^1 \operatorname{sign}(\xi - \eta)[\varphi(\eta) - f_0(\eta)]\,d\eta, \quad (\varkappa_0(\xi) = \varkappa(a\xi),\ 0 < \xi < 1), \tag{34}$$
$$T_0(\xi) = \frac{1}{\xi^2}\int_0^\xi \eta^2[\varphi(\eta) - f_0(\eta)]\,d\eta \quad \left(T_0(\xi) = \frac{T(a\xi)}{aG_1}\right), \tag{35}$$
$$\int_0^1 \eta^2\varphi(\eta)\,d\eta = Q_0 + S_0 \quad \left(Q_0 = \frac{M_0}{2\pi a^3 G_1},\ S_0 = \frac{T_0 h}{G_1 a},\ M_0 = 2\pi\int_0^a s^2\tau_0(s)\,ds\right). \tag{36}$$

Thus, after solving the governing IE (33), the characteristics of the elastic washer are determined by formulas (34) and (35), and with the help of (36) the dependence between the angle of rotation of the rigid washer ω and the torque $Q_0 + S_0$ is established.

The solution of IE (33) can be constructed by the method of spectral relationships for the kernel $M_0(\xi,\eta)$ [15, 16]. These relationships contain associated Legendre polynomials or equivalent Jacobi polynomials. The solution of IE (33) is represented by an infinite series in these polynomials with unknown coefficients with a distinguishing singularity for $\xi = 1$. An infinite system of linear algebraic equations is obtained for unknown coefficients. The solution of IE (33) can also be constructed by the collocation method at Chebyshev nodes. For this, we represent the solution by a finite series in the above-mentioned polynomials and use the Gaussian quadrature formulas to calculate the integrals. Then the kernel $M_0(\xi,\eta)$ can be replaced by the hypergeometric function from (32).

In the simplest particular case, when we have an elastic homogeneous half-space ($\gamma = 0$) and an absolutely rigid washer ($G_0 = \infty$ and hence $\Lambda = 0$), IE (33) takes the form

$$\int_0^1 M_0(\xi,\eta)\eta\varphi(\eta)\,d\eta = \omega\xi \quad (0 < \xi < 1), \tag{37}$$

which describes the known Reissner–Sagotzi contact problem [7]. The corresponding spectral relationship in [15, 16] is represented by the formula

$$\int_0^1 M_0(\xi,\eta)\frac{\eta^2\,d\eta}{\sqrt{1-\eta^2}} = \frac{\pi}{4}\xi \quad (0 < \xi < 1).$$

From which it follows that IE (37) has the solution

$$\varphi(\xi)=\frac{4\omega}{\pi}\frac{\xi}{\sqrt{1-\xi^2}}\quad (0<\xi<1), \tag{38}$$

that coincides with the known result of [7]. Substituting (38) into the moment equilibrium condition of the washer in (36), we come to the following relation between the hard angle of rotation of the washer and the torque:

$$8\omega=3\pi(Q_0+S_0).$$

4. On the contact interaction of an elastic continuous circular cylinder with a thin elastic cylindrical shell

First, as above, let us consider the physical model of a thin elastic circular cylindrical shell. For this, for a hollow elastic cylinder $D_0=\{a\le r\le b;\ 0\le\vartheta<2\pi;\ -\infty<z<\infty\}$ infinite in both directions with the axis Oz and the shear modulus G_0, we consider the first boundary-value problem in torsion. Let torsional tangential forces of intensity $\tau(z)$ act on the inner surface $r=a$ of the cylinder, and let torsional tangential forces of intensity $\tau_0(z)$ act on its outer surface $r=b$. With the help of (2)–(4), this problem for the circular displacements $v_0=v_0(r,z)$ is mathematically formulated as the boundary-value problem

$$\begin{cases}\dfrac{\partial^2 v_0}{\partial r^2}+\dfrac{1}{r}\dfrac{\partial v_0}{\partial r}-\dfrac{v_0}{r^2}+\dfrac{\partial^2 v_0}{\partial z^2}=0\quad (a<r<b,\ -\infty<z<\infty),\\ \left(\dfrac{\partial v_0}{\partial r}-\dfrac{v_0}{r}\right)\Big|_{r=a}=\dfrac{\tau(z)}{G_0},\quad \left(\dfrac{\partial v_0}{\partial r}-\dfrac{v_0}{r}\right)\Big|_{r=b}=\dfrac{\tau_0(z)}{G_0}\quad (-\infty<z<\infty),\\ \tau_{r\vartheta}^2+\tau_{\vartheta z}^2\to 0\quad \text{when } r^2+z^2\to\infty.\end{cases} \tag{39}$$

Now we apply the integral Fourier transform with respect to the variable z to the boundary-value problem (39) by setting

$$\{\bar v_0=\bar v_0(r,\lambda),\ \bar\tau_0(\lambda),\ \bar\tau(\lambda)\}=\int_{-\infty}^{\infty}\{v_0(r,z),\ \tau_0(z),\ \tau(z)\}e^{i\lambda z}\,dz.$$

As a result, we arrive at the one-dimensional boundary-value problem

$$\begin{cases}\dfrac{d^2\bar v_0}{dr^2}+\dfrac{1}{r}\dfrac{d\bar v_0}{dr}-\dfrac{\bar v_0}{r^2}-\lambda^2\bar v_0=0\quad (a<z<b,\ -\infty<\lambda<\infty),\\ \left(\dfrac{d\bar v_0}{dr}-\dfrac{\bar v_0}{r}\right)\Big|_{r=a}=\dfrac{\bar\tau(\lambda)}{G_0},\quad \left(\dfrac{d\bar v_0}{dr}-\dfrac{\bar v_0}{r}\right)\Big|_{r=b}=\dfrac{\bar\tau_0(\lambda)}{G_0}.\end{cases} \tag{40}$$

The general solution of differential equation (40) is given by the formula

$$\bar v_0(r,\lambda)=AI_1\left(\frac{\lambda}{r}\right)+BK_1\left(\frac{\lambda}{r}\right)\quad (a\le r\le b),$$

where $I_1(x)$ and $K_1(x)$ are modified Bessel functions of the first and third kind of order 1, respectively. Constants A and B are determined from the boundary conditions of problem (40).

IOP Conf. Series: Journal of Physics: Conf. Series **991** (2018) 012037 doi:10.1088/1742-6596/991/1/012037

Omitting the intermediate calculations, we give the final expressions for $\bar{v}_0(a,\lambda)$ and $\bar{v}_0(b,\lambda)$:

$$\begin{aligned}
G_0\Delta(\lambda)\bar{v}_0(a,\lambda) &= ab|\lambda|[I_0(a|\lambda|)K_1(a|\lambda|)+K_0(a|\lambda|)I_1(a|\lambda|)]\bar{\tau}_0(\lambda)\\
&- ab|\lambda|[I_0(b|\lambda|)K_1(a|\lambda|)+K_0(b|\lambda|)I_1(a|\lambda|)]\bar{\tau}(\lambda)\\
&+ 2a[I_1(b|\lambda|)K_1(a|\lambda|)-K_1(b|\lambda|)I_1(a|\lambda|)]\bar{\tau}(\lambda) \quad (-\infty<\lambda<\infty), \qquad (41)\\
G_0\Delta(\lambda)\bar{v}_0(b,\lambda) &= -ab|\lambda|[I_1(b|\lambda|)K_0(b|\lambda|)+I_0(b|\lambda|)K_1(b|\lambda|)]\bar{\tau}(\lambda)\\
&+ ab|\lambda|[I_1(b|\lambda|)K_0(a|\lambda|)+I_0(a|\lambda|)K_1(b|\lambda|)]\bar{\tau}_0(\lambda)\\
&+ 2b[I_1(b|\lambda|)K_1(a|\lambda|)-I_1(a|\lambda|)K_1(b|\lambda|)]\bar{\tau}_0(\lambda), \qquad (42)\\
\Delta(\lambda) &= ab\lambda^2[I_0(b|\lambda|)K_0(a|\lambda|)-I_0(a|\lambda|)K_0(b|\lambda|)]-2a|\lambda|[I_1(b|\lambda|)K_0(a|\lambda|)+K_1(b|\lambda|)I_0(a|\lambda|)]\\
&+ 2b|\lambda|[I_0(b|\lambda|)K_1(a|\lambda|)+I_1(a|\lambda|)K_0(b|\lambda|)]+2I_1(a|\lambda|)K_1(b|\lambda|)-2I_1(b|\lambda|)K_1(a|\lambda|). \qquad (43)
\end{aligned}$$

Here $I_0(x)$ and $K_0(x)$ are modified Bessel functions of order 0.

Next, we write the well-known representations of the modified Bessel functions [17]:

$$\begin{aligned}
&I_0(x)=\sum_{m=0}^{\infty}\frac{x^{2m}}{2^{2m}(m!)^2}, \quad I_1(x)=\sum_{m=0}^{\infty}\frac{x^{2m+1}}{2^{2m+1}m!l(m+1)!},\\
&K_0(x)=-I_0(x)\ln\frac{x}{2}+\sum_{m=0}^{\infty}\frac{x^{2m}\psi(m+1)}{2^{2m}(m!)^2}, \quad \psi(x)=\frac{\Gamma'(x)}{\Gamma(x)}, \qquad (44)\\
&K_1(x)=I_1(x)\ln\frac{x}{2}+\frac{1}{x}-\frac{1}{2}\sum_{m=0}^{\infty}\frac{x^{2m+1}[\psi(m+2)+\psi(m+1)]}{2^{2m+1}m!(m+1)!}, \quad (x>0),
\end{aligned}$$

where $\Gamma(x)$ is the well-known Euler gamma function. Now the functions appearing in (41)–(43) can be expanded in power series in λ, using representations (44) and restricting ourselves to terms of order λ^2. In addition, we set $b=a+h$ and, in the corresponding expansions in h, confine ourselves to terms of order h. After necessary transformations and calculations, we have

$$\begin{aligned}
-G_0h(2+a^2\lambda^2)\bar{v}_0(a,\lambda) &= -a(a+h)\left\{1+\frac{a^2}{4}[1+\psi(1)-\psi(2)]\lambda^2\right\}\bar{\tau}_0(\lambda)\\
&+a\left\langle a-h+\frac{a^2(a+h)}{4}\{1+[\psi(1)-\psi(2)]\}\lambda^2\right\rangle\bar{\tau}(\lambda), \qquad (45)\\
-G_0h(2+a^2\lambda^2)\bar{v}_0(b,\lambda) &= a\left\{a+\frac{a^2}{4}(a+2h)[1+\psi(1)-\psi(2)]\lambda^2\right\}\bar{\tau}(\lambda)\\
&-a\left\langle a+2h+\frac{a^2}{4}\{a+(a+2h)[\psi(1)-\psi(2)]+2h\}\lambda^2\right\rangle\bar{\tau}_0(\lambda). \qquad (46)
\end{aligned}$$

Returning to the originals in (45) and (46), we obtain

$$\begin{aligned}
G_0h\left(a^2\frac{d^2v_0}{dz^2}-2v_0\right) &= -a(a+h)\tau_0(z)+\frac{1}{4}a^3(a+h)[1+\psi(1)-\psi(2)]\tau_0''(z)\\
&+a(a-h)\tau(z)-\frac{a^3}{4}(a+h)[1+\psi(1)-\psi(2)]\tau''(z) \quad (-\infty<z<\infty), \qquad (47)\\
G_0h\left(a^2\frac{d^2v_0}{dz^2}-2v_0\right) &= a^2\tau(z)-a(a+h)\tau_0(z)-\frac{a^3}{4}(a+2h)[1+\psi(1)-\psi(2)]\tau''(z)\\
&+\frac{a^3}{4}\{a+(a+2h)[\psi(1)-\psi(2)]+2h\}\tau_0''(z) \quad (-\infty<z<\infty). \qquad (48)
\end{aligned}$$

Thus, in the framework of the accepted accuracy, the mechanical behavior of the inner thin cylindrical layer ($r = a$) is described by equation (47), and the outer thin cylindrical layer ($r = b$), by equation (48).

If in (47), (48) we neglect the terms conditioned by the second-order derivatives of the surface loads, then these relations simplify and take the forms

$$G_0 h\left(a^2\frac{d^2 v_0}{dz^2} - 2v_0\right) = -a(a+h)\tau_0(z) + a(a-h)\tau(z) \quad (-\infty < z < \infty), \tag{49}$$

$$G_0 h\left(a^2\frac{d^2 v_0}{dz^2} - 2v_0\right) = a^2\tau(z) - a(a+h)\tau_0(z) \quad (-\infty < z < \infty), \tag{50}$$

respectively. Now we construct the solution of the problem of torsion of a solid infinite circular cylinder $D_1 = \{0 \le r \le a,\ 0 \le \vartheta < 2\pi,\ -\infty < z < \infty\}$ with the shear modulus G_1, when torsional tangential forces of intensity $\tau(z)$ act on its surface $r = a$. This boundary-value problem is formulated mathematically as follows:

$$\begin{cases} \dfrac{\partial^2 v_1}{\partial r^2} + \dfrac{1}{r}\dfrac{\partial v_1}{\partial r} - \dfrac{v_1}{r^2} + \dfrac{\partial^2 v_1}{\partial z^2} = 0 \quad (0 < r < a,\ -\infty < z < \infty), \\ \left(\dfrac{\partial v_1}{\partial r} - \dfrac{v_1}{r}\right)\Big|_{r=a} = \dfrac{\tau(z)}{G_1} \quad (-\infty < z < \infty), \\ \tau_{r\vartheta}^2 + \tau_{\vartheta z}^2 \to 0 \quad \text{when } |z| \to \infty, \end{cases} \tag{51}$$

where $v_1 = v_1(r, z)$ are circular displacements of cylinder points. The solution of the boundary-value problem (51) can also be easily obtained by means of an integral Fourier transform with respect to the variable z. Proceeding as above, after separating the principal singular part of the kernel and its slowly convergent part, we have

$$\frac{dv_1(a,z)}{dz} = \frac{1}{\pi G_1}\int_{-\infty}^{\infty}\left[\frac{1}{u-z} + \frac{3\pi}{4a}\operatorname{sign}(u-z) + N(u-z)\right]\tau(u)\,du \quad (-\infty < z < \infty), \tag{52}$$

$$N(z) = \frac{1}{2a}\int_0^\infty \frac{2a^2\lambda^2 R(a\lambda) - 3}{\lambda}\sin(\lambda z)\,d\lambda, \quad R(a\lambda) = \frac{a\lambda[I_1(a\lambda) - I_0(a\lambda)] + 2I_1(a\lambda)}{a\lambda[a\lambda I_0(a\lambda) - 2I_1(a\lambda)]} \quad (\lambda > 0),$$

$$R(a\lambda) \sim 4/a^2\lambda^2 \quad (\lambda \to +0); \quad R(a\lambda) \sim 3/2a^2\lambda^2 \quad (\lambda \to \infty).$$

Now let us formulate the contact problem. Let the thin elastic cylindrical shell $D_0^{(c)} = \{a \le r \le a+h,\ 0 \le \vartheta < 2\pi,\ -c \le z \le c\}$ $(h \ll a)$ of finite length $2c$ and the shear modulus G_0, whose mechanical behavior is described by equation (49) (or (50)), be adhered to the cylinder along the segment $-c \le z \le c$ of its surface. We further assume that torsional tangential forces of intensity $\tau_0(z)$ act on the shell lateral surface $r = a + h$ and, in addition, the torques $M_{\pm c}$ act in the end face sections $z = \pm c$ of the shell $D_0^{(c)}$. It is required to determine the distribution of contact tangential stresses $\tau(z)$ in the contact area $-c \le z \le c$ and the torques M_z in any section z of shell.

We reduce solving the posed contact problem to solving the singular integral equation (SIE). For this, we write differential equation (49) in the form

$$\frac{d^2 v_0}{dz^2} - k^2 v_0 = f(z) \quad (-c < z < c),$$

$$k = \frac{\sqrt{2}}{a}, \quad f(z) = \frac{1}{a^2 h G_0}[a(a-h)\tau(z) - a(a+h)\tau_0(z)].$$

It has the following general solution

$$v_0(z) = A\cosh(kz) + B\sinh(kz) - \frac{1}{2k}\int_{-c}^{c} e^{-k|z-u|} f(u)\,du \quad (-c \leq z \leq c). \tag{53}$$

To determine the constants A and B, we should find the torque M_z in the shell section z. We can write

$$M_z = \int_0^{2\pi}\int_a^{a+h} \tau_{\vartheta z} r^2\,dr\,d\vartheta = 2\pi G_0 \int_a^{a+h} e_{\vartheta z} r^2\,dr = 2\pi G_0 \frac{dv_0}{dz}\int_a^{a+h} r^2\,dr = \frac{2\pi G_0}{3}\frac{dv_0}{dz}[(a+h)^3 - h^3].$$

After necessary simplifications, we obtain

$$M_z = 2\pi a^2 h G_0 \frac{dv_0}{dz}. \tag{54}$$

On the other hand, considering the moment equilibrium condition of the shell part $[-c, z]$, after some simplifications we obtain

$$M_z = M_{-c} - 2\pi a(a+2h)\int_{-c}^{z} \tau_0(u)\,du + 2\pi a^2 \int_{-c}^{z} \tau(u)\,du \quad (-c \leq z \leq c), \tag{55}$$

which implies the moment equilibrium condition of the shell for $z = c$:

$$\int_{-c}^{c} \tau(u)\,du = \left(1 + \frac{2h}{a}\right)\int_{-c}^{c} \tau_0(u)\,du + \frac{M_c - M_{-c}}{2\pi a^2}. \tag{56}$$

Now from (54) we get the boundary condition

$$\left.\frac{dv_0}{dz}\right|_{z=\pm c} = \frac{M_{\pm c}}{2\pi a^2 h G_0}. \tag{57}$$

Proceeding from (53) and satisfying boundary conditions (57), we determine the constants A and B. As a result, after simple manipulations, we have

$$\frac{dv_0}{dz} = \frac{(M_c - M_{-c})\sinh(kz)}{4\pi a^2 h G_0 \sinh(kc)} + \frac{(M_c + M_{-c})\cosh(kz)}{4\pi a^2 h G_0 \cosh(kc)} + \frac{a-h}{4ahG_0}\left\langle \int_{-c}^{c}\left\{\mathrm{sign}(z-u)e^{-k|z-u|}\right.\right.$$
$$\left. - e^{-kc}\left[\frac{\sinh(kz)}{\sinh(kc)}\cosh(ku) + \frac{\cosh(kz)}{\cosh(kc)}\sinh(ku)\right]\right\}\tau(u)\,du - \frac{1+h_0}{1-h_0}\int_{-c}^{c}\left\{\mathrm{sign}(z-u)e^{-k|z-u|}\right.$$
$$\left.\left. - e^{-kc}\left[\frac{\sinh(kz)}{\sinh(kc)}\cosh(ku) + \frac{\cosh(kz)}{\cosh(kc)}\sinh(ku)\right]\right\}\tau_0(u)\,du\right\rangle \quad \left(h_0 = \frac{h}{a},\ \frac{1+h_0}{1-h_0} \approx 1 + 2h_0\right). \tag{58}$$

Then we substitute (52) and (58) into the condition of contact between the shell and the solid elastic cylinder:

$$\frac{dv_0}{dz} = \frac{dv_1}{dz} \quad (-c < z < c).$$

Hence, introducing the dimensionless quantities

$$\xi = \frac{z}{c},\quad \eta = \frac{u}{c},\quad \psi_0(\eta) = \frac{\tau(c\eta)}{G_1},\quad k_0 = \frac{\sqrt{2}c}{a} = kc,\quad M_{\pm c}^{(0)} = \frac{M_{\pm c}}{a^2 h G_0},\quad h_0 = \frac{h}{a},$$
$$X_0(\eta) = \frac{\tau_0(a\eta)}{G_1},\quad L_0(\xi) = cN(c\xi),\quad \lambda_0 = \frac{c(a-h)G_1}{2ahG_0} = \frac{(1-h_0)G_1}{2\delta_0 G_0},\quad \delta_0 = \frac{h}{c},$$

we arrive at the following governing SIE of the contact problem under consideration:

$$\frac{1}{\pi}\int_{-1}^{1}\left[\frac{1}{\eta-\xi}+\frac{3\pi}{4}\mathrm{sign}(\eta-\xi)+L_0(\eta-\xi)-\pi\lambda_0 H_0(\xi,\eta)\right]\psi_0(\eta)\,d\eta=f_0(\xi)\quad(-1<\xi<1),$$

$$H_0(\xi,\eta)=\mathrm{sign}(\xi-\eta)e^{-k_0|\xi-\eta|}-e^{-k_0}\left[\frac{\sinh(k_0\xi)}{\sinh k_0}\cosh(k_0\eta)+\frac{\cosh(k_0\xi)}{\cosh k_0}\sinh(k_0\eta)\right],\tag{59}$$

$$f_0(\xi)=-(1+2h_0)\lambda_0\int_{-1}^{1}H_0(\xi,\eta)\mathrm{X}_0(\eta)\,d\eta+\frac{(M_c^{(0)}-M_{-c}^{(0)})\sinh(k_0\xi)}{4\pi\sinh k_0}+\frac{(M_c^{(0)}+M_{-c}^{(0)})\cosh(k_0\xi)}{4\pi\cosh k_0},$$

where the function $L_0(\xi)$ is defined by the function $N(\eta)$ from (52).

In dimensionless quantities, condition (56) becomes

$$\int_{-1}^{1}\psi_0(\eta)\,d\eta=(1+2h_0)\int_{-1}^{1}\mathrm{X}_0(\eta)\,d\eta+\frac{M_c^{(0)}-M_{-c}^{(0)}}{2\pi}\chi\delta_0\quad\left(\chi=\frac{G_0}{G_1},\ \delta_0=\frac{h}{c}\right),\tag{60}$$

and the expression for the torque from (55) become

$$M_z^{(0)}(\xi)=M_{-c}^{(0)}-\frac{2\pi(1+h_0)}{\chi\delta_0}\int_{-1}^{\xi}\mathrm{X}_0(\eta)\,d\eta+\frac{2\pi}{\chi\delta_0}\int_{-1}^{\xi}\psi_0(\eta)\,d\eta\quad(-1\le\xi\le1).\tag{61}$$

The solution of SIE must satisfy condition (60), after which the dimensionless torques $M_z^{(0)}(\xi)=M_z/a^2hG_0$ in the shell sections are determined by formula (61). A sufficiently effective solution of SIE (59), (60) can be obtained by the known numerical-analytical method for solving the SIE [11].

Note that in a particular case where $\tau_0(z)\equiv0$, $M_{-c}=M_c$, and hence $\tau(-z)=-\tau(z)$, besides $\tau(0)=0$, it follows from (55) that

$$\left.\frac{dM_z}{dz}\right|_{z=0}=0.$$

This means that the greatest torque appears in the middle section of the shell $z=0$.

The results of numerical analysis of the problems considered here is beyond the scope of this contribution and will be presented as a separate paper.

Conclusions

The paper deals with nonclassical contact problems related to the interaction of thin-walled elements with massive deformable bodies under torsion. The governing equations of physical models of thin-walled elements are derived from the exact equations of mathematical theory of elasticity. On the one hand, they are quite simple to use in specific problems, and, on the other hand, they provide rather high practical accuracy. The results and approaches discussed in the paper can be used to study a rather wide class of problems of contact interaction of thin-walled elements and massive elastic bodies under torsion, which are of both theoretical and practical interest.

References

[1] Love A E H 1996 *Treatise on the Mathematical Theory of Elasticity* (New York: Dover Publications) p 643
[2] Arutjunyan N K and Abramyan B L 1963 *Torsion of Elastic Bodies* (Moscow: Fizmatgiz) [in Russian]
[3] Solyanik-Krassa K V 1949 *Torsion of Shafts of Variable Cross-Section* (Moscow: Gostekhizdat) p 167 [in Russian]
[4] Lekhnitskii S G 1971 *Torsion of Anisotropic and Non-Homogeneous Beams* (Moscow: Nauka) [in Russian]
[5] Muskhelishvili N I 1966 *Some Basic Problems of the Mathematical Theory of Elasticity* (Moscow: Nauka) [in Russian]

IOP Conf. Series: Journal of Physics: Conf. Series **991** (2018) 012037 doi:10.1088/1742-6596/991/1/012037

[6] Novojilov V V 1958 *Theory of Elasticity* (Moscow: Sudpromgiz) [in Russian]
[7] Reissner E and Sagoci H F 1944 Forced Torsional Oscillation of an Elastic Half-Space *J. Appl. Phys.* **15** 652–4
[8] Galin L A (Editor) 1976 *Development of the Theory of Contact Problems in the USSR* (Moscow: Nauka) p 493 [in Russian]
[9] Melan E 1932 Ein Beitrag zur Theorie geschweißter Verbindungen *Arch. Appl. Mech.* **3** (2) 123–9
[10] Bufler H 1964 Zur Krafteinleitung in Scheiben über geschweißte oder geklebte Verbindungen *Österr. Ing. Arch.* **18** (3-4) 284–92
[11] Erdogan F, Gupta G D, and Cook T S 1973 The Numerical Solutions of Singular Integral Equations In *Methods of Analysis and Solution of Crack Problems* (Leyden: Noordhoff Intern Publ) pp 368–425
[12] Sneddon I 1951 *Fourier Transforms* (New York-Totonto: McGraw-Hill)
[13] Bateman H and Erdelyi A 1954 *Tables of Integral Transforms* vol 2 (New York: McGraw-Hill)
[14] Bateman H and Erdelyi A 1953 *Higher Transcendental Functions* vol 1 (New York: McGraw-Hill)
[15] Popov G Ia 1964 Certain properties of classical polynomials and their applications in contact problems *J. Appl. Math. Mech.* **28** (3) 544–55
[16] Mkhitarian S M 1984 Spectral relationships for the integral operators generated by a kernel in the form of a Weber–Sonin integral, and their applications to contact problems *J. Appl. Math. Mech.* **48** (1) 67–74
[17] Bateman H and Erdelyi A 1953 *Higher Transcendental Functions* vol 2 (New York: McGraw-Hill)

About the possibility of obtaining cementitious soil composites of high strength on the basis of belozems of carbonate composition

K A Karapetyan[1*], S G Hayroyan[1,2], and E S Manukyan[1]**
[1]Institute of Mechanics of the National Academy of Sciences of Armenia, Yerevan, Armenia
[2]Yerevan State University, Yerevan, Armenia
E-mail: *koryan@mechins.sci.am, **hairoyan@ysu.am

Abstract. The problem of manufacturing high strength cementitious soils based on belozems of carbonate composition, which experience compression (no less than 10 MPa), without application of surface active substances is considered.

The portland cement of type 400 was used as a binding agent to develop compositions of cementitious soil composites, and the ordinary pipe water was used to obtain solutions of cementitious soils.

The chemical and mineralogical composition of the initial ingredients and the granulometric composition of belozems were determined.

The measurements showed that the upper and lower plasticity limits, the optimum moisture content, and the maximal density of the skeleton of belozems, as well as the considered compositions of cementitious soils, are insignificant, while the plasticity index of cementitious soils is less than one for belozems.

It is experimentally proved that an increase in the portland cement amount lead to an increase in the compressive strength of cementitious soils with a decreasing speed. But for the same amount of portland cement used in the cementitious soil compositions, the values of the strength ratio of the pieces tested at the age of 60 and 28 days remain the same and are approximately equal to 1.2.

A comparison of experimental data showed that it seems to be real to manufacture a cementitious soil on the basis of belozems of carbonate composition, which contain 10% of cement of the weight of dry mixture and have strength more than 10 MPa, without adding any surfactants to the material composition.

1. Introduction

Abroad, the cementitious soils (ground-concrete) are widely used as building materials for constructing roads and runways in airports (layers under external coverings), for covering the canals and dams, for manufacturing blocks and stones, and for constructing one- and two-storey houses as well as monolithic ground mixes for preparation for floors [1, 2].

Modified cement mortar compositions (cementitious soils) and technologies for their application in construction of underground parts of buildings have recently been developed up to a mark of ±0.00, including foundations of various types (belt, column, cushion-walls), cellar walls, and socles [1]. In this case, as a rule, preference is given to the prefabricated cement mortar (cementitious soils) as the most suitable material for constructing the underground parts of buildings.

Such a wide application of cementitious soil materials usually composed of clayey rocks rich in calcium compounds (such as clay, sandy loam, loams, etc. [2]) is due to the relatively low laboriousness of manufacturing the material, its relatively good physical and mechanical characteristics, and its cheapness, especially if the soil is taken from a pit excavated under the construction [1].

It is known that the structurally unstable, under-compacted subsidence grounds (soils) are widespread in almost all continents of the globe. According to geological engineers, such soils occupy about 10% of the total land area, and the thickness of their formation varies from 10 to 40 m [3]. In some regions of the globe, these formations reach the thickness of 100 m [3].

Because of structurally under-compacted soils in Armenia, much attention is paid to belozems, which are most common in some areas of Yerevan and in the districts of Armavir, Kotayk, Lori, etc. [4].

Belozem is the name of strongly carbonate, often gypsum, suffosion-unstable, subsidence, light dusty-sand loams and heavy dusty sandy loams. On the other hand, because of their suffosion stability and subsidence, the belozems are most unfavorable [5].

The low water resistance of belozems, the appearance and intensive development of suffosion phenomena in them were recorded already in the 1920s, when the irrigation systems were operated in Armenia [5]. Despite this, in some regions of the country, active construction was carried out on the territory of belozem distribution, and its negative consequences were manifested rather soon. As an example, we can point out some buildings in the residential area of Ajapnyak, Yerevan, whose normal exploitation is currently violated. During the exploitation, the sewage waters penetrated into the grounds of basements of these buildings and caused their intensive saturation. The resulting subsidence was accompanied by formation of large cracks in the buildings and, in some cases, by subsidence of their individual parts.

The observations show that similar phenomena develop especially intensively in hydraulic engineering structures, where the filtration losses in soils are of catastrophic character (slakening of the belozems occurs very quickly — within 3–5 minutes [5]) and the water is accumulated in developed large voids [5].

It obviously follows from the above that, due to their engineering-geological features, the building on the territory of under-compacted belozem distribution requires artificial reinforcement, i.e., it is necessary to increase the physical characteristics of buildings including their suffosion stability and mechanical strength.

The goal in this work is to develop, without using any plasticizers, compositions of cementitious soil composites of increased strength on the basis of belozems which satisfy the requirements for building materials of this type.

2. Experimental part

The cement soil usually contains ordinary bound soils (clay, loam, sandy loam), mineral binders (for example, portland cement of at least 400 grade [1]), and water. It should be noted that, to increase the technological characteristics of the cement mortar (for instance, its workability) and to increase the indices of its physical and mechanical properties, purposefully selected plasticizers are sometimes introduced into the material composition [1].

In the study outlined in the framework of the problems considered in this paper, the following starting materials were used:

- the belozems taken from the areas adjacent to the territories of the Institute of Physics located in the residential district of Ajapnyak, Yerevan;
- the portland cement (type 400) produced by the Ararat Cement Plant (Republic of Armenia);
- the ordinary pipe water.

Table 1. Chemical and mineralogical composition of the portland cement.

Chemical composition							Mineralogical composition					
CaO	SiO_2	Al_2O_3	Fe_2O_3	MgO	SO_3	Loss on ignition	Hydroxide moisture	Insoluble sediment	$3CaOSiO_2$	$2CaOSiO_2$	$3CaOAl_2O_3$	$4CaOAl_2O_3Fe_2O_3$
62.7	21.8	5.1	3.1	3.2	2.2	0.95	0.48	1.35	52	23	7–9	15

Table 2. Chemical composition of belozems.

SiO_2	Fe_2O_3	FeO	TiO_2	Al_2O_3	CaO	MgO	MnO	P_2O_5	SO_3	Na_2O	K_2O	H_2O	Loss on ignition
30.64	1.56	1.81	0.35	5.60	27.88	1.99	traces	0.15	no	1.70	1.20	1.92	24.82

Table 3. Characteristics of the portland cement.

Volume weight, g/cm^3	Specific weight, g/cm^3	Stiffening deadline (GOST 310.3-76)		Normal density of cement mortar (paste), % (GOST 310.3-76)	Ultimate strength for age of 28 days, MPa (GOST 310.4-76)	
		start (min.)	end (hour)		for bending	for compression
1.21	3.04	48–51	8.0	24–25	5.5	38.6

Table 4. Average granulometric composition of belozems.

Particle size, mm	2–1	1–0.5	0.5–0.25	0.25–0.1	0.1–0.05	0.05–0.01	0.01–0.005	< 0.005
%	2.0	7.0	6.2	13.1	15.0	7.2	40.0	9.5

It is known that one of the main stages of the process of developing the composite materials, including the cementitious soils, consists of certain analyzes of the main ingredients of these materials. Therefore, the chemical and mineralogical analyzes of the portland cement and belozems were carried out (tables 1 and 2). The parameters of the physical and mechanical characteristics of the portland cement (table 3) and the granulometric composition of belozems sieved through sieve No. 2 (table 4) were determined.

The data presented in table 2 show that the component SiO_2 contained in belozems has a property contributing to the process of concrete formation in the cementitious soil, and the component CaO has a property contributing to the formation of concrete and increasing the cementitious soil strength by carbonization. Also, we note that the components Al_2O_3, K_2O, and Na_2O of belozems (table 2) can significantly increase the cement mortar workability.

As a result of analysis of the salt composition of belozems, it was found that no medium-soluble salts are contained in the soil, and the amount of readily soluble and sparingly soluble salts (CaO_3) is 20.02% and 20.3% of the mass, correspondingly.

Table 5. Water-physical parameters of belozems and cementitious soil.

Cement content, %	Plasticity limit			Optimum moisture content, W_{opt}	Max. density of the skeleton, g/cm^3
	upper limit, W_L	lower limit, W_P	plasticity index, I_p		
0	0.245	0.190	0.055	0.212	1.58
5.0	0.230	0.187	0.043	0.200	1.585
6.0	0.244	0.196	0.048	0.211	1.62
8.0	0.248	0.198	0.50	0.217	1.62
10.0	0.230	0.188	0.042	0.209	1.61

According to the data in table 4, the belozems used in the cement mortar compositions are represented by dusty sandy loams.

The investigation stages considered here are the following ones:

- development of preliminary compositions of cement;
- obtaining dry mixtures from original ingredients;
- production of cementitious soil mixtures on the basis of dry mixtures and water;
- production of test pieces (cylinders of diameter of 5.0 cm and height of 20.0 cm) from cement mortar mixtures by direct compression in appropriate forms;
- determination of the average density and compressive strength of test pieces at the age of 28 and 60 days.

The investigation stages are summarized below.

In view of the above discussion, in the process of developing the cementitious soil compositions, the portland cement of type 400 (table 3) was used with an addition of cement in an amount of 5, 6, 8, and 10% of the weight of dry mixture of the main ingredients. This approach to the development of preliminary cementitious soil compositions was applied with the aim of revealing the laws of changes in the indices of the soil basic physical and mechanical characteristics depending on the content of cement in its composition.

A screw mixer was used to prepare both dry (up to 1600 g) and cementitious soil (up to 1900 g) mixtures in each case under study. The mixing time of dry and cement mortar mixtures was 15 min for the above-indicated amounts. The degrees of homogeneity of the mixtures were determined by a known method.

The cylindrical test pieces were manufactured from cementitious soil mixtures was by direct pressing with the value of pressure 4.8–5.3 MPa ensuring the production of a material with the maximum density of the skeleton.

After manufacturing, the test pieces were released from the molds after 14 days, and further, till the time of testing at the age of 28 and 60 days, they were stored in a laboratory room at an average temperature of 23°C and a relative humidity of 57%.

The strength of cylindrical test pieces was determined on a universal testing machine ZD 10/90 at a loading speed of 3 mm/min. For each case under study, 3–4 test pieces (twins) were tested. The maximum spread of strength indices in relation to their average arithmetic value was equal to +5.7% and −6.2%.

3. Discussion of the obtained results

The indices of main water-physical characteristics of belozems and cement mortar. determined by standard methods [6], are given in table 5.

According to the data in table 5, the difference between the indices of both the upper and lower limits of belozem plasticity and the considered compositions of the cementitious

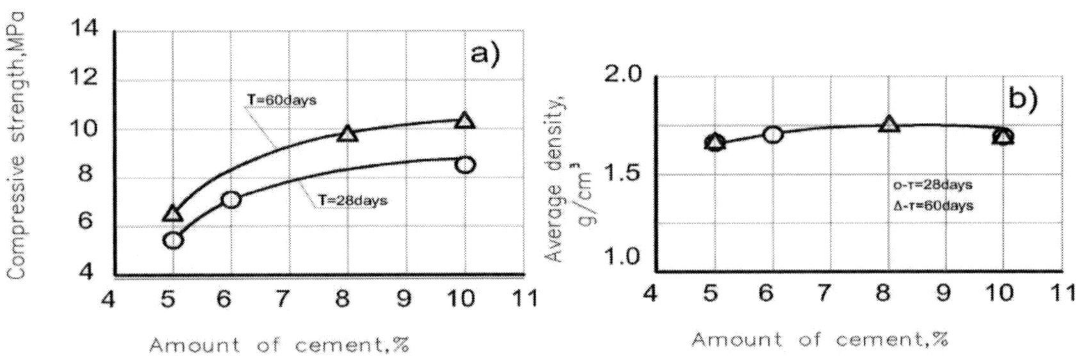

Figure 1. Curves of the strength (*a*) and the average density (*b*) of cementitious soil cylindrical test pieces depending on the amount of portland cement.

soil was insignificant. The values of these parameters vary within the limits, from 0.230 to 0.248 and from 0.188 to 0.198, correspondingly. A similar phenomenon occurs for the optimal humidity (from 0.200 to 0.217), as well as for the maximum density of the skeleton (from 1.58 to 1.62 g/cm^3) of these materials.

It also follows from the data in table 5 that the plasticity index of the considered cementitious soil composition is always lower than that of belozem. However, no experimentally determined clear pattern of variation in the plasticity index of cementitious soil in dependence on the amount of portland cement used in its composition has been observed.

The experimentally obtained results in the form of labels, whose ordinates correspond to the average arithmetic values of the measured characteristics of cementitious soil, are shown in figure 1.

Before analyzing the data given in figure 1, we note that all cementitious soil cylindrical test pieces acquired a light pink color at the time of the tests.

Comparing the data given in figure 1b, we see that the value of the average density of cementitious soil contained in the same amount of portland cement slightly depends on the age of the considered test pieces at the time of measurements. At the same time, an increase in the consumed portland cement amount from 5% to 6–8% of the weight of dry mixture leads to an increase in the average density of the cementitious soil by approximately 3%. In comparison with the average density of the material contained in the portland cement composition in the amount of 6–8%, the further increase in the portland cement amount from 8% to 10% leads to a decrease in the average density by approximately 1.2%.

From the data given in figure 1*a*, we note that the nature of changes in the compressive strength of cementitious soil cylindrical test pieces in dependence on the amount of portland cement in their compositions remains the same independently of the material age at the time of tests. In particular, an increase in the portland cement amount leads to an increase in the strength of test pieces with a significantly decreasing speed. Thus, an increase in the portland cement amount from 5% to 6%, from 6% to 8%, and from 8% to 10% leads to an increase in the strength of test pieces by approximately 30%, 17%, and 4%, correspondingly, for the composition age of 28 days, and by 25%, 18%, and 5% for the composition age of 60 days.

Comparing the data illustrated in the figure, we see that, for the same amount of portland cement in the cementitious soil composition, the values of the strength ratio of the test pieces tested at the age of 60 days and 28 days remain the same and are approximately equal to 1.2.

It should also be noted that the studies of the water resistance of the pieces formed in the compression testing of experimental cylinders showed that, during 3 weeks in the aqueous medium, the cementitious soil composites with different content of portland cement did not show any sign of slakening.

Conclusions

Under the normal conditions of hardening, the cementitious soil with the cement content of 120 kg per 1 m^3 attains the compressive strength of 1.6 MPa after 7 days, and 2.0 MPa after 28 days, but for the cement content of 180 kg per 1 m^3, the cementitious soil strength increases from 25% to 70% [7].

The cement suitable for operation in relatively few loaded parts of building structures (for example, foundations of various types) must have compressive strength of at least 10 MPa. To obtain a material with such a compressive strength and the cement consumption of at most 12% of the weight of dry mixture, several surfactants are usually introduced into its composition [1, 3, 8].

The cementitious soil grade for compressive strength is determined according to GOST 18105-06 on test pieces at the age of 90 days. This permits determining the indicated characteristics of cementitious soil at the age of 28 days multiplied by 1.5 [14]. In this case, one can obtain the compressive strength of cementitious soil using the test pieces, i.e., cylinders of diameter and height of 5.0 cm, with subsequent transition to the standard cubicle of dimensions $15.0 \times 15.0 \times 15.0$ cm with a coefficient of 0.7 [9].

The cementitious soil cylindrical test pieces of diameter of 5.0 cm and height of 20.0 cm, made on the basis of belozems and portland cement of grade 400 (the cement content in the material is 10% of the weight of dry mixture or 160 kg per 1 m^3), have compressive strength of 8.5 MPa at the age of 28 days.

Taking into account the above-mentioned recommendations about the dimensions of the experimental cylindrical test pieces whose use permits establishing the compressive strength level of cementitious soil and the experimentally confirmed patterns of variation in the compressive strength of experimental concrete test pieces [10] depending on the ratio of their height to the cross-sectional dimensions, one can draw the following conclusion: it seems to be real to manufacture a cementitious soil on the basis of belozems of carbonate composition, which contain 10% of cement of the weight of dry mixture and have strength more than 10 MPa, without adding any surfactants to the composition of the material.

References

[1] Ramzanov A A, Badaeva A D, Lanin E B, and Alnashash T A 2015 Soil-concrete in a foundation construction *Stroit. Unik. Zdanii Soorozh.* No. 3 (30) 111–28
[2] Minke G 2004 *Clay Concrete and Its Application* (Kaliningrad: Yantarny skaz) p 232
[3] Ananiev V P and Potapov A D 2005 *Engineering Geology* (Moscow: Vysshaya Shkola) p 400
[4] Sidorenko A V (Editor) 1968 *The Hydrogeology of the USSR* vol XI *Armyanskaya SSR* (Moscow: Nedra) p 352
[5] Boshnagyan P S 1956 *Questions of Geology and Hydrogeology* (Yerevan: Izd. AN ArmSSR)
[6] Trofimov V T, Korolev V A, Voznesensky E A, et al. 2005 *Soil Science* (Moscow: Izdat. MGU) p 1024
[7] Chudinov S A 2010 Research of influence of technology factors on soil cement durability *Vestnik Povolzhsk. Gos. Tekhn. Univ. Ser. Les. Ekol. Prirodopol;z.* No. 1 46–52
[8] Chudinov S A 2011 In-process Testing of the Grounds Reinforced with Portland Cement Containing Polyelectrolyte Additive *Izv. VUZov. Lesnoy Zh.* **6** 58–61
[9] Ivanov V V and Eremin V Ya 2014 Up-to-date technologies and materials for construction, repairs, and reconstruction of buildings' foundations *Gradostroit.* No. 1 (29) 78-81
[10] Karapetyan K S 1965 Influence of anisotropy on the creep of concrete during the compression as a function of the test piece height *Dokl. AN ArmSSR* **40** (5) 279–83

Modelling of mechanical and filtration processes near the well with regard to anisotropy

V I Karev*, D M Klimov, Yu F Kovalenko***, and K B Ustinov******

Ishlinsky Institute for Problems in Mechanics of the Russian Academy of Sciences, Moscow, Russia

E-mail: *klimov@ipmnet.ru, **wikarev@ipmnet.ru, ***perfolinkgeo@yandex.ru, ****ustinov@ipmnet.ru

Abstract. A geomechanical approach to modeling deformation and seepage is presented. Three stages of modeling are described: choice of an appropriate mechanical model and its adaptation to the case in question, experimental determination of parameters of the model, simulation of processes of seepage for particular configurations of the well. The applied model allows describing the main specific characteristics of mechanical behavior of the collector: the influence of the pore pressure on deformation; the influence of not only shear but also comprehensive stresses and pore pressure on the transition to inelastic behavior; the appearance of inelastic volumetric deformation and its nontrivial dependence on the stress state; the anisotropy of elastic, strength and seepage properties; non-obvious dependence of permeability on the stress strain state. The model unites essential characteristics of Hill's plastic flow theory for anisotropic materials and the Drucker–Prager theory for inelastic deformation of soils. The results of experimental determination of the involved parameters obtained using true triaxial loading system for the collector of Vladimir Filanovsky field in the Caspian Sea are presented.

1. Introduction

The process of hydrocarbon fluid seepage into productive wells is affected by many factors, one of which, and not the least significant, is the changes in the pore structure, and hence in permeability, due to deformation processes in collectors governed by the stress state. The situation is aggravated due to the back influence of seepage in rearrangement of the stress strain state, and hence it is necessary to solve the coupled elastic-plastic and filtration problem.

The approach of modeling used throughout the current research consists in the following key stages:

- choosing and, if necessary, developing an appropriate model of deformation and filtration;
- experimental determination of the parameters involved;
- solving the coupled mechanical (determining the stress strain state) and seepage (determining the fluid flow and the production rate) problem on the basis of the chosen model and experimentally obtained parameters.

The results of the research on these three stages are described below in the particular case of Filanovsky field collector.

2. Model of deformation and filtration

2.1. Generalities

One of the main goals of the current research, consists in choosing and developing an appropriate mechanical model of deformation and fracture of the material in the collector and the seepage processes in it. An adequate model should be able to describe the key characteristics of the process, among which we emphasize the following ones:

(1) the influence of pore pressure on deformation;
(2) the nonlinear stress-strain behavior with essential role of inelastic ("plastic") strains, and the influence of the level of not only shear but also comprehensive stress (and pore pressure) on the transition to inelastic behavior;
(3) the possible inelastic volumetric deformation and its nontrivial dependence on the stress state: at least the absence of proportionality between inelastic volumetric strain and volumetric stress (violation of the associative law);
(4) the possible anisotropy of elastic, strength and seepage properties;
(5) the non-obvious dependence of permeability on the stress strain state.

All of these characteristics have been well described by a number of suggested theories. Thus, the influence of pore pressure on deformation (p. 1) is described by the theory of poroelasticity [1–3]. The influence of comprehensive stress on the level of elastic/plastic transition (p. 2) is described by the Moor–Coulomb and Drucker–Prager criteria [4–6]. To describe the inelastic volumetric deformation nonproportional to the volumetric stress (p. 3), the concept of dilatancy was proposed [7–10], according to which the former is supposed to depend on the intensity of shear strain, with possible violation of the associative law. The anisotropy in the theory of plastic flow was taken into account by Hill [11, 12], and further the Hill theory was modified to account for the difference in the behavior in compression and in tension [13, 14]. The dependence of permeability on the stress state may vary in a wide range depending on the particular collector and is usually determined from experiments.

The above-mentioned models, which describe the listed characteristics of deformation and filtration, form a basis for solving geomechanical problems. Moreover, some of them have already been implemented to commercial codes. Meanwhile, describing some characteristics each of these models does not describe the others, which does not allow their direct application. Therefore, one of the main goals of the current study consists in adoption and further development of these models aiming at taking the whole set of the above-listed specific features into account.

2.2. Basic equation of the model

The fluid flow is described by Darcy's law

$$(\kappa p_{,i})_{,i} = 0, \tag{1}$$

where p is the pore pressure; κ is the permeability which may generally depend on the coordinate, pore pressure, the stress state and other factors; the indexes after comma stand for the derivative with respect to the corresponding coordinate.

The stress strain state is described by the laws of poroelasticity [2, 3]. Their generalization to the case of plastic deformations may be written as

$$\sigma_{ij,i} = 0, \tag{2}$$

$$s_{ij} = \sigma_{ij} + \alpha_P p \delta_{ij}, \tag{3}$$

$$s_{ij} = \Lambda_{ijkl} \varepsilon^E_{kl}, \tag{4}$$

$$\varepsilon^T_{ij} = \varepsilon^E_{ij} + \varepsilon^P_{ij} = \frac{1}{2}(u_{i,j} + u_{j,i}). \tag{5}$$

Here σ_{ij} and s_{ij} are components of the total and effective (acting on solid skeleton) stress tensors; ε^T_{ij}, ε^E_{ij}, and ε^P_{ij} are components of the total, elastic and plastic strain tensors; u_i are components of the displacement vector; Λ_{ijkl} are components of elasticity tensor (in the case of elastic isotropy, they are expressed in terms of two constants); $0 \le \alpha_P \le 1$ is Bio's coefficient characterizing the structure of porosity; for well permeable collectors, α_P approaches the unity from below [5]. For practical reasons, one can set $\alpha_P = 1$. Before reaching the criterion of elastic-plastic transition $\varepsilon^P_{ij} = 0$, equations (1)–(5) together with the boundary conditions for stresses (or displacements) and the pore pressure form a closed system.

To describe the plastic deformation, one needs to specify the plasticity criterion and the plasticity potential. Consider a criterion of the type of [13] modified for taking account of the effects of poroelasticity

$$F = [G^0_{(23)}(s_{22} - s_{33})^2 + G^0_{(13)}(s_{11} - s_{33})^2 + G^0_{(12)}(s_{11} - s_{22})^2 + 2L^0_{(23)}s^2_{23} + 2L^0_{(13)}s^2_{31} + 2L^0_{(12)}s^2_{12}]^{1/2} + (B^0_{(1)}s_{11} + B^0_{(2)}s_{22} + B^0_{(3)}s_{33}) - 1 > 0. \quad (6)$$

Here s_{ij} are components of the effective stress tensor in the coordinate frame related to the material axes of isotropy; $G^0_{(ij)}$, $L^0_{(ij)}$, $B^0_{(ij)}$ are material constants [11–14].

For a transverse isotropic medium with the isotropy plane with the normal n_3, the number of constants in (6) reduces to

$$G^0_{(13)} = G^0_{(23)}, \quad L^0_{(13)} = L^0_{(23)}, \quad L^0_{(12)} = G^0_{(13)} + 2G^0_{(12)}, \quad B^0_{(1)} = B^0_{(2)}. \quad (7)$$

In the isotropic case,

$$G^0_{(13)} = G^0_{(23)} = G^0_{(12)} = G, \quad L^0_{(13)} = L^0_{(23)} = L^0_{(12)} = 3G, \quad B^0_{(1)} = B^0_{(2)} = B^0_{(3)} = \tfrac{1}{3}B, \quad (8)$$

neglecting the effects related to the pore pressure (replacing effective stresses with the total ones) results in reducing (6) to the Druckep–Prager criterion [5].

On the other hand, for the vanishing influence of normal stresses on the elastic-plastic transition and effects related to pore pressure, equation (6) reduces to the Hill criterion [11, 12] of anisotropy plasticity.

According to the plastic flow theory (although modified for taking account of hydrostatic stresses), the actual state in the space of stresses during the active loading remains belonging to the yield surface (generalization of plasticity criterion (6))

$$F = [G^0_{(23)}(s_{22} - s_{33})^2 + G^0_{(13)}(s_{11} - s_{33})^2 + G^0_{(12)}(s_{11} - s_{22})^2 + 2L^0_{(23)}s^2_{23} + 2L^0_{(13)}s^2_{31} + 2L^0_{(12)}s^2_{12}]^{1/2} + (B^0_{(1)}s_{11} + B^0_{(2)}s_{22} + B^0_{(3)}s_{33}) - A(k), \quad (9)$$

where the dimensionless function $A(k)$ ($A(0) = 1$) serves as a characteristic of hardening, supposed to be isotropic for the model under consideration; k is the parameter of hardening and, for a widely used variant of the theory, k is associated with the work of plastic strains

$$dk = s_{ij}\, d\varepsilon^p_{ij}. \quad (10)$$

During active loading, the stress growth is accompanied by plastic strain growth. In the frame of the plastic flow theory, the increment ("rate") of plastic strain $d\varepsilon^p_{ij}$ is expressed in terms of the plastic potential Q as

$$d\varepsilon^P_{ij} = d\lambda \frac{\partial Q}{\partial \sigma_{ij}}. \quad (11)$$

Here $d\lambda$ is a still unknown coefficient. In the classical theory of plasticity, equating the plastic potential with the yield surface (such a relation is referred to as the associative law) allows elegant deducing to constructing constitutive laws capable of adequate description of the observed stress strain relations. However, using the associative law to describe the deformation of rocks and soils leads to a rather strong discrepancy from the phenomenon, namely, the predicted volumetric inelastic deformations are usually much greater than the actual ones. Therefore, the so-called non-associative laws with $Q \neq F$ are used to describe the deformation of rocks and soils [9, 10, 15].

Hereafter, the form of the plastic potential Q is chosen similarly to the form of the yield function F in equation (9) with dissimilar factors at the linear terms of stresses

$$Q = [G^0_{(23)}(s_{22} - s_{33})^2 + G^0_{(13)}(s_{11} - s_{33})^2 + G^0_{(12)}(s_{11} - s_{22})^2 \\ + 2L^0_{(23)}s^2_{23} + 2L^0_{(13)}s^2_{31} + 2L^0_{(12)}s^2_{12}]^{1/2} + (B^1_{(1)}s_{11} + B^1_{(2)}s_{22} + B^1_{(3)}s_{33}) - A(k). \quad (12)$$

In the case of vanishing or negligible inelastic volumetric strains (dilatancy), the following restriction takes place

$$B^1_{(1)} + B^1_{(2)} + B^1_{(2)} = 0. \quad (13)$$

For a number of rocks and soils, this equation, although not exact, serves as a rather accurate approximation.

To provide the deviator associativity, it is sufficient to set

$$B^1_{(i)} = B^0_{(i)} - B_0, \quad B_0 = \frac{1}{3}\sum_{j=1}^{3} B^0_{(j)}. \quad (14)$$

The value of $d\lambda$ is determined from equation (9). The differential of F should vanish

$$\frac{\partial F}{\partial s_{ij}} ds_{ij} + \frac{\partial F}{\partial k} dk = 0. \quad (15)$$

The substitution of dk from equation (10) into this relation yields

$$\frac{\partial F}{\partial s_{ij}} ds_{ij} + \frac{\partial F}{\partial k} s_{ij} d\varepsilon^p_{ij} = 0. \quad (16)$$

Expressing the value of plastic strain in terms of the plastic potential (11), one finds

$$d\lambda = \frac{\partial F}{\partial s_{ij}} ds_{ij} \left(H \frac{\partial Q}{\partial s_{ij}} s_{ij} \right)^{-1}. \quad (17)$$

Here $H \equiv -\partial F/\partial k$ is the material function determined from experiments. In the first approximation, it may be considered as the constant $H = E_p^{-1}$, where E_p is the plasticity modulus.

Thus, for given F, Q, and H, equations (11), (10), and (17) form a closed system.

3. Experimental determination of the model parameters

The experimental part of the study consists in determining the model parameters. Experiments were carried using the experimental unit IPMech RAS Triaxial Independent Loading Test System (TILTS) [16] for cubic specimens (40 × 40 × 40 mm) for one of lithotypes of the Filanovsky field collector. Two types of the loading program were used: uniaxial loading with lateral pressure and generalized shear [17–19].

According to the loading program of the first type, the specimens were subjected to comprehensive pressure up to a certain value (2, 10, or 20 MPa) and then loaded along one of the axis up to the level of elastic/plastic transition in the displacement controlled mode.

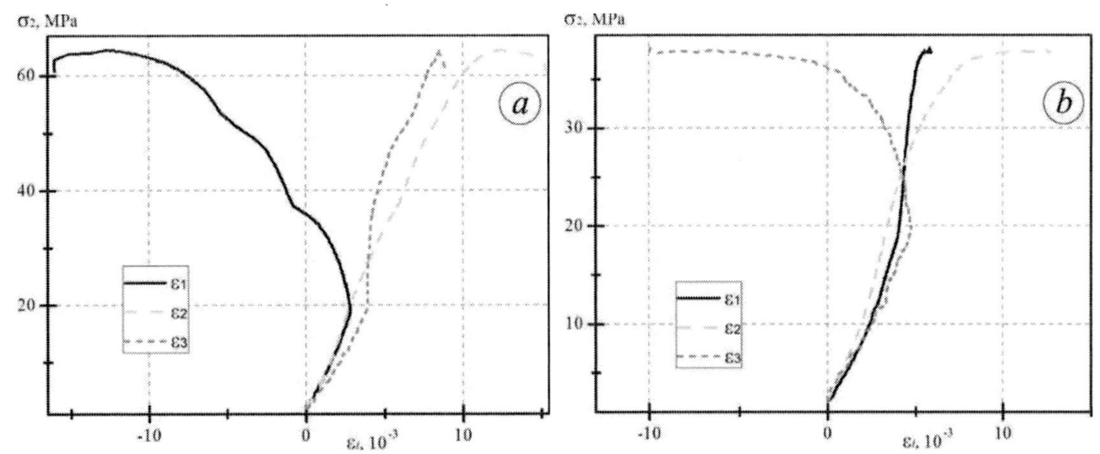

Figure 1. Stress-strain curves for the loading program of the second type; the maximal stress corresponding to the vertical (a) and horizontal (b) directions.

Table 1. Results of tests for the loading program of the first type (Specimen Pr-1).

Cycle	Lateral pressure, MPa	$E \times 10^{-3}$, MPa	$E_p \times 10^{-3}$, MPa	S_2^*, MPa
1	2	1.80	0.18	27
2	10	5.42	0.54	51
3	20	7.78	0.78	75

During the second loading program, the specimens experience volumetric compression up to the level corresponding to the in situ stress state, and then the loading continues along one of the axes accompanied with unloading along another axis so that the stress along the third axis remains constant; the average stress is conserved at this stage. The stress state constructed in this program corresponds to the state near the well.

The permeability was measured along the direction of the minimal stress during the whole period. The tests were carried out for two orientations of the minimal stresses: along the layers and normally to the layers, which corresponds to two different points at the contour of the horizontal well. As a result, the dependence of permeability on the stress and the yield stresses corresponding to these loading conditions was obtained.

The results of tests according to the loading programs of the first type are presented in table 1. Figure 1 presents the stress strain curves in the cases of the maximal stress aligned in vertical (figure 1 a) and one of the horizontal (figure 1 b) directions. Figure 2 presents the dependence of permeability on the stress intensity. Figure 3 presents experimental and calculated data about the critical stress according to criterion (9).

The parameters of yield criteria (6) were determined from the whole set of experiments of both types (the property corresponds to transverse isotropy (7)) by means of the least square method $G^0_{(12)} = 2.7 \times 10^{-3}$, $G^0_{(13)} = G^0_{(23)} = 1.27 \times 10^{-3}$, $B^0_{(1)} = B^0_{(2)} = 1.95 \times 10^{-2}$, $B^0_{(3)} = 3.28 \times 10^{-2}$, $L^0_{(13)} = 1.0 \times 10^{-2}$. It was impossible to determine the last parameter from the available experiments, its value was chosen by analogy with samples of similar lithotype [18]. The parameters of plastic potential (12) deviating from the corresponding parameters of the yield criteria are determined by formula (14): $B^1_{(1)} = -4.44 \times 10^{-3}$, $B^1_{(3)} = 8.88 \times 10^{-3}$.

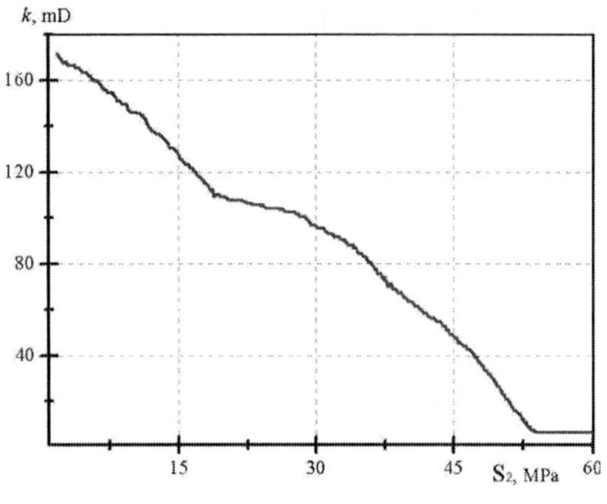

Figure 2. Dependence of permeability on the stress intensity.

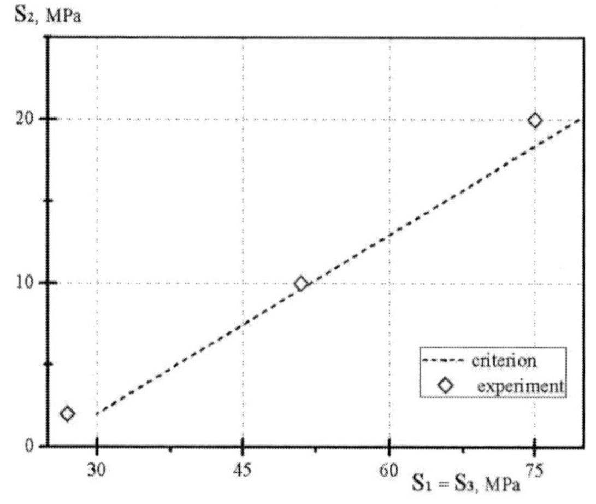

Figure 3. Value of critical stress S_{cr}: depending on the lateral pressure for tests of the second type. The dots correspond to experimental points, the line corresponds to modelling.

4. Simulation

Two variants of well faces are considered: simple unsupported well and unsupported well with perforation. The former case corresponds to a rather simple geometry, which allows comparing the obtained results of simulation with the analytical results. The latter case corresponds to a more complex situation. In both cases, the normal stresses and pressure at the well contour were assumed to be vanishing. The stress state at a far distance from the well was supposed to be the hydrostatic compression of magnitude corresponding to the weight of the overlying rocks; the pore pressure at a far distance from the well was set to be corresponding to hydrostatic.

For each configuration, the finite element (FEM) simulation was carried out as follows.

1. The problem of fluid seepage was solved to obtain the first iteration of the pore pressure.
2. The problem of poroelastoplasticity was solved to obtain the distribution of effective stresses, elastic and plastic strains. This stage consisted in the following substages.
   2.1. For the calculated field of pore pressure, the uncoupled problem of poroelasticity was solved.
   2.2. The parameters of plasticity for each element were modified according to the calculated fields of the attained stresses and pore pressure levels.
   2.3. The problem of poroelastoplasticity was solved for the modified properties of the elements.
3. In accordance with the experimentally obtained law of the permeability dependence on the stress strain state, the permeability of each element was modified in accordance with the calculated stress field.
4. The problem of fluid seepage was solved, the flows distribution and the production rate were obtained.

The calculations were performed in 3-D using meshes with 22356 nodes and 44001 elements corresponding (due to symmetry) to quarters of the initial areas. The geometrical properties were: the well radius, $R = 0.1\,\mathrm{m}$; the length of the cut $L = 0.46\,\mathrm{m}$; the thickness of the cut $h = 0.02\,\mathrm{m}$ (figure 4). The mechanical properties were listed in the preceding section.

The boundary conditions were the following: normal stress and pore pressure on the outer boundary were 31 MPa and 13 MPa, respectively; normal stress and pore pressure at the well

IOP Conf. Series: Journal of Physics: Conf. Series **991** (2018) 012039 doi:10.1088/1742-6596/991/1/012039

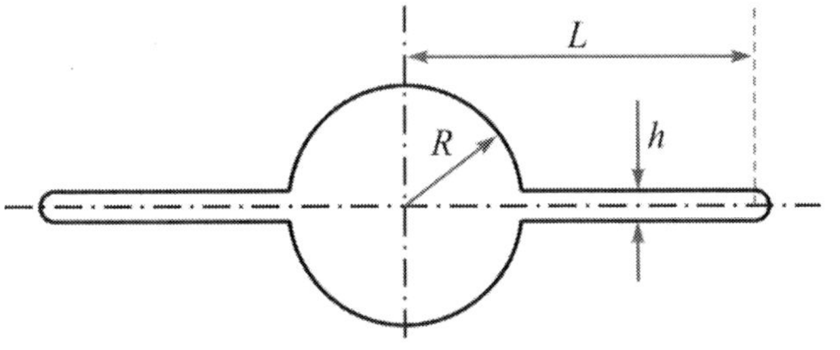

Figure 4. Geometry of the problems: the well perforated with cuts.

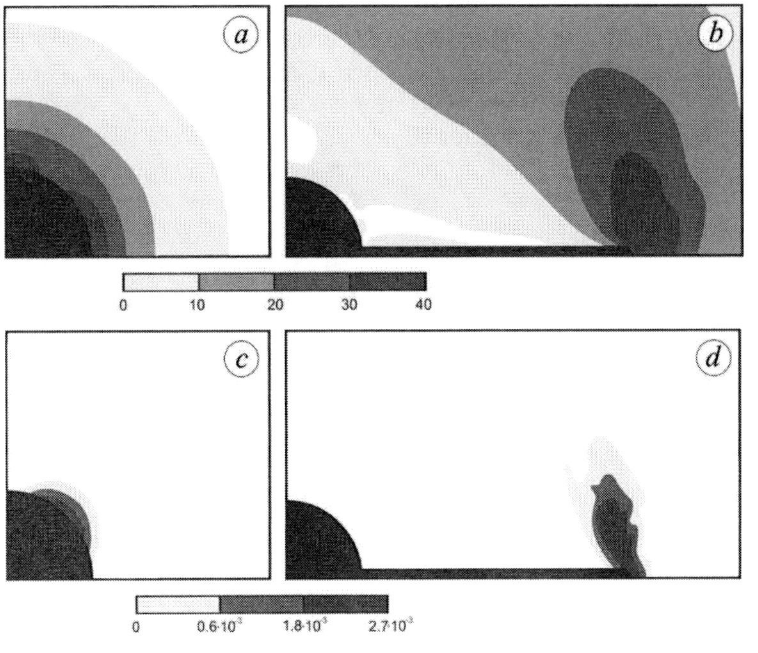

Figure 5. Distributions of the stress intensity (*a*, *b*) and plastic strains (*c*, *d*) for two configurations.

Table 2. production rates relative to the "ideal" well.

Geometry	No dependence	Elasticity	Anisotropic plasticity
Simple unsupported well	1	0.77	0.79
Perforated with cuts	1.27	0.87	0.89

and cut surfaces were zero. For the normal to the well surfaces, the zero displacements and zero fluid flow were set.

The production rates relative to the "ideal" well (for which the permeability is constant and equal to the initial permeability) are presented in table 2. The distributions of the stress intensity and plastic strains for both configurations are presented in figure 5.

Conclusion

The geomechanical approach to modeling the deformation and seepage is presented with regard to the anisotropy of elastic and plastic properties and dependence of the yield transition on volumetric stresses. Three stages of modeling are described, namely: choosing an appropriate mechanical model and its adaptation to the case in question, obtaining parameters of the model from direct experiments, computing the stress state and the parameters of the fluid flow for particular configurations of the well.

The used model allows describing the main specific characteristics of the mechanical behavior of the collector: the influence of pore pressure on deformation; the influence of not only shear but also comprehensive stress and pore pressure on the transition to inelastic behavior; the appearance of inelastic volumetric deformation and its nontrivial dependence on the stress state; the anisotropy of elastic, strength and seepage properties; the non-obvious dependence of the permeability on stress strain state. The model unites essential characteristics of Hill's plastic flow theory for anisotropic materials and the Drucker–Prager theory for inelastic deformation of soils.

The procedure of experimental determination of the involved parameters for the Filanovsky field using the experimental unit IPMech RAS Triaxial Independent Loading Test System (TILTS) is presented, and the experimental results are given.

The results of numerical simulation for the described model and the experimentally obtained parameters are presented. The cases of unsupported and perforated well ends are considered. The stress concentrations and the production rates were calculated.

The results of the study demonstrate that the proposed approach can be used to address geomechanical problems for optimizing technological processes.

Acknowledgments
This work is supported by the Russian Foundation for Basic Research, grant No. 16-11-10325.

References

[1] Terzaghi K 1925 *Erdbaumechanik auf Bodenphysikalischer Grundlage* (Leipzig & Wien: Franz Deuticke)
[2] Biot M A 1935 Le problème de la consolidation des matires argileuses sous une charge *Ann. Soc. Sc. de Brux Ser. B* **55** 110–3
[3] Hristianovich S A and Jeltov Yu P 1955 On hydraulic break of a petroliferous layer *Izv. AN SSSR. OTN.* **5** 3–41
[4] Coulomb C A 1776 Essai sur une application des règles des maximis et minimis à quelques problèmes de statique relatifs, à l'architecture *Mem. Acad. Roy. Div. Sav.* **7** 343–87
[5] Drucker D C and Prager W 1952 Soil mechanics and plastic analysis for limit design *Quart. of Appl. Math.* **10** (2) 157–65
[6] Goodman R E 1980 Introduction to rocks mechanics *New York: John Wiley and Sons*
[7] Reynolds O 1885 On the dilatancy of media composed of rigid particles in contact, with experimental illustrations *Phil. Mag. Ser. 5* **20** (127) 469–81
[8] Mead W J 1925 The geologic rôle of dilatancy *J. Geology* **33** (5) 685–98
[9] Nikolaevskii V N 1967 On interconnection of volume and shear strains and on shock waves in soft soils *Dokl. Akad. Nauk SSSR* **177** 542–3
[10] Nikolaevskii V N 1996 *Geomechanics and Fluid Dynamics* (Moscow: Nedra) p 448 [in Russian]
[11] Hill R 1948 A theory of the yielding and plastic flow of anisotropic metals *Proc. Roy. Soc. London A* **193** 281–97
[12] Hill R 1983 The Mathematical Theory of Plasticity *New York: Oxford University Press*
[13] Caddel R M, Raghava E S, and Atkins A G 1973 A yield criterion for anisotropic and pressure dependent solids such as oriented polymers *J. Mater. sci.* **8** 1641–6
[14] Deshpande V S, Fleck N A, and Ashby M F 2001 Effective properties of the octet-truss lattice material *J. Mech. Phys. Solids* **49** 1747–69
[15] Ustinov K B 2016 On application of models of plastic flow to description of inelastic behavior of anisotropic rocks *Processes in geomedia* **3** (7) 278–87
[16] Karev V I and Kovalenko Yu F 2013 Triaxial loading system as a tool for solving geotechnical problems of oil and gas production In *True Triaxial Testing of Rocks* (Leiden: CRC Press/Balkema) pp 301–10
[17] Karev V I, Klimov D M, Kovalenko Yu F, and Ustinov K B 2016 Fracture model of anisotropic rocks under complex loading *Phys. Mesomech.* **19** (6) 34–40
[18] Karev V I, Klimov D M, Kovalenko Yu F, and Ustinov K B 2016 Fracture of Sedimentary Rocks under a Complex Triaxial Stress State *Mech. Solids* **51** (5) 522–6

IOP Conf. Series: Journal of Physics: Conf. Series **991** (2018) 012040 doi:10.1088/1742-6596/991/1/012040

Contact interaction between a layered foundation and a system of annular punches with complex base shapes

K E Kazakov[1,2*] and S P Kurdina[2]

[1]Ishlinsky Institute for Problems in Mechanics of the Russian Academy of Sciences, Moscow, Russia

[2]Bauman Moscow State Technical University, Moscow, Russia

E-mail: *kazakov-ke@yandex.ru

Abstract. We study the contact interaction between a system of rigid annular punches and a viscoelastic two-layer foundation. The upper layer is thin compared with the punch width. We study the case where the punch shapes are described by a rapidly varying functions. We use special methods for constructing the solutions, because the standard methods are inefficient.

Introduction

Contact mechanics is a very important part of mechanics of solids, and it is of great practical interest. It can be used to solve many problems, for example, wheel-rail contact, in the calculation of couplings, brakes, tires, sliding and rolling bearings, internal combustion engines, hinges, seals; at stamping, metal working, ultrasonic welding, electric contacts, etc. The contact problems for bodies with coatings were first investigated in the second half of the twentieth century. Several contact problems related to the current paper were considered in [1–6]. These were plane and axisymmetric problems for conformal contact between layered foundations and one or several punches, contact problems for nonuniform foundations, rough punches, and bodies with complex rheology. Complex properties of coatings appears due to the peculiarity of their manufacture (see, for example, [7, 8]). In this paper, we for the first time solve the contact problem for a foundation with a homogeneous coating and a system of annular punches with complex base shapes.

1. Statement of the problem

A viscoelastic basement lies on a rigid basis. This basement consist of two layers: the upper layer made of viscoelastic material at time τ_2 and the lower layer made of a different viscoelastic material at time τ_1. Both layers are homogeneous, the height of the upper layer is h and the height of the lower layer is H. We consider the case of smooth layer-layer and layer-basis contacts. We assume that the upper layer rigidity is less than the rigidity of the lower layer or they are of the same order of magnitude, i.e., Young's modulus $E_1(t)$ of the upper layer is less than Young's modulus $E_2(t)$ of the upper layer.

At time $\tau_0 \geq \max\{\tau_1, \tau_2\}$, the forces $P_i(t)$ start to indent smooth rigid annular punches (figure 1) into the surface of such a foundation. The coating is assumed to be thin compared

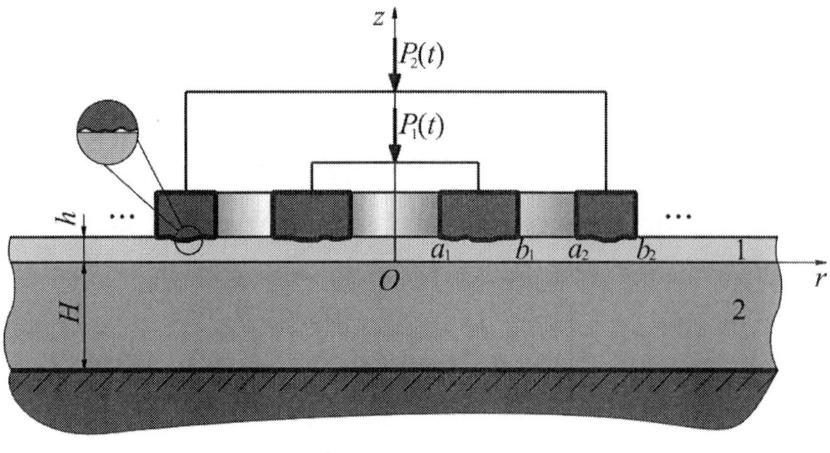

Figure 1. Multiple axisymmetric contact problem.

with the length of the contact area, i.e., its thickness satisfies the condition $h \ll (b_i - a_i)$ for all $i = 1, 2, \ldots, n$, where a_i and b_i are the interior and external radii of the ith punch and n is the quantity of punches. The base shape of the ith punch is described by a function $g_i(r)$ ($g_i(r) \geqslant 0$, $\exists r_{0i} \in [a_i, b_i] : g_i(r_{0i}) = 0$). We consider the axisymmetric problem.

As a result of this interaction, the punches are immersed in the base at the depths δ_i (δ_i is the settlement of the ith punch).

It is easy to show that the system of integral equations and additional conditions for the described problem has the form ($i = 1, 2, \ldots, n$)

$$(1-\nu_1^2)h\left[\frac{q_i(r,t)}{E_1(t-\tau_1)} - \int_{\tau_0}^{t} \frac{q_i(r,\tau)}{E_1(\tau-\tau_1)} K_1(t-\tau_1, \tau-\tau_1)\,d\tau\right]$$
$$+\frac{2(1-\nu_2^2)}{H}\sum_{j=1}^{n}\left[\int_{a_j}^{b_j} k_{\rm ax}\left(\frac{r}{H},\frac{\rho}{H}\right)\frac{q_j(\rho,t)}{E_2(t-\tau_2)}\rho\,d\rho\right.$$
$$\left.-\int_{\tau_0}^{t} K_2(t-\tau_2,\tau-\tau_2)\int_{a_j}^{b_j} k_{\rm ax}\left(\frac{r}{H},\frac{\rho}{H}\right)\frac{q_j(\rho,\tau)}{E_2(\tau-\tau_2)}\rho\,d\rho\,d\tau\right] = \delta_i(t) - g_i(r), \qquad (1)$$

$$2\pi\int_{a_i}^{b_i} q_i(\rho,t)\rho\,d\rho = P_i(t), \quad K_k(t,\tau) = E_k(\tau)\frac{\partial}{\partial\tau}\left[\frac{1}{E_k(\tau)} + C_k(t,\tau)\right], \qquad (2)$$

where $q^i(r,t)$ are contact pressures under the punches, ν_k, $K_k(t,\tau)$, $C_k(t,\tau)$ are the Poisson ratios, tensile creep kernels, and tensile creep functions of the upper ($k = 1$) and lower ($k = 2$) layers; $k_{\rm ax}(r/H, \rho/H)$ is the known kernel of the axisymmetric contact problem (see, e.g., [9]):

$$k_{\rm ax}(r,\rho) = \int_0^\infty L(u)J_0(ru)J_0(\rho u)\,du,$$

where

$$L(u) = \frac{\cosh 2u - 1}{\sinh 2u + 2u}$$

and $J_0(u)$ is the Bessel function of the first kind of order zero.

Let us change the variables in (1) and (2) by the formulas

$$
\begin{aligned}
&(r^*)^2 = \frac{r^2 - a_i^2}{b_i^2 - a_i^2}, \quad (\rho^*)^2 = \frac{\rho^2 - a_j^2}{b_j^2 - a_j^2}, \quad t^* = \frac{t}{\tau_0}, \quad \tau_k^* = \frac{\tau_k}{\tau_0},\\
&\lambda = \frac{H}{b_1 - a_1}, \quad \eta_i = \frac{a_i}{b_1 - a_1}, \quad \zeta_i^2 = \frac{b_i^2 - a_i^2}{(b_1 - a_1)^2}, \quad \delta^{i*}(t^*) = \frac{\delta_i(t)\zeta_i}{b_1 - a_1},\\
&g^{i*}(r^*) = \frac{g_i(r)\zeta_i}{b_1 - a_1}, \quad c^*(t^*) = \frac{E_2(t - \tau_2)}{E_1(t - \tau_1)} \frac{1 - \nu_1^2}{1 - \nu_2^2} \frac{h}{2(b_1 - a_1)},\\
&q^{i*}(r^*, t^*) = \frac{2q_i(r,t)(1 - \nu_2^2)\zeta_i}{E_2(t - \tau_2)}, \quad P^{i*}(t^*) = \frac{P_i(t)(1 - \nu_2^2)\zeta_i}{\pi E_2(t - \tau_2)(b_i^2 - a_i^2)},\\
&\mathbf{F}^{ij*} f(r^*) = \int_0^1 k^{ij}(r^*, \rho^*) f(\rho^*)\rho^*\, d\rho^*, \quad \mathbf{V}_k^* f(t^*) = \int_1^{t^*} K_k(t^*, \tau^*) f(\tau^*)\, d\tau^*,\\
&k^{ij}(r^*, \rho^*) = \frac{\zeta_i\zeta_j}{\lambda} k_{\mathrm{ax}}\left(\frac{r}{H}, \frac{\rho}{H}\right), \quad K^{1*}(t^*, \tau^*) = \frac{E_1(t - \tau_1)E_2(\tau - \tau_2)}{E_1(\tau - \tau_1)E_2(t - \tau_2)} K_1(t - \tau_1, \tau - \tau_1)\tau_0,\\
&K^{2*}(t^*, \tau^*) = K_2(t - \tau_2, \tau - \tau_2)\tau_0, \quad i, j = 1, 2, \ldots, n.
\end{aligned}
\tag{3}
$$

Then, omitting the asterisks, we obtain the system of mixed integral equation and additional conditions in dimensionless form

$$
c(t)(\mathbf{I} - \mathbf{V}_1)q^i(r,t) + (\mathbf{I} - \mathbf{V}_2)\sum_{j=1}^{n} \mathbf{F}^{ij} q^j(r,t) = \delta^i(t) - g^i(r),
$$

$$
\int_0^1 q^i(\rho, t)\rho\, d\rho = P^i(t), \quad r \in [0,1], \quad t \geq 1, \quad i = 1, 2, \ldots, n.
$$

It is easy to show that there exist three versions of mathematical statements for the contact problem with a system of punches in the axisymmetric case. We construct the solution in the case with known forces $P^i(t)$, unknown settlements $\delta^i(t)$, and contact pressures $q^i(r,t)$.

2. Operator representation

Assuming

$$
\begin{aligned}
&\mathbf{q}(r,t) = q^i(r,t)\mathbf{i}^i, \quad \boldsymbol{\delta}(t) = \delta^i(t)\mathbf{i}^i, \quad \mathbf{P}(t) = P^i(t)\mathbf{i}^i,\\
&\mathbf{k}(r,\rho) = k^{ij}(r,\rho)\mathbf{i}^i\mathbf{i}^j, \quad \mathbf{Ff}(r) = \int_0^1 \mathbf{k}(r,\rho)\cdot \mathbf{f}(\rho)\rho\, d\rho,
\end{aligned}
\tag{4}
$$

we can represent system with additional conditions (4) as

$$
c(t)(\mathbf{I} - \mathbf{V}_1)\mathbf{q}(r,t) + (\mathbf{I} - \mathbf{V}_2)\mathbf{Fq}(r,t) = \boldsymbol{\delta}(t) - \mathbf{g}(r), \tag{5}
$$

$$
\int_0^1 \mathbf{q}(\rho, t)\rho\, d\rho = \mathbf{P}(t), \quad r \in [0,1], \quad t \geq 1. \tag{6}
$$

Hereinafter, the summation will be over the repeated upper indices i and j from 1 to n if the left-hand side of the formula is independent of the index.

Thus, we study operator equation (5) with additional condition (6), where $\mathbf{q}(r,t)$ is a vector $L_2((0,1),V)$-function continuous in time t, $\mathbf{P}(t)$ and $\boldsymbol{\delta}(t)$ are vector functions continuous in time t, and $\mathbf{g}(r)$ is a vector function in $L_2((0,1),V)$ (it can be a rapidly varying function). The compact operator $\mathbf{F}$ is a self-adjoint and positive definite operator from $L_2((0,1),V)$ to $L_2((0,1),V)$.

3. Solution of the problem

We seek $\mathbf{q}(r,t)$ in the form

$$\mathbf{q}(r,t) = \mathbf{Q}(r,t) - \mathbf{g}(r)(\mathbf{I}-\mathbf{V}_1)^{-1}\frac{1}{c(t)},$$

where $\mathbf{Q}(r,t)$ is a new unknown vector function. Then the operator equation (5) and additional condition (6) become

$$c(t)(\mathbf{I}-\mathbf{V}_1)\mathbf{Q}(r,t) + (\mathbf{I}-\mathbf{V}_2)\mathbf{F}\mathbf{Q}(r,t) = \boldsymbol{\delta}(t) + \tilde{c}(t)\tilde{\mathbf{g}}(r), \tag{7}$$

$$\int_0^1 \mathbf{Q}(\rho,t)\rho\,d\rho = \tilde{\mathbf{P}}(t), \quad r\in[0,1], \quad t\geq 1, \tag{8}$$

$$\begin{aligned} &\tilde{c}(t) = (\mathbf{I}-\mathbf{V}_2)(\mathbf{I}-\mathbf{V}_1)^{-1}\frac{1}{c(t)}, \quad \tilde{\mathbf{g}}(r) = \int_0^1 k^{ij}(r,\rho)g^j(\rho)\rho\,d\rho\mathbf{i}^i, \\ &\tilde{\mathbf{P}}(t) = \mathbf{P}(t) + (\mathbf{I}-\mathbf{V}_1)^{-1}\int_0^1 \frac{\mathbf{g}(\rho)\rho\,d\rho}{c(t)}. \end{aligned} \tag{9}$$

We seek a solution of Eq. (7) under condition (8) in the class of vector functions continuous in time t in the Hilbert space $L_2((0,1),V)$. We use the basis $\{\mathbf{p}_k^i(r)\}_{\substack{k=0,1,2,\ldots\\ i=1,2,\ldots,n}} = \{p_k(r)\mathbf{i}^i\}_{\substack{k=0,1,2,\ldots,\\ i=1,2,\ldots,n}}$ where $\{p_k(r)\}_{k=0,1,2,\ldots}$ is a system of orthogonal polynomials with respect to the L_2-norm on the interval $[0,1]$ with weight r, i.e., $\int_0^1 p_{k_1}(r)p_{k_2}(r)r\,dr = \delta_{k_1k_2}$ ($\delta_{k_1k_2}$ is the Dirac delta function). So the system $\{\mathbf{p}_k^i(r)\}_{\substack{k=0,1,2,\ldots\\ i=1,2,\ldots,n}}$ satisfies the condition $\int_0^1 \mathbf{p}_{k_1}^{i_1}(r)\cdot\mathbf{p}_{k_2}^{i_2}(r)r\,dr = \delta_{k_1k_2}\delta_{i_1i_2}$.

The Hilbert space $L_2((0,1),V)$ can be represented as a direct sum of orthogonal subspaces $L_2((0,1),V) = L_2^{(0)}((0,1),V)\oplus L_2^{(1)}((0,1),V)$, where $L_2^{(0)}((0,1),V)$ is the Euclidean space with basis $\{\mathbf{p}_0^i(r)\}_{i=1,2,\ldots,n}$ and $L_2^{(1)}((0,1),V)$ is a Hilbert space with basis $\{\mathbf{p}_1^i(r),\mathbf{p}_2^i(r),\ldots\}_{i=1,2,\ldots,n}$. The integrand and the right-hand side of (7) can also be represented as an algebraic sum of vector functions continuous in time t and ranging in $L_2^{(1)}((0,1),V)$ and $L_2^{(2)}((0,1),V)$, respectively, i.e.,

$$\mathbf{Q}(r,t) = \mathbf{Q}_0(r,t) + \mathbf{Q}_1(r,t), \quad \boldsymbol{\delta}(t) + \tilde{c}(t)\tilde{\mathbf{g}}(r) = \boldsymbol{\Delta}_0(r,t) + \boldsymbol{\Delta}_1(r,t),$$

where $\mathbf{Q}_0(r,t)$, $\boldsymbol{\Delta}_0(r,t)\in L_2^{(0)}((0,1),V)$, $\mathbf{Q}_1(r,t)$, $\boldsymbol{\Delta}_1(r,t)\in L_2^{(1)}((0,1),V)$ and

$$\begin{aligned} &\mathbf{Q}_0(r,t) = z_0^i(t)\mathbf{p}_0^i(r), \quad \boldsymbol{\Delta}_0(r,t) = \left[\sqrt{\tfrac{1}{2}}\delta^i(t) + \tilde{g}_0^i\tilde{c}(t)\right]\mathbf{p}_0^i(r), \quad \boldsymbol{\Delta}_1(r,t) = \tilde{c}(t)\tilde{\mathbf{g}}_1(r), \\ &\tilde{g}_0^i = \sum_{l=0}^{\infty} K_{0l}^{ij}\int_0^1 p_l(\rho)g^j(\rho)\rho\,d\rho, \quad \tilde{\mathbf{g}}_1(r) = \tilde{\mathbf{g}}(r) - \tilde{g}_0^i\mathbf{p}_0^i(r), \\ &K_{ml}^{ij} = \int_0^1\!\!\int_0^1 k^{ij}(r,\rho)p_m(r)p_l(\rho)r\rho\,dr\,d\rho, \quad k^{ij}(r,\rho) = \sum_{m=0}^{\infty}\sum_{l=0}^{\infty} K_{ml}^{ij}p_m(r)p_l(\rho). \end{aligned} \tag{10}$$

Note that the representation for $\mathbf{Q}(r,t)$ contains the known term $\mathbf{Q}_0(r,t)$, which is determined by the additional conditions (8)

$$\mathbf{Q}_0(r,t) = z_0^i(t)\mathbf{p}_0^i(r), \quad z_0^i(t) = \sqrt{2}P^i(t),$$

and the term $\mathbf{Q}_1(r,t)$ to be found. Conversely, for the right-hand side, one should find $\boldsymbol{\Delta}_0(r,t)$, while $\boldsymbol{\Delta}_1(r,t)$ is known. These specific features permit classifying the obtained problem as a specific case of the generalized projection problem stated and solved in [10].

We can introduce the orthogonal projection operators, mapping the space $L_2((0,1),V)$ onto $L_2^{(0)}((0,1),V)$ and $L_2^{(0)}((0,1),V)$:

$$\mathbf{P}_0\mathbf{f}(r) = \int_0^1 \mathbf{f}(\rho)\cdot\mathbf{p}_0^i(\rho)\rho\,d\rho\;\mathbf{p}_0^i(r),\quad \mathbf{P}_1 = \mathbf{I} - \mathbf{P}_0.$$

In addition, the following relations hold: $\mathbf{P}_l\mathbf{Q}(r,t) = \mathbf{Q}_l(r,t)$, $\mathbf{P}_l\big[\boldsymbol{\delta}(t) + \tilde{c}(t)\tilde{\mathbf{g}}(r)\big] = \boldsymbol{\Delta}_l(r,t)$ $(l = 0,1)$.

We apply the orthogonal projection operator $\mathbf{P}_1$ to Eq. (7). As a result, we obtain the equation for determining $\mathbf{Q}_1(r,t)$ with a known right-hand side

$$c(t)(\mathbf{I}-\mathbf{V}_1)\mathbf{Q}_1(r,t) + (\mathbf{I}-\mathbf{V}_2)\mathbf{P}_1\mathbf{F}\mathbf{Q}_1(r,t) = -(\mathbf{I}-\mathbf{V}_2)\mathbf{P}_1\mathbf{F}\mathbf{Q}_0(r,t) + \tilde{c}(t)\tilde{\mathbf{g}}_1(r). \tag{11}$$

It is necessary to construct its solution in the form of a series in the eigenfunctions of the operator $\mathbf{P}_1\mathbf{F}$, which is a compact strongly positive self-adjoint operator from $L_2^{(1)}((0,1),V)$ to $L_2^{(1)}((0,1),V)$. The system of eigenfunctions of such an operator is a basis in the space $L_2^{(1)}((0,1),V)$, see [11]. The spectral problem for the operator $\mathbf{P}_1\mathbf{F}$ can be written as

$$\mathbf{P}_1\mathbf{F}\boldsymbol{\varphi}_k(r) = \gamma_k\boldsymbol{\varphi}_k(r),\quad \boldsymbol{\varphi}_k(r) = \sum_{m=1}^{\infty}\psi_{km}^i\mathbf{p}_m^i(r),\quad k = 1,2,3,\ldots$$

Now it is necessary to solve the system for the coefficients ψ_{km}^i and γ_k

$$\sum_{l=1}^{\infty} K_{ml}^{ij}\psi_{kl}^j = \gamma_k\psi_{km}^i,\quad k,m = 1,2,3,\ldots,\quad i = 1,2,\ldots,n,$$

where K_{ml}^{ij} are determined by formula (10) We expand the functions $\mathbf{Q}_1(r,t)$ and $\tilde{\mathbf{g}}_1(r)$ with respect to the basis $\{\boldsymbol{\varphi}_k(r)\}_{k=1,2,3,\ldots}$ in $L_2^{(1)}((0,1),V)$, i.e.,

$$\mathbf{Q}_1(r,t) = \sum_{k=1}^{\infty} z_k(t)\boldsymbol{\varphi}_k(r),\quad \tilde{\mathbf{g}}_1(r) = \sum_{k=1}^{\infty} g_k\boldsymbol{\varphi}_k(r),$$

substitute these representations, (9) and (10) into (11), and see that the unknown expansion functions $z_k(t)$ $(k = 1,2,3,\ldots)$ can be determined by the formula

$$z_k(t) = (\mathbf{I}+\mathbf{W}_k)\frac{(\mathbf{I}-\mathbf{V}_2)[g_k(\mathbf{I}-\mathbf{V}_1)^{-1}c^{-1}(t) - K_{0k}^i z_0^i(t)]}{c(t)+\gamma_k},$$

$$K_{0k}^i = \sum_{m=1}^{\infty} K_{0m}^{ij}\psi_{km}^j,\quad g_k = \sum_{m=1}^{\infty}\psi_{km}^i\sum_{l=0}^{\infty}K_{ml}^{ij}\int_0^1 p_l(\rho)g^j(\rho)\rho\,d\rho,\quad \mathbf{W}_k f(t) = \int_1^t R_k^*(t,\tau)f(\tau)\,d\tau,$$

where $R_k^*(t,\tau)$ $(k = 1,2,3,\ldots)$ are resolvents of the kernel $K_k^*(t,\tau) = [c(t)K^1(t,\tau) + \gamma_k K^2(t,\tau)]/[c(t)+\gamma_k]$.

Note that the obtained solution has the following structure

$$\mathbf{q}(r,t) = z_0^i(t)\mathbf{p}_0^i(r) + \sum_{m=1}^{\infty}\left[\sum_{k=1}^{\infty}\psi_{km}^i z_k(t)\right]\mathbf{p}_m^i(r) - \mathbf{g}(r)(\mathbf{I}-\mathbf{V}_1)^{-1}\frac{1}{c(t)},$$

i.e., one can explicitly distinguish the function $\mathbf{g}(r)$, and hence the punch base shape functions $g_i(r)$ related to it by the change of variables (3). The formulas thus obtained permit deriving efficient analytic solutions for contact problems for layers with thin coatings and a system of punches whose base shapes are described by rapidly varying functions. This can hardly be done by other well-known methods.

Determining the contact pressures under the punches, we can find the unknown settlements of punches. To this end, we must apply the operator $\mathbf{P}_0$ to Eq. (7):

$$c(t)(\mathbf{I}-\mathbf{V}_1)\mathbf{Q}_0(r,t)+(\mathbf{I}-\mathbf{V}_2)\mathbf{P}_0\mathbf{F}[\mathbf{Q}_0(r,t)+\mathbf{Q}_1(r,t)]=\mathbf{\Delta}_0(r,t),$$

The obtained relation readily implies the formulas for the settlements $\delta^i(t)$

$$\delta(t)=\sqrt{2}\Big\{c(t)(\mathbf{I}-\mathbf{V}_1)z_0^i(t)+(\mathbf{I}-\mathbf{V}_2)\Big[K_{00}^{ij}z_0^j(t)+\sum_{k=1}^{\infty}K_{0k}^i z_k(t)-g_0^i(\mathbf{I}-\mathbf{V}_1)^{-1}\frac{1}{c(t)}\Big]\Big\}.$$

Thus, we have constructed an analytic solution of the axisymmetric contact problem for a viscoelastic layer with a thin coating, which permits efficiently solving problems with complicated functions describing the shapes of the punch bases.

Conclusions

We pose and solve axisymmetric contact problems for viscoelastic aging basements with coatings and a system of rigid punches with complicated shapes of their bases. The solution of the problem is obtained analytically, and, in the expressions for the contact stresses, the punch base shape functions are distinguished explicitly, which allows one to perform computations for the real profiles of punch bases, which are described by rapidly varying functions. The explicit formulas were obtained for the settlements of the punches.

Acknowledgments

Author thanks A.V. Manzhirov for statement of the problem, useful discussions, and valuable advice. The research was financially supported by the Russian Science Foundation, project No. 17-19-01257.

References

[1] Manzhirov A V 1985 On a method of solving two-dimensional integral equations of axisymmetric contact problems for bodies with complex rheology *J. Appl. Math. Mech.* **49** (6) 777–82
[2] Kazakov K E and Manzhirov A V 2008 Conformal contact between layered foundations and punches *Mech. Solids.* **43** (3) 512–24
[3] Manzhirov A V and Kazakov K E 2016 Contact problem for a foundation with a rough coating *Lect. Notes Engng Comp. Sci.* **2224** (1) 877–82
[4] Kazakov K E and Kurdina S P 2017 Indentation of the regular system of punches into the foundation with rough coating In *Mechanics for Materials and Technologies. Advanced Structured Materials* vol 46, ed H Altenbach, R V Goldstein, and E Murashkin (Springer) pp 297–308
[5] Kazakov K E and Manzhirov A V 2017 Axisymmetric contact between an annular rough punch and a surface nonuniform foundation *Mech. Solids* **54** (4) 444–51
[6] Manzhirov A V and Kazakov K E 2017 Modeling of the contact interaction between a nonuniform foundation and rough punch *Mat. Model.* **29** (10) 95–104
[7] Manzhirov A V 2013 Mechanics of growing solids and phase transitions *Key Engng Mater.* **535–536** 89–93
[8] Manzhirov A V 2015 Mechanical design of viscoelastic parts fabricated using additive manufacturing technologies In *Lecture Notes in Engineering and Computer Science: Proceedings of the World Congress on Engineering 2015, 1–3 July, 2015, London, U.K.* Vol II (London: IAENG) pp 710–714
[9] Alexandrov V M and Mkhitaryan S M 1983 *Contact Problems for Bodies with Thin Coatings and Interlayers* (Moscow: Nauka) [in Russian]
[10] Manzhirov A V 2016 A mixed integral equation of mechanics and a generalized projection method of its solution *Dokl. Phys.* **61** (10) 489–93
[11] Kolmogorov A N and Fomin S V 1999 *Elements of the Theory of Functions and Functional Analysis* (New York: Dover Publications)

IOP Conf. Series: Journal of Physics: Conf. Series **991** (2018) 012041 doi:10.1088/1742-6596/991/1/012041

On the eigenfrequencies of elastic shear waves propagating in an inhomogeneous layer

V M Khachatryan
Institute of Mechanics of the National Academy of Sciences of Armenia, Yerevan, Armenia
E-mail: khachatryan-vazgen@inbox.ru

Abstract. In this work, we consider the problem of eigenfrequencies of elastic shear waves propagating in a layer whose Young's modulus and density are functions of the longitudinal coordinate. Taking into account the material inhomogeneity makes the problem of the eigenfrequencies of the waves propagating in the layer more complicated. In this paper, the problem of pure shear is considered. To solve the problem, we use an integral formula which allows us to represent the general solution of the original equation with variable coefficients in terms of the general solution of the accompanying equation with constant coefficients.

1. Formulation of the problem

Suppose that an elastic layer occupies the area $0 \leq x \leq a$, $0 \leq y \leq b$, $|z| < \infty$, for which the equation of motion of an elastic inhomogeneous medium in pure shear (antiplane problem) has the form [1, 2]

$$\frac{\partial}{\partial x}\left[G(x)\frac{\partial W}{\partial x}\right] + G(x)\frac{\partial^2 W}{\partial y^2} = \rho(x)\frac{\partial^2 W}{\partial t^2}, \tag{1}$$

where $G(x)$ is the shear modulus, $\rho(x)$ is the density of the medium material, which are variable, and $W = W(x, y, t)$ is the elastic displacement.

The boundary conditions are

$$W = 0 \quad \text{at } y = 0, b. \tag{2}$$

The solution of (1) satisfying boundary conditions (2) is represented as

$$W(x, y, t) = \sum_{n=1}^{\infty} F_n(x, t)\sin(\lambda_n y), \quad \lambda_n = \frac{\pi n}{b}. \tag{3}$$

Substituting (3) into (1) and equating the corresponding coefficients of the series, we obtain

$$\frac{\partial}{\partial x}\left[G(x)\frac{\partial F_n(x,t)}{\partial x}\right] - \lambda_n^2 G(x)F_n(x,t) = \rho(x)\frac{\partial^2 F_n(x,t)}{\partial t^2}. \tag{4}$$

The following three cases of boundary conditions at $x = 0, a$ are considered:

(1) $W = 0$ at $x = 0, a$. Then

$$F_n(0,t) = 0, \quad F_n(a,t) = 0, \quad n = 1, 2, \ldots . \tag{5}$$

(2) $W = 0$ at $x = 0$ and $\sigma_{13} = 0$ at $x = a$. Then

$$F_n(0,t) = 0, \quad \left.\frac{\partial F_n(x,t)}{\partial x}\right|_{x=a} = 0, \quad n = 1, 2, \dots . \tag{6}$$

(3) $\sigma_{13} = 0$ at $x = 0, a$. Then

$$\left.\frac{\partial F_n(x,t)}{\partial x}\right|_{x=0,a} = 0, \quad n = 1, 2, \dots . \tag{7}$$

Taking into account the inhomogeneity of the material makes the problem of the eigenfrequencies of the waves propagating in the layer more complicated.

2. Representation of the general solution of the original equation through the general solution of the accompanying equation

Let us average the boundary-value problem for equation (4). In [3] the averaging process for nonperiodic inhomogeneity is described in detail. On the same interval $0 \le x \le a$, we consider equation (4) with parameters $G_0 = \text{const}$ and $\rho_0 = \text{const}$:

$$\frac{\partial}{\partial x}\left[G_0 \frac{\partial P_n(x,t)}{\partial x}\right] - \lambda_n^2 G_0 P_n(x,t) = \rho_0 \frac{\partial^2 P_n(x,t)}{\partial t^2}. \tag{8}$$

This case corresponds to a homogeneous layer. It is assumed that $P_n(x,t)$ satisfies the same boundary conditions as $F_n(x,t)$. Formula (8) is the accompanying problem. The coefficients of equation (8) of the accompanying problem can be any positive constants, but it is expedient to associate them with the characteristics of the original problem [4]. Suppose that

$$G_0 = \frac{1}{\langle 1/G(x)\rangle}, \quad \rho_0 = \langle \rho(x)\rangle. \tag{9}$$

By averaging we mean the expression for the solution of original problem (4) in terms of the solution of accompanying problem (8).

We consider the case of steady-state harmonic oscillations. Suppose that

$$F_n(x,t) = f_n(x)e^{i\omega_n t}, \quad P_n(x,t) = p_n(x)e^{i\omega_n t}. \tag{10}$$

Substituting (10) into (4)–(7), we obtain

$$\frac{\partial}{\partial x}\left[G(x)\frac{df_n(x)}{dx}\right] + [\omega_n^2 \rho(x) - \lambda_n^2 G(x)]f_n(x) = 0 \tag{11}$$

with the boundary conditions

$$f_n(0) = 0, \quad f_n(a) = 0, \quad n = 1, 2, \dots \quad \text{(case 1)}, \tag{12}$$

$$f_n(0) = 0, \quad \left.\frac{df_n(x)}{dx}\right|_{x=a} = 0, \quad n = 1, 2, \dots \quad \text{(case 2)}, \tag{13}$$

$$\left.\frac{df_n(x)}{dx}\right|_{x=0,a} = 0, \quad n = 1, 2, \dots \quad \text{(case 3)}. \tag{14}$$

Substituting (10) into (8), we obtain

$$G_0 \frac{d^2 p_n(x)}{dx^2} + (\omega_n^2 \rho_0 - \lambda_n^2 G_0)p_n(x) = 0. \tag{15}$$

The boundary conditions for $p_n(x)$ are the same as for $f_n(x)$:

$$p_n(0) = 0, \quad p_n(a) = 0, \quad n = 1, 2, \ldots \quad \text{(case 1)}, \tag{16}$$

$$p_n(0) = 0, \quad \left.\frac{dp_n(x)}{dx}\right|_{x=a} = 0, \quad n = 1, 2, \ldots \quad \text{(case 2)}, \tag{17}$$

$$\left.\frac{dp_n(x)}{dx}\right|_{x=0,a} = 0, \quad n = 1, 2, \ldots \quad \text{(case 3)}. \tag{18}$$

The general solution of the accompanying equation (15) has the form

$$p_n(x) = Ae^{i\lambda_n\sqrt{\eta_n^2-1}x} + Be^{-i\lambda_n\sqrt{\eta_n^2-1}x}, \tag{19}$$

where η_n is a dimensionless characteristic, $V_{\Phi n}$ are the phase velocities of the corresponding modes, and c is the velocity of the elastic shear wave propagation

$$\eta_n = \frac{V_{\Phi n}}{c}, \quad V_{\Phi n} = \frac{\omega_n}{\lambda_n}, \quad c = \sqrt{\frac{G_0}{\rho_0}}. \tag{20}$$

The integral formula that permits determining the solution of the original equation (11) in terms of the solution of the accompanying equation (15) has the form

$$f_n(x) = p_n(x) + \int_0^a \frac{\partial K_n(s,x)}{\partial s}\tilde{G}(s)\frac{dp_n(s)}{ds}\,ds - \int_0^a [\omega_n^2\tilde{\rho}(s) - \lambda_n^2\tilde{G}(s)]p_n(s)K_n(s,x)\,ds, \tag{21}$$

where

$$\tilde{G}(s) = G_0 - G(s), \quad \tilde{\rho}(s) = \rho_0 - \rho(s), \tag{22}$$

and $K_n(s,x)$ is the continuous fundamental solution of (11) [5], which satisfies the equation

$$\frac{\partial}{\partial x}\left[G(x)\frac{\partial K_n(x,s)}{\partial x}\right] + \left[\omega_n^2\rho(x) - \lambda_n^2 G(x)\right]K_n(x,s) = -\delta(x-s), \tag{23}$$

where $x, s \in [0, a]$.

Substituting expression (21) into equation (11), we see that it is exactly satisfied if equations (15) and (23) are true.

Substituting (19) into (21), we obtain the general solution of the original equation

$$f_n(x) = A\left\{e^{\beta_n(\omega)x} + \beta_n(\omega)\int_0^a \frac{\partial K_n(s,x)}{\partial s}\tilde{G}(s)e^{\beta_n(\omega)s}\,ds - \int_0^a K_n(s,x)[\omega_n^2\tilde{\rho}(s) - \lambda_n^2\tilde{G}(s)]e^{\beta_n(\omega)s}\,ds\right\}$$
$$+ B\left\{e^{-\beta_n(\omega)x} - \beta_n(\omega)\int_0^a \frac{\partial K_n(s,x)}{\partial s}\tilde{G}(s)e^{-\beta_n(\omega)s}\,ds - \int_0^a K_n(s,x)[\omega_n^2\tilde{\rho}(s) - \lambda_n^2\tilde{G}(s)]e^{-\beta_n(\omega)s}\,ds\right\}, \tag{24}$$

where

$$\beta_n(\omega) = i\alpha_n(\omega), \quad \alpha_n(\omega) = \lambda_n\sqrt{\eta_n^2(\omega) - 1}. \tag{25}$$

3. Fundamental solution of the original equation

To find the Green's function satisfying equation (23), we use the method of successive approximations. For the zero approximation, we have

$$\frac{\partial}{\partial x}\left[G(x)\frac{\partial K_{n0}(x,s)}{\partial x}\right] = -\delta(x-s). \tag{26}$$

For the next approximations, we obtain the recurrent equations

$$\frac{\partial}{\partial x}\left[G(x)\frac{\partial K_{nm}(x,s)}{\partial x}\right] = -[\omega_n^2\rho(x) - \lambda_n^2 G(x)]K_{nm-1}(x,s) - \delta(x-s), \quad m \geq 1. \tag{27}$$

From (26) we obtain

$$K_{n0}(x,s) = -\int_0^x \frac{h(z-s)}{G(z)}dz = \begin{cases} 0, & x < s, \\ -\displaystyle\int_s^x \frac{dz}{G(z)}, & x \geq s. \end{cases} \tag{28}$$

From (27) we have

$$K_{nm}(x,s) = -\int_0^x \frac{dx_1}{G(x_1)}\int_0^{x_1}[\omega_n^2\rho(x_2) - \lambda_n^2 G(x_2)]K_{nm-1}(x_2,s)\,dx_2 - \int_0^x \frac{h(x_1-s)}{G(x_1)}\,dx_1, \tag{29}$$

$m \geq 1$.

Hence, if $\lim\limits_{m\to\infty} K_{nm}(x,s)$ converges, then

$$K_n(x,s) = \lim_{m\to\infty} K_{nm}(x,s). \tag{30}$$

Substituting (30) into (24) and satisfying boundary conditions (12), (13) or (14), we obtain the corresponding equations for the eigenfrequencies ω_n.

4. Frequency equations of the zero approximation

Considering only the zero approximation for Green's function, we obtain

$$\begin{aligned} &K_n(x,s) \approx -\int_s^x \frac{dz}{G(z)}, \quad x \geq s, \\ &K_n(x,s) = 0, \quad x < s. \end{aligned} \tag{31}$$

For the derivatives, we have

$$\frac{\partial K_n(s,x)}{\partial s} = \frac{1}{G(s)}, \quad \frac{\partial K_n(s,x)}{\partial x} = -\frac{1}{G(x)}, \quad \frac{\partial^2 K_n(s,x)}{\partial s \partial x} = 0 \tag{32}$$

and

$$\begin{aligned} \frac{\partial f_n(x)}{\partial x} = A\Bigg\{&\beta_n(\omega)\left[e^{\beta_n(\omega)x} + \int_0^a \frac{\partial^2 K_n(s,x)}{\partial s\partial x}\tilde{G}(s)e^{\beta_n(\omega)s}\,ds\right] - \int_0^a \frac{\partial K_n(s,x)}{\partial x}[\omega_n^2\tilde{\rho}(s) \\ &- \lambda_n^2\tilde{G}(s)]e^{\beta_n(\omega)s}\,ds\Bigg\} - B\Bigg\{\beta_n(\omega)\left[e^{-\beta_n(\omega)x} + \int_0^a \frac{\partial^2 K_n(s,x)}{\partial s\partial x}\tilde{G}(s)e^{-\beta_n(\omega)s}\,ds\right] \\ &+ \int_0^a \frac{\partial K_n(s,x)}{\partial x}[\omega_n^2\tilde{\rho}(s) - \lambda_n^2\tilde{G}(s)]e^{-\beta_n(\omega)s}\,ds\Bigg\}. \end{aligned} \tag{33}$$

Satisfying boundary conditions (12)–(14), we obtain

$$\begin{cases} A + B = 0, \\ A\left\{e^{a\beta_n(\omega)} + \displaystyle\int_0^a [\beta_n(\omega)Q(s) + R_n(s)L(s)]e^{\beta_n(\omega)s}\,ds\right\} \\ \qquad + B\left\{e^{-a\beta_n(\omega)} - \displaystyle\int_0^a [\beta_n(\omega)Q(s) - R_n(s)L(s)]e^{-\beta_n(\omega)s}\,ds\right\} = 0, \end{cases} \tag{34}$$

$$\begin{cases} A + B = 0, \\ A\left[\beta_n(\omega)e^{a\beta_n(\omega)} + \dfrac{1}{G(a)}\displaystyle\int_0^a R_n(s)e^{\beta_n(\omega)s}\,ds\right] \\ \qquad -B\left[\beta_n(\omega)e^{-a\beta_n(\omega)} - \dfrac{1}{G(a)}\displaystyle\int_0^a R_n(s)e^{-\beta_n(\omega)s}\,ds\right] = 0, \end{cases} \tag{35}$$

$$\begin{cases} \beta_n(\omega)(A - B) = 0, \\ A\left[\beta_n(\omega)e^{a\beta_n(\omega)} + \dfrac{1}{G(a)}\displaystyle\int_0^a R_n(s)e^{\beta_n(\omega)s}\,ds\right] \\ \qquad -B\left[\beta_n(\omega)e^{-a\beta_n(\omega)} - \dfrac{1}{G(a)}\displaystyle\int_0^a R_n(s)e^{-\beta_n(\omega)s}\,ds\right] = 0, \end{cases} \tag{36}$$

where

$$Q(s) = \frac{G_0}{G(s)} - 1, \quad R_n(s) = \omega_n^2[\rho_0 - \rho(s)] - \lambda_n^2[G_0 - G(s)], \quad L(s) = \int_s^a \frac{dz}{G(z)}. \tag{37}$$

Equating the determinants of systems (34)–(36) to zero, we obtain the following approximate frequency equations:

- for clamped ends,

$$\sin[a\alpha_n(\omega)] + \alpha_n(\omega)\int_0^a Q(s)\cos s\alpha_n(\omega)]\,ds + \int_0^a R_n(s)L(s)\sin[s\alpha_n(\omega)]\,ds = 0, \tag{38}$$

- for the clamped end $x = 0$ and the free end $x = a$,

$$G(a)\alpha_n(\omega)\cos[a\alpha_n(\omega)] + \int_0^a R_n(s)\sin[s\alpha_n(\omega)]\,ds = 0, \tag{39}$$

- and for free ends,

$$\alpha_n(\omega)\left\{G(a)\alpha_n(\omega)\sin[a\alpha_n(\omega)] - \int_0^a R_n(s)\cos[s\alpha_n(\omega)]\,ds\right\} = 0. \tag{40}$$

Specific examples of inhomogeneities will be considered in a subsequent paper.

Conclusions

Taking into account the material inhomogeneity makes the problem of the eigenfrequencies of the waves propagating in the layer more complicated. The above-mentioned integral formula allows us to solve the problem of pure shear by representing the general solution of the original equation with variable coefficients in terms of the general solution of the accompanying equation with constant coefficients.

References

[1] Brekhovskikh L M 1973 *Waves in Laminated Media* (Moscow: Nauka) [in Russian]
[2] Maugin G A 1988 *Continuum Mechanics of electromagnetic Solids* (Amsterdam: North Holland)
[3] Gorbachev V I 2001 Averaging of linear problems in the mechanics of composites with nonperiodic inhomogeneities *Mech. Solids* **36** (1) 24–9
[4] Gorbachev V I 2016 Natural frequencies of longitudinal oscillations for a nonuniform variable cross-section rod *Moscow Univ. Mech. Bull.* **71** (1) 7–15
[5] Vladimirov V S 1976 *Generalized functions in mathematical physics* (Moscow: Nauka) [in Russian]

IOP Conf. Series: Journal of Physics: Conf. Series **991** (2018) 012042 doi:10.1088/1742-6596/991/1/012042

Mathematical modeling of dynamic stability of shell structures under large elastoplastic deformations

K Z Khairnasov

Moscow Aviation Institute (National Research University), Moscow, Russia

E-mail: kamilh@mail.ru

Abstract. In this paper, a mathematical model of dynamic stability of a ring plate under the action of large elastoplastic strains is considered. Such problems are encountered in pressing, for example, by a pulse magnetic field, of ring plates supported on the inner contour. The obtained results show that the proposed mathematical model agrees well with experimental data.

1. Introduction

The manufacturing of thin-walled workpieces with bulging edges by traditional pressing methods leads to the loss of stability, requires a lot of hand work, and results in rather many defective pieces. The use of dynamic loading and, in particular, of pressing by a pulse magnetic field, permits significantly "hindering" the loss of stability of the workpiece, increasing the quality, and decreasing the cost of the workpieces. The problems of dynamic stability in the case of large displacements of plates and shells have been studied absolutely insufficiently and no physically justified criterion for the loss of stability has been proposed. In many operations of cold deformation, the stability is lost at the moment of a semi-finished workpiece shaping from a plane ring plate, and precisely this fact determined the choice of the computational model. Much attention has been paid to studying configurations under the action of momentary dynamic loads in [1–9]. At the same time, all numerous parameters of the dynamic loading, such as the rate of loading, the characteristics of the material dynamic behavior, etc., cannot be taken into account even in several papers.

2. Statement of the problem

We study the behavior of a ring plate supported on the inner contour under the action of highly intensive dynamic loads. The problem is solved under the assumption that the properties of the material are described by the theory of plastic flow and satisfy the Mises yield condition. It is required to construct a mathematical model of behavior of a ring plate under the action of intensive dynamic impact in a geometrically and physically nonlinear statement.

3. Method and construction of the solution

The problem is solved by the finite element method. As a finite element, we take a thin-walled axially symmetric finite element whose displacements in the normal, circular, and radial directions are approximated as polynomials in the radial distance and as Fourier series in the circular direction [5]. Let us consider a thin finite element (see figure 1). We

IOP Conf. Series: Journal of Physics: Conf. Series **991** (2018) 012042 doi:10.1088/1742-6596/991/1/012042

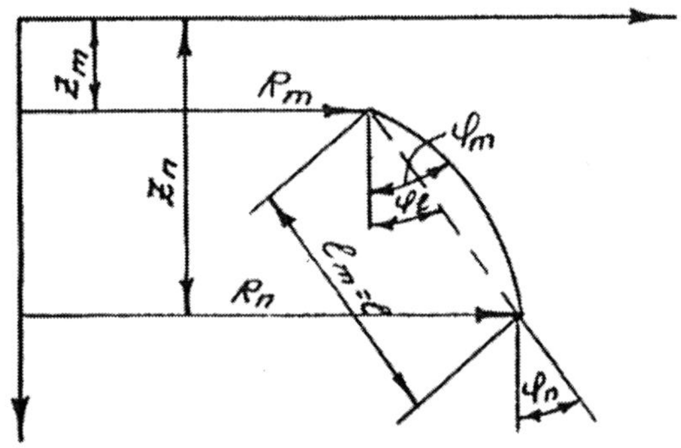

Figure 1. Axially symmetric element.

introduce the coordinate system r, s, θ related to the element. The angle φ is approximated as $\varphi = \alpha_1 + \alpha_2 s + \alpha_2 s^2$, where α_1, α_2, α_3 are the coefficients determined from the boundary conditions for $s = 0, l$:

$$\varphi_{s=0} = \varphi_m, \quad \varphi_{s=L} = \varphi_n, \quad \int_0^L \sin(\varphi - \varphi_L)\, ds = \int_0^L (\varphi - \varphi_L)\, ds = 0, \quad \int_0^L \cos(\varphi - \varphi_L)\, ds = L.$$

After several transformations, we obtain

$$\alpha_1 = \varphi_m, \quad \alpha_2 = \frac{6\varphi_L - 4\varphi_m - 2\varphi_n}{L}, \quad \alpha_3 = \frac{3m - 3\varphi_n - 6\varphi_L}{L^2},$$
$$\varphi_L = \arctan\frac{R_m - R_n}{z_m - z_n}, \quad R = R_m = \int_0^s \sin\varphi\, ds.$$

For the finite element shapes as a truncated cone, we have

$$\varphi = \varphi_L, \quad R = R_m + s\sin\varphi_L.$$

We assume that the field of displacements of the middle surface points inside a finite element has the form

$$\begin{Bmatrix} u(s,\theta) \\ v(s,\theta) \\ w(s,\theta) \end{Bmatrix} = \sum_{j=0}^{\infty} \begin{bmatrix} \cos(j\theta) & 0 & 0 \\ 0 & \sin(j\theta) & 0 \\ 0 & 0 & \cos(j\theta) \end{bmatrix} \begin{Bmatrix} u_s \\ v_s \\ w_s \end{Bmatrix}$$
$$+ \sum_{j=0}^{\infty} \begin{bmatrix} \sin(j\theta) & 0 & 0 \\ 0 & \cos(j\theta) & 0 \\ 0 & 0 & \sin(j\theta) \end{bmatrix} \begin{Bmatrix} u_s \\ v_s \\ w_s \end{Bmatrix}, \tag{1}$$
$$u(s,\theta) = \alpha_1 + \alpha_2 s, \quad v(s,\theta) = \alpha_7 + \alpha_8 s, \quad w(s,\theta) = \alpha_3 + \alpha_4 s + \alpha_5 s^2 + \alpha_6 s^3.$$

Such displacements $u(s,\theta)$, $v(s,\theta)$, and $w(s,\theta)$ are chosen to ensure the continuity of displacements and their first derivatives in transition from one element to another and the smoothness of the interface between elements.

We express the parameters α in terms of displacements at nodal points

$$\{\alpha^j\} = [C]\{q^j\}, \quad \{q^j\} = (u_m^j, w_m^j, \beta_m^j, v_m^j, u_n^j, w_n^j, \beta_m^j, v_m^j),$$
$$\{\alpha^j\} = (\alpha_1, \alpha_2, \ldots, \alpha_8), \quad \beta^j = \frac{dw^j}{ds} + u^j\frac{d\varphi}{ds}$$

IOP Conf. Series: Journal of Physics: Conf. Series **991** (2018) 012042 doi:10.1088/1742-6596/991/1/012042

The displacements $u(s,\theta)$, $v(s,\theta)$, and $w(s,\theta)$ expressed in terms of nodal displacements have the form

$$\{u\} = [A]\{\alpha\}, \quad \{u\} = \{u(s), v(s), w(s)\}, \quad \{u\} = [B]\{q^j\}.$$

To derive the mathematical model for studying the shell configurations made of homogeneous and multilayer composite materials under intensive momentary impacts with large displacements, we consider the Lagrange equations

$$\frac{d}{dt}\frac{\partial T}{\partial \dot{q}_k} + \frac{\partial U}{\partial q_k} = Q_k. \tag{2}$$

Here T is the kinetic energy, U is the potential energy, Q are external loads, $\dot{q}$ and q are the generalized velocities and displacements, and the k are the degrees of freedom.

We use the relations between strains and displacements for a moderate deflection

$$\begin{gathered}
\varepsilon_s = e_s + \frac{e_{13}^2 + e_{33}^2}{2}, \quad \varepsilon_\theta = e_\theta + \frac{e_{23}^2 + e_{33}^2}{2}, \quad \varepsilon_{s\theta} = e_{s\theta} + \frac{e_{13}^2 e_{23}^2}{2}, \quad \chi_s = \frac{\partial e_{13}}{\partial s}, \\
\chi_\theta = \frac{1}{r}\left(\frac{\partial e_{23}}{\partial \theta} + e_{13}\sin\varphi\right), \quad \chi_{s\theta} = \frac{1}{2}\left[\frac{\partial e_{23}}{\partial s} + \frac{1}{r}\frac{\partial e_{13}}{\partial \theta} - e_{23}\frac{\sin\varphi}{r} + \left(\frac{\partial \varphi}{\partial s} + \frac{\cos\varphi}{r}\right)e_{33}\right], \\
e_s = \frac{\partial u}{\partial s} - w\frac{\partial \varphi}{\partial s}, \quad e_\theta = \frac{1}{r}\left(u\sin\varphi + \frac{\partial v}{\partial \theta} + w\cos\varphi\right), \quad e_{s\theta} = \frac{1}{r}\left(\frac{\partial u}{\partial \theta} + r\frac{\partial v}{\partial s} - v\sin\varphi\right), \\
e_{13} = \left(\frac{\partial w}{\partial s} + u\frac{\partial \varphi}{\partial s}\right), \quad e_{23} = -\frac{1}{r}\left(\frac{\partial w}{\partial \theta} - v\sin\varphi\right), \quad e_{33} = -\frac{1}{2r}\left(r\frac{\partial v}{\partial s} + v\sin\varphi - \frac{\partial u}{\partial \theta}\right), \\
\varphi_s = -(w_s + u\sin\varphi), \quad \varphi_\theta = -\frac{w_\theta - v\cos\varphi}{r}.
\end{gathered} \tag{3}$$

In equations (3), we use the following notation: ε_s, ε_θ, and $\varepsilon_{s\theta}$ are the strains of the shell middle surface in the corresponding coordinates, χ_s, χ_θ, and $\chi_{s\theta}$ are the curvature strains, e_{13} and e_{23} are the angles of rotation about the coordinate lines, R is the shell radius, φ_s and φ_θ are the angle of the meridian inclination to the shell axis and the angle in the circular direction, and e_{33} is the "torsion" of the middle surface.

3.1. Equation of the stress-strain relationship

We consider the geometrical and physical dependencies for multilayer shells in the reduced rigidity method.

In the plane stress state, the matrices of elastic coefficients for an orthotropic material whose orthotropy axes coincide with the coordinate axes have the form

$$\{\sigma\} = [E]\{\varepsilon\}, \tag{4}$$

where

$$[E] = \begin{bmatrix} Q_{11} & Q_{12} & 0 \\ Q_{21} & Q_{22} & 0 \\ 0 & 0 & Q_{66} \end{bmatrix}, \quad \{\varepsilon\}^T = \{\varepsilon_s, \varepsilon_\theta, \varepsilon_{s\theta}\}, \quad \{\sigma\}^T = \{\sigma_s, \sigma_\theta, \sigma_{s\theta}\},$$

$$Q_{11} = \frac{E_s}{1 - v_{s\theta}v_{\theta s}}, \quad Q_{12} = \frac{v_{s\theta}E_s}{1 - v_{s\theta}v_{\theta s}}, \quad Q_{21} = \frac{v_{\theta s}E_s}{1 - v_{s\theta}v_{\theta s}}, \quad Q_{22} = \frac{E_\theta}{1 - v_{s\theta}v_{\theta s}}, \quad Q_{66} = G_{66}.$$

For the layer at the distance z from the middle surface, the strains become

$$\{\varepsilon\} = \{\varepsilon^o\} + z\{\chi^o\},$$

where $\{\varepsilon^o\}$ are the strains of the middle surface and $\{\chi^o\}$ are the curvature variations.

After substitution of these relations, equation (4) becomes

$$\{\sigma\} = [\bar{Q}]\{\varepsilon^o\} + z[\bar{Q}]\{\chi^o\}.$$

We take the relations

$$\begin{aligned} \{N\} &= \int_{-h/2}^{h/2} \{\sigma\}\,dz, \quad \{N\}^T = (N_s, N_\theta, N_{s\theta}), \\ \{M\} &= \int_{-h/2}^{h/2} \{\sigma\}z\,dz, \quad \{M\}^T = (M_s, M_\theta, M_{s\theta}), \end{aligned} \tag{5}$$

where N are membrane forces and M are bending moments, into account and integrate expressions (5) over the shell thickness. Then the forces and moments can be represented as

$$\left\{ \begin{array}{c} N \\ M \end{array} \right\} = [E] \left\{ \begin{array}{c} \varepsilon^o \\ \chi^o \end{array} \right\}, \quad [E] = \left[\begin{array}{cc} [A] & [B] \\ [B] & [D] \end{array} \right], \tag{6}$$

$$[A] = \begin{bmatrix} A_{11} & A_{12} & A_{16} \\ A_{21} & A_{22} & A_{26} \\ A_{61} & A_{62} & A_{66} \end{bmatrix}, \quad [B] = \begin{bmatrix} B_{11} & B_{12} & B_{16} \\ B_{21} & B_{22} & B_{26} \\ B_{61} & B_{62} & B_{66} \end{bmatrix}, \quad [D] = \begin{bmatrix} D_{11} & D_{12} & D_{16} \\ D_{21} & D_{22} & D_{26} \\ D_{61} & D_{62} & D_{66} \end{bmatrix},$$

$$\{A_{ij}, B_{ij}, D_{ij}\} = \int_{-h/2}^{h/2} Q_{ij}(1, z, z^2)\,dz \quad (i, j = 1, 2, 6).$$

If the coefficients of the matrix (6) are constant in each layer of the package, then we integrate (6) to obtain

$$A_{ij} = \sum_{k=1}^{n} \bar{Q}_{ij}(h_k - h_{k-1}), \quad B_{ij} = \sum_{k=1}^{n} \bar{Q}_{ij}(h_k^2 - h_{k-1}^2), \quad D_{ij} = \sum_{k=1}^{n} \bar{Q}_{ij}(h_k^3 - h_{k-1}^3), \quad i, j = 1, 2, 6,$$

where A_{ij}, B_{ij}, and D_{ij} are the membrane, bending-membrane, and bending rigidities.

The equations describing the behavior of shell configurations under dynamic loads were derived in detail by the author in paper "Mathematical modeling of shell configurations made of homogeneous and composite materials experiencing intensive short actions and large displacements" presented in this issue. The difference is that, in problems of large deformations, there is an additional term related to the accumulated strains.

3.2. Resolving equations

The mathematical model for studying configurations made of homogeneous and composite materials under the action of intensive momentary impacts in the case of large strains can be represented as

$$[M]\{\hat{q}\} + [K]\{q\} = \{Q\} + \{\Delta Q\} - [K_G]\{q\} - \{Q^{\rm nl}\} - \{Q^P\}, \tag{7}$$

where $[M] = \frac{\partial}{\partial \dot{q}} \iint (\dot{u}^2 + \dot{v}^2 + \dot{w}^2 + \dot{\beta}_s I_s + \dot{\beta}_\theta I_\theta)\,dA$ is the mass matrix, $[K] = \partial U_1^{(2)}/\partial q = \frac{\partial}{\partial q} \iint \sigma^{\rm l} \varepsilon^{\rm l}\,dA$ is the rigidity matrix, $\{Q\} = \frac{\partial}{\partial q} \iint (P_u u + P_v v + P_w w)\,dA$ is the vector of external forces, $\{Q^{\rm nl}\} = \partial U_{\rm n}^3/\partial q + \partial U_{\rm n}^4/\partial q = \frac{\partial}{\partial q} \iint (\varepsilon^{\rm l} A \varepsilon^{\rm n} + \varepsilon^{\rm n} B \chi + \varepsilon^{\rm n} A \varepsilon^{\rm n})\,dA$ is a geometrically nonlinear term, $K_G = \frac{\partial}{\partial q} \iint (\sigma_{\rm init}^{\rm l} + \sigma_{\rm init}^{\rm n}) \varepsilon^{\rm n}\,dA$ is the matrix of initial stresses, $\{\Delta Q\} = \frac{1}{2} \partial W^{(2)}/\partial q = \frac{1}{2} \frac{\partial}{\partial q} \iint [-P_w w(e_s + e_\theta) + P_v v e_{23} + P_u u e_{13}]\,dA$ is the variation in the potential of external forces in the case of nonconservative loads, and $\{Q^p\} = \frac{\partial}{\partial q}\iint(\sigma^{\rm l}+\sigma^{\rm n})\varepsilon^{\rm l}\,dA$ is the vector of accumulated plasticity. The matrix of elastoplastic strains is determined from the relation

$$\{\sigma\} = [E_{\rm pl}]\{\varepsilon\}.$$

Here $\{\sigma\} = \{\sigma_s, \sigma_\theta, \sigma_z, \tau_{s\theta}, \tau_{\theta z}, \tau_{sz}\}$ is the column vector of stresses, $\{\tau\} = \{\varepsilon_s, \varepsilon_\theta, \varepsilon_z, \varepsilon_{s\theta}, \varepsilon_{\theta z}, \varepsilon_{sz}\}$ is the column vector of strains, and $[E_{\rm pl}] = ([E] - [E]\{A\}\{A\}^T[E]^T)/(H + \{A\}^T[E]\{A\})$ is the elastoplastic modulus; $\{A\}^T = 6\{\sigma_s - \sigma_o, \sigma_\theta - \sigma_o, \sigma_z - \sigma_o, 2\tau_{s\theta}, 2\tau_{\theta z}, 2\tau_{sz}\}$, $\sigma_o = (\sigma_s + \sigma_\theta + \sigma_z)/3$, $H = EE^{\rm p}/(E - E^{\rm p})$, where E and $E^{\rm p}$ are elastic and plastic modules, respectively. To pass to the region of plastic flow, we use the Mises yield condition

$$\sqrt{0.5[(\sigma_s - \sigma_\theta)^2 + (\sigma_\theta - \sigma_z)^2 + (\sigma_s - \sigma_z)^2] + 6(\tau_{s\theta}^2 + \tau_{\theta z}^2 + \tau_{sz}^2)} \gg \sigma_T.$$

In formulas (7), we used the notation: the index "l" denotes the linear component, the index "n" denotes the nonlinear component, and index "init" denotes the initial stress (prestress), and the dot over a symbol denotes the differentiation with respect to time.

In the expression for the strain energy, we preserve the nonlinear terms up to the fourth order inclusively.

The process of obtaining exact solutions of nonlinear equations of shell and structure dynamics, especially for complex multilayer configurations, where the number of equations depends on the number of layers, encounters great mathematical difficulties. Therefore, such problems (differential equations with variable coefficients) are usually solved numerically.

In this method, the nonlinear terms are placed in the right-hand sides of equilibrium equations and are considered as additional generalized forces calculated from the values of generalized coordinates obtained at the preceding step of loading [1–9]. The convergence is usually improved by using iteration and extrapolation methods.

The equilibrium equations describing the nonlinear dynamic reaction, which are derived from the Lagrange equations (1), can be applied both to linear and nonlinear systems provided that the terms characterizing the strain energy and the work are expressed in terms of generalized coordinates, their time-derivatives, and variations.

Equations (7) were integrated by the Runge-Kutta method with automatic choice of the loading step with a prescribed accuracy under the following initial conditions:

$$\hat{q} = 0, \quad q = q_o \quad \text{for } t = 0,$$

where $\hat{q} = 0$ is the initial velocity, $q = q_o$ is the initial deflection, and t is the time.

The nonlinear terms were calculated for each harmonic from the values of generalized coordinates for the load increment at the preceding loading step.

At the first loading step, it was assumed that the system is linear and the nonlinearities are equal to zero; for the second loading increment, the linear interpolation was used:

$$q_j = 2q_{j-1} - q_{j-2}.$$

After the second and subsequent loading steps, the quadratic extrapolation was used:

$$q_j = \left(1 + \frac{3}{2}d + \frac{1}{2}d^2\right)q_{j-1} - (2d + d^2)q_{j-2} + \frac{1}{2}(d + d^2)q_{j-3},$$

where q_j are the extrapolated displacements, q_{j-1}, q_{j-2}, q_{j-3} are the known displacements at the preceding loading steps, and d is an extrapolation parameter. In the present paper, $d = 1$, which corresponds to the case of quadratic extrapolation.

IOP Conf. Series: Journal of Physics: Conf. Series **991** (2018) 012042 doi:10.1088/1742-6596/991/1/012042

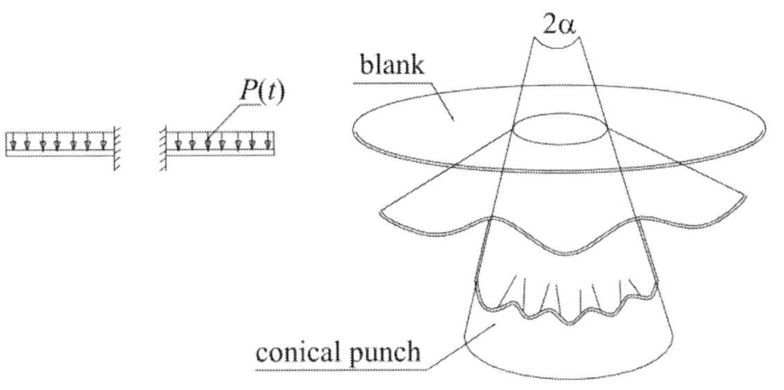

Figure 2. Ring plate deformation under pulse loading.

The shape of the loss of stability of the ring plate was taken as the initial deflection q_o. The maximal value (amplitude) of the initial deflection was 10^{-8} mm. In the calculations, the first three terms of the Fourier series (1) were preserved and the loading increment (workpiece conic punch) was set to be equal to 1×10^{-5}. In the further calculations, the program itself determined the step necessary to ensure the prescribed accuracy. The loss of accuracy at each step did not exceed $\Delta X = 1 \times 10^{-5}$. The solution with verified relative accuracy $\Delta X = 1 \times 10^{-7}$ was obtained to control the convergence of the solution. In both cases, the results were practically the same for the displacements. The discrepancy was at most 0.2%. The solution was stable until the relative deflection attained the value 0.1–0.2 for the time increment $\Delta t = 1 \times 10^{-6}$ s. To ensure the stability of the solution at the moment of sharp increase in the relative deflection, the increment was decreases to $\Delta t = 1 \times 10^{-7}$ s, and the time step was equal to $\Delta t = 1 \times 10^{-8}$ s in the supercritical domain. Figure 2 illustrates the variations in the ring plate shape under the action of rapidly increasing normal load.

The process of solving the problem of large displacements and strains splits into several stages. As the deflection becomes equal to half the plate thickness in the initial state, the mass and rigidity matrices and the vector of generalized external forces are recalculated with regard to the attained level of the stress strain state and the changed geometry of the plate. The calculations continue till the moment of the loss of stability of the plate. As the criterion for the loss of stability, we take the load at which a nonzero harmonic is suddenly excited. The obtained solutions were used to calculate the dynamic processes of flanging, drawing without pressing, and drawing with conic punch. A good agreement between the theoretical and experimental results was observed.

3.3. Experiment

The experiment was performed by using a magnetic pulse facility of energy capacity 30 kJ. The magnetic field pressure on the workpiece varied in time according to the damped sine law with maximal amplitude equal to $P_{\max} = 75$ MPa and duration of $t = (40\text{–}50)\times10^{-6}$ s. The workpiece deformation mechanism was investigated by using the high-speed filming with frequency varying from 83.330 to 5000 frames per second. The magnetic field topography was determined by the inductance transducers. The maximal pressure was attained at $t = 20\,\mu$s. A ring plate of diameter $d = 190$ mm and thickness $h = 1.5$ mm deformed into a ring matrix ($R = 190$ mm) was studied. Initially, the plate was pressed to the matrix by the force $P = 14.7$ kN. The material was the aluminum alloy Amg6M ($\sigma_s = 3.46 \times 10^8\,\mathrm{N/m^2}$, $\sigma_\theta = 715 \times 10^8\,\mathrm{N/m^2}$, the hardening modulus $3g = 4.35 \times 10^8\,\mathrm{N/m^2}$).

The plastic deformation starts at the matrix radius and propagates towards the grinding center as an extension wave and bending and shear waves. As a wave arrives, the speed of the pole part increases sharply, which often results in the fracture. For sufficiently thin workpieces ($h/R = 0.02$), the maximal strains and the fracture arise at the center independently of the fixation practice. In thicker rigidly fixed workpieces, the fracture occurs near the fixation. Any workpieces are similarly fractured under the action of a momentary powerful pulse.

The comparative experiments on static deformation of workpieces under the pressure of resin showed that the pulse loading permits increasing the limit degree of strain before the fracture by (50–90)%. The degree of strain before the loss of stability in operations of deep drawing was increased by (15–20)% compared with the static one, and in the operations of flanging, by 50%.

In the experiments, where the influence of the pulse value on the moment of the loss of stability was investigated, a ring workpiece of diameter $D = 100$ mm and thickness $h = 1.2$ mm manufactured from the D16M material ($\sigma_s = 3.0 \times 10^8$ N/m^2, $\sigma_\theta = 700 \times 10^8$ N/m^2, the hardening modulus $3g = 5.55 \times 10^8$ N/m^2) was considered. The central part of diameter $d = 45$ mm of the ring workpiece was supported by a steel punch. The dynamics of the workpiece motion depending on the pulse value and shape and the moment of the loss of stability were obtained by decoding the cinegrams. An analysis of the performed experiments showed that the plate limit degree of strain before the loss of stability increases by (15–40)% compared with the static loading.

As a rule, the discrepancy between all experimental data and the numerical data obtained by solving the problems did not exceed (10–20)%.

Conclusion

A mathematical model for studying the shell configurations made of homogeneous and composite materials under static and dynamic loads was derived in a geometrically and physically nonlinear statement. The main result of this study is the possibility of controlling the process of loss of stability of thin-walled plates and shells experiencing large elastoplastic deformations under the conditions of dynamic loading.

References

[1] Khairnasov K Z, Yudaev V B, and Favorin B N 1986 Dynamics and stability of plates under large strains In *Proc. II All-Union Symp. "Stability in Mechanics of Deformable Solids"* (Kalinin: Izdat. Kalinin. Politekh. Inst.) [in Russian]

[2] Yudaev V B, Khayrnassov K Z, and Favorin B N 1988 Behavior of pulse loaded plates at large deformations In *Mechanical Behaviour of Materials V. Proc. of the Fifth Int. Conf., Beijing, China, 3–6 June 1987* vol 1 ed M G Yan, S H Zhang, and Z M Zheng (Pergamon) pp 785–92

[3] Khairnasov K Z 2009 Modeling of behavior of ring plate under pulse loading, *Gorn. Inf.-Anal. Bull.* No. 7 52–63

[4] Khairnasov K Z 2013 Modeling of the aviation workpiece pressing *Vestnik MAI* No. 5 73–7

[5] Stricklin J, Navaratna D, and Pian T 1986 Improvements in Shell Computations by Matrix Displacement Method *Raketn. Tekhn. Kosmonavt.* **4** (6), 253–4

[6] Stricklin J, Haisler W, and Riesemann W. 1978 Evaluation of solution procedures for material and/or geometrically nonlinear structural analysis *Raketn. Tekhn. Kosmonavt.* **11** (3) 45–56

[7] Stricklin J, Martinez J, Tillerson J, et al. 1971 Nonlinear Dynamic Analysis of Shells of Revolution by Matrix Displacement Method *Raketn. Tekhn. Kosmonavt.* No. 4, 108–18

[8] Stricklin J, Haisler W, MacDougall H, and Stebbins F 1978 Nonlinear Analysis of Shells of Revolution by Matrix Displacement Method *Raketn. Tekhn. Kosmonavt.* **11** (3) 45–56

[9] Obraztsov I F, Volmir A S, and Khairnasov K Z 1982 Toroidal Shells: Delayed Catastrophes in Dynamic Loading *Dokl. Akad. Nauk SSSR* 266 (6), 1343–6

IOP Conf. Series: Journal of Physics: Conf. Series **991** (2018) 012043 doi:10.1088/1742-6596/991/1/012043

Mathematical modeling of shell configurations made of homogeneous and composite materials experiencing intensive short actions and large displacements

K Z Khairnasov

Moscow Aviation Institute (National Research University), Moscow, Russia

E-mail: kamilh@mail.ru

Abstract. The paper presents a mathematical model for solving the problem of behavior of shell configurations under the action of static and dynamic impacts. The problem is solved in geometrically nonlinear statement with regard to the finite element method. The composite structures with different material layers are considered. The obtained equations are used to study the behavior of shell configurations under the action of dynamic loads. The results agree well with the experimental data.

1. Introduction

Much attention has been paid to studying the structures under the action of short dynamic loads (see [1–3, 5, 6]). At the same time, it seems impossible to take into account all numerous parameters of the dynamic loading, such as the rate of loading, characteristics of the material dynamic behavior, etc., even in several papers.

2. Statement of the problem

The problem is to derive a mathematical model for studying the behavior of structures made of homogeneous and composite materials under static and dynamic actions in a geometrically nonlinear statement.

3. Method and construction of the solution

To derive the mathematical model of behavior of structures made of homogeneous and composite materials under intensive short actions in the case of large displacements, we consider the Lagrange equation

$$\frac{d}{dt}\frac{\partial T}{\partial \dot{q}_k}+\frac{\partial U}{\partial q_k}=Q_k, \tag{1}$$

where T is the kinetic energy, U is the potential energy of the system, Q_k are external loads, $\dot{q}_k$ and q_k are generalized velocity and displacements, and the k are the degrees of freedom.

We use the following strain-stress relations for a moderate deflection:

$$\begin{gathered}
\varepsilon_s = e_s + \frac{e_{13}^2 + e_{33}^2}{2}, \quad \varepsilon_\theta = e_\theta + \frac{e_{23}^2 + e_{33}^2}{2}, \quad \varepsilon_{s\theta} = e_{s\theta} + \frac{e_{13}^2 e_{23}^2}{2}, \quad \chi_s = \frac{\partial e_{13}}{\partial s}, \\
\chi_\theta = \frac{1}{r}\left(\frac{\partial e_{23}}{\partial \theta} + e_{13}\sin\varphi\right), \quad \chi_{s\theta} = \frac{1}{2}\left[\frac{\partial e_{23}}{\partial s} + \frac{1}{r}\frac{\partial e_{13}}{\partial \theta} - e_{23}\frac{\sin\varphi}{r} + \left(\frac{\partial \varphi}{\partial s} + \frac{\cos\varphi}{r}\right)e_{33}\right], \\
e_s = \frac{\partial u}{\partial s} - w\frac{\partial \varphi}{\partial s}, \quad e_\theta = \frac{1}{r}\left(u\sin\varphi + \frac{\partial v}{\partial \theta} + w\cos\varphi\right), \quad e_{s\theta} = \frac{1}{r}\left(\frac{\partial u}{\partial \theta} + r\frac{\partial v}{\partial s} - v\sin\varphi\right), \\
e_{13} = \left(\frac{\partial w}{\partial s} + u\frac{\partial \varphi}{\partial s}\right), \quad e_{23} = -\frac{1}{r}\left(\frac{\partial w}{\partial \theta} - v\sin\varphi\right), \quad e_{33} = -\frac{1}{2r}\left(r\frac{\partial v}{\partial s} + v\sin\varphi - \frac{\partial u}{\partial \theta}\right).
\end{gathered} \tag{2}$$

In equations (2), we use the following notation: ε_s, ε_θ, and $\varepsilon_{s\theta}$ are the strains of the middle surface of the shell in the corresponding coordinates, χ_s, χ_θ, and $\chi_{s\theta}$ are the curvature strains, e_{13} and e_{23} are the angles of rotation about the coordinate lines, r is the shell radius, φ_s and φ_θ are the angle of the meridian inclination to the shell axis and the angle in the circular direction, and e_{33} is the "torsion" of the middle surface.

3.1. Equations of the stress-strain relationship

We consider the geometrical and physical dependencies for multilayer shells in the reduced rigidity method. In the plane stress state, the matrices of elastic coefficients for an orthotopic material whose orthotropy axes coincide with the coordinate axes have the form

$$\{\sigma\} = [E]\{\varepsilon\},$$

where

$$\begin{gathered}
[E] = \begin{bmatrix} Q_{11} & Q_{12} & 0 \\ Q_{21} & Q_{22} & 0 \\ 0 & 0 & Q_{66} \end{bmatrix}, \quad \{\varepsilon\}^T = \{\varepsilon_s, \varepsilon_\theta, \varepsilon_{s\theta}\}, \quad \{\sigma\}^T = \{\sigma_s, \sigma_\theta, \sigma_{s\theta}\}, \\
Q_{11} = \frac{E_s}{1 - v_{s\theta}v_{\theta s}}, \quad Q_{12} = \frac{v_{s\theta}E_s}{1 - v_{s\theta}v_{\theta s}}, \quad Q_{21} = \frac{v_{\theta s}E_s}{1 - v_{s\theta}v_{\theta s}}, \quad Q_{22} = \frac{E_\theta}{1 - v_{s\theta}v_{\theta s}}, \quad Q_{66} = G_{66}.
\end{gathered} \tag{3}$$

As the coordinate axes rotate by an angle θ, the matrix of elastic coefficients becomes

$$[\bar{E}] = \begin{bmatrix} \bar{Q}_{11} & \bar{Q}_{12} & \bar{Q}_{16} \\ \bar{Q}_{21} & \bar{Q}_{22} & \bar{Q}_{26} \\ \bar{Q}_{61} & \bar{Q}_{62} & \bar{Q}_{66} \end{bmatrix}, \tag{4}$$

where

$$\begin{gathered}
\bar{Q}_{11} = c^4Q_{11} - s^4Q_{22} + 2(Q_{12} + 2Q_{66})s^2c^2, \quad \bar{Q}_{12} = (Q_{11} + Q_{22} - 4Q_{66})s^2c^2 + (s^2 + c^2)Q_{22}, \\
\bar{Q}_{16} = [c^2Q_{11} - s^2Q_{12} + (Q_{12} + 2Q_{66})(s^2 - c^2)]sc, \quad \bar{Q}_{22} = s^4Q_{11} - c^4Q_{22} + 2(Q_{12} + 2Q_{66})s^2c^2, \\
\bar{Q}_{26} = [s^2Q_{11} - c^2Q_{12} - (Q_{12} + 2Q_{66})(s^2 - c^2)]sc, \quad \bar{Q}_{66} = (Q_{11} - 2Q_{12} + Q_{22})s^2c^2 + (s^2 - c^2)Q_{66}, \\
s = \sin\theta, \quad c = \cos\theta.
\end{gathered} \tag{5}$$

Finally, the matrix of elastic coefficients of an arbitrary orthotropic layer has six components $\bar{Q}_{ij}$ depending on the six components Q_{ij}. In anisotropic materials, there are four independent constants Q_{ij} of the material.

For the layer at the distance z from the middle surface, the strains become

$$\{\varepsilon\} = \{\varepsilon^o\} + z\{\chi^o\},$$

where $\{\varepsilon^o\}$ are the strains of the middle surface and $\{\chi^o\}$ are the curvature variations.

After substitution of these relations, equation (4) becomes

$$\{\sigma\} = [\bar{Q}]\{\varepsilon^o\} + z[\bar{Q}]\{\chi^o\}.$$

We take the relations

$$\begin{aligned} \{N\} &= \int_{-h/2}^{h/2} \{\sigma\}\, dz, \quad \{N\}^T = \{N_s, N_\theta, N_{s\theta}\}, \\ \{M\} &= \int_{-h/2}^{h/2} \{\sigma\} z\, dz, \quad \{M\}^T = \{M_s, M_\theta, M_{s\theta}\}, \end{aligned} \tag{6}$$

where $\{N\}$ are membrane forces and $\{M\}$ are bending moments, into account and integrate expressions (6) over the shell thickness. Then the forces and moments can be represented as

$$\left\{ \begin{array}{c} N \\ M \end{array} \right\} = [E] \left\{ \begin{array}{c} \varepsilon^o \\ \chi^o \end{array} \right\}, \quad [E] = \left[\begin{array}{cc} [A] & [B] \\ [B] & [D] \end{array} \right], \tag{7}$$

$$\begin{aligned} &[A] = \left[\begin{array}{ccc} A_{11} & A_{12} & A_{16} \\ A_{21} & A_{22} & A_{26} \\ A_{61} & A_{62} & A_{66} \end{array} \right], \quad [B] = \left[\begin{array}{ccc} B_{11} & B_{12} & B_{16} \\ B_{21} & B_{22} & B_{26} \\ B_{61} & B_{62} & B_{66} \end{array} \right], \quad [D] = \left[\begin{array}{ccc} D_{11} & D_{12} & D_{16} \\ D_{21} & D_{22} & D_{26} \\ D_{61} & D_{62} & D_{66} \end{array} \right], \\ &\{A_{ij}, B_{ij}, D_{ij}\} = \int_{-h/2}^{h/2} Q_{ij}(1, z, z^2) dz \quad (i, j = 1, 2, 6). \end{aligned} \tag{8}$$

If the coefficients of the matrix (7) are constant in each layer of the package, then we integrate (8) to obtain

$$\begin{aligned} A_{ij} &= \sum_{k=1}^{n} \bar{Q}_{ij}(h_k - h_{k-1}), \quad i, j = 1, 2, 6, \\ B_{ij} &= \sum_{k=1}^{n} \bar{Q}_{ij}(h_k^2 - h_{k-1}^2), \quad i, j = 1, 2, 6, \\ D_{ij} &= \sum_{k=1}^{n} \bar{Q}_{ij}(h_k^3 - h_{k-1}^3), \quad i, j = 1, 2, 6, \end{aligned} \tag{9}$$

where A_{ij}, B_{ij}, and D_{ij} are the membrane, bending-membrane, and bending rigidities.

In the expression for Hooke's law, the linear variation in the transverse shear strains across the thickness supplements the matrices of elastic coefficients with the matrices of shear rigidities

$$\left\{ \begin{array}{c} \sigma_4 \\ \sigma_5 \end{array} \right\}^{(k)} = \left[\begin{array}{cc} \bar{Q}_{44} & 0 \\ 0 & \bar{Q}_{55} \end{array} \right] \left\{ \begin{array}{c} \bar{\varepsilon}_4 \\ \bar{\varepsilon}_5 \end{array} \right\}^{(k)},$$

where $\bar{Q}_{44} = G_{13}$, $\bar{Q}_{55} = G_{23}$, and G_{13}, G_{23} are the shear moduli.

Let us consider the kinematic model of multilayer shell in the case where the number of resolving equations depends on the number of the material layer. In this case, it is assumed that the displacements of the carrying layers must be determined, and they are independent parameters, while the displacements of the layers binding the carrying layers are determined from the displacements of the carrying layers according to some laws, and they are dependent parameters. Thus, the order of the system of equations with n carrying layers is n times greater than the order of the system of resolving equations in the reduced rigidity method. For simplicity, we consider a three-layer package of thickness t with two carrying layers of thicknesses t_1 and t_3

and a filler of thickness t_2 between the carrying layers. In the present paper, the displacements across the package thickness are distributed according to the polygonal line law.

The displacement of the neutral line in the filler and the angle of rotation of the normal in the deformation are determined by the relations

$$v_2 = \frac{\bar{v}_1 + \bar{v}_3}{2}, \quad \varphi_2 = \frac{\bar{v}_1 - \bar{v}_3}{t}, \quad \bar{v}_1 = v_1 - \frac{t_1 e_{13}}{2}, \quad \bar{v}_3 = v_3 + \frac{t_3 e_{23}}{2}$$

The displacements in the filler at the distance z from the neutral axis can be written as

$$v_2(z) = v_1 + z\varphi_2 = 0.5\left(v_1 - \frac{t_1 e_{13}}{2} + v_3 + \frac{t_3 e_{23}}{2}\right) + \frac{z(v_1 - t_1 e_{13}/2 - v_3 - t_3 e_{23}/2)}{t_2},$$

$$u_2(z) = u_1 + z\varphi_2 = 0.5\left(u_1 - \frac{t_1 e_{13}}{2} + u_3 + \frac{t_3 e_{23}}{2}\right) + \frac{z(u - t_1 e_{13}/2 - u_3 - t_3 e_{23}/2)}{t_2},$$

$$w_2(z) = w_2 + \frac{z(w_3 - w_1)}{t_2}.$$

The three-layer package model under study permits constructing a kinematic model of the shell composed of arbitrarily many layers.

3.2. Multilayer axially symmetric finite element

We consider the solution of equations of motion (1) for axially symmetric finite elements (see figure 1). The displacements for axially symmetric finite elements are represented as polynomials in the normal, circular, and radial directions and as Fourier series in the circular direction [4]:

$$u = \sum_{i=0} \left[\left(1 - \frac{s}{l}\right) q_1 + \frac{s}{l} q_5\right] \cos(i\theta), \quad v = \sum_{i=0} \left[\left(1 - \frac{s}{l}\right) q_1 + \frac{s}{l} q_5\right] \sin(i\theta),$$
$$w = \sum_{i=0} \left[\left(1 + \frac{3s^2}{l^2} + \frac{2s^3}{l^3}\right) q_2 + \left(s - \frac{2s^2}{l} + \frac{s^3}{l^2}\right) q_3 + \left(\frac{3s^2}{l^2} + \frac{2s^3}{l^3}\right) q_6 + \left(-\frac{s^2}{l} + \frac{s^3}{l^2}\right) q_7\right] \cos(i\theta), \tag{10}$$

where q_1, q_2, ..., q_8 are generalized displacements at nodal points.

To determine nonlinear components of potential energy we take the displacement w in the form

$$w = \sum_{i=0} \left[\left(1 - \frac{s}{l}\right) q_2 + \frac{s}{l} q_6\right] \cos(i\theta). \tag{11}$$

With this representation of the displacements taken into account, the equation of motion holds for any harmonic if there is a relationship between the harmonics contained in the last term.

3.3. Rigidity matrix

The method for direct determination of the rigidity matrix $[K]$ for structure elements consists in using the dependence between the strain energy of an element and the rigidity coefficients.

This dependence can be written as

$$[K]\{q\} = \frac{\partial U_1^{(2)}}{\partial q_i},$$

where $U_1^{(2)}$ is the linear part of the strain energy of an element and the q_i are displacements at the nodes.

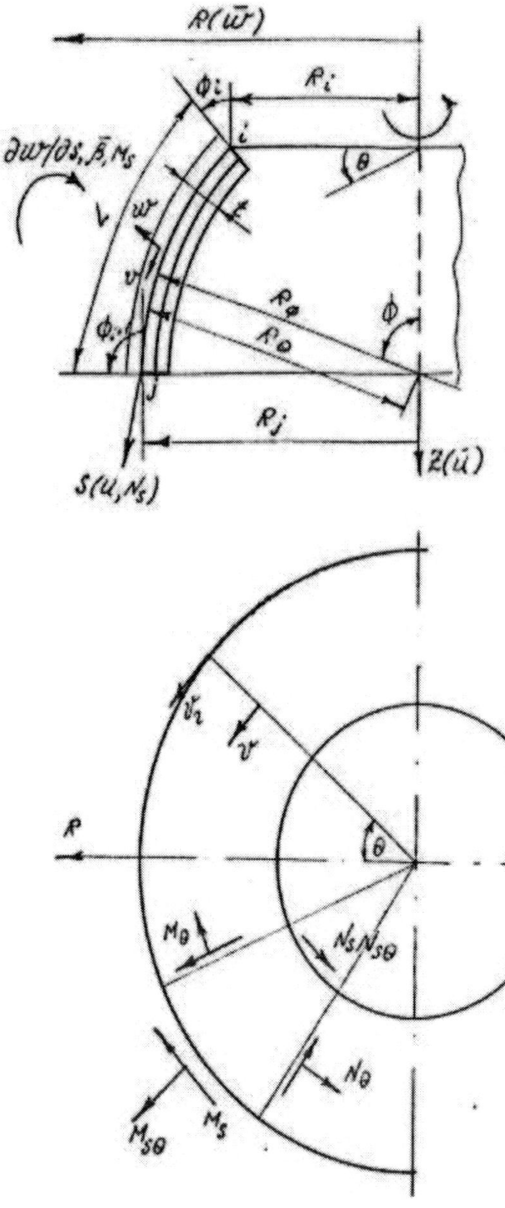

Figure 1. Multilayer axially symmetric finite element.

The strain energy of an element can be written as

$$U = 0.5 \iint \{\{N\}^T\{\varepsilon^o\} + \{M\}^T\{\chi^o\}\}\, dA, \qquad (12)$$

where we use the following notation: $\{N\}^T = \{N_s, N_\theta, N_{s\theta}\}$ are membrane forces, $\{M\}^T = \{M_s, M_\theta, M_{s\theta}\}$ are bending moments, $\{\varepsilon^o\} = (\varepsilon_s^o, \varepsilon_\theta^o, \varepsilon_{s\theta}^o)$ are strains in the middle surface plane, $\{\chi^o\} = \{\chi_s^o, \chi_\theta^o, \chi_{s\theta}^o\}$ are curvatures, and $dA = R\, ds\, d\theta$ is the area of integration.

IOP Conf. Series: Journal of Physics: Conf. Series **991** (2018) 012043 doi:10.1088/1742-6596/991/1/012043

For a layered shell, the stress-strain relation can be written as

$$\{N\} = [A]\{\varepsilon^o\} + [B]\{\chi^o\}, \quad \{M\} = [B]\{\varepsilon^o\} + [D]\{\chi^o\}, \tag{13}$$

where $[A]$, $[B]$, and $[D]$ are membrane, bending-membrane, and bending rigidity matrices of the layers.

Substituting (11) into (10), we obtain

$$U = 0.5 \iint (\{\varepsilon^o\}^T[A]\{\varepsilon^o\} + \{\chi^o\}^T[B]\{\varepsilon^o\} + \{\varepsilon^o\}^T[B]\{\chi^o\} + \{\chi^o\}^T[D]\{\chi^o\}])r\, ds\, d\theta. \tag{14}$$

After the substitution of strain expressions (10) and (11) into Eq. (14), the strain energy can be represented as

$$U = U_1^{(2)} + U_n^{(3)} + U_n^{(4)}, \tag{15}$$

where $U_1^{(2)}$ is the strain energy calculated for the corresponding strains e by the linear theory:

$$U_1^{(2)} = 0.5 \iint (\{\varepsilon\}^T[A]\{\varepsilon\} + \{\chi\}^T[B]\{\varepsilon\} + \{\varepsilon\}^T[B]\{\chi\} + \{\chi\}^T[D]\{\chi\})r\, ds\, d\theta, \tag{16}$$

where $U_n^{(3)}$ and $U_n^{(4)}$ are components of the strain energy due to the nonlinear terms in the relations between strains and displacements, which can be determined as

$$\begin{aligned} U_n^{(3)} &= 0.5 \iint (\{\varepsilon\}^T[A]\{\varepsilon^{(2)}\} + \{\chi\}^T[B]\{\varepsilon^{(2)}\} + \{\varepsilon^{(2)}\}^T[B]\{\chi\})r\, ds\, d\theta, \\ U_n^{(4)} &= 0.5 \iint (\{\varepsilon^{(2)}\}^T[A]\{\varepsilon^{(2)}\})r\, ds\, d\theta. \end{aligned} \tag{17}$$

For the layer of the filler, the strain energy can be written as

$$U = 0.5 \iint \{\sigma_3\}\{\varepsilon_3\}^T r\, ds\, d\theta, \quad \{\sigma_3\} = \{\tau_{13}, \tau_{23}\}, \quad \{\varepsilon_3\}^T = \{\chi_{13}, \chi_{23}\}. \tag{18}$$

3.4. Variations in the potential of external forces

In the case of hydrostatic load, the dependence between an increase in the volume bounded by the shell ΔV and the variation in the potential of external forces due to the shell deformation is determined by the relation $\Delta Q = P\Delta V$. Let us find the volume variation up to the squared bifurcation displacements and their derivatives. For this, a shell element shaped as a parallelepiped $ABCDL_1B_1C_1D_1$ (see figure 2) is divided into six elementary tetrahedrons so that the tetrahedrons of adjacent elements have the same boundaries AC_1BA, $AD_1C_1A_1$, ADC_1D, ABC_1B_1, ADC_1C, and $ACBC_1$. Then the volume of any shell element can be determined completely by the sum of these tetrahedrons.

It is well known that if one of the tetrahedron vertices coincides with the origin, then its volume is calculated by the formula

$$V = \frac{1}{6}\begin{pmatrix} x_1 & s_1 & w_1 \\ x_2 & s_2 & w_2 \\ x_3 & s_3 & w_3 \end{pmatrix},$$

where x_i, s_i, and w_i $(i = 1, 2, 3)$ are the coordinates of the vertices.

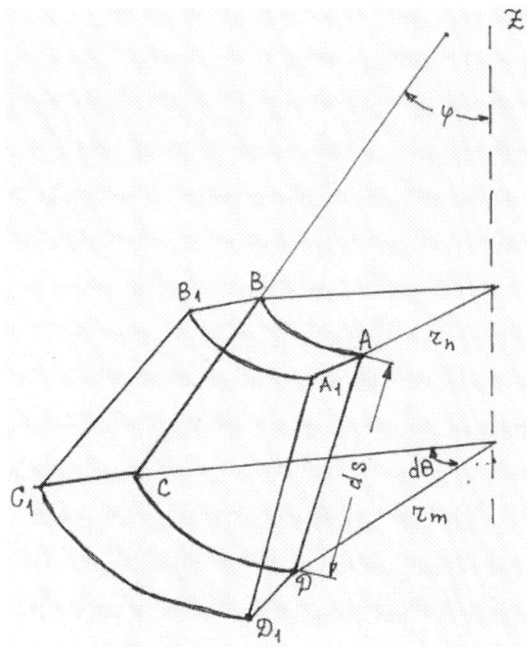

Figure 2. Ring plate deformation under pulse loading.

Let us calculate the volumes of the tetrahedrons AC_1BA, $AD_1C_1A_1$, ADC_1D, ABC_1B_1, ADC_1C, and $ACBC_1$ by the formulas

$$V_{ADC_1D_1} = \frac{1}{6}\begin{pmatrix} x_D & s_D & w_D \\ x_C & s_C & w_C \\ x_D & s_D & w_D \end{pmatrix}$$

$$= \frac{1}{6}\left(w + \frac{w}{r_n}\frac{\partial v}{\partial \theta} + \frac{w^2}{r_n}\cos\varphi - \frac{v}{r_n}\frac{\partial w}{\partial \theta} + \frac{v^2}{r_n}\cos\varphi + \frac{uw}{r_n}\sin\varphi\right) r_n\,d\theta\,ds,$$

$$V_{AC_1BA_1} = \frac{1}{6}\left(-u\frac{w}{ds} - \frac{v}{r_n}\frac{\partial w}{\partial \theta} + \frac{v^2}{r_n}\cos\varphi + w + \frac{w^2}{r_n}\frac{\partial v}{\partial \theta} + \frac{w^2}{r_n}\cos\varphi + v\frac{\partial u}{\partial s} + \frac{uw}{r_n}\sin\varphi\right) r_n\,d\theta\,ds,$$

$$V_{AD_1C_1A_1} = \frac{1}{6}\left(-u\frac{w}{ds} - \frac{v}{r_n}\frac{\partial w}{\partial \theta} + w + w\frac{\partial u}{\partial s} + \frac{w}{r_n}\frac{\partial v}{\partial \theta} + \frac{w^2}{r_n}\cos\varphi + \frac{uw}{r_n}\sin\varphi\right) r_n\,d\theta\,ds,$$

$$V_{ABC_1B_1} = \frac{1}{6}\left(w + w\frac{\partial w}{\partial s} - u\frac{\partial w}{\partial s}\right) r_n\,d\theta\,ds, \quad V_{ADCC_1} = \frac{1}{6}wr_n\,d\theta\,ds, \quad V_{ACBC_1} = \frac{1}{6}wr_n\,d\theta\,ds.$$

The values of x_D, s_D, and w_D are the coordinates of the tetrahedron vertices.

We sum the volumes of all tetrahedrons to obtain

$$d(\Delta V) = [w + 0.5w(e_s + e_\theta) - ve_{23} - ue_{13}]r_n\,ds\,d\theta.$$

Thus, the variation in the potential of external forces in the transition of the shell from the initial state into the deformed state becomes

$$W^{(2)} = -0.5\iint [P_w w(e_s + e_\theta) - P_v ve_{23} - P_u ue_{13}]r_n\,ds\,d\theta.$$

To solve the nonlinear equations, it is unnecessary to separate the problem of determining the initial state and the stability problem, as is usual in the case of static stability criterion. The critical loads can be determined from the limit points or from the branch points of the nonlinear solution (the points of the solution bifurcation).

It is admissible to use such a distinction between the displacements, because they satisfy the continuity condition for the $(n-1)$th derivative on the interface between the elements provided that the intrinsic energy is a function of the ith derivative of displacements [5, 6].

The nonlinear terms for an element are determined by calculating the partial derivatives of expression (12) for the strain energy, which corresponds to the nonlinearities in the generalized coordinates. For $U_{\text{n}}^{(3)} + U_{\text{n}}^{(4)}$, this implies

$$\frac{\partial U_{\text{n}}^{(3)}}{\partial q_m^n} = 0.5 \iint \Big[(C_1 e_{13}^2 + v_{s\theta} C_2 e_{23}^2) \frac{\partial e_s}{\partial q_m^n} + (C_2 e_{23}^2 + v_{s\theta} C_1 e_{13}^2) \frac{\partial e_\theta}{\partial q_m^n} + 2G_1 e_{13} e_{23} \frac{\partial e_{s\theta}}{\partial q_m^n} + 2(C_1 e_s e_{13} + v_{s\theta} C_1 e_s e_{13} + G_1 e_{s\theta} e_{23}) \frac{\partial e_{13}}{\partial q_m^n} + 2(2e_\theta e_{23} + v_{s\theta} C_1 e_s e_{23} + G_1 e_{s\theta} e_{13}) \frac{\partial e_{23}}{\partial q_m^n} \Big] r\, ds\, d\theta, \tag{19}$$

$$\frac{\partial U_{\text{n}}^{(4)}}{\partial q_m^n} = 0.5 \iint \Big\{ [C_1 e_{13}^3 + (v_{s\theta} C_1 + 2G_1) e_{13} e_{23}^2] \frac{\partial e_{13}}{\partial q_m^n} + [C_1 e_{13}^3 + (v_{s\theta} C_1 + 2G_1) e_{13} e_{23}^2] \frac{\partial e_{23}}{\partial q_m^n} \Big\} r\, ds\, d\theta,$$

where n is the harmonic number and $m = 1, 2, \ldots, 8$. The accepted functions of displacements permit separating the variables in equations (10) and (11) and satisfying the periodicity condition in the circular variable θ.

Substituting (11) into (3), we obtain

$$\begin{gathered} \varepsilon_s = \sum_{i=0} e_s^i \cos(i\theta), \quad \varepsilon_\theta = \sum_{i=0} e_\theta^i \cos(i\theta), \quad \varepsilon_{s\theta} = \sum_{i=0} e_{s\theta}^i \sin(i\theta), \\ \varepsilon_{13} = \sum_{i=0} e_{13}^i \cos(i\theta), \quad \varepsilon_{23} = \sum_{i=0} e_{23}^i \cos(i\theta). \end{gathered} \tag{20}$$

Here e_s, e_θ, $e_{s\theta}$, e_{13}, and e_{23} are linear strains, and the rotations for the ith harmonics are functions of only the meridian distance s. For example,

$$e_{23} = \left(-\frac{i}{r} \right) (\alpha_1^i + \alpha_2^i s + \alpha_3^i s^2 + \alpha_4^i s^3) - \cos \frac{\varphi}{r} (\alpha_7^i + \alpha_8^i s). \tag{21}$$

When calculating the partial derivatives, the integrals over the length are taken independently.

Because the products of trigonometric functions with wave numbers i, j, and k can be represented by a sum (difference) of trigonometric functions with wave numbers

$$k - j - i, \quad k - j + i, \quad k + j + i, \tag{22}$$

$$\cos(k\theta)\cos(j\theta)\cos(i\theta) = \frac{1}{4}[\cos(k-j-i)\theta + \cos(k-j+i)\theta + \cos(k+j-i)\theta + \cos(k+j+i)\theta],$$

$$\sin(k\theta)\sin(j\theta)\sin(i\theta) = \frac{1}{4}[\cos(k-j-i)\theta + \cos(k-j+i)\theta - \cos(k+j-i)\theta + \cos(k+j+i)\theta],$$

$$\cos(k\theta)\sin(j\theta)\sin(i\theta) = -\frac{1}{4}[\cos(k-j-i)\theta + \cos(k-j+i)\theta + \cos(k+j-i)\theta - \cos(k+j+i)\theta],$$

$$\sin(k\theta)\cos(j\theta)\sin(i\theta) = \frac{1}{4}[\cos(k-j-i)\theta - \cos(k-j+i)\theta + \cos(k+j-i)\theta - \cos(k+j+i)\theta],$$

the above relations are nonzero if relations (22) are equal to zero. Otherwise, they are identically zero.

IOP Conf. Series: Journal of Physics: Conf. Series **991** (2018) 012043 doi:10.1088/1742-6596/991/1/012043

When the fourth-order terms are calculated, expressions (8) form the following structures:

$$
\begin{aligned}
&\cos(k\theta)\cos(j\theta)\cos(i\theta)\cos(n\theta) = 0.25[H_1 + H_2 + H_3 + H_4 + H_5 + H_6 + H_7 + H_8],\\
&\sin(k\theta)\sin(j\theta)\sin(i\theta)\sin(n\theta) = 0.25[-H_1 - H_2 + H_3 + H_4 + H_5 + H_6 - H_7 - H_8],\\
&\sin(k\theta)\sin(j\theta)\sin(i\theta)\sin(n\theta) = 0.25[-H_1 - H_2 + H_3 + H_4 + H_5 + H_6 - H_7 - H_8],\\
&\cos(k\theta)\cos(j\theta)\sin(i\theta)\sin(n\theta) = 0.25[-H_1 + H_2 + H_3 - H_4 + H_5 - H_6 - H_7 + H_8],\\
&H_1 = \cos(k-j-i-n)\theta, \quad H_2 = \cos(k-j-i+n)\theta, \quad H_3 = \cos(k-j+i-n)\theta,\\
&H_4 = \cos(k-j+i+n)\theta, \quad H_5 = \cos(k+j-i+n)\theta, \quad H_6 = \cos(k-j+i-n)\theta,\\
&H_7 = \cos(k+j+i-n)\theta, \quad H_8 = \cos(k+j+i+n)\theta.
\end{aligned}
\tag{23}
$$

Expressions (23) are different from zero under the same conditions as the conditions for expressions (22), i.e., if $k \pm j \pm i \pm n = 0$ and expressions (23) are equal to zero otherwise.

Thus, the integrals over the angle θ can be calculated explicitly, while the integration over the length of an element consists of many operations and requires a greater volume of computer memory. Therefore, it was assumed in the computational program that the integrals over an element of meridian length are calculated by dividing this meridian into separate parts.

Substituting (10) and (11) into (12) and calculating the strain values at the center of the element, we obtain the linear strains e_s^i, e_θ^i, $e_{s\theta}^i$, e_{13}^i, and e_{23}^i in the form

$$
\begin{aligned}
&e_s^i = \frac{q_5^i - q_1^i}{l} - \frac{q_6^i - q_2^i}{l}\varphi_m, \quad e_\theta^i = \frac{1}{r_m}\left(i\frac{q_4^i - q_8^i}{2} - \frac{q_1^i - q_5^i}{2}\sin\varphi_m + \frac{q_2^i - q_6^i}{2}\cos\varphi_m\right),\\
&e_{s\theta}^i = -i\frac{q_5^i - q_1^i}{2r_m} - \frac{q_4^i + q_8^i}{2r_m}\sin\varphi_m, \quad e_{13}^i = \frac{q_2^i - q_6^i}{l} - \frac{q_1^i - q_5^i}{2}\varphi_m,\\
&e_{23}^i = \frac{1}{r_m}\left(i\frac{q_6^i - q_2^i}{2} + \frac{q_2^i - q_6^i}{2}\cos\varphi_m\right).
\end{aligned}
\tag{24}
$$

Substituting (24) into (19) and using the method of integration of a part over the coordinate varying in the meridional direction, we obtain

$$
\begin{aligned}
\frac{\partial U_{\mathrm{n}}^{(3)}}{\partial q_m^n} = \frac{r_m l}{2}\sum_{i=0}\sum_{j=0}\Big\{&(C_1^{ijn}e_{13}^i e_{13}^j + \nu_{s\theta}C_1^{\bar{i}\bar{j}n}e_{23}^i e_{23}^j)\frac{\partial e_s^n}{\partial q_m^n} + (C_1^{ijn}e_{23}^i e_{23}^j + \nu_{s\theta}C_1^{ijn}e_{13}^i e_{13}^j)\frac{\partial e_\theta^n}{\partial q_m^n}\\
&+ 2G_1^{\bar{i}\bar{j}\bar{n}}e_{13}^i e_{23}^j\frac{\partial e_{s\theta}^n}{\partial q_m^n} + (2C_1^{ijn}e_s^i e_{13}^i + 2\nu_{s\theta}C_1^{ijn}e_\theta^i e_{13}^j + 2G_1^{\bar{i}\bar{j}n}e_{s\theta}^i e_{23}^j)\frac{\partial e_{13}^n}{\partial q_m^n}\\
&+ (2C_1^{\bar{i}\bar{j}\bar{n}}e_\theta^i e_{23}^j + 2\nu_{s\theta}C_1^{i\bar{j}\bar{n}}e_s^i e_{13}^j + 2G_1^{\bar{i}\bar{j}\bar{n}}e_{s\theta}^i e_{13}^j)\frac{\partial e_{23}^n}{\partial q_m^n}\Big\},\\
\frac{\partial U_{\mathrm{n}}^{(4)}}{\partial q_m^n} = \frac{r_m l}{2}\sum_{i=0}\sum_{j=0}\sum_{k=0}\Big\{&[C_2^{\bar{i}\bar{j}\bar{k}\bar{n}}e_{13}^i e_{13}^j e_{13}^k + (\nu_{s\theta}C_1^{\bar{i}j\bar{k}n} + 2G_1^{\bar{i}\bar{j}k\bar{n}})e_{23}^i e_{23}^j e_{23}^k]\frac{\partial e_{13}^n}{\partial q_m^n}\\
&+ [C_2^{i\bar{j}\bar{k}n}e_{13}^i e_{13}^j e_{13}^k + (\nu_{s\theta}C_1^{\bar{i}j\bar{k}n} + 2G_1^{\bar{i}\bar{j}k\bar{n}})e_{23}^i e_{23}^j e_{23}^k]\frac{\partial e_{23}^n}{\partial q_m^n}\Big\}
\end{aligned}
$$

where r_m is the radius value in the middle of the element, and e_s^i, e_θ^i, $e_{s\theta}^i$, e_{13}^i, and e_{23}^i are the values of linear strains calculated at the middle point of the element (10).

The indices of the constants C_1, C_2, and G_1 show that a constant is multiplied by the integral of a trigonometric function. For example,

$$
\begin{aligned}
C_1^{ijkn} &= C_1\int_0^{2\pi}\cos(i\theta)\cos(j\theta)\cos(k\theta)\cos(n\theta)\,d\theta,\\
C_1^{ij\bar{k}\bar{n}} &= C_1\int_0^{2\pi}\cos(i\theta)\cos(j\theta)\sin(k\theta)\sin(n\theta)\,d\theta,
\end{aligned}
$$

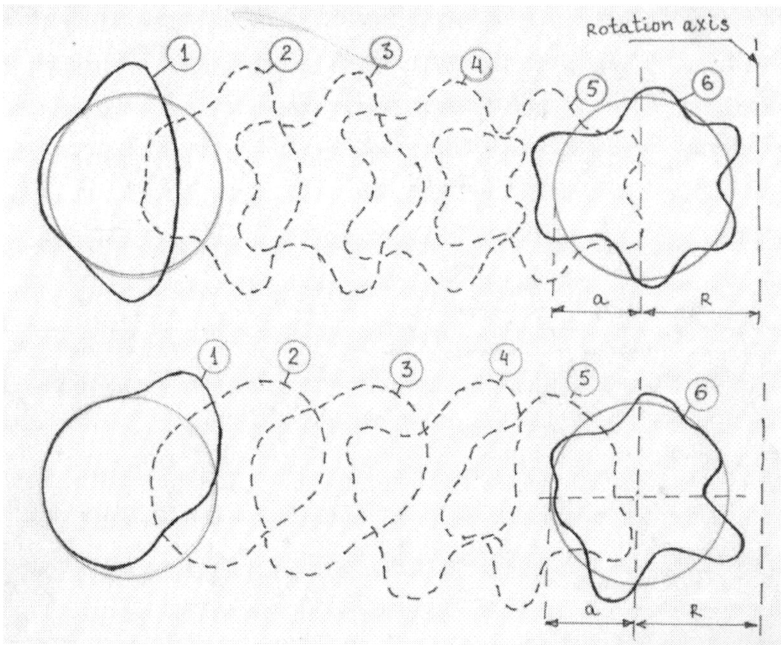

Figure 3. Bulging shapes of a toroidal shell with the parameters $R = a = 8$ and $s = h = 100$ in the process of loading by an external pressure $p = p_0 t$ which rapidly increases in time, where $p_0 = 1000\,\mathrm{MPa/s}$ is the loading rate and t is the time, for different values of $t^* = p = p_0$: (a) symmetric bulging, (b) skew-symmetric bulging.

where the bar over i, j, and k or n means that the cosine function must be replaced by the sine function. The other terms can be calculated similarly.

3.5. Resolving equations

The equilibrium equations describing the nonlinear dynamic reaction are obtained from the Lagrange equation (1), and these equations can be applied both to linear and nonlinear systems under the condition that the terms characterizing the strain energy and the work are expressed in terms of generalized coordinates and their time-derivatives and variations.

Substituting (12)–(19) into (1) and moving the quantities corresponding to the nonlinear terms into the right-hand side, we obtain the equations of motion, which form a mathematical model for studying the structures made of homogeneous and composite materials experiencing intensive short actions and large strains, in the form

$$[M]\{\hat{q}\} + [K]\{q\} = \{Q\} + \{\Delta Q\} - [K_G]\{q\} - \{Q^{\mathrm{nl}}\}, \tag{25}$$

where $[M] = \frac{\partial}{\partial q}\iint(\dot{u}^2 + \dot{v}^2 + \dot{w}^2 + \dot{\varphi}_s I_s + \dot{\varphi}_\theta I_\theta)\,dA$ is the mass matrix, $[K] = \partial U_{\mathrm{l}}^{(2)}/\partial q = \frac{\partial}{\partial q}\iint \sigma^{\mathrm{l}}\varepsilon^{\mathrm{l}}\,dA$ is the rigidity matrix, $\{Q\} = \frac{\partial}{\partial q}\iint(P_u u + P_v v + P_w w)\,dA$ is the vector of external forces, $\{Q^{\mathrm{nl}}\} = \partial U_{\mathrm{n}}^3/\partial q + \partial U_{\mathrm{n}}^4/\partial q = \frac{\partial}{\partial q}\iint(\varepsilon^{\mathrm{l}}A\varepsilon^{\mathrm{n}} + \varepsilon^{\mathrm{n}}B\chi + \varepsilon^{\mathrm{n}}A\varepsilon^{\mathrm{n}})\,dA$ is the geometrically nonlinear term, $[K_G] = \frac{\partial}{\partial q}\iint(\sigma^{\mathrm{l}}_{\mathrm{init}} + \sigma^{\mathrm{n}}_{\mathrm{init}})\varepsilon^{\mathrm{n}}\,dA$ is the matrix of initial stresses, and $\{\Delta Q\} = \frac{1}{2}\partial W^{(2)}/\partial q = \frac{1}{2}\frac{\partial}{\partial q}\iint[-P_w w(e_s + e_\theta) + P_v v e_{23} + P_u u e_{13}]\,dA$ is the variation in the potential of external forces in the case of nonconservative loads.

In formulas (25), we use the notation: the index "l" denotes the linear component, the index "n" denotes the nonlinear component, the index "init" denotes the initial stress (prestress), and the dot over a symbol denotes the differentiation with respect to time.

The process of exact solution of nonlinear equations of motion for shells and structures, especially for complex multilayer configurations, where the number of equations depends on the number of layers, encounters great mathematical difficulties. Therefore, such problems (differential equations with variable coefficients) are usually solved numerically.

In this method, the nonlinear terms are placed in the right-hand sides of equilibrium equations and are considered as additional generalized forces calculated from the values of generalized coordinates obtained at the preceding step of loading [5, 6]. The convergence is usually improved by iteration and extrapolation methods.

4. Analysis of the results

The above-developed program together with application of axially symmetric finite elements was used in geometrically nonlinear statement to study the behavior of a toroidal shell under the action of a sudden rapidly increasing external normal pressure [1]. The results of computations show that the behavior of a cross-section of the toroidal shell is independent of the initial deflection shape but is determined by its symmetry and skew-symmetry with respect to the horizontal axis of the cross-section (see figure 3).

Conclusion

A mathematical model for studying the shell configurations made of homogeneous and composite materials under static and dynamic loads was derived in a geometrically and physically nonlinear statement.

References

[1] Obraztsov I F, Volmir A S, and Khairnasov K Z 1982 Toroidal Shells: Delayed Catastrophes in Dynamic Loading *Dokl. Akad. Nauk SSSR* 266 (6), 1343–6
[2] Khairnasov K Z 2009 Modeling of behavior of ring plate under pulse loading, *Gorn. Inf.-Anal. Bull.* No. 7 52–63
[3] Khairnasov K Z 2013 Modeling of the aviation workpiece pressing *Vestnik MAI* No. 5 73–7
[4] Stricklin J, Navaratna D, and Pian T 1986 Improvements in Shell Computations by Matrix Displacement Method *Raketn. Tekhn. Kosmonavt.* **4** (6), 253–4
[5] Stricklin J, Martinez J, Tillerson J, et al. 1971 Nonlinear Dynamic Analysis of Shells of Revolution by Matrix Displacement Method *Raketn. Tekhn. Kosmonavt.* No. 4, 108–18
[6] Stricklin J, Haisler W, and Riesemann W 1978 Evaluation of solution procedures for material and/or geometrically nonlinear structural analysis *Raketn. Tekhn. Kosmonavt.* **11** (3) 45–56

Collaboration of polymer composite reinforcement and cement concrete

V G Khozin* **and A R Gizdatullin****

Kazan State University of Architecture and Civil Engineering, Kazan, Russia

E-mail: *khozin@kgasu.ru, **antonchiks@mail.ru

Abstract. The results of experimental study of bond strength of cement concrete of different types with fiber reinforcing polymer (FRP) bars are reported. The reinforcing bars were manufactured of glass fibers and had a rebar with different types of the surface relief formed by winding a thin strip impregnated with a binder or by "sanding". The pullout tests were carried out simultaneously for the steel reinforcing ribbed bars A400. The impact of friction, adhesion and mechanical bond on the strength of bonds between FRP and concrete was studied. The influence of the concrete strength and different operation factors on the bond strength of concrete was evaluated.

1. Introduction

In the recent years, in addition to the conventional steel reinforcing bars, polymer composite reinforcement (FRP) is increasingly used in the construction industry. FRP is manufactured from basalt, glass or carbon fibers and polymer binders based on epoxy and (less frequently) vinyl-ether resins. At that, a difference in molding techniques translates into the surface relief of FRP bar and exert a direct impact on its bond strength with concrete.

The recently published investigations have revealed specific characteristics of FRP interaction with cement concrete; the influence of anchorage length, bar diameter, and strength of concrete on the bond strength value has been studied. The majority of experiments have been carried out for rebar specimens with a "sandy" coating, helically-deformed (depressed) profile or machined annular "grooves" on the surface. At the same time, FRP manufactured by the needle-trusion method is popular in Russia because of simplicity of the technique. This FRP bar has a spirally wound roving which is pasted on the surface and impregnated by the same binder; this is an imitation of a periodical profile of steel reinforcing bars (ribbed bars).

In the present work, we report the data on the bond strength between FRP with different type of the surface profile and the cement concrete, which were obtained by the direct pullout test method under different operational conditions. The bond stress distributions were determined; the character of stress distribution along the rebar was described.

2. Materials and Methods

The FRP specimens of four types with a diameter ranging from 7.5 to 8 mm, which had different surface profiles (confinement), were studied. All samples were manufactured from glass fibers and epoxy binder. To compare the results, the tests were carried out for the A400-class steel rebar together with FRP specimens.

Table 1. Characterization of FRP.

No.	Type of fiber and surface relief	Geometry of rod, dimensions (mm)	External view of specimens (photograph)
1	Fiber – glass; Rebar surface is helically deformed by depressed winding (strip)	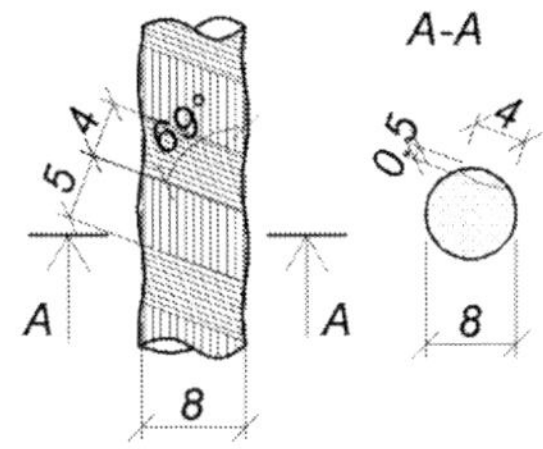	
2	Fiber – glass; Double helical roving is pasted on the rebar surface	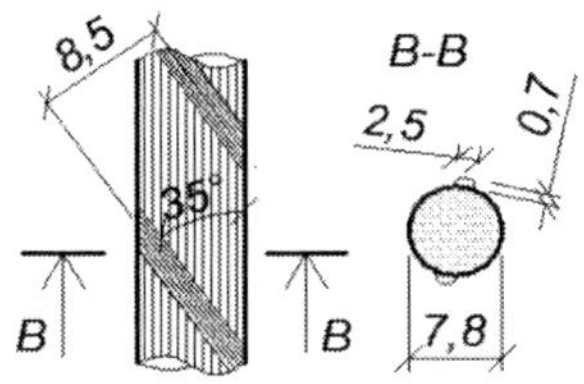	
3	Fiber – glass; Single helical roving is pasted on the rebar surface	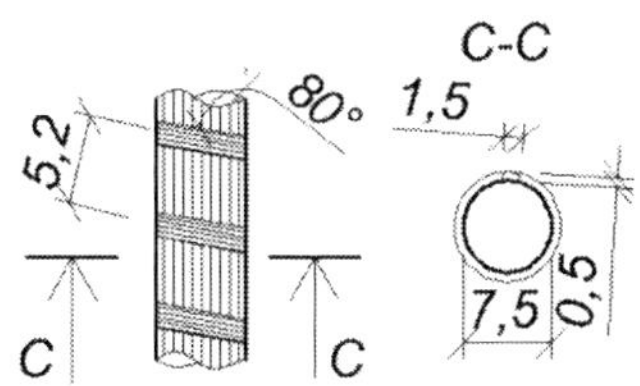	
4	Fiber – glass; Surface is coated by quartz sand (sandy)	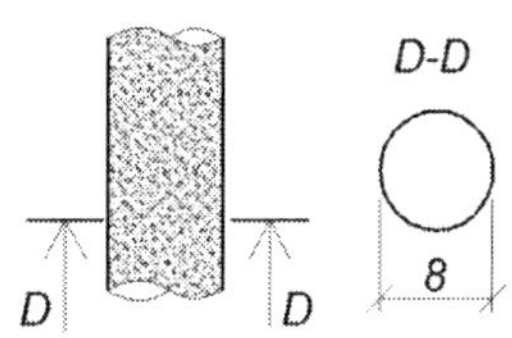	
3	Steel A400		

A fine-grained concrete with the strength compression of 30 MPa was used to fabricate the samples. The molding of concrete cylinders (110 mm in diameter and 100 mm in length) was carried out in a polyethylene molding matrix with a FRP bar embedded coaxially. The embedment length (length of the contact between a rebar and concrete) was 50 mm. The solidification of specimens proceeded under typical conditions (20°C and 100% humidity) within 28 days; and one experiment was carried out upon the steam curing processing (SCP) (2 + 6 + 2 hours)

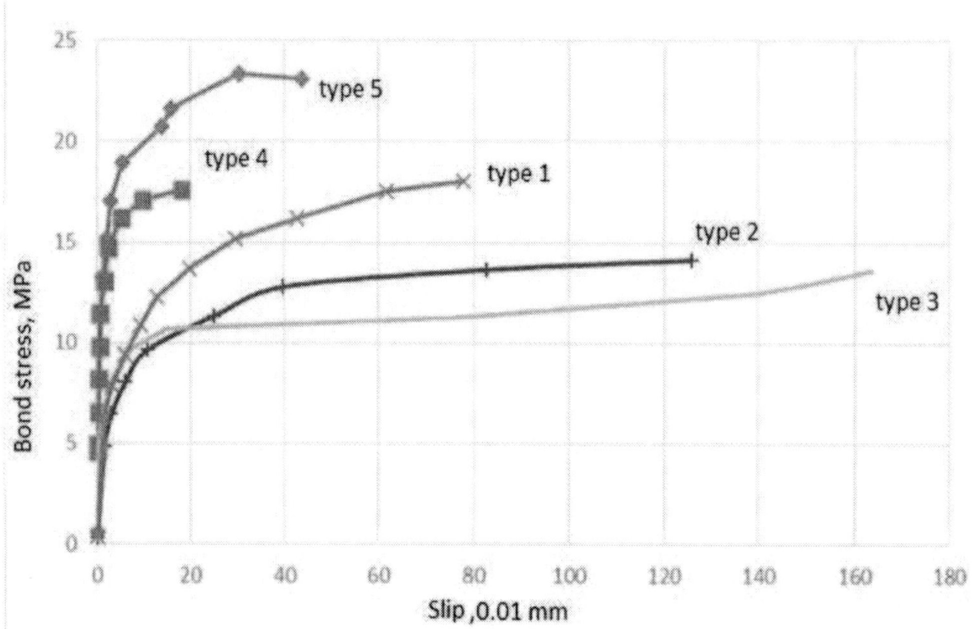

Figure 1. "Adhesion-slip" curves ($\tau - \Delta$) of FRP and steel specimens obtained in the reference tests.

The method of direct pulling out of the rebar from concrete was applied. The slip of rebar tips in concrete under the load of 200 kg-sec was determined from the tracer displacement. The loads that caused 0.1, 0.3 mm and the maximum displacement of the unloaded end were considered as the reference values. To evaluate the external technological and exploitation factors, which are related to adhesion, the tests were conducted under the conditions described in table 1.

As a part of study, the following tests were carried out:

- tests of reference samples;
- tests after steam curing processing;
- tests with regard to high (+40°C/+80°C) and low (−40°C) temperatures;
- tests after temperature oscillations;
- tests after exposure in water;
- tests after exposure in alkaline medium;
- pullout tests with concrete specimens of different strength.

3. Results

The results of investigations are listed in table 2; the results of pullout tests for the reference samples (most typical) are represented as "stress-slip" ($\tau - \Delta$) curves.

The plots in figure 1 show that the character of variations in the tangent stress "τ" on the interfacial "FRP–concrete" boundary in the course of the pullout tests carried out for the reference samples significantly depends on the type of FRP surface profile. The growth of "τ" up to the values about 15 MPa was observed for a FRP#1 specimen with depressed winding (deformed bar) at a FRP slip of less than 0.3 mm after which the FRP slip in concrete attained 1–3 mm upon a slight growth in "τ" attaining the maximum values ≈ 20 MPa and was characterized by an inflection of the "$\tau - \Delta$" curves in the figures. Then, the rebar occurred to

Table 2. Test results.

Test		Cohesion strength, MPa, for different samples at the reference values FRP#1 at 0.1 mm	FRP#2 at 0.1 mm	FRP#3 at 0.1 mm	FRP#4 at 0.1 mm	Steel at 0.1 mm
Reference samples		10.8 100.0%	9.6 100.0%	10.1 100.0%	19.01 100.0%	19.7 100.0%
Samples after SCP		10.9 100.9%	9.4 97.9%	8.8 87.1%	20.6 108.4	
Temperature action upon tests	Heating up to +40°C	10.8 100.0%	10.9 113.5%	10.6 105.0%	21.7 114.2%	18.8 95.4%
	Heating up to +80°C	10.2 94.4%	5.9 61.5%	4.8 47.5%	16.6 87.3%	17.6 89.3%
	Cooling down to −40°C	14 129.6%	12.5 130.2%	13.5 133.7%	23 121.0%	20.2 102.5%
Thermal cycling oscillations	20 cycles	13.8 127.8%	9 93.8%	10.2 101.0%	16.2 85.2%	22.2 112.7%
	40 cycles	12.4 114.8%	10.2 106.3%	10.69 105.8%	21.6 113.6%	20.4 103.6%
Exposure in water	30 cycles	12.7 117.6%	10.5 109.4%	11.43 113.2%	19.3 101.5%	22.7 115.2%
	60 cycles	12.6 116.7%	10.4 108.3%	10.8 106.9%	18.1 95.2%	21.9 111.2%
Exposure in alkaline medium	Exposure in concrete	11.5 106.5%	10.5 109.4%	6 59.4%	18.9 99.4%	
	Exposure of rods		3 31.3%		4.1 21.6%	
Influence of the strength of concrete	B12, 5	12.7 90.7%	10.5 84.0%	11.43 84.7%	19.3 83.9%	22.7 112.4%
	B35	12.6 90.0%	10.4 83.2%	10.8 80.0%	18.1 78.7%	21.9 108.4%
	B50	12.7 102.4%	10.5 102.9%	11.43 106.9%	19.3 89.4%	22.7 111.3%
	B60	12.6 101.6%	10.4 102.0%	10.8 101.0%	18.1 83.8%	21.9 107.4%
	B70	12.7 100.8%	10.5 101.0%	11.43 105.8%	19.3 106.6%	22.7 103.7%

be pulled out that was accompanied by a consequential wave-like decrease and increase in the load caused by inclusion of coils distant from the loaded end of FRP into the collaboration work. The pulled out FRP bars had partial surface damage between the depressed winding (cut-off of external layers of epoxy coating) and, generally, the destruction pattern at the boundary could be assessed as cohesion (via the material of the FRP bar).

Similarly to the previous samples, for the FRP#2 and 3 with roving pasted on the core, the

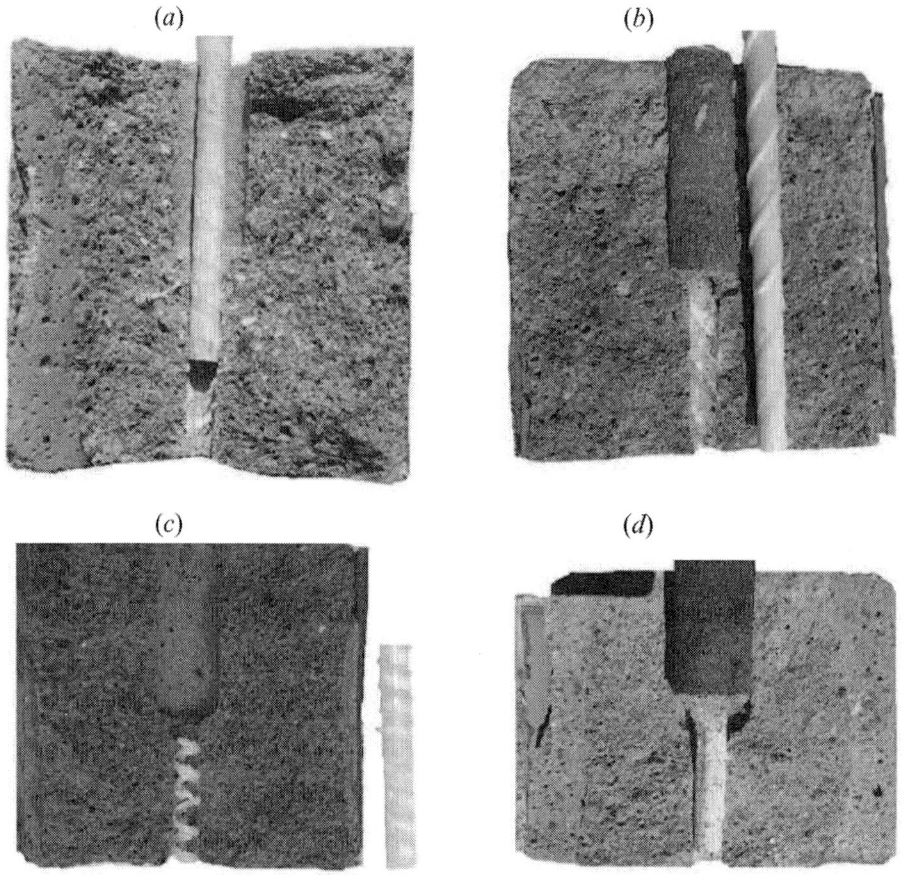

Figure 2. Cross-sections of concrete cylinder after the pullout test on FRP#1 (*a*), FRP#2 (*b*), FRP#3 (*c*), and FRP#4 (*d*).

main growth of "τ" up to 12–13 MPa was observed at a insignificant slip of 0.3 mm, then the growth of stresses up to 14–15 MPa was accompanied by FRP bond failure in concrete and the significant slip attained 2 mm (for FRP#2) and 4.5 mm (for FRP#3). The load dramatically dropped after it attained the maximal value, and the sample was pulled out of concrete by the minimal efforts caused only by friction of the rebar core on the concrete surface. Figures 2*b* and 2*c* show the fracture pattern of the FRP#2 and 3 specimens. Only the core could be seen in the FRP specimens which were pulled out of the concrete cylinders, the roving were cut away and partially remained in concrete (that is clearly seen in figure 2*c*).

Upon the tests on the reference samples, the FRP#4 samples with sandy surface demonstrated the following features of the pullout: the growth of the tangent stresses up to the maximal values $\approx$ 16–20 MPa occurred without any slip, after which the stresses dramatically decreased and the rebar was pulled out at minimal efforts. As figure 2*d* shows, the pulled out rebar were partially cut off including the sandy coating and external layers of epoxy binder over an area up to 50% of the contact zone, the cut of concrete on the rest of the area was fixed by the grains of sandy coating. No splitting of any concrete cylinders was observed upon the pullout test on FRP#4.

During the pullout test on the steel rebar, the tangent stresses "τ" increased up to $\approx$ 20 MPa almost without any slip; afterwards the "$\tau - \Delta$" plot bent and a gradual slip of the rebar began.

IOP Conf. Series: Journal of Physics: Conf. Series **991** (2018) 012044 doi:10.1088/1742-6596/991/1/012044

The maximal values of "τ" attained ≈ 22–26 MPa upon the slip on the free end of rebar of 0.3–0.5 mm; then the rebar specimens were gradually pulled out of the concrete cylinder, which was accompanied by a reduction in the tangent stresses. The rebar specimens, which were pulled out of concrete, showed no damage: both the rebar itself and the crescent-shaped coils remained undamaged; the bond failed at the "rebar-concrete" interfacial boundary due to the cut-off of concrete between the coils.

The strongly pronounced dynamics of the increase in the bond strength with an increase in the strength of concrete was established on the basis of the test results on the rebar specimens. However, this tendency developed differently for the specimens with unequal type of the surface relief.

For the FRP specimens with helically-deformed profile (depressed winding) — FRP#1, as the strength of concrete increased from 20.3 to 46.7 MPa, the maximal bond strength increased from 12.6 to 21.7 MPa (72.2%), but the further increase in the strength of concrete up to 98.9 MPa resulted in the maximal bond strength for FRP of only 24.7 MPa (an increase by 13.8%).

A similar tendency was found for the variations in the bond strength in the FRP#4 specimen with sandy surface: the main growth of the maximal bond strength from 14.3 to 22.7 MPa (increase by 58.7%) was observed when the strength of concrete increased from 20.3 to 46.7 MPa. The further increase in the strength of concrete up to 98.9 MPa led to the growth of the bond strength only by 14%.

The FRP#3 specimen with the winding pasted on the main bar showed an increase in the bond strength up to 59.2 MPa: the bond strength increased from 10.8 to 21.2 MPa (increase by 96.3%). A significant growth in the bond strength from 11.3 to 19.5 MPa was observed for the samples with winding (FRP#2) when the strength of concrete increased up to 79.5 MPa.

The maximal bond strength of the steel rebar varied directly with the strength of concrete. For example, when the class of concrete changed from B15 to B80, the maximal bond strength for the steel rebar increased from 19.4 to 35.7 MPa (increase by 84%).

The destruction pattern of FRP specimens also varied with the changes in the class of concrete. The minimal damages of the surface of specimens were observed when they were pulled out of concrete of B15 class and even B25: the sandy specimens were pulled out with the cut-off of concrete between the sand grains, the samples #1 with depressed winding also were pulled out without any damage with cut-off of concrete between the coils. The tests on the FRP#2 samples resulted in the cut-off of the lowest coils (distant from the plane of loading).

When the samples with winding were pulled out of concrete of B35 and B45, the cut-off was observed on the whole contact length. The damage of the surface epoxy coating was found for specimens with depressed winding (FRP#1). The pullout test on the FRP#4 specimen with sandy coating revealed the bond failure over an area up to 50% in the form of cut-off of the surface layer of epoxy and sand coatings, concrete was cut off between the sand grains. No damage was found after the pullout tests on the steel rebar. Upon the pullout tests for concrete of B60 and B80 classes, the rebar with depressed winding were extensively damaged; there was a cut-off of the epoxy coating and some segments with surface fibers. After the pullout tests, the specimens with sandy coating moved out with significant (up to 100%) damages (cut-off) of the sandy coating and surface layer of epoxy coating. The damage of the FRP#2 and 3 with winding has "classical" form — with the total cut-off of coils of the winding.

The tests on the steel rebar revealed an almost proportional growth of the bond strength with the strength of concrete. The damage of all specimens was in the form of the cut-off of concrete between the coils of rebar.

Thus, a limiting factor for the bond strength between concrete and FRP is the resistance to shear of the FRP surface profile, and, in the end, the resistance to shear of epoxy binder. This factor also determines the tensile strength of the polymer composite reinforcement.

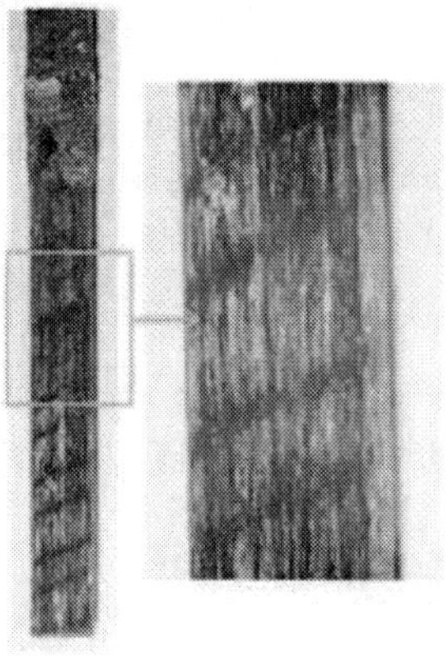

Figure 3. Damage pattern of FRP#4 after pullout tests for concrete of B60 class at ($\tau_{\max} = 25.9$ MPa.)

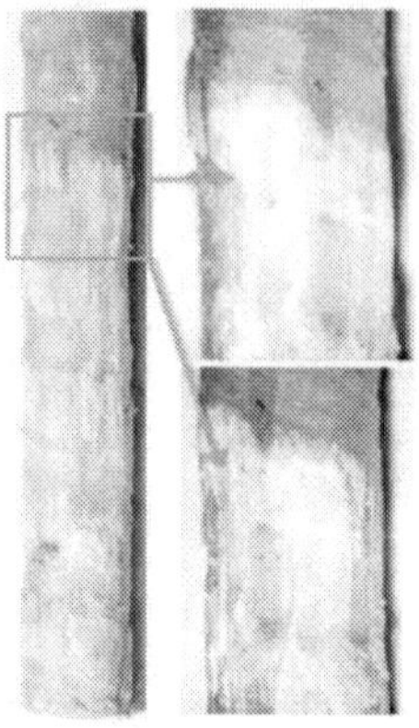

Figure 4. Damage pattern of FRP#1 after pullout tests for concrete of B60 class at ($\tau_{\max} = 24.3$ MPa.)

4. Analysis of data on the pullout tests

It is well known that the bond between a rebar and concrete is a continuous binding over the entire surface contact between them. It has been recently established that three factors are responsible for the bond of the composite rebar with concrete:

- adhesion (pasting) of cement to the FRP surface (F_d);
- mechanical interaction between the rebar surface relief and concrete (F_b);
- friction (F_f).

The adhesion factor and mechanical bond work at the initial stage of loading, and after the slip starts (in our experiments, a slit of 0.1 mm at the unloaded end of a polymer composite rebar was considered as a criteria of the adhesion failure), the adhesion failed, and the resistance against pulling out is provided by friction and mechanical bond of the rebar profile.

Adhesion (pasting) of cement to the FRP surface (F_d) arises during the set and hardening of concrete and is determined by chemical and physical processes which lead to intermolecular (physical) bonds on the "cement–epoxy polymer" interfacial surface. The adhesion forces are inoperative at relatively inconsiderable bond stresses and displacements of a rebar. It is known that these (physical) bonds cause the polymer susceptibility to higher temperatures and its susceptibility to aggressive liquids.

Mechanical bond strength of the rebar with concrete (F_b) is determined by the protruding elements of rebar profile: coils and sandy coverage.

In the case of the rebar with a deformed profile (depressed winding), the value of mechanical bond strength is taken as the minimal value: the cut-off of concrete between the raised grooves of the rebar and bearing (cut-off) of spiral "oval" ribs, and can be described by the equations

$$F_b = R_{sh}\frac{\pi dbn}{\sin\alpha}, \quad (1)$$

$$F_b = f_{sh}\frac{\pi dsn}{\sin\alpha}, \quad (2)$$

where R_{sh} is the shear resistance of concrete, d is the rebar diameter, b is the distance between the coils "in the clear", n is the number of coils along the contact length of the rebar in concrete, f_{sh} is the shear strength of epoxy polymer, s is the width of winding, and α is the angle of slope of coils to the longitudinal axis of the rebar.

IOP Conf. Series: Journal of Physics: Conf. Series **991** (2018) 012044 doi:10.1088/1742-6596/991/1/012044

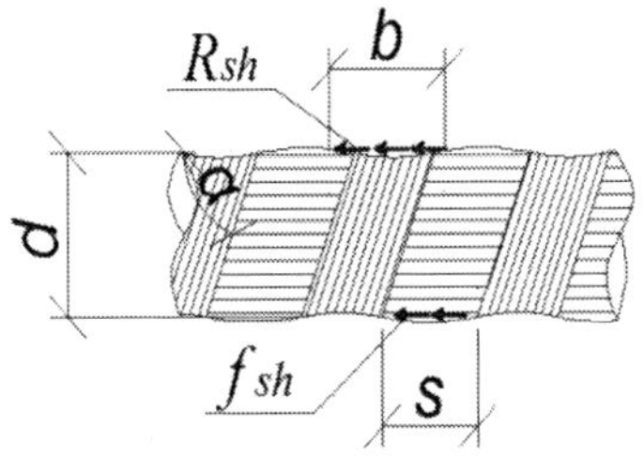

Figure 5. Fragment of the FRP specimen with depressed profile.

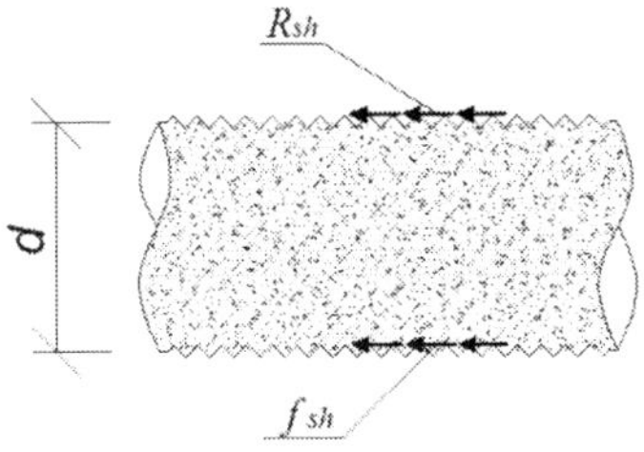

Figure 6. Fragment of the FRP speciment with sandy coating.

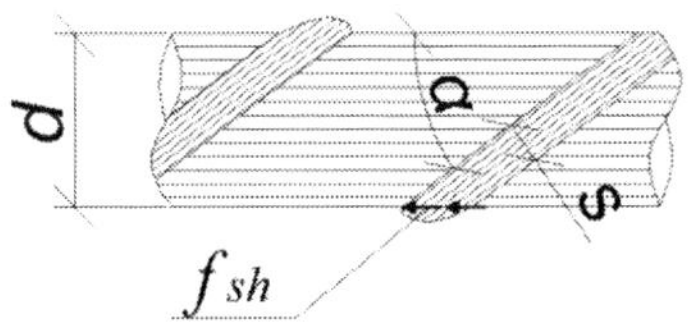

Figure 7. Fragment of the FRP specimen with winding pasted on the rebar core.

In the pullout tests on specimens with sandy coverage, the bond failure resulted from the cut-off of cement (C) between the sand grains or of epoxy coating together with the sand grains. Upon the pullout of specimens with sandy coating, the destruction resulted from the cut-off of cement between the sand grains or of the epoxy coverage with sand grains. Therefore, the contribution from mechanical bond for the sandy rebar is taken as the minimal value of the shear strength of epoxy coating and cement between the sand grains (or a sum of two unequal values), that is approximated by the equations

$$F_b = R_{sh}\frac{\pi dl}{2}, \tag{3}$$

$$F_b = f_{sh}\pi dl, \tag{4}$$

where R_{sh} is the shear strength of concrete, d is the rebar diameter, l is the contact length (contact between a rebar and concrete), and f_{sh} is the shear strength of epoxy polymer.

For the rebar specimens with the pasted single- and double-helical winding, the value of mechanical bond is limited by the strength of pasting of winding and described by the equation

$$F_b = f_{sh}\frac{\pi dsn}{\sin\alpha}, \tag{5}$$

where d is the rebar diameter, n is the number of coils in the zone of embedment, f_{sh} is the shear strength of epoxy polymer, s is the width of winding, and α is the angle of slope of coils to the longitudinal axis of the rebar.

The above-mentioned equations indicate that the mechanical bond for the FRP#1 and #4 with developed surface profile and sandy coating depends on both the strength of concrete and the shear strength of epoxy coating. For the FRP#2 and 3 with winding, the mechanical bond is constant and depends only on how firmly the winding is pasted to the core, but for low strength of pasting, the mechanical bond strength calculated by Eq. (5) is insignificant, and it can be neglected.

Friction (F_f) between the FRP surface and concrete. Several authors divide the friction into the static and dynamic terms. In our studies, due to low values of static friction, we include only the dynamic sliding friction in adhesion. The dynamic sliding friction arises after the failure of adhesion component at the moment of initiation of the rebar specimen slip in concrete.

The friction can be described by the equation

$$F_f = \pi dl\mu N, \tag{6}$$

where μ is the friction coefficient for "epoxy coating–concrete" or "concrete–concrete" contacts, d is the diameter of rebar, l is the length of embedment, and N is the edging draft of the rebar specimen during the test.

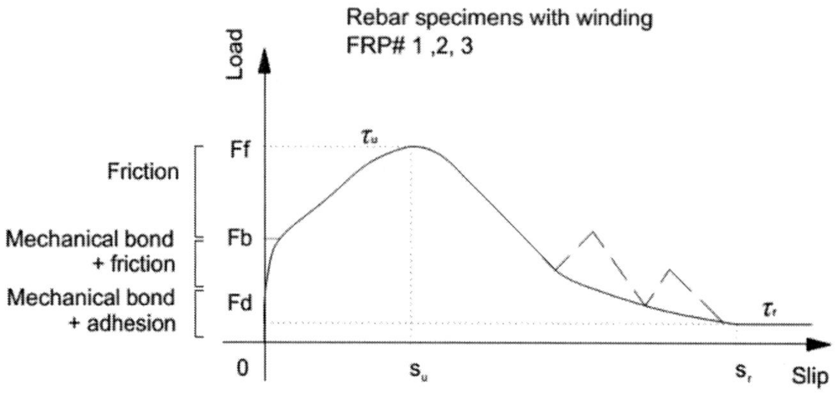

Figure 8. Scheme of changes in stresses upon the pullout of FRP#1, 2, and 3.

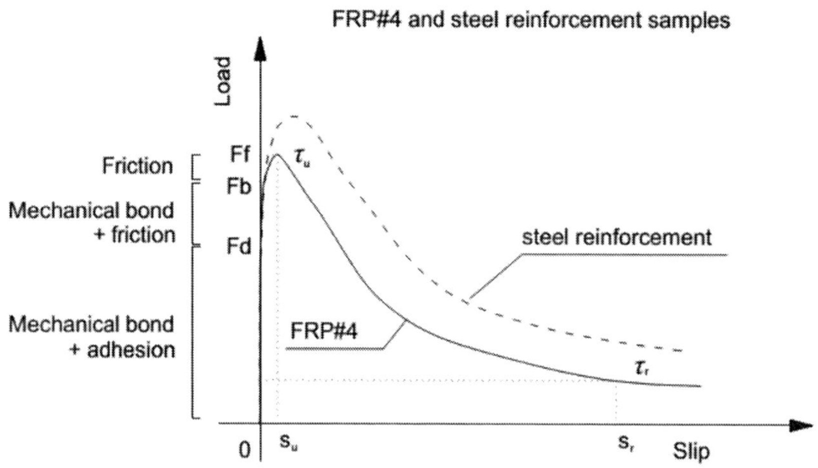

Figure 9. Scheme of changes in stresses upon the pullout of FRP#4 and steel reinforcement.

It should be noted that, upon friction, the normal pressure on the rebar remains constant in reciprocal friction as against the mechanical bond at which the normal stresses from the arising thrust turn out to be alternating quantities which depend on the type of profile.

The bond stress "τ" is determined as a sum of three above-mentioned factors:

$$F_p = F_d + F_b + F_f. \tag{7}$$

Therefore, the changes in the tangent stresses, which were determined in the testing procedures, are attributed to the changes of one of the factors F_d, F_b, and F_f.

Summarizing the above results, the contribution of each factor throughout all stages of the pullout of FRP from concrete can be represented as the following plots shown in figures 8 and 9.

The foregoing plots can be fitted using the equations, which have been recently suggested in [48]:

$$\frac{\tau}{\tau_u} = B_1\frac{s}{s_u} + B_2\left(\frac{s}{s_u}\right)^2,$$
$$\tau = \frac{s}{P_1 s - P - 2} \quad (s_u < s \le s_r),$$
$$\tau = \tau_r \quad (s > s_r),$$

where B_1 and B_2 are the coefficients determined from the results of tests for each type of FRP; P_1 and P_2 are the coefficients calculated from the equations

$$P_1 = \frac{s_u\tau_r - s_r\tau_u}{\tau_u\tau_r(s_u - s_r)}, \quad P_2 = \frac{\tau_r - \tau_u}{s_u - s_r}\frac{s_u s_r}{\tau_r\tau)u}.$$

5. Estimation of the tangent stress distributions over the embedment length

In our work, we carried out the numerical simulations using an ANSYS software in order to assess the irregularity degree of tangent stress distribution along the embedment length and to examine the stresses arising both in the concrete sample and rebar. The obtained results showed that the tangent stress distribution along the embedment length is irregular to a wide extent. Therefore, the internal stresses in the embedded part of the rebar are also irregularly distributed depending on the stage of loading and the bond, which transmits the load from the bar to concrete. The analysis of pullout test results enabled us to indicate three main stages of stress strain behavior at the "FRP–concrete cylinder" contact:

1st stage (figure 10): elastic state in which the deformation of the loaded rebar is proportional to the applied force and, mainly, reversible. The stresses in the rebar decrease rapidly with distance from the loaded end, as well as the bond stresses whose maximal value is observed near the loaded end of the rebar over the sections no more than 30 mm with relatively low bond strength elsewhere. At this stage, the bond strength is determined by adhesion and mechanical bond.

2nd stage (figure 11): partially elastic state in which the linearity and reversibility of deformation are interrupted, but no displacements of free (unloaded) end of the rebar occur. At this stage, the angle of slope of the "$\tau - \Delta$" curve changes abruptly. With an increase in the applied forces, the peak of maximal bond stress is shifted towards the unloaded end of the rebar, which indicates nonsimultaneous inclusion of tangent stresses into collaboration and distribution over the surface. This fact is caused by a low elastic modulus of the composite reinforcement leading to elongation of the rebar along the embedment length under tensile loading and, as a consequence, to adhesion failure.

The further increase in the loading leads to an increase in the elongation meaning the failure of physical bonds of the rebar with concrete in this section and the redistribution of tangent stresses over the next sections, which were formerly unloaded. At this stage, the mechanism of transmitting the stresses from the rebar to concrete is irregular: the elastic action of bond forces remains near the unloaded end of rebar, the concrete in the contact layer in the middle of the sample is most loaded, and the zone of contact is entirely destroyed near the loaded end.

3rd stage (figure 12): slip state when the rebar shifts relatively to concrete along the total embedment length. The maximal bond stresses are drawn towards the unloaded face of the sample. At the starting of embedment, the concrete in the contact layer is destroyed, but the particles of broken concrete still provide friction and splitting stresses. At this stage, an increase in the loads applied to the rebar leads to an intensive displacement of the loaded end. The embedment breakdown is the result of disintegration of the contact layer in concrete or epoxy binder on the FRP surface, or it occurs after the splitting of the sample.

IOP Conf. Series: Journal of Physics: Conf. Series **991** (2018) 012044 doi:10.1088/1742-6596/991/1/012044

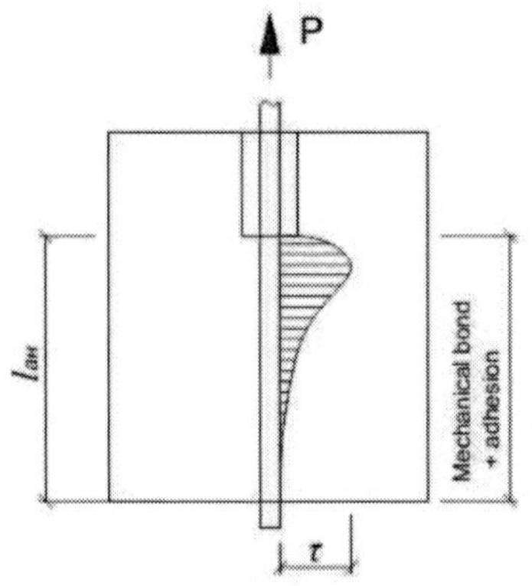

Figure 10. Tangent stress distribution at the first stage.

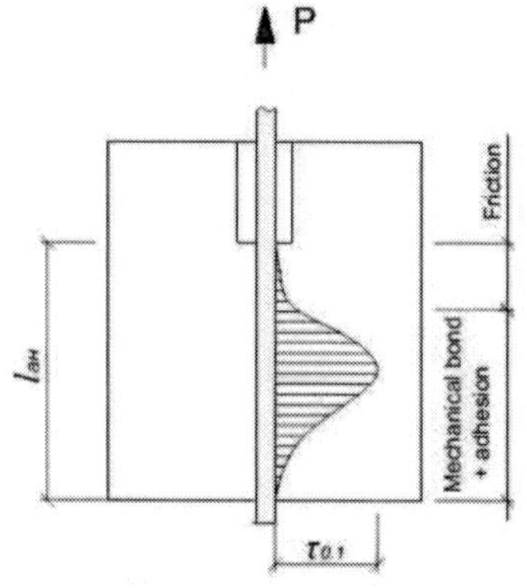

Figure 11. Tangent stress distribution at the second stage.

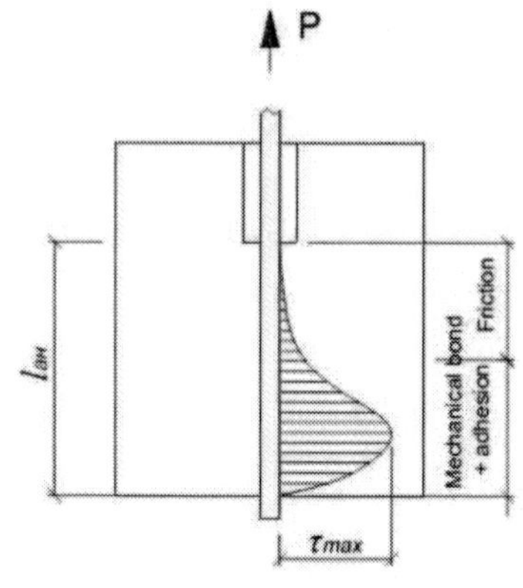

Figure 12. Tangent stress distribution at the third stage.

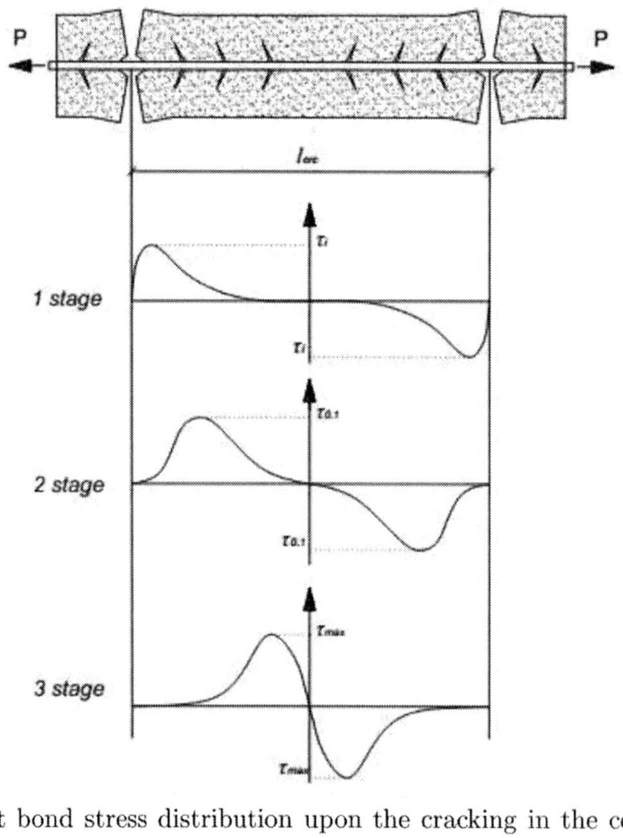

Figure 13. Tangent bond stress distribution upon the cracking in the concrete construction subjected to bending.

In regard to the concrete construction subjected to bending, which is reinforced by FRP, the tangent stress distribution can be described as is shown in figure 13. In the case of concrete beams reinforced by FRP, this manifests an increase in the number of cracks and a decrease in the spaces between them for the same loading and reinforcement ratio as in steel reinforcement.

Conclusions

1. Following the results of pullout test, it was established that regardless the type of surface profile, the bond strength is most sensitive to temperature: when heating up to +80°C, the bond strength is reduced by 49%, and conversely, when it is cooled down to −40°C, the bond strength increases by 21–36%. The exposure of the rebar in the caustic solution followed by the concrete casting also results in a decrease of the bond strength. The maximal drop of the strength by 79% was observed for specimens. The effect of other operational and technological factors on the bond strength is insignificant for all types of FRP regardless the surface profile.

2. An increase in the strength of concrete provides a proportional increase in the bond strength of all types of FRP. The maximal bond strength is limited by the strength of composite material, the shear resistance of the polymer binder, and its adhesion to the fiber filler and grains of sandy coverage. Therefore, when the limiting values (different for each FRP) are attained, the increase in the strength of concrete has almost no effect on the bond strength, in contrast to the direct growth of the bond strength in steel reinforcement.

3. As a result of study, the pattern of the stressed state of concrete and FRP is suggested, as well as their contact boundary upon the rebar pulling out. It was established that the tangent stress distribution along the pulled out rods was irregular and had a pronounced peak at the length of 30 mm, which moved in loading from the loaded end of FRP to its unloaded end. This peak of stresses was attributed to the lower elastic modulus of polymer composite reinforcement compared to concrete, and, as a consequence, to significant elongations of the rebar under loading and nonuniform inclusion into their collaboration.

4. Physical models of the bond between concrete and a series of FRP of four types were theoretically represented and examined as a sum of three factors: adhesion of concrete to polymer surface, mechanical bond with protruding profile elements, and friction.

References

[1] Khozin V G, Piskunov A A, Gizdatullin A R, and Kuklin A N 2013 Adhesion fiber-reinforced polymer bars with cement concrete *Izv. KGASU* No. 1 214–20

[2] Frolov N P 1980 *Glass-fiber plastic rebars and glass-fiber plastic concrete structures* (Moscow: Stroyizdat) p 102

[3] *Guide Test Methods for Fiber-Reinforced Polymers (FRPs) for Reinforcing or Strengthening Concrete Structures. ACI 440.3R-04* 2004 (American Concrete Institute) p 40

[4] *Design and Construction of Building Components with Fibre-Reinforced Polymers. CAN/CSA-S806-02 (R2007)* 2012 (Canadian Standards Association) p 206

[5] Trejo D, Gardoni P, Kim J J, and Zidek J 2009 *Long-Term Performance of GFRP reinforcement: Technical Report No. FHWA/TX-09/0/6069-1* p 124

[6] Muñoz M B 2010 *Study of Bond Behavior between FRP Reinforcement and Concrete* Technology Doctorate Program (Universitat de Girona) p 308

[7] Benmokrane B, El-Salakawy E, Cherrak Z, and Wiseman A 2004 Fibre-reinforced polymer composite bars for the structural concrete slabs of a Public Works and Government Services Canada parking garage *Canadian J. Civil Engng* **31** (5) 732–48

[8] El-Salakawy E, Kassem C, and Benmokrane B 2002 Field application of FRP composite bars as reinforcement for bridge decks In *Proc. of the Canadian Society for Civil Engineers 30th Annual Conference, 4th Structural Specialty Conference, Montréal, Québec, 5–8 June 2002* (Montréal, Québec: Canadian Society for Civil Engineers) S-23

[9] Trejo D, Aguiiga F, Yuan R L, et al. 2003 *Characterization of Design Parameters for Fiber Reinforced Polymer Composite Reinforced Concrete Systems. Report No.FHWA/TX-5/9-1520-3* URL `http://tti.tamu.edu/documents/9-1520-3.pdf`

[10] Ametrano D 2011 *Bond Characteristics of Glass Fibre-Reinforced Polymer Bars Embedded in High Performance and Ultra-High Performance Concrete* MASc (Toronto: Dept. Engng Arch. Sci. Ryerson Univ.)

Controllability of semi-infinite rod heating by a point source

A Khurshudyan
Institute of Mechanics of the National Academy of Sciences of Armenia, Yerevan, Armenia
E-mail: khurshudyan@mechins.sci.am

Abstract.
The possibility of control over heating of a semi-infinite thin rod by a point source concentrated at an inner point of the rod, is studied. Quadratic and piecewise constant solutions of the problem are derived, and the possibilities of solving appropriate problems of optimal control are indicated. Determining of the parameters of the piecewise constant solution is reduced to a problem of nonlinear programming. Numerical examples are considered.

1. Introduction

Controllability in unbounded domains, i.e., the ability to change the state of an a system defined in an unbounded domain as desired in given finite time, is one of the difficultly implementable features of control systems. However, in investigations of specific control systems, it turns out that, for a given initial state, it is impossible to find a control function providing (generally, approximately) the desired state in a finite time. More often, it is more difficult to determine the set of implementable terminal states rather than the control functions implementing them.

In [1], it is shown that, for arbitrary initial temperature, there do not exist L^2-boundary heating regimes u for unbounded three-dimensional bodies providing the null distribution of temperature in the body, i.e., the system

$$\frac{\partial\Theta(\mathbf{x},t)}{\partial t} - \Delta\Theta(\mathbf{x},t) = 0, \quad \mathbf{x}\in\Omega, \quad t\in(0,T), \tag{1}$$

$$\Theta(\mathbf{x},t) = u(t)\chi_{\partial\Omega_0}(\mathbf{x}), \quad (x,y,z)\in\partial\Omega, \quad t\in[0,T], \tag{2}$$

$$\Theta(\mathbf{x},0) = \Theta_0(\mathbf{x}), \quad \mathbf{x}\in\bar{\Omega}, \tag{3}$$

is not exact null-controllable in a given finite time T. Here $\Omega\subset\mathbb{R}^3$ is a semi-infinite domain of the body, $\partial\Omega$ is its boundary, Θ characterizes the heat distribution in the body, and $\chi_{\partial\Omega_0}$ is the characteristic function of $\partial\Omega_0\subseteq\partial\Omega$. It is important to note that heat is propagating in space with an *infinite* velocity. We also refer to [2–4] and the references therein.

In [5], a necessary and sufficient condition on the initial and terminal states is derived, for which there exists an L^1-control providing formal exact controllability of

$$\frac{\partial}{\partial x}\left[N(x)\frac{\partial w(x,t)}{\partial x}\right] - \rho(x)\frac{\partial^2 w(x,t)}{\partial t^2} = 0, \quad x \in (0,\infty), \quad t \in (0,T),$$
$$w(0,t) = u(t), \quad \lim_{x\to\infty} w(x,t) = 0, \quad t \in [0,T],$$
$$w(x,0) = w_0(x), \quad \left.\frac{\partial w(x,t)}{\partial t}\right|_{t=0} = w_0^1(x), \quad x \in [0,\infty),$$

in given finite T, for $0 \le N \in C_p^{(1)}[0,\infty)$, $0 < \rho \in C_p[0,\infty)$. A numerical analysis allows one to find the corresponding initial and terminal data. In this case, the *finiteness* of wave speed plays the key role.

For most of the systems defined in unbounded domains, it is generally impossible to provide given terminal state in given finite time exactly, i.e.,

$$\mathbf{x}(T) = \mathbf{x}_T,$$

where $\mathbf{x}$ is the state of the system and $\mathbf{x}_T$ is the desired terminal state, even if the set of admissible controls is large. In other words, the systems defined in unbounded domains are often not exactly controllable. More often, the terminal state is implemented approximately. In such cases, it is required to provide a terminal state, which is in a small neighborhood of the desired state [8, 9]:

$$|\mathbf{x}(T) - \mathbf{x}_T| \le \epsilon$$

for a given tolerance $\epsilon > 0$. Then, the system is called approximately controllable.

The aim of the paper is the approximate implementation of the desired terminal heat distribution in a semi-infinite insulated rod in given finite time, by means of a heat source concentrated at an internal point of the rod.

2. Statement of the problem

Let a semi-infinite rod of sufficiently small thickness[1] be heated by a point source with controllable intensity u, bounded by $|u| \le u_0$, and concentrated at a given point $x = x_0 \in (0,\infty)$ of the rod. The heat propagation in the rod obeys the one-dimensional heat equation

$$\frac{c_p\rho}{\kappa}\frac{\partial\Theta(x,t)}{\partial t} = \frac{\partial^2\Theta(x,t)}{\partial x^2} + \frac{u(t)}{\kappa}\delta(x-x_0), \quad x > 0, \quad t > 0, \tag{4}$$

where Θ is the temperature, c_p is the specific heat capacity, ρ is the density, κ is the thermal conductivity of the rod, and δ is the Dirac delta function.

Let, for simplicity, the $x = 0$ section and infinitely far sections of the rod be insulated, i.e.,

$$\Theta(0,t) = 0, \quad \lim_{x\to\infty}\Theta(x,t) = 0, \quad t \ge 0. \tag{5}$$

The initial distribution of temperature in the rod is known:

$$\Theta(x,0) = \Theta_0(x), \quad x \ge 0. \tag{6}$$

Our aim is to find a heating regime u such that, in for given finite T, the state

$$\Theta(x,T) = \Theta_T(x), \quad x \ge 0, \tag{7}$$

[1] It is basically assumed that all points of each cross-section of the rod have the same temperature at any fixed time instant.

IOP Conf. Series: Journal of Physics: Conf. Series **991** (2018) 012045 doi:10.1088/1742-6596/991/1/012045

is implemented. For simplicity, we assume that Θ_0 and Θ_T are continuous functions in $[0, \infty)$.

We introduce dimensionless variables and functions

$$\tilde{x} = \frac{x}{x_0}, \quad \tilde{t} = \frac{c_p \rho}{\kappa x_0^2} t, \quad \tilde{\Theta} = \frac{\Theta}{\Theta^0}, \quad \tilde{u} = \frac{x_0}{\Theta^0 \kappa} u,$$

where Θ^0 is the intensity of the initial temperature distribution in the rod, i.e., $\Theta_0(x) = \Theta^0 \tilde{\Theta}_0(\tilde{x})$. Note that $\tilde{T}$ then characterizes the ratio of the control time to the time required for the heat propagation at the distance x_0 from the source. It is taken into account that $x_0 \delta(x - x_0) = \delta(\tilde{x} - 1)$. We do not introduce new symbols and simply omit $\tilde{\cdot}$ over the corresponding characters.

The boundary and initial data are supposed to be consistent, i.e., the conditions

$$\Theta_0(0) = \Theta_T(0) = 0, \quad \Theta_0(1) = u(0), \quad \Theta_T(1) = u(T), \quad \lim_{x \to \infty} \Theta_0(x) = \lim_{x \to \infty} \Theta_T(x) = 0,$$

are satisfied.

3. Solution of the problem

Then, the solution of the problem is given by the following theorem.

Theorem 1. *For distributed controllability of* (4)–(6), *i.e., for the fulfilment of* (7), *in a finite* T, *it is necessary and sufficient that*

$$\int_0^T K(x, T - \tau) u(\tau) \, d\tau = F(x) \quad \textit{for all } x \geq 0, \tag{8}$$

where $K(x, t) = G(x, 1, t)$,

$$F(x) = \Theta_T(x) - \int_0^\infty G(x, \xi, T) \Theta_0(\xi) \, d\xi,$$

$$G(x, \xi, t) = \frac{1}{\sqrt{4\pi t}} \left\{ \exp\left[-\frac{(x - \xi)^2}{4t} \right] - \exp\left[-\frac{(x + \xi)^2}{4t} \right] \right\}.$$

At this, the control function satisfies

$$\int_0^T K(1, T - \tau) u(\tau) \, d\tau = F(1), \tag{9}$$

$$u(0) = \Theta_0(1), \quad u(T) = \Theta_T(1). \tag{10}$$

Proof. The proof is straightforward. Since the general solution of (4)–(6) is given by [6]

$$\Theta(x, t) = \int_0^t G(x, 1, t - \tau) u(\tau) \, d\tau + \int_0^\infty G(x, \xi, t) \Theta_0(\xi) \, d\xi, \quad x \geq 0, \quad t \geq 0, \tag{11}$$

we arrive at (8) evaluating it at $t = T$. Further, (9) is a consequence of the assumption that the point x_0 of the rod is heated by the source within $[0, T]$, and (10) are the consistency conditions between the boundary and initial and terminal data. □

In general, it is impossible exactly to satisfy condition (7) for all $x \geq 0$. Indeed, if a control function satisfying (9), (10) is found, then (8) is satisfied exactly only for $x = 1$. Evidently, (8) is satisfied for $x = 0$ and as $x \to \infty$.

IOP Conf. Series: Journal of Physics: Conf. Series **991** (2018) 012045 doi:10.1088/1742-6596/991/1/012045

It is also evident that the solution of (9), (10) is non-unique. Indeed, it can be represented in quadratic form $u(t) = \Theta_0(1) + [\Theta_T(1) - \Theta_0(1)]\frac{t}{T} + u_1 t(T-t)$, where u_1 is determined by substituting this form into (9) and by satisfying the equality. Another explicit form of the control function is

$$u(t) = F(1)\frac{L(t)}{K(1, T-t)},$$

where

$$L(0) = \frac{\Theta_0(1)}{F(1)}K(1,T), \quad L(T) = \frac{\Theta_T(1)}{F(1)}K(1,0), \quad \int_0^T L(t)\,dt = 1.$$

The non-uniqueness of the solution allows one to choose the following controls optimizing some criteria: the heating time, the error between the implemented and desired states (in the case of approximate controllability), etc.

4. Piecewise constant solution

The non-uniqueness of solution of (9), (10) allows one to consider piecewise constant intensities [5, 7, 10, 11]:

$$u(t) = \Theta_0(1) + \sum_{k=1}^{n} u_k \cdot \theta(t - t_k), \quad t \in [0, T], \tag{12}$$

where $\theta(t)$ is the Heaviside function, u_k, t_k ($t_0 = 0$, $t_n = T$) are determined by substituting (12) into (9):

$$\sum_{k=1}^{n} u_k[1 - \mathcal{K}_k(t_k)] = \int_0^\infty G(1, \xi, T)\Theta_0(\xi)\,d\xi, \tag{13}$$

$$u(T) = \Theta_0(1) + \sum_{k=1}^{n} u_k = \Theta_T(1), \quad \sum_{k=1}^{n} |u_k| \le u_0, \quad 0 \le t_k < t_{k+1} \le T, \tag{14}$$

where

$$\mathcal{K}_k(t_k) = \int_{t_{k-1}}^{t_k} K(1, 1-\tau)\,d\tau,$$

which can be computed exactly. It follows from (13) that $|\Theta_T(1)| \le |\Theta_0(1)| + u_0$.

If we require minimization of some functional like [12]

$$\max_k |u_k|, \quad \int_0^1 |u(t)|\,dt, \quad \sum_{k=1}^{n} |u_k|,$$

then the methods of nonlinear programming can be involved to determine u_k and t_k.

5. Numerical analysis

It is evident that it is practically impossible to satisfy (8) exactly for all $x \ge 0$. Therefore, our aim is to establish its approximate fulfillment.

Example 1. Let us begin with the quadratic control $u(t) = \Theta_0(1) + [\Theta_T(1) - \Theta_0(1)]\frac{t}{T} + u_1 t(T-t)$. It is given that $T = 8$, $\Theta_0(x) = x\exp(-x)$, and $\Theta_T(x) = x^2\exp(-0.5x)$. Then

$$u_1 = \frac{1}{\kappa_1 T - \kappa_2}\left[F(1) - \Theta_0(1)\kappa_0 - \frac{\Theta_T(1) - \Theta_0(1)}{T}\kappa_1\right], \quad \kappa_k = \int_0^T K(1, T-\tau)\tau^k\,d\tau.$$

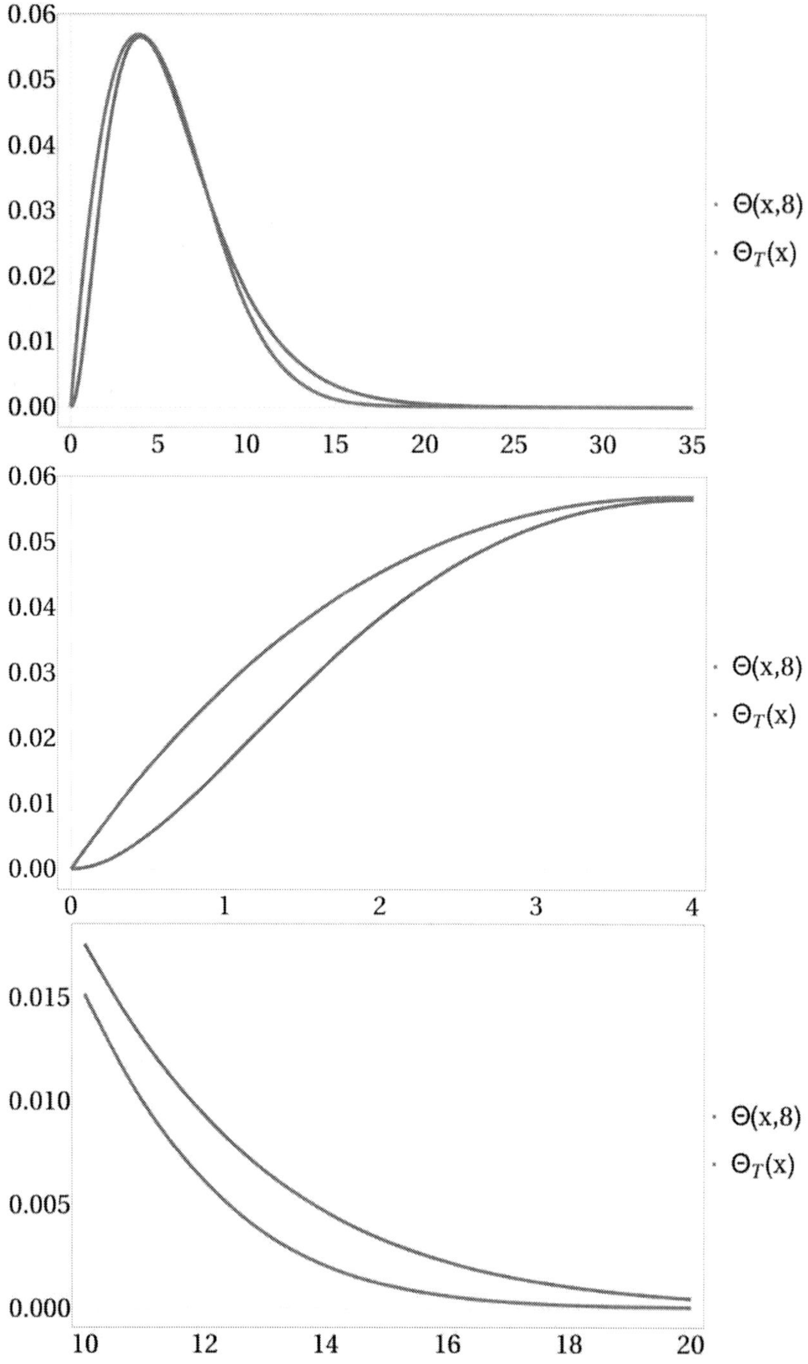

Figure 1. Realized and desired terminal states: quadratic control.

Figure 1 shows the graphs of the implemented and desired terminal states $\Theta(x,T)$ and $\Theta_T(x)$ for $u_1 = -0.086$. It turned out that the maximal distance is $\max\limits_{0\leq x\leq 35}|\Theta(x,8)-\Theta_T| \leq 0.005$. For larger x, $|\Theta(x,8)-\Theta_T| \to 0$.

Example 2. Now we consider the case where the heating regime is piecewise constant.

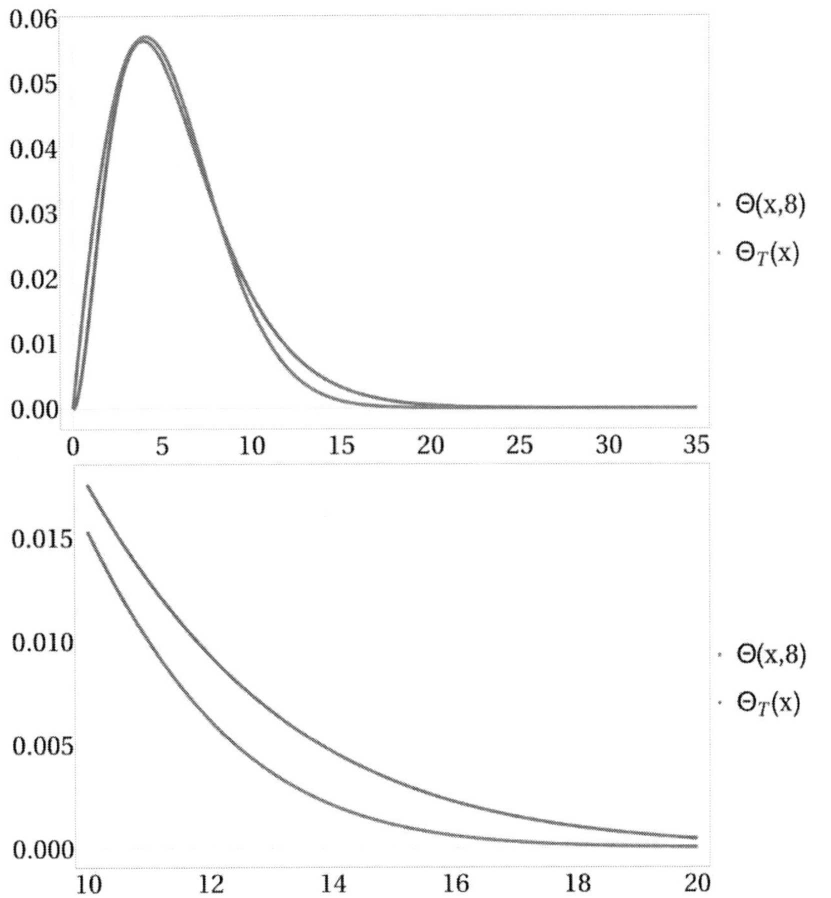

Figure 2. Realized and desired terminal states: piecewise constant control.

Let (12) contain only 3 terms. Then, from (13), (14) we obtain $u_1 = 0$, $u_2 = 0.1$, $u_2 = 0.138$, and $t_2 = 1$ ($t_1 = 0$, $t_3 = 8$).

Figure 2 presents the graphs of the implemented state $\Theta(x,T)$ and the desired state $\Theta_T(x)$. Evidently, the maximal distance is $\max\limits_{0\le x\le 35} |\Theta(x,8) - \Theta_T| \le 0.005$. For larger x, $|\Theta(x,8) - \Theta_T| \to 0$.

It is evident that, in the case of quadratic control, the two states have significant gap for $0 \le x \le 2$ and $10 \le x \le 15$; however, in the case of piecewise constant control, they differ only for $10 \le x \le 15$. Adding more terms into (12) or considering minimization problem

$$\max_{x\ge 0} |\Theta(x,T) - \Theta_T| \to \min$$

allow us to decrease the distance between the two states.

Example 3. It turns out that there is another possibility for improving the approximation rate. It is done by considering the case

$$u_s(t) = u(t)\chi_{[0,t_0]}(t),$$

where the source is switched off at $0 < t_0 < T$. The improvement of approximation is shown in figure 3 for the quadratic control with $u_1 = 0.41$ and $t_0 = 2$. In this case, $\max\limits_{0\le x\le 35} |\Theta(x,8) - \Theta_T| \le 10^{-4}$.

IOP Conf. Series: Journal of Physics: Conf. Series **991** (2018) 012045 doi:10.1088/1742-6596/991/1/012045

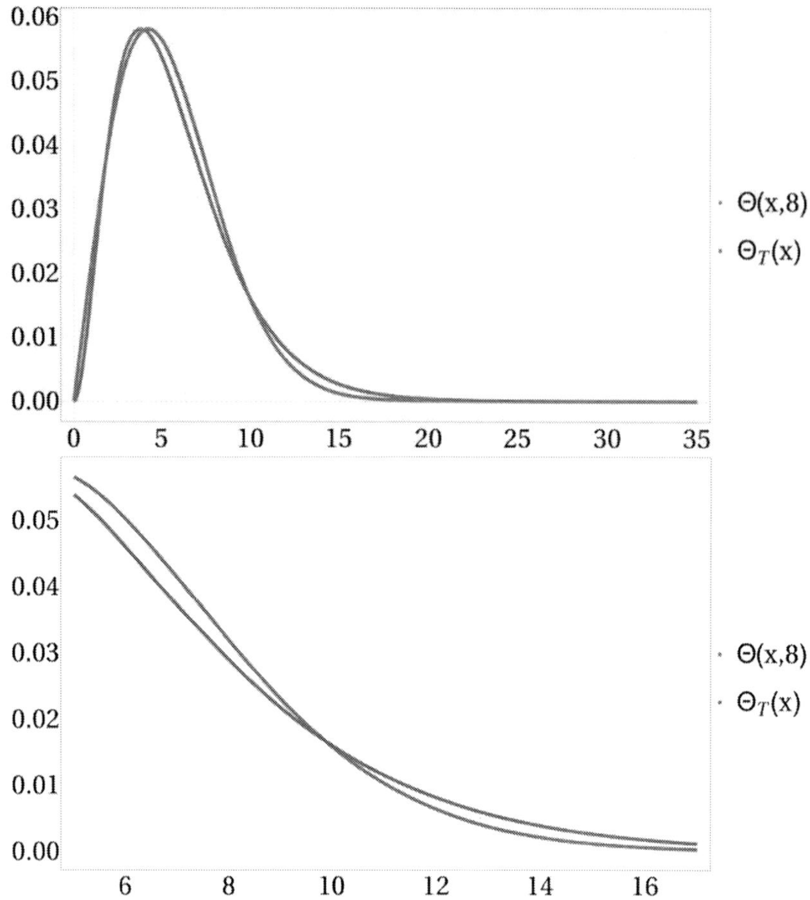

Figure 3. Realized and desired terminal states: switching intensity.

Conclusions

Under some assumptions, it is shown that sufficiently thin semi-infinite rod can be heated from some initial temperature distribution to some final one using a point source of heat. By the method of Green's function, the problem of finding a suitable heating regime is reduced to the problem of solving an integral equation with a singular kernel. Several regular solutions satisfying necessary restrictions are derived. A piecewise constant solution of the integral equation whose the parameters are determined by solving nonlinear programming problem with constraints of equality and inequality type is obtained. the non-uniqueness of the solution of the integral equation allows considering the optimization problems.

The results can be generalized to other PDEs, coupled systems of differential equations, etc., using the Green's function of which is known or can be easily found.

References

[1] Micu S and Zuazua E 2004 Null-controllability of the heat equation in unbounded domains In *Unsolved Problems in Mathematical Systems & Control Theory* ed V D Blondel and Al Megretski (Princeton: Princeton University Press) pp 163–8

[2] Zuazua E 2007 Controllability and observability of partial differential equations: some results and open problems In *Handbook of Differential Equations: Evolutionary Equations* vol 3 ed C M Dafermos and E Feireisl (North Holland) pp 527–621

[3] Barbu V 2014 Exact null internal controllability for the heat equation on unbounded convex domain *ESAIM: Control Opt. Calc. Variat.* **20** (1) 222–35
[4] Miller L 2005 On the null-controllability of the heat equation in unbounded domains *Bull. Sci. Math.* **129** (2) 175–85
[5] Khurshudyan As Zh 2015 Generalized control with compact support for systems with distributed parameters *Arch. Control Sci.* **25** (1) 5–20
[6] Butkovskii A G and Pustyl'nikov L M 1993 *Characteristics of Distributed-Parameter Systems* (Kluwer Academic Publishers)
[7] Sarkisyan S V, Jilavyan S H, and Khurshudyan As Zh 2015 Structural optimization of an inhomogeneous infinite layer in problems on propagation of periodic waves *Mech. Compos. Mater.* **51** (3) 277–84
[8] Avetisyan A S and Khurshudyan As Zh 2016 Green's function approach in approximate controllability problems *Izv. Nats. Akad. Nauk Armenii. Mekh.* **69** (2) 3–22
[9] Avetisyan A S and Khurshudyan As Zh 2017 Green's function approach in approximate controllability for nonlinear physical processes *Modern Phys. Lett. A* **32** (21) 1730015
[10] Khurshudyan As Zh 2015 The Bubnov-Galerkin method in control problems for bilinear systems *Automat. Remote Contr.* **76** (8) 1361–8
[11] Arakelyan Sh Kh and Khurshudyan As Zh 2015 The Bubnov-Galerkin procedure in mobile (scanning) control problems for systems with distributed parameters *Izv. Nats. Akad. Nauk Armenii. Mekh.* **68** (3) 54–75
[12] Kaczorek T and Borawski K 2016 Minimum energy control of descriptor discrete-time linear systems by the use of Weierstrass-Kronecker decomposition *Arch. Control Sci.* **26** (2) 177–87

Vibrations of geometrically nonlinear beam subjected to two oppositely moving loads and supported by three equidistant visco-elastic dampers

A Khurshudyan* **and S Ohanyan****
Institute of Mechanics of the National Academy of Sciences of Armenia, Yerevan, Armenia
E-mail: *khurshudyan@mechins.sci.am, **sergohanyan@mail.ru

Abstract. The bending vibrations of a geometrically nonlinear Euler–Bernoulli beam supported by three equidistant visco-elastic dampers and subjected to two oppositely moving point loads are studied. The Bubnov–Galerkin procedure is used for solving the strong formulation of the problem, and the finite element method is involved for numerical analysis.

List of symbols

u_0 and w_0	longitudinal and transverse displacements of the beam
ε_{xx}, ε_{xz}, ε_{zz}	von Kármán strains
N_{xx}	axial force
M_{xx}	bending moment
l	length of the beam
E and ρ	Young's modulus and density of the beam
S and J	cross-section area and moment of inertia of the beam
P_1 and P_2	intensities of the moving forces
v_1 and v_2	velocities of the moving forces
t_0	time instant when the second load starts to move
δ	Dirac delta function

1. Introduction

Simply supported beams subjected to moving loads are the most common engineering models of bridges [1–3]. The dependence of displacements of the beam on the load parameters is analyzed to estimate the areas of occurrence of critical residual stresses having accumulative nature [4]. To compensate the influence of the load on the beam, different types of dampers and vibration absorbers are usually involved. A tremendous body of literature is devoted to analysis of beams under moving loads supported by various kinds of dampers: elastic [5], viscous [6], linear visco-elastic [7], nonlinear visco-elastic [8, 9]. Linear beam theories of Euler–Bernoulli [3] and Timoshenko [10] are mainly covered. The two-dimensional non-classical theories are also involved [11]. The Fourier method of variable separation, the Bubnov–Galerkin procedure, and Green's function approach are the main tools for analysis of such problems.

The cases of material or geometrical nonlinearities are considered relatively less (see [12–15] and the references therein). The load-displacement relation is much too complicated to be

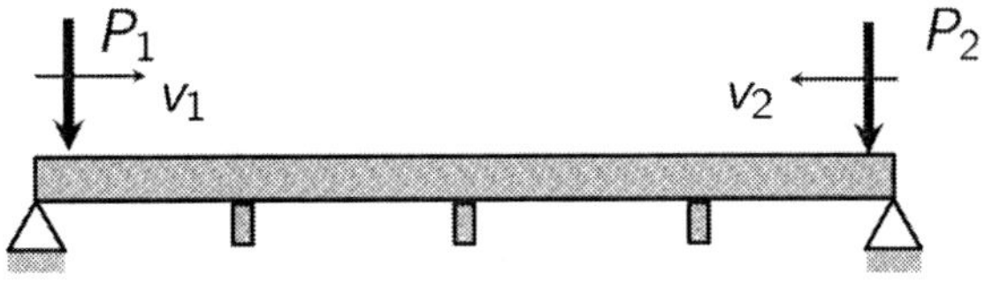

Figure 1. Simply supported beam subjected to two oppositely moving point loads and supported by three equidistant visco-elastic dampers.

derived in a closed form, and therefore approximation methods are involved. However, not all approximation methods can be useful in this case, especially because the mathematical model of the problem is formulated in terms of distributions. Due to its variational formulation, the Bubnov–Galerkin procedure [16] is usually applied efficiently.

In this paper, we analyze the load-displacement relation for isotropic, homogeneous, elastic, geometrically nonlinear Euler–Bernoulli beam, which is subjected to point loads of constant intensity uniformly moving along the upper surface of the beam in directions opposite to each other. The beam is supported by three equidistant similar visco-elastic dampers. Numerical analysis reveals the influence of dampers on the longitudinal and transverse vibrations of the beam.

2. Statement of the problem: strong formulation

Let an elastic isotropic homogeneous beam of length l be simply supported at the ends $x = 0$ and $x = l$ and be supported by three visco-elastic dampers placed at $x_1 = l/4$, $x_2 = l/2$, and $x_3 = 3l/4$. The beam is subjected to two constant loads with intensities P_1 and P_2 perpendicular to the beam axis and moving with constant velocities v_1 and v_2, respectively, in opposite directions (see figure 1).

The Euler–Bernoulli assumptions are accepted for the beam, and the Kelvin–Voigt model of visco-elastic body is assumed to hold for the dampers. However, the Cauchy relations between the strain tensor and the displacement vector components are assumed to be nonlinear, and the von Kármán strains are introduced. In this case [20],

$$\begin{aligned} u(x,t) &= u_0(x,t) - z\frac{\partial w_0}{\partial x}, \\ v(x,t) &\equiv 0, \\ w(x,t) &= w_0(x,t), \end{aligned} \tag{1}$$

and the only non-zero von Kármán strains are [21]

$$\begin{aligned} \varepsilon_{xx} &= \frac{\partial u_0}{\partial x} - z\frac{\partial^2 w_0}{\partial x^2} + \frac{1}{2}\left[\left(\frac{\partial u_0}{\partial x} - z\frac{\partial^2 w_0}{\partial x^2}\right)^2 + \left(\frac{\partial w_0}{\partial x}\right)^2\right], \\ \varepsilon_{xz} &= -\frac{1}{2}\left(\frac{\partial u_0}{\partial x} - z\frac{\partial^2 w_0}{\partial x^2}\right)\frac{\partial w_0}{\partial x}, \quad \varepsilon_{zz} = -\frac{1}{2}\left(\frac{\partial w_0}{\partial x}\right)^2. \end{aligned} \tag{2}$$

Since the beam material is supposed to be linear elastic, with neglecting rotary and longitudinal inertia, as well as shearing deformations, the balance of forces and moments gives

$$\begin{aligned} &\frac{\partial N_{xx}}{\partial x} = 0, \\ &\frac{\partial^2 M_{xx}}{\partial x^2} + \frac{\partial}{\partial x}\left(N_{xx}\frac{\partial w_0}{\partial x}\right) - \rho S\frac{\partial^2 w_0}{\partial t^2} - D(x,t) + F(x,t) = 0, \end{aligned} \tag{3}$$

where F characterizes the influence of the loads, and D characterizes that of the dampers. Here

$$N_{xx} \approx ES\left[\frac{\partial u_0}{\partial x} + \frac{1}{2}\left(\frac{\partial w_0}{\partial x}\right)^2\right],$$
$$M_{xx} \approx -EJ\frac{\partial^2 w_0}{\partial x^2}.$$

Substituting the obtained relations into (3), we obtain

$$\begin{aligned}&\frac{\partial}{\partial x}\left[\frac{\partial u_0}{\partial x} + \frac{1}{2}\left(\frac{\partial w_0}{\partial x}\right)^2\right] = 0,\\ &EJ\frac{\partial^4 w_0}{\partial x^4} - ES\left\{\left[\frac{\partial u_0}{\partial x} + \frac{1}{2}\left(\frac{\partial w_0}{\partial x}\right)^2\right]\frac{\partial^2 w_0}{\partial x^2}\right\} + \rho S\frac{\partial^2 w_0}{\partial t^2} = F(x,t) - D(x,t).\end{aligned} \tag{4}$$

Since there are no any longitudinal loadings acting on the beam, the boundary conditions for u_0 are

$$u_0(0,t) = \left.\frac{\partial u_0}{\partial x}\right|_{x=l} = 0. \tag{5}$$

Integration of the first equation gives

$$\frac{\partial u_0}{\partial x} + \frac{1}{2}\left(\frac{\partial w_0}{\partial x}\right)^2 = f(t), \tag{6}$$

which together with the second boundary condition in (5) leads to

$$f(t) = \frac{1}{2}\left.\left(\frac{\partial w_0}{\partial x}\right)^2\right|_{x=l}.$$

Integration of (6) once more gives

$$u_0(x,t) = -\frac{1}{2}\int_0^x \left(\frac{\partial w_0}{\partial \xi}\right)^2 d\xi + \frac{1}{2}\left.\left(\frac{\partial w_0}{\partial x}\right)^2\right|_{x=l} \cdot x, \tag{7}$$

where the first boundary condition in (5) is taken into account.

Thus,

$$\frac{\partial u_0}{\partial x} = -\frac{1}{2}\left(\frac{\partial w_0}{\partial x}\right)^2 + \frac{1}{2}\left.\left(\frac{\partial w_0}{\partial x}\right)^2\right|_{x=l}, \tag{8}$$

whose substitution into the second equation in (4) reduces to a single fourth order nonlinear ordinary differential equation with respect to w_0

$$EJ\frac{\partial^4 w_0}{\partial x^4} - \frac{ES}{2}\left.\left(\frac{\partial w_0}{\partial x}\right)^2\right|_{x=l}\frac{\partial^2 w_0}{\partial x^2} + \rho S\frac{\partial^2 w_0}{\partial t^2} = F(x,t) - D(x,t). \tag{9}$$

2.1. Influence of the loads and dampers

Involving the theory of generalized functions, the influence of moving loads can be described by the potential [18, 19, 22]

$$F(x,t) = P_1\delta(x - v_1 t) + P_2\delta[x - l + v_2(t - t_0)], \tag{10}$$

at this (for generality), we assume that the load of intensity P_2 starts moving from the end $x = l$ of the beam at time moment $t_0 \geq 0$. (The case $t_0 = 0$ corresponds to the case of simultaneous start of motion of the loads.)

The influence of the dampers, again within the theory of generalized functions, can be expressed by [18, 19, 22]:

$$D(x,t) = \left(\alpha w_0 + \beta \frac{\partial w_0}{\partial t}\right) \sum_{k=1}^{3} \delta(x - x_k), \quad x_k = \frac{kl}{4}. \tag{11}$$

2.2. Boundary and initial conditions

As was supposed above, the beam is simply supported, therefore the corresponding boundary conditions read as

$$\begin{aligned} &w_0(0,t) = w_0(l,t) \equiv 0, \\ &\left.\frac{\partial^2 w_0}{\partial x^2}\right|_{x=0} = \left.\frac{\partial^2 w_0}{\partial x^2}\right|_{x=l} \equiv 0, \quad t \geq 0. \end{aligned} \tag{12}$$

To derive a unique solution, the following initial conditions are given:

$$\begin{aligned} &w_0(x,0) = w_i(x), \\ &\left.\frac{\partial w_0}{\partial t}\right|_{t=0} = w_i^1(x), \quad 0 \leq x \leq l. \end{aligned} \tag{13}$$

We assume that w_i and w_i^1 are continuous on $[0,l]$. For simplicity, we can take $w_i = w_i^1 \equiv 0$, i.e., before P_1 starts to act on the beam, it was in equilibrium.

2.3. Bubnov–Galerkin procedure

To derive w_0 from $(4)_2$ (since u_0 is well defined in terms of w_0 from $(4)_1$), we involve the Bubnov–Galerkin procedure [16]. Because of its weak or variational nature, it is not required to pass to the weak formulation of the problem, as it is usually done [17]. An approximate solution of $(4)_2$, (12) is sought in the form

$$w_{0N}(x,t) = \sum_{n=1}^{N} \lambda_n(t)\mu_n(x), \quad \mu_n(x) := \sin(\pi n x). \tag{14}$$

Following the Bubnov–Galerkin procedure, λ_n, $n = 1, \ldots, N$, is determined from the system of nonlinear ordinary differential equations of second order

$$\begin{aligned} \rho S \ddot{\lambda}_\nu(t) + \sum_{n=1}^{N} \left\{ {}_kA_n^\nu \beta \dot{\lambda}_n(t) + ({}_kA_n^\nu \alpha + B_n^\nu EJ)\lambda_n(t) + \frac{ES}{2}\pi^4 n^2 C_n^\nu \lambda_n(t) \left[\sum_{k=1}^{N} k\cos(\pi k l)\lambda_k(t)\right]^2 \right\} \\ = P_1 \mu_\nu(v_1 t) + P_2 \mu_\nu[l - v_2(t - t_0)], \quad \nu = 1, \ldots, N, \end{aligned} \tag{15}$$

where

$${}_kA_n^\nu = \mu_n(x_k)\mu_\nu(x_k), \quad B_n^\nu = \int_0^l \mu_n^{IV}(x)\mu_\nu(x)\,dx, \quad C_n^\nu = \int_0^l \mu_n''(x)\mu_\nu(x)\,dx.$$

Note that (13) provides the initial conditions on λ_n:

$$\lambda_n(0) = \int_0^l w_i(x)\sin(\pi n x)\,dx, \quad \dot{\lambda}_n(0) = \int_0^l w_i^1(x)\sin(\pi n x)\,dx, \quad n = 1, \ldots, N. \tag{16}$$

Thus, determining the beam displacements is reduced to solving the Cauchy problem (15), (16). In this study, we involve a finite difference scheme to determine the expansion coefficients λ_n, $n = 1, \ldots, N$, without linearizing (15), as is usually the case [12–14].

IOP Conf. Series: Journal of Physics: Conf. Series **991** (2018) 012046 doi:10.1088/1742-6596/991/1/012046

3. Numerical results and discussions

Numerical analysis is carried out when the dampers are equally spaced, however, in [18, 19], it is shown that, in that way, one can attain the maximal damping rate and the minimal value of absolute value of the displacement.

We assume, that at initial moment, the beam is in equilibrium, i.e., $w_i = w_i^1 \equiv 0$. Let the cross-section of the beam be a solid rectangle of unit area, and let the beam material be steel with $E = 2 \times 10^{11}$ kg/m^2 and $\nu = 0.3$. The transverse and longitudinal displacements of the beam are examined for different combinations of the load parameters. We conventionally consider the following two cases. In the first case, we study the transverse and longitudinal displacements of the point $x_0 = 12$ m (the system of reference is at the left end of the beam, and $l = 100$ m), while in the second case, we study the same displacements of the point at which the loads meet.

The parameters are chosen in the following ranges: $P_1, P_2 \in \{10^5, 2 \times 10^5, 4 \times 10^5\}$ kg, $v_1, v_2 \in \{4, 8, 32, 64\}$ m/s, and $t_0 \in \{0, 2, 4\}$ s. The cases considered below are numbered as $m.n.k.l$, where m corresponds to P_1, n — to P_2, k — to v_1, and l — to v_2. So, when we write Case 1.2.3.4, we mean $P_1 = 10^5$ kg, $P_2 = 2 \times 10^5$ kg, $v_1 = 32$ m/s, and $v_2 = 64$ m/s, and t_0 is usually fixed.

Figures 2 and 3 indicate the influence of the right load on the beam horizontal and vertical deflection at point A for fixed values of the parameters of the left load: more precisely, the cases 1.1.1.1, 1.1.1.4, 1.1.2.4, 1.1.4.1, respectively.

Figure 2 indicates a certain shift of u_0 displacement in the horizontal direction depending on t_0: the greater t_0, the greater the shift. The upper two graphs show that an increase in the velocity of the right load leads to an increase in (absolute) u_0 for $t_0 = 2$ s and to a decrease in (absolute) u_0 for $t_0 = 4$ s (relative to $t_0 = 0$ s). An increase in the velocity of the left load leads to an increase in the (absolute) u_0 for $t_0 = 0$ s and to a decrease in (absolute) u_0 for both $t_0 = 2$ s and $t_0 = 4$ s.

The same dependence of (absolute) w_0 on v_1 and v_2 is observed in figure 3.

Figures 4 and 5 show the influence of the right load on the beam horizontal and vertical deflection at point B for fixed values of the parameters of the left load: more precisely, the cases 1.1.1.1, 1.1.1.4, 1.1.2.4, 1.1.4.1, respectively. Figure 4 indicates the same behavior of (absolute) u_0 on v_1 and v_2, while figure 5 reveals a different picture. An increase in v_2 leads to and increase in (absolute) w_0 for $t_0 = 4$ s only (relative to $t_0 = 0$ s and $t_0 = 2$ s). On the other hand, an increase in v_1 increases the maximal (absolute) value of w_0.

Figures 6, 7 and 8, 9 graphically express the dependence of u_0 and w_0 of points A and B, respectively, on P_1 and P_1 (v_1 and v_2 are fixed: the same behavior is seen for all v_1 and v_2). A shift depending on t_0 is detected in all cases. Unlike the previous case, neither P_1 nor P_2 lead to an increase or a decrease in u_0 and w_0 for different t_0.

It is planned to study the possibility of optimal location of the dampers under the beam in order to reduce its vibrations in the minimal possible time. The results will be reported in subsequent papers.

Conclusions

The influence of two different point loads moving on a beam in opposite directions is considered. The beam is simply supported at its ends and is also supported by three equidistant visco-elastic dampers. The von Kármán theory is accepted with respect to the beam, and it is assumed that the dampers are filled with Kelvin–Voight visco-elastic material. Assuming that the loads start to move at different time moments, the dependence of the beam vertical displacements on external parameters (loads intensities, velocities and starting time moment) is studied. Fully nonlinear problem is studied and strongly nonlinear behavior is observed. Future goals are set up.

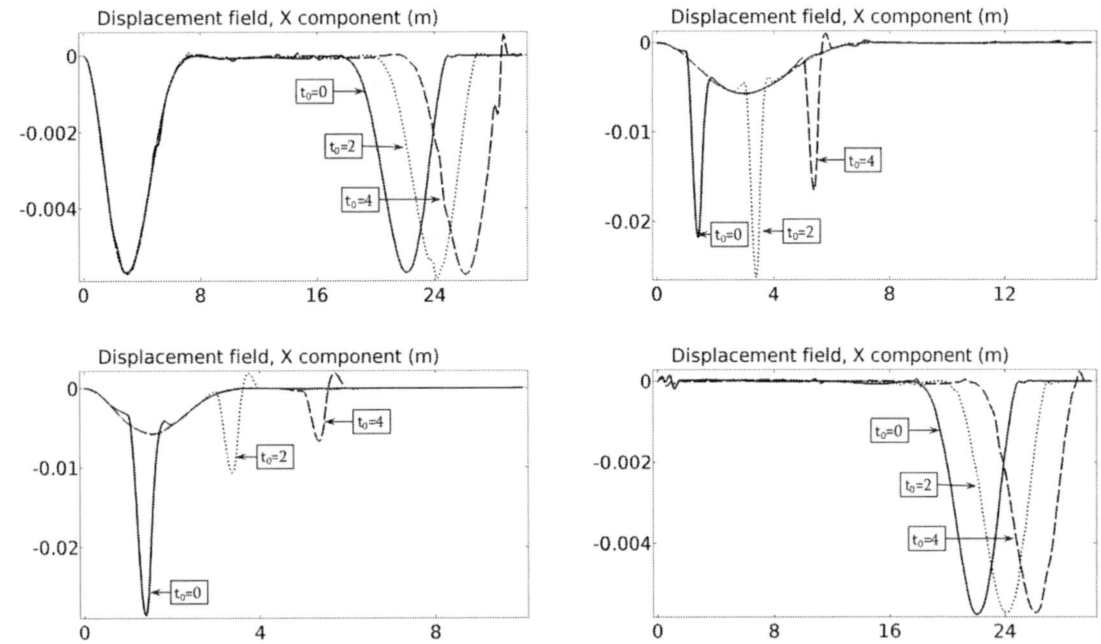

Figure 2. Horizontal deflection of point A in the cases 1.1.1.1, 1.1.1.4, 1.1.2.4, 1.1.4.1, respectively.

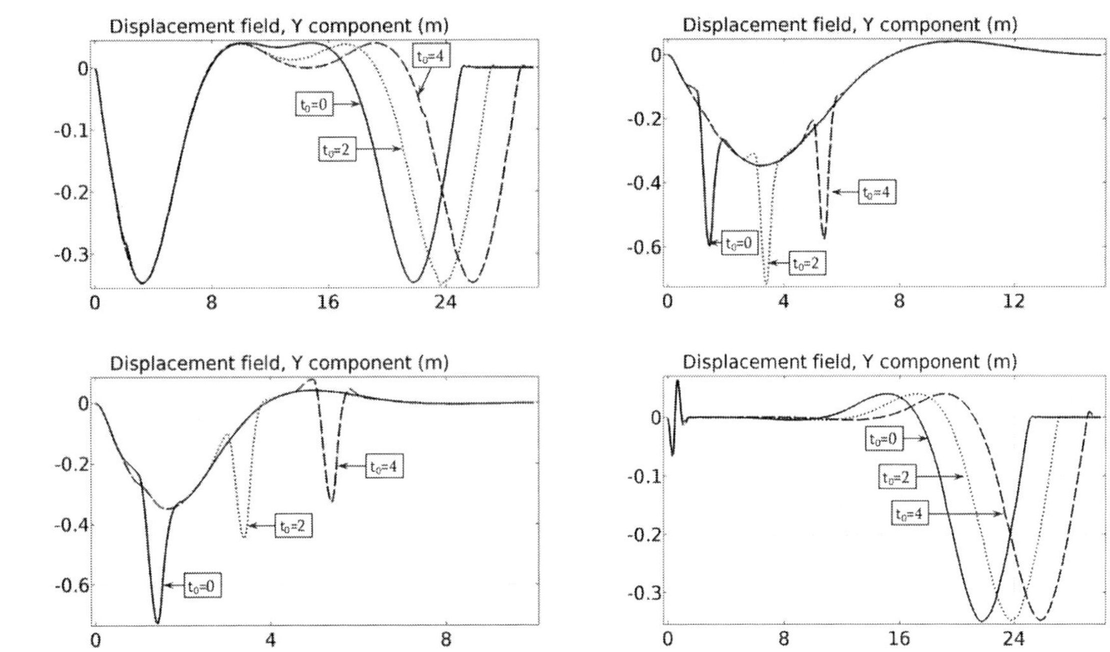

Figure 3. Vertical deflection of point A in the cases 1.1.1.1, 1.1.1.4, 1.1.2.4, 1.1.4.1, respectively.

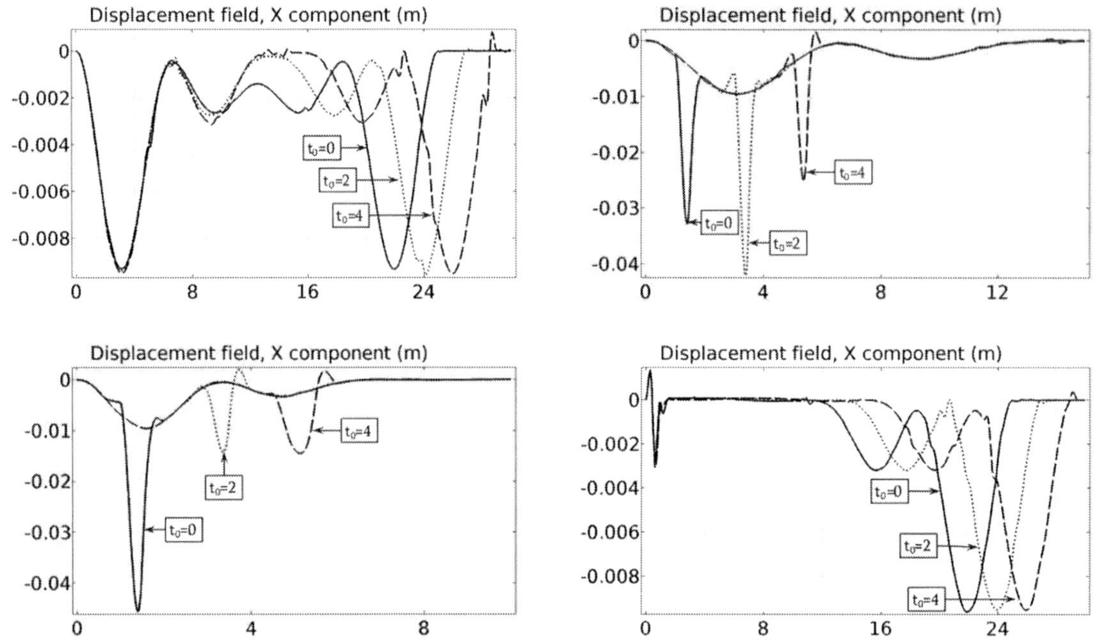

Figure 4. Horizontal deflection of point B in the cases 1.1.1.1, 1.1.1.4, 1.1.2.4, 1.1.4.1, respectively.

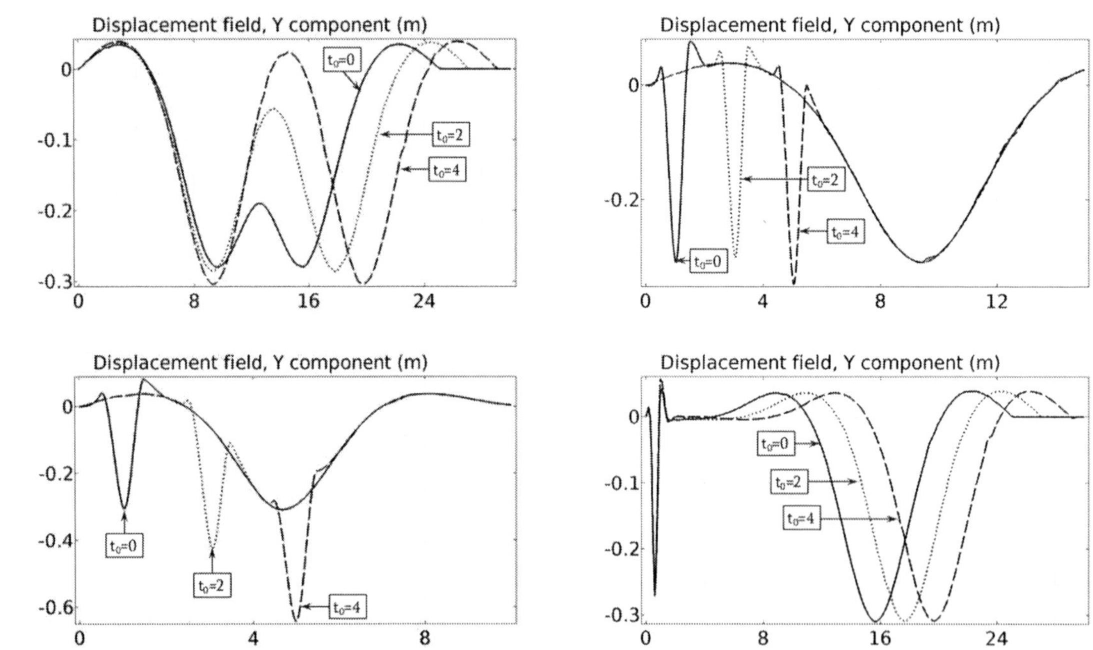

Figure 5. Vertical deflection of point B in the cases 1.1.1.1, 1.1.1.4, 1.1.2.4, 1.1.4.1, respectively.

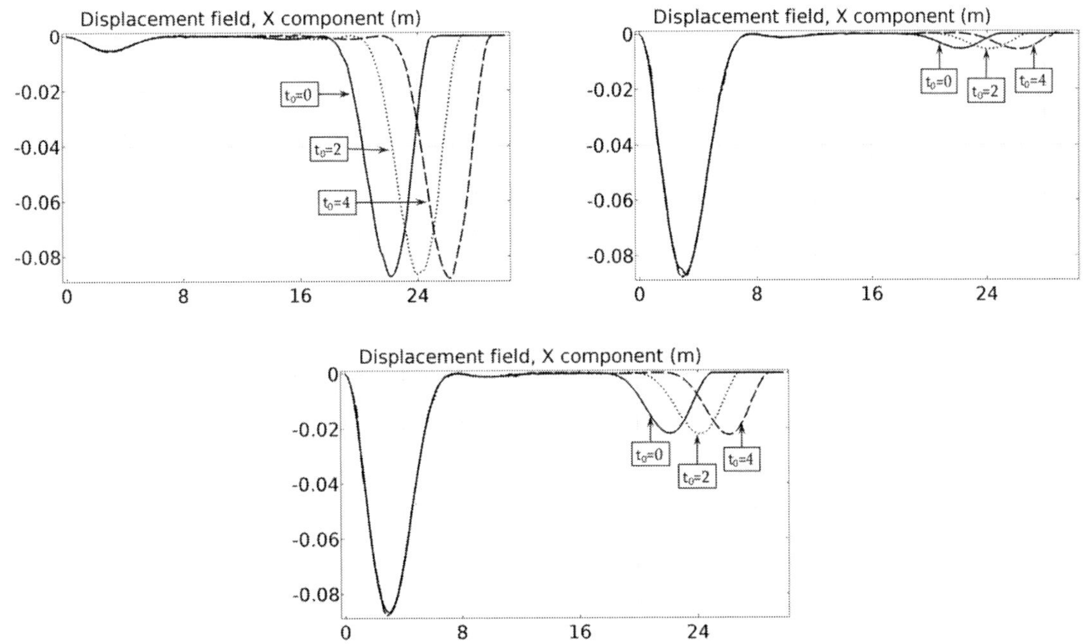

Figure 6. Horizontal deflection of point A in the cases 1.4.1.1, 4.1.1.1, 4.2.1.1, respectively.

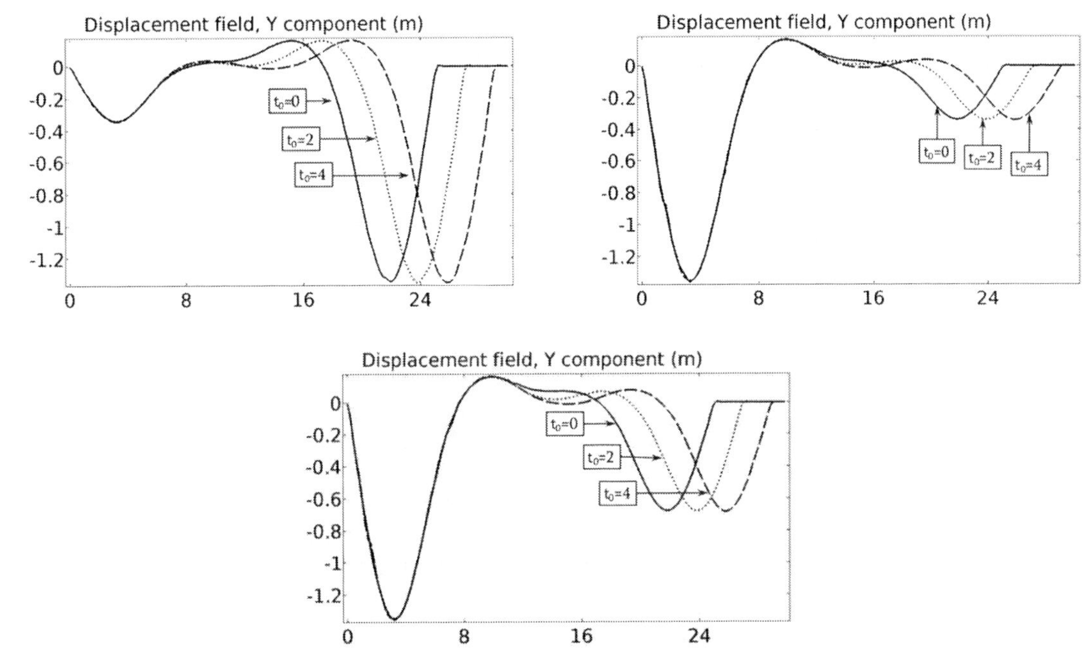

Figure 7. Vertical deflection of point A in the cases 1.4.1.1, 4.1.1.1, 4.2.1.1, respectively.

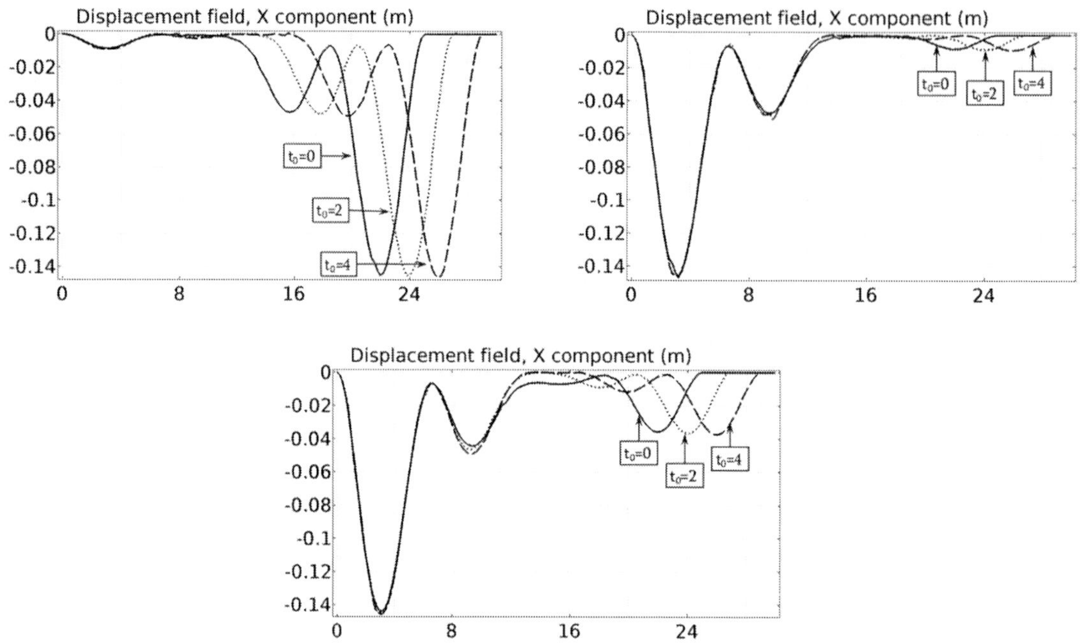

Figure 8. Horizontal deflection of point B in the cases 1.4.1.1, 4.1.1.1, 4.2.1.1, respectively.

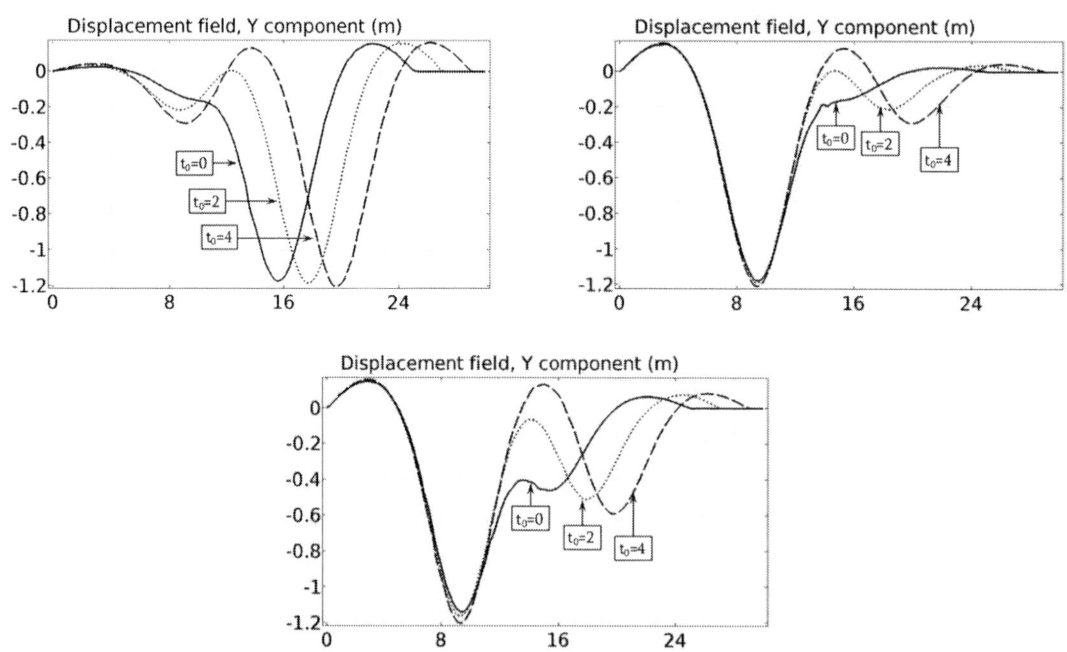

Figure 9. Vertical deflection of point B in the cases 1.4.1.1, 4.1.1.1, 4.2.1.1, respectively.

Acknowledgments
Acknowledgements to our teachers Prof. Dr. Eduard Kh. Grigoryan (1937–2016) and Dr. Hamlet V. Hovhannisyan (1956–2016), who unexpectedly passed away before realizing their great scientific ideas.

References
[1] Frýba L 1999 *Vibration of Solids and Structures under Moving Loads* (London: Thomas Telford)
[2] Bajer C and Dyniewicz B 2012 Numerical analysis of vibrations of structures under moving inertial load (Berlin - Heidelberg: Springer)
[3] Ouyang H 2011 Moving-load dynamic problems: A tutorial (with a brief overview) *Mech. Syst. Signal Process.* **25** 2039–60
[4] Savin E 2001 Dynamic amplification factor and response spectrum for the evaluation of vibrations of beams under successive moving loads *J. Sound Vibr.* **248** (2) 267–88
[5] Thambiratnam D and Zhuge Y 1996 Dynamic analysis of beams on an elastic foundation subjected to moving loads *J. Sound Vibr.* **198** (2) 149–69
[6] Museros P and Martinez-Rodrigo M D 2007 Vibration control of simply supported beams under moving loads using fluid viscous dampers *J. Sound Vibr.* **300** (1-2) 292–315
[7] Di Lorenzo S, Di Paola M, Failla G, and Pirrotta A 2016 On the moving load problem in Euler–Bernoulli uniform beams with viscoelastic supports and joints *Acta Mech.* **228** (3) 805–21
[8] Samani F S and Pellicano F 2009 Vibration reduction on beams subjected to moving loads using linear and nonlinear dynamic absorbers *J. Sound Vibr.* **325** (4-5) 742–54
[9] Samani F S and Pellicano F 2012 Vibration reduction of beams under successive traveling loads by means of linear and nonlinear dynamic absorbers *J. Sound Vibr.* **331** (10) 2272–90
[10] Sapountzakis E J and Dourakopoulos J A 2009 Nonlinear dynamic analysis of Timoshenko beams by BEM. Part I: Theory and numerical implementation *Nonlin. Dyn.* **58** 295–306
[11] Dyniewicz B, Pisarski D, and Bajer C I 2017 Vibrations of a Mindlin plate subjected to a pair of inertial loads moving in opposite directions *J. Sound Vibr.* **386** 265–82
[12] Yoshimura T, Hino J, and Anantharayana N 1986 Vibration analysis of a non-linear beam subjected to moving loads by using the Galerkin method *J. Sound Vibr.* **104** (2) 179–86
[13] Chang T-P and Liu Y-N 1996 Dynamic finite element analysis of a nonlinear beam subjected to a moving load *Int. J. Solids Struct.* **33** (12) 1673–88
[14] Yanmeni Wayou A N, Tchoukuegno R, and Woafo P 2004 Non-linear dynamics of an elastic beam under moving loads *J. Sound Vibr.* **273** (4-5) 1101–8
[15] Mamandi A, Kargarnovin M H, and Younesian D 2010 Nonlinear dynamics of an inclined beam subjected to a moving load *Nonlin. Dyn.* **60** (3) 277–93
[16] Mikhlin S M 1991 *Error Analysis in Numerical Processes* (Chichester - New York: Wiley & Sons)
[17] Liu C-S, Liu D, and Jhao W-S 2017 Solving a singular beam equation by using a weak-form integral equation method *Appl. Math. Lett.* **64** 51–8
[18] Khurshudyan Am Zh and Khurshudyan As Zh 2014 Optimal distribution of viscoelastic dampers under elastic finite beam under moving load Izv. Nats. Akad. Nauk Armenii. Mekh. **67** (3) 56–67
[19] Khurshudyan As Zh 2015 The Bubnov–Galerkin method in control problems for bilinear systems *Automat. Remote Contr.* **76** (8) 1361–8
[20] Reddy J N 2015 *An Introduction to Nonlinear Finite Element Analysis: with Applications to Heat Transfer, Fluid Mechanics, and Solid Mechanics. 2nd Edition* (Oxford: Oxford University Press) p 768
[21] Ciarlet F G 1997 *Mathematical Elasticity, Vol. II: Theory of Plates* (Amsterdam: North-Holland)
[22] Avetisyan A S and Khurshudyan As Zh 2016 Green's function approach in approximate controllability problems Izv. Nats. Akad. Nauk Armenii. Mekh. **69** (2) 3–20
[23] Avetisyan A S and Khurshudyan As Zh 2017 Green's function approach in approximate controllability for nonlinear physical processes *Modern Phys. Lett. A* **32** (21) 1730015

Wave propagation problem for a micropolar elastic waveguide

V A Kovalev[1], **E V Murashkin**[2,3*], **Y N Radayev**[2,4**]
[1]Moscow City Government University of Management, Sretenka Str 28, Moscow, 107045, Russia
[2]Ishlinsky Institute for Problems in Mechanics of the Russian Academy of Sciences, Vernadsky Ave 101 Bldg 1, Moscow, 119526, Russia
[3]Bauman Moscow State Technical University, 2nd Baumanskaya Str 5/1, Moscow, 105005, Russia
[4]Department of Energy Conversion Science, Graduate School of Energy Science, Kyoto University, Yoshida Honmachi, Sakyo-ku, Kyoto, 606-8501, Japan

E-mail: *murashkin@ipmnet.ru, **radayev@ipmnet.ru, **radaev.iurii.8e@kyoto-u.ac.jp

Abstract. A propagation problem for coupled harmonic waves of translational displacements and microrotations along the axis of a long cylindrical waveguide is discussed at present study. Microrotations modeling is carried out within the linear micropolar elasticity frameworks. The mathematical model of the linear (or even nonlinear) micropolar elasticity is also expanded to a field theory model by variational least action integral and the least action principle. The governing coupled vector differential equations of the linear micropolar elasticity are given. The translational displacements and microrotations in the harmonic coupled wave are decomposed into potential and vortex parts. Calibrating equations providing simplification of the equations for the wave potentials are proposed. The coupled differential equations are then reduced to uncoupled ones and finally to the Helmholtz wave equations. The wave equations solutions for the translational and microrotational waves potentials are obtained for a high-frequency range.

Introduction

Modelling of a specific mechanical and multi-physical behaviour of solids (including metamaterials) is an actual problem of modern Continuum Mechanics. The metamaterials exhibit physical properties not usually found in nature. For example, there are materials with negative Poissons ratio (auxetic materials), negative thermal expansion, negative electric permittivity and the magnetic permeability. These paradoxical physical phenomena cannot be described in the frameworks of the conventional Continuum Mechanics. In such a case, microstructure continuum theories are to be usually employed [1–5]. Even now the micropolar continuum models are intensively used in applied mechanics in order to resolve various shortcomings and difficulties arising while employing the classical theory of elasticity. Asymmetric stress and strain tensors are intrinsic to the micropolar elasticity modeling (see discussion in [4] for the subject). The problems of propagating surfaces of discontinuities of displacements and microrotations are previously discussed in a number of studies [8–18]. It is also true for the wave problems formulated for circular micropolar waveguides (see [19]).

The aim of the present paper is to give a method which provides determination of the translational and microrotation waves propagating along an infinite waveguide with circular

IOP Conf. Series: Journal of Physics: Conf. Series **991** (2018) 012047 doi:10.1088/1742-6596/991/1/012047

cross-section. Microrotation waves in an elastic solid are modelled within the linear micropolar elasticity frameworks by introducing in the governing equations of the Continuum Mechanics additional microrotation degree of freedom. Such a modelling can be regularly carried out in terms of the classical field theory [7], starting from the action integral and the least action variational principle.

The present paper is arranged as follows. In Secs. 1, 2 microrotation waves modelling is realized within the frameworks of linear micropolar elasticity by introducing the axial microrotation vector thus representing additional independent degrees of freedom of the elastic continuum. The coupled vector differential equations of the conventional micropolar elasticity are given. Actually Sec. 2 is to be considered as a script of the equations of the linear micropolar elasticity. Sec. 1 includes a breaf discussion of the conventional micropolar elasticity as a field theory based on the least action principle. In Sec. 3 the translational displacements and microrotations vectors in the coupled wave are decomposed into potential and vortex parts. Scalar and vector wave potentials of displacements and microrotations are introduced. The calibrating equations for the vector wave potentials are presented. Coupled vector equations involving the vector wave potentials are obtained. These equations are then uncoupled for harmonic waves of high-frequency range, when the governing differential equations belong to the hyperbolic analytical type. In Sec. 4 the uncoupled equations are reduced to the Helmholtz wave equations. The latter are resolved for the translational and microrotation waves of high-frequency range.

1. Micropolar elasticity as a field theory. The least action principle

The conventional model of micropolar elastic continuum (see [4] for details) is used throughout the paper. Such continua based on the existence of comlementary rotational degrees of freedom (reperesented by the rotation vector $\boldsymbol{\phi}$) along with translational degrees of freedom (described by the translational displacement vector $\mathbf{u}$). The model of micropolar elastic continuum can be expanded as a field theory if the "natural" action density is taken in the following form

$$\mathcal{L} = \frac{1}{2}\dot{\mathbf{u}} \cdot \boldsymbol{\rho} \cdot \dot{\mathbf{u}} + \frac{1}{2}\dot{\boldsymbol{\phi}} \cdot \mathfrak{J} \cdot \dot{\boldsymbol{\phi}} - \psi(\boldsymbol{e}, \boldsymbol{\gamma}), \tag{1}$$

where $\boldsymbol{\rho}$ is the mass density tensor, $\mathfrak{J}$ is the tensor of microinertia, ψ is the Helmholtz free energy per unit volume, $\boldsymbol{\nabla}$ is the three-dimensional Hamiltonian differential operator (the nabla symbol), the superimposed dot denotes partial differentiation with respect to time variable. For tensors $\boldsymbol{\rho}$ and $\mathfrak{J}$ the symmetry conditions are assumed to be valid.

The asymmetric strain tensor $\boldsymbol{e}$ and bending–torsion tensor $\boldsymbol{\gamma}$ are associated with translational displacements $\mathbf{u}$ and microrotations $\boldsymbol{\phi}$ by the formulae

$$\begin{aligned} \boldsymbol{e} &= \boldsymbol{\nabla} \otimes \mathbf{u} - \boldsymbol{\phi} \cdot \boldsymbol{\varepsilon}, \\ \boldsymbol{\gamma} &= \boldsymbol{\nabla} \otimes \boldsymbol{\phi}. \end{aligned} \tag{2}$$

We employ the notation $\boldsymbol{\varepsilon}$ for the three-dimensional permutation tensor. The equations (2) are represented in a rectangular co-ordinate system as

$$\begin{aligned} e_{ji} &= \partial_j u_i - \varepsilon_{jik}\phi_k, \\ \gamma_{ji} &= \partial_j \phi_i. \end{aligned} \tag{3}$$

In general, the variational action integral with "natural" action density (1) is given by following equation ($d\tau$ is an elementary spatial volume):

$$\Im = \int \mathcal{L}(\mathbf{u}, \boldsymbol{\phi}, \dot{\mathbf{u}}, \dot{\boldsymbol{\phi}}, \boldsymbol{\nabla} \cdot \boldsymbol{\phi}, \boldsymbol{\nabla} \cdot \mathbf{u}) d\tau dt. \tag{4}$$

IOP Conf. Series: Journal of Physics: Conf. Series **991** (2018) 012047 doi:10.1088/1742-6596/991/1/012047

The least action principle states that the actual field is realized in such a way that the action of (4) is minimum, i.e. for any admissible variations of the physical fields $\{\mathbf{u}, \boldsymbol{\phi}\}$ and non-variable spatial coordinates and the time variable the following equation is valid

$$\delta\Im = 0. \tag{5}$$

The differential field equations is simply obtained from the equation (5) as

$$\begin{aligned} &\partial_\alpha S^{\alpha\cdot}_{\cdot j} - \dot{P}_j = -\frac{\partial\mathcal{L}}{\partial u^j} && (\alpha = 1,2,3;\ j = 1,2,3), \\ &\partial_\beta \mathcal{M}^{\beta\cdot}_{\cdot j} + \mathcal{A}_j - \dot{\mathcal{Q}}_j = 0 && (\beta = 1,2,3;\ j = 1,2,3), \end{aligned} \tag{6}$$

wherein the following notations are used

$$P_j = \frac{\partial\mathcal{L}}{\partial\dot{u}^j}, \quad \mathcal{Q}_j = \frac{\partial\mathcal{L}}{\partial\dot{\phi}^j}, \quad S^{\alpha\cdot}_{\cdot j} = -\frac{\partial\mathcal{L}}{\partial(\partial_\alpha u^j)}, \quad \mathcal{M}^{\alpha\cdot}_{\cdot j} = -\frac{\partial\mathcal{L}}{\partial(\partial_\alpha \phi^j)}, \quad \mathcal{A}_j = \frac{\partial\mathcal{L}}{\partial\phi^j}. \tag{7}$$

2. Reminder of the linear micropolar elasticity

For linear micropolar elastic continuum we can expand the Helmholtz free energy ψ into the Taylor series in the vicinity of the natural state $\boldsymbol{e} = \boldsymbol{\gamma} = \mathbf{0}$, while disregarding the terms of higher order than the second ones. In a rectangular coordinate system the following form can be obtained for isotropic, homogeneous and centrosymmetric continua [4]:

$$\psi = \frac{\mu+\eta}{2}e_{ji}e_{ji} + \frac{\mu-\eta}{2}e_{ji}e_{ij} + \frac{\lambda}{2}e_{kk}e_{nn} + \frac{\gamma+\varepsilon}{2}\gamma_{ji}\gamma_{ji} + \frac{\gamma-\varepsilon}{2}\gamma_{ji}\gamma_{ij} + \frac{\beta}{2}\gamma_{kk}\gamma_{nn}. \tag{8}$$

where λ, μ, γ, β, ε, η are the constitutive constants of micropolar elastic continuum.

The constitutive equations for the force stress tensor $\boldsymbol{\sigma}$ and the moment stress tensor $\boldsymbol{\mu}$ are formulated in virtue of equations (7) as follows

$$\begin{aligned} \boldsymbol{\sigma} &= (\mu+\eta)\boldsymbol{e} + (\mu-\eta)\boldsymbol{e}^{\mathrm{T}} + \lambda\mathbf{I}\,\mathrm{tr}\boldsymbol{e}, \\ \boldsymbol{\mu} &= (\gamma+\varepsilon)\boldsymbol{\gamma} + (\gamma-\varepsilon)\boldsymbol{\gamma}^{\mathrm{T}} + \beta\mathbf{I}\mathrm{tr}\boldsymbol{\gamma}. \end{aligned} \tag{9}$$

In the above equations $\mathbf{I}$ denotes the three-dimensional unit tensor, $\boldsymbol{\gamma}$ is the strain tensor, $\boldsymbol{\kappa}$ is the bending-torsion tensor.

The linear vector partial differential equations determining the coupled translational and microrotation waves in a micropolar elastic continuum read (see [4]):

$$\begin{cases} (\lambda+2\mu)\boldsymbol{\nabla}\boldsymbol{\nabla}\cdot\mathbf{u} - (\mu+\eta)\boldsymbol{\nabla}\times(\boldsymbol{\nabla}\times\mathbf{u}) + 2\eta\boldsymbol{\nabla}\times\boldsymbol{\phi} = \rho\ddot{\mathbf{u}}, \\ (\beta+2\gamma)\boldsymbol{\nabla}\cdot\boldsymbol{\phi} - (\gamma+\varepsilon)\boldsymbol{\nabla}\times(\boldsymbol{\nabla}\times\boldsymbol{\phi}) + 2\eta\boldsymbol{\nabla}\times\mathbf{u} - 4\eta\boldsymbol{\phi} = \Im\ddot{\boldsymbol{\phi}}. \end{cases} \tag{10}$$

In view of equations (9) and (2) after chain of transformations the following equations for the traction vector $\mathbf{t}$ and the moment vector $\mathbf{m}$ can be obtained:

$$\begin{aligned} \mathbf{t} &= \lambda(\boldsymbol{\nabla}\cdot\mathbf{u})\mathbf{n} + 2\mu(\mathbf{n}\cdot\boldsymbol{\nabla})\mathbf{u} + (\mu-\eta)(\mathbf{n}\times\boldsymbol{\nabla})\times\mathbf{u} + 2\eta\mathbf{n}\times\boldsymbol{\phi}, \\ \mathbf{m} &= \beta(\boldsymbol{\nabla}\cdot\boldsymbol{\phi})\mathbf{n} + 2\gamma(\mathbf{n}\cdot\boldsymbol{\nabla})\boldsymbol{\phi} + (\gamma-\varepsilon)(\mathbf{n}\times\boldsymbol{\nabla})\times\boldsymbol{\phi}, \end{aligned} \tag{11}$$

where $\mathbf{n}$ is the unit normal vector determining the spatial orientation of the plane surface element.

The latter equations are required in order to formulate boundary conditions on the free from tractions and moments sidewall of the waveguide:

$$\mathbf{t} = \mathbf{0}, \quad \mathbf{m} = \mathbf{0}. \tag{12}$$

IOP Conf. Series: Journal of Physics: Conf. Series **991** (2018) 012047 doi:10.1088/1742-6596/991/1/012047

3. Wave potentials of translational displacements and microrotations

The vector partial differential equations (10) are coupled and can be uncoupled following a scheme described in this Section.

At the beginning we introduce wave vector potentials corresponding to the translational displacements and microrotations. The vector fields $\mathbf{u}$ and $\boldsymbol{\phi}$ are decomposed into potential and vortex parts according to the Helmholtz representations:

$$
\begin{aligned}
\mathbf{u} &= \boldsymbol{\nabla}\Phi + \boldsymbol{\nabla} \times \boldsymbol{\Psi}, \\
\boldsymbol{\phi} &= \boldsymbol{\nabla}\Sigma + \boldsymbol{\nabla} \times \mathbf{H},
\end{aligned}
\tag{13}
$$

where Φ, Σ are the scalar wave potentials, and $\boldsymbol{\Psi}$, $\mathbf{H}$ are the vector wave potentials of translational displacements and microrotations respectively. Discussion on the completeness and relations between such potentials and stress functions in micropolar elasticity see in [20].

Second, in the present study we shall impose the usual calibration conditions for the wave potentials $\boldsymbol{\Psi}$, $\mathbf{H}$. Those read the two scalar equations

$$
\begin{aligned}
\boldsymbol{\nabla} \cdot \boldsymbol{\Psi} &= 0, \\
\boldsymbol{\nabla} \cdot \mathbf{H} &= 0.
\end{aligned}
\tag{14}
$$

Then by substituting equations (13) in (10) and taking into consideration the calibrating equations (14) it is seen that the system (10) is satisfied if the scalar and the vector wave potentials are to be solutions of the uncoupled scalar equations

$$
\begin{aligned}
&\Delta\Phi - \frac{1}{c_{\|}^2}\ddot{\Phi} = 0, \\
&\Delta\Sigma - \frac{1}{{}_{\mu}c_{\|}^2}\ddot{\Sigma} - \frac{\Omega^2}{{}_{\mu}c_{\|}^2}\Sigma = 0,
\end{aligned}
\tag{15}
$$

and the following system of coupled vector partial differential equations:

$$
\begin{cases}
\Delta\boldsymbol{\Psi} - \dfrac{1}{{}''c_{\perp}^2}\ddot{\boldsymbol{\Psi}} + 2d_{\perp}^2\,\boldsymbol{\nabla} \times \mathbf{H} = 0, \\[2ex]
\Delta\mathbf{H} - \dfrac{1}{{}_{\mu}c_{\perp}^2}\ddot{\mathbf{H}} - \dfrac{\Omega^2}{{}_{\mu}c_{\perp}^2}\mathbf{H} + \dfrac{\Omega^2}{2\,{}_{\mu}c_{\perp}^2}\boldsymbol{\nabla} \times \boldsymbol{\Psi} = 0,
\end{cases}
\tag{16}
$$

wherein we have employed the notations for given below set of constitutive constants:

$$
\begin{aligned}
&\Omega^2 = \frac{4\eta}{\Im}, \quad c_{\|}^2 = \frac{\lambda + 2\mu}{\rho}, \quad {}_{\mu}c_{\|}^2 = \frac{\beta + 2\gamma}{\Im}, \quad {}_{\mu}c_{\perp}^2 = \frac{\mu + \eta}{\Im}, \\
&{}''c_{\perp}^2 = \frac{\mu + \eta}{\rho}, \, {}'c_{\perp}^2 = \frac{\eta}{\rho}, d_{\perp}^2 = \frac{{}'c_{\perp}^2}{{}''c_{\perp}^2}.
\end{aligned}
\tag{17}
$$

In the present study only coupled high-frequency waves of thranslational displacements and microrotaions are considered. That means we assume $\omega > \Omega$. In such a case the following differential equations for the wave potentials can be obtained

$$
\begin{cases}
(\Delta + \alpha_{\|}^2)\Phi = 0, \\
(\Delta + \beta_{\|}^2)\Sigma = 0; \\
(\Delta + \alpha_{\perp}^2)\boldsymbol{\Psi} + 2d_{\perp}^2\,\boldsymbol{\nabla} \times \mathbf{H} = \mathbf{0}, \\
(\Delta + \beta_{\perp}^2)\mathbf{H} + \dfrac{\Omega^2}{2\,{}_{\mu}c_{\perp}^2}\boldsymbol{\nabla} \times \boldsymbol{\Psi} = \mathbf{0}.
\end{cases}
\tag{18}
$$

IOP Conf. Series: Journal of Physics: Conf. Series **991** (2018) 012047 doi:10.1088/1742-6596/991/1/012047

In the above equations we have introduced the constants defined according to

$$\alpha_{\parallel}^2 = \frac{\omega^2}{c_{\parallel}^2}, \quad \beta_{\parallel}^2 = \frac{\omega^2 - \Omega^2}{\mu c_{\parallel}^2}, \quad \alpha_{\perp}^2 = \frac{\omega^2}{c_{\perp}^2}, \quad \beta_{\perp}^2 = \frac{\omega^2 - \Omega^2}{\mu c_{\perp}^2}. \tag{19}$$

The two vector equations of system (18) related to the vortex potentials $\mathbf{\Psi}$ and $\mathbf{H}$ are still coupled. Uncoupled equations involved them can be obtained however this requires higher order differential operators.

At last, for the vortex wave potentials $\mathbf{\Psi}$ and $\mathbf{H}$ the two separate equations are derived in the following identical forms

$$\begin{gathered}(\Delta + K_1^2)(\Delta + K_2^2)\mathbf{\Psi} = \mathbf{0}, \\ (\Delta + K_1^2)(\Delta + K_2^2)\mathbf{H} = \mathbf{0},\end{gathered} \tag{20}$$

wherein both the constants K_1, K_2 are real for the harmonic wave of the high-frequency range. They are defined by

$$\begin{gathered} K_{1,2}^2 = -\Delta_{1,2}\,, \\ \Delta_{1,2} = \frac{-(\alpha_{\perp}^2 + \beta_{\perp}^2 + \sigma_{\perp}^2) \pm \sqrt{(\alpha_{\perp}^2 - \beta_{\perp}^2 + \sigma_{\perp}^2)^2 + 4\beta_{\perp}^2\sigma_{\perp}^2}}{2}, \\ \sigma_{\perp}^2 = \frac{d_{\perp}^2\Omega^2}{\mu c_{\perp}^2}. \end{gathered} \tag{21}$$

4. Displacement and microrotation fields in propagating wave

We proceed the discussion to determination of the displacement and microrotaion vectors in the coupled propagating wave. The coupled translational and microrotation wave fields in a cylindrical waveguide are determined in the cylindrical coordinates r, φ, z by the separation of variables method (see details in [2]). This method provides investigation of waves of an arbitrary azimuthal number n propagating along the infinite waveguide.

Before all it is necessary to underline that the differential operators

$$(\Delta + K_1^2)$$

and

$$(\Delta + K_2^2)$$

commute. This fact enables us to expand the vector wave potential $\mathbf{\Psi}$ into the sum

$$\mathbf{\Psi} = \mathbf{\Psi}_{(1)} + \mathbf{\Psi}_{(2)}$$

with the summands satisfying the standard Helmholtz wave equations

$$(\Delta + K_1^2)\mathbf{\Psi}_{(1)} = \mathbf{0}, \quad (\Delta + K_2^2)\mathbf{\Psi}_{(2)} = \mathbf{0},$$

and it is also true for the vector wave potential $\mathbf{H}$, i.e.

$$\mathbf{H} = \mathbf{H}_{(1)} + \mathbf{H}_{(2)},$$

where

$$(\Delta + K_1^2)\mathbf{H}_{(1)} = \mathbf{0}, \quad (\Delta + K_2^2)\mathbf{H}_{(2)} = \mathbf{0}.$$

For the scalar wave potentials Φ, Σ of the coupled wave of displacevents and microrotations the following representations are obtained:

$$\begin{aligned} \Phi &= C_1 I_n(p_1 r) \left\{ \begin{array}{c} \cos n\varphi \\ -\sin n\varphi \end{array} \right\} e^{\pm ikz}, \\ \sum &= -C_2 I_n(p_2 r) \left\{ \begin{array}{c} \sin n\varphi \\ \cos n\varphi \end{array} \right\} e^{\pm ikz}, \end{aligned} \tag{22}$$

wherein the time harmonic exponent multipliers

$$e^{i\omega t}$$

have been omitted; k denotes the wavenumber of the propagating wave; C_1 and C_2 are arbitrary constants; $I_n(\cdot)$ is the standard Bessel function of the first kind of an imaginary argument;

$$p_1^2 = k^2 - \alpha_{\|}^2, \quad p_2^2 = k^2 - \beta_{\|}^2.$$

The vortex wave potentials of translational displacements and microrotations are obtainable in a more compicated forms given by the following formulae for their physical components in the cylindrical coordinate net:

$$\left\{ \begin{aligned} \Psi_r &= [C_3' I_{n-1}(q_1 r) + C_4' I_{n+1}(q_1 r) + C_3'' I_{n-1}(q_2 r) + C_4'' I_{n+1}(q_2 r)] \left\{ \begin{array}{c} \sin n\varphi \\ \cos n\varphi \end{array} \right\} e^{\pm ikz}, \\ \Psi_\varphi &= [C_3' I_{n-1}(q_1 r) - C_4' I_{n+1}(q_1 r) + C_3'' I_{n-1}(q_2 r) - C_4'' I_{n+1}(q_2 r)] \left\{ \begin{array}{c} \cos n\varphi \\ -\sin n\varphi \end{array} \right\} e^{\pm ikz}, \\ \Psi_z &= [C_5' I_n(q_1 r) + C_5'' I_n(q_2 r)] \left\{ \begin{array}{c} \sin n\varphi \\ \cos n\varphi \end{array} \right\} e^{\pm ikz}; \end{aligned} \right. \tag{23}$$

$$\left\{ \begin{aligned} H_r &= [L_3' I_{n-1}(q_1 r) + L_4' I_{n+1}(q_1 r) + L_3'' I_{n-1}(q_2 r) + L_4'' I_{n+1}(q_2 r)] \left\{ \begin{array}{c} -\cos n\varphi \\ \sin n\varphi \end{array} \right\} e^{\pm ikz}, \\ H_\varphi &= [L_3' I_{n-1}(q_1 r) - L_4' I_{n+1}(q_1 r) + L_3'' I_{n-1}(q_2 r) - L_4'' I_{n+1}(q_2 r)] \left\{ \begin{array}{c} \sin n\varphi \\ \cos n\varphi \end{array} \right\} e^{\pm ikz}, \\ H_z &= [L_5' I_n(q_1 r) + L_5'' I_n(q_2 r)] \left\{ \begin{array}{c} -\cos n\varphi \\ \sin n\varphi \end{array} \right\} e^{\pm ikz}, \end{aligned} \right. \tag{24}$$

wherein the time harmonic exponent multipliers have been omitted; $C_3' - C_5'$, $C_3'' - C_5''$, $L_3' - L_5'$ and $L_3'' - L_5''$ are arbitrary constants and

$$q_1^2 = k^2 - K_1^2, \quad q_2^2 = k^2 - K_2^2.$$

The coordinate representations (23) and (24) of the vortex wave potentials are derived from a formal solution

$$\left\{ \begin{aligned} \Gamma_r &= [C_1 I_{n-1}(q_* r) + C_2 I_{n+1}(q_* r)] \left\{ \begin{array}{c} \sin n\varphi \\ \cos n\varphi \end{array} \right\} e^{\pm ikz}, \\ \Gamma_\varphi &= [C_1 I_{n-1}(q_* r) - C_2 I_{n+1}(q_* r)] \left\{ \begin{array}{c} \cos n\varphi \\ -\sin n\varphi \end{array} \right\} e^{\pm ikz}, \\ \Gamma_z &= C_3 I_n(q_* r) \left\{ \begin{array}{c} \sin n\varphi \\ \cos n\varphi \end{array} \right\} e^{\pm ikz}. \end{aligned} \right. \tag{25}$$

of the vector Helmholtz wave equation

$$(\Delta + k_*^2)\mathbf{\Gamma} = \mathbf{0}, \tag{26}$$

IOP Conf. Series: Journal of Physics: Conf. Series **991** (2018) 012047 doi:10.1088/1742-6596/991/1/012047

which can be obtained by separating variables in the cylindrical coordinates r, φ, z.

We should also note that the constant q_* coming into the solution (25) of the Helmholtz wave equation (26) is defined according to

$$q_*^2 = k^2 - k_*^2.$$

In view of equations (22), (23) and (24) the physical components of displacement and microrotation vectors in the propagating coupled wave can be obtained. We give these components for any azimuthal number n and omitting as previously the time harmonic exponent multipliers:

$$u_r = \left[\begin{array}{l} C_1 \left(p_1 I_{n+1}(p_1 r) + \frac{n}{r} I_n(p_1 r) \right) + \frac{n}{r} \left(C_5' I_n(q_1 r) + C_5'' I_n(q_2 r) \right) + \\ +(\mp ik) \left(C_3' I_{n-1}(q_1 r) - C_4' I_{n+1}(q_1 r) + C_3'' I_{n-1}(q_2 r) - C_4'' I_{n+1}(q_2 r) \right) \end{array} \right] \times$$
$$\times \left\{ \begin{array}{c} \cos n\varphi \\ -\sin n\varphi \end{array} \right\} e^{\pm ikz}, \tag{27}$$

$$u_\varphi = \left[\begin{array}{l} -\frac{n}{r} C_1 I_n(p_1 r) - C_5' \left(q_1 I_{n+1}(q_1 r) + \frac{n}{r} I_n(q_1 r) \right) + \\ +(\pm ik) \left(C_3' I_{n-1}(q_1 r) + C_4' I_{n+1}(q_1 r) + C_3'' I_{n-1}(q_2 r) + C_4'' I_{n+1}(q_2 r) \right) - \\ -C_5'' \left(q_2 I_{n+1}(q_2 r) + \frac{n}{r} I_n(q_2 r) \right) \end{array} \right] \times$$
$$\times \left\{ \begin{array}{c} \sin n\varphi \\ \cos n\varphi \end{array} \right\} e^{\pm ikz}, \tag{28}$$

$$u_z = \left(C_1 (\pm ik) I_n(p_1 r) + (C_3' - C_4') q_1 I_n(q_1 r) + (C_3'' - C_4'') q_2 I_n(q_2 r) \right) \times$$
$$\times \left\{ \begin{array}{c} \cos n\varphi \\ -\sin n\varphi \end{array} \right\} e^{\pm ikz}; \tag{29}$$

$$\phi_r = \left[\begin{array}{l} -C_2 \left(p_2 I_{n+1}(p_2 r) + \frac{n}{r} I_n(p_2 r) \right) + \frac{n}{r} \left(L_5' I_n(q_1 r) + L_5'' I_n(q_2 r) \right) + \\ +(\mp ik) \left(L_3' I_{n-1}(q_1 r) - L_4' I_{n+1}(q_1 r) + L_3'' I_{n-1}(q_2 r) - L_4'' I_{n+1}(q_2 r) \right) \end{array} \right] \times$$
$$\times \left\{ \begin{array}{c} \sin n\varphi \\ \cos n\varphi \end{array} \right\} e^{\pm ikz}, \tag{30}$$

$$\phi_\varphi = \left[\begin{array}{l} -\frac{n}{r} C_2 I_n(p_2 r) - L_5' \left(q_1 I_{n+1}(q_1 r) + \frac{n}{r} I_n(q_1 r) \right) + \\ +(\pm ik) \left(L_3' I_{n-1}(q_1 r) + L_4' I_{n+1}(q_1 r) + L_3'' I_{n-1}(q_2 r) + L_4'' I_{n+1}(q_2 r) \right) - \\ -L_5'' \left(q_2 I_{n+1}(q_2 r) + \frac{n}{r} I_n(q_2 r) \right) \end{array} \right] \times$$
$$\times \left\{ \begin{array}{c} -\cos n\varphi \\ \sin n\varphi \end{array} \right\} e^{\pm ikz}, \tag{31}$$

$$\phi_z = \left(-C_2 (\pm ik) I_n(p_2 r) + (L_3' - L_4') q_1 I_n(q_1 r) + (L_3'' - L_4'') q_2 I_n(q_2 r) \right) \times$$
$$\times \left\{ \begin{array}{c} \sin n\varphi \\ \cos n\varphi \end{array} \right\} e^{\pm ikz}. \tag{32}$$

Arbitrary constants in the formulae (27)–(32) are related by the linear equations

$$\begin{array}{l} (C_3' + C_4') q_1 + (\pm ik) C_5' = 0, \\ (C_3'' + C_4'') q_2 + (\pm ik) C_5'' = 0, \\ (L_3' + L_4') q_1 + (\pm ik) L_5' = 0, \\ (L_3'' + L_4'') q_2 + (\pm ik) L_5'' = 0. \end{array} \tag{33}$$

which follows from the calibration conditions (14). Additional linear algebraic equations for them are derived from the boundary conditions (12) on sidewall of the waveguide. This conditions are to be expanded in virtue of formulae (11), representing the traction and moment vectors via the displacement and microrotation vectors, and the Helmholtz decompositions (13).

Note that the calibration equations (14) actually may not enter the script of the requsite equations related to the problem under consideration. In general they might be replaced by

$$\begin{aligned} \nabla \cdot \boldsymbol{\Psi} &= F, \\ \nabla \cdot \mathbf{H} &= G \end{aligned} \tag{34}$$

with arbitrary functions F and G of spatial coordinates and the time variable. This immediately leads to the conclusion that the equations (33) in fact may not be taken into consideration.

Conclusion

- The constitutive and governing equations of the linear micropolar elasticity have been furnished as a tool for mathematical modeling of microrations wave.
- The mathematical model of the linear (or even nonlinear) micropolar elasticity has been expanded to a field theory model by variational least action integral the least action principle.
- Propagation problem for coupled harmonic waves of translational displacements and microrotations along the axis of a long circular waveguide has been resolved.
- The translational displacements and microrotations in the coupled wave have been decomposed into potential and vortex parts and the coupled differential equations of motion has been reduced to uncoupled ones.
- The wave problem solution has been obtained for a coupled wave of the high-frequency range providing the hyperbolicity of the requisite equations.
- An excluded case of the coupled wave of the low-frequency range has been shown need a further analysis be carried out.

Acknowledgments

The present study is supported by the Federal Agency for Scientific Organizations (State Registration Number AAAA-A17-117021310381-8) and by the Russian Foundation for Basic Research (project No. 18-01-00844).

References

[1] Cosserat E et F 1909 *Theorie des corps deformables* (Paris: Librairie Scientifique A. Hermann et Fils)
[2] Mindlin R D and Tiersten H F 1962 *Archive for Rational Mechanics and Analysis* **11(1)** 415
[3] Toupin R A 1964 *Arc. Rat. Mech. and Anal.* **17(2)** 85
[4] Nowacki W 1986 *Theory of Asymmetric Elasticity* (Oxford: Pergamon Press)
[5] Eringen 1999 *Microcontinuum field theories. Vol. 1. Foundations and Solids* (Berlin; Heidelberg; N.Y.: Springer)
[6] Kovalev V A, Radayev Y N and Semenov D A 2009 *Izvestiya of Saratov University. New Series. Series Mathematics. Mechanics. Informatics* **9(5)** 94 (in Russian)
[7] Kovalev V A and Radayev Y N 2010 *Wave Problems of the Field Theory and Thermomechanics* (Saratov: Saratov University Press) (in Russian)
[8] Kovalev V A and Radayev Y N 2012 *Izvestiya of Saratov University. New Series. Series Mathematics. Mechanics. Informatics* **12(4)** 71 (in Russian)
[9] Murashkin E V and Radayev Y N 2014 *J. Samara State Tech. Univ., Ser. Phys. Math. Sci.* **4(37)** 85
[10] Kovalev V A, Murashkin E V and Radayev Y N 2014 *Izvestiya of Saratov University. New Series. Series Mathematics. Mechanics. Informatics* **14(1)** 78 (in Russian)
[11] Murashkin E V and Radayev Y N 2015 *Mat. Phys. and Mech.* **23** 10
[12] Kovalev V A, Murashkin E V and Radayev Y N 2015 *Bulletin of the Yakovlev Chuvash State Pedagogical University. Series: Mechanics of Limit State* (Cheboksary: ChGPU Publishing) **3** 61 (in Russian)

IOP Conf. Series: Journal of Physics: Conf. Series **991** (2018) 012047 doi:10.1088/1742-6596/991/1/012047

[13] Kovalev V A, Murashkin E V and Radayev Y N 2015 *Izvestiya of Saratov University. New Series. Series Mathematics. Mechanics. Informatics* **15(1)** 79
[14] Murashkin E V and Radaev Yu 2015 *Proceedings of the International conference Topical Problems of Continuum Mechanics, 21-26 September 2015, Tsaghkadzor, Armenia.* (Erevan: National University of Architecture and Construction of Armenia Publishing) 444
[15] Murashkin E V *Proceedings of the XXIV ICTAM, 21-26 August 2016, Montreal, Canada* PO.SM07-1.12.275
[16] Kovalev V A, Murashkin E V and Radayev Y N 2016 *Mechanics: Proceedings of the International School-Conference of Young Scientists, 3-7 October 2016, Tsakhkadzor, Armenia* (Erevan: National university of architecture and construction of Armenia) 160
[17] Murashkin E V and Radaev Yu N 2017 *On Thermodynamics of Wave Processes of Heat Transport, In: Mechanics for Materials and Technologies. Volume 6 of the series Advanced Structured Materials* (Cham: Springer)
[18] Kovalev V, Murashkin E and Radayev Y 2017 *JPCS* **788(1)** 012043
[19] Smith A C 1967 *Int. J Eng. Sc.* **5(10)** 741
[20] Nowacki W 1969 *On the completeness of potentials in micropolar elasticity(Completeness and relations between potentials and stress functions in micropolar elasticity and thermoelasticity)* Arch. Mech. Stos. **21(2)** 107

Numerical simulation of deformation and failure processes of a complex technical object under impact loading

E I Kraus*, **I I Shabalin, and T I Shabalin**

Khristianovich Institute of Theoretical and Applied Mechanics, Siberian Branch of the Russian Academy of Sciences, Novosibirsk, Russia

E-mail: *kraus@itam.nsc.ru

Abstract. The main points of development of numerical tools for simulation of deformation and failure of complex technical objects under nonstationary conditions of extreme loading are presented. The possibility of extending the dynamic method for construction of difference grids to the 3D case is shown. A 3D realization of discrete-continuum approach to the deformation and failure of complex technical objects is carried out. The efficiency of the existing software package for 3D modelling is shown.

1. Introduction

Numerical modeling of the matter behavior under the influence of shock load requires taking into account the laws of elastoplastic flow and, therefore, of stress and strain. Such problems are solved using the equations of mass, momentum, and energy conservation [1–4]. The physics of the processes at high-speed collision of solid bodies can be studied in one-dimensional case, for example, construction of equations describing the state of materials. However, there is a need for both two-dimensional and spatial calculations when modeling the processes of deformation and destruction of complex technical objects that do not fit into one-dimensional schemes. Thus, the two-dimensional numerical tools cannot be used to describe the phenomena of a nutating cylinder rebound off an obstacle in collision at an angle or the impact of a complex technical object, for example, a model reactor at its center line due to its multiply connected structure. Therefore, there is an urgent need for further development of the existing two-dimensional numerical tools [5] to be used in the spatial case.

The software package is designed to solve the time-dependent problems of deformation and fracture of materials at loading of complex technical objects. The package implements a continuous-discrete approach, combining Lagrangian coordinates for the solid material and discrete particles of finite size simulating fragments of the fractured material. The behavior of the materials is described by an elastoplastic model with a few-parameter equation of state. The difference ratios are put on a tetrahedral grid, which is dynamically constructed in complex multiply connected computational domains. All this allows one to simulate the processes of impact loading of complex technical objects to "the end," i.e., till the complete absorption of the kinetic energy of the projectile body and/or splitting of the object into separate fragments.

The software package developed for the spatial modeling includes:

IOP Conf. Series: Journal of Physics: Conf. Series **991** (2018) 012048 doi:10.1088/1742-6596/991/1/012048

- equations of mass, momentum, and energy balance [6, 7];
- the initial data and boundary conditions for all elements of the object [3, 6, 7];
- equations of state and equations of the process for the reactor materials [8–12];
- necessary criteria for the destruction of materials covering a variety of mechanisms of their destruction [2, 3, 5, 13];
- algorithm for replacing cells containing the fractured material by discrete particles of finite size [3, 5, 7];
- extension methods for dynamic construction of the difference grid [5, 14] to the spatial case;
- finite-difference equations for the balance equations on a tetrahedral grid [7];
- symmetric algorithm for calculating the contact boundaries and condition for obtaining stable solutions of the system of difference equations [5, 7].

Let us consider the above steps in more detail.

2. Formulation of the problem

We start from the Lagrangian method for describing the environment as the most appropriate way to describe the interaction of deformable solids [1]. The model consists of equations of motion, mass, momentum, and energy balance, as well as the equations of state and elastoplastic Prandtl–Reuss flow:

equations of the trajectory of motion of material particles

$$\dot{x}_i = u_i; \tag{1}$$

equation of the medium continuity

$$V_0\rho_0 = V\rho; \tag{2}$$

law of change in the material particle momentum

$$\rho\dot{u}_i = \sigma_{ij,j}; \tag{3}$$

change of the internal energy of a particle

$$\rho\dot{e}_i = \sigma_{ij}\dot{\varepsilon}_{ij}; \tag{4}$$

the strain rate tensor has the form

$$\dot{\varepsilon}_{ij} = 0.5(u_{i,j} + u_{j,i}); \tag{5}$$

the stress tensor can be presented in standard form

$$\sigma_{ij} = -\delta_{ij}P + s_{ij}, \tag{6}$$

where s_{ij} is deviator of the stress tensor responsible for the reaction to the shear deformation of the material particles; δ_{ij} is the Kronecker delta; P is a function of pressure in Mie–Grüneisen form.

The process equations are taken in the Prandtl–Reuss form

$$\bar{s}_{ij} + d\lambda' s_{ij} = 2G\varepsilon'_{ij}, \tag{7}$$

where $\dot{\varepsilon}'_{ij} = \dot{\varepsilon}_{ij} - \frac{1}{3}\dot{\varepsilon}_{kk}$ provided the Huber–Mises plasticity

$$s_{ij}s_{ij} \leq \frac{2Y_0^2}{3}, \tag{8}$$

where Y_0^2 is the dynamic yield strength, and the scalar factor $d\lambda'$ is determined by the known procedure of reducing to the yield circle. The above equations use the standard notation: each of the indices i, j takes values of 1 and 2; the repeated indices assume summation; the dot over a symbol is the time-derivative; the index after comma is the derivative with respect to the corresponding coordinate; x_i, u_i are components of the vectors of position and velocity of a particle, respectively; ρ is the current density; G is the shear modulus.

The initial conditions for the ith body at $t = 0$ in the domain $D(x,t)$ often have the form

$$\rho^i(\mathbf{x},0) = \rho^{i0}(\mathbf{x}), \quad u_j^i(\mathbf{x},0) = u_j^{i0}(\mathbf{x}), \quad s_{ij} = P = e = 0, \tag{9}$$

where $\rho^{i0}(\mathbf{x})$, $u_j^{i0}(\mathbf{x})$ are specified initial distributions of the material density and the velocity vector throughout the region $D(\mathbf{x},t)$.

3. Equation of state

We consider a three-term equation of state with the solid-phase free energy determined as

$$F(V,T) = E_x(V) + c_{v,l}T\ln\frac{\theta(V)}{T} - \frac{1}{2}c_{v,e0}T^2\left(\frac{V}{V_0}\right)^{2/3}, \tag{10}$$

where V is the specific volume, $E_x(V)$ is the "cold" energy, T is the temperature, $c_{v,l} = 3R/A$ is the specific heat of the lattice at constant volume, A is the mean atomic weight, R is the gas constant, $\theta(V)$ is the Debye temperature, and $c_{v,e0}$ is the experimental value of the electron heat capacity under standard conditions. The elastic (cold) component of energy $E_x(V)$ is related exclusively to interaction forces between the body atoms and is equal (including the energy of zero vibrations) to the specific internal energy at the absolute zero temperature.

The thermodynamic model of a few-parameter equation of state is based on the dependence of the Grüneisen coefficient on the volume

$$\gamma(V) = \frac{2}{3} - \frac{2}{1 - aV_0/V}, \quad a = 1 + \frac{2}{\gamma_s - 2/3} + \frac{2P_{t,0}}{K_s}, \tag{11}$$

where $\gamma_s = \beta K_s V_0/c_v$, K_s is the adiabatic modulus of volume compression, c_v is the specific heat at constant volume, and $P_{t,0}$ is the thermal pressure in the initial state.

The general expression for the volume dependence of the Grüneisen coefficient has the form

$$\gamma(V) = -\left(\frac{2-t}{3}\right) - \frac{V}{2}\frac{d^2(P_xV^{2t/3})/dV^2}{d(P_xV^{2t/3})/dV}. \tag{12}$$

In equation (12), the situation value corresponds to the Landau–Slater theory [15, 16] at $t = 0$, to the Dugdale–McDonald theory [17] at $t = 1$, and to the free-volume theory [18] at $t = 2$.

To determine the zero isotherm, we equate the expression for the Grüneisen coefficient (11) at zero temperature ($T = 0$K) to the expression for the generalized Grüneisen coefficient (12)

$$\frac{2}{3} - \frac{2}{1 - a_xV_0/V} = -\left(\frac{2-t}{3}\right) - \frac{V}{2}\frac{d^2(P_xV^{2t/3})/dV^2}{d(P_xV^{2t/3})/dV}. \tag{13}$$

Here, a_x is the value of the parameter $a\big|_{T=0}$ at the zero temperature in equation (11), which can be $a_x = a(0) = 1 + 2/(\gamma_s - 2/3)$ as the first approximation.

Differential equation (13) has an analytical solution for "cold" pressure and energy:

$$P_x(V) = C_1 V^{-2t/3} + C_2 H_2(V), \quad E_x(V) = -\left[C_1 \frac{V^{1-2t/3}}{1-2t/3} + C_2 H_1(V)\right] + C_3, \qquad (14)$$

where C_1, C_2, and C_3 are constants of integration and $H_1(V)$, $H_2(V)$ are polynomials computed by the formulas [10]

$$H_1(V) = V^{-2/3}\left(-1.5k_1\xi^4 - 12k_2\xi^3 V + 4.5k_3\xi^2 V^2 - \frac{12k_4\xi V^3}{7} + 0.3k_5 V^4\right),$$
$$H_2(V) = V^{-5/3}(k_1\xi^4 - 4k_2\xi^3 V + 6k_3\xi^2 V^2 - 4k_4\xi V^3 + 3k_5 V^4).$$

The following replacements are made here:

$$\xi = aV_0, \quad k_1 = t_2 t_{12} t_1 t_{72}, \quad k_2 = t_2 t_{12} t_{52} t_{72}, \quad k_3 = t_2 t_1 t_{52} t_{72}, \quad k_4 = t_{12} t_1 t_{52} t_{72},$$
$$k_5 = t_2 t_{12} t_1 t_{52}, \quad t_2 = t + 2, \quad t_{12} = t + \frac{1}{2}, \quad t_1 = t - 1, \quad t_{72} = t + \frac{7}{2}, \quad t_{52} = t - \frac{5}{2}.$$

We use the definition of the Grüneisen coefficient in the Debye approximation $\gamma = -(d\ln\theta/d\ln V)\big|_T$ and equation (11) to obtain the characteristic Debye temperature dependence on the volume:

$$\theta(V) = \theta_0 \left(\frac{a - V/V_0}{a-1}\right)^2 \left(\frac{V_0}{V}\right)^{2/3}$$

where $L_0 = L(V_0)$ is the Debye temperature under the initial conditions.

We also demonstrated there that the set of semi-empirical relations (10)–(13) describes the behavior of thermodynamic properties of solids within 5–10% in a wide range of pressures and temperatures. For the equation of state to be applied, it is sufficient to know only six constants V_0, β, K_t, c_v, L_0, and $c_{v,e0}$ corresponding to the values of these quantities under standard conditions, which can be found in reference books on physical and mechanical properties of substances.

The above-constructed equation of state was used to construct the principal Hugoniot for various materials. Figure 1 shows the principal Hugoniot and zero isotherms of aluminum, copper, and lead calculated by the method proposed by the author. For comparison, the results of experiments of various research groups are shown, whose data are put together in [19], as well as the results of calculation of the "cold" pressure [20] (these calculations are classical in the field of creating the equation of state). As one can see, the differences between the "cold" curves calculated in [20] and the authors' dependence are insignificant up to the compression ratios $V_0/V = 1.5$. At the compression ratios $V_0/V > 1.5$, the differences are more significant (but they do not exceed 7%) due to the method used to construct the model, where the reference state for the construction of equations is the normal state of the condensed matter.

The calculations were carried out for the parameter value $t = 0$ in equation (16), i.e., according to the theory of Landau–Slater. The calculations for the uranium dioxide UO_2 showed that there are no values of the fitting parameter and/or the initial data, which determine the initial course of the principal Hugoniot, that would allow approximating the experimental data from [20] at $t = 0$. For $t > 1$, the approximation is possible, but the derivative $\partial P/\partial\rho$ at the starting point for these experiments is quite small. Figure 2 shows the calculation of the shock Hugoniots at $t = 0$ by the Landau–Slater method and at $t = 0.6$ in equation (16).

Figures 3 and 4 show how the few-parameter equation of state is successfully used to calculate the behavior of ceramic materials (B_4C and Al_2O_3) at high pressures and temperatures.

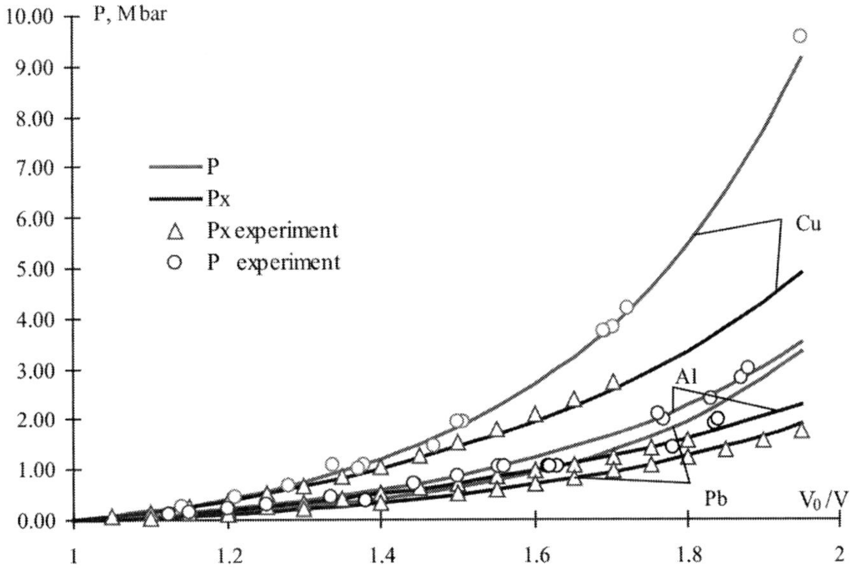

Figure 1. Principal Hugoniot and curves of "cold" compression of aluminum (Al), copper (Cu), and lead (Pb) [19].

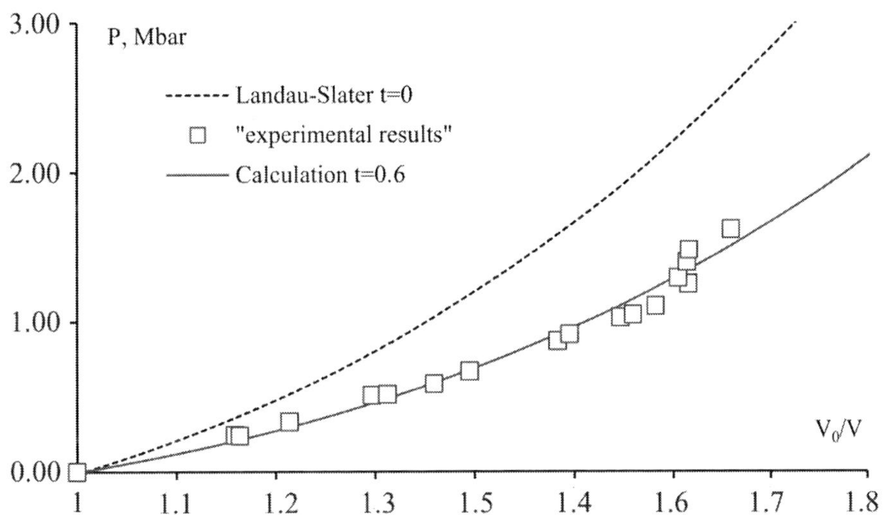

Figure 2. Principal Hugoniot of UO_2 [21].

4. Verification of the tool for high-velocity interaction and damage of solids

A test simulation was made to verify the program. The initial conditions and experiment results [34] are shown in figure 5. The experiments were carried out with tungsten alloy rods of equal length-to-diameter ratio $L/D = 20$ and equal mass $m = 74.2\,\text{g}$ at a nominal impact velocity of 1.615 km/s for the cylindrical rod penetrator with angles $\alpha_1/\alpha_2 = +2.4°/-0.7°$.

An example of conversion computations is shown in figure 6, where a tungsten rod impacts and perforates a steel plate. There is a good general agreement between the computed and

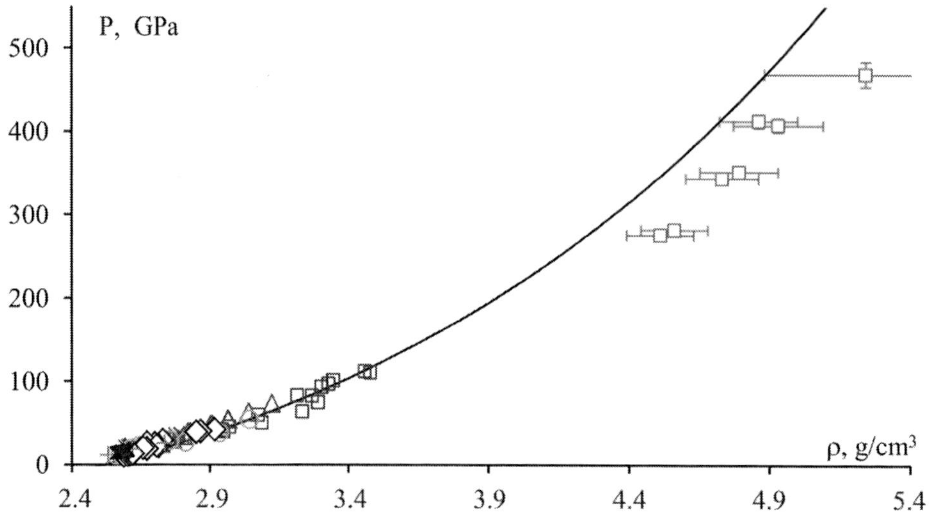

Figure 3. Principal Hugoniot of ceramic B_4C. Experimental data: + – [22], ◇ – [23], ○ – [24, 25], □ – [26], × – [27], △ – [28], × – [29], □ – [30].

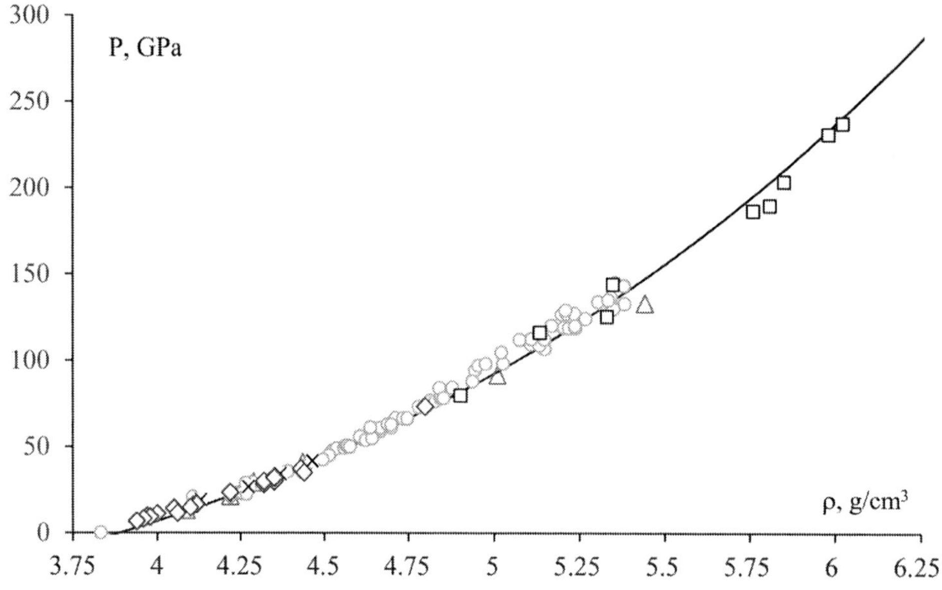

Figure 4. Principal Hugoniot of ceramic Al_2O_3. Experimental data: ○ – [21], △ – [22], □ – [31], × – [32]; ◇ – [33].

experimental results. The nose of the rod is eroded, the fragment distribution is similar (on both sides of the plate), and the hole in the plate is similar.

The failed elements are replaced with discrete particles during the computation. If an element is at the boundary of the computational domain and the parameters of the deformation reach a certain limit value, then the element is replaced by a discrete spherical particle. The radius of the particle is calculated from the condition of lying inside the element. The mass of the element is transferred to the particle. Only one layer of boundary elements can be converted into discrete particles in one time step, so that the velocity of the destruction wave front does

IOP Conf. Series: Journal of Physics: Conf. Series **991** (2018) 012048 doi:10.1088/1742-6596/991/1/012048

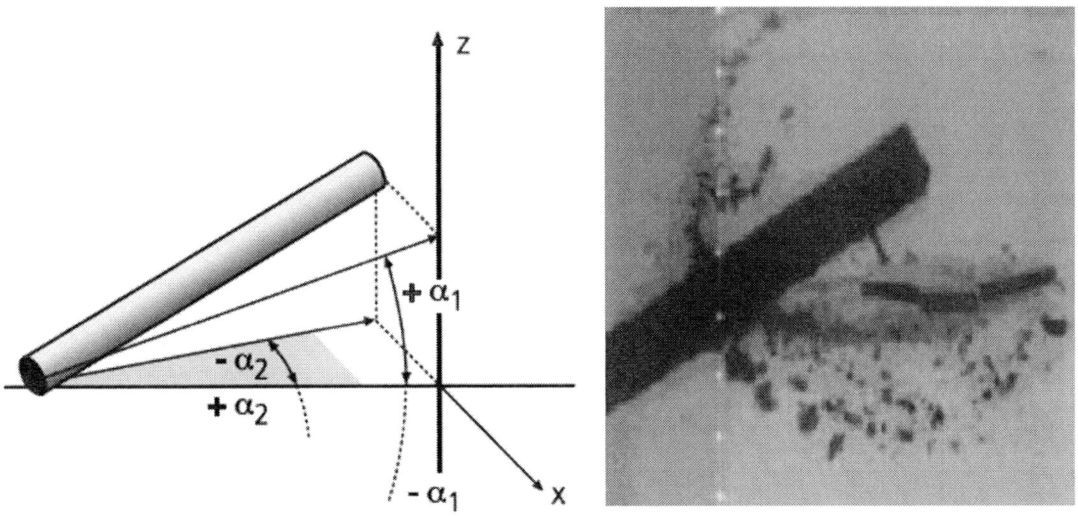

Figure 5. Bending and breaking of a cylindrical rod during perforation of a NATO 60° single-plate target dependent on the pitch and yaw angles α_1/α_2 [34].

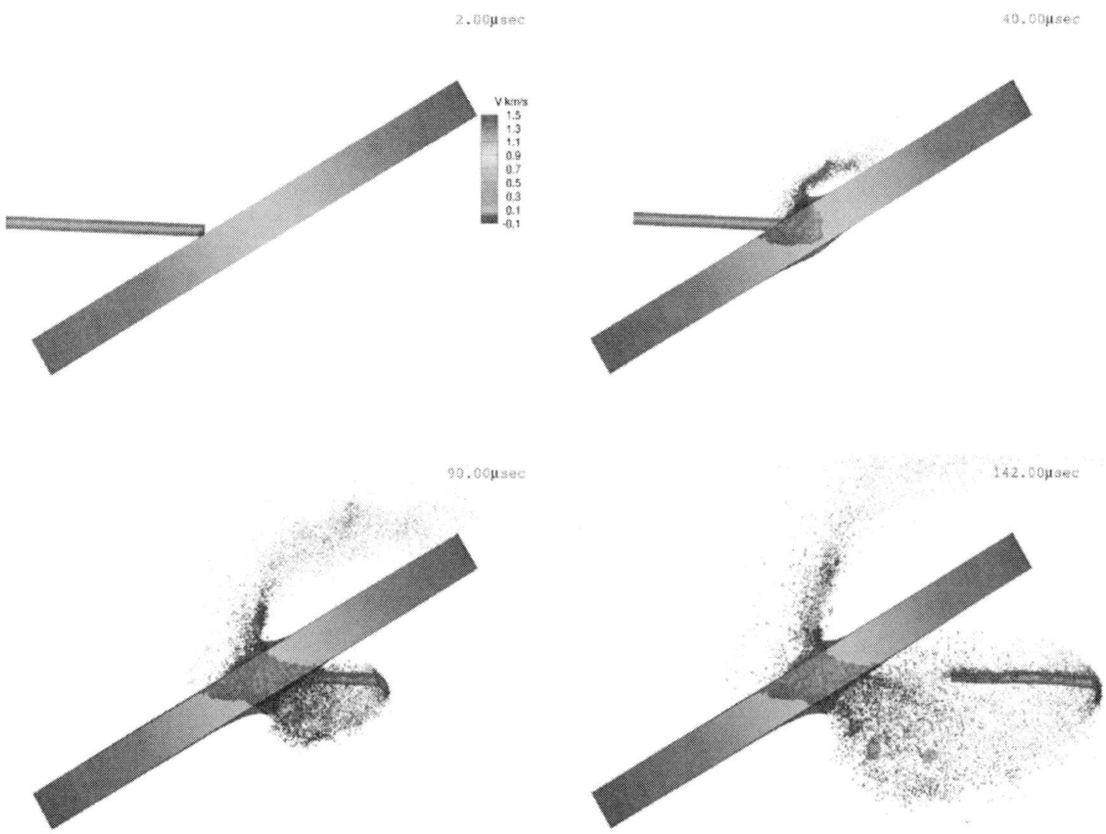

Figure 6. Frame sequence of a tungsten rod impacting and perforating a steel plate.

IOP Conf. Series: Journal of Physics: Conf. Series **991** (2018) 012048 doi:10.1088/1742-6596/991/1/012048

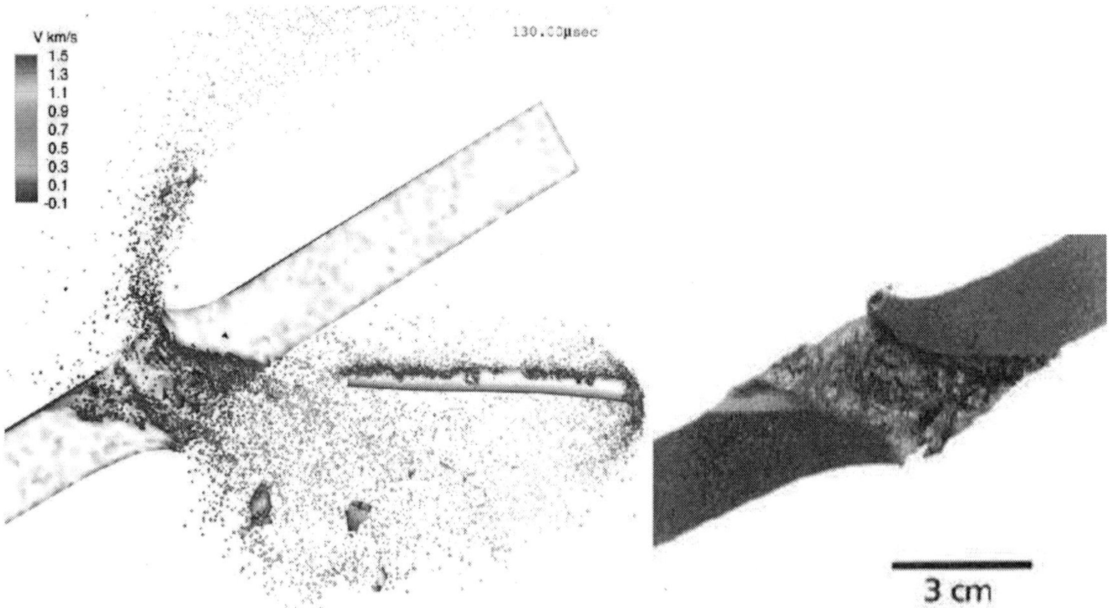

Figure 7. Simulation results (left) and experimental data (right).

not exceed the speed of disturbances in the medium. The conversion of triangle elements to particles is presented [3, 7].

For visualization purposes, the particles from the tungsten alloy rod are shown in red in figure 6, and the particles from the steel plate are shown in blue. When the rod hits the plate at an angle, the deformation process starts with formation of a cylindrical track ahead of the front edge of the rod. A "wave" of displaced plate material is formed ahead of the blunt tip of the rod, which then splashes out when the tip plunges into the target (figure 6, 40 μs time).

At the same time, the stresses increase and spread as a wave from the contact surface and transfer the momentum to the target material. When the compression wave hits the backside of the plate, the unloading wave is formed and the speed of the plate material increases in the normal direction. This normalizes the motion of the impacting rod and bends it. This effect is weak when the plate is thin and can only be seen near the tip of the rod in figure 7. The rod makes a through hole in the plate and forms a cloud of fragments. The simulation results are compared with the experimental data in figure 7 and a good qualitative and quantitative agreement is shown.

5. Examples of application of the tool for high-velocity interaction and damage of solids

To illustrate the efficiency of the numerical 3D tools, we consider the interaction of a model reactor of thermionic power plant with 12 regulating cylinders in the side reflector of neutrons and concentric rows of power-generating channels in the moderator of zirconium hydride. The cylindrical system is formed by fuel elements with integrated thermionic converter of thermal energy of nuclear reaction to electric energy. A great many contact surfaces and various materials show that the software package allows one to solve the problem "completely," i.e., till the complete failure and shutdown of the model reactor fragments. Figure 8 is an illustration of the geometric model of the reactor.

For a heavy tungsten rod, it is easy to penetrate a light and brittle beryllium cap (figure 9,

IOP Conf. Series: Journal of Physics: Conf. Series **991** (2018) 012048 doi:10.1088/1742-6596/991/1/012048

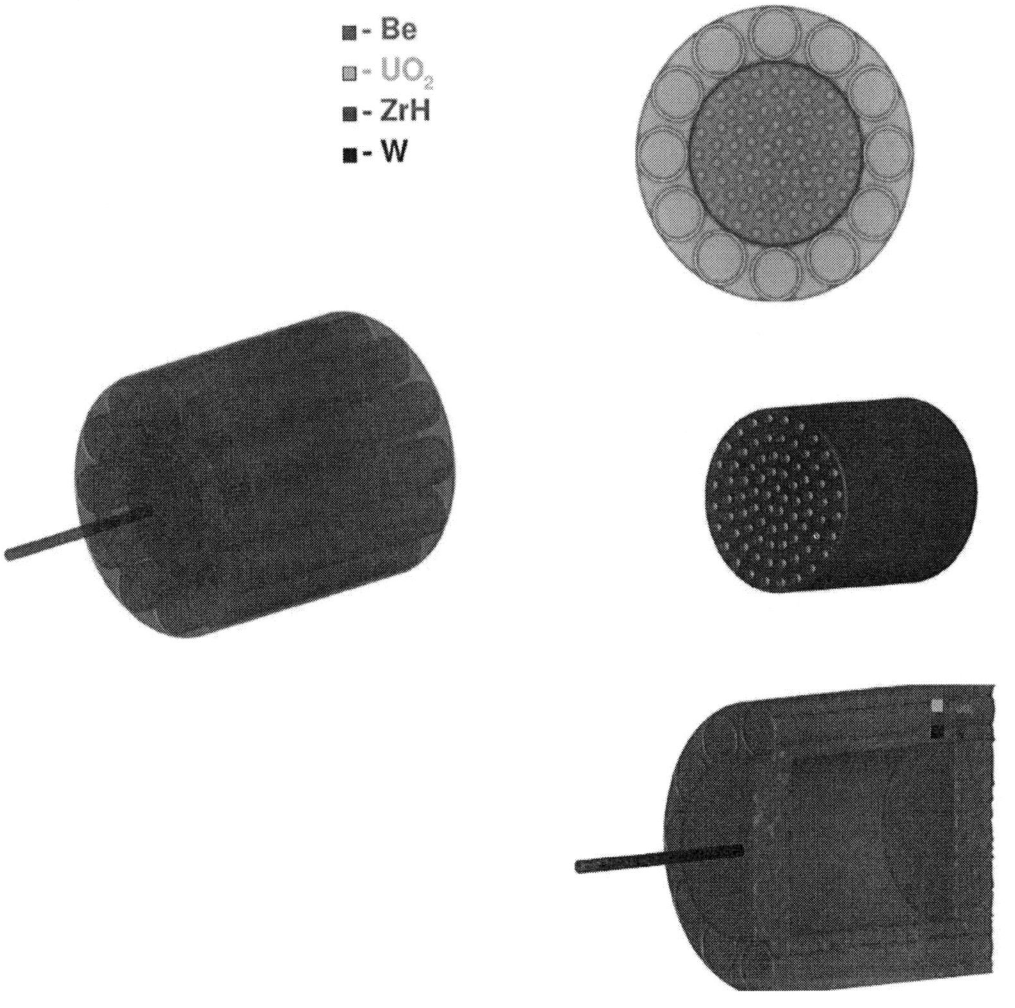

Figure 8. 3D Geometry construction presented by a set of elementary and multiply connected bodies.

25 μs). The rod has gone through the cap and its wearing out has barely started. However, when it contacts with heavier cylindrical systems, it quickly loses mass and its length is reduced to a third of its initial value (figure 9, 65 μs time). There are many contacts between different parts of the reactor, which give rise to a complicated pattern of wave interaction consisting of loading and unloading zones. Both the materials of zirconium hydride and homogenized cylindrical systems have low strength which makes the failed material zone spread in the direction of the rod motion, which can be seen in figure 9 (from 25 μs to 100 μs). It should be noted that the impact of the rod on the reactor leads to the catastrophic outcome for the latter.

Conclusions

It is shown that the existing software package REACTOR3D has evolved and is suitable to solve the impact problems in the 3D case. A method for dynamic construction of a tetrahedral mesh for complex, multiply connected objects is presented. An algorithm for modeling a material fractured by discrete particles of finite size is exposed. The key moments of interaction of the fractured material fragments with the boundaries of a solid material and with each other

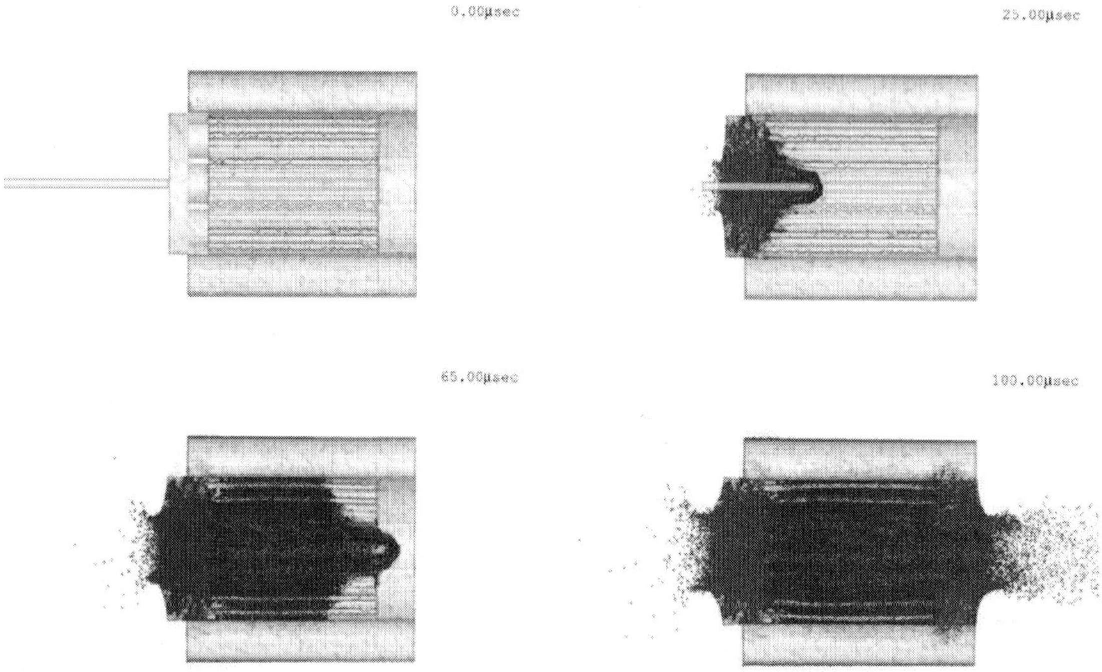

Figure 9. Frame sequence of a tungsten rod impacting a model reactor.

are shown. Examples of the efficiency of the computational modules for both regions of solid material and numerous fragments of the fractured material are given.

Acknowledgments

The work was supported by the Russian Science Foundation, grant No. 16-19-10300.

References

[1] Radchenko A, Radchenko P, and Batuev S 2015 Numerical study of the influence of anisotropy of physical-mechanical properties on the impact fracture of orthotropic composites *Russ. Phys. J.* **58** (3) 319–29
[2] Buzyurkin A, Gladky I, and Kraus E 2015 Determination and verification of Johnson-Cook model parameters at high-speed deformation of titanium alloys *Aerosp. Sci. Technol.* **45** 121–7
[3] Kraus E and Shabalin I 2013 Impact loading of a space nuclear powerplant *Frattura ed Integrità Strutturale* **24** 138–60
[4] Radchenko P, Batuev S, Radchenko A, and Tukaev A 2017 Simulation of interaction of damage agents of different shape with shaped-charge munition *AIP Conf. Proc.* **1800** 040001
[5] Kraus E and Shabalin I 2016 Reactor2D: A tool for simulation of shock deformation *AIP Conf. Proc.* **1770** 030092
[6] Wilkins M L 1999 *Computer simulation of dynamic phenomena* (Berlin - Heidelberg: Springer)
[7] Fomin V, Gulidov A, and Sapozhnikov G 1999 *High-Velocity Solids Interaction* (Novosibirsk: SO RAN)
[8] Fomin V, Kraus E, and Shabalin I 2004 An equation of state for condensed matter behind intense shockwaves *Mater. Phys. Mech.* **7** (1) 23–28
[9] Kraus E, Fomin V, and Shabalin I 2014 Determination of the shear module behind strong shock wave *Vestnik Yuzhno-Ural. Gos. Univ. Ser. Mat. Mod. Programm.* **7** (1) 49–61
[10] Kraus E I and Shabalin I I 2016 A few-parameter equation of state of the condensed matter *J. Phys.: Conf. Ser.* **774** (1) 012009
[11] Fomin V, Kraus E, and Shabalin I 2010 Theoretical and experimental investigation of shock wave stressing of metal powders by an explosion *EPJ Web Conf.* **10** 00027

[12] Kraus E and Shabalin I 2015 Calculation of elastic modules behind strong shock wave *J. Phys.: Conf. Ser.* **653** 012085
[13] Buzyurkin A, Kraus E, and Gladky I 2015 Determination of parameters of the Johnson–Cook model for the description of deformation and fracture of titanium alloys *J. Appl. Mech. Tech. Phys.* **56** (2) 330–6
[14] Shimada K and Gossard D C 1995 Bubble mesh: automated triangular meshing of non-manifold geometry by sphere packing In *Proc. of the third ACM Symp. on Solid Modeling and Applications* ed C Hoffmann and J Rossignac (New York: ACM) pp 409–19
[15] Slater I 1935 *Introduction in the Chemical Physics* (New York - London: McGraw Book Company Inc.)
[16] Landau L and Stanyukovich K 1945 On study of a detonation of condensed explosive substances *Dokl. Akad. Nauk SSSR* **46** (9) 399–402
[17] Dugdale J and McDonald D 1953 The thermal expansion of solids *Phys. Rev.* **89** (4) 832–4
[18] Vaschenko V and Zubarev V 1963 Concerning the Gruneisen constant *Sov. Phys. Solid State* **5** (3) 653–5
[19] Trunin R F, Gudarenko L F, Zhernokletov M V, and Simakov G V 2001 *Experimental Data on Shock Compression and Adiabatic Expansion of Condensed Matter* (Sarov: RFYaTs-VNIEEF)
[20] Al'tshuler L V, Kormer S B, Brazhnik M I, et al 1960 The isentropic compressibility of aluminum, copper, lead, and iron at high pressures *J. Exp. Theor. Phys.* **11** (4) 766–75
[21] Marsh S (Editor) 1979 *LASL Shock Hugoniot Data* (Berkeley: Univ. California Press)
[22] Pavlovskii M N 1971 *Sov. Phys. Solid State* **12** 1736–7
[23] Wilkins M L 1968 *Third Progress Report of Light Armor Program. Los Alamos National Laboratory Report UCRL-50460* (LLNL, University of California, CA)
[24] Grady D E 2015 Hugoniot equation of state and dynamic strength of boron carbide *J. Appl. Phys.* **117** (16) 165904
[25] Grady D E 1994 Shock-wave strength properties of boron carbide and silicon carbide *J. Phys. IV* **4** (C8) 385–91
[26] McQueen R, Marsh S, Taylor J, et al. 1970 The equation of state of solids from shock wave studies In *High-Velocity Wave Phenomena* (New York: Academic Press) pp 293–417
[27] Zhang Y, Mashimo T, Uemura Y, et al. 2006 Shock compression behaviors of boron carbide (B_4C) *J. Appl. Phys.* **100** (11) 113536
[28] Vogler T, Reinhart W, and Chhabildas L 2004 Dynamic behavior of boron carbide *J. Appl. Phys.* **95** (8) 4173–83
[29] Gust W and Royce E 1971 Dynamic yield strengths of B_4C, BeO, and Al_2O_3 Ceramics *J. Appl. Phys.* **42** (1) 276–95
[30] Fratanduono D E, Celliers P M, Braunet D G, et al. 2016 Equation of state, adiabatic sound speed, and Grüneisen coefficient of boron carbide along the principal Hugoniot to 700 GPa *Phys. Rev. B.* **94** (18) 184107
[31] Erskine D 1994 High pressure Hugoniot of sapphire *AIP Conf. Proc.* **309** (1) 141–3
[32] Grady D E 1995 *Dynamic Properties of Ceramic Materials. Sendia National Laboratories Report SAND 94-3266, Albuquerque, NM* (Albuquerque: Sendia Natl. Lab.)
[33] Mashimo T, Hanaoka Y, and Nagayama K 1988 Elastoplastic properties under shock compression of Al_2O_3 single crystal and polycrystal *J. Appl. Phys.* **63** (2) 327–36
[34] Heider N, Weber K, and Weidemaier P 2004 Experimental and numerical simulation analysis of the impact process of structured KE penetrators onto semi-infinite and oblique targets *Proc. of the 21st Int. Symp. on Ballistics, Adelaide, South Australia* 97–103

Healing of damaged metal by a pulsed high-energy electromagnetic field

K V Kukudzhanov[1,2*] and A L Levitin[1]**
[1]Ishlinsky Institute for Problems in Mechanics of the Russian Academy of Sciences, Moscow, Russia
[2]Moscow Institute of Physics and Technology, Dolgoprudny, Moscow Region, Russia
E-mail: *kconstantin@mail.ru, **alex_lev@ipmnet.ru

Abstract. The processes of defect (intergranular micro-cracks) transformation are investigated for metal samples in a high-energy short-pulsed electromagnetic field. This investigation is based on a numerical coupled model of the impact of high-energy electromagnetic field on the pre-damaged thermal elastic-plastic material with defects. The model takes into account the melting and evaporation of the metal and the dependence of its physical and mechanical properties on the temperature. The system of equations is solved numerically by finite element method with an adaptive mesh using the arbitrary Euler–Lagrange method. The calculations show that the welding of the crack and the healing of micro-defects under treatment by short pulses of the current takes place. For the macroscopic description of the healing process, the healing and damage parameters of the material are introduced. The healing of micro-cracks improves the material healing parameter and reduces its damage. The micro-crack shapes practically do not affect the time-dependence of the healing and damage under the treatment by the current pulses. These changes are affected only by the value of the initial damage of the material and the initial length of the micro-crack. The time-dependence of the healing and the damage is practically the same for all different shapes of micro-defects, provided that the initial lengths of micro-cracks and the initial damages are the same for these different shapes of defects.

1. Introduction

The micro-defects of linear dimensions of order $10\,\mu$m present the most widely spread dimension of defects in polycrystalline metals. They are mostly formed near the surfaces of the neighboring monocrystals (grains). A number of such defects are always formed between the grains after casting, and they can also arise and develop in metals in the process of their plastic deformation. From the standpoint of the further macro-fracture of a piece of metal, the most dangerous defects are the plane micro-cracks.

The hypothesis that the defects in conducting materials can be healed (changed) under the action of short pulses of a high-energy electromagnetic field (HEMF) was suggested by several authors [1–5] who tried to explain the phenomenon of electroplasticity of metals.

The hypotheses of the defect healing in the material can currently be treated as an experimentally observed phenomenon [6–10].

In this cases, the experimental data testify that not only the compressing stresses arise with crater (pore) melting at the defect tips, but the defect shape itself is changed and the continuous structure of the material is reconstructed, which is accompanied by variations in the bulk content of micro-defects (up to the complete disappearance of some micro-defects). This results in a

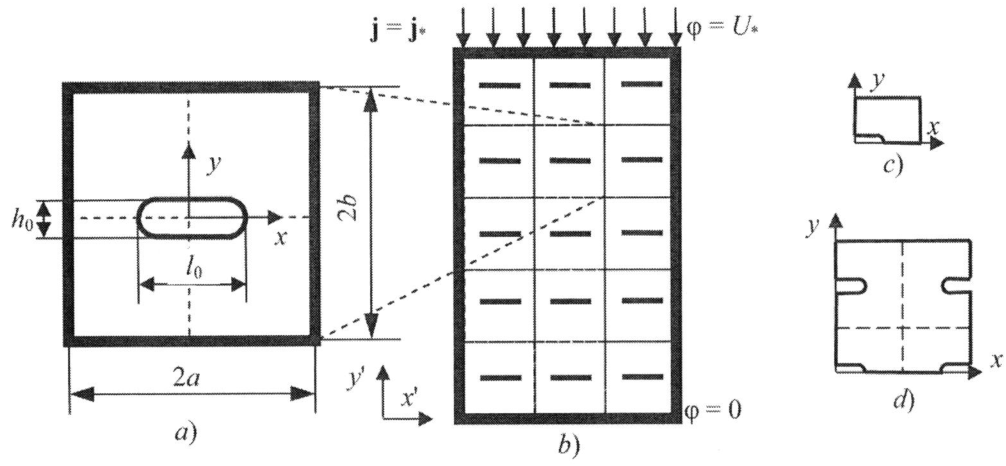

Figure 1. (a) Representative cell with micro-defect $h_0 \ll l_0$; (b) the ordered structure of micro-defects, the arrows show the direction of the electric current density vector at the sample boundary; (c–d) domains of integration used in the calculations.

decrease in the material damage and in an increase in the limit plastic strain till the fracture. But the mathematical models [3–5, 11, 12], which have been proposed until recently, cannot explain these experimental facts. Naturally, under such conditions, the problems of damage variations and, moreover, of the laws of these variations under the HEMF action on the material remained unstudied.

To study the physical processes which occur near micro-cracks under the action of intensive current pulses and to explain the healing effect, a model of the pulse HEMF action on a preliminary damaged material with defects was proposed in [13]. This model allowed one mathematically to describe the experimentally observed process of micro-defect transportation and the decrease in the damage of the metal. The problems of choosing the domain of integration and revealing the influence of the distance between the cracks and their mutual location on the processes of deformation and micro-defect healing were investigated in [14].

In the present paper, we study the influence of the micro-defect shape on the physical fields near them and on the processes of healing of micro-defects. We consider how the healing and damage of material vary in time under the action of a pulse high-energy electromagnetic field.

2. Electro-thermo-mechanical model

We consider the model of electromagnetic field action on a pre-damaged material with defects [13, 14] and use the finite element method for calculations.

We consider a damaged electro-conductive material containing uniform (in shape, size, and orientation) defects like plane thin ($h_0 \ll l_0$) micro-cracks with rounded tips (figure 1a). We assume that the defects in the material are arranged at the nodes of a rectangular lattice (figure 1b). For this arrangement, we can easily identify the representative volume cell (periodicity cells) as is shown in figure 1a.

The material is subjected to the effect of a short-pulse high-energy electromagnetic field by applying a potential difference to the outer boundaries of the sample, which, on these boundaries, induces an electric current with the density vector perpendicular to the xz-plane of micro-cracks (figure 1b). We consider high electromagnetic fields inducing an electric current with density varying from 10^8 to 10^{11} A/m^2 and duration 10^{-5}–10^{-4} s in the sample.

The solution is sought in the domains of integration (figure 1c–d) containing from one to four micro-cracks or from one to four parts of representative cells. This choice of the domains of integration is stipulated by the following results. In [14], we showed that the use of other domains of integration (and with more representative cells) does not increase the accuracy of the solution near the micro-crack (centered at $x = 0$, $y = 0$) and does not affect the healing processes.

During the considered treatment, the following physical processes occur in the material: electromagnetic, mechanical, and thermal. The characteristic time of each of these processes is approximately inversely proportional to the speed of propagation of the respective perturbations. Therefore, the time required for steady-state electromagnetic and mechanical processes considered in the present paper amounts to 10^{-13}–10^{-12} s and 10^{-8}–10^{-7} s, respectively, which is substantially smaller than the time of external action of the source of the electric current (10^{-5}–10^{-4} s). Therefore, the electric potential φ in the conducting material is obtained using the charge conservation law (assuming that the current in the sample is steady-state) written in the variation statement [15]. In this case, Ohm's law is considered to hold for the conductor. The displacement field $\mathbf{u}$ is determined from the equilibrium equations written in the form of the virtual work principle [15].

Strains are assumed to be finite. The rates of the elastic, plastic, and temperature strains are assumed to be additive. We also assume that Hooke's law for an isotropic body and the associated flow rule with the von Mises plasticity condition hold for the rates of elastic and plastic strains. The relationship between the rates of temperature strains and the time-derivative of the temperature is assumed to be linear.

The temperature field T is determined from the energy conservation law. The time of electromagnetic action on the material is small (10^{-5}–10^{-4} s) and the temperature gradients near the crack tips are very high (10^{7}°C/m) [16], and we ignore the heat conduction and consider the process to be adiabatic.

The complete system of equations for the displacement vector $\mathbf{u}$, the electric potential φ, and the temperature T in the considered electro-thermo-mechanical model has the form [13]

$$\int_V \nabla\,\delta\varphi\,\sigma^{\mathrm{E}}(T)\,\nabla\,\varphi\,dV = \int_S \delta\varphi\,j\,dS, \qquad \mathbf{j} = \sigma^{\mathrm{E}}\,\mathbf{E} = -\sigma^{\mathrm{E}}\frac{\partial\varphi}{\partial\mathbf{x}}, \qquad j = -\mathbf{j}\cdot\mathbf{n}, \tag{1}$$

$$\int_V \boldsymbol{\sigma}:\boldsymbol{\varepsilon}\,dV = \int_S \mathbf{t}\cdot\delta\mathbf{u}\,dS + \int_V \mathbf{f}\cdot\delta\mathbf{u}\,dV, \qquad \dot{\boldsymbol{\varepsilon}} = \dot{\boldsymbol{\varepsilon}}^{\mathrm{el}} + \dot{\boldsymbol{\varepsilon}}^{\mathrm{pl}} + \dot{\boldsymbol{\varepsilon}}^{\mathrm{th}}, \tag{2}$$

$$\dot{\boldsymbol{\sigma}} = \lambda(T)\,\dot{\boldsymbol{\varepsilon}}^{\mathrm{el}}:\mathbf{I} + 2\,\mu(T)\,\dot{\boldsymbol{\varepsilon}}^{\mathrm{el}}, \ \dot{\boldsymbol{\varepsilon}}^{\mathrm{pl}} = \dot{\Lambda}\frac{\partial\Phi}{\partial\boldsymbol{\sigma}} = \dot{\Lambda}\,\mathbf{s}, \ \bar{\sigma} = \sigma_{\mathrm{Y}}(T), \ \bar{\sigma} = \sqrt{\tfrac{3}{2}\mathbf{s}:\mathbf{s}}, \ d\boldsymbol{\varepsilon}^{\mathrm{th}} = \alpha(T)\,\mathbf{I}\,dT, \tag{3}$$

$$\rho(T)\,c(T)\,\dot{T} = r^{\mathrm{E}} + r^{\mathrm{pl}} + r^{\mathrm{melt}} + r^{\mathrm{eval}}, \quad r^{\mathrm{E}} = \eta^{\mathrm{E}}\,\mathbf{j}\cdot\mathbf{E} = \eta^{\mathrm{E}}\,\nabla\varphi\cdot\sigma^{\mathrm{E}}\cdot\nabla\varphi, \quad r^{\mathrm{pl}} = \eta^{\mathrm{pl}}\,\boldsymbol{\sigma}:\dot{\boldsymbol{\varepsilon}}^{\mathrm{pl}}, \tag{4}$$

$$T = T_{\mathrm{melt}}, \qquad t_{\mathrm{sol}} \le t \le t_{\mathrm{liq}}, \qquad \int_{t_{\mathrm{sol}}}^{t_{\mathrm{liq}}} \left(r^{\mathrm{E}} + r^{\mathrm{pl}}\right) dt = \rho\,\Lambda_{\mathrm{melt}}, \tag{5}$$

$$T = T_{\mathrm{evap}}, \qquad t_{\mathrm{eliq}} \le t \le t_{\mathrm{evap}}, \qquad \int_{t_{\mathrm{eliq}}}^{t_{\mathrm{evap}}} \left(r^{\mathrm{E}} + r^{\mathrm{pl}}\right) dt = \rho\,\Lambda_{\mathrm{evap}}, \tag{6}$$

where V is a arbitrary volume bounded by the piecewise-smooth surface S, $\mathbf{n}$ is the outer normal to S, T is the temperature, $\mathbf{E} = -\partial\varphi/\partial\mathbf{x}$ is the vector of the electric field intensity determined as the negative gradient of the electric potential, $\sigma^{\mathrm{E}}(T)$ is the electrical conductivity, $j = -\mathbf{j}\cdot\mathbf{n}$ is the current density along the normal to the surface S, $\delta\varphi$ is the variation in the electric potential satisfying the boundary conditions, $\mathbf{u}$ is the displacement vector, $\boldsymbol{\sigma}$ is the stress tensor, $\boldsymbol{\varepsilon}$ is the total strain tensor, $\mathbf{f}$ is the body force vector, $\mathbf{t}$ is the surface force vector, $\delta\mathbf{u}$ and $\delta\boldsymbol{\varepsilon}$ are the variations in displacements and their associated total strains, $\dot{\boldsymbol{\varepsilon}}^{\mathrm{el}}$, $\dot{\boldsymbol{\varepsilon}}^{\mathrm{pl}}$ and $\dot{\boldsymbol{\varepsilon}}^{\mathrm{th}}$ are tensors of the rates of elastic, plastic, and temperature strains, respectively, $\lambda(T)$ and $\mu(T)$ are Lamé's moduli of elasticity, $\sigma_{\mathrm{Y}}(T)$ is the yield strength, $\mathbf{s} = \boldsymbol{\sigma} - \frac{1}{3}\boldsymbol{\sigma}:\mathbf{I}$ is the stress tensor deviator, Φ is the

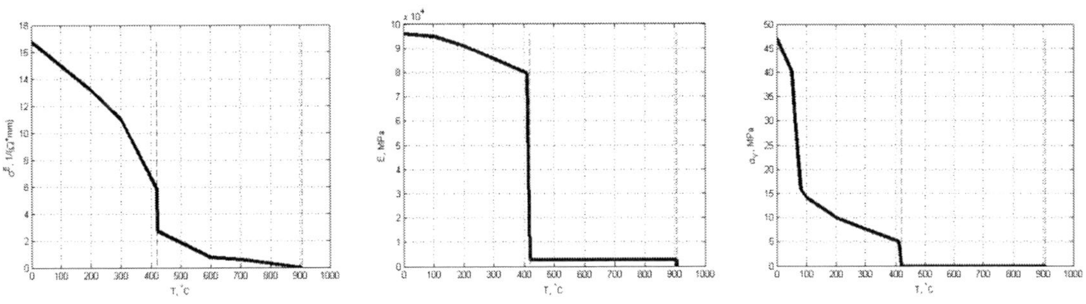

a) *b*) *c*)

Figure 2. Dependence of the electrical conductivity σ^{E} (a), Young's modulus E (b), and the yield stress σ_{Y} (c) on the temperature T (vertical dashed lines are the melting point $T_{\mathrm{melt}} = 419°\mathrm{C}$ and the evaporation point $T_{\mathrm{evap}} = 906°\mathrm{C}$) for a zinc monocrystal.

yield function, $\mathbf{I}$ is the unit tensor, $\alpha(T)$ is the coefficient of thermal expansion, $\rho(T)$ is the density, $c(T)$ is the specific heat capacity, $\dot{T}$ is the material derivative of the temperature, r^{k} is the heat released per volume unit in the current configuration of the body per unit time (r^{E} is the heat released due to the electric current flow, r^{pl} is the heat released during the plastic deformation, r^{melt} is the heat absorbed in the process of melting, and r^{evap} is the heat absorbed in the process of evaporation), η^{E} is an empirical coefficient defined as the fraction of power of electric current per volume unit dissipated as heat, η^{pl} is an empirical coefficient determined as the fraction of plastic power per volume unit dissipated as heat, T_{melt} is the temperature of the material melting, t_{sol} is the time at which the material begins to melt, t_{liq} is the time at which the material is totally melted, Λ_{melt} is the latent heat of melting, T_{evap} is the temperature of the material evaporation, t_{eliq} is the time at which the material begins to evaporate, t_{evap} is the time at which the material is totally evaporated, and Λ_{evap} is the latent heat of evaporation.

The evolution equation for the temperature takes into account the heat generated per unit volume in the current configuration of the body per unit time due to the electric current (Joule heating), the heat generated during plastic deformation, and the latent heat absorbed in the processes of melting and evaporation. To calculate the temperature field, we supplement the evolution equation for the temperature with additional equations which are the conditions of latent heat absorption during the matter transition from one aggregate state into another (during melting and evaporation).

The temperature in these processes varies from room temperature to the material evaporation temperature [16]. Therefore, all physical and mechanical properties of the material in the proposed model (density, specific heat, electrical conductivity, thermal expansion coefficient, elastic moduli, yield stress, etc.) depend on the temperature (until the evaporation temperature is attained).

At the points where the material is melted, its physical properties are changed sharply, including the conductivity, specific heat, density, linear thermal expansion coefficient, and all other mechanical characteristics of the material. Such a change in the properties of the material is consistent with the available experimental data [17–19]. Figure 2 shows the temperature-dependence of some of these characteristics for zinc.

Thus, after attaining the melting point, the material does not lose its ability to conduct the electric current (figure 2*a*), and the further heating of the melt occurs. The further fall in the elastic moduli and the yield stress vanishing (figure 2*b*, *c*) allow us to describe the melted material behavior at all subsequent moments by constitutive relations which degenerate into thermo-viscoelastic ones with a nonlinear viscosity.

Figure 3. Shapes of microdefects: (*a*) horizontal microcracks with rounded tips; (*b*) narrow microcracks with rounded tips; (*c*) elliptical microcracks; (*d*) dumbbell-shaped microdefects.

At the points where the material has evaporated entirely, at all subsequent moments, it is assumed that the current density $\mathbf{j} = \mathbf{0}$, the stress tensor $\boldsymbol{\sigma} = \mathbf{0}$, and the temperature is constant $T = T_{\text{evap}}$. Thus, in the framework of the proposed model, when the material attains the evaporation temperature, it loses its ability to conduct electric current, and it is not heated further. The metal loses the properties of a viscous fluid and is considered as a rarefied gas.

Because of the symmetry of representative cells, the used integration domains consisting of half or quarters of representative cells (figure 1*c*, *d*). On the horizontal and vertical boundaries of the integration domains, we posed the conditions that the normal component of the displacement vector and the tangential components of the stress tensor were absent and that the potential was unperturbed by the presence of a defect. The crack did not conduct a current, so the normal derivative of the potential at its boundary was assumed to be zero. The crack surface was assumed to be free of stresses. The initial fields of temperature, displacements and electric potential were assumed to be homogeneous ($T_0 = 20°\text{C}$, $\mathbf{u}_0 = \mathbf{0}$, $\varphi_0 = 0$).

In [13, 14], it was in particular shown that, under the action considered in this model, the volume of micro-defects decreases with time. Following the first works of the theory of damaged media [20], we associate the damage with the formation of voids in the material (porosity) due to the formation and development of microcracks and microvoids in the process of loading. We introduce the notation: $V(t)$ is the volume of the unit micro-defect, $V_0 = V(0)$ is the initial volume of the defect, and V_{re} is the volume of the representation cell containing such a micro-defect (figure 1). We define the damage $f = f(t)$ and healing $\chi = \chi(t)$ parameters as follows:

$$f(t) = \frac{V(t)}{V_{\text{re}}}, \qquad \chi(t) = \frac{V(0) - V(t)}{V(0)}. \tag{7}$$

The initial damage of the material at time $t = 0$ is equal to $f_0 = f(0) = V_0/V_{\text{re}}$. Thus, the damage of the material decreases and the healing increases due to the action of the electromagnetic field.

3. Results of numerical simulation

The coupled equations of the model are solved numerically together with boundary, contact, and initial conditions. The computations were performed for the plane strain using liner four-node isoparametric and three-node finite elements. The mesh was reconstructed on the basis of the arbitrary Euler–Lagrange method. The temperature is not a degree of freedom of the problem. The temperature evolution equation was solved directly at each point of integration using the temperature derivative approximation by the inverse difference in time.

The modeling was performed for samples manufactured of zinc, it physical and mechanical properties were temperature dependent [17–19].

The dimensions (a and b) of the representative cells varied from 7.5 to 37.5 μm due to the spread of the grain dimensions of the polycrystalline zinc.

The considered forms of microcracks are shown in figure 1*a*–*c*. To complete the picture, a dumbbell-shaped microdefect (figure 3*d*) was also considered (figure 1*d*). All defects had the same initial length (the size on the axis x) $l_0 = 10\,\mu$m and the same area (in the plane xy). Thus, the initial damage f_0 for all forms of microdefects was the same. In the calculations, it varied in the range $f_0 = 0.27$–2.45%. It seems that, for large initial damage, it is already impossible to treat the material as undamaged in the sense of macro- (meso-) fracture.

The potential difference (per unit length) used in the computations was $534.3\,\mathrm{mV/mm}$, which corresponds to the current of density $8.95\,\mathrm{kA/mm^2}$ in materials without defects. The potential difference was assumed to be constant during the entire time of the pulse action $\tau_0 = 90\,\mu\mathrm{s}$.

In [13, 14], it was shown that when the electric current flows through a sample with micro-cracks, there arise large gradients of the electric field near the cracks and this leads to a significant increase in the current density near the crack tips compared with the density of the current applied to the sample. This results in fast inhomogeneous local heating near the micro-crack tip accompanied by its thermal expansion (at the same time, there is no heating near the crack center and on its shores). The inhomogeneous thermal expansion generates large compression stresses (the pressure can exceed 100 MPa) near the crack and, as a consequence, results in simultaneous healing of the crack (narrowing the crack) thus decreasing the crack length and releasing the melted metal into the crack (figures 4 and 5).

The computations showed that, in the whole range of distances between the defects (or equivalently, for any initial damage), the healing $\chi(t)$ and the damage $f(t)$ vary identically in time under the HEMF action. Their variation is independent of whether the domains of integration, where these functions are calculated, consist of one (figure 1*c*) or four (figure 1*d*) representative cells. The maximal deviation of the obtained curves $\chi(t)$ and $f(t)$ for the domains of integration with one, four, and six representative cells was equal to 0.09%.

Thus, when modeling the process of micro-crack healing, one can, without loss of generality, consider only one quarter of the representative cell (figure 1*c*) as the domain of integration with the above-described boundary conditions posed on its boundaries.

Further, we study how the micro-defect shape affects the variations in the healing $\chi(t)$ and damage $f(t)$ of the material under the action of the current. Figure 6 shows the dependence of the healing $\chi(t)$ and damage $f(t)$ on the time t (μs) for squared representative cells of different dimensions with micro-defect shaped as horizontal, narrowing, and elliptic micro-cracks and as a dumbbell (the corresponding shapes are shown in figure 3*a*–*c* and figure 3*d*).

Figure 6 shows that, in the entire considered range of distances between the micro-cracks (initial damage), the curves of variation in the healing $\chi(t)$ and damage $f(t)$ practically coincide with each other for the same initial damage f_0. This effect ($\chi(t)$ and $f(t)$ are independent of the micro-defect shape) also takes place even for micro-defects of a cardinally different shape even for the dumbbell-shaped micro-defect.

The change of the representative cell shape from square to rectangle for the same f_0 does not practically change the dependence $\chi(t)$ and $f(t)$ shown in figure 6. In this case, the variations in the length and width of the representative cells did not exceed 25% of the dimensions shown in figure 6.

Now we consider the physical fields generated by the action of the current near the micro-defects.

Figure 4 presents the temperature fields induced by the current flow in the representative cell of size $a = b = 10\,\mu\mathrm{m}$ with defects of different shapes at time $t = 25.4\,\mu\mathrm{s}$. This figure shows that the temperature field is practically independent of the defect shape in the whole domain of integration except for the extreme points of the micro-defects on the axis x (which we call "micro-defect tips" for brevity). The temperature fields (figure 4) induce similar states in the entire domain of integration (except for the neighborhoods of the micro-defect tips).

Figure 7 illustrate the results of computation of the displacement fields u_y in the representative cell of size $a = b = 10\,\mu\mathrm{m}$ with defects of different shapes at time $t = 25.4\,\mu\mathrm{s}$.

At the other times, the obtained fields of temperature, stress, and displacement are also quantitatively close to one another for different shapes of defects. The considered stress-strain states show that the variations in the defect volume in the course of time depend on their shape

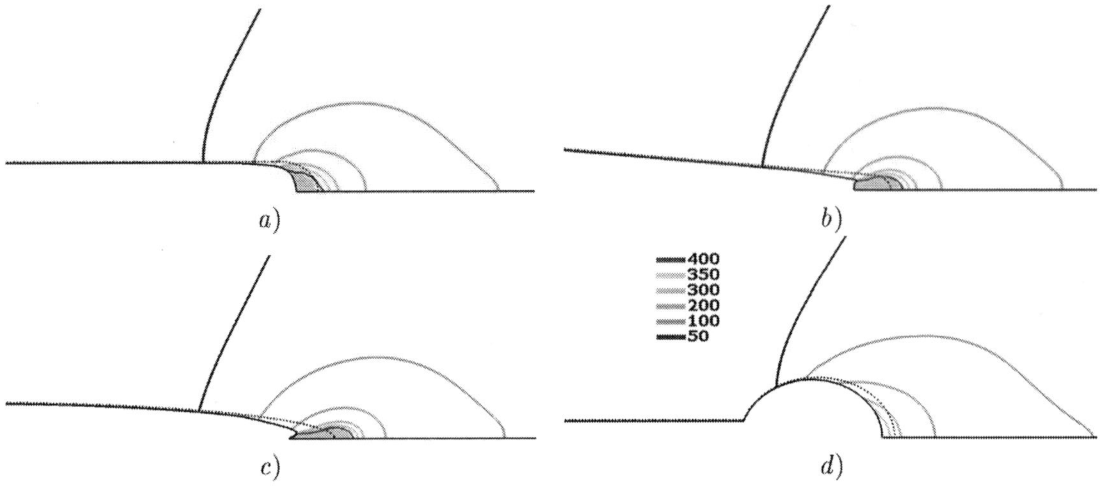

Figure 4. Healing of various micro-defects at time $t = 25.4\,\mu s$ (the dashed line shows the initial boundaries of the defects) and the temperature contours (°C), the area of melted metal is colored gray.

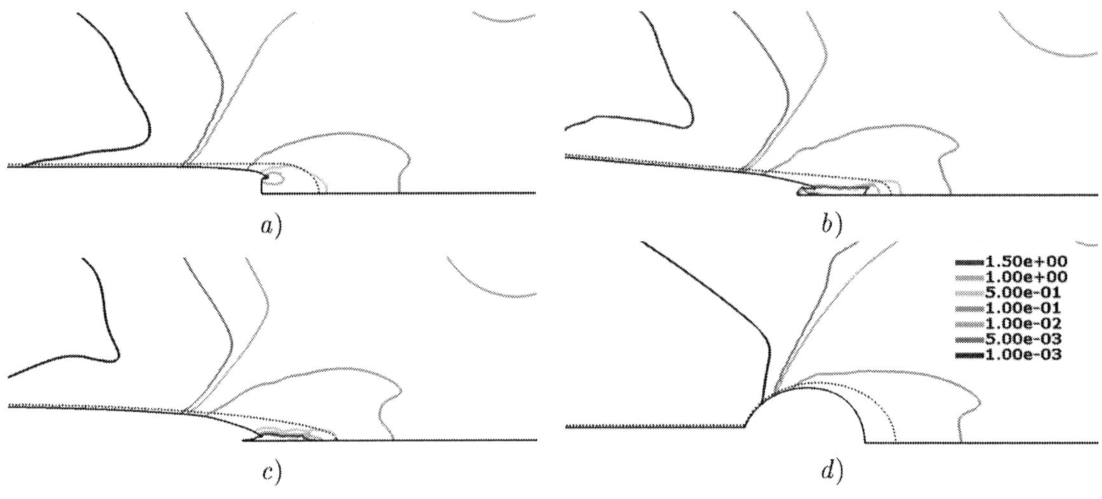

Figure 5. Boundaries of various micro-defects (the dashed line shows the initial boundaries of the defects) at time $t = 48.0\,\mu s$ and the contours of the equivalent plastic strain.

very weakly (for the same initial volumes and lengths of the defects).

The above-described deformation process is accompanied by very intensive plastic flow near the tips of micro-defects. Figure 5 shows the equivalent plastic strain (the second invariant of the plastic strain tensor) at time $t = 48.0\,\mu s$.

The residual plastic strains in these regions attain hundred percents. This fact allows us to believe that the shape of the micro-defect cannot be reconstructed in its initial dimensions.

There may be no viscous fracture of the material and formation of a new discontinuity (defect) at the site of the healed micro-crack, because the plastic deformation occurs under the conditions of strong uniform compression.

As was already noted, the experiments, see [1, 2, 6–10], confirmed that the continuity of the material structure is reconstructed in the material and this phenomenon is accompanied

IOP Conf. Series: Journal of Physics: Conf. Series **991** (2018) 012049 doi:10.1088/1742-6596/991/1/012049

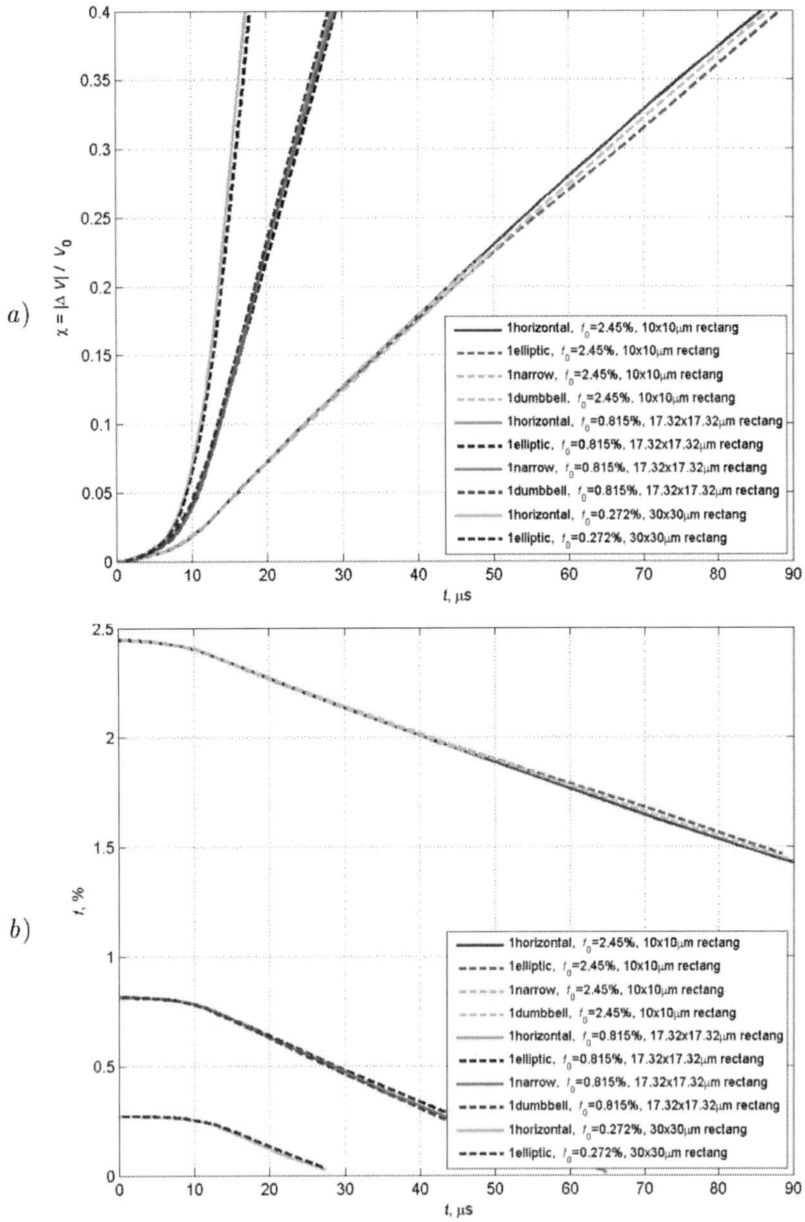

Figure 6. Dependence (a) healing χ and (b) damage f on time t (μs) for microdefects of different shapes (horizontal, narrow, elliptical microcracks and dumbbells) and various initial damage f_0.

by variations in the bulk content of micro-defects (till the total disappearance of several micro-defects). The results obtained above qualitatively agree with these experimental data.

Conclusions

In the modeling of the process of micro-crack healing, one can, without loss of generality, consider only one quarter of the representative cell (figure 1c) as the domain of integration. In this case, in the entire range of distances between the defects (or equivalently, for any initial damage),

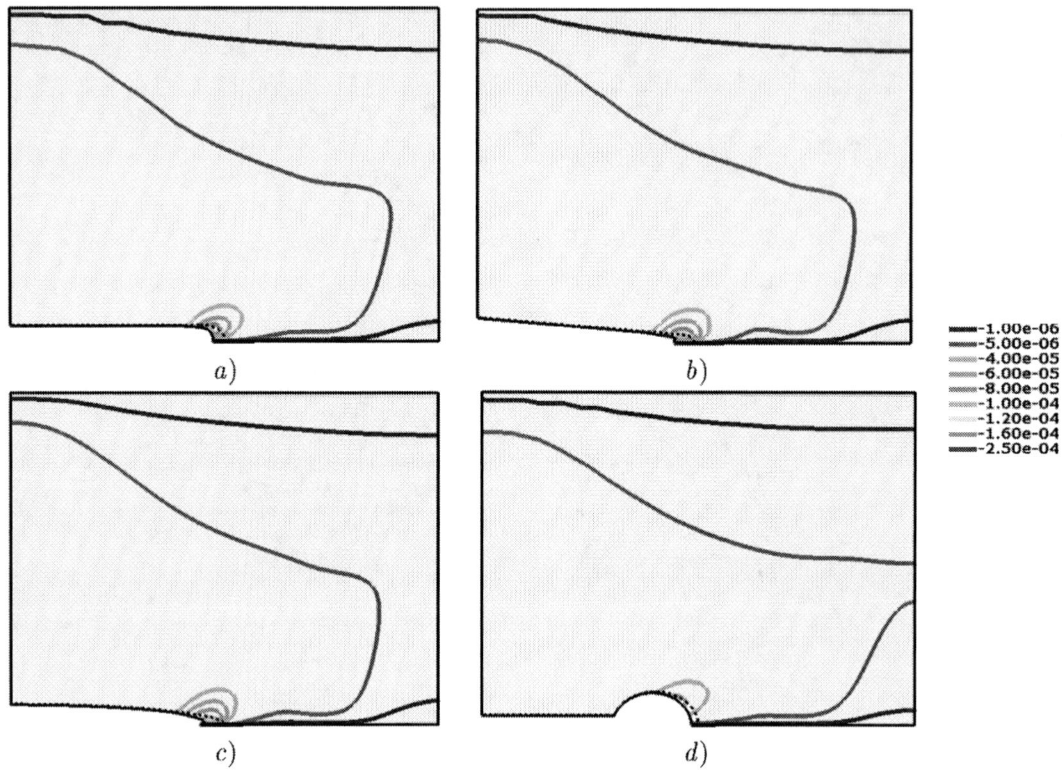

Figure 7. Contours of displacements u_y (mm) at time $t = 25.4\,\mu$s for defects of different shapes (initial damage $f_0 = 2.45\%$).

the healing $\chi(t)$ and the damage $f(t)$ vary identically in time under the action of current and their variation is independent of whether the domains of integration, where these functions are calculated, consist of one or several representative cells.

The decisive factor for variations in the healing and damage of the material in the course of time under the action of a current pulse is not the micro-defect shape (for the same length of the micro-defect) but the parameter of the material damage introduced above. The modeling shows that, for the same equal initial damage of the material f_0, no significant variations in the shape of the micro-defect can practically affect the dependence $\chi(t)$ and $f(t)$.

The obtained fields of temperature, stress, and displacement near micro-defects of different shapes turn out to be quantitatively close to one another.

Acknowledgments
This work was supported by the Russian Foundation for Basic Research (RFBR) under Grant No. 15-08-08693 and No. 18-08-00958.

References

[1] Beklemishev N N, Koryagin N I, and Shapiro G C 1984 Influence of locally inhomogeneous pulsed electromagnetic field on ductility and strength of conductive materials *Izv. Akad. Nauk SSSR. Metally* No. 4 184–7

[2] Beklemishev N N, Kukudzhanov V N, Porokhov V A et al 1989 *Plasticity and Strength of Metallic Materials with the Pulse Action of a High-Energy Electromagnetic Field Taken into Account* Preprint 372 (Moscow: IPM AN SSSR) p 56 [in Russian]

[3] Klyushnikov V D and Ovchinnikov I V 1988 Plane problem of effect of an instantaneous point heat source *Mech. Solids* **23** (4) 113–7
[4] Ovchinnikov I V 1993 Determination of the reserve of plasticity under the action of a current *Strength of Materials* **25** (6) 437–41
[5] Kukudzhanov V N and Kolomiets-Romanenko A V 2010 Study of the influence of electric current dynamical action on mechanical properties of materials with ordered structure of defects *Mech. Solids* **45** (3) 465–75 doi: 10.3103/S0025654410030167
[6] Song Hui, Wang Zhong-jin and Gao Tie-jun 2007 Effect of high density electropulsing treatment on formability of TC4 titanium alloy sheet *Trans. Nonferrous Soc. China* **17** (1) 87–92 doi: 10.1016/S1003-6326(07)60053-3
[7] Troitskii O A, Baranov Yu V, Avraamov Y S and Shlyapin A D 2004 *Physical Fundamentals and Technologies of Processing Advanced Materials (Theory, Technology, Structure, and Properties)* vol 1 (Moscow-Izhevsk: Inst. Komp. Issled.) p 590 [in Russian]
[8] Conrad H 1989 A study into the mechanism(s) for the electroplastic effect in metals and its application to metalworking, processing and fatigue (Final Report ARO Proposal Number 23090-MS, ARO Funding Document DAAL03-86-K-0015, U.S. Army Research Office, North Carolina State University, Raleigh, NC 27695) p 52
[9] Zuev L B, Tsellermaer V Ya, Gromov V E, and Murav'ev V V 1997 Ultrasonic monitoring of the accumulation of aging damage and recovery of the useful lifetime of industrial parts *Tech. Phys.* **49** (2) 1094–6 doi: 10.1134/1.1258774
[10] Zuev L B, Sosnin O V, Chirakadze D Z, Gromov V E, and Murav'ev V V 1998 Acoustic evaluation of the endurance of steel specimens and recovery of their serviceability *J. Appl. Mech. Techn. Phys.* **39** (4) 639–41
[11] Liu T J-C 2010 Effects of temperature-dependent material properties on stress and temperature in cracked metal plate under electric current load *Int. J. Mech., Aerospace, Industrial, Mechatronic and Manufact. Engng* **4** (5) 418–23
[12] Yu J, Zhang H, Deng D, Hao S, and Iqbal A 2014 Numerical calculation and experimental research on crack arrest by detour effect and joule heating of high pulsed current in remanufacturing *Chinese J. Mech. Engng* **27** (4) 745–53 doi: 10.3901/CJME.2014.0414.075
[13] Kukudzhanov K V and Levitin A L 2015 Modeling the healing of microcracks in metal stimulated by a pulsed high-energy electromagnetic field. Part I *Nanomech. Sci. Technol.: Int. J.* **6** (3) 233–50 doi: 10.1615/NanomechanicsSciTechnolIntJ.v6.i3.60
[14] Kukudzhanov K V and Levitin A L 2016 Modeling the healing of microcracks in metal stimulated by a pulsed high-energy electromagnetic field. Part II *Nanomech. Sci. Technol.: Int. J.* **7** (2) 123–48 doi: 10.1615/NanomechanicsSciTechnolIntJ.v7.i2.30
[15] Zienkiewicz O C, Taylor R L, and Zhu J Z 2005 *The Finite Element Method: Its Basis and Fundamentals* (Elsevier)
[16] Finkel' V M, Golovin Yu I, and Sletkov A A 1977 Fracture of crack tip by high-energy electromagnetic field *Sov. Phys. Dokl.* **237** (2) 325–7
[17] Gavrilin I V 2000 *Melting and Crystallization of Metals and Alloys* (Vladimir: Vladimir. Gos. Univ.) p 260 [in Russian]
[18] Pikunov M V 2005 *Metallurgy of the Melts. A Course of Lectures* (Moscow: MISiS) p 286 [in Russian]
[19] Pikunov M V 1997 *Metal Smelting. Alloy Crystallization. The Solidification of Castings* (Moscow: MISiS) p 374 [in Russian]
[20] Kachanov L M 1958 On the time of fracture in creep *Izv. Akad. Nauk SSSR. OTN* No. 8 26–31

IOP Conf. Series: Journal of Physics: Conf. Series **991** (2018) 012050 doi:10.1088/1742-6596/991/1/012050

Mathematical model of the heat transfer process taking into account the consequences of nonlocality in structurally sensitive materials

G N Kuvyrkin, I Y Savelyeva*, and D A Kuvshynnikova
Bauman Moscow State Technical University, Moscow, Russia
E-mail: *inga.savelyeva@gmail.com

Abstract. Creation of new materials based on nanotechnology is an important direction of modern materials science development. Materials obtained using nanotechnology can possess unique physical-mechanical and thermophysical properties, allowing their effective use in structures exposed to high-intensity thermomechanical effects. An important step in creation and use of new materials is the construction of mathematical models to describe the behavior of these materials in a wide range of changes under external effects. The model of heat conduction of structural-sensitive materials is considered with regard to the medium nonlocality effects. The relations of the mathematical model include an integral term describing the spatial nonlocality of the medium. A difference scheme, which makes it possible to obtain a numerical solution of the problem of nonstationary heat conduction with regard to the influence of the medium nonlocality on space, has been developed. The influence of the model parameters on the temperature distributions is analyzed.

1. Introduction

Contemporary structural and functional materials, which are combinations of micro- or nanostructural elements, are often called structurally sensitive materials. An important characteristic of such materials is the qualitative variation in the physical properties compared to massive materials [1–5]. For example, for large fluctuations of physical and mechanical characteristics of micro- or nanostructural elements of the material, the material cannot already be regarded as a simple material, because these characteristics of elements of such a material experience the influence of other surrounding structural elements, i.e., this is the so-called nonlocal medium.

It is well known that the use of methods of continuum mechanics for materials with microstructure is limited by the scale and boundary effects. The direct application of continuum mechanics methods is incorrect for composites modified by nanostructural inclusions [6–10]. Therefore, it is of interest to consider the theory, where, on the one hand, the presence of microstructure is taken into account and, on the other hand, the equations look as the usual equations of continuum mechanics which are generally integrodifferential and which can be solved by methods of continuum mechanics. The method where the concepts of classical continuum mechanics are generalized to media with micro- and nanostructure is called the continuous approximation method [8]. The field of science where the behavior of materials with micro- and nanostructure is studied by continuous approximation methods is sometimes called generalized

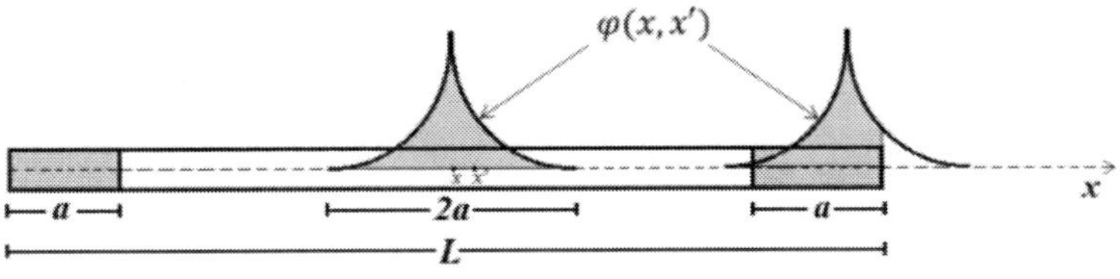

Figure 1. The bar and the influence function.

continuum mechanics. The main problem in these methods is to determine the relationship between the characteristics of the micro- (nano-) and macrolevels with regard to the effects of spatial nonlocality of the medium.

2. Mathematical model

There are comparatively few studies where models of behavior of a nonlocal medium are developed with regard to specific characteristics of the structure. In the present paper, we consider the surface heating problem for a bar in the one-dimensional setting without taking into account the relation between the temperature and stress fields. Under the assumption that the temperature depends only on time and the coordinate x directed along the normal inside the body, we write the heat equation as [11, 12]

$$\rho c \frac{\partial T}{\partial t} = -\frac{\partial q}{\partial x}, \quad x \in (0, L), \quad t > 0, \tag{1}$$

where ρ is the density, c is the heat capacity, q is the heat flux.

The model proposed by Eringen is based on the idea that the long-range forces responsible for the nonlocal behavior of the material at a given point x of space can adequately be described using a distance function $\varphi(|x' - x|)$ decreasing with increasing $|x' - x|$ [10, 13–15]. The function $\varphi(|x' - x|)$ is called the influence function.

We use this idea to take into account the effect of spatial nonlocality in the model of heat transfer in structure-sensitive materials and assume that the heat flux q satisfies the relation

$$q(x) = -p_1 \lambda \frac{\partial T}{\partial x} - p_2 \lambda \int_{x-a}^{x+a} \varphi(|x' - x|) \frac{\partial T}{\partial x'}\, dx', \tag{2}$$

where a is the radius of the nonlocality influence zone filled by the continuum conceived as an aggregate of material particles linked to one another by cohesive bonds (between adjacent particles) and long range forces, and the parameters p_1, p_2 are proportions of the influence of local and nonlocal effects, $p_1 + p_2 = 1$.

The influence function is chosen as follows (figure 1) [10, 16]:

$$\varphi(|x' - x|) = \frac{1}{2a} e^{-|x'-x|/a}, \quad |x' - x| < a. \tag{3}$$

Substituting expression (2) into the equation (1), we obtain the heat equation with nonlocal effects taken into account:

$$\rho c \frac{\partial T}{\partial t} = p_1 \lambda \frac{\partial^2 T}{\partial x^2} + p_2 \lambda \frac{\partial}{\partial x} \int_{x-a}^{x+a} \varphi(|x' - x|) \frac{\partial T}{\partial x'}\, dx', \quad x \in (0, L), \quad t > 0. \tag{4}$$

The boundary conditions for equation (4) are written in the form:

$$-p_1\lambda\frac{\partial T}{\partial x}-p_2\lambda\int_0^a\varphi(|x'-x|)\frac{\partial T}{\partial x'}\,dx'\bigg|_{x=0}=q_0(t),\tag{5}$$

$$p_1\lambda\frac{\partial T}{\partial x}+p_2\lambda\int_{L-a}^{L}\varphi(|x'-x|)\frac{\partial T}{\partial x'}\,dx'\bigg|_{x=L}=q_1(t),\tag{6}$$

where $q_0(t)$ and $q_1(t)$ are given heat fluxes at the left and right ends of the bar, respectively.

The initial condition is

$$T(x,0)=T_0=\text{const}.\tag{7}$$

3. Numerical solution

To solve problem (4)–(7) numerically, we construct its discrete analogue. We introduce a grid in the domain $\Omega_T=\{0\le x\le L, 0\le t\le t_0\}$ as follows: $\omega_h=\{x_i=ih, i=0,1,\dots,n,\ h=L/n\}$ is a uniform grid with step h on the interval $0\le x\le L$; $\omega_\tau=\{t_j=j\tau, j=0,1,\dots,m,\ \tau=t_0/m\}$ is a uniform grid with step τ on the interval $0\le t\le t_0$. Then $\omega_{\tau h}=\omega_h\times\omega_\tau=\{(x_i,t_j),x_i\in\omega_h,t_j\in\omega_\tau\}$ is a grid in Ω_T. The temperature value is referred to the cell center. Integrating relation (4) over the cell, we obtain

$$\rho c\frac{T_i^{j+1}-T_i^j}{\tau}h=p_1\lambda\frac{\partial T}{\partial x}\bigg|_{x_i}^{x_{i+1}}+p_2\lambda\left[\int_{x-a}^{x+a}\varphi(|x'-x|)\frac{\partial T}{\partial x'}\,dx'\right]\bigg|_{x_i}^{x_{i+1}}.\tag{8}$$

The first term on the right-hand side is responsible for local effects of the medium, and the second term, for nonlocal ones. Let us consider the second term in more detail:

$$\left[\int_{x-a}^{x+a}\varphi(|x'-x|)\frac{\partial T}{\partial x'}\,dx'\right]\bigg|_{x_i}^{x_{i+1}}=\int_{x_{i+1}-a}^{x_{i+1}+a}\varphi(|x'-x_{i+1}|)\frac{\partial T}{\partial x'}\,dx'-\int_{x_i-a}^{x_i+a}\varphi(|x'-x_i|)\frac{\partial T}{\partial x'}\,dx'.\tag{9}$$

We choose a spatial step h so that an integer number of steps h fit into the radius a of nonlocality. In this case, the boundaries of the influence of the nonlocality function fall on the boundaries of the cells.

The integrals in the right-hand side of relation (9) are computed by the quadrature formulas of trapezium. The heat flows are replaced by difference analogs by approximation of the Fourier thermal conductivity law. Then in the case $a=2h$, the difference analogue of the heat equation taking into account the nonlocality of the medium has the form

$$\begin{aligned}\rho c\frac{T_i^{j+1}-T_i^j}{\tau}h&=-p_1\left(\lambda\frac{T_{i+2}^{j+1}-T_{i+1}^{j+1}}{h}-\lambda\frac{T_{i+1}^{j+1}-T_i^{j+1}}{h}\right)\\&+\frac{1}{2}hp_2\left[\lambda\frac{c_1(T_{i-1}^{j+1}-T_{i-2}^{j+1})}{h}+\lambda\frac{2c_2(T_i^{j+1}-T_{i-1}^{j+1})}{h}\right]\\&-\frac{1}{2}hp_2\left[\lambda\frac{c_1(T_i^{j+1}-T_{i-1}^{j+1})}{h}+\lambda\frac{2c_2(T_{i+1}^{j+1}-T_i^{j+1})}{h}+\lambda\frac{2c_3(T_{i+2}^{j+1}-T_{i+1}^{j+1})}{h}\right]\\&+\frac{1}{2}hp_2\left[\lambda\frac{2c_3(T_{i+1}^{j+1}-T_i^{j+1})}{h}+\lambda\frac{2c_2(T_{i+2}^{j+1}-T_{i+1}^{j+1})}{h}+\lambda\frac{c_1(T_{i+3}^{j+1}-T_{i+2}^{j+1})}{h}\right]\\&-\frac{1}{2}hp_2\left[\lambda\frac{2c_2(T_{i+3}^{j+1}-T_{i+2}^{j+1})}{h}+\lambda\frac{c_1(T_{i+4}^{j+1}-T_{i+3}^{j+1})}{h}\right],\quad i=4,\dots,n-5,\quad j=0,\dots,m,\end{aligned}\tag{10}$$

where c_i is the value of the nonlocality influence function at the point x_i. Because of the symmetry of the function $\varphi(|x'-x|)$, the parameter range is $j\in\{1,2,3\}$.

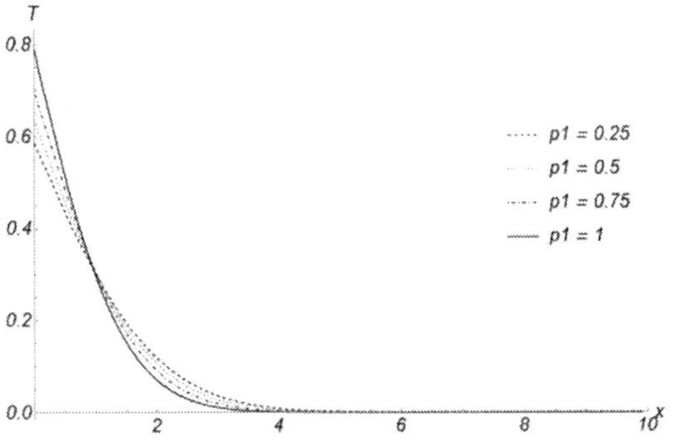

Figure 2. Influence of nonlocality parameters on the temperature distribution in the bar.

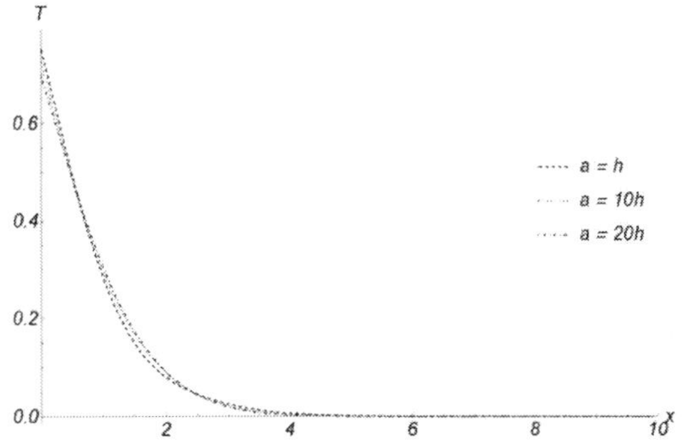

Figure 3. Influence of the radius of nonlocality a on the temperature distribution in the bar.

The expression for the heat flux at the left boundary of the bar has the form

$$p_1 q(0) - p_2 \int_0^a \varphi(|x'|) \frac{\partial T}{\partial x'}\, dx'$$
$$= -p_1 \frac{\partial T}{\partial x} - p_2 h \frac{1}{2} \left(c_3 q(0) + 2c_2 \lambda \frac{T_2^{j+1} - T_1^{j+1}}{h} + c_1 \lambda \frac{T_3^{j+1} - T_2^{j+1}}{h} \right) = q_0(t). \quad (11)$$

Similarly, we arrive at an approximation of the boundary condition on the right boundary of the bar. It should be noted that setting the heat flux in the form (11) leads to the influence of the boundary conditions of the second kind not only on the outermost cells but also on the cells adjacent to them, thus falling into the segment of the influence of the nonlocality.

The resulting difference problem leads to a system of linear algebraic equations with a band matrix whose tape width is greater by 1 than the number of cells that fall in the segment of the nonlocality effect (in the case $2a = 4h$, the matrix is five-diagonal, $2a = 6h$ is a seven-diagonal matrix, etc.). The solution of the resulting system of equations can be obtained by any known method.

4. Analysis of the results

We performed calculations for the surface heating in the case where a high-intensity heat flux is given on the left boundary and has the form:

$$q(t) = \frac{m^m e^{-mt} t^m}{(m-1)!}.$$

Figure 2 shows the temperature distribution in the bar of length $L = 10$ at time $t = 2$ for $a = 1 = 10h$, $h = 0.1$, $m = 2$, and for different values of the parameter p_1. Obviously, with an increase in the share of accounting for nonlocal effects, the temperature on the boundary increases.

Fix $p_1 = 0.5$, $h = 0.1$, and consider the effect of the nonlocality radius a on the temperature distribution in the bar at time $t = 2$. The graphs in Figure 3 show that the temperature increases with the radius of nonlocality.

Let us consider the influence of the heat flow intensity at different times on the temperature distribution in the bar of length $L = 5$ with parameters $a = 0.2$ and $p_1 = 0.5$.

The graphs presented in Figs. 4–7 show the temperature distribution in the bar at the times corresponding to different stages of high-intensity heating. The times $t = 0.5$, $t = 1$, $t = 2.5$ correspond to the initial stage and the peak and final stages of heating, respectively. The graphs show that the solution corresponds to the given boundary conditions.

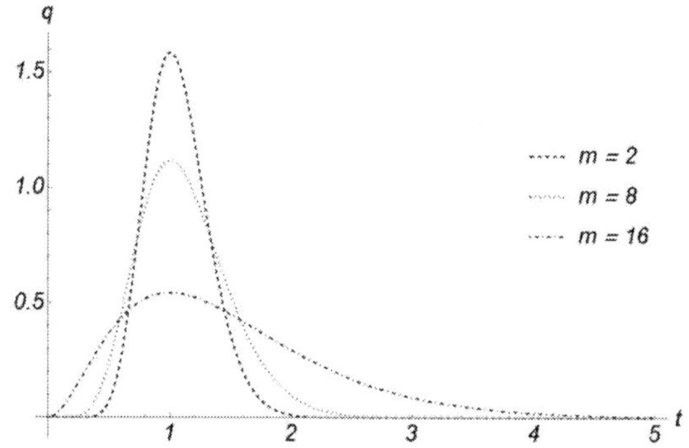

Figure 4. Graphs of the function of a high-intensity heat flux for different parameters m

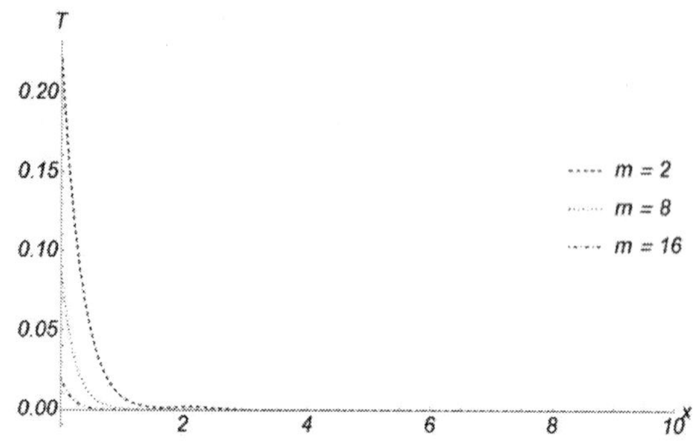

Figure 5. Comparison of the results of calculation of the intensity for different parameters m at time $t = 0.5$

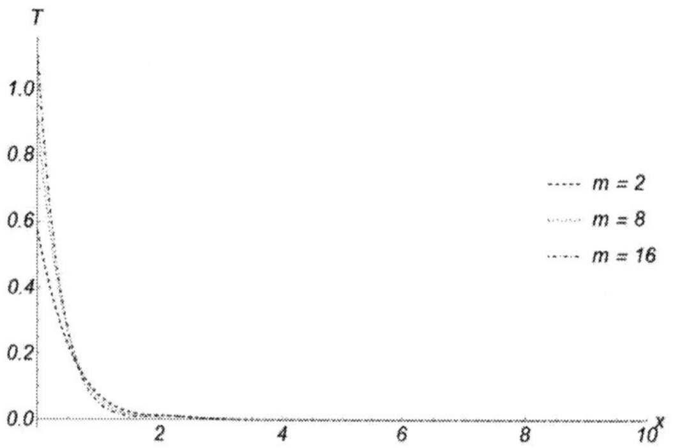

Figure 6. Comparison of the results of calculation of the intensity for different parameters m at time $t = 1$

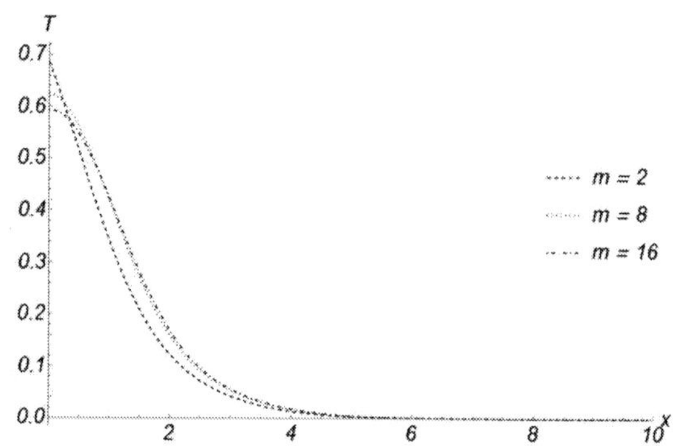

Figure 7. Comparison of the results of calculation of the intensity for different parameters m at time $t = 2.5$

Conclusions

The difference analogue of the one-dimensional heat conduction equation was constructed with regard to the medium nonlocality. The temperature distribution in time was determined and the dependence on radius and parameters of nonlocal nature of the environment was studied. It should be noted that the solution of equation (6) with nonlocality parameter $p_1 = 1$, i.e., where the influence of the nonlocal component is missing, coincides with the solution of the heat equation, where the contribution of the structure nonlocality is not taken into account.

Acknowledgments

The research was supported by the Russian Federation Ministry of Education and Science (grant No. 9.2422.2017).

References

[1] Andrievskii R A and Ragulya A V 2005 *Nanostructural Materials* (Moscow: Akademiya) [in Russian]
[2] Gusev A N 2005 *Nanomaterials, Nanostructures, and Nanotechnologies* (Moscow: Fizmatlit) [in Russian]
[3] Kobayashi H 2005 *Introduction to Nanotechnology* (Moscow: BINOM) [in Russian]
[4] Poole Ch P Jr and Owens F J 2003 *Introduction to Nanotechnology* (Wiley).
[5] Kelsall R W (Editor) 2005 *Nanoscale Science and Technology* (Wiley)

[6] Peddieson J, Buchanan G R, and McNitt R P 2003 Application of nonlocal continuum models to nanotechnology *Int. J. Engng Sci.* **41** 305–12
[7] Krivtsov A M 2007 *Deformation and Failure of Rigid Bodies with Microstructure* (Moscow: Fizmatlit) [in Russian]
[8] Kunin I A 1975 *Theory of Elastic Media with Microstructure. Nonlocal Theory of Elasticity* (Moscow: Nauka) [in Russian]
[9] Onemi M, Iwasimidzu S, Genka K, et al. 1987 *Introduction to Micromechanics* (Moscow: Metallurgiya) [in Russian]
[10] Eringen A C 2002 *Nonlocal continuum field theories* (New York - Berlin - Heidelberg: Springer-Verlag) p 393
[11] Zarubin V S and Kuvyrkin G N 2008 *Mathematical Models of Continuum Mechanics and Electrodynamics* (Moscow: Izdat. MGTU im. Baumana) [in Russian]
[12] Zarubin V S and Kuvyrkin G N (2003) Mathematical modeling of thermomechanical processes under intense thermal effect *Teplofiz. Vysokikh Temp.* **41** (2), 300–9 [*High Tempr.* (Engl. Transl.) **41** (2), 257–65].
[13] Pisano A A and Fuschi P 2003 Closed form solution for a nonlocal elastic bar in tension *Int. J. Solids Struct.* **40** 13–23
[14] Polizzotto C 2001 Nonlocal elasticity and related variational principles *Int. J. Solids Struct.* **38** 7359–80
[15] Zarubin V S, Kuvyrkin G N, and Savelieva I Yu (2013) Mathematical model of a nonlocal medium with internal state parameters, *Inzh.-Fiz. Zh.* **86** (4), 768–73 [*J. Engng Phys. Thermophys.* (Engl. Transl.) **86** (4), 820–6].
[16] Kuvyrkin G N and Savelieva I Yu (2016) Thermomechanical model of of nonlocal deformation of a solid *Mech. Solids* **51** (3) 256–62

Harmonic oscillations of a longitudinal shear infinite hollow cylinder arbitrary cross-section with a tunnel crack

O I Kyrylova and V G Popov*

National University "Odessa Maritime Academy," Odessa, Ukraine

E-mail: *`dr.vg.popov@gmail.com`

Abstract. An effective analytical-numerical method for determining the dynamic stresses in a hollow cylindrical body of arbitrary cross-section with a tunnel crack under antiplane strain conditions is proposed. The method allows separately solving the integral equations on the crack faces and satisfying the boundary conditions on the body boundaries. It provides a convenient numerical scheme. Approximate formulas for calculating the dynamic stress intensity factors in a neighborhood of the crack are obtained and the influence of the crack geometry and wave number on these quantities is investigated, especially from the point of view of the resonance existence.

1. Introduction

Research of the stress conditions of finite bodies with cracks under harmonic loads remains relevant nowadays. This actuality is stipulated by the need for determining the conditions of fracture through evaluation of the stress intensity factor (SIF) of dynamic stresses near the cracks and diagnosing defects by analyzing their influence on the resonance frequency [1]. Preliminary results in this area mainly appertain to infinite bodies with cracks or thin inclusions [2–10]. The situations where the body occupies a finite area is considered much less. This is primarily due to the fact that the application of the method of boundary integral equations for the source of the problems reduces the problem to solving systems of integral equations and satisfying conditions on the surface defects and on the boundary of the body [11–15], which significantly complicates the numerical implementation of the method. We propose an approach that allows solving separately integral equations on the surface cracks and satisfying the loading conditions on the boundary of the body.

2. Statements of the problem

Consider an elastic cylinder with generators parallel to the Oz axis (figure 1) whose intersection by the plane xOy is a doubly-connected domain bounded by arbitrary closed smooth curves. In the polar coordinates $Or\varphi$, whose center is located at the origin of coordinates, these curves are determined by the equations

$$r = r_0\psi_0(\varphi), \quad r = r_1\psi_1(\varphi), \quad 0 \leq \varphi < 2\pi.$$

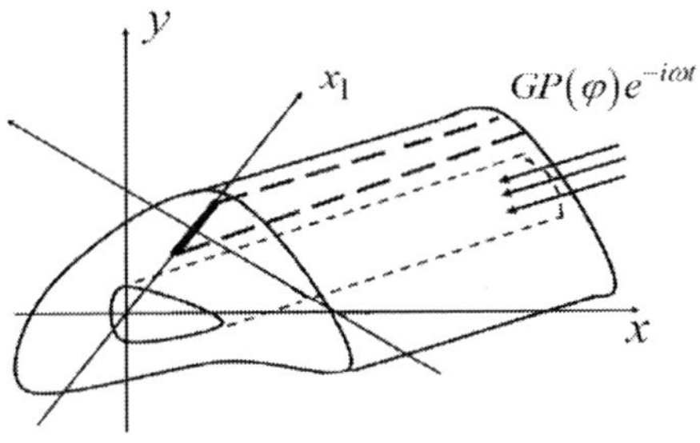

Figure 1. Hollow cylinder with a tunnel crack.

The cylinder contains a through crack. In the plane xOy, the defect does not fall outside the boundary of cross-section and occupies a segment of length $2a$. The oscillations of longitudinal shear go on in the lateral surface of the cylinder due to the action of a harmonic load $P(\varphi)\exp(-i\omega t)$, which is directed along the axis Oz. We omit the multiplier $\exp(-i\omega t$, which determines the time dependence, and consider only the amplitude values below. Under such conditions, only the z-component of displacement vector, which satisfies the Helmholtz equation [16], is different from zero. This equation in the polar coordinates has the form

$$\Delta w + \kappa_2^2 w = 0, \quad \Delta = \frac{\partial^2}{\partial r^2} + \frac{1}{r}\frac{\partial}{\partial r} + \frac{1}{r^2}\frac{\partial^2}{\partial \varphi^2}, \tag{1}$$

where $\kappa_2^2 = \omega^2 c_2^{-2}$, $c_2^2 = G\rho^{-1}$, G and ρ are the shear module and the density of the cylinder. Now we formulate conditions on the outer and inner surfaces of the cylinder. According to the conditions of loading of the outer surface, the stress must satisfy the equality on it

$$\tau_{\bar{n}z}(r_0\psi_0(\varphi), \varphi) = GP(\varphi), \quad 0 \le \varphi < 2\pi, \tag{2}$$

where $\bar{n}$ is the vector of normal on the cylinder surface. The inner surface of the cylinder is considered to be unmovable

$$w(r_1\psi_1(\varphi), \varphi) = 0, \quad 0 \le \varphi < 2\pi. \tag{3}$$

To formulate the boundary conditions on the defect, we connect a coordinate system $O_1x_1y_1$ with its center, as is shown in figure 1. The relation between the system $O_1x_1y_1$ and the polar coordinates is given by

$$x_1 = r\cos(\varphi - \alpha) - c\cos\alpha - d\sin\alpha, \quad y_1 = r\sin(\varphi - \alpha) - d\cos\alpha + c\sin\alpha, \tag{4}$$

where α is the angle between the axes O_1x_1 and Ox.

Let $w_1^{(1)}(x_1, y_1)$ be the z component of the displacement vector $w(r, \varphi)$ in passing from the polar coordinates to local coordinates, according to (4). Considering a crack, we assume that its surface is free from stresses:

$$\tau_{zy_1}(x_1, 0) = 0, \quad |x_1| < a. \tag{5}$$

In addition, the displacement $w_1^{(1)}(x_1, y_1)$ on the crack surface has a discontinuity, whose jump is denoted by $\chi(x_1)$:

$$w_1^{(1)}(x_1, +0) - w_1^{(1)}(x_1, -0) = \chi(x_1), \quad \chi(\pm a) = 0. \tag{6}$$

Thus, determining of the stress state of the considered body with crack under conditions of oscillations is reduced to solving equation (1) with conditions (2), (3), (5), and (6).

3. Satisfaction of the conditions on the crack surface

To solve our problem, we first, in the system $O_1x_1y_1$, construct a discontinuous solution of the Helmholtz equation [2, 17] with a jump $\chi(x_1)$ determined by (6):

$$w_1^{(1)}(x_1, y_1) = \frac{\partial}{\partial y_1}\int_{-a}^{a}\chi(\eta)r(\eta - x_1, y_1)\,d\eta, \quad r(\eta - x_1, y_1) = -\frac{i}{4}H_0^{(1)}\left(\kappa_2\sqrt{(\eta - x_1)^2 + y_1^2}\right), \quad (7)$$

where $H_0^{(1)}(z)$ is the Hankel function. The stresses corresponding to displacement (7) are determined as

$$\tau_{y_1z}^{(1)} = G\frac{\partial w_1^{(1)}(x_1, y_1)}{\partial y_1} = -G\left[\int_{-a}^{a}\chi(\eta)\frac{\partial^2}{\partial\eta^2}r(\eta - x_1, y_1)\,d\eta + \kappa_2^2\int_{-a}^{a}\chi(\eta)r(\eta - x_1, y_1)\,d\eta\right]. \quad (8)$$

Then, in the system $Or\varphi$, we represent the displacement as

$$w(r, \varphi) = w_0(r, \varphi) + w_1(r, \varphi), \quad (9)$$

where $w_1(r, \varphi)$ is the discontinuous solution (7) after passage to the polar coordinates according to (4) and $w_0(r, \varphi)$ is a unknown solution of the Helmholtz equation (1) such that the boundary conditions (2), (3) are satisfied for (9). We represent this unknown function as a linear combination of particular solutions of equation (1):

$$w_0(r, \varphi) = r_0\sum_{k=1}^{N}[A_kg_k(r, \varphi) + B_kh_k(r, \varphi)], \quad (10)$$
$$g_{2m-1}(r, \varphi) = J_{m-1}(\kappa_2 r)\cos[(m-1)\varphi], \quad g_{2m}(r, \varphi) = J_m(\kappa_2 r)\sin(m\varphi),$$
$$h_{2m-1}(r, \varphi) = H_{m-1}^{(1)}(\kappa_2 r)\cos[(m-1)\varphi], \quad h_{2m}(r, \varphi) = H_m^{(1)}(\kappa_2 r)\sin(m\varphi),$$

where $J_m(z)$, $H_m^{(1)}(z)$ are cylindrical functions, A_k and B_k are unknown coefficients.

The functions $g_k(r, \varphi)$, $h_k(r, \varphi)$ are linearly independent and form a complete closed system of functions in the domain of cross-section [18].

After substitution of expressions (9) and (10) for displacements, the boundary condition (5) on the crack becomes

$$\tau_{y_1z}^{(1)}(x_1, 0) + \tau_{y_1z}^{(0)}(x_1, 0) = 0, \quad (11)$$

where the first term $\tau_{y_1z}^{(1)}(x_1, 0)$ is given by (8), and

$$\tau_{y_1z}^{(0)}(x_1, 0) = Gr_0\sum_{k=1}^{N}\left(A_k\frac{\partial g_k^{(1)}(x_1, y_1)}{\partial y_1}\bigg|_{y_1=0} + B_k\frac{\partial h_k^{(1)}(x_1, y_1)}{\partial y_1}\bigg|_{y_1=0}\right),$$

where $g_k^{(1)}(x_1, y_1)$, $h_k^{(1)}(x_1, y_1)$ are the function $g_k(r, \varphi)$ and $h_k(r, \varphi)$ in the coordinate system $O_1x_1y_1$. We use relation (11) to obtain an integral equation, which, after integrating by parts and isolating the singular part in its kernel, can be written as

$$(2\pi)^{-1}\int_{-1}^{1}\chi'(a\tau)[(\tau - \varsigma)^{-1} + Q(\tau - \varsigma)]\,d\tau = f(\varsigma), \quad -1 < \zeta < 1. \quad (12)$$

In equation (12), we used the notation

$$\eta = a\tau, \quad x_1 = a\varsigma, \quad \kappa_0 = \kappa_2 r_0, \quad ar_0^{-1} = \gamma, \quad \kappa_2 a = \gamma\kappa_0, \quad r_1r_0^{-1} = \lambda.$$

IOP Conf. Series: Journal of Physics: Conf. Series **991** (2018) 012051 doi:10.1088/1742-6596/991/1/012051

The right-hand side of equation (12) has the form

$$f(\varsigma) = \sum_{k=1}^{N} [A_k f_k^1(\varsigma) + B_k f_k^2(\varsigma)], \quad f_k^1(\varsigma) = -\frac{\partial g_k^{(1)}(a\varsigma, 0)}{\partial y_1}, \quad f_k^2(\varsigma) = -\frac{\partial h_k^{(1)}(a\varsigma, 0)}{\partial y_1} \tag{13}$$

and its kernel has the asymptotics

$$Q(Z) = O(|Z| \ln |Z|), \quad Z \to 0.$$

We should supplement equation (12) with the equality

$$\int_{-1}^{1} \chi'(a\tau)\, d\tau = 0 \tag{14}$$

which follows from condition (6) of the crack closure.

Since equations (12) and (14) are linear, we can write the solution of equation (12) with right-hand side (13) in the form

$$\chi'(a\tau) = \sum_{k=1}^{N} [A_k S_k^1(\tau) + B_k S_k^2(\tau)]. \tag{15}$$

This representation allows us to transform equation (12) into a sequence of equations for the functions $S_k^1(\tau)$ and $S_k^2(\tau)$:

$$(2\pi)^{-1} \int_{-1}^{1} S_k^i(\tau)[(\tau - \varsigma)^{-1} + Q(\tau - \varsigma)]\, d\tau = f_k^i(\varsigma), \quad i = 1, 2, \quad k = \overline{1, N}. \tag{16}$$

As can be seen, these equations differ only in the right-hand sides and do not depend on the boundaries of the cross-section domain. We should supplement equations (16) with the equality arising from (14):

$$\int_{-1}^{1} S_k^i(\tau)\, d\tau = 0, \quad i = 1, 2, \quad k = \overline{1, N}. \tag{17}$$

To construct an approximate solutions of equations (16), we represent the unknown functions $S_k^i(\tau)$ as

$$S_k^i(\tau) = \psi_k^i(\tau)(1 - \tau^2)^{-1/2}, \quad i = 1, 2, \quad k = \overline{1, N}, \tag{18}$$

where each function $\psi_k^i(\tau)$ is assumed to satisfy the Hölder condition on $[-1, 1]$. To determine them, we apply the method of mechanical quadratures [19], choosing the roots of the second-kind Chebyshev polynomial $U_{n-1}(\varsigma)$ as the collocation points $\varsigma_j = \cos(\pi j/n)$, $j = 1, 2, \ldots, n-1$. Then, from (16) and (17), we obtain systems of linear algebraic equations for the values of the function $\psi_k^i(\tau)$ at the interpolation nodes:

$$\begin{aligned} &(2\pi)^{-1} \sum_{m=1}^{n} a_m \psi_{km}^i [(\tau_m - \varsigma_j)^{-1} + Q(\tau_m - \varsigma_j)] = f_k^i(\varsigma_j), \quad i = 1, 2, \quad k = \overline{1, n-1}; \\ &(2\pi)^{-1} \sum_{m=1}^{n} a_m \psi_{km}^i = 0, \end{aligned} \tag{19}$$

where $\psi_{km}^i = \psi_k^i(\tau_m)$ and $\tau_m = \cos(\pi(2m-1)/2n)$, $m = 1, \ldots, n$, are the roots of the first-kind Chebyshev polynomial $T_n(\tau)$. If system (19) is solved, then we obtain an approximation of each function $\psi_k^i(\tau)$ by the interpolation polynomial

$$\psi_k^i(\tau) \approx \sum_{m=1}^{n} \psi_{km}^i \frac{T_n(\tau)}{(\tau - \tau_m) T_n'(\tau_m)}. \tag{20}$$

IOP Conf. Series: Journal of Physics: Conf. Series **991** (2018) 012051 doi:10.1088/1742-6596/991/1/012051

4. Satisfaction of the conditions on the cylinder surfaces

The unknown coefficients A_k and B_k in equality (10) are determined from condition (2), (3) on the cylinder surfaces. For this, we write the stresses and the displacement as

$$\begin{aligned} \tau_{\bar{n}z}(r_0\psi_0(\varphi),\varphi) &= \tau_{xz}(r_0\psi_0(\varphi),\varphi)c_x + \tau_{yz}(r_0\psi_0(\varphi),\varphi)c_y, \\ w(r_1\psi_1(\varphi),\varphi) &= w_0(r_1\psi_1(\varphi),\varphi) + w_1(r_1\psi_1(\varphi),\varphi), \end{aligned} \tag{21}$$

where c_x, c_y are the direction cosines of the normal vector $\bar{n} = (c_x, c_y)$ of the outer surfaces of the cylinder. Substituting these expressions (21) for the stresses and the displacement in (2), (3) and using formulas (7)–(10), we obtain

$$\begin{aligned} &\sum_{k=1}^{N} A_k\left[\int_{-1}^{1}\Phi_k^1(\tau)G(\tau,\varphi)\,d\tau + F_k^1(\varphi)\right] \\ &\qquad + \sum_{k=1}^{N} B_k\left[\int_{-1}^{1}\Phi_k^2(\tau)G(\tau,\varphi)\,d\tau + F_k^2(\varphi)\right] = P(\varphi), \quad 0 \le \varphi < 2\pi, \\ &\sum_{k=1}^{N} A_k\int_{-1}^{1}\Phi_k^1(\tau)E(\tau,\varphi)\,d\tau + \sum_{k=1}^{N} B_k\int_{-1}^{1}\Phi_k^2(\tau)E(\tau,\varphi)\,d\tau \\ &\qquad + \sum_{k=1}^{N} A_k g_k(r_1(\varphi),\varphi) + \sum_{k=1}^{N} B_k h_k(r_1(\varphi),\varphi) = 0. \end{aligned} \tag{22}$$

The functions $G(\tau,\varphi)$, $E(\tau,\varphi)$ are here expressed through the values of the function $r(\eta - x_1, y_1)$ from (7) and its derivatives on the cylinder surfaces, and the functions $F_k^1(\varphi)$, $F_k^2(\varphi)$ are expressed in terms of the values of function $g_k(r,\varphi)$, $h_k(r,\varphi)$ on these surfaces given by formulas (10). In derivation of equations (22), the relation

$$\chi(a\tau) = a\sum_{k=1}^{N}[A_k\Phi_k^1(\tau) + B_k\Phi_k^2(\tau)], \quad [\Phi_k^i(\tau)]' = S_k^i(\tau), \quad i = 1,2.$$

follows from formula (15). For calculating the integrals in (22), we use the following approximation [10] for the functions $\Phi_k^i(\tau)$, which is obtained from (18) and (20):

$$\Phi_k^i(\tau) \approx (1-\tau^2)^{1/2}S_{kn}^i(\tau), \quad S_{kn}^i(\tau) = -\frac{2}{n}\sum_{m=1}^{n}\psi_{km}^i\sum_{p=1}^{n-1}p^{-1}T_p(\tau_m)U_{p-1}(\tau).$$

Here, $U_{p-1}(\tau)$, $T_p(\tau)$ are the first- and second-kind Chebyshev polynomials, respectively. Now we can calculate the integral in (22) by the quadrature formula [20]

$$\int_{-1}^{1}\Phi_k(\tau)K(\tau,\varphi)\,d\tau = \sum_{m=1}^{n}a_m\psi_{km}\sum_{l=1}^{n}D_{lm}K(Z_l,\varphi),$$

$$D_{lm} = -\frac{2}{n+1}\sin\frac{l\pi}{n+1}\sum_{p=1}^{n-1}p^{-1}\cos\frac{p(2m-1)\pi}{2n}\sin\frac{\pi lp}{n+1}, \quad Z_l = \cos\frac{\pi l}{n+1}.$$

Finally, the boundary condition (22) becomes

$$\begin{aligned}
&\sum_{k=1}^{N} A_k \left[\sum_{m=1}^{n} a_m \psi_{km}^1 \sum_{l=1}^{n} D_{lm} G(Z_l, \varphi) + F_k^1(\varphi) \right] \\
&\qquad + \sum_{k=1}^{N} B_k \left[\sum_{m=1}^{n} a_m \psi_{km}^2 \sum_{l=1}^{n} D_{lm} G(Z_l, \varphi) + F_k^2(\varphi) \right] = P(\varphi), \\
&\sum_{k=1}^{N} A_k \left[\sum_{m=1}^{n} a_m \psi_{km}^1 \sum_{l=1}^{n} D_{lm} E(Z_l, \varphi) + g_k(\varphi) \right] \\
&\qquad + \sum_{k=1}^{N} B_k \left[\sum_{m=1}^{n} a_m \psi_{km}^2 \sum_{l=1}^{n} D_{lm} E(Z_l, \varphi) + h_k(\varphi) \right] = 0.
\end{aligned} \tag{23}$$

Now we obtain a system of linear equations for determining the coefficients A_k and B_k by applying the collocation method to (23) and taking the quantities $\sigma_l = 2\pi l/N$, $l = 1, \ldots, N$, as the nodes:

$$\begin{aligned}
&\sum_{k=1}^{N} A_k \left[\sum_{m=1}^{n} a_m \psi_{km}^1 \sum_{l=1}^{n} D_{lm} G(Z_l, \sigma_l) + F_k^1(\sigma_l) \right] \\
&\qquad + \sum_{k=1}^{N} B_k \left[\sum_{m=1}^{n} a_m \psi_{km}^2 \sum_{l=1}^{n} D_{lm} G(Z_l, \sigma_l) + F_k^2(\sigma_l) \right] = P(\sigma_l), \\
&\sum_{k=1}^{N} A_k \left[\sum_{m=1}^{n} a_m \psi_{km}^1 \sum_{l=1}^{n} D_{lm} E(Z_l, \sigma_l) + g_k(\sigma_l) \right] \\
&\qquad + \sum_{k=1}^{N} B_k \left[\sum_{m=1}^{n} a_m \psi_{km}^2 \sum_{l=1}^{n} D_{lm} E(Z_l, \sigma_l) + h_k(\sigma_l) \right] = 0, \quad l = 1, \ldots, N.
\end{aligned} \tag{24}$$

As is well known, the possibility of crack propagation is determined by the stress intensity factors (SIF), which in our case are given by

$$K^{\pm} = \sqrt{a} \lim_{\varsigma \to \pm 1 \pm 0} (\varsigma^2 - 1)^{1/2} \tau_{y_1 z}(a\tau, 0).$$

If systems (19) and (24) are solved, then we obtain the following relations for the dimensionless SIF value:

$$\begin{aligned}
k^{\pm} &= \frac{K^{\pm}}{G\sqrt{a}}, \\
k^{+} &= \frac{1}{2n} \left[\sum_{k=1}^{N} A_k \sum_{m=1}^{n} (-1)^m \psi_{km}^1 \cot \frac{\gamma_m}{2} + \sum_{k=1}^{N} B_k \sum_{m=1}^{n} (-1)^m \psi_{km}^2 \cot \frac{\gamma_m}{2} \right], \\
k^{-} &= \frac{(-1)^{n+1}}{2n} \left[\sum_{k=1}^{N} A_k \sum_{m=1}^{n} (-1)^m \psi_{km}^1 \tan \frac{\gamma_m}{2} + \sum_{k=1}^{N} B_k \sum_{m=1}^{n} (-1)^m \psi_{km}^2 \tan \frac{\gamma_m}{2} \right].
\end{aligned}$$

5. Numerical analysis of the intensity of dynamic stresses and resonance phenomena

As an example, we consider a cylinder with the cross-section bounded by two ellipses with a common center, as is shown in figure 2.

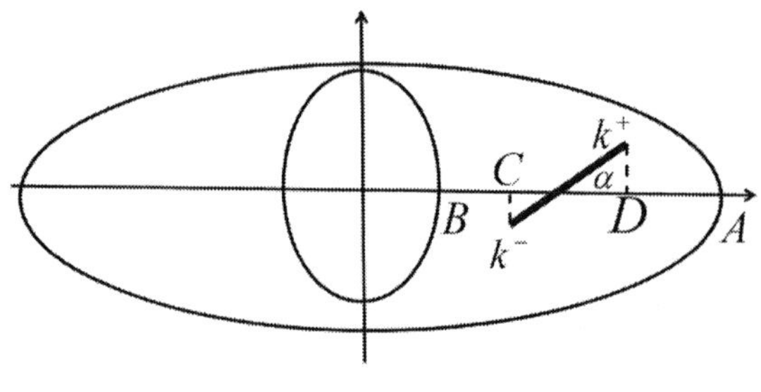

Figure 2. Cross-section of the cylinder with a crack.

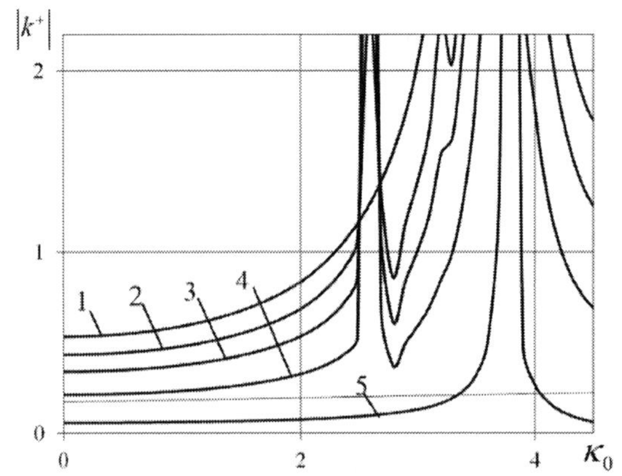

Figure 3. Dependence of absolute values on the wave number for different angles of crack inclination.

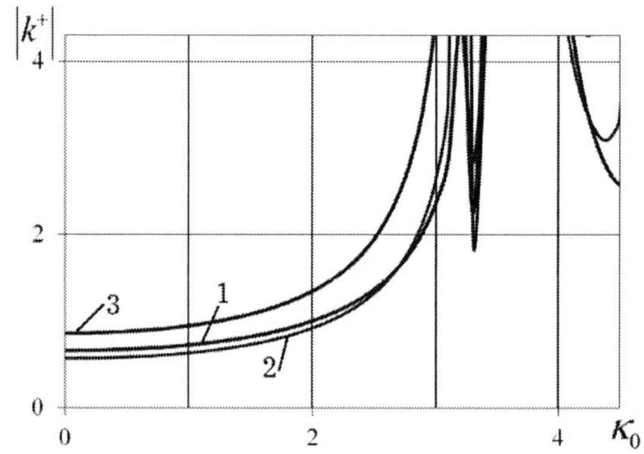

Figure 4. Dependence of absolute values of the dimensionless SIF on the wave number for increasing crack length.

The eccentricities of the both ellipses are the same $\varepsilon = 0.5$ and the main semiaxes are related as $r_1/r_0 = 0.5$. The load acting on the outer surface is determined by the formula. The center of the crack is located on the main axis of the outer ellipse. The results of numerical investigation of the SIF behavior in the frequency range and the transition to the resonance mode are shown in figures 3 and 4. The graphs in figure 3 are constructed in the case of a crack of fixed length inclined at an angle α.

The length of crack is equal to one third of the distance AB between the vertices of the ellipse and its center is located on its axis. The graphs correspond to the angles of the crack inclination: $1 — \alpha = 0°$, $2 — \alpha = 30°$, $3 — \alpha = 45°$, $4 — \alpha = 60°$, $5 — \alpha = 90°$. An analysis of these graphs shows that, at frequencies less than the frequency of the first resonance when the angle of the crack inclination increases, the value of SIF decreases. The SIF attains maximum values in this frequency range in the case where the crack is located on the axis of the ellipse ($\alpha = 0°$). At all values of the crack inclination angle, the resonance occurs at the frequency with $\kappa_0 \approx 3.8$.

The graphs in figure 4 are constructed for the angle $\alpha = 0°$ and a crack of variable length. The length varied so that the left end C of the crack was fixed, and the right end D approached to the outer boundary of the cross-section. This approximation is attained when the parameter $\gamma = a/r_0$ changes from 0.0945 to 0.189. At the last value of this parameter, the crack touches the outer boundaries of the cross-section. Graphs 1–3 in figure 4 correspond to the next parameter values: $1 — \gamma = 0.0946$, $2 — \gamma = 0.1412$, $3 — \gamma = 0.188$. The absence of influence of the boundary on the SIF value k^- at the internal end of the crack is established. The SIF value k^+ at the opposite cracks end increases when it approaches the boundary as the crack length increases. The resonance phenomena are observed in the frequency range $3 < \kappa_0 < 4$.

Conclusions

The presence of cracks in an elastic hollow cylinder under harmonic load is accompanied both by the intensity of the dynamic stresses in a neighborhood of the defect and the resonant nature of their changes as a result of generation of a wave process in a bounded area. The most convenient for such an analysis is to consider the longitudinal shear strain describing the contribution of the crack SIFs. The SIF determination in the frequency range is based on the SIF relation to the function of the dynamic displacement jump at the crack, while this function is determined by separately solving the integral equations for the crack and satisfying the boundary conditions of dynamic load of the cylindrical body on the lateral surfaces. In this frequency range, one or two resonances can occur depending on the crack angle made with the body boundaries. The changes in the angle, as well as the crack approach to the outer surface, significantly influence the SIF values and the speed of their exit to the resonant mode in the low-frequency region.

References

[1] Nazarchuk Z T and Skalskij V R 2009 *Acoustic Emission Diagnostics of Structural Elements* vol 3 (Kyev: Naukova dumka) [in Ukrainian]
[2] Popov V G 1993 Comparison of the fields of displacements and stresses at the diffraction of shear elastic waves on different defects: a crack and a thin rigid inclusion *Dinam. Sist.* **12** 35–41
[3] Ang D D and Knopoff L 1964 Diffraction of scalar elastic waves by a finite strip *Proc. Natl. Sci. USA* **51** (4) 593–8
[4] Loeber J F and Sih G C 1968 Diffraction of antiplane waves by a finite crack *J. Acoust. Soc. Amer.* **44** (1) 90–8
[5] Zi G, Chen H, Xu J, and Belytschko T 2005 The extended finite element method for dynamic fractures *Shock Vibrat.* **12** (1) 9–23
[6] Mal A K 1970 Interaction of elastic waves with a Griffith crack *Int. J. Engng Sci.* **8** (9) 763–6
[7] Mykhas'kiv V V and Zhbadynskyi I Ja 2007 Solving nonstationary for composite body with crack method integral equations *Fiz.-Khim. Mekh. Mater.* **43** (1) 33–42
[8] Mykhas'kiv V, Zhbadynskyi I, and Zhang Ch 2010 Elastodynamic analysis of multiple crack problem in 3-D bi-materials by a BEM *Int. J. Num. Meth. Biomed. Engng* **26** (12) 1934–46
[9] Popov V G 2013 Harmonic vibrations of a half-space with a surface-breaking crack under conditions of out-of-plane deformation *Mech. Solids* **48** (2) 194–202
[10] Popov V G 1998 Interaction of elastic waves of longitudinal shear with radially situated cracks *Prikl. Mekh.* **34** (2) 60–66
[11] Nistor I, Pantale O, and Caperaa S 2008 Numerical implementation of the extended finite element method for dynamic crack analysis *Adv. Engng Software* **39** (7) 573–87
[12] Hosseini-Tehrani P, Hosseini-Godarzi A, and Tavangar M 2005 Boundary element analysis of stress intensity factor K_I in some two-dimensional dynamic thermoelastic problems *Engng Anal. Bound. Elem.* **29** (3) 232–40
[13] Chirino F and Domingues J 1989 Dynamic analysis of cracks using boundary element method *Engng. Fract. Mech.* **34** (5-6) 1051–61
[14] Bobylev L L and Dobrova Ju L 2003 Application of the boundary element method to the numerical analysis of forced vibrations of the elastic bodies of finite sizes with crack *Visn. Kharkiv Nats. Univ. Ser. Mat. Model. Inform. Tehnol. Avtom. Sist. Upravl.* **590** (1) 49–54
[15] Zhang Ch 2002 A 2D hypersingular time-domain traction BEM for transient elastodynamic crack analysis *Wave Motion* **35** (1) 17–40
[16] Poruchikov V E 1986 *Methods of the Dynamic Theory of Elasticity* (Moscow: Nauka) p 328
[17] Popov V G 1992 A study of the fields of displacements and stresses at the diffraction of shear elastic waves on a thin rigid exfoliated inclusion *Izv. Ross. Akad. Nauk. Mekh. Tverdogo Tela* No. 3 139–46
[18] Vekua N P 1970 *Systems of singular integral equations and some boundary-value problems* (Moscow: Nauka) p 252
[19] Belotserkovskii S M and Lifanov I K 1985 *Numerical Methods in Singular Integral Equations and Their Application in Aerodynamics, Theory of Elasticity, and Electrodynamics* (Moscow: Nauka) p 256
[20] Krylov V I 1967 *Approximate Calculation of Integrals* (Moscow: Nauka) p 500

IOP Conf. Series: Journal of Physics: Conf. Series **991** (2018) 012052 doi:10.1088/1742-6596/991/1/012052

Determination of necking time in tensile test specimens, under high-temperature creep conditions, subjected to distribution of stresses over the cross-section

A Lokoshchenko[1*] **and W Teraud**[1,2**]

[1]Institute of Mechanics, Lomonosov Moscow State University, Moscow, Russia

[2]Trapeznikov Institute of Control Sciences of the Russian Academy of Sciences, Moscow, Russia

E-mail: *loko@imec.msu.ru, **ldrnww@gmail.com

Abstract. The work describes an experimental research of creep of cylindrical tensile test specimens made of aluminum alloy D16T at a constant temperature of 400°C. The issue to be examined was the necking at different values of initial tensile stresses. The use of a developed noncontacting measuring system allowed us to see variations in the specimen shape and to estimate the true stress in various times. Based on the obtained experimental data, several criteria were proposed for describing the point of time at which the necking occurs (necking point). Calculations were carried out at various values of the parameters in these criteria. The relative interval of deformation time in which the test specimen is uniformly stretched was also determined.

1. Introduction

It is known that the gradual fracture of tensile test specimens of constant cross-section occurs with instability such as necking, which results from inhomogeneity of the material microstructure. Upon necking, the material loses its capacity to resist a given load, or its load-bearing capacity; therefore, it is of critical importance to be able to estimate this point of time.

In [1], it is stated that, in 1885, Considére suggested that when an elastoplastic material is examined, the maximum force criteria should be used as a necking condition. According to this condition, the thinning occurs as soon as the stress attains its maximum value in the specimen exposed to the uniaxial tension (beginning in the area, where the dependence of the tensile force P on the deformation ε decreases). In the following, this "maximum force approach" [1] was improved (Swift [2], Hill [3], and Hora [4]).

A more refined criterion for determining the necking time was obtained by Michel [5]. To study the creep, one of the classical methods for determining the value of necking time τ is the time approach according to which it is assumed that $\tau \approx (0.9 \div 0.99) \cdot t^*$, where t^* is the time prior to the fracture. To all appearance, this approach is true for high-strength metals, e.g., such as Cr-Mo-V of steel [6]. Other criteria also reveal the necking shortly before fracture, such as those published in [7] and [8].

Table 1. Chemical composition of the alloy under study.

Fe	Si	Mn	Cr	Ti	Cu	Mg	Zn	other
< 0.5	< 0.5	0.3–0.9	< 0.1	< 0.15	3.8–4.9	1.2–1.8	< 0.25	0.15

A. Malyigin [9] proposed a deformation approach in which the homogeneous deformation ε_d is limited by the value $1/n$ ($\varepsilon_d < 1/n$), where n is the exponent in an exponential law of the creep formation.

A specific feature of all known criteria is that a real shape of the specimen is not taken into account. The creep tests are usually carried out in a closed oven, and therefore the only characteristic of the deformed state to be measured in the course of real experiments is the dependence of the specimen elongation l on the time t. In the room temperature tests, it is possible to use an approach, where the tests are periodically interrupted for measurements of the specimen shape [10]. But the use of this approach is not acceptable for studying the high-temperature creep. The development of known and generalized criteria [1–10] is stipulated exactly by this. The method of noncontacting measurements, developed by the author of [11], enables us to measure the real shape of specimens exposed to tensile stresses at a high temperature and, thereby, to register the necking occurrence.

To know the point of time at which homogeneous deformation of solids becomes inhomogeneous is of utmost interest for the correct description of material resistance to a given load. In the paper, various criteria are proposed for objective determining this point of time on the basis of real experiments carried out.

2. Setup and performance of real experiments

To carry out experiments, specimens of a circular cross-section were selected, made of D16T (Al-Cu-Mg) alloy whose chemical composition is shown in table 1. The specimen shape was selected according to the international standard ISO 783-99. The initial length of the test portion was $l_0 = 23$–31 mm, and the initial diameter was $d_0 = 4$ or 5 mm. All specimens were cut out from the same bar. For each value of the initial tensile stress σ_0, the tests were carried out with different amounts of specimens; k_3, k_4, k_5 were given constant values. Table 2 presents all details. In the bottom line, the values of $\bar{\tau}_m$ (%) are given calculated as average values of $\bar{\tau}$ except for one maximum and one minimum value in each column.

The experiments were carried out at the Institute of Mechanics of Lomonosov Moscow State University, using the "IMEX-5" test facility. First, the specimen fixed on rods was heated up to an operation temperature, then the camera set to shoot the specimen elongation at given intervals was switched on. The specimen was quickly loaded up to a given level of initial stress $\sigma_{0,i}$ ($i = 1, \ldots, 4$) using a weight $P_{0,i}$ in the lower part of the test facility. The further deformation was caused under creep conditions with a constant tensile force up to the fracture.

For experiments, one used a noncontacting system to measure the geometrical parameters of the specimen. The system was developed by the author of [11]. The specimen geometry was measured remotely, with pictures taken during the experiment.

The detailed data on experimental research, used in the present paper, are given in [12].

2.1. Accuracy of measurements

In addition to the optical system of measurements [11], a contact inductive sensor is fixed to measure the elongation of the specimen between the grips of the test facility. The comparison of curves corresponding to experiments 1 and 4 is illustrated in figure 1. The maximum deviation

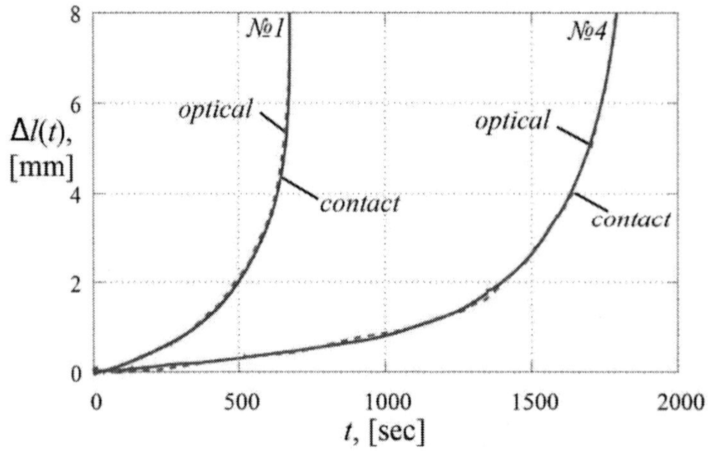

Figure 1. Elongation $\Delta l(t)$ of specimens 1 and 4 measured by a contact method (solid line) and an optical method (dashed line).

Table 2. Necking times τ in specimens of a circular cross-section with various parameters k_3, k_4, and k_5 for various criteria.

No.	σ_0 [MPa]	t^* [sec]	l_0^* [mm]	τ_3 [sec] ($\bar{\tau}_3$ [%]) for $k_3 = 0.1$ mm	τ_3 [sec] ($\bar{\tau}_3$ [%]) for $k_3 = 0.2$ mm	τ_4 [sec] ($\bar{\tau}_4$ [%]) for $k_4 = 0.3$ MPa	τ_4 [sec] ($\bar{\tau}_4$ [%]) for $k_4 = 1.4$ MPa	τ_5 [sec] ($\bar{\tau}_5$ [%]) for $k_5 = 0.3$ MPa	τ_5 [sec] ($\bar{\tau}_5$ [%]) for $k_5 = 1.4$ MPa
1	24.5	692	35.5	540 (78)	610 (88)	530 (77)	580 (84)	500 (72)	530 (77)
2	24.5	843	29	—	350 (46)	200 (26)	440 (57)	80 (10)	220 (29)
3	24.5	965	32.5	750 (78)	850 (88)	770 (80)	820 (85)	760 (79)	820 (85)
4	19.6	1733	42.5	1600 (92)	1660 (95)	1500 (86)	1720 (99)	1500 (86)	1720 (99)
5	19.6	1174	33.4	—	820 (72)	800 (70)	990 (86)	630 (55)	850 (74)
6	19.6	1575	32.9	1310 (82)	1400 (88)	1260 (79)	1430 (90)	1100 (69)	1420 (89)
7	19.6	1129	36.4	900 (80)	900 (80)	610 (54)	915 (81)	610 (54)	885 (78)
8	19.6	1407	40.8	850 (59)	1400 (98)	780 (55)	1400 (98)	—	1380 (96)
9	19.6	1707	37.2	1250 (73)	1500 (87)	1250 (73)	1500 (87)	1100 (64)	1400 (82)
10	17.6	2865	33.6	2520 (88)	2700 (94)	700 (24)	2640 (92)	700 (24)	2600 (91)
11	17.6	1772	28.3	1580 (90)	1650 (94)	—	1450 (83)	—	1400 (80)
			$\bar{\tau}_m$ [%]	81.3	87.3	64.2	87.4	59.7	83.5

between the methods for measuring the elongation, which are shown by the solid blue and dashed red lines, correspondingly, was 0.3 mm.

2.2. Tangent circle radius

To consider the stress distribution over the specimen cross-section according to [13], it is required to know the values of radii of longitudinal tangent circles R on the smallest cross-section of the neck at an arbitrary time. The radii of tangent circles were calculated by the least-square method for small ($\sim$ 1–3 mm of length) longitudinal parts of the specimen contours, measured

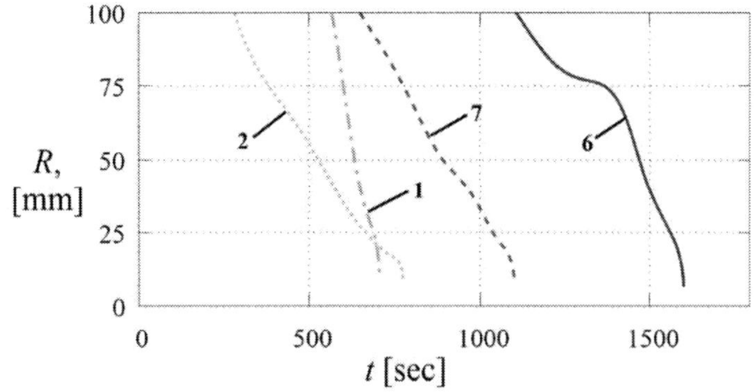

Figure 2. Time-dependence of the tangent circle radius R on the smallest cross-section. Specimens 1, 2, 6, and 7.

on the smallest cross-section, in each picture taken by the camera. The standard dependencies of $R(t)$ are presented in figure 2. From the necking occurrence and up to the point of fracture, the radius value decreases to 5–10 mm, while almost linear fracture occurs (figure 2).

3. Determination of the necking time

3.1. General points

To determine the necking time, one uses the Heaviside function

$$\Psi(t) = H[\varphi(t, l(t), d(t), k_i, \ldots)], \tag{1}$$

where φ is a function of parameters of mechanical tests, l is the length of the specimen, and d is its diameter. The $\Psi(t)$ functions take the values 0 or 1; the value 0 describes no deformation, and the value 1 characterizes the necking in the specimen. At $t = \tau$, stepwise changes in the function $\Psi(t)$ take place, from 0 to 1. Further, the $\Psi(t)$ functions with various types of dependence of φ are called criteria for the necking occurrence. Additionally, we introduce the relative time $\bar{t} = t/t^*$, and from here, the relative necking time $\bar{\tau} = \tau/t^*$.

Let us consider a specimen under homogenous deformation, i.e., a virtual specimen of diameter $d^0(t)$ subjected to deformation on the basis of the same pattern of changes $l(t)$ (to be observed in a "living" experiment), and this homogenous deformation of the whole test section lasts throughout the tension time. Based on the assumption that the volume of the cylindrical specimen is constant, it follows that

$$d^0(t) = d_0\sqrt{\frac{l_0}{l(t)}}. \tag{2}$$

As for the virtual specimen, its tensile stress is constant over the cross-section and the longitudinal coordinate y, because there is no distorting effect of the neck. Then, for each longitudinal cross-section, we obtain

$$\sigma^0(t) = \sigma_0\sqrt{\frac{l_0}{l(t)}}. \tag{3}$$

With no initial defects of the specimen structure, the real specimen deformation proceeds in the same way as that of the virtual one, up to the necking time. At a moment $t = \tau$, the deformation of the real specimen begins to differ from that of the virtual one.

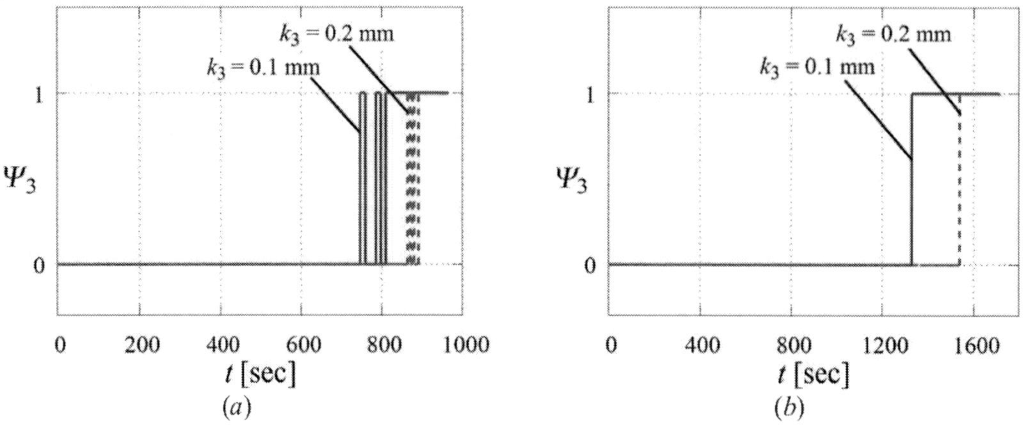

Figure 3. Dependence $\Psi_3(t)$ for $k_3 = 0.1\,\text{mm}$ (solid line) and $k_3 = 0.2\,\text{mm}$ (dashed line) for cylindrically shaped specimen (experiments 3 (a) and 9 (b)).

In [14], the Ψ_1 and Ψ_2 criteria are described with various types the dependence of φ on the arguments. Here, their further development is given.

3.2. Geometrical criterion

The experiments show that the deformation processes in all the experiments are axially symmetric. At an arbitrary point of time t, the circular specimen is characterized by the length $l(t)$ and the distribution of diameters $d(t, y)$. Let us assume that the necking criterion has the form

$$\Psi_3(t) = H[d^0(t) - \min_y(d(t, y)) - k_3]. \tag{4}$$

When the tests results were processed for each experiment according to (4), the time τ was calculated at which the value of Ψ_3 varies stepwise from 0 to 1. The dependence $\Psi_3(t)$ is illustrated in figure 3 for specimens 3 and 9. If the dependence $\Psi_3(t)$ fluctuates (i.e., is unstable), then the time τ is selected as an average value. The time values of necking according to (4) in all the experiments are shown in table 2. Their statistical processing allows us to concluded the following: the mathematical expectation of the necking time $\bar{\tau}_3$ for the Ψ_3 criterion is 81% for $k_3 = 0.1\,\text{mm}$ and 87% for $k_3 = 0.2\,\text{mm}$.

3.3. Kinematic criteria

The necking is related to a local increase in the tensile stress in a specimen. Therefore, to according to (3), during the necking, the maximum active tension $\sigma_{\max}(t)$ in the real specimen and the stress $\sigma^0(t)$ in the virtual one start to differ. A similar kinematic approach was described in [14]. If we assume that the stress is constant over the cross-section of the specimen, then $\sigma = \sigma_{m1}(t, y) = P_0/S(t, y) = 4P_0[\pi d^2(t, y)]^{-1}$, and then Ψ_4 becomes

$$\begin{aligned} \sigma_{m1}(t, y_{\min}) &= \max_y(\sigma_{m1}(t, y)), \\ \Psi_4(t) &= H[\sigma_{m1}(t, y_{\min}) - \sigma^0(t) - k_4]. \end{aligned} \tag{5}$$

Table 2 shows calculations of the necking time for the Ψ_4 criterion. The average value of the necking time $\bar{\tau}_4$, on the basis of Ψ_4, is 64% for $k_4 = 0.3\,\text{MPa}$ and 87% for $k_4 = 1.4\,\text{MPa}$.

On the basis of [13, 15], we conclude that, on the occurrence of a state of inhomogeneous deformation, the distribution of axial stress over the cross-section $y_{\min}$ of the minimum area is

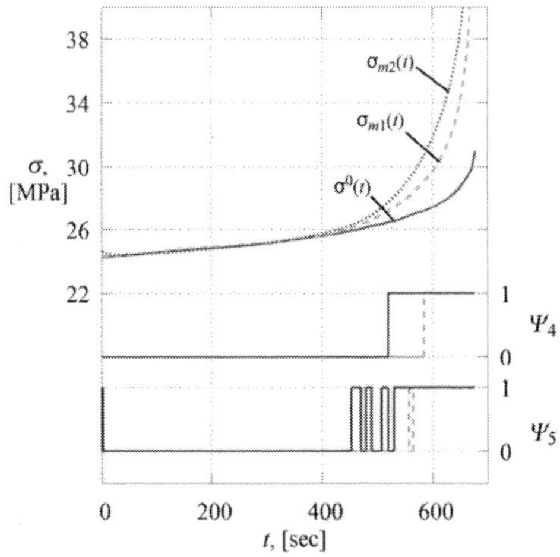

(a)

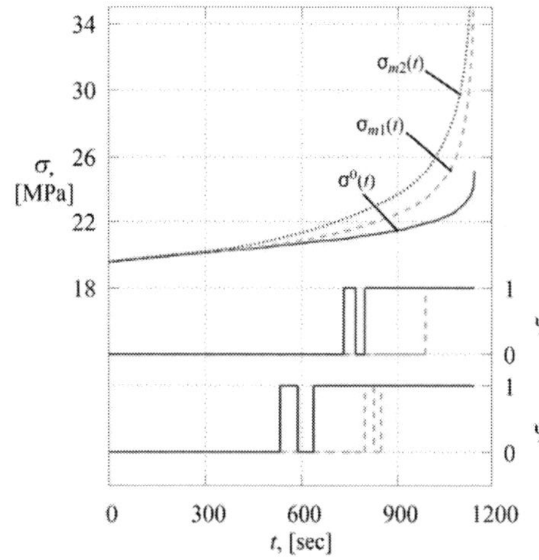

(b)

Figure 4. Dependence $\sigma^0(t)$ (solid line), $\sigma_{m1}(t)$ (dashed line), and $\sigma_{m2}(t)$ (dotted line) for the cross-section $y_{\min}$ and $\Psi_4(t)$ and $\Psi_5(t)$ criteria for $k_4 = k_5 = 0.3\,\text{MPa}$ (solid line) and $k_4 = k_5 = 1.4\,\text{MPa}$ (dotted line) (experiments 1 (a) and 5 (b)).

not constant and can be obtained from the parabolic expression

$$\sigma(t, y_{\min}, r) = \sigma_{m1}(t, y_{\min})\left[1 + \frac{0.25d^2(t, y_{\min}) - r^2}{Rd(t, y_{\min})}\right], \tag{6}$$

where R is the radius of neck curvature on the center line, r is the distance from an arbitrary point of the minimum cross-section to the specimen axis. This implies that the maximum stress value over the smallest cross-section, with the inhomogeneity taken into account, can be obtained at $r = 0$ as

$$\max_r(\sigma(t, y_{\min}, r)) = \sigma_{m2}(t, y_{\min}) = \sigma_{m1}(t, y_{\min})\left[1 + \frac{d(t, y_{\min})}{4R}\right]. \tag{7}$$

Therefore, replacing the criterion of constant stress over the specimen cross-section in the Ψ_4 criterion with an approach with a correcting expression according to [7] and taking $d_{\min}(t) = d(t, y_{\min})$, we obtain

$$\Psi_5(t) = H\left[\sigma_{m1}(t, y_{\min})\left(1 + \frac{d_{\min}}{4R}\right) - \sigma^0(t) - k_5\right]. \tag{8}$$

Therefore, Ψ_5 and Ψ_4 differ in the additional additive component $P_0/[\pi R d_{\min}(t)]$, which describes the increase in the stress at the specimen center relative to the average value due to the necking effect.

Figure 4 illustrates Experiments 1 and 5, where the curves of smoothed values of $\sigma_0(t)$, $\sigma_{m1}(t)$ and $\sigma_{m2}(t)$ are given for the cross-section $y_{\min}$, as well as values of Ψ_4 and Ψ_5 are given for the parameters $k_4 = k_5 = 0.3\,\text{MPa}$ and $k_4 = k_5 = 1.4\,\text{MPa}$. The deviations of the Ψ_4 and Ψ_5 curves can be explained by the corresponding fluctuations of the initial (chaotic) dependence $\sigma(t)$, which is not presented in the graph.

According to Ψ_5, the average necking time $\bar{\tau}_5$ is 60% for $k_5 = 0.3\,\text{MPa}$ and 83% for $k_5 = 1.4\,\text{MPa}$. Using the visual method for determining the time, we obtain the necking time equal to 61% (average value) of the time prior to fracture (curve $\sigma_{m1}(t)$) and 50.6% (curve $\sigma_{m2}(t)$). The Ψ_4 and Ψ_5 criteria, based on the experimental data, must be considered as basic ones to be used to determine the necking point.

Conclusions

The use of complex criteria (5) and (6) based on the introduction of a homogenous (virtual) specimen, allows one to conclude that the uniform deformation may take only 60% of the specimen life cycle. This must be taken into account in calculations of the load bearing capacity. As for geometrical criterion Ψ_3 (the difference between the diameter of a homogenous specimen and the minimum diameter of the specimen from the experiment) with $k_3 = 0.1$ mm, the necking time is 81%, and this is 87% for $k_3 = 0.2$ mm. Based on the Ψ_4 kinematic criteria (the difference between the stresses within a homogenous specimen and a real one) and based on the assumption that the stress value is constant over the cross-section, the necking time is 64% for $k_4 = 0.3$ MPa and 87% for $k_4 = 1.4$ MPa. For the kinematic criterion Ψ_5 (the difference between the stresses within a homogenous specimen and a real one) with the nonuniform character of stresses over the cross-section taken into account, the necking time is 60% for $k_5 = 0.3$ MPa and 83% for $k_5 = 1.4$ MPa. To estimate the necking time, it is most expedient to use the force criterion Ψ_5.

Acknowledgments

The work was supported by the Russian Foundation for Basic Research under grants Nos. 16-38-60200 and 17-08-00210.

References

[1] Altmeyer G 2011 *Modélisation théoretique et numérique des critères d'instabilité plastique. Application à la prédiction des phénomènes de striction et de localisation lors d'operation demise en forme par emboutissage.* PhD thesis (Françai: Arts et Métiers ParisTech)

[2] Swift H 1952 Plastic instability under plane stress *J. Mech. Phys. Solids* **1** (1) 1–18

[3] Hill R 1952 On discontinuous plastic states, with special reference to localized necking in thin sheets *J. Mech. Phys. Solids* **1** (1) 19–30

[4] Hora P, Tong L, and Reissner J 1996 Prediction method for ductile sheet metal failure in FE-simulation In *Proc. of the 3rd Int. Conf. NUMISHEET'96 "Numerical Simulation of 3-D Sheet Metal Forming Processes — Verification of Simulations with Experiments." Deaborn, Michigan, September 29 – October 3, 1996* ed J K Lee, G L Kinzel, and R H Wagoner (The Ohio State University) pp 252–6

[5] Brunet M and Clerc P 2007 Predictions of necking with analytical criteria and comparisons with experimental *AIP Conf. Proc.* **908** 81–9

[6] Wilshire B and Burt H 2006 Long-term creep design data for forged 1Cr–1Mo–0.25V steel *Strength, Fract. Compl.* **4** (2) 65–73

[7] Srinivas Naik, Janaki Ramulu P, and Ganesh Narayanan R 2010 Application of a few necking criteria in predicting the forming limit of unwelded and tailor-welded blanks *J. Strain Anal. Engng Des.* **45** (2) 79–96

[8] Radchenko V, Nebogina E, and Andreeva E 2009 Structural model of material softening at creep under complex stress conditions *Vestnik. Samar. GTU. Ser. Fiz.-Mat. Nauki* No. 1(18) 75–84

[9] Malygin G 2011 Influence of the grain size on the resistance of micro and nanocrystalline metals against the neck like localization of plastic deformation *Phys. Solid State* **53** (2) 363–8

[10] Nirmal K 2006 Constant-load tertiary creep in nickel-base single crystal superalloys *Mater. Sci. Engng* **432** (1-2) 129–41

[11] Teraud W 2013 Stretching of the cylinder specimen under high temperature measured based on the camera *Vestnik Dvigotelestr.* No. 2 236–9

[12] Lokoshchenko A and Teraud W 2013 Investigation of the localization of deformation in cylindrical samples by the method of photographic registration for high-temperature creep *Deform. Razrush. Mater.* **289** (11) 43–6

[13] Davidenkov N and Spiridonova N 1945 An analysis of the state of stress in the neck of the stretched sample *Zavod. Lab. Diagnost. Mater.* No. 6 583–93

[14] Teraud W 2017 Localization of the creep in rectangular samples at high temperature *Russ. Engng Res.* **37** (10) 850–6

[15] Bridgman P 1952 *Studies in Large Plastic Flow and Fracture with Special Emphasis on the Effects of Hydrostatic Pressure* (New York: McGraw-Hill) p 444

IOP Conf. Series: Journal of Physics: Conf. Series **991** (2018) 012053 doi:10.1088/1742-6596/991/1/012053

A stationary bulk planar ideal flow solution for the double shearing model

E A Lyamina[1*], N V Kalenova[2], and P P Date[3]

[1]Ishlinsky Institute for Problems in Mechanics of the Russian Academy of Sciences, Moscow, Russia
[2]Moscow Aviation Institute, Moscow, Russia
[3]Department of Mechanical Engineering, Indian Institute of Technology Bombay, Mumbai, India

E-mail: *lyamina@ipmnet.ru

Abstract. This paper provides a general ideal flow solution for the double shearing model of pressure-dependent plasticity. This new solution is restricted to a special class of stationary planar flows. A distinguished feature of this class of solutions is that one family of characteristic lines is straight. The solution is analytic. The mapping between Cartesian and principal lines based coordinate systems is given in parametric form with characteristic coordinates being the parameters. A simple relation that connects the scale factor for one family of coordinate curves of the principal lines based coordinate system and the magnitude of velocity is derived. The original ideal flow theory is widely used as the basis for inverse methods for the preliminary design of metal forming processes driven by minimum plastic work. The new theory extends this area of application to granular materials.

1. Introduction

The theory of stationary and non-stationary bulk ideal flow has been mainly developed for rigid perfectly plastic solids satisfying Tresca's yield condition and its associated flow rule. The first ideal flow solution has been derived in [1]. In this paper, a die profile for maximum efficiency in plane strain strip drawing has been determined. The existence of stationary three-dimensional ideal flows in Tresca's solids has been demonstrated in [2]. This result has been extended to non-stationary flows in [3]. A comprehensive overview on the ideal flow theory has been provided in [4]. This theory has been used as the basis for inverse methods for the preliminary design of bulk metal forming processes driven by minimum plastic work [1, 5, 6]. It has been shown in [7] that elasticity can be incorporated in the theory for stationary planar flow. A ductile fracture criterion has been incorporated in the ideal flow theory in [8]. The existence of stationary bulk planar ideal flows for anisotropic rigid plastic materials obeying the model proposed in [9] has been proven in [10]. All available results on the ideal flow theory have been obtained for pressure-independent materials. In the present paper, a general stationary ideal flow solution is given for the double shearing model proposed in [11]. This model is based on the Coulomb – Mohr yield criterion.

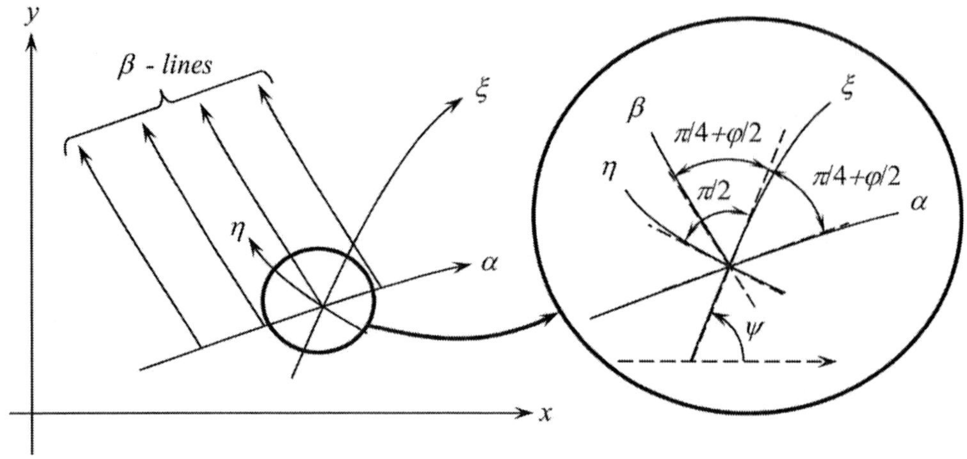

Figure 1.

2. System of equations and its properties

The constitutive equations of the double shearing model under plane strain conditions have been given in [11]. These equations comprise the Mohr-Coulomb yield criterion and the flow rule. Let $(\mu,\ \nu)$ be an arbitrary orthogonal curvilinear coordinate system. Then, the Mohr-Coulomb yield criterion and the flow rule read

$$(\sigma_{\mu\mu} + \sigma_{\nu\nu})\sin\varphi + \sqrt{(\sigma_{\mu\mu} - \sigma_{\nu\nu})^2 + 4\sigma_{\mu\nu}^2} = 2k\cos\varphi, \tag{1}$$

$$\varepsilon_{\mu\mu} + \varepsilon_{\nu\nu} = 0, \quad \sin(2\psi)(\varepsilon_{\mu\mu} - \varepsilon_{\nu\nu}) - 2\cos(2\psi)\varepsilon_{\mu\nu} - 2\sin\varphi\left(\omega_{\mu\nu} + \frac{d\psi}{dt}\right) = 0, \tag{2}$$

respectively. In equations (1) and (2), $\sigma_{\mu\mu}$, $\sigma_{\nu\nu}$, and $\sigma_{\mu\nu}$ are the components of the stress tensor, $\varepsilon_{\mu\mu}$, $\varepsilon_{\nu\nu}$, and $\varepsilon_{\mu\nu}$ are the components of the strain rate tensor and $\omega_{\mu\nu}$ is the only non-zero spin (vorticity) component in the (μ, ν) coordinate system. Also, ψ is the orientation of the major principal stress axis relative to the μ-direction, measured anti-clockwise positive, k is the cohesion, and φ is the angle of internal friction. Moreover, $d\psi/dt$ denotes the convected derivative. Equations (1) and (2) should be supplemented with the equilibrium equations.

Let us introduce a principal lines based coordinate system (ξ, η) where ξ-lines are trajectories of the major principal stress σ_ξ and η-lines are trajectories of the other principal stress in planes of flow σ_η. The system of stress equations comprising the Mohr-Coulomb yield criterion and the equilibrium equations is hyperbolic and the angle between each of the characteristic directions and the principal axis corresponding to the stress σ_ξ is equal to $\pi/4 + \varphi/2$. The (μ, ν) coordinate system can be chosen as a Cartesian coordinate system (x, y). The orientation of the principal lines based coordinate system and the characteristic coordinate system (α, β) relative to the Cartesian coordinate system is shown in figure 1.

It has been shown in [12] that the scale factors for the ξ- and η-coordinate lines, h_ξ and h_η, satisfy the relation

$$h_\xi h_\eta^b = 1, \tag{3}$$

where $b = (1 + \sin\varphi)/(1 - \sin\varphi)$.

IOP Conf. Series: Journal of Physics: Conf. Series **991** (2018) 012053 doi:10.1088/1742-6596/991/1/012053

3. Mapping between the Cartesian, principal lines based and characteristic coordinate systems

It follows from the geometry of figure 1 that

$$\frac{\partial x}{h_\xi \partial \xi} = \cos\psi, \quad \frac{\partial x}{h_\eta \partial \eta} = -\sin\psi, \quad \frac{\partial y}{h_\xi \partial \xi} = \sin\psi, \quad \frac{\partial y}{h_\eta \partial \eta} = \cos\psi.$$

Using (3) these equations can be transformed to

$$\frac{\partial x}{\partial \xi} = h^{-b}\cos\psi, \quad \frac{\partial x}{\partial \eta} = -h\sin\psi, \quad \frac{\partial y}{\partial \xi} = h^{-b}\sin\psi, \quad \frac{\partial y}{\partial \eta} = h\cos\psi \tag{4}$$

where $h \equiv h_\eta$. Substituting (4) into the compatibility equations

$$\frac{\partial^2 x}{\partial \xi \partial \eta} = \frac{\partial^2 x}{\partial \eta \partial \xi}, \quad \frac{\partial^2 y}{\partial \xi \partial \eta} = \frac{\partial^2 y}{\partial \eta \partial \xi}$$

gives

$$\begin{aligned} &-bh^{-b-1}\cos\psi\frac{\partial h}{\partial \eta} - h^{-b}\sin\psi\frac{\partial \psi}{\partial \eta} + \sin\psi\frac{\partial h}{\partial \xi} + h\cos\psi\frac{\partial \psi}{\partial \xi} = 0, \\ &-bh^{-b-1}\sin\psi\frac{\partial h}{\partial \eta} + h^{-b}\cos\psi\frac{\partial \psi}{\partial \eta} - \cos\psi\frac{\partial h}{\partial \xi} + h\sin\psi\frac{\partial \psi}{\partial \xi} = 0. \end{aligned} \tag{5}$$

Rotating the Cartesian coordinate system so that its x-axis coincides with principal axis corresponding to the stress σ_ξ at a given point transforms (5) to the following system of equations for h and ψ in the (ξ, η) coordinate system

$$bh^{-b-1}\frac{\partial h}{\partial \eta} - h\frac{\partial \psi}{\partial \xi} = 0, \quad h^{-b}\frac{\partial \psi}{\partial \eta} - \frac{\partial h}{\partial \xi} = 0. \tag{6}$$

Using a standard technique it is possible to show that this system of equations is hyperbolic. In particular, the characteristic curves are given by

$$\frac{d\eta}{d\xi} = -\sqrt{b}h^{-b-1}, \quad \frac{d\eta}{d\xi} = \sqrt{b}h^{-b-1}. \tag{7}$$

Here the first equation determines the α-lines and the second the β-lines. The characteristic relations along these lines are

$$\begin{aligned} &\sqrt{b}d\ln h + d\psi = 0 \quad \text{on } \alpha\text{-line}, \\ &-\sqrt{b}d\ln h + d\psi = 0 \quad \text{on } \beta\text{-line}. \end{aligned} \tag{8}$$

In what follows, it is assumed that β-lines are straight. In this case ψ is independent of β. Then, it is seen from the second equation in (8) that h is independent of β. The equations in (8) can be immediately integrated. Following [13] it is convenient to write the solution of these equations as

$$\sqrt{b}\ln\frac{h}{h_0} + \psi - \psi_0 = 0, \quad -\sqrt{b}\ln\frac{h}{h_0} + \psi - \psi_0 = 2\alpha\cos\varphi. \tag{9}$$

Here h_0 and ψ_0 are constant. Solving (9) for h and ψ yields

$$\sqrt{b}\ln\frac{h}{h_0} = -\alpha\cos\varphi, \quad \psi - \psi_0 = \alpha\cos\varphi. \tag{10}$$

Using (10) equations (7) can be rewritten as

$$\frac{\partial \eta}{\partial \alpha} = -\sqrt{b}h_0^{-b-1}\exp(2\alpha)\frac{\partial \xi}{\partial \alpha}, \quad \frac{\partial \eta}{\partial \beta} = \sqrt{b}h_0^{-b-1}\exp(2\alpha)\frac{\partial \xi}{\partial \beta}. \tag{11}$$

It is convenient to introduce the new variable ω by

$$\xi = \omega\exp(-2\alpha). \tag{12}$$

Eliminating ξ in (11) by means of this equation yields

$$\frac{\partial \eta}{\partial \alpha} = -\sqrt{b}h_0^{-b-1}\left(\frac{\partial \omega}{\partial \alpha} - 2\omega\right), \quad \frac{\partial \eta}{\partial \beta} = \sqrt{b}h_0^{-b-1}\frac{\partial \omega}{\partial \beta}. \tag{13}$$

The second equation can be immediately integrated to give

$$\eta = \sqrt{b}h_0^{-1-b}\omega + 2\sqrt{b}h_0^{-1-b}\Phi_1(\alpha), \tag{14}$$

where $\Phi_1(\alpha)$ is an arbitrary function of α. Eliminating η in the first equation in (13) by means of (14) results in

$$\frac{\partial \omega}{\partial \alpha} - \omega = -\frac{d\Phi_1}{d\alpha}.$$

The general solution of this equation is

$$\omega = -\Phi_1(\alpha) - e^{\alpha}\int \Phi_1(\alpha)e^{-\alpha}\,d\alpha + e^{\alpha}\Phi_2(\beta), \tag{15}$$

where $\Phi_2(\beta)$ is an arbitrary function of β. Eliminating ω in (12) and (14) by means of (15) yields

$$\begin{aligned} \xi &= -\Phi_1(\alpha)e^{-2\alpha} - e^{-\alpha}\int \Phi_1(\alpha)e^{-\alpha}\,d\alpha + e^{-\alpha}\Phi_2(\beta), \\ \frac{\eta h_0^{1+b}}{\sqrt{b}} &= -e^{\alpha}\int \Phi_1(\alpha)e^{-\alpha}\,d\alpha + e^{\alpha}\Phi_2(\beta) + \Phi_1(\alpha). \end{aligned} \tag{16}$$

These equations supply the mapping between the principal line based and characteristic coordinates. The infinitesimal arc lengths of the α- and β- characteristic curves are

$$ds_\alpha = \sqrt{\left(\frac{\partial x}{\partial \alpha}\right)^2 + \left(\frac{\partial y}{\partial \alpha}\right)^2}\,d\alpha, \quad ds_\beta = \sqrt{\left(\frac{\partial x}{\partial \beta}\right)^2 + \left(\frac{\partial y}{\partial \beta}\right)^2}\,d\beta,$$

respectively. Using the chain rule these relations can be transformed to

$$\begin{aligned} ds_\alpha &= \sqrt{\left(\frac{\partial x}{\partial \xi}\frac{\partial \xi}{\partial \alpha} + \frac{\partial x}{\partial \eta}\frac{\partial \eta}{\partial \alpha}\right)^2 + \left(\frac{\partial y}{\partial \xi}\frac{\partial \xi}{\partial \alpha} + \frac{\partial y}{\partial \eta}\frac{\partial \eta}{\partial \alpha}\right)^2}\,d\alpha, \\ ds_\beta &= \sqrt{\left(\frac{\partial x}{\partial \xi}\frac{\partial \xi}{\partial \beta} + \frac{\partial x}{\partial \eta}\frac{\partial \eta}{\partial \beta}\right)^2 + \left(\frac{\partial y}{\partial \xi}\frac{\partial \xi}{\partial \beta} + \frac{\partial y}{\partial \eta}\frac{\partial \eta}{\partial \beta}\right)^2}\,d\beta. \end{aligned} \tag{17}$$

Substituting (4), (10), and (11) into (17) gives

$$\begin{aligned} ds_\alpha &= -\sqrt{1+bh_0^{-b}}\exp[-\alpha(1-\sin\varphi)]\left(\frac{d\Phi_1}{d\alpha} + \omega\right)d\alpha, \\ ds_\beta &= \sqrt{1+bh_0^{-b}}\exp(\alpha\sin\varphi)\frac{d\Phi_2}{d\beta}\,d\beta. \end{aligned} \tag{18}$$

IOP Conf. Series: Journal of Physics: Conf. Series **991** (2018) 012053 doi:10.1088/1742-6596/991/1/012053

It is seen from this equation that it is possible to put with no loss of generality that $\Phi_2(\beta) = \beta$. Then, (16) and (18) become

$$\begin{aligned} \xi &= -\Phi_1(\alpha)e^{-2\alpha} - e^{-\alpha}\int \Phi_1(\alpha)e^{-\alpha}\,d\alpha + \beta e^{-\alpha}, \\ \frac{\eta h_0^{1+b}}{\sqrt{b}} &= -e^{\alpha}\int \Phi_1(\alpha)e^{-\alpha}\,d\alpha + \beta e^{\alpha} + \Phi_1(\alpha), \end{aligned} \tag{19}$$

$$\begin{aligned} ds_\alpha &= -\sqrt{1+bh_0^{-b}}\exp[-\alpha(1-\sin\varphi)]\left[\frac{d\Phi_1}{d\alpha} - \Phi_1(\alpha) - e^{\alpha}\int \Phi_1(\alpha)e^{-\alpha}\,d\alpha + e^{\alpha}\beta\right]d\alpha, \\ ds_\beta &= \sqrt{1+bh_0^{-b}}\exp(\alpha\sin\varphi)\,d\beta. \end{aligned} \tag{20}$$

It follows from the geometry of figure 1 that

$$\begin{aligned} \frac{\partial x}{\partial s_\alpha} &= \cos\left(\psi - \frac{\pi}{4} - \frac{\varphi}{2}\right), \quad \frac{\partial x}{\partial s_\beta} = \cos\left(\psi + \frac{\pi}{4} + \frac{\varphi}{2}\right), \\ \frac{\partial y}{\partial s_\alpha} &= -\cos\left(\psi + \frac{\pi}{4} - \frac{\varphi}{2}\right), \quad \frac{\partial y}{\partial s_\beta} = \cos\left(\psi - \frac{\pi}{4} + \frac{\varphi}{2}\right). \end{aligned} \tag{21}$$

Equations (20) and (21) combine to give

$$\begin{aligned} \frac{\partial x}{\partial \alpha} &= -\sqrt{1+bh_0^{-b}}\exp[-\alpha(1-\sin\varphi)]\left[\frac{d\Phi_1}{d\alpha} - \Phi_1(\alpha) - e^{\alpha}\int \Phi_1(\alpha)e^{-\alpha}\,d\alpha + e^{\alpha}\beta\right]\cos\left(\psi - \frac{\pi}{4} - \frac{\varphi}{2}\right), \\ \frac{\partial x}{\partial \beta} &= \sqrt{1+bh_0^{-b}}\exp(\alpha\sin\varphi)\cos\left(\psi + \frac{\pi}{4} + \frac{\varphi}{2}\right), \\ \frac{\partial y}{\partial \alpha} &= \sqrt{1+bh_0^{-b}}\exp[-\alpha(1-\sin\varphi)]\left(\frac{d\Phi_1}{d\alpha} - \Phi_1(\alpha) - e^{\alpha}\int \Phi_1(\alpha)e^{-\alpha}\,d\alpha + e^{\alpha}\beta\right)\cos\left(\psi + \frac{\pi}{4} - \frac{\varphi}{2}\right), \\ \frac{\partial y}{\partial \beta} &= \sqrt{1+bh_0^{-b}}\exp(\alpha\sin\varphi)\cos\left(\psi - \frac{\pi}{4} + \frac{\varphi}{2}\right). \end{aligned} \tag{22}$$

Integrating the second equation yields

$$x = \sqrt{1+bh_0^{-b}}\exp(\alpha\sin\varphi)\cos\left(\psi + \frac{\pi}{4} + \frac{\varphi}{2}\right)\beta + \sqrt{1+bh_0^{-b}}X(\alpha). \tag{23}$$

Here $X(\alpha)$ is an arbitrary function of α. Substituting this solution into the first equation in (22) results in

$$\frac{dX}{d\alpha} = -\left(\frac{d\Phi_1}{d\alpha} - \Phi_1(\alpha) - e^{\alpha}\int \Phi_1(\alpha)e^{-\alpha}\,d\alpha\right)\exp[\alpha(\sin\varphi - 1)]\cos\left(\psi - \frac{\pi}{4} - \frac{\varphi}{2}\right). \tag{24}$$

Analogously, integrating the forth equation in (22) yields

$$y = \sqrt{1+bh_0^{-b}}\exp(\alpha\sin\varphi)\cos\left(\psi - \frac{\pi}{4} + \frac{\varphi}{2}\right)\beta + \sqrt{1+bh_0^{-b}}Y(\alpha). \tag{25}$$

Here $Y(\alpha)$ is an arbitrary function of α. Substituting this solution into the third equation in (22) results in

$$\frac{dY}{d\alpha} = \left[\frac{d\Phi_1}{d\alpha} - \Phi_1(\alpha) - e^{\alpha}\int \Phi_1(\alpha)e^{-\alpha}\,d\alpha\right]\exp[\alpha(\sin\varphi - 1)]\cos\left(\psi + \frac{\pi}{4} - \frac{\varphi}{2}\right). \tag{26}$$

Eliminating ψ in (24) and (26) by means of (10) and integrating determine the functions $X(\alpha)$ and $Y(\alpha)$. Thus equations (19), (23), and (25) supply the mapping between the Cartesian and principal lines based coordinate systems in parametric form with α and β being the parameters.

4. Ideal flow solution

The solutions (19), (23), and (25) satisfy the equilibrium equations and Coulomb-Mohr yield criterion. In the case of ideal flow, the ξ-coordinate lines are streamlines. In order to prove the existence of ideal flow, it is necessary to show that this condition is compatible with the velocity equations of the double shearing model. Let us choose the (μ,ν)-coordinate system such that the μ-coordinate lines are the ξ-coordinate lines. In this case the angle ψ involved in (2) is the angle between the ξ-coordinate lines and the principal stress direction corresponding to σ_1. Therefore, $\psi = 0$ everywhere in the plastic region. Then, equation (2) becomes

$$\varepsilon_{\xi\xi} + \varepsilon_{\eta\eta} = 0, \quad \varepsilon_{\xi\eta} + \sin\varphi\omega_{\xi\eta} = 0. \tag{27}$$

It has been taken into account here that $\varepsilon_{\xi\xi} \equiv \varepsilon_{\mu\mu}$, $\varepsilon_{\eta\eta} \equiv \varepsilon_{\nu\nu}$, $\varepsilon_{\xi\eta} \equiv \varepsilon_{\mu\nu}$, and $\omega_{\xi\eta} \equiv \omega_{\mu\nu}$. Since the ξ-coordinate lines are streamlines, the velocity vector has one non-zero component $u_\xi = u$. In this case [14]

$$\varepsilon_{\xi\xi} = \frac{1}{h_\xi}\frac{\partial u}{\partial \xi}, \quad \varepsilon_{\eta\eta} = \frac{u}{h_\xi h_\eta}\frac{\partial h_\eta}{\partial \xi}, \quad 2\varepsilon_{\xi\eta} = \frac{1}{h_\eta}\frac{\partial u}{\partial \eta} - \frac{u}{h_\xi h_\eta}\frac{\partial h_\xi}{\partial \eta}, \quad 2\omega_{\xi\eta} = \frac{1}{h_\eta}\frac{\partial u}{\partial \eta} + \frac{u}{h_\xi h_\eta}\frac{\partial h_\xi}{\partial \eta}.$$

Using (3) these relations can be transformed to

$$\varepsilon_{\xi\xi} = h^b\frac{\partial u}{\partial \xi}, \quad \varepsilon_{\eta\eta} = uh^{b-1}\frac{\partial h}{\partial \xi}, \quad 2\varepsilon_{\xi\eta} = \frac{1}{h}\frac{\partial u}{\partial \eta} + \frac{bu}{h^2}\frac{\partial h}{\partial \eta}, \quad 2\omega_{\xi\eta} = \frac{1}{h}\frac{\partial u}{\partial \eta} - \frac{bu}{h^2}\frac{\partial h}{\partial \eta}. \tag{28}$$

Substituting (28) into (27) yields

$$\frac{\partial u}{\partial \xi} + \frac{u}{h}\frac{\partial h}{\partial \xi} = 0, \quad \frac{\partial u}{\partial \eta} + \frac{u}{h}\frac{\partial h}{\partial \eta} = 0.$$

These equations can be immediately integrated to give

$$u = \frac{C}{h}. \tag{29}$$

Here C is constant. Therefore, once any solution satisfying equations (19), (23), and (25) has been found, the magnitude of the velocity vector is determined from (29). The direction of this vector is tangent to the ξ-coordinate lines.

Conclusions

It has been shown that non-trivial stationary planar ideal flow solutions exist in pressure-dependent plasticity if the double shearing model based of the Coulomb–Mohr yield criterion is adopted. The solution found is based on the assumption that one family of characteristic lines is straight. In this case, the solution is practically analytic. In particular, the mapping between the principal lines based and characteristic coordinate systems is given by equation (19) and the mapping between the Cartesian and characteristic coordinate systems by equation (23) and (25). Thus the mapping between the principal lines based and Cartesian coordinate systems is given in parametric form. The function $\Phi_1(\alpha)$ involved in equations (19), (24), and (26) should be determined from boundary conditions. Once a solution for the mappings between different coordinate systems has been found, the magnitude of the velocity vector is determined from (29). The direction of this vector is tangent to the ξ-coordinate lines.

Acknowledgments

The research described was supported by the grants RSCF-16-49-02026 (Russia) and INT/RUS/RFBR/P-214 (India).

IOP Conf. Series: Journal of Physics: Conf. Series **991** (2018) 012053 doi:10.1088/1742-6596/991/1/012053

References

[1] Richmond O and Devenpeck M L 1962 A die profile for maximum efficiency in strip drawing *Proc. 4th U.S. National Congr. of Applied Mechanics (Berkeley)* vol 2, ed R M Rosenberg (New York: ASME) p 1053
[2] Hill R 1967 Ideal forming operations for perfectly plastic solids *J. Mech. Phys. Solids* **15** 223–7
[3] Richmond O and Alexandrov S 2002 The theory of general and ideal plastic deformations of Tresca solids *Acta Mech.* **158** (1) 33–42
[4] Chung K and Alexandrov S 2007 Ideal flow in plasticity *Appl. Mech. Rev.* **60** 316–35
[5] Richmond O and Morrison H L 1967 Streamlined wire drawing dies of minimum length *J. Mech. Phys. Solids.* **15** 195–203
[6] Richmond O 1968 Theory of streamlined dies for drawing and extrusion In *Mechanics of the Solid State* eds F P J Rimrott and J Schwaighofer (Toronto: University of Toronto Press) pp 154–67
[7] Hill R 1985 On the kinematics of steady plane flows in elastoplastic media In *Metal Forming and Impact Mechanics* ed S R Reid (Oxford: Pergamon Press) pp 3–17
[8] Alexandrov S 2003 Fracture prediction in steady ideal plastic flows *Acta Mech.* **163** 127–38
[9] Collins I F and Meguid S A 1977 On the influence of hardening and anisotropy on the plane-strain compression of thin metal strip *ASME J. Appl. Mech.* **44** (2) 271–8
[10] Alexandrov S, Mustafa Y, and Lyamina E 2016 Steady planar ideal flow of anisotropic materials *Meccanica* **51** 2235–41
[11] Spencer A J M 1964 A theory of the kinematics of ideal soils under plane strain conditions *J. Mech. Phys. Solids* **12** 337–51
[12] Alexandrov S and Harris D 2017 Geometry of principal stress trajectories for a. Mohr-Coulomb material under plane strain *ZAMM* **97** (4) 473–6
[13] Alexandrov S 2015 Geometry of plane strain characteristic fields in pressure-dependent plasticity *ZAMM* **95** (11) 1296–301
[14] Malvern L E 1969 *Introduction to the Mechanics of a Continuous Medium* (Englewood Cliffs: Prentice-Hall)

Mechanical and thermal properties of MoS_2 reinforced epoxy nanocomposites

S R Madeshwaran[1], R Jayaganthan[2], R Velmurugan[1*], N K Gupta[3], and A V Manzhirov[4,5]

[1]Department of Aerospace Engineering, Indian Institute of Technology Madras, Chennai, India
[2]Department of Engineering Design, Indian Institute of Technology Madras, Chennai, India
[3]Department of Applied Mechanics, Indian Institute of Technology Delhi, Delhi, India
[4]Ishlinsky Institute for Problems in Mechanics of the Russian Academy of Sciences, Moscow, Russia
[5]Bauman Moscow State Technical University, Moscow, Russia

E-mail: *mvs@ramanv@iitm.ac.in

Abstract. The effects of molybdenum disulfide (MoS_2) on thermal expansion and mechanical properties of epoxy composites were investigated. MoS_2 nanosheets were exfoliated by ultra-sonication and reinforced into epoxy as nanofiller by mechanical stirring. Transmission electron microscopy observations demonstrated that MoS_2 exhibited better dispersion in epoxy matrix. Thermal expansion measured by dilatometer has revealed that increasing MoS_2 fraction in epoxy matrix significantly reduced the coefficient of thermal expansion (CTE). The 0.5wt% MoS_2 incorporated epoxy composites shows 35.8% reduction in CTE as compared to neat epoxy. The addition of small fraction of MoS_2(0.1wt%) in the composites increased the tensile and flexural strength 39.2% and 9.0% respectively. The glass transition temperature (T_g) of 0.1wt% MoS_2 incorporated epoxy composites shows 7.39% increase in T_g.

1. Introduction

Polymer nanocomposites is one of the emerging fields in materials research because of their novel applications ranging from electrochemical energy storage, light weight structures in aerospace and automotive, and flexible substrate in electronics. Among various fillers, carbon based nanofillers such as carbon nanotubes and graphene exhibit excellent electrical, mechanical and thermal properties, which make them as potential reinforcing filler for polymer composites [1–7]. However, carbon nanotubes are expensive and limits its industrial applications. On the other hand, graphene is electrically conducting, which often restricts its application in power electronics, electric motors, packing etc. [8]. Recently, two-dimensional (2D) transition metal dichalcogenides such as molybdenum disulfide (MoS_2) and tungsten disulfide (WS_2) have emerged as an ideal material to replace carbon nanotubes and graphene because for their unique functional properties and wide range of applications including lubricants, electro catalyst, hydrogen storage etc. [9–13].

The detailed investigation of thermal and mechanical behavior of epoxy nanocomposites fabricated by using MoS_2 as filler is limited. Therefore, the present work has been focused to fabricate epoxy nanocomposites with varying addition of MoS_2 and subsequently investigated

IOP Conf. Series: Journal of Physics: Conf. Series **991** (2018) 012054 doi:10.1088/1742-6596/991/1/012054

for thermal and mechanical properties. The fabrication of MoS_2 reinforced epoxy composites was made by combination of ultra-sonication and mechanical shear mixing techniques.

2. Experimental studies

2.1. Materials

MoS_2 was procured from Loba Chemie Pvt. Ltd., Mumbai, India. N, N'-dimethylformamide (DMF) and tetrahydrofuran (THF) were purchased from Merck Life Sciences Pvt. Ltd., Mumbai, India. Diglycidyl ether of bisphenol A (DGEBA Araldite LY556) based epoxy resin was obtained from Huntsman, India. All the chemicals were used without further purification.

2.2. Synthesis of Exfoliated MoS_2 (ex-MoS_2) and Fabrication of epoxy-MoS_2 nanocomposites

First, 400 mg of MoS_2 powder was sonicated in 80 ml of DMF using a horn-type sonicator for 12 hours. Then, DMF was removed by filtration and the resultant exfoliated MoS_2 (ex-MoS_2) powder was dried in oven at 60°C for 12 hours. The as-obtained ex-MoS_2 powder was redispersed in a small amount of THF with a bath-type sonicator for 30 minutes. The resultant ex-MoS_2/THF dispersion was added to diluted epoxy resin and mechanically stirred using shear mixer for 4 hours at 700 rpm. Finally, calculated amount of curing agent (triethylenetetramine) was added and the reaction mixture was further stirred for 5 minutes. The resultant homogenous solution was poured in a silicone mold and cured at room temperature for 24 hours to obtain epoxy-MoS_2 nanocomposites. Finally, the samples were post cured at 60°C for 3 hours. The loadings of ex-MoS_2 in epoxy composites were varied (0.05wt%, 0.1wt%, 0.25wt% and 0.5wt%), and the resulting composites were named as Epoxy-MoS_2(0.05%), Epoxy-MoS_2(0.1%), Epoxy-MoS_2(0.25%), and Epoxy-MoS_2(0.5%) respectively.

2.3. Microstructural, Thermal, and Mechanical Characterization

Transmission electron microscopy (TEM) images of the ex-MoS_2 and epoxy-MoS_2 composites were obtained using a (JEOL JEM 2100 instrument). A small amount of ex-MoS_2 powder sample was dispersed in THF by bath sonication and then dropped on 200 mesh copper grids with supporting carbon films. For epoxy-Mos_2 composites, the TEM samples were prepared by ultramicrotoming technique. Scanning electron microscopy (SEM) images of bulk MoS_2 and ex-MoS_2 powder were obtained by using M/s FEI Quanta 3D FEG. Thermal expansion studies of nanocomposites were carried out using a Push-Rod Dilatometer, M/s VB ceramic consultants, India. Differential Scanning Calorimetry (DSC) measurements were performed using an M/s Netzsch DSC 200 F3 Maia instruments. The tensile properties of epoxy and EP-MoS_2 nanocomposites were using a UTM M/s Kalpak Instruments and Controls, Pune India. The flexural strength of the prepared nanocomposites was measured using UTM Instron 3365.

3. Results and discussion

The morphology of bulk MoS_2 powder and exfoliated MoS_2 (ex-MoS_2) was observed by SEM microscopy and the images are shown in figure 1. The figure 1 *a* and 1 *b* shows that bulk MoS_2 powder exhibits stacked layer-like morphology with thickness of several micrometers. However, ultra-sonication in DMF results in drastic exfoliation of layer structure, as evident from figure 1 *c* and 1 *d*.

To further understand the exfoliation state and morphology of ex-MoS_2 sheets, TEM analysis was carried out and the results are shown in figure 2. The ex-MoS_2 exhibits thin nano sheets morphology with several layers, suggesting that MoS_2 underwent effective exfoliation during ultrasonication process.

Flexural strength of epoxy and epoxy-MoS_2 nanocomposites were measured using a three point bend testing method and results are displayed in figure 3. As compared to neat

IOP Conf. Series: Journal of Physics: Conf. Series **991** (2018) 012054 doi:10.1088/1742-6596/991/1/012054

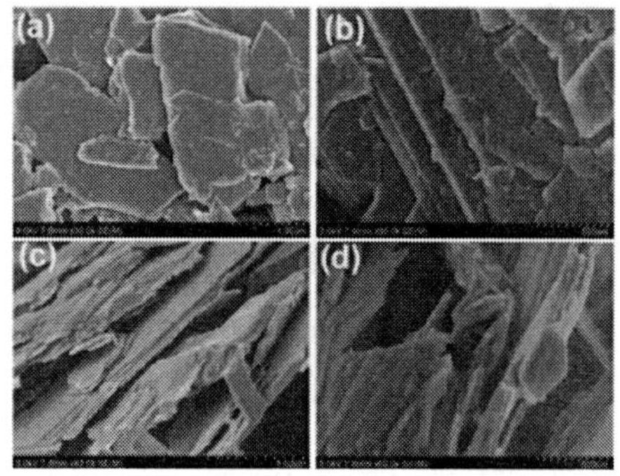

Figure 1. SEM images of (*a*, *b*) bulk MoS_2 and (*c*, *d*) ex-MoS_2.

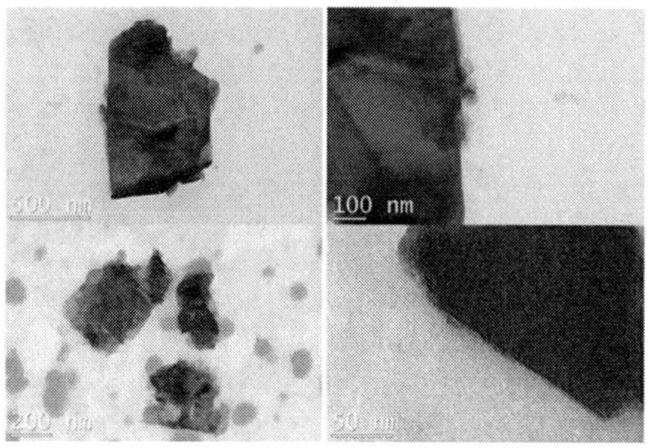

Figure 2. TEM image of ex-MoS_2.

epoxy (89.46 MPa), flexural strength of composites with 0.05wt% (113.49 MPa) and 0.1wt% (97.58 MPa) ex-MoS_2 content have increased 26.86% and 9.07%, respectively. However, the composites with 0.25wt% (64.90 MPa) and 0.5wt% (72.55 MPa) weight fractions of ex-MoS_2, the flexural strength has decreased to 27.45% and 18.90% with respect to neat epoxy (fig 3 *b*). The results indicate that the flexural strength increases for composites with lower nanofiller content ($\sim$ 0.1wt%), which can be attributed to good interfacial interaction between ex-MoS_2 and epoxy chains, as evidenced by TEM characterization (figure 2). However, flexural strength decreases for composites with higher nanofiller loadings (beyond 0.1wt%) which could be due to agglomeration of ex-MoS_2 results in lesser interfacial interaction between epoxy and nanofiller. Figure 3 *c* displays the flexural modulus values of neat epoxy and epoxy-MoS_2 composites. It has increased from 2.43 to 3.39, 3.46, 2.54, and 2.98 GPa when ex-MoS_2 content is increased from 0 to 0.05wt%, 0.1wt%, 0.25wt% and 0.5wt%, respectively. The reduction in flexural modulus beyond 0.1wt% nanofiller content in the composites was attributed to the agglomeration of MoS_2. The flexural strain at break values of neat epoxy and epoxy-MoS_2 composites is shown in figure 3 *d*.

Figure 4 *a* shows the tensile stress-strain plots for the neat epoxy and epoxy-MoS_2 composites with different weight fractions of ex-MoS_2 (0.05wt% to 0.5wt%). The test results demonstrate that tensile strength increases with increasing filler content (ex-MoS_2) up to 0.1wt%. However, beyond 0.1wt%, reducing trend is observed. The tensile strength of neat epoxy, epoxy-MoS_2(0.05wt%), epoxy-MoS_2(0.1wt%), epoxy-MoS_2(0.25%), and epoxy-MoS_2(0.5wt%) nanocomposites was found to be 51.03, 60.80, 71.04, 65.62, and 59.76 MPa respectively. The test results revealed that the tensile strength increased 19.15%, 39.21%,

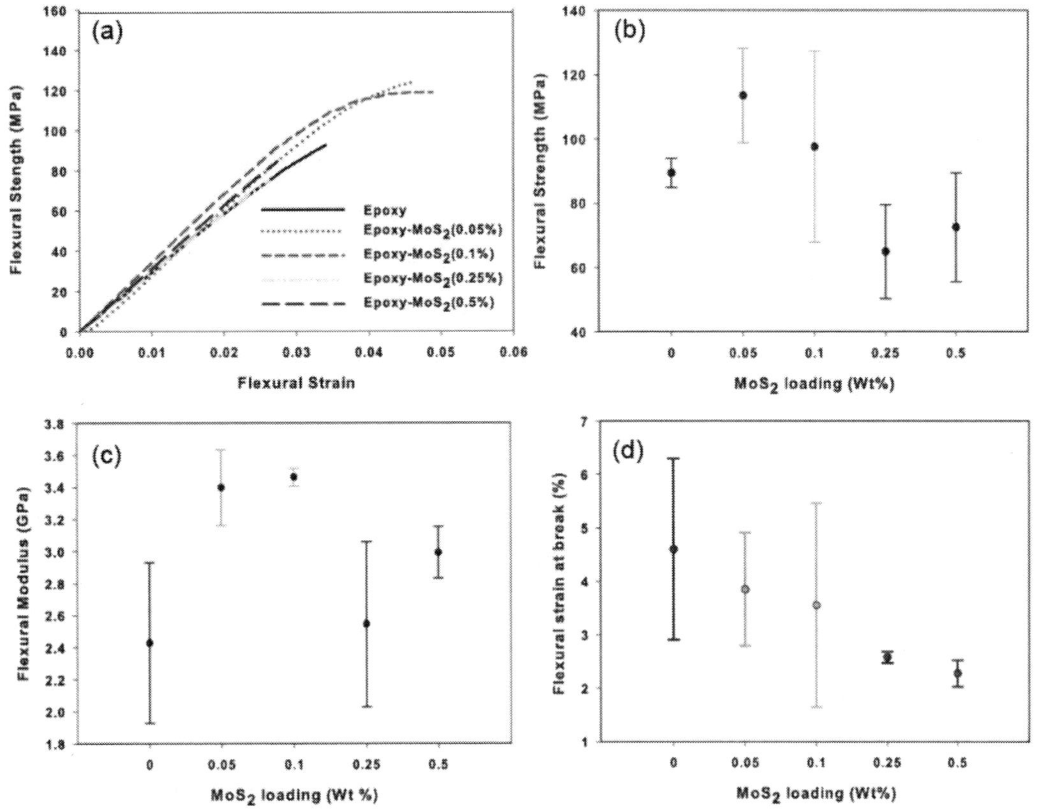

Figure 3. Flexural properties of epoxy-MoS_2 nanocomposites. (*a*) Flexural stress-strain curves, (*b*) Flexural strength with MoS_2 loading, (*c*) Flexural modulus with MoS_2 loading, (*d*) Flexural strain at break with MoS_2 loading.

28.59%, and 12.09%, respectively as compared to neat epoxy. The increase in tensile strength of the composites at ultralow nanofiller content (0.1wt%) was attributed to the good dispersion of MoS_2 and high interfacial interaction between nanofiller and polymer molecules, which is also evident from the TEM images (figure 2). However, at higher nanofiller content (beyond 0.1wt%), the ex-MoS_2 nanosheets are extremely close, as a result, the nanosheets tend to agglomerate and restack together due to their van der Waals forces. The Young's modulus of neat epoxy and epoxy-MoS_2 composites with different weight fractions of ex-MoS_2 ranging from 0.05wt%, 0.1wt%, 0.25wt%, and 0.5wt% to 0.5wt% was found to be 1.10, 1.15, 1.17, 1.10, and 1.0 GPa, respectively. The mechanical test results clearly indicate the effective reinforcement effect of ex-MoS_2 at ultralow filler content ($\sim$ 0.1wt%).

The thermal expansion studies of neat epoxy and its composites were carried out and coefficient of thermal expansion (CTEs) results are displayed in figure 5 *b*. The neat epoxy in this study exhibits CTE value of $6.7 \times 10^{-5}/°C$, whereas, epoxy-MoS_2(0.05wt%), epoxy-MoS_2(0.1wt%), epoxy-MoS_2(0.25wt%), and epoxy-MoS_2(0.5wt%) composites exhibits CTEs value of $6.0 \times 10^{-5}/°C$, $5.6 \times 10^{-5}/°C$, $4.8 \times 10^{-5}/°C$, and $4.3 \times 10^{-5}/°C$, respectively. The test results revealed that the CTEs decreased with increase in MoS_2 content in the composites. A similar trend was observed for the graphene based epoxy composites as reported in the literature. In order to understand the glass-transition temperature (T_g) of neat epoxy and its composites, differential scanning calorimetric studies were carried out (figure 5 *a*). The T_g of neat epoxy was found to be 67.6°C, however, incorporation of 0.05wt%, 0.1wt%, 0.25wt%, and 0.5wt% of

IOP Conf. Series: Journal of Physics: Conf. Series **991** (2018) 012054 doi:10.1088/1742-6596/991/1/012054

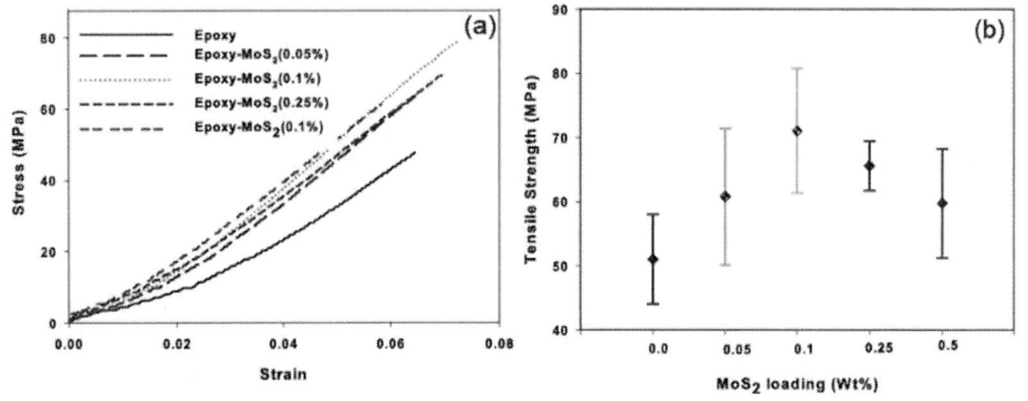

Figure 4. (*a*) Tensile stress-strain curves of epoxy and epoxy-MoS_2 nanocomposites, (*b*) Young's modulus of composites.

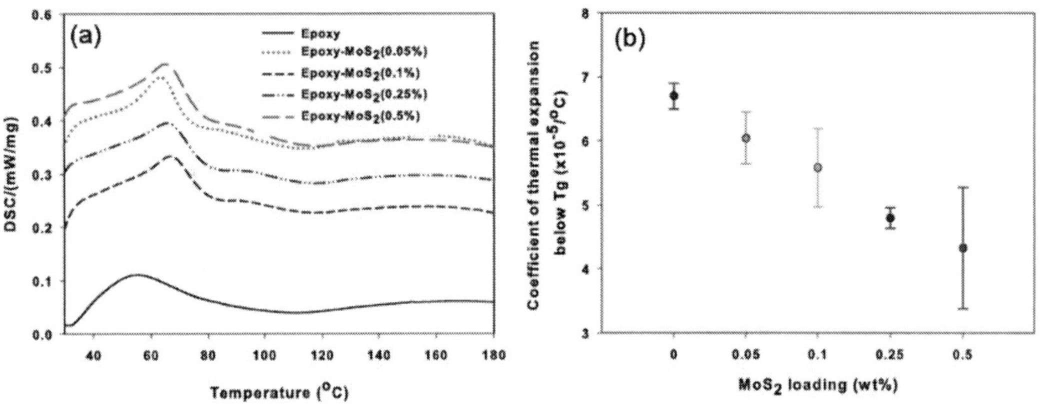

Figure 5. (*a*) DSC thermo grams of epoxy and epoxy-MoS_2 nanocomposites, (*b*) Coefficient of thermal expansion of epoxy-MoS_2 nanocomposites.

ex-MoS_2 in the composites increased the T_g to 68.0°C, 72.6°C, 72.3°C, and 71.1°C, respectively, which can be attributed to the restricted polymer chain mobility due to addition of nanofiller. The T_g result clearly indicates the good dispersion and interfacial interaction between nanofiller and polymer matrix.

Conclusions

MoS_2 reinforced epoxy nanocomposites were prepared by simple ultra-sonication and shear mixing techniques. The SEM observations revealed that the MoS_2 nanosheets were exfoliated after ultrasonication process. The TEM observations confirmed the significant interfacial interaction between MoS_2 and epoxy. The CTE of composites significantly reduced with addition of ex-MoS_2 in the composites. The incorporation of ultralow content (0.1% by weight) of ex-MoS_2 in the epoxy matrix has significantly increased the flexural (9.0%) and tensile strength (39.2%) properties of composites. The improvement in mechanical properties of epoxy nanocomposites can be attributed to the significant interfacial interaction between epoxy and MoS_2.

Acknowledgments
This study was financially supported by the Department of Science and Technology (DST) of India and the Russian Foundation for Basic Research (project No. 17-51-45054) through the National Post Doctoral Fellowship (N-PDF) and Indo-Russian collaborative project scheme and by the Federal Agency for Scientific Organizations (State Registration Number AAAA-A17-117021310381-8).

References
[1] Punetha V D, Rana S, Yoo H J, et al. 2017 Functionalization of carbon nanomaterials for advanced polymer nanocomposites: A comparison study between CNT and graphene *Prog. Polym. Sci.* **67** 1–47
[2] Kuilla T, Bhadra S, Yao D, et al. 2010 Recent advances in graphene based polymer composites *Prog. Polym. Sci.* **35** (11) 1350–75
[3] Sahoo N G, Rana S, Cho J W, et al. 2010 Polymer nanocomposites based on functionalized carbon nanotubes *Prog. Polym. Sci.* **35** (7) 837–67
[4] Kim H, Abdala A A, and Macosko C W 2010 Graphene/Polymer Nanocomposites *Macromolecules* **43** (16) 6515–30
[5] Stankovich S, Dikin D A, Dommett G H B, et al. 2006 Graphene-based composite materials *Nature* **442** 282–6
[6] Bak J M, Lee T, Seo E, et al. 2012 Thermoresponsive graphene nanosheets by functionalization with polymer brushes *Polymer* **53** (2) 316–23
[7] Ramasamy M S, Mahapatra S S, and Cho J W 2015 Functionalization of graphene with self-doped conducting polypyrrole by click coupling *J. Colloid Interface Sci.* **455** 63–70
[8] Eksik O, Gao J, Shojaee S A, et al. 2014 Epoxy nanocomposites with two-dimensional transition metal dichalcogenide additives *ACS Nano* **8** (5) 5282–9
[9] Li Y, Wang H, Xie L, et al. 2011 MoS_2 nanoparticles grown on graphene: an advanced catalyst for the hydrogen evolution reaction *J. Am. Chem. Soc.* **133** (19) 7296–9
[10] Han S, Ai Y, Tang Y, J. Jiang, D. Wu 2015 Carbonized polyaniline coupled molybdenum disulfide/graphene nanosheets for high performance lithium ion battery anodes *RSC Adv.* **5** 96660–4
[11] Meng F, Li J, Cushing S K, et al. 2013 Solar hydrogen generation by nanoscale p-n junction of p-type molybdenum disulfide/n-type nitrogen-doped reduced graphene oxide *J. Am. Chem. Soc.* **135** (28) 10286–9
[12] Xiao J, Choi D, Cosimbescu L, et al. 2010 Exfoliated MoS_2 nanocomposite as an anode material for lithium ion batteries *Chem. Mater.* **22** (16) 4522–4
[13] Tsai M-L, Su S-H, Chang J-K, et al. 2014 Monolayer MoS_2 heterojunction solar cells *ACS Nano* **8** (8) 8317–22

IOP Conf. Series: Journal of Physics: Conf. Series **991** (2018) 012055 doi:10.1088/1742-6596/991/1/012055

Influence of grain size and second phase particles on the process of void initiation

D K Magomedova[1*] and M Yu Murashkin[1,2]**

[1]Saint Petersburg State University, Saint Petersburg, Russia
[2]Ufa State Aviation Technical University, Ufa, Russia

E-mail: *magmedva.dasha@mail.ru, **m.murashkin@gmail.com

Abstract. Experimental results and analysis of pore formation in Al-6101 alloy in various structural states are presented. Experiments on tensile failure of samples of various geometries and different structural states of the commercial alloy Al-6101 were carried out. The surface of longitudinal section was examined for pore presence.

1. Introduction

Conditions for pore formation in metals in the frame of problem solving of material fracture are of particular interest in contemporary science. At the moment, there are no studies of pore formation in bulk nanomaterials. This process was previously described [1, 2] in both experimental and theoretical directions, but in coarse-grained metals; the methodology of this study was based on papers [1, 2]. In the present paper, the experimental data of the coarse and fine-grained Al alloy are analyzed and compared. A numerical calculation and an analysis of this problem will be carried out in subsequent works.

Within the framework of this problem, the loading conditions and the ability to observe the pore formation were of special interest. That is, it was necessary to deform a sample so that pores were formed in it without its fracture.

2. Experiment

Tensile tests were carried out on Shimadzu AG-50kNX testing machine with a loading rate of 0.0014 mm/s. The machine is equipped with a video extensometer for accurate measurement of deformation of small samples.

The samples and grips of special geometry for better fixation and control of the experiment were made according to All-Union State Standard (figures 1 and 2).

Before the experiment, samples of a coarse-grained alloy were placed in the annealing furnace for 2 hours at a temperature of 550°C and quenched with water.

Then one part of the samples was left for six days for natural aging, and the second part was placed in the furnace for 12 hours at a temperature of 170°C for artificial aging.

After annealing and quenching, tensile tests to failure were performed.

Samples of the fine-grained alloy were obtained by the ECAP method (Equal-Channel Angular Pressing) of two types: 4 and 6 passes/cycles.

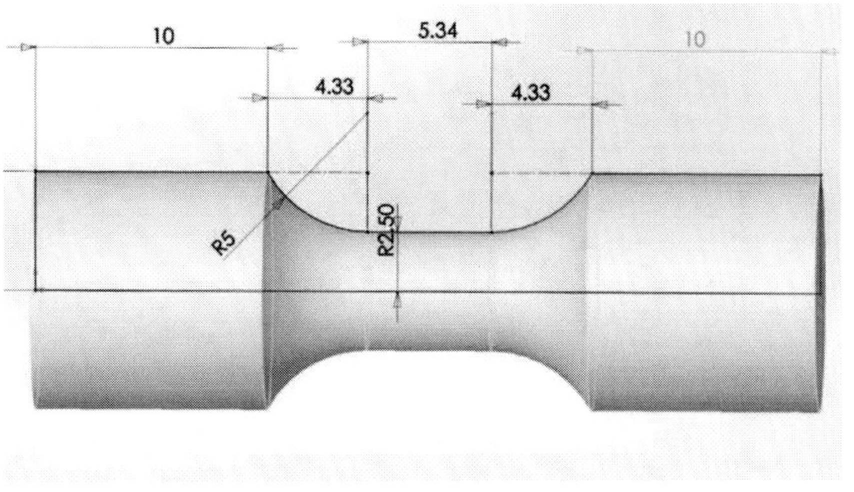

Figure 1. An example of a sample under study.

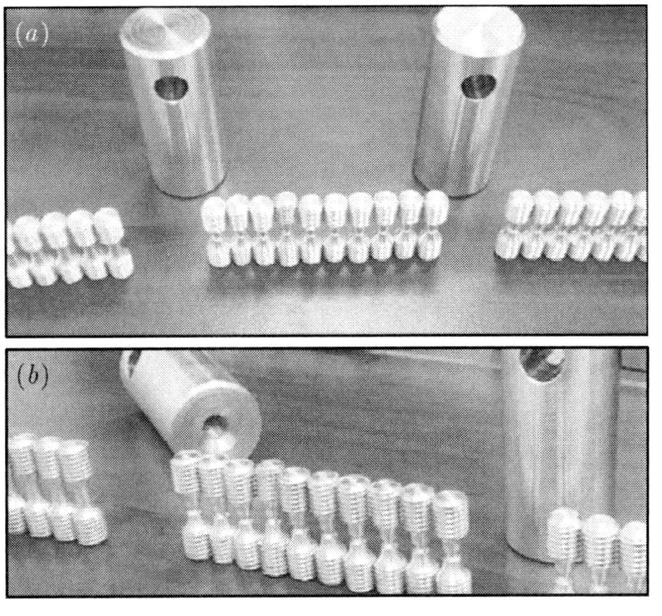

Figure 2. Images of samples and specialized grips.

After tensile testing, the samples were cut along the axis on the ARTA 123-PRO electroerosion machine with a precision accuracy up to 1–2 microns. The section surface was polished to a roughness of 50 nm to investigate the pore presence and location on the SEM.

3. Results

In this section, tensile diagrams of the described samples are presented.

3.1. Coarse-grained alloy

As can be seen from figure 3, the artificial aging shows great plasticity compared to the natural one, since during the artificial aging, β-particles, which lead to hardening the material, are

IOP Conf. Series: Journal of Physics: Conf. Series **991** (2018) 012055 doi:10.1088/1742-6596/991/1/012055

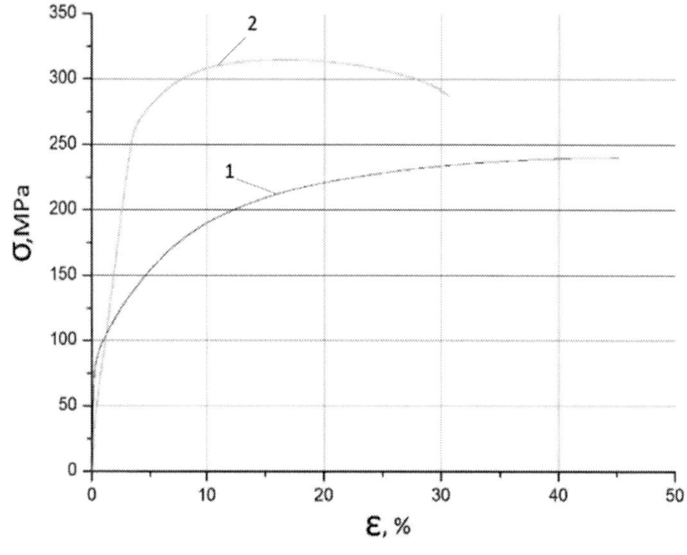

Figure 3. Tensile diagrams for the samples obtained by natural (curve *1*) and artificial (curve *2*) aging.

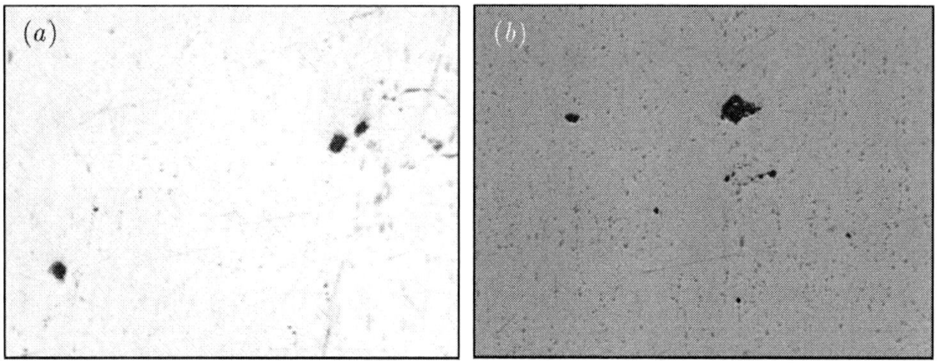

Figure 4. Image of the axial surface of a cut-off obtained with a microscope ((*a*) – 10-fold zoom, (*b*) – 40-fold zoom) for samples with natural (*a*) and artificial (*b*) aging.

precipitated.

In both cases, the pores formed during tension are clearly visible on the surface of the axial sections (figure 4).

3.2. Fine-grained alloy

As can be seen from the diagrams shown in figure 5, the difference between 4 and 6 cycles is practically indistinguishable in this loading regime.

The pore images on the axial surface of the sample section are presented on figures 6 and 7.

Conclusion

In figure 8, the comparative stress-strain diagram is presented for all structures of the Al-6101 alloy considered in this paper.

The main purpose of this work is the develop an experimental technique for search and

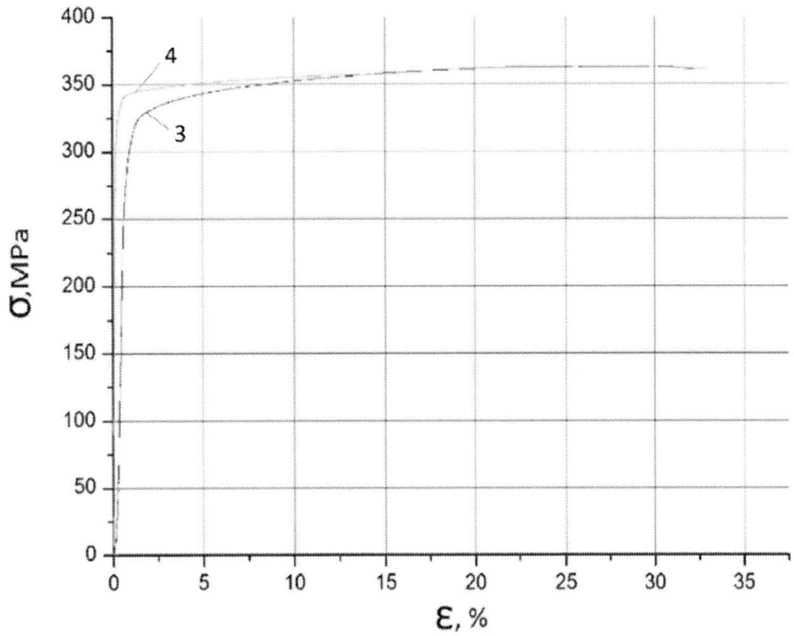

Figure 5. Stress-strain diagram for the fine-grained alloy: curve *3* — 4-cyclic, *4* — 6-cyclic.

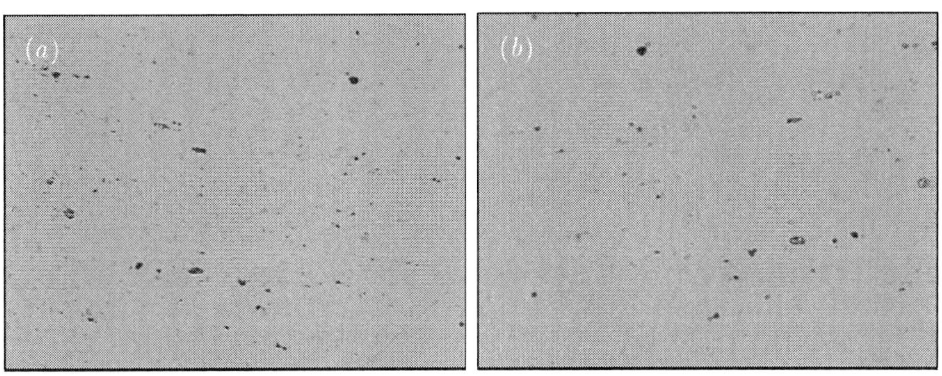

Figure 6. Images of the axial section of the 4-cyclic fine-grained alloy obtained with an optical microscope at the 10-fold zoom.

analysis of pore formation in the presented material. The special grips, geometry of the samples, the procedure of experiment and processing the samples for investigation are developed. A numerical analysis and calculation of the pore formation regions will be carried out in the subsequent work.

Acknowledgments
Useful discussions with Dr. I. Sabirov are gratefully acknowledged.

The authors would like to thank the Russian Science Foundation for financial support of this research by project No. 17-19-01311.

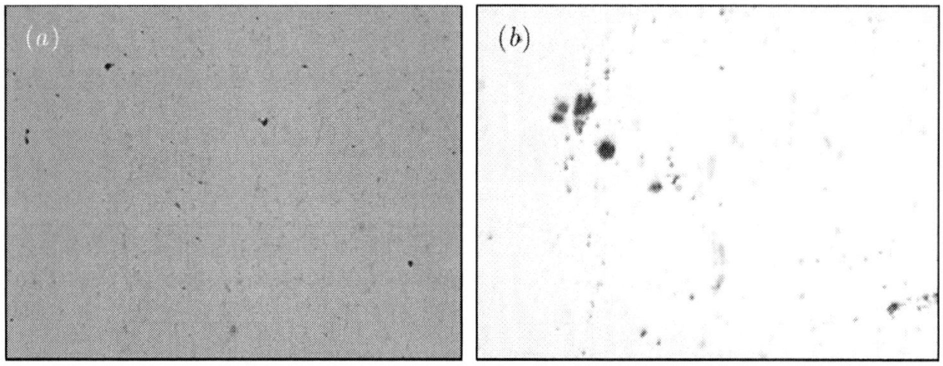

Figure 7. Images of the axial section of the 6-cyclic fine-grained alloy obtained with an optical microscope at the 10-fold (*a*) and 40-fold zoom (*b*).

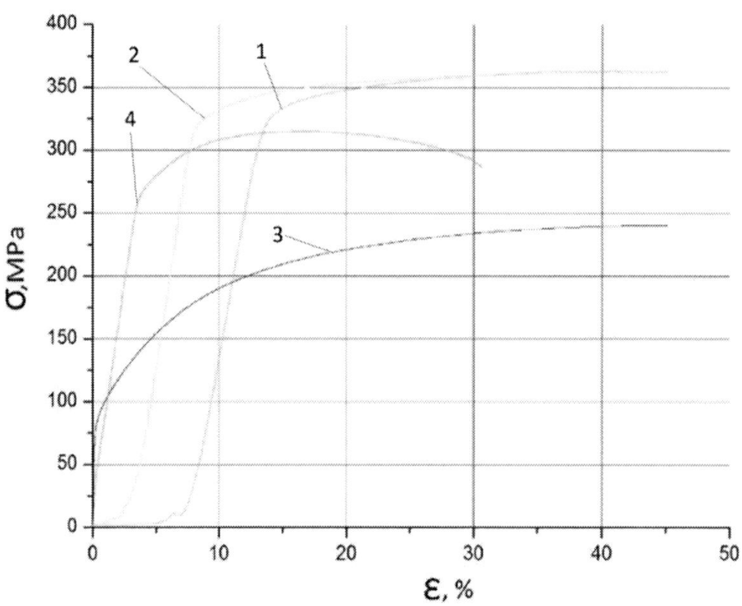

Figure 8. Stress-strain diagrams for samples obtained by 4-cyclic ECAP (curve *1*), 6-cyclic ECAP (curve *2*), natural (curve *3*), and artificial (curve *4*) aging.

References

[1] Argon A S and Im J 1975 Separation of second phase particles in spheroidized 1045 steel, Cu-0.6pct Cr alloy, and maraging steel in plastic straining *Met. Trans. A* **6** 839–51

[2] Argon A S, Im J, and Needleman A 1975 Distribution of plastic strain and negative pressure in necked steel and copper bars *Met. Trans. A* **6** 815–24

[3] Thomson R D and Hancock J W 1984 Ductile failure by void nucleation, growth and coalescence *Int. J. Fract.* **26** (2) 99–112

[4] Thomason P F 1990 *Ductile Fracture of Metals* (Oxford: Pergamon Press)

IOP Conf. Series: Journal of Physics: Conf. Series **991** (2018) 012056 doi:10.1088/1742-6596/991/1/012056

Some problems in mechanics of growing solids with applications to AM technologies

A V Manzhirov

Ishlinsky Institute for Problems in Mechanics of the Russian Academy of Sciences, Moscow, Russia

Bauman Moscow State Technical University, Moscow, Russia

E-mail: manzh@inbox.ru

Abstract. Additive Manufacturing (AM) technologies are an exciting area of the modern industrial revolution and have applications in engineering, medicine, electronics, aerospace industry, etc. AM enables cost-effective production of customized geometry and parts by direct fabrication from 3D data and mathematical models. Despite much progress in AM technologies, problems of mechanical analysis for AM fabricated parts yet remain to be solved. This paper deals with three main mechanical problems: the onset of residual stresses, which occur in the AM process and can lead to failure of the parts, the distortion of the final shape of AM fabricated parts, and the development of technological solutions aimed at improving existing AM technologies and creating new ones. An approach proposed deals with the construction of adequate analytical model and effective methods for the simulation of AM processes for fabricated solid parts.

1. Specific features of growing solids

A continuously growing solid is a solid whose composition, mass or volume varies because of a continuous addition of material to its surface. The process of adding new material to the solid is called growth as well as in particular cases accretion, erection, building-up, healing, etc. For continuous growth, the following basic stages of its deformation are strictly followed: before, during, and after the growth process. Each of these stages is characterized by instants when it starts and ends. The first is characterized by the instant of application of a load to the solid and the instant when growth starts. The second by the instant when growth starts and the instant when it ends. Conversely, the third is characterized by the instant when growth ends and the instant when it starts. The solid on whose surface new material is deposited starting from the instant when accretion starts is called the basic or original solid. Note that growth can also occur without the basic solid, starting from an infinitesimal material element. The part of the surface where infinitesimal pieces of the material are deposited at the actual instant is called the growth surface. The growth surface may be disconnected, in general. In particular, it can be the whole surface of the solid. Finally, the part of the surface of the original or the growing solid that coincides with the growth surface at the instant when growth starts will be called the base surface. The base surface is clearly the part of the surface of the solid on which material is to be deposited during the next stage of continuous growth. At different stages, it coincides, as a rule, with the surface between the basic solid and the additional solid as well as with the surfaces between the sub-solids [1–3].

We assume that the basic solid, which is made from a viscoelastic ageing material (e.g., see [3]), occupies a domain Ω_0 with surface S_0 and is free of stresses up to the time τ_0 of application of the load. From τ_0 up to time τ_1 when growth starts the classical boundary conditions are given on S_0, the specific form of which is stated below. At τ_1 the continuous growth of a solid begins due to the addition of material particles to the growth surface $S^*(t)$. As it grows, the solid occupies a domain $\Omega(t)$ with surface $S(t)$. It is obvious that $S^*(t) \subseteq S(t)$. The time when a particle characterized by a position vector x is deposited on the solid will be denoted by $\tau^*(\mathbf{x})$ and called the time of deposition of the particle on the growing solid. The configuration of the accreted solid is completely determined by the function $\tau^*(\mathbf{x})$ depending on the spatial coordinates. Boundedness and piecewise continuity are the general conditions usually imposed on $\tau^*(\mathbf{x})$.

We denote by $\tau_1^*(\mathbf{x})$ the time when an element of the growing solid is formed and by $\tau_0(\mathbf{x})$ the time when a load is applied to it. Naturally, $\tau_1^*(\mathbf{x}) \leq \tau_0(\mathbf{x}) = \tau_0$ for the elements of the basic solid ($\mathbf{x} \in \Omega_0$).

The vector equilibrium equation is obviously satisfied in the domain occupied by the growing solid at each instant of time. For quasistatic processes, it has the form

$$\mathbf{x} \in \Omega(t) \colon \nabla \cdot \mathbf{T} = \mathbf{O},$$

where T is the stress tensor and ∇ is the Hamilton operator. (Here and below we use the conventional notation of tensor calculus.)

The Cauchy conditions and the compatibility equations for deformations are always satisfied in the domain occupied by the basic solid, i.e.,

$$\mathbf{x} \in \Omega_0 \colon \mathbf{E} = \frac{1}{2}[\nabla \mathbf{u} + (\nabla \mathbf{u})^T], \quad \nabla \times (\nabla \times \mathbf{E})^T = \mathbf{O},$$

where $\mathbf{E}$ is the strain tensor and $\mathbf{u}$ is the displacement vector. But in the domain $\Omega^*(t)$ occupied by the additional solid ($\Omega^*(t) = \Omega(t)/\Omega_0$) only their analogs involving the rates of change of the corresponding variables are satisfied,

$$\mathbf{x} \in \Omega^*(t) \colon \mathbf{D} = \frac{1}{2}[\nabla \mathbf{v} + (\nabla \mathbf{v})^T], \quad \nabla \times (\nabla \times \mathbf{D})^T = \mathbf{O},$$
$$\mathbf{D} = \frac{\partial \mathbf{E}}{\partial t}, \quad \mathbf{v} = \frac{\partial \mathbf{u}}{\partial t};$$

i.e. the strains are usually incompatible.

This reflects the fact that the deposited elements may be subject to deforming actions prior to the time of deposition on the basic solid independently of the processes taking place in the solid itself.

To study the stress-strain state of a growing solid one must know the laws of deformation of the basic solid from the instant τ_0 when the load is applied up to the instant τ_1 when growth starts and of the deposited material from the instant $\tau_0(\mathbf{x})$ when a load is applied to this material up to the instant $\tau^*(\mathbf{x})$ of their deposition on the growing solid. The state of the original solid is determined from the solution of the problem with fixed boundary. The initial state of new elements which represent deposited surfaces as well as the boundary condition on the moving surface of the growing body can be determined by solving an additional contact problem of interaction between the solid and the surface.

The traditional boundary conditions for the displacement vector and the vector of surface forces are given on the stationary sections of the surface of the growing solid.

Currently a great number of AM fabricated part are made from viscoelastic materials with complex properties. To describe the behavior of these materials one can use the constitutive equations for a viscoelastic-ageing solid (see, e.g., [1, 2]).

The description of the process of continuous growth of a viscoelastic ageing solid involves three characteristic instants: the instant $\tau_1^*(\mathbf{x})$ when the element with coordinate $\mathbf{x}$ is formed, the instant $\tau_0(\mathbf{x})$ when a load is applied to this element, and the instant $\tau^*(\mathbf{x})$ when the element is deposited on the growing solid. These three instants are different, in general.

The deposition process is largely determined by specifying these three instants. If the processes of continuous concrete casting, ice formation, crystal growth, etc. are studied, then $\tau_1^*(\mathbf{x}) = \tau_0(\mathbf{x}) = \tau^*(\mathbf{x})$; i.e. the elements are deposited at the same instant as they are formed and a load is applied to them. If spray deposition or erection of a structure from a large number of blocks is modeled by a continuous growth process, then, as a rule, $\tau_0(\mathbf{x}) = \tau^*(\mathbf{x})$, and the instant $\tau_1^*(\mathbf{x})$ when the elements are formed is arbitrary. If the deformation of elements begins as soon as they are formed and they are being added to the basic solid only over some time interval, then $\tau_1^*(\mathbf{x}) = \tau_0(\mathbf{x}) \neq \tau^*(\mathbf{x})$, and so on.

Suppose that a homogeneous viscoelastic ageing solid occupying a domain Ω_0 with surface S_0 ($\mathbf{x} \in \Omega_0$) is formed at instant $\tau_1^*(\mathbf{x}) = 0$ and is free of stresses up to the instant $\tau_0 \geq 0$, when a load is applied. Starting from the latter instant, consider two kinds of boundary conditions on the surface of the solid (surface forces on $S_1(t)$ and displacements on $S_2(t)$).

The sections of the surface on which different boundary conditions are given do not intersect one another and cover the whole surface of the solid. The dependence of S_i on t enables us to take into account the possible evolution of the system of loads, punches, etc. on S_0, and is assumed to be piecewise constant. Unless the solid surface is closed, the behavior of stresses or strains at infinity is prescribed.

Continuous deposition of material formed simultaneously with the solid ($\tau_1^*(\mathbf{x}) = 0$) starts at $\tau_1 \geq \tau_0$. The solid occupies a domain $\Omega(t)$ with surface $S(t)$ during its growth. The growth surface $S^*(t)$ ($S^*(\tau_1) \subset S_0$) moves in space. In the general case the positions of $S(t)$ and $S^*(t)$ are unknown. The sections $S_i(t)$ ($i = 1, 2$) on which the common boundary conditions are given can vary because of the loading of the stationary surface of the additional solid. We assume that the growing surface is always free of outer loads and the newly deposited surfaces are loaded at the instant of their deposition.

At the instant $\tau_2 > \tau_1$, the growth of the solid ceases, and starting from this instant four kinds of boundary conditions are given on the sections $S_i(t)$ of the surface $S_1 = S(\tau_2)$ of the solid occupying the domain $\Omega_1 = \Omega(\tau_2)$. After some time, at the instant $\tau_3 > \tau_2$ the solid growth may start again.

2. Basic concepts in the theory of continuous surface growth

Recent research has shown that solids formed by growth processes differ in their properties essentially from solids in the traditional view. Moreover, the classical approaches of solid mechanics fail when modeling the growing solid behavior. They should be replaced by new ideas and methods of modern mechanics, mathematics, physics, and engineering sciences. The approach proposed here deals with the construction of an adequate model of surface growth processes of solids. This approach is based on the following statements:

- The surface growth of a solid is modeled by the motion of its boundary due to the influx of new material to the surface of the solid.
- The boundary conditions on the moving boundary (the growth surface) are found from an additional solid-surface contact problem depending on the specific features of the growth process.
- The strain rate tensor (or the stretch rate tensor) of a growing solid is compatible, while the strain tensor is, as a rule, incompatible. Accordingly, it is absolutely natural to take the stress rate tensor, the strain rate tensor, and the velocity vector to be the main variables in the problem.

In general, a boundary value problem for a growing solid contains three dependent controlled groups of values, namely, the surface and bulk loads, the stresses on the accretion surfaces, and the material influx rate on the accretion surfaces. We have consider the case in which the velocities of the boundary particles of the growing solid are much smaller than the influx rate of new particles and strains are small.

The process of accretion or deposition of new material to a solid is studied in the fundamental scientific area called Mechanics of Growing Solids. This area deals with all sorts of solid materials including elastic, viscoelastic, plastic, composite and graded materials.

3. Mechanical analysis of growing solids

Consider the stress-strain state of a viscoelastic ageing solid Ω_0 in the time interval $[\tau_0, \tau_1]$. We write out the equilibrium equation in the form ,

$$\nabla \cdot \mathbf{T} = \mathbf{0}. \tag{1}$$

We represent the boundary conditions described above as follows

$$\mathbf{x} \in S_1(t)\colon \mathbf{n} \cdot \mathbf{T} = \mathbf{p}_0, \quad \mathbf{x} \in S_2(t)\colon \mathbf{u} = \mathbf{u}_0, \tag{2}$$

where $\mathbf{p}_0$ and $\mathbf{u}_0$ are given vectors of surface forces and strains and $\mathbf{n}$ is the unit vector normal to the solid surface. The Cauchy conditions are written as follows:

$$\mathbf{E} = \frac{1}{2}[\nabla\mathbf{u} + (\nabla\mathbf{u})^T]. \tag{3}$$

We take the constitutive equations in the form in which the transverse contraction (Poisson's) ratios of the instantaneous elastic strain and the creep strain of the ageing material are identical and are equal to ν. Then we have (see [1–3])

$$\mathbf{T} = G(\mathbf{I} + \mathbf{N}(\tau_0, t))[2\mathbf{E} + (K-1)I_1(\mathbf{E})\mathbf{1}], \tag{4}$$

where

$$\begin{aligned}
&(\mathbf{I} + \mathbf{N}(\tau_0, t))^{-1} = (\mathbf{I} - \mathbf{L}(\tau_0, t)),\\
&2G = E(1+\nu)^{-1}, \quad K = (1-2\nu)^{-1},\\
&\mathbf{L}(\tau_0, t) f(t) = \int_{\tau_0}^{t} f(\tau) K(t,\tau)\, d\tau,\\
&\omega(t,\tau) = 2C(t,\tau)(1+\nu),\\
&K(t,\tau) = E(\tau)\frac{\partial}{\partial\tau}[E^{-1}(\tau) + C(t,\tau)] =\\
&\qquad = K_1(t,\tau) = G(\tau)\frac{\partial}{\partial\tau}[G^{-1}(\tau) + \omega(t,\tau)],
\end{aligned}$$

where $E = E(t)$ and $G = G(t)$ are the elastic moduli under tension and shear, $C(t,\tau)$ and $\omega(t,\tau)$ are the creep measures under tension and shear, $K(t,\tau)$ is the creep function under tension, and $\mathbf{1}$ is the unit tensor. The arguments are omitted in a number of obvious cases above. They will also be omitted in what follows and will be used only in those cases when their absence may be misleading.

Thus, Eqs. (1)–(4) constitute the boundary value problem of the linear theory of elasticity for a homogeneous ageing basic solid, the stress-strain state of which can be described by the solution of the system for $t \in [\tau_0, \tau_1]$.

IOP Conf. Series: Journal of Physics: Conf. Series **991** (2018) 012056 doi:10.1088/1742-6596/991/1/012056

We transform the boundary value problem for the basic solid. Let us introduce the notation

$$\begin{aligned}&\mathbf{N}^0 = \mathbf{H}(\tau_0, t)\mathbf{N}G^{-1}, \mathbf{a}^0 = \mathbf{H}(\tau_0, t)\mathbf{a}G^{-1},\\&\mathbf{H}(\psi, t) = (\mathbf{I} - \mathbf{L}(\psi, t)),\end{aligned} \tag{5}$$

where $\mathbf{N}$ and $\mathbf{a}$ are an arbitrary tensor and arbitrary vector, respectively. We apply the operator $\mathbf{H}(\tau_0, t)$ to the relations in (1)–(4) containing $\mathbf{T}$ after dividing them by G. Then, since $\mathbf{H}(\tau_0, t)$ commutes with the Hamilton operator, we obtain the following boundary value problem using (5) ($\tau_0 \le t \le \tau_1$):

$$\begin{aligned}&\nabla\cdot\mathbf{T}^0 = \mathbf{O},\\&\mathbf{x}\in S_1(t):\quad \mathbf{n}\cdot\mathbf{T}^0 = \mathbf{p}_0^0,\\&\mathbf{x}\in S_2(t):\quad \mathbf{u} = \mathbf{u}_0,\\&\mathbf{E} = \frac{1}{2}[\nabla\mathbf{u} + (\nabla\mathbf{u})^T],\\&\mathbf{T}^0 = 2\mathbf{E} + (K-1)I_1(\mathbf{E})\mathbf{1}.\end{aligned} \tag{6}$$

Unlike (1)–(4), time occurs in the boundary value problem (6) as a parameter. The latter is mathematically equivalent to the boundary value problem of the theory of elasticity with a parameter t. All analytic and numerical methods of the theory of elasticity can be used when constructing the solution of such a problem, which undoubtedly lends itself better to investigation than the problem of the theory of viscoelasticity.

To obtain the solution $\mathbf{T}$, $\mathbf{E}$, $\mathbf{u}$ of the boundary value problem (1)–(4), one must find the solution $\mathbf{T}^0$, $\mathbf{E}$, $\mathbf{u}$ of the boundary value problem (6) and use the relation as follows:

$$\mathbf{T}(\mathbf{x}, t) = G(t)[\mathbf{T}^0(\mathbf{x}, t) + \int_{\tau_0}^{t}\mathbf{T}^0(\mathbf{x}, \tau)R(t, \tau)\,d\tau], \tag{7}$$

Here $R(t,\tau)$ is the resolvent of the kernel $K(t,\tau)$.

Now consider the process of continuous growth of a solid ($\tau_1 \le t \le \tau_2$). For a growing solid, we have the equilibrium equation in the form

$$\nabla\cdot\mathbf{T} = \mathbf{O}. \tag{8}$$

The boundary conditions on the stationary part of the surface can be written as follows:

$$\begin{aligned}&\mathbf{x}\in S_1(t):\quad \mathbf{n}\cdot\mathbf{T} = \mathbf{p}_0,\\&\mathbf{x}\in S_2(t):\quad \mathbf{u} = \mathbf{u}_0.\end{aligned} \tag{9}$$

What regards the growth surface, we formulate the quasistatic and kinematic boundary conditions on it. The quasistatic boundary condition on the growth surface can be obtained from the solution of a contact problem for a solid and a surface provided that they interact without friction (see figure 1). Taking into account the following relations for small angles α_1 and α_2:

$$\begin{aligned}&2\sigma_1\,dh\,dS_2\,\sin\frac{d\alpha_1}{2} + 2\sigma_2\,dh\,dS_1\,\sin\frac{d\alpha_2}{2} + d\sigma\,dS_2\,dS_1 = 0,\\&\frac{d\sigma}{dt} = -s_n\left(\sigma_1\frac{d\alpha_1}{dS_1} + \sigma_2\frac{d\alpha_2}{dS_2}\right),\quad \frac{d\alpha_1}{dS_1} = \frac{1}{\rho_1},\quad \frac{d\alpha_2}{dS_2} = \frac{1}{\rho_2},\\&\frac{d\sigma}{dt} = -s_n\left(\frac{\sigma_1}{\rho_1} + \frac{\sigma_2}{\rho_2}\right) = -s_n\,(\mathcal{T}_s : \mathbf{L}),\quad dh = s_n\,dt,\quad s_n = \mathbf{v}\cdot\mathbf{n},\end{aligned} \tag{10}$$

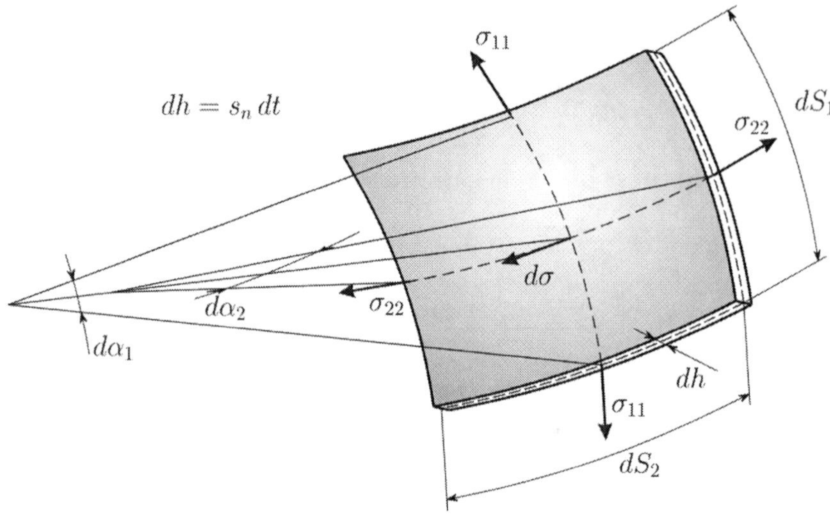

Figure 1. Scheme of an auxiliary contact problem.

one can obtain this condition in the form

$$\begin{aligned} &\mathbf{x} \in S^*(t)\colon \mathbf{n}\cdot\mathbf{S} = p\,\mathbf{n}, \quad \mathbf{S} = \frac{\partial(\mathbf{HT})}{\partial t}, \\ &p = -\frac{s_n}{G}(\mathcal{T}_s : \mathbf{L}), \quad \mathbf{v} = \frac{\partial \mathbf{u}}{\partial t} \quad (t = \tau^*(\mathbf{x})), \end{aligned} \tag{11}$$

where σ_{11} and σ_{11} are the 2D principal stresses, $\mathcal{T}_s$ is the 2D tensor of the deposited elastic surface tension, ρ_1 and ρ_2 are the principal radii of curvature of the contacting surface, and $\mathbf{L}$ is the 2D tensor of curvature of this surface.

The kinematic boundary condition is a form for the composition conservation law for a solid. It can be written in the form

$$\mathbf{x} \in S^*(t)\colon \quad \mathbf{v} = \mathbf{v}_{\text{def}} + \mathbf{v}_{\text{dep}}, \tag{12}$$

where $\mathbf{v}$ is the velocity of growth surface, $\mathbf{v}_{\text{def}}$ is the is the velocity of growth surface due to the deformation of a solid under the action of surface and volumetric forces as well as under the tension of new elements, and $\mathbf{v}_{\text{dep}}$ is the velocity of deposition of new elements to the growth surface. Here we suppose that $\mathbf{v}_{\text{def}}$ and $\mathbf{v}_{\text{dep}}$ are independent of each other. Hereinafter we study the case in which $\mathbf{v}_{\text{def}} \ll \mathbf{v}_{\text{dep}}$. In this case we can set $\mathbf{v}_{\text{def}} = 0$ in (11) with high accuracy and use the equation of the growth surface in the form $t = \tau_*(\mathbf{x})$. The latter equation gives us the value of s_n from (10). One can note that for the chosen case of problems statement we obtain the single boundary condition on the growth surface in the form (11).

The relation between the strain and displacement rate tensors is given by the equality

$$\mathbf{D} = \frac{1}{2}[\nabla\mathbf{v} + (\nabla\mathbf{v})^T], \tag{13}$$

The constitutive equation can be written as follows:

$$\mathbf{T} = G(\mathbf{I} + \mathbf{N}(\tau_0(\mathbf{x}), t))[2\mathbf{E} + (K-1)I_1(\mathbf{E})\mathbf{1}], \tag{14}$$

$$\tau_0(\mathbf{x}) = \begin{cases} \tau_0, & \mathbf{x} \in \Omega_0, \\ \tau^*(\mathbf{x}), & \mathbf{x} \in \Omega^*(t). \end{cases}$$

Relations (8)–(14) form a general noninertial initial–boundary value problem for a continuously growing solid, where the operator $(\mathbf{I} - \mathbf{L}(\tau_0(\mathbf{x}), t)) = \mathbf{H}(\tau_0(\mathbf{x}), t))$ and its inverse operator $(\mathbf{I}+\mathbf{N}(\tau_0(\mathbf{x}), t))$ can be determined from (4) with τ_0 replaced by $\tau_0(\mathbf{x})$. Let us transform the initial–boundary value problem for a continuously accreted viscoelastic ageing solid into a problem with time parameter that has the same form as the boundary value problem of the theory of elasticity. Omitting technical details we present the final result in the form of a boundary value problem as follows

$$
\begin{aligned}
&\nabla \cdot \mathbf{S} = \mathbf{O}, \\
&\mathbf{x} \in S_1(t): \quad \mathbf{n} \cdot \mathbf{S} = \mathbf{R}\mathbf{p}_0, \\
&\mathbf{x} \in S_2(t): \quad \mathbf{v} = \mathbf{v}_0, \\
&\mathbf{x} \in S^*(t): \quad \mathbf{n} \cdot \mathbf{S} = p\,\mathbf{n} \quad (t = \tau^*(\mathbf{x})), \\
&\mathbf{D} = \frac{1}{2}[\nabla \mathbf{v} + (\nabla \mathbf{v}^T], \\
&\mathbf{S} = 2\mathbf{D} + (K-1)I_1(\mathbf{D})\mathbf{1},
\end{aligned}
\tag{15}
$$

where $\mathbf{R}$ acts on an arbitrary vector $\mathbf{a}(\mathbf{x}, t)$ by the rule

$$
\mathbf{R}\mathbf{a}(\mathbf{x}, t) = \frac{1}{G(t)} \frac{\partial \mathbf{a}(\mathbf{x}, t)}{\partial t} + \int_{\tau_0(\mathbf{x})}^{t} \frac{\partial \mathbf{a}(\mathbf{x}, \tau)}{\partial \tau} \frac{\partial \omega(t, \tau)}{\partial t} d\tau + \mathbf{a}(\mathbf{x}, \tau_0(\mathbf{x})) \frac{\partial \omega(t, \tau_0(\mathbf{x}))}{\partial t}. \tag{16}
$$

Note that the conditions on $S_1(t)$ and $S^*(t)$ are identical.

Relations (15)–(16) supplemented with the initial conditions for the basic solid at $t = \tau_1$ form a boundary value problem with t as a parameter.

To obtain the solution $\mathbf{T}$, $\mathbf{E}$, and $\mathbf{u}$ of the initial–boundary value problem (8)–(14) one must find the solution $\mathbf{S}$, $\mathbf{D}$, and $\mathbf{v}$ of (15)–(16) and use the relations as follows

$$
\begin{aligned}
\mathbf{T}(\mathbf{x}, t) = G(t) &\left\{ \frac{\mathbf{T}(\mathbf{x}, \tau_0(\mathbf{x}))}{G(\tau_0(\mathbf{x}))} \left[1 + \int_{\tau_0(\mathbf{x})}^{t} R(t, \tau)\, d\tau \right] \right. \\
&\left. + \int_{\tau_0(\mathbf{x})}^{t} \left[\mathbf{S}(\mathbf{x}, \tau) + \int_{\tau_0(\mathbf{x})}^{\tau} \mathbf{S}(\mathbf{x}, \zeta)\, d\zeta R(t, \tau) \right] d\tau \right\}, \\
\mathbf{u}(\mathbf{x}, t) = \mathbf{u}&(\mathbf{x}, \tau_0(\mathbf{x})) + \int_{\tau_0(\mathbf{x})}^{t} \mathbf{v}(\mathbf{x}, \tau)\, d\tau.
\end{aligned}
\tag{17}
$$

Hence the solution of the growth problem for a viscoelastic ageing solid can be obtained by the solution of the mathematically identical problems with a parameter t, which have the same form as the boundary value problem of the classical theory of elasticity. Then the true stresses and displacements in the growing solid can be reconstructed using (17).

Suppose that the solid ceases to grow at instant τ_2. At this instant, it occupies a domain Ω_1 with surface S_1 on which two kinds of boundary conditions are specified, as in the case of the problem for the basic solid. Moreover, $S^*(\tau_2) = S_1^* \subseteq \cup_i S_i(t)$ for $i = 1, 2$. In this case the problem for the invariable solid occupying Ω_1 is similar to (8)–(14) without the initial–boundary condition on $S^*(t)$,

$$
\begin{aligned}
&\nabla \cdot \mathbf{T} + \mathbf{f} = \mathbf{O}, \\
&\mathbf{x} \in S_1(t): \mathbf{n} \cdot \mathbf{T} = \mathbf{p}_0, \quad \mathbf{x} \in S_2(t): \mathbf{u} = \mathbf{u}_0, \\
&\mathbf{D} = \frac{1}{2}[\nabla \mathbf{v} + (\nabla \mathbf{v})^T], \\
&\mathbf{T} = G(\mathbf{I} + \mathbf{N}(\tau_0(\mathbf{x}), t))[2\mathbf{E} + (K-1)I_1(\mathbf{E})\mathbf{1}],
\end{aligned}
\tag{18}
$$

with $\tau^*(\mathbf{x}) = \tau_2$ for $\mathbf{x} \in S_1^*$. The stresses, strains and displacements at $t = \tau_2$ found by solving the growth problem at the previous step serve as the initial conditions.

One can obtain the following boundary value problem:

$$\begin{aligned} &\nabla \cdot \mathbf{S} = \mathbf{O}, \\ &\mathbf{x} \in S_1(t)\colon \mathbf{n} \cdot \mathbf{S} = \mathbf{R}\mathbf{p}_0, \quad \mathbf{x} \in S_2(t)\colon \mathbf{v} = \mathbf{v}_0, \\ &\mathbf{D} = \frac{1}{2}[\nabla \mathbf{v} + (\nabla \mathbf{v})^T], \\ &\mathbf{S} = 2\mathbf{D} + (K-1)I_1(\mathbf{D})\mathbf{1}, \end{aligned} \tag{19}$$

where the initial conditions remain as before.

For $\mathbf{T}$, $\mathbf{E}$, and $\mathbf{u}$ to be solutions of the initial–boundary value problem (18) it is necessary and sufficient that $\mathbf{S}$, $\mathbf{D}$, and $\mathbf{v}$ form a solution of the boundary value problem (19) and that relations (17) be satisfied.

Thus, to construct the solution of the problem over the time interval $[\tau_0, \tau_3]$ one has to construct the solutions of the following three identical problems (having the same form as the boundary value problem of the theory of elasticity with a parameter t): problem (6) for $t = \tau_0$ as well as problems (13)–(14) and (17). The stress-strain state of the growing solid can then be reconstructed for any $t \in [\tau_0, \tau_3]$ from (7) and (15).

Among a great number of papers on AM technologies written by technologists, chemists, and physicists (see, e.g., [4–26]). there are few devoted to mechanical aspects and analisys of AM fabricated parts and similar problematics (see, e.g., [27–42]). In what follows we consider some mechanical problems of AM technologies in civil and structural engineering.

4. Additive manufacturing of a concrete dam

We consider a problem for a growing dam in the form of wedge which is manufactured from concrete. We use standard material moduli and functions for concrete and some plastics (e.g., see [1–3]). These problems differ one from each other by the rate as well as by the type of additive process. We study tree types of growth for a wedge: angle growth (figure 2 *a*), translational growth (figure 2 *b*)), and complex growth (figure 2 *c*). We have found that stress states for growing dams sufficiently differ from each other and from the classical case of fixed dam.

4.1. Translational type of building up

To be definite let us consider a problem for a dam, which in the process of its manufacturing, change the shape from a quarter-plane up to a plane (figure 2 *d*). Let us consider the problem of translational growth on a quarter-plane of non-stressed elements at a constant rate. Let a constant indenting force $P(t)$ and zero tilt angle of a punch which can be treated as an engineering block of a dam be given, and lets assume that the base of the punch is flat. The configuration of the body at any instant of time is specified by the coordinate of the tip of the quarter $x = -f(t)$ ($f(t) = 0$, when $\tau_0 \le t \le \tau_1$). We assume that the process of growth does not stop ($\tau_2 = \infty$), i.e. in the limit the body becomes a half-plane.

Let us determine the contact pressure q and the eccentricity e of the point of application of the force, ensuring that there is no a tilt of the punch at different rates of translational growth. In what follows, we use dimensionless variables and formulate general qualitative conclusions. We set $P(t) = 1$, $\alpha = \pi/2$. We consider 1) slow translational growth: $f(t) = t - 2$, $f'(t) = 1$ ($t \le \tau_1$); 2) rapid translational growth $f(t) = 6(t-2)$, $f'(t) = 6$ ($t \le \tau_1$); (the rate of growth $f'(t)$ is 6 times as high as in the case 1)).

Figures 3 *a* shows the limit distribution of the contact stresses at various rates $f'(t)$. The solid asymmetric line shows the initial distribution of the contact stresses when the dam has the shape of quarter-plane. The solid symmetric line demonstrates the contact stresses distribution

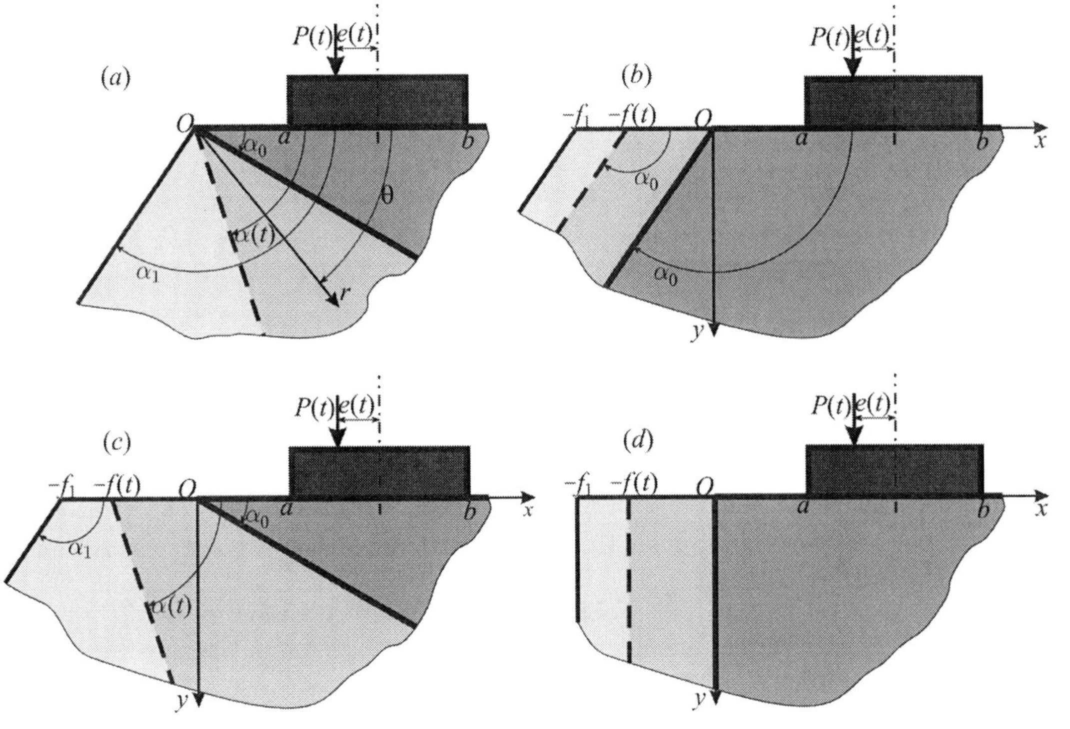

Figure 2. Types of additive manufacturing for a dam.

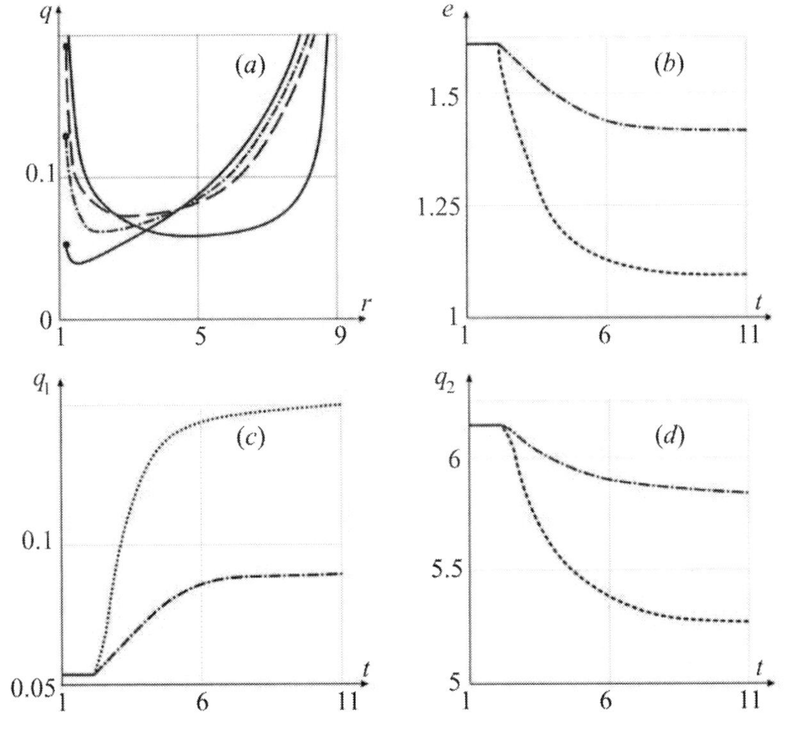

Figure 3. Contact stresses for the translational type of building up.

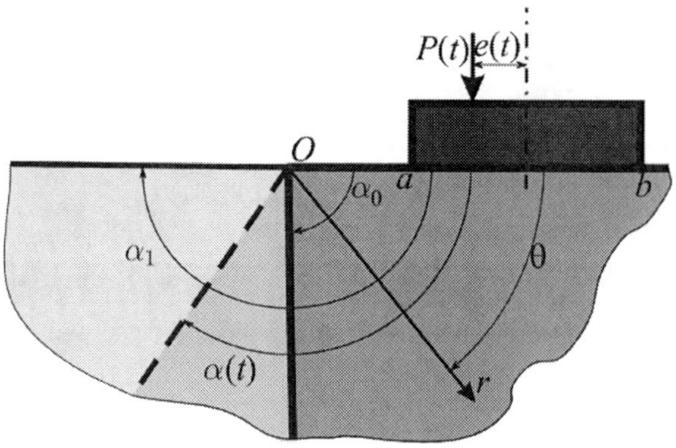

Figure 4. Angle building up of a dam.

for the classical problem of the interaction between a flat punch and a half-plane. The dashed-and-dot and dashed lines show the limiting distribution of contact stresses for slow and rapid translational growth respectively.

Figure 3 *b* shows the limiting values of the force eccentricity at various rates $f'(t)$. The dash-dot and dashed lines describe the behavior of the force eccentricity on time for slow and rapid translational growth respectively.

Figures 3 *c* and 3 *d* show the variation with time of the contact stresses q_1 at the left edge of the punch ($x = 0.1$) and q_2 at the right edge ($x = 8.9$) for slow (dash-dot lines) and rapid (dashed lines) growth respectively.

We note that in case 1) the stresses, e.g. at the left edge, increase by a factor of 1.4 and the eccentricity is reduced by a factor of 1.2. In case 2) these relations are equal to 2.7 and 1.5, respectively. When $t = t^\circ \approx 8$ and the rate of growth is arbitrary, the process becomes a steady-state process and further growth of the body does not affect the contact stresses and the eccentricity.

4.2. Angle type of building up

Let us now consider the problem of angle growth on a quarter-plane of non-stressed elements at a constant rate. The initial aperture angle of the wedge is $\alpha_0 = \pi/2$, becoming $\alpha_1 = \pi$ at the instant of termination. It is necessary to find contact stresses and the eccentricity of the indenting force which ensure that the punch is not misaligned. In the case of constant growth rate the instants of the onset τ_1 and termination τ_2 of growth completely determine the function

$$\alpha(t) = \tfrac{1}{2}\pi \frac{t + \tau_2 - 2\tau_1}{\tau_2 - \tau_1}$$

and thus configuration of the body at any instant of time. We chose concrete as the material of the wedge and use dimensionless quantities (see figure 4). Let us consider the case of a slow ($\tau_1 = 2$, $\tau_2 = 10$) and rapid ($\tau_1 = 2$, $\tau_2 = 2$) side growth of the wedge. From now on the dot-dash lines in the figures will present the basic characteristics during the slow growth, and dashed lines during the rapid growth (at a rate four times as high as that of the slow growth).

Figure 5 *a* shows the limit distributions of contact pressures under the punch q as $t \to \infty$ at different growth rates (dashed and dash-dot lines), and the distribution from the instant of the loading $\tau_0 = 1$ to the instant of onset of growth τ_1, constant over this time interval (solid line). The distribution functions have singularities at the punch edges, but in order to make the graphs easier to construct, we have eliminated the region $r < 0.1$ and indicated the values of the stresses at $r = 0.1$, by dots.

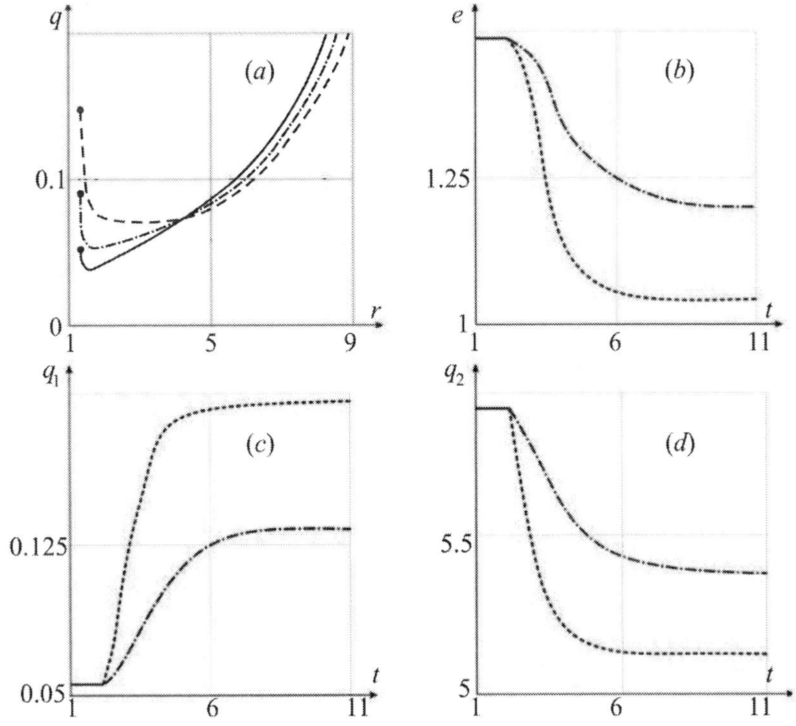

Figure 5. Contact stresses for the angle type of building up.

Figure 5 *b* shows the dependence of the eccetricity e of the point of application of the force on time t.

Figures 5 *c* and 5 *d* shows the variation with time of the contact stresses q_1 at the left edge of the punch ($r = 0.1$) and q_2 at the right edge ($r = 8.9$) for the two processes in question. We see that the processes characterized by the rate of growth show substantial qualitative and quantitative differences between each other. During the rapid growth the most intense change in the stresses and the eccentricity occurs in the interval $t \in [\tau_1, \tau_2]$ and continues after the growth has ceased. The stresses, e.g. at the left edge, increase by a factor of 3.5 and the eccentricity decreases by almost a half.

When the growth is slow, the characteristics change more smoothly and, beginning at a certain instant of time, the stress-strain state of the body becomes practically indifferent to the process of growth, i.e. under the conditions of limited creep the constant force acting on the punch becomes exhausted and further growth or its termination does not alter the already established values of the stresses and the eccentricity. We note for comparison that in case of a slow growth the stresses at the left edge increase by a factor of 2.4, and the eccentricity decreases by a factor of 1.5. By studying the process of slow growth we also broaden our understanding of the law governing the formation of the displacement and stress fields in a growing viecoelastic body acted upon by stationary forces.

Irrespective of the fact that the body has the form of a halfplane at the instant of termination, the distribution of contact pressures under the punch is not symmetrical at any $t \geq \tau_2$, and the eccentricity of the point of application of the force is very far from zero, i.e. the idea of a body which has grown to a half-plane may lead directly to incommensurable values of the fundamental characteristics.

IOP Conf. Series: Journal of Physics: Conf. Series **991** (2018) 012056 doi:10.1088/1742-6596/991/1/012056

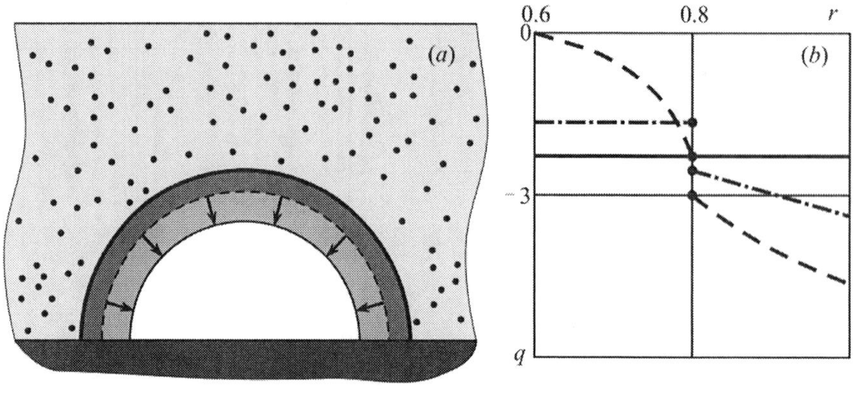

Figure 6. Additive manufacturing of a buried arch structure.

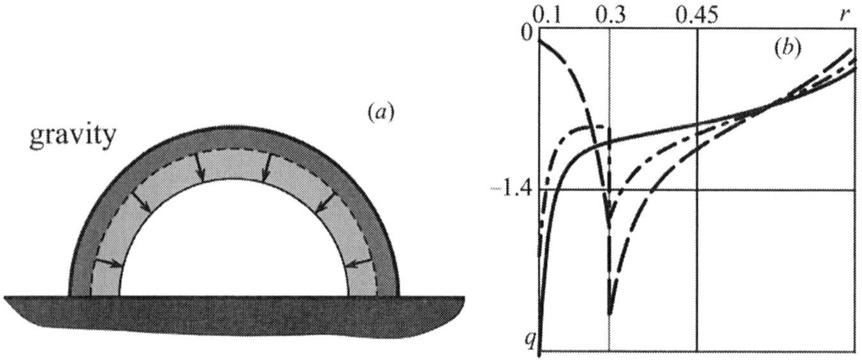

Figure 7. An arch structure under gravity.

5. Additive manufacturing of an arch structures

Some numerical results for AM fabricated concrete structures are demonstrated in this section. We suppose that concrete is a viscoelastic aging material, i.e., the material with time dependent mechanical properties (e.g., see [1–3]).

5.1. A Buried Arch

Consider now an arch structure manufactured from concrete and buried in a loose soil backfill (see figure 6 *a*). In order to strengthen this structure the process of growth inside it is initiated. During the growth the thickness of the arch wall doubles. Figure 6 *b* shows the limiting contact stress distribution ($t \to \infty$) on the end faces of the arch in the processes of slow (dashed line) and rapid (dash-dotted line) growth. Solid line shows the contact stress distribution on the end faces for a fixed arch with doubled wall thickness. One can easily see the fundamental difference in the behavior pattern of mechanical characteristics of growing solid as compared with a fixed one.

5.2. An Arch under Gravity

Consider one more AM process for an arch structure under gravity (see figure 7 *a*). This process is close to the previous one so we save all designations for variables and plots lines, i.e., in figure 7 *b* we use dashed and dash-dotted lines for the limiting contact stresses on the end faces of the arch for the modes of slow and rapid growth respectively and solid line for the analogous contact stresses in the mode of fixed arch.

One can see a very strong difference between the patterns of arch mechanical behavior for both problems and chosen modes.

It is quite clear that one can also obtain the solution for a buried arch under gravitation by using the superposition principle.

6. Conclusions

- Fundamentals of mechanical analysis for AM fabricated parts have been developed.
- New general nonclassical boundary value problems have been formulated includind a pair of specific quasistatic and kinematic conditions on the growth surface.
- The results obtained can also serve as a basis for the numerical research of AM processes in the case in which the velocity of growth surface due to the deformation of a solid under the action of surface and volumetric forces as well as under the tension of new elements is much smaller than the velocity of deposition of new elements to the solid's surface.
- New mechanical effects for AM fabricated parts and structures have been found out. Such effects can never arise in the case of another manufacturing processes.

Acknowledgments

The work was supported by the Russian Science Foundation under grant No. 17-19-01257.

References

[1] Aruyunyan N Kh, Manzhirov A V, and Naumov V E 1991 *Contact Problems in Mechanics of Growing Bodies* (Moscow: Nauka) p 172 [in Russian]

[2] Aruyunyan N Kh and Manzhirov A V 1999 *Contact Problems in the Theory of Creep* (Moscow: Nauka) p 320 [in Russian]

[3] Manzhirov A V 1995 The general non-inertial initial-boundary value problem for a viscoelastic ageing solid with piecewise-continuous accretion *J. Appl. Math. Mech.* **59** (5) 805–16

[4] Gibson I, Rosen D W, and Stucker B 2010 *Additive Manufacturing Technologies: Rapid Prototyping to Direct Digital Manufacturing* (New York: Springer Verlag, LLC) p XXI+498

[5] Gebhardt A and Hoetter J S 2016 *Additive Manufacturing: 3D Printing for Prototyping and Manufacturing* (Munich: Carl Hanser Verlag) p 611

[6] Guo N and Leu M C 2013 Additive manufacturing: technology, applications and research needs *Frontiers of Mech. Engng* **8** (3) 215–43

[7] Bikas H, Stavropoulos P, and Chryssolouris G 2016 Additive manufacturing methods and modelling approaches: a critical review *Int. J. Adv. Manufact. Techn.* **83** (1-4) 389–405

[8] Denlinger E R, Irwin J, and Michaleris P 2014 Thermomechanical modeling of additive manufacturing large parts *J. Manufact. Sci. Engng* **136** (6), 061007

[9] Frazier W E 2014 Metal additive manufacturing: A review *J. Mater. Engng Perform.* **23** (6), 1917–28

[10] Huang Y, Leu M C, Mazumder J, and Donmez A 2015 Additive manufacturing: current state, future potential, gaps and needs, and recommendations *J. Manufact. Sci. Engng* **137** (1), 014001

[11] Gibson I, Rosen D, and Stucker B (2010) *Additive Manufacturing Technologies* (New York: Springer) p 238

[12] Heigel J C, Michaleris P, and Reutzel E W 2015 Thermo-mechanical model development and validation of directed energy deposition additive manufacturing of Ti–6Al–4V *Addit. Manufact.* **5**, 9–19

[13] Chua C K and Leong K F 2014 *3D Printing and Additive Manufacturing: Principles and Applications (with Companion Media Pack). Fourth Edition of Rapid Prototyping* (World Scientific Publishing Co Inc)

[14] Biswas A, Shapiro V, and Tsukanov I 2004 Heterogeneous material modeling with distance fields *Comp. Aided Geom. Des.* **21** (3) 215–42

[15] Roberts I A, Wang, C J, Esterlein R, et al. 2009 A three-dimensional finite element analysis of the temperature field during laser melting of metal powders in additive layer manufacturing *Int. J. Machine Tools Manufact.* **49** (12) 916–23

[16] Melchels F P W, Domingos M A N, Klein T J, et al. 2012 Additive manufacturing of tissues and organs *Progress in Polymer Sci.* **37** (8) 1079–104

[17] Zhou J, Zhang Y, and Chen J K 2009 Numerical simulation of random packing of spherical particles for powder-based additive manufacturing *J. Manufact. Sci. Engng* **131** (3) 031004

[18] Parteli E J R 2013 DEM simulation of particles of complex shapes using the multisphere method: application for additive manufacturing *AIP Conf. Proc.* **1542** (1) 185–8
[19] Parteli E J R and Pöschel T 2016 Particle-based simulation of powder application in additive manufacturing *Powder Technol.* **288** 96–102
[20] Khajavi S H, Partanen J, and Holmstrom J 2014 Additive manufacturing in the spare parts supply chain *Computers in Industry* **65** (1) 50–63
[21] Patil N, Pal D, and Stucker B 2013 A new finite element solver using numerical Eigen modes for fast simulation of additive manufacturing processes In *Solid Freeform Fabrication Proceedings, August 2013* (The University of Texas at Austin) pp 535–43
[22] Kalpakjian S and Schmid S R 2014 *Manufacturing Engineering and Technology* (Upper Saddle River, NJ: Pearson)
[23] Hu D and Kovacevic R 2003 Modelling and measuring the thermal behaviour of the molten pool in closed-loop controlled laser-based additive manufacturing *Proc. Inst. Mech. Engineers. Part B: J. Engng Manufact.* **217** (4) 441–52
[24] Salmi M, Paloheimo K S, Tuomi J, et al. 2013 Accuracy of medical models made by additive manufacturing (rapid manufacturing) *J. Cranio-Maxillofacial Surgery* **41** (7) 603–9
[25] Li Y and Gu D 2014 Parametric analysis of thermal behavior during selective laser melting additive manufacturing of aluminum alloy powder *Mater. Des.* **63** 856–67
[26] Turner B N, Strong R, and Gold S A 2014 A review of melt extrusion additive manufacturing processes: I. Process design and modeling *Rapid Prototyping J.* **20** (3) 192–204
[27] Manzhirov A V 2013 Mechanics of growing solids and phase transitions *Key Engng Mater.* **535-536** 89–93
[28] Manzhirov A V Mechanics of growing solids: New track in mechanical engineering 2014 In *ASME 2014 Int. Mechanical Engineering Congr. and Expos. Volume 9: Mechanics of Solids, Structures and Fluids. Montreal, Quebec, Canada, November 14–20, 2014* (ASME) IMECE2014-36712, pp. V009T12A039
[29] Manzhirov A V and Lychev S A 2014 Mathematical modeling of additive manufacturing technologies In*Lecture Notes in Engineering and Computer Science: World Congr. on Engineering 2014, WCE 2014, 2–4 July, 2014, London, U.K.* (London: IAENG) pp 1404–9
[30] Manzhirov A V 2015 Mechanical design of viscoelastic parts fabricated using additive manufacturing technologies In *Lecture Notes in Engineering and Computer Science: Proc. of The World Congr. Engineering 2015, WCE 2015, 1–3 July, 2015, London, U.K.* (London: IAENG) pp 710–4
[31] Manzhirov A V and Parshin D A 2015 Influence of the erection regime on the stress state of a viscoelastic arched structure erected by an additive technology under the force of gravity *Mech. Solids* **50** (6) 657–75
[32] Manzhirov A V and Parshin D A 2016 Application of prestressed structural elements in the erection of heavy viscoelastic arched structures with the use of an additive technology *Mech. Solids* **51** (6) 692–700
[33] Manzhirov A V and Parshin D A 2016 Accretion of spherical viscoelastic objects under self-gravity *Lect. Notes Engng Comp. Sci.* **2224** (1) 1131–5
[34] Manzhirov A V 2016 A method for mechanical design of AM fabricated viscoelastic parts In *Transactions on Engineering Technologies* (Singapore: Springer) pp 223–35
[35] Manzhirov A V 2016 Fundamentals of mechanical design and analysis for AM fabricated parts *Proc. Manufact.* **7** 59–65
[36] Manzhirov A V 2017 Advances in the theory of surface growth with applications to additive manufacturing technologies *Proc. Engng* **173** 11-6
[37] Manzhirov A V 2017 Mechanical analysis of an AM fabricated viscoelastic shaft under torsion by rigid disks In *Lecture Notes in Engineering and Computer Science: Proc. of The World Congr. Engineering 2017, WCE 2017, 5–7 July, 2017, London, U.K., Vol. II* (London: IAENG) pp 856–60
[38] Manzhirov A V, Parshin D A, and Romanov A A 2017 Additive manufacturing of conical viscoelastic parts under axial tension-compression disks In *Lecture Notes in Engineering and Computer Science: Proc. of The World Congr. Engineering 2017, WCE 2017, 5–7 July, 2017, London, U.K., Vol. II* (London: IAENG) pp 934–939
[39] Manzhirov A V 2017 Mechanical design of AM fabricated prismatic rods under torsion *MATEC Web of Conf.* **95** 12002
[40] Manzhirov A V and Gupta N K 2017 Fundamentals of continuous growth processes in technology and nature *Proc. IUTAM* **23** 1–12
[41] Parshin D A and Manzhirov A V 2017 Quasistatic problems for piecewise-continuously growing solids with integral force conditions on surfaces expanding due to additional material influx *J. Phys.: Conf. Ser.* **937** 012038
[42] Bychkov P S, Kozintsev V M, Manzhirov A V, and Popov A L 2017 Determination of residual stresses in products in additive production by the layer-by-layer photopolymerization method *Mech. Solids* **52** (5) 524–9

IOP Conf. Series: Journal of Physics: Conf. Series **991** (2018) 012057 doi:10.1088/1742-6596/991/1/012057

2D problems of surface growth theory with applications to additive manufacturing

A V Manzhirov[1*] **and M N Mikhin**[2,3**]

[1]Ishlinsky Institute for Problems in Mechanics of the Russian Academy of Sciences, Moscow, Russia
[2]Branch of the Russian State University for the Humanities, Moscow Region, Domodedovo, Russia
[3]Moscow Aviation Institute (National Research University), Moscow, Russia

E-mail: *manzh@inbox.ru, **mmikhin@inbox.ru

Abstract. We study 2D problems of surface growth theory of deformable solids and their applications to the analysis of the stress-strain state of AM fabricated products and structures. Statements of the problems are given, and a solution method based on the approaches of the theory of functions of a complex variable is suggested. Computations are carried out for model problems. Qualitative and quantitative results are discussed.

Introduction

Traditional methods for the manufacturing of parts of complex shape involve various technological processing operations of material removal such as cutting, milling, drilling etc. There is an alternative class of technological processes where parts are manufactured by successive application of the material onto a substrate or a surface of arbitrary shape (e.g., by local polymerization, electrochemical reaction, welding deposition, local sintering, etc.). Such technologies include stereolithography, electrolytic deposition, winding, laser and thermal 3D printing, manufacturing of 3D electronic circuits, and many others (e.g., see [1, 2]).

Nowadays, additive manufacturing technologies undergo intensive development. The use of such technologies theoretically permits manufacturing 3D parts of very complex shape quickly and relatively cheaply from any material. However, although technological setup elements can be positioned with high precision, the deviations in the geometrical shape of the object to be manufactured from the design parameters can be high primarily owing to heat deformation, shrinkage, and other types of deformation of the object itself. Associated with these deformations are residual stresses, which inevitably arise in the object produced by additive manufacturing technologies. The residual stress distributions and magnitudes may result in buckling of thin-walled parts and destruction of massive bodies, which may happen as early as already in the manufacturing process. Thus, the development of mathematical models and theoretical methods for the analysis of internal technological stresses and strains in complex modern products is a topical scientific problem from the viewpoint of basic research as well as numerous applications. The basic scientific problems of surface growth are the subject of mechanics of growing bodies (e.g., see [3–5]). In what follows, we consider some 2D problems of surface growth theory with applications to the analysis and design of parts and structures produced by additive manufacturing technologies.

1. Plane problem of surface growth

Let a viscoelastic homogeneous ageing body occupy a plane multiply connected domain Ω_1 with boundary L_1 and be stress-free prior to the loading time instant τ_0. From the loading instant to the instant τ_1 of additive modification on a part $L_\sigma \subset L_1$ of the boundary, the body is subjected to a self-balanced load, and the part L^* of the boundary where new material is to be added remains load-free.

The continuous additive manufacturing of the body starts at time $\tau_1 \geq \tau_0$ by adding elements produced simultaneously with the body. In the course of the growth process, the body occupies a domain $\Omega(t)$ with boundary $L(t)$ consisting of two parts, $L(t) = L_\sigma \bigcup L^*(t)$, where $L^*(t)$ is the growth boundary or curve where the material inflow occurs at a given time ($L^*(t) = L^*$ for $\tau < \tau_1$) and $L_\sigma(t)$ is the boundary on which the load is given.

We assume that the growth boundary $L^*(t)$ remains load-free during the process of additive modification of the body. We also assume that the time $\tau_0 = \tau_0(x_1, x_2)$ at which the load is applied to the newly added elements coincides with the time $\tau^* = \tau^*(x_1, x_2)$ at which they are added to the growing body.

The AM fabrication process terminates at time $\tau_2 \geq \tau_1$, and after that the body occupies the domain $\Omega_2 = \Omega(\tau_2)$ bounded by the contour $L_2 = L(\tau_2) = L_\sigma(\tau_2) \bigcup L^*(\tau_2)$.

To be definite, consider the case of a plane stress state. In what follows, we deal with sufficiently slow processes such that one can neglect inertial terms in the equilibrium equations. We assume the bulk forces to be zero.

The boundary value problem for the main (original) viscoelastic homogeneous ageing body on the time interval $[\tau_0, \tau_1]$ is a traditional problem of viscoelasticity theory.

The initial-boundary value problem for the continuously growing body on the time interval $t \in [\tau_1, \tau_2]$ consists of the equilibrium equations

$$\frac{\partial \sigma_{11}}{\partial x_1} + \frac{\partial \sigma_{12}}{\partial x_2} = 0, \quad \frac{\partial \sigma_{12}}{\partial x_1} + \frac{\partial \sigma_{22}}{\partial x_2} = 0, \tag{1}$$

the Cauchy relations

$$D_{11} = \frac{\partial v_1}{\partial x_1}, \quad D_{22} = \frac{\partial v_2}{\partial x_2}, \quad D_{12} = \frac{1}{2}\left(\frac{\partial v_1}{\partial x_2} + \frac{\partial v_2}{\partial x_1}\right) \tag{2}$$

between the strain rates $D_{ij} = \partial \varepsilon_{ij}/\partial t$ and the displacement rates $v_i = \partial u_i/\partial t$, the state equation in the form

$$\begin{aligned}
&\sigma_{11} = 2G(\mathbf{I} + \mathbf{N}_{\tau_0})\left(\frac{2K}{K+1}\varepsilon_{11} + \frac{K-1}{K+1}\varepsilon_{22}\right), \\
&\sigma_{22} = 2G(\mathbf{I} + \mathbf{N}_{\tau_0})\left(\frac{K-1}{K+1}\varepsilon_{11} + \frac{2K}{K+1}\varepsilon_{22}\right), \\
&\sigma_{12} = 2G(\mathbf{I} + \mathbf{N}_{\tau_0})\varepsilon_{12}, \\
&\tau_0(x_1, x_2) = \begin{cases} \tau_0, & (x_1, x_2) \in \Omega_1, \\ \tau^*(x_1, x_2), & (x_1, x_2) \in \Omega^*(t), \end{cases} \\
&(\mathbf{I} + \mathbf{N}_{\tau_0})^{-1} = (\mathbf{I} - \mathbf{L}_{\tau_0}), \quad 2G = \frac{E}{1+v}, \quad K = \frac{1}{1-2v}, \\
&\mathbf{L}_s f(t) = \int_s^t f(\tau) K_1(t, \tau)\, d\tau, \quad K_1(t, \tau) = G(\tau)\frac{\partial}{\partial \tau}[G^{-1}(\tau) + \omega(t, \tau)],
\end{aligned} \tag{3}$$

the boundary condition

$$n_1\sigma_{11} + n_2\sigma_{12} = p_1, \quad n_1\sigma_{12} + n_2\sigma_{22} = p_1 \tag{4}$$

on the immovable part $L_\sigma(t)$ of the boundary, and the 3D-body-2D-surface contact condition on the growth boundary $L^*(t)$ (e.g., see [6–8]). For the zero prestress in the added elements, the last condition has the form

$$n_1\frac{\partial\sigma_{11}}{\partial t} + n_2\frac{\partial\sigma_{12}}{\partial t} = 0, \quad n_1\frac{\partial\sigma_{12}}{\partial t} + n_2\frac{\partial\sigma_{22}}{\partial t} = 0, \tag{5}$$

where $n = \{n_1, n_2\}$ is the unit outward normal vector to the lateral surface of the body, $p = \{p_1, p_2\}$ is the surface force vector, $\Omega^*(t) = \Omega(t)\backslash\Omega_1$ is the part of the body formed in the AM fabrication process (the additional body), $E = E(t)$ and $G = G(t)$ are the Young modulus and the shear modulus, respectively, $\omega(t,\tau)$ is the shear creep measure, $K_1(t,\tau)$ is the creep kernel, the elastic and creep Poisson ratios coincide and are equal to v, and $\mathbf{I}$ is the identity operator. The values of all functions for $\tau_0 \le t \le \tau_1$ are known from the solution of the problem for the main body.

The initial-boundary value problem (1)–(5) for an AM fabricated body has the following specific features, which take it beyond the framework of classical problems of mechanics of deformable solids: the strain consistency condition is violated in the domain occupied by the additional body, and only an analog of this condition and an analog of the Cauchy condition for the rates of the corresponding variables hold (this permits one to take into account the fact that, prior to the accretion to the main body, the added elements can undergo deforming actions independent of the processes occurring in the body itself); the constitutive relations depend on the function $\tau_0 = \tau_0(x_1, x_2)$, which may have jump discontinuities.

We use the notation $\sigma^0_{ij} = (\mathbf{I}-\mathbf{L}_{\tau_0})\sigma_{ij}G^{-1}$ and transform the problem of additive modification of a viscoelastic body with the constitutive relations (1)–(5) into the problem of additive modification of an elastic body described by Hooke's law,

$$\begin{gathered}
\frac{\partial\sigma^0_{11}}{\partial x_1} + \frac{\partial\sigma^0_{12}}{\partial x_2} = 0, \quad \frac{\partial\sigma^0_{12}}{\partial x_1} + \frac{\partial\sigma^0_{22}}{\partial x_2} = 0;\\
D_{11} = \frac{\partial v_1}{\partial x_1}, \quad D_{22} = \frac{\partial v_2}{\partial x_2}, \quad D_{12} = \frac{1}{2}\left(\frac{\partial v_1}{\partial x_2} + \frac{\partial v_2}{\partial x_1}\right),\\
\sigma^0_{11} = 2G\left(\frac{2K}{K+1}\varepsilon_{11} + \frac{K-1}{K+1}\varepsilon_{22}\right), \quad \sigma^0_{22} = 2G\left(\frac{K-1}{K+1}\varepsilon_{11} + \frac{2K}{K+1}\varepsilon_{22}\right), \quad \sigma^0_{12} = 2G\varepsilon_{12};\\
(x_1,x_2)\in L_\sigma: \quad n_1\sigma^0_{11} + n_2\sigma^0_{12} = (\mathbf{I}-\mathbf{L}_{\tau_0})p_1G^{-1} = p^0_1,\\
n_1\sigma^0_{12} + n_2\sigma^0_{22} = (\mathbf{I}-\mathbf{L}_{\tau_0})p_2G^{-1} = p^0_2;\\
(x_1,x_2)\in L^*(t): \quad n_1\frac{\partial\sigma^0_{11}}{\partial t} + n_2\frac{\partial\sigma^0_{12}}{\partial t} = 0, \quad n_1\frac{\partial\sigma^0_{12}}{\partial t} + n_2\frac{\partial\sigma^0_{22}}{\partial t} = 0.
\end{gathered} \tag{6}$$

We differentiate the equilibrium equations, the boundary condition on $L_\sigma(t)$, and the state equation in (6) with respect to t and reduce the boundary value problem for a growing body to the form

$$\begin{gathered}
\frac{\partial S_{11}}{\partial x_1} + \frac{\partial S_{12}}{\partial x_2} = 0, \quad \frac{\partial S_{12}}{\partial x_1} + \frac{\partial S_{22}}{\partial x_2} = 0,\\
S_{11} = 2\left(\frac{2K}{K+1}D_{11} + \frac{K-1}{K+1}D_{22}\right), \quad S_{22} = 2\left(\frac{K-1}{K+1}D_{11} + \frac{2K}{K+1}D_{22}\right),\\
S_{12} = 2D_{12}, \quad S_{ij} = \frac{\partial\sigma^0_{ij}}{\partial t},\\
(x_1,x_2)\in L_\sigma: \quad n_1S_{11} + n_2S_{12} = \mathbf{R}p_1 = q_1, \quad n_1S_{12} + n_2S_{22} = \mathbf{R}p_2 = q_2,\\
(x_1,x_2)\in L^*(t): \quad n_1S_{11} + n_2S_{12} = 0, \quad n_1S_{12} + n_2S_{22} = 0,
\end{gathered} \tag{7}$$

IOP Conf. Series: Journal of Physics: Conf. Series **991** (2018) 012057 doi:10.1088/1742-6596/991/1/012057

where the operator $\mathbf{R}$ is defined by the expression

$$\mathbf{R}f(x_1,x_2,t) = \frac{1}{G(t)}\frac{\partial f(x_1,x_2,t)}{\partial t} + \int_{\tau_0(x_1,x_2)}^{t} \frac{\partial f(x_1,x_2,\tau)}{\partial \tau}\frac{\partial \omega(t,\tau)}{\partial t}d\tau \\ + f(x_1,x_2,\tau_0(x_1,x_2))\frac{\partial \omega(t,\tau_0(x_1,x_2))}{\partial t}.$$

The boundary value problem (7) coincides in form with the boundary value problem of the elasticity theory with a parameter t. Its solution can be constructed by any analytical or numerical method efficient in the elasticity theory. The solution of the original initial-boundary problem of additive modification of a viscoelastic ageing body for $t \in [\tau_1, \tau_2]$ can be reconstructed by the formulas

$$\sigma_{ij}(x_1,x_2,t) = G(t)\left\{\frac{\sigma_{ij}(x_1,x_2,\tau_0(x_1,x_2))}{G(\tau_0(x_1,x_2))}\left[1 + \int_{\tau_0(x_1,x_2)}^{t} R(t,\tau)\,d\tau\right]\right. \\ \left. + \int_{\tau_0(x_1,x_2)}^{t}\left[S_{ij}(x_1,x_2,\tau) + \int_{\tau_0(x_1,x_2)}^{\tau} S_{ij}(x_1,x_2,\varsigma)\,d\varsigma R(t,\tau)\right]d\tau\right\}, \tag{8}$$
$$u_i(x_1,x_2,t) = u_i(x_1,x_2,\tau_0(x_1,x_2)) + \int_{\tau_0(x_1,x_2)}^{t} v_i(x_1,x_2,\tau)\,d\tau.$$

The constitutive relations of the problem for a body whose AM fabrication has been finished have the form (1)–(5), where the condition on the growth boundary is missing. This problem can be reduced to the form (7), and the formulas reconstructing the true characteristics of the stress-strain state preserve the form (8).

The exposition of the main new part of the method for solving the plane problem for a growing body is complete. The boundary value problems for all the main stages of the body evolution have been reduced to boundary value problems coinciding in form with the boundary value problems of the elasticity theory with some parameter. To study the latter, we use the methods of the theory of functions of a complex variable.

Since the boundary value problems at all stages are mathematically the same, we only consider the boundary value problem (7). To solve it, we express the stresses via the Airy stress function $U(x_1,x_2,t)$ (e.g., see [9]),

$$S_{11} = \frac{\partial^2 U}{\partial x_2^2}, \quad S_{22} = \frac{\partial^2 U}{\partial x_1^2}, \quad S_{12} = \frac{\partial^2 U}{\partial x_1 \partial x_2},$$

which is a biharmonic function.

It is well known that every biharmonic function can be represented via analytic functions of the complex variable,

$$U = \frac{1}{2}\left[\bar{z}\varphi_1(z,t) + z\overline{\varphi_1(z,t)} + \chi_1(z,t) + \overline{\chi_1(z,t)}\right].$$

Then the complex representation of the functions S_{ij} has the form

$$S_{11} + S_{22} = 4Re[\varphi_1(z,t)]',$$
$$S_{22} - S_{11} + 2iS_{12} = 2[\bar{z}\varphi_1''(z,t) + \psi'(z,t)],$$
$$\psi(z,t) = \chi_1'(z,t),$$

where the prime stands for differentiation with respect to z.

For the variables v_i, we write the complex representation in the form

$$2(v_1 + iv_2) = \frac{5K+1}{3K-1}\varphi_1(z,t) - z\overline{\varphi_1(z,t)} - \overline{\psi(z,t)}.$$

Thus, S_{ij} and v_i can be expressed via two analytic functions, which are commonly known as the Kolosov-Muskhelishvili functions [10].

The boundary condition can be represented in the form

$$\varphi_1(z,t) + z\overline{\varphi_1'(z,t)} + \overline{\psi(z,t)} = f,$$

where $f = i(Q_1 + Q_2)$ and Q_1 and Q_2 are given functions on the contour.

In many applications, one needs to have expressions for the desired functions in the orthogonal coordinate system defined via a conformal mapping $z = \omega(\varsigma, t)$ ($\varsigma = \rho e^{i\theta}$). In this case, we have

$$\begin{aligned}
&S_{\rho\rho}(\varsigma,t) + S_{\theta\rho}(\varsigma,t) = 4Re[\varphi(\varsigma,t)]',\\
&S_{\theta\rho}(\varsigma,t) - S_{\rho\rho}(\varsigma,t) + 2iS_{\theta\rho}(\varsigma,t) = \frac{2\varsigma^2}{p^2\overline{\omega'(\varsigma,t)}}\left[\overline{\omega(\varsigma,t)}\varphi''(\varsigma,t) + \omega'(\varsigma,t)\psi'(\varsigma,t)\right],\\
&2[v_\rho(\varsigma,t) + iv_\theta(\varsigma,t)] = \frac{\overline{\varsigma\omega'(\varsigma,t)}}{|\omega'(\varsigma,t)|}\left[\frac{5K+1}{3K-1}\varphi(\varsigma,t) - \frac{\omega(\varsigma,t)}{\overline{\omega'(\varsigma,t)}}\overline{\varphi'(\varsigma,t)} - \overline{\psi(\varsigma,t)}\right].
\end{aligned}$$

Since the domain in question is often mapped onto the unit disk (or onto the plane with the unit disk deleted) in many problems, we present the following useful formulas:

$$\begin{aligned}
\varphi(\varsigma,t) &= -\frac{Q_1 + iQ_2}{2\pi}\frac{3K-1}{8K}\ln\varsigma + \frac{\Gamma c(t)}{\varsigma} + \varphi_0(\varsigma,t),\\
\psi(\varsigma,t) &= -\frac{Q_1 - iQ_2}{2\pi}\frac{5K+1}{8K}\ln\varsigma + \frac{\Gamma' c(t)}{\varsigma} + \psi_0(\varsigma,t),
\end{aligned}$$

where $\varphi_0(\varsigma,t)$ and $\psi_0(\varsigma,t)$ are functions analytic in the interior and continuous up to the boundary of the unit disk. The functions $\varphi_0(\varsigma,t)$ and $\psi_0(\varsigma,t)$ satisfy the same equations as $\varphi(\varsigma,t)$ and $\psi(\varsigma,t)$ with the function f everywhere replaced by the function

$$f_0 = f - \frac{T_1 + iT_2}{2\pi}\ln\sigma - \frac{\Gamma c(t)}{\sigma} - \frac{\omega(\sigma,t)}{\overline{\omega'(\sigma,t)}}\left(\frac{T_1 - iT_2}{2\pi}\frac{3K-1}{8K}\sigma - \Gamma\overline{c(t)}\sigma^2\right) - \Gamma'\overline{c(t)}\sigma.$$

Once the stress components in the orthogonal coordinate system are known, one can find the components S_{ij} by the formulas

$$\begin{aligned}
&S_{\rho\rho}(\varsigma,t) + S_{\theta\rho}(\varsigma,t) = S_{11}(\varsigma,t) + S_{22}(\varsigma,t),\\
&S_{\theta\rho}(\varsigma,t) - S_{\rho\rho}(\varsigma,t) + 2iS_{\theta\rho}(\varsigma,t) = [S_{22}(\varsigma,t) - S_{11}(\varsigma,t) + 2iS_{12}(\varsigma,t)]e^{2i\alpha}.
\end{aligned} \tag{9}$$

Thus, having found two harmonic functions from the boundary conditions, one can find the desired functions S_{ij} and v_i and then use formulas (8) to reconstruct the true values of the stresses and displacements in the growing body.

The solution of each additive manufacturing problem for deformable bodies is a stand-alone laborious task (e.g., see [11–15].) However, the form of the mathematical relations obtained above permits one right away to predict such phenomena intrinsically typical of growing bodies as the onset of residual stresses after the load is removed, the occurrence of stress discontinuity surfaces in AM fabricated bodies, and the dependence of the stress-strain state of viscoelastic bodies on the additive manufacturing rate and method. As an example of possible applications of the theory, we note problems on the occlusion of various holes.

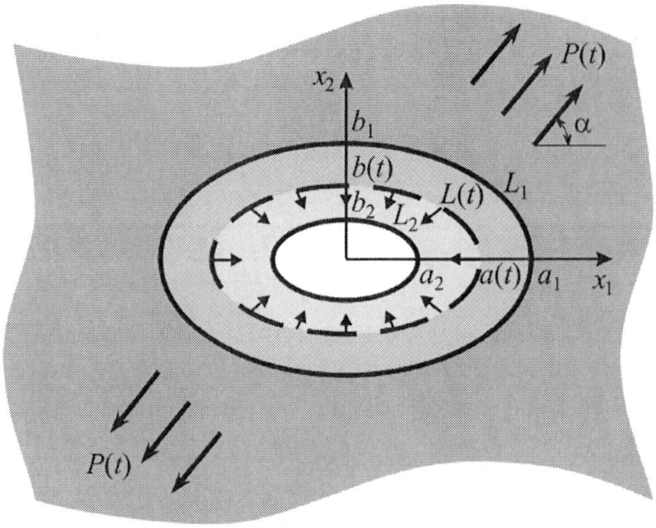

Figure 1. AM fabrication of a hole of the desired elliptic shape.

2. Stress concentration near AM fabricated holes of desired shape

By way of example, consider the problem on the stress concentration near a hole that must take a given shape in the AM fabrication process in a stressed medium. The medium material is viscous and ageing; i.e., its properties depend on time. The medium is in a uniaxial stress state. The additive manufacturing process should transform a large elliptic hole into a smaller elliptic hole.

Assume that an infinite plate Ω_1 weakened by an elliptic hole L_1 is manufactured at time zero from an ageing viscoelastic material. Let the contour L_1 be free of internal stresses, and let the stress state at infinity be characterized by a uniform extending stress $P(t)$ directed at an angle α with the axis Ox_1 (figure 1).

The continuous additive manufacturing of the hole by elements of the same age as the medium begins at time $\tau_1 \geq \tau_0$. In the growth process, the medium occupies a domain $\Omega(t)$ with boundary $L(t)$ that is an ellipse for each t. We consider the process of additive occlusion of the elliptic hole in which the ratio of the semiaxes $a(t)$ and $b(t)$ remains constant; i.e.,

$$\frac{a_1}{b_1} = \frac{a(t)}{b(t)}.$$

The additive occlusion of the hole terminates at time $\tau_2 \geq \tau_1$ and since then the hole occupies the domain $\Omega_2 = \Omega(\tau_2)$ bounded by the contour $L_2 = L(\tau_2)$.

We assume that the newly added elements are not prestressed. Since the boundary value problems obtained at all the considered stages of additive manufacturing are mathematically equivalent, it suffices to consider the growth stage. Then the initial-boundary value problem for the AM fabrication of the elliptic hole acquires the form

$$\begin{aligned}
&\frac{\partial\sigma_{11}}{\partial x_1}+\frac{\partial\sigma_{12}}{\partial x_2}=0,\quad \frac{\partial\sigma_{12}}{\partial x_1}+\frac{\partial\sigma_{22}}{\partial x_2}=0;\\
&D_{11}=\frac{\partial v_1}{\partial x_1},\quad D_{22}=\frac{\partial v_2}{\partial x_2},\quad D_{12}=\frac{1}{2}\left(\frac{\partial v_1}{\partial x_2}+\frac{\partial v_2}{\partial x_1}\right);\quad D_{ij}=\frac{\partial\varepsilon_{ij}}{\partial t},\quad v_i=\frac{\partial u_i}{\partial t};\\
&\sigma_{11}=2G(\mathbf{I}+\mathbf{N}_{\tau_0})\left(\frac{2K}{K+1}\varepsilon_{11}+\frac{K-1}{K+1}\varepsilon_{22}\right),\\
&\sigma_{22}=2G(\mathbf{I}+\mathbf{N}_{\tau_0})\left(\frac{K-1}{K+1}\varepsilon_{11}+\frac{2K}{K+1}\varepsilon_{22}\right),\\
&\sigma_{12}=2G(\mathbf{I}+\mathbf{N}_{\tau_0})\varepsilon_{12},\\
&(x_1,x_2)\in L(t):\quad n_1\frac{\partial\sigma_{11}}{\partial t}+n_2\frac{\partial\sigma_{12}}{\partial t}=0,\quad n_1\frac{\partial\sigma_{12}}{\partial t}+n_2\frac{\partial\sigma_{22}}{\partial t}=0.
\end{aligned}\tag{10}$$

The boundary value problem (7) becomes

$$\begin{aligned}
&\frac{\partial S_{11}}{\partial x_1}+\frac{\partial S_{12}}{\partial x_2}=0,\quad \frac{\partial S_{12}}{\partial x_1}+\frac{\partial S_{22}}{\partial x_2}=0,\\
&S_{11}=2\left(\frac{2K}{K+1}D_{11}+\frac{K-1}{K+1}D_{22}\right),\quad S_{22}=2\left(\frac{K-1}{K+1}D_{11}+\frac{2K}{K+1}D_{22}\right),\\
&S_{12}=2D_{12},\quad S_{ij}=\frac{\partial\sigma^0_{ij}}{\partial t},\quad \sigma^0_{ij}=(\mathbf{I}-\mathbf{L}_{\tau_0(x_1,x_2)})\sigma_{ij}G^{-1},\\
&(x_1,x_2)\in L(t):\quad n_1S_{11}+n_2S_{12}=0,\quad n_1S_{12}+n_2S_{22}=0.
\end{aligned}\tag{11}$$

Since the contour $L(t)$ is stress-free, it follows that the principal vector components are zero,

$$Q_1=0,\quad Q_2=0.$$

The principal stresses at infinity are given in the form

$$N_1=P^0=(\mathbf{I}-\mathbf{L}_{\tau_0})PG^{-1},\quad N_2=0.$$

Then we have

$$\operatorname{Re}\Gamma=\frac{1}{4}(N_1+N_2)=\frac{P^0}{4}\quad(\Gamma=\bar{\Gamma}),\quad \Gamma'=-\frac{1}{2}(N_1-N_2)e^{-2i\alpha}.$$

The function mapping the exterior of the disk onto the exterior of the ellipse has the form

$$z=\omega(\varsigma,t)=R(t)\left[\varsigma+\frac{m(t)}{\varsigma}\right],$$

where $R(t)=[a(t)+b(t)]/2$, $m(t)=[a(t)-b(t)]/[a(t)+b(t)]$.

The formulas for the components S_{rr}, $S_{\theta\theta}$, and $S_{\theta r}$ read

$$S_{rr} = \frac{q}{2}[A(r,\theta,t) - B(r,\theta,t)], \quad S_{\theta\theta} = \frac{q}{2}[A(r,\theta,t) + B(r,\theta,t)],$$

$$S_{\theta r} = \frac{qr^2}{2}\left(\frac{1}{[r^4 - 2mr^2\cos(2\theta) + m^2]^2}\{2(r^4 - r^2)\sin[2(\alpha-\theta)] + m(m^2 + r^4 - r^2)\sin(2\theta)\right.$$
$$- m^2\sin[2(\alpha+\theta)] + m^2r^2(1-m^2)\sin(2\theta)\cos(2\alpha) + [2m - m(r^2 - m^2)]\sin(2\alpha)\}$$
$$\left.+ \frac{(1-r^4)\sin[2(\theta-\alpha)] - mr^2\sin(2\alpha)}{r^2[r^4 - 2mr^2\cos(2\theta) + m^2]}\right),$$

$$A(r,\theta,t) = \frac{r^4 - 2mr^2\cos[2(\theta-\alpha)] - m^2 + 2m\cos(2\alpha)}{r^4 - 2mr^2\cos(2\theta) + m^2},$$

$$B(r,\theta,t) = \frac{r^2}{(r^4 - 2mr^2\cos(2\varphi) + m^2)^2}\{2[m - r^2\cos(2\theta)]\cos(2\alpha)$$
$$+ [\cos(2\alpha) - m](r^4 - m^2)[2\cos(2\theta) - m] + r^4 - m^2$$
$$+ (2r^4 + m^2 + 2m^3r^2)\sin(2\theta)\sin(2\alpha) + 2mr^2\sin(2\alpha)\cos(2\theta)\}$$
$$- \frac{\cos[2(\alpha-\theta)] + r^2\{r^2\cos[2(\theta-\alpha)] - m\cos(2\alpha)\}}{r^4 - 2mr^2\cos(2\theta) + m^2}.$$

We use Kolosov's formulas (9) to compute S_{11}, S_{12}, and S_{22}. The true stresses are determined by formulas (8).

For the material, we take concrete with instantaneous elastic shear modulus

$$G(t) = G_0(1 - e^{-\alpha_0 t}) \tag{12}$$

and with shear creep measure in the form

$$\omega(t,\tau) = (D_0 + Fe^{-\beta\tau})(1 - e^{-\gamma(t-\tau)}). \tag{13}$$

Let us pass to dimensionless variables using the limit shear modulus G_0, the coefficient γ in the approximation to the creep measure, and the final length $a_2 = a(\tau_2)$ of the major semiaxis of the ellipse. We make a change of variables by the formulas

$$t^* = \gamma t, \quad \tau_1^* = \gamma\tau_1, \quad \tau_2^* = \gamma\tau_2, \quad a^*(t^*) = \frac{a(t)}{a_2}, \quad b^*(t^*) = \frac{b(t)}{a_2}, \quad a_1^* = \frac{a_1}{a_2}, \quad a_2^* = \frac{a_2}{a_2} = 1,$$
$$b_1^* = \frac{b_1}{a_2}, \quad b_2^* = \frac{b_2}{a_2}, \quad \alpha_0^* = \frac{\alpha_0}{\gamma}, \quad \beta^* = \frac{\beta}{\gamma}, \quad \gamma^* = \frac{\gamma}{\gamma} = 1, \quad D_0^* = D_0G_0, \quad F^* = FG_0,$$

and, omitting asterisks in the notation, specify the following values of functions and parameters:

$$a(t) = a_1\left(\tau_2 - \frac{1}{3}\tau_1 - \frac{2}{3}t\right)(\tau_2 - \tau_1)^{-1}, \quad b(t) = b_1\left(\tau_2 - \frac{1}{3}\tau_1 - \frac{2}{3}t\right)(\tau_2 - \tau_1)^{-1},$$
$$a_1 = 3, \quad a_2 = 1, \quad b_1 = 1.5, \quad b_2 = 0.5, \quad \gamma = 0.06, \quad \alpha_0 = 2, \quad \beta = \frac{0.031}{0.06},$$
$$A = 0.5, \quad D_0 = 0.5522, \quad F = 4, \quad D_0^* = D_0G_0, \quad F^* = FG_0.$$

Obviously, each of the semiaxes of the ellipse becomes three times smaller after the accretion; i.e., $a_2 = a_1/3$ and $b_2 = b_1/3$.

Throughout the following, dashed lines in the figures stand for the shear stress intensity distribution at the beginning of the additive manufacturing process, dot-and-dash lines stand

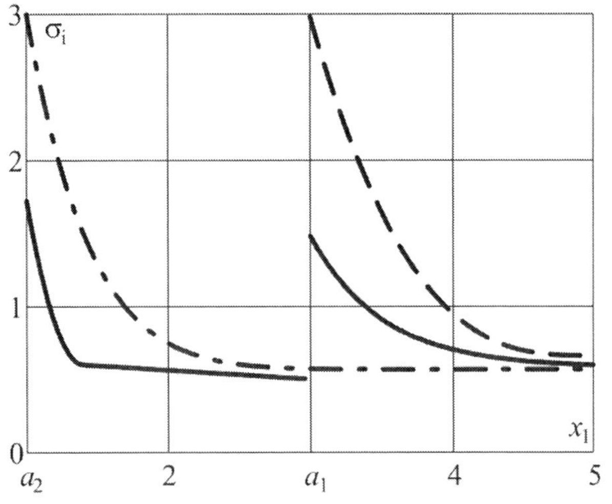

Figure 2. Shear stress intensity in the vicinity of the elliptic hole for the first mode of the fast AM fabrication.

for that in the final body without considering the additive manufacturing process, and solid lines stand for the shear stress intensity distribution at large times.

Let the stress state at infinity be characterized by an extending stress $P(t)$ directed at the angle $\alpha = \pi/2$ to the axis Ox_1. We study the shear stress intensity distribution along the major semiaxis of the hole B depending on the extending stress law $P(t)$ at infinity.

For the case of a generalized plane stress state, we compute the shear stress intensity by the formula $\sigma_i = \sqrt{1/3}(\sigma_{11}^2 + \sigma_{22}^2 - \sigma_{11}\sigma_{22} + 3\sigma_{12}^2)^{1/2}$.

Consider the case of a constant extending stress at infinity, $P(t) = 1$. For the fast AM fabrication starting shortly after the application of the load $P(t) = 1$ (the first mode of fast AM fabrication: $\tau_0 = 0.1$, $\tau_1 = 0.3$, $\tau_2 = 0.5$), the maximum of the shear stress intensity is attained on the boundary of the hole (figure 2).

If the AM fabrication starts much later than the load $P(t) = 1$ is applied (the second mode of the fast AM fabrication: $\tau_0 = 0.1$, $\tau_1 = 1.3$, $\tau_2 = 1.5$), then the maximum of the shear stress intensity is attained on the interface between the original and additional bodies (figure 3).

For slow AM fabrication, the maximum of the shear stress intensity is attained on the interface between the original and additional bodies (figure 4).

Now consider the case of stepwise variation of the extending stress on the time interval $t \in [1, 5]$ according to the formula

$$P(t) = \begin{cases} 0.1 & t \le T, \\ 1.1 & t > T. \end{cases}$$

For this AM fabrication mode, the maximum of the shear stress intensity is attained inside the additional body. Figure 5 shows the variations in the shear stress intensities for $T = 2$.

3. Torsion of prismatic rods under surface growth

Consider a homogeneous viscoelastic ageing body manufactured at time zero and occupying some cylindrical domain Π_1 whose cross-section Ω_1 has the boundary L_1. At time τ_0, the faces of the cylindrical body are subjected to forces statically equivalent to a couple with torque $M(t)$. The lateral surface of the body Π_1 is stress-free.

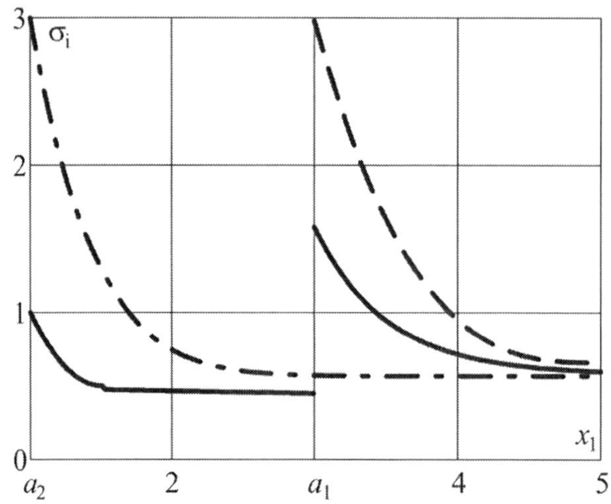

Figure 3. Shear stress intensity in the vicinity of the elliptic hole for the second mode of the fast AM fabrication.

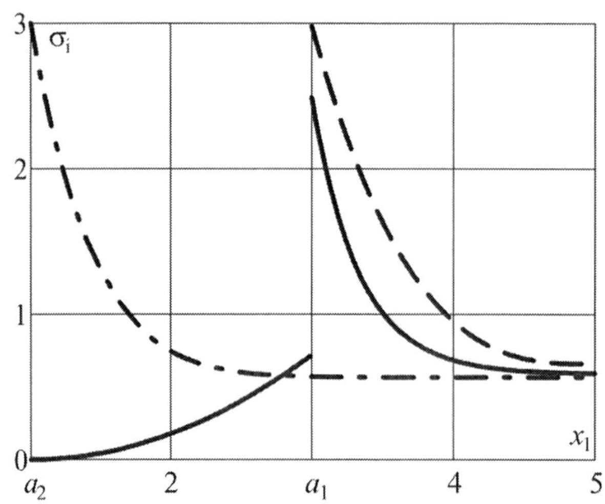

Figure 4. Shear stress intensity in the vicinity of the elliptic hole for the slow AM fabrication.

The continuous AM fabrication of the body by accretion of elements manufactured simultaneously with the body starts at time $\tau_1 \geq \tau_0$. The newly added elements are stress-free. Let $L(t)$ be the boundary of the cross-section $\Omega(t)$, which depends on time; we have $L(\tau_1) = L_1$ and $\Omega(\tau_1) = \Omega_1$. The boundary $L(t)$ of $\Omega(t)$ consists of two parts, $L(t) = L^*(t) \bigcup L_\sigma(t)$, where $L^*(t)$ is the growth boundary where the material inflow occurs at the current time ($L^*(t) = L^*$ for $\tau \leq \tau_1$) and the boundary $L_\sigma(t)$ is stress-free.

We assume that the time $\tau_0 = \tau_0(x_1, x_2)$ at which the load is applied to the newly added elements coincides with the time $\tau^* = \tau^*(x_1, x_2)$ at which they are added to the growing body.

The AM fabrication of the body terminates at time $\tau_2 \geq \tau_1$, and after that the body occupies the domain $\Pi_2 = \Pi(\tau_2)$ whose cross-section $\Omega_2 = \Omega(\tau_2)$ has the boundary $L_2 = L(\tau_2)$. Note that throughout the following we consider sufficiently slow processes, so that the inertial terms in the equilibrium equations can be neglected.

The boundary value problem for the main (original) viscoelastic ageing body on the time

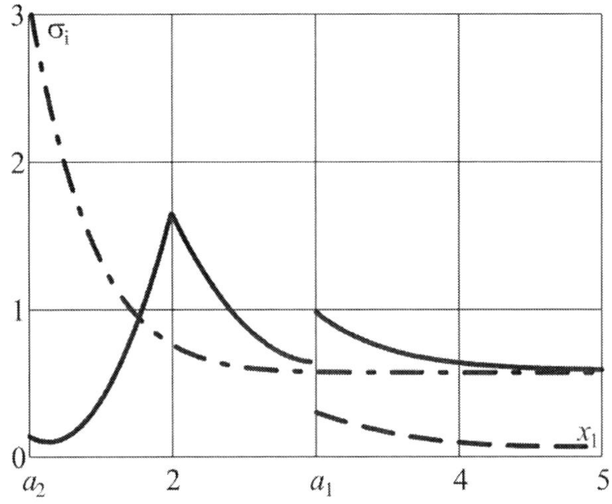

Figure 5. Shear stress intensity in the vicinity of the elliptic hole for the mode of AM fabrication with a step-function tension.

interval $[\tau_0, \tau_1]$ is a traditional torsion problem of viscoelasticity theory.

The initial-boundary value problem for the continuously growing body on the time interval $t \in [\tau_1, \tau_2]$ consists of the equilibrium equations

$$\frac{\partial \sigma_{13}}{\partial x_3} = 0, \quad \frac{\partial \sigma_{23}}{\partial x_1} = 0, \quad \frac{\partial \sigma_{13}}{\partial x_1} + \frac{\partial \sigma_{23}}{\partial x_2} = 0, \tag{14}$$

the Cauchy relations (see (1) and (2))

$$\begin{gathered} D_{11} = \frac{\partial v_1}{\partial x_1} = 0, \quad D_{22} = \frac{\partial v_2}{\partial x_2} = 0, \quad D_{33} = \frac{\partial v_3}{\partial x_3} = 0, \\ D_{12} = \frac{1}{2}\left(\frac{\partial v_2}{\partial x_1} + \frac{\partial v_1}{\partial x_2}\right) = 0, \quad D_{13} = \frac{1}{2}\left(\frac{\partial v_1}{\partial x_3} + \frac{\partial v_3}{\partial x_1}\right), \quad D_{23} = \frac{1}{2}\left(\frac{\partial v_2}{\partial x_3} + \frac{\partial v_3}{\partial x_2}\right) \end{gathered} \tag{15}$$

between the strain and displacement rates, the state equations (see (3))

$$\sigma_{13} = 2G(\mathbf{I} + \mathbf{N}_{\tau_0})\varepsilon_{13}, \quad \sigma_{23} = 2G(\mathbf{I} + \mathbf{N}_{\tau_0})\varepsilon_{23}, \tag{16}$$

the boundary condition (see condition (4))

$$(x_1, x_2) \in L_\sigma(t): \quad n_1\sigma_{13} + n_2\sigma_{23} = 0 \tag{17}$$

on the immovable part of the boundary, the boundary condition (see condition (5))

$$(x_1, x_2) \in L^*(t): \quad n_1\frac{\partial \sigma_{13}}{\partial t} + n_2\frac{\partial \sigma_{23}}{\partial t} = 0 \tag{18}$$

on the growth boundary $L^*(t)$, and the equilibrium conditions

$$M(t) = \iint_{\Omega(t)} (x_1\sigma_{23} - x_2\sigma_{13})\, dx_1\, dx_2, \quad \iint_{\Omega(t)} \sigma_{13}\, dx_1\, dx_2 = \iint_{\Omega(t)} \sigma_{23}\, dx_1\, dx_2 = 0 \tag{19}$$

for the face cross-sections $\Omega(t)$, where we use the notation defined after relations (5). The values of all functions at time $\tau_0 \le t \le \tau_1$ are known from the solution of the problem for the main body.

By analogy with the preceding, we transform the boundary value problem (14)–(19) of AM fabrication of a viscoelastic rod into the problem of AM fabrication of an elastic body,

$$\begin{aligned}
&\frac{\partial \sigma^0_{13}}{\partial x_1} + \frac{\partial \sigma^0_{23}}{\partial x_2} = 0,\\
&D_{13} = \frac{1}{2}\left(\frac{\partial v_1}{\partial x_3} + \frac{\partial v_3}{\partial x_1}\right), \quad D_{23} = \frac{1}{2}\left(\frac{\partial v_2}{\partial x_3} + \frac{\partial v_3}{\partial x_2}\right),\\
&\sigma^0_{13} = 2\varepsilon_{13}, \quad \sigma^0_{23} = 2\varepsilon_{23},\\
&(x_1, x_2) \in L_\sigma(t): \quad n_1\sigma^0_{13} + n_2\sigma^0_{23} = 0,\\
&(x_1, x_2) \in L^*(t): \quad n_1\frac{\partial \sigma^0_{13}}{\partial t} + n_2\frac{\partial \sigma^0_{23}}{\partial t} = 0,\\
&M(t) = \iint_{\Omega(t)} (x_1\sigma^0_{23} - x_2\sigma^0_{13})\, dx_1\, dx_2.
\end{aligned} \tag{20}$$

Let us transform the initial-boundary value problem (20) into a boundary value problem for the strain rates, displacement rates, and operator stress rates. To this end, we differentiate the equilibrium equation and the state equation with respect to t. As a result, we obtain the boundary value problem

$$\begin{aligned}
&\frac{\partial S_{13}}{\partial x_1} + \frac{\partial S_{23}}{\partial x_2} = 0,\\
&D_{13} = \frac{1}{2}\left(\frac{\partial v_3}{\partial x_1} + \frac{\partial v_1}{\partial x_3}\right), \quad D_{23} = \frac{1}{2}\left(\frac{\partial v_3}{\partial x_2} + \frac{\partial v_2}{\partial x_3}\right),\\
&S_{13} = 2D_{13}, \quad S_{23} = 2D_{23}, \quad S_{ij} = \frac{\partial \sigma^0_{ij}}{\partial t},\\
&(x_1, x_2) \in L(t): \quad n_1 S_{13} + n_2 S_{23} = 0,\\
&\frac{dM^0(t)}{dt} = \iint_{\Omega(t)} (x_1 S_{23} - x_2 S_{13})\, dx_1\, dx_2 + \int_{L^*(t)} (x_1\sigma^*_{23} - x_2\sigma^*_{13})\, dl.
\end{aligned} \tag{21}$$

Being supplemented with the initial conditions for the main body for $t = \tau_1$, relations (21), which also contain the initial-boundary condition on the growth boundary, form an initial-boundary value problem with time parameter t.

For the variables S_{ij} and v_i we have the formulas

$$\begin{aligned}
&v_1 = -\theta'_t(t)x_2x_3, \quad v_2 = \theta'(t)x_1x_3, \quad v_3 = \theta'(t)\varphi_t,\\
&S_{13} = \theta'_t(t)\left(\frac{\partial \varphi_t}{\partial x_1} - x_2\right), \quad S_{23} = \theta'_t(t)\left(\frac{\partial \varphi_t}{\partial x_2} + x_1\right),
\end{aligned} \tag{22}$$

where $\varphi_t = \varphi(x_1, x_2, t)$ is the torsion function to be determined, $\theta(t)$ is the torsion angle (twist), and $\theta'_t(t)$ is the time derivative of the twist. The torsion function φ_t is harmonic in the domain $\Omega(t)$. It must satisfy the boundary condition

$$(x_1, x_2) \in L(t): \quad \frac{\partial \varphi_t}{\partial n} = x_2 n_1 - x_1 n_2.$$

Thus, the solution of the torsion problem is reduced with the use of the torsion function φ_t to determining a harmonic function in the cross-section domain $\Omega(t)$ from a given value of its normal derivative on the contour $L(t)$ (the Neumann problem).

Obviously, in view of (21) one obtains

$$\begin{aligned}\frac{dM^0(t)}{dt} &= \theta_t'(t)D(t) + \int_{L^*(t)} (x_1\sigma_{23}^* - x_2\sigma_{13}^*)\, dl,\\ D(t) &= \iint_{\Omega(t)} \left(x_1^2 + x_2^2 + x_1\frac{\partial\varphi}{\partial x_2} - x_2\frac{\partial\varphi}{\partial x_1}\right) dx_1\, dx_2,\end{aligned} \tag{23}$$

where $D(t)$ is the variable torsional rigidity of the growing body.

Consider two possible versions of the statement of the problem.

1. The torque $M(t)$ is given, and the stresses σ_{ij}, the displacements u_i, and the twist $\theta(t)$ must be determined.

2. The twist $\theta(t)$ is given, and σ_{ij}, u_i, and $M(t)$ must be determined.

The solution for the first version of the statement (where the torque $M(t)$ is given) can be constructed as follows:

- The boundary condition is used to determine the function φ_t.
- Formulas (23) are used to find $\theta(t)$.
- Formulas (22) give v_i and S_{ij}.
- Finally, the true stresses σ_{ij} are reconstructed by formula (8).

For the second version of the statement,

- The torsion function $\varphi(x_1, x_2, t)$ is determined.
- The functions v_i and S_{ij} are found by formulas (22).
- The true stresses σ_{ij} are reconstructed by formula (8).
- The torque $M(t)$ is determined from the equilibrium (20) for the face cross-sections.

Now consider the stage of torsion of the body after the termination of growth. Let the AM fabrication of the body terminate at time τ_2. By then, the body occupies the domain Π_2 whose cross-section Ω_2 is bounded by the contour L_2. Here we obtain a problem similar to (21). In this case, the variables v_i and S_{ij} can be obtained by formulas (22), where one sets $t = \tau_2$. The stresses, displacements, and twist are given by formulas (8).

As a result, we have reduced the nonclassical boundary value problems arising in the analysis of torsion of AM fabricated bodies to well-known boundary value problems containing some parameter. The stress-strain state of the body can be completely reconstructed from the solutions of the latter problems by the deciphering formulas (8).

To solve the classical boundary value problems with a parameter, one can apply two methods of the theory of functions of a complex variable.

The first method is based on the reduction of the equation of the boundary to a special form, and the second method relies on the application of a conformal mapping. Consider the first method.

Consider the complex torsion function

$$F(z,t) = \varphi + i\psi,$$

where φ is the torsion function and ψ is the conjugate harmonic function.

In view of the formulas

$$F(z,t) + \overline{F(z,t)} = 2\mathrm{Re}\, F(z,t), \quad F(z,t) - \overline{F(z,t)} = 2\mathrm{Im}\, F(z,t).$$

The boundary condition in complex form reads

$$z \in L(t): \quad F(z,t) - \overline{F(z,t)} = z\bar{z}.$$

IOP Conf. Series: Journal of Physics: Conf. Series **991** (2018) 012057 doi:10.1088/1742-6596/991/1/012057

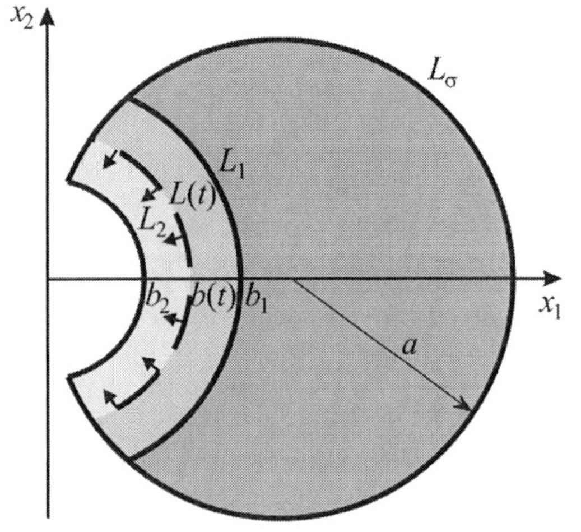

Figure 6. Torsion of an AM fabricated shaft with a recess.

Assume that the relationship between z and $\bar{z}$ for the points of the contour can be reduced to the form

$$z \in L(t): \quad z\bar{z} = h(z,t) + \overline{h(z,t)},$$

where $\overline{h(z,t)}$ is a function analytic inside the cross-section contour. Then one has

$$F(z,t) = ih(z,t) + C(t), \tag{24}$$

where $C(t)$ is an arbitrary function, which can be taken to be zero.

Having determined the complex torsion function $F(z,t)$, we obtain the torsion function

$$\varphi = \operatorname{Re}\{F(z,t)\}. \tag{25}$$

This method can be used to solve problems such as the torsion of an ellipse, of an equilateral triangle, and of a circular shaft with a longitudinal recess.

4. Torsion of an am fabricated shaft

By way of example, consider the problem on the torsion of an AM fabricated shaft with a longitudinal recess. Assume that a circular shaft Π_1 with a longitudinal recess is manufactured at time zero from an ageing viscoelastic material. The cross-section $\Omega(t)$ is the intersection of two circles (figure 6),

$$x_1^2 + x_2^2 \geq b(t), \quad (x_1 - a)^2 + x_2^2 \leq a^2 \quad (b(t) < a).$$

The boundary $L(t)$ of the section $\Omega(t)$ consists of two parts, $L(t) = L^*(t) \bigcup L_\sigma(t)$, where $L^*(t)$ is the growth boundary (the contour corresponding to the small circle of variable radius $b(t)$) and $L_\sigma(t)$ is the stress-free boundary. The shaft growth law is completely determined by the function $b(t)$.

We assume that the newly added elements are stress-free. Since the problems obtained for each stage of AM fabrication are mathematically equivalent, it suffices to consider the stage where material is being added. Then the initial-boundary value problem for the shaft acquires the form (21).

The torsion function $\varphi(x_1, x_2, t)$ is given by the formula

$$\varphi(x_1, x_2, t) = -ax_2 - \frac{ab^2(t)x_2}{x_1^2 + x_2^2}.$$

Having determined the torque $M(t)$, we find

$$\frac{dM^0(t)}{dt} = \frac{M_t'(t)}{G(t)} + \int_{\tau_0(x_1,x_2)}^{t} \frac{\partial M(t)}{\partial \tau}\frac{\partial \omega(t,\tau)}{\partial t}\,d\tau + M(\tau_0(x_1,x_2))\frac{\partial \omega(t, \tau_0(x_1,x_2))}{\partial t}$$

and then the twist rate

$$\theta_t'(t) = \frac{1}{2a_1 D(t)}\frac{dM^0(t)}{dt}, \quad D(t) = \frac{1}{24}\{\sin[4\alpha(t)] + 8\sin[2\alpha(t)] + 12\alpha(t)\},$$

$$-\frac{b^2(t)}{2a^2}\{\sin[2\alpha(t)] + 2\alpha(t)\} + \frac{4b^3(t)}{3a^3}\sin\alpha(t) - \frac{b^4(t)}{4a^4}\alpha(t), \quad \frac{b(t)}{a} = 2\cos\alpha(t).$$

The displacement rate and the variables S_{13} and S_{23} are found by the formulas

$$v_1 = -\theta_t'(t)x_2x_3, \quad v_2 = \theta_t'(t)x_1x_3, \quad v_3 = \theta_t'(t)\left[-ax_2 - \frac{ab^2(t)x_2}{x_1^2+x_2^2}\right],$$

$$S_{13} = \frac{1}{2a^2D_1(t)}\left[\frac{2ab^2(t)x_1x_2}{(x_1^2+x_2^2)^2} - x_2\right]\frac{dM^0(t)}{dt},$$

$$S_{23} = \frac{1}{2a^2D_1(t)}\left[-\frac{2ab^2(t)(x_1^2-x_2^2)}{(x_1^2+x_2^2)^2} + x_1 - a\right]\frac{dM^0(t)}{dt}.$$

The true characteristics of the stress-strain state are reconstructed by the deciphering formulas (8). To construct the solution at the per- and post-growth stages, its suffices to set $t = \tau_1$ and $t = \tau_2$, respectively, in the torsion function.

Let us carry out the computations for the shaft in view of the fact that it is manufactured from a viscoelastic ageing material with the mechanical characteristics given above. Consider the AM fabrication (recovery) of a shaft subjected to a torque for the case in which the recess radius $b(t)$ decreases fivefold from $b_1 = 0.5$ to $b_1 = 0.1$. All notation of the curves is the same as in the preceding example.

Let us study the shear stress intensity distribution along the diameter of the cross-section of a shaft with a longitudinal recess depending on how the torque varies in time. The shear stress intensity under torsion is given by the formula

$$\sigma_i = \sqrt{\sigma_{13}^2 + \sigma_{23}^2}.$$

Assume that $M(t) = 1$. Consider AM fabrication processes with $\tau_0 = 0.1$ and $\tau_1 = 0.3$. For rapid AM fabrication ($\tau_2 = 0.6$), the shear stress intensity is maximal on the boundary of the fabricated body at the base of the recess (figure 7).

Slow AM fabrication ($\tau_2 = 6$) is characterized by the fact that the stress state of the main body remains practically unchanged, while the additional body remains practically stress-free (figure 8).

Consider also AM fabrication with $\tau_0 = 0.1$, $\tau_1 = 1$, and $\tau_2 = 2$. Let the torque be given by

$$M(t) = \begin{cases} 0.1 & t \le T, \\ 1.8 & t > T. \end{cases}$$

If $T \in (\tau_1, \tau_2)$, then the maximum shear stress intensity is attained at an interior point of the AM fabricated body. Figure 9 shows the variations in the shear stress intensity for $T = 2$.

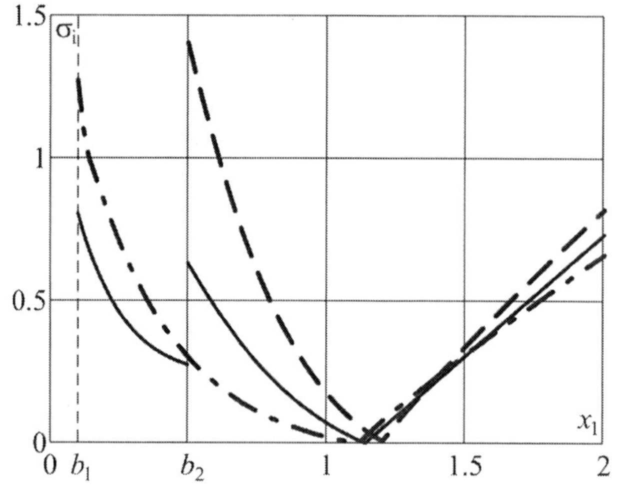

Figure 7. Shear stress intensity for the process of rapid AM fabrication of the shaft under constant torque.

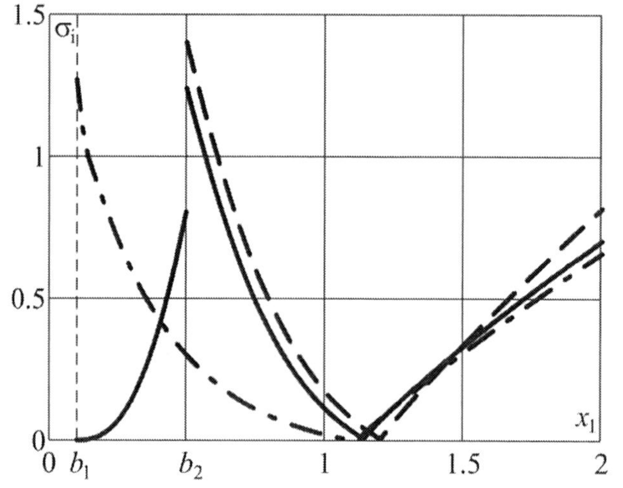

Figure 8. Shear stress intensity for the process of slow AM fabrication of the shaft under constant torque.

Conclusions

- We develop surface growth theory for studying 2D problems of AM fabrication of deformable bodies for the case in which the surface strain rate of the bodies due to forces and prestress of the added elements can be neglected compared with the new material inflow rate on the surface.
- We give the statement of classical and nonclassical initial-boundary value problems arising in the theory. Solution methods are proposed based on the reduction of nonclassical problems of AM fabrication of viscoelastic ageing bodies to elasticity problems with a parameter, the use of the theory of analytic functions for solving the latter problems, and on the reconstruction of true characteristics of the stress-strain state of the bodies with the help of deciphering formulas obtained.
- It is discovered that the maximum shear stress intensity in the manufactured object in 2D

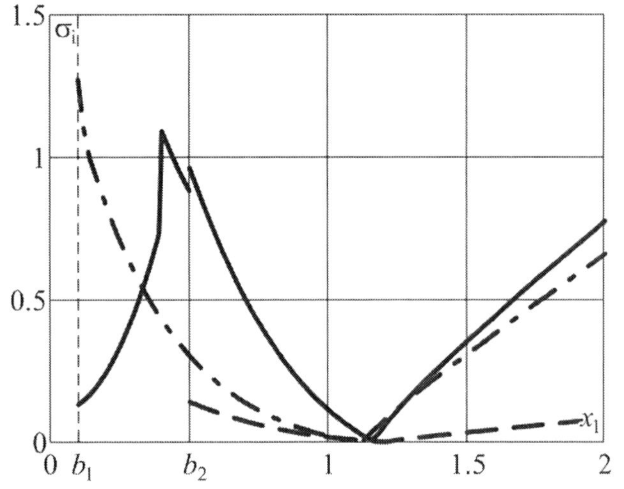

Figure 9. Shear stress intensity for the AM fabricated shaft under a step function torque.

problems on stress concentration near holes and on torsion is attained on the boundary of the body if the AM manufacturing process is not taken into account. For additive manufacturing, the maximum shear stress intensity can be attained on the interface between the main body and the added part, on the boundary of the AM fabricated body, and at an arbitrary point of the AM fabricated part of the body.

- The results can serve as a basis for the solution of important applied problems for parts and structural elements fabricated by additive manufacturing technologies.

Acknowledgments

The work was financially supported by the Russian Science Foundation under the Project 17-19-01257.

References

[1] Gebhardt A and Hoetter J S 2016 *Additive Manufacturing: 3D Printing for Prototyping and Manufacturing* (Munich: Carl Hanser Verlag) p 611

[2] Gibson I, Rosen D W, and Stucker B 2010 *Additive Manufacturing Technologies: Rapid Prototyping to Direct Digital Manufacturing* (New York: Springer Verlag, LLC) p XXI+498

[3] Hu D and Kovacevic R 2003 Modelling and measuring the thermal behaviour of the molten pool in closed-loop controlled laser-based additive manufacturing *Proc. Inst. Mech. Engineers. Part B: J. Engng Manufact.* **217** (4) 441–52

[4] Roberts I A, Wang C J, Esterlein R, et al. 2009 A three-dimensional finite element analysis of the temperature field during laser melting of metal powders in additive layer manufacturing *Int. J. Machine Tools Manufact.* **49** (12) 916–23

[5] Melchels F P W, Domingos M A N, Klein T J, et al. 2012 Additive manufacturing of tissues and organs *Progress in Polymer Sci.* **37** (8) 1079–104

[6] Zhou J, Zhang Y, and Chen J K 2009 Numerical simulation of random packing of spherical particles for powder-based additive manufacturing *J. Manufact. Sci. Engng* **131** (3) 031004

[7] Parteli E J R 2013 DEM simulation of particles of complex shapes using the multisphere method: application for additive manufacturing *AIP Conf. Proc.* **1542** (1) 185–8

[8] Khajavi S H, Partanen J, and Holmstrom J 2014 Additive manufacturing in the spare parts supply chain *Computers in Industry* **65** (1) 50–63

[9] Patil N, Pal D, and Stucker B 2013 A new finite element solver using numerical Eigen modes for fast simulation of additive manufacturing processes In *Solid Freeform Fabrication Proceedings, August 2013* (The University of Texas at Austin) pp 535–43

[10] Kalpakjian S and Schmid S R 2014 *Manufacturing Engineering and Technology* (Upper Saddle River, NJ: Pearson)

[11] Chua C K and Leong K F 2014 3D printing and additive manufacturing: Principles and Applications (with Companion Media Pack) of Rapid Prototyping (World Scientific Publishing Co Inc)

[12] Heigel J C, Michaleris P, and Reutzel E W 2015 Thermo-mechanical model development and validation of directed energy deposition additive manufacturing of Ti-6Al-4V *Addit. Manufact.* **5** 9–19

[13] Aruyunyan N Kh, Manzhirov A V, and Naumov V E 1991 *Contact Problems in Mechanics of Growing Bodies* (Moscow: Nauka) p 172 [in Russian]

[14] Aruyunyan N Kh and Manzhirov A V 1999 *Contact Problems in the Theory of Creep* (Moscow: Nauka) p 320 [in Russian]

[15] Manzhirov A V 1995 The general non-inertial initial-boundary value problem for a viscoelastic ageing solid with piecewise-continuous accretion *J. Appl. Math. Mech.* **59** (5) 805–16

[16] Manzhirov A V Mechanics of growing solids: New track in mechanical engineering 2014 In *ASME 2014 Int. Mechanical Engineering Congr. and Expos. Volume 9: Mechanics of Solids, Structures and Fluids. Montreal, Quebec, Canada, November 14–20, 2014* (ASME) IMECE2014-36712, pp. V009T12A039

[17] Manzhirov A V 2015 Mechanical design of viscoelastic parts fabricated using additive manufacturing technologies In *Lecture Notes in Engineering and Computer Science: Proc. of The World Congr. Engineering 2015, WCE 2015, 1–3 July, 2015, London, U.K.* (London: IAENG) pp 710–4

[18] Manzhirov A V 2016 A method for mechanical design of AM fabricated viscoelastic parts In *Transactions on Engineering Technologies* (Singapore: Springer) pp 223–35

[19] Arutyunyan N Kh and Abramyan B L 1963 *Torsion of Elastic Solids* (Moscow: Fizmatgiz) p 686 [in Russian]

[20] Muskhelishvili N I 1977 *Some Basic Problems of the Mathematical Theory of Elasticity* (Netherlands: Springer) p 732

[21] Manzhirov A V and Lychev S A 2014 Mathematical modeling of additive manufacturing technologies In*Lecture Notes in Engineering and Computer Science: World Congr. on Engineering 2014, WCE 2014, 2–4 July, 2014, London, U.K.* (London: IAENG) pp 1404–9

[22] Manzhirov A V and Lychev S A 2015 An approach to modeling of additive manufacturing technologies In *Transactions on Engineering Technologies: World Congr. on Engineering 2014* (Netherlands: Springer) pp 99–115

[23] Manzhirov A V 2016 Fundamentals of mechanical design and analysis for AM fabricated parts *Proc. Manufact.* **7** 59–65

[24] Manzhirov A V 2017 Advances in the theory of surface growth with applications to additive manufacturing technologies *Proc. Engng* **173** 11-6

[25] Manzhirov A V 2017 Mechanical analysis of an AM fabricated viscoelastic shaft under torsion by rigid disks In *Lecture Notes in Engineering and Computer Science: Proc. of The World Congr. Engineering 2017, WCE 2017, 5–7 July, 2017, London, U.K., Vol. II* (London: IAENG) pp 856–60

IOP Conf. Series: Journal of Physics: Conf. Series **991** (2018) 012058 doi:10.1088/1742-6596/991/1/012058

Investigation of problems of closing of geophysical cracks in thermoelastic media in the case of flow of fluids with impurities

A N Martirosyan*, **A V Davtyan****, **A S Dinunts*****, **and H A Martirosyan******

Goris State University, Goris, Armenia

E-mail: *ashot.martirosyan.14@gmail.com, **davtyananush@gmail.com, ***dinuntsas@gmail.com, ****hayk.martirosyan.75@mail.ru

Abstract. The purpose of this article is to investigate a problem of closing cracks by building up a layer of sediments on surfaces of a crack in an infinite thermoelastic medium in the presence of a flow of fluids with impurities. The statement of the problem of closing geophysical cracks in the presence of a fluid flow is presented with regard to the thermoelastic stress and the influence of the impurity deposition in the liquid on the crack surfaces due to thermal diffusion at the fracture closure. The Wiener–Hopf method yields an analytical solution in the special case without friction. Numerical calculations are performed in this case and the dependence of the crack closure time on the coordinate is plotted. A similar spatial problem is also solved. These results generalize the results of previous studies of geophysical cracks and debris in rocks, where the closure of a crack due to temperature effects is studied without taking the elastic stresses into account.

1. Introduction

The problems of increasing the impurity layer contained in a liquid entering a crack located in an infinite thermoelastic plane are considered. These problems were investigated in [1] as problems of closing cracks by injecting a hot hydrothermal mixture with crystallization of silica and their deposition on the crack surfaces under the assumption that the liquid flow and temperature at the point of entry into the crack are constant in time. In [2], the problem of closing a crack is considered when the liquid temperature and flow at the entry point into the crack are functions of time.

In [1, 2] and in numerous works of other foreign authors, the problems of closing a geophysical crack by deposition of crystals from a hydrothermal fluid due to thermal and thermal diffusion effects are studied without consideration of elastic stresses. In [3], the change of pressure and permeability of a fluid in a fracture in a rock array is studied with regard to elastic stresses and thermal diffusion, i.e., the effects of thermal stress and precipitation/dissolution of silicon dioxide are taken into account. The heat exchange between a crack and a rock array is modeled by taking only the thermal conductivity into account. The fluid pressure can vary with time, while the flow velocity is assumed to be constant. The experimental curves for closing bands filled with a fluid with inclusions in the Earth's medium are given in [4].

Numerous numerical experiments have been performed to simulate the fluid injection/liquid production. A temporary change in the crack aperture is considered in response to the individual and combined effects of thermal stress and silica dissolution/deposition. The results show that for smaller initial fracture ruptures, a significant increase in the fracture permeability and the associated pressure drop at the injection point are mainly due to thermoelastic effects, while an increase in the rupture aperture near the injection point is mainly the result of silica dissolution/deposition. On the other hand, for large initial holes, the effect of precipitation/dissolution of silicon dioxide is minimal, and the thermoelastic effects predominate, since the intensity of the connection between fractures with high permeability and mass of rocks with low permeability becomes weaker. As follows from these results, it is necessary to take into account the thermoelastic stresses and the effects of thermal diffusion, which is the main subject of this work.

In the present paper based on the Wiener–Hopf [5] and integral transforms methods, the mathematical solutions are obtained for the important applied problems of closing geophysical cracks by increasing the sediment layer on the crack surfaces in an infinite thermoelastic medium in the presence of a flow of fluids with impurities. The problem of closing the geophysical crack in the presence of a fluid flow is formulated with regard to the thermoelastic stress and influence of the impurity deposition in the liquid on the crack surfaces due to thermal diffusion at the fracture closure. The Wiener–Hopf method yields an analytic solution in the special case without friction. The solution of the nonstationary mixed plane boundary-value problem for thermoelasticity based on the Laplace integral transforms with respect to time and the Fourier integral transforms with respect to the coordinate is reduced to the Wiener–Hopf equation with continuous coefficients. After the inverse integral transformation, the solution for displacements is given in Smirnov-Sobolev form [6]. In this case, for some values of the parameters, numerical calculations were made and the crack closing time versus the coordinate was plotted. A similar spatial problem is also solved. The solution results are compared with the results obtained by the Wiener–Hopf method in the case of stationary crack, and the coincidence of these approaches is shown. These results generalize the previously performed work for geophysical cracks and faults in rocks, where the closure of a crack due to temperature effects is studied without consideration of elastic stresses.

2. Statement of the problem of crack in thermoelastic plane with a fluid flow

In this paper, the problem of closing cracks by building up sediments on crack surfaces is considered in infinite thermoelastic media in the case of flow of a fluid with impurities entering the crack at initial time $t = 0$.

In the plane x_1, x_2, the equation of crack profiles is $x_2 = \pm h(x_1, t)$, where $h(x_1, t) = h_0 + u_2(x_1, x_2, t)$, $x_2 \approx 0$, and the thickness $2h_0$ is small. Due to symmetry of the problem, we later take only the upper sign and solve the problem for $x_2 \geq 0$; here u_1, u_2 are displacement components in elastic media.

Following [1, 2, 7], we write the following approximate relation for $y = 0$:

$$\rho_s \frac{\partial h}{\partial t} = q\gamma \frac{\partial T}{\partial x} - K\sigma_{22} + \lambda_1 \frac{\partial C}{\partial y}, \tag{1}$$

where the left-hand side represents the mass of sediment deposited per unit time and proportional to the closure rate of the crack, the first term on the right-hand side is the mass flow density of inclusions in the fluid along the axis x_2 [2] transmitted by the fluid per unit time due to the temperature gradient along the x_1, the second term is the tribological term with stresses [8], the third term corresponds to the flux density of impurity concentration along the axis x_2, where $K =$ const is a tribological experimental constant of the crack boundary or an increase in its surface due to symmetry, where one can approximately include the term from barodiffusion, assuming that $\sigma_{x_2x_2} = -P$, $\partial P/\partial x_2 \approx -\sigma_{x_2x_2}/h_0$, ρ_s is the density of inclusions, C is the concentration of impurities in the fluid, $\gamma = \partial C/\partial T$, $\gamma = \text{const}$ [9], $q = \rho_f v h_0$ denotes the flow of entering fluid, ρ_f is the density of fluid, v is the fluid velocity along the axis x_1, $\lambda_1 = -\rho_s D$, and D is the diffusion coefficient. The temperature difference $T(x,y,t) - T_0$ is assumed to be the same in the elastic medium and fluid, where it is approximately assumed that $x_2 \approx 0$, and the notation $T(x,t) = T(x,0,t) - T_0$ is used.

The equations of motion in displacements for a thermoelastic medium [9] have the form

$$\begin{aligned} a^2\frac{\partial^2 u_1}{\partial x_1^2} + b^2\frac{\partial^2 u_1}{\partial x_2^2} + (a^2-b^2)\frac{\partial^2 u_2}{\partial x_1 \partial x_2} + \bar{\delta}\frac{\partial}{\partial x_1}\left(\frac{\partial u_1}{\partial x_1} + \frac{\partial u_2}{\partial x_2}\right) = \frac{\partial^2 u_1}{\partial t^2}, \\ a^2\frac{\partial^2 u_2}{\partial x_2^2} + b^2\frac{\partial^2 u_2}{\partial x_1^2} + (a^2-b^2)\frac{\partial^2 u_1}{\partial x_1 \partial x_2} + \bar{\delta}\frac{\partial}{\partial x_2}\left(\frac{\partial u_1}{\partial x_1} + \frac{\partial u_2}{\partial x_2}\right) = \frac{\partial^2 u_2}{\partial t^2}, \end{aligned} \tag{2}$$

where u_i $(i = 1,2,3)$ are displacement components, a and b are the velocities of longitudinal and transverse waves, $\bar{\delta} = K_1/\rho \cdot (C_\rho - C_v)/C_v$, $K_1 = \lambda + \frac{2}{3}\mu$ is the bulk modulus, λ and μ are the elastic Lame constants, t is the time, ρ is the density of thermoelastic medium, and C_ρ and C_v are the specific thermal capacities [9]. These formulas are obtained by neglecting the thermal conductivity in the temperature equation

$$\frac{\partial T}{\partial t} + \frac{C_\rho - C_v}{\alpha C_v}\frac{\partial}{\partial t}\left(\frac{\partial u_1}{\partial x_1} + \frac{\partial u_2}{\partial x_2}\right) = k\left(\frac{\partial^2 T}{\partial x_1^2} + \frac{\partial^2 T}{\partial x_2^2}\right), \tag{3}$$

where k is the coefficient of thermal conductivity and α is the coefficient of thermal expansion.

For the stress components, we have the relations

$$\begin{aligned} \frac{\sigma_{x_2x_2}}{\rho} &= (a^2 - 2b^2)\left(\frac{\partial u_1}{\partial x_1} + \frac{\partial u_2}{\partial x_2}\right) + 2b^2\frac{\partial u_2}{\partial x_2} + \bar{\delta}\left(\frac{\partial u_1}{\partial x_1} + \frac{\partial u_2}{\partial x_2}\right), \\ \frac{\sigma_{x_1x_2}}{\rho} &= b^2\left(\frac{\partial u_1}{\partial x_2} + \frac{\partial u_2}{\partial x_1}\right). \end{aligned} \tag{4}$$

After linearization, the boundary conditions of contact between the medium and the liquid, i.e., for $x_1 > 0$ and $x_2 = 0$, with the mechanical upbuilding taken into account, can also be written as

$$\begin{aligned} \frac{\partial u_1}{\partial t} &= \frac{K_2}{\rho_s}\sigma_{x_1x_2}, \\ \frac{\partial u_2}{\partial t} &= -\frac{q\gamma\rho(C_\rho - C_v)}{\rho_s \alpha C_v l}\left(\frac{\partial u_x}{\partial x} + \frac{\partial u_y}{\partial y}\right) - \frac{K}{\rho_s}\sigma_{x_2x_2} - \frac{i_0}{\rho_s}H(t)H(vt - x_1), \end{aligned} \tag{5}$$

where the diffusion flux $\lambda_1 \partial C/\partial y$ is assumed to be known and approximately equal to $-i_0 H(vt-x)H(t)$, i_0 is the constant diffusion flow of inclusions along the axis x_2 at the entrance to the crack, and approximately is taken $\partial T/\partial x_1 \approx (T - T_0)/l$, where l is a constant, $H(t)$ is the Heaviside unit step function, K_2 is the coefficient of the horizontal wear of the medium. In the linear approximation, the value q and i_0 are the same as in the initial section $x_1 = 0$, the term in the boundary condition containing i_0 corresponds to the flux density of the impurity concentration along the axis x_2, which occurs when the fluid enters the fracture.

In the general case, we have to add the diffusion equation for the impurity concentration, as well as the thermal diffusion equations for the fluid inside the crack $|x_2| \leq h(x_1,t)$. The above assumptions here we made to simplify the problem.

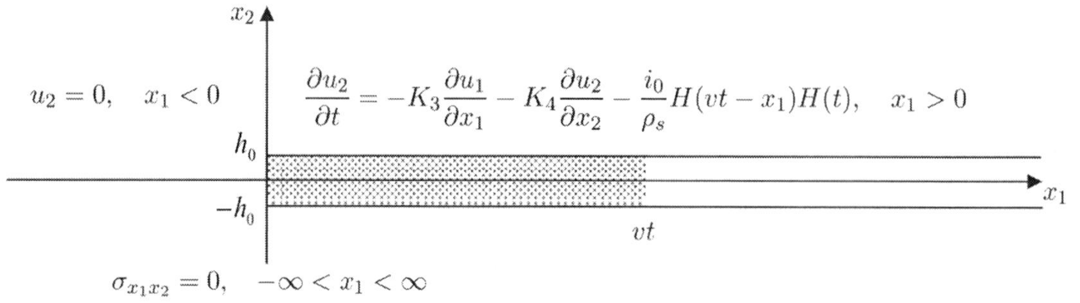

Figure 1. Crack profile and the boundary conditions of the problem.

3. Solution of problem of building up the crack in the absence of friction

Consider the following simplified mixed problem for a thermoelastic isotropic medium with the boundary conditions of the form (figure 1):

$$\begin{aligned} &\sigma_{x_1x_2}\big|_{x_2=0} = 0, \quad -\infty < x_1 < \infty, \\ &u_2\big|_{x_2=0} = 0, \quad x_1 < 0, \\ &\left(\frac{\partial u_2}{\partial t} + K_3\frac{\partial u_1}{\partial x_1} + K_4\frac{\partial u_2}{\partial x_2}\right)\bigg|_{x_2=0} = -\frac{i_0}{\rho_s}H(vt - x_1)H(t), \quad x_1 > 0, \end{aligned} \tag{6}$$

where $\bar{a}^2 = a^2 + \bar{\delta}$, $\xi = (C_p - C_v)/(C_v l) \cdot v h_0 \gamma \rho/(\alpha \rho_s)$, $K_3 = \rho/\rho_s \cdot K(\bar{a}^2 - b^2) + \xi$, $K_4 = \rho/\rho_s \cdot K\bar{a}^2 + \xi$.

In the plane case, the equations of motion for a thermoelastic isotropic medium in displacements have the form (2). The initial conditions are zero:

$$u_i\big|_{t=0} = 0, \quad \frac{\partial u_i}{\partial t}\bigg|_{t=0} = 0, \quad i = 1, 2.$$

We use the Laplace integral transformations with respect to t and the Fourier integral transformations with respect to x_1, denote the Laplace integral transformations of u_i by u_i^L and the Fourier integral transformations of u_i^L by u_i^{LF}, and seek the solution of the problem in the form

$$u_i^L = \sum_{n=1}^{2} \int_{-\infty}^{\infty} e^{i\bar{\alpha}x_1 + i\bar{\beta}_n x_2} u_{in}^{LF}\, d\bar{\alpha}, \quad \bar{\beta}_n = \sqrt{\frac{\omega^2}{c_n^2} - \bar{\alpha}^2}, \quad c_1 = \bar{a}, \quad c_2 = b, \quad i = 1, 2, \tag{7}$$

where $s = -i\omega$ is a parameter of plane Laplace transformation. Then problem (2), (6) is reduced to solving the Wiener–Hopf equation

$$\Omega_2^+ - \frac{i_0}{2\pi s \rho_s (s/v + i\bar{\alpha})} = \frac{R(\bar{\alpha})}{i\omega^2/b^2 \beta_1} V^-, \tag{8}$$

where

$$R(\bar{\alpha}) = \bar{\beta}_1 \frac{\omega^3}{b^2} - 2\bar{\alpha}^2\bar{\beta}_1\bar{\beta}_2(K_4 - K_3) - \left[K_4\frac{\omega^2}{\bar{a}^2} - (K_4 - K_3)\bar{\alpha}^2\right]\left(\frac{\omega^2}{b^2} - 2\bar{\alpha}^2\right), \tag{9}$$

$$\begin{aligned} &\Omega_2^+ = \frac{1}{2\pi}\int_{-\infty}^{0} \left(s u_2^L + K_3\frac{\partial u_1^L}{\partial x_1} + K_4\frac{\partial u_2^L}{\partial x_2}\right)\bigg|_{x_2=0} e^{-i\bar{\alpha}x_1}\, dx_1, \\ &V^- = \frac{1}{2\pi}\int_{0}^{\infty} u_2^L\big|_{y=0} e^{-i\bar{\alpha}x_1}\, dx_1. \end{aligned} \tag{10}$$

Here the index (+) stands for analytic functions of $\bar{\alpha}$ in the upper half-plane, and the index (−) stands for analytic functions in the lower half-plane $\bar{\alpha}$. Applying the Noble method [5], factorizing the function (9), solving equation (8), and making the inverse transformations, we obtained the solution of problem (2), (6) and the displacement component along the axis x_2 in the following form:

$$u_2 = \frac{C_0 i_0 \bar{a}\sqrt{1+\bar{a}/v}}{\pi\rho_s(\alpha_0+\bar{a}/v)D_+(\bar{a}/v)}\left\{\frac{(t-\alpha_0 x_1/\bar{a})\sqrt{1-\alpha_0}}{(\alpha_0-\bar{a}/v)D_-(\alpha_0)}H\left(t-\frac{\alpha_0 x}{\bar{a}}\right)\right.$$
$$-\frac{2x_1}{\bar{a}}H\left(\frac{\bar{a}t}{x}-1\right)\left[\frac{(\bar{a}/v-\bar{a}t/x_1)\sqrt{\bar{a}/v-1}}{2D_-(\bar{a}/v)(\bar{a}/v-\alpha_0)}\ln\left|\frac{\sqrt{\bar{a}t/x_1-1}-\sqrt{\bar{a}/v-1}}{\sqrt{\bar{a}t/x_1-1}+\sqrt{\bar{a}/v-1}}\right|\right.$$
$$+\sqrt{\frac{\bar{a}t}{x_1}-1}-\frac{(\bar{a}t/x_1-\alpha_0)\sqrt{1-\alpha_0}}{(\bar{a}/v-\alpha_0)D_-(\alpha_0)}\arctan\sqrt{\frac{\bar{a}t/x_1-1}{1-\alpha_0}}$$
$$\left.\left.+\int_1^{\bar{a}/b}\frac{F_2(u)\sqrt{u-1}(u-\bar{a}t/x_1)}{2(\bar{a}/v-u)(u-\alpha_0)}\ln\left|\frac{\sqrt{\bar{a}t/x_1-1}-\sqrt{u-1}}{\sqrt{\bar{a}t/x_1}-1}+\sqrt{u-1}\right|du\right]\right\}, \tag{11}$$

where

$$F_2(u) = -\mu(u)\exp[-\chi(u)],$$
$$\chi(u) = \frac{1}{2\pi i}\int_{1/\bar{a}}^{1/b}\ln\frac{R(\zeta)}{\overline{R(\zeta)}}\frac{d\zeta}{\zeta-u}$$
$$= \frac{1}{\pi}\int_{1/\bar{a}}^{1/b}\arctan\frac{\frac{1}{b}\sqrt{\zeta^2-\frac{1}{\bar{a}^2}}+2(K_3-K_4)\zeta^2\sqrt{\zeta^2-\frac{1}{\bar{a}^2}}\sqrt{\frac{1}{b^2}-\zeta^2}}{[(K_3-K_4)\zeta^2+\frac{K_4}{\bar{a}^2}](\frac{1}{b^2}-2\zeta^2)}\frac{d\zeta}{\zeta-u},$$
$$\mu(u) = \frac{\frac{1}{\pi}\left[-\frac{1}{b^2}\sqrt{u^2-\frac{1}{\bar{a}^2}}-2(K_3-K_4)u^2\sqrt{u^2-\frac{1}{\bar{a}^2}}\sqrt{\frac{1}{b^2}-u^2}\right]}{\sqrt{\left[\frac{1}{b^2}\sqrt{u^2-\frac{1}{\bar{a}^2}}+2(K_3-K_4)u^2\sqrt{u^2-\frac{1}{\bar{a}^2}}\sqrt{\frac{1}{b^2}-u^2}\right]^2+\left[(K_3-K_4)u^2+\frac{K_4}{\bar{a}^2}\right]^2\left(\frac{1}{b^2}-2u^2\right)^2}},$$
$$D_\pm(\alpha) = \exp\left\{\frac{1}{\pi}\int_1^{\bar{a}/b}\arctan\frac{\frac{\bar{a}^2}{b^2}\sqrt{\zeta^2-1}-\frac{2(K_4-K_3)}{\bar{a}}\zeta^2\sqrt{\zeta^2-1}\sqrt{\frac{\bar{a}^2}{b^2}-\zeta^2}}{\left(\frac{\bar{a}^2}{b^2}-2\zeta^2\right)\left[\frac{K_3}{\bar{a}}\zeta^2+\frac{K_4}{\bar{a}}(1-\zeta^2)\right]}\frac{d\zeta}{\zeta\pm\alpha}\right\};$$

here $\pm\alpha_0\frac{\omega}{\bar{a}}$ stand for zeros of the function $R(\bar{\alpha})$.

Calculating the function (12) for $\frac{\bar{a}}{b} = \sqrt{3}$, $\frac{K_3}{\bar{a}} = 0.8$, $\frac{K_4}{\bar{a}} = 0.3$, $\frac{i_0}{\rho_s\bar{a}} = 1/(2\times10^5)$, $\bar{a} = 1000$ cm/s, $\frac{\bar{a}}{v} = 100$, we obtain ±0.981431 for $\pm\alpha_0$ and the graph (figure 2) of the dependence of $u_2/\bar{a}t$ on $x_1/\bar{a}t$. As a result of calculations, it is shown that for $x_1/\bar{a}t \le 2.1\times10^{-7}$, $u_2/\bar{a}t > 0$, and for $x_1/\bar{a}t \ge 2.1\times10^{-7}$, $u_2/\bar{a}t < 0$, i.e., the crack width $h = h_0 + u_2$ decreases. Taking a very thin microcrack ($h_0 = 10^{-5}$ cm), it is possible to construct the crack closure time t dependence on the coordinate x_1 (figure 3). At the times $t = 4$ s and $t = 6$ s, the profile of the upper face of the fracture in the section 10^{-7} cm $< x_1 < 0.008$ cm has the shape shown in figure 4.

4. Solution of a spatial problem of closing a geophysical crack in a thermoelastic medium in the presence of a fluid flow

In this section, we consider a spatial problem similar to the plane problem solved in Sec. 3, where the main attention was paid to the effects of thermal diffusion and thermoelasticity in the medium. A crack at the initial time is an infinite half-strip confined by half-planes $x_3 = \pm h_0$, $x_1 > 0$, i.e., it is a gap of solid elastic medium of uniform width $2h_0$. The equations of crack profiles are $x_3 = \pm h(x_1, x_2, t)$, $x_1 > 0$. The fluid with impurities is injected into the crack and

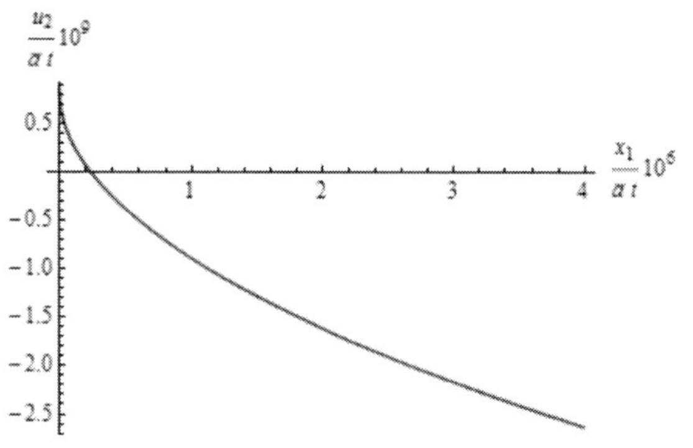

Figure 2. Graph of $u_2/(\bar{a}t)$ dependence on $x_1/(\bar{a}t)$.

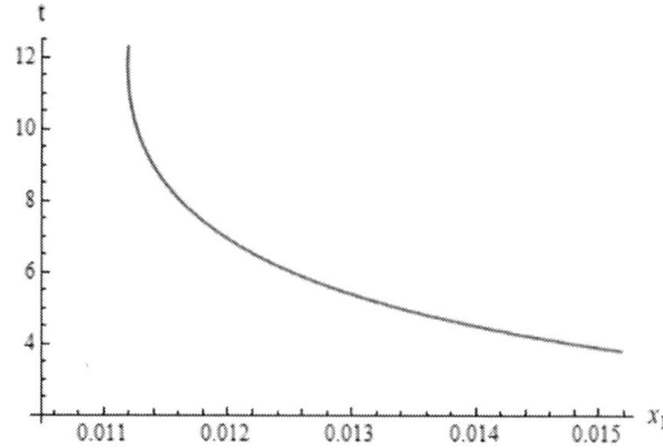

Figure 3. Graph of the dependence of time t of crack closing on x_1.

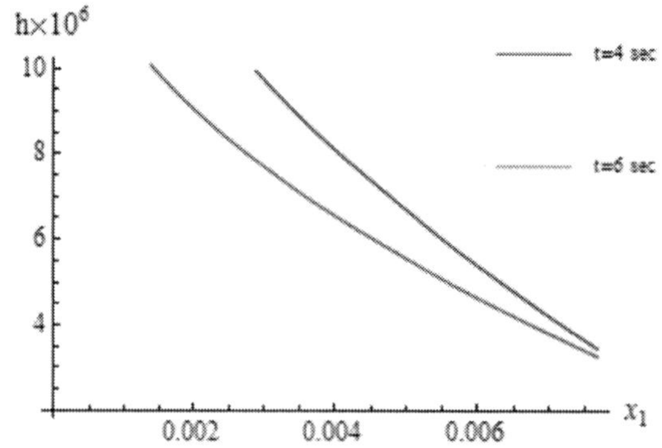

Figure 4. Profile of the upper face of the crack in the range $10^{-7}\,\text{cm} < x_1 < 0.008\,\text{cm}$ at times $t = 4\,\text{s}$ and $t = 6\,\text{s}$.

moves at a constant speed in the direction of the axis x_1. Due to cooling and thermal diffusion, as well as mechanical buildup, the precipitation of dissolved components occurs, which leads to a change in the crack profile.

In the three-dimensional problem, the equations of thermoelasticity in displacements for an isotropic medium, similarly to the equations obtained in the plane case, have the form

$$(\bar{a}^2 - b^2)\frac{\partial}{\partial x_i}\nabla + b^2\Delta u_i = \frac{\partial^2 u_i}{\partial t^2}, \quad i = 1, 2, 3, \quad \nabla = \operatorname{div} u, \quad \Delta = \sum_{i=1}^{3}\frac{\partial^2}{\partial x_i^2}, \tag{12}$$

where u_i $(i = 1, 2, 3)$ are the displacement components and the other notation is the same. The initial conditions are zero:

$$u_i\big|_{t=0} = 0, \quad \left.\frac{\partial u_i}{\partial t}\right|_{t=0} = 0, \quad i = 1, 2, 3.$$

With the crack build up taken into account, the boundary conditions on the crack $(x_1 > 0)$ and beyond it $(x_1 < 0)$ have the form $(|x_2| < \infty)$

$$\sigma_{x_1x_3}\big|_{x_3=0} = \rho b^2\left(\frac{\partial u_1}{\partial x_3} + \frac{\partial u_3}{\partial x_1}\right)\bigg|_{x_3=0} = 0 \quad \text{for } -\infty < x_1 < \infty, \tag{13}$$

$$\sigma_{x_2x_3}\big|_{x_3=0} = \rho b^2\left(\frac{\partial u_2}{\partial x_3} + \frac{\partial u_3}{\partial x_2}\right)\bigg|_{x_3=0} = 0 \quad \text{for } -\infty < x_1 < \infty, \tag{14}$$

$$u_3\big|_{x_3=0} = 0 \quad \text{for } x_1 < 0, \tag{15}$$

$$\left[\frac{\partial u_3}{\partial t} + K_3\left(\frac{\partial u_1}{\partial x_1} + \frac{\partial u_2}{\partial x_2}\right) + K_4\frac{\partial u_3}{\partial x_3}\right]\Bigg|_{x_3=0} = -\frac{i_0}{\rho_s}H(vt - x_1)H(c - x_2)H(x_2)H(t) \quad \text{for } x_1 > 0, \tag{16}$$

where $u_{1,2,3} = o(R_1^{1/2})$, $R_1 = \sqrt{x_1^2 + x_2^2} \to 0$ (the condition on the edge), $c = \text{const}$, i_0 is the constant diffusion flow of inclusions along the axis x_3 at the entrance to the crack, v is fluid velocity along the axis x_2, and the other notation is the same. By virtue of the boundary conditions, the problem can be considered as symmetric one with respect to the plane Ox_1x_2; moreover $u_{1,2}$ are even functions and u_3 is an odd function of x_3, and therefore we can formulate the problem in the half-space $x_3 \geq 0$.

The solution of the problem is constructed by combining [6] the Wiener–Hopf methods, the Laplace integral transformations with respect to t, the Fourier integral transformations with respect to x_1, x_2, and analytic methods of the theory of functions of a complex variable. After cumbersome calculations, the exact solution of the problem is obtained in polar coordinates in Smirnov–Sobolev form [6], and it is shown that there is no singularity at $x_1 \to +0$, $x_3 = 0$.

Conclusions

The study of closing of cracks is very actual and practical geophysical, technological and biological problem. In numerous works the problems of closing a geophysical crack due to thermal diffusion effects are studied without consideration of elastic stresses. In the present paper the statement of the problem of closing geophysical cracks in the presence of a fluid flow is presented with regard to the thermoelastic stress and the thermal diffusion effects. The nonstationary mixed plane boundary-value problem in the special case without friction based on the integral transforms and the Wiener–Hopf methods is solved. A similar spatial problem is also solved. The developed in the present paper analytical methods are rather general and obtained formulas are easy for calculations, which can be carried out in wide variety of engineering and medical, practical problems of closing of cracks.

References

[1] Lowell R P, Capperren P V, and Germanovich L N 1993 Silica precipitation in fracture and the evolution of permeability in hydrothermal upflow zones *Science* **260** 192–4
[2] Gliko A O 2008 Influence of process of sediment of solid phase from hydrothermal mixture on healing of system of plane-parallel cracks In *Proc. 6th Int. Conf. "The Problems of Dynamics of Interaction of Deformable Media," September 21–26, Goris - Stepanakert, 2008* (Goris - Stepanakert: Inst. Mekh. NAN RA) pp 170–7 [in Russian]
[3] Ghassemi A and Kumar S 2007 Changes in fracture aperture and fluid pressure due to thermal stress and silica dissolution (precipitation induced by heat extraction from subsurface craks) *Geothermics* **36** (2) 115–40
[4] Bhattacharia S, Blakeborough A, and Dush S 2008 Learning from collapse of piles in liquefiable soils *Proc. Inst. Civil Engineers* **161** 54–60
[5] Noble B 1958 *Methods Based on the Wiener–Hopf Technique for Solving Partial Differential Equations* (New York: Pergamon)
[6] Martirosyan A N 2007 *Mathematical Investigations of Nonstationary Linear Boundary Value Problems for Continuous Media* (Yerevan: Zangak-97), p 244 [in Russian]
[7] Bagdoev A G, Martirosyan G A, Martirosyan A N, et al. 2011 The investigation of problems of healing of fractures by injection of fluids with inclusions in various thermoelastic media In *The Problems of Dynamics of Interaction of Deformable Media. Proc. of the VII Int. Conf. September 19–23, 2011, Goris - Stepanakert* (Yerevan: Inst. Mekh. NAN RA) pp 442–7 [in Russian]
[8] Goryacheva I G and Dobychin M N 1988 *The Contact Problems in Tribology* (Moscow: Mashinostroenie) p 253
[9] Landau L D 1954 *Mechanics of Continua* (Moscow: TTL) p 795 [in Russian]

IOP Conf. Series: Journal of Physics: Conf. Series **991** (2018) 012059 doi:10.1088/1742-6596/991/1/012059

On a class of problems on interaction of stress concentrators of different types with an elastic semi-infinite plate

S M Mkhitaryan

Institute of Mechanics of the National Academy of Sciences of Armenia, Yerevan, Armenia

E-mail: smkhitaryan39@rambler.ru

Abstract. A class of mixed boundary-value problems of mathematical theory of elasticity dealing with interaction between stress concentrators of different types (such as cracks, absolutely rigid thin inclusions, punches, and stringers) and an elastic semi-infinite plate is considered. The method of Mellin integral transformation is used to reduce solving these problems to solving singular integral equations (SIE). After the governing SIE are solved, the following characteristics of the problem are determined: tangential contact stresses under stringers, dislocation density on the crack edges, breaking stresses outside the cracks on their line of location, the stress intensity factor (SIF), crack openings, jumps of contact stresses on the edges of inclusions.

1. Introduction

In many contact and mixed boundary-value problems of mechanics of deformable solids, the deformable foundations are often simulated as the classical foundation of half-plane type, for example, as an elastic half-plane or an elastic semi-infinite plate. In fracture mechanics, numerous problems on stress strain state of an elastic half-plane containing cracks of various geometric forms and in various combinations are investigated. In theoretical and practical aspects, the problems for an elastic half-plane containing a collinear system of cracks and a vertical edge crack are especially important [1–4]. In [5], a more general problem where a piecewise homogeneous elastic half-plane contains a crack, issuing perpendicularly to the direct section of the material, is considered. In [1, 2, 4, 5], the Wiener–Hopf method is used to construct a closed solution of the problem and to obtain exact explicit expressions of SIF in the cases of constant or polynomial loadings acting on the crack edges. The main results of studying this range of problems are presented sufficiently fully in monographs [6, 7] and in handbooks [8, 9].

The problems of interaction of absolutely rigid thin inclusions with massive deformable bodies of different shapes were considered in [10, 11]. In [12, 13], the crack development near the apex of a rigid inclusion is studied and the SIFs near the crack on the continuation of a linear rigid inclusion are determined.

The problems on stress state of an elastic half-plane, strip or half-plane reinforced by stringers and weakened by cracks were considered in [14–18].

In the present paper, a class of mixed boundary-value problems of elasticity dealing with interaction of stress concentrators (such as cracks, absolutely thin inclusions, punches, and stringers) with an elastic semi-infinite plate is considered. The following problems are considered:

(i) A problem on stress strain state of an elastic semi-infinite plate occupying the lower half-plane in the right rectangular system of coordinates when it contains arbitrarily many cracks on the vertical axis. It is assumed that the crack edges and the plate on its boundary are loaded by arbitrary asymmetric non-self-balanced distributed normal and tangential forces. The formulation of the problem is here somewhat wider than that in [3, 4, 7]. On the basis of the first boundary-value problem for the quarter-plane wedge with a right angle of opening, constructed by using the Mellin integral transformation, solving the given problem is reduced to solving the SIE for the unknown complex combination of dislocation densities on the crack edges. The governing SIE is solved by the well-known numeral-analytical method [7, 19–20] after which the following basic characteristics of the problem are determined: the dislocation density on the crack edges, the crack openings, the breaking stresses outside the cracks system on their lines of location, and SIF. The following important particular cases of the posed problem are discussed.
   (a) The plate contains only one inner crack located on the vertical axis at a distance from its boundary. It is noted that, in this case, the effective solution to the governing SIE of the problem can be obtained by using the Lagrange interpolation polynomial at Chebyshev nodes [21] or the Multhopp method [22]. It is also shown that when the upper tip of the crack approaches the plate boundary, the SIF at this tip increases infinitely taking also the value of the brittle fracture limit for the given material.
   (b) As a result, it is necessary to consider another important specific case where the crack vertically approaches the plate boundary. In this case, it is shown that the dislocation density at that upper tip of the crack has a finite value, and a simple dependence is established between this value and the load normal to the crack edges at the same point. Finally, the governing SIE is solved in the class of functions unbounded at the inner tip of the crack and bounded at its upper tip on the boundary.
   (c) The problem where the plate boundary is reinforced by an absolutely rigid broken infinite stringer and the crack vertically comes to the plate boundary at the point of the stringer fracture. In this case, it is shown that the dislocation density at the upper boundary point of the crack has a logarithmic singularity if, at the same point, the forces normal to the crack edges are different from zero.
   (d) Problem (b) where the crack edges near its inner tip are reinforced by a thin inclusion, which is simulated by continuously distributed linearly elastic springs [23]. The presence of such an inclusion decreases SIF and, therefore, prevents the crack from propagation.

(ii) Problem (i) where the system of cracks is replaced by a system of absolutely rigid thin inclusions. Particular cases, as well as the case of one inclusion approaching the plate boundary (reinforcement problem), are discussed.

(iii) The symmetric problem on stress state of an elastic infinite plate or a half-space under a plane deformation when, on the plate boundary, there is a system of arbitrarily many indented punches and the plate contains a system of cracks on the vertical axis.

(iv) The previous problem when the system of cracks is replaced by a system of absolutely rigid thin inclusions.

The results of investigations of the last two problems can used in calculations of the foundations. They are generalizations of classical contact problems of the theory of elasticity.

All problems are mathematically formulated as governing SIE and are solved exactly or by the numerical-analytical method mentioned above.

2. Solution of the first boundary-value problem for an elastic quarter-plane

First, we write the basic equations of elasticity in the case of plane deformation in cylindrical coordinates r, ϑ, z under the assumption that the polar axis Or coincides with the positive axis Ox and is directed vertically down and the axis Oy is horizontally directed to the right. In this system of coordinates, the Lame equations in the absence of volume forces have the form [24]

$$(\lambda+2\mu)\frac{\partial e}{\partial r}-\frac{\mu}{r^2}\frac{\partial\chi}{\partial\vartheta}=0,\quad (\lambda+2\mu)\frac{1}{r}\frac{\partial e}{\partial\vartheta}+\mu\frac{\partial}{\partial r}\left(\frac{\chi}{r}\right)=0,$$
$$e=\frac{\partial u_r}{\partial r}+\frac{1}{r}\left(u_r+\frac{\partial u_\vartheta}{\partial\vartheta}\right),\quad \chi=r\frac{\partial u_\vartheta}{\partial r}+u_\vartheta-\frac{\partial u_r}{\partial\vartheta}=2r\omega_z, \tag{1}$$

and Hook's law becomes

$$\sigma_r=2\mu\frac{\partial u_r}{\partial r}+\lambda e,\quad \sigma_\vartheta=\frac{2\mu}{r}\left(\frac{\partial u_\vartheta}{\partial\vartheta}+u_r\right)+\lambda e,\quad \tau_{r\vartheta}=\mu\left(\frac{\partial u_\vartheta}{\partial r}-\frac{u_\vartheta}{r}+\frac{1}{r}\frac{\partial u_r}{\partial\vartheta}\right). \tag{2}$$

Here σ_r, σ_ϑ, and $\tau_{r\vartheta}$ are the stress components, u_r and u_ϑ are correspondingly the components of point displacements of an elastic body in the radical and circular directions, e is the relative volume expansion, ω_z is the angle of rotation about the axis Oz perpendicular to the elastic plane, and λ and μ are Lame constants of the elastic body expressed in terms of the Young module E and Poisson ratios by the formulas

$$\lambda=\frac{E\nu}{(1+\nu)(1-2\nu)},\quad \mu=G=\frac{E}{2(1+\nu)}.$$

Now for the right elastic quarter-plane wedge $\Omega_+=\{0<r<\infty,\ 0<\vartheta<\pi/2\}$ with a right angle of opening, we consider the first boundary-value problem

$$\sigma_\vartheta\big|_{\vartheta=+0}=-q_0^+(r),\quad \tau_{r\vartheta}\big|_{\vartheta=+0}=-\tau_0^+(r),$$
$$\sigma_\vartheta\big|_{\vartheta=\pi/2-0}=q^+(r),\quad \tau_{r\vartheta}\big|_{\vartheta=\pi/2-0}=\tau^+(r)\quad (0<r<\infty), \tag{3}$$

where $q_0^+(r)$, $q^+(r)$, $\tau_0^+(r)$, $\tau^+(r)$ are functions of the load given on the wedge faces Ω_+. They are considered to be a prespecified integrable function.

The solution of the boundary-value problem (1)–(3) will be constructed by the method of Mellin integral transformation for which we introduce the Mellin transformants:

$$\{\bar\sigma_r(\vartheta,p);\bar\sigma_\vartheta(\vartheta,p);\bar\tau_{r\vartheta}(\vartheta,p);\bar e(\vartheta,p)\}=\int_0^\infty\{\sigma_r(r,\vartheta);\sigma_\vartheta(r,\vartheta);\tau_{r\vartheta}(r,\vartheta);e(r,\vartheta)\}r^p dr,$$
$$\{\bar u_r(\vartheta,p);\bar u_\vartheta(\vartheta,p);\bar\chi(\vartheta,p)\}=\int_0^\infty\{u_r(r,\vartheta);u_\vartheta(r,\vartheta);\chi(r,\vartheta)\}r^{p-1}dr \tag{4}$$
$$(-1+\varepsilon<\mathrm{Re}p<0,\ r>0).$$

Applying the Mellin transformation to equations (1), after simple manipulations, we obtain the general solution of these equations in Mellin transformants represented by the formulas

$$\bar e_+=A_1^+\sin(p+1)\vartheta+A_2^+\cos(p+1)\vartheta\quad\left(0\le\vartheta\le\frac{\pi}{2}\right),$$
$$\bar u_r^+=-A^*[A_1^+\sin(p+1)\vartheta+A_2^+\cos(p+1)\vartheta]+A_3^+\sin(p-1)\vartheta+A_4^+\cos(p-1)\vartheta,$$
$$\bar u_\vartheta^+=B^*[A_1^+\cos(p+1)\vartheta-A_2^+\sin(p+1)\vartheta]+A_4^+\sin(p-1)\vartheta-A_3^+\cos(p+1)\vartheta, \tag{5}$$
$$A^*=\frac{1}{4\mu p}[(\lambda+\mu)(p+1)+2\mu],\quad B^*=\frac{1}{4\mu p}[(\lambda+\mu)(p-1)-2\mu].$$

IOP Conf. Series: Journal of Physics: Conf. Series **991** (2018) 012059 doi:10.1088/1742-6596/991/1/012059

Further, the Mellin transformation is also applied to Hook's law (2) and boundary conditions (3). From them, with the help of (5), we determine the constants A_j^+ $(j=\overline{1,4})$:

$$
\begin{aligned}
A_1^+ &= \frac{1}{(\lambda+\mu)D(p)}\Big\{[2p-1+\cos(\pi p)]\bar{\tau}_0^+(p)-\sin(\pi p)\bar{q}_0^+(p) \\
&\quad -2p\cos\left(\frac{\pi p}{2}\right)\bar{q}^+(p)+2(p-1)\sin\left(\frac{\pi p}{2}\right)\bar{\tau}^+(p)\Big\}; \\
A_2^+ &= -\frac{1}{(\lambda+\mu)D(p)}\Big\{\sin(\pi p)\bar{\tau}_0^+(p)-[2p+1-\cos(\pi p)]\bar{q}_0^+(p) \\
&\quad -2p\cos\left(\frac{\pi p}{2}\right)\bar{\tau}^+(p)-2(p+1)\sin\left(\frac{\pi p}{2}\right)\bar{q}^+(p)\Big\}, \quad D(p)=2p^2-1+\cos(\pi p), \\
A_3^+ &= -\frac{1}{4\mu p D(p)}\Big\{(p-1)[2p+1-\cos(\pi p)]\bar{\tau}_0^+(p) \\
&\quad +(p+1)\left[\sin(\pi p)\bar{q}_0^+(p)-2(p-1)\sin\left(\frac{\pi p}{2}\right)\bar{\tau}^+(p)+2p\cos\left(\frac{\pi p}{2}\right)\bar{q}^+(p)\right], \\
A_4^+ &= -\frac{1}{4\mu p D(p)}\Big\{(p+1)[2p-1+\cos(\pi p)]\bar{q}_0^+p) \\
&\quad +(p-1)\left[\sin(\pi p)\bar{\tau}_0^+(p)-2p\cos\left(\frac{\pi p}{2}\right)\bar{\tau}^+(p)-2(p+1)\sin\left(\frac{\pi p}{2}\right)\bar{q}^+(p)\right]\Big\}.
\end{aligned}
\tag{6}
$$

Here, according to (4), $\bar{\tau}_0^+(p)$, $\bar{q}_0^+(p)$, $\bar{\tau}^+(p)$, and $\bar{q}^+(p)$ are Mellin transformants of the corresponding functions from boundary conditions (3).

Substituting the constants (6) into (5), we finally obtain $\bar{u}_r(\vartheta,p)$ and $\bar{u}_\vartheta(\vartheta,p)$. In particular, we have

$$
\begin{aligned}
\bar{u}_r^+(0,p) &= \frac{1}{2\mu(\lambda+\mu)pD(p)}\Big\langle(\lambda+2\mu)\sin(\pi p)\bar{\tau}_0^+(p)-\{2(\lambda+\mu)p(p+1) \\
&\quad +\mu[2p+1-\cos(\pi p)]\}\bar{q}_0^+(p)-2(\lambda+2\mu)p\cos\left(\frac{\pi p}{2}\right)\bar{\tau}^+(p)-2(\lambda+2\mu)(p+1)\sin\left(\frac{\pi p}{2}\right)\bar{q}^+(p)\Big\rangle, \\
\bar{u}_\vartheta^+(0,p) &= \frac{1}{2\mu(\lambda+\mu)pD(p)}\Big\langle\{2(\lambda+\mu)p(p-1)-\mu[2p-1+\cos(\pi p)]\}\bar{\tau}_0^+(p) \\
&\quad +(\lambda+2\mu)\left[\sin(\pi p)\bar{q}_0^+(p)-2(p-1)\sin\left(\frac{\pi p}{2}\right)\bar{\tau}^+(p)+2p\cos\left(\frac{\pi p}{2}\right)\bar{q}^+(p)\right]\Big\rangle.
\end{aligned}
\tag{7}
$$

The first boundary-value problem will also be considered for the left elastic quarter-plane, i.e. for the wedge $\Omega_-=\{0<r<\infty,\ -\pi/2<\vartheta<0\}$, assuming that

$$
\begin{aligned}
&\sigma_\vartheta\big|_{\vartheta=-0}=-q_0^-(r), \quad \tau_{r\vartheta}\big|_{\vartheta=-0}=-\tau_0^-(r), \\
&\sigma_\vartheta\big|_{\vartheta=-\pi/2+0}=-q^-(r), \quad \tau_{r\vartheta}\big|_{\vartheta=-\pi/2+0}=-\tau^-(r) \quad (0<r<\infty).
\end{aligned}
\tag{8}
$$

In this case, using the Mellin integral transformation, we again obtain equalities of the type of (5), where all "+" sign must be replaced by "-" signs. The constants A_j^- $(j=\overline{1,4})$ are determined by the above-mentioned method based on Hook's law (2) and boundary conditions (8) in Mellin transformants. They are quite similar to (6). We have

$$
\begin{aligned}
\bar{u}_r^-(0,p) &= -\frac{1}{2\mu(\lambda+\mu)pD(p)}\Big\langle \{2(\lambda+\mu)p(p+1)+\mu[2p+1-\cos(\pi p)]\}\bar{q}_0^-(p) \\
&\quad + (\lambda+2\mu)\Big[\sin(\pi p)\bar{\tau}_0^-(p) + 2p\cos\left(\frac{\pi p}{2}\right)\bar{\tau}^-(p) - 2(p+1)\sin\left(\frac{\pi p}{2}\right)\bar{q}^-(p)\Big]\Big\rangle, \\
\bar{u}_\vartheta^-(0,p) &= \frac{1}{2\mu(\lambda+\mu)pD(p)}\Big\langle \{2(\lambda+\mu)p(p-1)-\mu[2p-1+\cos(\pi p)]\}\bar{\tau}_0^-(p) \\
&\quad - (\lambda+2\mu)\Big[\sin(\pi p)\bar{q}_0^-(p) - 2(p-1)\sin\left(\frac{\pi p}{2}\right)\bar{\tau}^-(p) - 2p\cos(\frac{\pi p}{2})\bar{q}^-(p)\Big]\Big\rangle,
\end{aligned} \tag{9}
$$

where the notation is similar to that introduced above.

3. Stress state of an elastic half-plane with a collinear system of vertical cracks

Let an elastic half-plane or an elastic semi-infinite plate $x \geq 0$, on its vertical axis Ox (Or), contain a collinear system of cracks $L = \bigcup_{k=1}^{n}[a_k, b_k]$ consisting of arbitrarily many cracks along the segments $[a_k, b_k]$ $(k = \overline{1,n})$. Denote the set of the right and left edges of this system of cracks by $L_\pm$, respectively. Later, adhering to the general case, let us consider that the edges of the system of cracks $L_\pm$ are loaded by asymmetric non-self-balanced normal and tangential forces, i.e., according to (3) and (8), we have

$$(\tau_{r\vartheta} + i\sigma_\vartheta)\big|_{L_\pm} = -[\tau_0^\pm(r) + iq_0^\pm(r)] \quad (r \in L).$$

In addition, we assume that the boundary of the half-plane is also loaded by forces of this nature, i.e., according to (3) and (8), we have

$$(\tau_{r\vartheta} + i\sigma_\vartheta)\Big|_{\vartheta=\pm\pi/2\mp 0} = \pm[\tau^\pm(r) + iq^\pm(r)] \quad (0 < r < \infty).$$

It is required to determine the following characteristics of the problem: the dislocation density on the edges of the system of cracks L, the breaking stresses outside the system of cracks on their line of location, SIF, and the crack openings.

For their determination, we derive the governing SIE of the problem. With this aim in mind, we first introduce the notation:

$$
\begin{aligned}
&\bar{\Phi}_\pm(p) = -p\bar{u}_r^\pm(0,p), \quad \bar{\Psi}_\pm(p) = -p\bar{u}_\vartheta^\pm(0,p), \quad \bar{\mathrm{X}}_\pm(p) = \frac{1}{2}[\bar{\tau}_0^+(p) \pm \bar{\tau}_0^-(p)], \\
&\bar{\Omega}_\pm(p) = \frac{1}{2}[\bar{q}_0^+(p) \pm \bar{q}_0^-(p)], \quad \bar{\varphi}_\pm(p) = \frac{1}{2}[\bar{\Phi}_+(p) \pm \bar{\Phi}_-(p)], \quad \bar{\psi}_\pm(p) = \frac{1}{2}[\bar{\Psi}_+(p) \pm \bar{\Psi}_-(p)].
\end{aligned}
$$

Then we have

$$
\begin{aligned}
&\bar{\Phi}_\pm(p) = \bar{\varphi}_+(p) \pm \bar{\psi}_-(p), \quad \bar{\Psi}_\pm(p) = \bar{\psi}_+(p) \pm \bar{\psi}_-(p), \\
&\bar{\tau}_0^\pm(p) = \bar{\mathrm{X}}_+(p) \pm \bar{\mathrm{X}}_-(p), \quad \bar{q}_0^\pm(p) = \bar{\Omega}_+(p) \pm \bar{\Omega}_-(p).
\end{aligned}
$$

IOP Conf. Series: Journal of Physics: Conf. Series **991** (2018) 012059 doi:10.1088/1742-6596/991/1/012059

Later, we write formulas (7) and (9) in the new notation as follows:

$$\bar{\varphi}_+(p) + \bar{\varphi}_-(p) = \frac{1}{\Delta(p)}\Big\langle -(\lambda+2\mu)\sin(\pi p)[\bar{\mathrm{X}}_+(p) + \bar{\mathrm{X}}_-(p)]$$
$$+ \{2(\lambda+\mu)p(p+1) + \mu[2p+1-\cos(\pi p)]\}[\bar{\Omega}_+(p) + \bar{\Omega}_-(p)]$$
$$+ 2(\lambda+2\mu)p\cos\left(\frac{\pi p}{2}\right)\bar{\tau}^+(p) + 2(\lambda+2\mu)(p+1)\sin\left(\frac{\pi p}{2}\right)\bar{q}^+(p)\Big\rangle, \tag{10}$$

$$\bar{\psi}_+(p) + \bar{\psi}_-(p) = \frac{1}{\Delta(p)}\Big\langle -\{2(\lambda+\mu)p(p-1) - \mu[2p-1+\cos(\pi p)]\}[\bar{\mathrm{X}}_+(p) + \bar{\mathrm{X}}_-(p)]$$
$$-(1+2\mu)\Big\{\sin(\pi p)[\bar{\Omega}_+(p) + \bar{\Omega}_-(p)] - 2(p-1)\sin\left(\frac{\pi p}{2}\right)\bar{\tau}^+(p) + 2p\cos\left(\frac{\pi p}{2}\right)\bar{q}^+(p)\Big\}\Big\rangle, \tag{11}$$

$$\bar{\varphi}_+(p) - \bar{\varphi}_-(p) = \frac{1}{\Delta(p)}\Big\langle \{2(\lambda+\mu)p(p+1) + \mu[2p+1-\cos(\pi p)]\}[\bar{\Omega}_+(p) - \bar{\Omega}_-(p)]$$
$$+(\lambda+2\mu)\Big\{\sin(\pi p)[\bar{\mathrm{X}}_+(p) - \bar{\mathrm{X}}_-(p)] + 2p\cos\left(\frac{\pi p}{2}\right)\bar{\tau}^-(p) - 2(p+1)\sin\left(\frac{\pi p}{2}\right)\bar{q}^-(p)\Big\}\Big\rangle, \tag{12}$$

$$\bar{\psi}_+(p) - \bar{\psi}_-(p) = -\frac{1}{\Delta(p)}\Big\langle \{2(\lambda+\mu)p(p-1) - \mu[2p-1+\cos(\pi p)]\}[\bar{\mathrm{X}}_+(p) - \bar{\mathrm{X}}_-(p)]$$
$$-(\lambda+2\mu)\Big\{\sin(\pi p)[\bar{\Omega}_+(p) - \bar{\Omega}_-(p)] - 2(p-1)\sin\left(\frac{\pi p}{2}\right)\bar{\tau}^-(p) - 2p\cos\left(\frac{\pi p}{2}\right)\bar{q}^-(p)\Big\}\Big\rangle, \tag{13}$$

$$\Delta(p) = 2\mu(\lambda+\mu)D(p) = 2\mu(\lambda+\mu)[2p^2 - 1 + \cos(\pi p)].$$

Now from (12) we determine $\bar{\varphi}_+(p)$, from (13) we determine $\bar{\psi}_+(p)$, and substitute their expressions, correspondingly, in (11) and (10). After some calculations and transformations, we come to the equation

$$\bar{\mathrm{X}}_+(p) + i\bar{\Omega}_+(p) = -\frac{2\mu(\lambda+\mu)[2p^2-1+\cos(\pi p)]}{(\lambda+2\mu)\sin(\pi p)}[\bar{\varphi}_-(p) + i\bar{\psi}_-(p)] + \frac{1}{(\lambda+2\mu)\sin(\pi p)}$$
$$\times\Big\langle\Big\{2(\lambda+\mu)p^2 + \mu[1-\cos(\pi p)]\Big\}[\bar{\Omega}_-(p) - i\bar{\mathrm{X}}_-(p)] + 2(\lambda+2\mu)p[\bar{\Omega}_-(p) + i\bar{\mathrm{X}}_-(p)]\Big\rangle$$
$$+\frac{1}{\sin(\pi p)}\Big\{p\Big[\cos\left(\frac{\pi p}{2}\right) + i\sin\left(\frac{\pi p}{2}\right)\Big][\bar{\tau}^+(p) - i\bar{q}^+(p)] - p\Big[\cos\left(\frac{\pi p}{2}\right) - i\sin\left(\frac{\pi p}{2}\right)\Big]$$
$$\times[\bar{\tau}^-(p) - i\bar{q}^-(p)] - i\sin\left(\frac{\pi p}{2}\right)[\bar{\tau}^+(p) + i\bar{q}^+(p)] - i\sin\left(\frac{\pi p}{2}\right)[\bar{\tau}^-(p) + i\bar{q}^-(p)]\Big\}. \tag{14}$$

Equations (14) in the whole form the key equation of the posed problem for cracks in Mellin transformants. To write it in original form, we apply the inverse Mellin transform formula to (14). The line of integration over the variable p lies in the strip of regularity $-1+\varepsilon < \mathrm{Re}p < 0$. However, in our case, the strip of regularity of the Mellin transformation is actually wider than the above-indicated strip and includes the imaginary axis in the plane of the complex variable $p = u + iv$, since the functions in (14) do not have poles on the imaginary axis and are analytic on it. Transferring the line of integration in the Mellin inverse transform to the imaginary axis and calculating the required integrals, we obtain the following key equation of the problem under study:

$$\begin{aligned}
\mathrm{X}_+(r) + i\Omega_+(r) = &-\frac{16\mu(\lambda+\mu)}{\pi(\lambda+2\mu)}\int_L \frac{s^2 r[\varphi_-(s)+i\psi_-(s)]\,ds}{(s^2-r^2)(s+r)^2} \\
&+\frac{2}{\pi}\int_L\left[\frac{1}{s+r}-\frac{\mu}{\lambda+2\mu}\frac{1}{s-r}-\frac{\lambda+\mu}{\lambda+2\mu}\frac{s-r}{(s+r)^2}\right]\frac{s\Omega_-(s)\,ds}{s+r} \\
&+\frac{2i}{\pi}\int_L\left[\frac{1}{s+r}+\frac{\mu}{\lambda+2\mu}\frac{1}{s-r}+\frac{\lambda+\mu}{\lambda+2\mu}\frac{s-r}{(s+r)^2}\right]\frac{s\mathrm{X}_-(s)\,ds}{s+r} \\
&+\frac{2}{\pi}\int_0^\infty \frac{r-is}{(s^2+r^2)^2}s^2\tau^+(s)\,ds+\frac{2}{\pi}\int_0^\infty\frac{r-is}{(s^2+r^2)^2}rsq^+(s)\,ds \\
&-\frac{2}{\pi}\int_0^\infty \frac{r+is}{(r^2+s^2)^2}s^2\tau^-(s)\,ds+\frac{2}{\pi}\int_0^\infty\frac{r+is}{(r^2+s^2)^2}rsq^-(s)\,ds \quad (0<r<\infty). \qquad (15)
\end{aligned}$$

Note that, according to the above notation,

$$\begin{aligned}
&\mathrm{X}_\pm(r) = \frac{1}{2}[\tau_0^+(r)\pm\tau_0^-(r)], \quad \Omega_\pm(r)=\frac{1}{2}[q_0^+(r)\pm q_0^-(r)], \\
&\varphi_-(r) = \frac{1}{2}\frac{d}{dr}[u_r^+(r,0)-u_r^-(r,0)], \quad \psi_-(r)=\frac{1}{2}\frac{d}{dr}[u_\vartheta^+(r,0)-u_\vartheta^-(r,0)],
\end{aligned}$$

and equation (15) was derived with the stresses and displacements continuity outside the system of cracks on $L' = [0,\infty)\backslash L$ taken into account. Under these conditions, we have

$$\mathrm{X}_-(r) = \Omega_-(r) \equiv 0, \quad \varphi_-(r) = \psi_-(r) \equiv 0 \quad (r \in L').$$

The functions $\varphi_-(r)$ and $\psi_-(r)$ are halves of the dislocation density components on the cracks edges.

Now, considering the key equation (15) on the system of cracks L, we obtain the following governing SIE for the complex combination of the dislocation density components:

$$\begin{aligned}
&\frac{2\mu(\lambda+\mu)}{\pi(\lambda+2\mu)}\int_L K(r,s)[\varphi_-(s)+i\psi_-(s)]\,ds = f(r) \quad (r\in L), \qquad (16)\\
&K(r,s) = \frac{8s^2r}{(s^2-r^2)(s+r)^2} = \frac{1}{s-r}-\frac{1}{s+r}+\frac{6r}{(s+r)^2}-\frac{4r^2}{(s+r)^3}, \\
&f(r) = -[\mathrm{X}_+(r)+i\Omega_+(r)]+g(r), \\
&g(r) = \frac{2}{\pi}\int_L\left[\frac{1}{s+r}-\frac{\mu}{\lambda+2\mu}\frac{1}{s-r}-\frac{\lambda+\mu}{\lambda+2\mu}\frac{s-r}{(s+r)^2}\right]\frac{s\Omega_-(s)\,ds}{s+r} \\
&\quad +\frac{2i}{\pi}\int_L\left[\frac{1}{s+r}+\frac{\mu}{\lambda+2\mu}\frac{1}{s-r}+\frac{\lambda+\mu}{\lambda+2\mu}\frac{s-r}{(s+r)^2}\right]\frac{s\mathrm{X}_-(s)\,ds}{s+r} \\
&\quad +\frac{2}{\pi}\int_0^\infty \frac{r-is}{(s^2+r^2)^2}s^2\tau^+(s)\,ds+\frac{2}{\pi}\int_0^\infty\frac{r-is}{(s^2+r^2)^2}rsq^+(s)\,ds \\
&\quad -\frac{2}{\pi}\int_0^\infty \frac{r+is}{(r^2+s^2)^2}s^2\tau^-(s)\,ds+\frac{2}{\pi}\int_0^\infty\frac{r+is}{(r^2+s^2)^2}rsq^-(s)\,ds \quad (r\in L), \\
&\mathrm{X}_\pm(r) = \frac{1}{2}[\tau_0^+(r)\pm\tau_0^-(r)], \quad \Omega_\pm(r)=\frac{1}{2}[q_0^+(r)\pm q_0^-(r)] \quad (r\in L).
\end{aligned}$$

The solution of SIE (16) for $a_1 \neq 0$ must satisfy the conditions

$$\int_{a_k}^{b_k}[\varphi_-(s)+i\psi_-(s)]\,ds = 0 \quad (k=\overline{1,n}), \qquad (17)$$

which express the continuity conditions for displacements at the end points (tips) of the cracks.

The solution of SIE (16)–(17) can be constructed by the well-known numerical-analytical method for solving SIE [7, 19, 20], where the Gauss quadrature formulas at Chebyshev nodes are used to calculate the integrals. To this end, we must first introduce dimensionless quantities in (16), (17) and transform each integral on (a_k, b_k) in the system L into an integral on $(-1, 1)$. As a result, SIE (16), (17) is reduced to a system of linear algebraic equations.

However, if the key equation (16) is considered outside L on $L' = [0, \infty)\backslash L$, then we obtain a complex combination of breaking normal and tangential forces out of L on L'.

$$\begin{aligned} \mathrm{X}_+(r) + i\Omega_+(r) &= \tau(r) + iq(r) = \frac{1}{2}[\tau_0^+(r) + \tau_0^-(r)] + \frac{i}{2}[q_0^+(r) + q_0^-(r)] \\ &= -\frac{2\mu(\lambda+\mu)}{\pi(\lambda+2\mu)} \int_L K(r,s)[\varphi_-(s) + i\psi_-(s)]\,ds + g(r) \quad (r \in L' = [0,\infty)\backslash L), \end{aligned} \tag{18}$$

where we assumed that the solution of SIE (16)–(17) has already been found. From (18) and also from the expression for $\varphi_-(r) + i\psi_-(r)$, we can determine SIF by the well-known formula [8, 9] and find the crack openings.

3.1. Particular cases

Consider a particular case where there is one crack along the segment $[a, b]$. In this case, the Multhopp method [22] or the method based on application of Lagrange interpolation polynomials at Chebyshev nodes [21] can successfully be applied to solve SIE (16), (17). At the same time, it turns out that SIF infinitely increases at the endpoint (tip) $r = a$ of the crack as $a \to +0$ exceeding the brittle fracture limit for the given material. As a result, the crack vertically approaches the boundary of the elastic half-plane. To prevent the crack from propagation, it is possible to connect its edges together near the endpoint $r = a$ by an elastic inclusion which is modelled as continuously distributed linear springs [23]. Let us consider the case of an edge crack $[0, b]$ in more detail. For simplicity, we here assume that the boundary of the elastic half-plane is free from external forces, i.e., $\bar{\tau}_0^\pm(r) = \bar{q}_0^\pm(r) \equiv 0$ $(0 < r < \infty)$, and identical normal forces of intensity $p(r)$ (opening displacement) act on the crack edges, i.e., $\bar{\tau}_0^\pm(r) = 0$, $\bar{q}_0^+(r) = \bar{q}_0^-(r) = p(r)$ $(0 \le r < b)$. Then SIE (16) has the form

$$\begin{aligned} &\frac{1}{\pi}\int_0^b K(r,s)\psi(s)\,ds = -\Lambda p(r) \quad (0 < r < b), \\ &\psi(r) = 2\psi_-(r) = 2\frac{d}{dr}[u_\vartheta^+(r,0)] = -2\frac{d}{dr}u_\vartheta^-(r,0), \end{aligned} \tag{19}$$

where

$$\Lambda = \begin{cases} \dfrac{4(1-\nu^2)}{E} & \text{under the plane deformation,} \\ \dfrac{4}{E} & \text{in generalized plane stress state.} \end{cases}$$

Due to its theoretical and practical importance, the problem on the edge crack in an elastic half-plane, as was noted above, was discussed in many papers, and the Wiener–Hopf rigorous analytical method was applied to solve this problem. At the same time, for a detailed numerical analysis of the problem by simpler computational means, it is necessary to construct effective approximate but sufficiently accurate solutions of the problem. In [7], an approximate solution of the above-discussed problem is constructed by passing to the limit in the solution of a similar problem for the inner crack when the crack tip close to the boundary of the half-plane approaches it in the limit. But though very effective, such a solution does not sufficiently reflect the true behavior of the body stresses and deformations near the boundary point of the crack.

Here the solution of governing SIE (19) is constructed by the well-known numerical-analytical method for solving SIE [7, 19, 20]. With this aim in mind, we first note that, according to well-known results [25], the solution of SIE (19) is bounded at the point $r = 0$, i.e., at the upper boundary point of the edge crack. To establish the dependence between $\psi(0)$ and the external load $p(r)$, we use the well-known results on the behavior of Cauchy-type integrals near the endpoints of the integration interval [26]. In this way, we obtain

$$\psi(0) = -\frac{\pi\Lambda}{4}p(+0). \tag{20}$$

Hence it follows that $\psi(0) \neq 0$ not always, as was stated in [27].

Now, on the basis of (20), SIE (19) is transformed as

$$\frac{1}{\pi}\int_0^b K(r,s)\psi_0(s)\,ds = -\Lambda p(r) + \frac{\Lambda}{4}p(+0)\left[\ln\frac{b-r}{b+r} - \frac{6r}{b+r} + \frac{2r^2}{(b+r)^2} + 4\right] \quad (0 < r < b), \tag{21}$$
$$\psi_0(r) = \psi(r) - \psi(0).$$

Then in (21), we pass from the interval $(0, b)$ to the interval $(-1, 1)$ assuming that

$$r = \frac{b}{2}(\xi+1), \quad s = \frac{b}{2}(\eta+1) \quad (-1 < \xi, \eta < 1).$$

As a result, SIE (21) transforms into the SIE

$$\frac{1}{\pi}\int_{-1}^1\left[\frac{1}{\eta-\xi} + K_0(\xi,\eta)\right]\chi_0(\eta)\,d\eta = f_0(\xi) \quad (-1 < \xi < 1), \tag{22}$$
$$K_0(\xi,\eta) = \frac{1}{(\xi+\eta+2)^2}(\xi^2 - \eta^2 + 4\xi\eta + 6\xi + 2\eta + 4), \quad \chi_0(\xi) = \psi_0\left(\frac{b}{2}(\xi+1)\right),$$
$$\chi_0(-1) = 0, \quad f_0(\xi) = -p_0(\xi) + \frac{1}{4}p_0(+0)\left[\ln\frac{1-\xi}{3+\xi} + \frac{4(\xi+5)}{(\xi+3)^2}\right], \quad p_0(\xi) = \Lambda p\left(\frac{b}{2}(\xi+1)\right).$$

Note that the function $f_0(\xi)$ from (22) contains a logarithmic term $\ln(1-\xi)$, which becomes unbounded as $\xi \to 1-0$. In order to deal with SIE with bounded right-hand side, the solution of SIE (22) is represented as

$$\chi_0(\xi) = \varphi_0(\xi) + \omega_0(\xi) \quad (-1 < \xi < 1), \tag{23}$$

where $\omega_0(\xi)$ is the solution of SIE

$$\frac{1}{\pi}\int_{-1}^1\frac{\omega_0(\eta)}{\eta-\xi}d\eta = \frac{1}{4}p_0(+0)\ln(1-\xi) \quad (-1 < \xi < 1).$$

The solution of this SIE, bounded at the point $\xi = -1$ and unbounded at the point $\xi = 1$, is given by the formula [26]

$$\omega_0(\xi) = -\frac{1}{4\pi}p_0(+0)\sqrt{\frac{1+\xi}{1-\xi}}v_0(\xi), \quad v_0(\xi) = \int_{-1}^1\sqrt{\frac{1-\tau}{1+\tau}}\frac{\ln(1-\tau)\,d\tau}{\tau-\xi}. \tag{24}$$

Taking into account (23) and (24), we can rewrite SIE (22) in the form

$$\int_{-1}^1\left[\frac{1}{\eta-\xi} + K_0(\xi,\eta)\right]\varphi_0(\eta)\,d\eta = h_0(\xi) \quad (-1 < \xi < 1), \tag{25}$$
$$h_0(\xi) = h_1(\xi) + h_2(\xi), \quad h_1(\xi) = \frac{1}{4}p_0(+0)\left[\frac{4(\xi+5)}{(\xi+3)^2} - \ln(\xi+3)\right] - p_0(\xi),$$
$$h_2(\xi) = \frac{1}{4\pi^2}p_0(+0)\int_{-1}^1 K_0(\xi,\eta)\sqrt{\frac{1+\eta}{1-\eta}}v_0(\eta)\,d\eta, \quad v_0(\eta) = \int_{-1}^1\frac{(1-\tau)\ln(1-\tau)\,d\tau}{\sqrt{1-\tau^2}(\tau-\eta)}.$$

IOP Conf. Series: Journal of Physics: Conf. Series **991** (2018) 012059 doi:10.1088/1742-6596/991/1/012059

Now the solution of SIE (25) is given by the formula

$$\varphi_0(\xi) = \sqrt{\frac{1+\xi}{1-\xi}}\Phi_0(\xi) \quad (-1 < \xi < 1), \tag{26}$$

where $\Phi_0(\xi)$ is a Hölder function on the interval $-1 \le \xi \le 1$. From (26), the Jacobi polynomials $P_n^{(-1/2;1/2)}(\xi)$ with the weight function $w(\xi) = \sqrt{(1+\xi)/(1-\xi)}$ are orthogonal on the interval $(-1, 1)$ and besides [28]

$$P_n^{(-1/2;1/2)}(\xi) = \frac{\Gamma(n+\frac{1}{2})}{\sqrt{\pi}n!}\frac{T_{n+1}(\xi)+T_n(\xi)}{1+\xi} \quad (-1 < \xi < 1,\ n = 0, 1, \ldots),$$

where $\Gamma(x)$ is the Euler gamma function and $T_n(x)$ are Chebyshev polynomials of the first kind. The integral relationship with the Cauchy kernel, necessary in this case to implement the procedure outlined in [7, 19, 20], has the form

$$\frac{1}{\pi}\int_{-1}^{1}\frac{T_{n+1}(\eta)+T_n(\eta)}{1+\eta}\frac{1}{\eta-\xi}\sqrt{\frac{1+\eta}{1-\eta}}d\eta = U_n(\xi) + U_{n-1}(\xi)$$
$$(-1 < \xi < 1,\ n = 0, 1, 2, \ldots,\ U_{-1}(\xi) \equiv 0),$$

where $U_n(\xi)$ are Chebyshev polynomials of the second kind. This relationship immediately follows from the well-known relationship with the Cauchy kernel, which transforms Chebyshev polynomials of the first kind to Chebyshev polynomials of the second kind [28]. In the notation of [7], in our case, we have

$$P_M(\xi) = \frac{T_{M+1}(\xi)+T_M(\xi)}{1+\xi}, \quad Q_M(\xi) = -\frac{\pi}{2}[U_M(\xi) + U_{M-1}(\xi)],$$

where M is an arbitrary natural number. The roots of the equations $P_M(\eta) = 0$ and $Q_M(\xi) = 0$, are, respectively,

$$\eta_m = \cos\left(\frac{2m-1}{2M+1}\pi\right), \quad \xi_r = \cos\left(\frac{2\pi r}{2M+1}\right) \quad (m, r = \overline{1, M}).$$

Further, following the well-known procedure [7], solving SIE (25) is reduced to solving the final system of linear algebraic equations

$$\frac{4}{2M+1}\sum_{m=1}^{M}\cos^2\left[\frac{(2m-1)\pi}{2(2M+1)}\right]\left[\frac{1}{\eta_m-\xi_r} + K_0(\eta_m, \xi_r)\right]\Phi_0(\eta_m) = h_0(\xi_r) \quad (r = \overline{1, M}) \tag{27}$$

for the unknowns $\Phi_0(\eta_m)$. Note that the function $v_0(\eta)$ in (25) is calculated at Chebyshev ordinary nodes by the formula

$$v_0(\eta_r^*) = \frac{\pi}{M}\sum_{m=1}^{M}\frac{(1-\tau_m)\ln(1-\tau_m)}{\tau_m-\eta_r^*}, \quad (r = \overline{1, M-1}), \tag{28}$$

$$\tau_m = \cos\left[\frac{(2m-1)\pi}{2M}\right], \quad (m = \overline{1, M}), \quad \eta_r^* = \cos\left(\frac{\pi r}{M}\right), \quad (r = \overline{1, M-1}).$$

The values $v_0(\eta_m)$ in the expression for $h_2(\xi_r)$ are calculated by the Lagrange interpolation polynomial at Chebyshev nodes using (28). For dimensionless SIF $K_I^{(0)}$, we have

$$K_I^{(0)} = \frac{\Lambda}{\sqrt{2\pi b}} K_I = -\left[\Phi_0(1) - \frac{p_0(+0)}{4\pi} v_0(1)\right].$$

The crack opening and the breaking stresses outside the crack can also be calculated easily.

Let us consider another type of the edge crack in an elastic half-plane whose boundary is reinforced by an infinite broken stringer.

Let an elastic half-plane or an elastic semi-infinite plate $x \geq 0$ be reinforced on its boundary $x = 0$ by an infinite absolutely rigid stringer broken at the origin. On its vertical axis Ox (Or), the plane contains an edge crack coinciding with the segment $0 \leq x \leq l$. Further, let the edges of the crack be loaded only by normal forces of identical intensities $p(r)$, i.e.,

$$\sigma_\vartheta\big|_{\vartheta=\pm 0} = -p(r), \quad \tau_{r\vartheta}\big|_{\vartheta=\pm 0} = 0 \quad (0 < r < l).$$

It is assumed that the stringer does not resist bending in the vertical direction, i.e., it is an absolutely flexible thin-walled element, and in the horizontal direction, it behaves as an absolutely rigid element with respect to tension or compression.

Under these assumptions, due to symmetry, for the right quarter-plane Ω_+, we come to the mixed boundary-value problem

$$\begin{gathered}\tau_{r\vartheta}\big|_{\vartheta=+0} = 0 \quad (0 < r < \infty); \quad \sigma_\vartheta\big|_{\vartheta=+0} = -q_0(r) \quad (0 < r < \infty, \ q_0(r) = p(r) \text{ for } (0 < r < l); \\ u_\vartheta\big|_{\vartheta=+0} = 0 \quad (l < r < \infty); \quad \sigma_\vartheta\big|_{\vartheta=\pi/2-0} = 0; \quad u_r\big|_{\vartheta=\pi/2-0} = 0 \quad (0 < r < \infty).\end{gathered} \tag{29}$$

The solution of the problem is again obtained using the Mellin integral transformation based on the concepts (7). Here we take into account that the condition $\bar{u}_r(\vartheta, p)\big|_{\vartheta=\pi/2} = 0$ implies the relation

$$\bar{\tau}^+(p) = -\frac{p+1}{\cos(\pi p/2)} q_0^+(p) \quad (-1 < \text{Re} p < 1). \tag{30}$$

Further, let us introduce the Mellin transformant of dislocation density on the crack edges

$$\bar{\varphi}(p) = -2p\bar{u}_\vartheta\big|_{\vartheta=+0}$$

and express the transformant $\bar{q}_0(p)$ from the second boundary condition of problem (29) in terms of $\bar{\varphi}(p)$ as follows:

$$\bar{q}_0(p) = -\frac{1}{\Lambda}\frac{1+\cos(\pi p)}{\sin(\pi p)}\bar{\varphi}(p) \quad (-1 < \text{Re} p < 0). \tag{31}$$

Now we apply the Mellin inverse transformation to (31), taking the imaginary axis of complex plane p as the line of integration over the variable p and subtracting the half-residue of the integrand at the zero point $p = 0$. As a result, after some simple transformations, we obtain the key equation of problem (29)

$$q_0(r) = -\frac{1}{\Lambda\pi}\int_0^l \left(\frac{1}{s-r} - \frac{1}{s+r}\right)\varphi(s)\,ds \quad (0 < r < \infty). \tag{32}$$

Considering the key equation (32) on the crack, we come to the governing SIE of the mixed boundary-value problem (29):

$$\frac{1}{\pi}\int_0^l \left(\frac{1}{s-r} - \frac{1}{s+r}\right)\varphi(s)\,ds = -\Lambda p(r) \quad (0 < r < l). \tag{33}$$

IOP Conf. Series: Journal of Physics: Conf. Series **991** (2018) 012059 doi:10.1088/1742-6596/991/1/012059

The solution of SIE (33) must satisfy the continuity condition for displacements at the crack tips:

$$\int_0^l \varphi(s)\,ds = 0. \tag{34}$$

Let us investigate the behavior of the solution $\varphi(r)$ of SIE (33) as $r \to +0$. For this we use the well-known results on behavior of a Cauchy-type integral near the endpoints of the interval of integration [27] (pp. 74–75).

In application to this case, we consider the Cauchy-type integral

$$\Phi(z) = \frac{1}{2\pi i}\int_0^l \frac{\varphi(\tau)\,d\tau}{\tau - z}$$

and assume that the density of this integral at the end point $\tau = 0$ of the interval $0 < \tau < l$ has a logarithmic singularity: $\varphi(\tau) = \varphi^*(\tau)\ln\tau$, where the function $\varphi^*(\tau)$ satisfies the Hölder condition on $[0, l]$. Then near the zero point, we have the representations

$$\Phi(z) = \varphi^*(0)\omega_2(z,0) + \Phi_0(z), \quad \Phi(t) = \frac{1}{2}\varphi^*(0)[\omega_2^+(t,0) + \omega_2^-(t,0)] + \Phi_0(t), \tag{35}$$

where $\Phi_0(z)$ is a single-valued bounded function near 0 and tends to a certain limit as $z \to 0$. Here

$$\omega_1(z,0) = -\frac{1}{2\pi i}\ln z, \quad \omega_2(z,0) = -\frac{1}{4\pi i}\ln^2 z + \frac{1}{2}\ln z.$$

We put

$$\omega_1^+(t,0) = -\frac{1}{2\pi i}\ln t \quad (t \in (0,l)),$$

where any single-valued branch of the logarithmic function is taken in the complex plane z cut along $[0, l]$. Then

$$\omega_1^-(t,0) = -\frac{1}{2\pi i}\ln t - \frac{1}{2\pi i}2\pi i = \omega_1^+(t,0) - 1,$$

which implies

$$\omega_1^+(t,0) - \omega_1^-(t,0) = 1, \quad \omega_1^+(t,0) + \omega_1^-(t,0) = -\frac{1}{\pi i}\ln t - 1 \quad (t \in (0,l)).$$

In the same way, we obtain

$$\omega_2^+(t,0) - \omega_2^-(t,0) = \ln t, \quad \omega_2^+(t,0) + \omega_2^-(t,0) = -\frac{1}{2\pi i}\ln^2 t \quad (t \in (0,l)),$$

whence we have

$$\omega_2^-(t,0) = -\frac{1}{4\pi i}\ln^2 t - \frac{1}{2}\ln t \quad (t \in (0,l)).$$

Therefore,

$$\omega_2^-(-t,0) = -\frac{1}{4\pi i}\ln^2(-t) - \frac{1}{2}\ln(-t) = -\frac{1}{4\pi i}\ln^2 t + \frac{i\pi}{4}$$

and from (35), it follows that

$$\begin{aligned}\Phi(-t) = \frac{1}{2\pi i}\int_0^l \frac{\varphi^*(\tau)\ln\tau\,d\tau}{\tau + t} &= \varphi^*(0)\omega_2^-(-t,0) + \Phi_0(-t)\\ &= \left(-\frac{1}{4\pi i}\ln^2 t + \frac{i\pi}{4}\right)\varphi^*(0) + \Phi_0(-t) \quad (t \to +0).\end{aligned} \tag{36}$$

IOP Conf. Series: Journal of Physics: Conf. Series **991** (2018) 012059 doi:10.1088/1742-6596/991/1/012059

Substituting (35) and (36) into SIE (33), we now get an important relation

$$\varphi^*(0) = -\frac{2}{\pi}\Lambda p(+0), \tag{37}$$

which implies that if $p(+0) \neq 0$, then the function $\varphi(r)$ is the solution of SIE (33) and has a logarithmic singularity at the point $r = 0$.

On the basis of just stated results, as in [15], it is possible to show that the solution of SIE (33) cannot have power singularity at the point $r = 0$.

The SIE (33), (34) allow one to obtain an exact solution which obviously contains a term with logarithmic singularity. Actually, in (33), we continue the function $\varphi(r)$ to the interval $(-l, 0)$ in the even way, while the right-hand side $p(r)$ is continued in the odd way. Then (33) transforms as

$$\frac{1}{\pi}\int_{-l}^{l}\frac{\varphi(s)\,ds}{s-r} = -\Lambda q(r) \quad (q(r) = p(|r|)\mathrm{sgn}\, r), \tag{38}$$

and condition (34) takes the form

$$\int_{-l}^{l}\varphi(s)\,ds = 0. \tag{39}$$

Now SIE becomes [26]

$$\varphi(r) = \frac{\Lambda}{\pi\sqrt{l^2-r^2}}\int_{-l}^{l}\frac{\sqrt{l^2-s^2}q(s)\,ds}{s-r} + \frac{C}{\sqrt{l^2-r^2}} \quad (-l<r<l). \tag{40}$$

Integrating both sides of this equality with respect to r over the interval $(-l, l)$ and taking into account the expression of the indefinite integral with the Cauchy kernel from [29] (p. 111), we obtain $C = 0$. Then (40), where $C = 0$, becomes

$$\varphi(r) = \frac{\Lambda}{\pi\sqrt{l^2-r^2}}\int_{-l}^{l}\frac{\sqrt{l^2-s^2}q_0^*(s)\,ds}{s-r} + \frac{\Lambda p(+0)}{\pi\sqrt{l^2-r^2}}\int_{-l}^{l}\frac{\sqrt{l^2-s^2}\mathrm{sgn}\, s\,ds}{s-r} \quad (q_0^*(r) = [p(|r|) - p(+0)]\mathrm{sgn}\, r,\ -l<r<l).$$

After calculating the second integral in this relation, we have

$$\varphi(r) = -\frac{2l\Lambda p(+0)}{\pi\sqrt{l^2-r^2}} + \frac{\Lambda p(+0)}{\pi}\ln\left(\frac{l+\sqrt{l^2-r^2}}{l-\sqrt{l^2-r^2}}\right) + \frac{\Lambda}{\pi\sqrt{l^2-r^2}}\int_{-l}^{l}\frac{\sqrt{l^2-s^2}q_0^*(s)\,ds}{s-r} \quad (-l<r<l). \tag{41}$$

The second term in (41) has a logarithmic singularity at the point $r = 0$. After simple transformations, (41) can be represented as

$$\varphi(r) = \varphi^*(r)\ln r \quad (r \to +0),$$
$$\varphi^*(r) = -\frac{2\Lambda p(+0)}{\pi} - \frac{2\Lambda l p(+0)}{\pi\ln r\sqrt{l^2-r^2}} + \frac{2\Lambda p(+0)}{\pi\ln r}\ln(r+\sqrt{l^2-r^2}) + \frac{2\Lambda}{\pi\ln r\sqrt{l^2-r^2}}\int_{0}^{l}\frac{s\sqrt{l^2-s^2}q_0^*(s)\,ds}{s^2-r^2} \quad (0<r<l).$$

From here, again relationship (37) is immediately obtained.

Further, the Mellin inverse transformation is applied to (30) as well. Taking into account (32), after some simple transformations and calculations of necessary integrals, we have

$$\tau^+(r) = -\frac{4r^2}{\pi}\int_0^l \frac{sp(s)\,ds}{(s^2+r^2)^2} + \frac{4r^3}{\Lambda\pi^2}\int_0^l \left[\frac{1}{l^2+r^2} + \frac{1}{s^2+r^2}\ln\left(\frac{l^2-s^2}{l^2+r^2}\right)\right]\frac{\varphi(s)\,ds}{s^2+r^2} \quad (0 < r < \infty).$$

Whence, using the above-described results on the behavior of Cauchy-type integrals, we obtain from [26]

$$\tau^+(+0) = -\frac{2}{\pi}p(+0).$$

4. On the stress state of an elastic half-plane containing a collinear system of absolutely rigid thin inclusions on the vertical axis

Let an elastic half-space or an elastic semi-infinite plate $x \geq 0$ be loaded on its boundary according to boundary conditions (3) and (8) and, on the vertical axis Ox (Or), contain a collinear system of absolutely rigid thin inclusions along the system of the segments $L = \bigcup_{k=1}^{n}[a_k, b_k]$. Starting from (10)–(13) and using a method similar to that applied to the problem for a system of cracks, we reduce solving the considered problem to solving the following governing SIE:

$$\frac{1}{2(\lambda+2\mu)}\int_L \left\{\frac{2(\lambda+\mu)}{\mu}K_1(r,s) - \frac{2[3\mu(\lambda+\mu)+\lambda^2]}{\mu(\lambda+\mu)}K_2(r,s) - 2K_3(r,s)\right.$$
$$\left. + \frac{\mu}{\lambda+\mu}K_4(r,s) + \frac{\lambda^2+4\lambda\mu+3\mu^2}{2\mu(\lambda+\mu)}K_5(r,s)\right\}[\mathrm{X}_-(s) + i\Omega_-(s)]\,ds = rf(r) \quad (r \in L), \quad (42)$$

$$K_1(r,s) = \frac{1}{\pi}\int_0^\infty \frac{\alpha^4 \sin(\alpha a_0)\,d\alpha}{[\cosh(\pi\alpha) - 1 - 2\alpha^2]\sinh(\pi\alpha)} \quad \left(a_0 = \ln\frac{s}{r}\right);$$

$$K_2(r,s) = \frac{1}{\pi}\left\{\int_0^\infty \frac{\alpha^2\sin(\alpha a_0)\,d\alpha}{[2\alpha^2+1-\cosh(\pi\alpha)]\sinh(\pi\alpha)} + \frac{1}{\pi^2-4}\right\},$$

$$K_3(r,s) = \frac{1}{\pi}\left\{\int_0^\infty \frac{\alpha^2\cosh(\pi\alpha)\sin(\alpha a_0)\,d\alpha}{[2\alpha^2+1-\cosh(\pi\alpha)]\sinh(\pi\alpha)} + \frac{1}{\pi^2-4}\right\},$$

$$K_4(r,s) = \frac{1}{\pi}\left[\int_0^\infty \frac{\tanh(\pi/2)\sin(\alpha a_0)\,d\alpha}{2\alpha^2+1-\cosh(\pi\alpha)} + \frac{\pi^2}{2(\pi^2-4)}\right],$$

$$K_5(r,s) = -\frac{\sqrt{sr}}{\pi(s-r)} - \frac{1}{\pi}\left[\int_0^\infty \frac{(1+2\alpha^2)\tanh(\pi\alpha)\sin(\alpha a_0)\,d\alpha}{\cosh(\pi\alpha)-1-2\alpha^2} - \frac{\pi^2}{\pi^2-4}\right],$$

where the function $f(r)$ depends on the loads acting on the boundary of the elastic half-plane and the angles of rotation of the inclusions. Besides, here the unknown functions $\mathrm{X}_-(r)$ and $\Omega_-(r)$ characterize the jumps of the tangential normal stresses on the edges of the inclusions:

$$\mathrm{X}_-(r) = -[\tau_{r\vartheta}|_{\vartheta=+0} - \tau_{r\vartheta}|_{\vartheta=-0}], \quad \Omega_-(r) = -[\sigma_\vartheta|_{\vartheta=+0} - \sigma_\vartheta|_{\vartheta=-0}] \quad (r \in L).$$

The solution of SIE (42) must satisfy the equilibrium conditions for the inclusions

$$\int_{a_k}^{b_k}[\mathrm{X}_-(r) + i\Omega_-(r)]\,dr = 0, \quad \int_{a_k}^{b_k} r[\mathrm{X}_-(r) + i\Omega_-(r)]\,dr = 0. \quad (43)$$

Writing SIE (42), (43) in dimensionless variables and then transforming each interval (a_k, b_k) $(k = \overline{1,n})$ into the interval $(-1,1)$, we can apply the above-mentioned numerical-analytical method for solving SIE [7, 19, 20].

IOP Conf. Series: Journal of Physics: Conf. Series **991** (2018) 012059 doi:10.1088/1742-6596/991/1/012059

5. Indentation of a system of punches into an elastic half-plane containing a collinear system of absolutely rigid thin inclusions

Let the system of $2n+1$ punches be indented into an elastic half-plane $x \geq 0$ along the segments on the half-plane boundary defined as $L = [-a, a] \cup \bigcup_{k=-n,k\neq 0}^{n} [a_k, b_k]$ $(k = \overline{1,n})$. Further, let the elastic half-plane contain a collinear system of absolutely rigid thin inclusions on its vertical axis Ox (Or) on the set of segments $l = \bigcup_{k=1}^{m} [c_k, d_k]$. We assume that the punches are located symmetrically with respect to the origin of coordinates and a centrally applied vertical concentrated force P_k acts on the kth punch contacting with the elastic foundation along the segment $[a_k, b_k]$ $(k = \overline{-n,n},\ a_0 = a,\ b_0 = a,\ a_{-k} = -b_k,\ b_{-k} = -a_k)$; furthermore, $P_0 = P$, $P_{-k} = P_k$ $(k = \overline{1,n})$. The bases of the punches are geometrically characterized by the function $f(r)$ $(r = y)$. This problem generalizes the classical contact problem of elasticity for indenting a system of punches into an elastic half-plane and is encountered in foundation engineering in construction of buildings on elastic foundations reinforced by piles.

To reduce the above problem to the system of integral equations, we consider the following mixed boundary-value problem for the elastic right quarter-plane Ω_+:

$$\begin{gathered} \sigma_\vartheta\big|_{\vartheta=\pi/2-0} = -p(r) \quad (0<r<\infty,\ r=y); \quad \tau_{r\vartheta}\big|_{\vartheta=\pi/2-0} = 0 \quad (0<r<\infty); \\ u_r\big|_{\vartheta=+0} = 0, \quad u_\vartheta\big|_{\vartheta=+0} = 0 \quad (r \in l); \quad \tau_{r\vartheta}\big|_{\vartheta=+0} = u_\vartheta\big|_{\vartheta=+0} = 0 \quad (r \in [0,\infty)\backslash l). \end{gathered} \tag{44}$$

Here the function $p(r)$ is the contact pressure under the punches and is considered as a specified function.

The solution of the problem (44) is again constructed by the Mellin integral transformation which permits reducing the above-formulated contact problem to solving the following system of two SIEs:

$$\begin{gathered} \int_l K_{11}(r,s)\tau_0^+(s)\,ds + \int_{L_1} K_{12}(r,s)p(s)\,ds = 0 \quad (r \in l), \\ \int_l K_{21}(r,s)\tau_0^+(s)\,ds + \int_{L_1} K_{22}(r,s)p(s)\,ds = rf'(r) \quad \left(r \in L_1,\ L_1 = [0,a] \cup \bigcup_{k=1}^{n} [a_k, b_k]\right), \end{gathered} \tag{45}$$

$$\begin{aligned} K_{11}(r,s) &= -\frac{\mu}{\pi(\lambda+\mu)(\lambda+2\mu)}\frac{r}{s+r} \\ &+ \frac{2}{\pi\mu(\lambda+2\mu)}\left\{\frac{\lambda+3\mu}{\pi^2-4} + \int_0^\infty \frac{\alpha^2[\lambda+2\mu+\mu\cosh(\pi\alpha)]\sin(\alpha a_0)\,d\alpha}{[2\alpha^2+1-\cosh(\pi\alpha)]\sinh(\pi\alpha)}\right\} \\ &- \frac{\lambda+3\mu}{2\mu(\lambda+2\mu)\pi}\left[\frac{\pi^2}{\pi^2-4} - \frac{\sqrt{sr}}{s-r} + \int_0^\infty \frac{(1+2\alpha^2)\tanh(\pi\alpha)\sin(\alpha a_0)\,d\alpha}{2\alpha^2+1-\cosh(\pi\alpha)}\right] \\ &- \frac{2(\lambda+\mu)}{\pi\mu(\lambda+2\mu)}\int_0^\infty \frac{\alpha^4\sin(\alpha a_0)\,d\alpha}{[\cosh(\pi\alpha)-1-2\alpha^2]\sinh(\pi\alpha)} \quad \left(a_0 = \ln\frac{s}{r}\right), \\ K_{12}(r,s) &= -\frac{2}{\pi\mu}\int_0^\infty \frac{\alpha^3\cosh(\pi\alpha/2)\cos(\alpha a_0)\,d\alpha}{[\cosh(\pi\alpha)-1-2\alpha^2]\sinh(\pi\alpha)} \\ &+ \frac{2(\lambda+2\mu)}{\pi\mu(\lambda+\mu)}\left\{\frac{1}{\pi^2-4} + \int_0^\infty \frac{\alpha^2\cosh(\pi\alpha/2)\sin(\alpha a_0)\,d\alpha}{[2\alpha^2+1-\cosh(\pi\alpha)]\sinh(\pi\alpha)}\right\} \\ &+ \frac{1}{\pi(\lambda+\mu)}\int_0^\infty \frac{\alpha[\cosh(\pi\alpha)-1]\cosh(\pi\alpha/2)\cos(\alpha a_0)\,d\alpha}{[2\alpha^2+1-\cosh(\pi\alpha)]\sinh(\pi\alpha)} \\ &- \frac{\lambda+2\mu}{\pi\mu(\lambda+\mu)}\left[\int_0^\infty \frac{\alpha\sinh(\pi\alpha/2)\cos(\alpha a_0)\,d\alpha}{2\alpha^2+1-\cosh(\pi\alpha)} + \frac{\pi^2}{2(\pi^2-4)} + \int_0^\infty \frac{\sinh(\pi\alpha/2)\sin(\alpha a_0)\,d\alpha}{2\alpha^2+1-\cosh(\pi\alpha)}\right], \end{aligned}$$

$$
\begin{aligned}
K_{21}(r,s) &= \frac{2}{\pi\mu}\int_0^\infty \frac{\alpha^3\cosh(\pi\alpha/2)\cos(\alpha a_0)\,d\alpha}{[2\alpha^2+1-\cosh(\pi\alpha)]\sinh(\pi\alpha)} \\
&- \frac{2(\lambda+2\mu)}{\pi\mu(\lambda+\mu)}\left\{\frac{1}{\pi^2-4}+\int_0^\infty \frac{\alpha^2\cosh(\pi\alpha/2)\sin(\alpha a_0)\,d\alpha}{[2\alpha^2+1-\cosh(\pi\alpha)]\sinh(\pi\alpha)}\right\} \\
&+ \frac{1}{\pi\mu(\lambda+\mu)}\int_0^\infty \frac{\{\mu\cosh(\pi\alpha/2)[1-\cosh(\pi\alpha)]+(\lambda+2\mu)\sinh(\pi\alpha/2)\sinh(\pi\alpha)\}\alpha\cos(\alpha a_0)\,d\alpha}{[\cosh(\pi\alpha)-1-2\alpha^2]\sinh(\pi\alpha)} \\
&+ \frac{\lambda+2\mu}{\pi\mu(\lambda+\mu)}\left\{\frac{\pi^2}{2(\pi^2-4)}+\int_0^\infty \frac{\sinh(\pi\alpha/2)\sinh(\pi\alpha)\sin(\alpha a_0)\,d\alpha}{[2\alpha^2+1-\cosh(\pi\alpha)]\sinh(\pi\alpha)}\right\}, \\
K_{22}(r,s) &= -\frac{\lambda+2\mu}{\mu(\lambda+\mu)\pi}\frac{r^2}{s^2-r^2}.
\end{aligned}
$$

The solution of the system of SIEs (45) must satisfy the equilibrium conditions for the punches and inclusions

$$
\int_0^a p(s)\,ds = \frac{P}{2}, \quad \int_{a_k}^{b_k} p(s)\,ds = P_k \quad (k=\overline{1,n}), \quad 2\int_{c_k}^{d_k} \tau_0^+(s)\,ds = Q_k \quad (k=\overline{1,m}). \tag{46}
$$

Here $\tau_0^+(r)$ are tangential contact stresses acting on the edges of the kth inclusion and Q_k is the axial force applied to the inclusion.

To simplify the study and the procedure of solving the system of SIEs (45), (46), we set $m=n=1$, $L=[0,a]$, $L_0=[b,c]$. Assuming that $p(r)=P\delta(r)/2$, where $\delta(r)$ is the Dirac delta function, and passing to the limit as $a\to+0$, we obtain the following SIE from the system of SIEs (45):

$$
\int_b^c K_{11}(r,s)\tau_0^+(s)\,ds = \frac{\lambda+2\mu}{2\pi\mu(\lambda+\mu)}P, \tag{47}
$$

which describes the contact problem for the reinforcement in an elastic matrix. In this connection, the case $b=0$ is practically important.

Equations (45), (46), and (47) can efficiently be solved by the above-mentioned method proposed in [7, 19, 20].

Note that after solving SIEs (45), (46), the distribution of normal stresses on the vertical axis of the elastic half-plane is determined by the formula

$$
\begin{aligned}
q_0^+(r) &= \frac{4r}{\pi}\int_0^\infty \frac{s^2p(s)\,ds}{(s^2+r^2)^2} + \frac{16(\lambda+\mu)}{\pi(\lambda+2\mu)}r\int_0^\infty \frac{s^2(s^2-r^2)\tau_0^+(s)\,ds}{(s^2+r^2)^3} \\
&+ \frac{8r}{\pi}\int_0^\infty \frac{s^2\tau_0^+(s)\,ds}{(s^2+r^2)^2} + \frac{2\mu}{\pi(\lambda+2\mu)}\int_0^\infty \frac{s\tau_0^+(s)\,ds}{s^2-r^2} \quad (0<r<\infty).
\end{aligned}
$$

Equations similar to (45), (46), and (48) can describe the contact problem, where a system of $2n+1$ punches is indented into an elastic half-plane which contains a system of collinear cracks on its vertical axis, i.e., the case where, in the contact problem described above, the system of inclusions is replaced by a system of cracks.

Conclusions

In this paper, in a fairly general formulation, a class of boundary-value problems related to the interaction of stress concentrators (such as cracks, absolutely rigid thin inclusions and punches) with an elastic semi-infinite plate is studied by a unified method of SIE. Some results of other authors concerning the problems of an edge crack in an elastic plate, when the plate boundary is free from external forces or when it is reinforced by an infinite absolutely rigid broken stringer, are simultaneously clarified. The classical contact problems of indenting a punch system into an elastic half-plane, which are of both theoretical and practical interest, are generalized. The results and approaches discussed in the paper can be used to study related boundary-value problems of the theory of elasticity.

References

[1] Kojter W T 1959 An infinite row of collinear cracks in an infinite elastic sheet *Ingenieur-Arch.* **28** 168–72
[2] Kojter W T 1965 Discussion of the paper of Bovi "Stretching of a rectangular plate with symmetrical cracks on the edges" *ASME J. Appl. Mech.* **32** (1), 279–80
[3] Wigglesworth L A 1957 Stress distribution in a notched plate *Mathematika* **4** (1) 76–96
[4] Stallybrass M P 1970 A crack perpendicular to an elastic half-plane *Int. J. Engng Sci.* **8** (5) 351–62
[5] Khrapkov A A 1968 First fundamental problem for a piecewise homogeneous plane with a slit perpendicular to the line of separation *J. Appl. Math. Mech.* **32** (4) 666–78
[6] Cherepanov G P 1974 *Mechanics of Brittle Fracture* (Moscow: Nauka) p 640 [in Russian]
[7] Panasyuk V V, Savruk M P, and Datsyshin A P 1976 *Stress Distribution near Cracks in Plates and Shells* (Kiev: Naukova dumka) p 443 [in Russian]
[8] Murakami Y (Editor) 1987 *Stress Intensity Factors Handbook* (Oxford: Pergamon Press) p 1456
[9] Savruk M P 1988 *Stress Intensity Factors for the Bodies with Cracks. Handbook in 4 volumes* vol 2 (Kiev: Nauka Dumka) p 619 [in Russian]
[10] Berezhnitskii L T, Panasyuk V V, and Stazhchuk N G 1983 *Interaction of Rigid Linear Inclusions and Cracks in Deformable Body* (Kiev: Naukova Dumka) p 288 [in Russian]
[11] Hakobyan V N 2011 *Stress Concentration near Defects in Homogeneous and Compound Bodies* (Lap Lambert Academic Publishing) p 148
[12] Berezhnitskii L T and Stashchuk N G 1981 On the Development of Cracks in the Vicinity of the Top of a Rigid Inclusion *Doklady AN Ukr.SSR. Ser. A* No. 12 29–33
[13] Berezhnitskii L T and Stashchuk N G 1981 The Stress Intensity Coefficients near the Crack on the Continuation of a Linear Rigid Inclusion *Doklady AN Ukr.SSR. Ser. A* No. 11 49–53
[14] Greif R and Sanders J L Jr 1965 The effect of a stringer on the stress in a cracked sheet *ASME J. Appl. Mech.* **32** (1) 59–66
[15] Delale F and Erdogan F 1982 The crack problem for a half-plane stiffened by elastic cover plates *Int. J. Solids Struct.* **18** (3) 381–95
[16] Mkhitaryan S M and Aghayan K L 1978 On a periodic contact problem for an elastic strip weakened by cracks and reinforced by elastic stringers *Izv. Akad. Nauk. ArmSSR. Mekh.* **31** (3) 3–17
[17] Mkhitaryan S M 2014 Contact interaction between an infinite or finite stringers and elastic semi-infinite plate with a finite vertical crack coming out to the stringer by one end *Advanced Materials Research. Contemporary Problems in Architecture and Construction* vol 1020 (Trans Tech Publ) 253–257.
[18] Hoakbyan V N, Mirzoyan S E, and Mkhitaryan S M 2015 The problem of the contact between a broken stringer and an elastic infinite strip containing a vertical edge crack *Int. Appl. Mech.* **51** (2) 176–86
[19] Erdogan F, Gupta G D, and Cook T S 1973 The numerical solutions of singular integral equations. *Methods of Analysis an Solution of Crack Problems* (Leyden: Noordhoff Intern. Publ.) pp 368–425
[20] Theocaris P S and Ioakimidis N I 1977 Numerical integration methods for the solution of singular integral equations *Quart. Appl Math.* **XXXV** (1) 173–85
[21] Mkhitaryan S M, Mkrtchyan M S, and Kanetsyan E G (2017) On a method for solving Prandtl's integro-differential equation applied to problems of continuum mechanics using polynomial approximations *ZAMM* **97** (6) 639–54
[22] Multhopp H 1938 Die Berechnung der Auftriebsverteilung von Tragflugeln *Luftfahrtforschung* **15** (4) 153–66
[23] Rose R L 1987 Crack reinforcement by distributed springs *J. Mech. Phys. Solids* **35** (4) 363–405
[24] Novozhilov V V 1958 *Theory of Elasticity* (Leningrad: Sudpromgiz) p 372 [in Russian]
[25] Wiliams M L 1952 Stress singularities resulting from various boundary conditions in angular corners of plates in extension *J. Appl. Mech.* **19** (4) 526–8
[26] Gakhov F D 1977 *Boundary Value Problems* (Moscow: Nauka) p 640 [in Russian]
[27] Gupta G D and Erdogan F 1974 The problem of edge cracks in an infinite strip *J. Appl. Mech* **41** (4) 1001–6
[28] Bateman H and Erdalyi A 1953 *Higher Transcendental Functions* vol II (New York: McGraw–Hill Book Co.) p 295
[29] Alexandrov V M and Mkhitaryan S M 1983 Contact Problems for Bodies with Thin Coatings and Interlayers (Moscow: Nauka) p 483 [in Russian]

IOP Conf. Series: Journal of Physics: Conf. Series **991** (2018) 012060 doi:10.1088/1742-6596/991/1/012060

Coupled thermal stresses analysis in the composite elastic-plastic cylinder

E V Murashkin[1,2*], **E P Dats**[3**]

[1]Ishlinsky Institute for Problems in Mechanics of the Russian Academy of Sciences, Moscow, Russia
[2]Bauman Moscow State Technical University, Moscow, Russia
[3]Vladivostok State University of Economics and Service, Vladivostok, Russia

E-mail: *evmurashkin@gmail.com, **dats@dvo.ru

Abstract. The present study is devoted to the set of boundary value problems in the frameworks of coupled thermoelastoplasticity under axial symmetry conditions for a composite circular cylinder. Throughout the paper the conventional Prandtl–Reuss elastic–plastic model generalised on the thermal effects is used. The yield stress is assumed by linear function of the temperature. The plastic potential is chosen in the form of Tresca yield criterion and the associated plastic flow rule is derived. The adding process of a heated cylinder to another is simulated. The coupled thermal stresses are calculated during processes of cooling and material unloading. The elastic-plastic borders positions are calculated and plastic flow domains are localized. Numerical results are graphically analysed.

Introduction

The modern mechanical engineering deals with the additive manufacturing technologies based on the adding new material at the high temperature gradient. The process of adding new parts of the material can be considered as a process of discrete material growth used in the technology of additive manufacturing. The mechanics of growing bodies [1–4] can be considered as a theoretical basis for solving such problems. In [5, 6], boundary value problems of the growth of heavy viscoelastic bodies were solved with the gravitational forces presence. The thermal state of a growing viscoelastic sphere was discussed in [7, 8]. The Assemblages made by the method of hot fitting have become the most widespread in technological practice for the stressed joints [9–13]. Usually such joints are carried out on the cylindrical surfaces of the assembly elements, when the outer part is heated before fitting and the inner one is cooled or remains at room temperature. Simplicity of the method of connection is coupled with the possibility to transmit significant loading pressure with different magnitudes and directions. The effective approximate engineering approaches and numerical simulation were developed to prescribe the stress strain state and material behaviour during the fitting process in the assembled nodes. The main drawback of the existing approaches for fitting process simulations is the insufficiently consistent consideration of the plastic flow and irreversible deformations in the assembled materials. The accumulation of irreversible deformations during plastic flow and subsequent unloading and cooling is significantly affect to the formation of the residual stresses and the final tightness of assembled part.

IOP Conf. Series: Journal of Physics: Conf. Series **991** (2018) 012060 doi:10.1088/1742-6596/991/1/012060

The solutions of the boundary value problem of stresses computations in circular compounds during the hot fitting process taking into account the plastic properties of the material was considered in [10–13]. The present study deals with a new solution obtained in the frameworks of plastic flow theory and under conditions of complete plasticity and repeated plastic flows during the cooling of the assembled material.

1. Boundary value problem statement. Thermoelastic equilibrium

Consider the boundary value problem in the frameworks of the thermal stresses theory during an assembly of two infinitely long hollow cylinders made from the same thermoelastoplastic material. At the referential time $t = 0$ the inner circular cylinder with inner and outer radii R_0 and R_1 respectively is heated under referential temperature T_1. The outer circular cylinder under temperature $T_2 > T_1$ with inner and outer radii R_1 and R_2 respectively. The referential displacements in the material of the cylinders are assumed to be zero. The inner cold cylinder is inserted into the heated outer cylinder. The stresses are accumulated providing tightness in the assembled parts as a result of the thermal conductivity process through the contact surface $r = R_1$. It is necessary to calculate the temperature field arising from heat conduction to determine the stress-strain states in the cylinders during the fitting. The heat conduction equation in the cylindrical coordinate system under conditions of axial symmetry can read by

$$r\dot{T} = \kappa(T_{,r} + rT_{,rr}). \tag{1}$$

wherein κ is the thermal diffusivity, overdot denotes the time derivative, index after the comma denotes partial derivative with respect to the spatial coordinate. We numerically integrate equation (1) under boundary and initial conditions

$$\begin{aligned} &T(r,0) = \begin{cases} T_1 & R_0 \le r \le R_1, \\ T_2 & R_1 < r \le R_2, \end{cases} \\ &T_{,r}(R_0,t) = 0, \\ &T_{,r}(R_2,t) = 0. \end{aligned} \tag{2}$$

Fig. 1 illustrates the temperature field distribution at different times in the assembled cylinders. Note that the conditions for an ideal thermal contact between the connected cylinders, i.e. the equality of temperatures and heat fluxes at the surface $r = R_1$ are automatically satisfied in the numerical solution of the heat equation (1) under boundary and initial conditions (2).

Deformations d_{ij} in the cylinders material are assumed infinitesimal and are separated into reversible part e_{ij} and irreversible p_{ij} one additively by formula

$$d_{ij} = e_{ij} + p_{ij}, \quad d_{rr} = u_{r,r}, \quad d_{\varphi\varphi} = r^{-1}u_r. \tag{3}$$

The stresses are fully determined by reversible deformations according to the Duhamel–Neumann law [14–16]

$$\begin{aligned} \sigma_{rr} &= (\lambda + 2\mu)e_{rr} + \lambda(e_{\varphi\varphi} + e_{zz}) - (3\lambda + 2\mu)\Delta, \\ \sigma_{\varphi\varphi} &= (\lambda + 2\mu)e_{\varphi\varphi} + \lambda(e_{rr} + e_{zz}) - (3\lambda + 2\mu)\Delta, \\ \sigma_{zz} &= (\lambda + 2\mu)e_{zz} + \lambda(e_{\varphi\varphi} + e_{rr}) - (3\lambda + 2\mu)\Delta. \end{aligned} \tag{4}$$

Here λ, μ are the Lame parameters. The thermal expansion function Δ for considered problem can be furnished by

$$\Delta(r,0) = \alpha(T(r,t) - T(r,0)) = \begin{cases} \alpha(T(r,t) - T_1) & R_0 \le r \le R_1, \\ \alpha(T(r,t) - T_2) & R_1 < r \le R_2. \end{cases} \tag{5}$$

IOP Conf. Series: Journal of Physics: Conf. Series **991** (2018) 012060 doi:10.1088/1742-6596/991/1/012060

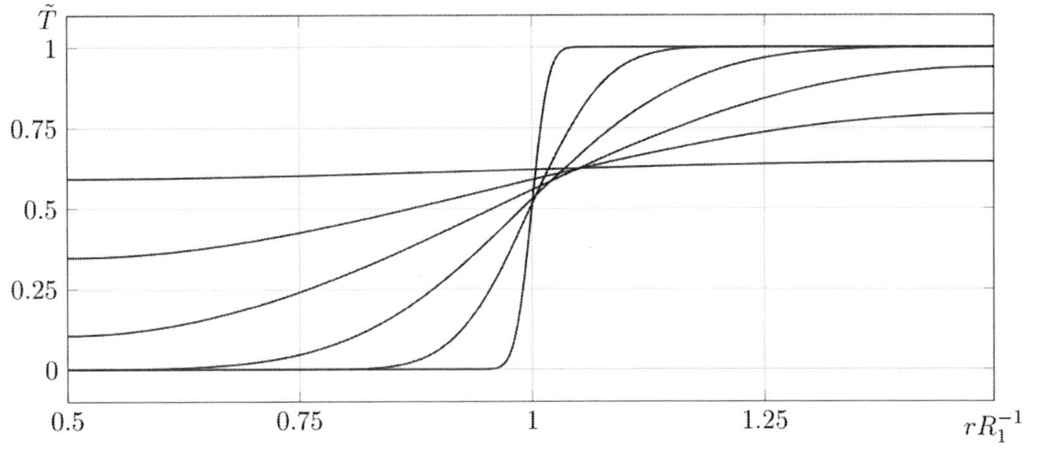

Figure 1. Temperature field in the assembled cylinders. $\tilde{T} = (T - T_1)(T_2 - T_1)^{-1}$.

Equation (5) gives the zero deformations and discplacements at time of the cylinder connection (fitting time) $t = 0$.

The equilibrium equation in the cylindrical coordinate system under conditions of axial symmetry reads as

$$\sigma_{rr,r} + r^{-1}(\sigma_{rr} - \sigma_{\varphi\varphi}) = 0. \tag{6}$$

The solution of the coupling equations (1), (4), (3), (6) under conditions (2), (5) and $p_{ij} = 0$ describe the thermoelastic equilibrium of assembled cylinders.

$$\begin{aligned}
\sigma_{rr}^{(i)} &= -\frac{2\omega}{r^2}\int_{R_0}^{r}\Delta(\rho,t)\rho d\rho + A_i(t) + \frac{B_i(t)}{r^2},\\
\sigma_{\varphi\varphi}^{(i)} &= \frac{2\omega}{r^2}\int_{R_0}^{r}\Delta(\rho,t)\rho d\rho + A_i(t) - \frac{B_i(t)}{r^2} - 2\omega\Delta(r,t),\\
\sigma_{zz}^{(i)} &= \frac{\lambda A_i(t)}{(\lambda+\mu)} - 2\omega\Delta(r,t).
\end{aligned} \tag{7}$$

Index (i) denote the inner (1) or outer (2) cylinders. Unknown time dependent functions $A_i(t)$, $B_i(t)$ can be derived from the system of boundary conditions

$$\begin{aligned}
&\sigma_{rr}^{(1)}(R_0,t) = 0, \quad \sigma_{rr}^{(1)}(R_1,t) = \sigma_{rr}^{(2)}(R_1,t),\\
&\sigma_{rr}^{(2)}(R_2,t) = 0, \quad u_r^{(1)}(R_1,t) = u_r^{(2)}(R_1,t).
\end{aligned} \tag{8}$$

Finally the solution of equations (8) can be obtained in the following form

$$A_i = \frac{2\omega}{(R_2^2 - R_0^2)}\int_{R_0}^{R_2}\Delta(\rho,t)\rho d\rho, \quad B_i = -\frac{2\omega R_0^2}{(R_2^2 - R_0^2)}\int_{R_0}^{R_2}\Delta(\rho,t)\rho d\rho. \tag{9}$$

2. Plastic flow in the outer cylinder

The plastic flow occurs on the contact surface of the outer cylinder and is propagated with the temperature field aligning inside the assembled material. We note that in some cases under a sufficiently high initial temperature decreasing a plastic flow can occur on the contact surface of the inner cylinder, but numerous simulations have shown that the level of plastic deformation in these cases is small in compared with deformations arising from the contact pressure and does not cause a noticeable influence on the distribution of residual stresses and deformations.

Let the time $t = t_p^{(2)}$ is the time of the plastic flow beginning on the inner surface of the outer cylinder. The plastic potential in the form of Tresca yield criterion is more preferable for considered problem

$$\sigma_{\varphi\varphi}^{(2)} - \sigma_{rr}^{(2)} = 2k(r,t). \tag{10}$$

At a time $t > t_p^{(2)}$ the plastic flow domain $R_1 < r < a_2(t)$ in the outer cylinder, where $a_2(t)$ denotes the elastic plastic border. Equation for stresses computation can be obtained by integration of the equilibrium equation (6) under condition (10):

$$\begin{aligned}
\sigma_{rr}^{(2)} &= 2\int_{R_0}^{r} \frac{k(\rho,t)}{\rho} d\rho + C_2(t),\\
\sigma_{\varphi\varphi}^{(2)} &= 2\int_{R_0}^{r} \frac{k(\rho,t)}{\rho} d\rho + C_2(t) + 2k(r,t),\\
\sigma_{zz}^{(2)} &= \frac{\lambda}{(\lambda+\mu)}\left(C_2(t) + 2\int_{R_0}^{r} \frac{k(\rho,t)}{\rho} d\rho + k(r,t)\right) - \gamma\Delta(r,t).
\end{aligned} \tag{11}$$

Note that the lower limit of integration in the equations (7), (11) which valid for the outer cylinder is the inner radius ($R_0 < R_2$) of the inner cylinder. For this case the functions $A_i(t)$, $B_i(t)$, $C_i(t)$... have a simplest form [17–20].

Equations for radial displacement and plastic deformation in the plastic flow domain $R_1 < r < a_2(t)$ read by

$$\begin{aligned}
u_r^{(2)} &= \frac{C_2(t)r}{2(\lambda+\mu)} + \frac{D_2(t)}{r} + \frac{r}{(\lambda+\mu)}\int_{R_0}^{r} \frac{k(\rho,t)}{\rho} d\rho + \frac{\gamma}{\mu r}\int_{R_0}^{r} \Delta(\rho,t)\rho d\rho,\\
p_{rr}^{(2)} &= \frac{(\lambda+2\mu)k(r,t)}{2\mu(\lambda+\mu)} + \frac{\gamma\Delta(r,t)}{2\mu} - \frac{\gamma}{\mu r^2}\int_{R_0}^{r} \Delta(\rho,t)\rho d\rho - \frac{D_2(t)}{r^2},\\
p_{zz}^{(2)} &= 0, \quad p_{\varphi\varphi}^{(2)} = -p_{rr}^{(2)}.
\end{aligned} \tag{12}$$

Unknown time dependent functions in virtue of the condition (8) and continuity conditions for radial stresses and displacement on the elastic plastic border a_2. Hereafter solutions for these functions are not shown because of their cumbersomeness. The position of the elastic plastic border is calculated by a numerical solution of equation $p_{rr}^{(2)}(a_2,t) = 0$, for a certain time $t > t_p^{(2)}$.

3. Plastic flow in the inner cylinder

The further temperature field rearranging in the assembled cylinders can lead to plastic flow on the inner surface of the inner cylinder at time $t = t_p^{(1)}$ [21–24]. Then the stresses keep the following form of the Tresca yield criterion [17, 18, 21]:

$$\sigma_{rr}^{(1)} - \sigma_{zz}^{(1)} = 2k(r,t). \tag{13}$$

Consequently for a time $t > t_p^{(1)}$ it is possible a plastic flow domain development $R_0 < a_1(t)$ in the material of the inner cylinder. The elastic domain is separated from the domain of irreversible deformation by the elastic plastic border $a_1(t)$. We can derived the plastic incompressibility conditions from the plastic flow rule associated with the yield criterion (13)

$$p_{zz} + p_{rr} = 0, \quad p_{\varphi\varphi} = 0; \tag{14}$$

The relations between the radial components of the reversible and irreversible strain tensors and the radial displacement are obtained from equations (3), (4), (14), (13) in following form

$$e_{rr} = \frac{1}{2}\left(u_{r,r} + \frac{k}{\mu}\right), \quad p_{rr} = \frac{1}{2}\left(u_{r,r} - \frac{k}{\mu}\right), \tag{15}$$

Then the differential equations for the radial displacement furnishes taking into account equilibrium equation (6) and equation (15) as follows

$$(ru_{r,r})_{,r} - \frac{\eta^2 u_r}{r} + \frac{(rk)_{,r}}{(\lambda+\mu)} - \frac{\gamma r \Delta_{,r}}{\mu} = 0, \quad \eta = \sqrt{\frac{(\lambda+2\mu)}{(\lambda+\mu)}}. \tag{16}$$

The function of the radial displacement is the solution of the equation (16) for the plastic flow domain

$$\begin{gathered} u_r = \frac{\gamma}{2\mu\eta}\left(\frac{(\eta+1)}{r^\eta}\int_{R_1}^{r}\rho^\eta\Delta(\rho,t)d\rho + (\eta-1)r^\eta\int_{R_1}^{r}\frac{\Delta(\rho,t)}{\rho^\eta}d\rho\right) - \\ -\frac{1}{2(\lambda+\mu)}\left(r^\eta\int_{R_1}^{r}\frac{k(\rho,t)}{\rho^\eta}d\rho + \frac{1}{r^\eta}\int_{R_1}^{r}\rho^\eta k(\rho,t)d\rho\right) + C(t)r^\eta + \frac{D(t)}{r^\eta}. \end{gathered} \tag{17}$$

The plastic deformation then is found by virtue of equations (15) as

$$\begin{gathered} p_{rr} = \frac{\eta}{4(\lambda+\mu)}\left(\frac{1}{r^{(\eta+1)}}\int_{R_1}^{r}\rho^\eta k(\rho,t)d\rho - r^{(\eta-1)}\int_{R_1}^{r}\frac{k(\rho,t)}{\rho^\eta}d\rho\right) + \frac{\gamma\Delta(r,t)}{2\mu} + \\ +\frac{\gamma}{4\mu}\left(r^{(\eta-1)}\int_{R_1}^{r}\frac{\Delta(\rho,t)}{\rho^\eta}d\rho - \frac{(\eta+1)}{r^{(\eta+1)}}\int_{R_1}^{r}\rho^\eta\Delta(\rho,t)d\rho\right) - \frac{\eta^2 k(r,t)}{2\mu} + \\ +\frac{\eta r^{(\eta-1)}C(t)}{2} - \frac{\eta D(t)}{2r^{(\eta+1)}}, \\ p_{zz} = -p_{rr}, \quad p_{\varphi\varphi} = 0. \end{gathered} \tag{18}$$

Let derive the equations for stresses (4) in the plastic flow domain by using equations (15), (17) and (18)

$$\begin{gathered} \sigma_{rr} = \sigma_{zz} + 2k(r,t) = \nu_1 r^{(\eta-1)}C(t) - \frac{\nu_2 D(t)}{r^{(\eta+1)}} - \\ -\frac{1}{2(\lambda+\mu)}\left(\nu_1 r^{(\eta-1)}\int_{R_1}^{r}\frac{k(\rho,t)}{\rho^\eta}d\rho - \frac{\nu_2}{r^{(\eta+1)}}\int_{R_1}^{r}\rho^\eta k(\rho,t)d\rho\right) + \\ +\frac{\gamma}{2\eta\mu}\left((\eta-1)\nu_1 r^{(\eta-1)}\int_{R_1}^{r}\frac{\Delta(\rho,t)}{\rho^\eta}d\rho - \frac{(\eta+1)\nu_2}{r^{(\eta+1)}}\int_{R_1}^{r}\rho^\eta\Delta(\rho,t)d\rho\right), \\ \nu_1 = (\eta\lambda + \lambda + \eta\mu), \quad \nu_1 = (\eta\lambda - \lambda + \eta\mu), \\ \sigma_{\varphi\varphi} = \upsilon_1 r^{(\eta-1)}C(t) + \frac{\upsilon_2 D(t)}{r^{(\eta+1)}} - \gamma\Delta(r,t) - \frac{\lambda k(r,t)}{(\lambda+\mu)} - \\ -\frac{1}{2(\lambda+\mu)}\left(\upsilon_1 r^{(\eta-1)}\int_{R_1}^{r}\frac{k(\rho,t)}{\rho^\eta}d\rho + \frac{\upsilon_2}{r^{(\eta+1)}}\int_{R_1}^{r}\rho^\eta k(\rho,t)d\rho\right) + \\ +\frac{\gamma}{2\eta\mu}\left((\eta-1)\upsilon_1 r^{(\eta-1)}\int_{R_1}^{r}\frac{\Delta(\rho,t)}{\rho^\eta}d\rho + \frac{(\eta+1)\upsilon_2}{r^{(\eta+1)}}\int_{R_1}^{r}\rho^\eta\Delta(\rho,t)d\rho\right), \\ \upsilon_1 = (\lambda + \eta\lambda + 2\mu), \quad \upsilon_2 = (\lambda - \eta\lambda + 2\mu). \end{gathered} \tag{19}$$

4. Complete plasticity in the inner cylinder

Note also that there is the possibility of the complete plasticity state in the frameworks of Tresca yield criterion [21, 22, 25, 26]. For this case we have two valid plastic flow conditions

$$\begin{cases} \sigma_{rr}^{(1)} - \sigma_{zz}^{(1)} = 2k(r,t), \\ \sigma_{rr}^{(1)} - \sigma_{\varphi\varphi}^{(1)} = 2k(r,t). \end{cases} \tag{20}$$

The stresses in the complete plasticity domain ($R_0 < r < b_1(t)$) are furnished by following relations

$$\begin{aligned} \sigma_{rr}^{(1)} &= -2\int_{R_1}^{r} \frac{k(\rho,t)}{\rho} d\rho + E(t), \\ \sigma_{\varphi\varphi}^{(1)} &= \sigma_{zz}^{(1)} - 2\int_{R_1}^{r} \frac{k(\rho,t)}{\rho} d\rho - 2k(r,t) + E(t). \end{aligned} \tag{21}$$

The plastic incompressibility condition $p_{rr} + p_{zz} + p_{\varphi\varphi} = 0$ implies the coupling between the total and reversible deformations as follows

$$u_{r,t} + \frac{u_r^{(1)}}{r} = e_{rr} + e_{\varphi\varphi} + e_{zz}. \tag{22}$$

Substitute the elastic deformation according to (4) and (21) in equation (22) and integrate obtained equation in the complete plasticity domain

$$\begin{aligned} u_r^{(1)} = &-\frac{1}{(3\lambda+2\mu)}\left(\frac{1}{r}\int_{R_1}^{r} k(\rho,t)\rho d\rho + 3r\int_{R_1}^{r} \frac{k(\rho,t)}{\rho} d\rho - \frac{3rE(t)}{2}\right) + \\ &+\frac{F(t)}{r} + \frac{3}{r}\int_{R_1}^{r} \Delta(\rho,t)\rho d\rho. \end{aligned} \tag{23}$$

Plastic deformations functions taking into account equation (23) in the domain ($R_1 < r < b_1(t)$) reads by

$$\begin{aligned} p_{rr}^{(1)} = &\frac{1}{(3\lambda+2\mu)}\left(\frac{1}{r^2}\int_{R_1}^{r} k(\rho,t)\rho d\rho - \int_{R_1}^{r} \frac{k(\rho,t)}{\rho} d\rho + \frac{E(t)}{2}\right) - \\ &-\frac{2k(r,t)}{\gamma} - \frac{3}{r^2}\int_{R_1}^{r} \Delta(\rho,t)\rho d\rho + 2\Delta(r,t) - \frac{F(t)}{r^2}, \\ p_{\varphi\varphi}^{(1)} = &-\frac{1}{(3\lambda+2\mu)}\left(\frac{1}{r^2}\int_{R_1}^{r} k(\rho,t)\rho d\rho + \int_{R_1}^{r} \frac{k(\rho,t)}{\rho} d\rho - \frac{E(t)}{2}\right) + \\ &+\frac{k(r,t)}{\gamma} + \frac{3}{r^2}\int_{R_1}^{r} \Delta(\rho,t)\rho d\rho - \Delta(r,t) + \frac{F(t)}{r^2}, \\ p_{zz}^{(1)} = &\frac{1}{(3\lambda+2\mu)}\left(2\int_{R_1}^{r} \frac{k(\rho,t)}{\rho} d\rho - E(t)\right) + \frac{k(r,t)}{\gamma} - \Delta(r,t). \end{aligned} \tag{24}$$

The satisfaction of the present Tresca yield criterion forms (10), (13) and (20) depends on the cylinder size and referential temperature gradient. Fig. 2 illustrates the thermal stresses in the assembled cylinders after full temperature equalization with the complete plasticity domain in the inner cylinder.

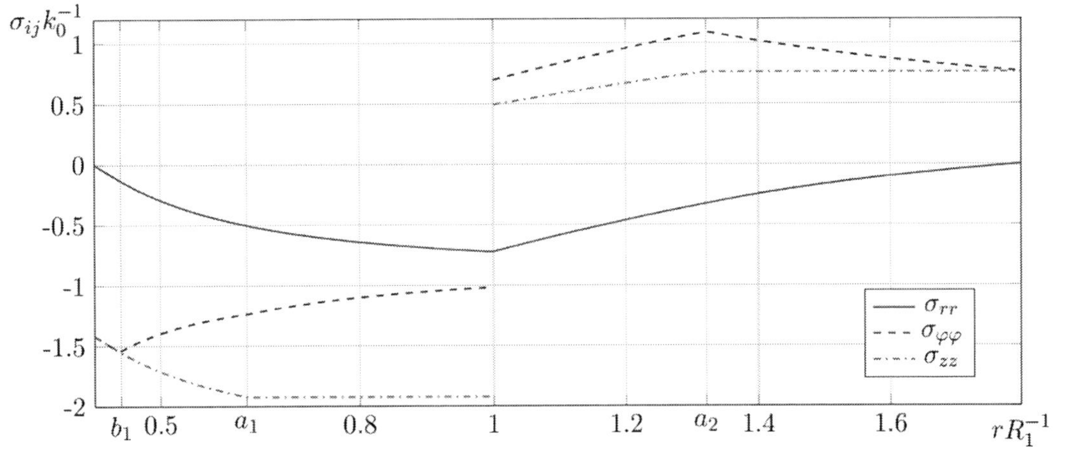

Figure 2. Thermal stresses in the composite cylinder under plastic flow. $R_0R_1^{-1} = 0.4$, $R_2R_1^{-1} = 1.8$.

5. Cooling to the referential temperature

Let consider the cooling process of the cylinders. We assume that the heat conduction with the external continuum is slow and the temperature field in the material of the cylinders corresponds to a uniform distribution slowly decreasing to the referential temperature of the inner cylinder. Thus the compressive stress $\sigma_{zz}^{(1)}$ during the cooling should decrease. Along with this, the value of the yield stress in the material of both cylinders increases. Consequently, the unloading occurs in the material of the inner cylinder during cooling (similar cases were discussed in [27–30]). The stress strain state of the outer cylinder due to increasing yield stress is also satisfies the unloading conditions for a certain temperature diapason. However a repeated plastic flow causing can begin as a cause of a high stress level $\sigma_z^{(2)}z$ corresponding to a stretching of the outer cylinder along the z axis.

Introduce notations $p_{ij}^{(1)}$ and $p_{ij}^{(2)}$ for the irreversible deformations of the cylinders computing at the time of final temperature equalization. These functions are vanished in elastic domains [31, 32]. Differential equation for the radial components of the stress tensor can be obtained from equations (4), (3), (6) in the form

$$\frac{\tau}{r}\left(p_{rr}^{(i)} - p_{\varphi\varphi}^{(i)} - rp_{\varphi\varphi,r}^{(i)}\right) - \frac{2\mu\lambda p_{zz,r}^{(i)}}{(\lambda + 2\mu)} - 2\omega\Delta_{,r} - 3\sigma_{rr,r}^{(i)} - r\sigma_{rr,rr}^{(i)} = 0, \qquad (25)$$

where $\tau = \dfrac{4\mu(\lambda + \mu)}{(\lambda + 2\mu)}$.

Integration of the equation (25) under condition $p_{zz}^{(1)} = -(p_{rr}^{(1)} + p_{\varphi\varphi}^{(1)})$, allows us to express the stresses in the inner cylinder by formulae

$$\sigma_{rr}^{(1)} = \frac{\tau}{2}\left(\int_{R_0}^{r} \frac{p_{rr}^{(1)}(\rho)}{\rho}d\rho - \int_{R_0}^{r} \frac{p_{\varphi\varphi}^{(1)}(\rho)}{\rho}d\rho\right) - \frac{2\omega}{r^2}\int_{R_0}^{r} \Delta(\rho,t)\rho d\rho -$$
$$-\frac{2\mu^2}{(\lambda+2\mu)r^2}\left(\int_{R_0}^{r} p_{rr}^{(1)}(\rho)\rho d\rho + \int_{R_0}^{r} p_{\varphi\varphi}^{(1)}(\rho)\rho d\rho\right) + X_1(t) + \frac{Y_1(t)}{r^2},$$

$$\sigma_{\varphi\varphi}^{(1)} = \frac{\tau}{2}\left(\int_{R_0}^{r} \frac{p_{rr}^{(1)}(\rho)}{\rho}d\rho - \int_{R_0}^{r} \frac{p_{\varphi\varphi}^{(1)}(\rho)}{\rho}d\rho\right) + \frac{2\omega}{r^2}\int_{R_0}^{r} \Delta(\rho,t)\rho d\rho +$$
$$+\frac{2\mu^2}{(\lambda+2\mu)r^2}\left(\int_{R_0}^{r} p_{rr}^{(1)}(\rho)\rho d\rho + \int_{R_0}^{r} p_{\varphi\varphi}^{(1)}(\rho)\rho d\rho\right) + X_1(t) - \frac{Y_1(t)}{r^2} + \quad (26)$$
$$+\frac{2\lambda\mu p_{rr}^{(1)}(r)}{(\lambda+2\mu)} - 2\omega\Delta(r,t) - 2\mu p_{\varphi\varphi}^{(1)}(r),$$

$$\sigma_{zz}^{(1)} = \frac{2\lambda\mu}{(\lambda+2\mu)}\left(\int_{R_0}^{r} \frac{p_{rr}^{(1)}(\rho)}{\rho}d\rho - \int_{R_0}^{r} \frac{p_{\varphi\varphi}^{(1)}(\rho)}{\rho}d\rho\right) + \tau p_{rr}^{(1)}(r) +$$

$$+2\mu p_{\varphi\varphi}^{(1)} + \frac{\lambda X_1(t)}{(\lambda+\mu)} - 2\omega\Delta(r,t).$$

The radial displacement in this case is calculated as

$$u_r^{(1)} = \frac{\omega}{\mu r}\int_{R_0}^{r} \Delta(\rho,t)\rho d\rho + \frac{rX_1(t)}{2(\lambda+\mu)} - \frac{Y_1(t)}{2\mu r} + \frac{\mu r}{(\lambda+2\mu)}\left(\int_{R_0}^{r} \frac{p_{rr}^{(1)}(\rho)}{\rho}d\rho -\right.$$
$$\left. - \int_{R_0}^{r} \frac{p_{\varphi\varphi}^{(1)}(\rho)}{\rho}d\rho + \frac{1}{r^2}\int_{R_0}^{r} p_{rr}^{(1)}(\rho)\rho d\rho + \frac{1}{r^2}\int_{R_0}^{r} p_{\varphi\varphi}^{(1)}(\rho)\rho d\rho\right). \quad (27)$$

The stresses in the outer cylinder in virtue of conditions $p_{rr}^{(2)} = -p_{\varphi\varphi}^{(2)}$ and $p_{zz}^{(2)} = 0$ reads by

$$\sigma_{rr}^{(2)} = \tau\int_{R_1}^{r} \frac{p_{rr}^{(2)}(\rho)}{\rho}d\rho - \frac{2\omega}{r^2}\int_{R_0}^{r} \Delta(\rho,t)\rho d\rho + X_2(t) + \frac{Y_2(t)}{r^2},$$

$$\sigma_{\varphi\varphi}^{(2)} = \tau\left(\int_{R_1}^{r} \frac{p_{rr}^{(2)}(\rho)}{\rho}d\rho + p_{rr}^{(2)}(r)\right) + X_2(t) -$$

$$-\frac{2\omega}{r^2}\left(\int_{R_0}^{r} \Delta(\rho,t)\rho d\rho - r^2\Delta(r,t)\right) - \frac{Y_2(t)}{r^2}, \quad (28)$$

$$\sigma_{zz}^{(2)} = \frac{2\lambda\mu}{(\lambda+2\mu)}\left(2\int_{R_1}^{r} \frac{p_{rr}^{(2)}(\rho)}{\rho}d\rho + p_{rr}^{(2)}(r)\right) + \frac{\lambda X_2(t)}{(\lambda+\mu)} - 2\omega\Delta(r,t).$$

The displacement in the outer cylinder in the time is satisfied to following equation

$$u_{rr}^{(2)} = \frac{2\mu r}{(\lambda+2\mu)}\int_{R_1}^{r} \frac{p_{rr}^{(2)}(\rho)}{\rho}d\rho + \frac{\omega}{r}\int_{R_0}^{r} \Delta(\rho,t)\rho d\rho + \frac{rX_2(t)}{2(\lambda+\mu)} - \frac{Y_2(t)}{2\mu r}. \quad (29)$$

Unknown time dependent functions $X_i(t)$, $Y_i(t)$ can be found by boundary conditions (8).

IOP Conf. Series: Journal of Physics: Conf. Series **991** (2018) 012060 doi:10.1088/1742-6596/991/1/012060

6. Repeated plastic flow

The stress strain state at a time $t = t_r$ can rich the yield surface on the inner surface of the outer cylinder as the assembled construction is cooling down. In this case Tresca yield criterion is valid in the following form

$$\sigma_{zz}^{(2)} - \sigma_{rr}^{(2)} = 2k. \tag{30}$$

From this time $t > t_r$ the repeated plastic flow domain is propagated in the outer cylinder ($R_1 < r < b_2(t)$). This repeated irreversible deformations are caused by cylinder stretching under cooling [23, 28].

To compute the stress-strain state parameters in this domain it is necessary to take into account the accumulated repeated irreversible deformations $p_{rr}^{(2)}$. Thus the equilibrium equation in terms of displacement is furnished by

$$(ru_{r,r})_{,r} - \frac{\eta^2 u_r}{r} - \frac{(rk + rp_{rr}^{(2)})_{,r}}{(\lambda+\mu)} - \frac{\gamma r \Delta_{,r}}{\mu} = 0, \tag{31}$$

We can solve the equation (31) and rewrite the displacement function by

$$\begin{aligned}
\tilde{u}_r = {} & \frac{\gamma}{2\mu\eta}\left(\frac{(\eta+1)}{r^\eta}\int_{R_1}^{r}\rho^\eta\Delta(\rho,t)d\rho + (\eta-1)r^\eta\int_{R_1}^{r}\frac{\Delta(\rho,t)}{\rho^\eta}d\rho\right) - \\
& + \frac{1}{2(\lambda+\mu)}\left(r^\eta\int_{R_1}^{r}\frac{k(\rho,t)}{\rho^\eta}d\rho + \frac{1}{r^\eta}\int_{R_1}^{r}\rho^\eta k(\rho,t)d\rho\right) + \\
& + \frac{\mu}{2(\lambda+\mu)}\left(r^\eta\frac{(\eta+2)}{\eta}\int_{R_1}^{r}\frac{p_{rr}^{(2)}(\rho)}{\rho^\eta}d\rho + \frac{1}{r^\eta}\frac{(\eta-2)}{\eta}\int_{R_1}^{r}\rho^\eta p_{rr}^{(2)}(\rho)d\rho\right) + \\
& + P_2(t)r^\eta + \frac{Q_2(t)}{r^\eta}.
\end{aligned} \tag{32}$$

Hereafter the tilde denote the functions determining in the repeated plastic flow domain. The plastic deformations by virtue of the solution (32) are calculated by formula

$$\begin{aligned}
\tilde{p}_{rr} = {} & \frac{\eta}{4(\lambda+\mu)}\left(r^{(\eta-1)}\int_{R_1}^{r}\frac{k(\rho,t)}{\rho^\eta}d\rho - \frac{1}{r^{(\eta+1)}}\int_{R_1}^{r}\rho^\eta k(\rho,t)d\rho\right) + \frac{\gamma\Delta(r,t)}{2\mu} + \\
& + \frac{\gamma}{4\mu}\left(r^{(\eta-1)}\int_{R_1}^{r}\frac{\Delta(\rho,t)}{\rho^\eta}d\rho - \frac{(\eta+1)}{r^{(\eta+1)}}\int_{R_1}^{r}\rho^\eta\Delta(\rho,t)d\rho\right) + \frac{\eta^2 k(r,t)}{2\mu} + \\
& + \frac{\mu\eta}{4(\lambda+\mu)}\left(r^{(\eta-1)}\frac{(\eta+2)}{\eta}\int_{R_1}^{r}\frac{p_{rr}^{(2)}(\rho)}{\rho^\eta}d\rho - \frac{1}{r^{(\eta+1)}}\frac{(\eta+2)}{\eta}\int_{R_1}^{r}\rho^\eta p_{rr}^{(2)}(\rho)d\rho\right) + \\
& + \frac{\eta r^{(\eta-1)}P_2(t)}{2} - \frac{\eta Q_2(t)}{2r^{(\eta+1)}} + \frac{\eta^2 p_{rr}^{(2)}(r)}{2},
\end{aligned} \tag{33}$$

$$\tilde{p}_{zz} = -\tilde{p}_{rr}(r,t) + p_{rr}^{(2)}(r), \quad \tilde{p}_{\varphi\varphi} = -p_{rr}^{(2)}(r).$$

The components of the stress tensor are derived by the known displacement (32) and plastic deformations (33) as

$$\begin{aligned}
&\tilde{\sigma}_{rr} = -\tilde{\sigma}_{zz} - 2k(r,t) = \nu_1 r^{(\eta-1)} P_2(t) - \frac{\nu_2 Q_2(t)}{r^{(\eta+1)}} - \\
&+\frac{1}{2(\lambda+\mu)}\left(\nu_1 r^{(\eta-1)}\int_{R_1}^{r}\frac{k(\rho,t)}{\rho^\eta}d\rho - \frac{\nu_2}{r^{(\eta+1)}}\int_{R_1}^{r}\rho^\eta k(\rho,t)d\rho\right)+ \\
&+\frac{\gamma}{2\eta\mu}\left((\eta-1)\nu_1 r^{(\eta-1)}\int_{R_1}^{r}\frac{\Delta(\rho,t)}{\rho^\eta}d\rho - \frac{(\eta+1)\nu_2}{r^{(\eta+1)}}\int_{R_1}^{r}\rho^\eta\Delta(\rho,t)d\rho\right)+ \\
&+\frac{\mu}{2(\lambda+\mu)}\left(\nu_1 r^{(\eta-1)}\frac{(\eta+2)}{\eta}\int_{R_1}^{r}\frac{p_{rr}^{(2)}(\rho)}{\rho^\eta}d\rho - \frac{\nu_2}{r^{(\eta+1)}}\frac{(\eta-2)}{\eta}\int_{R_1}^{r}\rho^\eta p_{rr}^{(2)}(\rho)d\rho\right), \\
&\nu_1 = (\eta\lambda+\lambda+\eta\mu), \quad \nu_1 = (\eta\lambda-\lambda+\eta\mu), \\
&\tilde{\sigma}_{\varphi\varphi} = \upsilon_1 r^{(\eta-1)} P_2(t) + \frac{\upsilon_2 Q_2(t)}{r^{(\eta+1)}} - \gamma\Delta(r,t) + \frac{\lambda k(r,t)}{(\lambda+\mu)} + \gamma p_{rr}^{(2)}(r) + \\
&+\frac{1}{2(\lambda+\mu)}\left(\upsilon_1 r^{(\eta-1)}\int_{R_1}^{r}\frac{k(\rho,t)}{\rho^\eta}d\rho + \frac{\upsilon_2}{r^{(\eta+1)}}\int_{R_1}^{r}\rho^\eta k(\rho,t)d\rho\right)+ \\
&\frac{\mu}{2(\lambda+\mu)}\left(\nu_1 r^{(\eta-1)}(\eta+2)\int_{R_1}^{r}\frac{p_{rr}^{(2)}(\rho)}{\rho^\eta}d\rho + \frac{\nu_2}{r^{(\eta+1)}}(\eta-2)\int_{R_1}^{r}\rho^\eta p_{rr}^{(2)}(\rho)d\rho\right)+ \\
&+\frac{\gamma}{2\eta\mu}\left((\eta-1)\upsilon_1 r^{(\eta-1)}\int_{R_1}^{r}\frac{\Delta(\rho,t)}{\rho^\eta}d\rho + \frac{(\eta+1)\upsilon_2}{r^{(\eta+1)}}\int_{R_1}^{r}\rho^\eta\Delta(\rho,t)d\rho\right), \\
&\upsilon_1 = (\lambda+\eta\lambda+2\mu), \quad \upsilon_2 = (\lambda-\eta\lambda+2\mu).
\end{aligned} \tag{34}$$

Unknown time dependent function $X_i(t)$, $Y_i(t)$, $P_2(t)$, $Q_2(t)$ can be found from the linear system of the boundary conditions (8) and continuity conditions of the radial stress and displacement at the boundary of the repeated plastic flow domain $b_2(t)$. The location of this border is numerically calculated as the solution of the equation $\tilde{p}_{rr}(b_2,t) = p_{rr}^{(2)}(r)$. The repeated plastic flow is propagated until the full cooling assembled cylinders to the referential temperature. After that the stress strain state riches the neutral loading state. On the Figure 3 the residual stresses are shown at the time of full cooling down. The neutral loading state as we can see on the Fig. 3 occupies the domain $R_1 < r < b_2$ with valid yield criterion $\tilde{\sigma}_{zz} - \tilde{\sigma}_{rr} = 2k_0$ and constant irreversible deformations.

Concluding remarks

The sequence of the considered boundary value problems may be violated by the cylinder size, material properties and referential temperature gradient. The extremal influence on the final contact pressure is induced by the irreversible deformation rate under yield criterion in form $\sigma_{zz} - \sigma_{rr} = 2k$. The dependence of the yield stress on temperature makes it possible to predict lower values of the contact pressure contrary to studies without the plastic properties accounting or with the constant yield stress. Therefore, to increase the value of the contact pressure it is necessary to reduce the initial temperatures in both parts of the assembled construction. Obviously, in this case the yield stress have the greatest value and, consequently, the level of the contact pressure is higher.

Acknowledgments

The authors would like to thank Professor A. Burenin for his helpful comments, suggestions and encouragement. The present study was financially supported by RSF (project No. 17-19-01257).

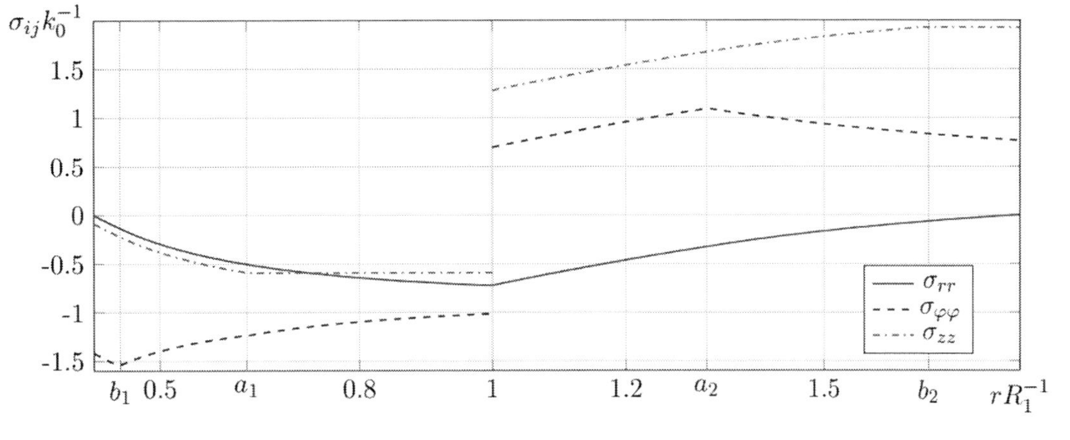

Figure 3. Thermal stresses in composite cylinder under cooling. $R_0R_1^{-1} = 0.4$, $R_2R_1^{-1} = 1.8$.

References

[1] Manzhirov A 1985 *Journal of Applied Mathematics and Mechanics* **49** 777–782
[2] Manzhirov A 1995 *Journal of Applied Mathematics and Mechanics* **59** 805–816
[3] Manzhirov A and Lychev S 2012 *Doklady Physics* **57** 160–163
[4] Manzhirov A V 2013 *Key Engineering Materials* **535** 89–93
[5] Manzhirov A 2015 *Lecture Notes in Engineering and Computer Science* **2218(1.1)** 710–714
[6] Joubert S, Shatalov M and Manzhirov A 2013 *Journal of Sound and Vibration* **332** 6169–6176
[7] Kuznetsov S, Manzhirov A and Fedotov I 2011 *Mech. Solids* **46**(6) 929–936
[8] Lychev S, Manzhirov A and Joubert S 2010 *Mechanics of solids* **45** 610–623
[9] Burenin A A, Dats E P and Tkacheva A V 2014 *Journal of Applied and Industrial Mathematics* **8** 493–499
[10] Dats E P and Tkacheva A V 2016 *Journal of Applied Mechanics and Technical Physics* **57** 569–576
[11] Lippmann H 1992 *Int. J. of Plasticity* **8** 567–582
[12] Mack W 1993 *Arch. of App. Mech.* **63** 42–50
[13] Kovacs A 1996 *Periodica Polytechnica: Mech. Eng.* **40** 103–112
[14] Boley B and Weiner J 1964 *Theory of Thermal Stresses* (New York: Wiley)
[15] Parkus G 1963 *Unsteady Thermal Stresses* (Moscow: Fizmatgiz)
[16] Odeno H 1969 *Journal Mechnical Engineering Science* **2** 384–391
[17] Orcan Y and Gamer U 1991 *Acta Mechanica* **90** 61–80
[18] Orcan Y 1995 *International Journal of Mechanical Sciences* **33** 1689–1698
[19] Raniecki B 1966 *Rozpr. Inz.* **14** 479
[20] Gamer U 1988 *Journal of Thermal Stresses* **13** 159–173
[21] Bland D 1956 *Journal of the Mechanics and Physics of Solids* **4** 209–229
[22] Dats E P, Murashkin E V and Gupta N K 2017 *Procedia IUTAM* **23** 187–200
[23] Dats E, Mokrin S and Murashkin E 2015 *Lecture Notes in Engineering and Computer Science* **2218** 1043–1046
[24] Dats E, Mokrin S and Murashkin E 2016 *Key Engineering Materials* **685** 37–41
[25] Dats E, Murashkin E and Stadnik N 2017 *Procedia Manufacturing* **7** 427–434
[26] Dats E and Murashkin E 2016 *Lecture Notes in Engineering and Computer Science* **2223** 69–72
[27] Dats E, Murashkin E and Stadnik N 2017 *Procedia Engineering* **173** 891–896
[28] Burenin A A, Dats E P, Mokrin S and Murashkin E V 2013 *Bulletin of Yakovlev Chuvash State Pedagogical University. Series: Mechanics of Limit State* **16** 22–28
[29] Burenin A A, Dats E P and Murashkin E V 2014 *mechanics of Solids* **49** 218–224
[30] Dats E, Murashkin E and Stadnik N 2017 *IAENG International Journal of Applied Mathematics* **47** 261–264
[31] Dats E P, Murashkin E V and Velmurugan R 2015 *Bulletin of Yakovlev Chuvash State Pedagogical University. Series: Mechanics of Limit State* **25** 168–175
[32] Gorshkov S A, Dats E P and Murashkin E V 2015 *Bulletin of Yakovlev Chuvash State Pedagogical University. Series: Mechanics of Limit State* **21** 169–175

IOP Conf. Series: Journal of Physics: Conf. Series **991** (2018) 012061 doi:10.1088/1742-6596/991/1/012061

Full thermomechanical coupling in modelling of micropolar thermoelasticity

E V Murashkin[1,2*], **Y N Radayev**[1,3**]

[1]Ishlinsky Institute for Problems in Mechanics of the Russian Academy of Sciences, Moscow, Russia
[2]Bauman Moscow State Technical University, Moscow, Russia
[3]Department of Energy Conversion Science, Graduate School of Energy Science, Kyoto University, Kyoto, Japan

E-mail: *murashkin@ipmnet.ru; **radayev@ipmnet.ru, radaev.iurii.8e@kyoto-u.ac.jp

Abstract. The present paper is devoted to plane harmonic waves of displacements and microrotations propagating in fully coupled thermoelastic continua. The analysis is carried out in the framework of linear conventional thermoelastic micropolar continuum model. The reduced energy balance equation and the special form of the Helmholtz free energy are discussed. The constitutive constants providing fully coupling of equations of motion and heat conduction are considered. The dispersion equation is derived and analysed in the form bi-cubic and bi-quadratic polynoms product. The equation are analyzed by the computer algebra system *Mathematica*. Algebraic forms expressed by complex multivalued square and cubic radicals are obtained for wavenumbers of transverse and longitudinal waves. The exact forms of wavenumbers of a plane harmonic coupled thermoelastic waves are computed.

Preliminaries & backgrounds

The concept of micropolar continua take its origin from the classical E. and F. Cosserat's study [1]. Micropolar continuum theories include not only translational displacements but also microrotations. Contrary to conventional elasticity a micropolar continuum is described by the asymmetric strain and stress tensors known from many previous discussions and surveys [2–7]. Thus the asymmetric elastic theory is characterized by a comparatively large number of constitutive elastic constants need to be determined from the experimental observations. By now there is no fully coupled model of micropolar thermoelastic behaviour taking accounts of torsion–bending processes of heat conduction.

The conventional thermoelastic micropolar continuum may be described in terms of field formalism, for example, from positions of the Green–Naghdi thermoelasticity [8, 9]. They are based on different modifications of the classical Fourier law of heat conduction. The refinements aim at derivations of *hyperbolic* partial differential equations of coupled thermoelasticity. Those are to simultaneously fulfill the following conditions:

(i) Finiteness of the heat signal propagation velocity;
(ii) The ability of the spatial propagation of the thermoelastic waves without attenuation;
(iii) Existence of distortionless wave forms akin to the classical d'Alembert type waves.

CTEM theory is based on the classical Fourier law of heat conduction with an infinite velocity of propagation of an exponentially decaying heat signal. Hyperbolic thermoelasticity theory is characterised by the energy conservation and the finite propagation velocity of thermal waves known as *second sound waves*.

In [4, 10–13] a non-linear mathematical model of thermoelastic micropolar continuum has been presented in terms of 4-covariant field theoretical formalism. In-depth study of weak and strong discontinuities in macropolar thermoelastic continua is given in [14–19]. In [20] problems concerning plane harmonic wavenumbers of coupled type-III thermoelastic waves are discussed. In [4] the linear symmetrical thermoelasticity is employed. A non-linear mathematical model of thermoelastic micropolar continuum is developed.

1. Requisites equations of micropolar thermoelasticity

A linear thermoelastic micropolar continuum model (see [21] for details) is used throughout the paper. For such continuum, equations of motion of micropolar media are written in direct tensor representation for the case of the absence of mass forces and mass moments

$$\begin{cases} \boldsymbol{\nabla} \cdot \boldsymbol{\sigma} = \rho \ddot{\mathbf{u}}, \\ \boldsymbol{\nabla} \cdot \boldsymbol{m} + \boldsymbol{\epsilon} \cdot\cdot \boldsymbol{\sigma} = \mathfrak{J} \ddot{\boldsymbol{\phi}}. \end{cases} \tag{1}$$

where $\boldsymbol{\nabla}$ is the three-dimensional Hamiltonian differential operator (the nabla symbol), ρ is the mass density, $\mathfrak{J}$ is a scalar dynamic characteristic of continuum (the rotational inertia), superimposed dot denotes partial differentiation with respect to time at fixed spatial coordinates, $\mathbf{u}$ is the translational displacement vector and $\boldsymbol{\phi}$ rotation vector, $\boldsymbol{\epsilon}$ is the three-dimensional Levi–Civita symbol (permutation symbol, antisymmetric symbol, or alternating symbol).

Displacements $\mathbf{u}$ and microrotations $\boldsymbol{\phi}$ are associated with asymmetric strain tensor $\boldsymbol{e}$ and bending–torsion tensor $\boldsymbol{\Gamma}$ by formulae

$$\begin{aligned} \boldsymbol{e} &= \boldsymbol{\nabla} \otimes \mathbf{u} - \boldsymbol{\epsilon} \cdot \boldsymbol{\phi}, \\ \boldsymbol{\Gamma} &= \boldsymbol{\nabla} \otimes \boldsymbol{\phi}, \end{aligned} \tag{2}$$

The equation (2) are represented in a rectangular co-ordinate system as

$$\begin{aligned} e_{ji} &= \partial_j u_i - \epsilon_{jik} \phi_k, \\ \Gamma_{ji} &= \partial_j \phi_i. \end{aligned} \tag{3}$$

The reduced energy balance equation can be furnished by

$$-(\dot{\psi} + s\dot{\theta}) + \operatorname{tr}(\boldsymbol{\sigma} \cdot \dot{\boldsymbol{e}}) + \operatorname{tr}(\boldsymbol{m} \cdot \dot{\boldsymbol{\Gamma}}) - \mathbf{h} \cdot \frac{\boldsymbol{\nabla}\theta}{\theta} = \theta\xi \tag{4}$$

where $\psi = \psi(\boldsymbol{e}, \boldsymbol{\Gamma}, \theta)$ is the Helmholtz free energy per unit of volume, s denotes the entropy referred to the unit of volume, $\mathbf{h}$ is the heat flux vector, θ is the actual temperature, ξ denotes the internal entropy product.

According to the postulate of the thermodynamics of irreversible processes the equation and inequality is valid

$$\xi = -\mathbf{h} \cdot \frac{\boldsymbol{\nabla}\theta}{\theta^2} \geq 0. \tag{5}$$

The inequality (5) is satisfied by using the Fourier's law of heat conduction, which states the proportionality of the heat flux vector $\mathbf{h}$ and negative spatial temperature gradient θ

$$\mathbf{h} = -\Lambda_* \boldsymbol{\nabla}\theta, \tag{6}$$

where Λ_* is the thermal conductivity coefficient (thermal diffusion coefficient), $\Lambda_* > 0$.

The following equation may be derived by transforming (4)

$$\dot{\psi} = \sigma_{ji}\dot{e}_{ji} + \mu_{ji}\dot{\Gamma}_{ji} - \dot{\theta}s. \tag{7}$$

Since the free energy is the function of the independent variables Γ_{ji}, e_{ji}, θ then

$$\dot{\psi} = \frac{\partial\psi}{\partial e_{ji}}\dot{e}_{ji} + \frac{\partial\psi}{\partial\Gamma_{ji}}\dot{\Gamma}_{ji} + \frac{\partial\psi}{\partial\theta}\dot{\theta}. \tag{8}$$

Thus following relations are valid

$$\sigma_{ji} = \frac{\partial\psi}{\partial e_{ji}}, \qquad \mu_{ji} = \frac{\partial\psi}{\partial\Gamma_{ji}}, \qquad s = -\frac{\partial\psi}{\partial\theta}. \tag{9}$$

Let us expand the free energy ψ into the Taylor series in the vicinity of the natural state $\boldsymbol{e} = \boldsymbol{\Gamma} = \boldsymbol{0}$, $\theta = 0$, disregarding the terms of higher order than the second one. The following form of the expansion is obtained for isotropic, homogeneous and centrosymmetric bodies

$$\begin{aligned}\psi = \frac{\mu+\eta}{2}e_{ji}e_{ji} + \frac{\mu-\eta}{2}e_{ji}e_{ij} + \frac{\lambda}{2}e_{kk}e_{nn} + \frac{\gamma+\varepsilon}{2}\Gamma_{ji}\Gamma_{ji} + \frac{\gamma-\varepsilon}{2}\Gamma_{ji}\Gamma_{ij} + \\ + \frac{\beta}{2}\Gamma_{kk}\Gamma_{nn} - \alpha e_{kk}\theta - \varsigma\Gamma_{kk}\theta - \frac{\Lambda_*}{2}\theta^2.\end{aligned} \tag{10}$$

Asymmetric tensors of the force stresses $\boldsymbol{\sigma}$ and the moment stresses $\boldsymbol{m}$ are determined according to the constitutive law of GNI/-thermoelasticity

$$\begin{aligned}\boldsymbol{\sigma} &= (\mu+\eta)\boldsymbol{e} + (\mu-\eta)\boldsymbol{e}^{\mathrm{T}} + (\lambda\mathrm{tr}\boldsymbol{e} - \alpha\theta)\mathbf{I},\\ \boldsymbol{m} &= (\gamma+\varepsilon)\boldsymbol{\Gamma} + (\gamma-\varepsilon)\boldsymbol{\Gamma}^{\mathrm{T}} + (\beta\mathrm{tr}\boldsymbol{\Gamma} - \varsigma\theta)\mathbf{I},\\ s &= \alpha\mathrm{tr}\boldsymbol{e} + \varsigma\mathrm{tr}\boldsymbol{\Gamma} + \Lambda_*\theta.\end{aligned} \tag{11}$$

wherein θ denotes the temperature increment over the referential temperature; $\lambda, \mu, \eta, \gamma, \beta, \varepsilon$ are isothermal constitutive constants of type-I micropolar thermoelastic continuum; α, ς are constitutive constants providing coupling of equations of motion and heat conduction. Constants α, ς depend not only on the mechanical properties of the continuum, but also depend on the thermal properties.

Dynamic equations (1) should be supplemented by heat conduction equation. After substituting the stress tensors $\boldsymbol{\sigma}$ and $\boldsymbol{m}$ from the formulas (11) in the equations of motion (1) and taking account of (2) the system of coupled partial differential equations of motion and heat conduction for a linear isotropic type-I micropolar thermoelastic continuum in the absence of mass forces, moments, and heat sources can be written as [?, 21]:

$$\begin{cases}(\lambda+\mu-\eta)\nabla\nabla\cdot\mathbf{u} + (\mu+\eta)\nabla\cdot\nabla\mathbf{u} + 2\eta\nabla\times\boldsymbol{\phi} - \alpha\nabla\theta - \rho\ddot{\mathbf{u}} = \mathbf{0},\\ (\beta+\gamma-\varepsilon)\nabla\nabla\cdot\boldsymbol{\phi} + (\gamma+\varepsilon)\nabla\cdot\nabla\boldsymbol{\phi} - 4\eta\boldsymbol{\phi} + 2\eta\nabla\times\mathbf{u} - \varsigma\nabla\theta - \mathfrak{J}\ddot{\boldsymbol{\phi}} = \mathbf{0},\\ \nabla^2\theta - \frac{\kappa}{\Lambda_*}\dot{\theta} - \frac{\alpha}{\Lambda_*}\nabla\cdot\dot{\mathbf{u}} - \frac{\varsigma}{\Lambda_*}\nabla\cdot\dot{\boldsymbol{\phi}} = 0.\end{cases} \tag{12}$$

Hereafter κ is the heat capacity (per unit volume) at constant (zero) strains.

As pointed out the nonzero constitutive constants α, ς provide coupling of micropolar thermoelasticity equations. It is usually assumed that $\varsigma = 0$ (see for example [21]). This constitutive constant is kept in all further considerations.

The scalar equation in the system (12) is called a generalized heat conduction equation conjugate to the equations of motion (the first and the second equations in (12)).

IOP Conf. Series: Journal of Physics: Conf. Series **991** (2018) 012061 doi:10.1088/1742-6596/991/1/012061

2. Plane harmonic waves propagation

Plane harmonic coupled micropolar thermoelastic wave manifests a fairly simple analytical structure:

$$\begin{aligned} \mathbf{u} &= \mathbf{A}e^{i(\mathbf{k}\cdot\mathbf{r}-\omega t)}, \\ \boldsymbol{\phi} &= \mathbf{S}e^{i(\mathbf{k}\cdot\mathbf{r}-\omega t)}, \\ \theta &= Be^{i(\mathbf{k}\cdot\mathbf{r}-\omega t)}, \end{aligned} \tag{13}$$

where $\mathbf{k}$ is a wave vector, ω is a cyclic frequency; $\mathbf{A}$, $\mathbf{S}$ are the polarization wave vectors; B is an amplitude of the deviation of temperature from the referential temperature. $\mathbf{A}$, $\mathbf{S}$, B are the values of weak discontinuities of displacements, microrotations and temperature, respectively. Wavenumber k (the modulus of the wave vector $\mathbf{k}$ or the waves propagation constant) can be real as well as complex variable.

Substituting the expression (13) into the differential equation (12), and given that

$$\nabla = i\mathbf{k}, \quad \partial/\partial t = -i\omega,$$

The following coupled system of equations of wave vector $\mathbf{k}$, the cyclic frequency ω, the polarization vectors of the plane wave $\mathbf{A}$, $\mathbf{S}$ and the amplitude B can be obtained:

$$\begin{cases} (\rho\omega^2 - (\mu+\eta)k^2)\mathbf{A} - (\lambda+\mu-\eta)\mathbf{k}(\mathbf{k}\cdot\mathbf{A}) + 2\eta i\mathbf{k}\times\mathbf{S} - \alpha i\mathbf{k}B = \mathbf{0}, \\ (\Im\omega^2 - 4\eta - (\gamma+\varepsilon)k^2)\mathbf{S} - (\beta+\gamma-\varepsilon)\mathbf{k}(\mathbf{k}\cdot\mathbf{S}) + 2\eta i\mathbf{k}\times\mathbf{A} - \varsigma i\mathbf{k}B = \mathbf{0}, \\ \left(\dfrac{\kappa}{\Lambda_*}i\omega - k^2\right)B - \dfrac{\alpha}{\Lambda_*}\omega\mathbf{k}\cdot\mathbf{A} - \dfrac{\varsigma}{\Lambda_*}\omega\mathbf{k}\cdot\mathbf{S} = 0. \end{cases} \tag{14}$$

The dispersion equation for the system (14) furnishes

$$\begin{aligned} &\left((c_\perp^{\mu 2}k^2 - \omega^2)(c_\perp^{\mu\mu 2}k^2 - \omega^2 + \Omega^2) - \frac{\Omega^2 k^2 \eta}{\rho}\right)\times \\ &\times\left((\rho\omega^2 - (\lambda+2\mu)k^2)(\Im\omega^2 - 4\eta - (\beta+2\gamma)k^2)\left(i\frac{\omega}{l_*^2} - k^2\right) - \right. \\ &\left.-\frac{i\omega\alpha^2k^2}{\Lambda_*}(\Im\omega^2 - 4\eta - (\beta+2\gamma)k^2) - \frac{i\varsigma^2k^2\omega}{\Lambda_*}(\rho\omega^2 - (\lambda+2\mu)k^2)\right) = 0, \end{aligned} \tag{15}$$

where

$$\Omega^2 = \frac{4\eta}{\Im}.$$

Hereafter we use following dimensionless notation:
wavenumbers of the transverse waves

$$k_\perp^{\mu\mu} = \frac{\omega}{c_\perp^{\mu\mu}}, \quad k_\perp^{\mu} = \frac{\omega}{c_\perp^{\mu}};$$

ratios of the transverse waves velocities

$$\widetilde{k_\perp}^2 = \frac{c_\perp^{\mu 2}}{c_\perp^{\mu\mu 2}}, \quad \widetilde{k_\perp^{\mu}}^2 = \frac{c_\perp^2}{c_\perp^{\mu 2}};$$

wavenumbers of the longitudinal waves

$$k_\parallel^{\mu\mu} = \frac{\omega}{c_\parallel^{\mu\mu}}, \quad k_\parallel = \frac{\omega}{c_\parallel};$$

ratios of the longitudinal waves velocities

$$\widetilde{k_\parallel} = \frac{k_\parallel^{\mu\mu}}{k_\parallel} = \frac{c_\parallel}{c_\parallel^{\mu\mu}};$$

constants elucidating the coupling of the mechanical and thermal features

$$s_1^2 = \frac{\alpha^2}{\omega \Lambda_* \rho}, \quad s_1^{\varsigma 2} = \frac{\varsigma^2}{\omega \Lambda_* \mathfrak{J}},$$

specific dimensionless parameters

$$\begin{aligned}
&h_1^2 = \left(\frac{\Omega}{\omega}\right)^2, \quad h_2^2 = h_1^2 \widetilde{k_\perp^\mu}^2 - 1 - \widetilde{k_\perp}^2. \\
&\widetilde{k_1}^2 = \widetilde{k_\parallel}^2 (1 - h_1^2), \quad Q_1^2 = \frac{\omega}{l_*^2 k_\parallel^2}, \\
&Q_2^2 = Q_1^2 + \widetilde{k_\parallel}^2 s_1^{\varsigma 2} + s_1^2, \quad Q_3^2 = Q_1^2(\widetilde{k_1}^2 + 1) + \widetilde{k_\parallel}^2 (s_1^{\varsigma 2} + s_1^2(1 - h_1^2)).
\end{aligned}$$

We can solve the dispersion equation separately as the bi-quadratic and bi-cubic equations and obtain the wavenumbers for the transverse waves

$$\sqrt{2}\frac{k_{1,2;3,4}}{k_\perp^\mu} = \pm\sqrt{-h_2^2 \pm \sqrt{h_2^4 - 4(1 - h_1^2)\widetilde{k_\perp}^2}};$$

and for the longitudinal waves

$$\begin{aligned}
&\frac{k_{1,2,3,4}}{k_\parallel} = \pm\sqrt{-\frac{1}{2}(a+b) \pm i\frac{\sqrt{3}}{2}(a-b) + \frac{1 + \widetilde{k_1}^2 + iQ_2^2}{3}}. \\
&\frac{k_{5,6}}{k_\parallel} = \pm\sqrt{a + b + \frac{1 + \widetilde{k_1}^2 + iQ_2^2}{3}}.
\end{aligned} \tag{16}$$

Here we use the following notations

$$\begin{aligned}
&a = \sqrt[3]{-\frac{q}{2} + \sqrt{\frac{q^2}{4} + \frac{p^3}{27}}}, \quad b = \sqrt[3]{-\frac{q}{2} - \sqrt{\frac{q^2}{4} + \frac{p^3}{27}}}, \\
&\mathrm{Re}p = \widetilde{k_1}^2 - \frac{1}{3}\left(\left(1 + \widetilde{k_1}^2\right)^2 - Q_2^4\right), \\
&\mathrm{Im}p = Q_3^2 - \frac{2}{3}Q_2^2\left(1 + \widetilde{k_1}^2\right), \\
&\mathrm{Re}q = \left(-\frac{2}{27}\widetilde{k_1}^4 + \frac{1}{9}\widetilde{k_1}^2 + \frac{1}{9} + \frac{2}{9}Q_2^4\right)\widetilde{k_1}^2 - \frac{2}{27} - \frac{1}{3}Q_3^2 Q_2^2 + \frac{2}{27}Q_2^4, \\
&\mathrm{Im}q = \left(\frac{1}{3}Q_3^2 - Q_1^2 - \frac{1}{9}Q_2^2 - \frac{2}{9}\widetilde{k_1}^2 Q_2^2\right)\widetilde{k_1}^2 - \frac{2}{9}Q_2^2 + \frac{1}{3}Q_3^2 + \frac{2}{27}Q_2^6.
\end{aligned}$$

3. Conclusions

(i) Thermomechanical coupling modelling has been developed within the frameworks of the conventional micropolar thermoelasticity taking account of full thermomechanical coupling.

(ii) The reduced energy balance equation have been discussed.

(iii) Constitutive equations and conservation laws have been derived from balance equations and the thermodynamics principles.

(iv) A plane harmonic coupled thermoelastic displacements and microrotations waves have been discussed.

(v) The dispersion equation have been derived and analysed in the form bi-cubic and bi-quadratic polynoms product.

(vi) Wavenumbers have been obtained and discriminated for a transverse and longitudinal thermoelastic waves.

Acknowledgments
The present study is supported by the Federal Agency for Scientific Organizations (State Registration Number AAAA-A17-117021310381-8) and by the Russian Foundation for Basic Research financially (project No. 18-01-00844).

References

[1] Cosserat E and Cosserat F 1909 *Théorie des corps déformables* (Paris: Librairie Scientifique A. Hermann et Fils)
[2] Felippa C A 1994 *Computer Methods in Applied Mechanics and Engineering* **113** 109–139
[3] Jasiuk I and Ostoja-Starzewski M 2004 *Micromechanics and Nanoscale Effects* (Springer) pp 175–212
[4] Kovalev V A and Radayev Y N 2010 *Wave Problems of Field Theory and Thermomechanics* (Saratov: Saratov University Publisher) (In Russian)
[5] Maugin G A and Metrikine A V 2010 *Advances in mechanics and mathematics* **21**
[6] Altenbach J, Altenbach H and Eremeyev V A 2010 *Archive of Applied Mechanics* **80** 73–92
[7] Hassanpour S and Heppler G R 2017 *Mathematics and Mechanics of Solids* **22** 224–242
[8] Green A E and Naghdi P M 1992 *J. Therm. Stress* **15** 253–264
[9] Green A E and Naghdi P M 1993 *J. Elasticity* **31** 189–208
[10] Kovalev V A and Radayev Y N 2009 *Elements of the Field Theory: Variational Symmetries and Geometric Invariants* (Moscow: FIZMATLIT) (In Russian)
[11] Kovalev V A, Murashkin E V and Radayev Y N 2016 *Mechanics: Proceedings of the International School-Conference of Young Scientists, 3-7 October 2016, Tsakhkadzor, Armenia* (Erevan: National University of Architecture and Construction of Armenia Publishing) pp 160–163
[12] Murashkin E V and Radayev Y N 2017 *On Thermodynamics of Wave Processes of Heat Transport* (Cham: Springer) pp 363–376 (*Advanced Structured Materials* no 46)
[13] Kovalev V A, Murashkin E V and Radayev Y N 2017 *JPCS* **788(1)** 012043
[14] Kovalev V A and Radayev Y N 2012 *Izvestiya of Saratov University. New Series. Series Mathematics. Mechanics. Informatics* **12**(4) 71–79 (In Russian)
[15] Murashkin E V and Radayev Y N 2014 *J. Samara State Tech. Univ., Ser. Phys. Math. Sci.* **4(37)** 85–97 (In Russian)
[16] Murashkin E V and Radayev Y N 2015 *Mat. Phys. and Mech.* **23** 10–13
[17] Kovalev V A, Murashkin E V and Radayev Y N 2015 *Bulletin of the Yakovlev Chuvash State Pedagogical University. Series: Mechanics of Limit State* **3(25)** 6165 (In Russian)
[18] Kovalev V A, Murashkin E V and Radayev Y N 2015 *Izvestiya of Saratov University. New Series. Series Mathematics. Mechanics. Informatics* **15**(1) 7989
[19] Murashkin E V 2016 *Contributions to the Foundations of Multidisciplinary Research in Mechanics. Papers presented during the 24th International Congress of Theoretical and Applied Mechanics ICTAM2016, Montreal 22-26, 2016* ed Floryan J (Montreal, Canada: ICTAM) pp 2271–2272
[20] Kovalev V A, Radayev Y N and Semenov D A 2009 *Izvestiya of Saratov University. New Series. Series Mathematics. Mechanics. Informatics* **9**(5) 94–127 (In Russian)
[21] Nowacki W 1986 *Theory of asymmetric elasticity* (Oxford: Pergamon Press)

IOP Conf. Series: Journal of Physics: Conf. Series **991** (2018) 012062 doi:10.1088/1742-6596/991/1/012062

Light radiation pressure upon a wrinkled membrane – parametrization of an optically orthotropic model

N A Nerovny* and V N Zimin**

Bauman Moscow State Technical University, Moscow, Russia

E-mail: *nick.nerovny@bmstu.ru, **zimin@bmstu.ru

Abstract. In this paper, the problem of representing the light pressure force upon the surface of a thin wrinkled film is discussed. The common source of wrinkles is the shear deformation of the membrane sample. The optical model of such a membrane is assumed to be optically orthotropic and an analytic equation for infinitesimal light pressure force is written. A linear regression model in the case of wrinkle geometry, where a surface element can have different optical parameters, is constructed and the Bayesian approach is used to calculate the parameters of this model.

1. Model of optically orthotropic surface

In this section, we describe the model of light radiation pressure upon a surface with wrinkles. We assume that the optical parameters of such a surface can be represented in as follows: there are two perpendicular directions in which the optical parameters are different.

Let us introduce the local coordinate frame $Ox_1x_2x_3$ on an infinitesimal element dA of the surface whose optical parameters are assumed to be orthotropic. The unit vectors of the local coordinate frame are $\hat{\mathbf{e}}_1$, $\hat{\mathbf{e}}_2$, $\hat{\mathbf{e}}_3$. Let $\hat{\mathbf{n}}$ be a unit vector of local normal to the surface element dA, $\hat{\mathbf{n}} = \hat{\mathbf{e}}_3$. The area element dA is located on the plane Ox_1x_2. For any vector, e.g., $\hat{\mathbf{r}}$, we use the direction angles (β, θ) in the local frame as follows:

- $\beta \in [0, \pi/2]$ is the angle between the vectors $\hat{\mathbf{r}}$ and $+x_3$.
- $\theta \in [0, 2\pi]$ is the angle between the axis Ox_1 and the projection of $\hat{\mathbf{r}}$ on the plane Ox_1x_2, counterclockwise around Ox_3.

Let $\hat{\mathbf{s}}$ be a unit vector from the light source to dA. In the formulas given below, q_0 is a light flux, c is the light speed in a vacuum, σ is the Stephan–Boltzmann constant, and T is the temperature of dA.

Let $\epsilon'_m(\beta, \theta)$ be the directional emissivity in the orthotropic model [1]:

$$\epsilon'_m(\beta, \theta) = [\epsilon_1 \cos^2(\theta - \theta_m) + \epsilon_2 \sin^2(\theta - \theta_m)] \cos \beta,$$

where θ_m is an angle in the plane Ox_1x_2, which determines the direction in which the emissivity is equal to ϵ_1. In the perpendicular direction, the emissivity is equal to ϵ_2.

IOP Conf. Series: Journal of Physics: Conf. Series **991** (2018) 012062 doi:10.1088/1742-6596/991/1/012062

The equation for the emission pressure from the orthotropic surface is given by the formula

$$d\mathbf{F}^S = -\frac{\epsilon_1 B_m \sigma T^4}{c}\hat{\mathbf{n}}, \tag{1}$$

where the modified Lambertian coefficient B_m is

$$B_m = \frac{\epsilon_1 + \epsilon_2}{3\epsilon_1}.$$

The equation for the light pressure of the absorbed light has the form

$$d\mathbf{F}^A = -\frac{q_0}{c}(\hat{\mathbf{n}}\cdot\hat{\mathbf{s}})\hat{\mathbf{s}}\,dA. \tag{2}$$

The bidirectional reflectivity for the diffuse reflection is

$$\rho''_{1m}(\beta,\theta,\beta^R,\theta^R) = 2\pi(1-s)[\rho_1\cos^2(\theta_m-\theta^R) + \rho_2\sin^2(\theta_m-\theta^R)]\cos\beta^R,$$

where β^R, θ^R is the orientation vector for the reflected light flux.
The optically orthotropic model for the specularity s becomes

$$s = s_1\cos^2(\theta_m-\theta^R) + s_2\sin^2(\theta_m-\theta^R).$$

The equation for the light pressure force of the light diffusely reflected from an orthotropic surface can be written in vector form

$$d\mathbf{F}^{R1} = \frac{q_0}{c}\rho(1-s)B_\rho(\hat{\mathbf{n}}\cdot\hat{\mathbf{s}})\hat{\mathbf{n}}, \tag{3}$$

where

$$B_\rho = \frac{(4-3s_1-s_2)\rho_1 + (4-s_1-3s_2)\rho_2}{12(1-s_1)\rho_1}.$$

The bidirectional reflectivity for the specular reflection is

$$\rho''_2 = 2\pi\delta(\beta^R-\beta_0)\delta(\theta^R-\theta_0-\pi)[\rho_1\cos^2(\theta_m-\theta^R)+\rho_2\sin^2(\theta_m-\theta^R)][s_1\cos^2(\theta_m-\theta^R)+s_2\sin^2(\theta_m-\theta^R)].$$

The unit vector of orientation of the reflection axes in local frame is

$$\hat{\mathbf{m}} = \cos\theta_m\hat{\mathbf{e}}_1 + \sin\theta_m\hat{\mathbf{e}}_2.$$

The light radiation pressure upon dA from a specularly reflected light flux is

$$d\mathbf{F}^{R2m} = \frac{q_0}{c}\frac{(\hat{\mathbf{n}}\cdot\hat{\mathbf{s}})\hat{\mathbf{s}} - 2(\hat{\mathbf{n}}\cdot\hat{\mathbf{s}})^2\hat{\mathbf{n}}}{[1-(\hat{\mathbf{n}}\cdot\hat{\mathbf{s}})^2]^2}\{\rho_1(\hat{\mathbf{m}}\cdot\hat{\mathbf{s}})^2 + \rho_2[(\hat{\mathbf{m}}\times\hat{\mathbf{s}})\cdot\hat{\mathbf{n}}]^2\}\{s_1(\hat{\mathbf{m}}\cdot\hat{\mathbf{s}})^2 + s_2[(\hat{\mathbf{m}}\times\hat{\mathbf{s}})\cdot\hat{\mathbf{n}}]^2\}. \tag{4}$$

The bidirectional reflectivity for the back reflection is

$$\rho''_{2b} = 2\pi\delta(\beta^R-\beta_0)\delta(\theta^R-\theta_0-\pi)[\rho_1\cos^2(\theta_m-\theta^R)+\rho_2\sin^2(\theta_m-\theta^R)]\frac{s_1\cos^2(\theta_m-\theta^R)+s_2\sin^2(\theta_m-\theta^R)}{\cos\beta^R}k,$$

where k is an empirical parameter.
The light pressure from the back reflection is

$$d\mathbf{F}^{R2b} = \frac{q_0}{c}\frac{k\hat{\mathbf{s}}}{[1-(\hat{\mathbf{n}}\cdot\hat{\mathbf{s}})^2]^2}\{\rho_1(\hat{\mathbf{m}}\cdot\hat{\mathbf{s}})^2 + \rho_2[(\hat{\mathbf{m}}\times\hat{\mathbf{s}})\cdot\hat{\mathbf{n}}]^2\}\{s_1(\hat{\mathbf{m}}\cdot\hat{\mathbf{s}})^2 + s_2[(\hat{\mathbf{m}}\times\hat{\mathbf{s}})\cdot\hat{\mathbf{n}}]^2\}. \tag{5}$$

IOP Conf. Series: Journal of Physics: Conf. Series **991** (2018) 012062 doi:10.1088/1742-6596/991/1/012062

Table 1. Summary of the posterior distribution for different simulations for surface with waves with different optical parameters [1].

Case	ρ_1	ρ_2	s_1	s_2	k
$\rho = 0.5,\ s = 0$	0.041	0.121	0.806	0.524	3.480
$\rho = 0.5,\ s = 0.5$	0.057	0.074	0.218	0.448	4.450
$\rho = 0.5,\ s = 1$	0.047	0.211	0.094	0.296	2.330
$\rho = 1,\ s = 0$	0.104	0.369	0.677	0.331	2.440
$\rho = 1,\ s = 0.5$	0.291	0.476	0.179	0.222	1.870
$\rho = 1,\ s = 1$	0.749	0.729	0.803	0.822	0.186

2. Parametrization of optical parameters

As has already been shown by many authors, both analytically [2, 3], numerically [4–6], and experimentally [7], the surface of a wrinkled thin membrane can be represented according to the sine or cosine law. The common source of wrinkles in practical applications is the shear deformation of the membrane. Another source is the uniaxial stretching. The magnitude of wrinkles can be calculated by the formula [2]

$$x'_3 = \delta \sin \frac{2\pi x'_2}{\lambda},$$

where $\delta = \delta(x'_1, x'_2)$ is a smooth field of wrinkle amplitude over the wrinkled domain $\mathcal{D}$ and $\lambda = \lambda(x'_1, x'_2)$ is a smooth field of wrinkle wavelength over the wrinkled domain $\mathcal{D}$.

In [1], we used the following parameters of the wrinkled surface: $x'_1,\ x'_2 \in [-0.5, 0.5]$, $x'_3 = \delta[\cos(8\pi x'_2) - 1]$. There were four waves of wavelength $\lambda = \text{const} = 0.25\,\text{m}$, and amplitude $\delta = const = 0.2\,\text{m}$, and the global frame was shifted along $+x'_3$ by $\delta/2$. The optical parameters of this surface were uniform and could be specular, diffuse or specular-diffuse with the corresponding specularity coefficient: $\rho_0 \in \{0.5, 1\}$ is the surface reflectivity, $s_0 \in \{0, 0.5, 1\}$ is the surface specularity, $B_0 = 2/3$ is the Lambertian coefficient of the front side; there are six combinations in the whole. We calculated the light radiation pressure vector for 189 combinations of the incident light flux and obtained parameterized values for the optical parameters using the Bayesian approach by Markov Chain Monte Carlo simulation [8] based on Metropolis-Hastings sampling [9] for statistical toolkit JAGS [10] for *R* programming language. The summary of the posterior distribution for optical parameters is represented in table 1.

In this paper, we introduce the following linear regression model for optical parameters of initially optically homogeneous wrinkled surface:

$$\begin{aligned} \rho_1 &\sim A_1\rho + A_2 s + A_3, \\ \rho_2 &\sim B_1\rho + B_2 s + B_3, \\ s_1 &\sim C_1\rho + C_2 s + C_3, \\ s_2 &\sim D_1\rho + D_2 s + D_3, \\ k &\sim E_1\rho + E_2 s + E_3. \end{aligned}$$

For this model, the inverse link function is a normal distribution with corresponding standard deviation σ_A, σ_B, σ_C, σ_D, and σ_E.

The prior distribution for all parameters of the model is a normal distribution with mean 0 and deviation 1. For standard deviation of the link function, the prior distribution is exponential with parameter 0.1.

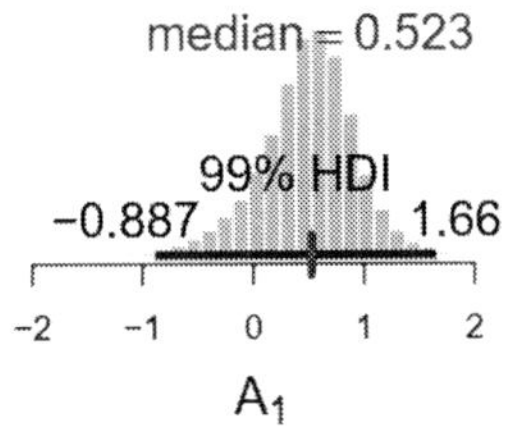

Figure 1. Posterior distribution for the parameter A_1 of linear regression model.

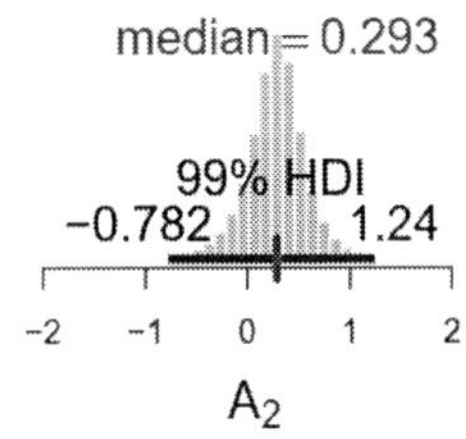

Figure 2. Posterior distribution for the parameter A_2 of linear regression model.

Table 2. Summary of posterior distribution for linear regression model of parameters of optically orthotropic surface. Limits are 99% Higher Density Interval.

Parameter	Value	Note
A_1	$0.523^{+1.137}_{-1.410}$	figure 1
A_2	$0.293^{+0.947}_{-1.075}$	figure 2
A_3	$-0.319^{+1.202}_{-0.971}$	figure 3
σ_A	$0.260^{+0.830}_{-0.179}$	figure 4
B_1	$0.726^{+0.764}_{-1.098}$	
B_2	$0.217^{+0.656}_{-0.684}$	
B_3	$-0.322^{+0.908}_{-0.665}$	
σ_B	$0.134^{+0.541}_{-0.095}$	
C_1	$0.311^{+1.529}_{-1.601}$	
C_2	$-0.216^{+1.396}_{-1.264}$	
C_3	$0.324^{+1.306}_{-1.374}$	
σ_C	$0.443^{+1.107}_{-0.289}$	
D_1	$0.096^{+1.374}_{-1.326}$	
D_2	$0.131^{+1.089}_{-1.181}$	
D_3	$0.299^{+1.171}_{-1.162}$	
σ_D	$0.330^{+0.890}_{-0.224}$	
E_1	$0.424^{+2.256}_{-2.204}$	
E_2	$0.085^{+2.346}_{-2.175}$	
E_3	$1.200^{+2.100}_{-2.440}$	
σ_E	$2.090^{+4.200}_{-1.407}$	

We calculated the posterior distribution using the previously mentioned MCMC approximation. The summary of this approximation is represented in table 2. The examples of posterior distribution for the parameters A_1, A_2, A_3, and σ_A are represented in figures 1–4, where HDI stands for the higher probability density interval, the distribution median is represented by a short vertical red line. The exact values of the probability density function are irrelevant since we use only relative probabilities and likelihoods.

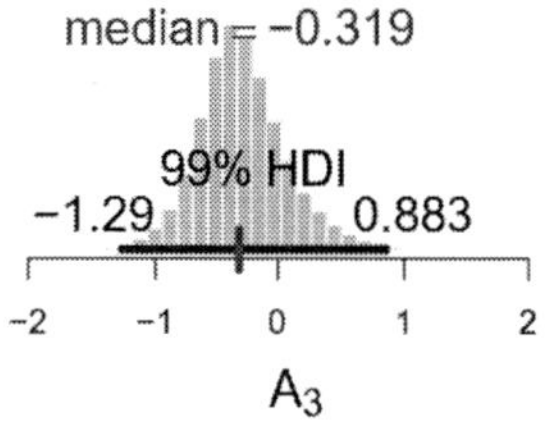

Figure 3. Posterior distribution for the parameter A_3 of linear regression model.

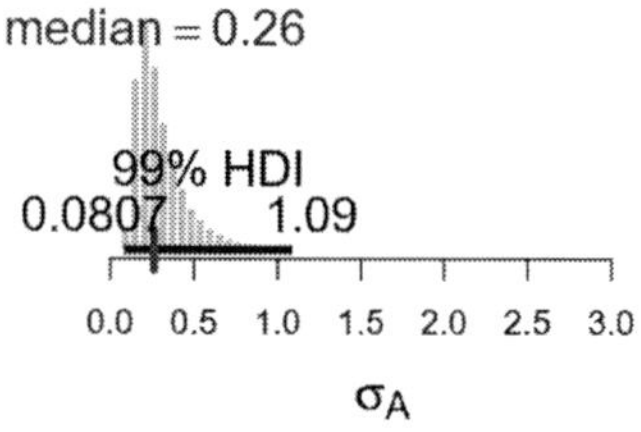

Figure 4. Posterior distribution for the parameter σ_A of linear regression model.

Colclusions

We presented an optically orthotropic model of optical parameters for the surface of thin film structure with wrinkles. The variation of wrinkle parameters leads to the variation of parameters of reflected flux. This dependence can be parametrized using linear regression for an optically orthotropic surface parameters.

References

[1] Nerovny N, Lapina I, and Grigorjev A 2017 Light radiation pressure upon an optically orthotropic surface *J. Quant. Spectrosc. Radiat. Transf.* **202** 64–73

[2] Epstein M 2003 Differential equation for the amplitude of wrinkles *AIAA J.* **41** (2) 327–9

[3] Wong W and Pellegrino S 2006 Wrinkled membranes II: analytical models *J. Mech. Mater. Struct.* **1** (1) 27–61

[4] Wong W and Pellegrino S 2006 Wrinkled membranes III: numerical simulations *J. Mech. Mater. Struct.* **1** (1) 63–95

[5] Wang C G, Du X W, Tan H F, and He X D 2009 A new computational method for wrinkling analysis of gossamer space structures *Int. J. Solids Struct.* **46** (6) 1516–26

[6] Xiao W-w, Chen W-j, and Fu G-y 2011 Wrinkle analysis of the space inflatable paraboloid antenna *J. Shanghai Jiaotong Univ. (Science)* **16** (1) 24–33

[7] Wong W and Pellegrino S 2006 Wrinkled Membranes I: Experiments *J. Mech. Mater. Struct.* **1** (1) 3–25

[8] Gilks W R, Richardson S, and Spiegelhalter D 1995 *Markov Chain Monte Carlo in Practice* (London: CRC Press)

[9] Chib S and Greenberg E 1995 Understanding the Metropolis-Hastings Algorithm *The American Statistician* **49** 327–35

[10] Plummer M 2013 *JAGS Version 3.4.0 user manual* URL `http://www.stats.ox.ac.uk/nicholls/MScMCMC15/jags_user_manual.pdf`

IOP Conf. Series: Journal of Physics: Conf. Series **991** (2018) 012063 doi:10.1088/1742-6596/991/1/012063

The mechanical problems on additive manufacturing of viscoelastic solids with integral conditions on a surface increasing in the growth process

D A Parshin[1,2*] **and A V Manzhirov**[1,2**]

[1]Institute for Problems in Mechanics of the Russian Academy of Sciences, Moscow, Russia
[2]Bauman Moscow State Technical University, Moscow, Russia

E-mail: *parshin@ipmnet.ru, **manzh@inbox.ru

Abstract. Quasistatic mechanical problems on additive manufacturing aging viscoelastic solids are investigated. The processes of piecewise-continuous accretion of such solids are considered. The consideration is carried out in the framework of linear mechanics of growing solids. A theorem about commutativity of the integration over an arbitrary surface increasing in the solid growing process and the time-derived integral operator of viscoelasticity with a limit depending on the solid point is proved. This theorem provides an efficient way to construct on the basis of Saint-Venant principle solutions of nonclassical boundary-value problems for describing the mechanical behaviour of additively formed solids with integral satisfaction of boundary conditions on the surfaces expanding due to the additional material influx to the formed solid. The constructed solutions will retrace the evolution of the stress-strain state of the solids under consideration during and after the processes of their additive formation. An example of applying the proved theorem is given.

1. Introduction

One can find a wide variety of processes of solids additive formation in nature and technology. Many of them can be considered as continuous accretion — when an infinitely thin layer of additional material joins to the surface of the solid being formed every infinitely small time period, — or as piecewise-continuous accretion — when the stages of continuous accretion alternate with pauses during which the additional material does not attach to the solid. In the course of additive processes different factors influence on solids being formed and cause their deformation. The traditional in solid mechanics kinematical description of deformation processes is not suitable for accreted solids. However, as particles of the additional material after its attaching to the accreted solid surface move in the composition of a solid continuum, the enough smooth field of the material particles velocities is uniquely determined in the time-varying space region occupied by the whole growing solid at the current time instant. Therefore, development of the stress-strain state of accreted solids can be described in terms of rates of stress-strain characteristics. Such approach requires knowing the whole history of changing the state of the additional material elements up to their inclusion in the composition of the considered solid.

In the proposed study we follow the indicated approach and consider the situation when the solid being additively formed exhibits the properties of deformation heredity (viscoelasticity) and aging (weakening the deformation properties over time regardless stresses existing in the solid), and therefore, during pauses in the accreting process as well as after the final cessation of accretion the solid continues to change its stress-strain state. This situation is quite difficult to simulate as rheological manifestations in the deformation response of the material continuously interact with mechanical reactions of the solid on the process of adding new material elements to it running in time.

Remark that different approaches to description of mechanical behavior of additively formed solids deforming in processes of their continuous or piecewise-continuous accretion are being actively developed nowadays in the framework of growing solids mechanics [1–12].

2. Defining relations of the material used

We consider a homogeneously aging isotropic linearly viscoelastic material with the common and constant Poisson's ratio $\nu = \text{const}$ for elastic and creep strain. The material is described by the following equation of state [13, 14]:

$$\mathcal{H}_{\tau_0(\mathbf{r})}\mathbf{T}(\mathbf{r},t) = 2\mathbf{E}(\mathbf{r},t) + \frac{2\nu}{1-2\nu}\mathbf{1}\,\mathrm{tr}\,\mathbf{E}(\mathbf{r},t) \tag{1}$$

were $\mathbf{T}$ and $\mathbf{E}$ are the stress and the small strain tensors at the point $\mathbf{r}$ at the time instant t; $\mathbf{1}$ is the unit tensor of the second rank. The linear time-operator $\mathcal{H}_s$ with the real parameter $s \geqslant 0$ is defined by the expressions

$$\begin{gathered}\mathcal{H}_s f(t) = \frac{f(t)}{G(t)} - \int_s^t \frac{f(\tau)}{G(\tau)} K(t,\tau)\,d\tau,\\ K(t,\tau) = G(\tau)\,\frac{\partial\Delta(t,\tau)}{\partial\tau}, \quad \Delta(t,\tau) = \frac{1}{G(\tau)} + \omega(t,\tau).\end{gathered} \tag{2}$$

Here $G(t)$ is the elastic shear modulus; $K(t,\tau)$, $\Delta(t,\tau)$, and $\omega(t,\tau)$ are respectively the kernel of creep, the specific strain function, and the creep measure for pure shear ($t \geqslant \tau \geqslant 0$). It is accepted by definition that $\omega(\tau,\tau) \equiv 0$.

The quantity $\tau_0(\mathbf{r})$ is the time instant at which nonzero stresses appear at the point corresponding to the radius-vector $\mathbf{r}$ in the solid under consideration. The start of timing t is taken be the moment of the material nucleation.

The material defining relations written out above were developed originally for the description of creep processes in concretes. But they are also well suited to describe the analogous processes in rocks, polymers, soils, ice.

3. The investigated problems

The state equation (1) is used in the present work to describe the mechanical behaviour of *growing solids* which are built up additively by attaching the additional material to the current solid surface. We suppose that the additional material is being loaded in such accreting process simultaneously with its attaching to the solid. In this case, the function $\tau_0(\mathbf{r})$ in (1) is to be determined in the following way.

The *original part* V_0 of the growing solid was initially formed stress-free and loaded then before the accreting process start. In the original part the function $\tau_0(\mathbf{r})$ should be identically equal to the time moment t_0 of loading of this part. In the formed due to accreting process *additional part* V_{A} of the solid the function $\tau_0(\mathbf{r})$ should coincide with the distribution $\tau_*(\mathbf{r})$ of moments of attaching particles $\mathbf{r}$ of the additional material to the solid. So,

$$\tau_0(\mathbf{r}) = \begin{cases} t_0, & \mathbf{r} \in V_0,\\ \tau_*(\mathbf{r}), & \mathbf{r} \in V_{\mathrm{A}}.\end{cases} \tag{3}$$

If the accreting process starts at the time instant $t = t_1$ then, obviously, $\tau_*(\mathbf{r}) \geqslant t_1 \geqslant t_0$.

We assume that the considered process of adding the material to the solid can be adequately modelled as a process of *piecewise-continuous accretion*. This means that the process consists of N *stages of continuous accretion* $t \in [t_{2k-1}, t_{2k})$, $k = 1, \ldots, N$, during which an infinitely thin layer of the additional material adheres to the growing solid surface every infinitely small time period. Before the first stage of continuous accretion when $t \in [t_0, t_1)$, between the stages when $t \in [t_{2k}, t_{2k+1})$, $k = 1, \ldots, N-1$, and after the last stage when $t \in [t_{2N}, +\infty)$, the additional material influx is absent, the solid does not grow. The part $\Sigma(t)$ of the growing solid surface to which the additional material continuously inflows at the current time instant t during any stage of continuous accretion is named the *(current) growth surface*. It is clear that the growth surface $\Sigma(t)$ is the t-level surface of the function $\tau_*(\mathbf{r})$.

We investigate mechanical problems for growing solids in quasistatic statement and in the approximation of small strains and displacements. The latter let us consider the time-variable space domain $V(t) = V_0 \cup V_*(t)$ occupied with the whole growing solid to the current time instant t to be known at any time instant and prescribed by the specific simulated growing process; here the domain $V_*(t) \subseteq V_{\mathrm{A}}$ is the piece of the additional part already formed to the time instant t. So the growth surface $\Sigma(t)$ of the growing solid moves in the space in a known manner, and its motion forms the domain V_{A}.

We denote for the notation conciseness

$$g^\circ(\mathbf{r}, t) = \mathcal{H}_{\tau_0(\mathbf{r})}\, g(\mathbf{r}, t) \tag{4}$$

for arbitrary function $g(\mathbf{r}, t)$ of solid point $\mathbf{r}$ and time t, and

$$h^\circ(t) = \mathcal{H}_{t_0}\, h(t) \tag{5}$$

for arbitrary function of time $h(t)$ which is not associated with specific points of considered solid.

4. Theorem about commutativity of the time-derived integral operator of viscoelasticity and the integration over an arbitrary surface increasing due to growing the solid

Let the mechanical problem to be solved for a solid growing in the above described manner contain integral conditions (e.g., integral conditions for the force or the moment of force value) on some parts of the solid surface expanding due to the influx of additional material. We can prove the following proposition giving an efficient way to solve the problem in question on the basis of Saint-Venant principle.

Theorem. *Let Ω_0 and Ω_{A} be two arbitrary surfaces inside or on the boundary of a solid subordinated to the state equation* (1) *and formed in a process of piecewise-continuous accretion in N stages of continuous growth $t \in [t_{2k-1}, t_{2k})$, $k = 1, \ldots, N$, with arbitrary long pauses $t_{2k+1} - t_{2k}$, $k = 1, \ldots, N-1$, between the stages. The surface Ω_0 lies entirely within the original (existing before accreting) part V_0 of the solid considered. The surface Ω_{A} lies entirely in the additional (formed in the accreting process) part V_{A} of the solid and is obtained by motion in space of an arbitrary curve $\Gamma(t)$, $t \in [t_1, +\infty)$, which belongs to the current growth surface $\Sigma(t)$ of the solid at every moment t of its continuous accreting and is fixed in the pauses between the stages of continuous accretion and after the accreting process end, i.e. on the time periods $t \in [t_{2k}, t_{2k+1})$, $k = 1, \ldots, N$, where $t_{2N+1} = +\infty$: $\Omega_{\mathrm{A}} = \{\Gamma(t) \subset \Sigma(t) \,|\, t_1 \leqslant t < +\infty\}$. Let $g(\mathbf{r}, t)$ be an arbitrary function defined at the points $\mathbf{r}$ of the both surfaces Ω_0 and Ω_{A} for $t \geqslant \tau_0(\mathbf{r})$. Assume that*

$$g\big(\mathbf{r}, \tau_*(\mathbf{r})\big) \equiv 0, \quad \mathbf{r} \in \Omega_{\mathrm{A}}. \tag{6}$$

Then, when $t > t_1$ the formula

$$\frac{d}{dt}\left[\int_{\Omega(t)} g(\mathbf{r},t)\,dS\right]^{\circ} = \int_{\Omega(t)} \frac{\partial g^{\circ}(\mathbf{r},t)}{\partial t}\,dS \tag{7}$$

will be fair, where the expanding in time (disconnected in general) surface $\Omega(t)$ combines the surface Ω_0 and that part of the surface Ω_{A}, which has already been formed by the time $t \geqslant t_0$:

$$\Omega(t) = \begin{cases} \Omega_0, & t \in [t_0, t_1], \\ \Omega_0 \cup \Omega_*(t), & t \in (t_1, +\infty), \end{cases} \qquad \Omega_*(t) = \big\{\Gamma(\tau)\,\big|\, t_1 \leqslant \tau \leqslant t\big\} \subseteq \Omega_{\mathrm{A}}. \tag{8}$$

We are to emphasize that the surfaces Ω_0 and Ω_{A} considered in the Theorem may have arbitrary curvatures. Meanwhile the boundaries of these surfaces may not have common points. It is also possible that $\Omega_0 = \varnothing$. Forming a surface Ω_{A} curves $\Gamma(t)$ can be unclosed or closed. In the latter case the surface Ω_{A} may "circle" the original part of the solid or form a "tube" enveloping only the material of the additional part of the having been finally formed solid.

5. Proof of the Theorem

Introduce on the surface Ω_{A} the parameterization $\mathbf{r} = \mathbf{r}(\xi, \eta)$ which is induced by the ongoing growth program, in the following way. Let us select a certain general parameter $\eta \in [A, B]$, where A and B — are some constants, for all the curves $\Gamma(t)$, $t \geqslant t_1$, the surface Ω_{A} is composed from. This means that curves $\Gamma(t)$ form a family of η-lines on the surface Ω_{A}. The choice of the parameter η for all the curves $\Gamma(t)$ is to ensure that through each point of the surface Ω_{A} it goes one and only one line from the second family — the family of lines consisting of all those points on different curves $\Gamma(t)$, which correspond to the same value η.

Let the geometric position of the curve $\Gamma(t)$ on the surface Ω_{A} at a particular value of time t be in one-to-one correspondance to a particular value of some quantity ξ, namely the value $\xi = \Xi(t)$: if at different t' and t'' the curves $\Gamma(t')$ and $\Gamma(t'')$ geometrically coincide, then $\Xi(t') = \Xi(t'')$; and vice versa, if at different t' and t'' the curves $\Gamma(t')$ and $\Gamma(t'')$ do not geometrically coincide, then $\Xi(t') \neq \Xi(t'')$. The function $\Xi(t)$ we consider continuous for all $t \geqslant t_1$ and monotonically non-decreasing on the time intervals $t \in [t_{2k-1}, t_{2k})$, $k = 1, \ldots, N$ corresponding to the stages of the continuous growth of the solid. Beyond these intervals, that is, when $t \in [t_{2k}, t_{2k+1})$, where $t_{2N+1} = +\infty$, when the influx of additional material to the solid is absent and therefore the movement of the curve $\Gamma(t)$ over the surface Ω_{A} is temporarily (when $k = 1, \ldots, N-1$) or ultimately (when $k = N$) completed the function $\Xi(t)$ takes obviously constant values:

$$\Xi(t) \equiv \xi_k, \quad t \in [t_{2k}, t_{2k+1}),$$

where $\xi_k = \Xi(t_{2k})$. Due to continuity of the function $\Xi(t)$ the value ξ_k is the value of the parameter ξ corresponding to the position of the curve $\Gamma(t)$ at the end of the k-th stage of continuous growth.

If the value $\xi_0 = \Xi(t_1)$ corresponds to the position of the curve $\Gamma(t)$ at the initial time instant of the process of growing the solid, then due to monotonicity of the function $\Xi(t)$ we will have

$$\xi_0 \leqslant \xi_1 \leqslant \ldots \leqslant \xi_{N-1} \leqslant \xi_N, \tag{9}$$

and the parameter ξ over the surface Ω_{A} will vary on the interval $\xi \in [\xi_0, \xi_N]$. An example of dependence of $\xi = \Xi(t)$ in the case $N = 3$ is illustrated on the graph in figure 1.

Note that a possible non-strict monotonicity (non-decreasing) of the function $\Xi(t)$ on the intervals of continuous growth $t \in [t_{2k-1}, t_{2k})$ and, consequently, possible non-strict signs in

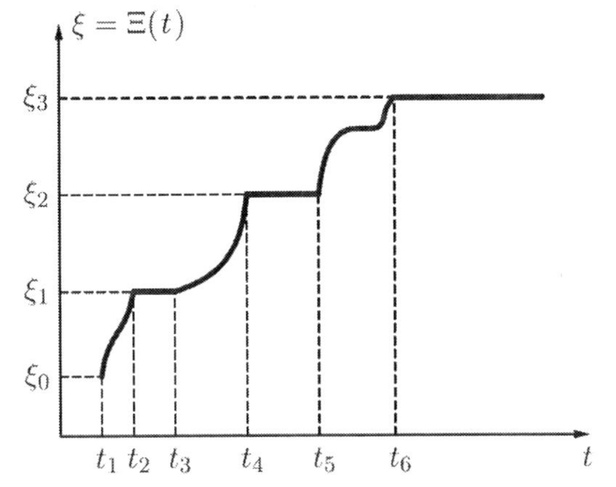

Figure 1. Example of dependence of the parameter ξ on the surface Ω_{A} on time t in the case of $N = 3$ stages of continuous accretion of the solid

chain (9), arise from the possibility of pursuing such a variant of accreting the solid when a region of space that it occupies at the stage of continuous growth is constantly expanding due to the influx of additional material to the part $\Sigma(t)$ of boundary surface of the growing solid — the current surface of growth, — but herewith the trace $\Gamma(t)$ of the moving in space growth surface $\Sigma(t)$ on the surface Ω_{A} selected by us inside or on the boundary of the growing solid remains for some time stationary.

The family of lines consisting of points corresponding to the same value of the parameter η on different curves $\Gamma(t)$ which was discussed above is obviously the family of ξ-lines on the surface Ω_{A}.

The couple of parameters (ξ, η) will be considered as curvilinear coordinates on the surface Ω_{A}. The convenience of introducing such coordinates is explained for us by the following key fact. Since at each value t the curve $\Gamma(t)$ lies entirely on the current growth surface $\Sigma(t)$, then at the coordinates (ξ, η) the time instant $\tau_0(\mathbf{r})$ of occurrence of stresses at the points $\mathbf{r}$ of the surface Ω_{A} that coincides with time instant $\tau_*(\mathbf{r})$ of the inclusion of these points in the composition of the growing solid (see (3)) depends only on the coordinate ξ:

$$\tau_0(\mathbf{r}) \equiv \tau_0(\xi), \tag{10}$$

moreover, in accordance with the definition of the function $\Xi(t)$ it is true the identity

$$\Xi\big(\tau_0(\xi)\big) \equiv \xi \tag{11}$$

for all $\xi \in [\xi_0, \xi_N]$.

By using the curvilinear coordinates (ξ, η) the integral over the surface $\Omega_*(t) \subseteq \Omega_{\mathrm{A}}$ of the arbitrary function $f(\mathbf{r}, t)$ of the point and time is written as follows:

$$\int_{\Omega_*(t)} f(\mathbf{r}, t)\, dS = \int_A^B d\eta \int_{\xi_0}^{\Xi(t)} f(\xi, \eta, t)\, J(\xi, \eta)\, d\xi. \tag{12}$$

Here the value $J(\xi, \eta)$ defines an element of the area of the surface Ω_{A} in the curvilinear coordinates (ξ, η) and is equal to

$$J(\xi, \eta) = \left| \frac{\partial \mathbf{r}}{\partial \xi} \times \frac{\partial \mathbf{r}}{\partial \eta} \right|.$$

By choosing $f(\mathbf{r},t) = \partial g^\circ(\mathbf{r},t)/\partial t$ and using the rule of differentiation of the integral by parameter, we can transform the inner integral in (12):

$$\int_{\xi_0}^{\Xi(t)} \frac{\partial g^\circ(\xi,\eta,t)}{\partial t} J(\xi,\eta)\,d\xi = \frac{\partial}{\partial t}\int_{\xi_0}^{\Xi(t)} g^\circ(\xi,\eta,t)\,J(\xi,\eta)\,d\xi - g^\circ\big(\Xi(t),\eta,t\big)\,J\big(\Xi(t),\eta\big)\,\Xi'(t). \quad (13)$$

Let us reveal the symbolic notation $(\,)^\circ$ in the integral standing on the right in (13) (see definition (4), (2) and formula (10)):

$$\int_{\xi_0}^{\Xi(t)} g^\circ(\xi,\eta,t)\,J(\xi,\eta)\,d\xi = \int_{\xi_0}^{\Xi(t)} \frac{g(\xi,\eta,t)}{G(t)} J(\xi,\eta)\,d\xi - \int_{\xi_0}^{\Xi(t)} J(\xi,\eta)\,d\xi \int_{\tau_0(\xi)}^{t} \frac{g(\xi,\eta,\tau)}{G(\tau)} K(t,\tau)\,d\tau. \quad (14)$$

Let us change in (14) the order of integration in the repeated integral (see figure 1 and identity (11)):

$$\int_{\xi_0}^{\Xi(t)} J(\xi,\eta)\,d\xi \int_{\tau_0(\xi)}^{t} \frac{g(\xi,\eta,\tau)}{G(\tau)} K(t,\tau)\,d\tau = \int_{t_1}^{t} K(t,\tau)\,d\tau \int_{\xi_0}^{\Xi(\tau)} \frac{g(\xi,\eta,\tau)}{G(\tau)} J(\xi,\eta)\,d\xi. \quad (15)$$

We consider now the term beyond the integrals in (13). We have (see definition (4), (2) and formula (10))

$$g^\circ\big(\Xi(t),\eta,t\big) = \frac{g\big(\Xi(t),\eta,t\big)}{G(t)} - \int_{\tau_0(\Xi(t))}^{t} \frac{g\big(\Xi(t),\eta,\tau\big)}{G(\tau)} K(t,\tau)\,d\tau.$$

On all pieces of the strict monotonicity of the function $\Xi(t)$ this function has a reverse one, which is the function $\tau_0(\xi)$ in accordance with identity (11). Therefore at the lower limit in the last integral it will be $\tau_0\big(\Xi(t)\big) \equiv t$ and this integral will be identically equal to zero. Herewith we also have

$$g\big(\Xi(t),\eta,t\big) = g(\xi,\eta,t)\big|_{\xi=\Xi(t)} = g(\xi,\eta,t)\big|_{t=\tau_0(\xi)} = g(\mathbf{r},t)\big|_{t=\tau_0(\mathbf{r})} \equiv 0$$

in accordance with condition (6) and representation (3).

Thus, on pieces of the strict monotonicity of the function $\Xi(t)$ the term beyond the integrals in (13) becomes zero. On pieces of the function $\Xi(t)$ constancy (in particular, in the pauses between the stages of continuous accretion of the solid and after its ultimate growth completing) it will be $\Xi'(t) \equiv 0$, therefore this term vanishes as well.

So, from (13)–(15) we get

$$\int_{\xi_0}^{\Xi(t)} \frac{\partial g^\circ(\xi,\eta,t)}{\partial t} J(\xi,\eta)\,d\xi = \frac{\partial}{\partial t}\left[\int_{\xi_0}^{\Xi(t)} \frac{g(\xi,\eta,t)}{G(t)} J(\xi,\eta)\,d\xi - \int_{t_1}^{t} K(t,\tau)\,d\tau \int_{\xi_0}^{\Xi(\tau)} \frac{g(\xi,\eta,\tau)}{G(\tau)} J(\xi,\eta)\,d\xi\right].$$

Hence, due to representation (12) we have

$$\int_{\Omega_*(t)} \frac{\partial g^\circ(\mathbf{r},t)}{\partial t}\,dS = \frac{d}{dt}\left[\int_{\Omega_*(t)} \frac{g(\mathbf{r},t)}{G(t)}\,dS - \int_{t_1}^{t} K(t,\tau)\,d\tau \int_{\Omega_*(\tau)} \frac{g(\mathbf{r},\tau)}{G(\tau)}\,dS\right]. \quad (16)$$

To complete the proof of formula (7) it remains to present the integral of the function $f(\mathbf{r},t) = \partial g^\circ(\mathbf{r},t)/\partial t$ over the surface Ω_0 supplementing the surface $\Omega_*(t)$ up to $\Omega(t)$ in a form similar to (16). As the surface Ω_0 is unchanged in time, then the time derivative can be taken outside the sign of the integral over this surface:

$$\int_{\Omega_0} \frac{\partial g^\circ(\mathbf{r},t)}{\partial t}\,dS = \frac{d}{dt}\int_{\Omega_0} g^\circ(\mathbf{r},t)\,dS. \quad (17)$$

IOP Conf. Series: Journal of Physics: Conf. Series **991** (2018) 012063 doi:10.1088/1742-6596/991/1/012063

We then reveal the symbolic notation $(\,)^\circ$ in the integral on the right (see definition (4), (2)) by reversing the order of integration over time and over the surface Ω_0 that does not dependent on time:

$$\int_{\Omega_0} g^\circ(\mathbf{r},t)\,dS = \int_{\Omega_0} \frac{g(\mathbf{r},t)}{G(t)}\,dS - \int_{\Omega_0} dS \int_{t_0}^{t} \frac{g(\mathbf{r},\tau)}{G(\tau)} K(t,\tau)\,d\tau$$
$$= \int_{\Omega_0} \frac{g(\mathbf{r},t)}{G(t)}\,dS - \int_{t_0}^{t} K(t,\tau)\,d\tau \int_{\Omega_0} \frac{g(\mathbf{r},\tau)}{G(\tau)}\,dS. \quad (18)$$

It is taken here into account that $\tau_0(\mathbf{r}) \equiv t_0$ for $\mathbf{r} \in \Omega_0$ as the surface Ω_0 lies entirely in the originally existing part of the solid, stresses at all points of which arose at the same moment of time $t = t_0$ (see (3)).

When $t > t_1$ we can split the time integral in (18) into two integrals — from t_0 to t_1 and from t_1 to t, — given that $\Omega(t) \equiv \Omega_0$ for $t \in [t_0, t_1]$ (see (8)):

$$\int_{\Omega_0} g^\circ(\mathbf{r},t)\,dS = \int_{\Omega_0} \frac{g(\mathbf{r},t)}{G(t)}\,dS - \int_{t_0}^{t_1} K(t,\tau)\,d\tau \int_{\Omega(\tau)} \frac{g(\mathbf{r},\tau)}{G(\tau)}\,dS - \int_{t_1}^{t} K(t,\tau)\,d\tau \int_{\Omega_0} \frac{g(\mathbf{r},\tau)}{G(\tau)}\,dS. \quad (19)$$

As a result, due to (17) we will have

$$\int_{\Omega_0} \frac{\partial g^\circ(\mathbf{r},t)}{\partial t}\,dS = \frac{d}{dt}\left[\int_{\Omega_0} \frac{g(\mathbf{r},t)}{G(t)}\,dS - \int_{t_1}^{t} K(t,\tau)\,d\tau \int_{\Omega_0} \frac{g(\mathbf{r},\tau)}{G(\tau)}\,dS - \int_{t_0}^{t_1} K(t,\tau)\,d\tau \int_{\Omega(\tau)} \frac{g(\mathbf{r},\tau)}{G(\tau)}\,dS\right]$$

which together with (16) with regard to (8) and (5) gives the formula (7) being proved.

6. An example of applying the Theorem to solving boundary-value problems for growing solids on the basis of Saint-Venant principle

Applying the above proved Theorem to solving the growing solids mechanics problems with integral force conditions is demonstrated in the paper [15] on the example of modelling additive processes of formation of the relatively long in the axial direction conical solids of rotation of the isotropic homogeneously aging linearly viscoelastic material subordinated to the constitutive equation (1).

It is assumed that in the process of formation of the solid its end surfaces are acted with a loading statically equivalent to the axial central tension–compression force P which can arbitrarily vary with time t. Forming the solid under consideration is carried out by means of its axisymmetric thickening in the radial direction due to attaching the additional material to the conic side surface. This process is piecewise-continuous in time, i.e. consists of arbitrary number of consecutive stages of continuous accretion alternating with pauses of arbitrary duration. Gradual thickening occurs in such a way that in each time moment the accreted body maintains the shape of a circular truncated cone of fixed length.

The cone lateral surface moving due to the influx of additional material represents the current growth surface $\Sigma(t)$ in the considered example. The trace of its passing in the space forms an additional part V_{A} of the growing cone.

The material used for the cone thickening is supposed to be initially free from stresses, so the initial condition

$$\mathbf{T}(\mathbf{r}, \tau_*(\mathbf{r})) \equiv 0, \quad \mathbf{r} \in V_{\mathrm{A}}, \quad (20)$$

should take place in the additional part of the considered growing solid.

With the help of the above proved Theorem we can solve on the basis of the Saint-Venant principle the boundary-value problem for the proposed mechanical model of the conical solid growth. As a surface $\Omega(t)$ expanding over time t due to the additional material influx it is necessary to consider in this problem the flat surface constituting one end side of the growing cone for every $t \geqslant t_0$. On this surface the integral boundary conditions for the components of the force P and its moment are to be formulated. In this case the surface Ω_A is annular, and its forming curves $\Gamma(t)$ are concentric circles. The surface Ω_0 is the disc of the radius being equal to the initial radius of the corresponding cone end.

The condition (20) in the considered example acts as the requirement (6) in the Theorem.

A closed form analytical solution of the described example problem is constructed in [15]. The time evolution of the stress-strain state of the conical solid before the start, during and upon the completion of the process of its additive manufacturing under specified conditions of loading is investigated.

Acknowledgments
This work was supported by the Russian Science Foundation under grant No. 17-19-01257.

References
[1] Manzhirov A V 1995 The general non-inertial initial-boundaryvalue problem for a viscoelastic ageing solid with piecewise-continuous accretion *J. Appl. Math. Mech.* **59** (5) 805–16
[2] Manzhirov A V 2013 Mechanics of growing solids and phase transitions *Key Engineering Materials* **535–536** 89–93
[3] Manzhirov A V and Lychev S A 2014 Mathematical modeling of additive manufacturing technologies In *Lecture Notes in Engineering and Computer Science: Proc. World Congr. on Engineering 2014, WCE 2014, 2-4 July 2014 London* pp 1404–9
[4] Manzhirov A V and Parshin D A 2006 Accretion of a viscoelastic ball in a centrally symmetric force field *Mech. Solids* **41** (1) 51–64
[5] Manzhirov A V and Parshin D A 2006 Modeling the accretion of cylindrical bodies on a rotating mandrel with centrifugal forces taken into account *Mech. Solids* **41** (6) 121–34
[6] Manzhirov A V and Parshin D A 2015 Arch structure erection by an additive manufacturing technology under the action of the gravity force *Mech. Solids* **50** (5) 559–70
[7] Manzhirov A V and Parshin D A 2015 Influence of the erection regime on the stress state of a viscoelastic arched structure erected by an additive technology under the force of gravity *Mech. Solids* **50** (6) 657–75
[8] Manzhirov A V and Parshin D A 2016 Accretion of spherical viscoelastic objects under self-gravity *Lecture Notes in Engineering and Computer Science* **2224** (1) 1131–5
[9] Manzhirov A V and Parshin D A 2016 Application of prestressed structural elements in the erection of heavy viscoelastic arched structures with the use of an additive technology *Mech. Solids* **51** (6) 692–700
[10] Parshin D A and Hakobyan V N 2017 Additive manufacturing of a cylindrical arch of viscoelastic aging material under gravity action at various modes of the process *Procedia IUTAM* **23** 66–77.
[11] Parshin D A 2017 The technological stresses in a vaulted structure built up on a falsework *Mechanics for Materials and Technologies* ed H Altenbach, R V Goldstein and E Murashkin *Advanced Structured Materials* vol 46 (Springer) pp 377–86
[12] Parshin D A 2017 Analytic solution of the problem of additive formation of an inhomogeneous elastic spherical body in an arbitrary nonstationary central field of forces *Mech. Solids* **52** (5) doi: 10.3103/S0025654417050089
[13] Arutyunyan N Kh 1966 *Some Problems in the Theory of Creep* (Oxford: Pergamon Press)
[14] Arutyunyan N Kh and Manzhirov A V 1999 *Contact Problems of the Theory of Creep* (Erevan: Izdat. Inst. Mekhaniki NAN RA) [in Russian]
[15] Manzhirov A V, Parshin D A, and Romanov A A 2017 Additive manufacturing of conical viscoelastic parts under axial tension-compression *Lecture Notes in Engineering and Computer Science: Proc. World Congr. on Engineering 2017, WCE 2017, 5–7 July 2017 London* pp 934–9

IOP Conf. Series: Journal of Physics: Conf. Series **991** (2018) 012064 doi:10.1088/1742-6596/991/1/012064

Analysis of syntactic foam – GFRP sandwich composites for flexural loads

Daniel Paul[1], R Velmurugan[1*], R Jayaganthan[2], N K Gupta[3], and A V Manzhirov[4,5]

[1]Department of Aerospace Engineering, Indian Institute of Technology Madras, Chennai, India
[2]Department of Engineering Design, Indian Institute of Technology Madras, Chennai, India
[3]Department of Applied Mechanics, Indian Institute of Technology Delhi, Delhi, India
[4]Ishlinsky Institute for Problems in Mechanics of the Russian Academy of Sciences, Moscow, Russia
[5]Bauman Moscow State Technical University, Moscow, Russia

E-mail: *ramanv@iitm.ac.in

Abstract. The use of glass microballoon (GMB) — epoxy syntactic foams as a sandwich core material is studied. The skins and foam core are fabricated and joined instantaneously unlike the procedures followed in the previous studies. Each successive layer of the sandwich is fabricated when the previous layer is in a semi-gelled state. These sandwich samples are characterized for their properties under flexural loading. The failure modes and mechanical properties are carefully investigated. The change in fabrication technique results in a significant increase in the load bearing pattern of the sandwich. In earlier studies, debonding was found to occur prematurely since the bonding between the skins and core is the weakest plane. Using the current technique, core cracking occurs first, followed by skin fiber breaking and debonding happens at the end. This ensures that the load carrying phase of the structure is extended considerably. The sandwich is also analytically studied using Reddy's higher order shear deformation theory. A higher order theory is selected as the sandwich can no longer be considered as a thin beam and thus shear effects also need to be considered in addition to bending effects.

1. Introduction

A syntactic foam is a material system composed of a matrix and hollow particles which are added to improve weight reduction and certain other properties. While most of the initial focus was on polymer matrices such as epoxy or vinyl ester [1,2], studies have recently been done even on metal matrices [3]. The lower weight of these syntactic foams enables us to use them in various applications, mainly in the aerospace and marine sectors. Their mechanical properties have been extensively studied. Another attractive use of syntactic foams is as the core in sandwich structures. These structures have been proposed as early as the 1990s [4]. Such sandwich composites have been fabricated and studied in the past both experimentally [5] and numerically [6]. The foam core is found to reduce the overall density of the composite while providing considerable stiffness under bending. Most sandwich structure cores studied have very low stiffness compared to the skin and models to analyse them have been discussed in literature [7]. Syntactic foam cores on the other hand, have considerably higher stiffness values.

In most of the studies on syntactic foam sandwich structures the skins and core are fabricated separately and combined at the end using a resin system. This caused premature debonding,

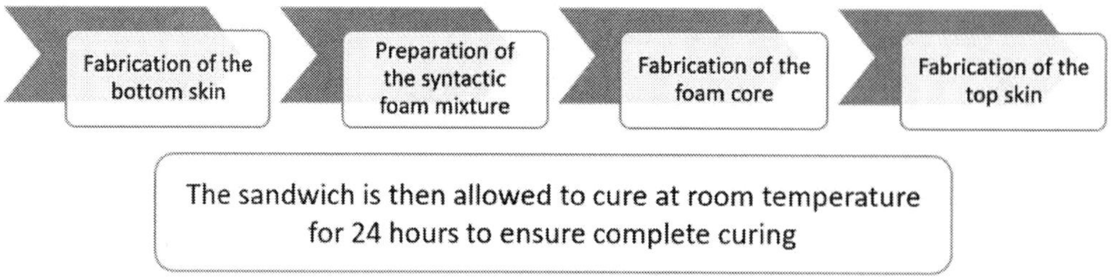

Figure 1. The stages involved in flexural failure of syntactic foam – GFRP sandwich composites.

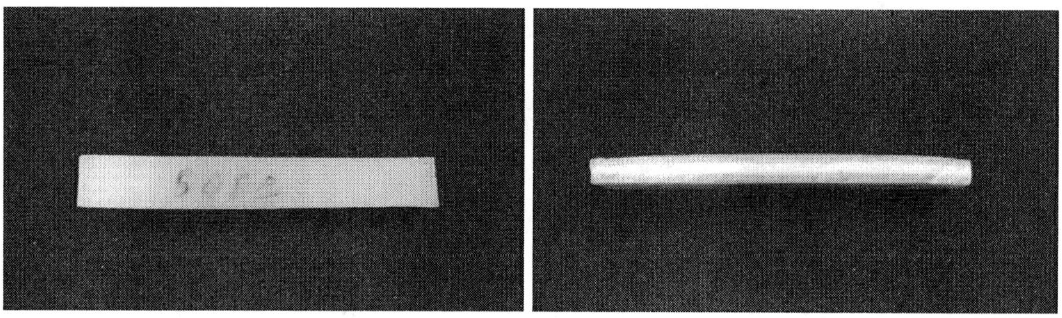

Figure 2. Different views of a sandwich sample after preparation.

leading to failure, under bending. In this study, the skins and core were fabricated together, thereby obtaining maximum load carrying capability from the sandwich before debonding takes place. Flexural tests were performed experimentally and the modes of failure were carefully observed. Analytical study was carried out and results were compared.

2. Experimental study

2.1. Fabrication of the sandwich composites

The materials used in the fabrication of the sandwich composites were 3M Scotchlite K15 type hollow glass microballoons, unidirectional stitched glass fiber mats, and a DGEBA-based epoxy resin system, Araldite LY556. A TETA-based hardener, Aradur HY951, was used as the curing agent. In previous studies, the skins were pre-fabricated and the foam core was then bonded separately. In this study, all the layers were fabricated simultaneously to ensure better bonding at the interfaces.

A predetermined quantity of GMB was mixed with epoxy system using a mechanical stirrer. The mixing was done at lower speeds and the mixture was heated at regular intervals while being stirred to prevent the breaking of the GMB and to reduce its viscosity. At the same time, the lower skin was fabricated by using a hand layup technique. As soon as the epoxy in the skin started gelling, the hardener was added to the GMB-epoxy mixture being stirred. The GMB-epoxy-hardener mixture was poured over the skin and allowed to gel into a semi-solid state. The top skin was then fabricated over the foam core. Both the upper and lower skins were comprised of three layers of glass fiber mats. Care was taken that the subsequent layers were fabricated before the layer preceding them gelled completely. The entire sandwich structure was allowed to cure for around 24 hours at room temperature. A sample with 3mm core thickness with a volume fraction of 0.5 is shown in figure 2.

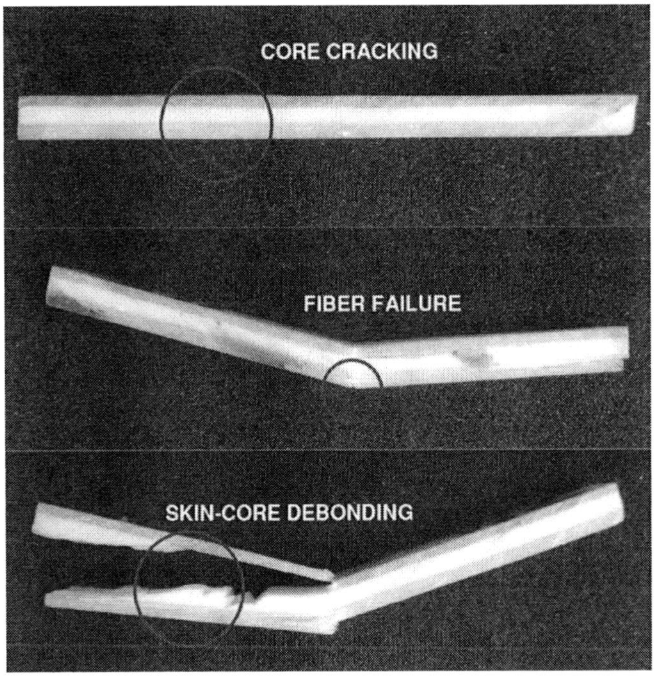

Figure 3. The stages involved in flexural failure of syntactic foam - GFRP sandwich composites.

2.2. Mechanical tests
Flexural tests were performed on a 5 kN Instron Universal Testing Machine (UTM) fitted with a three-point bending fixture. Rectangular samples were used for the tests. The gauge length was fixed at 50 mm.

2.3. Results and discussion
To compare the method of fabrication used in this study, samples were also fabricated by preparing the core and the skins separately and bonding them. Both were subjected to bending loads in order to understand the modes of failure in each section. The stages of failure observed in a sandwich specimen prepared instantly as described in this study are shown in figure 3. There were three easily distinguishable modes of failure in the sandwich structure: core failure (cracking), skin fiber breakage, and skin-core debonding. Initially, both the skin and the core carried the load applied. The core failed first as it had lower strength compared to the skin. This is the first, sudden drop seen in the stress-strain curve. Even though cracks developed in the core, it was held intact by the skins and the structure was able to take further loads. As the specimen was bent further, the skin fibers on the tensile side of the specimen started breaking. This resulted in sudden drops in the load carried alternated by regions where the load could still be carried by the remaining intact specimen. The final failure happened when the skin and core debonded due to shear failure. It was observed that the core surface after debonding was not smooth which indicated that the bonding between the foam and the skin was very strong. This was a result of the fabrication technique used where the entire specimen was allowed to cure as one single unit instead of fabricating each layer of the sandwich and bonding them at the end. In specimens prepared by bonding together the individual sandwich layers at the end, it is seen that the first mode of failure is the debonding of the skin and the core, which led to very low overall strength as the debonding occurred before the maximum load-bearing capacity of the skin had been reached.

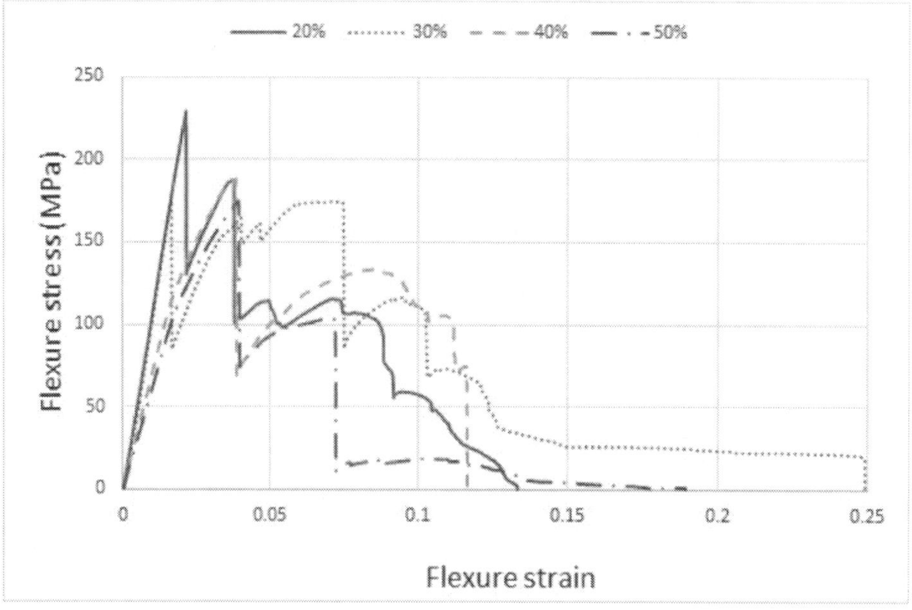

Figure 4. The flexural stress-strain curves of sandwich composites with varying foam volume fraction (core thickness equal 3 mm).

The results of the flexural tests for various core volume fractions of the sandwich composites are shown in figure 4. It can be seen that as the core volume fraction increases, the modulus and strength of the foam showed a slight decrease. Figure 5 compares the sandwich flexural moduli to those of the pure syntactic foams. The increase in modulus compared to the pure foam is due to the higher stiffness of the glass-fiber reinforcement. Despite the decrease in properties, the foam core helps in reducing the density of the composite significantly. The density of the fiber reinforced epoxy alone was found to be 1.646 g/cc while a sandwich containing a core having a GMB volume fraction of 50% had a reduced density of 1.155 g/cc. This helps in reducing the weight of the overall structure being fabricated.

3. Analytical study

Analytical studies were carried out using the higher order shear deformation theory proposed by Reddy [8]. A higher order theory was chosen since the sandwich composite being studied cannot be considered as a thin beam and thus will have shear effects in addition to pure bending. This is visibly noticed during the experimental bending tests where the core and the skin shear off at higher loads.

The displacement field is considered to be of the form

$$\mathbf{u}_1(x,y,z) = \mathbf{u}(x,y) + z\Psi_x(x,y) + z^2\xi_x(x,y) + z^3\zeta_x(x,y), \tag{1}$$
$$\mathbf{u}_2(x,y,z) = \mathbf{v}(x,y) + z\Psi_y(x,y) + z^2\xi_y(x,y) + z^3\zeta_y(x,y), \tag{2}$$
$$\mathbf{u}_3(x,y,z) = \mathbf{w}(x,y), \tag{3}$$

where the last two terms in (1) and (2) are the higher order terms which account for the shear effect which is not considered in first-order theories; Ψ_x and Ψ_y are the rotations of the normal to the mid-plane about the y and x axes, respectively. Using the constitutive and equilibrium equations leads us to form a stiffness matrix which involves a large number of coupling terms other than the usual A, B, and D matrices obtained in simpler theories. After simplifying the

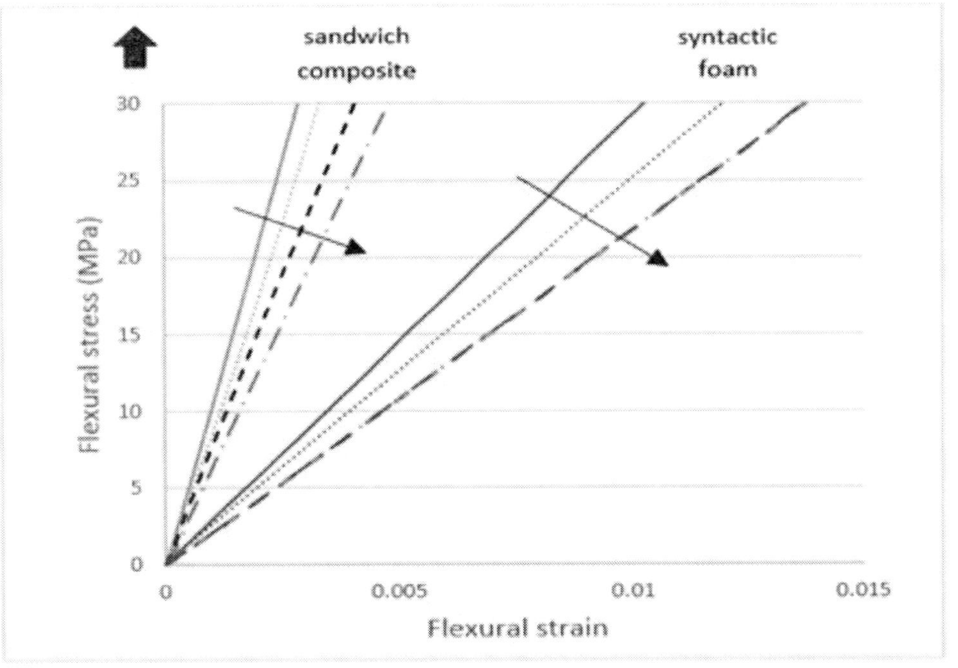

Figure 5. The initial portion of the flexural stress-strain curve compared with pure syntactic foam curves. The arrows depict increasing core volume fraction.

equations for symmetry and special material properties, assumed displacements and rotations are substituted into the equations in Fourier series fashion. Care is taken that these satisfy the boundary conditions. The unknowns in our study are the vertical displacement, w, and the rotations, Ψ_x and Ψ_y. The set of equations obtained is then solved using any equation solving software to obtain the displacement field at a given load. This can be used to calculate the modulus of the sandwich beam. The results obtained for specific volume fractions are shown in figure 6. They were found to be following the same trend as the experimental results.

Conclusions

Syntactic foam – GFRP sandwich composites were fabricated using an instantaneous method. Flexural tests were performed, which showed failure modes different from specimens made by the normal method. The modes of failure were studied in detail and it showed three modes of failure in the following order of occurrence: core failure (cracking), skin fiber breakage, and skin-core debonding. The effect of increasing core GMB volume fraction on the overall properties was studied. Though the stiffness and initial failure of the sandwich samples decreases with an increase in GMB volume fraction, the drawback in compensated by a corresponding decrease in overall sandwich density.

Analytical study was done using a higher order shear deformation theory to model the sandwich composite studied which ensured the shear effect in the thicker composite was taken into account. The values for the skins and core properties individually were obtained from experiments and from analysis. The analytical results were very close to the experimental valued obtained.

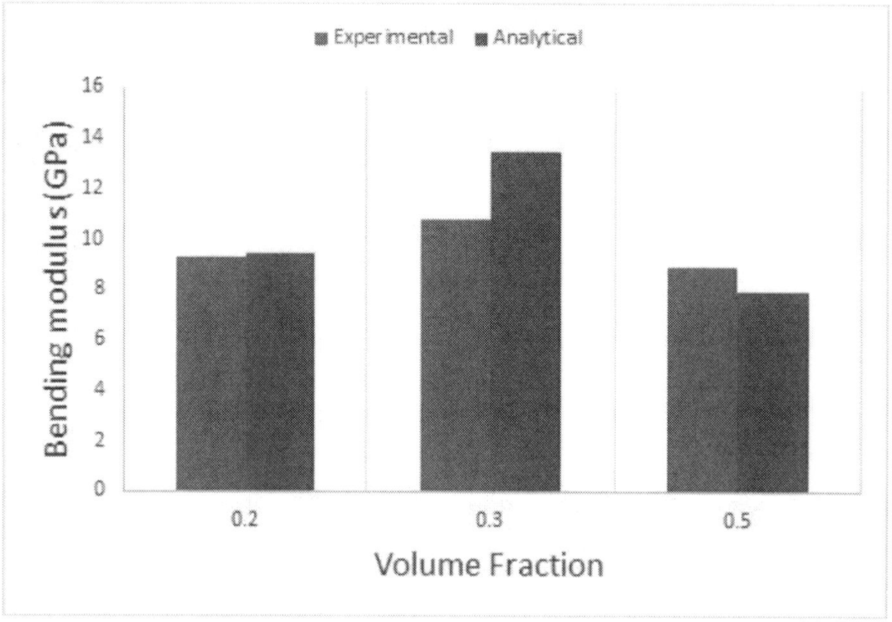

Figure 6. Comparison of analytical results with experimental values.

Acknowledgments
This study was financially supported by the Federal Agency for Scientific Organizations (State Registration Number AAAA-A17-117021310381-8) and by the Department of Science and Technology (DST) of India and the Russian Foundation for Basic Research (project No. 17-51-45054) through the Indo-Russian collaborative project scheme.

References
[1] Gupta N, Ye R, and Porfiri M 2010 Comparison of tensile and compressive characteristics of vinyl ester/glass microballoon syntactic foams *Compos. Part B: Engng* **41** (3) 236–45
[2] Wouterson E M, Boey F Y, Hu X, and Wong S C 2005 Specific properties and fracture toughness of syntactic foam: Effect of foam microstructures *Compos. Sci. Technol.* **65** (11) 1840–50
[3] Omar M Y, Xiang C, Gupta N, et al. 2015 Syntactic foam core metal matrix sandwich composite: compressive properties and strain rate effects *Mater. Sci. Engng: A* **643** 156–68
[4] Hiel C, Dittman D, and Ishai O 1993 Composite sandwich construction with syntactic foam core: a practical assessment of post-impact damage and residual strength *Composites* **24** (5) 447–50
[5] Gupta N and Woldesenbet E 2005 Characterization of flexural properties of syntactic foam core sandwich composites and effect of density variation *J. Compos. Mater.* **39** (24) 2197–212
[6] Corigliano A, Rizzi E, and Papa E 2000 Experimental characterization and numerical simulations of a syntactic-foam/glass-fibre composite sandwich *Compos. Sci. Technol.* **60** (11) 2169–80
[7] Hu H, Belouettar S, and Potier-Ferry M 2008 Review and assessment of various theories for modeling sandwich composites *Compos. Struct.* **84** (3) 282–92
[8] Reddy J N 1984 A simple higher-order theory for laminated composite plates *J. Appl. Mech.* **51** (4) 745–52

IOP Conf. Series: Journal of Physics: Conf. Series **991** (2018) 012065 doi:10.1088/1742-6596/991/1/012065

Polariton resonances in multilayered piezoelectric superlattices

D Piliposyan

Institute of Mechanics of the National Academy of Sciences of Armenia, Yerevan, Armenia

E-mail: piliposyan@mechins.sci.am

Abstract. Coupled electro-elastic SH waves propagating in a periodic piezoelectric finite-length superlattice with identical piezoelectric materials in a unit cell are considered in the framework of the full system of Maxwell's electrodynamic equations. In the long wavelength region, coupling between electro-magnetic and elastic waves creates frequency band gaps. It is shown that for piezoelectric superlattice at acoustic frequencies, acousto-optic coupling gives rise to polariton behavior at wavelengths much larger than the length of the unit cell. The results of the paper may be useful in design of narrow band filters or multi-channel piezoelectric filters.

1. Introduction

The interaction of waves with periodic structures, especially artificial superlattices, has recently attracted much attention. A periodic modulation of the dielectric or elastic material properties leads to absolute stop bands and ultimate control of propagation of waves in the structure. A piezoelectric or piezomagnetic superlattice made of a periodically domain-inverted dielectric crystal with periodically modulated piezoelectric or piezomagnetic coefficients, but a homogeneous refractive index, can be considered as a one-dimensional diatom ionic crystal with positive and negative ions arranged periodically [1]. The coupling between transverse lattice vibrations and electro-magnetic waves in an ionic crystal can lead to phonon polariton coupling with possible stop bands in the infrared region [2]. Analogously, piezoelectric and piezomagnetic periodic structures with periodicity of the lattice expanded from an atomic scale to microns can exhibit similar coupling and resonant band gap structure in the microwave region [3–7]. This coupling between the electromagnetic wave and the superlattice vibration takes place in the long wavelength region, where the superlattice can be considered as a deep subwavelength artificial material. Using the long wavelength approximation, it has been shown that the piezoelectric superlattice exhibits a new type of polariton, in which the resonance frequency is determined by the period of the superlattice and negative effective permittivity occurs near the high frequency side of resonance [8, 9]. Theoretical and experimental work has suggested that a different type of polariton is also possible in a piezoelectric superlattice that is coupling of electro-magnetic waves with longitudinal superlattice vibrations [10].

While the long-wave approximation only reveals the phonon-photon polariton at high acoustic frequencies in the middle of the Brillouin zone, the analytical solution shows that the coupling of photons and phonons is also possible at optical frequencies in the whole Brillouin zone [11, 12]. It also reveals a phonon-polariton gap in a piezoelectric phononic crystal with a unit cell made of

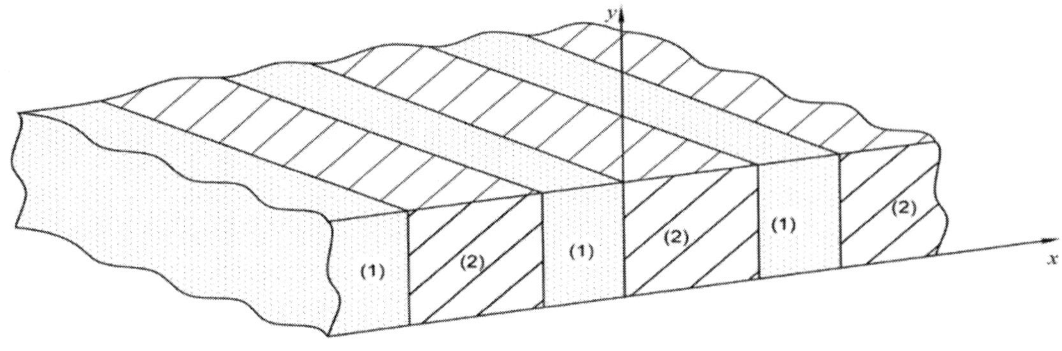

Figure 1. Periodic piezoelectric superlattice.

different constituent materials. The similarities and differences between artificial superlattices and real lattices suggest rich physics in artificial microstructures and gives a possibility to control and manipulate both photons and phonons simultaneously [10, 13].

The problem is more interesting in a periodic superlattice with full contact interfaces. The system in this case is described by two coupled electro-acoustic waves and exhibits a very interesting acousto-optic resonance (phonon-polariton) at high acoustic frequencies which cannot be observed in the previous problem. The dynamic setting for Maxwell's equation, where both the optical effect and the effect from the rotational part of the electric field are taken into account, permits investigating this problem as well.

2. Statement of the problem

We investigate the reflection/transmission properties of a finite stack of transversely isotropic hexagonal piezoelectric crystals (6 mm), a periodically stratified piezoelectric crystal with the crystallographic axes directed along the Oz direction (figure 1).

The anti-plane problem of interconnected elastic and electro-magnetic excitations can be written in terms of the variables $E_y(x,y)$, $u_z(x,y)$, $H_z(x,y)$, and $\sigma_{xz}(x,y)$ for which the interface boundary conditions are applied:

$$\frac{\partial u_z}{\partial x} = \frac{\sigma_{xz}}{G} + \frac{e_{15}}{i\omega G\varepsilon_{11}}\frac{\partial H_z}{\partial y}, \tag{1}$$

$$\frac{\partial E_y}{\partial x} = -\frac{e_{15}}{G\varepsilon_{11}}\frac{\partial \sigma_{xz}}{\partial y} + \left(\frac{c_{44}}{i\omega G\varepsilon_{11}}\frac{\partial^2 H_z}{\partial y^2} - i\omega\mu_{33}H_z\right), \tag{2}$$

$$\frac{\partial \sigma_{xz}}{\partial x} = e_{15}\frac{\partial E_y}{\partial y} - \left(c_{44}\frac{\partial^2 u_z}{\partial y^2} + \rho\omega^2 u_z\right), \tag{3}$$

$$\frac{\partial H_z}{\partial x} = -i\omega\varepsilon_{11}E_y - i\omega e_{15}\frac{\partial u_z}{\partial y}, \quad G = c_{44} + \frac{e^2}{\varepsilon}. \tag{4}$$

For the shear wave propagating in the XY plane and polarized along the Z axis, harmonic time dependence in the form $\exp(i\omega t)$ and plane-wave dependence in the form $\exp(ipy)$ for all time and space dependent variables is assumed with ω as the wave angular frequency and p as a wave number. After introducing $e_{15} = e$, $\varepsilon_{11} = \varepsilon$, $H_z = H$, $u_z = u$, the vector functions

$$\mathbf{v}(x,y) = \mathbf{v}(x)e^{ipy} \quad \text{and} \quad \mathbf{v}(x) = (-\omega u, -iE, i\sigma, iH)^T, \tag{5}$$

and the matrix

$$\hat{\mathrm{M}} = \begin{pmatrix} 0 & 0 & \dfrac{\omega}{G} & \dfrac{ep}{G\varepsilon} \\ 0 & 0 & \dfrac{ep}{G\varepsilon} & \mu\omega - \dfrac{c_{44}p^2}{G\varepsilon\omega} \\ \rho\omega - \dfrac{c_{44}p^2}{\omega} & -ep & 0 & 0 \\ -ep & \omega\varepsilon & 0 & 0 \end{pmatrix}, \tag{6}$$

the problem reduces to solving the following eigenvalue problem

$$\frac{1}{i}\frac{d}{dx}\mathbf{v}(x) = \hat{\mathrm{M}}\mathbf{v}(x). \tag{7}$$

The eigenvalues of (8) are

$$q = \pm\sqrt{\frac{\omega^2}{c_0^2} - p^2}, \quad s = \pm\sqrt{\frac{\omega^2}{c^2} - p^2}, \quad c_0 = \sqrt{\frac{G}{\rho}}, \quad c = \sqrt{\frac{1}{\varepsilon\mu}}, \tag{8}$$

and the eigenvectors are

$$\mathbf{v}_1 = \sqrt{\frac{Gq}{2\omega}}\left(\frac{\omega}{Gq}, \frac{ep}{Gq\varepsilon}, 1, 0\right)^T, \quad \mathbf{v}_2 = i\sqrt{\frac{Gq}{2\omega}}\left(\frac{-\omega}{Gq}, \frac{-ep}{Gq\varepsilon}, 1, 0\right)^T, \tag{9}$$

$$\mathbf{v}_3 = \sqrt{\frac{\varepsilon\omega}{2s}}\left(0, \frac{s}{\varepsilon\omega}, \frac{-ep}{\varepsilon\omega}, 1\right)^T, \quad \mathbf{v}_4 = i\sqrt{\frac{\varepsilon\omega}{2s}}\left(0, \frac{-s}{\varepsilon\omega}, \frac{-ep}{\varepsilon\omega}, 1\right)^T. \tag{10}$$

For a periodic structure described in figure 1, the solution for (7) in the nth unit cell can be written in the form

$$\mathbf{v}_n^{(j)}(x) = A_n^{(j)}\mathbf{v}_1^{(j)}e^{iq(x-n\beta)} + B_n^{(j)}\mathbf{v}_2^{(j)}e^{-iq(x-n\beta)} + C_n^{(j)}\mathbf{v}_3^{(j)}e^{is(x-n\beta)} + D_n^{(j)}\mathbf{v}_4^{(j)}e^{-is(x-n\beta)}, \quad j = 1, 2, \tag{11}$$

where $A_n^{(j)}$, $B_n^{(j)}$, $C_n^{(j)}$, and $D_n^{(j)}$ are the amplitudes of forward and backward travelling waves.

For perfectly bounded interfaces $x = n\beta + a_1$ and $x = (n+1)\beta$ between the layers $(n,1)/(n,2)$ and $(n,2)/(n+1,1)$, the unimodular propagator matrix $\hat{\mathrm{W}}$ coupling the forward and backward travelling wave amplitudes $A_n^{(1)}$, $C_n^{(1)}$, $B_n^{(1)}$, and $D_n^{(1)}$ in layers (1) in two neighboring cells (n) and ($n+1$) can be constructed directly as

$$\begin{pmatrix} A_n^{(1)} \\ B_n^{(1)} \\ C_n^{(1)} \\ D_n^{(1)} \end{pmatrix} = \hat{\mathrm{W}} \begin{pmatrix} A_{n+1}^{(1)} \\ B_{n+1}^{(1)} \\ C_{n+1}^{(1)} \\ D_{n+1}^{(1)} \end{pmatrix}. \tag{12}$$

For a superlattice with cells composed of identical but oppositely polarized piezoelectric material ($e_1 = e$, $e_2 = -e$) and with equal widths $a_1 = a_2 = a$, the propagator matrix $\hat{\mathrm{W}}$ has the form

$$\hat{W} = \begin{pmatrix} e^{-2iaq} - \xi_1 e^{-ia(q+s)} & i\xi_1 e^{-ia(q+s)} & \xi_3^* e^{-2iaq} & i\xi_4 e^{-2iaq} \\ i\xi_1 e^{ia(q-s)} & e^{2iaq} + \xi_1 e^{ia(q-s)} & -i\xi_4^* e^{2iaq} & \xi_3 e^{2iaq} \\ -\xi_3 e^{-2ias} & i\xi_4 e^{-2ias} & e^{-2ias} - \xi_2 e^{-ia(q+s)} & -i\xi e_2^{-ia(q+s)} \\ -i\xi_4^* e^{2ias} & -\xi_3^* e^{2ias} & -i\xi_2 e^{-ia(q-s)} & e^{2ias} + \xi_2 e^{-ia(q-s)} \end{pmatrix}, \tag{13}$$

where

$$\xi_1 = \gamma^2\theta^2(e^{2ias} - 1), \quad \xi_2 = \gamma^2\theta^2(e^{2iaq} - 1), \quad \xi_3 = \gamma\theta(e^{ia(s-q)} - 1), \quad \xi_4 = \gamma\theta(e^{-ia(s+q)} - 1),$$
$$\gamma = \frac{p^2}{qs}, \quad \theta = \frac{e^2}{G\varepsilon^2}, \tag{14}$$

θ is the electro-mechanical coupling coefficient, and $*$ in the superscript denotes the complex conjugate.

It now follows that the relationship between the amplitudes of forward and backward travelling wave fields in the first and last layers of the superlattice can be written as

$$\begin{pmatrix} A_0^{(1)} \\ B_0^{(1)} \\ C_0^{(1)} \\ D_0^{(1)} \end{pmatrix} = \hat{\mathrm{W}}^n \begin{pmatrix} A_n^{(1)} \\ B_n^{(1)} \\ C_n^{(1)} \\ D_n^{(1)} \end{pmatrix}. \tag{15}$$

Due to the symmetry of the problem, the propagator matrix $\hat{\mathrm{W}}$ is unimodular with eigenvalues $\lambda_{1,2} = e^{\pm ik_1\beta}$ and $\lambda_{3,4} = e^{\pm ik_2\beta}$, where $k_{1,2}$ are the Bloch–Floquet numbers [11]. The characteristic equation of the matrix $\hat{\mathrm{W}}$ gives the following dispersion equation for an elasto-electromagnetic wave propagation in a piezoelectric superlattice with $\beta = 2a$ [14]:

$$\cos(k_{1,2}\beta) = \frac{\cos(q\beta) + \cos(s\beta)}{2} - 2\gamma^2\theta^2\sin(aq)\sin(sa)$$
$$\pm\left[\cos^2(qa) - \cos^2(sa)\right]\sqrt{1 - 4\gamma^2\theta^2\frac{\sin(aq)\sin(sa)}{[\cos(qa) + \cos(sa)]^2}}. \tag{16}$$

It is clear from formula (16) that if the unit cell in the superlattice has oppositely polarised identical piezoelectric elements, it will allow the propagation of the Bloch–Floquet waves at acoustic frequencies. At optical frequencies, the opposite polarization does not generate band gaps, since the effect of piezoelectricity diminishes at such frequencies and the structure behaves as a homogeneous stricture with identical optical refractive indexes [11]. This means that only an acoustic incident wave will be considered for the investigation of the reflection/transmission properties in the superlattice. For an incident acoustic wave with amplitude A_I, the partial amplitudes of the wave fields in the first layer and at the exit are related as

$$\begin{pmatrix} A_I \\ B_R \\ 0 \\ D_R \end{pmatrix} = \hat{\mathrm{W}}^n \begin{pmatrix} A_T \\ 0 \\ C_T \\ 0 \end{pmatrix}, \tag{17}$$

where B_R and D_R are the amplitudes of the reflected elastic and electromagnetic waves and A_T and C_T are the amplitudes of the transmitted acoustic and electromagnetic waves. It follows from (17) that the reflection/transmission coefficients R_{ac}, R_{el}, T_{ac}, T_{el} for acoustic and electromagnetic waves in terms of the propagator matrix are

$$R_{\mathrm{ac}} = \frac{B_R}{A_I} = \frac{V_{23}V_{31} - V_{21}V_{33}}{V_{13}V_{31} - V_{11}V_{33}}, \quad T_{\mathrm{ac}} = \frac{A_T}{A_I} = \frac{V_{33}}{V_{11}V_{33} - V_{13}V_{31}}, \tag{18}$$
$$R_{\mathrm{el}} = \frac{D_R}{A_I} = \frac{V_{33}V_{41} - V_{31}V_{43}}{V_{11}V_{31} - V_{13}V_{31}}, \quad T_{\mathrm{el}} = \frac{C_T}{A_I} = \frac{V_{31}}{V_{13}V_{33} - V_{11}V_{33}}. \tag{19}$$

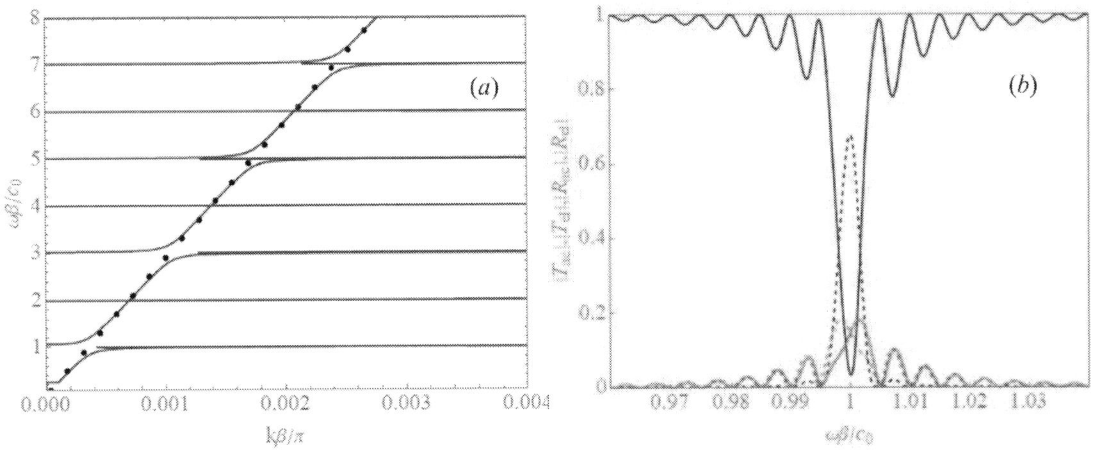

Figure 2. (*a*) Dispersion diagram of an infinite superlattice at acoustic frequencies near the center of the Brillouin zone. The horizontal lines correspond to the dimensionless phase velocity $k\beta/\pi$ of the acoustic wave enhanced by piezoelectricity, the vertical axis is the dimensionless frequency $\omega\beta/c_0$. The linear oblique dotted line corresponds to the phase velocity of a pure electromagnetic wave. (*b*) Transmitted acoustic (blue line), electromagnetic (red line), reflected acoustic (blue dashed line), and reflected electromagnetic (red dashed line) energy diagrams, $n = 200$.

3. Results and Discussion

Due to the electro-mechanical coupling in a piezoelectric superlattice, an incident acoustic wave generates a propagating electromagnetic wave. At specific frequencies in the long wavelength region, the elastic and generated electromagnetic waves resonate with each other creating a new form of a sub-wavelength band gap which is not associated with the Bragg reflection.

Since the dispersion curve of photons at acoustic frequencies is too close to the vertical axis, the coupling between the electromagnetic wave and the superlattice vibration can be seen in the long-wavelength region [12].

The solution of the dispersion equation (16) describes both coupled elastic and electromagnetic waves for the infinite piezoelectric superlattice. Here $k_1\beta$ describes the coupled acoustic wave field for frequencies lower than the pure electromagnetic wave frequency, and $k_2\beta$ describes the coupled acoustic and electromagnetic waves for frequencies higher than or equal to the frequencies of the pure electromagnetic wave. The resonant phonon polariton coupling appears when the square root in (16) becomes zero, i.e., $k_1\beta = k_2\beta$.

The first panel in figure 2 shows the subwavelength resonant bands described by dispersion curves (16) for an infinite superlattice at acoustic frequencies near the center of the Brillouin zone. The horizontal lines correspond to the phase velocity of the acoustic wave enhanced by the piezoelectric coupling. The linear oblique dotted line computed by the equation $k\beta/\omega = \sqrt{1/c^2 - \sin\phi^2/c_0^2}$ corresponds to the phase velocity of a pure electromagnetic wave uncoupled with the lattice vibrations. The region of crossover of these two lines is the resonance region, where neither an electromagnetic nor an acoustic wave can propagate for a very narrow range of the wave number.

As can be seen from the first panel in figure 2, the acousto-optic coupling for equal length layers in the cell takes place only for odd multiples of the fundamental resonance frequency $(\omega\beta/(c_0\pi) = n,\ n = 1, 3, 5, \ldots)$. For even n, the polariton is not excited. Here the red line corresponds to the $k_1\beta$ in (16), and the blue line corresponds to the $k_2\beta$.

Conclusions

Due to the electromechanical coupling, the piezoelectric superlattice exhibits unique reflection/transmission properties even when the unit cell is made of identical piezoelectric materials. When the interfaces in the supetlattice are perfectly bonded, the two identical elements in the unit cell can be oppositely polarized to make the structure behave as a phononic crystal. This however occurs only at acoustic frequencies, since the piezoelectric effect does not affect the band structure at optic frequencies where, in this case, the structure will act as a homogeneous material. The piezoelectric superlattice structure demonstrates an interesting coupling effect between the electro-magnetic wave and the superlattice vibration. The dispersion equation (16) contains information about this coupling which, in the long wavelength region, results in phonon-polariton gaps, where the resonance frequency is determined by the period of the superlattice.

Acknowledgments

This research was supported by the State Committee of Science of Armenia under grant No. SCS 115T-2C031.

References

[1] Lu Y-q, Zhu Y-y, Chen Y-f, et al. 1999 Optical properties of an ionic-type phononic crystal *Science* **284** (5421) 1822–4

[2] Huang K 1951 On the interaction between the radiation field and ionic crystals *Proc. Roy. Soc. Lond. A* **208** (1094) 352–65

[3] Zhang W, Liu Z, and Wang Z 2005 Band structures and transmission spectra of piezoelectric superlattices *Phys. Rev. B* **71** (19) 195114

[4] Liu H, Zhu S N, Dong Z G, et al. 2005 Coupling of electromagnetic waves and superlattice vibrations in a piezomagnetic superlattice: Creation of a polariton through the piezomagnetic effect *Phys. Rev. B* **71** (12) 125106

[5] Liu Z and Zhang W 2006 Study of the phonon–polaritons in piezomagnetic superlattices using a generalized transfer matrix method *J. Phys.: Condens. Matter* **18** (39) 9083

[6] Yang M-Y, Wu L-C, and Tseng J-Y 2008 Phonon–polariton in two-dimensional piezoelectric phononic crystals *Phys. Lett. A* **372** (26) 4730–5

[7] Liu H, Zhu S N, Dong Z G, et al. 2005 Coupling of electromagnetic waves and superlattice vibrations in a piezomagnetic superlattice: Creation of a polariton through the piezomagnetic effect *Phys. Rev. B* **71** (12) 125106

[8] Zhang X-J, Zhu R-Q, Zhao J C, et al 2004 Phonon-polariton dispersion and the polariton-based photonic band gap in piezoelectric superlattices *Phys. Rev. B* **69** (8) 085118

[9] Zhao J, Yin R-C, Fan T, et al. Coupled phonon polaritons in a piezoelectric-piezomagnetic superlattice *Phys. Rev. B* **77** (7) 075126

[10] Zhu Y, Zhang X, Lu Y, et al. New type of polariton in a piezoelectric superlattice *Phys. Rev. Lett.* **90** (5) 053903

[11] Piliposyan D G, Ghazaryan K B, and Piliposian G T 2014 Shear Bloch waves and coupled phonon–polariton in periodic piezoelectric waveguides *Ultrasonics* **54** (2) 644–54

[12] Piliposyan D G, Ghazaryan K B, and Piliposian G T 2014 Internal resonances in a periodic magneto-electro-elastic structure *J. Appl. Phys.* **116** (4) 044107

[13] Liu Z and Zhang W 2006 Multiband polaritonic filter made of piezoelectric superlattices *Phys. Lett. A* **351** (3) 192–7

[14] Shuvalov A L and Gorkunova A S 1999 Transverse acoustic waves in piezoelectric and ferroelectric antiphase superlattices *Phys. Rev. B* **59** (14) 9070–7

IOP Conf. Series: Journal of Physics: Conf. Series **991** (2018) 012066 doi:10.1088/1742-6596/991/1/012066

Problems of interaction longitudinal shear waves with V-shape tunnels defect

V G Popov

National University "Odessa Maritime Academy," Odessa, Ukraine

E-mail: dr.vg.popov@gmail.com

Abstract. The problem of determining the two-dimensional dynamic stress state near a tunnel defect of V-shaped cross-section is solved. The defect is located in an infinite elastic medium, where harmonic longitudinal shear waves are propagating. The initial problem is reduced to a system of two singular integral or integro-differential equations with fixed singularities. A numerical method for solving these systems with regard to the true asymptotics of the unknown functions is developed.

1. Introduction

At present, there exist numerous solutions of two-dimensional dynamical problems of the theory of elasticity for bodies with defects shaped as a segment of a straight line or an arc of a smooth curve. Examples of solving such problems can be found in [1–8]. However, the real defects can have corner points, be piecewise smooth, intersect one another, and bifurcate. Practically, the problems of determining the dynamical stress state in a neighborhood of such defects have not yet been solved. This is related to the difficulties arising when they are solved by using the method of boundary integral equations widely applied nowadays and the problems are reduced to singular integro-differential or hypersingular equations with fixed singularities. Most of the studies deal with static problems for bodies whose defects have corner points. First, we mention [9–12], where the exact solution was obtained by the Wiener–Hopf method, and an accurate value of the stress intensity factor was determined. These solutions and the results of [12] indicate that the presence of kernels with fixed singularities influences the singularity of the solutions in neighborhoods of the ends of integration intervals. The stress state near branched, broken, and edge cracks was also determined in [14–19], where the integral equations were numerically solved by the method of mechanical quadratures. This method is based on the application of the Gauss–Chebyshev quadrature formulas that give a root singularity of the solutions. In this case, the true asymptotics of the solution is not taken into account, or an additional condition leading to a singularity weaker than the root one is imposed on the solution. Another drawback of the numerical methods used in those works is the formal application of the Gauss–Chebyshev quadrature formulas to integrals with fixed singularities. As a result, the convergence is rather slow (in order to obtain results with an error less than 0.1%, several dozens of collocation points should be used). In the present paper, we solve the problem of determining the dynamical stress intensity factors for tunnel defects (cracks or thin rigid inclusions) which are V-shaped in a cross-section under the interaction with a harmonic longitudinal shear wave. The problem is reduced to a system of three singular integro-differential or integral equations solved by the

IOP Conf. Series: Journal of Physics: Conf. Series **991** (2018) 012066 doi:10.1088/1742-6596/991/1/012066

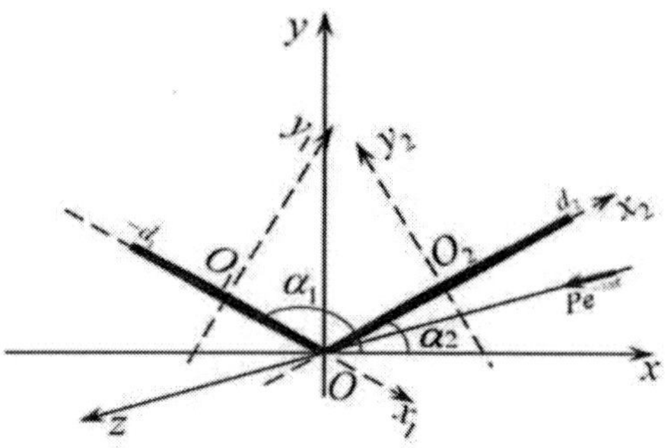

Figure 1. Cross-section of a defect on the coordinate plane.

method of collocations. Within this method, the true singularity of the solution is considered, and special quadrature formulas are used to calculate the integrals with fixed singularities.

2. Statement of the problem and its reduction to a system of singular integral equations

We consider the unbounded isotropic elastic medium under the conditions of antiplane deformation with a through V-shaped defect in a cross-section in the plane Oxy (figure 1).

It may be a crack or a thin rigid inclusion. The cross-section of the defect occupies two segments what are of length $2d_k$ and make angles α_k, $k = 1, 2$, with the axis Ox. The defect interacts with a plane longitudinal shear wave. Its front makes an angle θ_0 with the axis Ox and causes the displacements

$$W_0^{(i)}(x, y) = A_0 \exp[i\kappa_2(x \cos\theta_0 + y \sin\theta_0)], \quad \kappa_2^2 = \frac{\rho\omega^2}{G}, \tag{1}$$

in the medium along the axis Oz. In (1), the symbols G and ρ stand for the shear modulus and the medium density, and ω is the frequency of oscillations. The time dependence is given by the factor $\exp(-i\omega t)$ that is omitted here and below. Let the vector of displacements caused by the wave reflected from the defect have a single component $W(x, y)$ which is nonzero under an antiplane deformation. In the coordinate system Oxy, it satisfies the Helmholtz equation

$$\Delta W + \kappa_2^2 W = 0. \tag{2}$$

To specify the boundary conditions on the defects, we associate the coordinate system with each of their segments. Let these systems $O_k x_k y_k$, $k = 1, 2$ be centered at the middles of segments, and let its centers have coordinates $O_k(d_k \cos\alpha_k, d_k \sin\alpha_k)$. Let

$$W_k(x_k, y_k) = W(a_k + x_k \cos\alpha_k - y_k \sin\alpha_k, b_k + x_k \sin\alpha_k + y_k \cos\alpha_k) \tag{3}$$

be the displacement in the coordinate systems related to the kth segment of the defect. In the case of a thin rigid inclusion, in view of its small thickness, we formulate the corresponding boundary conditions on its median surface. Assume that the condition of perfect contact is realized between the body and the inclusion

$$W_k^1(x_k, 0) = c - W_k^{(i)}(x_k, 0), \quad k = 1, 2, \tag{4}$$

where

$$W_k^{(i)}(x_k, y_k) = W_0^{(i)}((d_k + x_k)\cos\alpha_k - y_k \sin\alpha_k, (d_k + x_k)\sin\alpha_k + y_k \cos\alpha_k).$$

IOP Conf. Series: Journal of Physics: Conf. Series **991** (2018) 012066 doi:10.1088/1742-6596/991/1/012066

On the surface of the inclusion, the tangential stresses are discontinues and their jumps are denoted by

$$\tau_{zy_k}(x_k,+0)-\tau_{zy_k}(x_k,-0)=\chi_{1k}(x_k),\quad -d_k<x_k<d_k,\quad k=1,2. \tag{5}$$

In formula (4), c is the amplitude of longitudinal oscillations of the inclusion in the equation of its motion as a solid body. In the case of harmonic oscillations, this equation becomes

$$-(m_1+m_2)\omega^2 c=P+\sum_{k=1}^{2}\int_{-d_k}^{d_k}\chi_{1k}(\eta)\,d\eta,\quad m_k=2\rho_0 d_k h, \tag{6}$$

where h is the inclusion thickness and ρ_0 is the inclusion density.

In the case of a crack, its surface is assumed to be free form stresses,

$$\tau_{zy_k}(x_k,0)=-\tau^{(i)}_{zy_k}(x_k,0),\quad -d_k<x_k<d_k,\quad k=1,2. \tag{7}$$

In addition, the displacement on the crack surface has a jump denoted by

$$W^1_k(x_k,+0)-W^1_k(x_k,-0)=\chi_{2k}(x_k),\quad -d_k<x_k<d_k,\quad k=1,2. \tag{8}$$

Also, the continuity of displacements along the edge yields

$$W_k(d_k,+0)=W_{k+1}(-d_{k+1},+0),\quad W_k(d_k,-0)=W_{k+1}(-d_{k+1},-0),\quad k=1,2.$$

We start solving the posed problem by constructing a discontinuous solution of the Helmholtz equation in the coordinate system $O_k x_k y_k$, $k=1,2$, for each link of the defect [6]. In the case of an inclusion, this is a solution with jumps (5):

$$W^1_{1l}(x_l,y_l)=\int_{-d_l}^{d_l}\frac{\chi_{1l}(\eta)}{G}r_2(\eta-x_l,y_l)\,d\eta,\quad l=1,2. \tag{9}$$

In the case of a crack, we construct the discontinuous solution with jumps (8):

$$W^d_k(x_k,y_k)=\frac{\partial}{\partial y_k}\int_{-d_k}^{d_k}\chi_k(\eta)r_2(\eta-x_k,y_k)\,d\eta. \tag{10}$$

In formulas (9), (10), we denote

$$r_2(\eta-x_l,y_l)=\frac{i}{4}H_0^{(1)}(\kappa_0\sqrt{(\eta-x_l)^2+y_l^2}),$$

where $H_0^{(1)}(z)$ is the Hankel function. Then the displacements of the reflected wave field can be given in the form

$$W(x,y)=\sum_{k=1}^{3}W^g_k(x,y), \tag{11}$$

where

$$W^g_k(x,y)=W^d_k((x-a_k)\cos\alpha_k+(y-b_k)\sin\alpha_k,-(x-a_k)\sin\alpha_k+(y-b_k)\cos\alpha_k).$$

In order to finally determine the displacements of the diffraction field, we should find the unknown jumps (5) or (8). To this end, we use conditions (4) or (7).

First, we consider the case of inclusion. To determinate jumps (5) from the boundary conditions (4), we obtain a system of singular integral equations. For the singular component of their kernels contain a singularity in the form of the Cauchy kernel, instead of (4), we should use [7] the conditions

$$\frac{\partial W_k(x_k,0)}{\partial x_k} = -\frac{\partial}{\partial x_k} W_k^{(i)}(x_k,0), \quad W_k(-d_k,0) = c, \quad k = 1,2. \tag{12}$$

The result of substitution of (3), (11), and (9) into (12) is the following singular integral equation with an auxiliary condition:

$$\begin{aligned}
&\int_{-1}^{1} \varphi_1(\tau)\left[\frac{1}{\tau-\xi} - R_{11}(\tau-\xi)\right] d\tau - \int_{-1}^{1} \varphi_2(\tau)[q_{12}(\tau,\xi) - R_{12}(\tau,\xi)]\, d\tau = f_1(\xi),\\
&\int_{-1}^{1} \varphi_2(\tau)\left[\frac{1}{\tau-\xi} - R_{22}(\tau-\xi)\right] d\tau - \int_{-1}^{1} \varphi_1(\tau)[q_{21}(\tau,\xi) - R_{21}(\tau,\xi)]\, d\tau = f_2(\xi),\\
&\frac{1}{2\pi}\sum_{l=1}^{2} \gamma_l \int_{-1}^{1} \varphi_l(\tau)[\ln|\tau + (-1)^l| + D_l(\tau)]\, d\tau = c_0, \quad c_0 = \frac{c}{d}.
\end{aligned} \tag{13}$$

In system (13), we introduce the notation

$$\begin{aligned}
&q_{kl}(\tau,\xi) = \frac{\gamma_l(\gamma_l\tau_l\cos\alpha_{lk} + \gamma_k\xi_k)}{s_{kl}^2(\tau,\xi)}, \quad s_{kl}^2(\tau,\xi) = \gamma_l^2\tau_l^2 + 2\gamma_1\gamma_2\tau_l\xi_k\cos\alpha_{lk} + \gamma_k^2\xi_k^2, \quad k,l = 1,2,\\
&\tau_1 = \tau - 1, \quad \tau_2 = \tau + 1, \quad \xi_1 = \xi - 1, \quad \xi_2 = \xi + 1,\\
&f_l(\zeta) = -i\kappa_0 C_0 \cos\theta_{0l} \exp[i\kappa_0\gamma_l(1+\zeta)\cos\theta_{0l}], \quad C_0 = \frac{A}{d}, \quad \kappa_0 = \kappa_2 d,\\
&\varphi_k(\tau) = G^{-1}\chi_{1k}(d_k\tau), \quad \eta = d_k\tau, \quad x_k = d_k\xi, \quad \gamma_k = d^{-1}d_k, \quad d = \max(d_1,d_2), \quad k = 1,2.
\end{aligned} \tag{14}$$

In the case of a crack, the following system of singular integro-differential equations for jumps (8) and their derivatives is obtained after substitution of (3), (10) and (11) into (7):

$$\begin{aligned}
&\frac{1}{2\pi}\int_{-1}^{1} \varphi_1'(\tau)\left[\frac{1}{\tau-\zeta} + R_{11}(\tau-\zeta)\right] d\tau + \frac{1}{2\pi}\int_{-1}^{1} \varphi_2'(\tau)[q_{12}(\tau,\zeta) + R_{12}(\tau,\zeta)]\, d\tau\\
&\quad + \frac{1}{2\pi}\int_{-1}^{1} \varphi_1(\tau)[-\gamma_1^2\kappa_0^2\ln|\tau-\zeta| + U_{11}(\tau-\zeta)]\, d\tau + \frac{1}{2\pi}\int_{-1}^{1} \varphi_2(\tau)U_{12}(\tau,\zeta)\, d\tau = f_1(\zeta),\\
&\frac{1}{2\pi}\int_{-1}^{1} \varphi_2'(\tau)\left[\frac{1}{\tau-\zeta} + R_{22}(\tau-\zeta)\right] d\tau + \frac{1}{2\pi}\int_{-1}^{1} \varphi_1'(\tau)[g_{21}(\tau,\zeta) + R_{21}(\tau,\zeta)]\, d\tau\\
&\quad + \frac{1}{2\pi}\int_{-1}^{1} \varphi_2(\tau)[-\gamma_2^2\kappa_0^2\ln|\tau-\zeta| + U_{22}(\tau-\zeta)]\, d\tau + \frac{1}{2\pi}\int_{-1}^{1} \varphi_2(\tau)U_{21}(\tau,\zeta)\, d\tau = f_2(\zeta).
\end{aligned} \tag{15}$$

We used the following notation in system (15):

$$\begin{aligned}
&q_{kl}(\tau,\zeta) = \frac{(-1)^l\{\gamma_l[1+(-1)^l\tau]\cos\alpha_{kl} + \gamma_k[1+(-1)^k\zeta]\}\gamma_l}{s_{kl}(\tau,\zeta)}, \quad \varphi_k(\tau) = d_k^{-1}\chi_k(d_k\tau),\\
&f_l(\zeta) = -i\kappa_0 C_0 \sin\theta_{0l} \exp[i\kappa_0\gamma_l(1+\zeta)\cos\theta_{0l}], \quad \theta_{0l} = \theta_0 - \alpha_l, \quad k,l = 1,2.
\end{aligned} \tag{16}$$

As is seen in (14) and (16), the functions $q_{kl}(\tau,\xi)$ have singularities at $\tau = \pm 1$, $\xi = \pm 1$. The functions R_{lk}, U_{lk}, $R_{lk}(\tau,\xi)$, $D_l(\tau)$, $k,l = 1,2$, are bounded for $-1 \le \tau$, $\xi \le 1$.

IOP Conf. Series: Journal of Physics: Conf. Series **991** (2018) 012066 doi:10.1088/1742-6596/991/1/012066

3. Solution of the problem in the case of an inclusion

The presence of the fixed singularity $\tau = \pm 1$, $\xi = \pm 1$ in the singular component of the system integral equations (13) affects the asymptotics of its solutions in neighborhoods of the points $\tau = \pm 1$. The singularity of the solutions in neighborhoods of these points can be determined by analyzing the asymptotic properties of singular integrals or by investigating the symbol of the singular kernel [13]. As a result, we established that the unknown function is to be sought in the form

$$\varphi_1(\tau) = (1+\tau)^{-1/2}(1-\tau)^{-\delta}\psi_1(\tau), \quad \varphi_2(\tau) = (1+\tau)^{-\delta}(1-\tau)^{-1/2}\psi_2(\tau), \tag{17}$$

where $\psi_k(\tau)$, $k = 1, 2$, satisfy the Hölder condition, and the exponents are

$$\delta = \begin{cases} \dfrac{\pi - 2\alpha}{2(\pi - \alpha)} & \text{for } 0 < \alpha \le \dfrac{\pi}{2}, \\ \dfrac{2\alpha - \pi}{2\alpha} & \text{for } \dfrac{\pi}{2} < \alpha < \pi, \end{cases} \qquad \alpha = |\alpha_2 - \alpha_1|.$$

Now, if we consider the function

$$\psi_k(\tau) = \psi_{0k}(\tau) + 0.5(1+\tau)\psi_k(1) + 0.5(1-\tau)\psi_k(-1) \tag{18}$$

then we obtain $\psi_{0k}(\pm 1) = 0$. This is why we can assume that

$$\psi_{0k}(\tau) = (1-\tau)^2 g_k(\tau), \quad k = 1, 2, \tag{19}$$

where $g_k(\tau)$ are new unknown functions. Substituting (19) and (18) into (17), we obtain the following representation of the solutions of integrals equation (13):

$$\varphi_k(\tau) = (1-\tau)^{\sigma_k}(1+\tau)^{\lambda_k}g_k(\tau) + 0.5(1+\tau)^{\lambda_k}(1-\tau)^{\sigma_k - 1}\psi_k(1) + 0.5(1+\tau)^{\lambda_k - 1}(1-\tau)^{\sigma_k}\psi_k(-1), \tag{20}$$

Here $\sigma_1 = \lambda_2 = 1 - \delta$, $\sigma_2 = \lambda_1 = 1/2$. Further, the approximate method of solution is based on approximation of the functions $g_k(\tau)$ by the interpolation polynomials of degree $(n-1)$:

$$g_k(\tau) \approx g_{k,n-1}(\tau) = \sum_{m=1}^{n} g_{mk}\frac{Q_{nk}(\tau)}{(\tau - \tau_{mk})Q'_{nk}(\tau_{mk})}, \quad g_{mk} = g_k(\tau_{mk}) \tag{21}$$

where $Q_{n1}(\tau) = P_n^{1-\delta,1/2}(\tau)$, $Q_{n2}(\tau) = P_n^{1/2,1-\delta}(\tau)$ are the Jacobi polynomials and τ_{m1}, τ_{m2} are roots of these polynomials. Then, for the singular integral with Cauchy kernel, the following quadrature formula holds [20]:

$$\int_{-1}^{1}\frac{\varphi_k(\tau)}{\tau - \xi_{jk}}d\tau = \sum_{m=1}^{n} g_{mk}\frac{A_{mk}}{\tau_{mk} - \xi_{jk}} + \frac{\psi_k(-1)}{2}b_{jk}^{-} + \frac{\psi_k(1)}{2}b_{jk}^{+}, \tag{22}$$

where ξ_{j1}, ξ_{j2} $(j = 1, \ldots, n+1)$ are roots of the Jacobi functions of the second kind $J_n^{1/2,1-\delta}(\xi)$, $J_n^{1-\delta,1/2}(\xi)$, and A_{mk} are the coefficients of the corresponding Gauss–Jacobi quadrature formula. The use of representation (20) requires to calculate the following integrals with Cauchy kernel:

$$b_{kj}^{+} = \int_{-1}^{1}\frac{(1-\tau)^{\sigma_k}(1+\tau)^{\lambda_k - 1}}{\tau - \xi_{kj}}d\tau, \quad b_{jk}^{-} = \int_{-1}^{1}\frac{(1-\tau)^{\sigma_k - 1}(1+\tau)^{\lambda_k}}{\tau - \xi_{jk}}d\tau. \tag{23}$$

We obtain their values by the method described in [21] and based on their transformation to the Mellin convolution. Further, applying the theorem on convolution, we can represent these integrals as the sum of residuals at the poles of the integrands.

IOP Conf. Series: Journal of Physics: Conf. Series **991** (2018) 012066 doi:10.1088/1742-6596/991/1/012066

Then we calculate the integrals with a fixed singularity

$$E_{kl}^{j} = \int_{-1}^{1} q_{kl}(\tau, \xi_{jk})(1-\tau)^{\sigma_l}(1+\tau)^{\lambda_l} g_l(\tau)\, d\tau. \tag{24}$$

Let $0 < r_1 < 1$ be a positive number. In the case where $1 \pm \xi_k > r_1$, the integral (24) is not singular, and it can be calculated by the Gauss–Jacobi quadrature formulas with the corresponding weight function [22]. If $1 \pm \xi_k \to 0$, then we should replace $\varphi_k(\tau)$ by expression (20) and use a method based on using the Mellin integral transformation [21]. We represent these integrals in terms of the following power series which converge for $0 \le 1 \pm \xi_k < r_1 < 1$:

$$\int_{-1}^{1} q_{kl}(\tau, \xi_{jk})\varphi_l(\tau)\, d\tau = \sum_{m=1}^{n} g_{m1} \frac{H_m^{kl}}{s_{kl}^2(\tau_{ml}, \xi_{jk})} + \frac{\psi_l(1)}{2} s_{jkl}^{+} + \frac{\psi_l(-1)}{2} s_{jkl}^{-}, \tag{25}$$

$$H_m^{kl} = A_{ml}\{\gamma_l[\tau_{ml} + (-1)^l]\cos\alpha_{kl} + \gamma_k[\xi_{jk} + (-1)^k]\} - \frac{\gamma_l^2[\tau_{ml} + (-1)^l]}{Q_n'(\tau_{ml})} h_{jl}^2 - \frac{\gamma_l\gamma_k[\xi_{jk} + (-1)^k]}{Q_n'(\tau_{ml})} h_{jl}^3,$$

$$h_{ml}^1 = J_n^{\sigma_l, \lambda_l}(\tau_{ml}) = A_{ml} Q_{nl}^l(\tau_{lm}),$$

$$h_{jl}^2 = \frac{\Gamma(\frac{3}{2}+n)(-1)^n}{\gamma_l} a_{kl}\left(\frac{1+(-1)^k\xi_{jk}}{2\beta_{kl}}\right), \quad h_{jl}^3 = \frac{\Gamma(\frac{3}{2}+n)(-1)^n}{\gamma_l} b_{kl}\left(\frac{1+(-1)^k\xi_{jk}}{2\beta_{kl}}\right), \quad \beta_{kl} = \frac{\gamma_l}{\gamma_k},$$

$$a_{kl}(y) = \frac{2^{3/2-\delta}}{n!}\left\{\frac{y^{1-\delta}}{\sin(\pi\delta)} \sum_{p=0}^{\infty} z_{1p} y^p \cos[(\pi - \alpha_{kl})(\delta - p - 2)] - \cot(\pi\delta) \sum_{p=0}^{\infty} z_{2p} y^p \cos[\alpha_{kl}(p+1)]\right\},$$

$$b_{kl}(y) = \frac{2^{3/2-\delta}}{n!}\left\{\frac{y^{1-\delta}}{\sin(\pi\delta)} \sum_{p=0}^{\infty} z_{1p} y^p \cos[(\pi - \alpha_{kl})(\delta - p - 1)] + \cot(\pi\delta) \sum_{p=0}^{\infty} z_{2p} \cos(\alpha_{kl} p)\right\};$$

$$z_{1p} = \frac{\Gamma(-0.5 - n + p)\Gamma(2 + n + p - \delta)(-1)^p}{p!\Gamma(2 + p - \delta)}, \quad z_{2p} = \frac{\Gamma(-1.5 - n + \delta + p)\Gamma(1 + n + p)}{p!\Gamma(\delta + p)}.$$

The integrals

$$s_{jkl}^{+} = \int_{-1}^{1} q_{kl}(\tau, \xi_{jk})(1-\tau)^{\sigma_l - 1}(1+\tau)^{\lambda_l}\, d\tau, \quad s_{jkl}^{-} = \int_{-1}^{1} q_{kl}(\tau, \xi_{jk})(1-\tau)^{\sigma_l}(1+\tau)^{\lambda_l - 1}\, d\tau$$

are computed by a method similar to that used to calculate integrals (23).

To calculate the integrals with a logarithmic singularity, we use representation (20) and approximate the function $g_k(\tau)$ by the interpolation polynomial (21), which we preliminarily transform according to the Darboux–Christoffel identity [22]. Thereafter we obtain the formula

$$\int_{-1}^{1} \varphi_l(\tau) \ln|\tau + (-1)^l|\, d\tau = \sum_{m=1}^{n} A_{ml} g_{ml} \theta_{ml} + \frac{\psi_l(1)}{2} E_l^{-} + \frac{\psi_l(-1)}{2} E_l^{+}, \tag{26}$$

where

$$\theta_{ml} = \sum_{s=0}^{n-1} \frac{u_{s1}}{\sigma_s^2} Q_{sl}(\tau_{ml}), \quad u_{s1} = -a_s, \quad u_{s2} = (-1)^{s+1} a_s, \quad a_s = \frac{2^{5/2-\delta}\Gamma(1.5+s)\Gamma(2-\delta)}{\Gamma(3.5-\delta+s)s!},$$

$$E_1^{\pm} = \frac{2^{1/2-\delta}\Gamma(\frac{2\pm 1}{2})\Gamma(\frac{3\mp 1}{2} - \delta)}{\Gamma(2.5-\delta)}\left[\ln 2 + \Psi\left(\frac{3\mp 1}{2} - \delta\right) - \Psi\left(\frac{5}{2} - \delta\right)\right].$$

Here $\Psi(x)$ is the logarithmic derivative for the Γ function.

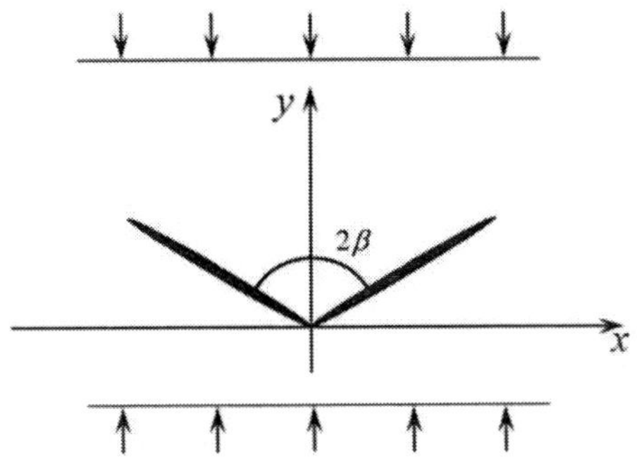

Figure 2. Shape of the defect for numerical analysis.

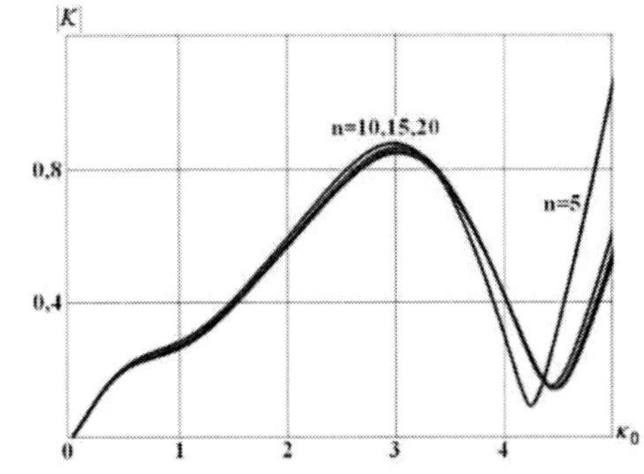

Figure 3. Numerical investigation of the practical convergence.

The obtained formulas for the singular integrals (22), (25), (26) and Gauss–Jacobi quadrature formulas applied to the integrals with regular kernels allow us to replace (13) and (6) by a system of linear algebraic equations. As a result of solving this system, we obtain $g_{m1} = g(\tau_{m1})$, $g_{m2} = g(\tau_{m2})$, $\psi_1(\pm 1)$, $\psi_2(\pm 1)$, c_0. Then the approximate solution of the system is determined by formulas (20) and (21). One of the important characteristics of the stress state near the inclusion is the stress intensity factor (SIF). It is determined from the known asymptotic representation [23] of stresses in neighborhoods of the end of the inclusion. We obtain the following approximate formulas for the dimensionless SIF:

$$K_1 = G\sqrt{d_1}2^{-\delta}\psi_1(-1), \quad K_2 = G\sqrt{d_2}2^{-\delta}\psi_2(1). \tag{27}$$

On the basis of the obtained formulas, the dependence of SIF on the dimensionless wave number $\kappa_0 = \kappa_2 d$ is investigated. For this purpose, the inclusion with links of equal length d symmetrical with respect to the axis Oy was considered (figure 2). The links of inclusion come from the origin of coordinates and make an angle 2β between them. Realizing the proposed method of solution, we first perform numerical investigation of its practical convergence. The calculation was carried out for the following data:

$$\bar{\rho} = \frac{\rho_k}{\rho} = 1, \quad \varepsilon = \frac{h_k}{d} = 0.05, \quad \theta_0 = 90^\circ, \quad \beta = 30^\circ.$$

Under these conditions, the equality $K_1 = K_2 = K$ is executed. As results of computation by formula (27), we construct graphs of the dependence of the absolute value of the dimensionless SIF $k = K/(G\sqrt{d})$ on the dimensionless wave number, which is shown in figure 3.

The curves correspond to the number of interpolation nodes $n = 5, 10, 15, 20$ in approximation (21). We can see that, in the region of low frequencies $\kappa_0 < 2$, to attain a fairly high accuracy (the relative error does not exceed 1%), it suffices to use $n = 5$ interpolation nodes. In the whole frequency range $0 \le \kappa_0 \le 5$, to obtain values of SIF with an error below 0.1%, it suffices to have $n = 20$ interpolation nodes in formula (21).

The graphs in figures 4 and 5 illustrate the influence of the angle β between the inclusion links on the dependence of SIF on the frequency κ_0. The curves correspond to the values: $1 — \beta = 5^\circ$, $2 — 45^\circ$, $3 — 60^\circ$, $4 — 87^\circ$. The graphs in figure 4 are constructed at the angle of wave propagation $\theta_0 = 90^\circ$, and in figure 5, at the angle $\theta_0 = 270^\circ$. We can see that a decrease in the angle between the inclusion links lead to a complication of frequency dependence of SIF. This complication is the appearance of local maxima and minima. The smallest values of SIF are observed for a rectilinear inclusion $\beta \approx 90^\circ$.

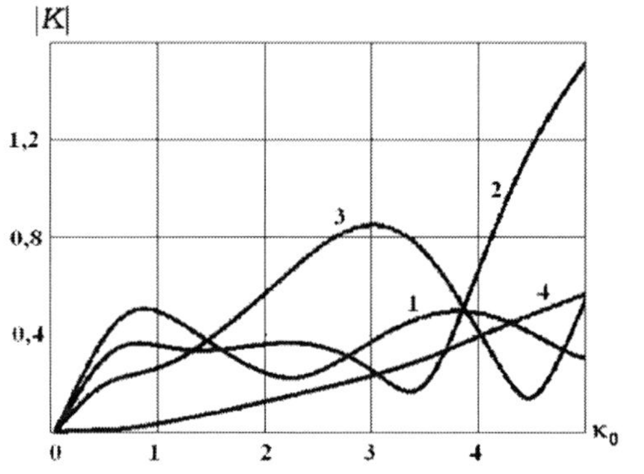

Figure 4. Dependence of SIF on the frequency ($\theta_0 = 90°$).

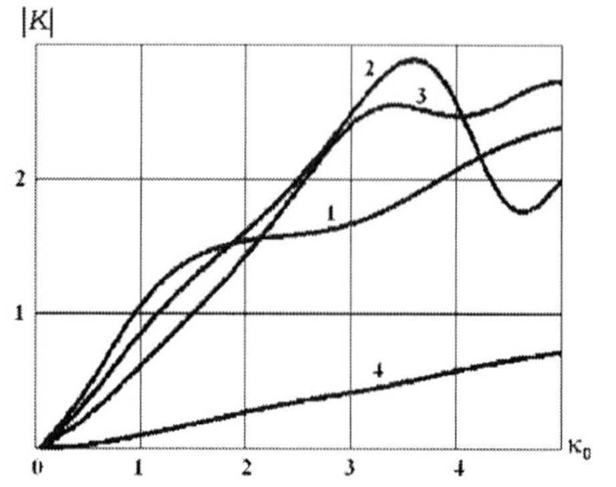

Figure 5. Dependence of SIF on the frequency ($\theta_0 = 270°$).

4. Solution of the problem in the case of a crack

In case of a crack, we represent the derivatives of solutions of system (15) in the form

$$\varphi_l'(\tau) = W_l(\tau) g_l(\tau), \quad l = 1,2, \quad W_1(\tau) = (1-\tau)^{-\sigma}(1+\tau)^{-1/2}, \quad W_2(\tau) = (1-\tau)^{-1/2}(1+\tau)^{-\sigma}, \tag{28}$$

where the functions $g_l(\tau)$ $(l = 1,2)$ satisfy the Hölder condition and the exponents are

$$\sigma = \frac{\pi - \alpha}{2\pi - \alpha}, \quad \alpha = |\alpha_2 - \alpha_1|, \quad 0 \le \alpha \le \pi.$$

Further, as in the previous case we approximate the function $g_l(\tau)$ by interpolation polynomials of degree $(n-1)$ by formula (21), where $Q_{1n}(\tau) = P_n^{-\sigma-1/2}(\tau)$, $Q_{2n}(\tau) = P_n^{-1/2-\sigma}(\tau)$ are the Jacobi polynomials and τ_{m1}, τ_{m2} are roots of these polynomials. Then, for the singular integral with Cauchy kernel, the following quadrature formula holds [20]:

$$\int_{-1}^{1} \frac{\varphi_l'(\tau)}{\tau - \zeta_{lj}}\, d\tau = \sum_{m=1}^{n} g_{lm} \frac{A_{lm}}{\tau_{lm} - \zeta_{lj}}, \quad l = 1,2, \quad j = 1,2,\ldots,n. \tag{29}$$

where ξ_{j1}, ξ_{j2} $(j = 1,2,\ldots,n)$ are roots of the Jacobi functions of the second kind $J_n^{-\sigma-1/2}(\tau)$, $J_n^{-1/2-\sigma}(\tau)$ and A_{lm} are the coefficients of the corresponding Gauss–Jacobi quadrature formula.

Further, we obtain similar formulas for the integrals with fixed singularities

$$E_{kl}^{j} = \int_{-1}^{1} q_{kl}(\tau, \xi_{jk}) \varphi_l'(\tau)\, d\tau. \tag{30}$$

If $1 \pm \xi > r_1 > 0$, then it follows from relations (16) that the functions $q_{kl}(\tau, \zeta)$ are infinitely differentiable. Therefore, we can apply the Gauss–Jacobi quadrature formulas to integrals (30). The principal difficulty is related to the calculation of these integrals as $1 \pm \xi \to 0$. Then, as in the previous case, we should replace $\varphi_l'(\tau)$ by expressions (28), (21) and again use the method based on the Mellin integral transformation [21]. The final formulas for calculating the integrals (30) become

$$E_{kl}^{j} = \sum_{m=1}^{n} g_{ml} \frac{H_m^{kl}}{s_{kl}^2(\tau_{ml}, \xi_{jk})}, \tag{31}$$

where the coefficients H_m^{kl} are represented in terms of the following power series similar to (25) and converging for $0 \le 1 \pm \xi < r_1 < 1$.

Since system (15) also contains the functions $\varphi_l(\tau)$, it is necessary to obtain the quadrature formulas for the corresponding integrals. First, we use the obvious formulas

$$\varphi_1(\tau) = \int_{-1}^{\tau} \varphi_1'(x)\,dx, \quad \varphi_2(\tau) = -\int_{\tau}^{1} \varphi_2'(x)\,dx. \tag{32}$$

Substituting expressions (28), (21) into formulas (32) and integrating, we obtain the representation

$$\varphi_l(\tau) = [1-(-1)^l\tau]^{1/2} \sum_{m=1}^{n} A_{lm}\psi_{lm}S_{lm}(\tau), \quad l=1,2, \tag{33}$$

$$S_{lm}(\tau) = F\left(\sigma, \frac{1}{2}, \frac{3}{2}, \frac{1-(-1)^l\tau}{2}\right)\frac{(-1)^l 2^{1-\sigma}}{\sigma_0} - [1+(-1)^l\tau]^{1-\sigma}\sum_{j=1}^{n}\frac{Q_{lj}(\tau_{lm})Q_{lj}^{(1)}(\tau)}{2j\sigma_j^2},$$

where $Q_{1j}(\tau) = P_j^{1-\sigma,1/2}(\tau)$, $Q_{2j}(\tau) = P_j^{1/2,1-\sigma}(\tau)$ are the Jacobi polynomials and F is a hypergeometric function. Representation (33) is the basis for the following quadrature formulas for integrals with unknown functions:

$$\begin{gathered}\int_{-1}^{1} \varphi_k(\tau)\begin{bmatrix} U_{kk}(\tau-\zeta_{kj}) \\ U_{lk}(\tau,\zeta_{kj})\end{bmatrix} d\tau = \sum_{m=1}^{n} A_{km}\psi_{km}\begin{bmatrix} U_{jm}^{kk} \\ U_{jm}^{lk}\end{bmatrix}, \\ \begin{bmatrix} U_{jm}^{kk} \\ U_{jm}^{lk}\end{bmatrix} = \sum_{p=1}^{n} B_{kp}S_{km}(z_{kp})\begin{bmatrix} U_{kk}(z_{kp}-\zeta_{kj}) \\ U_{lk}(z_{kp},\zeta_{kj})\end{bmatrix}\end{gathered} \tag{34}$$

where z_{kp} are roots of the Jacobi polynomials $P_n^{0\,1/2}(z_{1p}) = 0$, $P_n^{1/2\,0}(z_{2p}) = 0$ and $B_{1p} = A_p^{0\,1/2}$, $B_{2p} = A_p^{1/2\,0}$ are the coefficients of the corresponding Gauss–Jacobi quadrature formula.

Let us consider the integrals with logarithmic difference kernel from system (16). As a result of the integration by parts and the use of representations (28) and (21) for the derivatives of the unknown functions, we obtain the quadrature formulas

$$\int_{-1}^{1} \varphi_k(\tau)\ln|\tau-\zeta_{lj}|\,d\tau = \sum_{m=1}^{n} A_{lm}H_{jm}^{l}, \quad l=1,2, \tag{35}$$

$$H_{jm}^{(l)} = (-1)^l[1+(-1)^l\zeta_{lj}][\ln|1+(-1)^l\zeta_{lj}|-1] - (\tau_{lm}-\zeta_{lj})(\ln|\tau_{lm}-\zeta_{lj}-1).$$

Let us use quadrature formulas (29), (31), (34), (35), the Gauss–Jacobi quadrature formulas, and the roots of Jacobi polynomials as points of collocation. Then system (15) can be reduced to a system of linear algebraic equations for $g_{mk} = g_k(\tau_{mk})$. By solving this system, we obtain the following formulas for the approximate values of SIF:

$$K_l = -G\sqrt{d_l}2^{-(1+\sigma)}Q_{ln}((-1)^l)\sum_{m=1}^{n}\frac{g_{lm}}{P_{ln}^{l}(\tau_{lm})[1-(-1)^l\tau_{lm}]}.$$

The numerical realization of the proposed method is carried out for the crack shown in figure 3. The calculation was carried out for the same data as in the case of inclusion. The results of studying the practical convergence of the proposed method are given in figure 6 as graphs of the absolute value of SIF. Here $k = K/(G\sqrt{d})$ as a function of the dimensionless wave number $\kappa_0 = \kappa_2 d$. The curves correspond to the numbers of interpolation nodes

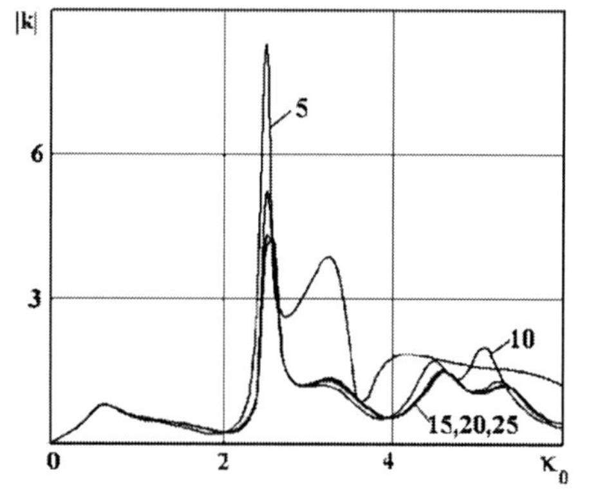

Figure 6. Numerical investigation of the practical convergence.

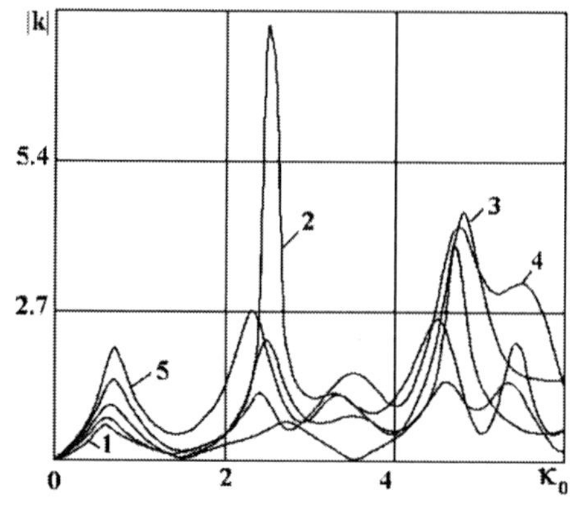

Figure 7. Dependence of SIF on the frequency ($\theta_0 = 90°$).

$n = 5, 10, 15, 20, 25$ in approximation (21). These results show that, in the region of low frequencies $\kappa_0 < 1$, to attain a fairly high accuracy (the relative error does not exceed 1%), it suffices to use $n = 5$ interpolation nodes. In the whole considered frequency range $0 \leq \kappa_0 \leq 6$, to obtain values of SIF with an error below 0.1%, it suffices to have $n = 20$ interpolation nodes in formula (21). We also numerically studied the influence of the angle β between the crack links on the dependence of SIF on the frequency. The results of calculations are given in figure 7. The curves correspond to values: 1 — $\beta = 15°$, 2 — 30°, 3 — 45°, 4 — 60°, 5 — 87° and are constructed at the angle of wave propagation $\theta_0 = 90°$. We can see that, at low frequencies $0 \leq \kappa_0 \leq 2$, as the angle increases, the value of SIF also increases. The largest values of SIF are observed as the crack is transformed in a segment of the straight line $\beta \approx 90°$. For the further increase in the frequency $2 \leq \kappa_0 \leq 6$, the dependence becomes more complicated and has many maxima and minima. This complication is related to the multiple reflections of waves from the sides of the angle made by the cracks links.

Conclusions

In the presence of angular points in cracks and thin inclusions the method of integral equations reduce boundary value problems to singular integral or integro-differential equations with fixed singularities. For the numerical solution of such equations one must take into account the real property of unknown functions and apply special quadrature formulas for singular integrals. It ensures rapid convergence of numerical method of solution. On the example of V-shaped defects it can be seen the shape and type of defect substantially influence on stress state in their region. In particular, the angle formed by the sides of the defect substantially influence on the SIF dependence on the frequency. So, in the low frequency region. In the case of crack, the largest values of the SIF are observed for strait lines cracks. If the defect is inclusion for this angle values of the SIF will be the smallest. The presence of maximus resonance type in the region of high frequencies caused by the multiple reflection of the waves from the defects sides is shown.

References

[1] Kit H S, Kunets Ya I, and Yemets V F 1999 Elastodynamic scattering from a thin-walled inclusion of low rigidity *Int. J. Engng Sci.* **37** 331–45
[2] Olsson P 1986 Elastodynamic scattering by fluid-filled nonplanar cracks *J. Nondestruct. Eval.* **5** (3) 161–8
[3] Olsson P, Datta S, and Boström A 1990 Elastodynamic scattering from inclusion surrounded by thin interface layers *J. Appl. Mech.* **57** (3) 672–6

[4] Kunets' Ya I, Matus Ya I, and Porokhovs'kyi V V 2000 Scattering of pulses of elastic SH-waves on a thin-walled elastic curvilinear inclusion *Mat. Met. Fiz.-Mekh. Polya* **43** (4) 150–4

[5] Takahuda Kuzuo, Tukizava Yasushi, Koizumi Tukashi, and Shibuya Toshikazu 1984 Dynamic interactions between cracks. Diffraction of SH waves being incident of Griffith cracks in an infinite body *Trans. Jpn. Soc. Mech. Eng. Ser. A* **50** (452) 799–804

[6] Popov V G 1992 Investigation of fields of displacements and stresses under diffraction of elastic shear waves on a thin rigid delaminated inclusion *Izv. Ross. Akad. Nauk. Mekh. Tverd. Tela* No. 3 139–46

[7] Popov V G 1993 Comparison of displacement fields and stress fields under diffraction of elastic shear waves on different defects: crack and thin rigid inclusion *Dinam. Sistemy* **12** 35–41

[8] Gross D and Zhang Ch 1988 Diffraction of SH waves by a system of cracks: Solution by integral equation method *Int. J. Solids Struct.* **24** (1) 41–9

[9] Afyan B A 1984 On the integral equations with fixed singularities in the theory of branched cracks *Dokl. Akad. Nauk Arm. SSR* **79** (4) 177–81

[10] Vitek V 1977 Plane strain stress intensity factors for branched cracks *Int. J. Fract.* **13** (4) 481–501

[11] Atkinson C 1973 Some ribbon-like inclusion problems *Int. J. Engng Sci.* **11** (2) 243–66

[12] Stullybrass M P 1970 A crack perpendicular to an elastic half-plane *Int. J. Engng Sci.* **8** (5) 351–62

[13] Duduchava R V 1979 *Integral Equations in Convolution with Discontinuous Presymbols, Singular Integral Equations with Fixed Singularities, and Their Applications to Some Problems of Mechanics* (Tbilisi: Metsniereba)

[14] Osiv P N and Savruk M P 1983 Determination of stresses in an infinite plate with broken or branching crack *J. Appl. Mech. Techn. Phys.* **24** (2) 266–71

[15] Savruk M P 1981 *Two-dimensional problems of elasticity for bodies with cracks* (Kiev: Naukova Dumka) p 323 [in Russian]

[16] Isida M and Noguchi H 1992 Stress intensity factors at tips of branched cracks under various loadings *Int. J. Fract.* **54** (4) 293–316

[17] Li Z and Achenbach I 1991 Interaction of rayleigh wave with a disband in a material interphase normal to a free surface *Ultrasonics* **29** (1) 45–52

[18] Gupta G D and Erdogan F E 1974 The problem of edge cracks in an infinity strip *Appl. Mech.* **41** (4) 1001–6

[19] Erdogan F, Gupta G, and Cook T 1973 Numerical solution of singular integral equations In *Methods of Analysis and Solutions of Crack Problems. Mechanics of Fracture* ed G C Sih vol 1 (Dordrecht: Springer) 368–425

[20] Andreev A V 2005 Direct numerical method for solving singular integral equations of the first kind with generalized kernels *Mech. Solids* **40** (1) 104–19

[21] Popov G Ya 1982 *Concentration of Elastic Stresses near Stamps, Cuts, Thin Inclusions and Reinforcements* (Moscow: Nauka) p 342 [in Russian]

[22] Krylov V I 1967 *The Approximate Calculation of Integrals* (Moscow: Nauka) p 500 [in Russian]

[23] Sulym H 2007 *Fundamentals of the Mathematical Theory of Thermoelastic Equilibrium of Deformable Solid Bodies with Thin Inclusions* (Lviv: Shevchenko Scientific Society) p 715 [in Ukrainian]

IOP Conf. Series: Journal of Physics: Conf. Series **991** (2018) 012067 doi:10.1088/1742-6596/991/1/012067

Modeling of thin-walled structures interacting with acoustic media as constrained two-dimensional continua

L N Rabinskiy[1] **and S I Zhavoronok**[2*]

[1]Moscow Aviation Institute (National Research University), Moscow, Russia
[2]Institute of Applied Mechanics of Russian Academy of Sciences, Moscow, Russia

E-mail: *zhavoronok@iam.ras.ru

Abstract. The transient interaction of acoustic media and elastic shells is considered on the basis of the transition function approach. The three-dimensional hyperbolic initial boundary-value problem is reduced to a two-dimensional problem of shell theory with integral operators approximating the acoustic medium effect on the shell dynamics. The kernels of these integral operators are determined by the elementary solution of the problem of acoustic waves diffraction at a rigid obstacle with the same boundary shape as the wetted shell surface. The closed-form elementary solution for arbitrary convex obstacles can be obtained at the initial interaction stages on the background of the so-called "thin layer hypothesis". Thus, the shell–wave interaction model defined by integro-differential dynamic equations with analytically determined kernels of integral operators becomes hence two-dimensional but nonlocal in time. On the other hand, the initial interaction stage results in localized dynamic loadings and consequently in complex strain and stress states that require higher-order shell theories. Here the modified theory of I.N.Vekua–A.A.Amosov-type is formulated in terms of analytical continuum dynamics. The shell model is constructed on a two-dimensional manifold within a set of field variables, Lagrangian density, and constraint equations following from the boundary conditions "shifted" from the shell faces to its base surface. Such an approach allows one to construct consistent low-order shell models within a unified formal hierarchy. The equations of the Nth-order shell theory are singularly perturbed and contain second-order partial derivatives with respect to time and surface coordinates whereas the numerical integration of systems of first-order equations is more efficient. Such systems can be obtained as Hamilton–de Donder–Weyl-type equations for the Lagrangian dynamical system. The Hamiltonian formulation of the elementary Nth-order shell theory is here briefly described.

1. Introduction

In general, a transient problem of interaction of a shell with a fluid can be solved using the finite element discretization of both the liquid and the deformable solid together with appropriate numerical algorithms, the so-called "fluid–structure interaction" [1–6]. Such an approach results in models with many thousands of degrees of freedom and leads to several computational difficulties [3, 4] especially for thin-walled and light-weight structures [7] that force to develop new computational approaches (see, e.g., [8, 9]). In particular, special finite volume schemes are required to suppress high-frequency oscillations near multiple wave fronts in liquids [10, 11].

Another way to study the transient dynamics of thin-walled structures interacting with liquids is based on the use of the elementary solution apparatus and leads to time-boundary

IOP Conf. Series: Journal of Physics: Conf. Series **991** (2018) 012067 doi:10.1088/1742-6596/991/1/012067

integral equations. The kernels of integral operators are determined by Green's functions of the corresponding unsteady problems of fluid dynamics with appropriate boundary conditions on the fluid-solid interfaces [12, 13]. Thus, the problem can be divided into two steps: at the first step, the initial boundary-value problem of wave diffraction at curvilinear obstacles must be solved, whereas the second step consists in solving the time-boundary integral equations. The diffraction problem can be linearized using the acoustic media approximation for liquids, so the solution can be finally determined as the superposition of incident waves, reflected waves, and waves radiated by the deformed solid [14, 15]. The closed-form definition of kernels of integral operators makes the time-boundary integral equations much more efficient. The appropriate analytical solutions of unsteady diffraction problems can be obtained on the basis of the integral transforms or integral equation [16–18] which are usually restricted by the obstacles on the surfaces of canonical geometry. In general, the solutions for arbitrary-shaped convex obstacles can be found using various simplifying assumptions estimated in [19–21]. The so-called "modified thin-layer hypothesis" [12, 13, 22–25] improving the known "plane reflection hypothesis" [26, 27] is used below; it consists in taking the medium curvature of the wetted surface of the obstacle into account and remains consistent with the wave–structure interaction for a short time after the contact initiation [27].

If the kernel of the integral operator is found, then the unsteady problem of shell dynamics is formulated as a system of integro-differential equations; in other words, the shell model is hence nonlocal [12, 13, 23–25]. The closed-form analytical solution for such problems can be constructed in the case of a shell with canonically shaped base surface, otherwise, numerical approaches [12, 13, 25, 28] or the eigenfunction expansion [13] can be useful. The numerical integration of singularly perturbed dynamic equations of various shell theories may be difficult due to high-frequency oscillatory solutions [10]; this drawback can be eliminated by using specific numerical methods [29, 30]. The eigenfunction expansion offers the possibility of reducing the problem by choosing an appropriate eigenfunction subset making a decisive effect on the dynamics of linear or nonlinear systems [31], but the shell dynamics at the initial diffraction stages is too complex to be adequately defined within a few eigenfunctions due to the local loading behavior. At the same time, the coupling of transverse shear and normal strains must be taken into account to compute higher-order eigenfunctions, moreover, the shell kinematics under high-frequency oscillations cannot be adequately defined within the assumptions of the classical shell model [32–37]. Thus, the higher-order shell theories are strongly required.

A wide range of nonclassical shell theories exists nowadays such as the well-known assumption-based approach [32, 38, 39], the asymptotic integration [40–42], the power or Fourier series expansion [43–36], etc. The asymptotic method allows a qualitative analysis of the shell dynamics [41, 47, 48]. On the other hand, the formal expansion approach offers the possibility of constructing a hierarchy of various-order shell theories using a unified formalism [43, 44, 49–51] and can be efficiently combined with numerical methods such as finite element modeling. This unified formalism consists in the Lagrangian continuum dynamics [52, 53]: a shell model is defined on the two-dimensional manifold within a set of field variables and Lagrangian density. The field variables are introduced as coefficients of a biorthogonal expansion for the three-dimensional displacement vector field [52], therefore the Legendre polynomials as compact functions can also be used as a base system. Such a theory shows its efficiency in steady problems of "thin-walled structure" dynamics [54–56] as in the transient dynamics [49]. An improved formalism of the so-called "extended shell theory" [57, 58] accounts for the boundary conditions on the shell faces as constraint equations for a variational problem of Lagrangian dynamics; moreover, it allows one to obtain the asymptotically consistent Kirchhoff model within a unified formal hierarchy without supplementary assumptions [59]. The extended Nth-order shell theory is used below to obtain an approximate model of transient fluid–shell interaction.

2. Approximate solution for the diffraction of a pressure wave at a rigid shell

2.1. Basics of the thin layer interaction model

Let us consider the diffraction of acoustic waves at a rigid shell which is a three-dimensional body $V \subset \mathrm{R}^3$, $\partial V = S_\pm \oplus S_B$, with faces $S_\pm \neq \emptyset$ and lateral surface S_B:

$$\forall M_* \in \bar{V} \quad \mathbf{R}(M_*) = \mathbf{R}(\xi^\alpha, \xi^3) = \xi^\alpha \mathbf{r}_\alpha + \xi^3 \mathbf{n}, \quad \mathbf{r}_\alpha(\xi^\alpha) = \partial_\alpha \mathbf{r}(\xi^\alpha) \equiv \partial_\alpha \mathbf{R}(\xi^\alpha, 0),$$

where $\xi^\alpha \in D_\xi \subseteq \mathrm{R}^2$, $\alpha = 1, 2$, are curvilinear coordinates on the wetted surface S_+ considered as the base, $\partial S = S \cup S_B$; $\xi^3 \in [-2h(\xi^\alpha), 0]$ is the normal coordinate, $\forall M \in S_+$ $\xi^3 = 0$; $\mathbf{r}(\xi^\alpha)$ are base vectors in the tangent fibration $T_M S$, and $\mathbf{n}$ is a normal unit vector; $\partial_i \equiv \partial / \partial \xi^i$ denote the partial differentiation with respect to the curvilinear coordinates ξ^i, $i = 1, 2, 3$.

The equations for acoustic media dynamics can be written as [12, 13]

$$\rho_0 \dot{\mathbf{v}} = -\nabla p, \quad \dot{p} = -\rho_0 c^2 \nabla \cdot \mathbf{v}, \tag{1}$$

where p is the pressure of the acoustic media, ∇ is the Nabla operator, and $\mathbf{v} = v^i \mathbf{R}_i$ is the velocity vector referred to the basis of the frame ξ^1, ξ^2, ξ^3 attached to the wetted surface S_+:

$$\mathbf{R}_\alpha = A_{\alpha\cdot}^{\cdot\beta} \mathbf{r}_\beta, \quad \mathbf{R}_3 \equiv \mathbf{n}, \quad A_{\alpha\cdot}^{\cdot\beta} = \delta_\alpha^\beta - \xi^3 b_\alpha^\beta, \quad |\mathbf{n}| = 1,$$

where $\mathbf{n}$ is the normal unit vector at a point $M_* \in S_+$, $A_{\alpha\cdot}^{\cdot\beta}$ are components of the parallel shift tensor [43], b_α^β is the curvature tensor of the surface S_+, and δ_α^β denotes the Kronecker delta.

Let us express equations (1) in terms of the physical components of the velocity vector v_{i0}:

$$\rho_0 \dot{v}_{k0} = -\partial_k p \sqrt{g^{kk}}, \quad \dot{p} = -\rho_0 c^2 g^{-1/2} \partial_i \left(v_{j0} g^{ij} \sqrt{\frac{g}{g^{jj}}} \right), \quad v_{k0} = v_k \sqrt{g^{kk}}. \tag{2}$$

Here c is the speed of sound, ρ_0 is the liquid mass density, $g_{ij} = \mathbf{R}_i \cdot \mathbf{R}_j$ is the covariant metric, g^{ij} is the contravariant metric, and $g = \det(g_{ij})$. Let us note that $g_{33} = g^{33} = 1$ and $g_{\alpha 3} \equiv 0$. Here the summation rule is used for both Latin and Greek indices except for the indices k and κ.

Hence let us consider the acoustic media dynamics in the layer of thickness δ near the wetted surface S_+, and let L be a characteristic dimension of the shell. The dimensionless coordinates $\tilde{\xi}^\alpha = \xi^\alpha / L^{m_\alpha}$, $\alpha = 1, 2$, $\zeta = \xi^3/\delta$, the time $\tau = ct/\delta$, the pressure $\tilde{p} = p/(\rho_0 c^2)$, and the velocity $\tilde{v}_{i0} = v_{i0}/c$ can be introduced. Thus, we have the following formulation of equations (2):

$$\dot{v}_{\kappa 0} = -\gamma \partial_\kappa p \sqrt{g^{\kappa\kappa}}, \quad \dot{v}_{30} = -\partial_3 p, \quad \dot{p} = -g^{-1/2} \left[\gamma \partial_\alpha \left(v_{\alpha 0} g^{\alpha\kappa} \sqrt{\frac{g}{g^{\kappa\kappa}}} \right) + \partial_\zeta (v_{30} \sqrt{g}) \right], \tag{3}$$

where $\gamma = \delta/L$, $\kappa = 1, 2$, and the partial differentiation is considered with respect to the dimensionless coordinates $\tilde{\xi}^\alpha$ and time τ. Here and below, the tilde symbol is omitted for convenience.

Let the acoustic layer be sufficiently thin, so that $\gamma \to 0$, then the corresponding terms in equation (3) vanish. This hypothesis becomes the first postulate of the thin layer theory. Of course, the introduced assumption fails if the time t increases; therefore, the developed model can be used to study the initial stage of dynamic interaction between acoustic waves and shells. As a result, we can simplify equations (3) by omitting the terms with γ [12]:

$$\dot{v}_{\alpha 0} = 0, \quad \dot{v}_{30} = -\partial_\zeta p, \quad \alpha = 1, 2; \quad \dot{p} = -g^{-1/2} \partial_\zeta (v_{30} \sqrt{g}). \tag{4}$$

Considering equations (4) and letting $\gamma \to 0$, we obtain the following approximate equation for the dimensionless pressure p (for more details, see [12] and [13]):

$$\ddot{p} = \partial_{\zeta\zeta} p + \frac{\partial_\zeta g}{4g} \partial_\zeta p, \quad g = \det(g_{ij}). \tag{5}$$

Let us consider the known expressions for the metric determinant g (see, e.g., [13, 52]):

$$\frac{\partial_\zeta g}{4g} = -\frac{H - \zeta K}{1 - 2\zeta H + \zeta^2 K} \approx -H, \quad 2H = b^\alpha_\alpha, \quad K = \det(b^\beta_\alpha), \tag{6}$$

where H is the medium curvature of the wetted surface S_+. This quantity is invariant under the frame $O\xi^1\xi^2$, we can write the final equation of the thin layer theory that depends only on the intrinsic geometry of the wetted surface S_+ [12, 13]:

$$\ddot{p} = \partial_{\zeta\zeta} p - 2H\partial_\zeta p. \tag{7}$$

2.2. Approximate transition functions based on the thin layer assumption
Let us pose the initial boundary-value problem for the transient function on the basis of (7):

$$\ddot{G}^\varsigma = \partial_{\zeta\zeta} G^\varsigma - 2H\partial_\zeta G^\varsigma, \tag{8}$$

$$G^\varsigma\big|_{\tau=0} = \dot{G}^\varsigma\big|_{\tau=0} = 0, \tag{9}$$

$$(\delta_{1\varsigma} v_n + \delta_{2\varsigma} p)\big|_{\zeta=0} = \delta(\tau)\delta(\xi^1 - \xi^1_0)\delta(\xi^2 - \xi^2_0), \tag{10}$$

$$G^\varsigma(\tau, r) = O(1), \quad r \to \infty, \quad r^2 = g_{ij}\xi^i\xi^j. \tag{11}$$

Here $G^\varsigma(r,\tau)$ corresponds to the acoustic medium pressure, $v_n = \mathbf{v}\cdot\mathbf{n}$ is the velocity normal to the wetted surface S_+, and $\delta(\tau)$, $\delta(\xi)$ are Dirac functions. The upper index $\varsigma = 1$ corresponds to the boundary-value problem of the first kind and $\varsigma = 2$ to the boundary-value problem of the second kind given by the general boundary condition (10) (for more details, see [12, 13]).

Let us apply the Laplace transform to problem (8)–(11) in the time domain:

$$\frac{d^2\bar{G}^\varsigma}{d\zeta^2} - 2H(\xi^\alpha)\frac{d\bar{G}^\varsigma}{d\zeta} - s^2\bar{G}^\varsigma = 0, \tag{12}$$

$$(\delta_{1\varsigma}\bar{v}_{30} + \delta_{2\varsigma}\bar{p})\big|_{\zeta=0} = \delta(\xi^1 - \xi^1_0)\delta(\xi^2 - \xi^2_0), \tag{13}$$

$$\bar{G}(r, s; \xi^j_0) = O(1), \quad r \to \infty. \tag{14}$$

Taking into account boundary conditions (13) and (14), we have the following general solution of linear equation (12) for the Laplace image $\bar{G}^\varsigma$ of the transition function G^ς:

$$\bar{G}^\varsigma = \frac{s e^{-\lambda\zeta}}{\delta_{1\varsigma}\lambda + \delta_{2\varsigma} s}\delta(\xi^1 - \xi^1_0)\delta(\xi^2 - \xi^2_0), \tag{15}$$

$$\lambda(\xi^\alpha, s) = -H + \sqrt{H^2 + s^2} \quad (\alpha = 1, 2), \tag{16}$$

where the root corresponds to $\Re\sqrt{\cdot} > 0$. Let us define the surface transient function image $\bar{G}^\varsigma_0$:

$$\bar{G}^\varsigma_0 = \bar{G}^\varsigma\big|_{\zeta=0} = s\frac{1}{\delta_{1\varsigma}\lambda + \delta_{2\varsigma} s}\delta(\xi^1 - \xi^1_0)\delta(\xi^2 - \xi^2_0). \tag{17}$$

With regard to (16), for the initial boundary-value problem of the first kind. we have

$$\bar{G}^1_0 = \frac{s}{-H + \sqrt{H^2 + s^2}}\delta(\xi^1 - \xi^1_0)\delta(\xi^2 - \xi^2_0), \tag{18}$$

whereas in case of the initial boundary-value problem of the second kind, we have

$$\bar{G}^2_0 = \delta(\xi^1 - \xi^1_0)\delta(\xi^2 - \xi^2_0). \tag{19}$$

The transient function G_0^1 can be obtained by the inverse transform of (18). Bearing in mind

$$\frac{H}{s^2} \div H\tau, \qquad \frac{1}{\sqrt{H^2+s^2}} \div J_0(H\tau),$$

and the Laplace image multiplication theorem which implies

$$\frac{H^2}{s^2\sqrt{H^2+s^2}} \div H^2\int_0^\tau (\tau-t)J_0(Ht)\,dt = \tau^2\left\{J_0(z) - \frac{\pi}{2}[J_0(z)\mathrm{H}_1(z) - J_1(z)\mathrm{H}_0(z)]\right\} + \frac{\tau}{H}J_1(z),$$

where $z = -H\tau$, $J_0(z)$, $J_1(z)$ are Bessel functions of the first kind, and $\mathrm{H}_0(z)$, $\mathrm{H}_1(z)$ are Struve functions of the zeroth and first orders (for more details, see [13]), we finally obtain

$$G_0^1(\xi^\alpha,\tau;\xi_0^\alpha) = \dot{F}^1(\xi^\alpha,\tau)\delta(\xi^1-\xi_0^1)\delta(\xi^2-\xi_0^2), \quad F^1(\xi^\alpha,\tau) = \Theta(\tau)\{Q(z) - z[1+J_1(z)]\}, \tag{20}$$
$$Q(z) = J_0(z) + z^2\left\{J_0(z) - \frac{\pi}{2}[J_0(z)\mathrm{H}_1(z) - J_1(z)\mathrm{H}_0(z)]\right\},$$

where $\Theta(t)$ is the Heaviside function. For the time derivative $\dot{F}^1(\xi^\alpha,\tau)$, we have

$$\dot{F}^1(\xi^\alpha,\tau) = \delta(\tau) - H[Q'(z) - zJ_0(z) - 1]\Theta(\tau), \tag{21}$$
$$Q'(z) = -J_1(z) + 2z\left\{J_0(z) - \frac{\pi}{4}[J_0(z)\mathrm{H}_1(z) - J_1(z)\mathrm{H}_0(z)]\right\}.$$

Thus, we can construct the following integral formula for the pressure p on the wetted surface S_+:

$$p(\xi^\alpha,\tau) = \int_0^\tau dt\int_{S_+} G_0^1(\xi^\alpha,\tau-t;\xi_0^\alpha)v_n(\xi_0^\alpha,t)\,dS_+ = v_n(\xi^\alpha,\tau) -$$
$$- H(\xi^\alpha)\int_0^\tau \{Q'[-H(\xi^\alpha)(\tau-t)] + H(\xi^\alpha)(\tau-t)J_0[-H(\xi^\alpha)(\tau-t)] - 1\}v_n(\xi^\alpha,t)\,dt. \tag{22}$$

Here the first term corresponds to the plane reflection assumption, whereas the second term accounts for the curvature of the wetted surface of the obstacle [12, 13]. As a result, we must know the normal velocity v_n on the wetted surface to approximate the pressure. In the case of elastic shell, taking the boundary condition on S_+ into account, we obtain

$$\dot{u}_3\big|_{S_+} = (\mathbf{v}_* \cdot \mathbf{n} + v_n)\big|_{S_+}, \tag{23}$$

where u_3 is the displacement of the wetted shell surface and $\mathbf{v}_*$ is the incident wave velocity. As a result, the shell is loaded by the summary pressure defined by the formula [13]

$$p(\xi^\alpha), t) = p_*(\xi^\alpha,t)\big|_{S_+} + G^1(\xi^\alpha,t) \star\star\star [\dot{u}_3 - \mathbf{v}_*\cdot\mathbf{n}]\big|_{S_+}, \tag{24}$$

where p_* is the pressure in the incident acoustic wave computed on the outer surface of the shell and the symbol "$\star$" denotes the integral convolution corresponding to (22).

3. Two-dimensional approximate model of the wave–shell interaction

3.1. Constrained Lagrangian model of the wave–shell interaction

Let the shell be elastic. Thus, let $\mathbf{s}$ be a stress tensor, $\mathbf{C}$ be an elasticity tensor, and $\mathbf{u} = u_\alpha \mathbf{r}^\alpha + u_3 \mathbf{n}$ be a displacement vector field. The higher-order shell model can be defined on the two-dimensional manifold corresponding to the base surface by the set of field variables $u_i^{(k)}(\xi^\alpha, t)$, $i = 1, 2, 3$, $k = 0, 1, \ldots, N$, $\alpha = 1, 2$ and the Lagrangian density $\mathcal{L}_S(u_i^{(k)}, \dot{u}_i^{(k)}, \bar{\nabla}_\beta u_i^{(k)})$ in terms of the dimensional reduction concept [52, 53]. The field variables are introduced by means of the biorthogonal expansion of the covariant displacement components $u_i(\xi^\alpha, z, t)$ with respect to the base system $\mathrm{p}_{(k)}(z)$; here z is the dimensionless thickness variable [52]. As a result, we have the following Lagrangian density formulation (for more details, see [52] and [53]):

$$\mathcal{L}_\mathcal{M} = \frac{1}{2}\rho_{(k)}^{(m)}(\dot{u}_{(m)}^{\alpha}\dot{u}_\alpha^{(k)} + \dot{u}_{(m)}^{3}\dot{u}_3^{(k)}) - \frac{1}{2}\Big[s_{(k)}^{\gamma\beta}(\bar{\nabla}_\gamma u_\beta^{(k)} - b_{\gamma\beta}u_3^{(k)}) + s_{(m)}^{33}D_{(k\cdot)}^{(\cdot m)}u_3^{(k)}$$
$$+ s_{(k)}^{\gamma 3}(\bar{\nabla}_\gamma u_3^{(k)} + b_\gamma^\alpha u_\alpha^{(k)}) + s_{(m)}^{3\beta}D_{(k\cdot)}^{(\cdot m)}u_\beta^{(k)}\Big] + P_{(k)}^\alpha u_\alpha^{(k)} + P_{(k)}^3 u_3^{(k)}, \qquad (25)$$

where $H_{\beta(k\cdot)}^{(\cdot m)}$ and $D_{(k\cdot)}^{(\cdot m)}$ are linear operators and $s_{(k)}^{ij}$ are generalized forces [52]:

$$s_{(k)}^{ij} = \bar{\bar{C}}_{(km)}^{ij\gamma\delta}\bar{\nabla}_\delta u_\gamma^{(m)} + \bar{\bar{C}}_{(km)}^{ij3\delta}\bar{\nabla}_\delta u_3^{(m)} + \bar{C}_{(km)}^{ij\gamma}u_\gamma^{(m)} + \bar{C}_{(km)}^{ij3}u_3^{(m)}, \quad i, j = 1, 2, 3, \quad \gamma, \delta = 1, 2. \quad (26)$$

The use of Legendre polynomials $\mathrm{p}_{(k)}(z)$ allows one to construct traditional shell theories [43, 44, 49, 54, 55], whereas the compact functions result in "finite layer" shell models. Let us note that the boundary conditions on the faces $S_\pm$ of a shell including the no-flow condition (23) can be satisfied only approximately as a result of convergence of the two-dimensional solution sequence as $N \to \infty$ [51], moreover, the first-order order shell theory given in [25] requires the plane stress assumption to be consistent [43, 45]. To eliminate this drawback of the dimensional reduction, the so-called extended shell theory of the Nth order [57–59] can be constructed.

Let us consider the boundary conditions on the faces $S_\pm$ given as

$$\mathbf{s}\big|_\pm \cdot \mathbf{n}_\pm \equiv [\mathbf{C} : (\nabla \otimes \mathbf{u})]\big|_\pm \cdot \mathbf{n}_\pm = \mathbf{q}_\pm, \qquad (27)$$

where $\mathbf{q}_\pm$ is the resultant force vector and $\mathbf{n}_\pm$ are normal unit vectors on $S_\pm$. The boundary conditions (27) can be "shifted" from $S_\pm$ to the base surface by representing the vector $\mathbf{u}$ in the basis $\mathbf{r}^\alpha, \mathbf{n}$ [57]. The wetted surface S_+ is considered below as the base surface according to the above-formulated statement of the diffraction problem, therefore, the no-flow condition does not require further transformations, and moreover $q_-^i \equiv 0$ and $q_+^\alpha \equiv 0$ because the acoustic media are nonviscous. Substituting the biorthogonal expansion of the displacement vector $(u_\alpha \mathbf{r}^\alpha + u_3\mathbf{n})\mathrm{p}_{(k)}$ into (27) and taking (24) into account, we obtain the equations

$$\bar{\bar{C}}_{+(k)}^{i\gamma\beta}(\bar{\nabla}_\delta u_\gamma^{(k)} + H_{\delta(m\cdot)}^{(\cdot k)}u_\gamma^{(m)} - b_{\gamma\delta}u_3^{(k)}) + \bar{\bar{C}}_{+(k)}^{i\gamma 3}(\bar{\nabla}_\delta u_3^{(k)} + H_{\delta(m\cdot)}^{(\cdot k)}u_3^{(m)} + b_\delta^\gamma u_\gamma^{(k)})$$
$$+ \bar{C}_{+(k)}^{i3j}h^{-1}D_{(m\cdot)}^{(\cdot k)}u_j^{(m)} - \delta_3^i\big[p_*\big|_{S_+} + G^1 \star\star\star (\dot{u}_3^{(k)}\mathrm{p}_{(k)} - \mathbf{v}_* \cdot \mathbf{n})\big|_{S_+}\big] = 0, \qquad (28)$$

$$\bar{\bar{C}}_{-(k)}^{i\gamma\beta}(\bar{\nabla}_\delta u_\gamma^{(k)} + H_{\delta(m\cdot)}^{(\cdot k)}u_\gamma^{(m)} - b_{\gamma\delta}u_3^{(k)}) + \bar{\bar{C}}_{-(k)}^{i\gamma 3}(\bar{\nabla}_\delta u_3^{(k)} + H_{\delta(m\cdot)}^{(\cdot k)}u_3^{(m)} + b_\delta^\gamma u_\gamma^{(k)})$$
$$+ \bar{C}_{-(k)}^{i3j}h^{-1}D_{(m\cdot)}^{(\cdot k)}u_j^{(m)} = 0, \quad i, j = 1, 2, 3. \qquad (29)$$

Here the "surface" values of the generalized elastic constants $\bar{C}^{ijpq}$ are introduced [57, 58] as

$$\bar{\bar{C}}_{\pm(k)}^{ij\delta} = \bar{C}_\pm^{i3j\delta}\mathrm{p}_{(k)}\big|_{S_\pm} + h_\beta^\pm \bar{\bar{C}}_\pm^{i\beta j\delta}\mathrm{p}_{(k)}\big|_{S_\pm}, \quad \bar{\bar{C}}_\pm^{ijpq} = \bar{\bar{C}}^{ijpq}\big|_{S_\pm}.$$

Equations (28), (29) become the constraint equations for the variational problem given by the Lagrangian (25); thus, the shell model becomes nonlocal due to the integral operators (22). The generalized Lagrange equations following from (25) and (28), (29) and their natural boundary conditions can be represented in the formulation [57, 58], where the pressure (24) is taken into account:

$$\rho_{(k)}^{(m)}\ddot{u}_{(m)}^{\alpha} = \bar{\nabla}_{\beta}\tilde{s}_{(k)}^{\alpha\beta} - H_{\beta(k\cdot)}^{(\cdot m)}\tilde{s}_{(m)}^{\alpha\beta} - b_{\beta}^{\alpha}\tilde{s}_{(k)}^{3\beta} - h^{-1}D_{(k\cdot)}^{(\cdot m)}\tilde{s}_{(m)}^{\alpha 3}, \tag{30}$$

$$\rho_{(k)}^{(m)}\ddot{u}_{(m)}^{3} = \bar{\nabla}_{\beta}\tilde{s}_{(k)}^{3\beta} - H_{\beta(k\cdot)}^{(\cdot m)}\tilde{s}_{(m)}^{3\beta} + b_{\alpha\beta}\tilde{s}_{(k)}^{\alpha\beta} - h^{-1}D_{(k\cdot)}^{(\cdot m)}\tilde{s}_{(m)}^{33}$$
$$+ p_*\big|_{S_+} + G^1 \star\star\star (\dot{u}_3^{(k)}\mathrm{p}_{(k)} - \mathbf{v}_*\cdot\mathbf{n})\big|_{S_+}, \tag{31}$$

$$(\tilde{s}_{(k)}^{i\beta}\nu_{\beta} - q_{B(k)}^{i})\delta u_i^{(k)}\big|_{\partial S_+} = 0, \quad i = 1,2,3, \quad \gamma,\beta = 1,2. \tag{32}$$

The shell model consists in (30)–(32), the initial conditions (for more details, see [52, 57]):

$$u_i^{(k)}\big|_{t=0} = U_i^{(k)}, \quad \dot{u}_i^{(k)}\big|_{t=0} = V_i^{(k)}, \quad i = 1,2,3, \quad k = 0,\ldots,N, \tag{33}$$

the constraint equations (28), (29), and the constitutive equations (34) [57] with (26) taken into account:

$$\tilde{s}_{(k)}^{ij} = \bar{\bar{C}}_{(km)}^{ij\gamma\delta}\bar{\nabla}_{\delta}u_{\gamma}^{(m)} + \bar{\bar{C}}_{(km)}^{ij3\delta}\bar{\nabla}_{\delta}u_3^{(m)} + \bar{C}_{(km)}^{ij\gamma}u_{\gamma}^{(m)} + \bar{C}_{(km)}^{ij3}u_3^{(m)} + \lambda_{\varepsilon}^{\pm}\bar{\bar{C}}_{\pm(k)}^{\varepsilon ij} + \lambda_3^{\pm}\bar{\bar{C}}_{\pm(k)}^{3ij}; \tag{34}$$

the forces $\tilde{s}_{(k)}^{ij}$ are consistent with the boundary conditions (27) by means of the Lagrange multipliers $\lambda_i^{\pm}$.

3.2. Hamilton–de Donder–Weyl shell model

The obtained system of equations (30), (31) contains the partial derivatives of the second order, whereas the efficient numerical integration requires systems of first-order partial differential equations [30]. The first-order equations can be obtained using the generalized Hamiltonian formalism [60].

Let us apply the Lagrange transform with respect to time and space derivatives to (25):

$$\mathcal{H}_S^{DW} = p_{(k)}^{\alpha}\dot{u}_{\alpha}^{(k)} + p_{(k)}^{3}\dot{u}_3^{(k)} - s_{(k)}^{\gamma\beta}\bar{\nabla}_{\gamma}u_{\beta}^{(k)} - s_{(k)}^{\gamma 3}\bar{\nabla}_{\gamma}u_3^{(k)} - \mathcal{L}_S, \tag{35}$$

where the quasicanonical polymomenta are defined as follows:

$$p_{(k)}^{\alpha} = \frac{\partial\mathcal{L}_S}{\partial\dot{u}_{\alpha}^{(k)}}, \quad p_{(k)}^{3} = \frac{\partial\mathcal{L}_S}{\partial\dot{u}_3^{(k)}}, \quad s^{\gamma\beta} = -\frac{\partial\mathcal{L}_S}{\partial(\bar{\nabla}_{\gamma}u_{\beta}^{(k)})}, \quad s^{\gamma 3} = -\frac{\partial\mathcal{L}_S}{\partial(\bar{\nabla}_{\gamma}u_3^{(k)})}. \tag{36}$$

Taking into account (25), (35), and (36), we obtain the system of equations

$$\dot{u}_{\alpha}^{(k)} = \frac{\partial\mathcal{H}_S^{DW}}{\partial p_{(k)}^{\alpha}}, \quad \dot{u}_3^{(k)} = \frac{\partial\mathcal{H}_S^{DW}}{\partial p_{(k)}^{3}}, \tag{37}$$

$$\dot{p}_{(k)}^{\alpha} - \bar{\nabla}_{\beta}s_{(k)}^{\alpha\beta} = -\frac{\partial\mathcal{H}_S^{DW}}{\partial u_{\alpha}^{(k)}}, \quad \dot{p}_{(k)}^{3} - \bar{\nabla}_{\beta}s_{(k)}^{\beta 3} = -\frac{\partial\mathcal{H}_S^{DW}}{\partial u_3^{(k)}}, \tag{38}$$

$$\bar{\nabla}_{\beta}u_{\alpha}^{(k)} = -\frac{\partial\mathcal{H}_S^{DW}}{\partial s_{(k)}^{\alpha\beta}}, \quad \bar{\nabla}_{\beta}u_3^{(k)} = -\frac{\partial\mathcal{H}_S^{DW}}{\partial s_{(k)}^{\beta 3}}. \tag{39}$$

Here equations (37) define the generalized velocities $\dot{u}_i^{(k)}$, (38) are dynamics equations, and (39) are constitutive equations of the elementary Nth-order shell theory. Expressing the covariant derivatives in terms of partial derivatives and Christoffel symbols and then transposing them to the right-hand sides, we obtain the quasicanonical formulation of system (37)–(39) corresponding to the Hamilton–de Donder–Weyl formalism of analytical continuum dynamics [60, 61].

Conclusions

Thus, the approximate theory of transient interaction between acoustic waves and elastic anisotropic shells is formulated for higher-order shell theories. The proposed theory allows one to model the unsteady shell dynamics at the initial stage of the shell–wave interaction, considers the coupling of the transverse shear and normal deformation and takes the higher degrees of freedom into account i n addition to the translation and rotation at a point of the base surface. Two formulations are shown, one is based on the Lagrangian formalism of analytical continuum mechanics and results in second-order partial differential equations, and the other is based on the Hamilton–de Donder–Weyl formalism and leads to dynamic equations with first-order partial derivatives.

Acknowledgments

This work was supported by the Russian Foundation for Basic Researches under grants Nos. 16-01-00751-a, 17-01-00837-a, and 17-08-01461-a.

References

[1] Wagner M. 2000 Hybride Randelementmethode in der Akustik und zur Struktur-Fluid-Interaktion *Berlin Inst. A. Mech.* No. 4 1–82

[2] Guriswamu Guru P 2002 A review of numerical fluids/structures interaction methods for computations using high-fidelity equations *Comput. Struct.* **80** (1) 31–41

[3] Farhat C 2004 CFD-based nonlinear computational aeroelasticity In *Encyclopaedia of Computational Mechanics* ed E Stein, R de Borst, and T J R Hughes (Chichester: John Wiley & Sons Ltd.) pp 459–80

[4] Golovanov A I, Tyuleneva O I, and Shigabutdinov A F 2006 *Finite Element Method in Statics and Dynamics of Thin-Walled Structures* (Moscow: FIZMATLIT) [in Russian]

[5] Liao K and Hu C 2013 A coupled FDM-FEM method for free surface flow interaction with thin elastic plate *J. Marine Sci. Technol.* **18** (1) 1–11

[6] Jith J and Sarkar S 2017 Acousto-elastic interactions in high-pressure CO_2 centrifugal compressors *J. Vibr. Acoust. Trans. ASME* **139** (6) 061013

[7] Wuchner R, Kupzok A, and Bletzingr K-U 2006 Simulation of fluid–structure interaction with free form membrane structures using an implicit coupling scheme with adaptive under relaxation In *Eur. Conf. on Computational Fluid Dynamics (ECOMMACS CFD 2006)* ed P Wesswling, E Onate, and J Periaux (Netherlands: TU Delft) pp 1–17

[8] Idelsohn S R, Marti J, Sonto-Iglesias A, and Onate E 2008 Unified Lagrangian formulation for elastic solids and incompressible fluids: application to fluid–structure interaction problem via PFEM *Comput. Methods Appl. Mech. Engng* **197** 1762–6

[9] Antoci S, Gallati M, and Cibilla S 2007 Numerical simulation of fluid-structure by SPH *Comput. Struct* **85** 879–90

[10] Bazhenov V G and Chekmaryov D T 2001 Numerical methods for solving problems of unsteady dynamics of thin-walled structures *Mech. Solids* **36** (5) 126–40

[11] Borisova N M and Ostapenko V V 2003 Accuracy of commutations of unsteady shock waves in shock-fitting methods *Comput. Math. Math. Phys.* **43** (10) 1437–58

[12] Medvedskii A L and Rabinskii L N 2007 *Method of Surface Influence Functions in Unsteady Diffraction Problems* (Moscow: MAI) p 256 [in Russian]

[13] Zhavoronok S I, Kuprikov M Yu, Medvedskii A L, and Rabinskii L N 2010 *Analytical and Numerical Methods of Solution for Problems of Diffraction of Acoustical Waves on Solids and Shells* (Moscow: FIZMATLIT) p 192 [in Russian]

[14] Grigolyuk E I and Gorshkov A G 2976 *Interaction of Elastic Structures and a Liquid* (Leningrad: Sudostroenie) p 200 [in Russian]

[15] Gorshkov A G and Tarlakovskii D V 1990 *Transient Aerohydroelasticity of Spherical Bodies* (Moscow: Nauka) p 264 [in Russian]

[16] Belov V E, Gorsky S M, Zalezsky A A, and Zinovyev A Y 1998 Application of the integral equation method to acoustic wave diffraction from elastic bodies in a fluid layer *J. Acoust. Soc. Amer.* **1998** (3) 1288–95

[17] Slepyan L I and Sorokin S V 1995 Analysis of structural-acoustic coupling problems by a two-level boundary integral method. Part 1: A general formulation and test problems *J. Sound Vibr.* **184** (2) 195–211

[18] Slepyan L I and Sorokin S V 1995 Analysis of structural-acoustic coupling problems by a two-level boundary

integral method. Part 2: Vibrations of a cylindrical shell of finite length in an acoustic medium *J. Sound Vibr.* **184** (2) 213–28

[19] Grigolyuk E I and Gorshkov A G 1971 Ineraction of weak shock waves with elastic structures In *Proc. of Institute of Mechanics of Moscow State Univ.* no. 13 (Moscow: MGU) pp 1–180 [in Russian]

[20] Grigolyuk E I and Gorshkov A G 1973 Determination of hydrodynamic loads during interaction of weak pressure waves and elastic shells In *Oscillations, Damping, and Radiation of Elastic Structures* (Moscow: Nauka) pp 3-11 [in Russian]

[21] Bedrosian B and Di Maggio F L 1972 Acoustic approximations in fluid–shell interactions *J. Engng Mech. Div. Proc. Amer. Soc. Civ. Engng* **98** (3) 731–42

[22] Gorshkov A G, Medvedsky A L, Rabinsky L N, and Zhavoronok S I 2003 Plane problem of diffraction of a pressure wave at a curvilinear obstacle *Mech. Solids* **38** (3) 118–23

[23] Gorshkov A G, Medvedsky A L, Rabinsky L N, and Zhavoronok S I 2004 Plane problems of diffraction of an acoustical wave at a thin orthotropic panel placed in a rigid shield *Mech. Solids* **39** (1) 161–70

[24] Gorshkov A G, Medvedskiy A L, Rabinskiy L N, and Zhavoronok S I 2007 A nonsteady problem of cylindrical pressure wave's diffraction at thin elliptical shell *Int. J. Comput. Civil Struct. Engng* **1** (2) 82–93

[25] Rabinskiy L N and Zhavoronok S I 2006 Axisymmetrical problem of transient interaction of an acoustical pressure wave with an elastic shell of revolution *Mekh. Kompoz. Mater. Konstr.* **12** (4) 541–54

[26] Gorshkov A G 1968 Interaction of plane acoustical shock waves with rigid and elastic shells *Mech. Solids* No. 1 157–8

[27] Gorshkov A G 1981 Unsteady interaction of plates and shells with continuum media *Mech. Solids* **16** (4) 170–89

[28] Egorova O V, Rabinskiy L N, and Zhavoronok S I 2010 Middle thickness shell's interaction with acoustical wave *Vestn. Mosk. Aviats. Inst.* **17** (2) 127–35

[29] Evseev E G and Semyonov A Yu 1990 Numerical method of solution of equations systems of thin shell dynamics *Mech. Solids* **25** (4) 785–8

[30] Kulikovskii A G, Pogorelov N V, and Semyonov A Yu 2001 Mathematical Problems of Numerical Integration of Hyperbolic Equations Systems (Moscow: FIZMATLIT) [in Russian]

[31] Amabili M 2013 Reduced-order models for nonlinear vibrations based on natural modes: the case of the circular cylindrical shell *Phil. Trans. Roy. Soc. A* **371** (1993) 2012.0474

[32] Grigolyuk E I and Selezov I T 1973 Nonclassical theories of vibrations of bars, plates, and shells In *Results of Science and Engineering. Mechanics of Deformable Solids* (Moscow: VINITI) [in Russian]

[33] Brischetto S and Carrera E 2010 Importance of higher-order modes and refined theories in free vibration analysis of composite plates *J. Appl. Math.* **77** (1) 011013

[34] Gupta A, Talha M, and Singh B H 2016 Vibration characteristics of functionally graded material plate with various boundary constraints using higher-order shear deformation theory *Compos. Part B: Engng* **94** (1) 64–74

[35] Gupta A and Talha M 2016 An assessment of a nonpolynomial based higher-order shear and normal deformation theory for vibration response of gradient plates with initial geometric imperfections *Compos. Part B: Engng* **107** (1) 141–61

[36] Kang J H and Leissa A W 2004 Three-dimensional vibration analysis of thick, complete conical shells *Trans. ASME* **71** (6) 502–7

[37] Sheikholeslami A and Saidi A R 2013 Vibration analysis of functionally graded rectangular plates resting on elastic foundation using higher-order shear and normal deformation theory *Compos. Struct.* **106** 350–61

[38] Ambartsumyan S A 1974 *General Theory of Anisotropic Shells* (Moscow: Nauka) [in Russian]

[39] Reddy J N and Arciniega R A 2004 Shear deformation plate and shell theories: form Stavsky to present *Mech. Adv. Mater. Struct.* **11** (6) 535–82

[40] Goldenveizer A L 1961 *Theory of Elastic Thin Shells* (Oxford: Pergamon Press)

[41] Kaplunov J D, Kossovich L Yu, and Nolde E V 1998 *Dynamics of Thin-Walled Elastic Bodies* (San Diego - London - Boston - New York - Sydney - Tokyo - Toronto: Academic Press)

[42] Aghalovyan L 2015 *Asymptotic Theory of Anisotropic Plates and Shells* (Singapore - London: World Scientific Publ.)

[43] Vekua I N 1985 *Shell Theory: General Methods of Construction* (Boston: Pitman Adv. Publ. Program)

[44] Khoma I Yu 1986 *General Anosotropic Shell Theory* (Kiev: Naukova Dumka) [in Russian]

[45] Kienzler R and Altenbach H 2004 *Theories of Plates and Shells: Critical Review and New Applications* (Berlin: Springer-Verlag)

[46] Annin B D and Volchkov Yu M 2016 Nonclassical models of the theory of plates and shells *J. Appl. Math. Tech. Phys.* **57** (5) 769–76

[47] Aghalovyan L 2017 On space dynamic problems of plates and shells *Izv. Nats. Akad. Nauk Armenii. Mekh.* **70** (1) 3–21

IOP Conf. Series: Journal of Physics: Conf. Series **991** (2018) 012067 doi:10.1088/1742-6596/991/1/012067

[48] Goldenveizer A L and Kaplunov J D 1988 Dynamic boundary layer in problems of vibrations of shells *J. Appl. Math. Mech.* **32** (4) 704–18

[49] Gulyaev V I, Bazhenov V A, and Lizunov P P 1978 *Nonclassical Shell Theory and Applications in Engineering* (Kiev: Vishcha Shkola) [in Russian]

[50] Gordeziani D, Avalishvili M, and Avalishvili G 2006 Hierarchical models for elastic shells in curvilinear coordinates *Comput. Math. Appl.* **51** 1789–808

[51] Zhavoronok S I, Leontiev A N, and Leontiev K A 2010 Analysis of thick-walled rotation shells based on Legendre polynomials *Int. J. Comput. Civil Struct. Engng* **6** (1-2) 105–11

[52] Zhavoronok S I 2014 A Vekua-type linear theory of thick elastic shells *ZAMM* **91** (1-2) 164–84

[53] Zhavoronok S I 2014 Variational formulations of Vekua-type shell theories and some their applications *Shell Structures: Theory and Applications* vol 3 ed W Pietraszkiewicz and J Gorski (Leiden: CRC Press / Balkema, Taylor & Francis Gr.) pp 341–4

[54] Zhavoronok S I 2010 Investigation of harmonic waves in elastic layer using Nth-order three-dimensional shell theory *Mekh. Kompoz. Mater. Konstr.* **16** (5) 693–701

[55] Zhavoronok S I 2012 Kinematics of normal modes in elastic layer for some wavenumber's investigation based on Nth-order three-dimensional shells'theory *Mekh. Kompoz. Mater. Konstr.* **18** (1) 45–56

[56] Zhavoronok S I 2012 A formulation of the three-dimensional approximated shells theory of Nth-order using generalized displacements and its application to steady dynamics *Mekh. Kompoz. Mater. Konstr.* **18** (3) 333–44

[57] Zhavoronok S I 2015 On the variational formulation of the extended thick anisotropic shells theory of I. N. Vekua type *Proc. Engng* **111** 888–95

[58] Zhavoronok S I 2015 The generalized Lagrange equations of the second kind for the extended three-dimensional Nth-order theory of anisotropic shells *Mekh. Kompoz. Mater. Konstr.* **21** (3) 370–81

[59] Zhavoronok S I 2016 The extended shell theory of Vekua–Amosov type and the low-order plate models *Int. J. Comput. Civil Struct. Engng* **12** (4) 27–35

[60] Kastrup H A 1983 Canonical theories of lagrangian dynamical systems in physics *Phys. Rep.* **101** (1-2) 1–167

[61] Kilchevskiy N A, Kilchinskaya G A, and Tkachenko N E *Fundamentals of the Analytical Mechanics of Continuous Systems* (Kiev: Naukova Dumka) [in Russian]

IOP Conf. Series: Journal of Physics: Conf. Series **991** (2018) 012068 doi:10.1088/1742-6596/991/1/012068

A spring system method for a mesh generation problem

A Romanov

Ishlinsky Institute for Problems in Mechanics of the Russian Academy of Sciences, Moscow, Russia

E-mail: malimo93@gmail.com

Abstract. A new direct method for the 2d-mesh generation for a simply-connected domain using a spring system is observed. The method can be used with other methods to modify a mesh for growing solid problems. Advantages and disadvantages of the method are shown. Different types of boundary conditions are explored. The results of modelling for different target domains are given. Some applications for composite materials are studied.

1. Introduction

Requirements to endurance and reliability of machines, constructions, and mechanisms are constantly increasing as the functionality of devices and their complexity are growing as a result of technical and technological progress. The methods for precise calculations of stresses, deformations, and distributions of different fields in machine parts are necessary for accepting new challenges. Various properties of devices, for instance, endurance and reliability, depend on the manufacturing process. One of the new and prospective processes is the additional manufacturing, so it is important to study its procedures [1–3]. Thus, we need to have methods for numerical solution of such problems. This is usually obtained by the finite element method (FEM) which consist in approximation of unknown displacements by a known function with parameters which are going to be determined. This approximation is based on a mesh in the solid under study. It is usually difficult to create a convenient mesh in 3d- and 2d-domains, especially for ones with complex geometry. There are a lot of methods for mesh generation methods and most of them provide a fine mesh either triangular or quadrangular. Composite materials [4] are also used in the growing process and, as a result, we can get composite solids (solids where the size of inclusions is comparable with the size of the body). Thus, we have to study such solids as well.

2. Description and comparison of the methods

All the methods can be divided into two large groups: methods for a triangular mesh and methods for a quadrangular mesh. The second ones are, in turn, divided into direct and indirect methods.

Triangular meshes. This class contains a lot of methods: Delauney triangulation [5], quad-tree-based method [6], LBWARP [7], and a bubble method [8]. They are not studied here because the quad-meshes are better than triangular ones in many cases for FEM.

Quadrangular meshes. Direct methods. In direct methods, the quadrilaterals are constructed at once. They use some kind of advancing front technique [9], regular grid-based methods, or quad-tree ones [10]. The advancing front methods for quads are considered to be nonrobust, and the quad-tree methods usually produce low quality elements close to the boundaries of the domain and are unable to fulfil general size constraints (anisotropy, strong variations).

Indirect methods. In indirect methods, a triangular mesh is built first. Then the triangle-merge methods use the triangles of the initial mesh and recombine them to form quadrangles [11, 12]. Other more sophisticated indirect methods use a mix of advancing front and triangle merge [13].

The methods described above do not suit us, because they generate a mesh from scratch every time, but in problems of growing bodies, this is improper because these problems are often solved in the quasi-state formulation [14, 15] and changing domains. So we have to find a method which only modified the mesh but does not regenerate. Our method is supposed to be as follows.

The method described in the paper is a direct method and can create meshes consisting only on quads for simply-connected domain. It was inspired by the theory of elasticity, the theory of conformal mapping [16], and the theory of neural net [17, 18].

The main point of the method is that we have a mesh in an initial domain. The initial domain can have a simple geometry, and it is easy to generate a mesh on it, or a mesh can be created for the initial state of the studied solid by using the techniques described above. Than if we have a mapping between the boundary of the initial domain and the boundary of the domain, where we would like to make a mesh, we have a mapping between the whole domains. And finally we can project the initial mesh on the manifold which we are interested in.

We do not have an exact mapping in the method, but only an approximate one. But if $\|\partial\Omega(t) - l(t)\| < \varepsilon$, where $\partial\Omega(t)$ is the boundary of the target domain, $l(t)$ is a curve which approximates the boundary, t is a parameter, and ε is the required precision, then one can say that we have a mapping between the initial and target boundaries.

3. Description of the method

We use a system of springs to obtain mapping inside the domains. Let us assume that we have such a system in the initial domain free of stresses and deformations. If we change the position of each point belonging to the initial boundary, i.e., they have some displacements, then the inner points change their positions correspondingly. We write the elastic energy of the system as

$$E = \sum_{i=1}^{K} \frac{k_i[(u^x_{i_1} - u^x_{i_2})^2 + (u^y_{i_1} - u^y_{i_2})^2]}{2} + \sum_{j=1}^{N} \vec{\lambda}_j(\vec{u}_j - \vec{b}_j)$$

where K is the number of springs, k_i is the stiffness of the ith spring, $u^k_{i_1}$ is the displacement at the first side of the spring along the axis k, $u^k_{i_2}$ is the displacement at the second side of the spring along the axis k, N is the number of the boundary points, $\vec{\lambda}_j$ are Lagrangian coefficients, and $\vec{b}_j$ is the value of the displacement of a point on the boundary. Only the boundary conditions for displacements are used, because the explicit position of the target boundary is known.

The displacement of each point is found from the statement

$$E[\vec{u}] \to \min,$$

which is equal to

$$\begin{cases} \dfrac{\partial E}{\partial u_i^k} = 0, \\ \dfrac{\partial E}{\partial \vec{\lambda}_j} = 0. \end{cases}$$

Thus

$$\begin{cases} \dfrac{\partial E}{\partial u_k^x} = \sum\limits_{i=1}^{K} k_i(u_{i_1}^x - u_{i_2}^x)(\delta_{i_1}^k - \delta_{i_2}^k) + \sum\limits_{j=1}^{N} \lambda_j^1 \delta_j^k, \\ \dfrac{\partial E}{\partial u_k^y} = \sum\limits_{i=1}^{K} k_i(u_{i_1}^y - u_{i_2}^y)(\delta_{i_1}^k - \delta_{i_2}^k) + \sum\limits_{j=1}^{N} \lambda_j^2 \delta_j^k, \\ \dfrac{\partial E}{\partial \vec{\lambda}_k} = \vec{u}_k - \vec{b}_k. \end{cases} \tag{1}$$

where δ_i^k is the Kronecker delta. The equations of system (1) can be rewritten in matrix form

$$A = \begin{pmatrix} u \\ \lambda \end{pmatrix} = f.$$

The matrix A is the following one:

$$A = \begin{pmatrix} B & \Lambda \\ \Lambda^T & 0 \end{pmatrix},$$

where B is the stiffness matrix, Λ corresponds to the Lagrangian coefficients and contains only ones and zeros. Ones are in the places corresponding to u_i^k belonging to the initial boundary. The right-hand side is

$$f = \begin{pmatrix} 0 \\ \vec{b}_i \end{pmatrix}.$$

In case where the mesh is regular, in a rectangular domain and at all inner points which have only four neighbors, we have the system of linear algebraic equations

$$\begin{cases} n_i k_i \vec{u}_i - \sum\limits_{j=1}^{n_i} k_{I(j)} \vec{u}_{I(j)} + \lambda_i = 0, \quad i = \overline{1, K}, \\ \vec{u}_k = \vec{b}_k, \quad k = \overline{1, N}. \end{cases}$$

where $n_i = 4$ if the point is inner one, $n_i = 3$ if the point belongs to the boundary but is not a corner point, $n_i = 2$ if the point is a corner point, I is the set of indexes of the neighbor points, and the cardinal number of I is n_i. Different types of points and their neighbors are shown on figure 1, red color – an inner point, blue – a boundary one, green – a corner one.

The mapping between the initial and target boundaries is obtained as the matching of two sets of points $J_1 \in \partial\Omega_1$ and $J_2 \in \Omega_2$, $|J_1| = |J_2|$, $\partial\Omega_1$ is the initial boundary, and $\partial\Omega_2$ is the target one. The set J_1 also represents the boundary points of the initial mesh, and the set J_2 gives their final positions. If $X_1 = \{(x, y)|\ x, y \in \mathbb{R}\}$ is a set of the positions of initial boundary points and $X_2 = \{(x, y)|\ x, y \in \mathbb{R}\}$ is a set of the positions of points in the set J_2, then the boundary conditions are $\vec{b}_i = X_2^i - X_1^i$.

As was said before, the approximation of the boundaries should be enough precise. And as before, $\|\partial\Omega_1(t) - l_1(t)\| < \varepsilon_1$, $\|\partial\Omega_2(t) - l_2(t)\| < \varepsilon_2$, $l_1(t)$ and $l_2(t)$ are linear approximations of the appropriate boundary based on the appropriate set J_i. This leads to the following statement.

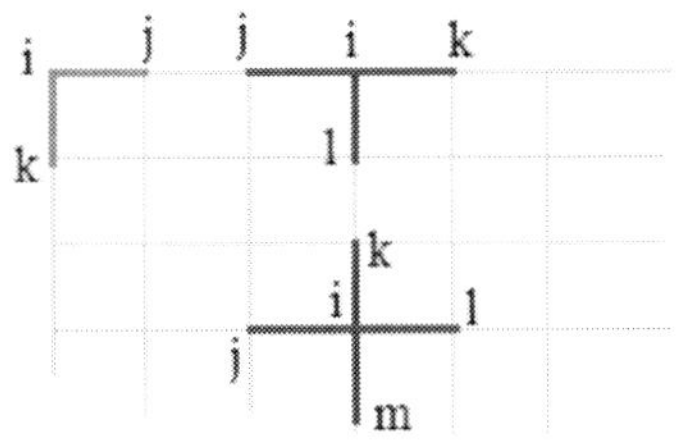

Figure 1. Numeration of neighbor points.

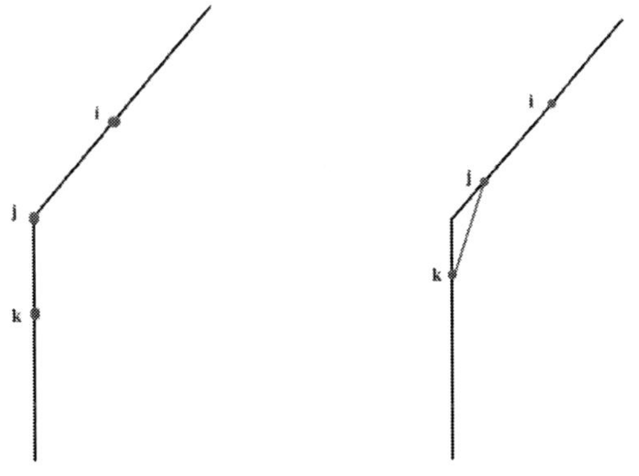

Figure 2. A precise (left) and imprecise (right) approximation.

Statement. All corner points of the boundary should be in the set for the compliance J. If not, then one can choose such an ε smaller than $\|\partial\Omega(t) - l(t)\|$, and thus the condition of precision is broken. This fact is demonstrated in figure 2.

Also an amount of points near parts of the boundary with high curvature should be considerable, otherwise the precision will be low, but this is a common rule for every mesh.

The authors of [19] stated that their algorithm is the fastest one among the methods for quadrangular meshing and gives an estimate of the time required for the mesh generation. The time t_1 is $\mathcal{O}(n_p(n_e + \ln n_p))$, where n_p is the number of points, n_e is the number of edges. The estimate can be slightly rewritten to compare with the time needed for our method, $t_1 \approx \mathcal{O}(n_p^2)$. The time t_2 for our method is $\mathcal{O}(n_p^2)$. So our method is not slower than the others, but we can improve the time using iteration methods to solve a system of linear algebraic equations.

Here is the algorithm for using this method in problems of growing [20–23].

1. Create a mesh for either a simple domain or the initial domain. If chosen, the initial one go to 3.
2. Project the mesh onto the initial domain.
3. Solve one step of the problem of a growing solid.
4. If there are cells of area larger than the maximum area $S_{\max}$, then create new elements on the growing surface and go to 5.
5. Stretch the mesh on the grown area. The displacement for each point is zero except for points belonging to the growing surface. Here the displacement is $v(\vec{x})\, dt$, where $\vec{x}$ belongs to the growing surface.
6. If this is the last time step, complete the program or else go to 3.

IOP Conf. Series: Journal of Physics: Conf. Series **991** (2018) 012068 doi:10.1088/1742-6596/991/1/012068

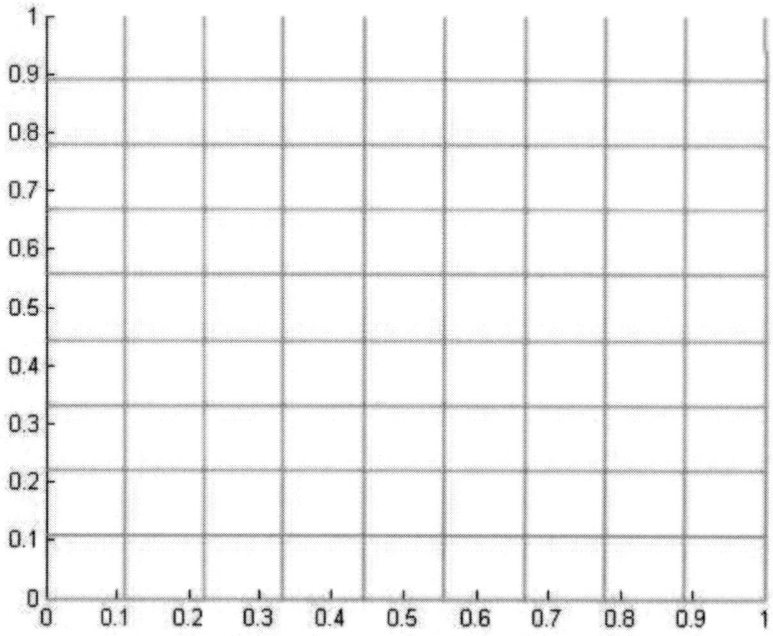

Figure 3. Initial domain and the mesh in it.

4. Numerical realization

The program was developed to demonstrate the method. It was written using the programming language C++ and QT libraries. One should enter an amount of points of the mesh, scales of the initial rectangle, and the boundary of the target manifold into the program. All springs are considered equal with the same stiffness k. The initial domain with the mesh is shown in figure 3. The result of this work is shown in figures 4 and 5. All results were obtained with the mesh of 10×10 cells. Both the initial and final boundaries were divided into the same amount of uniform parts, and a mapping was created between these divisions.

The resulting meshes consist of non-uniform cells, and the mesh in the second test near the peaks on the lateral surfaces is rough. Probably, it would be better to use different stiffness for each spring and mapping between the boundaries.

Conclusions

The method is valid only for simply-connected domains, but if other methods are used to create the initial mesh, then it is valid for any domain. There are two possible modifications of the method for multi-connected domains. The first one makes the initial manifold to be a multi-connected one. The second one first makes the mesh not to take the holes in the target domain into account, then deletes every point inside the holes except for the points which are the closest ones to the holes boundary. Then these points are projected onto the holes boundaries, and the mesh is recalculated. An additional requirement is that the mesh cell size should be smaller than the smallest one of the holes.

The same techniques can be used for composite solids, but it is not the optimal way. It is better to apply another method. Let us assume that the solid under study consists of several domains with very different physical properties and the precise boundary surfaces are given. Then one can generate meshes for each domain and modify the meshes near the boundaries to obtain one continuous mesh using a system of springs. Such a modification is required to

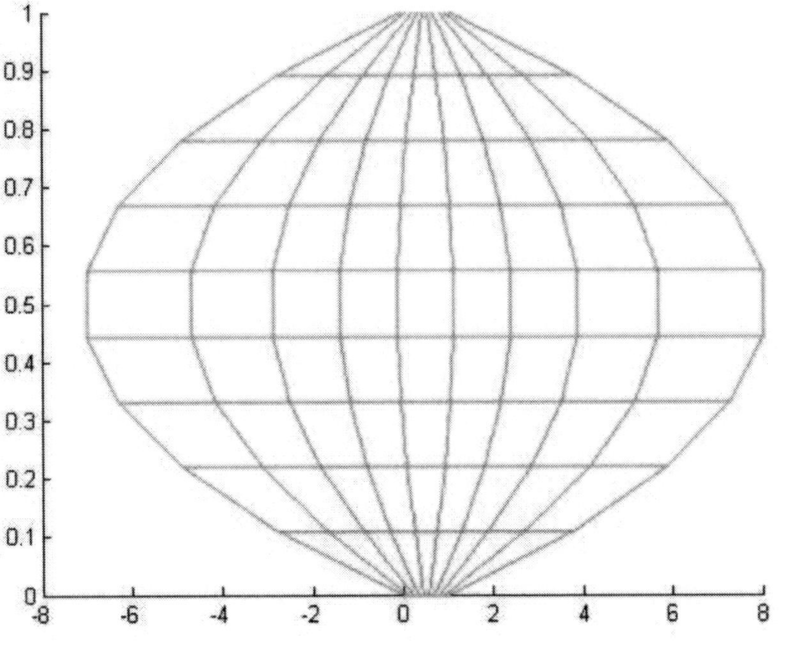

Figure 4. The first test.

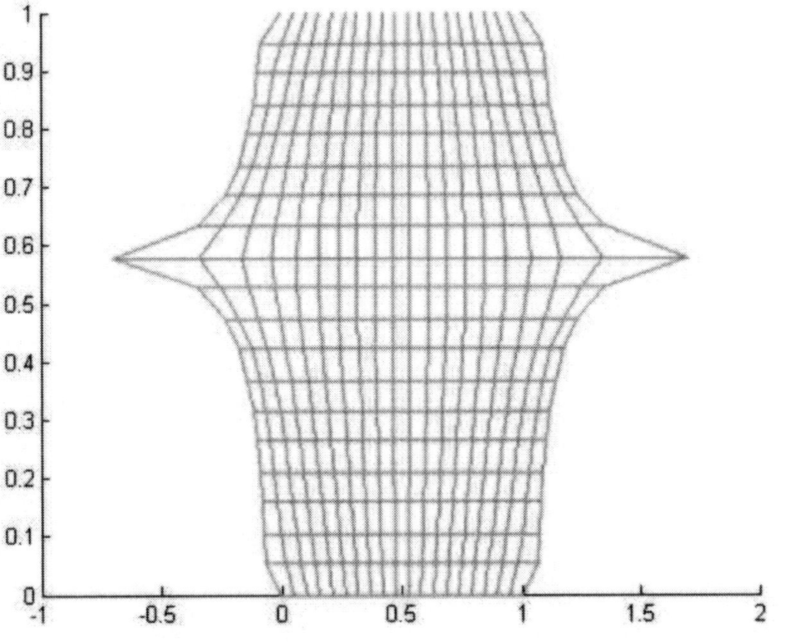

Figure 5. The second test.

avoid additional restrictions on the boundary points during the modelling, which simplifies the computational process.

Acknowledgments

The author thanks professor A.V. Manzhirov for the statement of the problem and helpful comments. The work was supported by the Russian Science Foundation (project No. 17-19-01257).

References

[1] Manzhirov A V 2016 A method for mechanical design of AM fabricated viscoelastic parts In *Transactions on Engineering Technologies* ed S Ao, G C Yang, and L Gelman (Singapore: Springer) 223–35
[2] Manzhirov A V 2015 Mechanical design of viscoelastic parts fabricated using additive manufacturing technologies *Lect. Notes Engng Comp. Sci.* **2** 710–4
[3] Manzhirov A V 2014 Mechanics of growing solids: new track in mechanical engineering *ASME 2014 Int. Mechanical Engineering Congr. and Exposition. Vol 9: Mechanics of Solids, Structures and Fluids Montreal, Quebec, Canada, November 14–20, 2014* (ASME) V009T12A039
[4] Harris P J 2004 Carbon nanotube composites *Int. Mater. Rev.* **49** (1) 31–43
[5] Chen L and Xu J C 2004 Optimal Delaunay triangulations *J. Comp. Math.* **22** (2) 299–308
[6] GVS P R, Montas H J, Samet H, and Shirmohammadi A 2001 Quad-tree-based triangular mesh generation for finite element analysis of heterogeneous spatial data In *The 2001 ASAE Annual Int. Meeting* (St. Joseph, Mich.: ASAE) Paper No. 01-3072.
[7] Shontz S M and Vavasis S A 2003 A mesh warping algorithm based on weighted Laplacian smoothing In *Proc. of the 12th Int. Meshing Roundtable* 147–58
[8] Shimada K and Gossard D C 1995 Bubble mesh: automated triangular meshing of non-manifold geometry by sphere packing In *Proc. of the Third ACM Symp. on Solid Modeling and Application* ed C Hoffmann and J Rossignac (New York: ACM) pp 409–19
[9] Blacker T D and Stephenson M B 1991 Paving: A new approach to automated quadrilateral mesh generation *Int. J. Num. Meth. Engng* **32** (4) 811–47
[10] Liang X, Ebeida M S, and Zhang Y 2010 Guaranteed-quality all-quadrilateral mesh generation with feature preservation *Comp. Meth. Appl. Mech. Engng* **199** (29) 2072–83
[11] Lee C K and Lo S H 1994 A new scheme for the generation of a graded quadrilateral mesh *Comput. Struct.* **52** 847–57 Borouchaki H and Frey P 1998 Adaptive triangular–quadrilateral mesh generation *Int. J. Num. Meth. Engng* **41** (5) 915–34
[12] Owen S J, Staten M L, Canann S A, and Saigal S 1999 Q-morph: An indirect approach to advancing front quad meshing *Int. J. Num. Meth. Engng* **44** (9) 1317–40
[13] Manzhirov A V 2017 Fundamentals of mechanical design and analysis for am fabricated parts *Proc. Manufact.* **7** 59–65
[14] Manzhirov A V 2017 Advances in the theory of surface growth with applications to additive manufacturing technologies *Proc. Engng* **173** 11-6
[15] Smith W and Volberg A 2011 A conformal mapping example *Comptes Rendus Math.* **349** (9-10) 511-4
[16] Hodge V J, O'Keefe S., and Austin J 2016 Hadoop neural network for parallel and distributed feature selection *Neural Netw.* **78** 24–35
[17] Kim J, Kim J, Jang G J, and Lee M 2017 Fast learning method for convolutional neural networks using extreme learning machine and its application to lane detection *Neural Netw.* 2017 **87** 109–21
[18] Remacle J F, Lambrechts J, Seny B, et al. 2012 Blossom-Quad: A non-uniform quadrilateral mesh generator using a minimum-cost perfect-matching algorithm *Int. J. Num. Meth. Engng* **89** (9) 1102–19
[19] Gabow H N 1990 Data structures for weighted matching and nearest common ancestors with linking In *Proc. of the 1st Annual ACM-SIAM Symp. on Discrete Algorithms* pp 434–43
[20] Manzhirov A V 1995 The general non-inertial initial-boundary value problem for a viscoelastic ageing solid with piecewise-continuous accretion *J. Appl. Math. Mech.* **59** (5) 805–16
[21] Lychev S A, Lycheva T N, and Manzhirov A V 2011 Unsteady vibration of a growing circular plate *Mech. Solids* **46** (2), 325–333
[22] Manzhirov A V and Lychev S A 2012 The mathematical theory of growing solids: finite deformations *Dokl. Phys.* **57** (4) 160–3
[23] Joubert S V, Shatalov M Y, and Manzhirov A V 2013 Bryan's effect and isotropic nonlinear damping *J. Sound. Vibr.* **332** (23) 6169–76

IOP Conf. Series: Journal of Physics: Conf. Series **991** (2018) 012069 doi:10.1088/1742-6596/991/1/012069

The fractional derivative Kelvin–Voigt model of viscoelasticity with and without volumetric relaxation

Yu A Rossikhin and M V Shitikova

Research Center on Dynamics of Solids and Structures, Voronezh State Technical University, Voronezh, Russia

E-mail: mvs@vgasu.vrn.ru

Abstract. The fractional derivative Kelvin–Voigt model of viscoelasticity involving the time-dependent Poisson's operator has been studied not only for the case of a time-independent bulk modulus, but also when the volumetric relaxation is taken into account. It has been shown that such a model could describe the features of auxetic materials.

1. Introduction

It is well known that each isotropic elastic material possesses only two independent constants, and all others are expressed in terms of two constants which should be preassigned [1]. Thus, if Young's modulus E and Poisson's ratio ν are known, then Lame constants λ and μ and bulk modulus K are expressed as

$$\mu = \frac{E}{2(1+\nu)}, \quad \lambda = \frac{E\nu}{(1-2\nu)(1+\nu)}, \quad K = \frac{E}{3(1-2\nu)}, \tag{1}$$

or in the case, when shear modulus μ and bulk modulus K are preassigned, other constants, E, ν, and λ, are defined as

$$E = \frac{9K\mu}{3K+\mu}, \quad \nu = \frac{3K-2\mu}{2(3K+\mu)}, \quad \lambda = \frac{3K-2\mu}{9}. \tag{2}$$

In the case of an isotropic viscoelastic material, material properties are time-dependent, and once again only two time-dependent viscoelastic operators should be known, while others could be expressed in terms of two preassigned operators utilizing the correspondence principle and relationships (1) or (2).

In the present paper, viscoelastic operators will be constructed for the fractional derivative Kelvin–Voigt model with and without volume relaxation, since this model is of frequent use in engineering applications [2].

2. Application of the simplest viscoelastic models for the description of dynamic response of current materials

Simplified approach of many researchers for solving dynamic viscoelastic problems is connected first of all with the ineptitude to decode intricate operator expressions involving fractional operators which appear during the solution of dynamic problems. The desire to solve a formulated problem by any means often leads to the fact that some simplifying assumptions are introduced into operator expressions, resulting in incorrectness of these expressions, and finally to the loss of their physical meaning.

That is why, despite that in different applications the utilization of fractional derivatives in problems of mechanics and physics is growing significantly during last two decays, the variety of rheological models used is rather restricted. During description of such viscoelastic bodies as beams, plates and shells, the most frequently used models for the Young operator are the fractional derivative Kelvin–Voigt relation [3–8], the fractional derivative Maxwell model [9, 10], and standard linear solid model [11, 12], in so doing the Poisson ratio of viscoelastic material is assumed as a constant.

However as experimental data have shown [13, 14], Poisson's ratio is always a time-dependent operator $\tilde{\nu}$ [15–17], and only the bulk extension-compression operator $\tilde{K}$ may be expressed as the time-independent value, which for the most viscoelastic materials weakly varies during deformation.

However, the fractional derivative Kelvin–Voigt model with time-independent Poisson's ratio is only acceptable for the description of the dynamic behavior of elastic bodies in a viscoelastic medium [2, 18–20] or on a viscoelastic foundation [21, 22].

The review of 'traditional' fractional calculus models ('traditional' in the sense that such models consider time-independent Poisson's ratios) could be found in [2, 23–25].

2.1. Modelling the Young operator $\tilde{E}$ using the fractional derivative Kelvin–Voigt model without volumetric relaxation

Let us first choose Young's operator as

$$\tilde{E} = E_0[1 + (\tau_\sigma^E)^\gamma D^\gamma], \tag{3}$$

where τ_σ^E is the retardation, or creep, time during longitudinal deformations, D^γ is the Riemann-Liouville fractional derivative

$$D^\gamma x(t) = \frac{d}{dt}\int_0^t \frac{(t-t')^{-\gamma}}{\Gamma(1-\gamma)} x(t')\, dt', \tag{4}$$

$\Gamma(1-\gamma)$ is the Gamma-function, and $x(t)$ is a certain function.

It should be noted that the choice of the Riemann-Liouville derivative of the fractional order is not random. Since during the solution of dynamic problems of viscoelasticity it is often needed to find resolvent operators and to decode intricate operator relationships [25, 26], then the utilization of other definitions of fractional derivatives tunes out to be inconvenient. Details could be found in [27].

The bulk extension-compression operator $\tilde{K}$ is assumed to be time-independent, i.e., volumetric relaxation is neglected (this assumption is due to the fact that for many viscoelastic materials volumetric relaxation is much smaller than the shear relaxation)

$$\tilde{K} = K_0, \tag{5}$$

resulting in

$$\frac{\tilde{E}}{1-2\tilde{\nu}} = 3K_0, \tag{6}$$

where K_0 is a certain constant.

Substituting (3) in (6), we could find Poisson's operator

$$\tilde{\nu} = \nu_0 - \frac{E_0}{6K_0}(\tau_\sigma^E)^\gamma D^\gamma, \tag{7}$$

and finally

$$\nu(t) = \tilde{\nu}H(t) = \nu_0 - \frac{E_0}{6K_0}(\tau_\sigma^E)^\gamma \frac{t^{-\gamma}}{\Gamma(1-\gamma)}, \tag{8}$$

whence it follows that

$$\lim_{t\to 0} \tilde{\nu}H(t) = -\infty, \quad \lim_{t\to\infty} \tilde{\nu}H(t) = \nu_0, \tag{9}$$

where $\nu_0 = (3K_0 - E_0)/(6K_0)$.

From (9) it is seen that this model lacks the physical meaning, since for real materials the low limiting value of Poisson's ratio $\nu(0)$ could not take on extremely negative value $-\infty$. The similar result for this model was reported in [28] but with reverse sign: $\nu(0) = +\infty$, and $\nu(\infty) = \nu_0$. But nevertheless this model is inappropriate for dealing with real viscoelastic materials.

2.2. Modelling the shear operator $\tilde{\mu}$ using the fractional derivative Kelvin–Voigt model without volumetric relaxation

The shear operator $\tilde{\mu}$ most frequently is preassigned using the fractional derivative Kelvin–Voigt model

$$\tilde{\mu} = \mu_0[1 + (\tau_\sigma^\mu)^\gamma D^\gamma], \tag{10}$$

where μ_0 is the relaxed shear modulus, τ_σ^μ is the retardation time during shear deformations, while the bulk operator is assumed to be constant according to (5).

In order to evaluate the dynamic response of viscoelastic bodies, it is necessary to calculate Young's operator. For this purpose using the Volterra correspondence principle, the following formula could be utilized:

$$\tilde{E} = \frac{9K_0\tilde{\mu}}{3K_0 + \tilde{\mu}}. \tag{11}$$

First we write the operator

$$3K_0 + \tilde{\mu} = (3K_0 + \mu_0)(1 + t_\sigma^\gamma D^\gamma), \tag{12}$$

where $t_\sigma^\gamma = \mu_0(\tau_\sigma^\mu)^\gamma(3K_0 + \mu_0)^{-1}$.

Then we find the operator reverse to (12), i.e.,

$$(3K_0 + \tilde{\mu})^{-1} = \frac{1}{3K_0 + \mu_0}\frac{1}{1 + t_\sigma^\gamma D^\gamma} = (3K_0 + \mu_0)^{-1} \ni_\gamma^* (t_\sigma^\gamma), \tag{13}$$

where

$$\ni_\gamma^* (t_\sigma^\gamma) = \frac{1}{1 + t_\sigma^\gamma D^\gamma} \tag{14}$$

is the dimensionless Rabotnov fractional operator [23, 25, 26].

Before calculating operator $\tilde{E}$, let us remind one useful formula [29]

$$t_\sigma^\gamma D^\gamma \cdot \ni_\gamma^* (t_\sigma^\gamma) = \frac{t_\sigma^\gamma D^\gamma}{1 + t_\sigma^\gamma D^\gamma} = 1 - \ni_\gamma^* (t_\sigma^\gamma). \tag{15}$$

Substituting (10) and (13) in (11) and considering (15) yield

$$\tilde{E} = 9K_0\left[1 - \frac{E_0}{3\mu_0} \ni_\gamma^* (t_\sigma^\gamma)\right]. \tag{16}$$

Now we could calculate Poisson's operator $\tilde{\nu}$ via the formula

$$\frac{\tilde{E}}{1 - 2\tilde{\nu}} = 3K_0. \tag{17}$$

Substituting (16) in (17), we have

$$\tilde{\nu} = -1 + \frac{E_0}{2\mu_0} \ni^*_\gamma (t^\gamma_\sigma). \tag{18}$$

If we study the relaxation process, i.e., consider the longitudinal deformation in a rod as constant, then operator $\tilde{E}$ will act on the unit Heaviside function $\tilde{E}H(t)$. With due account for

$$\int_0^\infty \ni_\gamma \left(-\frac{t-t'}{\tau_i} \right) H(t')\, dt' = 1, \tag{19}$$

where

$$\ni_\gamma \left(-\frac{t}{\tau_i} \right) = \frac{t^{\gamma-1}}{\tau_i^\gamma} \sum_{n=0}^{\infty} (-1)^n \frac{(t/\tau_i)^{\gamma n}}{\Gamma[\gamma(n+1)]} \tag{20}$$

is Rabotnov fractional exponential function [30] which at $\gamma = 1$ goes over into the traditional exponent, we obtain

$$\lim_{t\to 0} \tilde{E}H(t) = 9K_0, \quad \lim_{t\to\infty} \tilde{E}H(t) = \frac{9K_0\mu_0}{3K_0 + \mu_0} = E_0. \tag{21}$$

For operator $\tilde{\nu}H(t)$ we have [29]

$$\lim_{t\to 0} \tilde{\nu}H(t) = -1, \quad \lim_{t\to\infty} \tilde{\nu}H(t) = \frac{3K_0 - 2\mu_0}{2(3K_0 + \mu_0)} = \nu_0. \tag{22}$$

According to the classical theory of elasticity it has been considered that for the conventional materials the Poisson's ratio ν varies within the interval $0 < \nu \leq 0.5$. However, nowadays a wide variety of so-called auxetic materials has been fabricated, including polymeric and metallic foams, microporous polymers, carbonfibre laminates and honeycomb structures [31], which possess such unusual mechanical property as a negative Poisson's ratio, i.e., the Poisson's ratio ν could vary within the interval $-1 < \nu \leq 0.5$. Therefore, the proposed model (10) with $\tilde{K} = K_0 = \text{const}$ could describe the behavior of viscoelastic auxetic materials, in so doing Poisson's ratio could vary from -1 to its relaxed magnitude ν_0.

In [29], substituting (16) and (18) in formula

$$\tilde{d} = \frac{\tilde{E}}{1 - \tilde{\nu}^2} \tag{23}$$

the rigidity operator was calculated, which is needed when solving different dynamic problems of thin plates and shells.

In conclusion of this subsection note that Zhu *et al.* [32, 33] for studying the dynamic response of a Timoshenko beam considered shear operator $\tilde{\mu}$ (10) and time-independent volumetric modulus (5) with further interpretation of the Young operator in the form of (11). The same model was adopted in [34] for quasi-static analysis of a Bernoulli-Euler beam.

However, the authors were not able to decode operator (11), although in order to it they would have to know only one formula from the algebra of fractional operators, namely: (15), where $\ni^*_\gamma (t^\gamma_\sigma)$ is defined in (14), and this formula even does not need demonstration, since it is evident.

So Poisson's operator in [32] is given in the form

$$\tilde{\nu} = \frac{3K_0 - 2\mu_0[1 + (\tau^\mu_\sigma)^\gamma D^\gamma]}{6K_0 + 2\mu_0[1 + (\tau^\mu_\sigma)^\gamma D^\gamma]}. \tag{24}$$

Further the formulated problem was solved without expanding the expression (24), and deflection of the beam was found. However, expanding formula (24) and considering (14) yield

$$\tilde{\nu} = \nu_0(1 - \tau^{\gamma} D^{\gamma}) \ni^*_{\gamma} (t^{\gamma}_{\sigma}), \tag{25}$$

where ν_0 and t^{γ}_{σ} have been defined above, and

$$\tau^{\gamma} = \frac{2\mu_0(\tau^{\mu}_{\sigma})^{\gamma}}{3K_0 - 2\mu_0}.$$

Now removing parenthesis in (25) and considering (15), we have

$$\tilde{\nu} = \nu_0\left\{ \ni^*_{\gamma} (t^{\gamma}_{\sigma}) - \frac{\tau^{\gamma}}{t^{\gamma}_{\sigma}}[1- \ni^*_{\gamma} (t^{\gamma}_{\sigma})]\right\} = \nu_0\left[-\frac{1}{\nu_0} + \left(1 + \frac{1}{\nu_0}\right) \ni^*_{\gamma} (t^{\gamma}_{\sigma})\right],$$

that is

$$\tilde{\nu} = -1 + (1 + \nu_0) \ni^*_{\gamma} (t^{\gamma}_{\sigma}). \tag{26}$$

Since $1 + \nu_0 = E_0(2\mu_0)^{-1}$, then (26) coincides with (18). Thus, the fact that for this model Poisson's ratio could increase from -1 to 0 (what is characteristic for viscoelastic auxetic materials) and then it could increase further from 0 to ν_0 (what is valid for ordinary viscoelastic materials with positive magnitudes of Poisson's ratio) has not been revealed in [32–34].

Our reasoning is supported by [35], where it is noted that "$\nu(0) = -1$ indicates that such a material experiences lateral expansion under axial tension at the very beginning of the elongation process."

2.3. Models considering volumetric relaxation

In the state-of-the-art papers [2, 26] it has been emphasized that in the majority of articles utilizing the fractional calculus models the second, or bulk, viscosity is not taken into account, in spite the fact that there are numerous evidence of its existence in scientific literature [36–38]. It is more important for modern composite materials [37].

One of the first papers in the field was published by Meshkov and Pachevskaya [36], wherein the attempt for considering the influence of the bulk relaxation on the internal friction phenomenon was carried out on the example of longitudinal harmonic vibrations of a three-dimensional hereditary elastic rod under the conditions of homogeneous deformation. A fractional exponential function has been used as a hereditary kernel. Investigation of the frequency dependence of the tangent of the phase shift between the stress and deformation, i.e., mechanical loss tangent, has revealed two peaks, namely, shear and bulk, in so doing the peak due to the shear deformation is five times larger than that resulting from the bulk deformation.

However, the account for the volumetric relaxation results in more cumbersome calculations. Thus, if operators $\tilde{E}$ and $\tilde{\mu}$, which are defined, respectively, by (3) and (10), are known, then first it is necessary to find operator $\tilde{K}^{-1}$ utilizing the relationship

$$\tilde{J} = \tilde{E}^{-1} = \frac{3\tilde{K} + \tilde{\mu}}{9\tilde{K}\tilde{\mu}} = \frac{1}{3}\tilde{\mu}^{-1} + \frac{1}{9}\tilde{K}^{-1}, \tag{27}$$

whence it follows that

$$\tilde{K}^{-1} = 9\tilde{E}^{-1} - 3\tilde{\mu}^{-1} = a_1 \ni^*_{\gamma} (\tau^{E}_{\sigma})^{\gamma} - a_2 \ni^*_{\gamma} (\tau^{\mu}_{\sigma})^{\gamma}, \tag{28}$$

where $a_1 = 9E_0^{-1}$ and $a_2 = 3\mu_0^{-1}$.

Now we could calculate the ratio of operators considering formulas (3) and (28)

$$\frac{\tilde{E}}{\tilde{K}} = E_0[1 + (\tau_\sigma^E)^\gamma D^\gamma][a_1 \ni_\gamma^* (\tau_\sigma^E)^\gamma - a_2 \ni_\gamma^* (\tau_\sigma^\mu)^\gamma],$$

or

$$\frac{\tilde{E}}{\tilde{K}} = E_0\left[a_1 - a_2\left(\frac{\tau_\sigma^E}{\tau_\sigma^\mu}\right)^\gamma\right] - E_0 a_2\left[1 - \left(\frac{\tau_\sigma^E}{\tau_\sigma^\mu}\right)^\gamma\right] \ni_\gamma^* (\tau_\sigma^\mu)^\gamma. \tag{29}$$

Computing operator $\tilde{\nu}$ via formula

$$\frac{\tilde{E}}{\tilde{K}} = 3(1 - 2\tilde{\nu}) \tag{30}$$

and considering (29) yield

$$\tilde{\nu} = \frac{1}{2} - \frac{E_0}{6}\left[a_1 - a_2\left(\frac{\tau_\sigma^E}{\tau_\sigma^\mu}\right)^\gamma\right] + \frac{E_0}{6} a_2\left[1 - \left(\frac{\tau_\sigma^E}{\tau_\sigma^\mu}\right)^\gamma\right] \ni_\gamma^* (\tau_\sigma^\mu)^\gamma. \tag{31}$$

Considering the product $\nu(t) = \tilde{\nu} \cdot H(t)$, we find

$$\nu(0) = \frac{1}{2} - \frac{E_0}{6}\left[a_1 - a_2\left(\frac{\tau_\sigma^E}{\tau_\sigma^\mu}\right)^\gamma\right] = -1 + \frac{E_0}{2\mu_0}\left(\frac{\tau_\sigma^E}{\tau_\sigma^\mu}\right)^\gamma, \tag{32}$$

$$\nu(\infty) = \frac{1}{2} + \frac{E_0}{6}(a_2 - a_1) = -1 + \frac{E_0}{2\mu_0} = \nu_0. \tag{33}$$

This model with two preassigned operators $\tilde{E}$ (3) and $\tilde{\mu}$ (10) was used in [39] for the analysis of dynamic response of viscoelastic Timoshenko beam. Since the Timoshenko beam equation does not involve the Poisson's ratio in the explicit form, then there was no need for the authors of [39] to find operators $\tilde{K}$ and $\tilde{\nu}$.

Reference to relationship (32) shows that the low limiting magnitude of Poisson's ratio $\nu(0)$ depends on the ratio of two retardation times and fractional parameter γ, as distinct from the Kelvin–Voigt model without volumetric relaxation for which $\nu(0) = -1$ according to (22).

However note that when $(\tau_\sigma^E/\tau_\sigma^\mu)^\gamma \to 0$, then from (32) it follows that $\nu(0) \to -1$. It means that the model under consideration could be used for modelling the features of viscoelastic auxetic materials until $(\tau_\sigma^E/\tau_\sigma^\mu)^\gamma \to 2\mu_0/E_0$ resulting in $\nu(0) \to 0$. For magnitudes $(\tau_\sigma^E/\tau_\sigma^\mu)^\gamma \geq 2\mu_0/E_0$, $\nu(0) \geq 0$ what corresponds to traditional viscoelastic materials.

Note that the standard linear solid model with two preassigned Lame operators $\tilde{\mu}$ and $\tilde{\lambda}$

$$\tilde{\mu} = \mu_0 \frac{1 + (\tau_\sigma^\mu)^\gamma D^\gamma}{1 + (\tau_\varepsilon^\mu)^\gamma D^\gamma}, \quad \tilde{\lambda} = \lambda_0 \frac{1 + (\tau_\sigma^\lambda)^\gamma D^\gamma}{1 + (\tau_\varepsilon^\lambda)^\gamma D^\gamma} \tag{34}$$

has been studied in [29].

In this case, the account for bulk relaxation results in the viscoelastic Poisson's operator $\tilde{\nu}$ which could be computed via formula

$$\tilde{\nu} = \frac{\tilde{\lambda}}{2(\tilde{\lambda} + \tilde{\mu})} = \nu_\infty[1 + M_1 \ni_\gamma^* (T_1^\gamma) + M_2 \ni_\gamma^* (T_2^\gamma)], \tag{35}$$

where M_i and T_i $(i = 1, 2)$ are material's parameters, whence it follows that viscoelastic Poisson's ratio varies from $\nu(t)\big|_{t\to 0} = \nu_\infty$ to $\nu(t)\big|_{t\to\infty} = \nu_0$.

It is also very important to note that if there exists some 'hypothetical' viscoelastic material with retardation times equal in magnitude $\tau_\sigma^E = \tau_\sigma^\mu$, then from (32) and (33) it is seen that such a material should possess the time-independent Poisson's ratio $\tilde{\nu} = \nu = \text{const}$. As a result, all existing viscoelastic operators are proportional each other, what, generally speaking, does not refer with all real viscoelastic materials. Some authors based their research on such a 'hypothetical' viscoelastic material [7, 40] and tried to solve different intricate problems of Mechanics of Solids with such a so-called viscoelastic material, what, broadly speaking, is inappropriate from the point of view of Mechanics.

The critical review of the role of Poisson's ratio in linear viscoelasticity is presented by Tschoegl *et al.* [17] and Hilton [13, 16], wherein it has been shown that "no dynamic effects and no body forces, as well as no moving boundaries, i.e., no penetration or ablation problems, can be included" in the consideration adopting the assumption of time-independent Poisson's ratios.

Conclusions

For the fractional derivative Kelvin–Voigt model involving the time-dependent Poisson's operator it has been shown that such a model could describe the features of auxetic materials. For the model without bulk relaxation the Poisson's ratio varies from -1 to a positive magnitude ν_0, while for the model considering the volumetric relaxation the Poisson's ratio increases from a negative magnitude larger than -1 (which depends on the retardation times) to a positive magnitude ν_0.

Acknowledgments

This research has been supported by the Russian Foundation for Basic Research Project No. 17-01-00490.

References

[1] Landau L D and Lifshits E M 2010 *(Theoretical Physics. Vol 7. Theory of Elasticity* (Moscow: Nauka)
[2] Rossikhin Yu A and Shitikova M V 2010 Application of fractional calculus for dynamic problems of solid mechanics: Novel trends and recent results *Appl. Mech. Rev.* **63** (1) 010801
[3] Cajić M, Karličić D, and Lazarević M 2015 Nonlocal vibration of a fractional order viscoelastic nanobeam with attached nanoparticle *Theor. Appl. Mech.* **42** 167-90
[4] Demir D D, Bildik N, and Sinir B G 2014 Linear dynamical analysis of fractionally damped beams and rods *J. Engng Math.* **85** 131–47
[5] Freundlich J 2013 Vibrations of a simply supported beam with a fractional viscoelastic material model – supports movement excitation *Shock Vibr.* **20** 1103–12
[6] Ingman D and Suzdalnitsky J 2008 Response of viscoelastic plate to impact *ASME J. Vibr. Acous.* **130** 011010
[7] Katsikadelis J T 2012 Nonlinear dynamic analysis of viscoelastic membranes described with fractional differential models *J. Theor. Appl. Mech.* **50** (3) 743–53
[8] Di Paola M, Heuer R, and Pirrotta A 2013 Fractional visco-elastic Euler-Bernoulli beam *Int. J. Solids Struct.* **50** 3505–10
[9] Lewandowski R and Pawlak Z 2011 Dynamic analysis of frames with viscoelastic dampers modelled by rheological models with fractional derivatives *J. Sound Vibr.* **330** 923–36
[10] Pritchard R H and Terentjev E M 2017 Oscillations and damping in the fractional Maxwell materials *J. Rheology* **61** 187–203
[11] Sotoodeh Bahraini S M, Eghtesad M, Farid M, and Ghavanloo E 2013 Large deflection of viscoelastic beams using fractional derivative model *J. Mech. Sci. Tech.* **27** (4) 1063–70
[12] Galucio A C, Deü J F, and Ohayon R 2004 Finite element formulation of viscoelastic sandwich beams using fractional derivative operators *Comp. Mech.* **33** 282–91
[13] Hilton H H 2011 Clarifications of certain ambiguities and failings of Poisson's ratios in linear viscoelasticity *J. Elasticity* **104** 303–18
[14] Rabotnov Yu N 1980 *Elements of Hereditary Solid Mechanics* (Moscow: Nauka)
[15] Hilton H H 2001 Implications and constraints of time-independent Poisson ratios in linear isotropic and anisotropic viscoelasticity *J. Elasticity* **63** 221–51

[16] Hilton H H 2009 The elusive and fickle viscoelastic Poissons ratio and its relation to the elastic-viscoelastic correspondence principle *J. Mech. Mat. Struct.* **4**(7-8) 1341–64
[17] Tschoegl N W, Knauss W G, and Emri I 2002 Poisson's ratio in linear viscoelasticity – a critical review *Mech. Time-Depend. Mater.* **6** (1) 3–51
[18] Malara G and Spanos P D 2016 Nonlinear random vibrations of plates endowed with fractional derivative elements *Meccanica dei Materiali e delle Strutture* **VI** (1) 17–24
[19] Rossikhin Yu A, Shitikova M V, and Estrada Meza M G 2016 Modeling of the impact response of a beam in a viscoelastic medium *Appl. Math. Sci.* **10** (49-52) 2471–81
[20] Rossikhin Yu A, Shitikova M V, and Trung P T 2016 Application of the fractional derivative Kelvin–Voigt model for the analysis of impact response of a Kirchhoff-Love plate *WSEAS Trans. Math.* **15** 498–501
[21] Yan Q and Su Z 2012 Free vibration of elastic Timoshenko beam on fractional derivative Winkler viscoelastic foundation *Adv. Mater. Res.* **368-373** 1034–7
[22] Cajić M, Karličić D, and Lazarević M 2017 Damped vibration of a nonlocal nanobeam resting on viscoelastic foundation: fractional derivative model with two retardation times and fractional parameters *Meccanica* **52** (1-2) 363–82
[23] Rossikhin Yu A and Shitikova M V 2014 Centennial jubilee of Academician Rabotnov and contemporary handling of his fractional operator *Fract. Calculus Appl. Anal.* **17** 674–83
[24] Rossikhin Yu A and Shitikova M V 2014 The simplest models of viscoelasticity involving fractional derivatives and their connectedness with the Rabotnov fractional order operators *Int. J. Mech.* **8** 326–31
[25] Rossikhin Yu A and Shitikova M V 2015 Chapter 8 — Features of fractional operators involving fractional derivatives and their applications to the problems of mechanics of solids In *Fractional calculus, Applications* eds Zeid Daou R A and Moreau X (New York: Nova Publishers) pp 165–226
[26] Rossikhin Yu A and Shitikova M V 2013 Two approaches for studying the impact response of viscoelastic engineering systems: An overview *Comp. Math. Appl.* **66** 755–73
[27] Rossikhin Yu A, Shitikova M V, and Popov I I 2017 Dynamic response of a viscoelastic beam impacted by a viscoelastic sphere *Comp. Math. Appl.* **73** (6) 970–84
[28] Yin D, Duan X, and Zhou X 2013 Fractional time-dependent deformation component models for characterizing viscoelastic Poisson's ratio *Eur. J. Mech A/Solids* **42** 422–9
[29] Rossikhin Yu A, Shitikova M V, and Krusser A I 2016 To the question on the correctness of fractional derivative models in dynamic problems of viscoelastic bodies *Mech. Res. Com.* **77** (1) 44–9
[30] Rabotnov Yu N 2014 Equilibrium of an elastic medium with after-effect *Fract. Calculus Appl. Anal.* **17** 684–96
[31] Prawoto Y 2012 Seeing auxetic materials from the mechanics point of view: A structural review on the negative Poisson's ratio *Comp. Mat. Sci.* **58** 140–53
[32] Zhu Z Y, Li G G, and Cheng C J 2002 Quasi-static and dynamical analysis for viscoelastic Timoshenko beam with fractional derivative constitutive relation *Appl. Math. Mech.* **23** 1–12
[33] Zhu Z Y, Li G G, and Cheng C J 2003 A numerical method for fractional integral with applications *Appl. Math. Mech.* **24** 373–84
[34] Yao Q, Liu L, and Yan Q 2011 Quasi-static analysis of beam described by fractional derivative Kelvin viscoelastic model under lateral load *Adv. Mater. Res.* **189-193** 3391–4
[35] Makris N 1997 Three-dimensional constitutive viscoelastic laws with fractional order time derivatives *J. Rheology* **41** (5) 1007–20
[36] Meshkov S I and Pachevskaja G N 1967 Allowance for bulk relaxation by the method of internal friction *J. Appl. Mech. Tech. Phys.* **8** (2) 47–8
[37] Selivanov M F 2009 Effective properties of a linear viscoelastic composite *Int. Appl. Mech.* **45** 1084–91
[38] Selivanov M F and Chornoivan Y O 2012 Computational optimization of characteristics for composites of viscoelastic components *J. Engng Math.* **74** 91–100
[39] Li GG 2002 Dynamical behaviors of Timoshenko beam with fractional derivative constitutive relation *Int. J. Nonlin. Sci. Num. Sim.* **3** 67–73
[40] Zhu H H, Liu L C, Pei H F, and Shi B 2001 Settlement analysis of viscoelastic foundation under vertical line load using a fractional Kelvin–Voigt model *Geomech. Engng* **4** (1) 67–78

IOP Conf. Series: Journal of Physics: Conf. Series **991** (2018) 012070 doi:10.1088/1742-6596/991/1/012070

Method of mechanical quadratures for solving singular integral equations of various types

A V Sahakyan* **and H A Amirjanyan****

Institute of Mechanics of the National Academy of Sciences of the Republic of Armenia, Yerevan, Armenia

E-mail: *avsahakyan@gmail.com, **amirjanyan@gmail.com

Abstract. The method of mechanical quadratures is proposed as a common approach intended for solving the integral equations defined on finite intervals and containing Cauchy-type singular integrals. This method can be used to solve singular integral equations of the first and second kind, equations with generalized kernel, weakly singular equations, and integro-differential equations. The quadrature rules for several different integrals represented through the same coefficients are presented. This allows one to reduce the integral equations containing integrals of different types to a system of linear algebraic equations.

Introduction

The rapid development of computer technologies in the second half of the last century revived the interest of researchers in computational methods and their effective use, in particular, in methods for solving singular integral equations with Cauchy kernel. The importance of "bringing to a number" the solutions of singular integral equations had been repeatedly emphasized by well-known mathematicians, such as N. I. Mukhelishvili, M. A. Lavrentiev, S. G. Mikhlin, H. Multhopp, I. N. Vekua.

The first important results in this direction were obtained by M. A. Lavrentiev [1]. Noting the importance of this work, Academician N. I. Muskhelishvili noted in all editions of his monograph [2]: "The further development of this and similar methods for solving singular integral equations approximately is, I believe, one of the most important problems in the theory of these equations."

From the middle of the twentieth century, approximate methods for solving singular integral equations began to develop intensively. There are a huge number of papers pertaining to this problem, but the vast majority of them are related to the most widespread particular case, when the behavior of the singular integral density at the ends of the segment of integration is described by a root function. The number of papers devoted to approximate calculations of singular integrals whose densities contain the weight function of Jacobi polynomials with arbitrary admissible real exponents is much smaller. The works, where these exponents can be complex, are not known to the authors.

Of particular interest is the work of F. Erdogan, G. D. Gupta and T. S. Cook [3]. This work is important because of a wider range of the considered material and good presentation of the methods for investigating the behavior of solutions at singular points and for solving singular integral equations of various types. However, unfortunately, in this work, there is no single approach to the problem of reducing singular integral equations to a system of algebraic equations.

Despite such a large number of works devoted to approximate methods for solving singular integral equations, interest in this problem does not decrease and continues to be on the agenda. This is testified by a considerable number of publications in the recent years, for example, [4–11].

In this paper, we give a series of quadrature formulas for various integrals arising in singular integral or integro-differential equations of the first and second kind, and describe the method of mechanical quadratures for solving these equations.

1. Quadrature formulas

In this section, several quadrature formulas of Gauss type for calculating integrals containing the weight function of Jacobi polynomials are given. Some of them were obtained in [12–16].

Suppose we have a function $\Phi(x)$ defined on the interval $(-1, 1)$ and representable in the form

$$\Phi(x) = \varphi(x)(1-x)^{\alpha}(1+x)^{\beta} \quad (\operatorname{Re}\alpha, \operatorname{Re}\beta > -1) \tag{1}$$

where $\varphi(x)$ is a function that satisfies the Hölder condition on the interval $[-1, 1]$.

The application of these quadrature formulas for complex values of exponents α and β is explained in detail in [17].

Replacing the function $\varphi(x)$ by an interpolation polynomial

$$\varphi_n(x) = \frac{2}{n+\alpha+\beta+1} \sum_{j=1}^{n} \frac{\varphi(\xi_j) P_n^{(\alpha,\beta)}(x)}{(x-\xi_j) P_{n-1}^{(\alpha+1,\beta+1)}(\xi_j)},$$

we obtain the quadrature formulas

$$\int_{-1}^{1} \Phi(x)\, dx \approx \sum_{j=1}^{n} w_j \varphi(\xi_j), \tag{2}$$

$$\int_{-1}^{1} \frac{\Phi(x)}{x-z}\, dx \approx \sum_{j=1}^{n} \frac{w_j \varphi(\xi_j)}{\xi_j - z}\left[1 - \frac{R_n^{(\alpha,\beta)}(z)}{R_n^{(\alpha,\beta)}(\xi_j)}\right], \tag{3}$$

$$\int_{-1}^{1} \frac{\Phi'(x)}{x-z} dx \approx \sum_{j=1}^{n} \frac{w_i \varphi(\xi_j)}{\xi_j - z}\left[2(n+1)\frac{R_{n+1}^{(\alpha-1,\beta-1)}(z)}{R_n^{(\alpha,\beta)}(\xi_j)} + \frac{R_n^{(\alpha,\beta)}(z) - R_n^{(\alpha,\beta)}(\xi_j)}{R_n^{(\alpha,\beta)}(\xi_j)(z-\xi_j)}\right], \tag{4}$$

$$\int_{-1}^{y} \frac{\Phi(x)}{x-z} dx \approx \sum_{i=1}^{n} \frac{w_i \varphi(\xi_i)}{\xi_i - z}\left[K_1(y,\xi_i) - \sum_{j=1}^{n} \frac{a_j(z)}{a_1(\xi_i)} K_j(y,z) - I_0(y,z)\right],$$

$$\int_{-1}^{y} \Phi(x)\, dx \approx \sum_{i=1}^{n} w_i \varphi(\xi_i)\left[\frac{B_{\frac{1+y}{2}}(\beta+1,\alpha+1)}{B(\beta+1,\alpha+1)} - \sum_{m=1}^{n-1} P_m^{(\alpha,\beta)}(\xi_i) S_m^{(\alpha,\beta)}(y)\right], \tag{5}$$

$$\int_{-1}^{1} \operatorname{sign}(x-y)\Phi(x)\, dx \approx \sum_{i=1}^{n} w_i \varphi(\xi_i)\left[1 - \frac{2B_{(1+y)/2}(\beta+1,\alpha+1)}{B(\beta+1,\alpha+1)} + 2\sum_{m=1}^{n-1} P_m^{(\alpha,\beta)}(\xi_i) S_m^{(\alpha,\beta)}(y)\right], \tag{6}$$

$$\int_{-1}^{1} \ln\frac{1}{|x-y|}\Phi(x)\, dx \approx \sum_{i=1}^{n} w_i \varphi(\xi_i)\left[L_0^{(\alpha,\beta)}(y) + \sum_{m=1}^{n-1} P_m^{(\alpha,\beta)}(\xi_i) L_m^{(\alpha,\beta)}(y)\right]. \tag{7}$$

Here the points ξ_i $(i = \overline{1,n})$ are the roots of the Jacobi polynomial $P_n^{(\alpha,\beta)}(x)$, z can be any point of the complex plane except for the points ± 1, and y is a point in the interval $(-1, 1)$.

The weighting factors w_i $(i = \overline{1,n})$ are defined as

$$w_j = \frac{2}{n+\alpha+\beta+1}\frac{R_n^{(\alpha,\beta)}(\xi_j)}{P_{n-1}^{(\alpha+1,\beta+1)}(\xi_j)}$$
$$= \frac{2^{\alpha+\beta+3}}{1-\xi_j^2}\frac{\Gamma(\alpha+n+1)\Gamma(\beta+n+1)}{\Gamma(n+1)\Gamma(\alpha+\beta+n+1)}\left[\frac{1}{(\alpha+\beta+n+1)P_{n-1}^{(\alpha+1,\beta+1)}(\xi_j)}\right]^2, \quad (8)$$

$B(\alpha,\beta)$, $B_y(\alpha,\beta)$ are complete and incomplete beta functions, $F[a,b;c;x]$, ${}_3F_2[a,b,c;e,d;x]$ are hypergeometric and generalized hypergeometric functions, $\Gamma(x)$ is the Euler gamma function, $\psi(x) = \Gamma'(x)/\Gamma(x)$ is the digamma function,

$$R_n^{(\alpha,\beta)}(z) = -\left(\frac{2}{z-1}\right)^{n+1} 2^{\alpha+\beta} B(n+\alpha+1, n+\beta+1) F\left[n+1, n+\alpha+1; 2n+\alpha+\beta+2; \frac{2}{1-z}\right],$$
$$h_m^{(\alpha,\beta)} = \frac{2^{\alpha+\beta+1}\Gamma(m+\alpha+1)\Gamma(m+\beta+1)}{(2m+\alpha+\beta+1)\Gamma(m+1)\Gamma(m+\alpha+\beta+1)},$$
$$k_m^{(\alpha,\beta)} = \frac{\Gamma(2m+\alpha+\beta+1)}{2^m\Gamma(m+\alpha+\beta+1)\Gamma(m+1)},$$
$$a_j(z) = \frac{k_{n-j+1}^{(\alpha,\beta)} h_{n-j}^{(\alpha,\beta)} P_n^{(\alpha,\beta)}(z)}{k_{n-j}^{(\alpha,\beta)} P_{n-j}^{(\alpha,\beta)}(z) P_{n-j+1}^{(\alpha,\beta)}(z)},$$
$$K_j(y,z) = \frac{B_{(1+y)/2}(1+\beta, 1+\alpha)}{B(1+\alpha,1+\beta)} - \sum_{m=1}^{n-j} P_m^{(\alpha,\beta)}(z) S_m^{(\alpha,\beta)}(y),$$
$$S_m^{(\alpha,\beta)}(y) = (1-y)^{\alpha+1}(1+y)^{\beta+1}\frac{1}{2m h_m^{(\alpha,\beta)}} P_{m-1}^{(\alpha+1,\beta+1)}(y) \quad (m \geq 1),$$
$$L_0^{(\alpha,\beta)}(y) = -\frac{B_{(1-y)/2}(1+\alpha,1+\beta)}{B(1+\beta,1+\alpha)}\pi\cot(\pi\alpha) - \ln 2 - \psi(\alpha+1) + \psi(\alpha+\beta+2)$$
$$+ \frac{(1-y)(1+\beta+\alpha)}{2\alpha}{}_3F_2\left[1,1,-\alpha-\beta;2,1-\alpha;\frac{1-y}{2}\right],$$
$$L_m^{(\alpha,\beta)}(z) = \left(\frac{2}{z-1}\right)^m \frac{2^{\alpha+\beta+1}}{m} B(m+\alpha+1, m+\beta+1) F\left[m+\alpha+1, m; 2m+\alpha+\beta+2, \frac{2}{1-z}\right].$$

Note that the functions $R_n^{(\alpha,\beta)}(z)$ and $L_n^{(\alpha,\beta)}(z)$ are uniquely determined in the complex plane which is cut respectively along the segment $[-1,1]$ and the ray $[-1,\infty]$, when for the cut points, the half-sum of the values from above $(y+i0)$ and below $(y-i0)$ is used.

It is known that quadrature formula (2) is the formula of the highest algebraic accuracy, i.e., it is satisfied exactly if the function $\varphi(x)$ is a polynomial up to order $2n-1$. In the general case, i.e., at an arbitrary value z, formula (3) is not a formula of the highest algebraic accuracy, but it becomes so if z coincides with the roots of the function $R_n^{(\alpha,\beta)}(z)$. The remaining formulas are exact only in the case when the function $\varphi(x)$ is a polynomial of order less than n.

The main advantage of quadrature formulas (2)–(7) is that they are all expressed in terms of the same quantities $\varphi(\xi_j)$, which are the unknowns in the integral equations. This allows one to reduce an integral equation containing arbitrary combinations of the integrals shown above to a system of linear algebraic equations with respect to the unknowns $\varphi(\xi_j)$.

2. Method of mechanical quadratures

Below we step by step describe the basic procedures required to apply the method of mechanical quadratures for solving the singular integral equations.

IOP Conf. Series: Journal of Physics: Conf. Series **991** (2018) 012070 doi:10.1088/1742-6596/991/1/012070

Suppose we have an equation that, in addition to the singular integral (3), also contains integrals of some type from (2)–(7). We note that, in the integral equation, formula (2) appears in the form

$$\int_{-1}^{1} R(x,y)\Phi(x)\,dx \approx \sum_{j=1}^{n} w_j R(\xi_j, y)\varphi(\xi_j), \tag{9}$$

where $R(x, y)$ is a regular function in both variables.

It is obvious that, in the case of an arbitrary interval of integration, the equation can be reduced to the interval $(-1, 1)$ by a simple linear transformation. Further, the solution of the equation is sought in the form (1) with unknown coefficients α and β.

Step 1. We assume that the solution of the given equation has representation (1), where the unknown function $\varphi(x)$ is sufficiently smooth and bounded on the closed interval $[-1, 1]$. We will call it the regular part of the solution of the given equation.

Using the results of N. I. Mukhelishvili on the behavior of the Cauchy-type integral at the ends of the line of integration [2], we write the principal terms of the asymptotics of the behavior of each of the integrals in the equation, and of the right-hand side, the known part of the equation in a neighborhood of each of the ends of the interval of integration. Equating the coefficients of the corresponding terms of the asymptotics to each other, we obtain equations for determining the exponents α and β. Usually, these roots are in the interval $(-1, 1)$. In the case when the equations obtained have two roots in $(-1, 1)$, usually of opposite signs, the choice of the necessary root is dictated by the statement of the problem.

Step 2. Having found the exponents α and β and choosing a certain order of approximation n, we can replace each of the integrals of the equation, according to formulas (2)–(7), by a quadrature sum. As a result, we obtain a functional equation that is linear with respect to unknown coefficients, which are the values of the regular part of the solution of the integral equation in the roots of the Jacobi polynomial corresponding to the found exponents.

Step 3. To determine the unknown coefficients, it is necessary to equate the right- and left-hand sides of the obtained functional equation at a certain number of points, called collocation points. The effective choice of these points is dictated by the following considerations.

Suppose we have a singular integral equation of the first kind, which contains only the singular integral (3) and the regular integral (9). As was mentioned above, because, at the roots of the function $R_n^{(\alpha,\beta)}(z)$, the quadrature formulas for both integrals are quadrature formulas of the highest algebraic accuracy, it is obvious that these roots should be chosen as collocation points. This also fixes the number of collocation points necessary to obtain a closed system of linear algebraic equations. In the case under consideration, when the regular kernel and the right-hand side of the equation do not have singularities at the ends of the interval of integration, the exponents α and β can take the values $\alpha = \pm\beta = \pm 0.5$. As was indicated above, the choice of adequate values is dictated by the formulation of the problem. For clarity, we illustrate this with an example of the classical problem of indenting a rigid stamp into an elastic half-plane.

Let us have an elastic strip or an inhomogeneous half-plane into which a rigid stamp of a certain shape is pressed under the action of a force P (figure 1). It is assumed that there is a smooth contact and only normal stresses appear in the contact zone.

After reducing the contact zone intervals $(-a, a)$, $(-c, a)$, and $(-c, b)$ to the interval $(-1, 1)$ in all three cases, the problem reduces to solving the equation

$$\int_{-1}^{1} \frac{\Phi(\xi)}{\xi - \eta}\,d\xi + \int_{-1}^{1} K(\xi,\eta)\Phi(\xi)\,d\xi = f(\eta) \quad (-1 < \eta < 1) \tag{10}$$

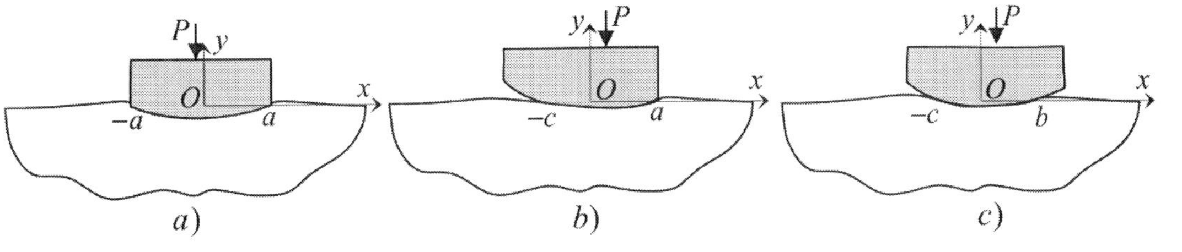

Figure 1. Contact interaction.

under the condition of the stamp equilibrium

$$\int_{-1}^{1} \Phi(\xi)\, d\xi = P_*. \tag{11}$$

Here $\Phi(\xi) = p(x(\xi))/\mu$ is the dimensionless contact pressure, $f(\eta)$ is a known dimensionless function determined by the shape of the base of the stamp, μ is the modulus of shear of the material of the base, and $P_* = P/(a\mu)$ is the dimensionless value of the pressing force.

The solution of equation (10) is sought in the form (1). According to step 1, we find that the exponents α and β can take the values $\alpha = \pm\beta = \pm 0.5$.

In case $a)$ in figure 1, at the ends of the contact zone, the contact pressure has a singularity and, therefore, the values $\alpha = \beta = -0.5$ should be selected. The function $R_n^{(-0.5,-0.5)}(y)$ becomes the Chebyshev polynomial of the second kind $U_{n-1}(y)$ which has $n-1$ roots, and therefore the number of collocation points is $n-1$. This is in complete agreement with the well-known fact that equation (10) does not have a unique solution and is determined up to the summand $A(1-\xi^2)^{-0.5}$ whose coefficient A must be found from the condition (11).

Thus, by combining the $n-1$ collocation equations (10) and condition (11), which is discretized, according to (2), for determining n unknown coefficients $\varphi(\xi_j)$ we obtain a closed system of linear algebraic equations

$$\begin{aligned} &\sum_{j=1}^{n} \frac{w_j \varphi(\xi_j)}{\xi_j - \zeta_k} + \sum_{j=1}^{n} w_j K(\xi_j, \zeta_k) \varphi(\xi_j) = f(\zeta_k) \quad (k = 1, 2, \ldots, n-1), \\ &\sum_{j=1}^{n} w_j \varphi(\xi_j) = P_*. \end{aligned} \tag{12}$$

In case $b)$ in figure 1, at the left end of the contact zone, due to continuity, the contact pressure must go to zero and, therefore, the values $\alpha = -0.5$, $\beta = 0.5$ should be selected. The function $R_n^{(-0.5,0.5)}(y)$ is the Jacobi polynomial $P_n^{(0.5,-0.5)}(y)$ and the number of collocation points is n. Therefore, the resulting system of linear algebraic equations will contain $n+1$ equations. However, it should be noted that in addition to unknown coefficients $\varphi(\xi_j)$, in this case the location of the left end of the contact zone c/a is also unknown. Since P_* appears only in condition (11), it is more expedient to solve the inverse problem, i.e., to set the value c/a, find the coefficients $\varphi(\xi_j)$ from the system of collocation equations, and, further, from equilibrium condition (11), to determine for what P_* this is the case. The system of linear algebraic equations in this case becomes

$$\begin{aligned} &\sum_{j=1}^{n} \frac{w_j \varphi(\xi_j)}{\xi_j - t_k} + \sum_{j=1}^{n} w_j K(\xi_j, t_k) \varphi(\xi_j) = f(t_k) \quad (k = 1, 2, \ldots, n), \\ &\sum_{j=1}^{n} w_j \varphi(\xi_j) = P_* \end{aligned} \tag{13}$$

In case c) in figure 1, the contact pressure is zero at both ends of the contact zone. The function $R_n^{(0.5,0.5)}(y)$ is a Chebyshev polynomial of the first kind $T_{n+1}(y)$, which has $n+1$ roots, that is, the number of collocation points is $n+1$. It is known that, in this class of functions, equation (10) has a solution only if the following condition is satisfied:

$$\int_{-1}^{1}\left[\int_{-1}^{1}\frac{\Phi(\xi)}{\xi-\eta}\,dx\right]\frac{d\eta}{\sqrt{1-\eta^2}}=0. \tag{14}$$

Since, unlike the preceding case, the location of the right end of the contact zone b/a is also unknown here, condition (14) is necessary for its determination.

In [18] it is shown that if an additional unknown γ_{0n}, with a unit coefficient, is introduced into the resulting system of linear algebraic equations, then $\lim_{n\to\infty}\gamma_{0n}=0$ if and only if condition (14) is satisfied. The unknown γ_{0n} is called the regularizing factor, and its behavior serves as a convenient indicator of the existence of the solution. As in the preceding case, it is convenient here to solve the inverse problem, i.e., to set, for example, the value c/a, then select a value b/a so that when the collocation system is solved, then the regularizing factor γ_{0n} is close to zero. Then the corresponding value P_* is determined from condition (11).

It is worth to note that, in this case, the unknown γ_{0n} can be interpreted as the angle of rotation of the stamp. Really, if both values c/a and b/a are set, and the value P_* is found, then γ_{0n} means the angle of the stamp rotation needed for the contact zone to coincide with the interval $(-c/a, b/a)$. When $\gamma_{0n}=0$, the stamp is pressed vertically into the half-plane.

In this case, the system of linear algebraic equations has the form

$$\begin{gathered}\gamma_{0n}+\sum_{j=1}^{n}\frac{w_j\varphi(\xi_j)}{\xi_j-t_k}+\sum_{j=1}^{n}w_jK(\xi_j,t_k)\varphi(\xi_j)=f(t_k)(k=1,2,\dots,n+1),\\ \sum_{j=1}^{n}w_j\varphi(\xi_j)=P_*.\end{gathered} \tag{15}$$

Unlike the equations of the first kind, other singular integral equations contain at least one term for which the approximation formula is exact only for polynomials of order at most $n-1$. Although the above justification for choosing the roots of $R_n^{(\alpha,\beta)}(z)$ as collocation points is not valid here, nevertheless, the experience shows that it is also acceptable in this case. Indeed, the number of roots of the function $R_n^{(\alpha,\beta)}(z)$, if any, coincides with the necessary number of collocation points. The choice of collocation points themselves can be arbitrary, since there are no restrictions on the external variable in formulas (2)–(7).

When solving the singular integral equations of the second kind, it is more convenient to use a slightly different representation of quadrature formula (3).

The function $R_n^{(\alpha,\beta)}(z)$ for the points of the interval of integration can be written as

$$R_n^{(\alpha,\beta)}(y)=\pi\cot(\alpha\pi)(1-y)^\alpha(1+y)^\beta P_n^{(\alpha,\beta)}(y)-2^{\alpha+\beta}B(\alpha,n+\beta+1)F\left[n+1,-n-\alpha-\beta;1-\alpha;\frac{1-y}{2}\right].$$

Since, in the case of a singular integral equation of the second kind, the index of the equation κ is equal to $\kappa=\alpha+\beta=\{-1,0,1\}$, the last expression becomes

$$R_n^{(\alpha,\beta)}(y)=\pi\cot(\alpha\pi)(1-y)^\alpha(1+y)^\beta P_n^{(\alpha,\beta)}(y)-\frac{2^\kappa\pi}{\sin(\alpha\pi)}P_{n+\kappa}^{(-\alpha,-\beta)}(y). \tag{16}$$

Substituting (16) into formula (3), after simple transformations, we obtain the quadrature formula

$$\int_{-1}^{1} \frac{\Phi(x)}{x-y}dx - \pi\cot(\alpha\pi)\Phi(y) \approx \sum_{j=1}^{n} \frac{w_j\varphi(\xi_j)}{\xi_j - z}\left[1 - \frac{P_{n+\kappa}^{(-\alpha,-\beta)}(y)}{P_{n+\kappa}^{(-\alpha,-\beta)}(y)}\right]. \tag{17}$$

Suppose we have a singular integral equation of the second kind

$$\int_{-1}^{1} \frac{\Phi(x)}{x-y}dx + \pi\lambda\Phi(y) + \int_{-1}^{1} K(x,y)\Phi(x)\,dx = f(y) \quad (-1 < y < 1) \tag{18}$$

Under the assumption that the regular kernel and the right-hand side of the equation have no singularity at the ends of the integration interval, the exponents α and β are determined from the equations

$$\cot(\alpha\pi) - \lambda = 0, \quad \cot(\beta\pi) + \lambda = 0.$$

Here, depending on the value of the index κ, all three cases mentioned above can occur.

We note that, using the quadrature formula (17) and choosing the zeros of the polynomial $P_{n+\kappa}^{(-\alpha,-\beta)}(y)$ as collocation points, equation (18) reduces to a system of linear algebraic equations of the form (12), (13) or (15).

A similar approach can be used for more complex singular integral equations as well.

3. Approbation of the method

The method of mechanical quadratures in application to singular integral equations of the first kind is well known, and there are many works using it. Below we cite the works where the method of mechanical quadratures is used to solve singular equations of other types. Note that in many of these papers, the method is called the method of discrete singularities.

In [19, 20], contact problems of theory of elasticity are solved in the case where the defining equations are singular integral equations of the first kind with a right-hand side containing a logarithmic singularity at the ends of the interval of integration.

The contact problems for moving stamps, which reduce to solving singular integral equations of the second kind with real coefficient of the free term and, consequently, real exponents of the Jacobi weight function, were solved in [21–26]. The problem of indenting a stamp into an elastic massive body with allowance for the bonding zone in the middle part of the contact zone and slip zones at its ends [27–29], as well as the mixed problems for a piecewise homogeneous plane with interphase cracks or rigid inclusions, considered in the formulation of the antiplane problem of the theory of elasticity [30], can also be reduced to a system of such equations.

A large number of papers are devoted to problems which can be reduced to systems of singular integral equations of the second kind with imaginary coefficient of the free term and, consequently, with complex exponents of the Jacobi weight function. In [31–35], problems for homogeneous massive bodies containing thin rigid inclusions are solved, when, on one side, the inclusion is connected to a massive body, and on the other side, a crack is formed. In [36–38], problems for piecewise-homogeneous massive bodies with cracks and thin rigid inclusions on the material separation line are solved.

In [39–48], problems of the theory of elasticity in plane and antiplane formulations for massive bodies containing stress concentrators of the types indicated above are solved, which reduce to solving systems of singular integral equations with a generalized Cauchy kernel. These equations are also known as equations with a fixed singularity. This type of singularities often appears when an end of a stress concentrator reaches the boundary of the homogeneous body or the line of separation of materials of the piecewise homogeneous body or another concentrator.

Contact problems for massive bodies, reinforced by thin elastic plates or containing thin elastic inclusions, reduce to solving singular integro-differential equations. In [49–55], such problems were solved by the method of mechanical quadratures. In [53] the solution of the Prandtl equation is presented in two ways using the method of mechanical quadratures.

In a number of works, the rate of convergence of the approximate solution to the exact one, depending on the order of approximation, is investigated by numerical analysis. The practice shows that, with the correct choice of the exponents of the singularity α and β and the corresponding quadrature formulas, at $n = 10 \sim 12$, it is already possible to obtain solutions with accuracy higher than four significant digits.

Conclusions

The paper presents the development of the method of mechanical quadratures, which is well known for solving singular integral equations of the first kind. The quadrature formulas given in the paper, which are written with respect to the same coefficients (values of the unknown functions at the same points), allow us considerably to expand the scope of the method. This is tested by a large number of problems solved by this method whose defining equations qualitatively differ from each other.

References

[1] Lavrent'ev M A 1932 On the construction of a stream flowing around an arc of a given form *Trudi TsAGI* **118** pp 3-56
[2] Muskhelishvili N I 1968 *Singular Integral Equations* (Moscow: Nauka) [in Russian]
[3] Erdogan F, Gupta G D, and Cook T S 1973 The numerical solutions of singular integral equations. *Methods of Analysis an Solution of Crack Problems* (Leyden: Noordhoff Intern. Publ.) pp 368–425
[4] Andreev A V 2005 Direct numerical method for solving singular integral equations of the first kind with generalized kernels *Mech. Solids* **40** (1) 104–19
[5] Okecha G E 2007 Solution of Cauchy-type singular integral equations of the first kind with zeros of Jacobi polynomials as interpolation nodes *Int. J. Math. Math. Sci.* **2007** 10957 doi: http://dx.doi.org/10.1155/2007/10957
[6] Capobianco M R, Criscuolo G, and Junghanns P 2008 On the numerical solution of a hypersingular integral equation with fixed singularities In *Recent Advances in Operator Theory and Applications* pp 95–116
[7] Concetta Laurita 2012 A quadrature method for Cauchy singular integral equations with index −1 *IMA J. Num. Anal.* **32** (3) 1071–95 doi: https://doi.org/10.1093/imanum/drr032
[8] Setia A 2014 Numerical solution of various cases of Cauchy type singular integral equation *Appl. Math. Comput.* **230** 200–7
[9] Criscuolo G 2014 Numerical evaluation of certain strongly singular integrals *IMA J. Num. Anal.* **34** (2) 651–74 doi: https://doi.org/10.1093/imanum/drt017
[10] Huang C and Stynes M 2017 Spectral Galerkin methods for a weakly singular Volterra integral equation of the second kind *IMA J. Num. Anal.* **37** (3) 1411–36 doi: https://doi.org/10.1093/imanum/drw034
[11] Setia A, Sharma V and Liu Y 2017 Numerical method to solve Cauchy type singular integral equation with error bounds *AIP Conf. Proc.* **1798** doi: http://dx.doi.org/10.1063/1.4972733
[12] Sahakyan A V 2002 Gauss type quadrature formulae for singular integrals In *Problems of Mechanics of Thin Deformable Bodies, Coll. of Scientific Papers Dedicated to 80-th Anniversary of Acad. S. A. Ambartsumian* (Yerevan: Gitutyun) pp 259–66 [in Russian]
[13] Sahakyan A V 2010 Quadrature formulas for calculation of the integral with a variable upper limit In *Topical Problems of Continuum Mechanics: Proc. of II Int. Conf. 4–8 October 2010, Dilijan, Armenia* vol 2 (Yerevan: EGUAS) 107–11 [in Russian]
[14] Sahakyan A V 2012 Quadrature formulae for Cauchy type integral with variable integration limit In *Problems of Mechanics of Thin Deformable Bodies, Coll. of Scientific Papers Dedicated to 80-th Anniversary of Acad. S. A. Ambartsumian* (Yerevan: Gitutyun) pp 240–4 [in Russian]
[15] Amirjanyan H A and Hakobyan L V 2010 Uniform movement of an absolutely rigid die over the inner surface of an infinite hollow cylinder In *Topical Problems of Continuum Mechanics: Proc. of II Int. Conf. 4–8 October 2010, Dilijan, Armenia* vol 1 (Yerevan: EGUAS) 81–5 [in Russian]
[16] Sahakyan A V 2011 Method of discrete singularities for solution of singular integral and integro-differential equations *Proc. of A. Razmadze Math. Inst.* **156** 101–11
[17] Saakyan A V 2012 Quadrature formulas of maximum algebraic accuracy for Cauchy type integrals with complex exponents of the Jacobi weight function *Mech. Solids* **47** (6) 695–9

[18] Lifanov I K 1980 On the ill-posedness and regularization of numerical solutions of a singular integral equation of the first kind *Dokl. Akad. Nauk SSSR* **255** (5) 1046–50

[19] Amirjanyan H A and Sahakyan A V 2012 About the indentation of a rigid punch with concave base into half plane, taking into account the influence of filler in cavity *Izv. Nats. Akad. Nauk Armenii. Mekh.* **65** (1) 7–16

[20] Amirjanyan A A and Sahakyan A V 2017 Mechanical quadrature method as applied to singular integral equations with logarithmic singularity on the right-hand side *Comput. Math. Math. Phys.* **57** 1285–93 doi: https://doi.org/10.1134/S0965542517080036

[21] Lifanov I K and Saakian A V 1982 Method of numerical solution of the problem of impressing a moving stamp into an elastic half-plane, taking heat generation into account *J. Appl. Math. Mech.* **46** (3) 388–94

[22] Sahakyan A V 1996 Ch. 12.3. On the indentation of a pair of uniformly moving punches into an elastic strip In Lifanov I K *Singular Integral Equations and Discrete Vortices* (Utrecht: VSP BVP)

[23] Hakobyan V N and Amirjanyan H A 2009 About indentation of the system of uniformly moving rigid punches in piecewise homogeneous half-plane *Izv. Nats. Akad. Nauk Armenii. Mekh.* **62** (3) 3–10

[24] Sahakyan A V 2009 The influence of friction heat generation on the contact stress distribution between uniformly moving punch and elastic half-plane In *Proc. of the 8th Int. Congr. on Thermal Stresses. TS2009. 1-4 July, Univ. of Illinois at Urbana-Champaing. Illinois, USA* vol 1 (Illinois) pp 117–20

[25] Amirjanyan H A and Hakobyan L V 2010 On uniform motion of absolutely rigid strips on inner surface of elastic infinite hollow cylinder In *Topical Problems of Continuum Mechanics: Proc. of II Int. Conf. 4-8 October 2010, Dilijan, Armenia* vol 1 (Yerevan: EGUAS) 81–5 [in Russian]

[26] Amirjanyan H A 2011 On uniform motion of a periodic system of absolutely rigid stamps, in the hollow cylinder *Izv. Nats. Akad. Nauk Armenii. Mekh.* **64**(1) 5–10

[27] Sahakyan A V 2013 Solution of a contact problem with contact and adhesion regions (Galin's problem) by the method of discrete singularities In *Development of L. A. Galin's Ideas in Mechanics. Collection of Research Papers* (Moscow - Izhevsk) pp 103–20

[28] Hakobyan V N and Sahakyan A V 2014 Stress state of elastic plane with partially detached from matrix absolutely rigid inclusion within Galins' model *Proc. of VIII Int. Conf. "The Problems of Dynamics of Interaction of Deformable Media" September 22-26, Goris - Stepanakert, Armenia, 2014* (Goris - Stepanakert: Inst. Mekh. NAN RA) pp 43–7 [in Russian]

[29] Amirjanyan A A and Sahakyan A V 2017 On indentation of a pair of rigid punches connected by an elastic beam into an elastic half-plane with regard to the friction and adhesion forces in the contact region *Mech. Solids* **52** (2) 161–71

[30] Hakobyan V N and Hakobyan L V 2016 Stress state of piece-wise homogeneous space with doubly-periodic system of interphase defects *Izv. Nats. Akad. Nauk Armenii. Mekh.* **69** (3) 3-15

[31] Akopyan V N and Saakyan A V 1999 The stress state of a homogeneous elastic plane with a star-shaped crack under mixed boundary conditions on the crack faces *Mech. Solids* **34** (3) 90–5

[32] Sahakyan A V 2005 The mixed boundary value problem for elastic wedge with thin rigid inclusion one side of which is separated from the matrix In *Collection of Papers, Dedicated to the Memory of Professor P. S. Theocaris* (Yerevan: Gitutyun NAN RA) pp 231–9

[33] Akopyan V N and Saakyan A V 1999 A mixed problem for an elastic wedge weakened by a crack *Mech. Solids* **34** (6) 56–65

[34] Hakobyan V N, Sahakyan A V, and Sargsyan A H 2014 The plane deformation state of elastic plane with finite rigid inclusion under harmonic loading In *Proc. of V Int. Conf. "The Problems of Dynamics of Interaction of Deformable Media" October 1-7, Goris, Armenia, 2014* pp 56–60

[35] Hakobyan V N and Sahakyan A V 2014 On the pressing of two smooth punches into an elastic half-plane containing a rigid inclusion of finite length, one face of which is detached from the matrix In *Proc. XVII Int. Conf. "Modern Problems of Continuum Mechanics" October 14-17, Rostov-on-Don, Russia, 2014* vol 1 pp 29–33 [in Russian]

[36] Sahakyan A V and Dashtoyan L L 2003 On one problem for compound plane with crack and inclusion with the existence of temperature field In *Proc. of the 5th Int. Congress on Thermal Stresses and Related Topics, TS2003, 8-11 June 2003, Blacksburg, VA* pp TM-4-1-1-4

[37] Hakobyan V and Dashtoyan L 2016 Doubly periodic problem for piecewise homogeneous plane with absolutely rigid inclusions In *Proc. of the 8th Int. Conf. Contemporary Problems of Architecture and Construction, Yerevan* (Yerevan) pp 125–8 [in Russian]

[38] Hakobyan V N 2016 Axisymmetric stress state of a piecewise homogeneous, layered space with parallel coin-like cracks In *Proc. XVIII Int. Conf. "Modern Problems of Continuum Mechanics" 7-10 November 2016, Rostov-on-Don, Russia* vol 1 35–9 [in Russian]

[39] Sahakyan A V 2000 Application of the method of discrete singularities for solving the singular integral equations with unmoved singularities *Izv. Nats. Akad. Nauk Armenii. Mekh.* **53** (3) 12–9

[40] Sahakyan A V 2004 The method of discrete singularities of solutions of singular integral equations with unmoved singularity In *Topics in Analysis and its Applications. NATO Science Series, II. Mathematics, Physics and Chemistry* vol 147 (The Netherlands, Kluwer Acad. Publ.) pp 327–32
[41] Aghayan K L, Amirjanyan H A, and Zakaryan V G 2016 Longitudinal shear of an elastic space containing cracks and inclusions *Nauchn. Trudy NUACA* **1** (60) 35–44
[42] Aghayan K L, Amirjanyan H A, and Zakaryan V G 2016 Anti-plane contact problem for a compound elastic half-space with a finite crack In *Proc. of the 8th Int. Conf. Contemporary Problems of Architecture and Construction, Yerevan* (Yerevan) pp 50–3
[43] Sahakyan A V 2010 Method of discrete singularities in use to solution of the problems for regions with boundary cracks *Proc. of the XIV Int. Conf. "Actual Problems of Mechanics of Solids," 19–24 June, 2010, Rostov-on-Don, Russia* pp 290–4 [in Russian]
[44] Sahakyan A V and Hakobyan V N 2007 On the stressed state of an elastic plane with a semi-infinite cut which perpendicularly terminates at the rigid inclusion of finite length In *Problems of Mechanics of Thin Deformable Bodies, Coll. of Scientific Papers Dedicated to 80-th Anniversary of acad. S. A. Ambartsumian* (Yerevan: Gitutyun) pp 16–25 [in Russian]
[45] Sahakyan A V 2011 A method of discrete singularities in application to a mixed problem for an elastic half-plane with boundary inclusion In *Proc. XV Int. Symp. "Method of Discrete Singularities in Problems of Mathematical Physics" Kharkov-Kherson* pp 330–3 [in Russian]
[46] Hakobyan L V and Sahakyan A V 2015 A refined solution of the mixed problem for an elastic space with a T-shaped crack at antiplane deformation In *Topical Problems of Continuum Mechanics: Proc. of IV Int. Conf. 21–26 September 2015, Tsaghkadzor, Armenia* (Yerevan: Nats. Univ. Arkh. Stroit. Armenii) pp 38–42 [in Russian]
[47] Hakobyan V N, Dashtoyan L L, and Hakobyan L V 2012 On the stressed state of an elastic plane with a semi-infinite cut which perpendicularly terminates at the rigid inclusion which is partially torn from the matrix In *Proc. III Int. Conf. "Topical Problems of Mechanics of Solids" Dedicated to the Century Anniversary of Acad. N. Kh. Arutyunyan, Tsaghkadzor, Armenia* vol 1 pp 57-61 [in Russian]
[48] Hakobyan V N and Amirjanyan H A 2015 Stress state of semi-infinite plane with absolutely rigid inclusion and crack *Izv. Nats. Akad. Nauk Armenii. Mekh.* **68** (1) 25–36
[49] Sahakyan A V 2008 Method of discrete singularities in use to solution of a singular integro-differential equations In *Proc. 6th Int. Conf. "The Problems of Dynamics of Interaction of Deformable Media," September 21–26, Goris - Stepanakert, 2008* (Goris - Stepanakert: Inst. Mekh. NAN RA) pp 383–7 [in Russian]
[50] Hakobyan V N and Amirjanyan H A 2012 Load-transfer from deformable inclusions to infinite visco-elastic plate with slits *Izv. Nats. Akad. Nauk Armenii. Mekh.* **65** (4) 6–21
[51] Hakobyan V N and Amirjanyan H A 2013 On the load transfer from elastic inclusions to an infinite elastic orthotropic plane with cuts *Mech. Solids* **48** (6) 628–35
[52] Sahakyan A V 2011 Method of discrete singularities for solution of singular integral and integro-differential equations *Proc. of A. Razmadze Math. Inst.* **156** 101–11
[53] Sahakyan A V and Shavlakadze N N 2014 Two methods for direct numerical integration of the Prandtl equation and comparative analysis between them *Comput. Math. Math. Phys.* **54** (8) 1244–50
[54] Sahakyan A V and Shavlakadze N N 2016 The contact problem for piecewise-homogeneous elastic plate reinforced by finite elastic stringer of variable stiffness *Abstr. of the 40-th Solid Mechanics Conf, Warsaw* URL `http://solmech2016.ippt.gov.pl/Abstracts/0109.pdf`
[55] Sahakyan A V and Shavlakadze N N 2016 Numerical analysis of the contact problem for an elastic stringer of variable rigidity that extends to the material separation line of a composite elastic plane In *Proc. of the XVIII Int. Conf. "Actual Problems of Mechanics of Solids," 7–10 November, 2016, Rostov-on-Don, Russia* pp 171–5 [in Russian]

IOP Conf. Series: Journal of Physics: Conf. Series **991** (2018) 012071 doi:10.1088/1742-6596/991/1/012071

Mixed boundary-value problem for an orthotropic rectangular strip with variable coefficients of elasticity

M Z Sargsyan* **and H M Poghosyan****

Institute of Mechanics of the National Academy of Sciences of Armenia, Yerevan, Armenia

E-mail: *messarg@gmail.com, **haikp@mail.ru

Abstract. A dynamical problem for a rectangular strip with variable coefficients of elasticity is solved by an asymptotic method. It is assumed that the strip is orthotropic, the elasticity coefficients are exponential functions of y, and mixed boundary conditions are posed. The solution of the inner problem is obtained using Bessel functions.

1. Basic equations and statement of the problem

In the case of an orthotropic rectangular strip, it is required to find the solution of a plane problem of elasticity in the domain $D = \{(x, y) \colon 0 \le x \le \ell, -h \le y \le h, h \ll l\}$ [1] (see figure 1) under the conditions

$$\begin{aligned} u_x(y = -h) &= 0, \\ u_y(y = -h) &= 0 \end{aligned} \tag{1}$$

and

$$\begin{aligned} \sigma_{yy}(y = h) &= P(\xi)\exp(i\Omega t), \quad \xi = \frac{x}{l}, \\ \sigma_{xy}(y = h) &= 0, \end{aligned} \tag{2}$$

where $P(\xi)$ is a given function and Ω is the frequency of the forcing effect.

It is required to find solutions of dynamic equations of the elasticity theory for an orthotropic body

$$\begin{aligned} \frac{\partial \sigma_{xx}}{\partial x} + \frac{\partial \sigma_{xy}}{\partial y} &= \rho \frac{\partial^2 u}{\partial t^2}, \\ \frac{\partial \sigma_{xy}}{\partial x} + \frac{\partial \sigma_{yy}}{\partial y} &= \rho \frac{\partial^2 v}{\partial t^2}, \end{aligned} \tag{3}$$

together with Hook's law

$$\begin{aligned} \frac{\partial u}{\partial x} &= a_{11}(y)\sigma_{xx} + a_{12}(y)\sigma_{yy}, \\ \frac{\partial v}{\partial y} &= a_{12}(y)\sigma_{xx} + a_{22}(y)\sigma_{yy}, \\ \frac{\partial u}{\partial y} + \frac{\partial v}{\partial x} &= a_{66}(y)\sigma_{xy}, \end{aligned} \tag{4}$$

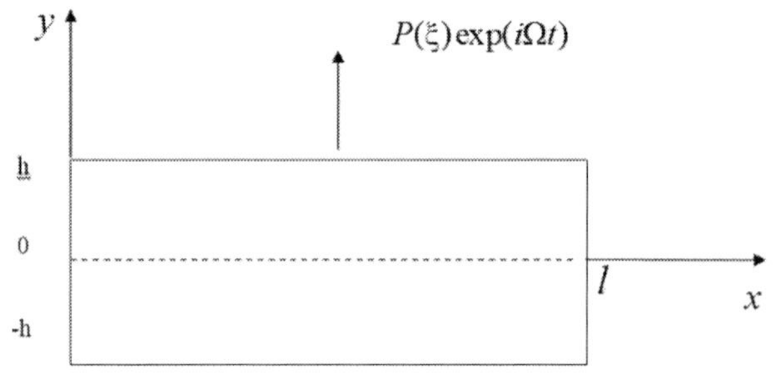

Figure 1.

where [2]

$$a_{11}(y) = \frac{1}{E_x}, \quad a_{12}(y) = -\frac{\nu_{xy}}{E_y}, \quad a_{22}(y) = \frac{1}{E_y}, \quad a_{66}(y) = \frac{1}{Ge^{-my}},$$
$$E_x = E_y = Ee^{-my}, \quad \nu_{xy} = \nu_{yx} \approx \nu,$$
$$a_{11} = a_{22} = \frac{1}{E}, \quad a_{12} = -\frac{\nu}{E}, \quad a_{66} = \frac{1}{G},$$

Here m describes the velocity of exponential variation in the elasticity coefficients.

2. Solution of the problem

Only the inner problem is solved, because no conditions on the lateral surface of the plate (which correspond to the appearance of boundary layers) are considered. The solution of boundary problem (1)–(4) is sought in the form

$$\begin{aligned} \sigma_{xx}(x,y,t) &= \sigma_{11}(x,y)\exp(i\Omega t),\\ \sigma_{xy}(x,y,t) &= \sigma_{12}(x,y)\exp(i\Omega t),\\ \sigma_{yy}(x,y,t) &= \sigma_{22}(x,y)\exp(i\Omega t),\\ u(x,y,t) &= u_x(x,y)\exp(i\Omega t),\\ v(x,y,t) &= u_y(x,y)\exp(i\Omega t). \end{aligned} \tag{5}$$

To solve the problem, we switch to dimensionless coordinates $\xi = x/l$, $\zeta = y/h$ and dimensionless displacement $U = u_x/l$, $V = u_y/l$ in the equations of elasticity theory and solve the obtained system which is singularly perturbed by a small parameter $\varepsilon = h/l$.

In this case, for the boundary conditions we have

$$\begin{aligned} U^{(s)}(\zeta = -1) &= 0,\\ V^{(s)}(\zeta = -1) &= 0, \end{aligned} \tag{6}$$

and

$$\begin{aligned} \sigma_{22}^{(s)}(\zeta = 1) &= P^{(0)}(\xi),\\ \sigma_{12}^{(s)}(\zeta = 1) &= 0, \end{aligned} \tag{7}$$

where $P^{(0)}(\xi) = \varepsilon P(\xi)$ and $P^{(s)}(\xi) = 0$, $s \neq 0$.

IOP Conf. Series: Journal of Physics: Conf. Series **991** (2018) 012071 doi:10.1088/1742-6596/991/1/012071

The solution of the inner problem is sought in the form

$$\begin{aligned}
\sigma_{11} &= \varepsilon^{-1+s}\sigma_{11}^{(s)}(\xi,\zeta),\\
\sigma_{12} &= \varepsilon^{-1+s}\sigma_{12}^{(s)}(\xi,\zeta),\\
\sigma_{22} &= \varepsilon^{-1+s}\sigma_{22}^{(s)}(\xi,\zeta),\\
(U,V) &= \varepsilon^{s}(U^{(s)},V^{(s)}), \quad s=\overline{0,N}.
\end{aligned} \tag{8}$$

Finally, we obtain equations to determine the functions $U^{(s)}$, $V^{(s)}$, $W^{(s)}$ [1, 3]:

$$\begin{aligned}
&\frac{\partial^2 U^{(s)}}{\partial\zeta^2} + a_{66}e^{n\zeta}\Omega_*^2 U^{(s)} = R_u^{(s)},\\
&\frac{\partial^2 V^{(s)}}{\partial\zeta^2} + \frac{\Delta}{a_{11}}e^{n\zeta}\Omega_*^2 V^{(s)} = R_v^{(s)},
\end{aligned} \tag{9}$$

where

$$\begin{aligned}
R_u^{(s)} &= -\left(a_{66}e^{n\zeta}\frac{\partial\sigma_{11}^{(s-1)}}{\partial\xi} + \frac{\partial^2 V^{(s-1)}}{\partial\xi\partial\zeta}\right),\\
R_v^{(s)} &= \frac{a_{12}}{a_{11}}\frac{\partial^2 U^{(s-1)}}{\partial\xi\partial\zeta} - \frac{e^{n\zeta}\Delta}{a_{11}}\frac{\partial\sigma_{12}^{(s-1)}}{\partial\xi},
\end{aligned}$$

The general solution of equations (9) is expressed in terms of Bessel functions as

$$\begin{aligned}
U^{(s)} &= J_0(g(\zeta))C_1(\xi) + 2Y_0(g(\zeta))C_2(\xi) + U_*^{(s)},\\
V^{(s)} &= J_0(f(\zeta))C_3(\xi) + 2Y_0(f(\zeta))C_4(\xi) + V_*^{(s)},
\end{aligned} \tag{10}$$

where the first part is the complete solution of homogeneous equations, the second part ($U_*^{(s)}$, $V_*^{(s)}$) is a particular solution of inhomogeneous equations (9), and J, Y are Bessel functions.

Satisfying boundary conditions (1), (2), we obtain the stress tensor components

$$\begin{aligned}
\sigma_{11}^{(s)} &= \frac{a_{12}\Omega_*}{\sqrt{a_{11}e^{n\zeta}\Delta}}[J_1(f(\zeta))C_3(\xi) + 2Y_1(f(\zeta))C_4(\xi)] + \sigma_{11*}^{(s)},\\
\sigma_{22}^{(s)} &= \frac{\Omega_*}{\sqrt{\frac{\Delta}{a_{11}}e^{n\zeta}}}[J_1(f(\zeta))C_3(\xi) + 2Y_1(f(\zeta))C_4(\xi)] + \sigma_{22*}^{(s)},\\
\sigma_{12}^{(s)} &= -\frac{\Omega_*}{\sqrt{a_{66}}}\sqrt{e^{n\zeta}}[J_1(g(\zeta))C_1(\xi) + 2Y_1(g(\zeta))C_2(\xi)] + \sigma_{12*}^{(s)},
\end{aligned} \tag{11}$$

where

$$\begin{aligned}
f(\zeta) &= \frac{2\Omega_*}{n}\sqrt{\frac{\Delta}{a_{11}}e^{n\zeta}},\\
g(\zeta) &= \frac{2\Omega_*}{n}\sqrt{a_{66}e^{n\zeta}},
\end{aligned} \tag{12}$$

and $\sigma_{11*}^{(s)}$, $\sigma_{22*}^{(s)}$, $\sigma_{12*}^{(s)}$ depend on the particular solution of inhomogeneous equations (9).

When the forcing effect is constant $P(\xi) = P = \text{const}$ and the iteration process stops at $s = 0$, we have $R_u^{(0)} = 0$, $R_v^{(0)} = 0$. Therefore $U_*^{(0)} = 0$, $V_*^{(s)} = 0$. From (10) we obtain the displacement vector components

$$\begin{aligned}
U^{(0)} &= J_0(g(\zeta))C_1(\xi) + 2Y_0(g(\zeta))C_2(\xi),\\
V^{(0)} &= J_0(f(\zeta))C_3(\xi) + 2Y_0(f(\zeta))C_4(\xi),
\end{aligned} \tag{13}$$

and from (9) we obtain the stress tensor components

$$\begin{aligned}
\sigma_{11}^{(0)} &= \frac{a_{12}2\Omega_*^2}{a_{11}nf(\zeta)}[J_1(f(\zeta))C_3(\xi) + 2Y_1(f(\zeta))C_4(\xi)],\\
\sigma_{22}^{(0)} &= \frac{2\Omega_*^2}{nf(\zeta)}[J_1(f(\zeta))C_3(\xi) + 2Y_1(f(\zeta))C_4(\xi)],\\
\sigma_{12}^{(0)} &= -\frac{n}{2a_{66}}g(\zeta)[J_1(g(\zeta))C_1(\xi) + 2Y_1(g(\zeta))C_2(\xi)].
\end{aligned} \tag{14}$$

Satisfying boundary conditions (6), (7), we obtain the coefficients C_1, C_2, C_3, C_4:

$$\begin{aligned}
C_1(\xi) &= 0,\\
C_2(\xi) &= 0,\\
C_3(\xi) &= \frac{Pf(1)Y_0(f(-1))}{2\Omega_*^2[J_1(f(1))Y_0(f(-1)) + J_0(f(-1))Y_1(f(1))]},\\
C_4(\xi) &= -\frac{Pf(1)J_0(f(-1))}{4\Omega_*^2[J_1(f(1))Y_0(f(-1)) + J_0(f(-1))Y_1(f(1))]}.
\end{aligned} \tag{15}$$

From (5), (8), (14), (15) we get the final solution in the form

$$\begin{aligned}
u &= 0,\\
\sigma_{xy} &= 0,\\
v &= \frac{lPf(1)[J_0(f(\zeta))Y_0(f(-1)) - J_0(f(-1))Y_0(f(\zeta))]}{2\Omega_*^2(J_1(f(1))Y_0(f(-1)) + J_0(f(-1))Y_1(f(1)))}\exp(i\Omega t),\\
\sigma_{xx} &= P\frac{la_{12}f(1)}{ha_{11}nf(\zeta)}\frac{J_1(f(\zeta))Y_0(f(-1)) - J_0(f(-1))Y_1(f(\zeta))}{J_1(f(1))Y_0(f(-1)) + J_0(f(-1))Y_1(f(1))}\exp(i\Omega t),\\
\sigma_{yy} &= P\frac{lf(1)}{hnf(\zeta)}\frac{J_1(f(\zeta))Y_0(f(-1)) - J_0(f(-1))Y_1(f(\zeta))}{J_1(f(1))Y_0(f(-1)) + J_0(f(-1))Y_1(f(1))}\exp(i\Omega t).
\end{aligned} \tag{16}$$

Solution (16) easily shows that resonance occurs if

$$J_1(f(1))Y_0(f(-1)) + J_0(f(-1))Y_1(f(1)) = 0.$$

Conclusions

It is shown that the solution of a dynamical problem for a rectangular strip in case of variable coefficients of elasticity contains Bessel functions. By the use of the properties of those functions, we can carry out numerical analysis of the solution of the inner problem. It is shown when the forcing effect is constant the iteration process is interrupted and mathematically exact solutions are received. Conditions to avoid resonance are obtained.

References

[1] Aghalovyan L A 2015 *Asymptotic Theory of Anisotropic Plates and Shells* (Singapore - London: World Scientific Publishing) p 375
[2] Aghalovyan L A 1982 On the structure of solution of one class of plane problems of elasticity theory of anisotropic body In *Mechanics: Interuniversity transactions* vol 2 (Yerevan: Yerevan University Publishing House) pp 7–12
[3] Aghalovyan L A and Aghalovyan M L On asymptotic theory of beams, plates, and shells *J. Curved Layered Struct.* **3** (1) 74–81

IOP Conf. Series: Journal of Physics: Conf. Series **991** (2018) 012072 doi:10.1088/1742-6596/991/1/012072

An analytical theory for a three-dimensional thick-disc thin-plate vibratory gyroscope

G T Sedebo*, **S V Joubert****, **and M Y Shatalov*****

Department of Mathematics and Statistics, Tshwane University of Technology, Pretoria, South Africa

E-mail: *SedeboGT@tut.ac.za, **JoubertSV@tut.ac.za; ***ShatalovM@tut.ac.za

Abstract. We consider a cylindrical vibratory gyroscope comprising a not necessarliy thin-shelled annular disc with small-plate thickness, vibrating in the m-th vibration mode in-plane and in the $(m+1)$st vibration mode out-of-plane. We derive the equations of motion for this contrivance in the "force-to-rebalance regime" and show how a slow (three-dimensional) inertial rotation rate of the gyroscope can be calculated in terms of amplitudes of vibration and other constants, all of which can be measured experimentally or calculated when the eigenfunctions and eigenvalues of the system are known. By means of a concrete example, a numerical experiment demonstrates how varying the inner radius of the annulus as well as the thickness of the plate allows us to "tune" the vibration frequencies of the in-plane and out-of-plane vibrations so that they coincide (for all practical purposes), eliminating any frequency split. Conventionally, an array of at least three thin-shelled hemispherical (or thin-ring) vibratory (resonator) gyroscopes is used to measure any three-dimensional rotation of the craft to which the gyroscopes are fixed. With the design proposed here, the array can be reduced to a solitary, tuned, annular thick-disc thin-plate vibratory gyroscope, reducing both size and cost.

1. Introduction

A gyroscope is a sensor that measures the angular rate or the angle of rotation of an object (see [1]). Rotating wheel gyroscopes have many disadvantages such as bearing friction and wear. The hemispherical vibratory gyroscopes (HVGs), ring vibratory gyroscopes (RVGs), disc vibratory gyroscopes (DVG), et cetera, eliminate mechanical rotation and avoid bearing problems. In 1890, G.H. Bryan [2] laid the foundation for the development of vibratory gyroscopes (VGs) based on the Coriolis effect. He observed what we call *Bryan's law* or *effect* [3]:

> *The vibration pattern of a revolving cylinder or bell revolves at a rate proportional to the inertial rotation rate of the cylinder or bell.*

Bryan's law has been used in the design and manufacture of numerous variations of VGs. Owing to their high performance, high reliability, and small size, VGs are used in many fields with great diversity. These include navigation of air, space and marine vehicles, tracking control and detection systems in motor vehicles.

Commercial three-dimensional VGs consist of "many body arrays" comprising at least, say, three HVGs, each with orthogonal axes of rotation. These arrays are expensive, and engineers have therefore designed various micro-VGs operating in single-axis, dual-axis and three-dimensional modes, with varying sensitivities and performance [4]. For instance, [5] used

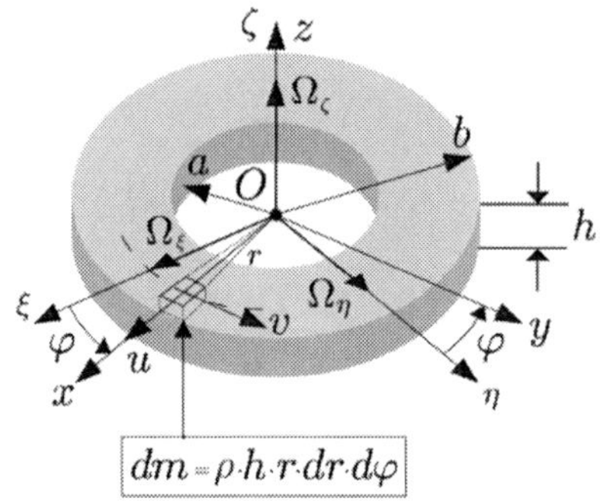

Figure 1. The polar coordinates r and φ of the position of rest of particle $dm = \rho hr\,dr\,d\varphi$ vibrating in-plane and out-of-plane within the annular disc of axial thickness h, inner radius a and outer radius b rotating slowly at a rate $(\Omega_\xi, \Omega_\eta, \Omega_\zeta)^T$ in three dimensions.

a MEMS device consisting of three orthogonal yaw-axis gyroscopes integrated into a single microchip, yielding a "yaw-sensing, pitch-sensing and roll-sensing gyroscope."

Below we tentatively discuss under various assumptions, the *first approximation mathematical principles* of a single "cylindrical" vibratory gyroscope comprising of a not necessarily thin-shelled annular disc for in-plane vibrations and a thin plate for out-of-plane vibrations that measures three-dimensional rotation (see figure 1). We call this gyroscope a *tuned disc vibratory gyroscope* (TDVG) for reasons that will become apparent. A TDVG eliminates the integration of separate proof masses for each orthogonal axis in the gyroscope body, replacing the need for an array of orthogonal symmetry axis VGs. Such a design should reduce both the size and cost of the gyroscope.

Similar to our TDVG is the three-dimensional VDG described in [6]. They use the FEM to verify their experimental observations, but "shell theory" (thin or otherwise) is never discussed and a complete system of gyroscopic equations of motion is not provided. In contrast, our tentative theoretical approach for the TDVG demonstrates that we are not limited to "thin-shell theory" for in-plane vibrations of the annulus, but that we are limited to "thin-plate theory" for out-of-plane vibrations. The natural frequencies of both modes of vibration are tuned and determined by selecting suitable in-plane and out-of-plane vibration modes and appropriate values of the plate thickness and radii of the disc. We analytically derive the gyroscopic equations of motion in terms of four generalised variables, defining Bryan's factor and two auxiliary Bryan's factors in terms of quotients of integrals of eigenfunctions. Because the eigenfunctions may be determined in terms of Bessel and Neumann functions, the three Bryan's factors may be calculated. Two of these factors enable us to show exactly how to calculate the three orthogonal components of the slow inertial rotation rate in terms of measurable quantities such as amplitudes of vibration and linear damping coefficients.

2. Preliminaries

We say that a vibratory gyroscope exhibits *ideal behaviour* if it operates according to Bryan's law. It is well known that a *frequency split* (vibration "*beats*") causes deviation from the ideal behaviour of a vibratory gyroscope. Indeed, various schemes have been introduced to eliminate the frequency split caused by mass and/or stiffness and/or shape imperfections and/or anisotropic damping (see [3] and the by no means exhaustive list of citations therein). We assume that the TDVG described below has been manufactured from a substance that exhibits isotropic damping with perfect axial symmetry and that mass and/or stiffness imperfections have been eliminated for both in-plane vibration and out-of-plane vibration by first applying an optimised

IOP Conf. Series: Journal of Physics: Conf. Series **991** (2018) 012072 doi:10.1088/1742-6596/991/1/012072

mass-balancing action (see [7]) and then an appropriate parallel-plate capacitor control array (see [3]). Hence, when we determine the Euler-Lagrange equations of motion for the vibrating system, we assume ideal conditions, taking only light *isotropic linear damping* into consideration (and deal with isotropic nonlinear damping in a future study).

Because of the nature of the TDVG described below, a frequency split between the resonant frequency of in-plane and the resonant frequency of out-of plane vibration might be inevitable. We describe below how to "*tune*" in-plane and out-of-plane resonant frequencies so that this cause of frequency split may be eliminated.

Consider two coordinate systems $0\xi\eta\zeta$ and $0xyz$ respectively, where $0\xi\eta\zeta$ is connected to the disc and rotates with respect to any inertial coordinate system, whereas $0xyz$ is located at an angle φ from the 0ξ-axis of the $0\xi\eta\zeta$ system and characterises the location of the infinitesimal element $dm = \rho h r dr d\varphi$ as shown in figure 1. Furthermore, assume that the $0\zeta-$axis of $0\xi\eta\zeta$ and the $0z$-axis of $0xyz$ system overlap and coincide with the axis of the disc, as shown in figure 1. Suppose the disc is subjected to *slow three-dimensional rotation* which has components Ω_ξ, Ω_η, and Ω_ζ with directions along the 0ξ-axis, 0η-axis, and 0ζ-axis, respectively. Assume that the annular disc (of axial plate thickness h) has an inner radius a and outer radius b with r the position of a vibrating particle in the disc, as shown in figure 1. Let E be Young's modulus, with ρ and μ the mass density and Poisson's ratio respectively of the material of the disc. We assume that $h \ll \min(a, b)$, and hence we use "thin-plate theory" to simplify matters. Furthermore, assume that the position vector of a point on the disc is given by $\vec{R} = (r + u, v, w)^T$, where u, v, and w are the displacements in the radial, tangential and axial directions, respectively. It is clear that

$$\vec{\Omega} = (\Omega_x, \Omega_y, \Omega_z)^T = \begin{pmatrix} \Omega_\xi \cos\varphi + \Omega_\eta \sin\varphi \\ -\Omega_\xi \sin\varphi + \Omega_\eta \cos\varphi \\ \Omega_\zeta \end{pmatrix}. \tag{1}$$

Thus it follows that the absolute linear velocity of the particle element dm in figure 1 is given by Euler's formula (see [8])

$$\vec{V} = (\dot{u}, \dot{v}, \dot{w})^T + \vec{\Omega} \times \vec{R}, \tag{2}$$

where

$$\vec{\Omega} \times \vec{R} = \begin{pmatrix} (\Omega_\eta \cos\varphi - \Omega_\xi \sin\varphi)w - \Omega_\zeta v \\ \Omega_\zeta (r + u) - (\Omega_\xi \cos\varphi + \Omega_\eta \sin\varphi)w \\ (\Omega_\xi \cos\varphi + \Omega_\eta \sin\varphi)v + (\Omega_\xi \sin\varphi - \Omega_\eta \cos\varphi)(r + u) \end{pmatrix}. \tag{3}$$

Assume that the components of displacement of the vibrating particle of figure 1 are given by

$$u = U(r)[C(t)\cos(m\varphi) + S(t)\sin(m\varphi)], \tag{4}$$

$$v = V(r)[C(t)\sin(m\varphi) - S(t)\cos(m\varphi)], \tag{5}$$

$$w = W(r)[A(t)\cos(m+1)\varphi + B(t)\sin(m+1)\varphi], \tag{6}$$

(7)

where m is a vibration mode number, $U(r)$ and $V(r)$ are eigenfunctions of the in-plane vibrations and $W(r)$ is the eigenfunction of the out-of-plane vibration. These eigenfuctions may be calculated in terms of known functions using well-known standard techniques. For instance, using the theory of thin shells to investigate out-of-plane vibration of the disc ([9] or [10]), if ω_{out} is the angular frequency of the nth mode of out-of-plane vibration, then

$$W(r) = c_1 J_n(\beta r) + c_2 Y_n(\beta r) + c_3 I_n(\beta r) + c_4 K_n(\beta r), \tag{8}$$

where $J_n(\beta r)$ (resp. $Y_n(\beta r)$) is a *Bessel* function of the *first* (resp. *second*) kind of order n, $I_n(\beta r)$ (resp. $K_n(\beta r)$) is a *modified Bessel* function of the *first* (resp. *second*) kind of order n,

$$\beta^4 = \frac{12\rho(1 - v^2)}{Eh^2}\omega_{\text{out}}^2$$

and the constants c_1, c_2, c_3, c_4 determine the mode shape and are determined using the chosen boundary conditions. For the in-plane vibration pattern, $C(t)$ and $S(t)$ may respectively be regarded as the "cosine" and "sine" channels of a vibratory gyroscope connected to a two-channel oscilloscope; the same applies to the A- and B-channels for out-of-plane vibration (see [11]). However, the functions of time $C(t)$, $S(t)$, $A(t)$, and $B(t)$ may readily be determined numerically once the equations of motion (see (16) to (19) below) have been established.

3. Energy

We assume that the size of the angular rate of rotation $\vec{\Omega}$ is substantially smaller than the lowest eigenvalue ω_0 of either in-plane or out-of-plane vibration. Consequently, we neglect terms containing Ω_ξ^2, Ω_η^2, Ω_ζ^2, $\Omega_\zeta\Omega_\eta$, $\Omega_\eta\Omega_\xi$, $\Omega_\xi\Omega_\zeta$. Neglecting the effects of rigid body motion, from (2) we obtain the the kinetic energy E_k of the vibrating disc as

$$E_k = \frac{\rho h}{2}\int_0^{2\pi}\int_a^b (\vec{V}\cdot\vec{V})r\,dr\,d\varphi \approx \pi\big[I_0(\dot{C}^2+\dot{S}^2) + 2\Omega_\zeta(\dot{C}S - C\dot{S})I_1 \\ + I_3\Omega_\eta(A\dot{C} - \dot{A}C + B\dot{S} - \dot{B}S) + I_3\Omega_\xi(A\dot{S} - \dot{A}S + \dot{B}C - B\dot{C}) + I_4(\dot{A}^2+\dot{B}^2)\big], \quad (9)$$

where

$$I_0 = \frac{\rho h}{2}\int_a^b (U^2+V^2)r\,dr, \quad I_1 = \rho h\int_a^b UVr\,dr, \\ I_3 = \frac{\rho h}{2}\int_a^b (U+V)Wr\,dr, \quad I_4 = \frac{\rho h}{2}\int_a^b W^2 r\,dr. \quad (10)$$

If respectively, σ_r, σ_φ, $\sigma_{r\varphi}$ (resp. ϵ_r, ϵ_φ, $\epsilon_{r\varphi}$) are the usual tensile radial, tensile tangential and shear stresses (resp. strains) for in-plane vibrations, then the potential energy of in-plane vibrations is (see [12] and [13] (8) and (17))

$$E_{p,\text{in}} = \frac{Eh}{2(1-\mu^2)}\int_0^{2\pi}\int_a^b [\sigma_r\epsilon_r + \sigma_\varphi\epsilon_\varphi + \sigma_{r\varphi}\epsilon_{r\varphi}]r\,dr\,d\varphi = \pi I_2(C^2+S^2), \quad (11)$$

where the definite integral I_2 is given by [13] (20). Using the assumptions of "thin-plate theory," with

$$\chi_1 = -\frac{\partial^2 w}{\partial r^2}, \quad \chi_2 = -\frac{1}{r}\frac{\partial w}{\partial r} - \frac{1}{r^2}\frac{\partial^2 w}{\partial \varphi^2}, \quad \tau = -\frac{1}{r}\frac{\partial^2 w}{\partial r\partial\varphi} + \frac{1}{r^2}\frac{\partial w}{\partial\varphi}, \quad (12)$$

the potential energy for out-of plane vibrations is (see Novozhilov [9] (9.12))

$$E_{p,\text{out}} = \frac{Eh^3}{24(1-\mu^2)}\int_0^{2\pi}\int_a^b [\chi_1^2 + \chi_2^2 + 2\mu\chi_1\chi_2 + 2(1-\mu)\tau^2]r\,dr\,d\varphi = \pi I_5(A^2+B^2), \quad (13)$$

where

$$I_5 = \frac{Eh^3}{24(1-\mu^2)}\int_a^b \left\{\frac{2\mu W'W''}{r} - \frac{[2(\mu-1)(m+1)^2-1](W')^2 - 2\mu(m+1)^2WW''}{r^2} \right. \\ \left. + \frac{(2\mu-3)(m+1)^2WW'}{r^3} + (W'')^2 + \frac{[m^2(m^2+4m+8-2\mu)-2(\mu-2)(2m+1)-1]W^2}{r^4}\right\} r\,dr. \quad (14)$$

Note that the definite integral I_2 ([13] (20)) is given similarly in terms of E, μ, and an integrand containing sums of products and/or powers of U, U', V, V'.

IOP Conf. Series: Journal of Physics: Conf. Series **991** (2018) 012072 doi:10.1088/1742-6596/991/1/012072

4. Equations of motion

We assume that there is ever-present light, isotropic, linear damping, with damping factor δ for both in-plane and out-of-plane damping and define Rayleigh's dissipation functions R (see [14])

$$R = \pi I_0 \delta(\dot{C}^2 + \dot{S}^2). \tag{15}$$

By "light damping" we mean that δ is substantially smaller than the lowest eigenvalue ω_0 of either in-plane or out-of-plane vibration. As demonstrated in [14], we assume that the rotation rates of vibrating patterns are not affected by light isotropic damping.

Using (9), (11), and (13), we determine Lagrangian $L = E_k - (E_{p,\text{in}} + E_{p,\text{out}}) + E_e$, where E_e is the electrical potential energy introduced to achieve electronic control of the gyroscope. We choose E_e in such a manner that it also produces a so-called *force-to-rebalance regime* (see [15] for a similar setup) that drives only the C- and S-channels (not the A- and B-channels) of the gyroscope at an operational angular frequency ω close to the natural angular frequencies, respectively, ω_{in} and ω_{out} of in-plane and out-of-plane vibrations. As in [3] (12.33), we apply the Euler-Lagrange equations to the generalised coordinates C, S, A, and B in the Lagrangian L, obtaining the *coupled system of ODE of gyroscopic motion* appropriate to our TDVG:

$$\ddot{C} + 2\delta\dot{C} + 2\beta_1\Omega_\zeta\dot{S} + \beta_1\dot{\Omega}_\zeta S + \omega_{\text{in}}^2 C + \beta_2[2(\Omega_\eta\dot{A} - \Omega_\xi\dot{B}) + (\dot{\Omega}_\eta A - \dot{\Omega}_\xi B)] = f\cos(\omega t + \psi), \tag{16}$$

$$\ddot{S} + 2\delta\dot{S} - 2\beta_1\Omega_\zeta\dot{C} - \beta_1\dot{\Omega}_\zeta C + \omega_{\text{in}}^2 S + \beta_2[2(\Omega_\xi\dot{A} + \Omega_\eta\dot{B}) + (\dot{\Omega}_\xi A + \dot{\Omega}_\eta B)] = F\cos(\omega t + \Psi), \tag{17}$$

$$\ddot{A} + 2\delta\dot{A} + \omega_{\text{out}}^2 A - \beta_3[2(\Omega_\eta\dot{C} + \Omega_\xi\dot{S}) + (\dot{\Omega}_\eta C + \dot{\Omega}_\xi S)] = 0, \tag{18}$$

$$\ddot{B} + 2\delta\dot{B} + \omega_{\text{out}}^2 B + \beta_3[2(\Omega_\xi\dot{C} - \Omega_\eta\dot{S}) + (\dot{\Omega}_\xi C - \dot{\Omega}_\eta S)] = 0. \tag{19}$$

Here ψ and Ψ are phase angles with

$$\beta_1 = \frac{I_1}{I_0}, \quad \beta_2 = \frac{I_3}{2I_0}, \quad \beta_3 = \frac{I_3}{2I_4}, \quad \omega_{\text{in}} = \sqrt{\frac{I_2}{I_0}}, \quad \omega_{\text{out}} = \sqrt{\frac{I_5}{I_4}}. \tag{20}$$

The constant β_1 is *Bryan's factor* (see [3] (12.8)), while β_2 and β_3 are *auxiliary Bryan's factors.* Recall that the quantity ω_{in} (resp. ω_{out}) is the eigenvalue or angular rate of the in-plane (resp. out-of-plane) vibration.

5. Bryan's factors and inertial rotation rates

The *control or excitation force* $f\cos(\omega t + \psi)$ in (16) is used to keep the *vibration amplitude of the C-channel constant*, with the phase shift ψ *chosen* in such a way that $C(t) = C_c\cos(\omega t)$. With an appropriate phase shift Ψ, the *rebalance force* $F\cos(\omega t + \Psi)$ in (17) is used to yield $S(t) = 0$.

To demonstrate how β_1 and β_3 are *related to rotation rates*, we use the force-to-rebalance electronic array discussed above and assume that a steady-state solution for (16) to (19) is of the form

$$C(t) = C_c\cos(\omega t), \tag{21}$$

$$S(t) = 0, \tag{22}$$

$$A(t) = A_c\cos(\omega t) + A_s\sin(\omega t), \tag{23}$$

$$B(t) = B_c\cos(\omega t) + B_s\sin(\omega t). \tag{24}$$

Owing to the slow inertial rotation rate $\vec{\Omega}$, the amplitudes A_c, A_s, B_c, and B_s are *slowly varying time-dependent quantities*, which we *assume are constant* in order to determine simplified expressions for angular rates. The exact nature of these amplitudes is left for future study, while good average estimations of these amplitudes should be measurable by experiment.

Substituting (21) to (24) into (16) to (19) yields equations for the angular velocities Ω_ζ, Ω_η, Ω_ξ and angular accelerations $\dot{\Omega}_\zeta$, $\dot{\Omega}_\eta$, $\dot{\Omega}_\xi$ as follows:

$$\Omega_\zeta = -\frac{F\sin\Psi}{2\omega C_c\beta_1}, \quad \Omega_\eta = \frac{1}{2}\frac{2A_c\delta\omega + A_s\omega^2 - A_s\omega_{\text{out}}^2}{C_c\beta_3\omega}, \quad \Omega_\xi = -\frac{1}{2}\frac{2B_c\delta\omega + B_s\omega^2 - B_s\omega_{\text{out}}^2}{C_c\beta_3\omega}, \tag{25}$$

$$\dot{\Omega}_\zeta = -\frac{F\cos\Psi}{C_c\beta_1}, \quad \dot{\Omega}_\eta = -\frac{A_c\omega^2 - A_c\omega_{\text{out}}^2 - 2A_s\delta\omega}{C_c\beta_3}, \quad \dot{\Omega}_\xi = \frac{B_c\omega^2 - B_c\omega_{\text{out}}^2 - 2B_s\delta\omega}{C_c\beta_3}. \tag{26}$$

In order to eliminate the frequency split introduced by a difference between the in-plane angular rate and the out-of-plane angular rates, the gyroscope may be "tuned" by varying the radial and/or axial thickness, enabling us to achieve $\omega_{\text{in}} = \omega_{\text{out}} = \omega$ for all practical purposes. We have verified that this "tuning" is possible by conducting a numerical experiment as discussed below. This "tuning" yields the following simplified equations

$$\Omega_\zeta = -\frac{F\sin\Psi}{2\omega C_c\beta_1}, \quad \Omega_\eta = \frac{A_c\delta}{C_c\beta_3}, \quad \Omega_\xi = -\frac{B_c\delta}{C_c\beta_3}, \tag{27}$$

$$\dot{\Omega}_\zeta = -\frac{F\cos\Psi}{C_c\beta_1}, \quad \dot{\Omega}_\eta = \frac{2A_s\delta\omega}{C_c\beta_3}, \quad \dot{\Omega}_\xi = -\frac{2B_s\delta\omega}{C_c\beta_3}. \tag{28}$$

The first equation in (27) demonstrates that the axial inertial rotation rate Ω_ζ is directly proportional to the amplitude F of the rebalance force. This result is *essentially* the same as that reported for the axial rotation rate of an HVG operating under the force-to-rebalance regime (see [15] (6)).

The amplitude F and phase angle Ψ of the rebalance force, the operational angular frequency ω as well as the amplitudes A_c, B_c, and C_c should all be experimentally measurable. Once the eigenfunctions $U(r)$, $V(r)$, and $W(r)$ have been calculated (as outlined immediately above (8)), Bryan's factors β_1 and β_3 can be calculated from the formulae in (20). Consequently the angular rate and accelerations of each orthogonal axis can be calculated from (27).

6. Tuning

Ultimately, we provide a numerical experiment, which for this first approximation theory, demonstrates that it may well be possible to "tune" the TDVG once it has been manufactured. Consider an aluminium disc with Young's modulus $E = 7 \times 10^10\,\text{N}\cdot\text{m}^{-2}$, density $\rho = 2700\,\text{kg}\cdot\text{m}^{-3}$, and Poisson's ratio $\mu = 0.33$. With reference to figure 1, the thickness and outer radius are kept constant at $h = 4.824\,\text{mm}$ and $b = 100\,\text{mm}$, respectively, while the inner radius a is varied.

Considering mode number $m = 2$ for in-plane vibration and mode number $m+1 = 3$ for out-of-plane vibration, and assuming free boundaries on the TDVG, the eigenfunctions are calculated as outlined immediately above (8). Tables 1 and 2 were generated using (20) and the computer algebra system *Maple 18* at standard working precision.

In table 1, with $h = 4.824\,\text{mm}$ and $b = 100\,\text{mm}$, the inner radius in each case is approximately $a = 40\,\text{mm}$. Hence the thickness T of the annular disc is

$$T = b - a \approx 60\,\text{mm}$$

for each inner radius, and so the value of the ratio

$$\frac{T}{h} \approx 12$$

lies in the interval defined for a "thin plate" in Section 1.1.3 of [10]. Consequently, the values used in table 1 are commensurate with the theory developed above. With an inner radius

IOP Conf. Series: Journal of Physics: Conf. Series **991** (2018) 012072 doi:10.1088/1742-6596/991/1/012072

Table 1. Eigenfrequencies and frequency split for an aluminium vibratory disc gyroscope for a fixed outer radius $b = 100\,\text{mm}$ and thickness $h = 4.824\,\text{mm}$ for various values of inner radius a.

a (mm)	$f_{\text{in}} = \frac{\omega_p}{2\pi}$ (kHz)	$f_{\text{out}} = \frac{\omega_p}{2\pi}$ (kHz)	$\Delta f = \lvert f_{\text{in}} - f_{\text{out}}\rvert$ (kHz)
39.51	6.1778	5.8329	0.3449
40.51	6.0341	5.8049	0.2293
41.51	5.8922	5.7777	0.1145
42.51	5.7521	5.7518	0.0003

Table 2. Bryan's factor and auxiliary Bryan's factors for an aluminium vibratory disc gyroscope for a fixed outer radius $b = 100\,\text{mm}$ and thickness $h = 4.824\,\text{mm}$, for various values of inner radius a.

a (mm)	β_1	β_2	β_3
39.51	−0.7513	−0.1037	−0.0060
40.51	−0.7513	−0.1057	−0.0050
41.51	−0.7514	−0.1067	−0.0040
42.51	−0.7516	−0.1064	−0.0032

of 42.51 mm, it appears that it is feasible to "tune" the gyroscope and achieve a frequency split $\Delta f \approx 0\,\text{kHz}$ (to four significant figures of accuracy).

In table 2, the negative Bryan's factor β_1 indicates that, for in-plane-vibrations, the vibration pattern is *rotating in the opposite direction* to the direction of the inertial axial rotation rate Ω_ζ. As the annulus decreases in thickness (inner radius increases), so Bryan's factor β_1 and the auxiliary Bryan's factor β_2 appear to increase in size, whereas the auxiliary Bryan's factor β_3 decreases in size.

Conclusions

We have developed a first approximation theory for an essentially ideal tuning disc vibratory gyroscope (TDVG) rotating at a slow inertial angular rate in three dimensions. Under various mild assumptions, we have derived the equations of motion of a TDVG. Under a force-to-rebalance regime, assuming a steady-state solution, we have derived approximate expressions for the angular rotation rate along each orthogonal rotation axis as well as the corresponding angular accelerations. Our expression for the axial inertial angular rate agrees with known results for a single axis of rotation, hemispherical vibratory gyroscope. We have conducted a numerical experiment using our first approximation theory, which illustrates how it might be feasible to build and tune a three-dimensional TDVG.

Acknowledgments

The authors thank the the National Research Foundation (NRF) of South Africa (grant reference numbers IFR160211157784 and 106308) as well as the Tshwane University of Technology (TUT) for financial support.

IOP Conf. Series: Journal of Physics: Conf. Series **991** (2018) 012072 doi:10.1088/1742-6596/991/1/012072

References

[1] Akar C and Shkel A 2009 *MEMS Vibratory Gyroscopes: Structural Approaches to Improve Robustness* (New York: Springer)
[2] Bryan G H 1890 On the beats in the vibrations of a revolving cylinder or bell *Proc. Camb. Phil. Soc. Math. Phys. Sci.* **7** 101–14
[3] Joubert S V, Shatalov M Y, and Spoelstra H 2017 On electronically restoring an imperfect vibratory gyroscope to an ideal state *Mechanics for Materials and Technologies* ed Altenbach H, Goldstein R V, and Murashkin E (Cham: Springer) pp 231–56
[4] Tsai N-C and Sue C-Y 2010 Experimental analysis and characterization of electrostatic-drive tri-axis micro-gyroscope *Sensors Actuat. A: Phys.* **158** (2) 231–9
[5] Sung W 2013 *High-frequency Tri-axial Resonant Gyroscopes* Ph.D. thesis Georgia Institute of Technology Atlanta-Georgia, USA
[6] Johari H, Shah J, and Ayazi F 2008 High frequency XYZ single-disk silicon gyroscope In *Proc. IEEE Conf. on MEMS, Tucson, AZ, USA* (2008) 856–9
[7] Shatalov M Y, Joubert S V, and Spoelstra H 2015 Modelling and controlling imperfections in vibratory gyroscopes In *Int. Conf. on Sound and Vibration (ICSV22), Florence, Italy* 1–8
[8] Spiegel M 1967 *Theoretical Mechanics* (New York: Schaum Publishing Co.)
[9] Novozhilov V 1970 *The Theory of Thin Shells* (Groningen: Wolters-Noordhoff.)
[10] Ventsel E and Krauthammer T 2001 *Thin Plates and Shells: Theory, Analysis and Applications* 1st ed (New York: Marcel Dekker, Inc.) ISBN 0-8247-0575-0
[11] Shatalov M Y, Joubert S V, and Coetzee C E 2011 The influence of mass imperfections on the evolution of standing waves in slowly rotating spherical bodies *J. Sound Vibr.* **330** 127–35
[12] Redwood M 1960 *Mechanical Waveguides* (Oxford: Pergamon Press)
[13] Joubert S V, Shatalov M Y, and Fay T H 2009 Rotating structures and Bryan's effect *Am. J. Phys.* **77** (6) 520–5
[14] Joubert S V, Shatalov M Y, and Manzhirov A V 2013 Bryan's effect and isotropic nonlinear damping *J. Sound. Vibr.* **332** (23) 6169–76
[15] Wang X, Wu W, Luo B, et al. 2011 Force to rebalance control of HRG and suppression of its errors on the basis of FPGA *Sensors* **11** 11761–73

IOP Conf. Series: Journal of Physics: Conf. Series **991** (2018) 012073 doi:10.1088/1742-6596/991/1/012073

Vibration of high-voltage electric machines with rotors on rolling bearings

H G Shekyan* **and A V Gevorgyan**

Institute of Mechanics of the National Academy of Sciences of Armenia, Yerevan, Armenia

E-mail: *hamlet@mechins.sci.am

Abstract. The paper presents an investigation of vibrational activity of electric machines due to high-harmonic vibrational loadings. It is shown that the vibrational loadings experienced by bearings may result in the interruption of their normal operation and even take them out of action. Therefore, the values of the vibrational speed-up leading to high harmonics are factors determining the admissible dynamic loading on the bearings. In the paper, an attempt is made to consider the factors which result in origination of high harmonics and to illustrate methods for their smoothing.

1. Introduction

Despite numerous literature sources devoted to calculations of the machine body vibrations, there is still no calculational method which can give a result differing from the experimental data by less than 50%. This can be explained by the fact that, in calculations, it is impossible to take into account all numerous factors influencing the body vibration (nonlinear rigidity, kinematic excitation of rolling bearings, rigid characteristics of the stator shields and beds, etc.). It suffices to say that numerous harmonic components of a machine have vibration spectra, which originate due to the nonlinear rigidity and kinetic perturbation of the bearings and which are to be determined when evaluating the dynamic situation and the working capacity of the bearings of electric machines,

Taking into account these factors would, of course, permit improving the existing methods of calculations [1, 2].

2. Statement of the problem

Let us consider the stationary motion of a rigid rotor on rolling bearings in a rigid shock-proof housing (figure 1).

The motion of the system is described by the differential equations

$$\begin{cases} m_1\ddot{y}_1 = P_1 + P_M - F_{y1} + P(t), \\ m_2\ddot{y}_2 = P_2 + F_{y1} - F_{y2} - P_M, \end{cases} \tag{1}$$

$$F_{y1} = C_1 B_0 (y_1 - y_2 + \lambda_{1st} - \lambda_{2st})^{3/2}, \tag{2}$$

$$P_M = C_M (y_1 - y_2 + \lambda_{1st} - \lambda_{2st}), \tag{3}$$

$$F_{y2} = C_2 (y_2 + \lambda_{2st}), \tag{4}$$

$$P_1 = C_1 B_0 (\lambda_{1st} - \lambda_{2st})^{3/2} - C_M (\lambda_{1st} - \lambda_{2st}), \tag{5}$$

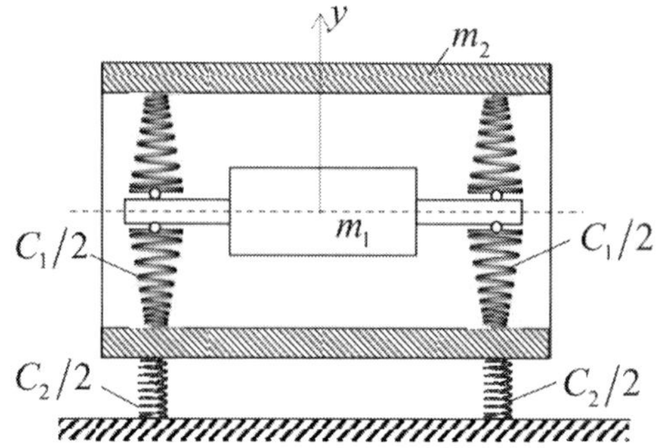

Figure 1. Rigid rotor on nonlinearly elastic supports in an amortized body.

$$P_2 = C_2(y_2 - \lambda_{2st}) - C_1 B(\lambda_{1st} - \lambda_{2st})^{3/2} + C_M(\lambda_{1st} - \lambda_{2st}), \tag{6}$$
$$B = B_0 - B_1 \cos(\omega_1 t), \tag{7}$$
$$P(t) = m_1 F \sin(\omega_1 t) - m_1 \ddot{s}, \quad F = \omega_1^2 e, \tag{8}$$

where y_1 and y_2 are the respective displacements of the rotor and stator along the axis y; F_{y1} and F_{y2} are the respective restoring forces due to the contact elasticity of the bearings and the rigidity of the shock absorbers; P_M is the force of the one-way magnetic traction, C_M is the conditional rigidity of the magnetic field; P_1 is the rotor and stator weight; C_1 is the Hertz coefficient, C_2 is the total rigidity of shock absorbers, B is a periodic function of time which is approximated by an expression [3, 4]; $P(t)$ is the perturbation force due to the rotor out-of-balance and the kinematic perturbation of the bearings; e is the rotor eccentricity, and s is a function of kinematic displacement [3, 4].

Substituting (2)–(7) into (1), we obtain

$$\begin{cases} m_1\ddot{y}_1 + C_1[B_0 - B_1\cos(\omega_1 t)](y_1 - y_2 + \lambda_{1st} - \lambda_{2st})^{3/2} - C_M(y_1 - y_2 + \lambda_{1st} - \lambda_{2st}) \\ \qquad = CB_0(\lambda_{1st} - \lambda_{2st})^{3/2} - C_M(\lambda_{1st} - \lambda_{2st}) + P(t), \\ m_2\ddot{y}_2 + C_2(y_2 + \lambda_{2st}) - C[B_0 - B_1\cos(\omega_1 t)](y_1 - y_2 + \lambda_{1st} - \lambda_{2st})^{3/2} \\ \qquad + C_M(y_1 - y_2 + \lambda_{1st} - \lambda_{2st}) = C_2\lambda_{2st} - CB_0(\lambda_{1st} - \lambda_{2st}) + C_m(\lambda_{1st} - \lambda_{2st}). \end{cases} \tag{9}$$

Expanding the function $(y_1 - y_2 + \lambda_{1st} - \lambda_{2st})^{3/2}$ into a Taylor series and preserving only the linear terms for small vibrations, we obtain

$$\begin{aligned} &\ddot{f} + \ddot{y}_2 + [h^2 - q\cos(\omega_1 t)]\xi = v\cos(\omega_1 t) + P'(t), \\ &\ddot{y}_2 + \omega_2^2 y_2 - d[h^2 - q\cos(\omega_1 t)]\xi = -dn\cos(\omega_1 t), \end{aligned} \tag{10}$$

where

$$\xi = y_1 - y_2, \quad h^2 = \frac{3}{2}\frac{C_1 B_0}{m_1}(\lambda_{1st} - \lambda_{2st})^{1/2} - \frac{C_M}{m_1}, \quad P'(t) = \frac{P(t)}{m_1},$$
$$q = \frac{3}{2}\frac{C_1 B_1}{m_1}(\lambda_{1st} - \lambda_{2st})^{1/2}, \quad \nu = \frac{C_1 B_1}{m_1}(\lambda_{1st} - \lambda_{2st})^{3/2}, \quad d = \frac{m_1}{m_2}, \quad \omega_2^2 = \frac{C_2}{m_2}.$$

The solution of the system of equations (10) shows that, at certain values of the rigid characteristics of the bearings and shock absorbers, there is a sharp increase in the amplitudes of separate vibrational components of spectrum, and the parametric resonance may occur. In

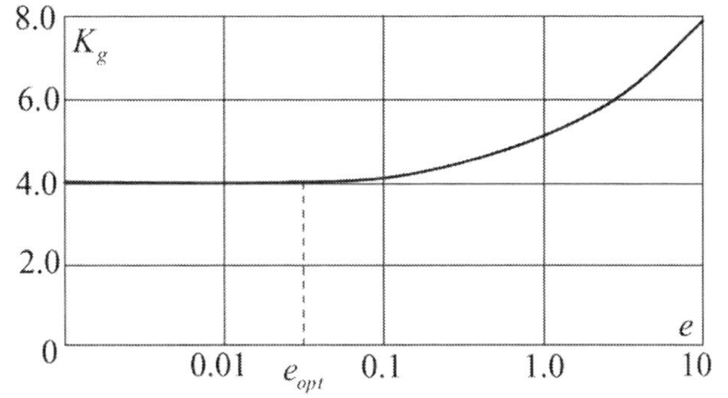

Figure 2. Dependence of the vibration overload coefficient on the rotor out-or-balance of generator C-75.

this case, it is necessary to change the rigid characteristics of the machine so as to satisfy the conditions

$$n^2\omega_1^2 > \omega_2^2 + (d+1)h^2 \quad \text{and} \quad [(n-1)\omega_1 \pm \omega_{ki}]^2 > \omega_2^2 + (d+1)h^2 \tag{11}$$

obtained by analyzing the solutions of system (10) by the method of a small parameter [5].

In (11), $\omega_k i$ is the frequency of the ith harmonic whose speed-up amplitude is the greatest. With the aim of evaluating the system dynamic state, the vibration overload coefficient is introduced in the form

$$k_g = \frac{P_y}{G},$$

where P_y is the centrifugal force of inertia due to the precession motion of the rotating rotor and G is the rotor weight. In the case of a high-speed rotor rotating on the rolling bearings, the static equilibrium state is unstable if the rotating rotor is precessing with the angular speed of rotation in a circle whose radius is equal to the bearing deformation.

If the rotor eccentricity is equal to e, then the centrifugal force is $P_y = m_1\omega_1^2(e + y_1 - y_2)/g$, where $y_1 - y_2 = [P_y/C \cdot B]^{3/2}$.

For $e = 0$, we have $P_y = m_1^3 \cdot \omega_1^6/C \cdot B$ and $k_g = m_1^2 \cdot \omega_1^6/(C \cdot B)^3$.

As we see, the dynamic factor increases significantly because of the nonlinear rigidity of the bearings.

If the rotor eccentricity is nonzero, then

$$k_g = \frac{\omega_1^2(e + y_1 - y_2)}{g},$$

and the dependence of the vibrational overload coefficient on the rotor out-of-balance is shown in figure 2. The graph is constructed for a bearing of type 160507 in the serial generator C-75 at $\omega = 800\,\text{rad/s}$.

The graph shows that, up to a certain eccentricity value ($e = 0.05\,\mu\text{m}$), the reaction between the shaft and the bearing is practically constant and only the further increase in the eccentricity leads to a sharp increase in the vibration overload coefficient. In figure 3, the dependence of k_g on the angular speed of rotation of a rigidly balanced rotor ($e = 0$) is depicted.

Figure 3 shows that, up to some values of the angular speed $\omega = 300\,\text{rad/s}$, the value of the vibration overload coefficient is practically insignificant. The further increase in the rotor angular speed leads to a sharp increase in the vibration overload coefficient. Therefore, for each type of the bearing carrying a rotor, there is a value of the angular speed at which the

IOP Conf. Series: Journal of Physics: Conf. Series **991** (2018) 012073 doi:10.1088/1742-6596/991/1/012073

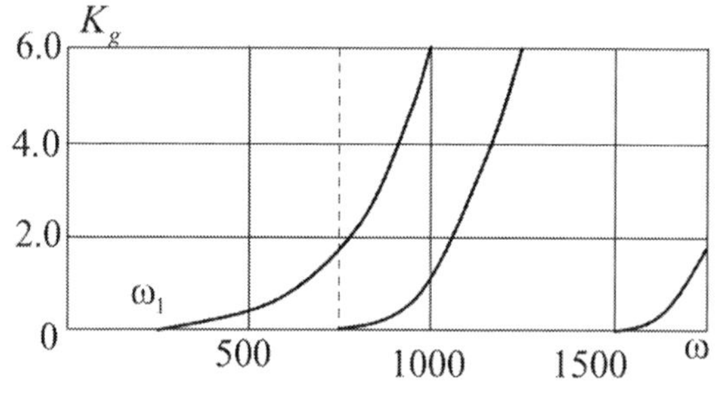

Figure 3. Dependence of the vibration overload coefficient on the angular speed of the rotor rotation: *1* – without taking the rotor ductility into account, *2* – with the rotor ductility taken into account, *3* – the rotor on linearly elastic supports.

Table 1. Spectral components of bearing harmonics of generator C-75.

Bearing type	Bearing number	Bearing spectrum					
		Outer circle			Inner circle		
		harmonic number	harmonic amplitude, μm	harmonic frequency, Hz	harmonic number	harmonic amplitude, μm	harmonic frequency, Hz
308	1	1	0.11	53	12	0.24	960
		4	0.2	212	21	0.13	1680
		10	0.16	530	33	0.1	2650
		16	0.09	850	47	0.15	3760
		24	0.06	1270	64	0.07	5150
		44	0.02	2420	93	0.03	7450
		86	0.01	4600	125	0.08	10000
		115	0.04	610	130	0.01	10050
308	2	2	0.28	107	15	0.18	1200
		5	0.13	268	26	0.14	2080
		15	0.31	810	33	0.22	2640
		47	0.08	2500	61	0.1	4860
		66	0.06	3520	82	0.07	6550
		92	0.02	4900	120	0.03	9600
		131	0.01	7000	160	0.02	10500

Comment. Bearing 1 is the generator bearing on the drive unit, and bearing 2, on the rear shield.

dynamic reactions begin to increase sharply. The dynamic reaction remains constant until a certain value of eccentricity is attained, and since the further balancing does not decrease the dynamic reaction, it makes no sense to decrease the eccentricity, as in the case $e = 0$, because the support deformation and the dynamic reaction do not vanish.

Generator C-75-V1 is considered as an example of calculations of the vibration spectrum. The basic parameters of the machine and its elements which are required to calculate the vibrations are given in tables 1 and 2, and the results of calculations the carcass vibrations and the experimental data are given in table 3.

It follows that the data obtained by calculations satisfactorily coincide with the experimental

Table 2. Basic parameters of machine C-75 and its elements.

Stator weight kg	Rotor weight kg	Turnover number, turn/min	Magnetic field conditional rigidity, N/m	Rotor eccentricity, μm	Bearing type	Bearing number	Hertz coefficient, $\frac{kg}{m^{3/2}}$	Parameter h^2, $\frac{N}{kg \cdot m}$	Parameter q, $\frac{N}{kg \cdot m}$	Parameter ν, $\frac{N}{kg \cdot m}$
100	57	8000	9.8×10^6	7	308	1	0.84×10^9	1.67×10^9	2.5×10^4	2.6×10^4
						2	0.84×10^9	1.3×10^9	2.5×10^4	2.6×10^4

Table 3. Spectral components of harmonics of the body of machine C-75.

Experimental			Calculated		Experimental			Calculated	
Harmonic frequency, Hz	Displacement amplitude, μm	Speed amplitude, m/sec^2	Displacement amplitude, μm	Speed amplitude, m/sec^2	Harmonic frequency, Hz	Displacement amplitude, μm	Speed amplitude, m/sec^2	Harmonic frequency, Hz	Displacement amplitude, μm
53	0.19	0.21	0.22	0.24	100	5.8	2.3	6	2.5
100	7	2.82	7.3	3	268	0.12	0.4	0.17	0.5
212	0.2	0.36	0.16	0.3	810	0.25	7.4	0.3	8
530	0.15	3.4	0.2	4.22	1200	0.1	9	0.15	10.5
850	0.075	2.3	0.07	2.4	2500	0.078	20	0.09	18
1270	0.1	9.4	0.18	11	2640	0.02	55.5	0.028	47
2420	0.025	7	0.04	7.6	3520	0.03	20	0.04	24
2650	0.03	11	0.05	16.8	4860	0.14	105	0.2	110
3760	0.013	6.7	0.02	15	5000	0.02	22	0.028	20
4600	0.01	10	0.021	20	6550	0.025	50	0.03	57
6100	0.03	43.3	0.04	48	7000	0.01	19	0.015	24
7450	0.02	48	0.028	60	9600	0.034	110	0.05	100
10050	0.07	36	0.01	39.8	10500	0.025	113	0.03	98

data. Besides, the maximum displacement amplitude of the machine carcass takes place at the rotation frequency. As the harmonic number increases, the displacement amplitudes decrease, while the speed-up amplitudes increase. At the frequency 10500 kg from the gear, the speed-up of the bearing shield is 114 m/s^2.

Therefore, the dynamic force transmitted from the rotor through the bearings to the carcass is equal to

$$F_g = m_2 \ddot{y}_2 = 100 \times 11.4 = 1140\,\text{kg},$$

which is almost 8–9 times greater than the static loading, which means that the state of the considered bearing is unsatisfactory. To improve the state of the bearing, it is necessary to minimize the speed-up of higher harmonics of the carcass either by selecting bearings with more plane characteristics of higher harmonic numbers of the spectrum or by localizing the rotor vibration from the carcase by using an elastic support (bearing), which permits cutting down the higher harmonics.

IOP Conf. Series: Journal of Physics: Conf. Series **991** (2018) 012073 doi:10.1088/1742-6596/991/1/012073

Conclusions

1. Considering the rotors vibrations of high-speed electric machines with the nonlinear rigidity and kinematic perturbation taken into account, one can explain the reasons of increase in the spectral components of higher harmonics and find ways for their smoothing.
2. To estimate of vibration activity of high-speed electric machines, it is necessary to estimate the vibrational speed-up, which is a force characteristic of the inertial system.
3. Under the technical conditions of high-speed machine operation determined by the durability of the bearings, rotation speed, and rotor mass, the vibration acceleration requirements should be formulated in the whole range of frequencies.
4. To improve the status of the bearings, the acceleration of higher harmonic carcasses should be minimized by selecting the bearings with more smooth characteristics of the spectrum of high harmonics or introducing elastic elements into the bearing construction so as to to cut off the higher harmonics.

References

[1] Shekyan H G 2004 *Dynamics of Rotor Machines* (Yerevan: Izdat. NAN RA "Gitutiun") p 329 [in Russian]
[2] Shekyan H G 1976 Stability of high-speed rotors, rotating on the bearings of roll *Trudy VNIIKE* **8** 136–47 [*Proc. HSII* (Engl. Transl.) **8** 62–77]
[3] Shekyan H G, Khalatyan R P, and Koshkaryan G N 1980 Dynamics of flexible rotors on bearings of roll *Izv. Akad. Nauk ArmSSR. Ser. Tekh. Nauk* **XXXIII** (2) 7–11
[4] Sheftell B T and Shanitsin A A 1973 Vibration of ball bearing with radical gap *Mashinoved.* No. 4 29–35
[5] Malkin I G 2004 *Lyapunov and Poincare Methods in Theory of Nonlinear Oscillations* (Moscow: Editorial URSS) p 248

IOP Conf. Series: Journal of Physics: Conf. Series **991** (2018) 012074 doi:10.1088/1742-6596/991/1/012074

Biomechanics of stem cells

A A Spector*, D Yuan, S Somers, and W L Grayson
Johns Hopkins University, Baltimore, USA
E-mail: *aspector@jhu.edu

Abstract. Stem cells play a key role in the healthy development and maintenance of organisms. They are also critically important in medical treatments of various diseases. It has been recently demonstrated that the mechanical factors such as forces, adhesion, stiffness, relaxation, etc. have significant effects on stem cell functions. Under physiological conditions, cells (stem cells) in muscles, heart, and blood vessels are under the action of externally applied strains. We consider the stem cell microenvironment and performance associated with their conversion (differentiation) into skeletal muscle cells. Two problems are studied by using mathematical models whose parameters are then optimized by fitting experiments. First, we present our analysis of the process of stem cell differentiation under the application of cyclic unidirectional strain. This process is interpreted as a transition through several (six) stages where each of them is defined in terms of expression of a set of factors typical to skeletal muscle cells. The stem cell evolution toward muscle cells is described by a system of nonlinear ODEs. The parameters of the model are determined by fitting the experimental data on the time course of expression of the factors under consideration. Second, we analyse the mechanical (relaxation) properties of a scaffold that serves as the microenvironment for stem cells differentiation into skeletal muscle cells. This scaffold (surrounded by a liquid solution) is composed of unidirectional fibers with pores between them. The relaxation properties of the scaffold are studied in an experiment where a long cylindrical specimen is loaded by the application of ramp displacement until the strain reaches a prescribed value. The magnitude of the corresponding load is recorded. The specimen is considered as transversely isotropic poroelastic cylinder whose force relaxation is associated with liquid diffusion through the pores. An analytical solution for the total force applied to the cylinder in terms of the mechanical properties of the scaffold (longitudinal and lateral Young's moduli, two Poisson's ratios, and typical time of liquid diffusion) is used. The number of constant is then reduced to three by estimating the longitudinal Young's modulus and one of Poisson's ratios from an earlier experiment. Finally, three remaining parameters are estimated by fitting the relaxation curve corresponding to strain rate of loading of $0.01\,s^{-1}$. The developed mathematical solution is then tested by comparing the theoretical and experimental results for another strain rate of $0.0025\,s^{-1}$. The scaffold relaxation properties can be important for differentiation of stem cells inside the pores.

1. Introduction

Stem cells, under certain conditions, are capable of conversion (differentiation) into different type(s) of cells. Among other (biochemical) factors, the mechanical factors play a significant role in stem cell differentiation. The externally applied forces (stresses) and strains may direct or accelerate the process of stem cell differentiation [1]. In the case of skeletal muscle, cyclic unidirectional strains affect stem cell differentiation [2].

Here we model of the process of stem cells differentiation into skeletal muscle cells under the action of cyclic strain of 10% magnitude and 0.5 Hz frequency. We describe this process as a transition through several stages and solve the corresponding system of nonlinear ODEs.

Another way to direct stem cell differentiation is to affect it by changing the surface or (and) bulk properties of the substrate. In the former case, the substrate surface has special topography that provides cell aligning in a prescribed direction [3]. The substrate/cell interaction can be changed by changing the bulk mechanical properties of the substrate. It has been demonstrated that stem cell differentiation can be directed toward brain cells, or muscle cells, or bone cells by switching from soft -to medium-stiff- to rigid substrates [4]. The viscous properties of the substrate have also a significant effect on stem cell differentiation which was recently demonstrated by using substrates of the same stiffness and different relaxation times [5].

Here we consider the relaxation properties of a novel porous scaffold developed for differentiation of stem cells [6]. We show that the relaxation properties of such scaffold can be associated with the diffusion of the liquid through the pores and explained by a biphasic poroelastic model. We simulate an experiment with ramp-displacement loading of a cylindrical specimen and estimate the parameters of the transversely isotropic poroelastic scaffold. We consider the modeling results for different strain rates and show that they are close to the experimental data.

2. Mechanical (strain) effects on stem cell differentiation

2.1. Mathematical model [7, 8]

The focus of our model of stem cell differentiation toward skeletal muscle cells is the kinetics of expression of five typical factors (called PAX3, desmin, MyoD, myogenin, and MHC) as well as the effect of the applied strain. We formulate the model in terms of the number of cells in the distinct stages of differentiation which are determined by particular combinations of the expressed factors (figure 1). The model is described to the following system of ODEs:

$$\frac{dn_0}{dt} = [(2r_0 - 1)p_0 n_0 - d_0 n_0] f(n_{tot}), \tag{1}$$

$$\frac{dn_i}{dt} = [(2r_i - 1)p_i n_i - d_i n_i + 2(1 - r_{i-1})p_{i-1} n_{i-1}] f(n_{tot}), \quad i = 1, 2, \tag{2}$$

$$\frac{dn_3}{dt} = [p_3 n_3 - d_3 n_3 + 2(1 - r_2)p_2 n_2 - D_3(n_3, \varepsilon)] f(n_{tot}), \tag{3}$$

$$\frac{dn_3}{dt} = [p_3 n_3 - d_3 n_3 + 2(1 - r_2)p_2 n_2 - D_3(n_3, \varepsilon)] f(n_{tot}), \tag{4}$$

$$\frac{dn_5}{dt} = [D_4(n_4, \varepsilon) - d_5 n_5] f(n_{tot}). \tag{5}$$

Here, the right-hand sides on equations (1)–(5) give the sums of fluxes that determine the rates of the cell numbers in stages 0, 1,…, 5, respectively, and n_0, …, n_5 are the cell numbers in these stages. The terms on the right-hand side of equation (1) are associated with division (proliferation rate, p_0, day^{-1}), self-renewal (self-renewal coefficient, r_0), and death (death rate, d_0, day^{-1}) in stage 0. The right-hand side in equations (2) are determined by division, (proliferation rates p_i, day^{-1}), self-renewal (self-renewal, coefficients r_i), and death (death rates, d_i) in the current (i) stage and by division (proliferation rate, p_i, day^{-1}) and differentiation (differentiation coefficient, $1 - r_i$) in the previous stage ($i - 1$). The right-hand side of equation (3) is determined by division and differentiation in previous stage 2, proliferation, direct (without division) differentiation, and death in current stage 3. The right-hand side of equation (4) is determined by direct differentiation in previous stage 3 and direct differentiation and death in current stage 4. Finally, the right-hand side of equation (5) is determined by direct differentiation in previous stage 4 and death in current stage 5. The function $f(n_{tot})$ (n_{tot} is the total cell number) describes the effects of stem cell density on their differentiation.

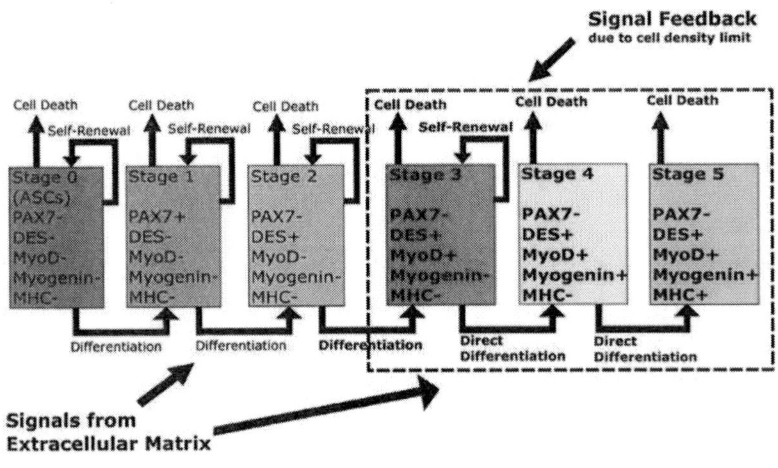

Figure 1. Multi-stage model of stem cell differentiation into skeletal muscle cells [8].

Direct differentiation in late stages 3, 4, and 5 is described by functions D_3 and D_4 that depend on the applied strains. Assuming that the applied strains initiate signalling that ultimately effect expression of the late factors, the direct differentiation functions take the following form:

$$D_3(n_3, \varepsilon) = \beta_3 \left(\frac{\varepsilon n_3}{\varepsilon + k_s^*} \right) \left(\frac{\varepsilon n_3}{\varepsilon + k_s^*} + k_3 \right)^{-1}, \tag{6}$$

$$D_4(n_4, \varepsilon) = \beta_4 \left(\frac{\varepsilon n_4}{\varepsilon + k_s^*} \right) \left(\frac{\varepsilon n_4}{\varepsilon + k_s^*} + k_4 \right)^{-1}, \tag{7}$$

where ε is the magnitude of the applied strain and β_3, β_4, k_s, k_3, and k_4 are parameters estimated by fitting the experimental data.

2.2. Optimization of the model parameters

We estimate the parameters of our model by optimally fitting the experimental data on the differentiation into skeletal muscle cells of adipose-derived stem cells (stem cells derived from fat tissue of a patient). In this experiment, the stem cells are seeded on a flexible membrane to which a unidirectional cyclic strain of a magnitude 10% and frequency 0.5 Hz is applied. The goal of the strain application is to maximize the outcome of differentiated skeletal muscle cells which is quite low without the mechanical stimulation. In the experiment, the numbers of cells expressing the groups of factors that correspond to 6 stages of our model are measured at several moments of time (3, 7, 14, and 21 days). The result of the parameter optimization are presented in figure 2 where the time course of the computed n_1, ..., n_5, and n_{tot} are shown vs. corresponding experimental points at the four moments of time

2.3. Prediction of the differentiation time course for different strains

By using the estimated model parameters we are able to predict stem cell differentiation results for strains different from the experimental 10% as well as for the times different from 21 days. Figures 3 *a–c* present the computed results of the time course (up to 33 days) of cell numbers in all six stages for external strain of 3%, 9%, and 15%.

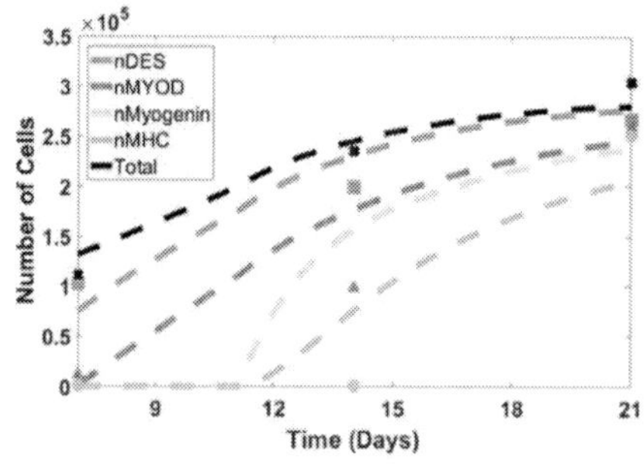

Figure 2. Optimization of the model parameters by fitting the experiment with the stem cell differentiation under the action of strain of 10% magnitude [8].

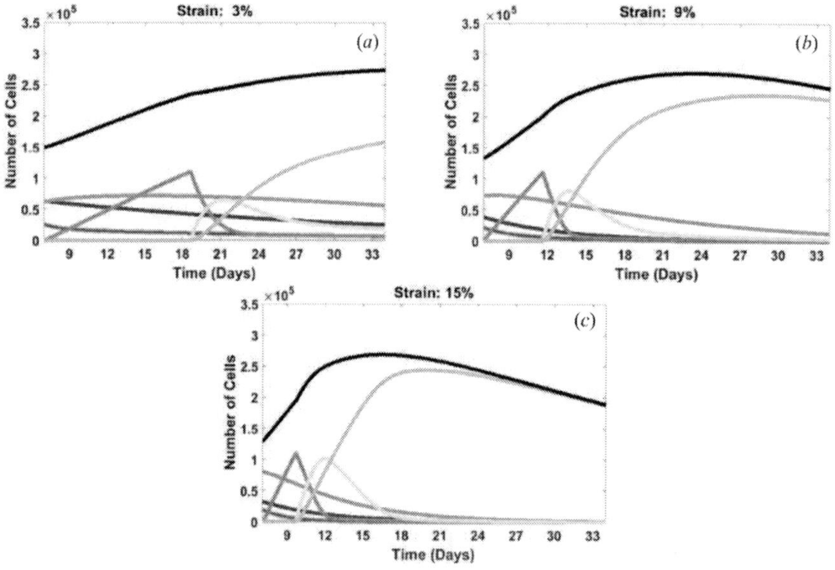

Figure 3. Predicted time course of stem cell differentiation for different strain magnitudes by using optimized model parameters 3% (a), 9% (b), and 15% (c).

3. Mechanics of a fibrous/porous scaffold for stem cell differentiation

3.1. Force relaxation experiment

A novel fibrous porous scaffold is designed for stem cells differentiation into skeletal muscle cells via cell alignment along the fiber direction, tuning by the scaffold mechanical properties, and application of the external strain [6]. The stem cells are located inside the pores of the scaffold. The relaxation properties of the scaffold have been demonstrated to be important to stem cell differentiation, and we focus on the experiment that helps estimate such properties. In this experiment, a cylindrical specimen is a subject to ramp displacement following by a regime of constant strain, and the corresponding horizontal force is recorded. The specimens are located in the liquid solution moving in the porous specimen in response to the applied tensile strain (figure 4).

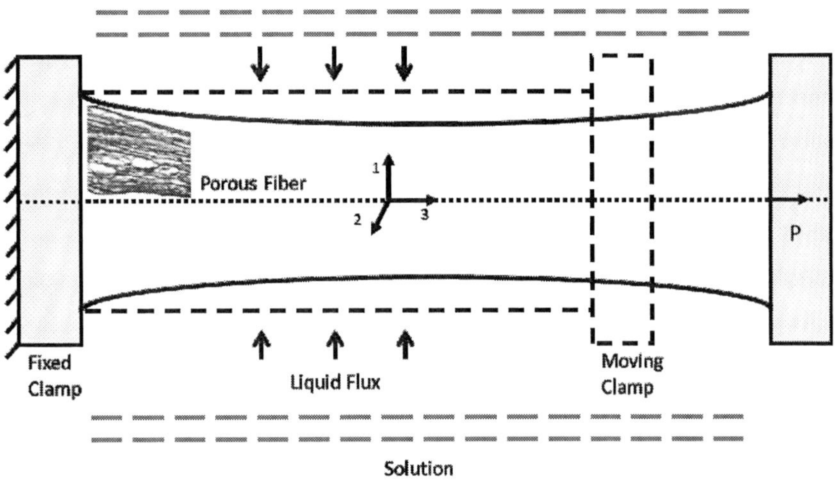

Figure 4. Sketch of the relaxation experiment with a cylindrical specimen of the fibrous/porous scaffold (insert) for stem cell differentiation.

3.2. A poroelastic model to interpret the force relaxation experiment

We assume transverse isotropy of the specimen properties and axisymmetry of the stress-strain distribution. We propose a poroelastic mechanism where the applied force relaxation is associated with the fluid diffusion inside the specimen in response to its tensile deformation. A solution for the compression of a short transversely isotropic poroelastic cylinder is available [9], and it can be adjusted to our case of the extension of a long poroelastc cylinder. Thus, we use the following function describing the time course of the force $P(t)$ applied to the cylinder:

$$P(t) = E_3\dot{\varepsilon}_0 t + E_1\dot{\varepsilon}_0 t_g F_1(t_g, E_1, E_3, \nu_{21}, \nu_{31}) \quad \text{for} \quad 0 \le t \le t_0, \tag{8}$$

$$P(t) = E_3\dot{\varepsilon}_0 t_g - E_1\dot{\varepsilon}_0 t_g F_2(t_g, E_1, E_3, \nu_{21}, \nu_{31}) \quad \text{for} \quad t \ge t_0, \tag{9}$$

where:

$$F_1 = \Delta_3\left(\frac{1}{8} - \sum_n \exp\left(-\alpha_n^2\frac{t}{t_g}\right)\{\alpha_n^2[\Delta_2^2\alpha_n^2 - \Delta_1(1+\nu_{21})]\}^{-1}\right), \tag{10}$$

$$F_2 = \Delta_3\sum_n\left[\exp\left(-\alpha_n^2\frac{t}{t_g}\right) - \exp\left(-\alpha_n^2\frac{t-t_0}{t_g}\right)\right]\{\alpha_n^2[\Delta_2^2\alpha_n^2 - \Delta_1(1+\nu_{21})]\}^{-1}, \tag{11}$$

$$\Delta_1 = 1 - \nu_{21} - \frac{2\nu_{31}^2 E_1}{E_3}, \quad \Delta_2 = 1 - \nu_{31}^2\frac{E_1}{E_3}, \quad \Delta_3 = 1 - \frac{2\nu_{31}^2\Delta_2}{\Delta_1}. \tag{12}$$

Here E_1 and E_3 are the transverse and longitudinal Young's moduli, ν_{21} and ν_{31} are in-transverse plane and transverse plane/out-of plane Poisson's ratios, t_g is a typical time of fluid filtration through the pores, and α_n are the roots of a characteristic equation [9].

3.3. Estimation of the scaffold mechanical properties

We reduce the number of parameters of the proposed poroelastic model to three by extracting Young's modulus, $E_3 = 30\,\text{kPa}$ and Poisson's ratio, $\nu_{31} = 0.3$ from the previously reported independent experiment [6]. The remaining three parameters are estimated by fitting the current relaxation experiment. Figure 5 shows the theoretical solution with optimal values of the parameters ($E_1 = 11.5\,\text{kPa}$, $t_g = 38.8\,\text{s}$, and $\nu_{21} = 0.77$) vs. the experimental data in the case of strain of 10% and strain rate of $0.01\,\text{s}^{-1}$. We finally test the model by checking its results against the data of relaxation experiment for a different strain rate of $0.0025\,\text{s}^{-1}$ (figure 6).

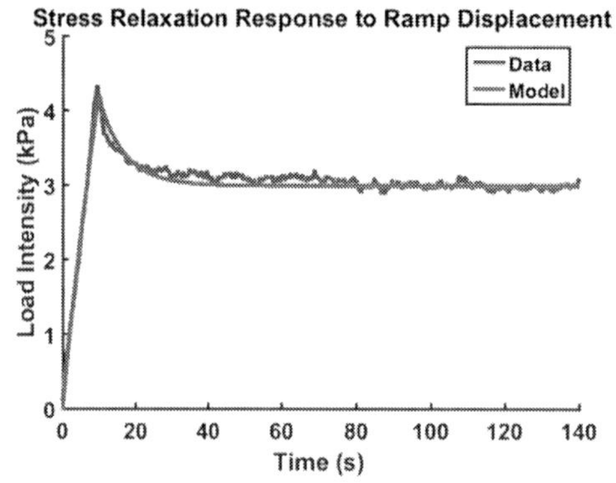

Figure 5. Optimization of the poroelastic model parameters by fitting the data of the force relaxation experiment for ramp displacement loading (strain rate of $0.01\,s^{-1}$) following by a constant strain regime (strain of 10%).

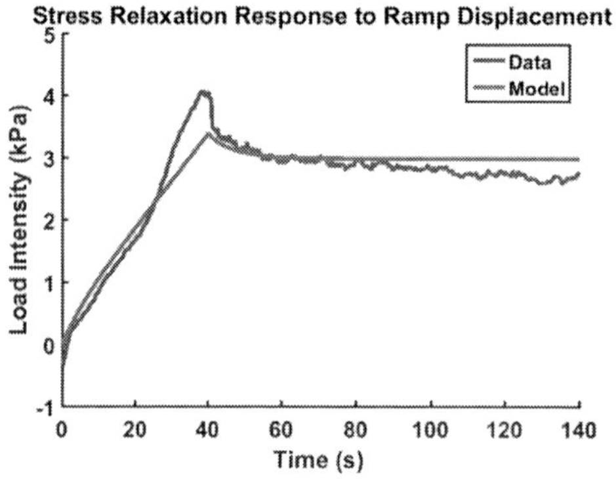

Figure 6. Modelling results for the optimal parameters vs. the data of the force relaxation experiment for the ramp displacement loading with the strain rate of $0.0025\,s^{-1}$.

References

[1] Dado D, Sagi M, Levenberg S, and Zemel A 2012 Mechanical control of stem cell differentiation *Cell. Mol. Bioeng.* **7** 101–16

[2] Huri P Y, Wang A, Spector A A, and Grayson W L 2014 Multistage adipose-derived stem cell myogenesis: An experimental and modeling study *Cell. Mol. Bioeng.* **7** 497–509

[3] McNamara L E, McMurray R J, Biggs M J P, et al. 2010 Nanotopographical control of stem cell differentiation *J. Tissue Engng* **1** 120623

[4] Engler A J, Sen S, Sweeney H L, and Discher D E 2006 Matrix elasticity directs stem cell lineage specification *Cell* **126** 677–89

[5] Chaudhuri O, Gu L, Klumpers D, et al. 2016 Hydrogels with tunable stress relaxation regulate stem cell fate and activity *Nature Mater.* **15** 326–34

[6] Cook C A, Huri P Y, Ginn B P, et al. 2016 Characterization of a novel bioreactor system for 3D cellular mechanobiology studies *Biotechnol. Bioeng.* **113** 1826–37

[7] Deshpande R S, Grayson W L, and Spector A A 2015 A modeling insight into adipose-derived stem cell myogenesis *PLoS ONE* **10** e0137918

[8] Deshpande R S and Spector A A 2017 Modeling stem cell myogenic differentiation *Sci. Rep.* **7** 40639

[9] Cohen B, Lai W M, and Mow V C 1998 A transverselty isotropic biphasic model for unconfined compression of growth plate and chondroepiphysis *J. Biomech. Engng* **120** 491–6

IOP Conf. Series: Journal of Physics: Conf. Series **991** (2018) 012075 doi:10.1088/1742-6596/991/1/012075

Continuum mathematical modelling of pathological growth of blood vessels

N E Stadnik[1*] and E P Dats[2]**

[1]Ishlinsky Institute for Problems in Mechanics of the Russian Academy of Sciences, Moscow, Russia
[2]Vladivostok State University of Economics and Service, Vladivostok, Russia

E-mail: *nik-122@mail.ru, **dats@dvo.ru

Abstract. The present study is devoted to the mathematical modelling of a human blood vessel pathological growth. The vessels are simulated as the thin-walled circular tube. The boundary value problem of the surface growth of an elastic thin-walled cylinder is solved. The analytical solution is obtained in terms of velocities of stress strain state parameters. The condition of thinness allows us to study finite displacements of cylinder surfaces by means of infinitesimal deformations. The stress-strain state characteristics, which depend on the mechanical parameters of the biological processes, are numerically computed and graphically analysed.

Preliminary Remarks

Diseases associated with the pathological growth of the walls of blood vessels are the one of the main causes of death of the population [1, 2]. At the heart of pathological growth of the vessel wall is atherosclerosis, growth processes in infectious and immune diseases, arteriolosclerosis and arteriosclerosis, hypertension, vasculitis. There are three types of vessels in the human circulatory system: arteries, veins and microcirculatory bed. The latter vessels type includes hemocapillaries, arterioles, venules and arteriolo-venular anastomoses. Arteries and veins are formed according to a similar scheme, Figure 1 [3–6].

The blood vessels growth is the thickening of the vessel wall by virtue of a pathological process. In the medical context, such growth is usually represented by an inflammatory process in the vessel wall and the process of dystrophy (accumulation of certain components in the vessel stroma). The vessel wall growth can also lead to the dystrophy process without an inflammatory reaction. Such widespread diseases as atherosclerosis and vasculitis are initiated by an inflammatory reaction leading to a chronic course and accompanying by the accumulations of lipids and proteins inside the vessel wall. Vessel remodelling process under arterial hypertension are frequently not accompanied by an inflammatory reaction, but thickening of the vessel wall occurs due to accumulation of the intercellular matrix components and active proliferation of smooth muscle cells.

Atherosclerosis is characterized by the defeats of arteries due to lipids and proteins deposition in the intima and reactive growth of connective tissue as the plaques.

The fat–protein detritus (atere) and the focal proliferation of connective tissue (sklerosis) occur due to atherosclerosis in the arteries intima and lead to the formation of an atherosclerotic

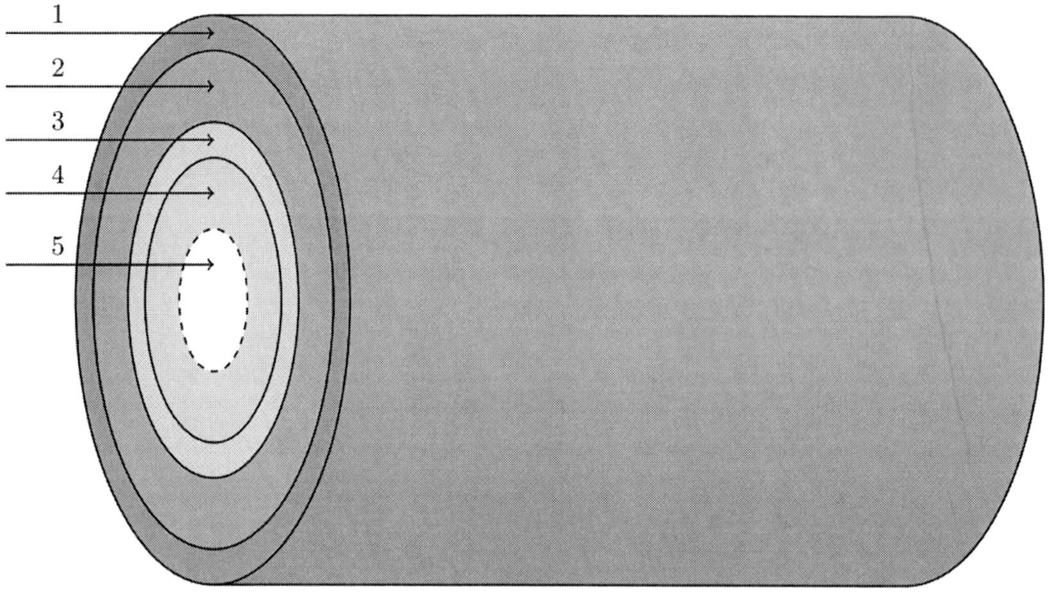

Figure 1. The scheme of a wall vessels (arteries and veins) structure. 1. Tunica adventitia is formed by a loose connective tissue. 2. Tunica media consists of muscle cells, elastic membranes and collagen. 3. Tunica intima consists of the endothelium and the elements of connective tissue located above it. 4. Growth (atherosclerosis) layer. 5. Vessel lumen.

plaques narrowing the vessel lumen. Arteries of large and medium calibre are more often affected by atherosclerosis then the small lumen arteries. The following stages of morphogenesis of atherosclerosis are discriminated:

(i) *Prelipoidosis stage* is not macroscopically determined.

(ii) *Lipoidosis stage* is characterized by focal impregnation of intima with lipids (cholesterol), lipoproteins, which leads to the formation of fat (lipid) spots and bands. Such fat spots macroscopically look like yellow areas, which sometimes can merge and form flat, elongated bands that do not rise above the surface of the intima.

(iii) *Liposclerosis stage* when the fibroblast cells are proliferated stimulating the growth of the young connective tissue in the intima. The subsequent maturation of this tissue is accompanied by the formation of a fibrous plaques. Macroscopically fibrous plaques are dense formation having round or oval forms of white or yellowish-white flowers rising above the intima surface.

(iv) *Atheromatosis stage.* Atheromatous masses are discriminated from the vessel lumen by a layer of mature, hyalineized connective tissue. The plaque cover is formed.

(v) *Ulceration stage* accompanied by the formation of an atheromatous ulcer. The intima defect is covered by the thrombotic overlays.

(vi) *Atherocalcinosis stage* is characterized by deposition in the fibrous plaques of calcium salts, i.e. their calcification (petrification). This is the final stage of atherosclerosis.

The elastosis and elastofibrosis are revealed due to hypertension in the arteries of large and medium calibre. Elastosis and elastofibrosis are sequential stages and represent hyperplasia and cleavage of the internal elastic membrane. Hyperplasia is compensatory developed in virtue of response to a persistent blood pressure increasing. Then the elastic fibres are destructed and replaced to collagen fibres. The vessel wall is thickened and the vessel lumen is narrowed. In small vessels and arterioles, the cells produce semi-transparent masses like a hyaline cartilage. Over time, these masses are becoming denser, are thickening the walls and are narrowing the lumen.

The vessel walls growth can be described using mathematical models of volume or surface growth. Let us focus only on the processes of surface growth of thin-walled vessels. We use the ideas of the mechanics of growing bodies [7–14]. Some one-dimensional problems in frameworks of the thermoelastoplasticity are studied under conditions of axial or central symmetries in [15–20]. The principal variables of the boundary value problem for a growing body are the stress rate tensor, the strain rate tensor and the velocity vector. On the surface of growth we set a specific boundary condition depending on the curvature tensor of the growth surface and the tension and inflow rates of the incremented elements.

Some model problems for an elastic thin-walled surface-growing cylinder are considered at present work. The condition of thinness allows us to study finite displacements of cylinder points under the condition of small deformations. This, in particular, makes it possible to solve the problem with exact boundary conditions on a moving surface. The behaviour features of the main strain–stress state characteristics depending on the pressure on the inner and outer surfaces of the cylinder.

1. Boundary Value Problems Statement and Solution

Consider an infinitely long hollow elastic cylinder with internal and external radii R_1 and R_2, respectively. The relations between the stress tensor σ_{ij} and the infinitesimal deformations e_{ij} for the isotropic elastic material are furnished by the Hooke's law

$$\begin{aligned} \sigma_{rr} &= (\lambda + 2\mu)e_{rr} + \lambda(e_{\varphi\varphi} + e_{zz}), \\ \sigma_{\varphi\varphi} &= (\lambda + 2\mu)e_{\varphi\varphi} + \lambda(e_{rr} + e_{zz}), \\ \sigma_{zz} &= (\lambda + 2\mu)e_{zz} + \lambda(e_{\varphi\varphi} + e_{rr}), \end{aligned} \tag{1}$$

where λ, μ are the Lame parameters.

The following components of the infinitesimal strain tensor are not vanished by virtue of the hypothesis of a plane strain state

$$e_{rr} = \frac{\partial u_r}{\partial r}, \qquad e_{\varphi\varphi} = \frac{u_r}{r}, \qquad e_{zz} = 0, \tag{2}$$

where u_r is the radial component of the displacement vector.

The components of the stress tensor satisfy the equation of equilibrium

$$\frac{\partial \sigma_{rr}}{\partial r} + \frac{\sigma_{rr} - \sigma_{\varphi\varphi}}{r} = 0. \tag{3}$$

One can transform the equilibrium equation (3) by determining the deformations from equation (1) in terms of the stresses by

$$\frac{\partial}{\partial r}\left(2\sigma_{rr} + r\frac{\partial \sigma_{rr}}{\partial r}\right) = 0. \tag{4}$$

A general solution of the equation (4) constitutes the radial stress depending on radius and undefined functions A, B

$$\sigma_{rr} = A + \frac{B}{r^2}. \tag{5}$$

The angular stress $\sigma_{\varphi\varphi}$, axis one σ_{zz} and radial displacement u_r define according to (1)–(5)

$$\sigma_{\varphi\varphi} = A - \frac{B}{r^2}, \quad \sigma_{zz} = \frac{\lambda A}{(\lambda+\mu)}, \quad u_r = \frac{Ar}{2(\lambda+\mu)} - \frac{B}{2\mu r}. \tag{6}$$

The lateral surface of the cylinder is subjected to a stationary loading pressure

$$\sigma_{rr}(R_1) = p_1, \qquad \sigma_{rr}(R_2) = -p_2. \tag{7}$$

The relations for an undefined functions A and B are derived by the solution of the boundary conditions system (7) in forms

$$A = -\frac{p_1 R_1^2 + p_2 R_2^2}{R_2^2 - R_1^2}, \qquad B = \frac{R_1^2 R_2^2 (p_1 + p_2)}{R_2^2 - R_1^2}. \tag{8}$$

The equations (5)–(8) determine the stress-strain state parameters before the growth. Suppose at time $t = 0$ on the inner surface of the cylinder the new material is adding with the velocity $v = v(t)$

$$\begin{aligned} &\dot{\sigma}_{rr}(R_1(t)) = -\frac{v(t)\tau(t)}{R_1(t)}, \\ &R_1(t) = R_1 - v(t)t, \\ &\dot{\sigma}_{rr}(R_2) = 0. \end{aligned} \tag{9}$$

Herein the dot denotes the speed of the considered variable, which can be determined as a partial derivative with respect to time within the frameworks of the infinitesimal deformations approach; $\tau(t)$ is the circumferential tension of the adding layer at the moment of joining to the cylinder surface. For a correct description of the stress-strain state evolution under the conditions of continuous growth $t > 0$ it is necessary to transform governing equations by replacing all variables on its velocities (1)–(3). After that we obtain following solution in terms of the velocities

$$\begin{aligned} &\dot{\sigma}_{rr} = X(t) + \frac{Y(t)}{r^2}, \quad \dot{\sigma}_{\varphi\varphi} = X(t) - \frac{Y(t)}{r^2}, \\ &\dot{\sigma}_{zz} = \frac{\lambda X(t)}{(\lambda+\mu)}, \qquad \dot{u}_r = \frac{X(t) r}{2(\lambda+\mu)} - \frac{Y(t)}{2\mu r}. \end{aligned} \tag{10}$$

Unknown functions $X(t)$ and $Y(t)$ can be computed by the solution of the system (9) in following form

$$X(t) = \frac{R_1(t) v(t) \tau(t)}{R_2^2 - R_1^2(t)}, \qquad Y(t) = -\frac{R_1(t) v(t) \tau(t) R_2^2}{R_2^2 - R_1^2(t)}. \tag{11}$$

The relations for stresses and displacements after integrating on time the equations (10) taking into account the boundary conditions (5)–(6) read by

$$\begin{aligned} &\sigma_{rr} = A + \frac{B}{r^2} + \int_0^t \left(X(s) + \frac{Y(s)}{r^2} \right) dt, \\ &\sigma_{\varphi\varphi} = A - \frac{B}{r^2} + \int_0^t \left(X(s) - \frac{Y(s)}{r^2} \right) dt, \\ &\sigma_{zz} = \frac{\lambda}{(\lambda+\mu)} \left(A + \int_0^t X(s) dt \right), \\ &u_r = \frac{Ar}{2(\lambda+\mu)} - \frac{B}{2\mu r} + \left(\int_0^t \frac{X(s) r}{2(\lambda+\mu)} - \frac{Y(s)}{2\mu r} \right) dt. \end{aligned} \tag{12}$$

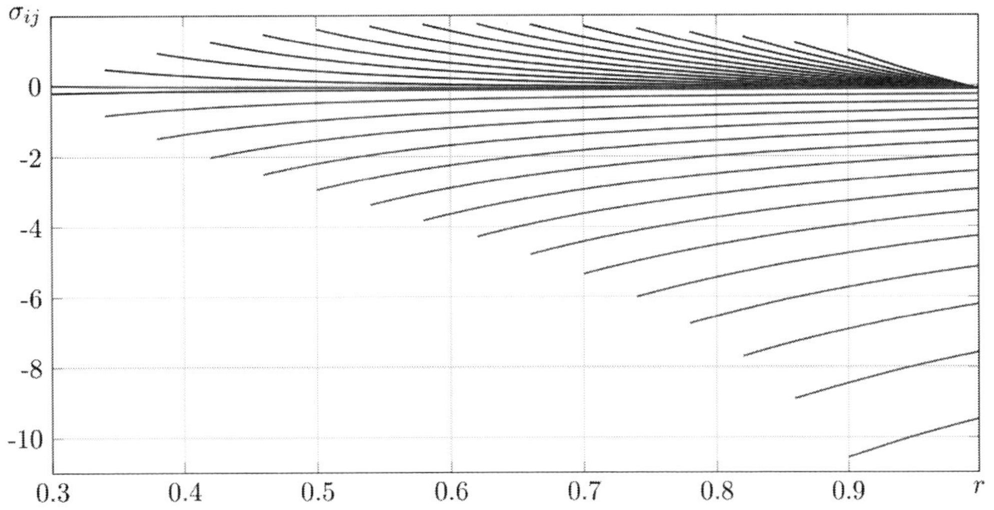

Figure 2. Stresses in the growth cylinder at different times

2. Numerical Results Discussion

The calculations were carried out for various values of the growing layer size $R_1(t)$ and the growth layer tension level $\tau(t)$. For the case of $v(t) = const$ and $\tau(t) = const$ it is established that the growth tension level τ level has a key influence on the formation of the final stress-strain state of the material. Figure 2 illustrates the radial stress field (blue) and circumferential one (red) during the growth. The stresses are vanished on the inner surface $R_1(t)$ for a certain value of growth layer tension τ during the inner cylinder wall growth. On Figure 3 is shown the radial displacement in the process of the inner cylinder wall growth. Calculations were carried out with the following material dimensionless parameters

$$\tau = 6, \quad \lambda = 4, \quad \mu = 2, \quad v = 0.1, \quad p_1 = 1, \quad p_2 = 0.1, \quad R_1 = 0.9, \quad R_2 = 1.$$

The stresses concentration caused by the influence of internal pressure can be computed in the absence of initial growth tension in the adding layer by formulae (5) and (8). The stresses concentration caused by the high level of the initial growth tension during the narrowing of the vessel lumen are calculated. Thus, it is possible to conclude, that on the one hand, with correct initial parameters of the growing process, it is possible to achieve a minimum effect of the internal pressure in the vessel on its walls deformation. To another hand, a certain growth regime is capable of causing irreversible deformation and fracture vessel wall.

Conclusions

- The constitutive and governing equations of the linear elasticity have been furnished as a tool for mathematical modeling of blood vessels growth.
- The boundary value problem of the surface growth of an elastic thin-walled cylinder has been solved.
- The stresses concentration caused by the different levels of the initial growth tension during the narrowing of the vessel lumen have been calculated.
- The stress-strain state characteristics, which depend on the mechanical parameters of the biological processes, are numerically computed and graphically analysed.

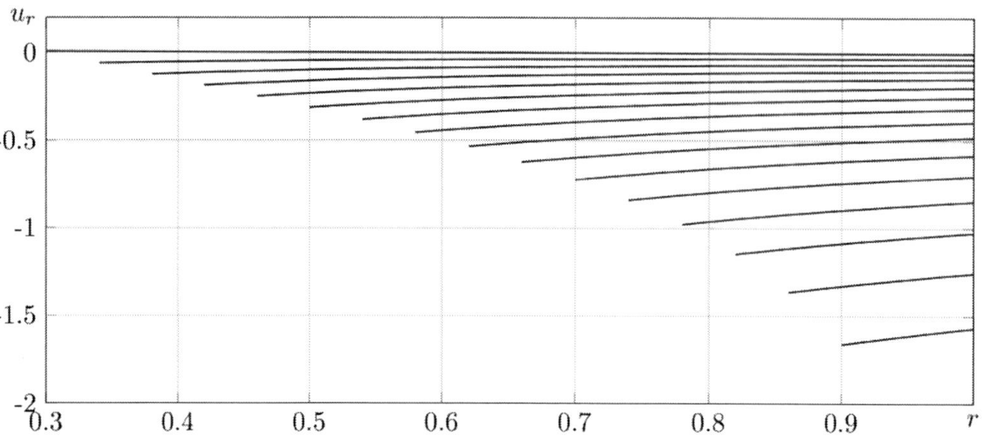

Figure 3. Displacement in the growth cylinder at different times

Acknowledgments
The authors would like to thank Professor A. V. Manzhirov for his helpful comments, suggestions and encouragement. The research was financially supported by RSF (project No. 17-19-01257).

References
[1] Kahlon R, Shapero J and Gotlieb A 1992 *The Canadian journal of cardiology* **8** 60–64
[2] Folkman J 2006 *Annu. Rev. Med.* **57** 1–18
[3] Folkman J 1995 *Nature medicine* **1** 27
[4] Risau W 1997 *Nature* **386** 671
[5] Carmeliet P and Jain R K 2000 *Nature* **407** 249–257
[6] Sluimer J C and Daemen M J 2009 *The Journal of pathology* **218** 7–29
[7] Manzhirov A 1985 *Journal of Applied Mathematics and Mechanics* **49** 777–782
[8] Manzhirov A 1995 *Journal of Applied Mathematics and Mechanics* **59** 805–816
[9] Lychev S, Lycheva T and Manzhirov A 2011 *Mechanics of Solids* **46** 325–333
[10] Manzhirov A and Lychev S 2012 *Doklady Physics* **57** 160–163
[11] Joubert S, Shatalov M and Manzhirov A 2013 *Journal of Sound and Vibration* **332** 6169–6176
[12] Manzhirov A 2015 *Lecture Notes in Engineering and Computer Science* **2218(1.1)** 710–714
[13] Manzhirov A 2017 *Procedia Engineering* **173** 11–16
[14] Manzhirov A 2017 *MATEC Web of Conferences* **95** 12002
[15] Burenin A A, Dats E P, Mokrin S and Murashkin E V 2013 *Bulletin of Yakovlev Chuvash State Pedagogical University. Series: Mechanics of Limit State* **16** 22–28
[16] Burenin A A, Dats E P and Murashkin E V 2014 *mechanics of Solids* **49** 218–224
[17] Dats E, Murashkin E and Stadnik N 2017 *Procedia Manufacturing* **7** 427–434
[18] Dats E, Murashkin E and Stadnik N 2017 *Procedia Engineering* **173** 891–896
[19] Dats E P, Murashkin E V and Gupta N K 2017 *Procedia IUTAM* **23** 187–200
[20] Dats E, Murashkin E and Stadnik N 2017 *IAENG International Journal of Applied Mathematics* **47** 261–264

IOP Conf. Series: Journal of Physics: Conf. Series **991** (2018) 012076 doi:10.1088/1742-6596/991/1/012076

Contact problem for a solid indenter and a viscoelastic half-space described by the spectrum of relaxation and retardation times

F I Stepanov

Ishlinsky Institute for Problems in Mechanics of the Russian Academy of Sciences, Moscow, Russia

E-mail: stepanov_ipm@mail.ru

Abstract. The mechanical properties of a material which is modeled by an exponential creep kernel characterized by a spectrum of relaxation and retardation times are studied. The research is carried out considering a contact problem for a solid indenter sliding over a viscoelastic half-space. The contact pressure, indentation depth of the indenter, and the deformation component of the friction coefficient are analyzed with respect to the case of half-space material modeled by single relaxation and retardation times.

1. Introduction

3D contact problems for a viscoelastic half-space interacting with hard bodies has been considered by many researchers in different statements of the problem [1–8]. Many of the problems are solved by the boundary element method (BEM) that allows one to consider any type of material model including a viscoelastic half-space described by a spectrum of relaxation and retardation times. Nevertheless, no analysis has still been made for the mechanical properties of a viscoelastic half-space described by a spectrum of relaxation times. The purpose of this research is to study the contact parameters of a solid indenter sliding over a viscoelastic half-space and to compare them with the case where the half-space is described by a single relaxation and retardation time.

2. Statement of the problem

We consider a solid indenter sliding over a viscoelastic half-space with a constant velocity V. The indenter is loaded by a vertical force Q and a tangential force T which ensure constant moving (figure 1). The material model is specified by the Volterra integral operator that defines

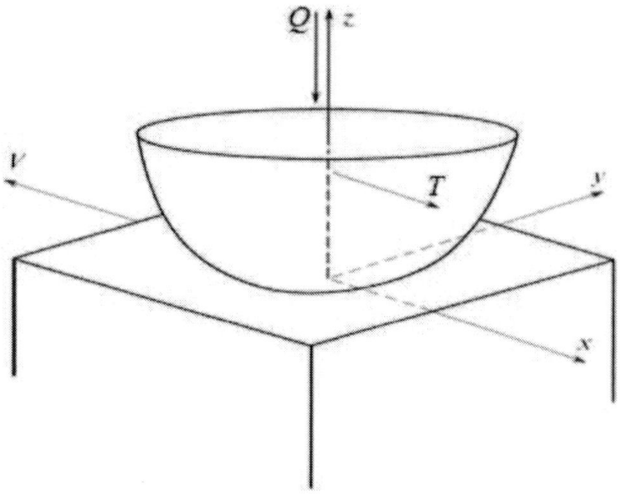

Figure 1. Scheme of contact.

the relationship between strains ($\gamma(t)$, $e(t)$) and stresses ($\tau(t)$, $\sigma(t)$) in the form

$$\begin{aligned}
&\gamma(t) = \frac{1}{G}\tau(t) + \frac{1}{G}\int_{-\infty}^{t} \tau(t)K(t-\tau)d\tau,\\
&e_x(t) = \frac{1}{E}\{\sigma_x(t) - \nu[\sigma_y(t) + \sigma_z(t)]\} + \frac{1}{E}\int_{-\infty}^{t} \{\sigma_x(t) - \nu[\sigma_y(t) + \sigma_z(t)]\}K(t-\tau)\,d\tau,\\
&e_y(t) = \frac{1}{E}\{\sigma_y(t) - \nu[\sigma_x(t) + \sigma_z(t)]\} + \frac{1}{E}\int_{-\infty}^{t} \{\sigma_y(t) - \nu[\sigma_x(t) + \sigma_z(t)]\}K(t-\tau)\,d\tau, \quad (1)\\
&e_z(t) = \frac{1}{E}\{\sigma_z(t) - \nu[\sigma_y(t) + \sigma_x(t)]\} + \frac{1}{E}\int_{-\infty}^{t} \{\sigma_z(t) - \nu[\sigma_y(t) + \sigma_x(t)]\}K(t-\tau)\,d\tau,\\
&K(t') = \sum_i k_i \exp\left(-\frac{t'}{\lambda_i}\right),
\end{aligned}$$

where G and E are the shear modulus and the modulus of elasticity, ν is the Poisson ratio, λ_i and $1/k_i$ are the retardation and relaxation times, and $K(t')$ is the creep kernel.

The following contact conditions are met:

$$\begin{aligned}
&z = 0\colon \tau_{xz}(x,y) = 0, \quad \tau_{yz} = 0, \quad w(x,y) = f(x,y) + D, \quad (x,y) \in \Omega,\\
&\sigma_z = 0, \quad \tau_{xz} = 0, \quad \tau_{yz} = 0, \quad (x,y) \notin \Omega, \quad (2)\\
&-\infty < x < +\infty, \quad -\infty < y < +\infty.
\end{aligned}$$

Here D is the indentation depth, $w(x,y)$ is the surface vertical displacement, and Ω is the unknown contact region to be determined by solving the contact problem. The indenter shape is described by the function

$$f(x,y) = \frac{x^2 + y^2}{2r}, \quad (3)$$

where r is the indenter radius. The equilibrium condition is considered in the form

$$Q = \iint_{\Omega} p(x,y)\,dx\,dy. \quad (4)$$

The relation between the contact pressure and the surface displacement is given by the equation [8]

$$w(x,y,0) = \frac{1-2\nu}{4\pi G}\iint_{\Omega}\tau_{xz}(\xi,\eta)\left[\frac{\xi-x}{R^2}+\frac{1}{V}\sum_{i=1}^{n}k_i I_1\left(\frac{\xi-x}{\lambda_i V},\frac{\eta-y}{\lambda_i V}\right)\right]d\xi\,d\eta -$$
$$-\frac{1-\nu}{2\pi G}\iint_{\Omega}q(\xi,\eta)\left[\frac{1}{R}+\frac{1}{V}\sum_{i=1}^{n}k_i I_2\left(\frac{\xi-x}{\lambda_i V},\frac{\eta-y}{\lambda_i V}\right)\right]d\xi\,d\eta, \quad (5)$$
$$I_1(\alpha,\beta) = e^{\alpha}\int_{\alpha}^{\infty}\frac{e^{-u}u\,du}{u^2+\beta^2}, \quad I_2(\alpha,\beta) = e^{\alpha}\int_{\alpha}^{\infty}\frac{e^{-u}\,du}{\sqrt{u^2+\beta^2}}.$$

The contact problem is solved by the BEM. The solution method is presented explicitly in [6, 7]. An arbitrary area containing the contact area is a priori divided into square elements of constant pressure. To obtain the unknown contact area, contact pressure, and the indentation D, the following system of equations based on (2), (4), (5) should be solved numerically in the iteration procedure:

$$\begin{bmatrix} \Delta^2 & \cdots & \Delta^2 & 0 \\ k_1^1 & \cdots & k_N^1 & -1 \\ \vdots & \ddots & \vdots & \vdots \\ k_1^N & \cdots & k_N^N & -1 \end{bmatrix} \cdot \begin{bmatrix} p_1 \\ \vdots \\ p_N \\ D \end{bmatrix} = \begin{bmatrix} Q \\ f_1 \\ \vdots \\ f_N \end{bmatrix}. \quad (6)$$

After each step of the procedure, the elements with negative pressure are eliminated and the system is solved again. The iteration procedure stops when the solution contains no elements with negative pressure. The deformation component of the friction coefficient arises due to asymmetric distribution of the contact pressure and can be obtained after solving the contact problem. It is determined by the equation

$$\mu' = \frac{\iint xp(x,y)\,dx\,dy}{r\iint p(x,y)\,dx\,dy}. \quad (7)$$

3. Calculation results

The numerical results are used to analyze the contact pressure distribution, the indentation depth, and the deformation component of the frictional force in the case where the material model is determined by three relaxation and retardation times. The following dimensionless parameters where used in the calculations:

$$(x',y',\xi',\eta') = \frac{1}{r}(x,y,\xi,\eta), \quad c = \lambda_i k_i = \frac{G}{G_l}, \quad A_i = \frac{r}{\lambda_i V}, \quad B_i = k_i\frac{r}{V},$$
$$Q' = \frac{Q}{G_l r^2}, \quad p'(x,y) = \frac{p(x,y)}{G_l}, \quad V' = \frac{V}{r}. \quad (8)$$

Figure 2 illustrates the normal displacement of the surface due to the impact of a moving load distributed inside the square element with center at $(0,0)$. The area of the square element is $S' = 1/900$, the pressure inside the element is constant ($p' = 1.0$). The lines in the graph represent the surface displacement considering $y' = 0$ for three different relaxation times and for the spectrum consisting of the above-mentioned relaxation times. Figure 2 clearly presents the effect of retardation (λ) and relaxation times ($1/k$). The less the retardation time (λ), the greater the surface displacement that occurs due to the moving load. In the case of a smaller value of the relaxation time ($1/k$), the surface displacement decreases more intensively after passing of the distributed load. Line *4* corresponds to the spectrum of relaxation times and shows the

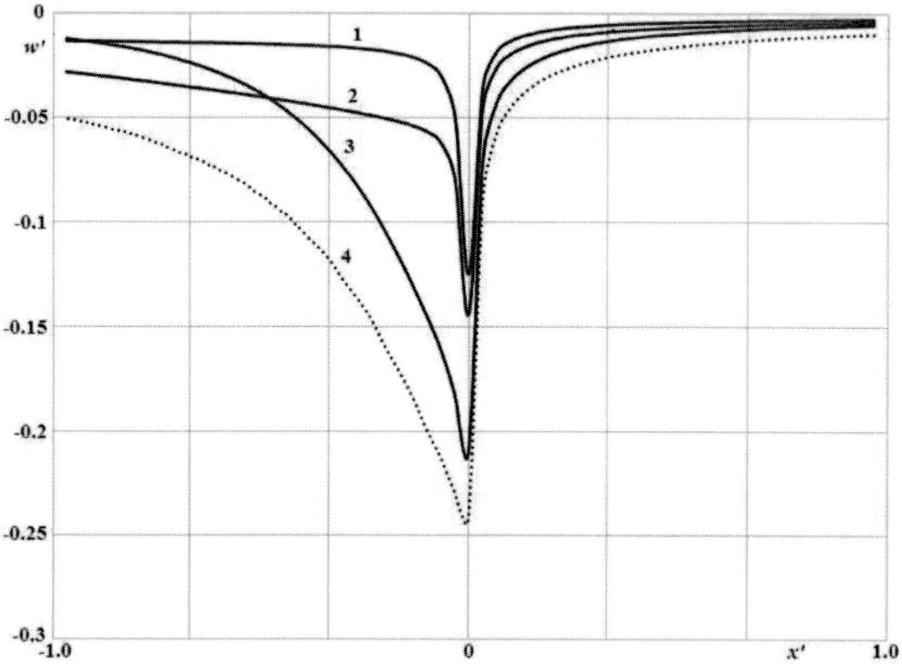

Figure 2. Normal displacement of the surface. Calculations for following parameters: $A = 5$, $P' = 1.0$, $\nu = 0.3$, $V' = 3$; $\lambda = 0.005$, $k = 1000$ for line *1*; $\lambda = 0.001$, $k = 5000$ for line *2*; $\lambda = 0.0002$, $k = 25000$ for line *3*; $\lambda_{1,2,3} = 0.0002, 0.001, 0.005$, $k_{1,2,3} = 25000, 5000, 1000$ for line *4*.

greatest surface displacement due to the moving load. The surface displacement decreases after passing of the distributed load more gradually than in the case of a single relaxation time of the smallest value.

The dependence of the mechanical component of the friction coefficient on the sliding velocity was studied. The parameter $c = 5$ which is the relation between the instant and longitudinal shear moduli is assumed to be constant. The calculations were carried out for the materials described by a creep kernel with a single relaxation time (figure 3, lines *1–3*) and for the spectrum of relaxation times (figure 3, line *4*). The maximum value of the mechanical component of the friction coefficient does not depend on the relaxation time, though it appears at different sliding velocities relevant to the relaxation time (lines *1–3*). In the case of a greater relaxation time, the maximum of the friction coefficient appears at a smaller sliding velocity. The friction coefficient attains its maximum and then decreases more gradually in the case of relatively small values of the relaxation time.

When the material is characterized by the spectrum of relaxation times, the value of the mechanical component of the friction coefficient is greater than that in the case of a material characterized by a single relaxation time.

The dependence of the indentation depth on the sliding velocity is presented in figure 4. The calculations were made using the same parameters as in figure 3. Lines *1–3* present the indentation of the indenter into the half-space described by a single relaxation time. Line *4* corresponds to the spectrum of relaxation times. The indentation of a body is the smallest in the case of a material characterized by the greatest relaxation time. In the case of the spectrum of relaxation times, the indentation is the greatest of all the cases under study. The indentation depth decreases more rapidly as the sliding velocity increases in the case of a greatest value of

IOP Conf. Series: Journal of Physics: Conf. Series **991** (2018) 012076 doi:10.1088/1742-6596/991/1/012076

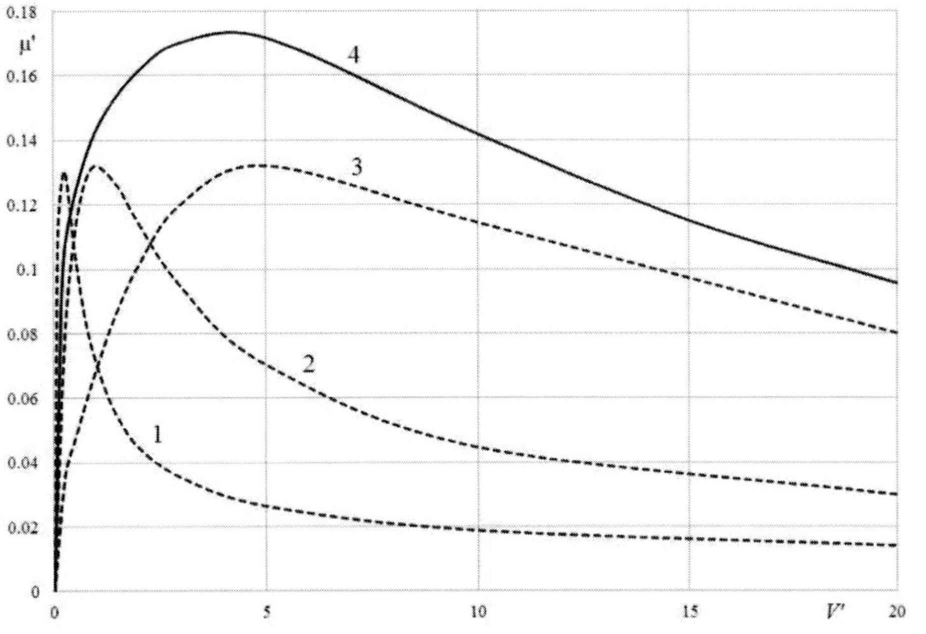

Figure 3. Mechanical component of friction. Calculations for following parameters: $A = 5$, $Q' = 0.06$, $\nu = 0.3$; $\lambda = 0.005$, $k = 1000$ for line *1*; $\lambda = 0.001$, $k = 5000$ for line *2*; $\lambda = 0.0002$, $k = 25000$ for line *3*; $\lambda_{1,2,3} = 0.0002, 0.001, 0.005$, $k_{1,2,3} = 25000, 5000, 1000$ for line *4*.

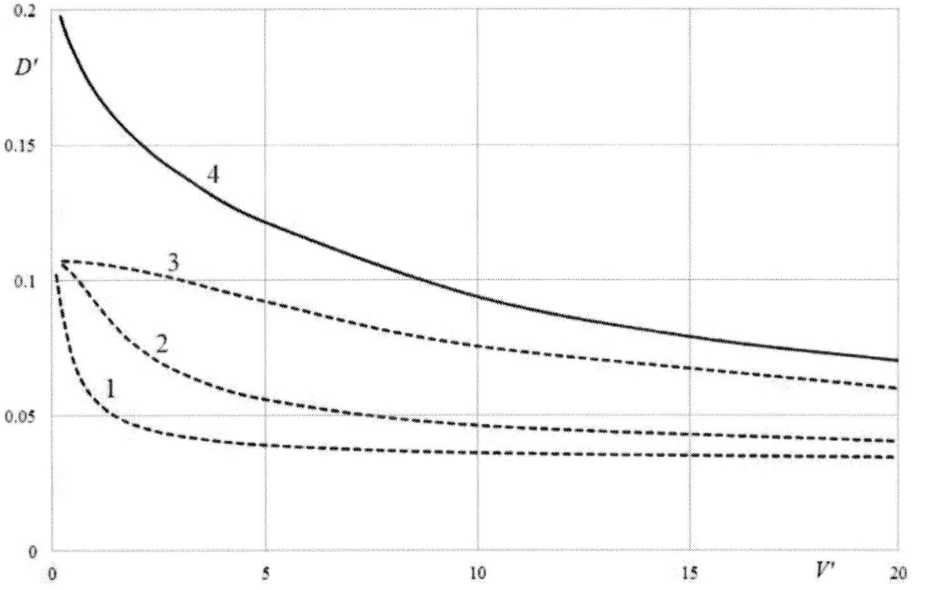

Figure 4. Indentation depth. Calculations for following parameters: $A = 5$, $Q' = 0.06$, $\nu = 0.3$; $\lambda = 0.005$, $k = 1000$ for line *1*; $\lambda = 0.001$, $k = 5000$ for line *2*; $\lambda = 0.0002$, $k = 25000$ for line *3*; $\lambda_{1,2,3} = 0.0002, 0.001, 0.005$, $k_{1,2,3} = 25000, 5000, 1000$ for line *4*.

the relaxation time.

Conclusions

The mechanical properties of viscoelastic half-space modeled by different relaxation times and the spectrum being a combination of the above-mentioned relaxation times have been studied. The calculations were carried out for the surface displacement due to the impact of a distributed load moving over a viscoelastic half-space and for a solid indenter sliding with a constant velocity over the viscoelastic half-space. The performed analysis allowed making the following conclusions:

- under the assumption that the relation between the instant and longitudinal shear moduli is constant, the maximum of the mechanical component of the friction coefficient is the same for different relaxation times, it occurs at different sliding velocities (the smaller the relaxation time, the greater the velocity);
- in the case where the spectrum of relaxation times is considered, the maximum value of the mechanical component of the friction coefficient is greater than that in the cases of a single relaxation time;
- the mechanical component of the friction coefficient decreases more rapidly with velocity when the value of the relaxation time is greater;
- the indentation depth is greater in the case of a smaller value of the relaxation time and it is the greatest in the case of a relaxation time spectrum;
- the indentation depth decreases more rapidly with the velocity in the case of a greater relaxation time.

Acknowledgments

This work was supported by the Russian Science Foundation (grant No. 14-29-00198).

References

[1] Flom D G and Bueche A M 1959 Theory of rolling friction for spheres *J. Appl. Phys.* **30** (11) 1725–30
[2] Panek C and Kalker J J 1977 A solution for the narrow rectangular punch *J. Elasticity* **7** (2) 213–8
[3] Panek C and Kalker J J 1980 Three-dimensional contact of a rigid roller traversing a viscoelastic half-space *IMA J. Appl. Math.* **26** (3) 299–313
[4] Kusche S 2016 Frictional force between a rotationally symmetric indenter and a viscoelastic half-space *ZAMM* **97** (2) 226–39
[5] Stepanov F I and Torskaya E V 2016 Study of stress state of viscoelastic half-space in sliding contact with smooth indenter *J. Frict. Wear* **37** (2) 101–6
[6] Goryacheva I G, Stepanov F I, and Torskaya E V 2015 Sliding of a smooth indentor over a viscoelastic half-space when there is friction *J. Appl. Math. Mech.* **79** (6) 596–603
[7] Goryacheva I G, Stepanov F I, and Torskaya E V 2016 Effect of friction in sliding contact of a sphere over a viscoelastic half-space In *Computational Methods in Applied Sciences* vol 40 *Mathematical Modeling and Optimization of Complex Structures* (Springer) pp 93–103
[8] Alexandrov V M and Goryacheva I G 2005 Mixed problems of mechanics of deformable solid In *Mixed Problems in Solid Mechanics: Proc. of V Russian Conf. with Int. Participation* ed N F Morozov (Saratov: Izdat. Sarat. Univ.) pp 23–5 [in Russian]

IOP Conf. Series: Journal of Physics: Conf. Series **991** (2018) 012077 doi:10.1088/1742-6596/991/1/012077

Strength analysis of aged polymer composites subjected to tensile loads

S Valesyan

Institute of Mechanics of the National Academy of Sciences of Armenia, Yerevan, Armenia

E-mail: valesyan@mail.ru

Abstract. It follows from the obtained data that the change of durability of the getinacks in stretching significantly depends on the pressing pressure value both at the age of 1 year and at the age of 4 years.

According to the data, in the case of samples of the first series, the ageing has not practically affected the durability of getinacks in stretching. In the case of samples of other series, the increase of age from 1 year to 4 years results in an increase of the getinacks durability, in particular, the increase is about 9% for the third series.

Comparing the values of failure tensile stresses given in the handbook of electrotechnics materials [1] with the data obtained by experimental investigation of aged glass textolite (GFRP composite-laminate) with the woven fiber orientation 0° and 90°, one can see a corresponding increase by approximately 25% and 35%.

The test results are approximated and compared with the experimental data. The corresponding figures are plotted on the basis of these data.

1. Introduction

The layered plastics, such as getinacks or textolites, have become more and more applicable in the electrotechnics as technologically convenient materials for the electro-isolation [1]. Depending on the chemical nature of the binding solution, filling, and a manufacturing technology, the layered plastics are suitable for long-term work at temperatures up to 180°C and for short-term work at higher temperatures [1].

The glass textolite (Glass Fibre Reinforced Polimer composite-laminate (GFRP composite-laminate)) is characterized by excellent thermal conductivity, moisture and biological resistance, and high strength, as well as it is lightweight, has good insulating properties, and is non-toxic. Because of its mechanical properties, heat resistance, and resistance to external environments, the glass textolite (GFRP composite–laminate) exceeds the PCB, and the maintenance of products manufactured from the glass textolite in the open air for 20 years is an important factor of production.

The polymer composites such as getinacks and glass textolite (GFRP composite–laminate) are easily liable to machine treatment such as cutting, drilling, turning and grinding, and for this reason, they are widely applied in aviation, metallurgy, oil and gas processing industry, shipbuilding, and in production of hardware and technical equipment of trolley buses and trams. In particular, the getinacks and glass textolite (GFRP composite-laminate) are used for production of items applicable in electrical engineering as structural and insulation materials.

A wide range of papers and books deal with investigation of mechanical properties of composites.

The influence of long-term storage on the mechanical properties of reinforced polymeric plastics was studied in [2]. The strength properties and ageing of sheet fiber glasses under various climatic conditions were studied in [3].

The influence of outer space conditions on parts of constructions subjected to tension, compression, and shear were investigated in [4]. The conditions of ageing both under the laboratory conditions and in the open space are compared.

The dissipative properties of plate reinforced composites cut out of composite at the angles $\varphi = 0°$, $5°$, $15°$, $30°$, $45°$, $60°$, and $90°$ to the reinforcement direction were investigated in [5].

The deformation behavior, strength and mechanical properties of unidirectional carbon fiber laminates are considered in [6]. The dissipative properties of the polymeric composites determined as a result of their repeated–static loading with regard to the technological factors and the environment temperature–moisture conditions are known from [7].

The first part of the book [8] reviews the processes and modeling of composite ageing including the physical and chemical ageing of polymeric composites, ageing of glass–ceramic matrix composites, chemical ageing mechanisms, stress corrosion cracking, thermo-oxidative ageing, spectroscopy of ageing composites, modeling physical and accelerated ageing, and ageing of silicon carbide composites. The second part examines the ageing of composites in transport applications including aircraft, vehicles and ships. The third part reviews the ageing of composites in nontransport applications such as implants in medical devices, oil and gas refining, construction, chemical processing, and underwater applications.

In the paper [9], the author analyzes the mechanical properties of plain-weave composites by employing different micromechanical models, and their in-plane and flexural properties are estimated by analytical approaches including the rule of mixtures and one-dimensional composite beam and two-dimensional mosaic models. Expressions for effective material properties are obtained in [9] for one-, two-, and three-ply woven composites.

It is of great interest to study the influence of "age" (ageing) on the mechanical properties of getinacks and glass textolite (GFRP composite-laminate), which are widely used in various branches of engineering.

2. Experimental part

Experimental samples of getinacks (figure 1) as double-sided blades whose sizes satisfy the requirements of GOST (State Standard Specification) 1199-78 were cut out from a sheet of getinacks of thickness 5 mm received by regulated thermo-pressing technology [10] for three different pressure values (1.9, 2.4, and 5.1 MPa).

In all three parties, the sheet getinacks cut out from the sulfate–cellulose paper saturated by the phenol–formaldehyde binding solution were cooled down after polymerization in the switched off furnace to avoid the temperature pressure.

The experimental samples of glass textolite (GFRP composite-laminate) (figure 2) as double-sided blades whose sizes satisfy the requirements of GOST (State Standard Specification) 1199-78 were cut out from a sheet of glass textolite (GFRP composite–laminate) of thickness ≈ 9 mm at the angles of the woven fiber orientation 0° and 90°.

The glass textolite (GFRP composite–laminate) is a laminate made of glass–cloth saturated by the thermosetting tar. The glass textolite (GFRP composite–laminate) is intended for manufacturing sheets and products for constructional purposes.

According to GOST (State Standard Specification) 25500-82, the getinacks and glass textolite (GFRP composite–laminate) test pieces were stored in horizontal position at a distance of at least 50 mm from the floor. The storage room was closed and the dry indoor air temperature was 20–25°T (the temperature interval stated by State Standard Specification from −10°C

IOP Conf. Series: Journal of Physics: Conf. Series **991** (2018) 012077 doi:10.1088/1742-6596/991/1/012077

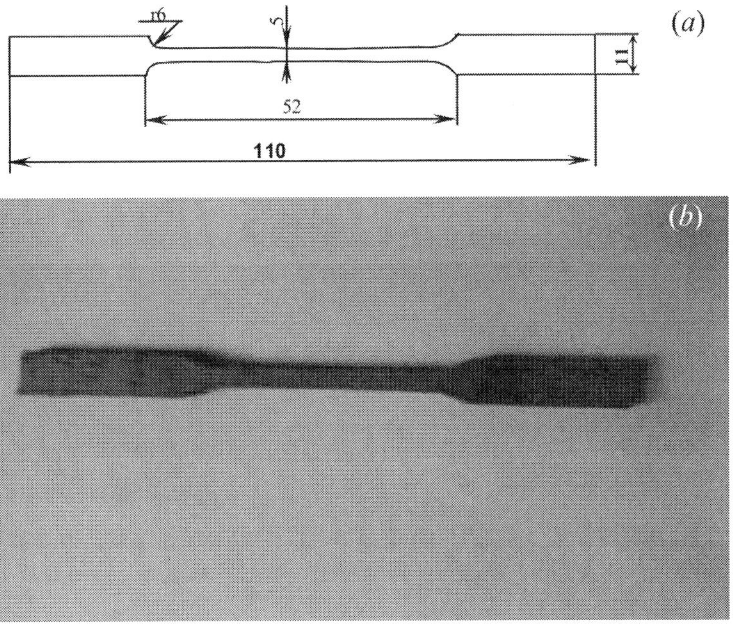

Figure 1. Experimental double-sided spattle-shaped getinacks sample.

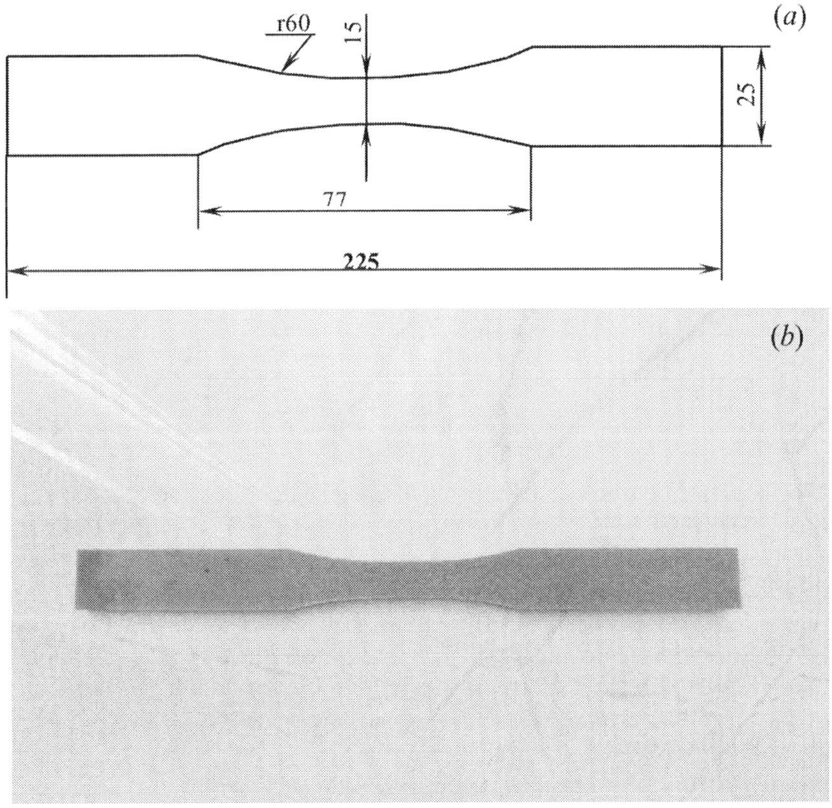

Figure 2. Experimental double-sided spattle-shaped GFRP composite–laminate sample.

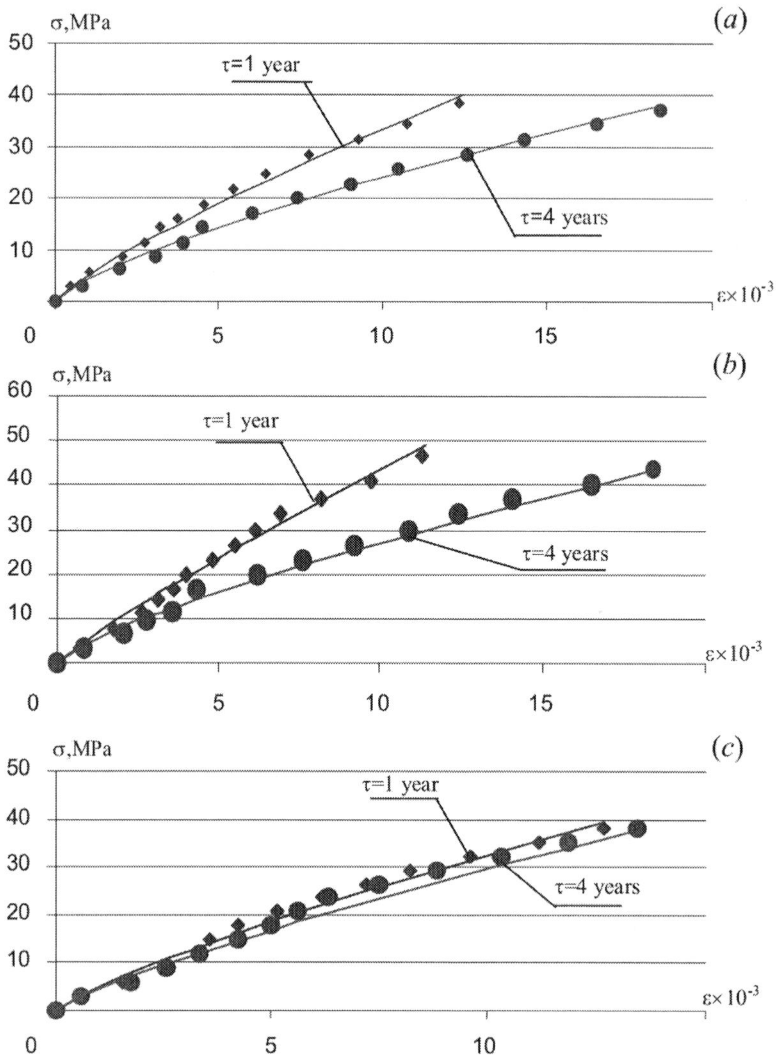

Figure 3. Curves of unidirectional tension of getinacks ((*a*) — series I, (*b*) — series II, (*c*) — series III)).

to +35°C), while the relative humidity did not exceed 80% (according to State Standard Specification).

The tests were performed by the testing machine ZD 10/90 at the loading speed 5 mm/min.

Experimental researches were carried out in 1 year and 4 years after the manufacturing of getinacks experimental samples, and in approximately 35 years after the manufacturing of glass textolite (GFRP composite–laminate) experimental samples.

During the unidirectional tensile loading, the samples were tested with the loading step $0.1\sigma_u$ (σ_u is ultimate strength of samples). The maximal value of the applied loading was $0.9\sigma_u$ and $0.8\sigma_u$ for getinacks and glass textolite (GFRP composite–laminate) experimental samples, respectively. The experimental data were recorded by a clock-type meter. The experimental data are evaluated and curves are plotted (figures 3 and 4).

The experimental data (σ and ε) are approximated by the equation

$$\sigma = \alpha\varepsilon + \beta\varepsilon^n,$$

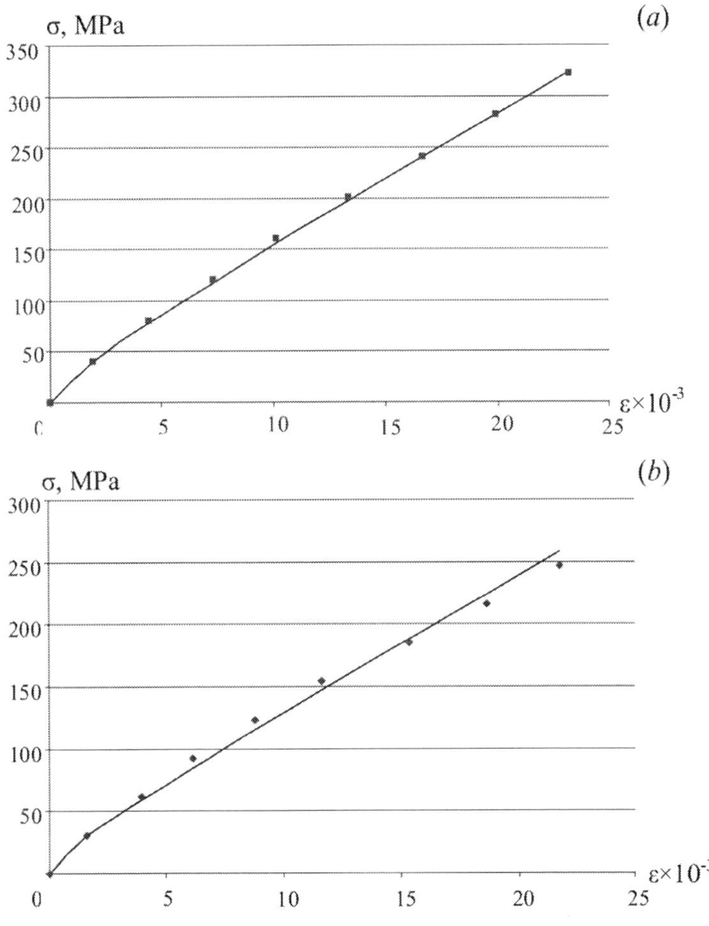

Figure 4. Curves of unidirectional tension of glass textolite ((*a*) — the angle of woven fiber orientation is 0°, (*b*) — the angle of woven fiber orientation is 90°).

where ε is the strain value determined from the permanent set value, and α, β, and n are approximation parameters.

3. Results and discussion

The results of testing are given in tables 1 and 2 and in figures 3 and 4. In figure 3 (*a*, *b*, and *c*) and figure 4 (*a* and *b*), the experimental data are plotted by dotted curves, and the theoretical values are plotted by solid lines.

It follows from the data given in table 1 that the change of durability of getinacks in stretching significantly depends on pressing pressure value both at the age of 1 year and at the age of 4 years. So, for $\tau = 1$ year, the durability of samples of the second series is by 16% greater than the durability of samples of the first series. The durability of samples of the third series is approximately by 20% greater than the durability of samples of the first series (see the table 1).

For τ=4 years, the durability of samples of the second and third series is approximately by 22% and 29% greater than the durability of samples of the first series, respectively.

According to the data given in table 1, in the case of samples of the first series, the ageing has not practically affected the durability of getinacks in stretching. In the case of samples of other series, the increase of the age from 1 year to 4 years results in an increase in the durability

Table 1. Durability of samples in stretching.

Series No.	Pressure of pressing, MPa	Age of samples τ at the time of experiment, year	Durability of samples during the stretching, MPa
I	1.9	1	117.3 ± 1.322
		4	118.4 ± 1.334
II	2.4	1	136.1 ± 14.997
		4	144.3 ± 15.903
III	5.1	1	140.4 ± 5.64
		4	152.8 ± 5.176

Table 2. Failure tensile stress.

	Failure tensile stress, MPa	
Woven fiber orientation	0°	90°
GFRP handbook values [1]	300	200
GFRP aged values	402	307

of getinacks; in particular, the increase is about 9% for the third series.

It follows from the diagrams in figure 3 that, under the influence of short-term forces, the ageing leads to an increase in the deformability of getinacks. The higher the level of the pressing pressure on the material, the less the deformability of getinacks.

Comparing the failure tensile stress values given in the handbook of electrotechnics materials [1] and obtained by the experimental investigation of aged glass textolite (GFRP composite–laminate) with the woven fiber orientation 0° and 90°, one can see an increase of approximately 25% and 35%, respectively (table 2).

It follows from the diagrams in figure 4 that, under the influence of short-term forces, the ageing leads to an increase in the deformability of getinacks. For glass textolite (GFRP composite–laminate) with the angles of the woven fiber orientation 0° and 90°, the experimental and theoretical curves coincide by 97% (the error is 3%) and 95.55% (the error is 4.45%), respectively.

Conclusions

The constructional elements from getinacks on the basis of sulfate–cellulose paper saturated by a phenol–formaldehyde binding solution should be manufactured by pressing under the pressure at least equal to 5.1 MPa, which ensures smaller deformation changes and the practical independence of the mechanical properties of the age.

The influence of ageing on the mechanical properties of glass textolite (GFRP composite–laminate) subjected to unidirectional tensile loads should be insignificant.

The applied equation allows plotting the theoretical curves close to the curves plotted on the basis of experimental data.

References

[1] Koritskii U V, Pasinkov V V and Tareev B M 1986 *Handbook of Electrotechnics Materials* (Moscow: Energoizdat) [in Russian]

[2] Ogibalov P M and Tuneeva I M 1969 Effect of extended storage on the mechanical properties of reinforced plastics *Polymer. Mech.* **5** (3) 478–80
[3] Smirnova Z A and Vlasov P V 1971 Strength properties and aging of sheet glass-reinforced plastics under various climatic conditions *Polymer. Mech.* **7** (3) 497–500
[4] Jegun E G, Ivonin U N, Shlica R P and Nikishin E F 1988 Effect of long-term exposure to extreme conditions on the properties of preloaded composites in tension, compression, and shear *Mech. Compos. Mater.* **23** (5) 567–72
[5] Limonov V A, Perevozchikov V G and Tamuzh V P 1988 Fatigue of laminated composites with various reinforcement systems. 1. Experimental results *Mech. Compos. Mater.* **24** (5) 585–94
[6] Kucher N K, Zemtsov M P and Zarazovskii M N 2006 Deformation behavior and strength of unidirectional carbon fiber laminates *Mech. Compos. Mater.* **42** (5) 407–18
[7] Karapetyan K A 2001 Deformation of glass-plastic tubes under repeated–static uniaxial and combined loadings *Dokl. Nats. Akad. Nauk Armenii* **101** (4) 317–23
[8] Martin R 2008 *Aging of Composites* (Boca Raton: CRC Press)
[9] Soykasap Ö 2011 Analysis of plain-weave composites *Mech. Compos. Mater.* **47** (2) 161
[10] Simonyan A M and Valesyan S Sh 1999 Design of rims for regulated self-pressing of composites during polymerization *Izv. Nats. Akad. Nauk Armenii. Ser. Tekhn. Nauk* **52** (1) 14–9

IOP Conf. Series: Journal of Physics: Conf. Series **991** (2018) 012078 doi:10.1088/1742-6596/991/1/012078

Investigation of stress concentration at corner points for orthotropic plate bending problem

N G Vasilyan

Institute of Mechanics of the National Academy of Sciences of Armenia, Yerevan, Armenia

E-mail: narine@mechins.sci.am

Abstract. This article deals with the bending problem for an orthotropic semi-infinite plate strip when three edges of the plate are hinged and the fourth edge goes to infinity. The plate is loaded with distributed load of intensity $q(y)$. A. Nadai's approach is applied, which says that to obtain the solution at a far distance from the edge, it is necessary to solve the problem of cylindrical bending. The generalized shearing forces on the fixed edge are investigated.

1. Introduction

In the Cartesian coordinate system, we explore a semi-infinite plate strip of constant thickness $2h$ occupying the area $0 \le x < \infty$, $0 \le y \le b$, $-h \le z \le h$. A distributed load of intensity $q(y)$ is acting on the plate [3].

We assume that the material of the plate, due to its elastic properties, has three planes of symmetry, i.e., the plate is orthotropic. If these planes are taken as coordinate axes, then the relationships between the components of stress and strain can be expressed by the equations

$$\sigma_x = E'_x\varepsilon_x + E''\varepsilon_y, \quad \sigma_y = E'_y\varepsilon_y + E''\varepsilon_x, \quad \sigma_{xy} = G\varepsilon_{xy}. \tag{1}$$

We assume that the perpendicular to the middle surface of the plate linear elements remains straight and normal to the curved surface of the plate after bending.

We use the following expressions of the strain components:

$$\varepsilon_x = \frac{\partial u}{\partial x} - z\frac{\partial^2 w}{\partial x^2}, \quad \varepsilon_y = \frac{\partial v}{\partial x} - z\frac{\partial^2 w}{\partial y^2}, \quad \varepsilon_{xy} = \frac{\partial u}{\partial y} + \frac{\partial v}{\partial x} - 2z\frac{\partial^2 w}{\partial x \partial y}. \tag{2}$$

When investigating the bending of plates, we consider only the perpendicular displacement [6]. Consequently, for the bending and twisting moments, we have [2]

$$\begin{gathered} M_x = \int_{-h}^{h} z\sigma_x\, dz = -D_x\frac{\partial^2 w}{\partial x^2} - D_1\frac{\partial^2 w}{\partial y^2}, \quad M_y = \int_{-h}^{h} z\sigma_y\, dz = -D_y\frac{\partial^2 w}{\partial y^2} - D_1\frac{\partial^2 w}{\partial x^2}, \\ H = -\int_{-h}^{h} z\sigma_{xy}\, dz = 2D_{xy}\frac{\partial^2 w}{\partial x \partial y}. \end{gathered} \tag{3}$$

IOP Conf. Series: Journal of Physics: Conf. Series **991** (2018) 012078 doi:10.1088/1742-6596/991/1/012078

where the coefficients, which characterizing the mechanical properties are:

$$E'_x = \frac{E_x}{1-\nu_{xy}\nu_{yx}}, \quad E'_y = \frac{E_y}{1-\nu_{xy}\nu_{yx}}, \quad E'' = \frac{E_y\nu_{xy}}{1-\nu_{xy}\nu_{yx}},$$
$$D_x = \frac{2E'_x h^3}{3}, \quad D_y = \frac{2E'_y h^3}{3}, \quad D_1 = \frac{2E'' h^3}{3}, \quad D_{xy} = \frac{2Gh^3}{3}. \tag{4}$$

Substituting (3) into the equations of equilibrium [8], we obtain the equation of bending of orthotropic plates

$$D_x\frac{\partial^4 w}{\partial x^4} + 2H_1\frac{\partial^4 w}{\partial x^2\partial y^2} + D_y\frac{\partial^4 w}{\partial y^4} = q(y), \tag{5}$$

where

$$H_1 = D_1 + 2D_{xy}.$$

In the specific case of isotropic plate, we have

$$E'_x = E'_y = \frac{E}{1-\vartheta^2}, \quad E'' = \frac{\vartheta E}{1-\vartheta^2}, \quad G = \frac{E}{2(1+\vartheta)},$$
$$D_x = D_y = H_1 = D = \frac{2Eh^3}{3(1-\vartheta^2)} \tag{6}$$

and obtain the equation of bending of plates in the well-known form $\Delta\Delta w = q/D$.

2. Satisfaction of the boundary conditions

It is assumed that the edges of the plate $y = 0, b$ are hinged which allows us to obtain the solution of equation (6) in the form

$$w = \sum_{n=1}^{\infty} f_n(x)\sin(\lambda_n y), \quad q = \sum_{n=1}^{\infty} q_n \sin(\lambda_n y), \quad \lambda_n = \frac{\pi n}{b}. \tag{7}$$

Substituting (7) into (5), we obtain the fourth-order inhomogeneous differential equation [7]

$$D_x f_n^{IV} - 2H_1\lambda_n^2 f_n^{II} + D_y\lambda_n^4 f_n = q_n. \tag{8}$$

We seek a homogeneous solution of the differential equation in the form

$$f_n(x) = Ae^{r\lambda_n x}. \tag{9}$$

The characteristic equation of (8) becomes

$$D_x r^4 - 2H_1 r^2 + D_y = 0, \tag{10}$$

$$r^2 = \frac{H_1 \pm \sqrt{H_1^2 - D_x D_y}}{D_x}. \tag{11}$$

We calculate its four roots and obtain the general solution of the problem of bending of an orthotropic semi-infinite plate strip with two hinged edges in the formn

$$f_n(x) = A_n e^{-r_1\lambda_n x} + B_n e^{-r_2\lambda_n x} + C_n e^{r_1\lambda_n x} + D_n e^{r_2\lambda_n x} + \frac{q_n}{D\lambda_n^4}. \tag{12}$$

Applying the Nadai approach, which says that, at a far distance from the edge $x = 0$, the solution must tend to the solution of the problem of cylindrical bending,

$$\lim_{x\to\infty} f_n(x) = \frac{q_n}{D\lambda_n^4}, \tag{13}$$

we obtain

$$f_n(x) = A_n e^{-r_1\lambda_n x} + B_n e^{-r_2\lambda_n x} + \frac{q_n}{D\lambda_n^4}. \tag{14}$$

The unknown coefficients A_n and B_n are determined from the boundary conditions on $x = 0$.

First, we examine the expression $H_1^2 - D_x D_y$, which is obtained for the root values of the solution of the differential equation. We consider the following three cases.

a. $H_1^2 - D_x D_y > 0$. Therefore

$$f_n(x) = A_n e^{-r_1\lambda_n x} + B_n e^{-r_2\lambda_n x}, \quad r_{1,2} = \sqrt{\frac{H_1 \pm \sqrt{H_1^2 - D_x D_y}}{D_x}}. \tag{15}$$

We assume that the edge $x = 0$ is hinged and have the boundary conditions: $w = 0$, $\frac{\partial^2 w}{\partial x^2} = 0$. Satisfying the boundary conditions, we find the unknown coefficients and determine the plate deflection

$$w(x,y) = \sum_{n=1}^{\infty} \frac{q_n}{D_y\lambda_n^4}\left[\frac{H_1 - \sqrt{H_1^2 - D_x D_y}}{2\sqrt{H_1^2 - D_x D_y}} \exp\left(-\sqrt{\frac{H_1 + \sqrt{H_1^2 - D_x D_y}}{D_x}}\lambda_n x\right)\right.$$
$$\left. - \frac{H_1 + \sqrt{H_1^2 - D_x D_y}}{2\sqrt{H_1^2 - D_x D_y}} \exp\left(-\sqrt{\frac{H_1 - \sqrt{H_1^2 - D_x D_y}}{D_x}}\lambda_n x\right) + 1\right] \sin(\lambda_n y). \tag{16}$$

The generalized shearing forces for orthotropic materials have the form

$$V_x(x,y) = -\frac{\partial}{\partial x}\left(D_x\frac{\partial^2 w}{\partial x^2} + H_2\frac{\partial^2 w}{\partial y^2}\right), \quad V_y(x,y) = -\frac{\partial}{\partial y}\left(D_y\frac{\partial^2 w}{\partial y^2} + H_2\frac{\partial^2 w}{\partial x^2}\right), \tag{17}$$

where

$$H_2 = H_1 + 2D_{xy} = D_1 + 4D_{xy}.$$

Therefore, for generalized shearing forces, we obtain

$$V_x(x,y) = \sum_{n=1}^{\infty} \frac{q_n}{2B\lambda_n}\frac{1}{D_y\sqrt{D_x}}\left\{\sqrt{H_1 + B}[D_x D_y - H_2(H_1 - B)]\exp\left(-\sqrt{\frac{H_1 + B}{D_x}}\lambda_n x\right)\right.$$
$$\left. + \sqrt{H_1 - B}[D_x D_y - H_2(H_1 + B)]\exp\left(-\sqrt{\frac{H_1 - B}{D_x}}\lambda_n x\right)\right\}\sin(\lambda_n y),$$
$$V_y(x,y) = \sum_{n=1}^{\infty} \frac{q_n}{2B\lambda_n}\left[2B - (B + 2D_{xy})\exp\left(-\sqrt{\frac{H_1 + B}{D_x}}\lambda_n x\right)\right.$$
$$\left. - (2H_1 + B + 2D_{xy})\exp\left(-\sqrt{\frac{H_1 - B}{D_x}}\lambda_n x\right)\right]\cos(\lambda_n y), \tag{18}$$

where we used the following notation: $B = \sqrt{H_1^2 - D_x D_y}$.

IOP Conf. Series: Journal of Physics: Conf. Series **991** (2018) 012078 doi:10.1088/1742-6596/991/1/012078

b. $H_1^2 - D_x D_y = 0$. This coincides with the case of isotropic material [4]. We seek $f_n(x)$ in the form $f_n(x) = A_n e^{-\lambda_n x}$. If the plate is hinged on the edge $x = 0$, then we obtain

$$w(x,y) = \sum_{n=1}^{\infty} \frac{q_n}{D\lambda_n^4}\left[1 - e^{-\lambda_n x} - \frac{\lambda_n x}{2} e^{-\lambda_n x}\right] \sin(\lambda_n y), \tag{19}$$

and for generalized shearing forces, we obtain

$$\begin{aligned} V_x(x,y) &= \sum_{n=1}^{\infty} \frac{q_n}{2\lambda_n}[(3-\nu)e^{-\lambda_n x} + (1-\nu)\lambda_n x e^{-\lambda_n x}] \sin(\lambda_n y), \\ V_y(x,y) &= \sum_{n=1}^{\infty} \frac{q_n}{2\lambda_n}[2(1 - e^{-\lambda_n x}) + (1-\nu)\lambda_n x e^{-\lambda_n x}] \cos(\lambda_n y). \end{aligned} \tag{20}$$

c. $H_1^2 - D_x D_y < 0$. The solution (15) has the form

$$f_n(x) = e^{-\alpha\lambda_n x}[A_n \sin(\beta\lambda_n x) + B_n \cos(\beta\lambda_n x)] + \frac{q_n}{D_y \lambda_n^4}. \tag{21}$$

where

$$\alpha = \frac{\sqrt{D_x D_y} + H_1}{2D_x}, \quad \beta = \frac{\sqrt{D_x D_y} - H_1}{2D_x}. \tag{22}$$

Satisfying the boundary conditions of hinge joining, we obtain

$$w(x,y) = \sum_{n=1}^{\infty} \frac{q_n}{D_y\lambda_n^4}\left\{1 - \left[\frac{\alpha^2-\beta^2}{2\alpha\beta}\sin(\beta\lambda_n x) + \cos(\beta\lambda_n x)\right] e^{-\alpha\lambda_n x}\right\} \sin(\lambda_n y). \tag{23}$$

For generalized shearing forces, we obtain the expression

$$\begin{aligned} V_x(x,y) &= \sum_{n=1}^{\infty} \frac{q_n}{\lambda_n}\frac{H_2}{D_y}\frac{\alpha^2+\beta^2}{2\alpha\beta}\left\{\left[1 + \frac{D_x}{H_2}(\alpha^2+\beta^2)\right]\beta\cos(\beta\lambda_n x)\right. \\ &\quad \left. + \left[1 - \frac{D_x}{H_2}(\alpha^2+\beta^2)\right]\alpha\sin(\beta\lambda_n x)\right\} e^{-\alpha\lambda_n x}\sin(\lambda_n y), \\ V_y(x,y) &= \sum_{n=1}^{\infty} \frac{q_n}{\lambda_n}\frac{1}{D_y}\frac{1}{2\alpha\beta}\left\{\alpha\beta D_y - 2\alpha\beta D_y e^{-\alpha\lambda_n x}\cos(\beta\lambda_n x)\right. \\ &\quad \left. + [H_2(\alpha^2+\beta^2)^2 - D_y(\alpha^2-\beta^2)]e^{-\alpha\lambda_n x}\sin(\beta\lambda_n x)\right\}\cos(\lambda_n y). \end{aligned} \tag{24}$$

3. Analysis of generalized shearing forces

The calculations for different materials showed that $H_1^2 - D_x D_y < 0$ for all materials, i.e., the third case. For generalized shearing forces in the middle of the fixed edge, we have

$$V_x\left(0, \frac{b}{2}\right) = \sum_{n=1}^{\infty} \frac{q_n}{\lambda_n}\frac{H_2}{D_y}\frac{\alpha^2+\beta^2}{2\alpha}\left[\frac{D_2}{H_2}(\alpha^2+\beta^2) + 1\right]\sin\frac{\pi n}{2}. \tag{25}$$

If we introduce a new coefficient γ equal to

$$\gamma = \frac{H_2}{D_y}\frac{\alpha^2+\beta^2}{2\alpha}\left[\frac{D_2}{H_2}(\alpha^2+\beta^2) + 1\right], \tag{26}$$

Table 1. Generalizing shearing force in the middle of the fixed edge.

(x,y)	$(0,0)$	$(0,b/2)$	$(0,b)$
$V_x(x,y)$	0	$\gamma\sum_{n=1}^{\infty}\frac{q_n}{\lambda_n}\sin\frac{\pi n}{2}$	0
$V_y(x,y)$	$-\sum_{n=1}^{\infty}\frac{q_n}{2\lambda_n}$	0	$-\sum_{n=1}^{\infty}\frac{q_n}{2\lambda_n}\cos(\pi n)$

Table 2. Values of γ for the investigated materials.

	Kevlar 49 [9]	Fiberglass AG-4 [1]	Fiberglass anisotropic materials [1]	Tornel 40 [10]	Plywood, 3 ply and 5 [10]
γ	1.87	1.22	1.26	0.59	2.33

Table 3. Values of V_x/q_0b for investigated materials.

Material	Mechanical characteristic coefficients	V_x/q_0b
Kevlar 49 [9]	$E_x = 81.8\times10^9$ Pa, $E_y = 5.1\times10^9$ Pa, $\nu_{xy} = 0.31$, $\nu_{yx} = 0.019$, $G = 1.82\times10^9$ Pa	0.46
Fiberglass AG-4 [1]	$E_x = 2.1\times10^9$ Pa, $E_y = 1.6\times10^9$ Pa, $\nu_{xy} = 0.07$, $\nu_{yx} = 0.092$, $G = 0.42\times10^9$ Pa	0.305
Fiberglass anisotropic materials [1]	$E_x = 3.05\times10^9$ Pa, $E_y = 1.88\times10^9$ Pa, $\nu_{xy} = 0.195$, $\nu_{yx} = 0.12$, $G = 0.49\times10^9$ Pa	0.315
Tornel 40: Carbon [10]	$E_r = 1.61\times10^9$ Pa, $E_z = 23\times10^9$ Pa, $\nu_{rz} = 0.021$, $\nu_{zr} = 0.3$, $G_{rz} = 2\times10^9$ Pa	0.15
Plywood, 3 and 5 plies [10]	$E'_x = 1.4\times10^9$ Pa, $E'_y = 0.117\times10^9$ Pa, $E'' = 0.054\times10^9$ Pa, $G = 0.12\times10^9$ Pa	0.58

then

$$V_x\left(0,\frac{b}{2}\right) = \gamma\sum_{n=1}^{\infty}\frac{q_n}{\lambda_n}\sin\frac{\pi n}{2}. \tag{27}$$

The solution of shearing forces depends on the coefficient γ (which depends only on the characteristic coefficients of the mechanical properties of the material) in the middle of the fixed edge (see table 1).

For the materials under study, the values of γ are given in table 2.

Table 3 shows the values of generalized shearing forces V_x for different materials with $q = q_0 = \text{const}$.

The case $H_1^2 - D_xD_y > 0$ will be obtained if we can find a material for which $E_2 > E_1$ and $G > E_2/6$.

The following two cases of the load were considered to calculate the generalized shearing forces:

$$1)\quad q = q_0 = \text{const},\quad \text{hence}\quad V_y(0,0) = -\frac{q_0 b}{2\pi}\sum_{n=1}^{\infty}\frac{1}{n},\quad V_y(0,b) = -\frac{q_0 b}{2\pi}\ln 2,$$

$$2)\quad q = q_0\sin(\lambda_1 y),\quad \text{hence}\quad V_y(0,0) = -\frac{q_1 b}{2\pi},\quad V_y(0,b) = \frac{q_1 b}{2\pi}.$$

Conclusions

In this article, the values of generalized shearing forces for orthotropic semi-infinite plate strip are obtained. At the corners of the fixed edge, we have $V_x = 0$, and at the middle points, we have values different from zero. Thus, for the bending problem of an orthotropic semi-infinite plate strip with three hinged edges, we obtain that the generalized shearing forces are concentrated on the fixed edge. For an isotropic semi-infinite plate strip ($b = 10a$), the equation of static equilibrium is obtained. Therefore, as compared to V. V. Vasil'ev's work [5] who studied a simply supported plate, the transformation of Kirchhoff and Thomson–Tait is true in this case.

References

[1] Ambartsumyan S A 1987 *Theory of Anisotropic Plates: Strength, Stability, and Vibrations* (Moscow: Nauka) p 360 [in Russian]
[2] Antonyan S S and Vasilyan N G 2011 The investigation of the stress-strain state of plate bending problem in vicinity of the edge of the plate boundary condition of sliding contact. In *The Problems of Dynamics of Interaction of Deformable Media. Proc. of the VII Int. Conf. September 19-23, 2011, Goris - Stepanakert* (Yerevan: Inst. Mekh. NAN RA) 44–8 [in Russian]
[3] Antonyan S S and Vasilyan N G 2013 The stress-strain state of bending plate around of hinged edges In *MECHANICS-2013. Proc. of the Int. Young Scientists School-Conference, 1-4 October 2013, Tsakhkadzor, Armenia* (Yerevan: EGUAS) 69–73
[4] Belubekyan M V 2002 On the equations of plate theory taking into account transverse sear stresses In *Problems of mechanics of thin deformable bodies: Collection of Papers, Devoted to the 80th Anniversary of Academician of NAS of Armenia S. A. Ambartsumyan)* (Yerevan: Inst. Mekh. NAN RA) 67–88
[5] Vasil'ev V V 2012 Kirchhoff and Thomson–Tait Transformations in the Classical Theory of Plates *Mech. Solids* No. 5 571–9
[6] Lehnitsky S G 1947 *Anisotropic Plates* (Moscow–Leningrad: OGIZ GITTL) p 355
[7] Meleshko V V 2004 Biharmonic problem for rectangular: History and modernity *Mat. Met. Fiz. Mekh. Polya* **47** (3) 45–67
[8] Timoshenko S P and Woinowsky-Krieger S 1959 *Theory of Plates and Shells* (New York: McGraw-Hill)
[9] Lubin G (Editor) 1982 *Handbook of Composites* (New York: Springer)
[10] *Encyclopaedia of Mechanical Engineering XXL* URL `http://mash-xxl.info`
[11] Nadai A 1925 Die elastischen Platten (Berlin: Verlag von Julius Springer)

IOP Conf. Series: Journal of Physics: Conf. Series **991** (2018) 012079 doi:10.1088/1742-6596/991/1/012079

Some remarks on recent developments in micropolar continuum theory

E N Vilchevskaya[1,2*] and W H Müller[3]**

[1]Institute for Problems in Mechanical Engineering of the Russian Academy of Sciences, St. Petersburg, Russia
[2]Peter the Great Saint-Petersburg Polytechnic University, St. Petersburg, Russia
[3]Institute of Mechanics, Chair of Continuum Mechanics and Constitutive Theory, Technische Universität Berlin, Berlin, Germany

E-mail: *vilchevska@gmail.com, **whmueller1000@gmail.com

Abstract. This paper considers micropolar media that can undergo structural changes and do not *a priori* consist of indestructible material particles. Initially the pertinent literature is reviewed. Then the necessary theoretical framework for a continuum of that type is presented. The standard macroscopic equations for mass, linear and angular momentum are complemented by a recently proposed balance for the moment of inertia tensor, which contains a production term. Two examples illustrate the effect of the production. In the first example, we study a continuous stream of matter on a conveyor belt going through a crusher so that the total number of particles will change. In context with this example, it is also clear that the traditional Lagrangian way of describing the motion of solids is no longer adequate and must be replaced by the Eulerian point of view known from fluid mechanics. The second example deals with hollow particles which rotate because of the presence of body couples. Now a transient temperature field is superimposed such that the moment of inertia field changes due to the thermal expansion of particles. This in turn results in rotational motion that is no longer constant but varies in space and time.

1. Introduction: Short review of the state-of-the-art

Generalized Continuum Theories (GCTs) have always been interesting to the materials science community in order to model and quantitatively describe high performance materials with inner degrees of freedom. Such materials can be used in large and small scale applications ranging from light-weight aerospace and automotive panels to micromechanics and microelectronic gadgets. A special GCT is the so-called theory of micropolar materials. It emphasizes the aspect of inner rotational degrees of freedom in a material. The introduction of such a feature is particularly useful for applications to soils, polycrystalline and composite matter, granular and powder-like materials, and even to porous media and foams.

It should be noted that the tensor field of the moment of inertia $\boldsymbol{J}$ traditionally plays an important role in context with rotational motion, specifically in combination with the angular velocity vector $\boldsymbol{\omega}$ describing, as a product, the local angular momentum in a continuum. Like the mass it is traditionally assigned to a material element. As such it is conserved and known *a priori*, see [1], pg. 14; [2], pg. 966; [3], Section 13; and [4], pg. 33. A different approach was suggested in [5], where it was assumed that the inertia of polar particles may change as the continuum deforms. This idea was further elaborated by [6] who clearly stated that the moment

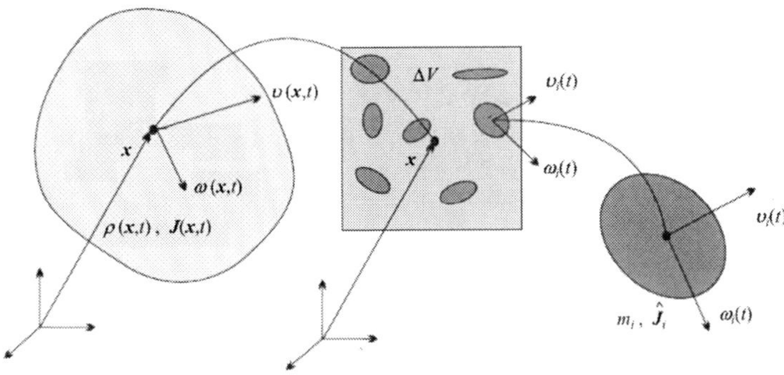

Figure 1. Macro- meso-, and microviewpoint (see text).

of inertia tensor should be treated as an independent field, just like the inertia associated with the linear momentum, namely the field of mass density, ρ. In fact, they proposed a balance equation for $\boldsymbol{J}$, which contains a production term χ_J of moment of inertia due to "structural transformations" as they called it. This means that the moment of inertia will change due to combination or fragmentation of particles during mechanical crushing or chemical reactions. In addition, a production may arise because of phase transitions or physical property changes, such as electric magnetization or polarization.

However, the departure from the idea of an indestructible material particle also necessitates us to abandon the Lagrangian description and turn to the Eulerian perspective (a.k.a. spatial description) instead. Originally, the Eulerian description stems from fluid mechanics. It does not impose strict constraints on the motion of "points." Rather it follows the idea of an open system *a priori* allowing for the exchange of mass, momentum, energy, moment of inertia, *etc.* between and within the cells of the Eulerian grid.

We therefore need to extend the original goals of micropolar theory. In fact, in what follows, we initially formulate the corresponding equations in a rather general manner, such that they are ready for future investigations of static or dynamic problems which involve linear and angular momentum simultaneously. However, we do not allow for a coupling between the two. The theoretical parts of this aspect are addressed (for example) in [4], Section 4.3; [6], or [7]. However, concrete examples were yet not provided. It is to be expected that the coupling will open up another degree of freedom in describing complex motion as it was shown for a point particle in [8], pp. 193. We now proceed to explain the relevant theory for structurally transforming micropolar materials.

2. Outline of the theory

In what follows, we consider the theoretical aspects of micropolar theory, first, from the macroscopic, *i.e.*, the continuum point of view, which results in a complete set of balances including one for the rotational inertia. Because the balance for the inertia tensor field is extended by the production term, we, second, present mesoscopic aspects of relevant fields in terms of microproperties of rigid particles. The situation is illustrated in figure 1.

2.1. The macroscopic aspect

Mathematically speaking, in a very general form, the objective of micropolar theory is to determine the following primary fields: (a) the scalar field of mass density $\rho(\boldsymbol{x},t)$; (b) the vector field of linear velocity $\boldsymbol{v}(\boldsymbol{x},t)$; (c) the symmetric, second rank, and positive definite specific moment of inertia tensor field $\boldsymbol{J}(\boldsymbol{x},t)$ in units of m^2; and (d) the spin (a.k.a. angular velocity) field $\boldsymbol{\omega}(\boldsymbol{x},t)$ at all points $\boldsymbol{x}$ and at all times t within a region of space $\mathcal{B}$, which can be either a material volume, *i.e.*, it consists of the same matter at all times, or a region through which the matter is flowing (see the left inset of figure 1).

The determination of these fields relies on field equations for the primary fields. The field equations are based on balance laws and need to be complemented by suitable constitutive relations. At regular points, these macroscopic balances read as follows:

- balance of mass:

$$\frac{\delta\rho}{\delta t}+\rho\boldsymbol{\nabla}\cdot\boldsymbol{v}=0, \tag{1}$$

- balance of momentum:

$$\rho\frac{\delta\boldsymbol{v}}{\delta t}=\boldsymbol{\nabla}\cdot\boldsymbol{\sigma}+\rho\boldsymbol{f}, \tag{2}$$

- balance of the moment of inertia tensor:

$$\frac{\delta\boldsymbol{J}}{\delta t}+\boldsymbol{J}\times\boldsymbol{\omega}-\boldsymbol{\omega}\times\boldsymbol{J}=\boldsymbol{\chi}_J, \tag{3}$$

- balance of spin:

$$\rho\frac{\delta\boldsymbol{J}\cdot\boldsymbol{\omega}}{\delta t}=\boldsymbol{\nabla}\cdot\boldsymbol{\mu}+\boldsymbol{\sigma}_\times+\rho\boldsymbol{m}+\rho\boldsymbol{\chi}_J\cdot\boldsymbol{\omega}. \tag{4}$$

We denote by

$$\frac{\delta(\cdot)}{\delta t}=\frac{\mathrm{d}(\cdot)}{\mathrm{d}t}+(\boldsymbol{v}-\boldsymbol{w})\cdot\boldsymbol{\nabla}(\cdot) \tag{5}$$

the substantial (a.k.a. material) derivative of a field quantity. Here $\mathrm{d}(\cdot)/\mathrm{d}t$ is the total derivative and $\boldsymbol{w}$ the mapping velocity of the observation point (see [9]). Moreover, $\boldsymbol{\sigma}$ is the (nonsymmetric) Cauchy stress tensor, $\boldsymbol{f}$ is the specific body force, $\boldsymbol{\chi}_J$ (a second-rank symmetric tensor) is the production related to the moment of inertia tensor $\boldsymbol{J}$; $\boldsymbol{\mu}$ is the couple stress tensor, $\boldsymbol{\sigma}_\times:=\boldsymbol{\epsilon}\cdot\cdot\boldsymbol{\sigma}$ is the Gibbsian cross applied to the (nonsymmetric) Cauchy stress tensor, $\boldsymbol{\epsilon}$ being the Levi-Civita tensor, and $\boldsymbol{m}$ are specific volume couples.

Equation (3) is worthy of a detailed discussion. It has already been mentioned that no need for the conservation of microinertia was seen in [5]. However, in the presented form, this equation can only be found in a recent paper by one of the authors [6]. There is a precedent to the balance for the inertia tensor $\boldsymbol{J}$, namely what is called "conservation of microinertia" in [1], pg. 15. However, no production term $\boldsymbol{\chi}_J$ can be found in there, because the particle $\boldsymbol{X}$ is material, *i.e.*, indestructible. Following [4], pg. 33 we may write if the translational and rotational inertia is conserved:

$$\rho=\frac{\rho_0}{\det\boldsymbol{F}},\quad \boldsymbol{F}:=\boldsymbol{\nabla}_{\boldsymbol{X}}\hat{\boldsymbol{x}}(\boldsymbol{X},t)\quad\text{for }\boldsymbol{J}=J\mathbf{I}:\quad J=J_0, \tag{6}$$

where all functions of the current configuration depend on $(\boldsymbol{X},t)$ and the ones in the reference placement (identified by the subscript 0) on $(\boldsymbol{X},t_0)$. Obviously the Lagrangian viewpoint is applicable with no problem whatsoever.

However, as indicated before, there is a catch. Consider our first example to be studied in the next section: A granular medium is milled. This affects the material particle, because its subunits will be crushed. They will change their individual masses and moments of inertia and, what is more, during the milling process there might even be an exchange of crushed subunits between neighboring material particles, which are then no longer material in the Lagrangian sense. Consequently, on a macroscopic scale the moments of inertia will change as well and all of this gives rise to the production term $\boldsymbol{\chi}_J$. On the macroscopic continuum level, this new term could be interpreted as a constitutive quantity and its concrete form depending on the primary fields should be found following the established principle of continuum theory. However, the mesoscopic point of view may give us some guidance as to how the production term should look like. We proceed to discuss this issue.

2.2. The mesoscopic aspect

In what follows, we summarize some essential relations on the mesoscopic level from paper [6]. In there and in [10], more information can be found.

Recall the (representative) volume element ΔV at a fixed position $\boldsymbol{x}$ in space (see figure 1, center) containing matter carrying inertial characteristics $\rho(\boldsymbol{x},t)$, $\boldsymbol{J}(\boldsymbol{x},t)$ (in units of m^2) moving with linear and angular velocities $\boldsymbol{v}(\boldsymbol{x},t)$ and $\boldsymbol{\omega}(\boldsymbol{x},t)$, respectively, all of which are fields. Note that ΔV does not refer to the volume occupied by a material point. Rather imagine it to be a small cell in a spatial Eulerian grid.

Within the volume element ΔV, there are sufficiently many "rigid particles," $i = 1, \ldots, N(\boldsymbol{x},t)$. The presence of a very large number of particles is required since establishing a continuous field theory would not be possible otherwise, and fluctuations would become dominant. Furthermore, note that the total particle number N can depend on space $\boldsymbol{x}$ and on time t.

One of the rigid particles, i, is depicted on the very right in figure 1. It carries known time-independent inertial characteristics m_i and $\hat{\boldsymbol{J}}_i$ (in units of kg·m^2, w.r.t. the center of ΔV). Moreover, the particle moves with individual independent linear and angular velocities $\boldsymbol{v}_i(t)$ and $\boldsymbol{\omega}_i(t)$, respectively, both of which can be time-dependent. Both are no fields but discrete quantities. We are now going to relate the macroscopic and particle worlds by forming mesoscopic averages. The idea is to replace the particles within an elementary volume by an ensemble of identical particles each having an average mass and an average tensor of inertia. To this end, the inertial characteristics of the elementary volume are assumed to coincide with those of the average particle. Moreover, the linear and angular momenta of the elementary volume consisting of original particles are required to equal those of the elementary volume consisting of average particles.

We start by defining average inertial characteristics within ΔV by

$$m(\boldsymbol{x},t) := \frac{\sum_{i=1}^{N(\boldsymbol{x},t)} m_i}{N(\boldsymbol{x},t)}, \quad \hat{\boldsymbol{J}}(\boldsymbol{x},t) := \frac{\sum_{i=1}^{N(\boldsymbol{x},t)} \hat{\boldsymbol{J}}_i}{N(\boldsymbol{x},t)}, \tag{7}$$

followed by the fields of particle and mass densities,

$$n(\boldsymbol{x},t) := \frac{N(\boldsymbol{x},t)}{\Delta V}, \quad \rho(\boldsymbol{x},t) := \frac{\sum_{i=1}^{N(\boldsymbol{x},t)} m_i}{\Delta V}, \tag{8}$$

which are then related by

$$\rho(\boldsymbol{x},t) = m(\boldsymbol{x},t)n(\boldsymbol{x},t). \tag{9}$$

For the rotational inertial characteristic, we have by definition:

$$\rho(\boldsymbol{x},t)\boldsymbol{J}(\boldsymbol{x},t) := \frac{\sum_{i=1}^{N(\boldsymbol{x},t)} \hat{\boldsymbol{J}}_i}{\Delta V}, \tag{10}$$

so that

$$\rho(\boldsymbol{x},t)\boldsymbol{J}(\boldsymbol{x},t) = n(\boldsymbol{x},t)\hat{\boldsymbol{J}}(\boldsymbol{x},t). \tag{11}$$

We shall now connect the linear and angular momenta of the macroscopic world with those of the particles. If we ignore coupling (see [4], Section 4.3; or [11], Sections 3.2.3ff for a discussion of coupling effects in general; and [7] and [10] in context with mesoscopic considerations of micropolar media in particular), then the linear and the angular momenta of one particle simply read

$$\hat{\boldsymbol{K}}_{1,i}(t) = m_i \boldsymbol{v}_i(t), \quad \hat{\boldsymbol{K}}_{2,i}(t) = \hat{\boldsymbol{J}}_i \cdot \boldsymbol{\omega}_i(t). \tag{12}$$

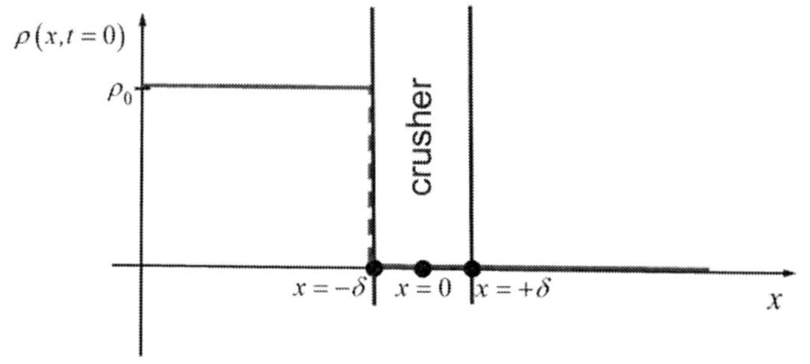

Figure 2. Idealization of the crusher problem (see text).

We define average linear and angular momenta fields within ΔV by

$$\hat{\boldsymbol{K}}_1(\boldsymbol{x},t) := \frac{\sum_{i=1}^{N(\boldsymbol{x},t)} \hat{\boldsymbol{K}}_{1,i}(t)}{N(\boldsymbol{x},t)}, \quad \hat{\boldsymbol{K}}_2(\boldsymbol{x},t) := \frac{\sum_{i=1}^{N(\boldsymbol{x},t)} \hat{\boldsymbol{K}}_{2,i}(t)}{N(\boldsymbol{x},t)}, \tag{13}$$

and postulate the following connection between the macroscopic linear momentum and the spin with the particle averages from equation (13):

$$\rho\boldsymbol{v} = n\hat{\boldsymbol{K}}_1, \quad \rho\boldsymbol{J}\cdot\boldsymbol{\omega} = n\hat{\boldsymbol{K}}_2, \tag{14}$$

where all quantities are fields depending on $\boldsymbol{x}$ and t. This in turn means that there are no simple relations between the particle velocities $\boldsymbol{v}_i$, $\boldsymbol{\omega}_i$ and the macroscopic fields $\boldsymbol{v}$, $\boldsymbol{\omega}$ and, in particular, we must conclude

$$\boldsymbol{v}(\boldsymbol{x},t) \neq \frac{\sum_{i=1}^{N(\boldsymbol{x},t)} \boldsymbol{v}_{1,i}(t)}{N(\boldsymbol{x},t)}, \quad \boldsymbol{\omega}(\boldsymbol{x},t) \neq \frac{\sum_{i=1}^{N(\boldsymbol{x},t)} \boldsymbol{\omega}_{2,i}(t)}{N(\boldsymbol{x},t)}. \tag{15}$$

3. Two examples

The effects and consequences of a balance for the moment of inertia with a production term are analyzed by two examples. In the first one, the crusher, an expression for the production is postulated phenomenologically, which allows us to mimic the milling of particles. In the second one, referred to as contracting and expanding hollow spheres, the production term is postulated based on a mesoscopic model of thermally sensitive hollow spheres in a thermal field varying in space and time. Here their state of rotation couples to the change of moment of inertia.

3.1. Analysis of a one-dimensional crusher

Consider the following problem in infinite one-dimensional space $-\infty < x < +\infty$ depicted in figure 2. A continuous flow of spherical particles with initial density $\rho(x, t = 0)$ is coming in from the left and keeps moving to the right at a constant prescribed speed v_0. On its way it enters a region $-\delta \leq 0 \leq +\delta$ symmetrically arranged around the position $x = 0$, where it is continuously crushed to form smaller and smaller particles. Hence we assume that the momentum balance shown in (2) is identically satisfied and the balance of mass (1) and of moment of inertia (3) read

$$\frac{\partial\rho}{\partial t} + v_0\frac{\partial\rho}{\partial x} = 0, \quad \frac{\partial J}{\partial t} + v_0\frac{\partial J}{\partial x} = \chi_J. \tag{16}$$

For the production of moment of inertia we postulate the following relationship:

$$\chi_J(x,t) = \begin{cases} 0, & \text{if } -\infty < x < -\delta, \\ -\alpha\langle\rho(x,t)\rangle^0\langle J(x,t) - J_*\rangle, & \text{if } -\delta \leq x \leq \delta, \\ 0, & \text{if } \delta < x < \infty, \end{cases} \tag{17}$$

where α is a positive rate constant, J_* indicates the moment of inertia pertinent to the minimum crushing size of the particles, and $\langle\bullet\rangle$ denotes the Macaulay bracket[1]. The density field is incorporated through a Heaviside function in order to ensure that there is no production if the density is zero.

A few more remarks regarding the constitutive character of equation (17) are in order. We may want to interpret the quantity J_* as a rotational inertial characteristic of the size of "glued" grains constituting the incoming ore. As such it is a characteristic of the material and not of the sieve size of the crusher (say), which has no inertial characteristics. Similarly the coefficient α could be interpreted as a measure of the dynamic resistance of the glue, *i.e.*, also as a constitutive property. On the other hand, it might also depend on the state of stress or on the stress rate in terms of an equivalent measure, just like yield stress, hence "converting" the action of the crusher blades to a material response. Thereby it is also related to the effectiveness or to the rotational speed of the crusher blades and transmission of its exerted forces. Hence, at least in the presented case the production term shows both, material as well as process, characteristics.

The field of the mass density can be obtained in closed form from equation $(16)_1$ in combination with the piecewise constant initial condition

$$\rho(x,t=0) =: \rho_0(x) = \begin{cases} \rho_0 = \text{const}, & \text{if } -\infty < x < -\delta, \\ 0, & \text{if } -\delta \le x \le +\delta, \\ 0, & \text{if } +\delta < x < +\infty, \end{cases} \tag{18}$$

and by using the method of characteristics for initial-value problems on an infinite domain (see [7] for details):

$$\rho(x,t) = \rho_0(x - v_0 t). \tag{19}$$

In other words, the solution is a step function of height ρ_0 moving steadily and uniformly from the left to the right. As expected, the solution for the rotational inertia shows a more complex behavior. By using the method of characteristics for equation $(16)_2$ in combination with the initial condition

$$J(x,t=0) =: J_0(x) = \begin{cases} J_0 = \text{const}, & \text{if } -\infty < x < -\delta, \\ 0 & \text{if } -\delta \le x \le +\delta, \\ 0 & \text{if } +\delta < x < +\infty. \end{cases} \tag{20}$$

we obtain

- to the left of the crusher region, *i.e.*, at positions $-\infty < x < -\delta$ and at times $0 \le t < \infty$:

$$J(x,t) = J_0; \tag{21}$$

- the crusher region at times $0 \le t < 2\delta/v_0$ for $-\delta \le x < x_s$:

$$J(x,t) = J_* + (J_0 - J_*)\exp\left[-\frac{\alpha}{v_0}(x+\delta)\right]; \tag{22}$$

and for $x_s \le x < +\delta$:

$$J(x,t) = 0, \tag{23}$$

and at times $2\delta/v_0 \le t < \infty$ for $-\delta \le x < +\delta$:

$$J(x,t) = J_* + (J_0 - J_*)\exp\left[-\frac{\alpha}{v_0}(x+\delta)\right], \tag{24}$$

whilst the shock front moves at a constant speed $\mathrm{d}x_s/\mathrm{d}t = v_0$;

[1] The Macaulay bracket is defined by $\langle x\rangle = \frac{1}{2}(|x| + x)$.

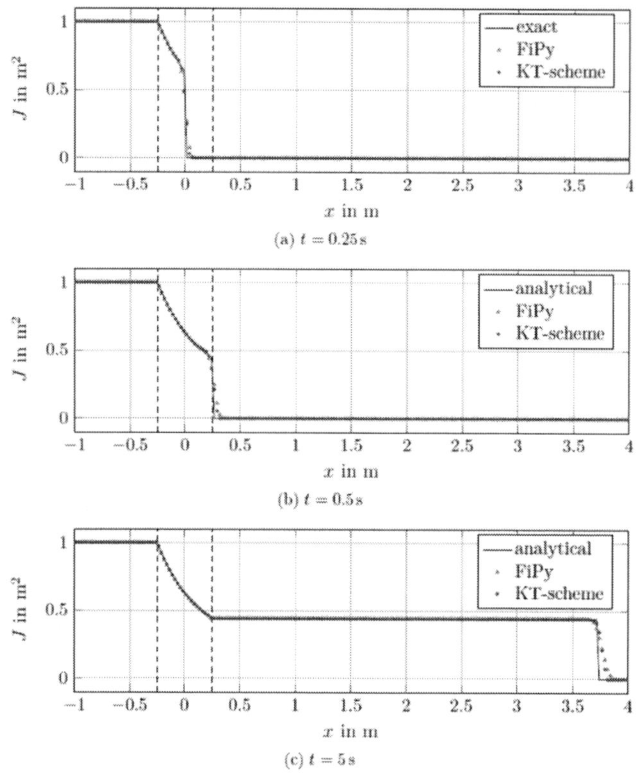

Figure 3. Evolving moment of inertia field for matter going through crusher region.

- to the right of the crusher region at times $0 \leq t < 2\delta/v_0$ for $\delta \leq x < \infty$:

$$J(x,t) = 0; \tag{25}$$

and at times $2\delta/v_0 \leq t < \infty$ for $\delta \leq x < v_0 t - \delta$:

$$J(x,t) = J_* + (J_0 - J_*) \exp\left(- \frac{2\alpha\delta}{v_0} \right), \tag{26}$$

and for $v_0 t - \delta \leq x < \infty$:

$$J(x,t) = 0, \tag{27}$$

and for $-\delta \leq x < \infty$:

$$J(x,t) = 0. \tag{28}$$

This result is graphically depicted in figure 3. The exponential decay in the crusher related to the milling and refinement of particles is visible. In fact, parameters were chosen such that the time the matter spends within the crusher was too short for milling down to the finest size J_*. Moreover, various finite volume schemes were used to solve the problem numerically, the FiPy method due to van Leer and the Kurganov–Tadmor scheme. The problem is to capture the abrupt decline of shock fronts accurately. This problem was analyzed in more detail in [12].

3.2. Analysis of rotating contracting and expanding hollow spheres

Consider a medium consisting of empty hollow elastic spheres homogeneously distributed within a one-dimensional region $x \in [0, l]$. Their initial inner and outer radii are R_i and R_o, respectively.

By positioning the medium in between two reservoirs kept at temperatures T_0 and T_l and attached at positions $x = 0$ and $x = l$ of the region, respectively, the temperature of this medium will gradually change from an initially constant value $T_{\rm ini}$. The development of temperature T is governed by Fourier heat conduction and the solution to the temperature problem is given by ([13], Section 3.4):

$$\bar{T}(\bar{x},\bar{t}) = \bar{T}_0 + (\bar{T}_l - \bar{T}_0) + \frac{2}{\pi}\sum_{n=1}^{\infty}\frac{\bar{T}_l\cos(n\pi) - \bar{T}_0}{n}\sin(n\pi\bar{x})\exp(-n^2\pi^2\bar{t})$$
$$+\frac{2}{\pi}\sum_{n=1}^{\infty}\frac{1-\cos(n\pi)}{n}\sin(n\pi\bar{x})\exp(-n^2\pi^2\bar{t}). \tag{29}$$

The bar on symbols refers to dimensionless quantities, namely,

$$\bar{x} := \frac{x}{l}, \quad \bar{t} := \frac{Dt}{l^2}, \quad \bar{T} := \frac{T}{T_{\rm ini}}, \tag{30}$$

where D is the thermal diffusivity. According to this temperature field, the hollow spheres do now contract and expand their inner and outer radii. Their moment of inertia will change accordingly. For simplicity, we assume that all of this happens instantaneously:

$$\boldsymbol{J} = J\mathbf{I}, \quad J = J_0[1 + \alpha T_{\rm ini}(\bar{T} - 1)]^2, \quad J_0 := \frac{2}{5}R_{\rm o}^2\frac{1-\beta^5}{1-\beta^3}, \quad \beta := \frac{R_{\rm i}}{R_{\rm o}}. \tag{31}$$

The production can now be found on the basis of equation (3):

$$\boldsymbol{\chi}_J = \chi_J\mathbf{I}, \quad \chi_J \equiv \frac{\partial J}{\partial t} = \chi_0[1 + \alpha T_{\rm ini}(\bar{T} - 1)]\frac{\partial \bar{T}}{\partial \bar{t}}, \quad \chi_0 := 2\alpha T_{\rm ini}J_0\frac{D}{l^2}. \tag{32}$$

Figure 4 shows the development of the (normalized) moment of inertia on the left ($\alpha T_{\rm ini}$=0.5). The blue and red curves depict the situation shortly after the heat conduction process has started. The green curve represents the situation corresponding to the less stationary temperature distribution (linear profile in $\bar{T}$). The picture on the right shows the temporal evolution of the production at various points, namely, at $\bar{x} = 0.1$ (blue), $\bar{x} = 0.5$ (red), $\bar{x} = 1/3$ (magenta, where $\bar{T} = 1$ for the choice of parameters), and $\bar{x} = 0.9$ (green). The production vanishes for long times when we move toward the stationary temperature profile.

We now turn to the problem of how the angular velocity of the hollow spheres will be influenced when subjected to the nonstationary temperature profile. To this end, we need to make some assumptions regarding the body couple $\boldsymbol{m}$. We consider the micro-model shown on the left in figure 5. Following the ideas presented in [14] we imagine the hollow spheres to be electrically polarized dipoles with zero net charge, $q^+ = -q^- = q$. Now we apply a constant external electric field $\boldsymbol{E}_0$ in the negative x-direction. The total Coulomb force, and therefore (after homogenization) the body force (in the x-direction), will then vanish. However, the moment couple acting on the sphere will not:

$$\boldsymbol{M} = (q^+ - q^-)\boldsymbol{R}_{\rm o} \times \boldsymbol{E}_0 = 2qR_{\rm o}E_0\cos\varphi(t)\boldsymbol{e}_z \neq \boldsymbol{0}. \tag{33}$$

Then the body moment couple becomes:

$$\boldsymbol{m} = m_0\cos\left[\int_{\bar{t}=0}^{\bar{t}=t}\omega(\bar{t})\mathrm{d}\bar{t}\right]\boldsymbol{e}_z, \quad m_0 = 2\frac{q}{m_{\rm p}}R_{\rm o}E_0 \tag{34}$$

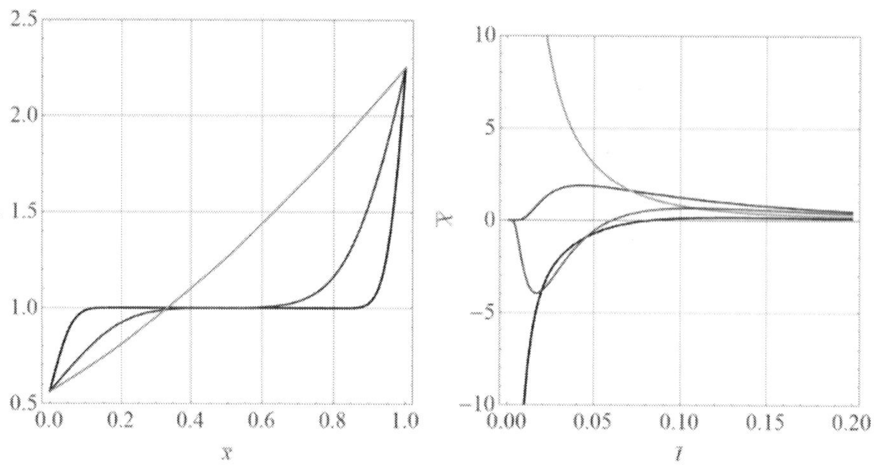

Figure 4. Moment of inertia and production fields for hollow spheres (see text).

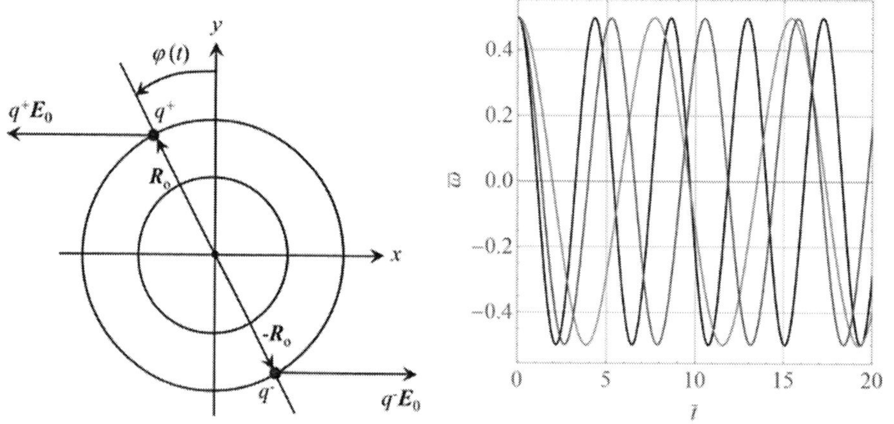

Figure 5. A simple model for inducing angular rotation (see text).

for $\varphi(0) = 0$ and m_p being the mass of one particle. We assume that the stress tensor and the couple stress tensor vanish (dust) and rewrite the spin balance (4) as

$$\frac{\partial}{\partial t}\left(\frac{J}{m_0}\frac{\partial \omega}{\partial t}\right) = -\omega\sqrt{1 - \left(\frac{J}{m_0}\frac{\partial \omega}{\partial t}\right)^2}. \tag{35}$$

A numerical solution of this ODE in time for each fixed position with $\omega(\bar{x}, \bar{t} = 0) = 0.4$, $\dot{\omega}(\bar{x}, \bar{t} = 0) = 0$ leads to oscillating behavior shown in figure 5 on the right. The different colors refer to different (normalized) positions ($\bar{x} = 0.1$ (blue), $\bar{x} = 1/3$ (magenta, where $\bar{T} = 1$). Noticeably the different speeds of temperature change influence the oscillation frequencies. More details are presented in [15].

Summary

The intention of this paper was to draw attention on some recent activities in the field of micropolar media capable of structural change. An extended set of balance equations was presented, where the extension consisted of an independent balance for the field of moment of inertia with production. Two examples were presented showing the effect of this new term. More research in this field is planned for the future in order to study fully coupled problems for micropolar media.

Acknowledgments

Support of this work by grants from the Russian Foundation for Basic Research (project No. 16-01-00815) as well as from Deutsche Forschungsgemeinschaft (MU-1752/43-1) is gratefully acknowledged.

References

[1] Eringen A C and Kafadar C B 1976 Polar field theories In *Continuum Physics IV* (London: Academic Press)
[2] Eringen A C 2000 A unified continuum theory for electrodynamics of polymeric liquid crystals *Int. J. Engng Sci.* **38** 959–87
[3] Eringen C 2002 *Nonlocal continuum field theories* (New York: Springer Science & Business Media)
[4] Eremeyev V A, Lebedev L P, and Altenbach H 2012 *Foundations of Micropolar Mechanics* (Heidelberg - New York - Dordrecht - London: Springer Science & Business Media)
[5] Dłużevski P H 1993 Finite deformations of polar elastic media *Int. J. Solids Struct.* **30** (16) 2277–85
[6] Ivanova E A and Vilchevskaya E N 2016 Micropolar continuum in spatial description *Continuum Mech. Thermodyn.* **28** (6) 1759–80
[7] Müller W H, Vilchevskaya E N, and Weiss W 2017 Micropolar theory with production of rotational inertia: A farewell to material *Phys. Mesomech.* **20** (3) 13–24
[8] Zhilin P A 2015 Dynamics of the rigid body) (St. Petersburg: Izdat. SPbGTU) [in Russian]
[9] Ivanova E, Vilchevskaya E, and Müller W H 2016 Time derivatives in material and spatial description — what are the differences and why do they concern us? In *Advanced Methods of Continuum Mechanics for Materials and Structures* ed Naumenko K and Aßmus M (Springer) pp 3–28
[10] Müller W H and Vilchevskaya E N 2017 Micropolar theory from the viewpoint of mesoscopic and mixture theories *Phys. Mesomech.* **20** (3) 263–79
[11] Zhilin P A 2012 *Rational Continuum Mechanics* (St. Petersburg: Izdat. Politekh. Univ) [in Russian]
[12] Glane S, Rickert W, Müller W H, and Vilchevskaya E 2017 Micropolar media with structural transformations: Numerical treatment of a particle crusher In *Proceedings Advanced Problems in Mechanics APM 2017* (St. Petersburg: Russian Academy of Sciences, IPME RAS, Peter the Great St. Petersburg Polytechnic University) pp 197–211
[13] Carslaw H S and Jaeger J C 1959 *Conduction of Heat in Solids, Second Edition* (Oxford: Clarandon Press)
[14] Zhilin P A and Kolpakov Y E 2005 A micro-polar theory for piezoelectric materials In *Lecture at XXXIII Summer School — Conference Advanced Problems in Mechanics* pp 250–61 URL `http://teormeh.spbstu.ru/Zhilin_New/pdf/Zhilin_Kolpakov2_eng.pdf`

IOP Conf. Series: Journal of Physics: Conf. Series **991** (2018) 012080 doi:10.1088/1742-6596/991/1/012080

Dependence of equivalent thermal conductivity coefficients of single-wall carbon nanotubes on their chirality

V S Zarubin[1*] **and E S Sergeeva**[2]
[1]Bauman Moscow State Technical University, Moscow, Russia
[2]JSC "Kompozit", Korolev, Moscow Region, Russia
E-mail: *zarubin@bmstu.ru

Abstract. Composite materials (composites) composed of a matrix and reinforcing components are currently widely used as structural materials for various engineering devices designed to operate under extreme thermal and mechanical loads. By modifying a composite with structure-sensitive inclusions such as single-wall carbon nanotubes, one can significantly improve the thermomechanical properties of the resulting material. The paper presents relationships obtained for the equivalent thermal conductivity coefficients of single-wall carbon nanotubes versus their chirality using a simulation model developed to simulate the heat transfer process through thermal conductivity in a transversely isotropic environment. With these coefficients, one can conventionally substitute a single-wall carbon nanotube with a continuous anisotropic fiber, thus allowing one to estimate the thermal properties of composites reinforced with objects of this sort by using the well-known models developed for fibered composites. The results presented here can be used to estimate the thermal properties of carbon nanotube-reinforced composites.

1. Introduction

Given their high thermoelastic performance, the nanoscale objects such as graphene flakes, fibers, single- and multi-wall carbon nanotubes (SWCNT / MWCNT) [1–9] have increasingly been used nowadays as modifying elements for composite materials. These objects are capable of aggregating into nanoclusters; however, the uniformly-dispersed reinforcement can also be attained. It is commonly known that the nanoscale graphene objects, even in a small proportion, can significantly improve the thermoelastic characteristics of a composite reinforced with them [10–13]. The development of a composite-based structure designed to operate under extreme mechanical and thermal loads requires correct estimates of the material elastic and thermal properties that are mainly governed by the corresponding characteristics of the reinforcing elements, which in turn depend on their geometry. There are plenty of publications addressing the dependence of the elastic properties of nanoscale elements on their geometry. However, the dependence of the thermal properties of these objects on their characteristic dimensions has not yet been studied in detail.

To construct a simulation model representing the SWCNT-reinforced composite thermal properties, a nanotube can be represented as an anisotropic circular cylinder, its length being

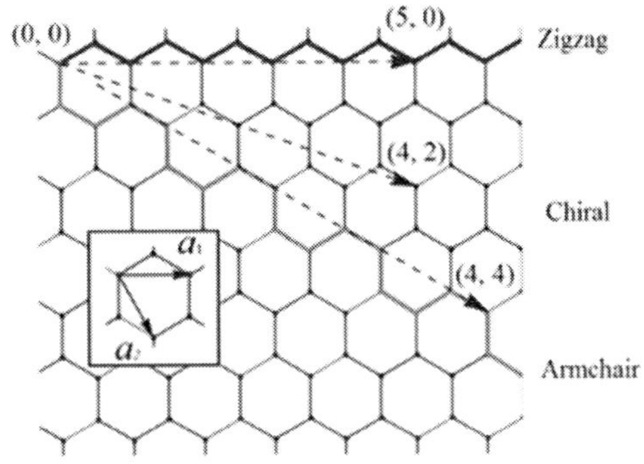

Figure 1. Several ways of rolling the graphene plane into a tube and their corresponding SWCNT configurations.

much larger than its radius. With such a representation, one can use the well-known models developed for estimating the thermal conductivity coefficient of fibered composites [14].

This paper focuses on the dependence of the equivalent thermal conductivity coefficients, both in the axial direction and in the cross-sectional plane, on the SWCNT diameter and configuration type determined by the chirality index [15].

2. Main relationships

A single-wall carbon nanotube (SWCNT) is a cylinder, up to several tens nanometer in diameter, representing a single graphene plane rolled in a seamless tube, namely, a surface composed of regular hexagons, with carbon atoms residing at their vertices [15]. The result of this rolling procedure depends on the graphene plane orientation angle as related to the nanotube axis. The orientation angle, in turn, governs the nanotube chirality, thus determining the nanotube characteristics. For the nanotube chirality, the notation (m, n) is used, indicating the coordinates of the hexagon to be aligned with the one residing at the origin point after the graphene plane is rolled into a tube (see figure 1).

The chirality indices (m, n) of a single-wall carbon nanotube uniquely determine its diameter D [15]

$$D = \frac{\sqrt{3}d_0}{\pi}\sqrt{m^2 + n^2 + mn}. \quad (1)$$

Of all possible ways of rolling the graphene plane into a tube, we set aside those for which the alignment of hexagon (m, n) with the origin point does not distort its structure. This condition is met, particularly, with angles $\alpha = 0°$ (Armchair-type configuration) and $\alpha = 30°$ (Zigzag-type configuration). These configurations correspond to chiralities (n, n) and $(m, 0)$, correspondingly. The remaining combinations of indices refer to the Chiral-type configuration.

For our simulation model, the nanotube is represented as a circular cylindrical shell formed by rolling a polyatomic graphene layer into a tube, with halves of fullerene molecules on the edges [10]. Assume that the thermal conductivity coefficient of graphene in its isotropy plane is equal to λ_0. In this case, the thermal conductivity coefficient is equal to λ_0 along any direction tangent to the middle surface of the shell [16]. However, given the conventional substitution of SWCNT with a continuous circular cylinder, the equivalent thermal conductivity coefficients $\lambda_\parallel$ and $\lambda_\perp$ along the longitudinal axis and in the plane perpendicular to it are, correspondingly, different. Since the SWCNT length is assumed to be much larger than its outer radius, the effect of the shell semispherical parts on the values of $\lambda_\parallel$ and $\lambda_\perp$ may be neglected.

3. Equivalent thermal conductivity coefficient

The dependence of the effective SWCNT thermal conductivity coefficient on the SWCNT diameter and configuration can be obtained using the simulation model representing the heat transfer through thermal conductivity in a nanotube [16].

The equivalent thermal conductivity coefficient $\lambda_{\parallel}$ along the SWCNT axis can be determined from the axial heat resistance equality between the circular cylindrical shell, its thermal conductivity coefficient equal to λ_0 in any tangential direction, and the continuous circular cylinder substituting the shell:

$$\frac{L}{2\pi R_0 h_0 \lambda_0} = \frac{L}{\pi (R_0 + h_0/2)^2 \lambda_{\parallel}},$$

where L is shell/cylinder length; R_0 and h_0 are the middle surface radius and the shell thickness, correspondingly. The latter value is assumed to be equal to 0.34 nm, i.e., the distance between two neighboring atomic planes in the graphite crystal [18]. From the above equality follows that

$$\lambda_{\parallel} = \frac{2R_0 h_0 \lambda_0}{\tilde{R}^2}, \tag{2}$$

where $\tilde{R} = R_0 + h_0/2$. This relation is meaningful for $R_0 \geq h/2$, which, given equation (1), for the Zigzag-type SWCNT corresponds to the condition $n \geq 5$, while for Armchair-type SWCNT, to the condition $n = m \geq 3$. For other chirality index combinations (Chiral-type SWCNT), the condition $n + m \geq 5$ shall be satisfied. Depending on the chirality indices or, more precisely, the their sum, the polygon, namely, the SWCNT cross section, tends to a circle. To obtain a first-order estimate, we assume that SWCNT can be represented by a cylinder satisfying the condition $n + m \geq 5$.

To find the equivalent thermal conductivity coefficient $\lambda_{\perp}$, we use the model simulating the interaction between a sufficiently long circular cylindrical shell and its environment of unlimited volume with thermal conductivity coefficient λ [16]. The temperature distribution established in the area $L/\tilde{R} \gg 1$ can be represented as a function $T(r, \varphi)$ under the assumption that it remains unchanged along the shell axis and satisfies the Laplace equation

$$\frac{1}{r}\frac{\partial}{\partial r}\left(r\frac{\partial T}{\partial r}\right) + \frac{1}{r^2}\frac{\partial^2 T}{\partial \varphi^2} = 0$$

in the polar coordinate system, with origin on shell axis.

If we assume that the temperature field gradient vector is equal to G in the environment at a distance r greater than $\tilde{R}$ in the direction of the polar angle φ, then the obtained temperature distribution in this environment, with regard to the temperature field distortion induced by the cylindrical shell, is described by the formula [16]

$$T(r, \varphi) = Gr\cos\varphi + \frac{B_0}{r}\cos\varphi, \quad r \geq \tilde{R}, \quad 0 \leq \varphi < 2\pi, \tag{3}$$

where B_0 is a constant coefficient which can be determined from the condition of thermal interaction between the environment and the cylindrical shell.

The temperature T_0 is assumed uniform through the shell thickness, depending on the polar angle φ alone. Under the assumption that the thermal contact between the shell and the environment to be perfect, from formula (3) for the temperature the cylindrical surface at $r = \tilde{R}$ we obtain

$$T_0(\varphi) = \left(G\tilde{R} + \frac{B_0}{\tilde{R}}\right)\cos\varphi. \tag{4}$$

Table 1. Chirality index for various SWCNT configurations.

Armchair		Zigzag		Chiral	
m	m	n	n	m	n
5	5	5	0	10	1
10	10	10	0	10	5
20	20	20	0	10	11
50	50	50	0	10	50
100	100	100	0	10	100
200	200	200	0	10	200

From (3) and (4), given the thermal balance in the shell, we have

$$\frac{B_0}{\tilde{R}^2} = G\frac{1-\beta_0}{1+\beta_0}, \quad \beta_0 = \frac{\lambda_0 h_0}{\lambda R_0}. \tag{5}$$

The replacement of the cylindrical shell by a continuous cylinder of radius $\tilde{R}$ and with thermal conductivity coefficient $\lambda_\perp$ will also distort the temperature field in the environment, which is described by the term $(\tilde{B}/r^2)\cos\varphi$ [17], where

$$\tilde{B} = G\tilde{R}^2\frac{\lambda-\lambda_\perp}{\lambda+\lambda_\perp}. \tag{6}$$

Assuming that the temperature distribution over the outer surface of the cylindrical shell is similar to that over the surface of the continuous cylinder substituting the shell, we have $B_0 = \tilde{B}$, from which, considering (5) and (6), we obtain the second SWCNT equivalent thermal conductivity coefficient:

$$\lambda_\perp = \frac{\lambda_0 h_0}{R_0}. \tag{7}$$

4. Analysis of the results

As the input data for SWCNT thermal conductivity coefficient λ_0, we use the value 3500 W/(m·K) [18]. The values of the chirality index chosen for various SWCNT configurations are given in table 1.

Using formula (1), the SWCNT diameter values were found for different chirality index combinations (see the Table below), with the previously established limits for these values taken into account.

From (2) and (7), we find the values of $\lambda_\parallel$ and $\lambda_\perp$, correspondingly. Thus we have determined the dependence of the SWCNT equivalent thermal conductivity coefficient on the nanotube diameter, and hence, on the chirality type. The dependence is shown in figure 2.

In figure 2, one can see that the values of the SWCNT equivalent thermal conductivity coefficient decrease as the nanotube diameter increases. Figure 3 shows the dependence of $\tilde{\lambda} = (\lambda_\parallel/\lambda_\perp)/(\lambda'_\parallel/\lambda'_\perp)$ on the ratio $\tilde{d} = d/d'$, where $\lambda'_\parallel$, $\lambda'_\perp$, and d' are maximum values of the equivalent thermal conductivity coefficient and the diameter for each SWCNT type.

It is clear from figure 3 that for a certain chirality index, the corresponding values of $\tilde{\lambda}$ are greater for the Armchair-type SWCNT than those obtained for the Zigzag-/Thiral-type SWCNT; however, when the dimensionless coefficient $\tilde{d}$ becomes equal to or larger than 0.5, the difference becomes almost unnoticeable. It should also be noted that the Chiral-type SWCNT and Zigzag-type SWCNT are almost identical by their values of $\tilde{\lambda}$ obtained with the selected chirality indices.

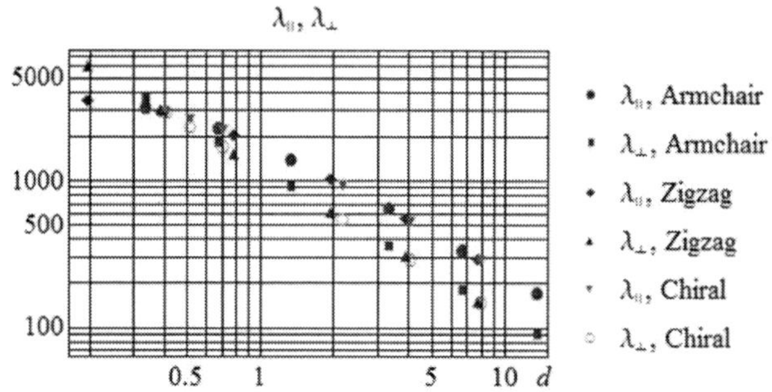

Figure 2. Dependence of the SWCNT equivalent thermal conductivity coefficient (W/(m·K)) on the diameter and chirality type.

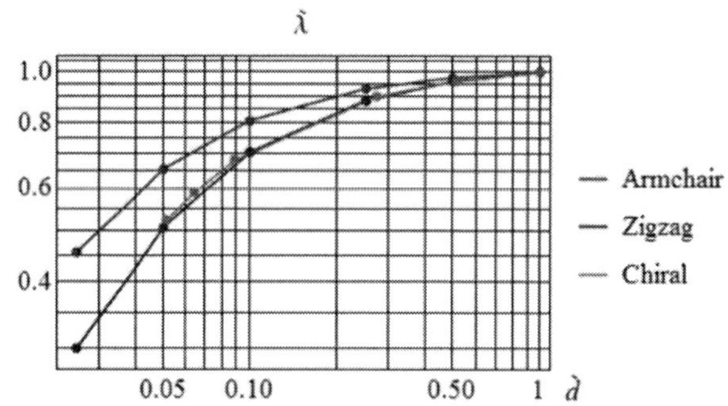

Figure 3. Dependence of $\tilde{\lambda}$ on $\tilde{d}$.

Conclusion

The relationships outlined above can be used to estimate the equivalent thermal conductivity coefficient of single-wall carbon nanotubes that are currently widely used as reinforcing inclusions for advanced structural materials in devices designed to operate under extreme loads of different nature. Based on these coefficients and the nanotube representation as an anisotropic continuous circular cylinder, one can predict the thermal properties of materials on the basis of simulation models developed for fibered composites.

Acknowledgments

The work was performed in the framework of implementation of the basic part of the governmental task of the Ministry of Education and Science of the Russian Federation (Project 9.7784.2017/BP).

References

[1] Palermo P 2015 Structural ceramic nanocomposites: A review of properties and powders' synthesis methods *Nanomater.* **5** 656–96

[2] Casati R and Vedani M 2014 Metal matrix composites reinforced by nano-particles — A review *Metals* **4** 65–83

[3] Alibeigloo A and Liew K M 2013 Thermoelastic analysis of functionally graded carbon nanotube-reinforced composite plate using theory of elasticity *Compos. Struct.* **106** 873–81

[4] Ayatollahi M R, Shadlou S, and Shokrieh M M 2011 Multiscale modeling for mechanical properties of carbon nanotube reinforced nanocomposites subjected to different types of loading *Compos. Struct.* **93** 2250–9

[5] Liew K M, Lei Z X, Zhang L W, et al. 2014 Mechanical analysis of functionally graded carbon nanotube reinforced composites: A review *Compos. Struct.* **120** 90–7

[6] Montinaro N and Pantano A 2014 Parameters influencing the stiffness of composites reinforced by carbon nanotubes — A numerical–analytical approach *Compos. Struct.* **109** 246–52
[7] Ürk D, Demir E, Bulut O, et al. 2016 Understanding the polymer type and CNT orientation effect on the dynamic mechanical properties of high volume fraction CNT polymer nanocomposites *Compos. Struct.* **155** 255–62
[8] Konstantinos G and Costas G 2012 Polymer–nanotube interaction in MWCNT/poly(vinyl alcohol) composite mats *Carbon* **50** 4291–301
[9] Jin G P, Qunfeng Ch, Jun L, et al. 2012 Thermal conductivity of MWCNT/epoxy composites: The effects of length, alignment and functionalization *Carbon* **50** 2083–90
[10] Katz A E 2008 *Fullerenes, Carbon Nanotubes And Nanoclusters: Genealogy of Forms and Ideas* (Moscow: LKI Publishers) p 296
[11] Zarubin V S, Sergeeva E S, and Shishkina S I 2016 Estimating the elastic properties of the carbon nanotube-reinforced composite matrix *Nauka Obraz. Nauchn. Izdanie* **9** 155–70
[12] Sergeeva E S 2016 Investigations of elastic properties of a composite containing ellipsoidal inclusions *Molodezhn. Nauchn.-Techn. Vestnik* **5** URL `http://sntbul.bmstu.ru/doc/839933.html`
[13] Sergeeva E S 2016 Investigations of elastic properties of nanocomposites *Molodezhn. Nauchn.-Techn. Vestnik* **8** URL `http://sntbul.bmstu.ru/doc/846958.html`
[14] Zarubin V S, Kuvyrkin G N, and Savelieva I Yu 2013 *Thermal Conductivity of Fiber Reinforced Composites: Derivation, Verification and Parametric Analysis of Calculation Formula* (Saarbrücken: LAP LAMBERT Academic Publishing) p 121
[15] Yeletski A V 2002 Carbon nanotubes and their emission properties *Usp. Fiz. Nauk* **172** (4) 401–38
[16] Zarubin V S 2013 Estimates for the equivalent thermal conductivity coefficients of fullerenes and single-wall carbon nanotubes *Inzh. Zh. Nauka Innovats.* **4(16)** URL `http://engjournal.ru/catalog/mathmodel/hidden/669.html`
[17] Carslaw H S and Jaeger J C 1964 *Conduction of Heat in Solids* (Moscow: Nauka) p 488
[18] Yeletski A V, Iskandarova I M, Knizhnik A A, and Krasikov D N 2011 Graphene: fabrication methods and thermophysical properties *Usp. Fiz. Nauk* **181** (3) 233–268